ANATOMY

DESCRIPTIVE AND SURGICAL

ANATOMY

DESCRIPTIVE AND SURGICAL

BY

HENRY GRAY, F.R.S.

LECTURER IN ANATOMY AT SAINT GEORGE'S HOSPITAL

THE DRAWINGS

BY H. V. CARTER, M.D.

LATE DEMONSTRATOR OF ANATOMY AT ST. GEORGE'S HOSPITAL

THE DISSECTIONS

JOINTLY BY THE AUTHOR AND DR. CARTER

This edition published and distributed by Parragon, 1998

Parragon
Unit 13–17, Avonbridge Trading Estate
Atlantic Road, Avonmouth
Bristol BS11 9QD

Produced by Magpie Books,
an imprint of Robinson Publishing Ltd, London

All rights reserved. This book is sold subject to the condition that it shall not, by way of trade or otherwise, be lent, resold, hired out or otherwise circulated in any form of binding or cover other than that in which it is published and without a similar condition including this condition being imposed on the subsequent purchaser.

ISBN 0-75252-409-7

A copy of the British Library Cataloguing-in-Publication Data
is available from the British Library.

Printed and bound in the E.C.

ANATOMY

DESCRIPTIVE
AND
SURGICAL

CONTENTS

Osteology.

The Skull.

The Thorax.

The Upper Extremity.

The Lower Extremity.

The Articulations.

Articulations of the Trunk.

Articulations of the Upper Extremity.

Muscles and Fasciæ.

The Arteries.

The Veins.

The Lymphatics.

Nervous System.

Organs of Sense.

VISCERA.

Organs of Digestion and their Appendages.

Organs of Voice and Respiration.

The Urinary Organs.

Male Generative Organs.

Female Organs of Generation.

Surgical Anatomy of Inguinal Hernia.

Surgical Anatomy of Femoral Hernia.

Surgical Anatomy of Perinæum and Ischio-Rectal Region.

LIST OF ILLUSTRATIONS

☞ The illustrations, when copied from any other work, have the author's name affixed. When no such acknowledgment is made, the drawing is to be considered original.

Osteology.

Articulations.

Muscles and Fasciæ.

Arteries.

Veins.

Lymphatics.

Nervous System.

Cranial Nerves.

Spinal Nerves.

Organs of Sense.

Organs of Digestion and their Appendages.

Organs of Circulation.

Organs of Voice and Respiration.

The Urinary and Generative Organs.

ANATOMY

DESCRIPTIVE AND SURGICAL.

ANATOMY

DESCRIPTIVE AND SURGICAL.

The Skeleton.

THE entire skeleton in the adult consists of 200 distinct bones. These are—

The Spine or vertebral column (sacrum and coccyx included) .	26
Cranium	8
Face	14
Os hyoides, sternum, and ribs	26
Upper extremities	64
Lower extremities	62
	200

In this enumeration, the patellæ are included as separate bones, but the smaller sesamoid bones, and the ossicula auditûs, are not reckoned. The teeth belong to the tegumentary system.

These bones are divisible into four classes: *Long*, *Short*, *Flat*, and *Irregular*.

The *Long Bones* are found in the limbs, where they form a system of levers, which have to sustain the weight of the trunk, and to confer the power of locomotion. A long bone consists of a lengthened cylinder or shaft, and two extremities. The *shaft* is a hollow cylinder, the walls consisting of dense compact tissue of great thickness in the middle, and becoming thinner towards the extremities; the spongy tissue is scanty, and the bone is hollowed out in its interior to form the *medullary canal*. The *extremities* are generally somewhat expanded for greater convenience of mutual connection, for the purposes of articulation, and to afford a broad surface for muscular attachment. Here the bone is made up of spongy tissue with only a thin coating of compact substance. The long bones are, the *humerus*, *radius*, *ulna*, *femur*, *tibia*, *fibula*, *metacarpal*, and *metatarsal* bones, and the *phalanges*. The *clavicle* is also usually reckoned as a long bone.

Short Bones. Where a part of the skeleton is intended for strength and compactness, and its motion is at the same time slight and limited, it is divided into a number of small pieces united together by ligaments, and the separate bones are short and compressed, such as the bones of the *carpus* and *tarsus*. These bones, in their structure, are spongy throughout, excepting at their surface, where there is a thin crust of compact substance.

Flat Bones. Where the principal requirement is either extensive protection, or the provision of broad surfaces for muscular attachment, we find the osseous structure expanded into broad flat plates, as is seen in the bones of the skull and the shoulder-blade. These bones are composed of two thin layers of compact tissue enclosing between them a variable quantity of cancellous tissue. In the cranial bones, these layers of compact tissue are familiarly known as the *tables* of the skull; the outer one is thick and tough; the inner one thinner, denser, and more brittle, and hence termed the *vitreous table*. The intervening cancellous tissue is called the *diploë*. The flat bones are, the *occipital*, *parietal*, *frontal*, *nasal*, *lachrymal*, *vomer*, *scapulæ*, *ossa innominata*, *sternum*, and *ribs*.

The *Irregular* or *Mixed* bones are such as, from their peculiar form, cannot be grouped under either of the preceding heads. Their structure is similar to that of other bones, consisting of a layer of compact tissue externally, and of spongy cancellous tissue within. The irregular bones are, the *vertebræ, sacrum, coccyx, temporal, sphenoid, ethmoid, superior maxillary, inferior maxillary, palate, inferior turbinated* and *hyoid.*

Surfaces of Bones. If the surface of any bone is examined, certain eminences and depressions are seen, to which descriptive anatomists have given the following names.

A prominent process projecting from the surface of a bone, which it has never been separate from, or moveable upon, is termed an *apophysis* (from ἀπόφυσις, *an excrescence*); but if such process is developed as a separate piece from the rest of the bone, to which it is afterwards joined, it is termed an *epiphysis* (from ἐπίφυσις, *an accretion.*

These eminences and depressions are of two kinds: *articular*, and *non-articular.* Well-marked examples of articular eminences are found in the heads of the humerus and femur; and of articular depressions, in the glenoid cavity of the scapula, and the acetabulum. Non-articular eminences are designated according to their form. Thus, a broad, rough, uneven elevation is called a *tuberosity*; a small rough prominence, a *tubercle*; a sharp, slender, pointed eminence, a *spine*; a narrow rough elevation, running some way along the surface, a *ridge*, or *line.*

The non-articular depressions are also of very variable form, and are described as fossæ, grooves, furrows, fissures, notches, etc. These non-articular eminences and depressions serve to increase the extent of surface for the attachment of ligaments and muscles, and are usually well marked in proportion to the muscularity of the subject.

THE SPINE.

The Spine is a flexuous and flexible column, formed of a series of bones called *Vertebræ.*

The Vertebræ are thirty-three in number, exclusive of those which form the skull, and have received the names *cervical, dorsal, lumbar, sacral,* and *coccygeal* according to the position which they occupy; seven being found in the cervical region, twelve in the dorsal, five in the lumbar, five in the sacral, and four in the coccygeal.

This number is sometimes increased by an additional vertebra in one region, or the number may be diminished in one region, the deficiency being supplied by an additional vertebra in another. These observations do not apply to the cervical portion of the spine, the number of bones forming which is seldom increased or diminished.

The Vertebræ in the three uppermost regions of the spine are separate throughout the whole of life; but those found in the sacral and coccygeal regions are, in the adult, firmly united, so as to form two bones—five entering into the formation of the upper bone or *sacrum*, and four into the terminal bone of the spine or *coccyx.*

General Characters of a Vertebra.

Each vertebra consists of two essential parts, an anterior solid segment or body, and a posterior segment or arch. The arch is formed of two pedicles and two laminæ, supporting seven processes; viz. four articular, two transverse, and one spinous process.

The Bodies of the vertebræ are piled one upon the other, forming a strong pillar, for the support of the cranium and trunk; the arches forming a hollow cylinder behind for the protection of the spinal cord. The different vertebræ are connected together by means of the articular processes, and the intervertebral cartilages; while the transverse and spinous processes serve as levers for the attachment of muscles which move the different parts of the spine. Lastly, between each pair of vertebræ, apertures exist through which the spinal nerves pass from the cord. Each of these constituent parts must now be separately examined.

The Body is the largest and most solid part of a vertebra. Above and below, it is slightly concave, presenting a rim around its circumference; and its upper and lower surfaces are rough, for the attachment of the intervertebral fibro-cartilages. In front, it is convex from side to side, concave from above downwards. Behind, it is flat from above downwards and slightly concave from side to side. Its anterior surface is perforated by a few small apertures, for the passage of nutrient vessels; whilst, on the posterior surface, is a single large irregular aperture, or occasionally more than one, for the exit of veins from the body of the vertebra, the *venæ basis vertebræ*.

The *Pedicles* project backwards, one on each side, from the upper part of the body of the vertebra, at the line of junction of its posterior and lateral surfaces. The concavities above and below the pedicles are the *intervertebral notches*; they are four in number, two on each side, the inferior ones being generally the deeper. When the vertebræ are articulated, the notches of each contiguous pair of bones form the intervertebral foramina which communicate with the spinal canal and transmit the spinal nerves.

The *Laminæ* are two broad plates of bone, which complete the vertebral arch behind, enclosing a foramen which serves for the protection of the spinal cord; they are connected to the body by means of the pedicles. Their upper and lower borders are rough, for the attachment of the *ligamenta subflava*.

The *Articular Processes*, four in number, two on each side, spring from the junction of the pedicles with the laminæ. The two superior project upwards, their articular surfaces being directed more or less backwards, the two inferior project downwards, their articular surfaces looking more or less forwards.*

The *Spinous Process* projects backwards from the junction of the two laminæ, and serves for the attachment of muscles.

The *Transverse Processes*, two in number, project one at each side from the point where the articular processes join the pedicle. They also serve for the attachment of muscles.

Characters of the Cervical Vertebræ (fig. 1).

The Body is smaller than in any other region of the spine, and broader from side to side than from before backwards. The anterior and posterior surfaces are

1.—A Cervical Vertebra.

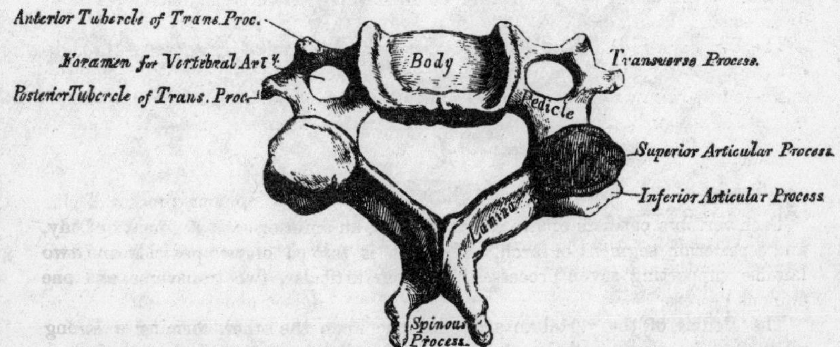

* It may, perhaps, be as well to remind the reader, that the direction of a surface is determined by that of a line drawn at right angles to it.

flattened and of equal depth; the former is placed on a lower level than the latter, and its inferior border is prolonged downwards so as to overlap the upper and fore part of the vertebra below. Its upper surface is concave transversely, and presents a projecting lip on each side; its lower surface being convex from side to side, concave from before backwards, and presenting laterally a shallow concavity, which receives the corresponding projecting lip of the adjacent vertebra. The *pedicles* are directed obliquely outwards, and the superior intervertebral notches are deeper, but narrower, than the inferior. The *laminæ* are narrow, long, thinner above than below, and overlap each other; enclosing the spinal foramen, which is very large, and of a triangular form. The *spinous processes* are short and bifid at the extremity, to afford greater extent of surface for the attachment of muscles, the two divisions being often of unequal size. They increase in length from the fourth to the seventh. The *transverse processes* are short, directed downwards, outwards, and forwards, bifid at their extremity, and marked by a groove along their upper surface, which runs downwards and outwards from the superior intervertebral notch, and serves for the transmission of one of the cervical nerves. The transverse processes are pierced at their base by a foramen, for the transmission of the vertebral artery, vein, and plexus of nerves. Each process is formed by two roots: the anterior root arises from the side of the body, and corresponds to the ribs: the posterior root springs from the junction of the pedicle with the lamina, and corresponds with the transverse processes in the dorsal region. It is by the junction of the two that the foramen for the vertebral vessels is formed. The extremities of each of these roots form the *anterior* and *posterior tubercles* of the transverse processes. The *articular processes* are oblique: the superior are of an oval form, flattened and directed upwards and backwards; the inferior downwards and forwards.

The peculiar vertebræ in the cervical region are the first or *Atlas*; the second or *Axis*; and the seventh or *Vertebra prominens.* The great modifications in the form of the atlas and axis are designed to admit of the nodding and rotatory movements of the head.

The *Atlas* (fig. 2) (so named from supporting the globe of the head). The chief

2.—1st Cervical Vertebra, or Atlas.

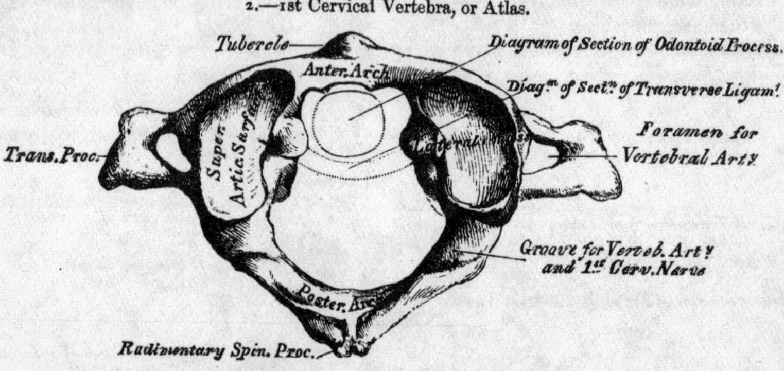

peculiarities of this bone are, that it has neither body nor spinous process. The body is detached from the rest of the bone, and forms the odontoid process of the second vertebra; while the parts corresponding to the pedicles pass in front, and join to form the anterior arch. The atlas consists of an anterior arch, a posterior arch, and two lateral masses. The *anterior* arch forms about one-fifth of the bone; its anterior surface is convex, and presents about its centre a tubercle, for the attachment of the Longus colli muscle; posteriorly it is concave, and marked by a smooth oval or circular facet, for articulation with the odontoid process of the axis. The *posterior* arch forms about two-fifths of the circumference of the bone; it terminates behind in a tubercle, which is the rudiment of a spinous process, and gives origin to the Rectus capitis posticus minor. The diminutive size of this process prevents any interference in the movements between it and the cranium. The posterior part of the arch presents, above, a rounded edge; whilst, in front, immediately behind each superior articular process, is a groove, sometimes converted into a foramen by a delicate bony spicula which arches backwards from the posterior extremity of the superior articular process. These grooves represent the superior intervertebral notches, and are peculiar from being situated behind the articular processes, instead of before them, as in the other vertebræ. They serve for the transmission of the vertebral artery, which, ascending through the foramen in the transverse process, winds round the lateral mass in a direction backwards and inwards. They also transmit the sub-occipital nerves. On the under surface of the posterior arch, in the same situation, are two other grooves, placed behind the lateral masses, and representing the inferior intervertebral notches of other vertebræ. They are much less marked than the superior. The *lateral masses* are the most bulky and solid parts of the atlas, in order to support the weight of the head; they present two articulating processes above, and two below. The two superior are of large size, oval, concave, and approach towards one another in front, but diverge behind; they are directed upwards, inwards, and a little backwards, forming a kind of cup for the condyles of the occipital bone, and are admirably adapted to the nodding movements of the head. Not unfrequently they are partially subdivided by a more or less deep indentation which encroaches upon each lateral margin. The inferior articular processes are circular in form, flattened, or slightly concave, and directed downwards, inwards, and a little backwards, articulating with the axis, and permitting the rotatory movements. Just below the inner margin of each superior articular surface is a small tubercle, for the attachment of a ligament which, stretching across the ring of the atlas, divides it into two unequal parts; the anterior or smaller segment receiving the odontoid process of the axis, the posterior allowing the transmission of the spinal cord and its membranes. This part of the spinal canal is of considerable size, to afford space for the spinal cord; and hence lateral displacement of the atlas may occur without compression of the spinal cord. The transverse processes are of large size, for the attachment of special muscles which assist in rotating the head—long, not bifid, perforated at their base by a canal for the vertebral artery, which is directed from below, upwards and backwards.

The *Axis* (fig. 3), (so named from forming the pivot upon which the head

3.—2nd Cervical Vertebra, or Axis.

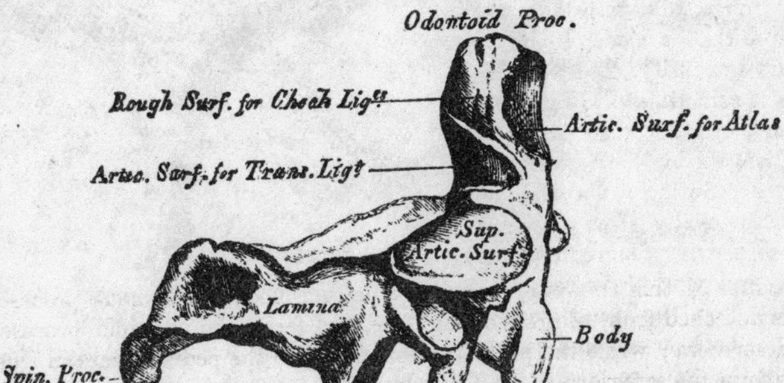

rotates). The most distinctive character of this bone is the strong prominent process, tooth-like in form (hence the name odontoid), which rises perpendicularly from the upper part of the body. The body is of a triangular form; deeper in front than behind, and prolonged downwards anteriorly so as to overlap the upper and fore part of the adjacent vertebra. It presents in front a median longitudinal ridge, separating two lateral depressions for the attachment of the Longus colli muscles of each side. The odontoid process presents two articulating surfaces: one in front of an oval form, for articulation with the atlas; another behind, for the transverse ligament; the latter frequently encroaching on the sides of the process; the apex is pointed. Below the apex, the process is somewhat enlarged, and presents on either side a rough impression for the attachment of the odontoid or check ligaments, which connect it to the occipital bone; the base of the process, where it is attached to the body, is constricted, so as to prevent displacement from the transverse ligament, which binds it in this situation to the anterior arch of the atlas. Sometimes, however, this process does become displaced, especially in children, in whom the ligaments are more relaxed: instant death is the result of this accident. The pedicles are broad and strong, especially their anterior extremities, which coalesce with the sides of the body and the root of the odontoid process. The laminæ are thick and strong, and the spinal foramen very large. The superior articular surfaces are round, slightly convex, directed upwards and outwards, and are peculiar in being supported on the body, pedicles, and transverse processes. The inferior articular surfaces have the same direction as those of the other cervical vertebræ. The superior intervertebral notches are very shallow, and lie behind the articular processes; the inferior in front of them, as in the other cervical vertebræ. The transverse processes are very small, not bifid, and perforated by the vertebral foramen, or foramen for the vertebral artery, which is directed obliquely upwards and outwards. The spinous process is of large size, very strong, deeply channeled on its under surface, and presents a bifid tubercular extremity for the attachment of muscles, which serve to rotate the head upon the spine.

Seventh Cervical (fig. 4). The most distinctive character of this vertebra is the existence of a very long and prominent spinous process; hence the name 'Vertebra prominens.' This process is thick, nearly horizontal in direction, not bifurcated, and has attached to it the ligamentum nuchæ. The transverse process is usually of large size, especially its posterior root; its upper surface has usually a shallow groove, and it seldom presents more than a trace of bifurcation at its extremity. The vertebral foramen is sometimes as large as in the other cervical vertebræ, usually smaller, on one or both sides, and sometimes wanting. On the left side it occasionally gives passage to the vertebral artery; more frequently the vertebral vein traverses it on both sides; but the usual arrangement is for both artery and vein to pass through the foramen in the transverse process of the sixth cervical.

4.—7th Cervical Vertebra, or Vertebra Prominens.

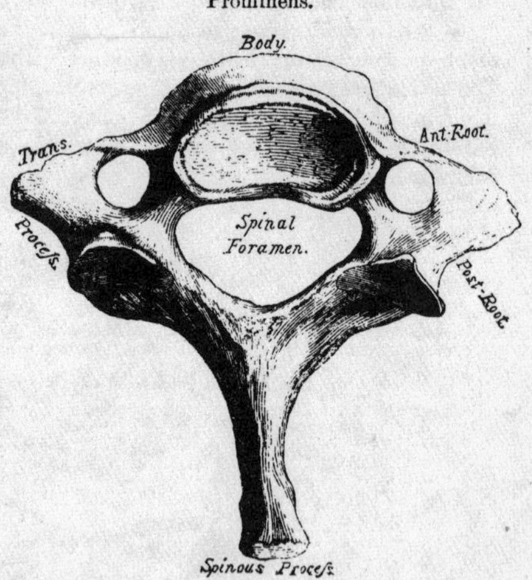

Characters of the Dorsal Vertebræ.

The bodies of the dorsal vertebræ resemble those in the cervical and lumbar regions at the respective ends of this portion of the spine; but in the middle of the dorsal region, their form is very characteristic, being heart-shaped, and broader in the antero-posterior than in the lateral direction. They are thicker behind than in front, flat above and below, convex and prominent in front, deeply concave behind, slightly constricted in front and at the sides, and marked on each side, near the root of the pedicle, by two demi-facets, one above, the other below. These are covered with cartilage in the recent state; and, when articulated with the adjoining vertebræ, form oval surfaces for the reception of the heads of the corresponding ribs. The pedicles are directed backwards, and the inferior intervertebral notches are of large size, and deeper than in any other region of the spine. The laminæ are broad and thick, and the spinal foramen small, and of a circular form. The articular processes are flat, nearly vertical in direction, and project from the upper and lower part of the pedicles, the superior being directed backwards and a little outwards and upwards, the inferior forwards and a little inwards and downwards. The transverse processes arise from the same parts of the arch as the posterior roots of the transverse processes in the neck; they are thick, strong, and of great length, directed obliquely backwards and outwards, presenting a clubbed extremity, which is tipped on its anterior part by a small concave surface, for articulation with the tubercle of a rib. Besides the articular facet for the rib, two indistinct tubercles may be seen rising from the extremity of the transverse processes, one near the upper, the other near the lower border. In man, they are comparatively of small size, and serve only for the attachment of muscles. But, in some animals, they attain considerable magnitude either for the purpose of more closely connecting the segments of this portion of the spine, or for muscular and ligamentous attachment. The spinous processes are long, triangular in form, directed obliquely downwards, and terminating by a tubercular margin. They overlap one another from the fifth to the eighth, but are less oblique in direction above and below.

5.—A Dorsal Vertebra.

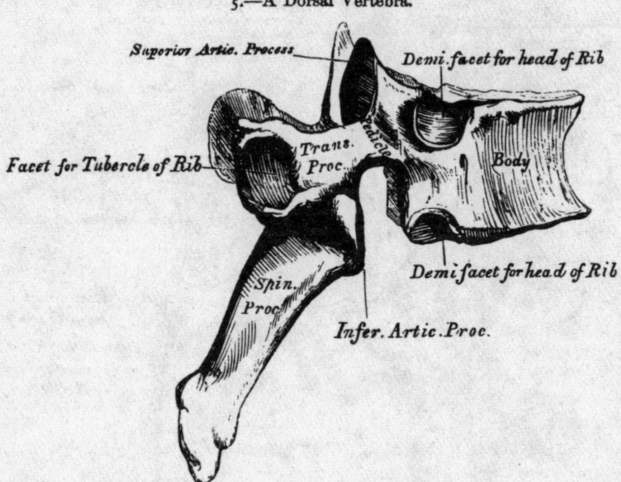

The peculiar dorsal vertebræ are the *first, ninth, tenth, eleventh, and twelfth* (fig. 6).

The *First Dorsal Vertebra* presents, on each side of the body, a single entire articular facet for the head of the first rib, and a half facet for the upper half of the second. The upper surface of the body is like that of a cervical vertebra, being broad transversely, concave, and lipped on each side. The *articular surfaces* are oblique and the *spinous process* thick, long, and almost horizontal.

The *Ninth Dorsal* has no demi-facet below. In some subjects, however, the ninth has two demi-facets on each side, then the tenth has a demi-facet at the upper part; none below.

The *Tenth Dorsal* has (except in the cases just mentioned) an entire articular facet on each side above; it has no demi-facet below.

In the *Eleventh Dorsal*, the body approaches in its form and size to the lumbar. The articular facets for the heads of the ribs, one on each side, are of large size,

6.—Peculiar Dorsal Vertebræ.

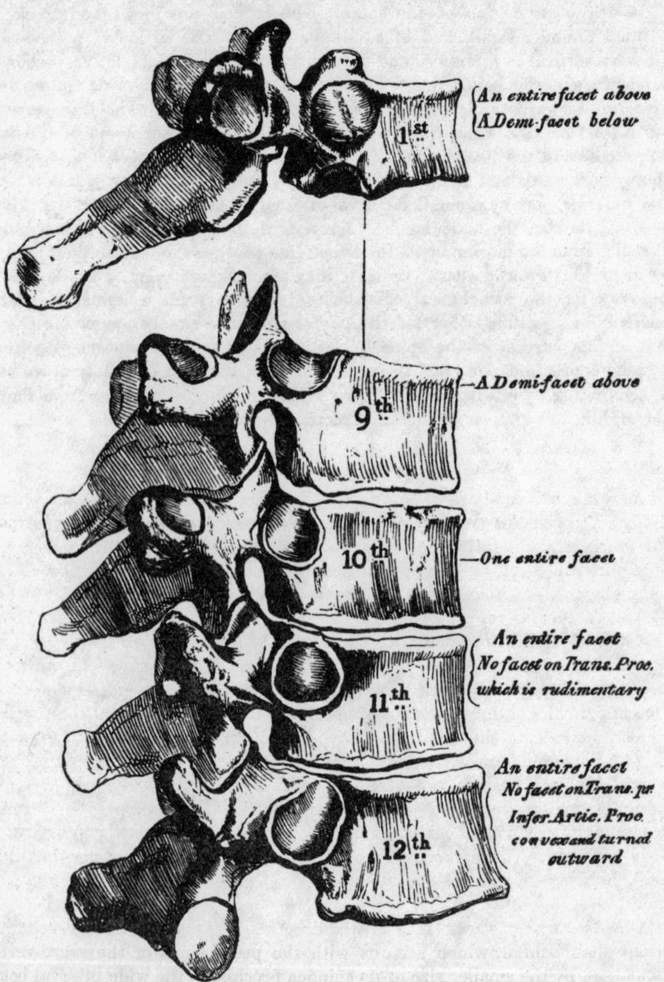

and placed chiefly on the pedicles, which are thicker and stronger in this and the next vertebra, than in any other part of the dorsal region. The *transverse* processes are very short, tubercular at their extremities, and have no articular facets for the tubercles of the ribs. The spinous process is short, nearly horizontal in direction, and presents a slight tendency to bifurcation at its extremity.

The *Twelfth Dorsal* has the same general characters as the eleventh; but may be distinguished from it by the inferior articular processes being convex and turned outwards, like those of the lumbar vertebræ; by the general form of the body, laminæ, and spinous process, approaching to that of the lumbar vertebræ; and by the transverse processes being shorter, and the tubercles at their extremities more marked.

Characters of the Lumbar Vertebræ.

The Lumbar Vertebræ (fig. 7) are the largest segments of the vertebral column. The body is large, broader from side to side than from before backwards, about equal in depth in front and behind, flattened or slightly concave above and below, concave behind, and deeply constricted in front and at the sides, presenting promi-

7.—Lumbar Vertebra.

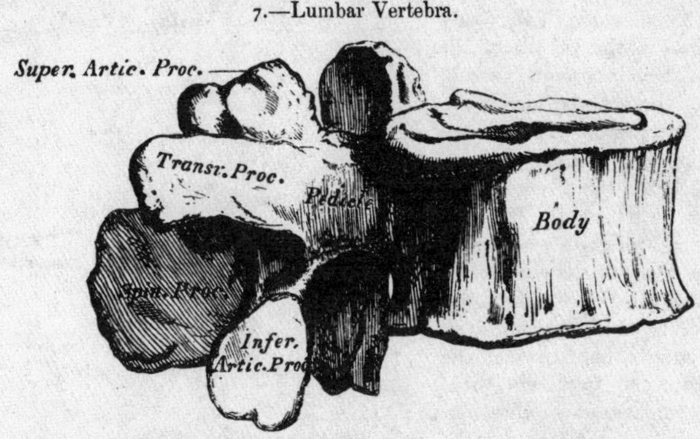

nent margins, which afford a broad basis for the support of the superincumbent weight. The pedicles are very strong, directed backwards from the upper part of the bodies; consequently the inferior intervertebral notches are of large size. The laminæ are short, but broad and strong; and the foramen triangular, larger than in the dorsal, smaller than in the cervical region. The superior articular processes are concave, and look almost directly inwards; the inferior, convex, look outwards and a little forwards; the former are separated by a much wider interval than the latter, embracing the lower articulating processes of the vertebra above. The transverse processes are long, slender, directed transversely outwards in the upper three lumbar vertebræ, slanting a little upwards in the lower two. By some anatomists they are considered homologous with the ribs. Of the two tubercles noticed in connection with the transverse processes in the dorsal region, the superior ones become connected in this region with the back part of the superior articular processes. Although in man they are comparatively small, in some animals they attain considerable size, and serve to lock the vertebræ more closely together. The spinous processes are thick and broad, somewhat quadrilateral, horizontal in direction, thicker below than above, and terminating by a rough uneven border.

The *Fifth Lumbar* vertebra is characterised by having the body much thicker in front than behind, which accords with the prominence of the sacro-vertebral articulation, by the smaller size of its spinous process, by the wide interval between the inferior articulating processes, and by the greater size and thickness of its transverse processes.

Structure of the Vertebræ. The structure of a vertebra differs in different parts. The body is composed of light spongy cancellous tissue, having a thin coating of compact tissue on its external surface perforated by numerous orifices, some of large size, for the passage of vessels; its interior is traversed by one or two large canals for the reception of veins, which converge towards a single large irregular or several small apertures at the posterior part of the body of each bone. The arch and processes projecting from it have, on the contrary, an exceedingly thick covering of compact tissue.

Development. Each vertebra is formed of three primary cartilaginous portions (fig. 8); one for each lamina and its processes, and one for the body. Ossification commences in the laminæ about the sixth week of fœtal life, in the situation where the transverse processes afterwards project, the ossific granules shooting backwards to the spine, forwards to the body, and outwards into the transverse and articular processes. Ossification in the body commences in the middle of the cartilage about the eighth week. At birth these three pieces are perfectly separate. During the first year the laminæ become united behind, by a portion of cartilage in which the spinous process is ultimately formed, and thus the arch is completed. About the third year the body is joined to the arch on each side, in such a manner that the body is formed from the three original centres of ossification, the amount contributed by the pedicles increasing in extent from below upwards. Thus the bodies of the sacral vertebræ are formed almost entirely from the central nuclei, the bodies of the lumbar segments are formed laterally and behind by the pedicles; in the dorsal region, the pedicles advance as far forwards as the articular depressions for the heads of the ribs, forming these cavities of reception; and in the neck the whole of the lateral portions of the bodies are formed by the advance of the pedicles. Before puberty, no other changes occur, excepting a gradual increase in the growth of these primary centres, the upper and under surface of the bodies, and the ends of the transverse and spinous processes, being tipped with cartilage, in which ossific granules are not as yet deposited. At sixteen years (fig. 9), four secondary centres appear, one for the tip of each transverse process, and two (sometimes

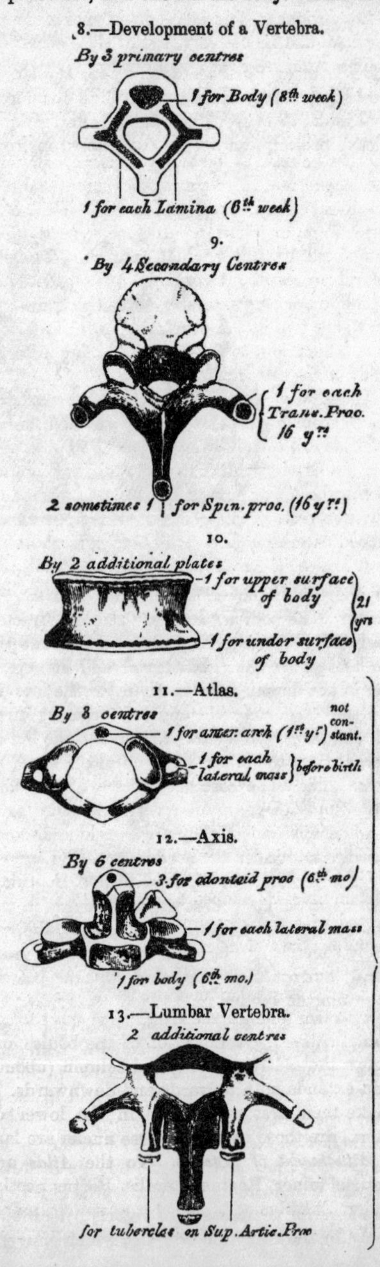

8.—Development of a Vertebra.

9.

10.

11.—Atlas.

12.—Axis.

13.—Lumbar Vertebra.

united into one) for the end of the spinous process. At twenty-one years (fig. 10), a thin circular plate of bone is formed in the layer of cartilage situated on the upper and under surface of the body, the former being the thicker of the two. All these become joined; and the bone is completely formed about the thirtieth year of life.

Exceptions to this mode of development occur in the first, second, and seventh cervical, and in the vertebræ of the lumbar region.

The *Atlas* (fig. 11) is developed by *two* primary centres, and by *one* or more epiphyses. The two primary centres are destined for the two lateral or neural masses, the ossification of which commences before birth, near the articular processes, and extends backwards: these portions of bone are separated from one another behind, at birth, by a narrow interval filled in with cartilage. Between the second and third years, they unite either directly or through the medium of an epiphysal centre, developed in the cartilage near their point of junction. The anterior arch, at birth, is altogether cartilaginous, and this portion of the atlas is completed by the gradual extension forwards and ultimate junction of the two neural processes. Occasionally, a separate nucleus is developed in the anterior arch, which, extending laterally, joins the neural processes in front of the pedicles; or, there are two nuclei developed in the anterior arch, one on either side of the median line, which join to form a single mass, afterwards united to the lateral portions in front of the articulating processes.

The *Axis* (fig. 12) is developed by *six* centres. The body and arch of this bone are formed in the same manner as the corresponding parts in the other vertebræ: one centre for the lower part of the body, and one for each lamina. The odontoid process consists originally of an extension upwards of the cartilaginous mass, in which the lower part of the body is formed. At about the sixth month of fœtal life, two osseous nuclei make their appearance in the base of this process: they are placed laterally, and join before birth to form a conical bi-lobed mass deeply cleft above; the interval between the cleft and the summit of the process, is formed by a wedge-shaped piece of cartilage; the base of the process being separated from the body by a cartilaginous interval, which gradually becomes ossified, sometimes by a separate epiphysal nucleus. Finally, as Dr. Humphry has demonstrated, the apex of the odontoid process has a separate nucleus.

The Seventh Cervical. The anterior or costal part of the transverse process of the seventh cervical, is developed from a separate osseous centre at about the sixth month of fœtal life, and joins the body and posterior division of the transverse process between the fifth and sixth years. Sometimes this process continues as a separate piece, and, becoming lengthened outwards, constitutes what is known as a cervical rib.

The Lumbar Vertebræ (fig. 13) have *two additional centres* (besides those peculiar to the vertebræ generally), for the tubercles, which project from the back part of the superior articular processes. The transverse process of the first lumbar is sometimes developed as a separate piece, which may remain permanently unconnected with the remaining portion of the bone; thus forming a lumbar rib, a peculiarity which is rarely met with.

Progress of Ossification in the Spine generally. Ossification of the laminæ of the vertebræ commences at the upper part of the spine, and proceeds gradually downwards; hence the frequent occurrence of spina bifida in the lower part of the spinal column. Ossification of the bodies, on the other hand, commences a little below the centre of the spinal column (about the ninth or tenth dorsal vertebræ), and extends both upwards and downwards. Although, however, the ossific nuclei make their first appearance in the lower dorsal vertebræ, the lumbar and first sacral are those in which these nuclei are largest at birth.

Attachment of Muscles. To the *Atlas* are attached the Longus colli, Rectus anticus minor, Rectus lateralis, Rectus posticus minor, Obliquus superior and inferior, Splenius colli, Levator anguli scapulæ, Interspinous, and Intertransverse.

To the *Axis* are attached the Longus colli, Obliquus inferior, Rectus posticus major, Semi-spinalis colli, Multifidus spinæ, Levator anguli scapulæ, Splenius colli, Transversalis colli, Scalenus posticus, Intertransversales, Interspinales.

To the remaining Vertebræ generally are attached, *anteriorly*, the Rectus anticus major, Longus colli, Scalenus anticus and posticus, Psoas magnus, Psoas parvus, Quadratus lumborum, Diaphragm, Obliquus internus and transversalis,—*posteriorly*, the Trapezius, Latissimus dorsi, Levator anguli scapulæ, Rhomboideus major and minor, Serratus posticus superior and inferior, Splenius, Sacro-lumbalis, Longissimus dorsi, Spinalis dorsi, Cervicalis ascendens, Transversalis collis, Trachelo-mastoid, Complexus, Semi-Spinalis dorsi and colli, Multifidus spinæ, Interspinales, Supraspinales, Intertransversales, Levatores costarum.

Sacral and Coccygeal Vertebræ.

The Sacral and Coccygeal Vertebræ consist, at an early period of life, of nine separate pieces, which are united in the adult, so as to form two bones, five entering into the formation of the sacrum, four into that of the coccyx. Occasionally, the coccyx consists of five bones.*

The Sacrum (fig. 14) is a large triangular bone, situated at the lower part

14.—Sacrum, Anterior Surface.

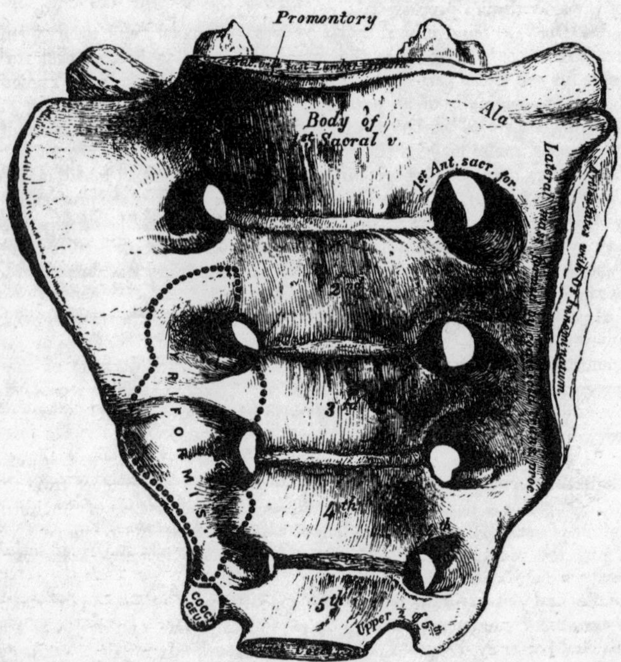

of the vertebral column, and at the upper and back part of the pelvic cavity, where it is inserted like a wedge between the two ossa innominata; its upper part, or base, articulating with the last lumbar vertebra, its apex with the coccyx. The sacrum is curved upon itself, and placed very obliquely, its upper extremity projecting forwards, and forming, with the last lumbar vertebra, a very prominent

* Dr. Humphry describes this as the usual composition of the Coccyx. — *On the Skeleton*, p. 456.

angle, called the *promontory* or *sacro-vertebral angle*, whilst its central part is directed backwards, so as to give increased capacity to the pelvic cavity. It presents for examination an anterior and posterior surface, two lateral surfaces, a base, an apex, and a central canal.

The Anterior Surface is concave from above downwards, and slightly so from side to side. In the middle are seen four transverse ridges, indicating the original division of the bone into five separate pieces. The portions of bone intervening between the ridges correspond to the bodies of the vertebræ. The body of the first segment is of large size, and in form resembles that of a lumbar vertebra; the succeeding ones diminish in size from above downwards, are flattened from before backwards, and curved so as to accommodate themselves to the form of the sacrum, being concave in front, convex behind. At each end of the ridges abovementioned, are seen the *anterior sacral foramina*, analogous to the intervertebral foramina, four in number on each side, somewhat rounded in form, diminishing in size from above downwards, and directed outwards and forwards; they transmit the anterior branches of the sacral nerves. External to these foramina is the *lateral mass*, consisting, at an early period of life, of separate segments, which correspond to the anterior transverse processes: these become blended, in the adult, with the bodies, with each other, and with the posterior transverse processes. Each lateral mass is traversed by four broad shallow grooves, which lodge the anterior sacral nerves as they pass outwards, the grooves being separated by prominent ridges of bone, which give attachment to the slips of the Pyriformis muscle.

15.—Vertical Section of the Sacrum.

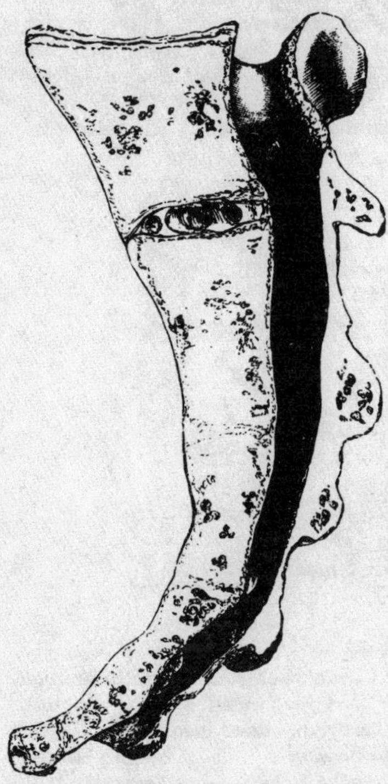

If a vertical section is made through the centre of the bone (fig. 15), the bodies are seen to be united at their circumference by bone, a wide interval being left centrally, which, in the recent state, is filled by intervertebral substance. In some bones, this union is more complete between the lower segments, than between the upper ones.

The *Posterior Surface* (fig. 16) is convex and much narrower than the anterior. In the middle line, are three or four tubercles, which represent the rudimentary spinous processes of the sacral vertebræ. Of these tubercles, the first is usually prominent, and perfectly distinct from the rest; the second and third, are either separate, or united into a tubercular ridge, which diminishes in size from above downwards; the fourth usually, and the fifth always, remaining undeveloped. External to the spinous processes on each side, are the *laminæ*, broad and well marked in the three first pieces; sometimes the fourth, and generally the fifth, being undeveloped; in this situation the lower end of the sacral canal is exposed. External to the laminæ are a linear series of indistinct tubercles representing the *articular processes*; the upper pair are large, well developed, and correspond in shape and direction to the superior articulating processes of a lumbar vertebra; the second and third are small; the fourth and fifth (usually blended together) are situated on each side of the sacral canal: they are called the *sacral cornua*, and articulate with the cornua of the coccyx. External to the articular processes are the four *posterior sacral foramina*: they are smaller in size, and less regular in form than the anterior, and transmit the posterior branches of the sacral nerves. On the outer side of the posterior sacral foramina are a series of tubercles, the rudimentary

16.—Sacrum, Posterior Surface.

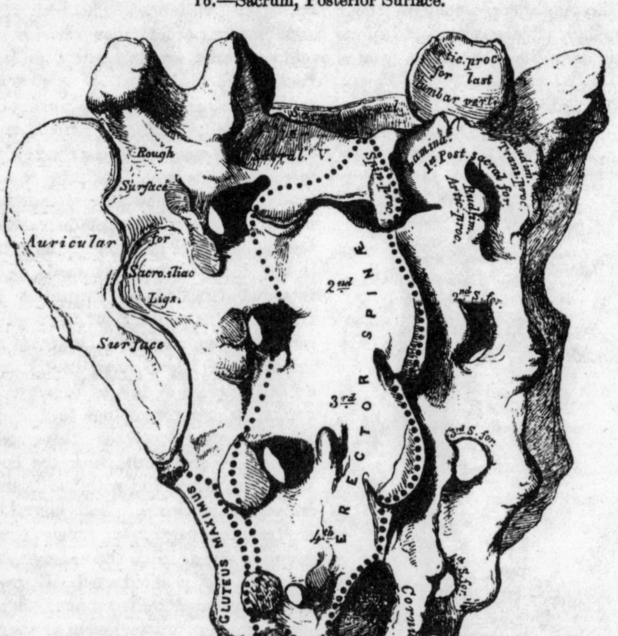

posterior *transverse processes* of the sacral vertebræ. The first pair of transverse tubercles are of large size, very distinct, and correspond with each superior angle of the bone; the second, small in size, enter into the formation of the sacro-iliac articulation; the third give attachment to the oblique sacro-iliac ligaments; and the fourth and fifth to the great sacro-ischiatic ligaments. The interspace between the spinous and transverse processes on the back of the sacrum, presents a wide shallow concavity, called the *sacral groove*; it is continuous above with the vertebral groove, and lodges the origin of the Erector spinæ.

The *Lateral Surface*, broad above, becomes narrowed into a thin edge below. Its upper half presents in front a broad ear-shaped surface for articulation with the ilium. This is called the *auricular* surface, and in the fresh state is coated with cartilage. It is bounded posteriorly by deep and uneven impressions, for the attachment of the posterior sacro-iliac ligaments. The lower half is thin and sharp, and gives attachment to the greater and lesser sacro-ischiatic ligaments, and to some fibres of the Gluteus maximus; below, it presents a deep notch, which is converted into a foramen by articulation with the transverse process of the upper piece of the coccyx, and transmits the anterior branch of the fifth sacral nerve.

The *Base* of the sacrum, which is broad and expanded, is directed upwards and forwards. In the middle is seen an oval articular surface, which corresponds with the under surface of the body of the last lumbar vertebra, bounded behind by the large triangular orifice of the sacral canal. This orifice is formed behind by the spinous process and laminæ of the first sacral vertebra, whilst projecting from it on each side are the superior articular processes; they are oval, concave, directed backwards and inwards, like the superior articular processes of a lumbar vertebra; and in front of each articular process is an intervertebral notch, which forms the lower half of the last intervertebral foramen. Lastly, on each side of the articular surface is a broad and flat triangular surface of bone, which extends outwards, and is continuous on each side with the iliac fossa.

The *Apex*, directed downwards and forwards, presents a small oval concave surface for articulation with the coccyx.

The *Sacral Canal* runs throughout the greater part of the bone; it is large and triangular in form above, small and flattened from before backwards below. In this situation, its posterior wall is incomplete, from the non-development of the laminæ and spinous processes. It lodges the sacral nerves, and is perforated by the anterior and posterior sacral foramina, through which these pass out.

17.—Development of Sacrum.

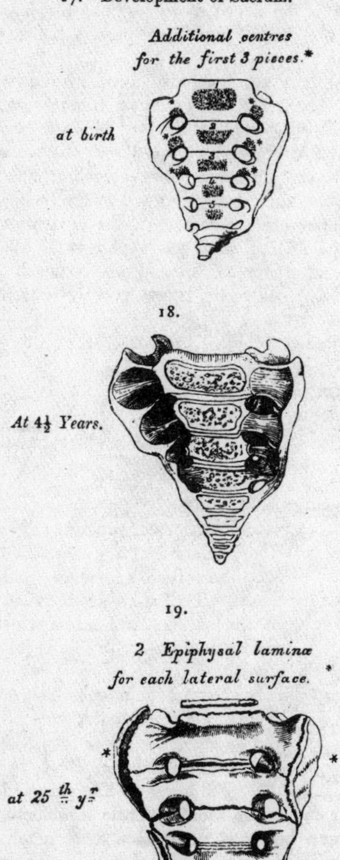

Structure. It consists of much loose spongy tissue within, invested externally by a thin layer of compact tissue.

Differences in the Sacrum of the Male and Female. The sacrum in the female is usually wider than in the male; and it is much less curved, the upper half of the bone being nearly straight, the lower half presenting the greatest amount of curvature. The bone is also directed more obliquely backwards; which increases the size of the pelvic cavity, and forms a more prominent sacro-vertebral angle. In the male, the curvature is more evenly distributed over the whole length of the bone, and is altogether greater than in the female.

Peculiarities of the Sacrum. This bone, in some cases, consists of six pieces; occasionally the number is reduced to four. Sometimes the bodies of the first and second segments are not joined, or the laminæ and spinous processes have not coalesced. Occasionally, the upper pair of transverse tubercles are not joined to the rest of the bone on one or both sides; and lastly, the sacral canal may be open for nearly the lower half of the bone, in consequence of the imperfect development of the laminæ and spinous processes. The sacrum, also, varies considerably with respect to its degree of curvature. From the examination of a large number of skeletons, it would appear, that, in one set of cases, the anterior surface of this bone was nearly straight, the curvature, which was very slight, affecting only its lower end. In another set of cases, the bone was curved throughout its whole length, but especially towards its middle. In a third set the degree of curvature was less marked, and affected especially the lower third of the bone.

Development (fig. 17). The sacrum, formed by the union of five vertebræ, has *thirty-five* centres of ossification.

The *bodies* of the sacral vertebræ have each three ossific centres; one for the central part, and one for the epiphysal plates on its upper and under surface.

The *laminæ* of the sacral vertebræ are each developed by two centres; these meet behind to form the arch, and subsequently join the body.

The *lateral masses* have six additional centres, two for each of the first three vertebræ. These centres make their appearance above and to the outer side of the anterior sacral foramina (fig. 17), and are developed into separate segments, which correspond with the anterior transverse processes (fig. 18); they are subsequently blended with each other, and with the bodies and the posterior transverse processes, to form the lateral mass.

Lastly, each *lateral surface* of the sacrum is developed by two epiphysal plates (fig. 19); one for the auricular surface, and one for the remaining part of the thin lateral edge of the bone.

Period of Development. At about the eighth or ninth week of fœtal life, ossification of the central part of the bodies of the first three vertebræ commences; and, at a somewhat later period, that of the last two. Between the sixth and eighth months ossification of the laminæ takes place; and, at about the same period, the characteristic osseous tubercles for the three first sacral vertebræ make their appearance. The laminæ join to form the arch, and are united to the bodies, first, in the lowest vertebræ. This occurs about the second year, the uppermost segment appearing as a single piece about the fifth or sixth year. About the sixteenth year the epiphyses for the upper and under surfaces of the bodies are formed; and, between the eighteenth and twentieth years, those for each lateral surface of the sacrum make their appearance. At about this period the last two segments are joined to one another; and this process gradually extending upwards, all the pieces become united, and the bone completely formed from the twenty-fifth to the thirtieth year of life.

Articulations. With four bones: the last lumbar vertebra, coccyx, and the two ossa innominata.

Attachment of Muscles. In front, the Pyriformis and Coccygeus; behind, the Gluteus maximus and Erector spinæ.

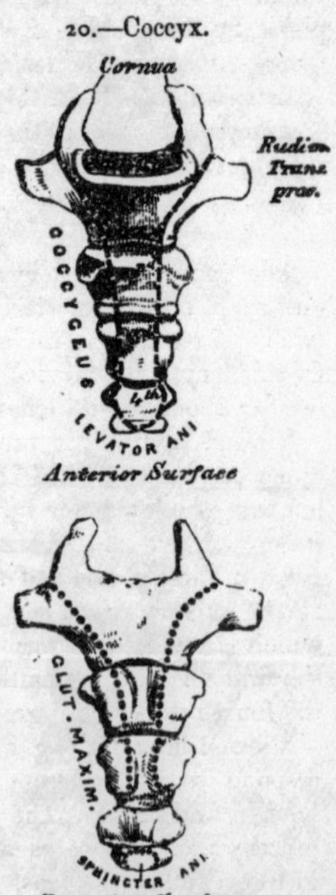

20.—Coccyx.

The Coccyx.

The Coccyx (κόκκυξ, *cuckoo*), so called from having been compared to a cuckoo's beak (fig. 20), is usually formed of four small segments of bone, the most rudimentary parts of the vertebral column. In each of the first three segments may be traced a rudimentary body, articular and transverse processes; the last piece (sometimes the third) is a mere nodule of bone, without distinct processes. All the segments are destitute of laminæ and spinous processes; and, consequently, of spinal canal and intervertebral foramina. The first segment is the largest; it resembles the lowermost sacral vertebra, and often exists as a separate piece; the last three, diminishing in size from above downwards, are usually blended together so as to form a single bone. The gradual diminution in the size of the pieces gives this bone a triangular form, the base of the triangle joining the end of the sacrum. It presents for examination an anterior and posterior surface, two borders, a base, and an apex. The *anterior surface* is slightly concave, and marked with three transverse grooves, indicating the points of junction of the different pieces. It has attached to it the anterior sacro-coccygeal ligament and Levator ani muscle, and supports the lower end of the rectum. The *posterior surface* is convex, marked by transverse grooves similar to those on the anterior surface; and presents on each side a lineal row of tubercles, the rudimentary articular processes of the coccygeal vertebræ. Of these, the superior pair are very large; and are called the *cornua of the coccyx*; they project upwards, and articulate with the cornua of the sacrum, the junction between these two bones completing the fifth sacral foramen for the transmission of the posterior branch of the fifth sacral nerve. The *lateral borders* are thin, and present a series of small eminences, which represent the transverse processes of the coccygeal vertebræ. Of these, the first on each side is of large size, flattened from before backwards; and often ascends to join the lower part of the thin lateral edge of the sacrum, thus completing the fifth sacral foramen: the others diminish in size from above downwards, and are often wanting. The borders of the coccyx are narrow, and give attachment on each side to the sacro-sciatic ligaments and Coccygeus muscle. The *base* presents an oval surface for articulation with the sacrum. The *apex* is rounded, and has attached to it the tendon of the external Sphincter muscle. It is occasionally bifid, and sometimes deflected to one or other side.

Development. The coccyx is developed by *four* centres, one for each piece. Occasionally, one of the first three pieces of this bone is developed by two centres, placed side by side. The ossific nuclei make their appearance in the following order: in the first segment, at birth; in the second piece, at from five to ten years; in the third, from ten to fifteen years; in the fourth, from fifteen to twenty years. As age advances, these various segments become united in the following order: the first two pieces join; then the third and fourth; and, lastly, the bone is completed by the union of the second and third. At a late period of life, especially in females, the coccyx often becomes joined to the end of the sacrum.

Articulation. With the sacrum.

Attachment of Muscles. On either side, the Coccygeus; behind, the Gluteus maximus; at the apex, the Sphincter ani; and in front, the Levator ani.

Of the Spine in general.

The spinal column, formed by the junction of the vertebræ, is situated in the median line, at the posterior part of the trunk: its average length is about two feet two or three inches, measured along the curved anterior surface of the column. Of this length the cervical part measures about five, the dorsal about eleven, the lumbar about seven inches, and the sacrum and coccyx the remainder.

Viewed in front, it presents two pyramids joined together at their bases, the upper one being formed by all the vertebræ from the second cervical to the last lumbar; the lower one by the sacrum and coccyx. When examined more closely, the upper pyramid is seen to be formed of three smaller pyramids. The uppermost of these consists of the six lower cervical vertebræ; its apex being formed by the axis or second cervical; its base, by the first dorsal. The second pyramid, which is inverted, is formed by the four upper dorsal vertebræ, the base being at the first dorsal, the smaller end at the fourth. The third pyramid commences at the fourth dorsal, and gradually increases in size to the fifth lumbar.

Viewed laterally (fig. 21), the spinal column presents several curves, which correspond to the different regions of the column, and are called *cervical*, *dorsal*, *lumbar*, and *pelvic*. The *cervical* curve commences at the apex of the odontoid process, and terminates at the middle of the second dorsal vertebra; it is convex in front, and is the least marked of all the curves. The *dorsal* curve, which is concave forwards, commences at the middle of the second, and terminates at the middle of the twelfth dorsal. Its most prominent point behind corresponds to the body of the seventh or eighth vertebra. The *lumbar* curve commences at the middle of the last dorsal vertebra, and terminates at the sacro-vertebral angle. It is convex anteriorly; the convexity of the lower three vertebræ being much greater than that of the upper ones. The *pelvic* curve commences at the sacro-vertebral articulation, and terminates at the point of the coccyx. It is concave anteriorly. These curves are partly due to the shape of the bodies of the vertebræ, and partly to the intervertebral substances, as will be explained in the *Articulations of the Spine.*

21.—Lateral View of the Spine.

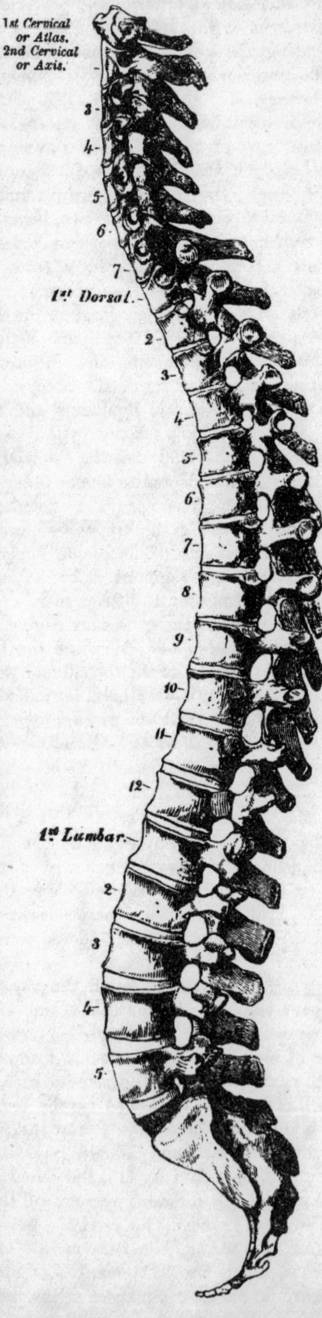

The spine has also a slight lateral curvature, the convexity of which is directed toward the right side. This is most probably produced, as Bichat first explained, chiefly by muscular action; most persons using the right arm in preference to the left, especially in making long-continued efforts, when the body is curved to the right side. In support of this explanation, it has been found, by Béclard, that in one or two individuals who were left-handed, the lateral curvature was directed to the left side.

The spinal column presents for examination an anterior, a posterior, and two lateral surfaces; a base, summit, and vertebral canal.

The *anterior surface* presents the bodies of the vertebræ separated in the recent state by the intervertebral discs. The bodies are broad in the cervical region, narrow in the upper part of the dorsal, and broadest in the lumbar region. The whole of this surface is convex transversely, concave from above downwards in the dorsal region, and convex in the same direction in the cervical and lumbar regions.

The *posterior surface* presents in the median line the spinous processes. These are short, horizontal, with bifid extremities in the cervical region. In the dorsal region, they are directed obliquely above, assume almost a vertical direction in the middle, and are horizontal below, as are also the spines of the lumbar vertebræ. They are separated by considerable intervals in the loins, by narrower intervals in the neck, and are closely approximated in the middle of the dorsal region. Occasionally one of these processes deviates a little from the median line, a fact to be remembered in practice, as irregularities of this sort are attendant also on fractures or displacements of the spine. On either side of the spinous processes, extending the whole length of the column, is the vertebral groove, formed by the laminæ in the cervical and lumbar regions, where it is shallow, and by the laminæ and transverse processes in the dorsal region, where it is deep and broad. In the recent state, these grooves lodge the deep muscles of the back. External to the vertebral grooves are the articular processes, and still more externally the transverse processes. In the dorsal region, the latter processes stand backwards, on a plane considerably posterior to the same processes in the cervical and lumbar regions. In the cervical region, the transverse processes are placed in front of the articular processes, and between the intervertebral foramina. In the lumbar, they are placed also in front of the articular processes, but behind the intervertebral foramina. In the dorsal region, they are posterior both to the articular processes and foramina.

The *lateral surfaces* are separated from the posterior by the articular processes in the cervical and lumbar regions, and by the transverse processes in the dorsal. These surfaces present in front the sides of the bodies of the vertebræ, marked in the dorsal region by the facets for articulation with the heads of the ribs. More posteriorly are the intervertebral foramina, formed by the juxtaposition of the intervertebral notches, oval in shape, smallest in the cervical and upper part of the dorsal regions, and gradually increasing in size to the last lumbar. They are situated between the transverse processes in the neck, and in front of them in the back and loins, and transmit the spinal nerves. The *base* of the vertical column is formed by the under surface of the body of the fifth lumbar vertebra; and the *summit* by the upper surface of the atlas. The *vertebral canal* follows the different curves of the spine; it is largest in those regions in which the spine enjoys the greatest freedom of movement, as in the neck and loins, where it is wide and triangular; and narrow and rounded in the back, where motion is more limited.

THE SKULL.

The Skull, or superior expansion of the vertebral column, is composed of four vertebræ, the elementary parts of which are specially modified in form and size, and almost immoveably connected, for the reception of the brain, and special organs of the senses. These vertebræ are the occipital, parietal, frontal, and nasal. Descriptive anatomists, however, divide the skull into two parts, the Cranium and the Face. The Cranium (κράνος, *a helmet*), is composed of *eight* bones: viz., the *occipital, two parietal, frontal, two temporal, sphenoid, and ethmoid.* The face is composed of *fourteen* bones: viz., the *two nasal, two superior maxillary, two lachrymal, two malar, two palate, two inferior turbinated, vomer, and inferior maxillary.* The *ossicula auditûs*, the *teeth*, and *Wormian bones*, are not included in this enumeration.

Skull, 22 bones.	*Cranium, 8 bones.*	Occipital.
		Two Parietal.
		Frontal.
		Two Temporal.
		Sphenoid.
		Ethmoid.
	Face, 14 bones.	Two Nasal.
		Two Superior Maxillary.
		Two Lachrymal.
		Two Malar.
		Two Palate.
		Two Inferior Turbinated.
		Vomer.
		Inferior Maxillary.

The Occipital Bone.

The *Occipital Bone* (fig. 22) is situated at the back part and base of the cranium, is trapezoid in form, curved upon itself, and presents for examination two surfaces, four borders, and four angles.

The *External Surface* is convex. Midway between the summit of the bone and the posterior margin of the foramen magnum is a prominent tubercle, the external occipital protuberance, for the attachment of the Ligamentum nuchæ ; and descending from it as far as the foramen, a vertical ridge, the external occipital crest. This tubercle and crest vary in prominence in different skulls. Passing outwards from the occipital protuberance on each side are two semi-circular ridges, the superior curved lines; and running parallel with these from the middle of the crest, are the two inferior curved lines. The surface of the bone above the superior curved lines is smooth on each side, and, in the recent state, is covered by the Occipito-frontalis muscle, whilst the ridges, as well as the surface of the bone

22.—Occipital Bone. Outer Surface.

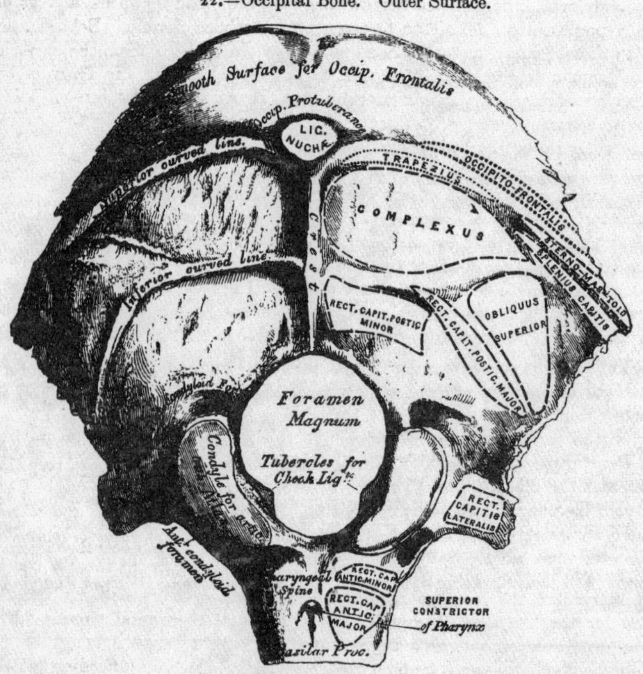

between them, serve for the attachment of numerous muscles. The superior curved line gives attachment internally to the Trapezius, externally to the Occipito-frontalis and Sterno-cleido-mastoid, to the extent shown in fig. 22; the depressions between the curved lines to the Complexus internally, the Splenius capitis and Obliquus capitis superior externally. The inferior curved line, and the depressions below it, afford insertion to the Rectus capitis posticus, major and minor.

The *foramen magnum* is a large oval aperture, its long diameter extending from before backwards. It transmits the spinal cord and its membranes, the spinal accessory nerves, and the vertebral arteries. Its back part is wide for the transmission of the cord, and the corresponding margin rough for the attachment of the dura mater enclosing the cord; the fore-part is narrower, being encroached upon by the condyles; it has projecting towards it from below the odontoid process, and its margins are smooth and bevelled internally to support the medulla oblongata. On each side of the foramen magnum are the condyles, for articulation with the atlas; they are convex, oblong, or reniform in shape, and directed downwards and outwards; they converge in front, and encroach slightly upon the anterior segment of the foramen. On the inner border of each condyle is a rough tubercle for the attachment of the ligaments (check) which connect this bone with the odontoid process of the axis; whilst external to them is a rough tubercular prominence, the transverse or jugular process (the representative of the transverse process of a vertebra (channelled in front by a deep notch, which forms part of the jugular foramen. The under surface of this process affords attachment to the Rectus capitis lateralis; its upper or cerebral surface presents a deep groove which lodges part of the lateral sinus, whilst its prominent extremity is marked by a quadrilateral rough surface, covered with cartilage in the fresh state, and articulating with a similar surface on the petrous portion of the temporal bone. On the outer side of each condyle, near its fore part, is a foramen, the anterior condyloid; it is directed downwards, outwards, and forwards, and transmits the hypoglossal nerve. This foramen is sometimes double. Behind each condyle is a fossa,* sometimes perforated at the bottom by a foramen, the posterior condyloid, for the transmission of a vein to the lateral sinus. In front of the foramen magnum is a strong quadrilateral plate of bone, the basilar process, wider behind than in front; its under surface, which is rough, presenting in the median line a tubercular ridge, the pharyngeal spine, for the attachment of the tendinous raphé and Superior constrictor of the pharynx; and, on each side of it, rough depressions for the attachment of the Recti capitis antici, major and minor.

The *Internal or Cerebral Surface* (fig. 23) is deeply concave. The posterior or occipital part is divided by a crucial ridge into four fossæ. The two superior fossæ receive the posterior lobes of the cerebrum, and present slight eminences and depressions corresponding to their convolutions. The two inferior, which receive the lateral lobes of the cerebellum, are larger than the former, and comparatively smooth; both are marked by slight grooves for the lodgment of arteries. At the point of meeting of the four divisions of the crucial ridge is an eminence, the internal occipital protuberance. It nearly corresponds to that on the outer surface, and is perforated by one or more large vascular foramina. From this eminence, the superior division of the crucial ridge runs upward to the superior angle of the bone; it presents occasionally a deep groove for the superior longitudinal sinus, the margins of which give attachment to the falx cerebri. The inferior division, the internal occipital crest, runs to the posterior margin of the foramen magnum, on the edge of which it becomes gradually lost; this ridge, which is bifurcated below, serves for the attachment of the falx cerebelli. It is usually marked by two small grooves, which commence on either side of the posterior margin of the foramen magnum, join together above, and run into the depression for the Torcular Herophili. They lodge the occipital sinuses. The transverse grooves pass outwards to the lateral angles; they are deeply channelled, for the lodgment of the lateral sinuses, their prominent margins affording attachment to the tentorium cerebelli.† At the point of meeting of these grooves is a depression, the 'Torcular Herophili,'‡ placed a little to one or the other side of

* This fossa presents many variations in size. It is usually shallow; and the foramen small; occasionally wanting, on one, or both sides. Sometimes both fossa and foramen are large, but confined to one side only; more rarely, the fossa and foramen are very large on both sides.

† Usually one of the transverse grooves is deeper and broader than the other; occasionally both grooves are of equal depth and breadth, or both equally indistinct. The broader of the two transverse grooves is nearly always continuous with the vertical groove for the superior longitudinal sinus, and occupies the corresponding side of the median line.

‡ The columns of blood coming in different directions were supposed to be *pressed* together at this point.

the internal occipital protuberance. More anteriorly is the foramen magnum, and on each side of it, but nearer its anterior than its posterior part, the internal openings of the anterior condyloid foramina; the internal openings of the posterior condyloid foramina being a little external and posterior to them, protected by a

23.—Occipital Bone. Inner Surface.

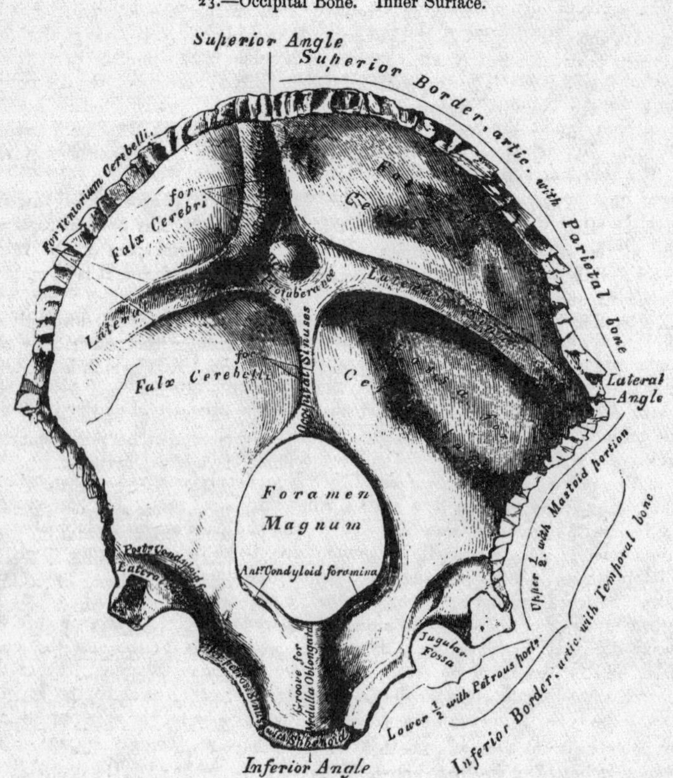

small arch of bone. At this part of the internal surface there is a very deep groove, in which the posterior condyloid foramen, when it exists, has its internal termination. This groove is continuous in the complete skull with that which separates the upper from the lower fossæ, and lodges the end of the same sinus, the lateral. In front of the foramen magnum is the basilar process, presenting a shallow depression, the basilar groove, which slopes from behind, upwards and forwards, and supports the medulla oblongata; and on each side of the basilar process is a narrow channel, which, when united with a similar channel on the petrous portion of the temporal bone, forms a groove, which lodges the inferior petrosal sinus.

Angles. The *superior* angle is received into the interval between the posterior superior angles of the two parietal bones: it corresponds with that part of the skull in the fœtus which is called the *posterior fontanelle.* The *inferior* angle is represented by the square-shaped surface of the basilar process. At an early period of life, a layer of cartilage separates this part of the bone from the sphenoid; but in the adult, the union between them is osseous. The *lateral angles* correspond to the outer ends of the transverse grooves, and are received into the interval between the posterior inferior angles of the parietal and the mastoid portion of the temporal.

Borders. The *superior* extends on each side from the superior to the lateral angle, is deeply serrated for articulation with the parietal bone, and forms, by this union, the lambdoid suture. The *inferior* border extends from the lateral to the inferior angle; its upper half is rough, and articulates with the mastoid portion of the temporal, forming the masto-occipital suture: the inferior half articulates with the petrous portion of the temporal, forming the petro-occipital suture; these two portions are separated from one another by the jugular process. In front of this process is a deep notch, which, with a similar one on the petrous portion of the temporal, forms the foramen lacerum posterius. This notch is occasionally subdivided into two parts by a small process of bone, and presents an aperture at its upper part, the internal opening of the posterior condyloid foramen.

Structure. The occipital bone consists of two compact laminæ, called the *outer* and *inner tables*, having between them the diploic tissue: this bone is especially thick at the ridges, protuberances, condyles, and anterior part of the basilar process; whilst at the bottom of the fossæ, especially the inferior, it is thin, semi-transparent, and destitute of diploë.

Development (fig. 24). The occipital bone has *four* centres of development; one for the posterior or occipital part, which is formed in membrane; one for the basilar portion, and one for each condyloid portion, which are formed in cartilage.

24.—Development of Occipital Bone.
By Four centres.

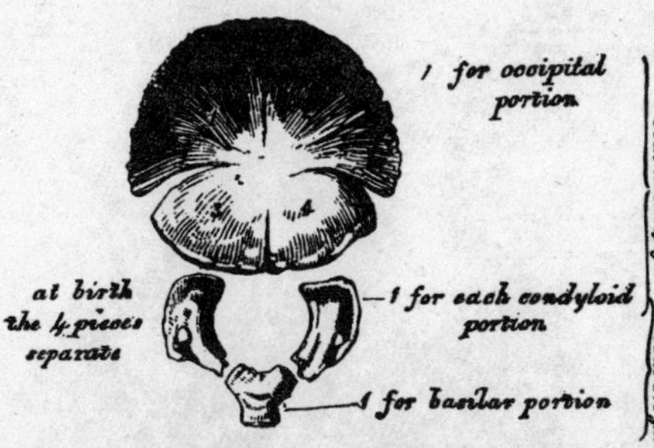

The centre for the occipital portion appears about the tenth week of fœtal life; and consists, according to Blandin and Cruvelhier, of a small oblong plate which appears in the situation of the occipital protuberance.* The condyloid portions then ossify, and lastly the basilar portion. At birth, the bone consists of four parts, separate from one another, the occipital portion being fissured in the direction indicated in the plate above. At about the fourth year, the occipital and the two condyloid pieces join; and about the sixth year the bone consists of a single piece. At a later period, between the eighteenth and twenty-fifth years, the occipital and sphenoid become united, forming a single bone.

Articulations. With six bones; two parietal, two temporal, sphenoid, and atlas.

Attachment of Muscles. To the superior curved line are attached the Occipito-frontalis, Trapezius, and Sterno-cleido-mastoid. To the space between the curved lines, the Complexus, Splenius capitis, and Obliquus superior; to the inferior curved line, and the space between it and the foramen magnum, the Rectus posticus major and minor; to the transverse process, the Rectus lateralis; and to the basilar process, the Rectus anticus major and minor, and Superior Constrictor of the pharynx.

The Parietal Bones.

The *Parietal Bones* (*paries*, a wall) form by their union the sides and roof of

* Béclard considers this segment to have four centres of ossification, arranged in pairs, two above and two below the curved lines, and Meckel describes eight, four of which correspond in situation with those above described: of the other four, two are placed in juxtaposition, at the upper angle of the bone, and the remaining two, one at each side, in the lateral angles.

the skull. Each bone is of an irregular quadrilateral form, and presents for examination two surfaces, four borders and four angles.

Surfaces. The *external surface* (fig. 25) is convex, smooth, and marked about its centre by an eminence, called the parietal eminence, which indicates the point where ossification commenced. Crossing the middle of the bone in an arched direction is a curved ridge, the temporal ridge, for the attachment of the temporal fascia. Above this ridge, the surface of the bone is rough and porous, and covered by the aponeurosis of the Occipito-frontalis; below it the bone is smooth, forms part of the temporal fossa, and affords attachment to the temporal muscle. At the back part of the superior border, close to the sagittal suture, is a small foramen, the parietal foramen, which transmits a vein to the superior longitudinal sinus. Its existence is not constant, and its size varies considerably.

The *internal surface* (fig. 26), concave, presents eminences and depressions for lodging the convolutions of the cerebrum, and numerous furrows for the ramifications of the meningeal arteries; the latter run upwards and backwards from the anterior inferior angle, and from the central and posterior part of the lower border of the bone. Along the upper margin is part of a shallow groove, which, when joined to the opposite parietal, forms a channel for the superior longitudinal sinus, the elevated edges of which afford attachment to the falx cerebri. Near the groove are seen several depressions; they lodge the Pacchionian bodies. The internal opening of the parietal foramen is also seen when that aperture exists.

25.—Left Parietal Bone. External Surface.

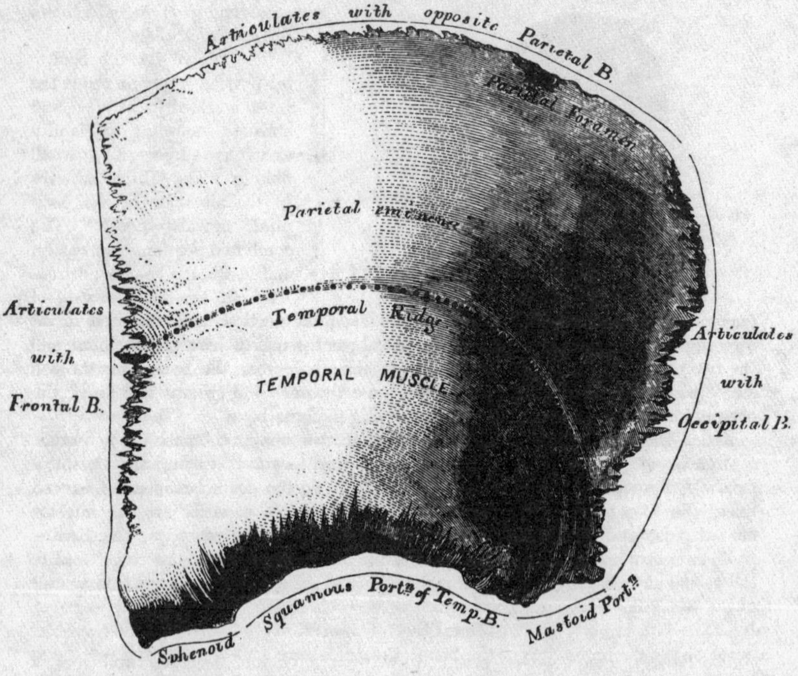

Borders. The *superior*, the longest and thickest, is dentated to articulate with its fellow of the opposite side, forming the sagittal suture. The *inferior* is divided into three parts; of these, the anterior is thin and pointed, bevelled at the expense of the outer surface, and overlapped by the tip of the great wing of the sphenoid; the middle portion is arched, bevelled at the expense of the outer surface, and overlapped by the squamous portion of the temporal; the posterior portion is thick and serrated for articulation with the mastoid portion of the temporal. The *anterior border*, deeply serrated, is bevelled at the expense of the outer surface above, and of the inner below; it articulates with the frontal bone, forming the coronal suture. The *posterior* border, deeply denticulated, articulates with the occipital, forming the lambdoid suture.

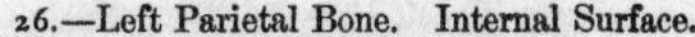

26.—Left Parietal Bone. Internal Surface.

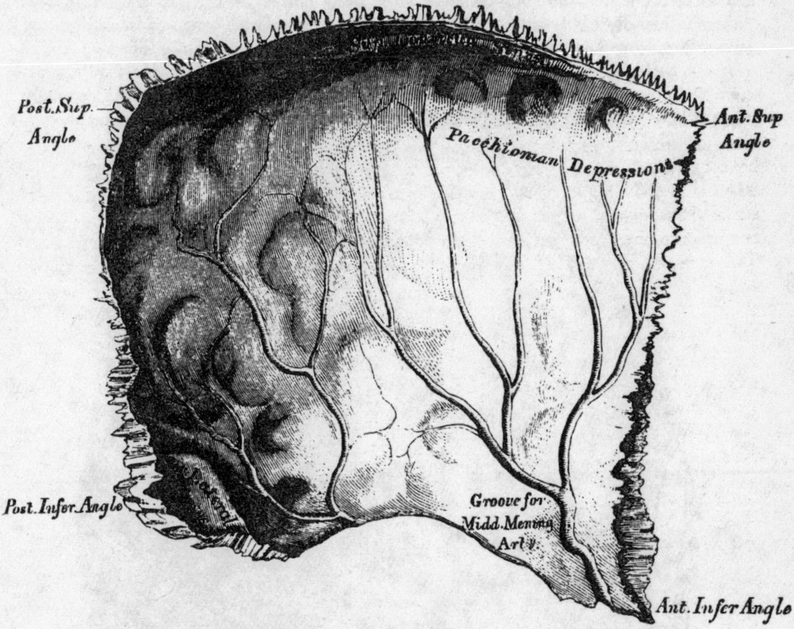

Angles. The *anterior superior*, thin and pointed, corresponds with that portion of the skull which in the fœtus is membranous, and is called the *anterior fontanelle*. The *anterior inferior angle* is thin and lengthened, being received in the interval between the great wing of the sphenoid and the frontal. This point will be found about one inch behind the upper and outer angle of the orbit. Its inner surface is marked by a deep groove, sometimes a canal, for the anterior branch of the middle meningeal artery. The *posterior superior angle* corresponds with the junction of the sagittal and lambdoid sutures. In the fœtus this part of the skull is membranous, and is called the *posterior fontanelle*. The *posterior inferior angle* articulates with the mastoid portion of the temporal bone, and generally presents on its inner surface a broad shallow groove for lodging part of the lateral sinus.

Development. The parietal bone is formed in membrane, being developed by *one* centre, which corresponds with the parietal eminence, and makes its first appearance about the fifth or sixth week of fœtal life. Ossification gradually extends from the centre to the circumference of the bone: the angles are consequently the parts last formed, and it is in their situation, that the fontanelles exist, previous to the completion of the growth of the bone.

Articulations. With five bones; the opposite parietal, the occipital, frontal, temporal, and sphenoid.

Attachment of Muscles. One only, the Temporal.

The Frontal Bone.

This bone, which resembles a cockle-shell in form, consists of two portions—a *vertical* or *frontal* portion, situated at the anterior part of the cranium, forming the forehead; and a *horizontal* or *orbito-nasal* portion, which enters into the formation of the roof of the orbits and nose.

Vertical Portion. External Surface (fig. 27). In the median line, traversing the bone from the upper to the lower part, is occasionally seen a slightly elevated ridge, and in young subjects a suture, which represents the line of union of the two lateral halves of which the bone consists at an early period of life: in the adult, this suture is usually obliterated, and the bone forms one piece: traces of the obliterated suture are, however, generally perceptible at the lower part. On either side of this ridge, a little below the centre of the bone, is a rounded eminence, the frontal eminence. These eminences vary in size in different individuals, and are occasionally unsymmetrical in the same subject. They are especially prominent in cases of well-

27.—Frontal Bone. Outer Surface.

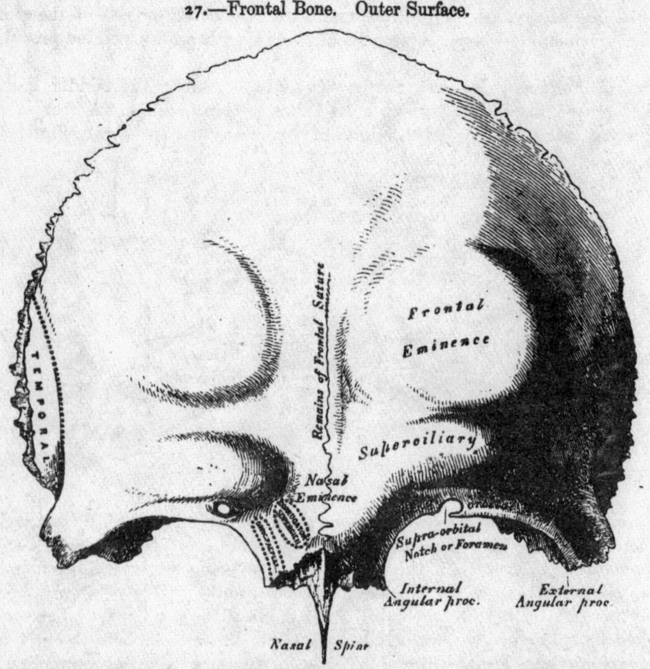

marked cerebral development. The whole surface of the bone above this part is smooth, and covered by the aponeurosis of the Occipito-frontalis muscle. Below the frontal eminence, and separated from it by a slight groove, is the superciliary ridge, broad internally where it is continuous with the nasal eminence, but less distinct as it arches outwards. These ridges are caused by the projection outwards of the frontal sinuses, and give attachment to the Orbicularis palpebrarum and Corrugator supercilii.* Beneath the superciliary ridge is the supra-orbital arch, a curved and

* Some confusion is occasioned to students commencing the study of anatomy, by the name 'sinuses' having been given to two perfectly different kinds of spaces connected with the skull. It may be as well, therefore, to state here, at the outset, that the 'sinuses' on the interior of the cranium, marked by grooves on the inner surface of the bones, are venous channels along which the blood runs in its passage back from the brain, while the 'sinuses' on the outside of the cranium (the frontal, ethmoidal, sphenoid, and maxillary) are hollow spaces in the bones themselves, which communicate with the nostrils, and contain air.

prominent margin, which forms the upper boundary of the orbit, and separates the vertical from the horizontal portion of the bone. The outer part of the arch is sharp and prominent, affording to the eye, in that situation, considerable protection from injury; the inner part is less prominent. At the inner third of this arch is a notch, sometimes converted into a foramen by a bony process, and called the *supra-orbital notch* or *foramen*. It transmits the supra-orbital artery, veins, and nerve. A small aperture is seen in the upper part of the notch, which transmits a vein from the diploë to join the ophthalmic vein. The supra-orbital arch terminates externally in the external angular process, and internally in the internal angular process. The external angular process is strong, prominent, and articulates with the malar bone: running upwards and backwards from it is a sharp curved crest, the temporal ridge, for the attachment of the temporal fascia; and beneath it a slight concavity, that forms the anterior part of the temporal fossa, and gives origin to the Temporal muscle. The internal angular processes are less marked than the external, and articulate with the lachrymal bones. Between the internal angular processes is a rough uneven interval, the *nasal notch*, which articulates in the middle line with the nasal bone, and on either side with the nasal process of the superior maxillary bone. The notch is continuous below with a long pointed process, the *nasal spine*.

Vertical Portion. Internal Surface (fig. 28). Along the middle line is a vertical groove, the edges of which unite below to form a ridge, the frontal crest; the groove lodges the superior longitudinal sinus, whilst its edges afford attach-

28.—Frontal Bone. Inner Surface.

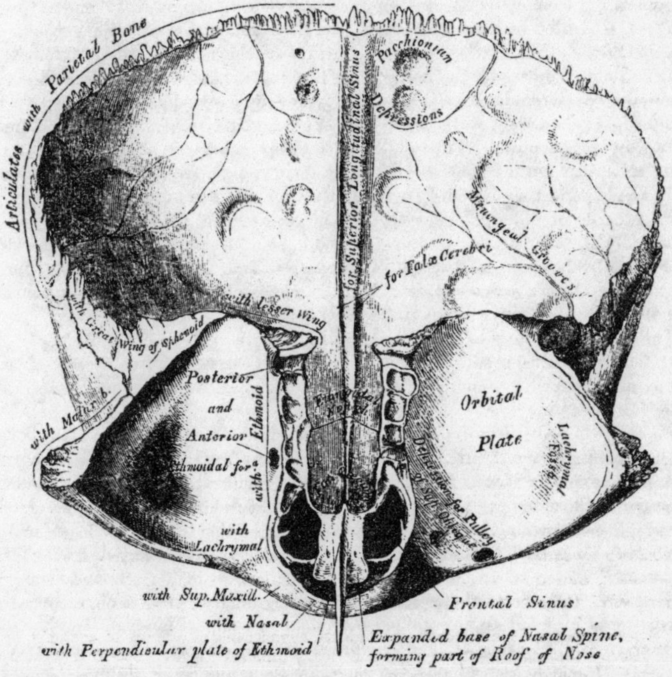

ment to the falx cerebri. The crest terminates below at a small opening, the foramen cæcum, which is generally completed behind by the ethmoid. This foramen varies in size in different subjects, is usually partially or completely impervious, lodges a process of the falx cerebri, and, when open, transmits a vein from the lining-membrane of the nose to the superior longitudinal sinus. On either side of the groove, the bone is deeply concave, presenting eminences and depressions for the convolutions of the brain, and numerous small furrows for lodging the ramifications of the anterior meningeal arteries. Several small, irregular fossæ are also seen on either side of the groove, for the reception of the Pacchionian bodies.

Horizontal Portion. External Surface. This portion of the bone consists of two thin plates, which form the vault of the orbits, separated from one another by the ethmoidal notch. Each orbital vault consists of a smooth, concave, triangular plate of bone, marked at its anterior and external part (immediately beneath the external angular process) by a shallow depression, the lachrymal fossa, for lodging the lachrymal gland; and at its anterior and internal part, by a depression (sometimes a small tubercle) for the attachment of the fibrous pulley of the Superior oblique muscle. The ethmoidal notch separates the two orbital plates; it is quadrilateral; and filled up, when the bones are united, by the cribriform plate of the ethmoid. The margins of this notch present several half-cells, which, when united with corresponding half-cells on the upper surface of the ethmoid, complete the ethmoidal cells; two grooves are also seen crossing these edges transversely; they are converted into canals by articulation with the ethmoid, and are called the *anterior* and *posterior* ethmoidal canals; they open on the inner walls of the orbit. The anterior one transmits the nasal nerve and anterior ethmoidal vessels, the posterior one the posterior ethmoidal vessels. In front of the ethmoidal notch is the nasal spine, a sharp-pointed eminence, which projects downwards and forwards, and articulates in front with the crest of the nasal bones; behind, it is marked by two grooves, separated by a vertical ridge; the ridge articulates with the perpendicular lamellæ of the ethmoid, the grooves form part of the roof of the nasal fossæ. On either side of the base of the nasal spine are the openings of the frontal sinuses. These are two irregular cavities, which extend upwards and outwards, a variable distance, between the two tables of the skull, and are separated from one another by a thin bony septum. They give rise to the prominences above the root of the nose, called the *nasal eminences* and *superciliary ridges.* In the child they are generally absent, and they become gradually developed as age advances. These cavities vary in size in different persons, are larger in men than in women, and are frequently of unequal size on the two sides, the left being commonly the larger. Occasionally they are subdivided by incomplete bony laminæ. They are lined by mucous membrane, and communicate with the nose by the infundibulum, and occasionally with each other by apertures in their septum.

The *Internal Surface* of the *Horizontal Portion* presents the convex upper surfaces of the orbital plates, separated from each other in the middle line by the ethmoidal notch, and marked by eminences and depressions for the convolutions of the anterior lobes of the brain.

Borders. The border of the vertical portion is thick, strongly serrated, bevelled at the expense of the internal table above, where it rests upon the parietal bones, and at the expense of the external table at each side, where it receives the lateral pressure of those bones: this border is continued below into a triangular rough surface, which articulates with the great wing of the sphenoid. The border of the horizontal portion is thin, serrated, and articulates with the lesser wing of the sphenoid.

Structure. The vertical portion and external angular processes are very thick, consisting of diploic tissue contained between two compact laminæ. The horizontal portion is thin, translucent, and composed entirely of compact tissue; hence the facility with which instruments can penetrate the cranium through this part of the orbit.

Development (fig. 29). The frontal bone is formed in membrane, being developed by *two* centres, one for each lateral half, which make their appearance, at an early period of fœtal life, in the situation of the orbital arches. From this point ossification extends, in a radiating manner, upwards into the forehead, and backwards over the orbit. At birth it consists of two pieces, which afterwards become united, along the middle line, by a suture which runs from the vertex to the root of the nose. This suture usually becomes obliterated within a few years after birth: but it occasionally remains throughout life.

29.—Frontal Bone at Birth.
Developed by two lateral Halves.

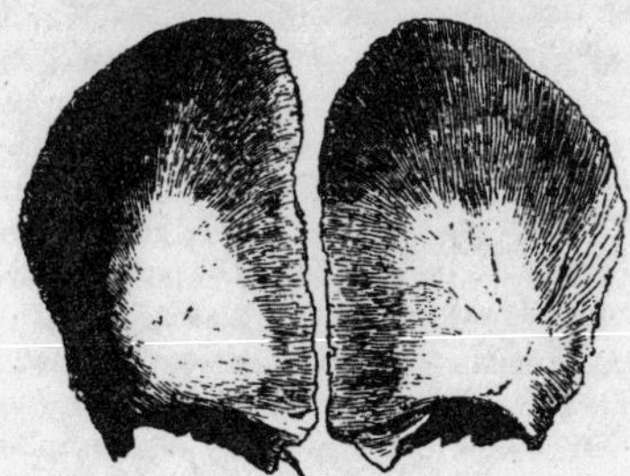

Articulations. With twelve bones: two parietal, sphenoid, ethmoid; two nasal, two superior maxillary, two lachrymal, and two malar.

Attachment of Muscles. The Corrugator supercilii, Orbicularis palpebrarum, and Temporal, on each side.

The Temporal Bones.

The Temporal Bones are situated at the side and base of the skull, and present for examination a *squamous*, *mastoid*, and *petrous* portion.

The *Squamous Portion* (*squama*, a scale), (fig. 30), the anterior and upper part of

30.—Left Temporal Bone. Outer Surface.

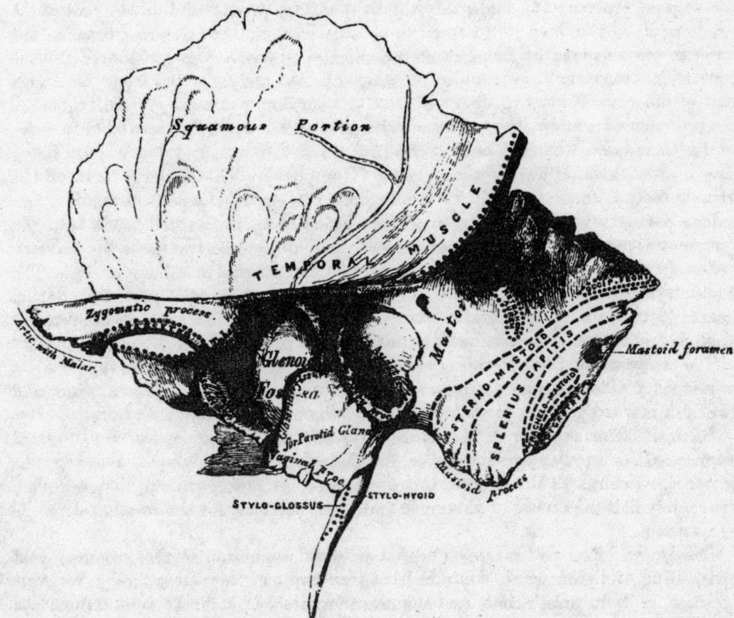

the bone, is scale-like in form, and thin and translucent in texture. Its outer surface is smooth, convex, and grooved at its back part for the deep temporal arteries;

it affords attachment to the Temporal muscle, and forms part of the temporal fossa. At its back part may be seen a curved ridge—part of the temporal ridge; it serves for the attachment of the temporal fascia, limits the origin of the Temporal muscle, and marks the boundary between the squamous and mastoid portion of the bone. Projecting from the lower part of the squamous portion is a long arched outgrowth of bone, the zygomatic process. This process is at first directed outwards, its two surfaces looking upwards and downwards; it then appears as if twisted upon itself, and runs forwards, its surfaces now looking inwards and outwards. The superior border of the process is long, thin, and sharp, and serves for the attachment of the temporal fascia. The inferior, short, thick, and arched, has attached to it some fibres of the Masseter muscle. Its outer surface is convex and subcutaneous; its inner is concave, and also affords attachment to the Masseter. The extremity, broad and deeply serrated, articulates with the malar bone. The zygomatic process is connected to the temporal bone by three divisions, called its *roots*—an anterior, middle, and posterior. The anterior, which is short but broad and strong, runs transversely inwards into a rounded eminence, the eminentia articularis. This eminence forms the front boundary of the glenoid fossa, and in the recent state is covered with cartilage. The middle root forms the outer margin of the glenoid cavity; running obliquely inwards, it terminates at the commencement of a well-marked fissure, the Glaserian fissure; whilst the posterior root, which is strongly marked, runs from the upper border of the zygoma, in an arched direction, upwards and backwards, forming the posterior part of the temporal ridge. At the junction of the anterior root with the zygoma is a projection, called the *tubercle*, for the attachment of the external lateral ligament of the lower jaw; and between the anterior and middle roots is an oval depression, forming part of the glenoid fossa (γλήνη, *a socket*), for the reception of the condyle of the lower jaw. This fossa is bounded, in front, by the eminentia articularis; behind, by the vaginal process; and, externally, by the auditory process and middle root of the zygoma; and is divided into two parts by a narrow slit, the Glaserian fissure. The anterior part, formed by the squamous portion of the bone, is smooth, covered in the recent state with cartilage, and articulates with the condyle of the lower jaw. This part of the glenoid fossa is separated from the auditory process by a small tubercle, the *post-glenoid process*, the representative of a prominent tubercle which, in some of the mammalia, descends behind the condyle of the jaw, and prevents it being displaced backwards during mastication (Humphry). The posterior part of the glenoid fossa is formed chiefly by the vaginal process of the petrous portion, and lodges part of the parotid gland. The Glaserian fissure, which leads into the tympanum, lodges the processus gracilis of the malleus, and transmits the Laxator tympani muscle and the tympanic branch of the internal maxillary artery. The chorda tympani nerve passes through a separate canal parallel to the Glaserian fissure (canal of Huguier), on the outer side of the Eustachian tube, in the retiring angle between the squamous and petrous portions of the temporal bone.

The *internal surface* of the squamous portion (fig. 31) is concave, presents numerous eminences and depressions for the convolutions of the cerebrum, and two well-marked grooves for the branches of the middle meningeal artery.

Borders. The superior border is thin, bevelled at the expense of the internal surface, so as to overlap the lower border of the parietal bone, forming the squamous suture. The anterior inferior border is thick, serrated, and bevelled, alternately at the expense of the inner and outer surfaces, for articulation with the great wing of the sphenoid.

The *Mastoid Portion* (μαστὸς, *a nipple* or *teat*) is situated at the posterior part of the bone; its outer surface is rough, and perforated by numerous foramina: one of these, of large size, situated at the posterior border of the bone, is termed the *mastoid foramen*; it transmits a vein to the lateral sinus and a small artery. The position and size of this foramen are very variable. It is not always present: sometimes it is situated in the occipital bone, or in the suture between the temporal and the occipital. The mastoid portion is continued below into a conical projection, the mastoid process, the size and form of which varies somewhat. This process serves for the attachment of the Sterno-mastoid, Splenius capitis, and Trachelo-mastoid muscles. On the inner side of the mastoid process is a deep groove, the digastric fossa, for the attachment of the Digastric muscle; and running parallel with it, but more internal, the occipital groove, which lodges the occipital artery. The internal surface of the mastoid portion presents a deep curved groove, which lodges part of the lateral sinus; and into it may be seen opening the mastoid foramen. A section of the mastoid process shows it to be hollowed out into a number of cellular spaces, communicating with each other, called the *mastoid cells*; they open by a single or double orifice into the back of the tympanum; are lined by a prolongation of its lining membrane; and, probably, form some secondary part of the organ of hearing. The mastoid cells, like the other sinuses of the cranium, are not developed until after puberty; hence the prominence of this process in the adult.

Borders. The superior border of the mastoid portion is broad and rough, its serrated edge sloping outwards, for articulation with the posterior inferior angle of

31.—Left Temporal Bone. Inner Surface.

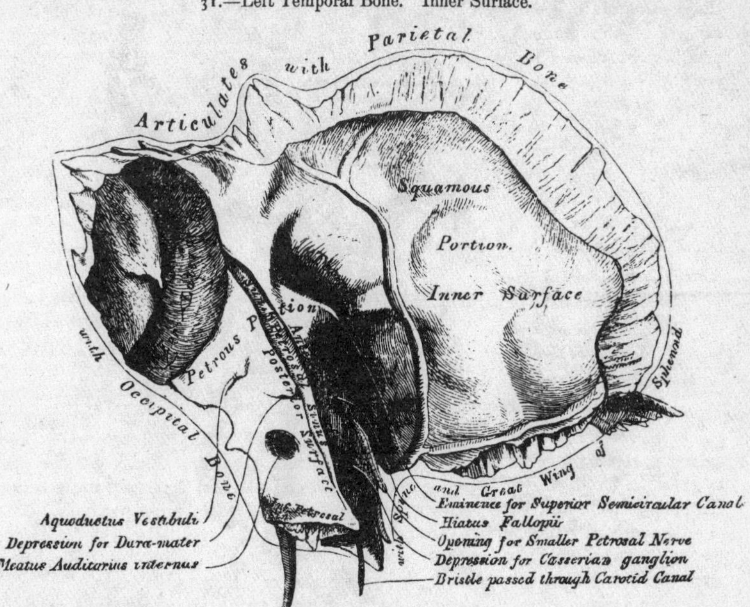

the parietal bone. The posterior border, also uneven and serrated, articulates with the inferior border of the occipital bone between its lateral angle and jugular process.

The *Petrous Portion* (*πέτρος, a stone*), so named from its extreme density and hardness, is a pyramidal process of bone, wedged in at the base of the skull between the sphenoid and occipital bones. Its direction from without is inwards, forwards, and a little downwards. It presents for examination a base, an apex, three surfaces, and three borders; and contains, in its interior, the essential parts of the organ of hearing. The *base* is applied against the internal surface of the squamous and mastoid portions, its upper half being concealed; but its lower half is exposed by the divergence of those two portions of the bone which brings into view the oval expanded orifice of a canal leading into the tympanum, the meatus auditorius externus. This canal is situated between the mastoid process and the posterior and middle roots of the zygoma; its upper margin is smooth and rounded, but the greater part of its circumference is surrounded by a curved plate of bone, the auditory process, the free margin of which is thick and rough, for the attachment of the cartilage of the external ear.

The *apex* of the petrous portion, rough and uneven, is received into the angular interval between the spinous process of the sphenoid and the basilar process of the occipital; it presents the anterior or internal orifice of the carotid canal, and forms the posterior and external boundary of the foramen lacerum medium.

The *anterior surface* of the petrous portion (fig. 31) forms the posterior boundary of the middle fossa of the skull. This surface is continuous with the squamous portion, to which it is united by a suture, the temporal suture, the remains of which are distinct even at a late period of life: it presents six points for examination: 1. an eminence near the centre, which indicates the situation of the superior vertical semicircular canal: 2. on the outer side of this eminence a depression, in-

32.—Petrous Portion. Inferior Surface.

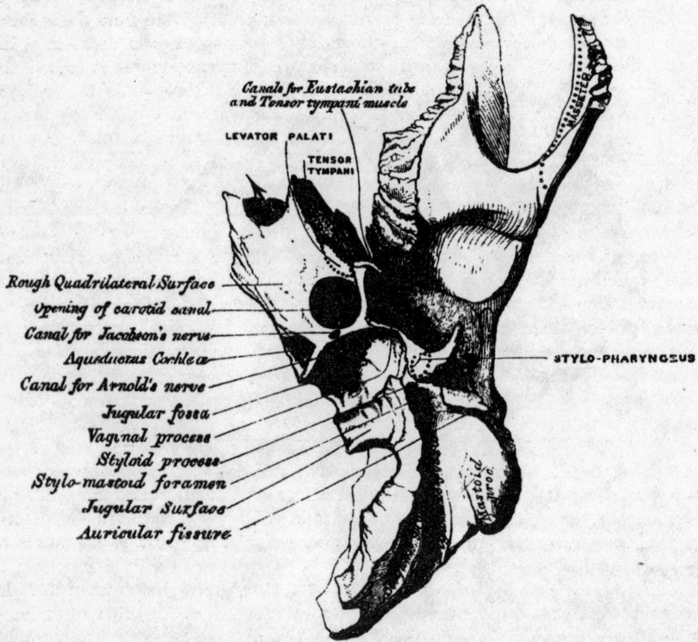

dicating the position of the tympanum, the layer of bone which separates the tympanum from the cranial cavity being extremely thin: 3. a shallow groove, sometimes double, leading backwards to an oblique opening, the hiatus Fallopii, for the passage of the petrosal branch of the Vidian nerve: 4. a smaller opening, occasionally seen external to the latter for the passage of the smaller petrosal nerve: 5. near the apex of the bone the termination of the carotid canal, the wall of which in this situation is deficient in front: 6. above this canal, a shallow depression for the reception of the Casserian ganglion.

The *posterior surface* forms the front boundary of the posterior fossa of the skull, and is continuous with the inner surface of the mastoid portion of the bone. It presents three points for examination: 1. about its centre, a large orifice, the meatus auditorius internus, whose size varies considerably; its margins are smooth and rounded; and it leads into a short canal, about four lines in length, which runs directly outwards, and is closed by a vertical plate, divided by a horizontal crest into two unequal portions: the canal transmits the auditory and facial nerves, and auditory artery: 2. behind the meatus auditorius, a small slit, almost hidden by a thin plate of bone, leading to a canal, the aquæductus vestibuli, which transmits a small artery and vein, and lodges a process of the dura mater: 3. in the interval between these two openings, but above them, an angular depression which lodges a process of the dura mater, and transmits a small vein into the cancellous tissue of the bone.

The *inferior* or *basilar surface* (fig. 32) is rough and irregular, and forms part of the base of the skull. Passing from the apex to the base, this surface presents eleven points for examination: 1. a rough surface, quadrilateral in form, which serves partly for the attachment of the Levator palati and Tensor tympani muscles: 2. the large circular aperture of the carotid canal, which ascends at first vertically, and then, making a bend, runs horizontally forwards and inwards; it transmits the internal carotid artery and the carotid plexus: 3. the aquæductus cochleæ, a small triangular opening, lying on the inner side of the latter, close to the posterior border of the petrous portion; it transmits a vein from the cochlea, which joins the internal jugular: 4. behind these openings a deep depression, the jugular fossa, which varies in depth and size in different skulls; it lodges the internal jugular vein, and, with a similar depression on the margin of the occipital bone, forms the foramen lacerum posterius: 5. a small foramen for the passage of Jacobson's nerve (the tympanic branch of the glosso-pharyngeal); this foramen is seen in front of the bony ridge dividing the carotid canal from the jugular fossa: 6. a small foramen on the inner wall of the jugular fossa, for the entrance of the auricular branch of the pneumogastric (Arnold's) nerve: 7. behind the jugular fossa, a smooth square-shaped facet, the jugular surface; it is covered with cartilage in the recent state, and articulates with the jugular process of the occipital bone: 8. the vaginal process, a very broad sheath-like plate of bone, which extends from the carotid canal to the mastoid process; it divides behind into two laminæ, receiving between them the 9th point for examination, the styloid process; a long sharp spine, about an inch in length, continuous with the vaginal process, between the laminæ of which it is received; it is directed downwards, forwards, and inwards, varies in size and shape, and sometimes consists of several pieces united by cartilage; it affords attachment to three muscles, the Stylo-pharyngeus, Stylo-glossus, and Stylo-hyoideus; and two ligaments, the stylo-hyoid and stylo-maxillary: 10. the stylo-mastoid foramen, a rather large orifice, placed between the styloid and mastoid processes; it is the termination of the aquæductus Fallopii, and transmits the facial nerve and stylo-mastoid artery: 11. the auricular fissure, situated between the vaginal and mastoid processes, for the exit of the auricular branch of the pneumogastric nerve.

Borders of the petrous portion. The *superior*, the longest, is grooved for the superior petrosal sinus, and has attached to it the tentorium cerebelli; at its inner extremity is a semilunar notch, upon which the fifth nerve lies. The *posterior* border is intermediate in length between the superior and the anterior. Its inner half is marked by a groove, which, when completed by its articulation with the occipital, forms the channel for the inferior petrosal sinus. Its outer half presents a deep excavation—the jugular fossa—which, with a similar notch on the occipital, forms the foramen lacerum posterius. A projecting eminence of bone occasionally stands out from the centre of the notch, and divides the foramen into two parts. The *anterior* border is divided into two parts—an outer joined to the squamous portion by a suture, the remains of which are distinct; an inner, free, articulating with the spinous process of the sphenoid. At the angle of junction of the petrous and squamous portions are seen two canals, separated from one another by a thin plate of bone, the processus cochleariformis: they both lead into the tympanum, the upper one transmitting the Tensor tympani muscle, the lower one the Eustachian tube.

Structure. The squamous portion is like that of the other cranial bones, the mastoid portion cellular, and the petrous portion dense and hard.

Development (fig. 33). The temporal bone is developed by *four* centres, exclusive of those for the internal ear and the ossicula, viz.:—one for the squamous portion including the zygoma, one for the petrous and mastoid parts, one for the styloid, and one for the auditory process (tympanic bone). The first traces of the development of this bone appear in the squamous portion, about the time when osseous matter is deposited in the vertebræ; the auditory process succeeds next; it consists of a curved piece of bone, forming about three-fourths of a circle, the deficiency being above; it is grooved along its concave surface for the attachment of the membrana tympani, and becomes united by its extremities to the squamous portion during the last months of intra-uterine life. The petrous and mastoid portions then become ossified, and lastly the styloid process, which remains separate a considerable period, and is occasionally never united to the rest of the bone. At birth, the temporal bone, excluding the styloid process, is formed of three pieces—the squamous and zygomatic, the petrous and mastoid, and the auditory. The auditory process joins with the squamous about the time of birth. The petrous and mastoid join with the squamous during the first year and the styloid process becomes united between the second and third years. The subsequent changes in this bone are, that the auditory process extends outwards, so as to form the meatus auditorius; the glenoid fossa becomes deeper; and the mastoid part, which at an early period of life is quite flat, enlarges from the development of the cellular cavities in its interior.

33.—Development of the Temporal Bone. By four Centres.

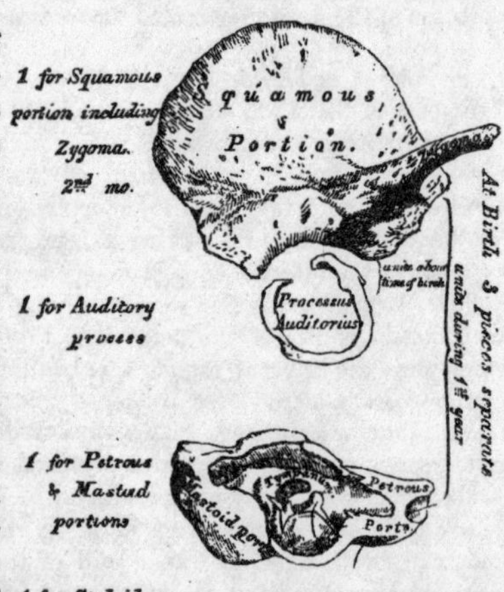

Articulations. With five bones—occipital, parietal, sphenoid, inferior maxillary, and malar.

Attachment of Muscles. To the squamous portion, the Temporal; to the zygoma, the Masseter; to the mastoid portion, the Occipito-frontalis, Sterno-mastoid, Splenius capitis, Trachelo-mastoid, Digastricus, and Retrahens aurem; to the styloid process, the Stylo-pharyngeus, Stylo-hyoideus, and Stylo-glossus; and to the petrous portion, the Levator palati, Tensor tympani, and Stapedius.

The Sphenoid Bone.

The Sphenoid bone (σφὴν, a *wedge*) is situated at the anterior part of the base of the skull, articulating with all the other cranial bones, which it binds firmly and solidly together. In its form it somewhat resembles a bat, with its wings extended; and is divided into a central portion or body, two greater and two lesser wings extending outwards on each side of the body; and two processes, the pterygoid processes, which project from it below.

The *Body* is of large size, quadrilateral in form, and hollowed out in its interior so as to form a mere shell of bone. It presents for examination *four* surfaces—a superior, an inferior, an anterior, and a posterior.

The *superior surface* (fig. 34). In front is seen a prominent spine, the ethmoidal spine, for articulation with the ethmoid; behind this a smooth surface presenting, in the median line, a slight longitudinal eminence, with a depression on each side, for lodging the olfactory nerves. A narrow transverse groove, the optic groove, bounds the above-mentioned surface behind; it lodges the optic commissure, and terminates on either side in the optic foramen, for the passage of the optic nerve and ophthalmic artery. Behind the optic groove is a small eminence, olive-like in shape, the olivary process; and still more posteriorly, a deep depression, the pituitary fossa, or 'sella Turcica,' which lodges the pituitary body. This fossa is perforated by numerous foramina, for the transmission of

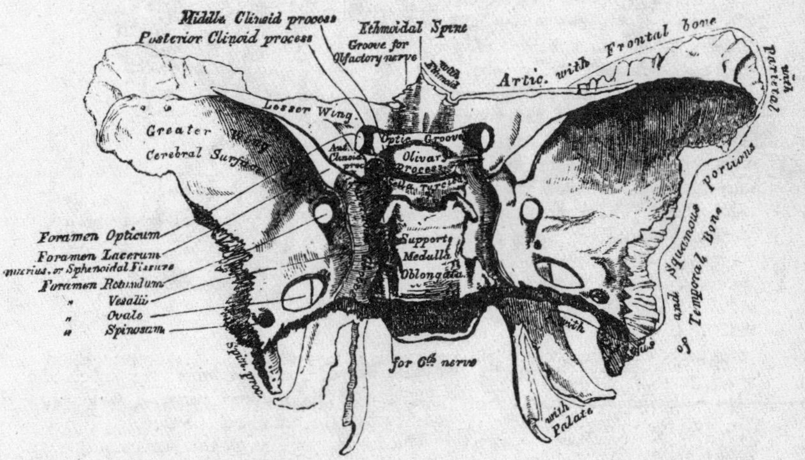

34.—Sphenoid Bone. Superior Surface.

nutrient vessels to the substance of the bone. It is bounded in front by two small eminences, one on either side, called the middle clinoid processes (κλίνη, *a bed*), and behind by a square-shaped plate of bone, terminating at each superior angle in a tubercle, the posterior clinoid processes, the size and form of which vary considerably in different individuals. These processes deepen the pituitary fossa, and serve for the attachment of prolongations from the tentorium cerebelli. The sides of the plate of bone supporting the posterior clinoid processes are notched for the passage of the sixth pair of nerves; and behind, this plate of bone presents a shallow depression, which slopes obliquely backwards, and is continuous with the basilar groove of the occipital bone; it supports the medulla oblongata. On either side of the body is a broad groove, curved something like the italic letter *f*; it lodges the internal carotid artery and the cavernous sinus, and is called the *cavernous groove.* The *posterior surface*, quadrilateral in form, articulates with the basilar process of the occipital bone. During childhood these bones are separated by a layer of cartilage; but in after-life (between the eighteenth and twenty-fifth years) this becomes ossified, ossification commencing above, and extending downward; and the two bones then form one piece. The *anterior surface* (fig. 35) presents, in the middle line, a vertical lamella of bone which articulates in front with the perpendicular plate of the ethmoid, forming part of the septum of the nose. On either side of it are the irregular openings leading into the sphenoid cells or sinuses. These are two large irregular cavities, hollowed out of the interior of the body of the sphenoid bone, and separated from one another by a more or less complete perpendicular bony septum. Their form and size vary considerably; they are seldom symmetrical, and are often partially subdivided by irregular osseous laminæ. Occasionally they extend into the basilar process of the occipital nearly as far as the foramen magnum. The septum is seldom quite vertical, being commonly bent to one or the other side. These sinuses do not exist in children, but they increase in size as age advances. They are partially

35.—Sphenoid Bone. Anterior Surface.*

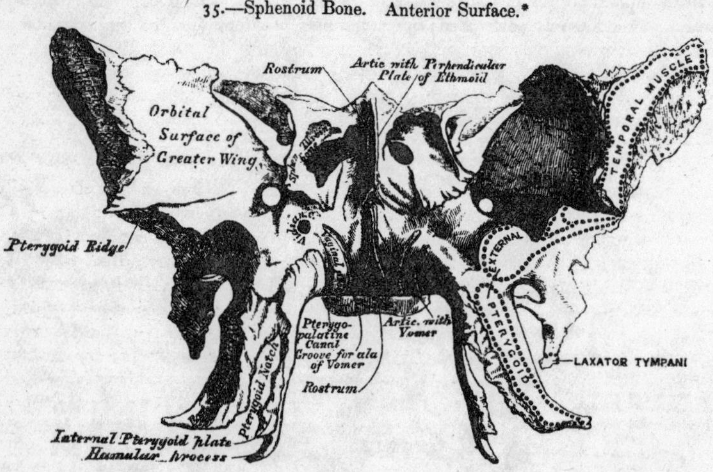

closed, in front and below, by two thin curved plates of bone; the sphenoidal turbinated bones, leaving a round opening at their upper parts, by which they communicate with the upper and back part of the nose, and occasionally with the posterior ethmoidal cells or sinuses. The lateral margins of this surface present a serrated edge, which articulates with the os planum of the ethmoid, completing the posterior ethmoidal cells; the lower margin, also rough and serrated, articulates with the orbital process of the palate bone; and the upper margin with the orbital plate of the frontal bone. The *inferior surface* presents, in the middle line, a triangular spine, the rostrum, which is continuous with the vertical plate on the anterior surface, and is received into a deep fissure between the alæ of the vomer. On each side may be seen a projecting lamina of bone, which runs horizontally inwards from near the base of the pterygoid process: these plates, termed the vaginal processes, articulate with the edges of the vomer. Close to the root of the pterygoid process is a groove, formed into a complete canal when articulated with the sphenoidal process of the palate bone; it is called the pterygo-palatine canal, and transmits the pterygo-palatine vessels and pharyngeal nerve.

The *Greater Wings* are two strong processes of bone, which arise from the sides of the body, and are curved in a direction upwards, outwards, and backwards; being prolonged behind into a sharp-pointed extremity, the *spinous process* of the sphenoid. Each wing presents three surfaces and a circumference. The *superior* or *cerebral* surface (fig. 34) forms part of the middle fossa of the skull; it is deeply concave, and presents eminences and depressions for the convolutions of the brain. At its anterior and internal part is seen a circular aperture, the foramen rotundum, for the transmission of the second division of the fifth nerve. Behind and external to this is a large oval foramen, the foramen ovale, for the transmission of the third division of the fifth nerve, the small meningeal artery, and the small petrosal nerve. At the inner side of the foramen ovale, a small aperture may

* In this figure, both the anterior and inferior surfaces of the body of the sphenoid bone are shown, the bone being held with the pterygoid processes almost horizontal.

occasionally be seen opposite the root of the pterygoid process; it is the foramen Vesalii, transmitting a small vein. Lastly, in the apex of the spine of the sphenoid is a short canal, sometimes double, the foramen spinosum; it transmits the middle meningeal artery. The *external* surface (fig. 35) is convex, and divided by a transverse ridge, the pterygoid ridge, into two portions. The superior or larger, convex from above downwards, concave from before backwards, enters into the formation of the temporal fossa, and attaches part of the Temporal muscle. The inferior portion, smaller in size and concave, enters into the formation of the zygomatic fossa, and affords attachment to the External pterygoid muscle. It presents, at its posterior part, a sharp-pointed eminence of bone, the spinous process, to which is connected the internal lateral ligament of the lower jaw, and the Laxator tympani muscle. The pterygoid ridge, dividing the temporal and zygomatic portions, gives attachment to part of the External pterygoid muscle. At its inner extremity is a triangular spine of bone, which serves to increase the extent of origin of this muscle. The *anterior* or *orbital* surface, smooth and quadrilateral in form, assists in forming the outer wall of the orbit. It is bounded above by a serrated edge, for articulation with the frontal bone; below, by a rounded border, which enters into the formation of the spheno-maxillary fissure; internally, it enters into the formation of the sphenoidal fissure; whilst externally it presents a serrated margin, for articulation with the malar bone. At the upper part of the inner border is a notch for the transmission of a branch of the ophthalmic artery; and at its lower part a small pointed spine of bone, which serves for the attachment of part of the lower head of the External rectus. One or two small foramina may occasionally be seen for the passage of arteries; they are called the *external orbital foramina*. *Circumference of the great wing* (fig. 34): commencing from behind, from the body of the sphenoid to the spine, the outer half of this margin is serrated, for articulation with the petrous portion of the temporal bone; whilst the inner half forms the anterior boundary of the foramen lacerum medium, and presents the posterior aperture of the Vidian canal. In front of the spine the circumference of the great wing presents a serrated edge, bevelled at the expense of the inner table below, and of the external above, which articulates with the squamous portion of the temporal bone. At the tip of the great wing a triangular portion is seen, bevelled at the expense of the internal surface, for articulation with the anterior inferior angle of the parietal bone. Internal to this is a broad serrated surface, for articulation with the frontal bone: this surface is continuous internally with the sharp inner edge of the orbital plate, which assists in the formation of the sphenoidal fissure.

The *Lesser Wings* (processes of Ingrassias) (fig. 34) are two thin triangular plates of bone, which arise from the upper and lateral parts of the body of the sphenoid; and, projecting transversely outwards, terminate in a sharp point. The superior surface of each is smooth, flat, broader internally than externally, and supports the anterior lobe of the brain. The inferior surface forms the back part of the roof of the orbit, and the upper boundary of the sphenoidal fissure or foramen lacerum anterius. This fissure is of a triangular form, and leads from the cavity of the cranium into the orbit; it is bounded internally by the body of the sphenoid—above, by the lesser wing; below, by the orbital surface of the great wing—and is converted into a foramen by the articulation of this bone with the frontal. It transmits the third, the fourth, the ophthalmic division of the fifth and the sixth nerves, and the ophthalmic vein. The anterior border of the lesser wing is serrated for articulation with the frontal bone; the posterior, smooth and rounded, is received into the fissure of Sylvius of the brain. The inner extremity of this border forms the anterior clinoid process. The lesser wing is connected to the side of the body by two roots, the upper thin and flat, the lower thicker, obliquely directed, and presenting on its outer side, near its junction with the body, a small tubercle, for the attachment of the common tendon of the muscles of the eye. Between the two roots is the optic foramen, for the transmission of the optic nerve and ophthalmic artery.

The *Pterygoid* processes (*πτέρυξ, a wing; εἶδος, likeness*), (fig. 36), one on each side, descend perpendicularly from the point where the body and great wing unite. Each process consists of an external and an internal plate, separated behind by an intervening notch,—the pterygoid fossa; but joined partially in front. The *external pterygoid plate* is broad and thin, turned a little outwards, and forms part of the inner wall of the zygomatic fossa. It gives attachment, by its outer surface, to the External pterygoid; its inner surface forms part of the pterygoid fossa, and gives attachment to the Internal pterygoid. The *internal pterygoid plate* is much narrower and longer, curving outwards, at its extremity, into a hook-like process of bone, the hamular process, around which turns the tendon of the Tensor palati muscle. At the base of this plate is a small, oval, shallow depression, the scaphoid fossa, from which arises the Tensor palati, and above which is seen the posterior orifice of the Vidian canal. The outer surface of this plate forms part of the pterygoid fossa, the inner surface forming the outer boundary of the posterior aperture of the nares. The Superior Constrictor of the pharynx is attached to its posterior edge. The two pterygoid plates are separated below by an angular interval, in which the pterygoid process, or tuberosity, of the palate-bone is received. The anterior surface of the pterygoid process is very broad at its base, and forms the posterior wall of the spheno-maxillary fossa. It supports Meckel's ganglion. It presents, above, the anterior orifice of the Vidian canal; and below, a rough margin, which articulates with the perpendicular plate of the palate-bone.

36.—Sphenoid Bone. Posterior Surface.

The *Sphenoidal Spongy Bones* are two thin curved plates of bone, which exist as separate pieces until puberty, and occasionally are not joined to the sphenoid in the adult. They are situated at the anterior and inferior part of the body of the sphenoid, an aperture of variable size being left in their anterior wall, through which the sphenoidal sinuses open into the nasal fossæ. They are irregular in form, and taper to a point behind, being broader and thinner in front. Their inner surface, which looks towards the cavity of the sinus, is concave; their outer surface convex. Each bone articulates in front with the ethmoid, externally with the palate; behind, its point is placed above the vomer, and is received between the root of the pterygoid process on the outer side, and the rostrum of the sphenoid on the inner.

Development. The sphenoid bone is developed by *ten* centres, six for the posterior sphenoidal division, and four for the anterior sphenoid. The six centres for the posterior sphenoid are—one for each greater wing and external pterygoid plate; one for each internal pterygoid plate; two for the posterior part of the body. The four for the anterior sphenoid are, one for each lesser wing and anterior part of the body, and one for each sphenoidal turbinated bone. Ossification takes

37.—Plan of the Development of Sphenoid. By ten Centres.

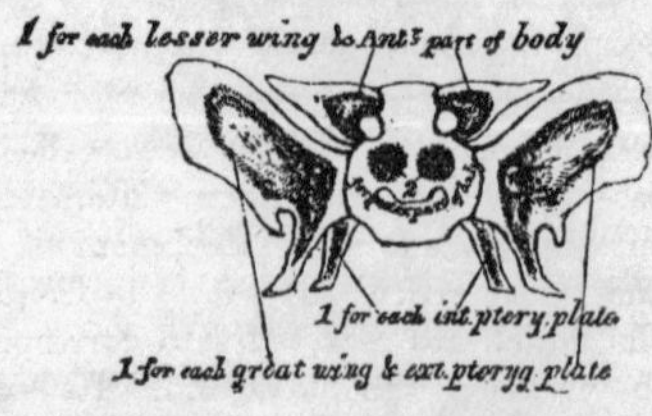

place in these pieces in the following order: the greater wing and external pterygoid plate are first formed, ossific granules being deposited close to the foramen rotundum on each side, at about the second month of fœtal life; from thence ossification spreads outwards into the great wing, and downwards into the external pterygoid plate. Each internal pterygoid plate is then formed, and becomes united to the external about the middle of fœtal life. The two centres for the posterior part of the body appear as separate nuclei, side by side, beneath the sella Turcica; they join, about the middle of fœtal life, into a single piece, which remains un-united to the rest of the bone until after birth. Each lesser wing is formed by a separate centre, which appears on the outer side of the optic foramen, at about the third month; they become united and join with the body at about the eighth month of fœtal life. At about the end of the third year, ossification has made its appearance in the sphenoidal spongy bones.

At birth the sphenoid consists of three pieces: viz. the greater wing and pterygoid processes on each side; the lesser wings and body united. At the first year after birth, the greater wings and body are united. From the tenth to the twelfth year the spongy bones are partially united to the sphenoid, their junction being complete by the twentieth year. Lastly, the sphenoid joins the occipital.

Articulations. The sphenoid articulates with *all* the bones of the cranium, and five of the face; the two malar, two palate, and vomer: the exact extent of articulation with each bone is shown in the accompanying figures.

Attachment of Muscles. The Temporal, External pterygoid, Internal pterygoid, Superior constrictor, Tensor palati, Laxator tympani, Levator palpebræ, Obliquus superior, Superior rectus, Internal rectus, Inferior rectus, External rectus.

The Ethmoid Bone.

The *Ethmoid* (*ἠθμὸς, a sieve*), is an exceedingly light spongy bone, of a cubical form, situated at the anterior part of the base of the cranium, between the two orbits, at the root of the nose, and contributing to form each of these cavities. It consists of three parts: a horizontal plate, which forms part of the base of the cranium; a perpendicular plate, which forms part of the septum nasi; and two lateral masses of cells.

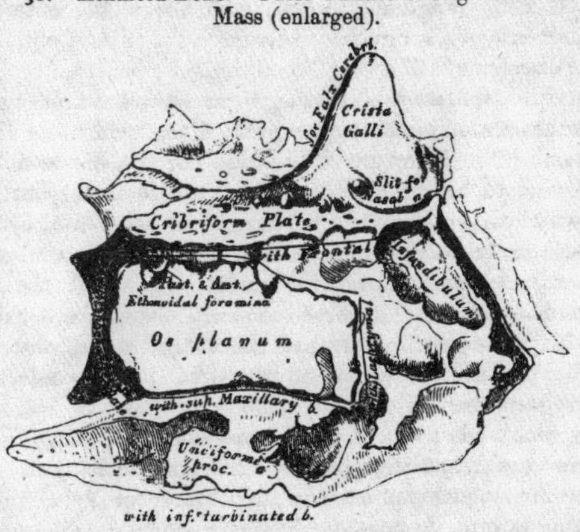

38.—Ethmoid Bone. Outer Surface of Right Lateral Mass (enlarged).

The *Horizontal* or *Cribriform Plate* (fig. 38) forms part of the anterior fossa of the base of the skull, and is received into the ethmoid notch of the frontal bone between the two orbital plates. Projecting upwards from the middle line of this plate, is a thick smooth triangular process of bone, the crista galli, so called from its resemblance to a cock's comb. Its base joins the cribriform plate. Its posterior border, long, thin, and slightly curved, serves for the attachment of the falx cerebri. Its anterior border, short and thick, articulates with the frontal bone, and presents two small projecting alæ, which are received into corresponding depressions in the frontal, completing the foramen cœcum behind. Its sides are smooth, and sometimes bulging; in which case it is found to enclose a small sinus. On each side of the christa galli, the cribriform plate is narrow, and deeply grooved, to support the bulb of the olfactory nerve, and perforated by foramina for the passage of its filaments. These foramina are arranged in three rows: the innermost, which are the largest and least numerous, are lost in grooves on the upper part of the septum; the foramina of the outer row are continued on to the surface of the upper spongy bone. The foramina of the middle row are the smallest; they perforate the bone, and transmit nerves to the roof of the nose. At the front part of the cribriform plate, on each side of the crista galli, is a small fissure, which transmits the nasal branch of the ophthalmic nerve; and at its posterior part a triangular notch, which receives the ethmoidal spine of the sphenoid.

39.—Perpendicular Plate of Ethmoid (enlarged). Shown by removing the right Lateral Mass.

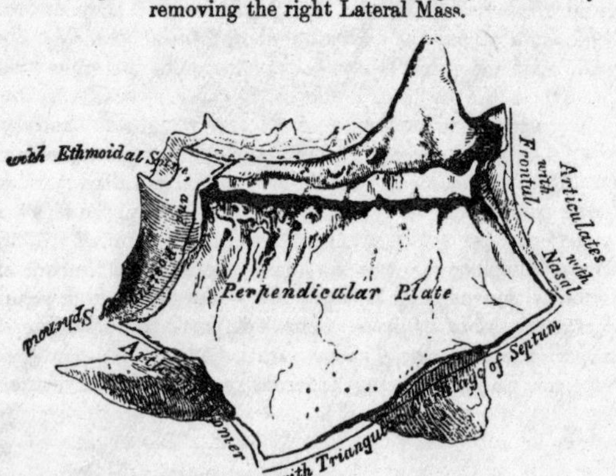

The *Perpendicular Plate* (fig. 39) is a thin flattened lamella of bone, which descends from the under-surface of the cribriform plate, and assists in forming the septum of the nose. It is much thinner in the middle than at the circumference, and is generally deflected a little to one side. Its anterior border articulates with the frontal spine and crest of the nasal bones. Its posterior, divided into two parts, is connected by its upper half with the rostrum of the sphenoid—by its lower half with the vomer. The inferior border serves for the attachment of the triangular cartilage of the nose. On each side of the perpendicular plate numerous grooves and canals are seen, leading from foramina on the cribriform plate; they lodge filaments of the olfactory nerves.

The *Lateral Masses* of the ethmoid consist of a number of thin-walled cellular cavities, the *ethmoidal cells*, interposed between two vertical plates of bone, the outer one of which forms part of the orbit, and the inner one part of the nasal fossa of the corresponding side. In the disarticulated bone many of these cells appear to be broken; but when the bones are articulated, they are closed in at every part. The upper surface of each lateral mass presents a number of apparently half-broken cellular spaces; these are closed in when articulated by the edges of the ethmoidal notch of the frontal bone. Crossing this surface are two grooves on each side, converted into canals by articulation with the frontal; they are the anterior and posterior ethmoidal foramina, and open on the inner wall of the orbit. The posterior surface also presents large irregular cellular cavities, which are closed in by articulation with the sphenoidal turbinated bones, and orbital process of the palate. The cells at the anterior surface are completed by the lachrymal bone and nasal process of the superior maxillary, and those below also by the superior maxillary. The outer surface of each lateral mass is formed of a thin smooth square plate of bone, called the *os planum*; it forms part of the inner wall of the orbit, and articulates above with the orbital plate of the frontal; below, with the superior maxillary and orbital process of the palate; in front, with the lachrymal; and behind, with the sphenoid.

From the inferior part of each lateral mass, immediately beneath the os planum, there projects downwards and backwards an irregular lamina of bone, called the *unciform process*, from its hook-like form: it serves to close in the upper part of the orifice of the antrum, and articulates with the ethmoidal process of the inferior turbinated bone. It is often broken in disarticulating the bones.

The inner surface of each lateral mass forms part of the outer wall of the nasal fossa of the corresponding side. It is formed of a thin lamella of bone, which descends from the under surface of the cribriform plate, and terminates below in a free convoluted margin, the middle turbinated bone. The whole of this surface is rough, and marked above by numerous grooves, which run nearly vertically downwards from the cribriform plate: they lodge branches of the olfactory nerve, which are distributed on the mucous membrane covering the bone. The back part of this surface is subdivided by a narrow oblique fissure, the superior meatus of the nose, bounded above by a thin curved plate of bone—the superior turbinated bone. By means of an orifice at the upper part of this fissure, the posterior ethmoidal cells open into the nose. Below, and in front of the superior meatus, is seen the convex surface of the middle turbinated bone. It extends along the whole length of the inner surface of each lateral mass; its lower margin is free and thick, and its concavity, directed outwards, assists in forming the middle meatus. It is by a large orifice at the upper and front part of the middle meatus, that the anterior ethmoidal cells, and through them the frontal sinuses, communicate with the nose, by means of a funnel-shaped canal, the infundibulum. The cellular cavities of each lateral mass, thus walled in by the os planum on the outer side, and by the other bones already mentioned, are divided by a thin transverse bony partition into two sets, which do not communicate with each other; they are termed the *anterior* and *posterior ethmoidal cells*, or *sinuses*. The former, smaller but more numerous, communicate with the frontal sinuses above, and the middle meatus below, by means of a long flexuous cellular canal, the *infundibulum*; the posterior, larger but less numerous, open into the superior meatus, and communicate (occasionally) with the sphenoidal sinuses.

40.—Ethmoid Bone. Inner Surface of Right Lateral Mass (enlarged).

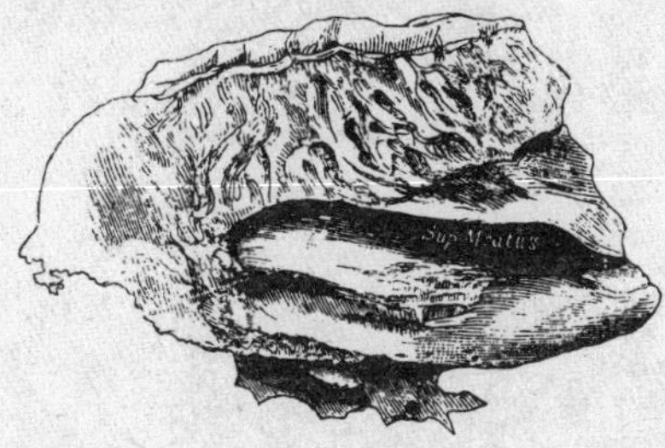

Development. By *three* centres—one for the perpendicular lamella, and one for each lateral mass.

The lateral masses are first developed, ossific granules making their first appearance in the os planum between the fourth and fifth months of fœtal life, and afterwards in the spongy bones. At birth, the bone consists of the two lateral masses, which are small and ill-developed; but when the perpendicular and horizontal plates begin to ossify, as they do about the first year after birth, the lateral masses become joined to the cribriform plate. The formation and increase in the ethmoidal cells, which complete the bone, take place about the fifth or sixth year.

Articulations. With fifteen bones: the sphenoid, two sphenoidal turbinated, the frontal, and eleven of the face—the two nasal, two superior maxillary, two lachrymal, two palate, two inferior turbinated, and the vomer.

Development of the Cranium.

The development of the cranium commences at a very early period, on account of the importance of the organ it is intended to protect. In its most rudimentary state, it consists of a thin membranous capsule, enclosing the cerebrum, and accurately moulded upon its surface. This capsule is placed external to the dura mater, and in close contact with it; its walls are continuous with the canal for the spinal cord, and the chorda dorsalis, or primitive part of the vertebral column, is continued forwards, from the spine, along the base, to its fore-part, where it terminates in a tapering point. The next step in the process of development is the formation of cartilage. This is deposited in the base of the skull, in two symmetrical segments, one on either side of the median line; these subsequently coalesce, so as to enclose the chorda dorsalis—the chief part of the cerebral capsule still retaining its membranous form. Ossification first takes place in the roof, and is preceded by the deposition of a membranous blastema upon the surface of the cerebral capsule, in which the ossifying process extends; the primitive membranous capsule becoming the internal periosteum, and being ultimately blended with the dura mater. Although the bones of the vertex of the skull appear before those at the base, and make considerable progress in their growth: at birth ossification is more advanced in the base, this portion of the skull forming a solid immovable groundwork.

41.—Skull at birth, showing the Anterior and Posterior Fontanelles.

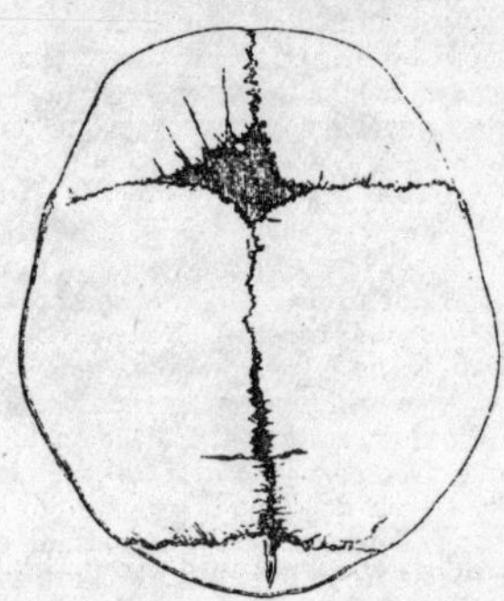

42.—The Lateral Fontanelles.

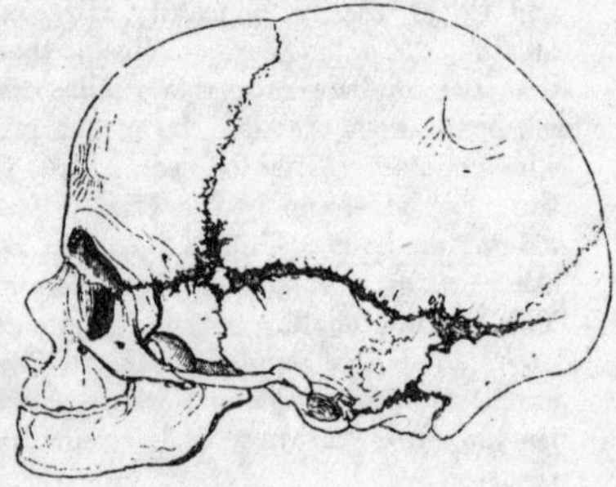

The Fontanelles (figs. 41, 42).

Before birth, the bones at the vertex and side of the skull are separated from each other by membranous intervals, in which bone is deficient. These intervals, at certain parts, are of considerable size, and are termed the *fontanelles,* so called from the pulsations of the brain, which are perceptible at the anterior fontanelle, and were likened to the rising of water in a fountain. The fontanelles are four in number, and correspond to the junction of the four angles of the parietal with the contiguous bones. The anterior fontanelle is the largest, and corresponds to the junction of the sagittal and coronal sutures; the posterior fontanelle, of smaller size, is situated at the junction of the sagittal and lambdoid sutures; the two remaining ones are situated at the inferior angles of the parietal bone. The latter are closed soon after birth; the two at the superior angles remain open longer: the posterior being closed in a few months after birth; the anterior remaining open until the first or second year. These spaces are gradually filled in by an extension of the ossifying process, or by the development of a Wormian bone. Sometimes the anterior fontanelle remains open beyond two years, and is occasionally persistent throughout life.

Supernumerary or Wormian* Bones.

When ossification of any of the tabular bones of the skull proves abortive, the membranous interval which would be left, is usually filled in by a supernumerary piece of bone. This is developed from a separate centre, and gradually extends until it fills the vacant space. These supernumerary pieces are called Wormian bones; they are called also, from their usual form, *ossa triquetra*; but they present much variation in situation, number, and size.

They occasionally occupy the situation of the fontanelles. Bertin, Cruvelhier, and Cuvier have each noticed the presence of one in the anterior fontanelle. There are two specimens in the Museum of St. George's Hospital, which present Wormian bones in this situation. In one, the skull of a child, the supernumerary piece is of considerable size, and of a quadrangular form.

They are occasionally found in the posterior fontanelle, appearing to replace the superior angle of the occipital bone. Not unfrequently, there is one replacing the extremity of the great wing of the sphenoid, or the anterior inferior angle of the parietal bone, in the fontanelle there situated.

They have been found in the different sutures on the vertex and side of the skull, and in some of those at the base. They are most frequent in the lambdoid suture. Mr. Ward mentions an instance 'in which one-half of the lambdoid suture was formed by large Wormian bones disposed in a double row, and jutting deeply into each other;' and refers to similar specimens described by Dumontier and Bourgery.

* Wormius, a physician in Copenhagen, is said to have given the first detailed description of these bones.

A deficiency in the ossification of the flat bones would appear in some cases to be *symmetrical* on the two sides of the skull; for it is not uncommon to find these supernumerary bones corresponding in form, size, and situation on each side. Thus, in several instances, I have seen a pair of large Wormian bones symmetrically placed in the lambdoid suture; in another specimen, a pair in the coronal suture, with a supernumerary bone in the spheno-parietal suture of both sides.

The size of these supernumerary pieces varies, they being in some cases not larger than a pin's head, and confined to the outer table; in other cases so large, that one pair of these bones may form the whole of the occipital bone above the superior curved lines, as described by Béclard and Ward. Their number is generally limited to two or three; but more than a hundred have been found in the skull of an adult hydrocephalic skeleton. In their development, structure, and mode of articulation, they resemble the other cranial bones.

Congenital Fissures and Gaps.

Dr. Humphry has called attention to the not unfrequent existence of *congenital fissures* in the cranial bones, the result of incomplete ossification. These fissures have been noticed in the frontal, parietal, and squamous portion of the temporal bones; they extend from the margin towards the middle of the bone; and are of great interest in a medico-legal point of view, as they are liable to be mistaken for fractures. An arrest of the ossifying process may also give rise to the *deficiencies* or *gaps* occasionally found in the cranial bones. Such deficiencies are said to occur most frequently when ossification is imperfect, and to be situated near the natural apertures for vessels. Dr. Humphry describes such deficiencies to exist in a calvarium, in the Cambridge Museum, where a gap sufficiently large to admit the end of the finger is seen on either side of the sagittal suture, in the place of the parietal foramen. There is a specimen precisely similar to this in the Museum of St. George's Hospital; and another, in which a small circular gap exists in the parietal bone of a young child, just above the parietal eminence. Similar deficiencies are not unfrequently met with in hydrocephalic skulls; being most frequent, according to Dr. Humphry, in the frontal bones; and, in the parietal bones, on either side of the sagittal suture.

Bones of the Face.

The Facial Bones are fourteen in number, viz., the

Two Nasal,	Two Palate,
Two superior Maxillary,	Two Inferior Turbinated,
Two Lachrymal,	Vomer,
Two Malar,	Inferior Maxillary.

Nasal Bones.

The Nasal are two small oblong bones, varying in size and form in different individuals; they are placed side by side at the middle and upper part of the face, forming, by their junction, 'the bridge,' of the nose. Each bone presents for examination two surfaces, and four borders. The *outer* surface is concave from above downwards, convex from side to side; it is covered by the Pyramidalis and Compressor nasi muscles, marked by numerous small arterial furrows, and perforated about its centre by a foramen, sometimes double, for the transmission of a small vein. Sometimes this foramen is absent on one or both sides, and occasionally the foramen cœcum opens on this surface. The *inner* surface is concave from side to side, convex from above downwards; in which direction it is traversed by a longitudinal groove (sometimes a canal), for the passage of a

43.—Right Nasal Bone.

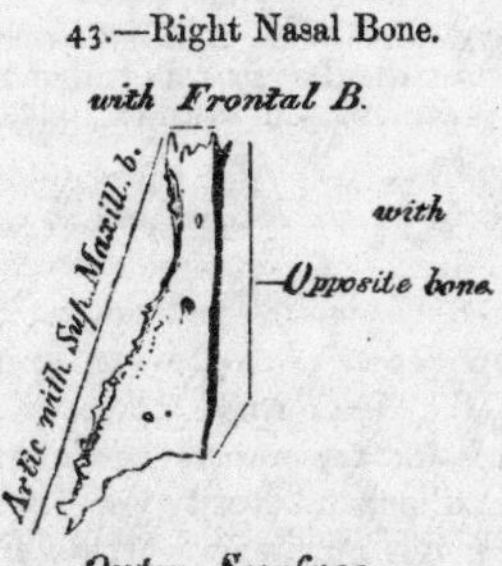

44.—Right Nasal Bone.

with Frontal Spine
crest
with Perpendicular Plate of Ethmoid
groove for nasal nerve
Inner Surface

branch of the nasal nerve. The superior border is narrow, thick, and serrated for articulation with the nasal notch of the frontal bone. The inferior border is broad, thin, sharp, directed obliquely downwards, outwards, and backwards, and serves for the attachment of the lateral cartilage of the nose. This border presents about its centre a notch, through which passes the branch of the nasal nerve above referred to; and is prolonged at its inner extremity into a sharp spine, which, when articulated with the opposite bone, forms the nasal angle. The external border is serrated, bevelled at the expense of the internal surface above, and of the external below, to articulate with the nasal process of the superior maxillary. The internal border, thicker above than below, articulates with its fellow of the opposite side, and is prolonged behind into a vertical crest, which forms part of the septum of the nose: this crest articulates with the nasal spine of the frontal above, and the perpendicular plate of the ethmoid below.

Development. By *one* centre for each bone, which appears about the same period as in the vertebræ.

Articulations. With four bones: two of the cranium, the frontal and ethmoid, and two of the face, the opposite nasal and the superior maxillary.

No muscles are directly attached to this bone.

Superior Maxillary Bone.

The Superior Maxillary is one of the most important bones of the face in a surgical point of view, on account of the number of diseases to which some of its parts are liable. Its minute examination becomes, therefore, a matter of considerable interest. It is the largest bone of the face, excepting the lower jaw; and forms, by its union with its fellow of the opposite side, the whole of the upper jaw. Each bone assists in the formation of three cavities, the roof of the mouth, the floor and outer wall of the nose, and the floor of the orbit; and also enters into the formation of two fossæ, the zygomatic and spheno-maxillary; and two fissures, the spheno-maxillary and pterygo-maxillary.

The bone presents for examination a body and four processes, malar, nasal, alveolar, and palatine.

The body is somewhat quadrilateral, and is hollowed out in its interior to form a large cavity, the antrum of Highmore. Its surfaces are four—an external or facial, a posterior or zygomatic, a superior or orbital, and an internal.

The *external* or *facial surface* (fig. 45) is directed forwards and outwards. In the median line of the bone, just above the incisor teeth, is a depression, the incisive or myrtiform fossa, which gives origin to the Depressor alæ nasi. Above and a little external to it, the Compressor nasi arises. More external, is another depression, the canine fossa, larger and deeper than the incisive fossa, from which it is separated by a vertical ridge, the canine eminence, corresponding to the socket of the canine tooth. The canine fossa gives origin to the Levator anguli oris. Above the canine fossa is the infra-orbital foramen, the termination of the infra-orbital canal; it transmits the infra-orbital nerve and artery. Above the infra-orbital foramen is the margin of the orbit, which affords partial attachment to the Levator labii superioris proprius.

The *posterior* or *zygomatic surface* is convex, directed backwards and outwards, and forms part of the zygomatic fossa. It presents about its centre several apertures leading to canals in the substance of the bone; they are termed the *posterior dental canals*, and transmit the posterior dental vessels and nerves. At the lower part of this surface is a rounded eminence, the maxillary tuberosity, especially prominent after the growth of the wisdom-tooth, rough on its inner side for articulation with the tuberosity of the palate bone. Immediately above the rough surface is a groove, which, running obliquely down on the inner surface of the bone, is converted into a canal by articulation with the palate bone, forming the posterior palatine canal.

The *superior* or *orbital surface* is thin, smooth, triangular, and forms part of the floor of the orbit. It is bounded internally by an irregular margin which articulates, in front, with the lachrymal; in the middle, with the os planum of the ethmoid; behind, with the orbital process of the palate bone; bounded externally by a smooth rounded edge which enters into the formation of the spheno-maxillary fissure, and which sometimes articulates at its anterior extremity with the orbital

45.—Left Superior Maxillary Bone. Outer Surface.

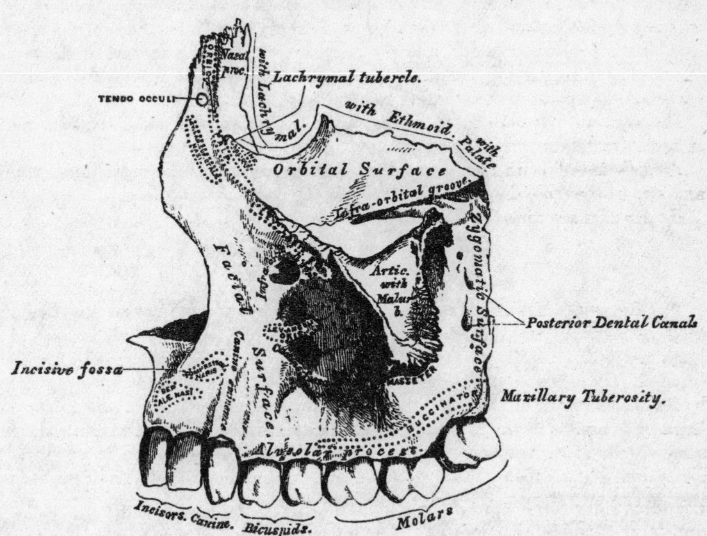

plate of the sphenoid; bounded, in front, by part of the circumference of the orbit, which is continuous, on the inner side with the nasal, on the outer side with the malar process. Along the middle line of the orbital surface is a deep groove, the infra-orbital, for the passage of the infra-orbital nerve and artery. This groove commences at the middle of the outer border of the surface, and, passing forwards, terminates in a canal which subdivides into two branches; one of the canals, the infra-orbital, opens just below the margin of the orbit; the other, which is smaller, runs into the substance of the anterior wall of the antrum; it is called the anterior dental canal, transmitting the anterior dental vessels and nerves to the front teeth of the upper jaw. At the inner and fore part of the orbital surface, just external to the lachrymal canal, is a minute depression, which gives origin to the Inferior oblique muscle of the eye.

The *internal surface* (fig. 46) is unequally divided into two parts by a horizontal projection of bone, the palate process; the portion above the palate process forms part of the outer wall of the nose; that below it forms part of the cavity of the mouth. The superior division of this surface presents a large irregular opening leading into the antrum of Highmore. At the upper border of this aperture are numerous broken cellular cavities, which, in the articulated skull, are closed in by the ethmoid and lachrymal bones. Below the aperture is a smooth concavity which forms part of the inferior meatus of the nose, traversed by a fissure, the maxillary fissure, which runs from the lower part of the orifice of the antrum obliquely downwards and forwards, and receives the maxillary process of the palate bone. Behind it is a rough surface which articulates with the perpendicular plate of the palate bone, traversed by a groove, which, commencing near the middle of the posterior border, runs obliquely downwards and forwards, and forms, when completed by its articulation with the palate bone, the posterior palatine canal. In front of the opening of the antrum is a deep groove, converted into a canal by the lachrymal and inferior turbinated bones, which is coated with mucous membrane, and called the nasal duct. More anteriorly is a well-marked rough ridge, the inferior turbinated crest, for articulation with the inferior turbinated bone. The concavity

46.—Left Superior Maxillary Bone. Inner Surface.

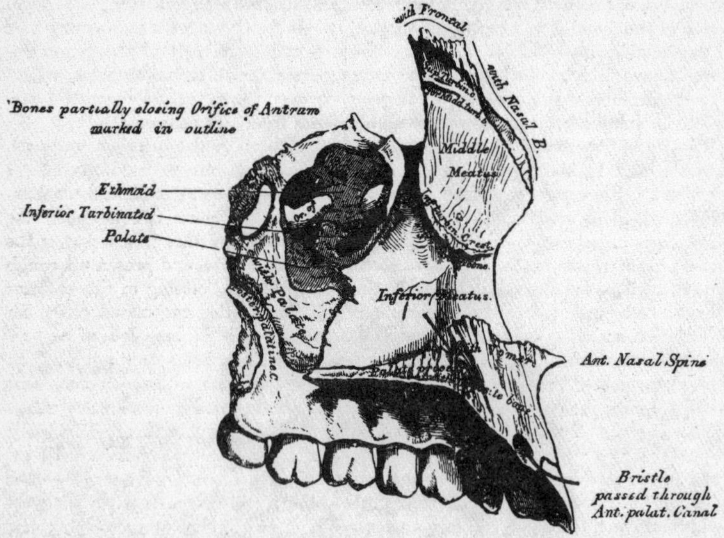

above this ridge forms part of the middle meatus of the nose; whilst that below it forms part of the inferior meatus. The inferior division of this surface is concave, rough, and uneven, and perforated by numerous small foramina for the passage of nutrient vessels.

The *Antrum of Highmore*, or Maxillary Sinus, is a large triangular-shaped cavity, hollowed out of the body of the maxillary bone; its apex, directed outwards, is formed by the malar process; its base, by the outer wall of the nose. Its walls are everywhere exceedingly thin, its roof being formed by the orbital plate, its floor by the alveolar process, its anterior wall by the facial, and its posterior by the zygomatic surface. Its inner wall, or base, presents, in the disarticulated bone, a large irregular aperture, which communicates with the nasal fossa. The margins of this aperture are thin and ragged, and the aperture itself is much contracted by its articulation with the ethmoid above, the inferior turbinated below, and the palate bone behind.* In the articulated skull, this cavity communicates with the middle meatus of the nose generally by two small apertures left between the above-mentioned bones. In the recent state, usually only one small opening exists, near the upper part of the cavity, sufficiently large to admit the end of a probe, the other being closed by the lining membrane of the sinus.

Crossing the cavity of the antrum, are often seen several projecting laminæ of bone, similar to those seen in the sinuses of the cranium; and on its posterior wall are the posterior dental canals, transmitting the posterior dental vessels and nerves to the teeth. Projecting into the floor are several conical processes, corresponding to the roots of the first and second molar teeth; † in some cases the floor is per-

* In some cases, at any rate, the lachrymal bone encroaches slightly on the anterior superior portion of the opening, and assists in forming the inner wall of the antrum.

† The number of teeth whose fangs are in relation with the floor of the antrum is variable. The antrum 'may extend so as to be in relation to all the teeth of the true maxilla, from the canine to the *dens sapientiæ*.'—See Mr. Salter on Abscess of the Antrum, in a *System of Surgery*, edited by T. Holmes, vol. iv. p. 25.

forated by the teeth in this situation. It is from the extreme thinness of the walls of this cavity, that we are enabled to explain how a tumour, growing from the antrum, encroaches upon the adjacent parts, pushing up the floor of the orbit, and displacing the eyeball, projecting inward into the nose, protruding forwards on to the cheek, and making its way backwards into the zygomatic fossa, and downwards into the mouth.

The *Malar Process* is a rough triangular eminence, situated at the angle of separation of the facial from the zygomatic surface. In front it is concave, forming part of the facial surface; behind, it is also concave, and forms part of the zygomatic fossa; above, it is rough and serrated for articulation with the malar bone; whilst below, a prominent ridge marks the division between the facial and zygomatic surfaces. A small part of the Masseter muscle arises from this process.

The *Nasal Process* is a thick triangular plate of bone, which projects upwards, inwards, and backwards, by the side of the nose, forming part of its lateral boundary. Its external surface is concave, smooth, perforated by numerous foramina, and gives attachment to the Levator labii superioris alæque nasi, the Orbicularis palpebrarum, and Tendo oculi. Its internal surface forms part of the outer wall of the nose; it articulates above with the frontal, and presents a rough uneven surface, which articulates with the ethmoid bone, closing in the anterior ethmoid cells; below this is a transverse ridge, the superior turbinated crest, for articulation with the middle turbinated bone of the ethmoid, bounded below by a smooth concavity which forms part of the middle meatus; below this again is the inferior turbinated crest (already described), for articulation with the inferior turbinated bone; and still more inferiorly, the concavity which forms part of the inferior meatus. The anterior border of the nasal process is thin, directed obliquely downwards and forwards, and presents a serrated edge for articulation with the nasal bone: its posterior border is thick, and hollowed into a groove for the nasal duct: of the two margins of this groove, the inner one articulates with the lachrymal bone, the outer one forms part of the circumference of the orbit. Just where the latter joins the orbital surface is a small tubercle, the lachrymal tubercle; this serves as a guide to the surgeon in the performance of the operation for fistula lachrymalis. The lachrymal groove in the articulated skull is converted into a canal by the lachrymal bone, and lachrymal process of the inferior turbinated; it is directed downwards, and a little backwards and outwards, is about the diameter of a goose-quill, slightly narrower in the middle than at either extremity, and lodges the nasal duct.

The *Alveolar Process* is the thickest and most spongy part of the bone, broader behind than in front, and excavated into deep cavities for the reception of the teeth. These cavities are eight in number, and vary in size and depth according to the teeth they contain. That for the canine tooth is the deepest; those for the molars are the widest, and subdivided into minor cavities; those for the incisors are single, but deep and narrow. The Buccinator muscle arises from the outer surface of this process, as far forward as the first molar tooth.

The *Palate Process*, thick and strong, projects horizontally inwards from the inner surface of the bone. It is much thicker in front than behind, and forms a considerable part of the floor of the nostril, and the roof of the mouth. Its upper surface is concave from side to side, smooth and forms part of the floor of the nose. In front is seen the upper orifice of the anterior palatine (incisor) canal, which leads into a fossa formed by the junction of the two superior maxillary bones, and situated immediately behind the incisor teeth. It transmits the anterior palatine vessels, the naso-palatine nerves passing through the intermaxillary suture. The inferior surface, also concave, is rough and uneven, and forms part of the roof of the mouth. This surface is perforated by numerous foramina for the passage of nutritious vessels, channelled at the back part of its alveolar border by a longitudinal groove, sometimes a canal, for the transmission of the posterior palatine vessels, and a large nerve, and presents little depressions for the lodgment of the palatine glands. This surface presents anteriorly the lower orifice of the anterior palatine fossa. In some bones, a delicate linear suture may be seen extending from the anterior palatine fossa to the interval between the lateral incisor and the canine tooth. This marks out the intermaxillary, or incisive, bone, which in some animals exists permanently as a separate piece. It includes the whole thickness of the alveolus, the corresponding part of the floor of the nose, and the anterior nasal spine, and contains the sockets of the incisor teeth. The outer border of the palate process is incorporated with the rest of the bone. The inner border is thicker in front than behind, and is raised above into a ridge, which, with the corresponding ridge in the opposite bone, forms a groove for the reception of the vomer. The anterior margin is bounded by the thin concave border of the opening of the nose, prolonged forwards internally into a sharp process, forming, with a similar process of the opposite bone, the anterior nasal spine. The posterior border is serrated for articulation with the horizontal plate of the palate bone.

47.—Development of Superior Maxillary Bone. By four Centres.

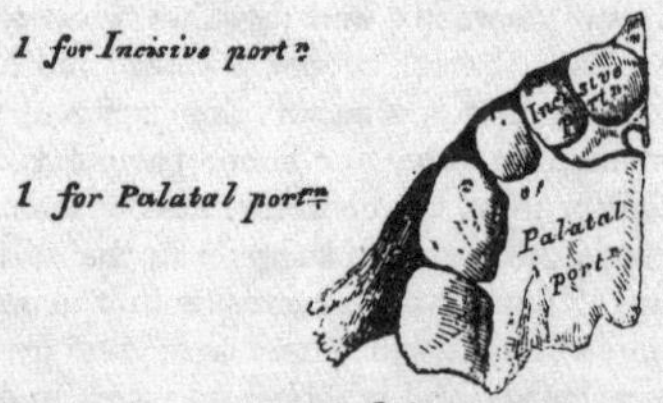

Development. This bone is formed at such an early period, and ossification proceeds in it with such rapidity, that it has been found impracticable hitherto to determine with accuracy its number of centres. It appears, however, probable that it has *four* centres of development, viz., one for the nasal and facial portions, one for the orbital and malar, one for the incisive, and one for the palatal portion, including the entire palate except the incisive segment. The incisive portion is indicated in young bones by a fissure, which marks off a small segment of the palate, including the two incisor teeth. In some animals, this remains permanently as a separate piece, constituting the intermaxillary bone; and in the human subject, where the jaw is malformed, as in cleft palate, this segment may be separated from the maxillary bone by a deep fissure extending backwards between the two into the palate. If the fissure be on both sides, both segments are quite isolated from the maxillary bones, and hang from the end of the vomer: they are not unfrequently much displaced, and the deformity is often accompanied by congenital fissure of the upper lip, either on one or both sides of the median line. The maxillary sinus appears at an earlier period than any of the other nasal sinuses, its development commencing about the fourth month of fœtal life.

Articulations. With *nine* bones; two of the cranium—the frontal and ethmoid, and seven of the face, viz., the nasal, malar, lachrymal, inferior turbinated, palate, vomer, and its fellow of the opposite side. Sometimes it articulates with the orbital plate of the sphenoid.

Attachment of Muscles. Orbicularis palpebrarum, Obliquus inferior oculi, Levator labii superioris alæque nasi, Levator labii superioris proprius, Levator anguli oris, Compressor nasi, Depressor alæ nasi, Masseter, Buccinator.

The Lachrymal Bones.

The *Lachrymal* are the smallest and most fragile bones of the face. They are situated at the front part of the inner wall of the orbit, and resemble somewhat in form, thinness, and size, a finger-nail; hence they are termed the *ossa unguis*. Each bone presents, for examination, two surfaces and four borders. The external (fig. 48) or orbital surface is divided by a vertical ridge into two parts. The portion of bone in front of this ridge presents a smooth, concave, longitudinal groove, the free margin of which unites with the nasal process of the superior maxillary bone, completing the lachrymal groove. The upper part of this groove lodges the lachrymal sac; the lower part assists in the formation of the lachrymal canal, and lodges the nasal duct. The portion of bone behind the ridge is smooth, slightly concave, and forms part of the inner wall of the orbit. The ridge, with a part of the orbital surface immediately behind it, affords attachment to the Tensor tarsi: the ridge terminates below in a small hook-like process, which articulates with the lachrymal tubercle of the superior maxillary bone and completes the upper orifice of the lachrymal canal. It sometimes exists as a separate piece, which is then called the *lesser lachrymal bone*. The internal or nasal surface presents a depressed furrow, corresponding to the ridge on its outer surface. The surface of bone in front of this forms part of the middle meatus; and that behind it articulates with the ethmoid bone, filling in the anterior ethmoidal cells. Of the *four borders*, the anterior is the longest, and articulates with the nasal process of the superior maxillary bone. The posterior, thin and uneven, articulates with the os planum of the ethmoid. The superior, the shortest and thickest, articulates with the internal angular process of the frontal bone. The inferior is divided by the lower edge of the vertical crest into two parts: the posterior part articulates with the orbital plate of the superior maxillary bone; the anterior portion is prolonged downwards into a pointed process, which articulates with the lachrymal process of the inferior turbinated bone, and assists in the formation of the lachrymal canal.

48.—Left Lachrymal Bone. External Surface. (Slightly enlarged.)

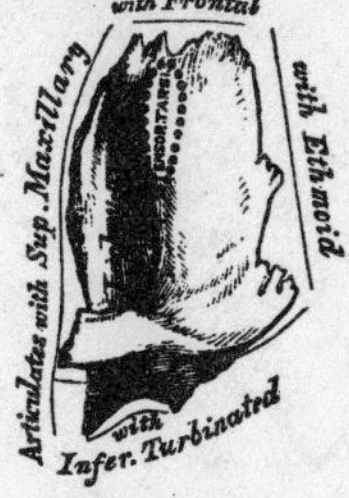

Development. By a single centre, which makes its appearance soon after ossification of the vertebræ has commenced.

Articulations. With four bones; two of the cranium, the frontal and ethmoid, and two of the face, the superior maxillary and the inferior turbinated.

Attachment of Muscles. The Tensor tarsi.

The Malar Bones.

The *Malar* are two small quadrangular bones, situated at the upper and outer part of the face: they form the prominence of the cheek, part of the outer wall and floor of the orbit, and part of the temporal and zygomatic fossæ. Each bone presents for examination an external and an internal surface; four processes, the frontal, orbital, maxillary, and zygomatic; and four borders. The external surface (fig. 49) is smooth, convex, perforated near its centre by one or two small apertures, the malar foramina, for the passage of nerves and vessels, covered by the Orbicularis palpebrarum muscle, and affords attachment to the Zygomaticus major and minor muscles.

The internal surface (fig. 50), directed backwards and inwards, is concave, presenting internally a rough triangular surface, for articulation with the superior maxillary bone; and externally, a smooth concave surface, which forms the anterior boundary of the temporal fossa above; and below, where it is wider, forms part of the zygomatic fossa. This surface presents, a little above its centre, the aperture of one or two malar canals, and affords attachment to part of two muscles, the Temporal above and the Masseter below. Of the four processes, the *frontal* is thick and serrated, and articulates with the external angular process of the frontal bone. The *orbital* process is a thick and strong plate, which projects backwards from the orbital margin of the bone. Its upper surface, smooth and concave, forms, by its junction with the great ala of the sphenoid, the outer wall of the orbit. Its under surface, smooth and convex, forms part of the temporal fossa. Its anterior margin is smooth and rounded, forming part of the circumference of the orbit. Its superior margin, rough, and directed horizontally, articulates with the frontal bone behind the external angular process. Its posterior margin is rough and serrated for articulation with the sphenoid; internally it is also serrated for articulation with the orbital surface of the superior maxillary. At the angle of junction of the sphenoidal and maxillary portions, a short rounded non-articular margin is generally seen: this forms the anterior boundary of the spheno-maxillary fissure; occasionally, no such non-articular margin exists, the fissure being completed by the direct junction of the maxillary and sphenoid bones, or by the interposition of a small Wormian bone in the angular interval between them. On the upper surface of the orbital process are seen the orifices of one or two temporo-malar canals; one of these usually opens on the posterior surface, the other (occasionally two), on the facial surface: they transmit filaments (temporo-malar) of the orbital branch of the superior maxillary nerve. The *maxillary* process is a rough triangular surface, which articulates with the superior maxillary bone. The *zygomatic* process, long, narrow, and serrated, articulates with the zygomatic process of the temporal bone. *Of the four borders*, the superior or orbital is smooth, arched, and forms a considerable part of the circumference of the orbit. The inferior, or zygomatic, is continuous with the lower border of the zygomatic arch, affording attachment by its rough edge to the Masseter muscle. The anterior or maxillary border is rough, and bevelled at the expense of its inner table, to articulate with the superior maxillary bone; affording attachment by its outer margin to the Levator labii superioris proprius, just at its point of junction with the superior maxillary. The posterior or temporal border, curved like an italic *ſ*, is continuous above with the commencement of the temporal ridge; below, with the upper border of the zygomatic arch: it affords attachment to the temporal fascia.

49.—Left Malar Bone. Outer Surface.

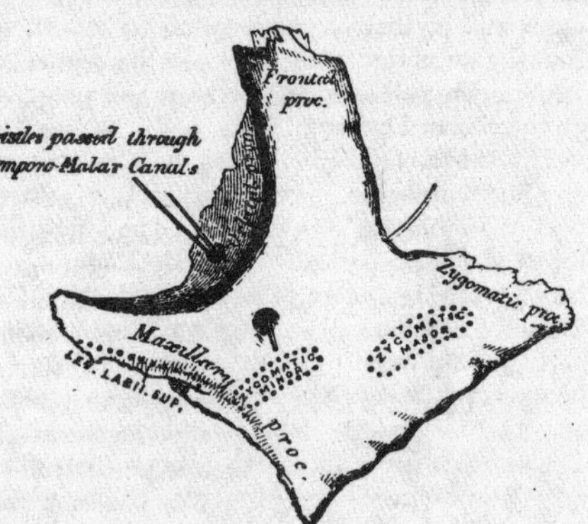

50.—Left Malar Bone. Inner Surface.

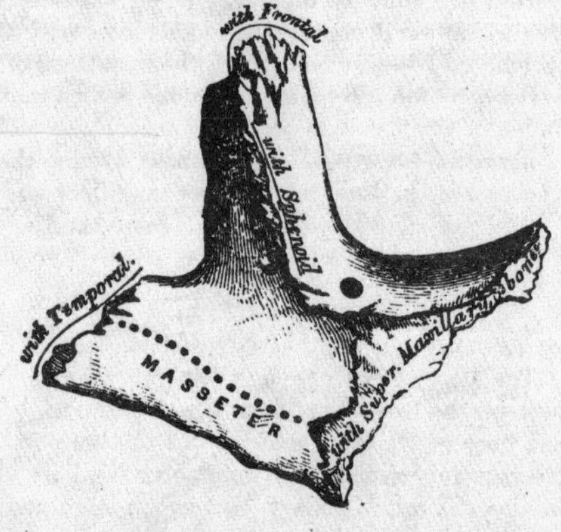

Development. By a single centre of ossification, which appears at about the same period when ossification of the vertebræ commences.

Articulations. With four bones: three of the cranium, frontal, sphenoid, and temporal; and one of the face, the superior maxillary.

Attachment of Muscles. Levator labii superioris proprius, Zygomaticus major and minor, Masseter, and Temporal.

The Palate Bones.

The Palate Bones are situated at the back part of the nasal fossæ; they are wedged in between the superior maxillary and the pterygoid process of the sphenoid. Each bone assists in the formation of three cavities; the floor and outer wall of the nose, the roof of the mouth, and the floor of the orbit; and enters into the formation of three fossæ: the zygomatic, spheno-maxillary, and pterygoid. In form the palate bone somewhat resembles the letter L, and may be divided into an inferior or horizontal plate, and a superior or vertical plate.

The *Horizontal Plate* is thick, of a quadrilateral form, and presents two surfaces and four borders. The superior surface, concave from side to side, forms the back part of the floor of the nostril. The inferior surface, slightly concave and rough, forms the back part of the hard palate. At its posterior part may be seen a transverse ridge, more or less marked, for the attachment of the aponeurosis of the Tensor palati muscle. At the outer extremity of this ridge is a deep groove, converted into a canal by its articulation with the tuberosity of the superior maxillary bone, and forming the posterior palatine canal. Near this groove, the orifices of one or two small canals, accessory posterior palatine, may frequently be seen. The anterior border is serrated, bevelled at the expense of its inferior surface, and articulates with the palate process of the superior maxillary bone. The posterior border is concave, free, and serves for the attachment of the soft palate. Its inner extremity is sharp and pointed, and, when united with the opposite bone, forms a projecting process, the posterior nasal spine, for the attachment of the Azygos uvulæ. The external border is united with the lower part of the perpendicular plate almost at right angles. The internal border, the thickest, is serrated for articulation with its fellow of the opposite side; its superior edge is raised into a ridge, which, united with the opposite bone, forms a crest in which the vomer is received.

The *Vertical Plate* (fig. 51) is thin, of an oblong form, and directed upwards and a little inwards. It presents two surfaces, an external and an internal, and four borders.

51.—Left Palate Bone. Internal View (enlarged).

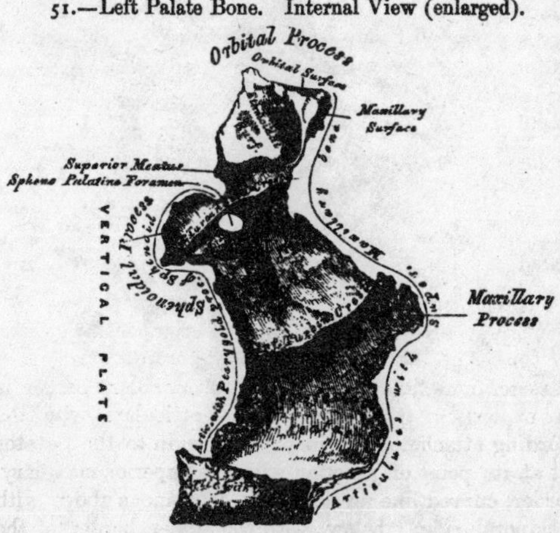

The *internal surface* presents at its lower part a broad shallow depression, which forms part of the inferior meatus of the nose. Immediately above this is a well-marked horizontal ridge, the inferior turbinated crest, for articulation with the inferior turbinated bone; above this, a second broad shallow depression, which forms part of the middle meatus, surmounted above by a horizontal ridge less prominent than the inferior, the superior turbinated crest, for articulation with the middle turbinated bone. Above the superior turbinated crest is a narrow horizontal groove, which forms part of the superior meatus.

The *external surface* is rough and irregular throughout the greater part of its extent, for articulation with the inner surface of the superior maxillary bone, its upper and back part being smooth where it enters into the formation of the spheno-maxillary fossa; it is also smooth in front, where it covers the orifice of the antrum. Towards the back part of this surface is a deep groove, converted into a canal, the posterior palatine, by its articulation with the superior maxillary bone. It transmits the posterior, or descending palatine vessels, and a large nerve.

The anterior border is thin, irregular, and presents opposite the inferior turbinated crest a pointed projecting lamina, the maxillary process, which is directed forwards, and closes in the lower and back part of the opening of the antrum, being received into a fissure that exists at the inferior part of this aperture. The posterior border (fig. 52) presents a deep groove, the edges of which are serrated for articulation with the pterygoid process of the sphenoid. At the lower part of this border is seen a pyramidal process of bone, the *pterygoid process* or tuberosity of the palate, which is received into the angular interval between the two pterygoid plates of the sphenoid at their inferior extremity. This process presents at its back part three grooves, a median and two lateral ones. The former is smooth, and forms part of the pterygoid fossa, affording attachment to the Internal pterygoid muscle; whilst the lateral grooves are rough and uneven, for articulation with the anterior border of each pterygoid plate. A few fibres of the External pterygoid muscle also arise from the tuberosity of the palate bone. The base of this process, continuous with the horizontal portion of the bone, presents the apertures of the accessory descending palatine canals; whilst its outer surface is rough for articulation with the inner surface of the body of the superior maxillary bone.

The superior border of the vertical plate presents two well-marked processes separated by an intervening notch or foramen. The anterior, or larger, is called the *orbital process;* the posterior, the *sphenoidal.*

The *Orbital Process,* directed upwards and outwards, is placed on a higher level than the sphenoidal. It presents five surfaces, which enclose a hollow cellular cavity, and is connected to the perpendicular plate by a narrow constricted neck. Of these five surfaces, three are articular, two non-articular, or free surfaces. The three articular are the anterior or *maxillary* surface, which is directed forwards, outwards, and downwards, is of an oblong form, and rough for articulation with the superior maxillary bone. The posterior or *sphenoidal* surface is directed backwards, upwards, and inwards. It ordinarily presents a small open cell, which communicates with the sphenoidal sinus, and the margins of which are serrated for articulation with the vertical part of the sphenoidal turbinated bone. The internal or *ethmoidal* surface is directed inwards, upwards, and forwards, and articulates with the lateral mass of the ethmoid bone. In some cases, the cellular cavity above-mentioned opens on this surface of the bone; it then communicates with the posterior ethmoidal cells. More rarely it opens on both surfaces, and then communicates both with the

52.—Left Palate Bone. Posterior View (enlarged).

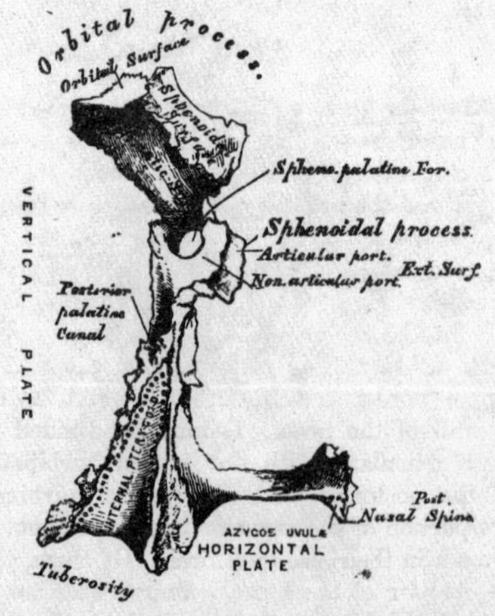

posterior ethmoidal cells, and the sphenoidal sinus. The non-articular or free surfaces are the superior or *orbital*, directed upwards and outwards, of triangular form, concave, smooth, and forming the back part of the floor of the orbit, and the external or *zygomatic* surface, directed outwards, backwards, and downwards, of an oblong form, smooth, lying in the spheno-maxillary fossa, and looking into the zygomatic fossa. The latter surface is separated from the orbital by a smooth rounded border, which enters into the formation of the spheno-maxillary fissure.

The *Sphenoidal Process* of the palate bone is a thin compressed plate, much smaller than the orbital, and directed upwards and inwards. It presents three surfaces and two borders. The superior surface, the smallest of the three, articulates with the horizontal part of the sphenoidal turbinated bone; it presents a groove which contributes to the formation of the pterygo-palatine canal. The internal surface is concave, and forms part of the outer wall of the nasal fossa. The external surface is divided into an articular and a non-articular portion; the former is rough for articulation with the inner surface of the pterygoid process of the sphenoid; the latter is smooth, and forms part of the zygomatic fossa. The anterior border forms the posterior boundary of the spheno-palatine foramen. The posterior border, serrated at the expense of the outer table, articulates with the inner surface of the pterygoid process.

The orbital and sphenoidal processes are separated from one another by a deep notch, which is converted into a foramen, the spheno-palatine, by articulation with the sphenoidal turbinated bone. Sometimes the two processes are united above, and form between them a complete foramen, or the notch is crossed by one or more spiculæ of bone, so as to form two or more foramina. In the articulated skull, this foramen opens into the back part of the outer wall of the superior meatus, and transmits the spheno-palatine vessels and nerves.

Development. From a single centre, which makes its appearance at the angle of junction of the two plates of the bone. From this point ossification spreads inwards to the horizontal plate, downwards into the tuberosity, and upwards into the vertical plate. In the fœtus, the horizontal plate is much longer than the vertical; and even after it is fully ossified, the whole bone is at first remarkable for its shortness.

Articulations. With six bones; the sphenoid, ethmoid, superior maxillary, inferior turbinated, vomer, and opposite palate.

Attachment of Muscles. The Tensor palati, Azygos uvulæ, Internal and External pterygoid.

The Inferior Turbinated Bones.

The *Inferior Turbinated* Bones are situated one on each side of the outer wall of the nasal fossæ. Each consists of a layer of thin spongy bone, curled upon itself like a scroll, hence its name 'turbinated;' and extends horizontally along the outer wall of the nasal fossa, immediately below the orifice of the antrum. Each bone presents two surfaces, two borders, and two extremities.

The *internal surface* (fig. 53) is convex, perforated by numerous apertures, and traversed by longitudinal grooves and canals for the lodgment of arteries and veins. In the recent state it is covered by the lining membrane of the nose. The *external surface* is concave (fig. 54), and forms part of the inferior meatus. Its upper border is thin, irregular, and connected to various bones along the outer wall of the nose. It may be divided into three portions; of these, the anterior articulates with the inferior turbinated crest of the superior maxillary bone; the posterior with the inferior turbinated crest of the palate bone; the middle portion of the superior border presents three well-marked processes, which vary much in their size and form. Of these, the anterior and smallest is situated at the junction of the anterior fourth with the posterior three-fourths of the bone; it is small and pointed, and is called the *lachrymal process*, for it articulates with the anterior inferior angle of the lachrymal bone, and by its margins, with the groove on the back of the nasal process of the superior maxillary, and thus assists in forming the lachrymal canal. At the junction of the two middle fourths of the bone, but encroaching on its posterior fourth, a broad thin plate, the *ethmoidal process*, ascends to join the unciform process of the ethmoid: from the lower border of this process a thin lamina of bone curves downwards and outwards, hooking over the lower edge of the orifice of the antrum, which it narrows below: it is called the *maxillary process*, and fixes the bone firmly on to the outer wall of the nasal fossa. The inferior border is free, thick and cellular in structure, more especially in the middle of the bone. Both extremities are more or less narrow and pointed. If the bone is held so that its outer concave surface is directed backwards (i. e. towards the holder), and its superior border, from which the lachrymal and ethmoidal processes project, upwards, the lachrymal process will be directed to the side to which the bone belongs.

53.—Right Inferior Turbinated Bone. Inner Surface.

54.—Right Inferior Turbinated Bone. Outer Surface.

Development. By a single centre which makes its appearance about the middle of fœtal life.

Articulations. With four bones; one of the cranium, the ethmoid, and three of the face, the superior maxillary, lachrymal and palate.

No muscles are attached to this bone.

The Vomer.

The *Vomer* is a single bone, situated vertically at the back part of the nasal fossæ, forming part of the septum of the nose. It is thin, somewhat like a ploughshare in form; but it varies in different individuals, being frequently bent to one or the other side; it presents for examination two surfaces and four borders. The lateral surfaces are smooth, marked by small furrows for the lodgment of blood-vessels, and by a groove on each side, sometimes a canal, the naso-palatine, which runs obliquely downwards and forwards to the intermaxillary suture between the two anterior palatine canals; it transmits the naso-palatine nerve. The superior border, the thickest, presents a deep groove, bounded on each side by a horizontal projecting ala of bone: the groove receives the rostrum of the sphenoid, whilst the alæ are overlapped and retained by laminæ (the vaginal processes) which project from the under surface of the body of the sphenoid at the base of the pterygoid processes. At the front of the groove a fissure is left for the transmission of blood-vessels to the substance of the bone. The inferior border, the longest, is broad and uneven

55.—Vomer.

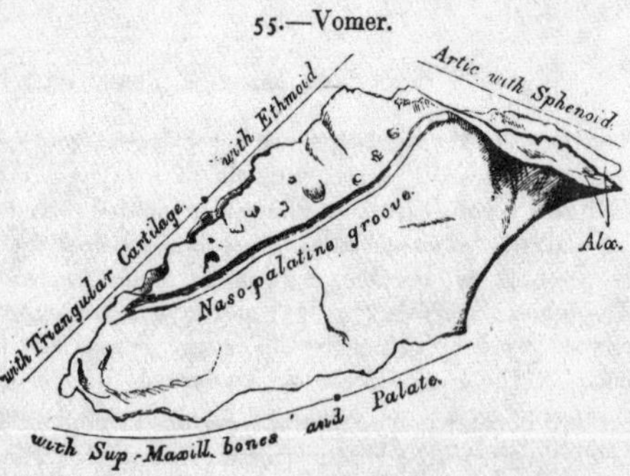

in front, where it articulates with the two superior maxillary bones; thin and sharp behind, where it joins with the palate bones. The upper half of the anterior border usually consists of two laminæ of bone, between which is received the perpendicular plate of the ethmoid, the lower half consisting of a single rough edge, also occasionally channelled, which is united to the triangular cartilage of the nose. The posterior border is free, concave, and separates the nasal fossæ behind. It is thick and bifid above, thin below.

Development. The vomer at an early period consists of two laminæ separated by a very considerable interval, and enclosing between them a plate of cartilage which is prolonged forwards to form the remainder of the septum. Ossification commences in it at about the same period as in the vertebræ (the coalescence of the laminæ taking place from behind forwards), but is not complete until after puberty.

Articulations. With six bones; two of the cranium, the sphenoid and ethmoid; and four of the face, the two superior maxillary and the two palate bones, and with the cartilage of the septum.

The vomer has no muscles attached to it.

The Inferior Maxillary Bone.

The *Inferior Maxillary* Bone, the largest and strongest bone of the face, serves for the reception of the lower teeth. It consists of a curved horizontal portion, the body, and two perpendicular portions, the rami, which join the back part of the body nearly at right angles.

The *Horizontal* portion, or body (fig. 56), is convex in its general outline, and curved somewhat like a horse-shoe. It presents for examination two surfaces

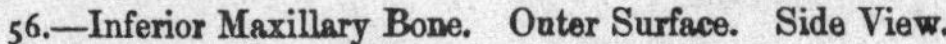

56.—Inferior Maxillary Bone. Outer Surface. Side View.

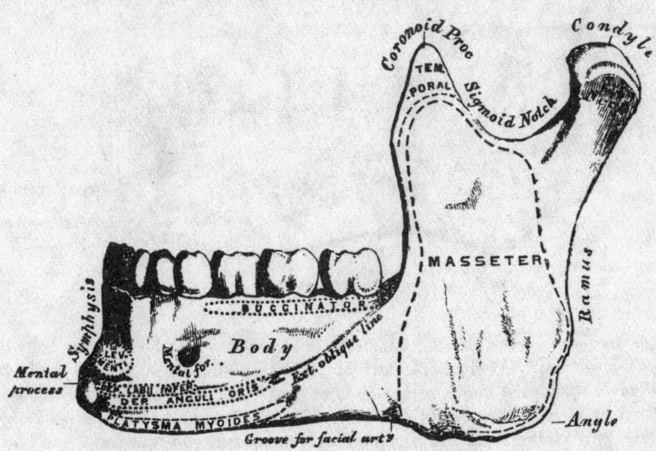

and two borders. The *external surface* is convex from side to side, concave from above downwards. In the median line is a vertical ridge, the symphysis, which extends from the upper to the lower border of the bone, and indicates the point of junction of the two pieces of which the bone is composed at an early period of life. The lower part of the ridge terminates in a prominent triangular eminence, the mental process. On either side of the symphysis, just below the roots of the incisor teeth, is a depression, the incisive fossa, for the attachment of the Levator menti; and still more externally, a foramen, the mental foramen, for the passage of the mental nerve and artery. This foramen is placed just below the root of the second bicuspid tooth. Running outwards from the base of the mental process on each side, is a well-marked ridge, the external oblique line. This ridge is at first nearly horizontal, but afterwards inclines upwards and backwards, and is continuous with the anterior border of the ramus; it affords attachment to the Depressor labii inferioris and Depressor anguli oris, below which the Platysma myoides is inserted. The external oblique line and the internal or mylo-hyoidean line (to be afterwards described), divide the body of the bone into a superior or alveolar, and an inferior or basilar portion.

The *internal surface* (fig. 57) is concave from side to side, convex from above downwards. In the middle line is an indistinct linear depression, corresponding to the symphysis externally; on either side of this depression, just below its centre, are four prominent tubercles, placed in pairs, two above and two below; they are called the *genial tubercles*, and afford attachment, the upper pair to the Genio-hyo-glossi muscles, the lower pair to the Genio-hyoidei muscles. Sometimes the tubercles on each side are blended into one, or they all unite into an irregular eminence of bone, or nothing but an irregularity may be seen on the surface of the bone at this part. On either side of the genial tubercles is an oval depression, the sublingual fossa, for lodging the sublingual gland; and beneath the fossa, a

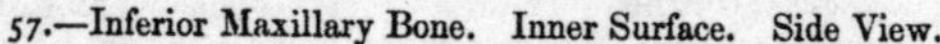

57.—Inferior Maxillary Bone. Inner Surface. Side View.

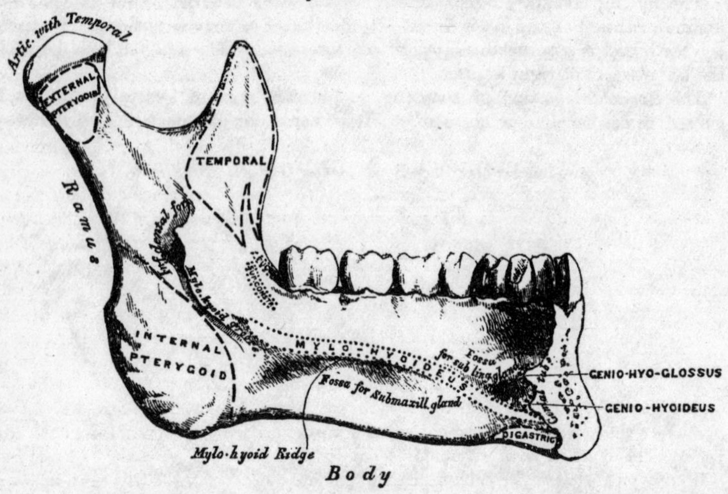

rough depression on each side, which gives attachment to the anterior belly of the Digastric muscle. At the back part of the sublingual fossa, the internal oblique line (mylo-hyoidean) commences; it is at first faintly marked, but becomes more distinct as it passes upwards and outwards, and is especially prominent opposite the last two molar teeth; it divides the lateral surface of the bone into two portions, and affords attachment throughout its whole extent to the Mylo-hyoid muscle, the Superior constrictor being attached above its posterior extremity, nearer the alveolar margin. The portion of bone above this ridge is smooth, and covered by the mucous membrane of the mouth: whilst that below it presents an oblong depression, the submaxillary fossa, wider behind than in front, for the lodgment of the submaxillary gland.

The *superior* or *alveolar border* is wider, and its margins thicker behind than in front. It is hollowed into numerous cavities, for the reception of the teeth; these cavities are sixteen in number, and vary in depth and size according to the teeth which they contain. To its outer side, the Buccinator muscle is attached as far forward as the first molar tooth. The *inferior border* is rounded, longer than the superior, and thicker in front than behind; it presents a shallow groove, just where the body joins the ramus, over which the facial artery turns.

The *Perpendicular Portions*, or *Rami*, are of a quadrilateral form. Each presents for examination two surfaces, four borders, and two processes. The *external surface* is flat, marked with ridges, and gives attachment throughout nearly the whole of its extent to the Masseter muscle. The *internal surface* presents about its centre the oblique aperture of the inferior dental canal, for the passage of the inferior dental vessels and nerve. The margin of this opening is irregular; it presents in front a prominent ridge, surmounted by a sharp spine, which gives attachment to the internal lateral ligament of the lower jaw; and at its lower and back part a notch leading to a groove, the mylo-hyoidean, which runs obliquely downwards to the back part of the submaxillary fossa; and lodges the mylo-hyoid vessels and nerve: behind the groove is a rough surface, for the insertion of the Internal pterygoid muscle. The inferior dental canal runs obliquely downwards and forwards in the substance of the ramus, and then horizontally forwards in the body; it is here placed under the alveoli, with which it communicates by small openings. On arriving at the incisor teeth, it turns back to communicate with the mental foramen, giving off two small canals, which run forward, to be lost in the cancellous tissue of the bone beneath the incisor teeth. This canal, in the posterior two-thirds of the bone, is situated nearer the internal surface of the jaw; and in the anterior third, nearer its external surface. Its walls are composed of compact tissue at either extremity, and of cancellous in the centre. It contains the inferior dental vessels and nerve, from which branches are distributed to the teeth through small apertures at the bases of the alveoli. The *upper border* of the ramus is thin, and presents two processes, separated by a deep concavity, the sigmoid notch. Of these processes, the anterior is the coronoid, the posterior the condyloid.

The *Coronoid Process* is a thin, flattened, triangular eminence of bone, which varies in shape and size in different subjects, and serves chiefly for the attachment of the Temporal muscle. Its *external surface* is smooth, and affords attachment to the Masseter and Temporal muscles. Its *internal surface* gives attachment to the Temporal muscle, and presents the commencement of a longitudinal ridge, which is continued to the posterior part of the alveolar process. On the outer side of this ridge is a deep groove, continued below on the outer side of the alveolar process; this ridge and part of the groove afford attachment, above, to the Temporal; below, to the Buccinator muscle.

The *Condyloid Process*, shorter but thicker than the coronoid, consists of two portions: the condyle, and the constricted portion which supports the condyle, the neck. The condyle is of an oblong form, its long axis being transverse, and set obliquely on the neck in such a manner that its outer end is a little more forward and a little higher than its inner. It is convex from before backwards, and from side to side, the articular surface extending further on the posterior than on the anterior surface. The neck of the condyle is flattened from before backwards, and strengthened by ridges which descend from the fore part and sides of the condyle. Its lateral margins are narrow, and present externally a tubercle for the external lateral ligament. Its posterior surface is convex; its anterior is hollowed out on its inner side by a depression (the pterygoid fossa) for the attachment of the External pterygoid.

The *lower border* of the ramus is thick, straight, and continuous with the body of the bone. At its junction with the posterior border is the angle of the jaw, which is either inverted or everted, and marked by rough oblique ridges on each side for the attachment of the Masseter externally, and the Internal pterygoid internally; the stylo-maxillary ligament is attached to the bone between these muscles. The *anterior border* is thin above, thicker below, and continuous with the external oblique line. The *posterior border* is thick, smooth, rounded, and covered by the parotid gland.

SIDE-VIEW OF THE LOWER JAW AT DIFFERENT PERIODS OF LIFE.

58.—At Birth.

59.—At Puberty.

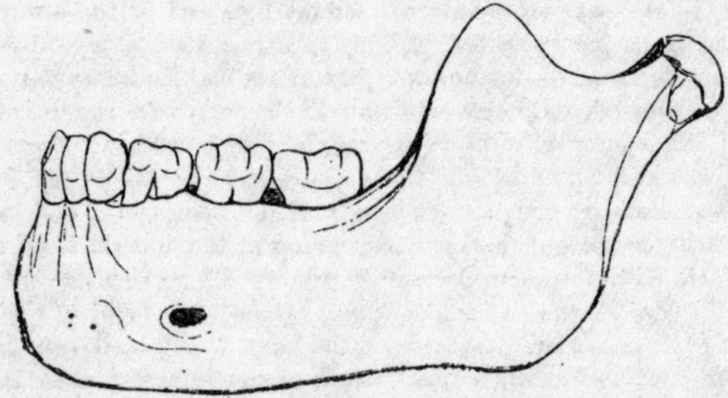

60.—In the Adult.

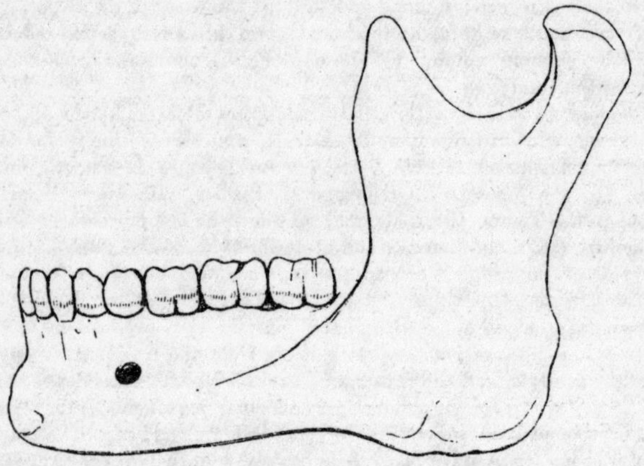

61.—In Old Age.

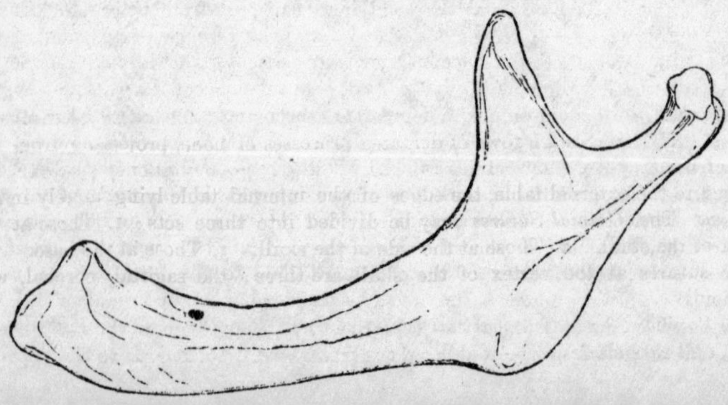

The *Sigmoid Notch*, separating the two processes, is a deep semilunar depression, crossed by the masseteric artery and nerve.

Development. This bone is formed at such an early period of life, before, indeed, any other bone except the clavicle, that it has been found impossible at present to determine its earliest condition. It appears probable, however, that it is developed by *two* centres, one for each lateral half, the two segments meeting at the symphysis, where they become united. Additional centres have also been described for the coronoid process, the condyle, the angle, and the thin plate of bone which forms the inner side of the alveolus.

Changes produced in the Lower Jaw by Age.

The changes which the Lower Jaw undergoes after birth, relate—1. To the alterations effected in the body of the bone by the first and second dentitions, the loss of the teeth in the aged, and the subsequent absorption of the alveoli. 2. To the size and situation of the dental canal; and, 3. To the angle at which the ramus joins with the body.

At birth (fig. 58), the bone consists of two lateral halves, united by fibro-cartilaginous tissue, in which one or two osseous nuclei are generally found. The body is a mere shell of bone containing the sockets of the two incisor, the canine, and the two temporary molar teeth, imperfectly partitioned from one another. The dental canal is of large size, and runs near the lower border of the bone, the mental foramen opening beneath the socket of the first molar. The angle is obtuse, from the jaws not being as yet separated by the eruption of the teeth.

After birth (fig. 59), the two segments of the bone become joined at the symphysis, from below upwards, in the first year; but a trace of separation may be visible in the beginning of the second year, near the alveolar margin. The body becomes elongated in its whole length, but more especially behind the mental foramen, to provide space for the three additional teeth developed in this part. The depth of the body becomes greater, owing to increased growth of the alveolar part, to afford room for the fangs of the teeth, and by thickening of the subdental portion which enables the jaw to withstand the powerful action of the masticatory muscles; but the alveolar portion is the deeper of the two, and, consequently, the chief part of the body lies above the oblique line. The dental canal, after the second dentition, is situated just above the level of the mylo-hyoid ridge; and the mental foramen occupies the position usual to it in the adult. The angle becomes less obtuse, owing to the separation of the jaws by the teeth.

In the adult (fig. 60), the alveolar and basilar portions of the body are usually of equal depth. The mental foramen opens midway between the upper and lower border of the bone, and the dental canal runs nearly parallel with the mylo-hyoid line. The ramus is almost vertical in direction, and joins the body nearly at right angles.

In old age (fig. 61), the bone becomes greatly reduced in size; for, with the loss of the teeth, the alveolar process is absorbed, and the basilar part of the bone alone remains; consequently, the chief part of the bone is *below* the oblique line. The dental canal, with the mental foramen opening from it, is close to the alveolar border. The rami are oblique in direction, and the angle obtuse.

Articulations. With the glenoid fossæ of the two temporal bones.

Attachment of Muscles. To its external surface, commencing at the symphysis, and proceeding backwards: Levator menti, Depressor labii inferioris, Depressor anguli oris, Platysma myoides, Buccinator, Masseter. To its internal surface, commencing at the same point: Genio-hyo-glossus, Genio-hyoideus, Mylo-hyoideus, Digastric, Superior constrictor, Temporal, Internal pterygoid, External pterygoid.

THE SUTURES.

The bones of the cranium and face are connected to each other by means of sutures. The sutures are rows of dentated processes of bone, projecting from the edge of either bone, and locking into each other: the dentations, however, are confined to the external table, the edges of the internal table lying merely in apposition. The *Cranial Sutures* may be divided into three sets: 1. Those at the vertex of the skull. 2. Those at the side of the skull. 3. Those at the base.

The sutures at the vertex of the skull are three: the sagittal, coronal, and lambdoid.

The *Sagittal Suture* (*interparietal*) is formed by the junction of the two parietal bones, and extends from the middle of the frontal bone, backwards to the superior angle of the occipital. In childhood, and occasionally in the adult, when the two halves of the frontal bone are not united, it is continued forwards to the root of the nose. This suture sometimes presents, near its posterior extremity, the parietal foramen on each side; and in front, where it joins the coronal suture, a space is occasionally left, which encloses a large Wormian bone.

The *Coronal Suture* (*fronto-parietal*) extends transversely across the vertex of the skull, and connects the frontal with the parietal bones. It commences at the extremity of the great wing of the sphenoid on one side, and terminates at the same point on the opposite side. The dentations of this suture are more marked at the sides than at the summit, and are so constructed that the frontal rests on the parietal above, whilst laterally the frontal supports the parietal.

The *Lambdoid Suture* (*occipito-parietal*), so called from its resemblance to the Greek letter Λ, connects the occipital with the parietal bones. It commences on each side at the mastoid portion of the temporal bone, and inclines upwards to the end of the sagittal suture. The dentations of this suture are very deep and distinct, and are often interrupted by several small Wormian bones.

The sutures at the side of the skull are also three in number: the spheno-parietal, squamo-parietal, and masto-parietal. They are subdivisions of a single suture, formed between the lower border of the parietal, and the temporal and sphenoid bones, and which extends from the lower end of the lambdoid suture behind, to the lower end of the coronal suture in front.

The *Spheno-parietal* is very short; it is formed by the tip of the great wing of the sphenoid, which overlaps the anterior inferior angle of the parietal bone.

The *Squamo-parietal*, or squamous suture, is arched. It is formed by the squamous portion of the temporal bone overlapping the middle division of the lower border of the parietal.

The *Masto-parietal* is a short suture, deeply dentated, formed by the posterior inferior angle of the parietal, and the superior border of the mastoid portion of the temporal.

The sutures at the base of the skull are, the basilar in the centre, and on each side, the petro-occipital, the masto-occipital, the petro-sphenoidal, and the squamo-sphenoidal.

The *Basilar Suture* is formed by the junction of the basilar surface of the occipital bone with the posterior surface of the body of the sphenoid. At an early period of life, a thin plate of cartilage exists between these bones; but in the adult they become fused into one. Between the outer extremity of the basilar suture, and the termination of the lambdoid, an irregular suture exists, which is subdivided into two portions. The inner portion, formed by the union of the petrous part of the temporal with the occipital bone, is termed the *petro-occipital.* The outer portion, formed by the junction of the mastoid part of the temporal with the occipital, is called the *masto-occipital.* Between the bones forming the petro-occipital suture, a thin plate of cartilage exists; in the masto-occipital is occasionally found the opening of the mastoid foramen. Between the outer extremity of the basilar suture and the spheno-parietal, an irregular suture may be seen, formed by the union of the sphenoid with the temporal bone. The inner and smaller portion of this suture is termed the *petro-sphenoidal*; it is formed between the petrous portion of the temporal and the great wing of the sphenoid; the outer portion, of greater length, and arched, is formed between the squamous portion of the temporal and the great wing of the sphenoid: it is called the *squamo-sphenoidal.*

The cranial bones are connected with those of the face, and the facial bones with each other, by numerous sutures, which, though distinctly marked, have received no special names. The only remaining suture deserving especial consideration, is the *transverse.* This extends across the upper part of the face, and is formed by the junction of the frontal with the facial bones; it extends from the external angular process of one side, to the same point on the opposite side, and connects the frontal with the malar, the sphenoid, the ethmoid, the lachrymal, the superior maxillary, and the nasal bones on each side.

The sutures remain separate for a considerable period after the complete formation of the skull. It is probable that they serve the purpose of permitting the growth of the bones at their margins; while their peculiar formation, and the interposition of the sutural ligament between the bones forming them, prevents the dispersion of blows or jars received upon the skull. Dr. Humphry remarks, 'that, as a general rule, the sutures are first obliterated at the parts in which the ossification of the skull was last completed, viz. in the neighbourhood of the fontanelles; and the cranial bones seem in this respect to observe a similar law to that which regulates the union of the epiphyses to the shafts of the long bones.'

THE SKULL.

The Skull, formed by the union of the several cranial and facial bones already described, when considered as a whole, is divisible into five regions: a superior region or vertex, an inferior region or base, two lateral regions, and an anterior region, the face.

Vertex of the Skull.

The *Superior Region*, or *Vertex*, presents two surfaces, an external and an internal.

The *External Surface* is bounded, in front, by the nasal eminences and superciliary ridges; behind, by the occipital protuberance and superior curved lines of the occipital bone; laterally, by an imaginary line extending from the outer end of the superior curved line, along the temporal ridge, to the external angular process of the frontal. This surface includes the vertical portion of the frontal, the greater part of the parietal, and the superior third of the occipital bone; it is smooth, convex, of an elongated oval form, crossed transversely by the coronal suture, and from before backwards by the sagittal, which terminates behind in the lambdoid. From before backwards may be seen the frontal eminences and remains of the suture connecting the two lateral halves of the frontal bone; on each side of the sagittal suture is the parietal foramen and parietal eminence, and still more posteriorly the smooth convex surface of the occipital bone.

The *Internal Surface* is concave, presents eminences and depressions for the convolutions of the cerebrum, and numerous furrows for the lodgment of branches of the meningeal arteries. Along the middle line of this surface is a longitudinal groove, narrow in front, where it terminates in the frontal crest; broader behind; it lodges the superior longitudinal sinus, and its margins afford attachment to the falx cerebri. On either side of it are several depressions for the Pacchionian bodies, and at its back part, the internal openings of the parietal foramina. This surface is crossed, in front, by the coronal suture; from before backwards, by the sagittal; behind, by the lambdoid.

Base of the Skull.

The *Inferior Region*, or *base* of the skull, presents two surfaces, an internal or cerebral, and an external or basilar.

The *Internal*, or *Cerebral Surface* (fig. 62), presents three fossæ on each side, called the *anterior*, *middle*, and *posterior* fossæ of the cranium.

The *Anterior Fossa* is formed by the orbital plate of the frontal, the cribriform plate of the ethmoid, the ethmoidal process and lesser wing of the sphenoid. It is the most elevated of the three fossæ, convex externally where it corresponds to the roof of the orbit, concave in the median line in the situation of the cribriform plate of the ethmoid. It is traversed by three sutures, the ethmoido-frontal, ethmo-sphenoidal, and fronto-sphenoidal; and lodges the anterior lobe of the cerebrum. It presents, in the median line, from before backwards, the commencement of the groove for the superior longitudinal sinus, and the crest for the attachment of the falx cerebri; the foramen cœcum, an aperture formed by the frontal bone and the crista galli of the ethmoid, which, if pervious, transmits a small vein from the nose to the superior longitudinal sinus; behind the foramen cœcum, the crista galli, the posterior margin of which affords attachment to the falx cerebri; on either side of the crista galli, the olfactory groove, which supports the bulb of the olfactory nerve, and is

62.—Base of Skull. Inner or Cerebral Surface.

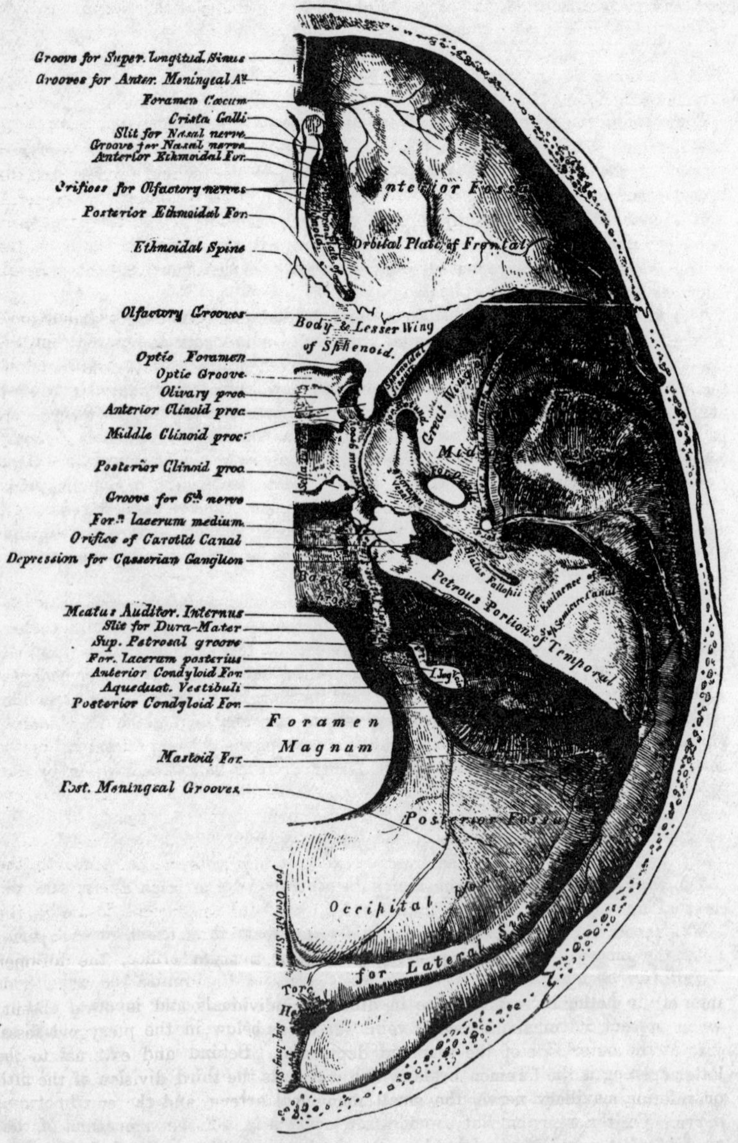

perforated by three rows of orifices for its filaments, and in front by a slit-like opening, for the nasal branch of the ophthalmic nerve. On the outer side of each olfactory groove are the internal openings of the anterior and posterior ethmoidal foramina; the former, situated about the middle of the outer margin of the olfactory groove, transmits the anterior ethmoidal artery and the nasal nerve, which runs in a depression along the surface of the ethmoid, to the slit-like opening above mentioned; whilst the posterior ethmoidal foramen opens at the back part of this margin under cover of the projecting lamina of the sphenoid, and transmits the posterior ethmoidal artery and vein to the posterior ethmoidal cells. Further back in the middle line is the ethmoidal spine, bounded behind by an elevated ridge, separating a longitudinal groove on each side which supports the olfactory nerve. The anterior fossa presents laterally eminences and depressions for the convolutions of the brain, and grooves for the lodgment of the anterior meningeal arteries.

The *Middle Fossa*, somewhat deeper than the preceding, is narrow in the middle, and becomes wider as it expands laterally. It is bounded in front by the posterior margin of the lesser wing of the sphenoid, the anterior clinoid process, and the anterior margin of the optic groove; behind, by the petrous portion of the temporal, and basilar suture; externally, by the squamous portion of the temporal, and anterior inferior angle of the parietal bone, and is separated from its fellow by the sella Turcica. It is traversed by four sutures, the squamous, spheno-parietal, spheno-temporal, and petro-sphenoidal.

In the middle line, from before backwards, is the optic groove, which supports the optic commissure, and terminates on each side in the optic foramen, for the passage of the optic nerve and ophthalmic artery; behind the optic groove is the olivary process, and laterally the anterior clinoid processes, to which are attached the folds of the dura mater, which form the cavernous sinuses. Separating the middle fossæ is the sella Turcica, a deep depression, which lodges the pituitary gland, bounded in front by a small eminence on either side, the middle clinoid process, and behind by a broad square plate of bone, surmounted at each superior angle by a tubercle, the posterior clinoid process; beneath the latter process is a groove, for the sixth nerve. On each side of the sella Turcica is the cavernous groove; it is broad, shallow, and curved somewhat like the italic letter *f*: it commences behind at the foramen lacerum medium, and terminates on the inner side of the anterior clinoid process. This groove lodges the cavernous sinus, the internal carotid artery, and the nerves of the orbit. The sides of the middle fossa are of considerable depth; they present eminences and depressions for the middle lobes of the brain, and grooves for the branches of the middle meningeal artery; the latter commence on the outer side of the foramen spinosum, and consist of two large branches, an anterior and a posterior; the former passing upwards and forwards to the anterior inferior angle of the parietal bone, the latter passing upwards and backwards. The following foramina may also be seen from before backwards. Most anteriorly is the foramen lacerum anterius, or sphenoidal fissure, formed above by the lesser wing of the sphenoid: below, by the greater wing; internally, by the body of the sphenoid; and completed externally by the orbital plate of the frontal bone. It transmits the third, fourth, the three branches of the ophthalmic division of the fifth, the sixth nerve, and the ophthalmic vein. Behind the inner extremity of the sphenoidal fissure is the foramen rotundum, for the passage of the second division of the fifth or superior maxillary nerve; still more posteriorly is seen a small orifice, the foramen Vesalii, an opening, situated between the foramen rotundum and ovale, a little internal to both; it varies in size in different individuals, and is often absent; when present, it transmits a small vein. It opens below in the pterygoid fossa, just at the outer side of the scaphoid depression. Behind and external to the latter opening is the foramen ovale, which transmits the third division of the fifth or inferior maxillary nerve, the small meningeal artery, and the small petrosal nerve. On the outer side of the foramen ovale is the foramen spinosum, for the passage of the middle meningeal artery; and on the inner side of the foramen ovale, the foramen lacerum medium. The lower part of this aperture is filled up with cartilage in the recent state. On the anterior surface of the petrous portion of the temporal bone is seen, from without inwards, the eminence caused by the projection of the superior semicircular canal, the groove leading to the hiatus Fallopii, for the transmission of the petrosal branch of the Vidian nerve; beneath it, the smaller groove, for the passage of the smaller petrosal nerve; and, near the apex of the bone, the depression for the Casserian ganglion, and the orifice of the carotid canal, for the passage of the internal carotid artery and carotid plexus of nerves.

The *Posterior Fossa*, deeply concave, is the largest of the three, and situated on a lower level than either of the preceding. It is formed by the occipital, the petrous and mastoid portions of the temporal, and the posterior inferior angle of the parietal bone; is crossed by three sutures, the petro-occipital, masto-occipital, and masto-parietal; and lodges the cerebellum, pons Varolii, and medulla oblongata. It is separated from the middle fossa in the median line by the basilar suture, and on each side by the superior border of the petrous portion of the temporal bone. This serves for the attachment of the tentorium cerebelli, is grooved externally for the superior petrosal sinus, and at its inner extremity presents a notch, upon which rests the fifth nerve. Its circumference is bounded posteriorly by the grooves for the lateral sinuses. In the centre of this fossa is the foramen magnum, bounded on either side by a rough tubercle, which gives attachment to the odontoid ligaments; and a little above these are seen the internal openings of the anterior condyloid foramina. In front of the foramen magnum is the basilar process, grooved for the support of the medulla oblongata and pons Varolii, and articulating on each side with the petrous portion of the temporal bone, forming the petro-occipital suture, the anterior half of which is grooved for the inferior petrosal sinus, the posterior half being encroached upon by the foramen lacerum posterius, or jugular foramen. This foramen is partially subdivided into two parts; the posterior and larger division transmitting the internal jugular vein, the anterior the eighth pair of nerves. Above the jugular foramen is the internal auditory foramen, for the auditory and facial nerves and auditory artery; behind and external to this is the slit-like opening leading into the aquæductus vestibuli; whilst between the two latter, and near the superior border of the petrous portion, is a small triangular depression which lodges a process of the dura mater, and occasionally transmits a small vein into the substance of the bone. Behind the foramen magnum are the inferior occipital fossæ, which lodge the lateral lobes of the cerebellum, separated from one another by the internal occipital crest, which serves for the attachment of the falx cerebelli, and lodges the occipital sinuses. The posterior fossæ are surmounted, above, by the deep transverse grooves for the lodgment of the lateral sinuses. These channels, in their passage outwards, groove the occipital bone, the posterior inferior angle of the parietal, the mastoid portion of the temporal, and the occipital just behind the jugular foramen, at the back part of which they terminate. Where this sinus grooves the mastoid part of the temporal bone, the orifice of the mastoid foramen may be seen; and, just previous to its termination, it has opening into it the posterior condyloid foramen.

The *External Surface* of the base of the Skull (fig. 63) is extremely irregular. It is bounded in front by the incisor teeth in the upper jaws; behind, by the superior curved lines of the occipital bone; and laterally, by the alveolar arch, the lower border of the malar bone, the zygoma, and an imaginary line, extending from the zygoma to the mastoid process and extremity of the superior curved line of the occiput. It is formed by the palate processes of the superior maxillary and palate bones, the vomer, the pterygoid, under surface of the great wing, spinous process and part of the body of the sphenoid, the under surface of the squamous, mastoid, and petrous portions of the temporal, and the under surface of the occipital bone. The anterior part of the base of the skull is raised above the level of the rest of this surface (when the skull is turned over for the purpose of examination), surrounded by the alveolar process, which is thicker behind than in front, and excavated by sixteen depressions for lodging the teeth of the upper

63.—Base of the Skull. External Surface.

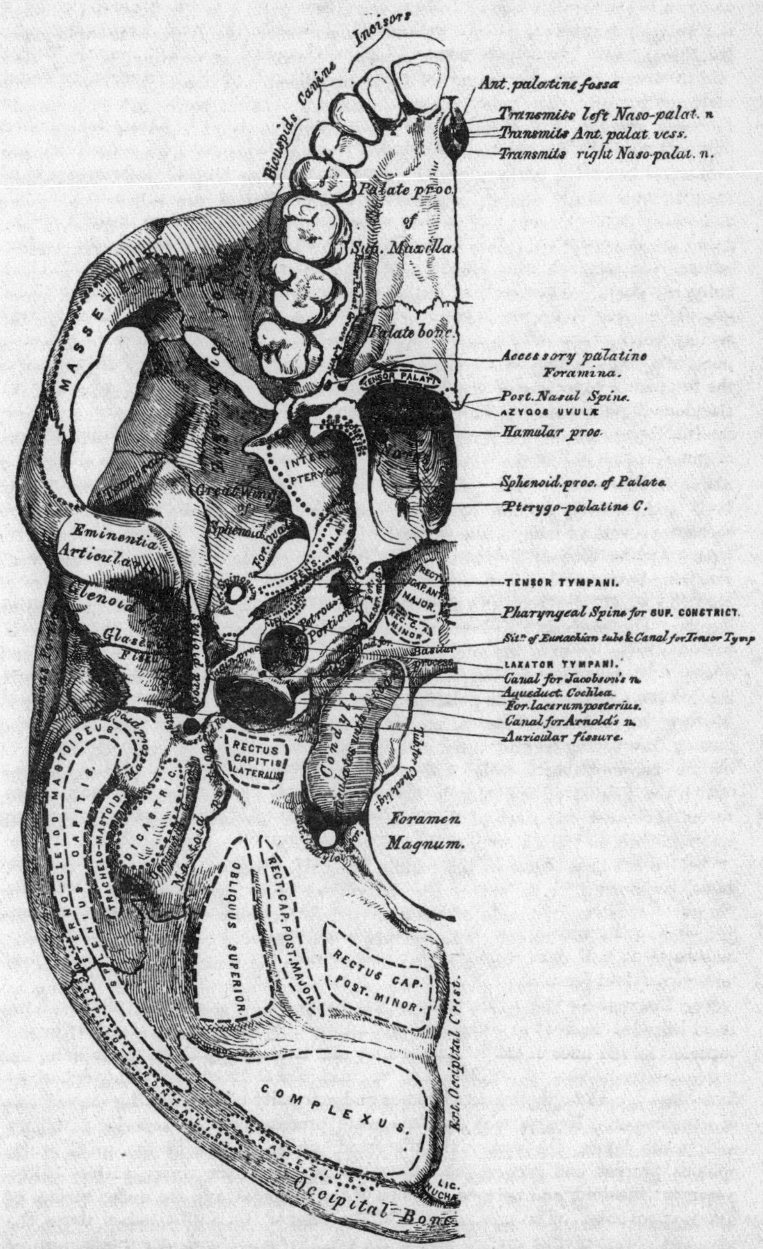

jaw; the cavities varying in depth and size according to the teeth they contain. Immediately behind the incisor teeth is the anterior palatine fossa. At the bottom of this fossa may usually be seen four apertures, two placed laterally, which open above, one in the floor of each nostril, and transmit the anterior palatine vessels, and two in the median line of the intermaxillary suture, one in front of the other, the anterior transmitting the left, and the posterior (the larger) the right naso-palatine nerve. These two latter canals are sometimes wanting, or they may join to form a single one, or one of them may open into one of the lateral canals above referred to. The palatine vault is concave, uneven, perforated by numerous foramina, marked by depressions for the palatal glands, and crossed by a crucial suture, formed by the junction of the four bones of which it is composed. One or two small foramina, in the alveolar margin behind the incisor teeth, occasionally seen in the adult, almost constant in young subjects, are called the *incisive foramina*; they transmit nerves and vessels to the incisor teeth. At each posterior angle of the hard palate is the posterior palatine foramen, for the transmission of the posterior palatine vessels and descending palatine nerve, and running forwards and inwards from it a groove, which lodges the same vessels and nerve. Behind the posterior palatine foramen is the tuberosity of the palate bone, perforated by one or more accessory posterior palatine canals, and marked by the commencement of a ridge, which runs transversely inwards, and serves for the attachment of the tendinous expansion of the Tensor palati muscle. Projecting backwards from the centre of the posterior border of the hard palate is the posterior nasal spine, for the attachment of the Azygos uvulæ. Behind and above the hard palate is the posterior aperture of the nares, divided into two parts by the vomer, bounded above by the body of the sphenoid, below by the horizontal plate of the palate bone, and laterally by the pterygoid processes of the sphenoid. Each aperture measures about an inch in the vertical, and half an inch in the transverse direction. At the base of the vomer may be seen the expanded alæ of this bone, receiving between them the rostrum of the sphenoid. Near the lateral margins of the vomer, at the root of the pterygoid processes, are the pterygo-palatine canals. The pterygoid process, which bounds the posterior nares on each side, presents near its base the pterygoid or Vidian canal, for the Vidian nerve and artery. Each process consists of two plates, which bifurcate at the extremity to receive the tuberosity of the palate bone, and are separated behind by the pterygoid fossa, which lodges the Internal pterygoid muscle. The Internal plate is long and narrow, presenting on the outer side of its base the scaphoid fossa, for the origin of the Tensor palati muscle, and at its extremity the hamular process, around which the tendon of this muscle turns. The external pterygoid plate is broad, forms the inner boundary of the zygomatic fossa, and affords attachment, by its outer surface, to the External pterygoid muscle.

Behind the nasal fossæ in the middle line is the basilar surface of the occipital bone, presenting in its centre the pharyngeal spine for the attachment of the Superior constrictor muscle of the pharynx, with depressions on each side for the insertion of the Rectus capitis anticus major and minor. At the base of the external pterygoid plate is the foramen ovale; behind this, the foramen spinosum, and the prominent spinous process of the sphenoid, which gives attachment to the internal lateral ligament of the lower jaw and the Laxator tympani muscle. External to the spinous process is the glenoid fossa, divided into two parts by the Glaserian fissure (p. 30), the anterior portion concave, smooth, bounded in front by the eminentia articularis, and serving for the articulation of the condyle of the lower jaw; the posterior portion rough, bounded behind by the vaginal process, and serving for the reception of part of the parotid gland. Emerging from between the laminæ of the vaginal process is the styloid process; and at the base of this process is the stylo-mastoid foramen, for the exit of the facial nerve, and entrance of the stylo-mastoid artery. External to the stylo-mastoid foramen is the auricular fissure for the auricular branch of the pneumogastric, bounded behind by the mastoid process. Upon the inner side of the mastoid process is a deep groove, the digastric fossa;

and a little more internally, the occipital groove, for the occipital artery. At the base of the internal pterygoid plate is a large and somewhat triangular aperture, the foramen lacerum medium, bounded in front by the great wing of the sphenoid, behind by the apex of the petrous portion of the temporal bone, and internally by the body of the sphenoid and basilar process of the occipital bone; it presents in front the posterior orifice of the Vidian canal, behind the aperture of the carotid canal. The basilar surface of this opening is filled up in the recent state by a fibro-cartilaginous substance; across its upper or cerebral aspect passes the internal carotid artery and Vidian nerve. External to this aperture, the petro-sphenoidal suture is observed, at the outer termination of which is seen the orifice of the canal for the Eustachian tube, and that for the Tensor tympani muscle. Behind this suture is seen the under surface of the petrous portion of the temporal bone, presenting, from within outwards, the quadrilateral rough surface, part of which affords attachment to the Levator palati and Tensor tympani muscles; external to this surface the orifices of the carotid canal and the aquæductus cochleæ, the former transmitting the internal carotid artery and the ascending branches of the superior cervical ganglion of the sympathetic, the latter serving for the passage of a small artery and vein to the cochlea. Behind the carotid canal is a large aperture, the jugular fossa, formed in front by the petrous portion of the temporal, and behind by the occipital; it is generally larger on the right than on the left side; and towards its cerebral aspect is divided into two parts by a ridge of bone, which projects usually from the temporal, the anterior, or smaller portion, transmitting the three divisions of the eighth pair of nerves; the posterior, transmitting the internal jugular vein and the ascending meningeal vessels, from the occipital and ascending pharyngeal arteries. On the ridge of bone dividing the carotid canal from the jugular fossa, is the small foramen for the transmission of the tympanic nerve; and on the outer wall of the jugular foramen, near the root of the styloid process, is the small aperture for the transmission of Arnold's nerve. Behind the basilar surface of the occipital bone is the foramen magnum, bounded on each side by the condyles, rough internally for the attachment of the alar ligaments, and presenting externally a rough surface, the jugular process, which serves for the attachment of the Rectus lateralis. On either side of each condyle anteriorly is the anterior condyloid fossa, perforated by the anterior condyloid foramen, for the passage of the hypoglossal nerve. Behind each condyle are the posterior condyloid fossæ, perforated on one or both sides by the posterior condyloid foramina, for the transmission of a vein to the lateral sinus. Behind the foramen magnum is the external occipital crest, terminating above at the external occipital protuberance, whilst on each side are seen the superior and inferior curved lines; these, as well as the surfaces of the bone between them, being rough for the attachment of the muscles, which are enumerated on page 20.

Lateral Region of the Skull.

The *Lateral Region* of the Skull is of a somewhat triangular form, the base of the triangle being formed by a line extending from the external angular process of the frontal bone along the temporal ridge backwards to the outer extremity of the superior curved line of the occiput: and the sides by two lines, the one drawn downwards and backwards from the external angular process of the frontal bone to the angle of the lower jaw, the other from the angle of the jaw upwards and backwards to the extremity of the superior curved line. This region is divisible into three portions, temporal, mastoid, and zygomatic.

The Temporal Fossa.

The *Temporal* fossa is bounded above and behind by the temporal ridge, which extends from the external angular process of the frontal upwards and backwards across the frontal and parietal bones, curving downwards behind to terminate at the root of the zygomatic process. In front, it is bounded by the frontal, malar, and great wing of the sphenoid: externally, by the zygomatic arch, formed conjointly by the malar and temporal bones; below, it is separated from the zygomatic fossa by the pterygoid ridge, seen on the outer surface of the great

64.—Side View of the Skull.

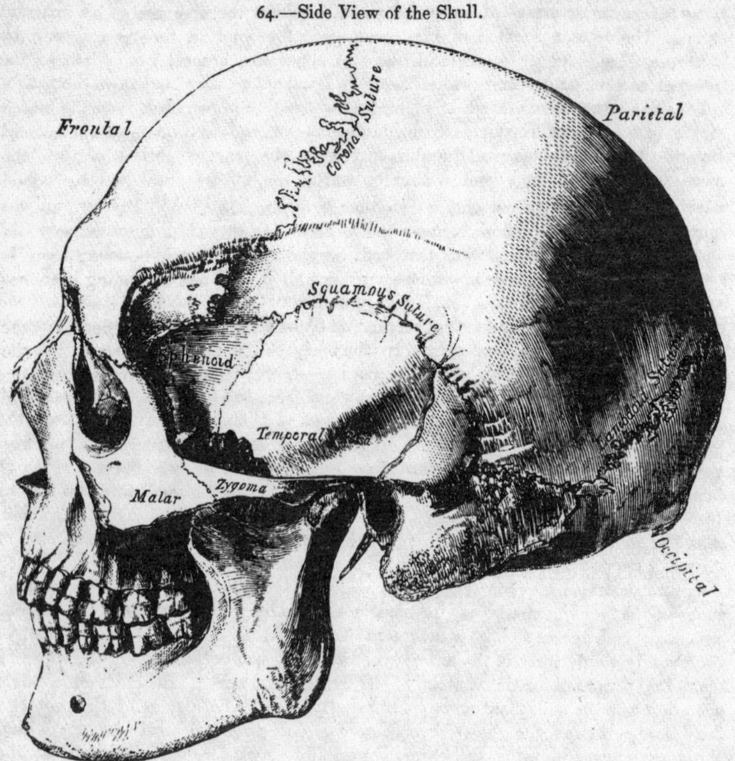

wing of the sphenoid. This fossa is formed by five bones, part of the frontal, great wing of the sphenoid, parietal, squamous portion of the temporal, and malar bones, and is traversed by five sutures, the transverse facial, coronal, spheno-parietal, squamo-parietal, and squamo-sphenoidal. It is deeply concave in front, convex behind, traversed by grooves which lodge branches of the deep temporal arteries, and filled by the Temporal muscle.

The *Mastoid Portion* of the side of the skull is bounded in front by the anterior root of the zygoma; above, by a line which runs from the posterior root of the zygoma to the end of the masto-parietal suture; behind and below, by the masto-occipital suture. It is formed by the mastoid and part of the squamous portion of the temporal bone; its surface is convex and rough for the attachment of muscles, and presents, from behind forwards, the mastoid foramen, the mastoid process, the external auditory meatus, surrounded by the auditory process, and, most anteriorly, the glenoid fossa, bounded in front by the eminentia articularis, behind by the vaginal process.

The Zygomatic Fossa.

The *Zygomatic* fossa is an irregularly-shaped cavity, situated below, and on the inner side of the zygoma; bounded, in front, by the tuberosity of the superior maxillary bone and the ridge which descends from its malar process; behind, by the posterior border of the pterygoid process; above, by the pterygoid ridge on the outer surface of the great wing of the sphenoid and squamous portion of the temporal; below, by the alveolar border of the superior maxilla; internally, by the external pterygoid plate; and externally, by the zygomatic arch and ramus of the jaw. It contains the lower part of the Temporal, the External, and Internal pterygoid muscles, the internal maxillary artery, the inferior maxillary nerve, and their branches. At its upper and inner part may be observed two fissures, the spheno-maxillary and pterygo-maxillary.

The *Spheno-maxillary* fissure, horizontal in direction, opens into the outer and back part of the orbit. It is formed above by the lower border of the orbital surface of the great wing of the sphenoid; below, by the external border of the orbital surface of the superior maxilla and a small part of the palate bone; externally, by a small part of the malar bone; internally, it joins at right angles with the pterygo-maxillary fissure. This fissure opens a communication from the orbit into three fossæ, the temporal, zygomatic, and spheno-maxillary; it transmits the superior maxillary nerve, infraorbital artery, and ascending branches from Meckel's ganglion.

The *Pterygo-maxillary* fissure is vertical, and descends at right angles from the inner extremity of the preceding; it is an elongated interval, formed by the divergence of the superior maxillary bone from the pterygoid process of the sphenoid. It serves to connect the spheno-maxillary fossa with the zygomatic, and transmits branches of the internal maxillary artery.

The Spheno-Maxillary Fossa.

The Spheno-maxillary fossa is a small triangular space situated at the angle of junction of the spheno-maxillary and pterygo-maxillary fissures, and placed beneath the apex of the orbit. It is formed above by the under surface of the body of the sphenoid; in front, by the superior maxillary bone: behind, by the pterygoid process of the sphenoid; internally, by the vertical plate of the palate. This fossa has three fissures terminating in it, the sphenoidal, spheno-maxillary, and pterygo-maxillary; it communicates with three fossæ, the orbital, nasal, and zygomatic, and with the cavity of the cranium, and has opening into it five foramina. Of these there are three on the posterior wall; the foramen rotundum above; below, and internal to this, the Vidian, and still more inferior and internal, the pterygo-palatine. On the inner wall is the spheno-palatine foramen by which the spheno-maxillary communicates with the nasal fossa, and below is the superior orifice of the posterior palatine canal, besides occasionally the orifices of two or three accessory posterior palatine canals.

Anterior Region of the Skull.

The Anterior Region of the Skull, which forms the face, is of an oval form, presents an irregular surface, and is excavated for the reception of the two principal organs of sense, the eye and the nose. It is bounded above by the nasal eminences and margins of the orbit; below, by the prominence of the chin; on each side, by the malar bone, and anterior margin of the ramus of the jaw. In the median line are seen from above downwards, the nasal eminences, which indicate the situation of the frontal sinuses; and diverging from which are the superciliary ridges which support the eyebrows. Beneath the nasal eminences is the arch of the nose, formed by the nasal bones, and the nasal processes of the superior maxillary. The nasal arch is convex from side to side, concave from above downwards, presenting in the median line the internasal suture, formed between the nasal bones, laterally the naso-maxillary suture, formed between the nasal bone and the nasal process of the superior maxillary bone, both these sutures terminating above in that part of the transverse suture which connects the nasal bones and nasal processes of the superior maxillary with the frontal. Below the nose is seen the opening of the anterior nares, which is heart-shaped, with the narrow end upwards, and presents laterally the thin sharp margins serving for the attachment of the lateral cartilages of the nose, and in the middle line below, a prominent process, the anterior nasal spine, bounded by two deep notches. Below this is the intermaxillary suture, and on each side of it the incisive fossa. Beneath this fossa is the alveolar process of the upper and lower jaw, containing the incisor teeth, and at the lower part of the median line, the symphysis of the chin, the mental eminence, and the incisive fossa of the lower jaw.

On each side, proceeding from above downwards, is the supraorbital ridge, terminating externally in the external angular process at its junction with the malar, and internally in the internal angular process; towards the inner third of

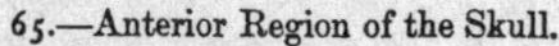

65.—Anterior Region of the Skull.

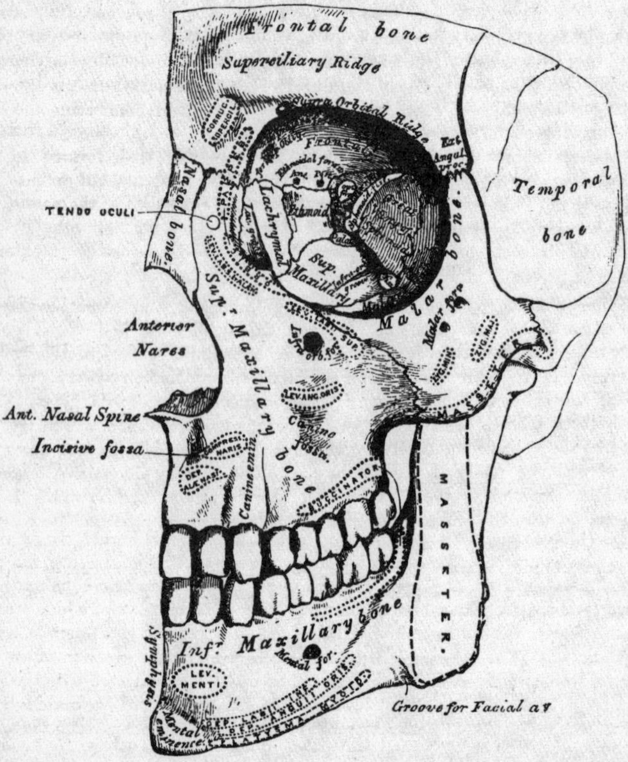

this ridge is the supraorbital notch or foramen, for the passage of the supraorbital vessels and nerve, and at its inner side a slight depression for the attachment of the pulley of the Superior oblique muscle. Beneath the supraorbital ridge is the opening of the orbit, bounded externally by the orbital ridge of the malar bone; below, by the orbital ridge formed by the malar, superior maxillary, and lachrymal bones; internally, by the nasal process of the superior maxillary, and the internal angular process of the frontal bone. On the outer side of the orbit, is the quadrilateral anterior surface of the malar bone, perforated by one or two small malar foramina. Below the inferior margin of the orbit, is the infraorbital foramen, the termination of the infraorbital canal, and beneath this, the canine fossa, which gives attachment to the Levator anguli oris; bounded below by the alveolar processes, containing the teeth of the upper and lower jaw. Beneath the alveolar arch of the lower jaw is the mental foramen for the passage of the mental nerve and artery, the external oblique line, and at the lower border of the bone, at the point of junction of the body with the ramus, a shallow groove for the passage of the facial artery.

The Orbits.

The Orbits (fig. 65) are two quadrilateral pyramidal cavities, situated at the upper and anterior part of the face, their bases being directed forwards and outwards, and their apices backwards and inwards. Each orbit is formed of *seven* bones, the frontal, sphenoid, ethmoid, superior maxillary, malar, lachrymal, and palate; but three of these, the frontal, ethmoid, and sphenoid, enter into the formation of *both* orbits, so that the two cavities are formed of *eleven* bones only. Each cavity presents for examination, a roof, a floor, an inner and an outer wall, four angles, a circumference or base, and an apex. The *Roof* is concave, directed downwards and forwards, and formed in front by the orbital plate of the frontal; behind, by the lesser wing of the sphenoid. This surface presents internally the depression for the fibro-cartilaginous pulley of the Superior oblique muscle; externally, the depression for the lachrymal gland, and posteriorly, the suture connecting the frontal and lesser wing of the sphenoid.

The *Floor* is nearly flat, and of less extent than the roof; it is formed chiefly by the orbital process of the superior maxillary; in front, to a small extent, by the orbital process of the malar, and behind, by the orbital surface of the palate. This surface presents at its anterior and internal part, just external to the lachrymal canal, a depression for the attachment of the Inferior oblique muscle; externally, the suture between the malar and superior maxillary bones; near its middle, the infraorbital groove; and posteriorly, the suture between the maxillary and palate bones.

The *Inner Wall* is flattened, and formed from before backwards by the nasal process of the superior maxillary, the lachrymal, os planum of the ethmoid, and a small part of the body of the sphenoid. This surface presents the lachrymal groove, and crest of the lachrymal bone, and the sutures connecting the ethmoid with the lachrymal bone in front, and the sphenoid behind.

The *Outer Wall* is formed in front by the orbital process of the malar bone; behind, by the orbital plate of the sphenoid. On it are seen the orifices of one or two malar canals, and the suture connecting the sphenoid and malar bones.

Angles. The *superior external angle* is formed by the junction of the upper and outer walls; it presents, from before backwards, the suture connecting the frontal with the malar in front, and with the orbital plate of the sphenoid behind; quite posteriorly is the foramen lacerum anterius, or sphenoidal fissure, which transmits the third, fourth, the ophthalmic division of the fifth and the sixth nerves, and the ophthalmic vein. The *superior internal angle* is formed by the junction of the upper and inner wall, and presents the suture connecting the frontal bone with the lachrymal in front, and with the ethmoid behind. This suture is perforated by two foramina, the anterior and posterior ethmoidal, the former transmitting the anterior ethmoidal artery and nasal nerve, the latter the posterior ethmoidal artery and vein. The *inferior external angle*, formed by the junction of the outer wall or floor, presents the spheno-maxillary fissure, which transmits the infraorbital vessels and nerve, and the ascending branches from the spheno-palatine ganglion. The *inferior internal angle* is formed by the union of the lachrymal and os planum of the ethmoid, with the superior maxillary and palate bones. The *circumference*, or base, of the orbit, quadrilateral in form, is bounded above by the supraorbital arch; below, by the anterior border of the orbital plate of the malar, superior maxillary, and lachrymal bones; externally, by the external angular process of the frontal and the malar bone; internally, by the internal angular process of the frontal, and the nasal process of the superior maxillary. The circumference is marked by three sutures, the fronto-maxillary internally, the fronto-malar externally, and the malo-maxillary below; it contributes to the formation of the lachrymal groove, and presents above, the supraorbital notch (or foramen), for the passage of the supraorbital artery, veins and nerve. The *apex*, situated at the back of the orbit, corresponds to the optic foramen, a short circular canal, which transmits the optic nerve and ophthalmic artery. It will thus be seen that there are *nine* openings communicating with each orbit, viz. the optic, foramen lacerum anterius, spheno-maxillary fissure, supra-orbital foramen, infraorbital canal, anterior and posterior ethmoidal foramina, malar foramina, and lachrymal canal.

The Nasal Fossæ.

The *Nasal Fossæ* are two large irregular cavities, situated in the middle line of the face, extending from the base of the cranium to the roof of the mouth, and separated from each other by a thin vertical septum. They communicate by two large apertures, the anterior nares, with the front of the face; and with the pharynx behind by the two posterior nares. These fossæ are much narrower above than below, and in the middle than at the anterior or posterior openings: their depth, which is considerable, is much greater in the middle than at either extremity. Each nasal fossa communicates with four sinuses, the frontal above, the sphenoidal behind, and the maxillary and ethmoidal on either side. Each fossa also communicates with four cavities: with the orbit by the lachrymal canal, with the mouth by the anterior palatine canal, with the cranium by the olfactory foramina, and with the spheno-maxillary fossa by the spheno-palatine foramen; and they occasionally communicate with each other by an aperture in the septum. The bones

66.—Roof, Floor, and Outer Wall of Nasal Fossa.

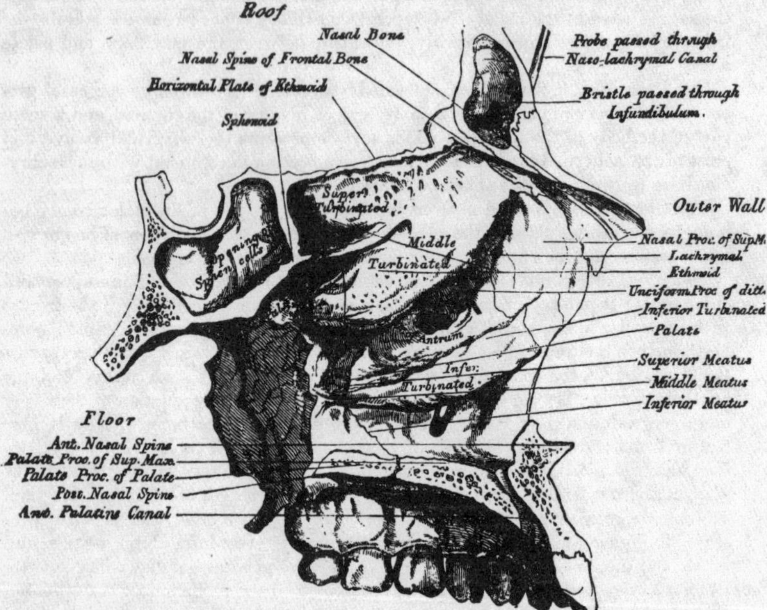

entering into their formation are fourteen in number: three of the cranium, the frontal, sphenoid, the ethmoid, and all the bones of the face, excepting the malar and lower jaw. Each cavity is bounded by a roof, a floor, an inner and an outer wall.

The *upper wall*, or roof (fig. 66), is long, narrow, and concave from before backwards; it is formed in front by the nasal bones and nasal spine of the frontal, which are directed downwards and forwards; in the middle, by the cribriform lamella of the ethmoid, which is horizontal; and behind, by the under surface of the body of the sphenoid, and sphenoidal turbinated bones, which are directed downwards and backwards. This surface presents, from before backwards, the internal aspect of the nasal bones; on their outer side, the suture formed between the nasal bone and the nasal process of the superior maxillary; on their inner side, the elevated crest which receives the nasal spine of the frontal, and the perpendicular plate of the ethmoid, and articulates with its fellow of the opposite side; whilst the surface of the bones is perforated by a few small vascular apertures, and presents the longitudinal groove for the nasal nerve: further back is the transverse suture, connecting the frontal with the nasal in front, and the ethmoid behind, the olfactory foramina and nasal slit on the under surface of the cribriform plate, and the suture between it and the sphenoid behind: quite posteriorly are seen the sphenoidal turbinated bones, the orifices of the sphenoidal sinuses and the articulation of the alæ of the vomer with the under surface of the body of the sphenoid.

The *floor* is flattened from before backwards, concave from side to side, and wider in the middle than at either extremity. It is formed in front by the palate process of the superior maxillary; behind, by the palate process of the palate bone. This surface presents, from before backwards, the anterior nasal spine; behind this, the upper orifice of the anterior palatine canal: internally, the elevated crest which articulates with the vomer; and behind, the suture between the palate and superior maxillary bones, and the posterior nasal spine.

The *inner wall*, or septum (fig. 67), is a thin vertical partition, which separates the nasal fossæ from one another; it is occasionally perforated, so that the fossæ

67.—Inner Wall of Nasal Fossæ, or Septum of Nose.

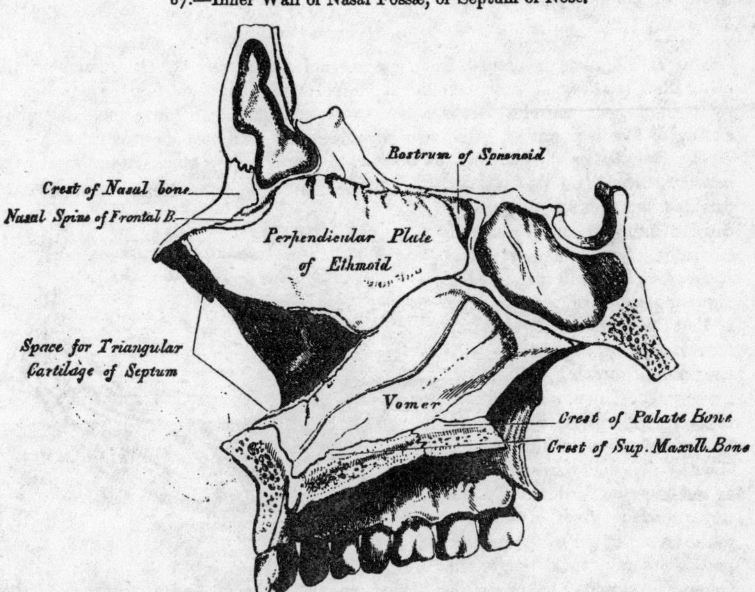

communicate, and it is frequently deflected considerably to one side. It is formed, in front, by the crest of the nasal bones and nasal spine of the frontal; in the middle, by the perpendicular lamella of the ethmoid; behind, by the vomer and rostrum of the sphenoid; below, by the crest of the superior maxillary and palate bones. It presents, in front, a large triangular notch, which receives the triangular cartilage of the nose; above, the lower orifices of the olfactory canals; and behind, the guttural edge of the vomer. Its surface is marked by numerous vascular and nervous canals and the groove for the naso-palatine nerve, and is traversed by sutures connecting the bones of which it is formed.

The *outer wall* (fig. 66) is formed, in front, by the nasal process of the superior maxillary and lachrymal bones; in the middle, by the ethmoid and inner surface of the superior maxillary and inferior turbinated bones; behind, by the vertical plate of the palate bone, and the internal pterygoid process of the sphenoid. This surface presents three irregular longitudinal passages, or meatuses, formed between three horizontal plates of bone that spring from it; they are termed the superior, middle, and inferior meatuses of the nose. The *superior meatus,* the smallest of the three, is situated at the upper and back part of each nasal fossa, occupying the posterior third of the outer wall. It is situated between the superior and middle turbinated bones, and has opening into it two foramina, the spheno-palatine at the back of its outer wall, the posterior ethmoidal cells at the front part of the upper wall. The opening of the sphenoidal sinuses is usually at the upper and back part of the nasal fossæ, immediately behind the superior turbinated bone. The *middle meatus* is situated between the middle and inferior turbinated bones, and occupies the posterior two-thirds of the outer wall of the nasal fossa. It presents two apertures. In front is the orifice of the infundibulum, by which the middle meatus communicates with the anterior ethmoidal cells, and through these with the frontal sinuses. At the centre of the outer wall is the orifice of the antrum, which varies somewhat as to its exact position in different skulls. The *inferior meatus,* the largest of the three, is the space between the inferior turbinated bone and the floor of the nasal fossa. It extends along the entire length of the outer wall of the nose, is broader in front than behind, and presents anteriorly the lower orifice of the lachrymal canal.

Os Hyoides.

The Hyoid bone is named from its resemblance to the Greek Upsilon; it is also called the *lingual bone,* because it supports the tongue, and gives attachment to its numerous muscles. It is a bony arch, shaped like a horse-shoe, and consisting of five segments, a body, two greater cornua, and two lesser cornua.

The *Body* forms the central part of the bone, and is of a quadrilateral form: its *anterior surface* (fig. 68) convex, directed forwards and upwards, is divided into two parts by a vertical ridge, which descends along the median line, and is crossed at right angles by a horizontal ridge, so that this surface is divided into four muscular depressions. At the point of meeting of these two lines is a prominent elevation, the tubercle. The portion above the horizontal ridge is directed upwards, and is sometimes described as the superior border. The anterior surface gives attachment to the Genio-hyoid in the greater part of its extent; above, to the Genio-hyo-glossus; below, to the Mylo-hyoid, Stylo-hyoid, and aponeurosis of the Digastric; and between these to part of the Hyo-glossus. The *posterior surface* is smooth, concave, directed backwards and downwards, and separated from the epiglottis by the thyro-hyoid membrane, and by a quantity of loose areolar tissue. The *superior border* is rounded, and gives attachment to the thyro-hyoid membrane, and part of the Genio-hyo-glossi muscles. The *inferior border* gives attachment, in front, to the Sterno-hyoid; behind, to part of the Thyro-hyoid, and to the Omo-hyoid at its junction with the great cornu. The *lateral surfaces* are small, oval, convex facets, covered with cartilage for articulation with the greater cornua.

68.—Hyoid Bone. Anterior Surface, (enlarged).

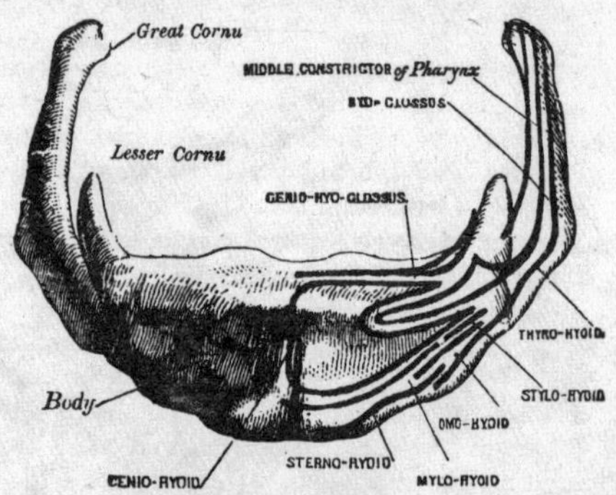

The *Greater Cornua* project backwards from the lateral surfaces of the body; they are flattened from above downwards, diminish in size from before backwards, and terminate posteriorly in a tubercle for the attachment of the thyro-hyoid ligament. Their outer surface gives attachment to the Hyo-glossus; their upper border, to the Middle constrictor of the pharynx; their lower border, to part of the Thyro-hyoid muscle.

The *Lesser Cornua* are two small conical-shaped eminences, attached by their bases to the angles of junction between the body and greater cornua, and giving attachment by their apices to the stylo-hyoid ligaments. In youth, the cornua are connected to the body by cartilaginous surfaces, and held together by ligaments; in middle life, the body and greater cornua usually become joined; and in old age, all the segments are united together, forming a single bone.

Development. By *five* centres; one for the body, and one for each cornu. Ossification commences in the body and greater cornua towards the end of fœtal life, those of the cornua first appearing. Ossification of the lesser cornua commences some months after birth.

Attachment of Muscles. Sterno-hyoid, Thyro-hyoid, Omo-hyoid, aponeurosis of the Digastricus, Stylo-hyoid, Mylo-hyoid, Genio-hyoid, Genio-hyo-glossus, Hyo-glossus, Middle Constrictor of the pharynx, and occasionally a few fibres of the Lingualis. It also gives attachment to the thyro-hyoidean membrane, and the stylo-hyoid, thyro-hyoid, and hyo-epiglottic ligaments.

THE THORAX.

The Thorax, or chest, is an osseo-cartilaginous cage, intended to contain and protect the principal organs of respiration and circulation. It is the largest of the three cavities connected with the spine, and is formed by the sternum and costal cartilages in front, the twelve ribs on each side, and the bodies of the dorsal vertebræ behind.

The Sternum.

The Sternum (figs. 69, 70) is a flat narrow bone, situated in the median line of the front of the chest, and consisting, in the adult, of three portions. It has been likened to an ancient sword: the upper piece, representing the handle, is termed the *manubrium*; the middle and largest piece, which represents the chief part of the blade, is termed the *gladiolus*; and the inferior piece, which is likened to the point of the sword, is termed the *ensiform* or *xiphoid appendix.* In its natural position, its inclination is oblique from above, downwards, and forwards. It is flattened in front, concave behind, broad above, becoming narrowed at the point where the first and second pieces are connected; after which it again widens a little, and is pointed at its extremity. Its average length in the adult is six inches, being rather longer in the male than in the female.

The *First Piece* of the sternum, or *Manubrium*, is of a somewhat triangular form, broad and thick above, narrow below at its junction with the middle piece. Its *anterior surface*, convex from side to side, concave from above downwards, is smooth, and affords attachment on each side to the Pectoralis major and sternal origin of the sterno-cleido-mastoid muscle. In well-marked bones, the ridges limiting the attachment of these muscles are very distinct. Its *posterior surface*, concave and smooth, affords attachment on each side to the Sterno-hyoid and Sterno-thyroid muscles. The *superior border*, the thickest, presents at its centre the interclavicular notch; and on each side, an oval articular surface, directed upwards, backwards, and outwards, for articulation with the sternal end of the clavicle. The *inferior border* presents an oval rough surface, covered in the recent state with a thin layer of cartilage, for articulation with the second portion of the bone. The *lateral borders* are marked above by an articular depression for the first costal cartilage, and below by a small facet, which, with a similar facet on the upper angle of the middle portion of the bone, forms a notch for the

69.—Sternum and Costal Cartilages

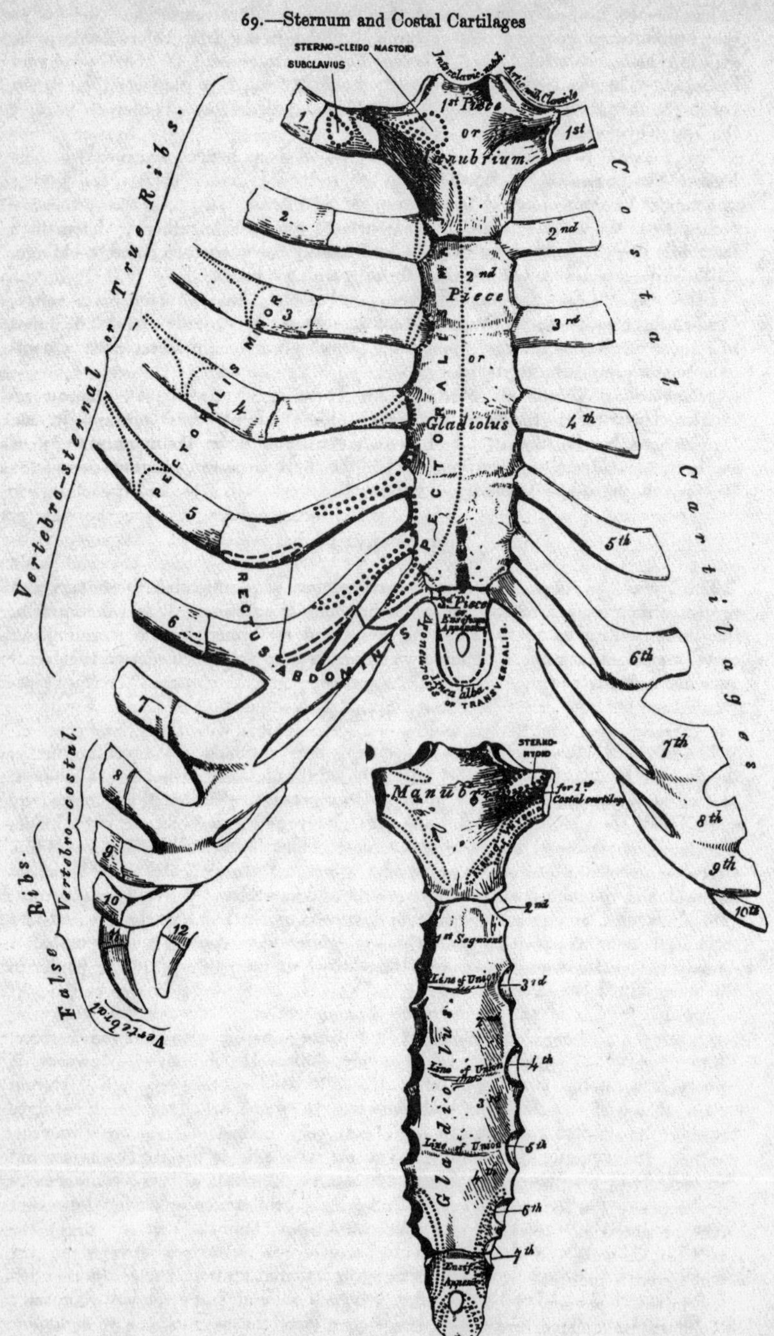

70.—Posterior Surface of Sternum.

reception of the costal cartilage of the second rib. These articular surfaces are separated by a narrow curved edge which slopes from above downwards and inwards.

The *Second Piece* of the sternum, or *gladiolus*, considerably longer, narrower, and thinner than the first piece, is broader below than above. Its *anterior surface* is nearly flat, directed upwards and forwards, and marked by three transverse lines which cross the bone opposite the third, fourth, and fifth articular depressions. These lines are produced by the union of the four separate pieces of which this part of the bone consists at an early period of life. At the junction of the third and fourth pieces, is occasionally seen an orifice, the sternal foramen; it varies in size and form in different individuals, and pierces the bone from before backwards. This surface affords attachment on each side to the sternal origin of the Pectoralis major. The *posterior surface*, slightly concave, is also marked by three transverse lines; but they are less distinct than those in front: this surface affords attachment below, on each side, to the Triangularis sterni muscle, and occasionally presents the posterior opening of the sternal foramen. The *superior border* presents an oval surface for articulation with the manubrium. The *inferior border* is narrow and articulates with the ensiform appendix. Each *lateral border* presents at each superior angle a small facet, which, with a similar facet on the manubrium, forms a cavity for the cartilage of the second rib; the four succeeding angular depressions receive the cartilages of the third, fourth, fifth, and sixth ribs, whilst each inferior angle presents a small facet, which, with a corresponding one on the ensiform appendix, forms a notch for the cartilage of the seventh rib. These articular depressions are separated by a series of curved inter-articular intervals, which diminish in length from above downwards, and correspond to the intercostal spaces. Most of the cartilages belonging to the true ribs, as will be seen from the foregoing description, articulate with the sternum at the line of junction of two of its primitive component segments. This is well seen in many of the lower animals, where the separate parts of the bone remain ununited longer than in man. In this respect a striking analogy exists between the mode of connection of the ribs with the vertebral column, and the connection of their cartilages with the sternal column.

The *Third Piece* of the sternum, the *ensiform* or *xiphoid appendix*, is the smallest of the three; it is thin and elongated in form, cartilaginous in structure in youth, but more or less ossified at its upper part in the adult. Its *anterior surface* affords attachment to the costo-xiphoid ligament; its *posterior surface*, to some of the fibres of the Diaphragm and Triangularis sterni muscles: its *lateral borders*, to the aponeurosis of the abdominal muscles. Above, it is continuous with the lower end of the gladiolus; below, by its pointed extremity, it gives attachment to the linea alba, and at each superior angle presents a facet for the lower half of the cartilage of the seventh rib. This portion of the sternum is very various in appearance, being sometimes pointed, broad and thin, sometimes bifid, or perforated by a round hole, occasionally curved, or deflected considerably to one or the other side.

Structure. The bone is composed of delicate cancellated texture, covered by a thin layer of compact tissue, which is thickest in the manubrium, between the articular facets for the clavicles.

Development. The sternum, including the ensiform appendix, is developed by *six* centres; one for the first piece, or manubrium, four for the second piece or gladiolus, and one for the ensiform appendix. Up to the middle of fœtal life, the sternum is entirely cartilaginous, and when ossification takes place, the ossific granules are deposited in the middle of the intervals between the articular depressions for the costal cartilages, in the following order (fig. 71). In the first piece, between the fifth and sixth months; in the second and third, between the sixth and seventh months; in the fourth piece, at the ninth month; in the fifth, within the first year, or between the first and second years after birth; and in the ensiform appendix, between the second and the seventeenth or eighteenth years, by a single centre which makes its appearance at the upper part, and proceeds gradually downwards. To these may be added the occasional existence, as described by Breschet, of two small episternal centres, which make their appearance one on each side of the interclavicular notch. These are regarded by him as the anterior rudiments of a rib, of which the posterior rudiment is the anterior lamina of the transverse process of the seventh cervical vertebra. It occasionally happens that some of the segments are formed from more than one centre, the number and position of which vary (fig. 73). Thus the first piece may have two, three, or even six centres. When two are present, they are generally situated one above the other, the upper one being the larger; the second piece has seldom more than one; the third, fourth, and fifth pieces are often formed from two centres placed laterally, the irregular union of which will serve to explain the occasional occurrence of the sternal foramen (fig. 74), or of the vertical fissure which occasionally intersects this part of the bone. Union of the various centres commences from below, and proceeds upwards, taking place in the following order (fig. 72). The fifth piece is joined to the fourth soon after puberty; the third to the third, between the twentieth and twenty-fifth years; the third to the second, between the thirty-fifth and fortieth years; the second is rarely joined to the first except in very advanced age.

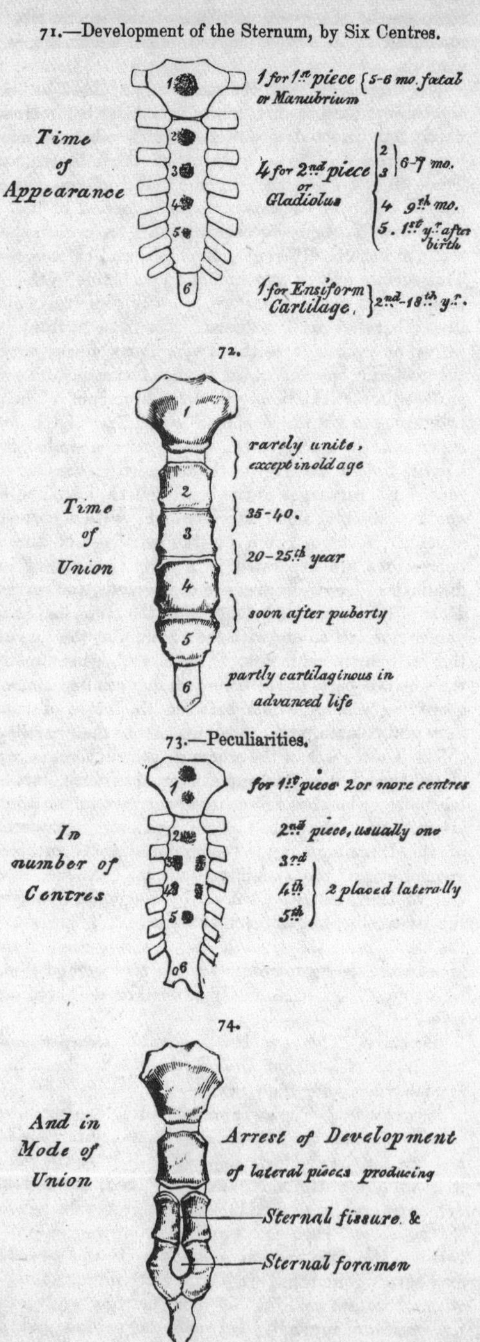

71.—Development of the Sternum, by Six Centres.

72.

73.—Peculiarities.

74.

Articulations. With the clavicles, and seven costal cartilages on each side.

75.—A Central Rib of Right Side. Inner Surface.

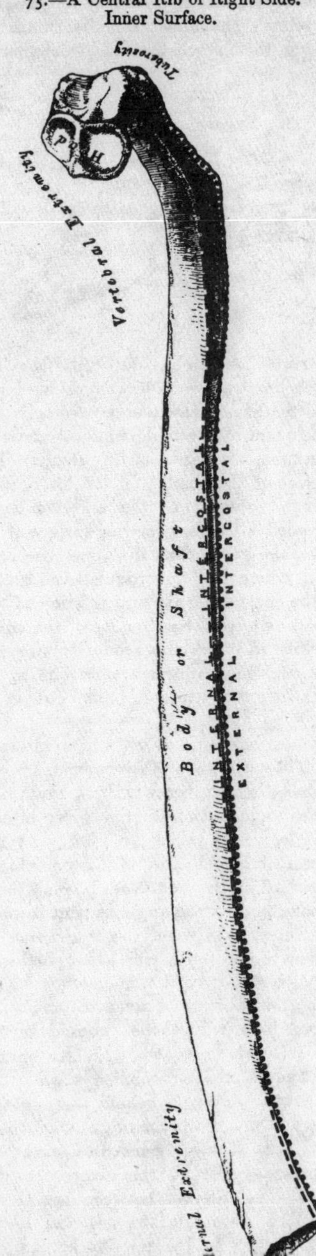

Attachment of Muscles. The Pectoralis major, Sterno-cleido-mastoid, Sterno-hyoid, Sterno-thyroid, Triangularis sterni, aponeurosis of the Obliquus externus, Obliquus internus, and Transversalis muscles, Rectus and Diaphragm.

The Ribs.

The Ribs are elastic arches of bone, which form the chief part of the thoracic walls. They are twelve in number on each side; but this number may be increased by the development of a cervical or lumbar rib, or may be diminished to eleven. The first seven are connected behind with the spine, and in front with the sternum, through the intervention of the costal cartilages; they are called *vertebro-sternal*, or true ribs. The remaining five are false ribs; of these the first three, being connected behind with the spine, and in front with the costal cartilages, are called the *vertebro-costal ribs*: the last two are connected with the vertebræ only, being free at their anterior extremities; they are termed *vertebral* or *floating ribs*. The ribs vary in their direction, the upper ones being placed nearly at right angles with the spine, the lower ones obliquely, so that the anterior extremity is lower than the posterior. The extent of obliquity reaches its maximum at the ninth rib, and gradually decreases from that rib to the twelfth. The ribs are situated one beneath the other in such a manner that spaces are left between them, which are called *intercostal spaces*. Their length corresponds to the length of the ribs, their breadth is more considerable in front than behind, and between the upper than between the lower ribs. The ribs increase in length from the first to the seventh, when they again diminish to the twelfth. In breadth they decrease from above downwards; in each rib the greatest breadth is at the sternal extremity.

Common characters of the Ribs (fig. 75). A rib from the middle of the series should be taken in order to study the common characters of the ribs.

Each rib presents two extremities, a posterior or vertebral, an anterior or sternal, and an intervening portion, the body or shaft. The *posterior* or *vertebral extremity* presents for examination a head, neck, and tuberosity. The *head* (fig. 76.) is marked by a kidney-shaped articular surface, divided by a horizontal ridge into two facets for articulation with the costal cavity formed by the junction of the bodies of two contiguous dorsal vertebræ; the upper facet is small, the inferior one of large size; the ridge separating them serves for the attachment of the inter-articular ligament.

76.—Vertebral Extremity of a Rib. External Surface.

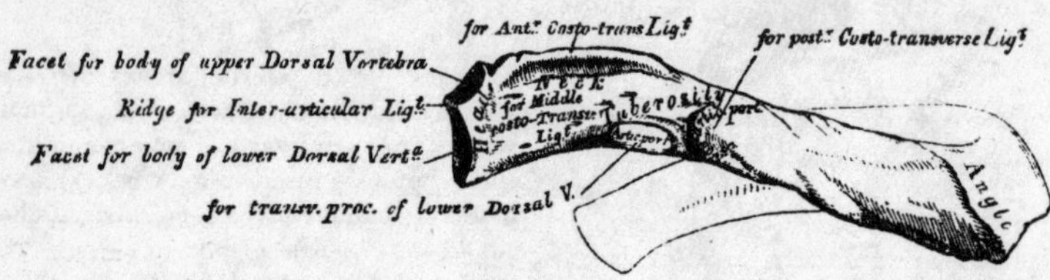

The *neck* is that flattened portion of the rib which extends outwards from the head; it is about an inch long, and rests upon the transverse process of the lower of the two vertebræ with which the head articulates. Its *anterior surface* is flat and smooth, its *posterior* rough, for the attachment of the middle costo-transverse ligament, and perforated by numerous foramina, the direction of which is less constant than those found on the inner surface of the shaft. Of its two borders the *superior* presents a rough crest for the attachment of the anterior costo-transverse ligament; its *inferior border* is rounded. On the posterior surface of the neck, just where it joins the shaft, and nearer the lower than the upper border, is an eminence—the tuberosity, or tubercle; it consists of an articular and a non-articular portion. The *articular portion*, the most internal and inferior of the two, presents a small oval surface, for articulation with the extremity of the transverse process of the lower of the two vertebræ to which the head is connected. The *non-articular portion* is a rough elevation, which affords attachment to the posterior costo-transverse ligament. The tubercle is much more prominent in the upper than in the lower ribs.

The *shaft* is thin and flat, so as to present two surfaces, an external and an internal; and two borders, a superior and an inferior. The *external surface* is convex, smooth, and marked, at its back part, a little in front of the tuberosity, by a prominent line, directed obliquely from above, downwards and outwards; this gives attachment to a tendon of the Sacro-lumbalis muscle, and is called the *angle*. At this point, the rib is bent in two directions. If the rib is laid upon its lower border, it will be seen that the anterior portion of the shaft, as far as the angle, rests upon this margin, while the vertebral end of the bone, beyond the angle, is bent inwards and at the same time tilted upwards. The interval between the angle and the tuberosity increases gradually from the second to the tenth rib. The portion of bone between these two parts is rounded, rough, and irregular, and serves for the attachment of the Longissimus dorsi. The portion of bone between the angle and sternal extremity is also slightly twisted upon its own axis, the external surface looking downwards behind the angle, a little upwards in front of it. This surface presents, towards its sternal extremity, an oblique line, the anterior angle. The *internal surface* is concave, smooth, directed a little upwards behind the angle; a little downwards in front of it. This surface is marked by a ridge which commences at the lower extremity of the head; it is strongly marked as far as the inner side of the angle, and gradually becomes lost at the junction of the anterior with the middle third of the bone. The interval between it and the inferior border is deeply grooved, to lodge the intercostal vessels and nerve. At the back part of the bone, this groove belongs to the inferior border, but just in front of the angle, where it is deepest and broadest, it corresponds to the internal surface. The superior edge of the groove is rounded; it serves for the attachment of the Internal intercostal muscle. The inferior edge corresponds to the lower margin of the rib, and gives attachment to the External intercostal. Within the groove are seen the orifices of numerous small foramina, which traverse the wall of the shaft obliquely from before backwards. The *superior border*, thick and rounded, is marked by an external and an internal lip, more distinct behind than in front; they serve for the attachment of the External and Internal intercostal muscles. The *inferior border*, thin and sharp, has attached the External intercostal muscle. The anterior or sternal extremity is flattened, and presents a porous oval concave depression, into which the costal cartilage is received.

Peculiar Ribs.

The ribs which require especial consideration are five in number, viz., the first, second, tenth, eleventh, and twelfth.

The *first rib* (fig. 77) is one of the shortest and the most curved of all the ribs; it is broad, flat, and placed horizontally at the upper part of the thorax, its surfaces looking upwards and downwards; and its borders inwards and outwards. The *head* is of small size, rounded, and presents only a single articular facet for articulation with the body of the first dorsal vertebra. The *neck* is narrow and rounded. The *tuberosity*, thick and prominent, rests on the outer border. There is no angle, and the shaft is not twisted on its axis. The upper surface of the shaft is marked by two shallow depressions, separated from one another by a ridge, which becomes more prominent towards the internal border, where it terminates in a tubercle: this tubercle and ridge serve for the attachment of the Scalenus anticus muscle, the groove in front of it transmitting the subclavian vein: that behind it, the subclavian artery. Between the groove for the subclavian artery and the tuberosity, is a depression for the attachment of the Scalenus medius muscle. The *under surface* is smooth, and destitute of the groove observed on the other ribs. The *outer border* is convex, thick, and rounded; the *inner*, concave, thin, and sharp, and marked about its centre by the tubercle before mentioned. The *anterior extremity* is larger and thicker than any of the other ribs.

The *second rib* (fig. 78) is much longer than the first, but bears a very considerable resemblance to it in the direction of its curvature. The non-articular portion of the tuberosity is occasionally only slightly marked. The *angle* is slight, and situated close to the tuberosity, and the shaft is not twisted, so that both ends touch any plane surface upon which it may be laid. The shaft is not horizontal, like that of the first rib; its *outer surface*, which is convex, looking upwards and a little outwards. It presents, near the middle, a rough eminence for the attachment of part of the second and third digitations of the Serratus magnus. The *inner surface*, smooth and concave, is directed downwards and a little inwards: it presents a short groove towards its posterior part.

The *tenth rib* (fig. 79) has only a single articular facet on its head.

The *eleventh* and *twelfth ribs* (figs. 80 and 81) have each a single articular facet on the head, which is of rather large size; they have no neck or tuberosity, and are pointed at the extremity. The eleventh has a slight angle and a shallow groove on the lower border. The twelfth has neither, and is much shorter than the eleventh.

Structure. The ribs consist of cancellous tissue, enclosed in a thin compact layer.

Development. Each rib, with the exception of the last two, is developed by *three* centres, one for the shaft, one for the head, and one for the tubercle. The last two have only *two* centres, that for the tubercle being wanting. Ossification commences in the body of the ribs at a very early period, before its appearance in the vertebræ. The epiphysis of the head, which is of a slightly angular shape, and that for the tubercle, of a lenticular form, make their appearance between the sixteenth and twentieth years, and are not united to the rest of the bone until about the twenty-fifth year.

Attachment of Muscles. The Intercostals, Scalenus anticus, Scalenus medius, Scalenus posticus, Pectoralis minor, Serratus magnus, Obliquus externus, Transversalis Quadratus lumborum, Diaphragm, Latissimus dorsi, Serratus posticus superior, Serratus posticus inferior, Sacro-lumbalis, Musculus accessorius ad sacro-lumbalem, Longissimus dorsi, Cervicalis ascendens, Levatores costarum.

Peculiar Ribs.

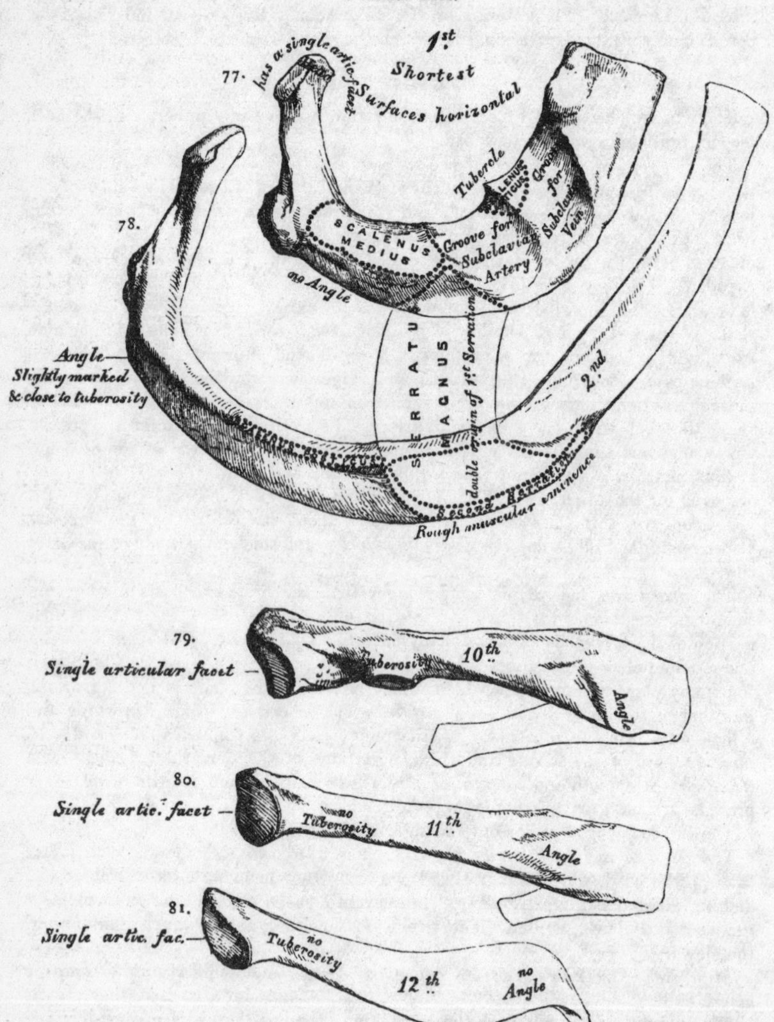

The Costal Cartilages.

The *Costal Cartilages* (fig. 69, p. 76) are white elastic structures, which serve to prolong the ribs forward to the front of the chest, and contribute very materially to the elasticity of its walls. The first seven are connected with the sternum, the next three with the lower border of the cartilage of the preceding rib. The cartilages of the last two ribs, which have pointed extremities, float freely in the walls of the abdomen. Like the ribs, the costal cartilages vary in their length, breadth, and direction. They increase in length from the first to the seventh, then gradually diminish to the last. They diminish in breadth, as well as the intervals between them, from the first to the last. They are broad at their attachment to the ribs, and taper towards their sternal extremities, excepting the first two, which are of the same breadth throughout, and the sixth, seventh, and eighth, which are enlarged where their margins are in contact. In direction they also vary; the first descends a little, the second is horizontal, the third ascends slightly, whilst all the rest follow the course of the ribs for a short extent, and then ascend to the sternum or preceding cartilage. Each costal cartilage presents two surfaces, two borders, and two extremities. The *anterior surface* is convex, and looks forwards and upwards; that of the first gives attachment to the costo-clavicular ligament; that of the first, second, third, fourth, fifth, and sixth, at their sternal ends, to the Pectoralis major. The others are covered by, and give partial attachment to, some of the great flat muscles of the abdomen. The *posterior surface* is concave, and directed backwards and downwards, the six or seven inferior ones affording attachment to the Transversalis muscle, and the Diaphragm. Of the two borders, the superior is concave; the inferior, convex; they afford attachment to the Intercostal muscles, the upper border of the sixth giving attachment to the Pectoralis major muscle. The contiguous borders of the sixth, seventh, and eighth, and sometimes the ninth and tenth costal cartilages present smooth oblong surfaces at the points where they articulate. Of the two extremities, the outer one is continuous with the osseous tissue of the rib to which it belongs. The inner extremity of the first is continuous with the sternum; the six succeeding ones have rounded extremities, which are received into shallow concavities on the lateral margins of the sternum. The inner extremities of the eighth, ninth, and tenth costal cartilages are pointed, and lie in contact with the cartilage above. Those of the eleventh and twelfth are free, and pointed.

The costal cartilages are most elastic in youth, those of the false ribs being more so than the true. In old age, they become of a deep yellow colour. Under certain diseased conditions, they are prone to ossify. Dr. Humphry's observations on this subject have led him to regard the ossification of the costal cartilages as a sign of disease rather than of age. 'The ossification takes place in the first cartilage sooner than in the others; and in men more frequently, and at an earlier period of life than in women.'

Attachment of Muscles. The Subclavius, Sterno-thyroid, Pectoralis major, Internal Oblique, Transversalis, Rectus, Diaphragm, Triangularis sterni, Internal and External Intercostals.

OF THE EXTREMITIES.

The extremities, or limbs, are those long-jointed appendages of the body, which are connected to the trunk by one end, and free in the rest of their extent. They are *four* in number: an *upper* or *thoracic pair*, connected with the thorax through the intervention of the shoulder, and subservient mainly to tact and prehension; and a *lower pair*, connected with the pelvis, intended for support and locomotion. Both pairs of limbs are constructed after one common type, so that they present numerous analogies; while at the same time certain differences are observed in each, dependent on the peculiar offices they perform.

OF THE UPPER EXTREMITY.

The upper extremity consists of the arm, the fore-arm, and the hand. Its continuity with the trunk is established by means of the shoulder, which is homologous with the innominate or haunch bone in the lower limb.

Of the Shoulder.

The Shoulder is placed upon the upper part and side of the chest, connecting the upper extremity to the trunk; it consists of two bones, the clavicle, and the scapula.

The Clavicle.

The *Clavicle* (*clavis*, a 'key'), or collar-bone, forms the anterior portion of the shoulder. It is a long bone, curved somewhat like the italic letter *ſ*, and placed nearly horizontally at the upper and anterior part of the thorax, immediately above the first rib. It articulates by its inner extremity with the upper border of the sternum, and, by its outer extremity, with the acromion process of the scapula; serving to sustain the upper extremity in the various positions which it assumes, whilst, at the same time, it allows of great latitude of motion in the arm. The clavicle is nearly horizontal. It presents a double curvature, when looked at in front; the convexity being forwards at the sternal end, and the concavity at the scapular end. Its outer third is flattened from above downwards, and extends in the natural position of the bone, from the coracoid process to the acromion. Its inner two-thirds are of a cylindrical form, and extend from the sternum to the coracoid process of the scapula.

External or Flattened Portion. The *outer third* is flattened from above downwards, so as to present two surfaces, an upper and a lower; and two borders, an anterior and a posterior. The *upper surface* is flattened, rough, marked by impressions for the attachment of the Deltoid in front, and the Trapezius behind: between these two impressions, externally, a small portion of the bone is subcutaneous. The *under surface* is flattened. At its posterior border, where the prismatic joins with the flattened portion, is a rough eminence, the *conoid tubercle*; this, in the natural position of the bone, surmounts the coracoid process of the scapula, and gives attachment to the conoid ligament. From this tubercle, an oblique line, occasionally a depression, passes forwards and outwards to near the outer end of the anterior border; it is called the *oblique line*, and affords attachment to the trapezoid ligament. The *anterior border* is concave, thin, and rough; it limits the attachment of the Deltoid, and occasionally presents, near the centre, a tubercle, the *deltoid tubercle*, which is sometimes distinct in the living subject. The *posterior border* is convex, rough, broader than the anterior, and gives attachment to the Trapezius.

Internal or Cylindrical Portion. The cylindrical portion forms the inner two-thirds of the bone. It is curved, so as to be convex in front, concave behind, and is marked by three borders separating three surfaces. The *anterior border* is continuous with the anterior margin of the flat portion. At its commencement it is smooth and corresponds to the interval between the attachment of the Pectoralis major and Deltoid muscles; about the centre of the clavicle it divides to enclose an elliptical space for the attachment of the clavicular portion of the Pectoralis major. This space extends inwards as far as the anterior margin of the sternal extremity. The *superior border* is continuous with the posterior margin of the flat portion, and separates the anterior from the posterior surface. At its commencement it is smooth and rounded, becomes rough towards the inner third for the attachment of the Sterno-mastoid muscle, and terminates at the upper angle of the sternal extremity. The *posterior* or *subclavian border* separates the posterior from the inferior surface, and extends from the conoid tubercle to the rhomboid impression. It forms the posterior boundary of the groove for the Subclavius muscle, and gives attachment to the fascia which encloses that muscle. The *anterior surface* is included between the superior and anterior borders. It is directed forwards and a little upwards at the sternal end, outwards and still more upwards at the acromial extremity, where it becomes continuous with the upper surface of the flat portion. Externally, it is smooth, convex, nearly subcutaneous, being covered only by the Platysma; but corresponding to the inner half of the bone, it is divided by a more or less prominent line into two parts: an anterior portion, elliptical in form, rough, and slightly convex, for the attachment of the Pectoralis major; and an upper part, which is rough behind, for the attachment of the Sterno-cleido-mastoid. Between the two muscular impressions is a small subcutaneous interval. The *posterior* or *cervical surface* is smooth, flat, directed vertically, and looks backwards towards the root of the neck. It is limited, above, by the superior border; below, by the subclavian border; internally, by the margin of the sternal extremity; externally, it is continuous with the posterior border of the flat portion. It is concave from within outwards, and is in relation, by its lower part, with the suprascapular vessels. It gives attachment, near the sternal extremity, to part of the Sterno-hyoid muscle; and presents, at or near the middle, a foramen, directed obliquely outwards, which transmits the chief nutrient artery of the bone. Sometimes, there are two foramina on the posterior surface, or one on the posterior, the other on the inferior surface. The *inferior* or *subclavian surface* is bounded, in front, by the anterior border; behind by the subclavian border. It is narrow internally,

82.—Left Clavicle. Anterior Surface.

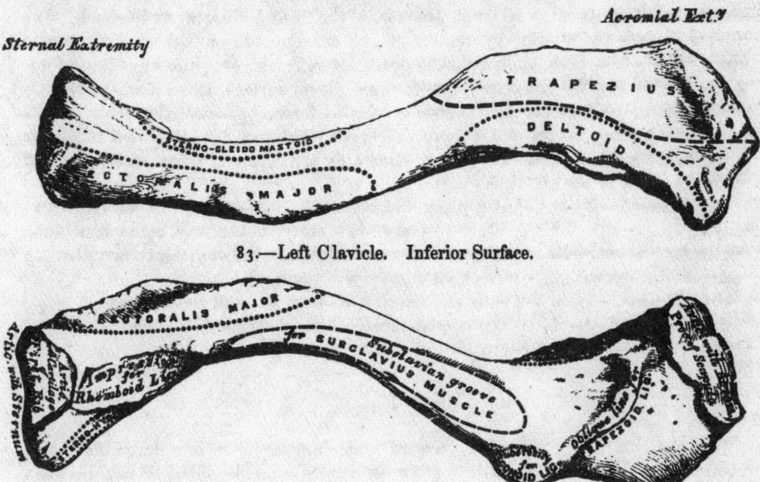

83.—Left Clavicle. Inferior Surface.

but gradually increases in width externally, and is continuous with the under surface of the flat portion. Commencing at the sternal extremity may be seen a small facet for articulation with the cartilage of the first rib. This is continuous with the articular surface at the sternal end of the bone. External to this is a broad rough impression, the rhomboid, rather more than an inch in length, for the attachment of the costo-clavicular (rhomboid) ligament. The remaining part of this surface is occupied by a longitudinal groove, the subclavian groove, broad and smooth externally; narrow and more uneven internally; it gives attachment to the Subclavius muscle, and, by its anterior margin, to the strong aponeurosis which encloses it. Not unfrequently this groove is subdivided into two parts, by a longitudinal line, which gives attachment to the intermuscular septum of the Subclavius muscle.

The *internal* or *sternal* end of the clavicle is triangular in form, directed inwards, and a little downwards and forwards; and presents an articular facet, concave from before backwards, convex from above downwards, which articulates with the sternum through the intervention of an inter-articular fibro-cartilage; the circumference of the articular surface is rough, for the attachment of numerous ligaments. This surface is continuous with the costal facet on the inner end of the inferior or subclavian surface, which articulates with the cartilage of the first rib.

The *outer* or *acromial extremity*, directed outwards and forwards, presents a small, flattened, oval facet, which looks obliquely downwards and inwards, for articulation with the acromion process of the scapula. The direction of this surface serves to explain the greater frequency of dislocation upwards rather than downwards, beneath the acromion process. The circumference of the articular facet is rough, especially above, for the attachment of the acromio-clavicular ligaments.

Peculiarities of the Bone in the Sexes and in Individuals. In the female, the clavicle is generally less curved, smoother, and more slender than in the male. In those persons who perform considerable manual labour, which brings into constant action the muscles connected with this bone, it acquires considerable bulk, becomes shorter, more curved, its ridges for muscular attachment become prominently marked, and its sternal end of a prismatic form. The right clavicle is generally heavier, thicker, and rougher, and often shorter, than the left.

Structure. The shaft, as well as the extremities, consists of cancellous tissue, invested in a compact layer much thicker in the centre than at either end. The clavicle is highly elastic, by reason of its curves. From the experiments of Mr. Ward, it has been shown that it possesses sufficient longitudinal elastic force to project its own weight nearly two feet on a level surface, when a smart blow is struck on it, and sufficient transverse elastic force, opposite the centre of its anterior convexity, to throw its own weight about a foot. This extent of elastic power must serve to moderate very considerably the effect of concussions received upon the point of the shoulder.

Development. By *two* centres: one for the shaft, and one for the sternal extremity. The centre for the shaft appears very early, before any other bone; the centre for the sternal end makes its appearance about the eighteenth or twentieth year, and unites with the rest of the bone a few years after.

Articulations. With the sternum, scapula, and cartilage of the first rib.

Attachment of Muscles. The Sterno-cleido mastoid, Trapezius, Pectoralis major, Deltoid, Subclavius, and Sterno-hyoid.

The Scapula.

The *Scapula* forms the back part of the shoulder. It is a large flat bone, triangular in shape, situated at the posterior aspect and side of the thorax, between the first and eighth ribs, its posterior border or base being about an inch from, and nearly parallel with, the spinous processes of the vertebræ. It presents for examination two surfaces, three borders, and three angles.

The *anterior surface*, or *venter* (fig. 84), presents a broad concavity, the subscapular fossa. It is marked, in the posterior two-thirds, by several oblique ridges, which pass from behind obliquely outwards and upwards, the anterior third being smooth. The oblique ridges give attachment to the tendinous intersections, and the surfaces between them, to the fleshy fibres, of the Subscapularis muscle. The anterior third of the fossa, which is smooth, is covered by, but does not afford attachment to, the fibres of this muscle. This surface is separated from the posterior border by a smooth triangular margin at the superior and inferior angles, and in the interval between these by a narrow edge which is often deficient. This marginal surface affords attachment throughout its entire extent to the Serratus magnus muscle. The subscapular fossa presents a transverse depression at its upper part, called the *subscapular angle*; it is in this situation that the fossa is deepest; so that the thickest part of the Subscapularis muscle lies in a line parallel with the glenoid cavity, and must consequently operate most effectively on the humerus which is contained in that cavity.

The *posterior surface*, or *dorsum* (fig. 85), is arched from above downwards, alternately convex and concave from side to side. It is subdivided unequally into two parts by the spine; the portion above the spine is called the supraspinous fossa, and that below it, the infraspinous fossa.

The *supraspinous fossa*, the smaller of the two, is concave, smooth, and broader, at the vertebral than at the humeral extremity. It affords attachment by its inner two-thirds to the Supraspinatus muscle.

84.—Left Scapula. Anterior Surface, or Venter.

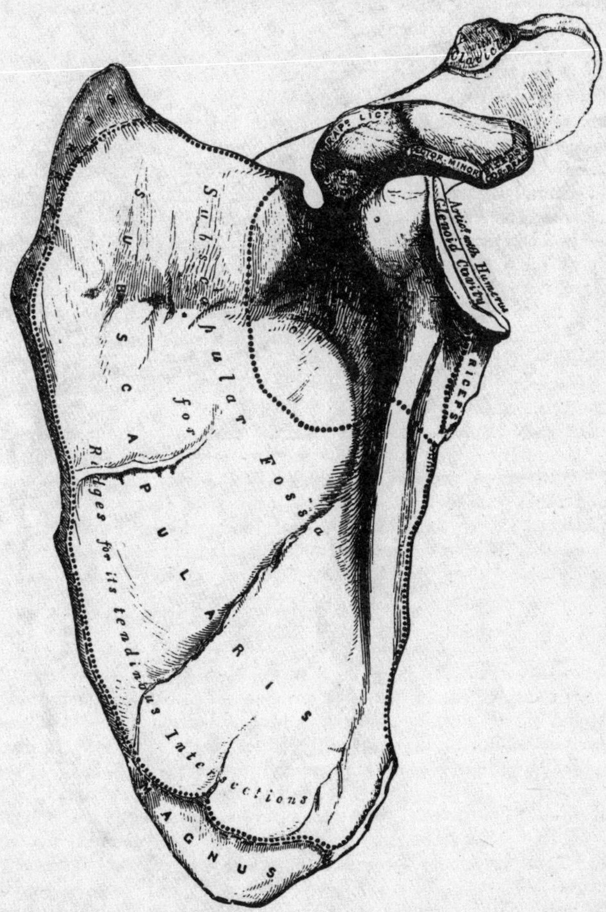

The *infraspinous fossa* is much larger than the preceding; towards its vertebral margin a shallow concavity is seen at its upper part; its centre presents a prominent convexity, whilst towards the axillary border is a deep groove, which runs from the upper towards the lower part. The inner three-fourths of this surface afford attachment to the infraspinatus muscle; the outer fourth is only covered by it, without giving origin to its fibres. This surface is separated from the axillary border by an elevated ridge, which runs from the lower part of the glenoid cavity, downwards and backwards to the posterior border, about an inch above the inferior angle. The ridge serves for the attachment of a strong aponeurosis, which separates the Infraspinatus from the two Teres muscles. The surface of bone between this line and the axillary border is narrow in the upper two-thirds of its extent, and traversed near its centre by a groove for the passage of the dorsalis scapulæ vessels; it affords attachment to the Teres minor. Its lower third presents a broader, somewhat triangular surface, which gives origin to the Teres major, and

85.—Left Scapula. Posterior Surface, or Dorsum.

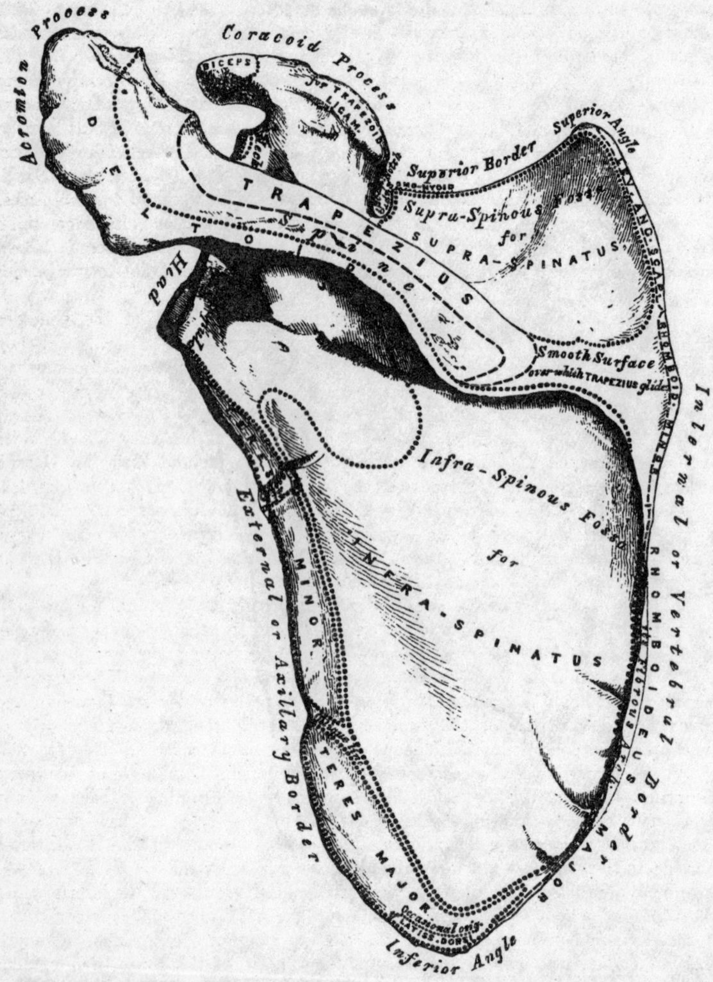

over which glides the Latissimus dorsi; sometimes the latter muscle takes origin by a few fibres from this part. The broad and narrow portions of bone above alluded to are separated by an oblique line, which runs from the axillary border, downwards and backwards: to it is attached the aponeurosis separating the two Teres muscles from each other.

The *Spine* is a prominent plate of bone, which crosses obliquely the inner four-fifths of the dorsum of the scapula at its upper part, and separates the supra- from the infraspinous fossa: it commences at the vertebral border by a smooth triangular surface, over which the Trapezius glides, separated from the bone by a bursa; and, gradually becoming more elevated as it passes forwards, terminates in the acromion process which overhangs the shoulder joint. The spine is triangular and flattened from above downwards, its apex corresponding to the posterior border; its base, which is directed outwards, to the neck of the scapula. It presents two surfaces and three borders. Its *superior surface* is concave, assists in forming the supraspinous fossa, and affords attachment to part of the Supraspinatus muscle. Its *inferior surface* forms part of the infraspinous fossa, gives origin to part of the Infraspinatus muscle, and presents near its centre the orifice of a nutrient canal. Of the three borders, the *anterior* is attached to the dorsum of the bone; the *posterior*, or *crest* of the spine, is broad, and presents two lips, and an intervening rough interval. To the superior lip is attached the Trapezius, to the extent shown in the figure. A very rough prominence is generally seen occupying that portion of the spine which receives the insertion of the middle and inferior fibres of this muscle. To the inferior lip, throughout its whole length, is attached the Deltoid. The interval between the lips is also partly covered by the fibres of these muscles. The *external border*, the shortest of the three, is slightly concave, its edges thick and round, continuous above with the under surface of the acromion process; below, with the neck of the scapula. The narrow portion of bone external to this border serves to connect the supra- and infraspinous fossæ.

The *Acromion process*, so called from forming the summit of the shoulder (ἄκρον, a summit; ὦμος, the shoulder), is a large and somewhat triangular process, flattened from behind forwards, directed at first a little outwards, and then curving forwards and upwards, so as to overhang the glenoid cavity. Its *upper surface*, directed upwards, backwards, and outwards, is convex, rough, and gives attachment to some fibres of the Deltoid. Its *under surface* is smooth and concave. Its *outer border*, which is thick and irregular, affords attachment to the Deltoid muscle. Its *inner margin*, shorter than the outer, is concave, gives attachment to a portion of the Trapezius muscle, and presents about its centre a small oval surface, for articulation with the scapular end of the clavicle. Its *apex*, which corresponds to the point of meeting of these two borders in front, is thin, and has attached to it the coraco-acromial ligament.

Of the three borders or costæ of the scapula, the *superior* is the shortest and thinnest; it is concave, terminating at its inner extremity at the superior angle, at its outer extremity at the coracoid process. At its outer part is a deep semicircular notch, the suprascapular, formed partly by the base of the coracoid process. This notch is converted into a foramen by the transverse ligament, and serves for the passage of the suprascapular nerve. The adjacent margin of the superior border affords attachment to the Omo-hyoid muscle. The *external*, or *axillary*, *border* is the thickest of the three. It commences above at the lower margin of the glenoid cavity, and inclines obliquely downwards and backwards to the inferior angle. Immediately below the glenoid cavity, is a rough depression about an inch in length, which affords attachment to the long head of the Triceps muscle; to this succeeds a longitudinal groove, which extends as far as its lower third, and affords origin to part of the Subscapularis muscle. The inferior third of this border, which is thin and sharp, serves for the attachment of a few fibres of the Teres major behind, and of the Subscapularis in front. The *internal* or *vertebral border*, also named the base, is the longest of the three, and extends from the superior to the inferior angle of the bone. It is arched, intermediate in thickness between the superior and the external borders, and the portion of it above the spine is bent considerably outwards, so as to form an obtuse angle with the lower part. The vertebral border presents an anterior lip, a posterior lip, and an intermediate space. The *anterior lip* affords attachment to the Serratus magnus; the *posterior lip*, to the Supraspinatus above the spine, the Infraspinatus below; the interval between the two lips, to the Levator anguli scapulæ above the triangular surface at the commencement of the spine; the Rhomboideus minor, to the edge of that surface; the Rhomboideus major being attached by means of a fibrous arch, connected above to the lower part of the triangular surface at the base of the spine, and below to the lower part of the posterior border.

Of the three angles, the *superior*, formed by the junction of the superior and internal borders, is thin, smooth, rounded, somewhat inclined outwards, and gives attachment to a few fibres of the Levator anguli scapulæ muscle. The *inferior* angle, thick and rough, is formed by the union of the vertebral and axillary borders, its outer surface affording attachment to the Teres major, and occasionally a few fibres of the Latissimus dorsi. The *anterior* angle is the thickest part of the bone, and forms what is called the *head* of the scapula. The head presents a shallow, pyriform, articular surface, the *glenoid cavity* (γλήνη, *a socket*), whose longest diameter is from above downwards, and its direction outwards and forwards. It is broader below than above; at its apex is attached the long tendon of the Biceps muscle. It is covered with cartilage in the recent state; and its margins, slightly raised, give attachment to a fibro-cartilaginous structure, the glenoid ligament, by which its cavity is deepened. The neck of the scapula is the slightly depressed surface which surrounds the head; it is more distinct on the posterior than on the anterior surface, and below than above. In the latter situation, it has, arising from it, a thick prominence, the coracoid process.

The *Coracoid process*, so called from its fancied resemblance to a crow's beak (κόραξ, *a crow*), is a thick curved process of bone, which arises by a broad base from the upper part of the neck of the scapula; it ascends at first upwards and inwards; then, becoming smaller, it changes its direction, and passes forwards and outwards. The ascending portion, flattened from before backwards, presents in front a smooth concave surface, over which passes the Subscapularis muscle. The horizontal portion is flattened from above downwards; its upper surface is convex and irregular; its under surface is smooth; its anterior border is rough, and gives attachment to the Pectoralis minor; its posterior border is also rough for the coraco-acromial ligament, while the apex is embraced by the conjoined tendon of origin of the short head of the Biceps and Coraco-brachialis muscles. At the inner side of the root of the coracoid process is a rough depression for the attachment of the conoid ligament, and, running from it obliquely forwards and outwards on the upper surface of the horizontal portion, an elevated ridge for the attachment of the trapezoid ligament.

Structure. In the head, processes, and all the thickened parts of the bone, it is cellular in structure, of a dense compact tissue in the rest of its extent. The centre and upper part of the dorsum, but especially the former, is usually so thin as to be semi-transparent; occasionally the bone is found wanting in this situation, and the adjacent muscles come into contact.

Development (fig. 86). By *seven* centres; one for the body, two for the coracoid process, two for the acromion, one for the posterior border, and one for the inferior angle.

Ossification of the body of the scapula commences about the second month of fœtal life, by the formation of an irregular quadrilateral plate of bone, immediately behind the glenoid cavity. This plate extends itself so as to form the chief part of the bone, the spine growing up from its posterior surface about the third month. At birth, the chief part of the scapula is osseous, only the coracoid and acromion processes, the posterior border, and inferior angle, being cartilaginous. About the first year after birth, ossification takes place in the middle of the coracoid process; which usually becomes joined with the rest of the bone at the time when the other centres make their appearance. Between the fifteenth and seventeenth years, ossification of the remaining centres takes place in quick succession, and in the following order: first, near the base of the acromion, and in the root of the coracoid process, the latter appearing in the form of a broad scale; secondly, in the inferior angle and contiguous part of the posterior border; thirdly, near the extremity of the acromion; fourthly, in the posterior border. The acromion process, besides being formed of two separate nuclei, has its base formed by an extension into it of the centre of ossification which belongs to the spine, the extent of which varies in different cases. The two separate nuclei unite, and then join with the extension carried in from the spine. These various epiphyses become joined to the bone between the ages of twenty-two and twenty-

86.—Plan of the Development of the Scapula. By Seven Centres.

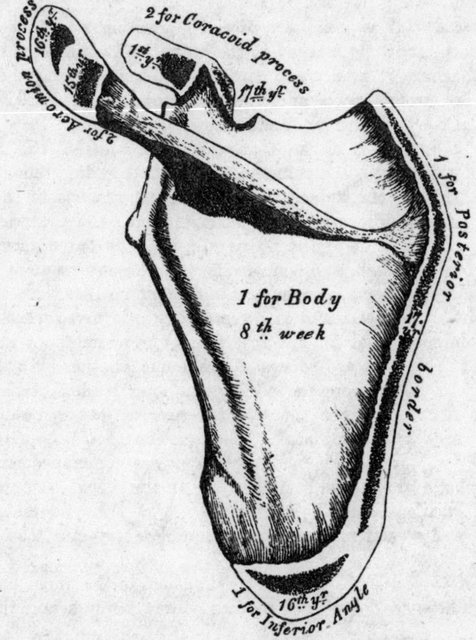

The epiphyses (except one for the Coracoid process) appear from fifteen to seventeen years, and unite between twenty-two and twenty-five years of age.

five years. Sometimes failure of union between the acromion process and spine occurs, the junction being effected by fibrous tissue, or by an imperfect articulation; in some cases of supposed fracture of the acromion with ligamentous union, it is probable that the detached segment was never united to the rest of the bone.

Articulations. With the humerus and clavicle.

Attachment of Muscles. To the anterior surface, the Subscapularis; posterior surface, Supraspinatus, Infraspinatus; spine, Trapezius, Deltoid; superior border, Omo-hyoid; vertebral border, Serratus magnus, Levator anguli scapulæ, Rhomboideus, minor and major; axillary border, Triceps, Teres minor, Teres major, glenoid cavity, long head of the Biceps; coracoid process, short head of the Biceps, Coraco-brachialis, Pectoralis minor; and to the inferior angle occasionally a few fibres of the Latissimus dorsi.

The Humerus.

The *Humerus* is the longest and largest bone of the upper extremity; it presents for examination a shaft and two extremities.

The *Upper Extremity* is the largest part of the bone; it presents a rounded head, joined to the shaft by a constricted part, called the neck, and two other eminences, the greater and lesser tuberosities (fig. 87).

The *head*, nearly hemispherical in form, is directed upwards, inwards, and a little backwards; its surface is smooth, coated with cartilage in the recent state, and articulates with the glenoid cavity of the scapula. The circumference of its articular surface is slightly constricted, and is termed the *anatomical neck*, in contradistinction to the constriction which exists below the tuberosities, and is called the *surgical neck*, from its often being the seat of fracture. It should be remembered, however, that fracture of the *anatomical neck* does sometimes, though rarely, occur.

The *anatomical neck* is obliquely directed, forming an obtuse angle with the shaft. It is more distinctly marked in the lower half of its circumference than in the upper half, where it presents a narrow groove, separating the head from the tuberosities. Its circumference affords attachment to the capsular ligament, and is perforated by numerous vascular foramina.

The *greater tuberosity* is situated on the outer side of the head and lesser tuberosity. Its upper surface is rounded and marked by three flat facets, separated by two slight ridges: the anterior facet gives attachment to the tendon of the Supraspinatus; the middle one to the Infraspinatus; the posterior facet, and the shaft of the bone below it, to the Teres minor. The outer surface of the great tuberosity is convex, rough, and continuous with the outer side of the shaft.

The *lesser tuberosity* is more prominent, although smaller than the greater: it is situated in front of the head, and is directed inwards and forwards. Its summit presents a prominent facet for the insertion of the tendon of the Subscapularis muscle. The tuberosities are separated from one another by a deep groove, the *bicipital groove*, so called from its lodging the long tendon of the Biceps muscle. It commences above between the two tuberosities, passes obliquely downwards and a little inwards, and terminates at the junction of the upper with the middle third of the bone. It is deep and narrow at its commencement, and becomes shallow and a little broader as it descends. In the recent state it is covered with a thin layer of cartilage, lined by a prolongation of the synovial membrane of the shoulder-joint, and receives part of the tendon of insertion of the Latissimus dorsi about its centre.

The *Shaft* of the humerus is almost cylindrical in the upper half of its extent, prismatic and flattened below, and presents three borders and three surfaces for examination.

The *anterior border* runs from the front of the great tuberosity above to the coronoid depression below, separating the internal from the external surface. Its upper part is very prominent and rough, forms the outer lip of the bicipital groove, and serves for the attachment of the tendon of the Pectoralis major. About its centre is seen the rough deltoid impression; below, it is smooth and rounded, affording attachment to the Brachialis anticus.

The *external border* runs from the back part of the greater tuberosity to the external condyle, and separates the external from the posterior surface. It is rounded and indistinctly marked in its upper half, serving for the attachment of the external head of the Triceps muscle; its centre is traversed by a broad but shallow oblique depression, the musculo-spiral groove; its lower part is marked by a prominent rough margin, a little curved from behind forwards, which presents an anterior lip for the attachment of the Supinator longus above and Extensor carpi radialis longior below, a posterior lip for the Triceps, and an interstice for the attachment of the external intermuscular aponeurosis.

The *internal border* extends from the lesser tuberosity to the internal condyle. Its upper third is marked by a prominent ridge, forming the inner lip of the bicipital groove, and gives attachment from above downwards to the tendons of the Latissimus dorsi, Teres major, and part of the origin of the inner head of the Triceps. About its centre is a rough ridge for the attachment of the Coraco-brachialis, and just below this is seen the entrance of the nutrient canal directed downwards. Sometimes there is a second canal higher up, which takes a similar direction. The inferior third of this border is raised into a slight ridge, which

87.—Left Humerus. Anterior View.

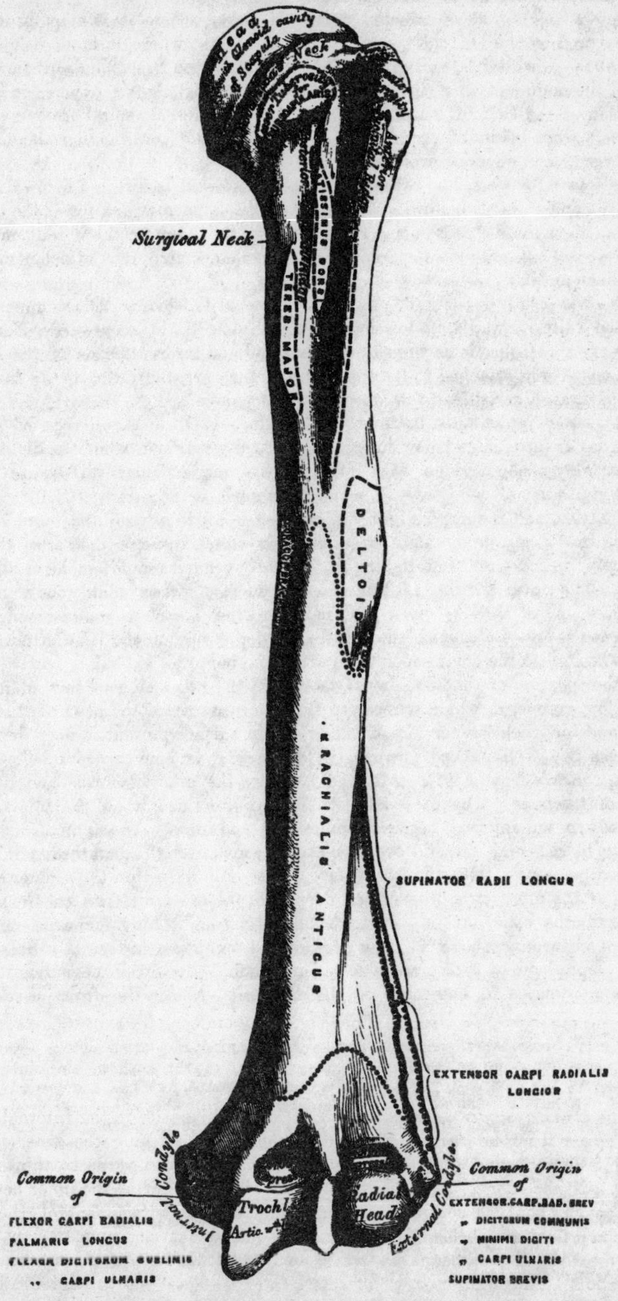

becomes very prominent below; it presents an anterior lip for the attachment of the Brachialis anticus, a posterior lip for the internal head of the Triceps, and an intermediate space for the internal intermuscular aponeurosis.

The *external surface* is directed outwards above, where it is smooth, rounded, and covered by the Deltoid muscle; forwards below, where it is slightly concave from above downwards, and gives origin to part of the Brachialis anticus muscle. About the middle of this surface is seen a rough triangular impression for the insertion of the Deltoid muscle, and below it the musculo-spiral groove, directed obliquely from behind, forwards, and downwards, and transmitting the musculo-spiral nerve and superior profunda artery.

The *internal surface*, less extensive than the external, is directed forwards above, forwards and inwards below: at its upper part it is narrow, and forms the bicipital groove. The middle part of this surface is slightly rough for the attachment of the Coraco-brachialis; its lower part is smooth, concave, and gives attachment to the Brachialis anticus muscle.*

The *posterior surface* (fig. 88) appears somewhat twisted, so that its upper part is directed a little inwards, its lower part backwards and a little outwards. Nearly the whole of this surface is covered by the external and internal heads of the Triceps, the former being attached to its upper and outer part, the latter to its inner and back part, at either side of the musculo-spiral groove.

The *Lower Extremity* is flattened from before backwards, and curved slightly forwards; it terminates below in a broad articular surface, which is divided into two parts by a slight ridge. On either side of the articular surface are the external and internal condyles. The articular surface extends a little lower than the condyles, and is curved slightly forwards, so as to occupy the more anterior part of the bone; its greatest breadth is in the transverse diameter, and it is obliquely directed, so that its inner extremity occupies a lower level than the outer. The outer portion of the articular surface presents a smooth rounded eminence, which has received the name of the *lesser* or *radial head* of the humerus; it articulates with the cup-shaped depression on the head of the radius, and is limited to the front and lower part of the bone, not extending as far back as the other portion of the articular surface. On the inner side of this eminence is a shallow groove, in which is received the inner margin of the head of the radius. The inner or trochlear portion of the articular surface presents a deep depression between two well-marked borders. This surface is convex from before backwards, concave from side to side, and occupies the anterior lower and posterior parts of the bone. The external border, less prominent than the internal, corresponds to the interval between the radius and ulna. The internal border is thicker, more prominent, and consequently of greater length than the external. The grooved portion of the articular surface fits accurately within the greater sigmoid cavity of the ulna; it is broader and deeper on the posterior than on the anterior aspect of the bone, and is directed obliquely from behind forwards, and from without inwards. Above the back part of the trochlear surface is a deep triangular depression, the olecranon fossa, in which is received the summit of the olecranon process in extension of the forearm. Above the front part of the

* A small hook-shaped process of bone, varying from $\frac{1}{10}$ to $\frac{3}{4}$ of an inch in length, is not unfrequently found projecting from the inner surface of the shaft of the humerus two inches above the internal condyle. It is curved downwards, forwards, and inwards, and its pointed extremity is connected to the internal border, just above the inner condyle, by a ligament or fibrous band; completing an arch, through which the median nerve and brachial artery pass, when these structures deviate from their usual course. Sometimes the nerve alone is transmitted through it, or the nerve may be accompanied by the ulnar-interosseous artery, in cases of high division of the brachial. A well-marked groove is usually found behind the process, in which the nerve and artery are lodged. This space is analogous to the supracondyloid foramen in many animals, and probably serves in them to protect the nerve and artery from compression during the contraction of the muscles in this region. A detailed account of this process is given by Dr. Struthers, in his 'Anatomical and Physiological Observations,' p. 202.

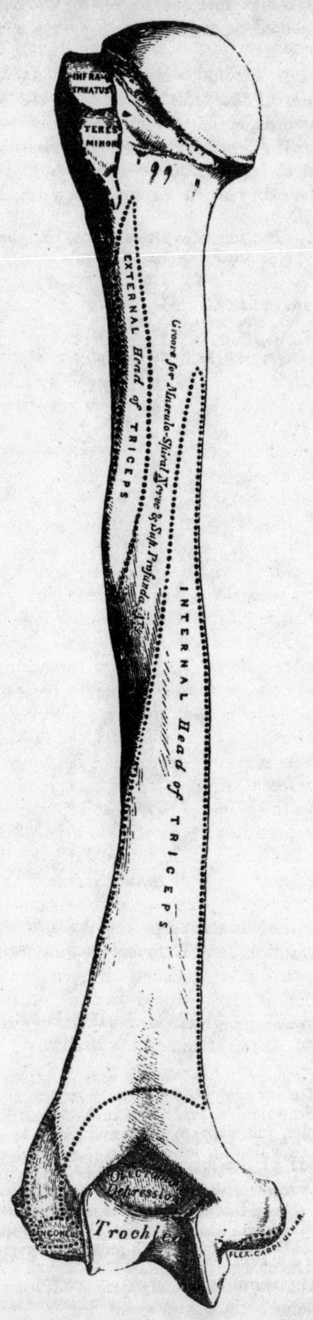

88.—Left Humerus. Posterior Surface.

trochlear surface is seen a smaller depression, the coronoid fossa, which receives the coronoid process of the ulna during flexion of the forearm. These fossæ are separated from one another by a thin transparent lamina of bone, which is sometimes perforated; their margins afford attachment to the anterior and posterior ligaments of the elbow-joint, and they are lined in the recent state by the synovial membrane of this articulation. Above the front part of the radial tuberosity is seen a slight depression, which receives the anterior border of the head of the radius when the forearm is strongly flexed. The external condyle is a small tubercular eminence, less prominent than the internal, curved a little forwards, and giving attachment to the external lateral ligament of the elbow-joint, and to a tendon common to the origin of some of the extensor and supinator muscles. The internal condyle, larger and more prominent than the external, is directed a little backwards: it gives attachment to the internal lateral ligament, and to a tendon common to the origin of some of the flexor muscles of the forearm. These eminences are directly continuous above with the external and internal borders. The greater prominence of the inner one renders it more liable to fracture.

Structure. The extremities consist of cancellous tissue, covered with a thin compact layer; the shaft is composed of a cylinder of compact tissue, thicker at the centre than at the extremities, and hollowed out by a large medullary canal, which extends along its whole length.

Development. By *seven* centres (fig. 89)—one for the shaft, one for the head, one for the greater tuberosity, one for the radial, one for the trochlear portion of the articular surface, and one for each condyle. The centre for the shaft appears very early, soon after ossification has commenced in the clavicle, and soon extends towards the extremities. At birth the humerus is ossified nearly in its whole length, the extremities remaining cartilaginous. Between the first and second years ossification commences in the head of the bone, and between the second and third years the centre for the tuberosities makes its appearance, usually by a single ossific point, but sometimes, according to Béclard, by one for each tuberosity, that for the lesser being small, and not appearing until after the fourth year. By the fifth year the centres for the head and tuberosities have enlarged and become joined, so as to form a single large epiphysis.

The lower end of the humerus is developed in the following manner: At the end of the second year ossification commences in the radial portion of the articular surface, and from this point extends inwards, so as to form the chief part of the articular end of the bone, the centre for the inner part of the articular surface not appearing until about the age of twelve. Ossification commences in the internal condyle about the fifth year, and in the external one not until between the thirteenth or fourteenth year. About sixteen or seventeen years, the outer condyle and both portions of the articulating surface (having already joined) unite with the shaft; at eighteen years the inner condyle becomes joined, whilst the upper epiphysis, although the first formed, is not united until about the twentieth year.

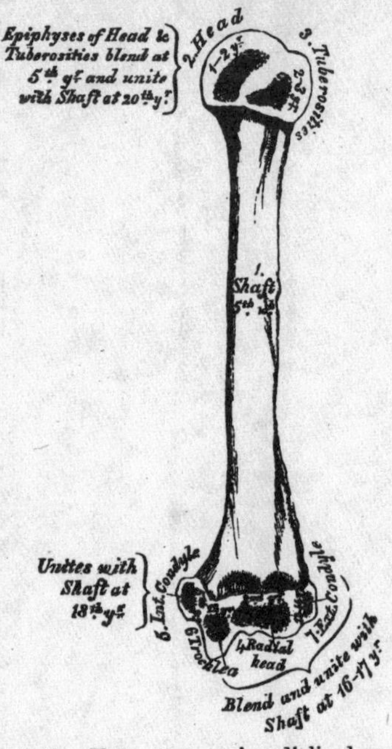

89.—Plan of the Development of the Humerus. By Seven Centres.

Articulations. With the glenoid cavity of the scapula, and with the ulna and radius.

Attachment of Muscles. To the greater tuberosity, the Supraspinatus, Infraspinatus, and Teres minor; to the lesser tuberosity, the Subscapularis; to the anterior bicipital ridge, the Pectoralis major; to the posterior bicipital ridge and groove, the Latissimus dorsi and Teres major; to the shaft, the Deltoid, Coraco-brachialis, Brachialis anticus, external and internal heads of the Triceps; to the internal condyle, the Pronator radii teres, and common tendon of the Flexor carpi radialis, Palmaris longus, Flexor digitorum sublimis, and Flexor carpi ulnaris; to the external condyloid ridge, the Supinator longus, and Extensor carpi radialis longior; to the external condyle, the common tendon of the Extensor carpi radialis brevior, Extensor communis digitorum, Extensor minimi digiti, and Extensor carpi ulnaris, the Anconeus, and Supinator brevis.

The *Forearm* is that portion of the upper extremity which is situated between the elbow and wrist. It is composed of two bones, the Ulna and the Radius.

The Ulna.

The *Ulna* (figs. 90, 91), so called from its forming the elbow (ὠλένη), is a long bone, prismatic in form, placed at the inner side of the forearm, parallel with the radius. It is the larger and longer of the two bones. Its upper extremity, of great thickness and strength, forms a large part of the articulation of the elbow-joint; it diminishes in size from above downwards, its lower extremity being very small, and excluded from the wrist-joint by the interposition of an interarticular fibro-cartilage. It is divisible into a shaft, and two extremities.

The *Upper Extremity*, the strongest part of the bone, presents for examination

90.—Bones of the Left Forearm. Anterior Surface.

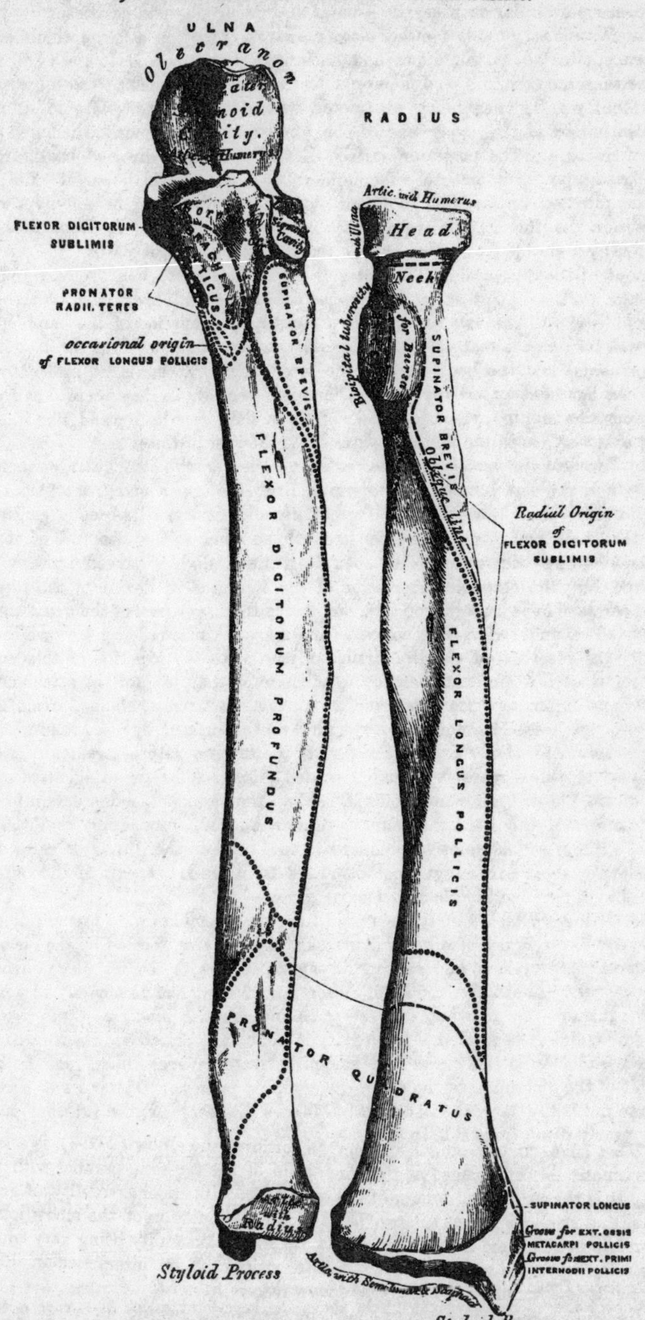

two large curved processes, the Olecranon process and the Coronoid process; and two concave articular cavities, the greater and lesser Sigmoid cavities.

The *Olecranon Process* (*ωλένη, elbow; κράνον, head*) is a large thick curved eminence, situated at the upper and back part of the ulna. It rises somewhat higher than the coronoid, and is curved forwards at the summit so as to present a prominent tip, its base being contracted where it joins the shaft. This is the narrowest part of the upper end of the ulna, and, consequently, the most usual seat of fracture. The posterior surface of the olecranon, directed backwards, is of a triangular form, smooth, subcutaneous, and covered by a bursa. Its upper surface, directed upwards, is of a quadrilateral form, marked behind by a rough impression for the attachment of the Triceps muscle; and in front, near the margin, by a slight transverse groove for the attachment of part of the posterior ligament of the elbow-joint. Its anterior surface is smooth, concave, covered with cartilage in the recent state, and forms the upper and back part of the great sigmoid cavity. The lateral borders present a continuation of the same groove that was seen on the margin of the superior surface; they serve for the attachment of ligaments, viz., the back part of the internal lateral ligament internally, the posterior ligament externally. The Olecranon process, in its structure as well as in its position and use, resembles the Patella in the lower limb; and, like it, sometimes exists as a separate piece, not united to the rest of the bone.*

The *Coronoid Process* (*κορώνη, a crow's beak*) is a rough triangular eminence of bone which projects horizontally forwards from the upper and front part of the ulna, forming the lower part of the great sigmoid cavity. Its base is continuous with the shaft, and of considerable strength, so much so that fracture of it is an accident of rare occurrence. Its apex is pointed, slightly curved upwards, and received into the coronoid depression of the humerus in flexion of the forearm. Its upper surface is smooth, concave, and forms the lower part of the great sigmoid cavity. The under-surface is concave, and marked internally by a rough impression for the insertion of the Brachialis anticus. At the junction of this surface with the shaft is a rough eminence, the tubercle of the ulna, for the attachment of the oblique ligament. Its outer surface presents a narrow, oblong, articular depression, the lesser sigmoid cavity. The inner surface, by its prominent free margin, serves for the attachment of part of the internal lateral ligament. At the front part of this surface is a small rounded eminence for the attachment of one head of the Flexor digitorum sublimis, behind the eminence, a depression for part of the origin of the Flexor profundus digitorum, and, descending from the eminence, a ridge, which gives attachment to one head of the Pronator radii teres. Occasionally the Flexor longus pollicis arises from the lower part of the Coronoid process by a rounded bundle of muscular fibres.

The *Greater Sigmoid Cavity*, so called from its resemblance to the old shape of the Greek letter Σ, is a semilunar depression of large size, formed by the olecranon and coronoid processes, and serving for articulation with the trochlear surface of the humerus. About the middle of either lateral border of this cavity is a notch, which contracts it somewhat, and serves to indicate the junction of the two processes of which it is formed. The cavity is concave from above downwards, and divided into two lateral parts by a smooth elevated ridge, which runs from the summit of the olecranon to the tip of the coronoid process. Of these two portions, the internal is the larger; it is slightly concave transversely, the external portion being nearly plane from side to side.

The *Lesser Sigmoid Cavity* is a narrow, oblong, articular depression, placed on the outer side of the coronoid process, and serving for articulation with the head of the radius. It is concave from before backwards; and its extremities, which are prominent, serve for the attachment of the orbicular ligament.

The *Shaft* is prismatic in form at its upper part, and curved from behind

* Professor Owen regards the olecranon as homologous not with the patella, but with an extension of the upper end of the fibula above the knee-joint, which is met with in the Ornithorynchus, Echidna, and some other animals. (OWEN, '*On the Nature of Limbs.*')

forwards, and from within outwards, so as to be convex behind and externally; its central part is quite straight; its lower part rounded, smooth, and bent a little outwards; it tapers gradually from above downwards, and presents for examination three borders, and three surfaces.

The *anterior border* commences above at the prominent inner angle of the coronoid process, and terminates below in front of the styloid process. It is well marked above, smooth and rounded in the middle of its extent, and affords attachment to the Flexor profundus digitorum: sharp and prominent in its lower fourth for the attachment of the Pronator quadratus. It separates the anterior from the internal surface.

The *posterior border* commences above at the apex of the triangular surface at the back part of the olecranon, and terminates below at the back part of the styloid process; it is well marked in the upper three-fourths, and gives attachment to an aponeurosis common to the Flexor carpi ulnaris, the Extensor carpi ulnaris, and the Flexor profundus digitorum muscles; its lower fourth is smooth and rounded. This border separates the internal from the posterior surface.

The *external border* commences above by two lines, which converge one from each extremity of the lesser sigmoid cavity, enclosing between them a triangular space for the attachment of part of the Supinator brevis, and terminates below at the middle of the head of the ulna. Its two middle fourths are very prominent, and serve for the attachment of the interosseous membrane; its lower fourth is smooth and rounded. This border separates the anterior from the posterior surface.

The *anterior surface*, much broader above than below, is concave in the upper three-fourths of its extent, and affords attachment to the Flexor profundus digitorum; its lower fourth, also concave, to the Pronator quadratus. The lower fourth is separated from the remaining portion of the bone by a prominent ridge, directed obliquely from above downwards and inwards; this ridge marks the extent of attachment of the Pronator above. At the junction of the upper with the middle third of the bone is the nutrient canal, directed obliquely upwards and inwards.

The *posterior surface*, directed backwards and outwards, is broad and concave above, somewhat narrower and convex in the middle of its course, narrow, smooth, and rounded below. It presents above an oblique ridge, which runs from the posterior extremity of the lesser sigmoid cavity, downwards to the posterior border; the triangular surface above this ridge receives the insertion of the Anconeus muscle, whilst the ridge itself affords attachment to the Supinator brevis. The surface of bone below this is subdivided by a longitudinal ridge, sometimes called the perpendicular line, into two parts: the internal part is smooth, concave, and gives origin to (occasionally is merely covered by) the Extensor carpi ulnaris; the external portion, wider and rougher, gives attachment from above downwards to part of the Supinator brevis, the Extensor ossis metacarpi pollicis, the Extensor secundi internodii pollicis, and the Extensor indicis muscles.

The *internal surface* is broad and concave above, narrow and convex below. It gives attachment by its upper three-fourths to the Flexor profundus digitorum muscle; its lower fourth is subcutaneous.

The *Lower Extremity* of the ulna is of small size, and excluded from the articulation of the wrist-joint. It presents for examination two eminences, the outer and larger of which is a rounded articular eminence, termed the head of the ulna; the inner, narrower and more projecting, is a non-articular eminence, the styloid process. The *head* presents an articular facet, part of which, of an oval form, is directed downwards, and plays on the surface of the triangular fibro-cartilage, which separates this bone from the wrist-joint; the remaining portion, directed outwards, is narrow, convex, and received into the sigmoid cavity of the radius. The *styloid process* projects from the inner and back part of the bone, and descends a little lower than the head, terminating in a rounded summit, which affords attachment to the internal lateral ligament of the wrist. The head is separated from the styloid process by a depression for the attachment of the triangular inter-

91.—Bones of the Left Forearm. Posterior Surface.

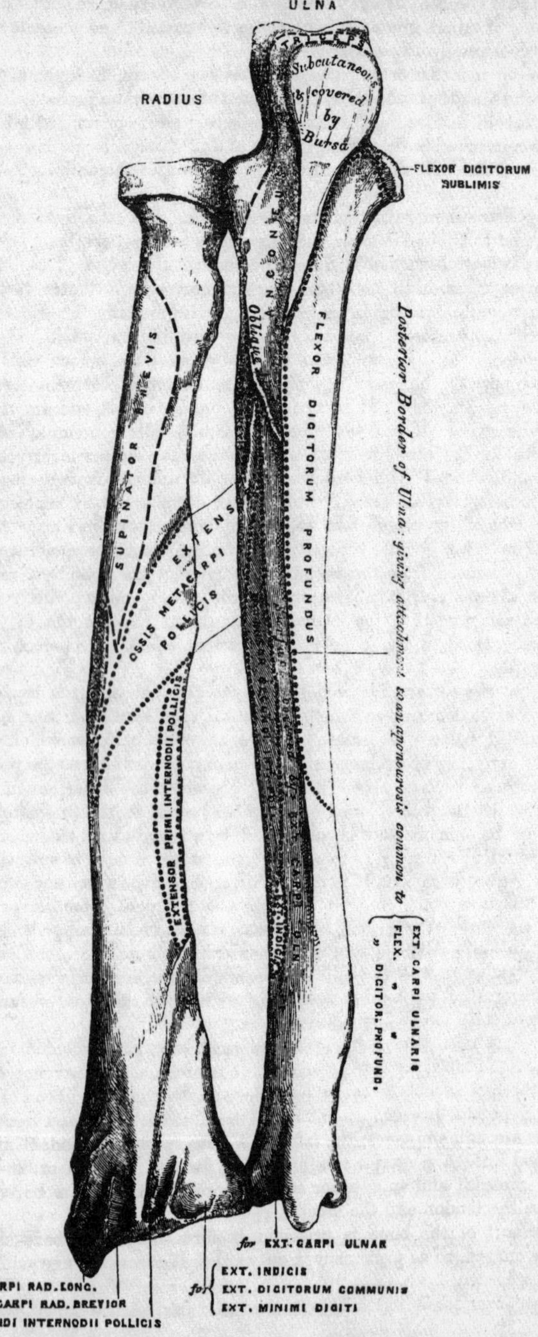

articular fibro-cartilage; and behind, by a shallow groove for the passage of the tendon of the Extensor carpi ulnaris.

Structure. Similar to that of the other long bones.

Development. By *three* centres: one for the shaft, one for the inferior extremity, and one for the olecranon (fig. 92). Ossification commences near the middle of the shaft about the fifth week, and soon extends through the greater part of the bone. At birth the ends are cartilaginous. About the fourth year, a separate osseous nucleus appears in the middle of the head, which soon extends into the styloid process. About the tenth year, ossific matter appears in the olecranon near its extremity, the chief part of this process being formed from an extension of the shaft of the bone into it. At about the sixteenth year, the upper epiphysis becomes joined, and at about the twentieth year the lower one.

Articulations. With the humerus and radius.

Attachment of Muscles. To the olecranon: the Triceps, Anconeus, and one head of the Flexor carpi ulnaris. To the coronoid process: the Brachialis anticus, Pronator radii teres, Flexor sublimis digitorum, and Flexor profundus digitorum, occasionally, also the Flexor longus pollicis. To the shaft: the Flexor profundus digitorum, Pronator quadratus, Flexor carpi ulnaris, Extensor carpi ulnaris, Anconeus, Supinator brevis, Extensor ossis metacarpi pollicis, Extensor secundi internodii pollicis, and Extensor indicis.

92.—Plan of the Development of the Ulna. By Three Centres.

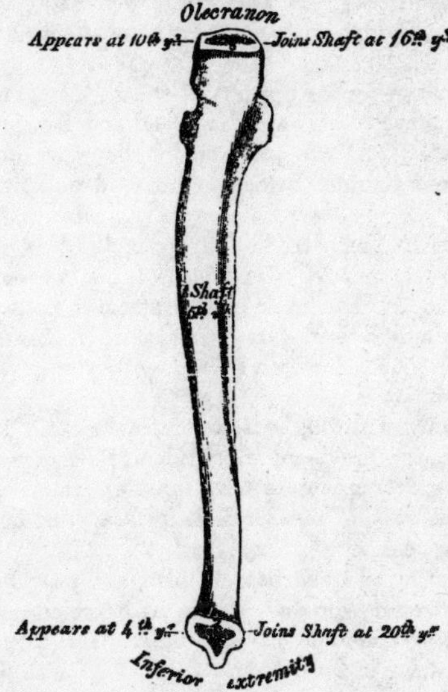

The Radius.

The *Radius* is situated on the outer side of the forearm, lying parallel with the ulna, which exceeds it in length and size. Its upper end is small, and forms only a small part of the elbowjoint; but its lower end is large, and forms the chief part of the wrist. It is one of the long bones, prismatic in form, slightly curved longitudinally, and like other long bones has a shaft and two extremities.

The *Upper Extremity* presents a head, neck, and tuberosity. The *head* is of a cylindrical form, depressed on its upper surface into a shallow cup, which articulates with the radial or lesser head of the humerus in flexion of the joint. Around the circumference of the head is a smooth articular surface, coated with cartilage in the recent state, broad internally where it rotates within the lesser sigmoid cavity of the ulna; narrow in the rest of its circumference, to play in the orbicular ligament. The head is supported on a round, smooth, and constricted portion of bone, called the *neck*, which presents, behind, a slight ridge, for the attachment of part of the Supinator brevis. Beneath the neck, at the inner and front aspect of the bone, is a rough eminence, the *tuberosity*. Its surface is divided into two parts by a vertical line—a posterior rough portion, for the insertion of the tendon of the Biceps muscle; and an anterior smooth portion, on which a bursa is interposed between the tendon and the bone.

The *Shaft* of the bone is prismoid in form, narrower above than below, and slightly curved, so as to be convex outwards. It presents three surfaces, separated by three borders.

The *anterior border* extends from the lower part of the tuberosity above, to the anterior part of the base of the styloid process below. It separates the anterior from the external surface. Its upper third is very prominent; and from its oblique direction, downwards and outwards, has received the name of the *oblique line of the radius*. It gives attachment, externally, to the Supinator brevis; internally, to the Flexor longus pollicis, and between these to the Flexor digitorum sublimis. The middle third of the anterior border is indistinct and rounded. Its lower fourth is sharp, prominent, affords attachment to the Pronator quadratus, and terminates in a small tubercle, into which is inserted the tendon of the Supinator longus.

The *posterior border* commences above, at the back part of the neck of the radius, and terminates below, at the posterior part of the base of the styloid process; it separates the posterior from the external surface. It is indistinct above and below, but well marked in the middle third of the bone.

The *internal* or *interosseous border* commences above, at the back part of the tuberosity, where it is rounded and indistinct, becomes sharp and prominent as it descends, and at its lower part bifurcates into two ridges, which descend to the anterior and posterior margins of the sigmoid cavity. This border separates the anterior from the posterior surface, and has the interosseous membrane attached to it throughout the greater part of its extent.

The *anterior surface* is narrow and concave for its upper two-thirds, and gives attachment to the Flexor longus pollicis muscle; below, it is broad and flat, and gives attachment to the Pronator quadratus. At the junction of the upper and middle thirds of this surface is the nutrient foramen, which is directed obliquely upwards.

The *posterior surface* is rounded, convex and smooth, in the upper third of its extent, and covered by the Supinator brevis muscle. Its middle third is broad, slightly concave, and gives attachment to the Extensor ossis metacarpi pollicis above, the Extensor primi internodii pollicis below. Its lower third is broad, convex, and covered by the tendons of the muscles, which subsequently run in the grooves on the lower end of the bone.

The *external surface* is rounded and convex throughout its entire extent. Its upper third gives attachment to the Supinator brevis muscle. About its centre is seen a rough ridge, for the insertion of the Pronator radii teres muscle. Its lower part is narrow, and covered by the tendons of the Extensor ossis metacarpi pollicis and Extensor primi internodii pollicis muscles.

The *Lower Extremity* of the radius is large, of quadrilateral form, and provided with two articular surfaces, one at the extremity for articulation with the carpus, and one at the inner side of the bone for articulation with the ulna. The carpal articular surface is of triangular form, concave, smooth, and divided by a slight antero-posterior ridge into two parts. Of these, the external is large, of a triangular form, and articulates with the scaphoid bone; the inner, smaller and quadrilateral, articulates with the semilunar. The articular surface for the ulna is called the *sigmoid cavity* of the radius; it is narrow, concave, smooth, and articulates with the head of the ulna. The circumference of this end of the bone presents three surfaces, an anterior, external, and posterior. The *anterior surface*, rough and irregular, affords attachment to the anterior ligament of the wrist-joint. The *external surface* is prolonged obliquely downwards into a strong conical projection, the styloid process, which gives attachment by its base to the tendon of the Supinator longus, and by its apex to the external lateral ligament of the wrist-joint. The outer surface of this process is marked by two grooves, which run obliquely downwards and forwards, and are separated from one another by an elevated ridge. The anterior one gives passage to the tendon of the Extensor ossis metacarpi pollicis, the posterior one to the tendon of the Extensor primi internodii pollicis. The *posterior surface* is convex, affords attachment to the posterior ligament of the wrist, and is marked by three grooves. Proceeding from without inwards, the first groove is broad, but shallow, and subdivided into two by a slightly elevated ridge: the outer of these two transmits the tendon of the Extensor carpi radialis longior, the inner the tendon of the Extensor carpi radialis brevior. The second, which is near the centre of the bone, is a deep but narrow groove, directed obliquely from above, downwards and outwards; it transmits the tendon of the Extensor secundi internodii pollicis. The third, lying most internally, is a broad groove, for the passage of the tendons of the Extensor communis digitorum, Extensor indicis and Extensor minimi digiti; the tendon of the last-named muscle passing through the groove at the point of articulation of the radius with the ulna.

Structure. Similar to that of the other long bones.

Development (fig. 93). By *three* centres, one for the shaft, and one for each extremity. That for the shaft makes its appearance near the centre of the bone, soon after the development of the humerus commences. At birth the shaft is ossified, but the ends of the bone are cartilaginous. About the end of the second year, ossification commences in the lower epiphysis; and about the fifth year, in the upper one. At the age of puberty, the upper epiphysis becomes joined to the shaft; the lower epiphysis becoming united about the twentieth year.

93.—Plan of the Development of the Radius. By Three Centres.

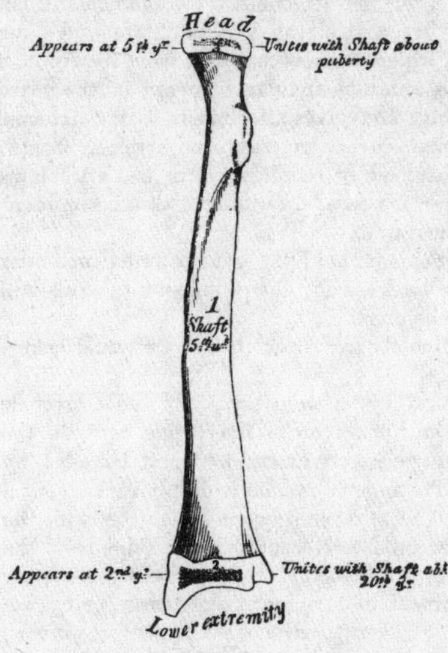

Articulations. With four bones—the humerus, ulna, scaphoid, and semilunar.

Attachment of Muscles. To the tuberosity, the Biceps; to the oblique ridge, the Supinator brevis, Flexor digitorum sublimis, and Flexor longus pollicis; to the shaft (its anterior surface), the Flexor longus pollicis and Pronator quadratus; (its posterior surface), the Extensor ossis metacarpi pollicis and Extensor primi internodii pollicis; (its outer surface), the Pronator radii teres; and to the styloid process, the Supinator longus.

THE HAND.

The Hand is subdivided into three segments—the Carpus or wrist, the Metacarpus or palm, and the Phalanges or fingers.

Carpus.

The bones of the Carpus, eight in number, are arranged in two rows. Those of the upper row, enumerated from the radial to the ulnar side, are the scaphoid, semilunar, cuneiform, and pisiform; those of the lower row, enumerated in the same order, are the trapezium, trapezoid, os magnum, and unciform.

Common Characters of the Carpal Bones.

Each bone (excepting the pisiform) presents six surfaces. Of these, the *anterior* or *palmar*, and the *posterior* or *dorsal*, are rough, for ligamentous attachment, the dorsal surface being generally the broader of the two. The *superior* and *inferior* are articular, the superior generally convex, the inferior concave; and the *internal* and *external* are also articular when in contact with contiguous bones, otherwise rough and tubercular. Their structure in all is similar, consisting within of cancellous tissue enclosed in a layer of compact bone. Each bone is also developed from a single centre of ossification.

Bones of the Upper Row. (Figs. 94, 95.)

The *Scaphoid* is the largest bone of the first row. It has received its name from its fancied resemblance to a boat, being broad at one end, and narrowed like a prow at the opposite. It is situated at the upper and outer part of the carpus, its direction being from above downwards, outwards, and forwards. The *superior surface* is convex, smooth, of triangular shape, and articulates with the lower end of the radius. The *inferior surface*, directed downwards, outwards, and backwards, is smooth, convex, also triangular, and divided by a slight ridge into two parts, the external of which articulates with the trapezium, the inner with the trapezoid. The *posterior* or *dorsal surface* presents a narrow, rough groove, which runs the entire breadth of the bone, and serves for the attachment of ligaments. The *anterior* or *palmar surface* is concave above, and elevated at its lower and outer part into a prominent rounded tubercle, which projects forwards from the front of the carpus, and gives attachment to the anterior annular ligament of the wrist. The *external surface* is rough and narrow, and gives attachment to the external lateral ligament of the wrist. The *internal surface* presents two articular facets: of these, the superior or smaller one is flattened, of semilunar form, and articulates with the semilunar; the inferior or larger is concave, forming with the semilunar bone, a concavity for the head of the os magnum.

To ascertain to which hand this bone belongs, hold the convex radial articular surface upwards, and the dorsal surface backwards; the prominent tubercle will be directed to the side to which the bone belongs.

Articulations. With five bones: the radius above, trapezium and trapezoid below, os magnum and semilunar internally.

The *Semilunar* bone may be distinguished by its deep concavity and crescentic outline. It is situated in the centre of the upper row of the carpus, between the scaphoid and cuneiform. The *superior surface*, convex, smooth, and bounded by four edges, articulates with the radius. The *inferior surface* is deeply concave, and of greater extent from before backwards than transversely; it articulates with the head of the os magnum, and by a long narrow facet (separated by a ridge from the general surface) with the unciform bone. The *anterior* or *palmar* and *posterior* or *dorsal surfaces* are rough, for the attachment of ligaments, the former being the broader, and of somewhat rounded form. The *external surface* presents a narrow, flattened, semilunar facet, for articulation with the scaphoid. The *internal surface* is marked by a smooth, quadrilateral facet, for articulation with the cuneiform.

To ascertain to which hand this bone belongs, hold it with the dorsal surface upwards, and the convex articular surface backwards; the quadrilateral articular facet will then point to the side to which the bone belongs.

Articulations. With five bones: the radius above, os magnum and unciform below, scaphoid and cuneiform on either side.

The *Cuneiform* (*l'Os Pyramidal*) may be distinguished by its pyramidal shape, and by its having an oval, isolated facet, for articulation with the pisiform bone. It is situated at the upper and inner side of the carpus. The *superior surface* presents an internal, rough, nonarticular portion; and an external or articular portion, which is convex, smooth, and separated from the lower end of the ulna by the interarticular fibro-cartilage of the wrist. The *inferior surface*, directed outwards, is concave, sinuously curved, and smooth for articulation with the unciform. The *posterior* or *dorsal surface* is rough, for the attachment of ligaments. The *anterior* or *palmar surface* presents, at its inner side, an oval facet, for articulation with the pisiform; and is rough externally, for ligamentous attachment. The *external surface*, the base of the pyramid, is marked by a flat, quadrilateral, smooth facet, for articulation with the semilunar. The *internal surface*, the summit of the pyramid, is pointed and roughened, for the attachment of the internal lateral ligament of the wrist.

94.—Bones of the Left Hand. Dorsal Surface.

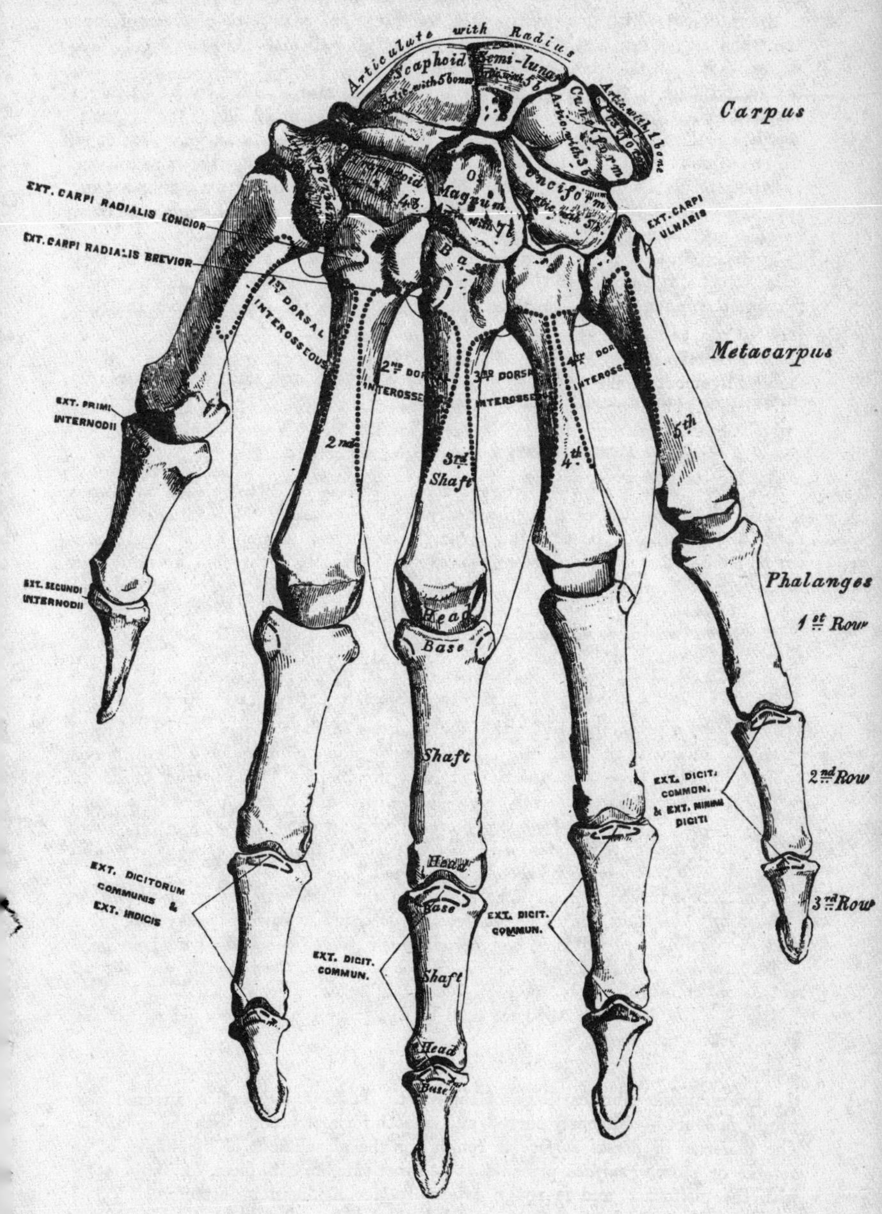

To ascertain to which hand this bone belongs, hold it so that the base is directed backwards, and the articular facet for the pisiform bone upwards; the concave articular facet will point to the side to which the bone belongs.

Articulations. With three bones: the semilunar externally, the pisiform in front, the unciform below, and with the triangular interarticular fibro-cartilage which separates it from the lower end of the ulna.

The *Pisiform* bone may be known by its small size, and by its presenting a single articular facet. It is situated at the anterior and inner side of the carpus, is nearly circular in form, and presents on its *posterior surface* a smooth oval facet, for articulation with the cuneiform bone. This facet approaches the superior, but not the inferior, border of the bone. The *anterior* or *palmar surface* is rounded and rough, and gives attachment to the anterior annular ligament. The *outer* and *inner surfaces* are also rough, the former being convex, the latter usually concave.

To ascertain to which hand it belongs, hold the bone with its posterior or articular facet downwards, and the nonarticular portion of the same surface backwards; the inner concave surface will then point to the side to which the bone belongs.

Articulations. With one bone, the cuneiform.

Attachment of Muscles. To two: the Flexor carpi ulnaris, and Abductor minimi digiti; and to the anterior annular ligament.

Bones of the Lower Row. (Figs. 94, 95.)

The *Trapezium* is of very irregular form. It may be distinguished by a deep groove, for the tendon of the Flexor carpi radialis muscle. It is situated at the external and inferior part of the carpus, between the scaphoid and first metacarpal bone. The *superior surface*, concave and smooth, is directed upwards and inwards, and articulates with the scaphoid. The *inferior surface*, directed downwards and outwards, is oval, concave from side to side, convex from before backwards, so as to form a saddle-shaped surface, for articulation with the base of the first metacarpal bone. The *anterior* or *palmar surface* is narrow and rough. At its upper part is a deep groove, running from above obliquely downwards and inwards; it transmits the tendon of the Flexor carpi radialis, and is bounded externally by a prominent ridge, the oblique ridge of the trapezium. This surface gives attachment to the Abductor pollicis, Flexor ossis metacarpi, and Flexor brevis pollicis muscles; and the anterior annular ligament. The *posterior* or *dorsal surface* is rough, and the *external surface* also broad and rough, for the attachment of ligaments. The *internal surface* presents two articular facets: the upper one, large and concave, articulates with the trapezoid; the lower one, narrow and flattened, with the base of the second metacarpal bone.

To ascertain to which hand it belongs, hold the bone with the grooved palmar surface upwards, and the external, broad, nonarticular surface backwards; the saddle-shaped surface will then be directed to the side to which the bone belongs.

Articulations. With four bones: the scaphoid above, the trapezoid and second metacarpal bones internally, the first metacarpal below.

Attachment of Muscles. Abductor pollicis, Flexor ossis metacarpi, and part of the Flexor brevis pollicis.

The *Trapezoid* is the smallest bone in the second row. It may be known by its wedge-shaped form, the broad end of the wedge forming the dorsal, the narrow end the palmar surface; and by its having four articular surfaces touching each other, and separated by sharp edges. The *superior surface*, quadrilateral in form, smooth and slightly concave, articulates with the scaphoid. The *inferior surface* articulates with the upper end of the second metacarpal bone; it is convex from side to side, concave from before backwards, and subdivided, by an elevated ridge, into two unequal lateral facets. The *posterior* or *dorsal* and *anterior* or *palmar surfaces* are rough, for the attachment of ligaments, the former being the larger of the two. The *external surface*, convex and smooth, articulates with

95.—Bones of the Left Hand. Palmar Surface.

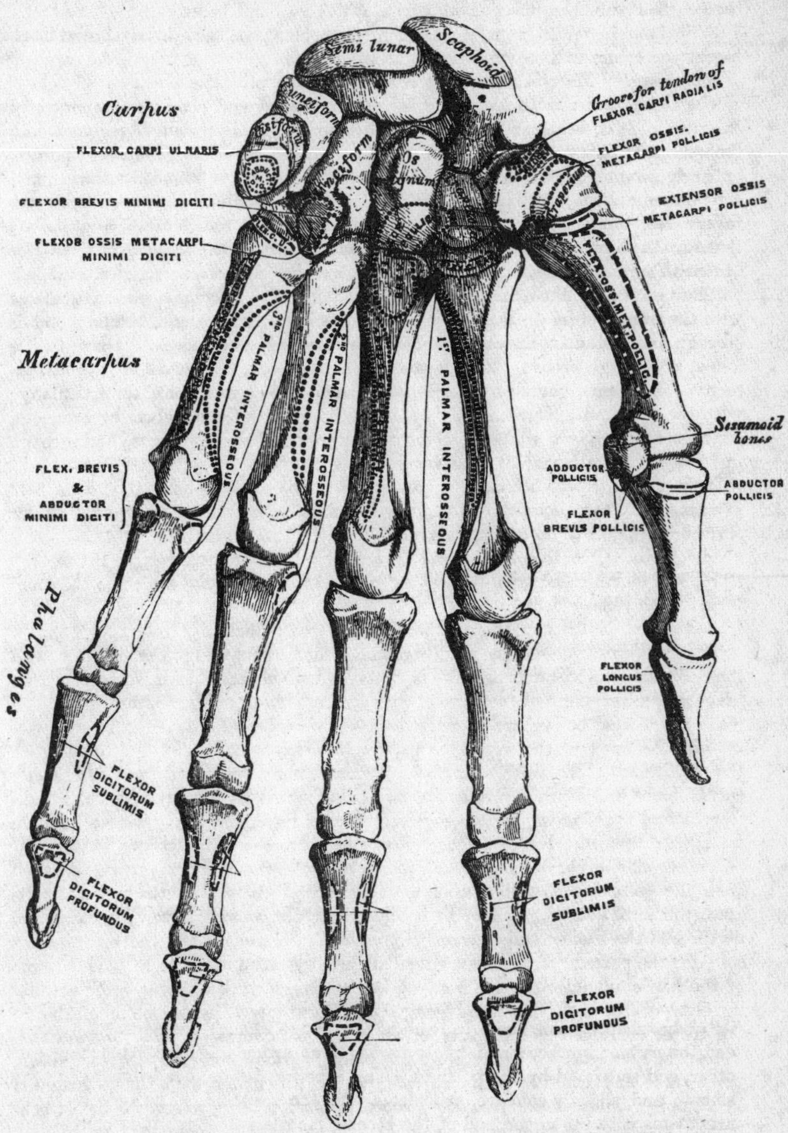

the trapezium. The *internal surface* is concave and smooth below, for articulation with the os magnum; rough above, for the attachment of an interosseous ligament.

To ascertain to which side this bone belongs, let the broad dorsal surface be held upwards, and the inferior concavo-convex surface forwards; the internal concave surface will then point to the side to which the bone belongs.

Articulations. With four bones: the scaphoid above, second metacarpal bone below, trapezium externally, os magnum internally.

Attachment of Muscles. Part of the Flexor brevis pollicis.

The *Os Magnum* is the largest bone of the carpus, and occupies the centre of the wrist. It presents above a rounded portion or head, which is received into the concavity formed by the scaphoid and semilunar bones; a constricted portion or neck; and, below, the body. The *superior surface* is rounded, smooth, and articulates with the semilunar. The *inferior surface* is divided by two ridges into three facets, for articulation with the second, third, and fourth metacarpal bones; that for the third (the middle facet) being the largest of the three. The *posterior* or *dorsal surface* is broad and rough; the *anterior* or *palmar*, narrow, rounded, and also rough, for the attachment of ligaments. The *external surface* articulates with the trapezoid by a small facet at its anterior inferior angle, behind which is a rough depression for the attachment of an interosseous ligament. Above this is a deep and rough groove, which forms part of the neck, and serves for the attachment of ligaments, bounded superiorly by a smooth convex surface, for articulation with the scaphoid. The *internal surface* articulates with the unciform by a smooth, concave, oblong facet, which occupies its posterior and superior parts; and is rough in front, for the attachment of an interosseous ligament.

To ascertain to which hand this bone belongs, the rounded head should be held upwards, and the broad dorsal surface forwards; the internal concave articular surface will point to its appropriate side.

Articulations. With seven bones: the scaphoid and semilunar above; the second, third, and fourth metacarpal below; the trapezoid on the radial side; and the unciform on the ulnar side.

Attachment of Muscles. Part of the Flexor brevis pollicis.

The *Unciform* bone may be readily distinguished by its wedge-shaped form, and the hook-like process that projects from its palmar surface. It is situated at the inner and lower angle of the carpus, with its base downwards, resting on the two inner metacarpal bones, and its apex directed upwards and outwards. The *superior surface*, the apex of the wedge, is narrow, convex, smooth, and articulates with the semilunar. The *inferior surface* articulates with the fourth and fifth metacarpal bones, the concave surface for each being separated by a ridge, which runs from before backwards. The *posterior* or *dorsal surface* is triangular and rough, for ligamentous attachment. The *anterior* or *palmar surface* presents, at its lower and inner side, a curved hook-like process of bone, the unciform process, directed from the palmar surface forwards and outwards. It gives attachment, by its apex, to the annular ligament; by its inner surface, to the Flexor brevis minimi digiti, and the Flexor ossis metacarpi minimi digiti; and is grooved on its outer side, for the passage of the Flexor tendons into the palm of the hand. This is one of the four eminences on the front of the carpus, to which the anterior annular ligament is attached; the others being the pisiform internally, the oblique ridge of the trapezium and the tuberosity of the scaphoid externally. The *internal surface* articulates with the cuneiform by an oblong surface, cut obliquely from above, downwards and inwards. The *external surface* articulates with the os magnum by its upper and posterior part, the remaining portion being rough, for the attachment of ligaments.

To ascertain to which hand it belongs, hold the apex of the bone upwards, and the broad dorsal surface backwards: the concavity of the unciform process will be directed to the side to which the bone belongs.

Articulations. With five bones: the semilunar above, the fourth and fifth metacarpal below, the cuneiform internally, the os magnum externally.

Attachment of Muscles. To two: the Flexor brevis minimi digiti, the Flexor ossis metacarpi minimi digiti; and to the anterior annular ligament.

The Metacarpus.

The Metacarpal bones are five in number; they are long cylindrical bones, presenting for examination a shaft, and two extremities.

Common Characters of the Metacarpal Bones.

The *shaft* is prismoid in form, and curved longitudinally, so as to be convex in the longitudinal direction behind, concave in front. It presents three surfaces: two lateral, and one posterior. The *lateral surfaces* are concave, for the attachment of the Interossei muscles, and separated from one another by a prominent line. The *posterior* or *dorsal surface* is triangular, smooth, and flattened below, and covered, in the recent state, by the tendons of the Extensor muscles. In its upper half it is divided by a ridge into two narrow lateral depressions, for the attachment of the Dorsal interossei muscles. This ridge bifurcates a little above the centre of the bone, and its branches run to the small tubercles on each side of the digital extremity.

The *carpal extremity*, or *base*, is of a cuboidal form, and broader behind than in front: it articulates above with the carpus, and on each side with the adjoining metacarpal bones; its *dorsal* and *palmar surfaces* being rough, for the attachment of tendons and ligaments.

The *digital extremity*, or *head*, presents an oblong surface, flattened at each side, for articulation with the first phalanx; it is broader, and extends farther forwards in front than behind, and is longer in the antero-posterior than in the transverse diameter. On either side of the head is a deep depression, surmounted by a tubercle, for the attachment of the lateral ligament of the metacarpo-phalangeal joint. The *posterior surface*, broad and flat, supports the Extensor tendons; the *anterior surface* presents a median groove, bounded on each side by a tubercle, for the passage of the Flexor tendons.

Peculiar Characters of the Metacarpal Bones.

The *metacarpal bone of the thumb* is shorter and wider than the rest, diverges to a greater degree from the carpus, and its *palmar surface* is directed inwards towards the palm. The *shaft* is flattened and broad on its dorsal aspect, and does not present the bifurcated ridge which is found on the other metacarpal bones; it is concave from before backwards on its palmar surface. The *carpal extremity*, or *base*, presents a concavo-convex surface, for articulation with the trapezium, and has no lateral facets. The *digital extremity* is less convex than that of the other metacarpal bones, broader from side to side than from before backwards, and terminates anteriorly in a small articular eminence on each side, over which play two sesamoid bones.

The *metacarpal bone of the index finger* is the longest, and its base the largest of the other four. Its *carpal extremity* is prolonged upwards and inwards. The dorsal and palmar surfaces of this extremity are rough, for the attachment of tendons and ligaments. It presents four articular facets: the first, at the end of the bone, is concave from side to side, convex from before backwards, and articulates with the trapezoid; the second, on the radial side, is a flat quadrilateral facet, for the trapezium; the third, which occupies the outer part of the ulnar side of the extremity, is a long narrow facet, for the os magnum; and the fourth, which occupies the inner part of the same side, is a considerably broader surface, for the third metacarpal bone.

The *metacarpal bone of the middle finger* is a little smaller than the preceding; it presents a pyramidal eminence on the radial side of its base (dorsal aspect), which extends upwards behind the os magnum. The carpal articular facet is concave behind, flat and horizontal in front, and corresponds to the os magnum. On the radial side is a smooth concave facet, for articulation with the second metacarpal bone; and on the ulnar side two small oval facets, for articulation with the fourth metacarpal.

The *metacarpal bone of the ring-finger* is shorter and smaller than the preceding, and its base small and quadrilateral; the carpal surface of the base presenting two facets, for articulation with the unciform and os magnum. On the radial side are two oval facets, for articulation with the third metacarpal bone; and on the ulnar side a single concave facet, for the fifth metacarpal.

The *metacarpal bone of the little finger* may be distinguished by the concavo-convex form of its carpal surface, which articulates with the unciform; and from having only one lateral articular facet, which corresponds with the fourth metacarpal bone. On its ulnar side is a prominent tubercle, for the insertion of the tendon of the Extensor carpi ulnaris. The dorsal surface of the shaft is marked by an oblique ridge, which extends from near the ulnar side of the upper extremity to the radial side of the lower. The outer division of this surface serves for the attachment of the fourth Dorsal interosseous muscle; the inner division is smooth, and covered by the Extensor tendons of the little finger.

Articulations. The first metacarpal bone articulates with the trapezium; the second with the trapezium, trapezoides, os magnum, and third metacarpal bones; the third with the os magnum, and second and fourth metacarpal bones; the fourth with the os magnum, unciform, and third and fifth metacarpal bones; and the fifth with the unciform and fourth metacarpal.

Attachment of Muscles. To the metacarpal bone of the thumb, three: the Flexor ossis metacarpi pollicis, Extensor ossis metacarpi pollicis, and first Dorsal interosseous. To the second metacarpal bone, five: the Flexor carpi radialis, Extensor carpi radialis longior, first and second Dorsal interosseous, and first Palmar interosseous. To the third, five: the Extensor carpi radialis brevior, Flexor brevis pollicis, Adductor pollicis, and second and third Dorsal interosseous. To the fourth, three: the third and fourth Dorsal and second Palmar interosseous. To the fifth, five: the Extensor carpi ulnaris, Flexor carpi ulnaris, Flexor ossis metacarpi minimi digiti, fourth Dorsal, and third Palmar interosseous.

Phalanges.

The Phalanges are the bones of the fingers; they are fourteen in number, three for each finger, and two for the thumb. They are long bones, and present for examination a shaft, and two extremities. The *shaft* tapers from above downwards, is convex posteriorly, concave in front from above downwards, flat from side to side, and marked laterally by rough ridges, which give attachment to the fibrous sheaths of the Flexor tendons. The *metacarpal extremity* or *base*, in the first row, presents an oval concave articular surface, broader from side to side than from before backwards; and the same extremity in the other two rows, a double concavity separated by a longitudinal median ridge, extending from before backwards. The *digital extremities* are smaller than the others, and terminate, in the first and second row, in two small lateral condyles, separated by a slight groove; the articular surface being prolonged farther forwards on the palmar than on the dorsal surface, especially in the first row.

The *Ungual phalanges* are convex on their dorsal, flat on their palmar surfaces; they are recognised by their small size, and by a roughened elevated surface of a horseshoe form on the palmar aspect of their ungual extremity, which serves to support the sensitive pulp of the finger.

Articulations. The first row with the metacarpal bones, and the second row of phalanges; the second row with the first and third; the third, with the second row.

Attachment of Muscles. To the base of the first phalanx of the thumb, four muscles: the Extensor primi internodii pollicis, Flexor brevis pollicis, Abductor pollicis, Adductor pollicis. To the second phalanx, two: the Flexor longus pollicis, and the Extensor secundi internodii. To the base of the first phalanx of the index finger, the first Dorsal and the first Palmar interosseous; to that of the middle finger, the second and third dorsal interosseous; to that of the ring-finger, the fourth Dorsal and the second Palmar interosseous; and to that of the little finger, the third Palmar interosseous, the Flexor brevis minimi digiti, and Abductor minimi digiti. To the second phalanges, the Flexor sublimis digitorum, Extensor communis digitorum; and, in addition, the Extensor indicis to the index finger, the Extensor minimi digiti to the little finger. To the third phalanges, the Flexor profundus digitorum and Extensor communis digitorum.

Development of the Bones of the Hand.

The *Carpal bones* are each developed by a *single* centre. At birth, they are all cartilaginous. Ossification proceeds in the following order (fig. 96): in the os magnum and unciform an ossific point appears during the first year, the former preceding the latter; in the cuneiform, at the third year; in the trapezium and semilunar, at the fifth year, the former preceding the latter; in the scaphoid, at the sixth year; in the trapezoid, during the eighth year; and in the pisiform, about the twelfth year.

96.—Plan of the Development of the Hand.

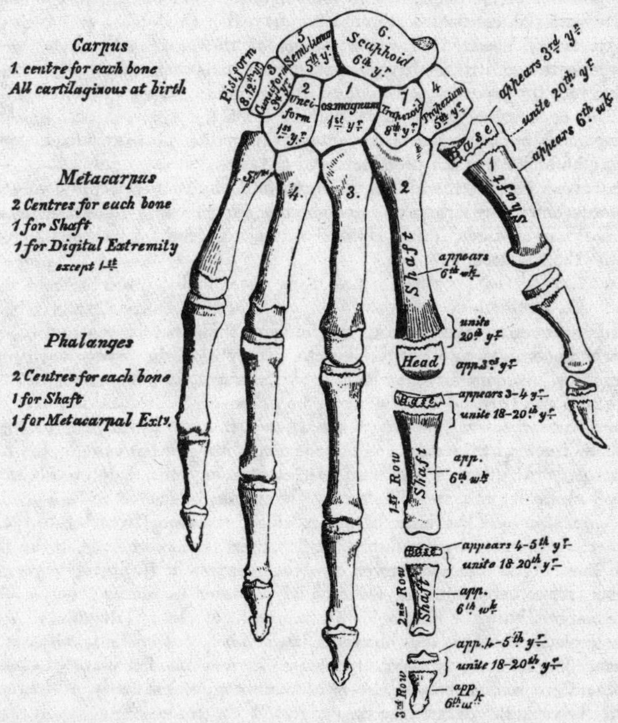

The *Metacarpal bones* are each developed by *two* centres: one for the shaft, and one for the digital extremity, for the four inner metacarpal bones; one for the shaft, and one for the base, for the metacarpal bone of the thumb, which in this respect resembles the phalanges. Ossification commences in the centre of the shaft about the sixth week, and gradually proceeds to either end of the bone; about the third year the digital extremities of the four inner metacarpal bones, and the base of the first metacarpal, commence to ossify, and they unite about the twentieth year.

The *Phalanges* are each developed by *two* centres: one for the shaft, and one for the base. Ossification commences in the shaft, in all three rows, at about the sixth week, and gradually involves the whole of the bone excepting the upper extremity. Ossification of the base commences in the first row between the third and fourth years, and a year later in those of the second and third rows. The two centres become united in each row, between the eighteenth and twentieth years.

Of the Lower Extremity.

The Lower Extremity consists of three segments, the *thigh*, *leg*, and *foot*, which correspond to the *arm*, *forearm*, and *hand* in the upper extremity. It is connected to the trunk through the os innominatum, or haunch, which is homologous with the shoulder.

The Os Innominatum.

The *Os Innominatum*, or nameless bone, so called from bearing no resemblance to any known object, is a large irregular-shaped bone, which, with its fellow of the opposite side, forms the sides and anterior wall of the pelvic cavity. In young subjects it consists of three separate parts, which meet and form the large cup-like cavity, situated near the middle of the outer side of the bone; and, although in the adult these have become united, it is usual to describe the bone as divisible into three portions—the ilium, the ischium, and the pubes.

The *ilium*, so called from its supporting the flank (ilia), is the superior broad and expanded portion which runs upwards from the upper and back part of the acetabulum, and forms the prominence of the hip.

The *ischium* (*ἰσχίον*, the hip) is the inferior and strongest portion of the bone; it proceeds downwards from the acetabulum, expands into a large tuberosity, and then, curving upwards, forms with the descending ramus of the pubes a large aperture, the obturator foramen.

The *pubes* is that portion which runs horizontally inwards from the inner side of the acetabulum for about two inches, then makes a sudden bend, and descends to the same extent: it forms the front of the pelvis, supports the external organs of generation, and has received its name from being covered with hair.

The *Ilium* presents for examination two surfaces, an external and an internal, a crest, and two borders, an anterior and a posterior.

External Surface or *Dorsum of the Ilium* (fig. 97). The back part of this surface is directed backwards, downwards, and outwards; its front part forwards, downwards, and outwards. It is smooth, convex in front, deeply concave behind; bounded above by the crest, below by the upper border of the acetabulum; in front and behind, by the anterior and posterior borders. This surface is crossed in an arched direction by three semicircular lines, the superior, middle, and inferior curved lines. The superior curved line, the shortest of the three, commences at the crest, about two inches in front of its posterior extremity; it is at first distinctly marked, but as it passes downwards and outwards to the upper part of the great sacro-sciatic notch, where it terminates, it becomes less marked, and is often altogether lost. The rough surface included between this line and the crest, affords attachment to part of the Gluteus maximus above, a few fibres of the Pyriformis below. The middle curved line, the longest of the three, commences at the crest, about an inch behind its anterior extremity, and, taking a curved direction downwards and backwards, terminates at the upper part of the great sacro-sciatic notch. The space between the middle and superior curved lines and the crest is concave, and affords attachment to the Gluteus medius muscle. Near the central part of this line may often be observed the orifice of a nutrient foramen. The inferior curved line, the least distinct of the three, commences in front at the upper part of the anterior inferior spinous process, and taking a curved direction backwards and downwards, terminates at the anterior part of the great sacro-sciatic notch. The surface of bone included between the middle and inferior curved lines is concave from above downwards, convex from before backwards, and affords attach-

97.—Right Os Innominatum. External Surface.

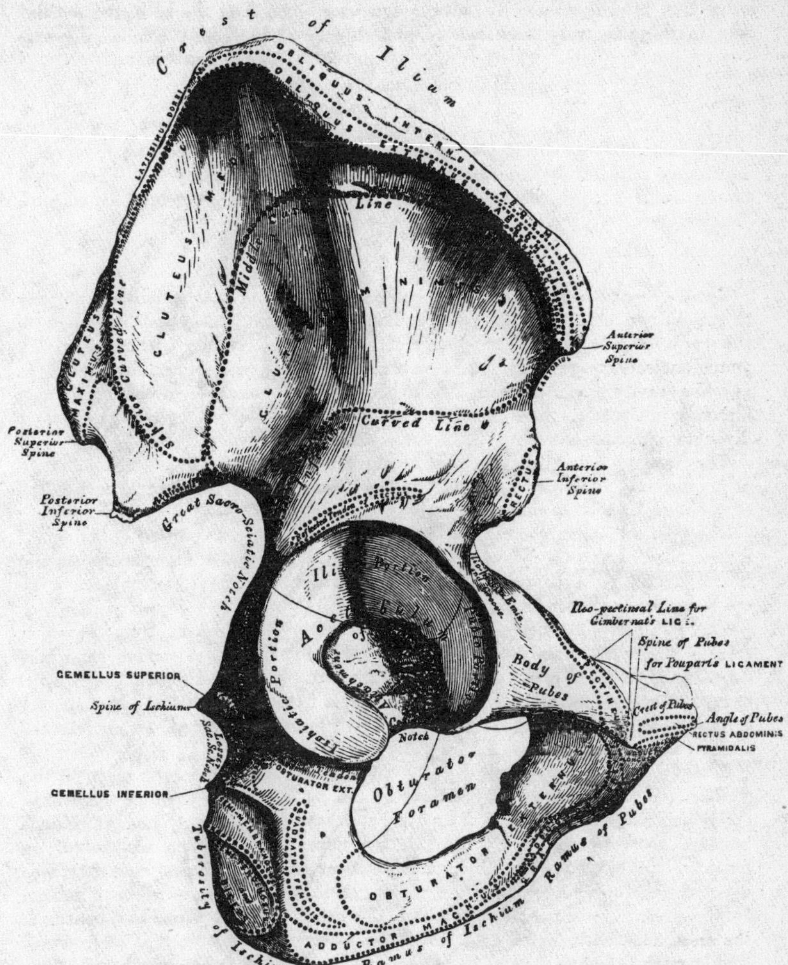

ment to the Gluteus minimus muscle. Beneath the inferior curved line, and corresponding to the upper part of the acetabulum, is a smooth eminence (sometimes a depression), to which is attached the reflected tendon of the Rectus femoris muscle.

The *Internal Surface* (fig. 98) of the ilium is bounded above by the crest, below by a prominent line, the linea ilio-pectinea, and before and behind by the anterior and posterior borders. It presents anteriorly a large smooth concave surface called the *internal iliac fossa*, or *venter of the ilium*, which lodges the Iliacus muscle, and presents at its lower part the orifice of a nutrient canal. Behind the iliac fossa is a rough surface, divided into two portions, a superior and an inferior.

The inferior or auricular portion, so called from its resemblance in shape to the ear, is coated with cartilage in the recent state, and articulates with a similar shaped surface on the side of the sacrum. The superior portion is concave and rough, for the attachment of the posterior sacro-iliac ligaments.

The crest of the ilium is convex in its general outline and sinuously curved, being bent inwards anteriorly, outwards posteriorly. It is longer in the female than in the male, very thick behind, and thinner at the centre than at the extremities. It terminates at either end in a prominent eminence, the anterior superior and posterior superior spinous process. The surface of the crest is broad, and divided into an external lip, an internal lip, and an intermediate space. To the external lip is attached the Tensor vaginæ femoris, Obliquus externus abdominis, and Latissimus dorsi, and by its whole length the fascia lata; to the interspace between the lips, the Internal oblique; to the internal lip, the Transversalis, Quadratus lumborum, and Erector spinæ.

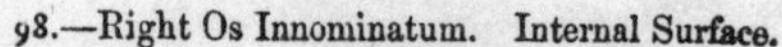
98.—Right Os Innominatum. Internal Surface.

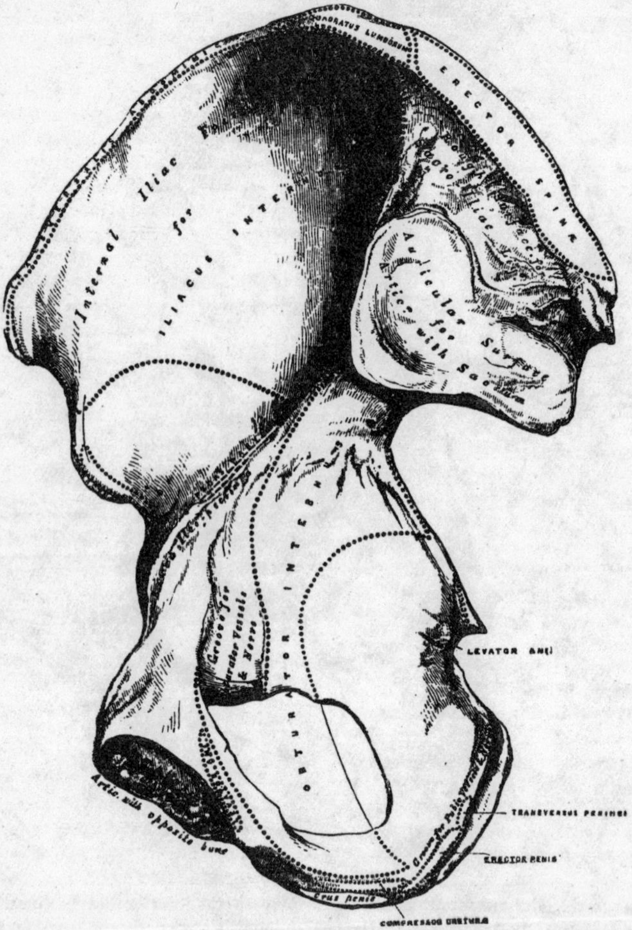

The anterior border of the ilium is concave. It presents two projections separated by a notch. Of these, the uppermost, situated at the junction of the crest and anterior border, is called the anterior superior spinous process of the ilium, the outer border of which gives attachment to the fascia lata, and the origin of the Tensor vaginæ femoris; its inner border, to the Iliacus internus; whilst its extremity affords attachment to Poupart's ligament, and the origin of the Sartorius. Beneath this eminence is a notch which gives attachment to the Sartorius muscle, and across which passes the external cutaneous nerve. Below the notch is the anterior inferior spinous process, which terminates in the upper lip of the acetabulum; it gives attachment to the straight tendon of the Rectus femoris muscle. On the inner side of the anterior inferior spinous process is a broad shallow groove over which passes the Iliacus muscle. The posterior border of the ilium, shorter than the anterior, also presents two projections separated by a notch, the posterior superior and the posterior inferior spinous processes. The former corresponds with that portion of the posterior surface of the ilium which serves for the attachment of the sacro-iliac ligaments; the latter to the auricular portion which articulates with the sacrum. Below the posterior inferior spinous process is a deep notch, the great sacro-sciatic.

The *Ischium* forms the lower and back part of the os innominatum. It is divisible into a thick and solid portion, the body; the tuberosity, a large rough eminence, on which the body rests in sitting; and a thin ascending part, the ramus.

The *body*, somewhat triangular in form, presents three surfaces, external, internal, and posterior. The *external surface* corresponds to that portion of the acetabulum formed by the ischium; it is smooth and concave above, and forms a little more than two-fifths of that cavity; its outer margin is bounded by a prominent rim or lip, to which the cotyloid fibro-cartilage is attached. Below the acetabulum, between it and the tuberosity, is a deep groove, along which the tendon of the Obturator externus muscle runs, as it passes outwards, to be inserted into the digital fossa of the femur. The *internal surface* is smooth, concave, and forms the lateral boundary of the true pelvic cavity; it is broad above, and separated from the venter of the ilium by the linea ilio-pectinea; narrow below, its posterior border is encroached upon, a little below its centre, by the spine of the ischium, above and below which are the greater and lesser sacro-sciatic notches; in front it presents a sharp margin, which forms the outer boundary of the obturator foramen. This surface is perforated by two or three large vascular foramina, and affords attachment to part of the Obturator internus muscle. The *posterior surface* is quadrilateral in form, broad and smooth above, narrow below where it becomes continuous with the tuberosity; it is limited, in front, by the margin of the acetabulum; behind, by the front part of the great sacro-sciatic notch. This surface supports the Pyriformis, the two Gemelli, and the Obturator internus muscles, in their passage outwards to the great trochanter. The body of the ischium presents three borders, posterior, inferior, and internal. The *posterior border* presents, a little below the centre, a thin and pointed triangular eminence, the spine of the ischium, more or less elongated in different subjects. Its external surface gives attachment to the Gemellus superior, its internal surface to the Coccygeus and Levator ani; whilst to the pointed extremity is connected the lesser sacro-sciatic ligament. Above the spine is a notch of large size, the great sacro-sciatic, converted into a foramen by the lesser sacro-sciatic ligament; it transmits the Pyriformis muscle, the gluteal vessels, and superior gluteal nerve passing out of the pelvis above the muscle; the sciatic artery, the greater and lesser sciatic nerves, the internal pudic vessels and nerve, and a small nerve to the Obturator internus muscle below it. Below the spine is a smaller notch, the lesser sacro-sciatic; it is smooth, coated in the recent state with cartilage, the surface of which presents numerous markings corresponding to the subdivisions of the tendon of the Obturator internus which winds over it. It is converted into a foramen by the sacro-sciatic ligaments, and transmits the tendon of the Obturator internus, the nerve which supplies that muscle, and the pudic vessels and nerve. The *inferior border* is thick and broad;

at its point of junction with the posterior is the tuberosity of the ischium. The *internal border* is thin, and forms the outer circumference of the obturator foramen.

The *tuberosity* presents for examination an external lip, an internal lip, and an intermediate space. The external lip gives attachment to the Quadratus femoris, and part of the Adductor magnus muscles. The inner lip is bounded by a sharp ridge, for the attachment of a falciform prolongation of the great sacro-sciatic ligament; it presents a groove on the inner side of this for the lodgment of the internal pudic vessels and nerve; and, more anteriorly, has attached the Transversus perinei and Erector penis muscles. The intermediate surface presents four distinct impressions. Two of these, seen at the front part of the tuberosity, are rough, elongated, and separated from each other by a prominent ridge; the outer one gives attachment to the Adductor magnus, the inner one to the great sacro-sciatic ligament. Two, situated at the back part, are smooth, larger in size, and separated by an oblique ridge: from the upper and outer arises the Semi-membranosus; from the lower and inner, the Biceps and Semi-tendinosus. The uppermost part of the tuberosity gives attachment to the Gemellus inferior.

The *ramus*, or *ascending ramus*, is the thin flattened part of the ischium, which ascends from the tuberosity upwards and inwards, and joins the ramus of the pubes—their point of junction being indicated in the adult by a rough eminence. The outer surface of the ramus is rough, for the attachment of the Obturator externus muscle, also some fibres of the Adductor magnus, and of the Gracilis; its inner surface forms part of the anterior wall of the pelvis. Its inner border is thick, rough, slightly everted, forms part of the outlet of the pelvis, and serves for the attachment of the crus penis. Its outer border is thin and sharp, and forms part of the inner margin of the obturator foramen.

The *Pubes* forms the anterior part of the os innominatum; it is divisible into a horizontal ramus or body, and a perpendicular ramus.

The *body*, or *horizontal ramus*, presents for examination two extremities, an outer and an inner, and four surfaces. The *outer extremity*, the thickest part of the bone, forms one-fifth of the cavity of the acetabulum; it presents, above, a rough eminence, the ilio-pectineal, which serves to indicate the point of junction of the ilium and pubes. The *inner extremity* is the symphysis; it is oval, covered by eight or nine transverse ridges, or a series of nipple-like processes arranged in rows, separated by grooves; they serve for the attachment of the interarticular fibro-cartilage, placed between it and the opposite bone. The *upper surface*, triangular in form, wider externally than internally, is bounded behind by a sharp ridge, the pectineal line, or linea ilio-pectinea, which, running outwards, marks the brim of the true pelvis. The surface of bone in front of the pubic portion of the linea ilio-pectinea, serves for the attachment of the Pectineus muscle. This ridge terminates internally at a tubercle, which projects forwards, and is called the *spine* of the pubes. The portion of bone included between the spine and inner extremity of the pubes is called the *crest*; it serves for the attachment of the Rectus, Pyramidalis, and conjoined tendon of the Internal oblique and Transversalis. The point of junction of the crest with the symphysis is called the *angle of the pubes*. The *inferior surface* presents, externally, a broad and deep oblique groove, for the passage of the obturator vessels and nerve; and, internally, a sharp margin, which forms part of the circumference of the obturator foramen. Its *external surface*, flat and compressed, serves for the attachment of muscles. Its *internal surface*, convex from above downwards, concave from side to side, is smooth, and forms part of the anterior wall of the pelvis.

The *descending ramus* of the pubes passes outwards and downwards, becoming thinner and narrower as it descends, and joins with the ramus of the ischium. Its *external surface* is rough, for the attachment of muscles; the Adductor longus above, the Adductor brevis below; the Gracilis along its inner border, the Compressor urethræ towards its internal aspect; and a portion of the Obturator externus where it enters into the formation of the foramen of that name. Its *inner surface* is smooth. Its *inner border* is thick, rough, and everted, especially in females. In the male it serves for the attachment of the crus penis. Its *outer border* forms part of the circumference of the obturator foramen.

The *cotyloid cavity*, or *acetabulum*, is a deep, cup-shaped, hemispherical depression; formed, internally, by the pubes, above by the ilium, behind and below by the ischium; a little less than two-fifths being formed by the ilium, a little more than two-fifths by the ischium, and the remaining fifth by the pubes. It is bounded by a prominent uneven rim, which is thick and strong above, and serves for the attachment of a fibro-cartilaginous structure which contracts its orifice, and deepens the surface for articulation. It presents on its inner side a deep notch, the cotyloid notch, which transmits the nutrient vessels into the interior of the joint, and is continuous with a circular depression at the bottom of the cavity: this depression is perforated by numerous apertures, lodges a mass of fat, and its margins serve for the attachment of the ligamentum teres. The notch is converted, in the natural state, into a foramen by a dense ligamentous band which passes across it. Through this foramen the nutrient vessels and nerves enter the joint.

The *obturator* or *thyroid foramen* is a large aperture, situated between the ischium and pubes. In the male it is large, of an oval form, its longest diameter being obliquely from above downwards; in the female it is smaller, and more triangular. It is bounded by a thin uneven margin, to which a strong membrane is attached; and presents, at its upper and outer part, a deep groove, which runs from the pelvis obliquely forwards, inwards, and downwards. This groove is converted into a foramen by the obturator membrane, and transmits the obturator vessels and nerve.

Structure. This bone consists of much cancellous tissue, especially where it is thick, enclosed between two layers of dense compact tissue. In the thinner parts of the bone, as at the bottom of the acetabulum and centre of the iliac fossa, it is usually semi-transparent, and composed entirely of compact tissue.

Development (fig. 99). By *eight* centres: three primary—one for the ilium, one for the ischium, and one for the pubes; and *five* secondary—one for the crest of the ilium its whole length, one for the anterior inferior spinous process (said to occur more frequently in the male than the female), one for the tuberosity of the ischium, one for the symphysis pubis (more frequent in the female than the male), and one for the Y-shaped piece at the bottom of the acetabulum. These various centres appear in the following order: First, in the ilium, at the lower part of the bone, immediately above the sciatic notch, at about the same period that the development of the vertebræ commences. Secondly, in the body of the ischium, at about the third month of fœtal life. Thirdly, in the body of the pubes, between the fourth and fifth months. At birth, the three primary centres are quite separate, the crest, the bottom of the acetabulum, and the rami of the ischium and pubes, being still cartilaginous. At about the sixth year, the rami of the pubes and ischium are almost completely ossified. About the thirteenth or fourteenth year, the three divisions of the bone have extended their growth into the bottom of the acetabulum, being separated from each other by a Y-shaped portion of cartilage, which now presents traces of ossification. The ilium and ischium then become joined, and lastly the pubes, through the intervention of this Y-shaped portion. At about the age of puberty, ossification takes place in each of the remaining portions, and they become joined to the rest of the bone about the twenty-fifth year.

Articulations. With its fellow of the opposite side, the sacrum and femur.

Attachment of Muscles. Ilium. To the outer lip of the crest, the Tensor vaginæ femoris, Obliquus externus abdominis, and Latissimus dorsi; to the internal lip, the Transversalis, Quadratus lumborum, and Erector spinæ; to the interspace between the lips, the Obliquus internus. To the outer surface of the ilium, the Gluteus maximus, Gluteus medius, Gluteus minimus, reflected tendon of Rectus, portion of Pyriformis; to the internal surface, the Iliacus; to the anterior border, the Sartorius and straight tendon of the Rectus. *Ischium.* To its outer surface, the Obturator externus; internal surface, Obturator internus and Levator ani.

To the spine, the Gemellus superior, Levator ani, and Coccygeus. To the tuberosity, the Biceps, Semi-tendinosus, Semi-membranosus, Quadratus femoris, Adductor magnus, Gemellus inferior, Transversus perinei, Erector penis. To the pubes, the Obliquus externus, Obliquus internus, Transversalis, Rectus, Pyramidalis, Psoas parvus, Pectineus, Adductor longus, Adductor brevis, Gracilis, Obturator externus and internus, Levator ani, Compressor urethræ, and occasionally a few fibres of the Accelerator urinæ.

99.—Plan of the Development of the Os Innominatum.

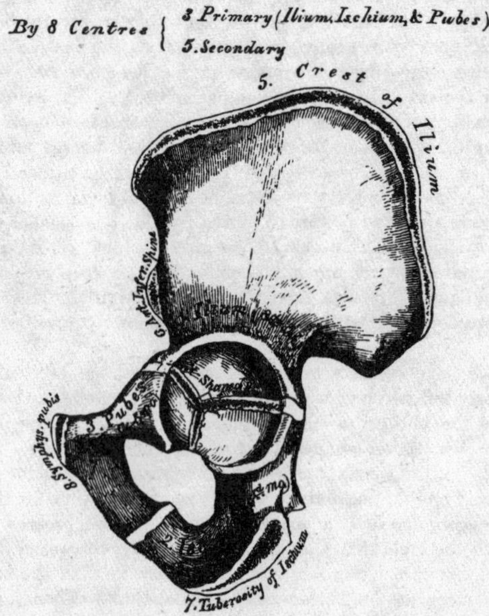

The Pelvis (figs. 100, 101).

The pelvis, so called from its resemblance to a basin, is stronger and more massively constructed than either the cranial or thoracic cavity; it is a bony ring, interposed between the lower end of the spine, which it supports, and the lower extremities, upon which it rests. It is composed of four bones—the two ossa innominata, which bound it on either side and in front; and the sacrum and coccyx, which complete it behind.

The pelvis is divided by a prominent line, the linea ilio-pectinea, into the false and true pelvis.

The *false pelvis* is all that expanded portion of the pelvic cavity which is situated above the linea ilio-pectinea. It is bounded on each side by the ossa ilii; in front it is incomplete, presenting a wide interval between the spinous processes of the ilia on either side, which is filled up in the recent state by the parietes of the abdomen; behind, in the middle line, is a deep notch. This broad shallow cavity is fitted to support the intestines, and to transmit part of their weight to the anterior wall of the abdomen.

The *true pelvis* is all that part of the pelvic cavity which is situated beneath the linea ilio-pectinea. It is smaller than the false pelvis, but its walls are more perfect. For convenience of description, it is divided into a superior circumference or inlet, an inferior circumference or outlet, and a cavity.

The *superior circumference* forms the margin or brim of the pelvis, the included space being called the *inlet*. It is formed by the linea ilio-pectinea, completed in front by the spine and crest of the pubes, and behind by the anterior margin of the base of the sacrum and sacro-vertebral angle.

100.—Male Pelvis (Adult).

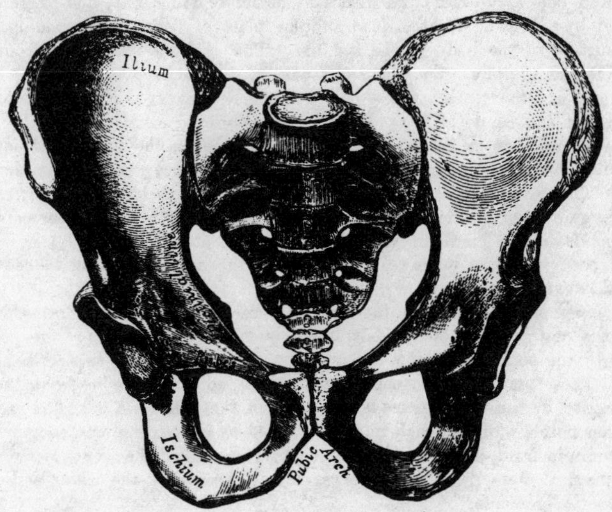

101.—Female Pelvis (Adult).

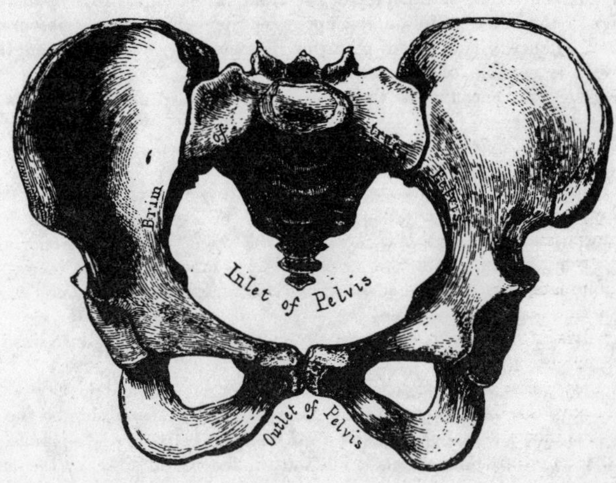

The *inlet* of the pelvis is somewhat heart-shaped, obtusely pointed in front, diverging on either side, and encroached upon behind by the projection forwards of the promontory of the sacrum. It has three principal diameters: antero-posterior (sacro-pubic), transverse, and oblique. The antero-posterior extends from the sacro-vertebral angle to the symphysis pubis; its average measurement is four inches. The transverse extends across the greatest width of the inlet, from the middle of the brim on one side to the same point on the opposite; its average measurement is five inches. The oblique extends from the margin of the pelvis, corresponding to the ilio-pectineal eminence on one side, to the sacro-iliac symphysis on the opposite side; its average measurement is also five inches.

The *cavity* of the true pelvis is bounded in front by the symphysis pubis; behind by the concavity of the sacrum and coccyx, which, curving forwards above and below, contracts the inlet and outlet of the canal; and laterally it is bounded by a broad, smooth, quadrangular plate of bone, corresponding to the inner surface of the body of the ischium. The cavity is shallow in front, measuring at the symphysis an inch and a half in depth, three inches and a half in the middle, and four inches and a half posteriorly. From this description, it will be seen that the cavity of the pelvis is a short curved canal, considerably deeper on its posterior than on its anterior wall, and broader in the middle than at either extremity, from the projection forwards of the sacro-coccygeal column above and below. This cavity contains, in the recent subject, the rectum, bladder, and part of the organs of generation. The rectum is placed at the back of the pelvis, and corresponds to the curve of the sacro-coccygeal column; the bladder in front, behind the symphysis pubis. In the female, the uterus and vagina occupy the interval between these parts.

The *lower circumference* of the pelvis is very irregular, and forms what is called the *outlet*. It is bounded by three prominent eminences: one posterior, formed by the point of the coccyx; and one on each side, the tuberosities of the ischia. These eminences are separated by three notches: one in front, the *pubic arch*, formed by the convergence of the rami of the ischia and pubes on each side. The other notches, one on each side, are formed by the sacrum and coccyx behind, the ischium in front, and the ilium above: they are called the *sacro-sciatic notches*; in the natural state they are converted into foramina by the lesser and greater sacro-sciatic ligaments.

The diameters of the outlet of the pelvis are two, antero-posterior and transverse. The *antero-posterior* extends from the tip of the coccyx to the lower part of the symphysis pubis, and the *transverse* from the posterior part of one ischiatic tuberosity to the same point on the opposite side: the average measurement of both is four inches. The antero-posterior diameter varies with the length of the coccyx, and is capable of increase or diminution, on account of the mobility of that bone.

102.—Vertical Section of the Pelvis, with lines indicating the Axes of the Pelvis.

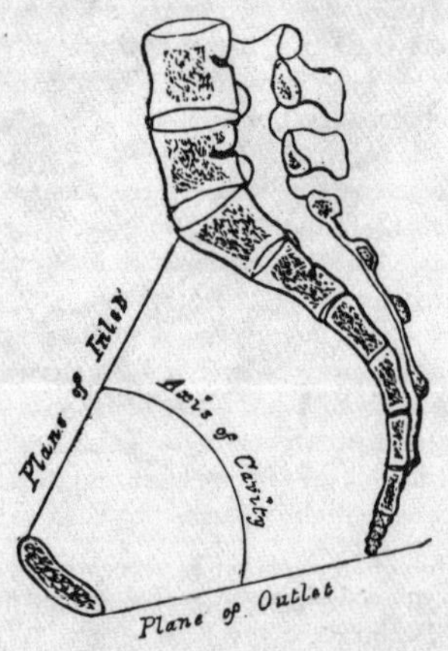

Position of the Pelvis. In the erect posture, the pelvis is placed obliquely with regard to the trunk of the body; the pelvic surface of the symphysis pubis looking upwards and backwards, the concavity of the sacrum and coccyx looking downwards and forwards; the base of the sacrum in well-formed female bodies, being nearly four inches above the upper border of the symphysis pubis, and the apex of the coccyx a little more than half an inch above its lower border. The obliquity is much greater in the fœtus, and at an early period of life, than in the adult.

Axes of the Pelvis (fig. 102). The

plane of the inlet of the true pelvis will be represented by a line drawn from the base of the sacrum to the upper margin of the symphysis pubis. A line carried at right angles with this at its middle, would correspond at one extremity with the umbilicus, and at the other with the middle of the coccyx; the axis of the inlet is therefore directed downwards and backwards. The axis of the outlet produced upwards would touch the base of the sacrum, and is therefore directed downwards and forwards. The axis of the cavity is curved like the cavity itself: this curve corresponds to the concavity of the sacrum and coccyx, the extremities being indicated by the central points of the inlet and outlet. A knowledge of the direction of these axes serves to explain the course of the fœtus in its passage through the pelvis during parturition. It is also important to the surgeon, as indicating the direction of the force required in the removal of calculi from the bladder, and as determining the direction in which instruments should be used in operations upon the pelvic viscera.

Differences between the Male and Female Pelvis. In the *male* the bones are thicker and stronger, and the muscular eminences and impressions on their surfaces more strongly marked. The male pelvis is altogether more massive; its cavity is deeper and narrower, and the obturator foramina of larger size. In the *female* the bones are lighter and more expanded, the muscular impressions on their surfaces are only slightly marked, and the pelvis generally is less massive in structure. The iliac fossæ are broad, and the spines of the ilia widely separated; hence the great prominence of the hips. The inlet and the outlet are larger; the cavity is more capacious, and the spines of the ischia project less into it. The promontory is less projecting, the sacrum wider and less curved,* and the coccyx more moveable. The arch of the pubes is wider, and its edges more everted. The tuberosities of the ischia and the acetabula are wider apart.

In the *fœtus* and for several years after birth, the pelvis is small in proportion to that of the adult. The cavity is deep, and the projection of the sacro-vertebral angle less marked. The antero-posterior and transverse diameters are nearly equal. *About puberty*, the pelvis in both sexes presents the general characters of the adult male pelvis, but *after puberty* it acquires its proper sexual characters.

The Femur or Thigh-Bone.

The Femur is the longest, largest, and strongest bone in the skeleton, and almost perfectly cylindrical in the greater part of its extent. In the erect posture it is not vertical, being separated from its fellow above by a considerable interval, which corresponds to the entire breadth of the pelvis, but inclining gradually downwards and inwards, so as to approach its fellow towards its lower part, for the purpose of bringing the knee-joint near the line of gravity of the body. The degree of this inclination varies in different persons, and is greater in the female than in the male, on account of the greater breadth of the pelvis. The femur, like other long bones, is divisible into a shaft, and two extremities.

The *Upper Extremity* presents for examination a head, a neck, and the greater and lesser trochanters.

The *head*, which is globular, and forms rather more than a hemisphere, is directed upwards, inwards, and a little forwards, the greater part of its convexity being above and in front. Its surface is smooth, coated with cartilage in the recent state, and presents, a little behind and below its centre, an ovoid depression, for the attachment of the ligamentum teres. The *neck* is a flattened pyramidal process of bone, which connects the head with the shaft. It varies in length and

* It is not unusual, however, to find the sacrum in the female presenting a considerable curve extending throughout its whole length.

103.—Right Femur. Anterior Surface.

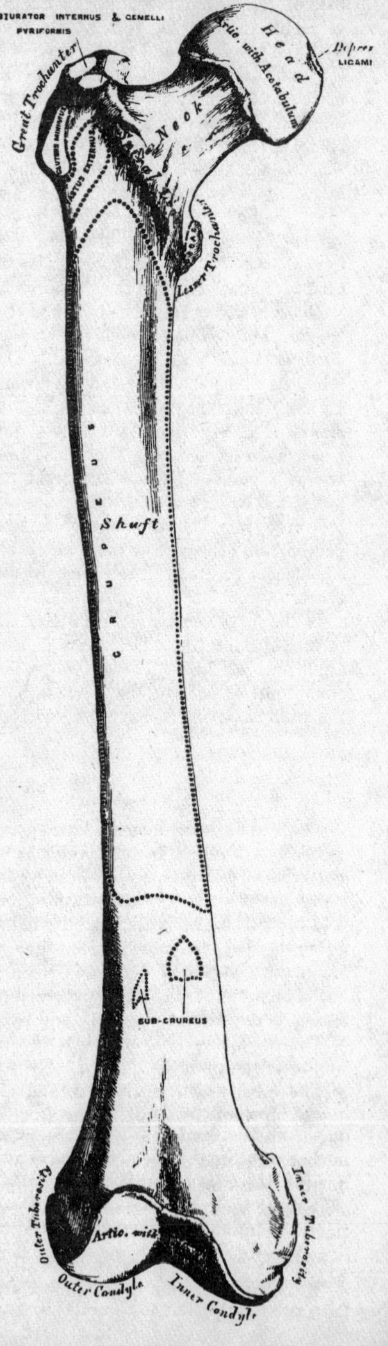

obliquity at various periods of life, and under different circumstances. Before puberty it is directed obliquely, so as to form a gentle curve from the axis of the shaft. In the adult male it forms an obtuse angle with the shaft, being directed upwards, inwards, and a little forwards. In the female it approaches more nearly a right angle. Occasionally, in very old subjects, and more especially in those greatly debilitated, its direction becomes horizontal; so that the head sinks below the level of the trochanter, and its length diminishes to such a degree, that the head becomes almost contiguous with the shaft. The neck is flattened from before backwards, contracted in the middle, and broader at its outer extremity, where it is connected with the shaft, than at its summit, where it is attached to the head. It is much broader in the vertical than in the antero-posterior diameter, and much thicker below than above, on account of the greater amount of resistance required in sustaining the weight of the trunk. The *anterior surface* of the neck is perforated by numerous vascular foramina. The *posterior surface* is smooth, and is broader and more concave than the anterior; it receives towards its outer side the attachment of the capsular ligament of the hip. The *superior border* is short and thick, bounded externally by the great trochanter, and its surface perforated by large foramina. The *inferior border*, long and narrow, curves a little backwards, to terminate at the lesser trochanter.

The Trochanters (τρόχαω, *to run* or *roll*) are prominent processes of bone which afford leverage to the muscles which rotate the thigh on its axis. They are two in number, the greater and the lesser.

The *Great Trochanter* is a large irregular quadrilateral eminence, situated at the outer side of the neck, at its junction with the upper part of the shaft. It is directed a little outwards and backwards, and, in the adult, is about three quarters of an inch lower than the head. It presents for examination two surfaces, and four borders. The *external surface*, quadrilateral in form, is broad, rough, convex, and marked by a prominent diagonal line, which extends from the posterior superior to the anterior inferior angle: this line serves for the attachment of the tendon of the Gluteus medius.

Above the line is a triangular surface, sometimes rough for part of the tendon of the same muscle, sometimes smooth for the interposition of a bursa between that tendon and the bone. Below and behind the diagonal line is a smooth triangular surface over which the tendon of the Gluteus maximus muscle plays, a bursa being interposed. The *internal surface* is of much less extent than the external, and presents at its base a deep depression, the digital or trochanteric fossa for the attachment of the tendon of the Obturator externus muscle. The *superior border* is free; it is thick and irregular, and marked by impressions for the attachment of the Pyriformis behind, the Obturator internus and Gemelli in front. The *inferior border* corresponds to the point of junction of the base of the trochanter with the outer surface of the shaft; it is rough, prominent, slightly curved, and gives attachment to the upper part of the Vastus externus muscle. The *anterior border* is prominent, somewhat irregular, as well as the surface of bone immediately below it; it affords attachment by its outer part to the Gluteus minimus. The *posterior border* is very prominent, and appears as a free rounded edge, which forms the back part of the digital fossa.

The *Lesser Trochanter* is a conical eminence, which varies in size in different subjects; it projects from the lower and back part of the base of the neck. Its base is triangular, and connected with the adjacent parts of the bone by three well-marked borders; of these, the *superior* is continuous with the lower border of the neck; the *posterior*, with the posterior intertrochanteric line; and the *inferior*, with the middle division of the linea aspera. Its summit, which is directed inwards and backwards, is rough, and gives insertion to the tendon of the Psoas magnus. The Iliacus is inserted into the shaft below the lesser trochanter, between the Vastus internus in front, and the Pectineus behind. A well-marked prominence, of variable size, which projects from the upper and front part of the neck, at its junction with the great trochanter, is called the *tubercle of the femur*; it is the point of meeting of three muscles, the Gluteus minimus externally, the Vastus externus below, and the tendon of the Obturator internus and Gemelli above. Running obliquely downwards and inwards from the tubercle is the spiral line of the femur, or anterior intertrochanteric line; it winds round the inner side of the shaft, below the lesser trochanter, and terminates in the linea aspera, about two inches below this eminence. Its upper half is rough, and affords attachment to the capsular ligament of the hip joint; its lower half is less prominent, and gives attachment to the upper part of the Vastus internus. The posterior intertrochanteric line is very prominent, and runs from the summit of the great trochanter downwards and inwards to the upper and back part of the lesser trochanter. Its upper half forms the posterior border of the great trochanter. A well-marked eminence commences about the middle of the posterior intertrochanteric line, and passes vertically downwards for about two inches along the back part of the shaft: it is called the *linea quadrati*, and gives attachment to the Quadratus femoris, and a few fibres of the Adductor magnus muscles.

The *Shaft*, almost perfectly cylindrical in form, is a little broader above than in the centre, and somewhat flattened from before backwards below. It is slightly arched, so as to be convex in front; concave behind, where it is strengthened by a prominent longitudinal ridge, the linea aspera. It presents for examination three borders separating three surfaces. Of the three borders, one, the linea aspera, is posterior; the other two are placed laterally.

The *linea aspera* (fig. 104) is a prominent longitudinal ridge or crest, presenting, on the middle third of the bone, an external lip, an internal lip, and a rough intermediate space. A little above the centre of the shaft, this crest divides into three lines;* the most external one becomes very rough, and is continued almost vertically upwards to the base of the great trochanter; the middle one, the least distinct, is continued to the base of the trochanter minor; and the internal one is lost above in the spiral line of the femur. Below, the linea aspera divides into

* Of these three lines, only the outer and inner are described by many anatomists: the linea aspera is then said to bifurcate above and below.

two bifurcations, which enclose between them a triangular space (the popliteal space), upon which rests the popliteal artery. Of these two bifurcations, the outer branch is the most prominent, and descends to the summit of the outer condyle. The inner branch is less marked, presents a broad and shallow groove for the passage of the femoral artery, and terminates in a small tubercle at the summit of the internal condyle.

To the inner lip of the linea aspera, along its whole length, is attached the Vastus internus; and to the whole length of the outer lip, the Vastus externus. The Adductor magnus is also attached to the whole length of the linea aspera, being connected with the outer lip above, and the inner lip below. Between the Vastus externus and the Adductor magnus are attached two muscles, viz., the Gluteus maximus above, and the short head of the Biceps below. Between the Adductor magnus and the Vastus internus four muscles are attached: the Iliacus and Pectineus above (the latter to the middle of the upper divisions); below these, the Adductor brevis and Adductor longus. The linea aspera is perforated a little below its centre by the nutrient canal, which is directed obliquely upwards.

The two lateral borders of the femur are only slightly marked, the outer one extending from the anterior inferior angle of the great trochanter to the anterior extremity of the external condyle; the inner one from the spiral line, at a point opposite the trochanter minor, to the anterior extremity of the internal condyle. The internal border marks the limit of attachment of the Crureus muscle internally.

The *anterior surface* includes that portion of the shaft which is situated between the two lateral borders. It is smooth, convex, broader above and below than in the centre, slightly twisted, so that its upper part is directed forwards and a little outwards, its lower part forwards and a little inwards. To the upper three-fourths of this surface the Crureus is attached; the lower fourth is separated from the muscle by the intervention of the synovial membrane of the knee-joint, and affords attachment to the Subcrureus to a small extent. The *external surface* includes the portion of bone between the

104.—Right femur. Posterior surface.

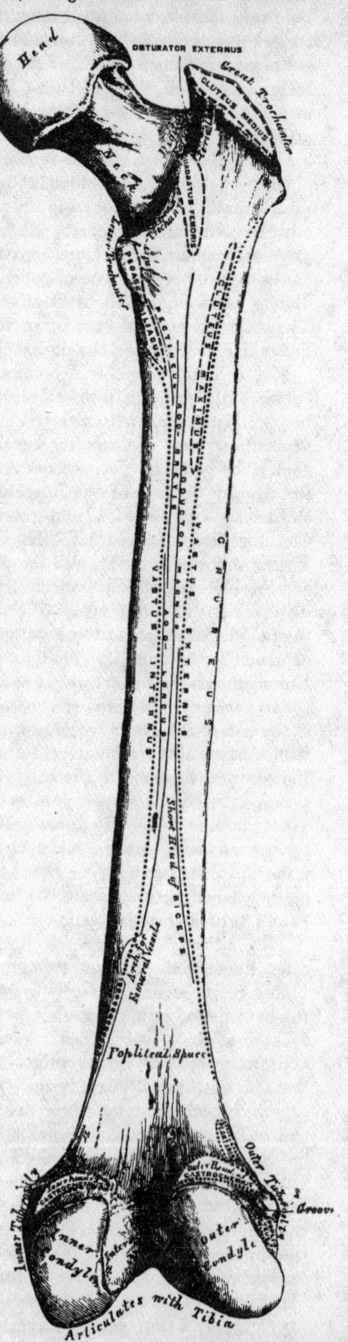

external border and the outer lip of the linea aspera; it is continuous above, with the outer surface of the great trochanter; below with the outer surface of the external condyle: to its upper three-fourths is attached the outer portion of the Crureus muscle. The *internal surface* includes the portion of bone between the internal border and the inner lip of the linea aspera; it is continuous, above, with the lower border of the neck; below, with the inner side of the internal condyle: it is covered by the Vastus internus muscle.

The *Lower Extremity*, larger than the upper, is of a cuboid form, flattened from before backwards, and divided by an interval presenting a smooth depression in front and a notch of considerable size behind, into two large eminences, the condyles (κόνδυλος, *a knuckle*). The interval is called the *intercondyloid notch*. The *external condyle* is the more prominent anteriorly, and is the broader both in the antero-posterior and transverse diameters. The *internal condyle* is the narrower, longer, and more prominent internally. This difference in the length of the two condyles is only observed when the bone is perpendicular, and depends upon the obliquity of the thigh-bones, in consequence of their separation above at the articulation with the pelvis. If the femur is held obliquely, the surfaces of the two condyles will be seen to be nearly horizontal. The two condyles are directly continuous in front, and form a smooth trochlear surface, the external border of which is more prominent, and ascends higher than the internal one. This surface articulates with the patella. It presents a median groove, which extends downwards and backwards to the intercondyloid notch; and two lateral convexities, of which the external is the broader, more prominent, and prolonged farther upwards upon the front of the outer condyle. The intercondyloid notch lodges the crucial ligaments; it is bounded laterally by the opposed surfaces of the two condyles, and in front by the lower end of the shaft.

Outer Condyle. The *outer surface* of the external condyle presents, a little behind its centre, an eminence, *the outer tuberosity*; it is less prominent than the inner tuberosity, and gives attachment to the external lateral ligament of the knee. Immediately beneath it, is a groove which commences at a depression a little behind the centre of the lower border of this surface: the depression is for the tendon of origin of the Popliteus muscle; the groove in which this tendon is contained, is smooth, covered with cartilage in the recent state, and runs upwards and backwards to the posterior extremity of the condyle. The *inner surface* of the outer condyle forms one of the lateral boundaries of the intercondyloid notch, and gives attachment, by its posterior part, to the anterior crucial ligament. The *inferior surface* is convex, smooth, and broader than that of the internal condyle. The posterior extremity is convex and smooth: just above the articular surface is a depression for the tendon of the outer head of the Gastrocnemius, above which is the origin of the Plantaris.

Inner Condyle. The *inner surface* of the inner condyle presents a convex eminence, the *inner tuberosity*, rough, for the attachment of the internal lateral ligament. Above this tuberosity, at the termination of the inner bifurcation of the linea aspera, is a tubercle, for the insertion of the tendon of the Adductor magnus; and behind and beneath the tubercle a depression, for the tendon of the inner head of the Gastrocnemius. The *outer side* of the inner condyle forms one of the lateral boundaries of the intercondyloid notch, and gives attachment, by its anterior part, to the posterior crucial ligament. Its *inferior* or *articular surface* is convex, and presents a less extensive surface than the external condyle.

Structure. The shaft of the femur is a cylinder of compact tissue, hollowed by a large medullary canal. The cylinder is of great thickness and density in the middle third of the shaft, where the bone is narrowest, and the medullary canal well formed; but above and below this, the cylinder gradually becomes thinner, owing to a separation of the layers of the bone into cancelli, which project into the medullary canal and finally obliterate it, so that the upper and lower ends of the shaft, and the articular extremities more especially, consist of cancellated tissue invested by a thin compact layer.

The arrangement of the cancelli in the ends of the femur is remarkable. In the upper end (fig. 105), they run in parallel columns *a a* from the summit of the head to the thick under wall of the neck, while a series of transverse fibres *b b* cross the parallel columns, and connect them to the thin upper wall of the neck. Another series of plates *c c* springs from the whole interior of the cylinder above the lesser trochanter; these pass upwards and converge to form a series of arches beneath the upper wall of the neck, near its junction with the great trochanter. This structure is admirably adapted to sustain, with the greatest mechanical advantage, concussion or weight transmitted from above, and serves an important office in strengthening a part especially liable to fracture.

105.—Diagram showing the Structure of the Neck of the Femur. (WARD.)

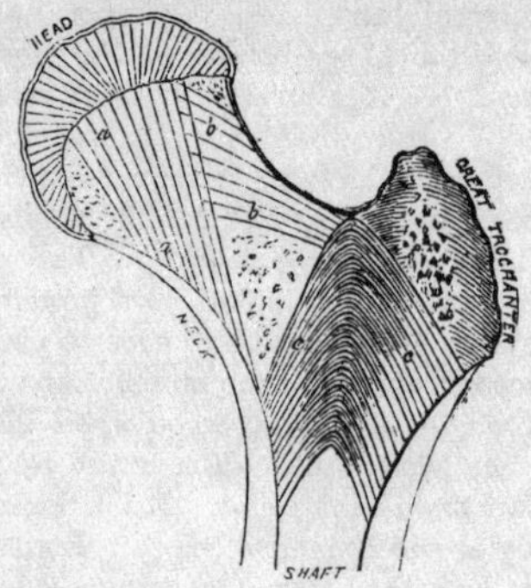

In the lower end, the cancelli spring on all sides from the inner surface of the cylinder, and descend in a perpendicular direction to the articular surface, the cancelli being strongest and having a more accurately perpendicular course, above the condyles.

Articulations. With three bones: the os innominatum, tibia, and patella.

Development (fig. 106). The femur is developed by *five* centres; one for the shaft, one for each extremity, and one for each trochanter. Of all the long bones, except the clavicle, it is the first to show traces of ossification; this commences in the shaft, at about the fifth week of fœtal life, the centres of ossification in the epiphyses appearing in the following order: First, in the lower end of the bone, at the ninth month of fœtal life; from this the condyles and tuberosities are formed; in the head, at the end of the first year after birth; in the great trochanter, during the fourth year; and in the lesser trochanter, between the thirteenth and fourteenth. The order in which the epiphyses are joined to the shaft, is the reverse of that of their appearance ; their junction does not commence until after puberty, the lesser trochanter being first joined, then the greater, than the head, and, lastly, the inferior extremity (the first in which ossification commenced), which is not united until the twentieth year.

106.—Plan of the Development of the Femur. By Five Centres.

Attachment of Muscles. To the great trochanter: the Gluteus medius, Gluteus minimus, Pyriformis, Obturator internus, Obturator externus, Gemellus superior, Gemellus inferior, and Quadratus femoris. To the lesser trochanter: the Psoas magnus, and the Iliacus below it. To the shaft, its posterior surface: the Vastus externus, Gluteus maximus, short head of the Biceps, Vastus internus, Adductor magnus, Pectineus, Adductor brevis, and Adductor longus; to its anterior surface: the Crureus, and Sub-crureus. To the condyles: the Gastrocnemius, Plantaris, and Popliteus.

THE LEG.

The skeleton of the Leg consists of three bones: the Patella, a large sesamoid bone, placed in front of the knee; and the Tibia, and Fibula.

The Patella. (Figs. 107, 108.)

The *Patella* is a flat, triangular bone, situated at the anterior part of the knee-joint. It resembles the sesamoid bones, from being developed in the tendon of the Quadriceps extensor, and in its structure, being composed throughout of dense cancellous tissue; but it is generally regarded as analogous to the olecranon process of the ulna, which occasionally exists as a separate piece, connected to the shaft of the bone by a continuation of the tendon of the Triceps muscle.* It serves to protect the front of the joint, and increases the leverage of the Quadriceps extensor by making it act at a greater angle. It presents an anterior and posterior surface, three borders, a base, and an apex.

107.—Right Patella. Anterior Surface.

108.—Posterior Surface.

The *anterior surface* is convex, perforated by small apertures, for the passage of nutrient vessels, and marked by numerous rough longitudinal striæ. This surface is covered, in the recent state, by an expansion from the tendon of the Quadriceps extensor, and separated from the integument by a bursa. It gives attachment below to the ligamentum patellæ. The *posterior surface* presents a smooth, oval-shaped, articular surface, covered with cartilage in the recent state, and divided into two facets by a vertical ridge, which descends from the superior towards the inferior angle of the bone. The ridge corresponds to the groove on the trochlear surface of the femur, and the two facets to the articular surfaces of the two condyles; the outer facet, for articulation with the outer condyle, being the broader and deeper. This character serves to indicate the side to which the bone belongs. Below the articular surface is a rough, convex, non-articular depression, the lower half of which gives attachment to the ligamentum patellæ; the upper half being separated from the head of the tibia by adipose tissue.

The *superior* and *lateral borders* give attachment to the tendon of the Quadriceps extensor; the *superior border*, to that portion of the tendon which is derived from the Rectus and Crureus muscles; and the *lateral borders*, to the portion derived from the external and internal Vasti muscles.

The *base*, or *superior border*, is thick, directed upwards, and cut obliquely at the expense of its outer surface; it receives the attachment, as already mentioned, of part of the Quadriceps extensor tendon.

The *apex* is pointed, and gives attachment to the ligamentum patellæ.

* Professor Owen states, that, 'in certain bats, there is a development of a sesamoid bone in the biceps brachii, which is the true homotype of the patella in the leg,' regarding the olecranon to be homologous, not with the patella, but with an extension of the upper end of the fibula above the knee-joint, which is met with in some animals. ('*On the Nature of Limbs*,' pp. 19, 24.)

Structure. It consists of dense cancellous tissue, covered by a thin compact lamina.

Development. By a single centre, which makes its appearance, according to Béclard, about the third year. In two instances, I have seen this bone cartilaginous throughout, at a much later period (six years). More rarely, the bone is developed by two centres, placed side by side.

Articulations. With the two condyles of the femur.

Attachment of Muscles. The Rectus, Crureus, Vastus internus, and Vastus externus. These muscles joined at their insertion constitute the Quadriceps extensor cruris.

109.—Bones of the Right Leg. Anterior Surface.

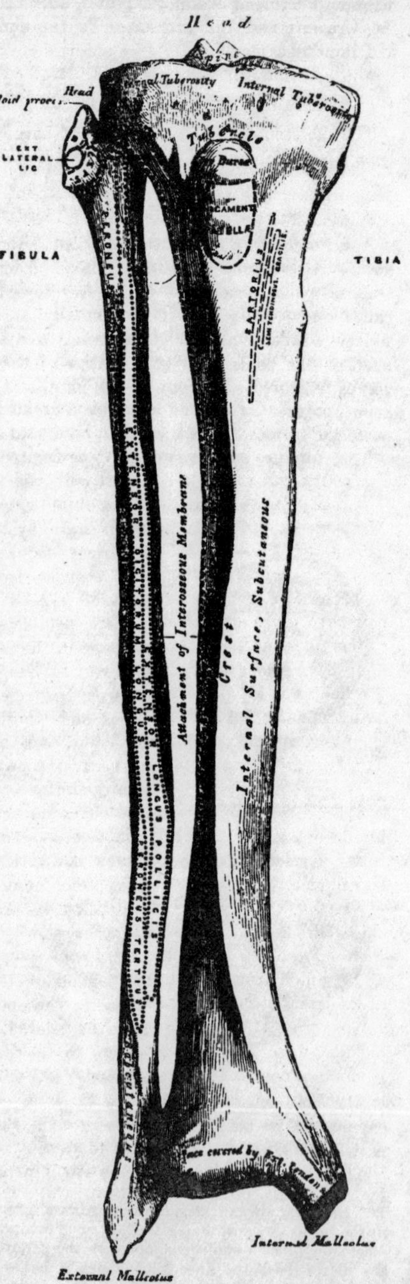

THE TIBIA. (Figs. 109, 110.)

The Tibia is situated at the front and inner side of the leg, and, excepting the femur, is the longest and largest bone in the skeleton. It is prismoid in form, expanded above, where it enters into the knee-joint, more slightly enlarged below. In the male, its direction is vertical, and parallel with the bone of the opposite side; but in the female it has a slight oblique direction downwards and outwards, to compensate for the oblique direction of the femur inwards. It presents for examination a shaft and two extremities.

The *Upper Extremity*, or head, is large and expanded on each side into two lateral eminences, the tuberosities. Superiorly, the tuberosities present two smooth concave surfaces, which articulate with the condyles of the femur; the internal articular surface is longer than the external, and oval from before backwards, to articulate with the internal condyle; the external one being broader, flatter, and more circular, to articulate with the external condyle. Between the two articular surfaces, and nearer the posterior than the anterior aspect of the bone, is an eminence, the spinous process of the tibia, surmounted by a prominent tubercle on each side, which gives attachment to the extremities of the semilunar fibro-cartilages; in front and behind the spinous process is a rough depression for the attachment of the anterior and posterior crucial ligaments and the semilunar cartilages. The anterior surfaces of the tuberosities are continuous with one another, forming a single large surface, which is somewhat flattened: it is triangular, broad above, and perforated by large vascular foramina, narrow below, where it terminates in a prominent oblong elevation of large size, the tubercle of the tibia; the lower half of this tubercle is rough, for the attachment of the ligamentum patellæ; the upper half is a smooth facet corresponding, in the recent state, with a bursa which separates the ligament from the bone. Posteriorly, the tuberosities are separated from each other by a shallow depression, the popliteal notch, which gives attachment to the posterior crucial ligament. The posterior surface of the inner tuberosity presents a deep transverse groove, for the insertion of the tendon of the Semimembranosus; and the posterior surface of the outer one, a flat articular facet, nearly circular in form, directed downwards, backwards, and outwards, for articulation with the fibula. The lateral surfaces are convex and rough: the internal one, the most prominent, gives attachment to the internal lateral ligament.

The *Shaft* of the tibia is of a triangular prismoid form, broad above, gradually decreasing in size to the commencement of its lower fourth, its most slender part where fracture most frequently occurs; it then enlarges again towards its lower extremity. It presents for examination three surfaces and three borders.

The *anterior border*, the most prominent of the three, is called the *crest of the tibia*, or, in popular language, the *shin*; it commences above at the tubercle, and terminates below at the anterior margin of the inner malleolus. This border is very prominent in the upper two-thirds of its extent, smooth and rounded below. It presents a very flexuous course, being curved outwards above, and inwards below; it gives attachment to the deep fascia of the leg.

The *internal border* is smooth and rounded above and below, but more prominent in the centre; it commences at the back part of the inner tuberosity, and terminates at the posterior border of the internal malleolus; its upper third gives attachment to the internal lateral ligament of the knee, and to some fibres of the Popliteus muscle; its middle third, to some fibres of the Soleus and Flexor longus digitorum muscles.

The *external border* is thin and prominent, especially its central part, and gives attachment to the interosseous membrane; it commences above in front of the fibular articular facet, and bifurcates below, to form the boundaries of a triangular rough surface, for the attachment of the interosseous ligament, connecting the tibia and fibula.

The *internal surface* is smooth, convex, and broader above than below; its upper third, directed forwards and inwards, is covered by the aponeurosis derived from the tendon of the Sartorius, and by the tendons of the Gracilis and Semitendinosus, all of which are inserted nearly as far forwards as the anterior border; in the rest of its extent it is subcutaneous.

The *external surface* is narrower than the internal; its upper two-thirds present a shallow groove for the attachment of the Tibialis anticus muscle; its lower third is smooth, convex, curves gradually forwards to the anterior part of the bone, and is covered from within outwards by the tendons of the following muscles: Tibialis anticus, Extensor proprius pollicis, Extensor longus digitorum, Peroneus tertius.

The *posterior surface* (fig. 110) presents, at its upper part, a prominent ridge, the oblique line of the tibia, which extends from the back part of the articular facet for the fibula, obliquely downwards, to the internal border, at the junction of its upper and middle thirds. It marks the limit for the insertion of the Popliteus muscle, and serves for the attachment of the popliteal fascia, and part of the Soleus, Flexor longus digitorum, and Tibialis posticus muscles; the triangular concave surface, above and to the inner side of this line, gives attachment to the Popliteus muscle. The middle third of the posterior surface is divided by a vertical ridge into two lateral halves: the ridge is well marked at its commencement at the oblique line, but becomes gradually indistinct below: the inner and broader half gives attachment to the Flexor longus digitorum, the outer and narrower to part of the Tibialis posticus. The remaining part of the bone is covered by the Tibialis posticus, Flexor longus digitorum, and Flexor longus pollicis muscles. Immediately below the oblique line is the medullary foramen, which is directed obliquely downwards.

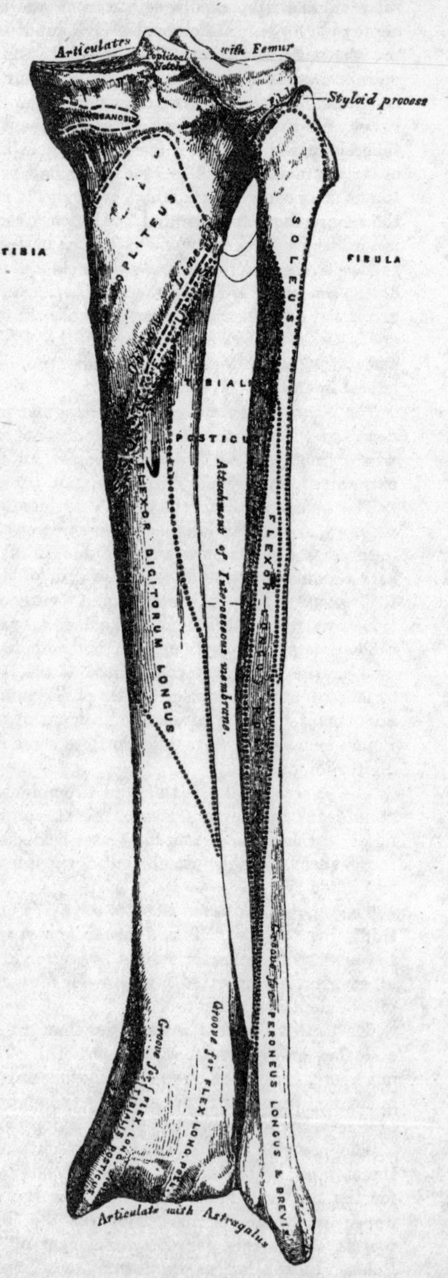

110.—Bones of the Right Leg. Posterior Surface.

The *Lower Extremity*, much smaller than the upper, presents five surfaces; it is prolonged downwards, on its inner side, into a strong process, the internal malleolus. The *inferior surface* of the bone is quadrilateral, and smooth, for articulation with the astragalus. This surface is narrow internally, where it becomes continuous with the articular surface of the inner malleolus, broader externally, and traversed from before backwards by a slight elevation, separating two lateral depressions. The *anterior surface* of the lower extremity is smooth and rounded above, and covered by the tendons of the Extensor muscles of the toes; its lower margin presents a rough transverse depression, for the attachment of the anterior ligament of the ankle joint: the *posterior surface* presents a superficial groove directed obliquely downwards and inwards, continuous with a similar groove on the posterior extremity of the astragalus, and serving for the passage of the tendon of the Flexor longus pollicis: the *external surface* presents a triangular rough depression for the attachment of the inferior interosseous ligament connecting it with the fibula; the lower part of this depression is smooth in some bones, covered with cartilage in the recent state, and articulating with the fibula. This surface is bounded by two prominent ridges, continuous above with the interosseous ridge; they afford attachment to the anterior and posterior tibio-fibular ligaments. The *internal surface* of the lower extremity is prolonged downwards to form a strong pyramidal process, flattened from without inwards, the inner malleolus. The *inner surface* of this process is convex and subcutaneous; its *outer surface*, smooth and slightly concave, deepens the articular surface for the astragalus; its *anterior border* is rough, for the attachment of ligamentous fibres; its *posterior border* presents a broad and deep groove, directed obliquely downwards and inwards, which is occasionally double: this groove transmits the tendons of the Tibialis posticus and Flexor longus digitorum muscles. The *summit* of the internal malleolus is marked by a rough depression behind, for the attachment of the internal lateral ligament of the ankle joint.

Structure. Like that of the other long bones.

Development. By three centres (fig. 111): one for the shaft, and one for each extremity. Ossification commences in the centre of the shaft about the same time as in the femur, the fifth week, and gradually extends towards either extremity. The centre for the upper epiphysis appears at birth; it is flattened in form, and has a thin tongue-shaped process in front, which forms the tubercle. That for the lower epiphysis appears in the second year. The lower epiphysis joins the shaft at about the twentieth year, and the upper one about the twenty-fifth year. Two additional centres occasionally exist, one for the tongue-shaped process of the upper epiphysis, the tubercle, and one for the inner malleolus.

111.—Plan of the Development of the Tibia. By Three Centres.

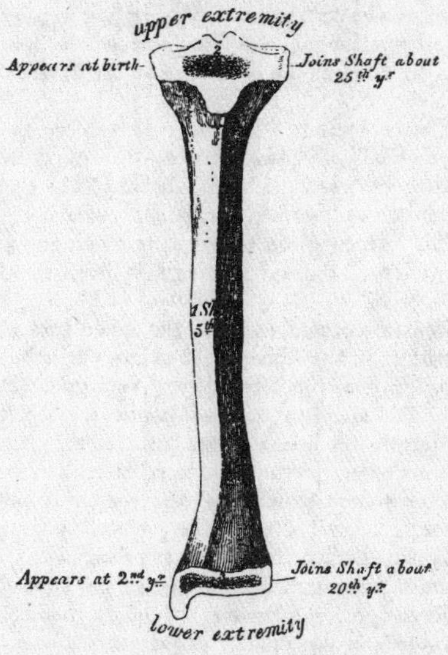

Articulations. With three bones: the femur, fibula, and astragalus.

Attachment of Muscles. To the inner tuberosity, the Semimembranosus: to the outer tuberosity, the Tibialis anticus and Extensor longus digitorum: to the shaft, its internal surface, the Sartorius, Gracilis, and Semitendinosus; to its external surface, the Tibialis anticus; to its posterior surface, the Popliteus, Soleus, Flexor longus digitorum, and Tibialis posticus; to the tubercle, the ligamentum patellæ.

The Fibula. (Figs. 109, 110.)

The Fibula is situated at the outer side of the leg. It is the smaller of the two bones, and, in proportion to its length, the most slender of all the long bones; it is placed nearly parallel with the tibia. Its upper extremity is small, placed below the level of the knee joint, and excluded from its formation; the lower extremity inclines a little forwards, so as to be on a plane anterior to that of the upper end, projects below the tibia, and forms the outer ankle. It presents for examination a shaft and two extremities.

The *Upper Extremity*, or *Head*, is of an irregular rounded form, presenting above a flattened articular facet, directed upwards and inwards, for articulation with a corresponding facet on the external tuberosity of the tibia. On the outer side is a thick and rough prominence, continued behind into a pointed eminence, the styloid process, which projects upwards from the posterior part of the head. The prominence gives attachment to the tendon of the Biceps muscle, and to the long external lateral ligament of the knee, the ligament dividing the tendon into two parts. The summit of the styloid process gives attachment to the short external lateral ligament. The remaining part of the circumference of the head is rough, for the attachment, in front, of the anterior superior tibio-fibular ligament, and the upper and anterior part of the Peroneus longus; and behind, to the posterior superior tibio-fibular ligament, and the upper fibres of the outer head of the Soleus muscle.

The *Lower Extremity*, or *external malleolus*, is of a pyramidal form, somewhat flattened from without inwards, and is longer, and descends lower, than the internal malleolus. Its *external surface* is convex, subcutaneous, and continuous with a triangular (also subcutaneous) surface on the outer side of the shaft. The *internal surface* presents in front a smooth triangular facet, broader above than below, and convex from above downwards, which articulates with a corresponding surface on the outer side of the astragalus. Behind and beneath the articular surface is a rough depression, which gives attachment to the posterior fasciculus of the external lateral ligament of the ankle. The *anterior border* is thick and rough, and marked below by a depression for the attachment of the anterior fasciculus of the external lateral ligament. The *posterior border* is broad and marked by a shallow groove, for the passage of the tendons of the Peroneus longus and brevis muscles. The *summit* is rounded, and gives attachment to the middle fasciculus of the external lateral ligament.

The *Shaft* presents three surfaces and three borders. The *anterior border* commences above in front of the head, runs vertically downwards to a little below the middle of the bone, and then, curving a little outwards, bifurcates below. The two lines so formed bound the triangular subcutaneous surface immediately above the outer side of the external malleolus. This border gives attachment to an intermuscular septum, which separates the muscles on the anterior surface from those on the external.

The *internal border*, or *interosseous ridge*, is situated close to the inner side of the preceding, and runs nearly parallel with it in the upper third of its extent, but diverges from it so as to include a broader space in the lower two-thirds. It commences above just beneath the head of the bone (sometimes it is quite indistinct for about an inch below the head), and terminates below at the apex of a rough triangular surface immediately above the articular facet of the external malleolus. It serves for the attachment of the interosseous membrane, and separates the extensor muscles in front from the flexor muscles behind. The portion of bone included between the anterior and interosseous lines forms the anterior surface.

The *posterior border* is sharp and prominent; it commences above at the base of the styloid process, and terminates below in the posterior border of the outer malleolus. It is directed outwards above, backwards in the middle of its course, backwards and a little inwards below, and gives attachment to an aponeurosis which separates the muscles on the outer from those on the inner surface of the shaft. The portion of bone included between this line and the interosseous ridge, and which includes more than half of the whole circumference of the fibula, is known as the internal surface. Its upper three-fourths are subdivided into two parts, an anterior and a posterior, by a very prominent ridge, the *oblique line of the fibula*, which commences above at the inner side of the head, and terminates by becoming continuous with the interosseous ridge at the lower fourth of the bone. The oblique line attaches an aponeurosis which separates the Tibialis posticus from the Soleus above, and the Flexor longus pollicis below. This line sometimes ceases just before approaching the interosseous ridge.

The *anterior surface* is the interval between the anterior and interosseous lines. It is extremely narrow and flat in the upper third of its extent; broader and grooved longitudinally in its lower third; it serves for the attachment of three muscles, the Extensor longus digitorum, Peroneus tertius, and Extensor longus pollicis.

The *external surface*, much broader than the preceding, and often deeply grooved, is directed outwards in the upper two-thirds of its course, backwards in the lower third, where it is continuous with the posterior border of the external malleolus. This surface is completely occupied by the Peroneus longus and brevis muscles.

The *internal surface* is the interval between the interosseous ridge and the posterior border, and occupies nearly two-thirds of the circumference of the bone. Its upper three-fourths are divided into an anterior and a posterior portion by a very prominent ridge already mentioned, the oblique line of the fibula. The anterior portion is directed inwards, and is grooved for the attachment of the Tibialis posticus muscle. The posterior portion is continuous below with the rough triangular surface above the articular facet of the outer malleolus; it is directed backwards above, backwards and inwards at its middle, directly inwards below. Its upper fourth is rough, for the attachment of the Soleus muscle; its lower part presents a triangular rough surface, connected to the tibia by a strong interosseous ligament, and between these two points the entire surface is covered by the fibres of origin of the Flexor longus pollicis muscle. At about the middle of this surface is the nutrient foramen, which is directed downwards.

In order to distinguish the side to which the bone belongs, hold it with the lower extremity downwards, and the broad groove for the Peronei tendons backwards, towards the holder: the triangular subcutaneous surface will then be directed to the side to which the bone belongs.

112.—Plan of the Development of the Fibula. By Three Centres.

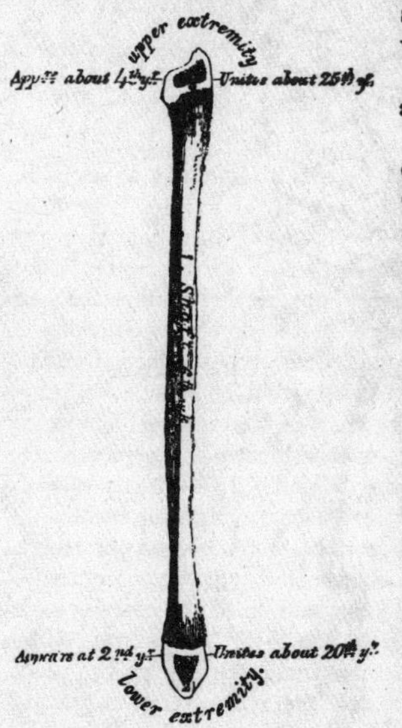

Articulations. With two bones; the tibia and astragalus.

Development. By three centres (fig. 112); one for the shaft, and one for each extremity. Ossification commences in the shaft about the sixth week of fœtal life, a little later than in the tibia, and extends gradually towards the extremities. At birth both ends are cartilaginous. Ossification commences in the lower end in the second year, and in the upper one about the fourth year. The lower epiphysis, the first in which ossification commences, becomes united to the shaft first, contrary to the law which appears to prevail with regard to the junction of the epiphyses with the shaft: this takes place about the twentieth year; the upper epiphysis is joined about the twenty-fifth year.*

Attachment of Muscles. To the head, the Biceps, Soleus, and Peroneus longus: to the shaft, its anterior surface, the Extensor longus digitorum, Peroneus tertius, and Extensor longus pollicis: to the internal surface, the Soleus, Tibialis posticus, and Flexor longus pollicis: to the external surface, the Peroneus longus and brevis.

THE FOOT. (Figs. 113, 114.)

The skeleton of the Foot consists of three divisions: the Tarsus, Metatarsus, and Phalanges.

THE TARSUS.

The bones of the Tarsus are seven in number: viz., the calcaneum, or os calcis, astragalus, cuboid, scaphoid, internal, middle, and external cuneiform bones.

* It will be observed that in the fibula, as in other long bones, the epiphysis towards which the nutrient artery is directed is the one first joined to the shaft.

113.—Bones of the Right Foot. Dorsal Surface.

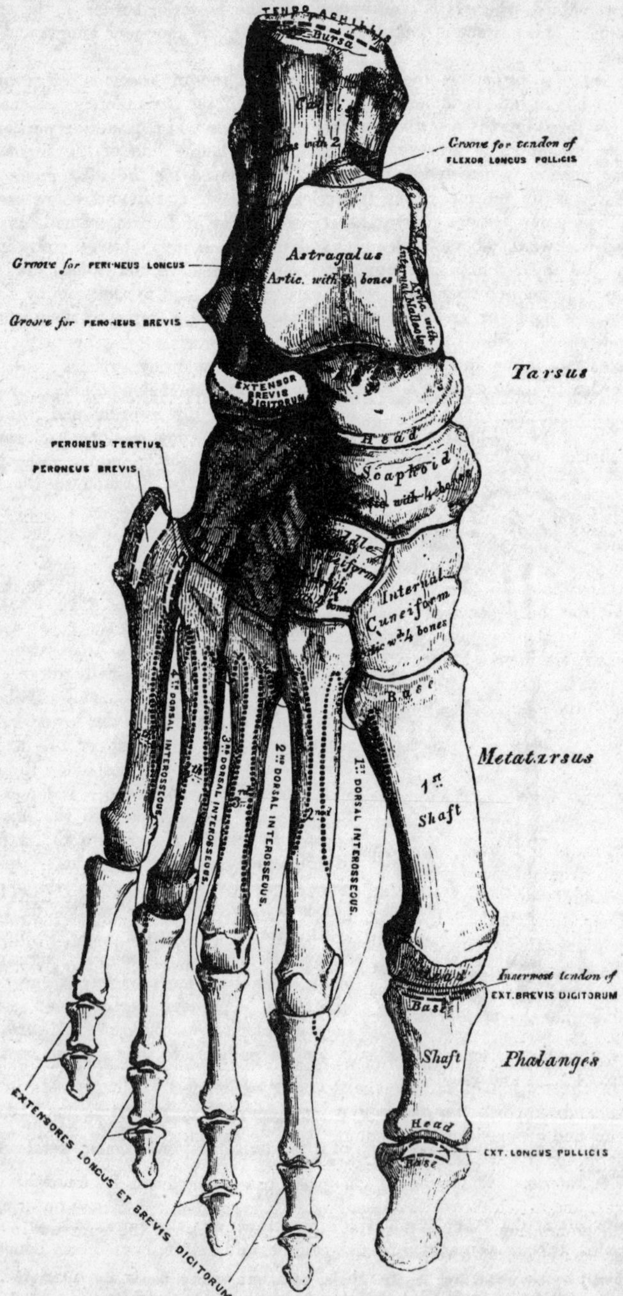

The Calcaneum.

The *Calcaneum*, or *Os Calcis*, is the largest and strongest of the tarsal bones. It is irregularly cuboidal in form, and situated at the lower and back part of the foot, serving to transmit the weight of the body to the ground, and forming a strong lever for the muscles of the calf. It presents for examination six surfaces; superior, inferior, external, internal, anterior, and posterior.

The *superior surface* is formed behind, of the upper aspect of that part of the os calcis which projects backwards to form the heel. It varies in length in different individuals; is convex from side to side, concave from before backwards, and corresponds above to a mass of adipose substance placed in front of the tendo Achillis. In the middle of the superior surface are two (sometimes three) articular facets, separated by a broad shallow groove, which is directed obliquely forwards and outwards, and is rough for the attachment of the interosseous ligament connecting the astragalus and os calcis. Of the two articular surfaces, the *external* is the larger, and situated on the body of the bone: it is of an oblong form, wider behind than in front, and convex from before backwards. The *internal articular surface* is supported on a projecting process of bone, called the *lesser process* of the calcaneum (sustentaculum tali); it is also oblong, concave longitudinally, and sometimes subdivided into two parts, which differ in size and shape. More anteriorly is seen the upper surface of the *greater process*, marked by a rough depression for the attachment of numerous ligaments, and the origin of the Extensor brevis digitorum muscle.

The *inferior surface* is narrow, rough, uneven, wider behind than in front, and convex from side to side; it is bounded posteriorly by two tubercles, separated by a rough depression; the *external*, small, prominent, and rounded, gives attachment to part of the Abductor minimi digiti; the *internal*, broader and larger, for the support of the heel, gives attachment, by its prominent inner margin, to the Abductor pollicis, and in front to the Flexor brevis digitorum muscles; the depression between the tubercles attaches the Abductor minimi digiti and plantar fascia. The rough surface in front of the tubercles gives attachment to the long plantar ligament, and to the outer head of the Flexor accessorius muscle; and to a prominent tubercle nearer the anterior part of the bone, as well as to a transverse groove in front of it, is attached the short plantar ligament.

The *external surface* is broad, flat, and almost subcutaneous; it presents near its centre a tubercle, for the attachment of the middle fasciculus of the external lateral ligament. Above the tubercle is a broad smooth surface, giving attachment, at its upper and anterior part, to the external astragalo-calcanean ligament; and in front of the tubercle a narrow surface marked by two oblique grooves, separated by an elevated ridge: the *superior groove* transmits the tendon of the Peroneus brevis; the *inferior*, the tendon of the Peroneus longus; the intervening ridge gives attachment to a prolongation from the external annular ligament.

The *internal surface* presents a deep concavity, directed obliquely downwards and forwards, for the transmission of the plantar vessels and nerves and Flexor tendons into the sole of the foot; it affords attachment to part of the Flexor accessorius muscle. This surface presents an eminence of bone, the *lesser process*, which projects horizontally inwards from its upper and fore part, and to which a slip of the tendon of the Tibialis posticus is attached. This process is concave above, and supports the anterior articular surface of the astragalus; below, it is convex, and grooved for the tendon of the Flexor longus pollicis. Its free margin is rough, for the attachment of ligaments.

The *anterior surface*, of a somewhat triangular form, is smooth, concavo-convex, and articulates with the cuboid. It is surmounted, on its outer side, by a rough prominence, which forms an important guide to the surgeon in the performance of Chopart's amputation.

The *posterior surface* is rough, prominent, convex, and wider below than above. Its lower part is rough, for the attachment of the tendo Achillis, and of the Plantaris muscle; its upper part is smooth, coated with cartilage, and corresponds to a bursa which separates that tendon from the bone.

Articulations. With two bones: the astragalus and cuboid.

Attachment of Muscles. Part of the Tibialis posticus, the tendo Achillis, Plantaris, Abductor pollicis, Abductor minimi digiti, Flexor brevis digitorum, Flexor accessorius, and Extensor brevis digitorum.

The Cuboid.

The *Cuboid* bone is placed on the outer side of the foot, in front of the os calcis, and behind the fourth and fifth metatarsal bones. It is of a pyramidal shape, its base being directed upwards and inwards, its apex downwards and outwards. It may be distinguished from the other tarsal bones by the existence of a deep groove on its under surface, for the tendon of the Peroneus longus muscle. It presents for examination six surfaces; three articular, and three non-articular.

The non-articular surfaces are the superior, inferior, and external. The *superior* or *dorsal surface*, directed upwards and outwards, is rough, for the attachment of numerous ligaments. The *inferior* or *plantar surface* presents in front a deep groove, which runs obliquely from without, forwards and inwards; it lodges the tendon of the Peroneus longus, and is bounded behind by a prominent ridge, terminating externally in an eminence, the tuberosity of the cuboid, the surface of which presents a convex facet, for articulation with the sesamoid bone of the tendon contained in the groove. The ridge and surface of bone behind it are rough, for the attachment of the long and short plantar ligaments. A few fibres of the Flexor brevis pollicis may be traced to this surface. The *external surface*, the smallest and narrowest of the three, presents a deep notch formed by the commencement of the peroneal groove.

The articular surfaces are the posterior, anterior, and internal. The *posterior surface* is smooth, triangular, concavo-convex, for articulation with the anterior surface of the os calcis. The *anterior*, of smaller size, but also irregularly triangular, is divided by a vertical ridge into two facets: the inner facet, quadrilateral in form, articulates with the fourth metatarsal bone; the outer one, larger and more triangular, articulates with the fifth metatarsal. The *internal surface* is broad, rough, irregularly quadrilateral, presenting at its middle and upper part a small oval facet, for articulation with the external cuneiform bone; and behind this (occasionally) a smaller facet, for articulation with the scaphoid; it is rough in the rest of its extent, for the attachment of strong interosseous ligaments.

To ascertain to which foot it belongs, hold the bone so that its under surface, marked by the peroneal groove, looks downwards, and the large concavo-convex articular surface backwards, towards the holder: the narrow non-articular surface, marked by the commencement of the peroneal groove, will point to the side to which the bone belongs.

Articulations. With four bones: the os calcis, external cuneiform, and the fourth and fifth metatarsal bones, occasionally with the scaphoid.

Attachment of Muscles. Part of the Flexor brevis pollicis.

The Astragalus.

The *Astragalus* (fig. 113) is the largest of the tarsal bones, next to the os calcis. It occupies the middle and upper part of the tarsus, supporting the tibia above, articulating with the malleoli on either side, resting below upon the os calcis, and joined in front to the scaphoid. This bone may easily be recognised by its large rounded head, by the broad articular facet on its upper convex surface, or by the two articular facets separated by a deep groove on its under concave surface. It presents six surfaces for examination.

The *superior surface* presents, behind, a broad smooth trochlear surface, for articulation with the tibia. The trochlea is broader in front than behind, convex from before backwards, slightly concave from side to side; in front of it is the

114.—Bones of the Right Foot. Plantar Surface.

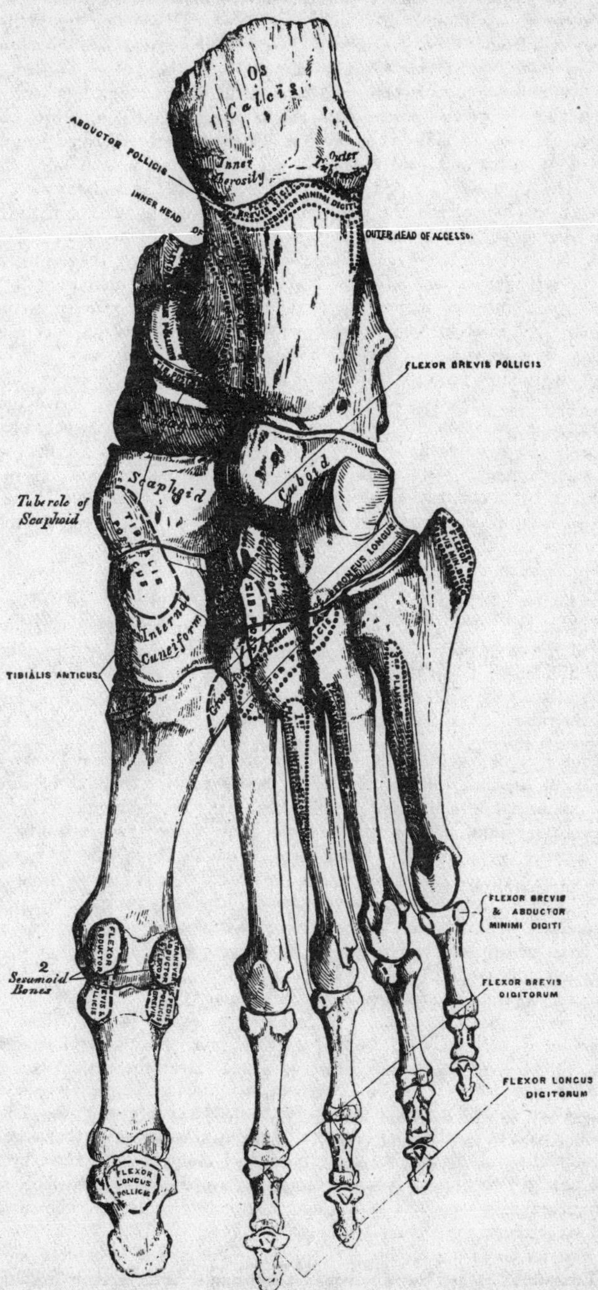

upper surface of the neck of the astragalus, rough for the attachment of ligaments. The *inferior surface* presents two articular facets separated by a deep groove. The groove runs obliquely forwards and outwards, becoming gradually broader and deeper in front: it corresponds with a similar groove upon the upper surface of the os calcis, and forms, when articulated with that bone, a canal, filled up in the recent state by the calcaneo-astragaloid interosseous ligament. Of the two articular facets, the posterior is the larger, of an oblong form, and deeply concave from side to side; the anterior, although nearly of equal length, is narrower, of an elongated oval form, convex longitudinally, and often subdivided into two by an elevated ridge; of these the posterior articulates with the lesser process of the os calcis; the anterior, with the upper surface of the calcaneo-scaphoid ligament. The *internal surface* presents at its upper part a pear-shaped articular facet for the inner malleolus, continuous above with the trochlear surface; below the articular surface is a rough depression, for the attachment of the deep portion of the internal lateral ligament. The *external surface* presents a large triangular facet, concave from above downwards, for articulation with the external malleolus; it is continuous above with the trochlear surface; and in front of it is a rough depression for the attachment of the anterior fasciculus of the external lateral ligament. The *anterior surface*, convex and rounded, forms the head of the astragalus; it is smooth, of an oval form, and directed obliquely inwards and downwards; it is continuous below with that part of the anterior facet on the under surface which rests upon the calcaneo-scaphoid ligament. The head is surrounded by a constricted portion, the neck of the astragalus. The *posterior surface* is narrow, and traversed by a groove, which runs obliquely downwards and inwards, and transmits the tendon of the Flexor longus pollicis.

To ascertain to which foot it belongs, hold the bone with the broad articular surface upwards, and the rounded head forwards; the lateral triangular articular surface for the external malleolus will then point to the side to which the bone belongs.

Articulations. With four bones: tibia, fibula, os calcis, and scaphoid.

The Scaphoid.

The *Scaphoid* or *Navicular* bone, so called from its fancied resemblance to a boat, is situated at the inner side of the tarsus, between the astragalus behind and the three cuneiform bones in front. This bone may be distinguished by its form, being concave behind, convex and subdivided into three facets in front.

The *anterior surface*, of an oblong form, is convex from side to side, and subdivided by two ridges into three facets, for articulation with the three cuneiform bones. The *posterior surface* is oval, concave, broader externally than internally, and articulates with the rounded head of the astragalus. The *superior surface* is convex from side to side, and rough for the attachment of ligaments. The *inferior* is somewhat concave, irregular, and also rough for the attachment of ligaments. The *internal surface* presents a rounded tubercular eminence, the tuberosity of the scaphoid, which gives attachment to part of the tendon of the Tibialis posticus. The *external surface* is broad, rough, and irregular, for the attachment of ligamentous fibres, and occasionally presents a small facet for articulation with the cuboid bone.

To ascertain to which foot it belongs, hold the bone with the concave articular surface backwards, and the convex dorsal surface upwards; the broad external surface will point to the side to which the bone belongs.

Articulations. With four bones: astragalus and three cuneiform; occasionally also with the cuboid.

Attachment of Muscles. Part of the Tibialis posticus.

The Cuneiform Bones have received their name from their wedge-like shape. They form with the cuboid the most anterior row of the tarsus, being placed between the scaphoid behind, the three innermost metatarsal bones in front, and the cuboid externally. They are called the *first*, *second*, and *third*, counting from the inner to the outer side of the foot, and, from their position, *internal*, *middle*, and *external*.

The Internal Cuneiform.

The *Internal Cuneiform* is the largest of the three. It is situated at the inner side of the foot, between the scaphoid behind and the base of the first metatarsal in front. It may be distinguished from the other two by its large size, and its more irregular wedge-like form. Without the others, it may be known by the large kidney-shaped anterior articulating surface, and by the prominence on the non-articular surface (or base of the wedge), for the attachment of a large tendon. It presents for examination six surfaces.

The *internal surface* is subcutaneous, and forms part of the inner border of the foot; it is broad, quadrilateral, and presents at its anterior inferior angle a smooth oval facet, over which the tendon of the Tibialis anticus muscle glides; in the rest of its extent it is rough, for the attachment of ligaments. The *external surface* is concave, presenting, along its superior and posterior borders, a narrow surface for articulation with the middle cuneiform behind, and second metatarsal bone in front: in the rest of its extent, it is rough for the attachment of ligaments, and prominent below, where it forms part of the tuberosity. The *anterior surface*, kidney-shaped, articulates with the metatarsal bone of the great toe. The *posterior surface* is triangular, concave, and articulates with the innermost and largest of the three facets on the anterior surface of the scaphoid. The *inferior* or *plantar surface* is rough, and presents a prominent tuberosity at its back part for the attachment of part of the tendon of the Tibialis posticus. It also gives attachment in front to part of the tendon of the Tibialis anticus. The *superior surface* is the narrow pointed end of the wedge, which is directed upwards and outwards; it is rough for the attachment of ligaments.

To ascertain to which side it belongs, hold the bone so that its superior narrow edge looks upwards, and the long articular surface forwards; the external surface, marked by its vertical and horizontal articular facets, will point to the side to which it belongs.

Articulations. With four bones: scaphoid, middle cuneiform, first and second metatarsal bones.

Attachment of Muscles. The Tibialis anticus and posticus.

The Middle Cuneiform.

The *Middle Cuneiform*, the smallest of the three, is of very regular wedge-like form, the broad extremity being placed upwards, the narrow end downwards. It is situated between the other two bones of the same name, and corresponds to the scaphoid behind, and the second metatarsal in front. It may be distinguished from the external cuneiform bone, which it much resembles in general appearance, by the articular facet, of angular form, which runs round the upper and back part of its inner surface.

The *anterior surface*, triangular in form, and narrower than the posterior, articulates with the base of the second metatarsal bone. The *posterior surface*, also triangular, articulates with the scaphoid. The *internal surface* presents an articular facet, running along the superior and posterior borders, for articulation with the internal cuneiform, and is rough below for the attachment of ligaments. The *external surface* presents posteriorly a smooth facet for articulation with the external cuneiform bone. The *superior surface* forms the base of the wedge; it is quadrilateral, broader behind than in front, and rough for the attachment of ligaments. The *inferior surface*, pointed and tubercular, is also rough for ligamentous attachment.

To ascertain to which foot the bone belongs, hold its superior or dorsal surface upwards, the broadest edge being towards the holder: the smooth facet (limited to the posterior border) will then point to the side to which it belongs.

Articulations. With four bones: scaphoid, internal and external cuneiform, and second metatarsal bone.

The External Cuneiform.

The *External Cuneiform*, intermediate in size between the two preceding, is of a very regular wedge-like form, the broad extremity being placed upwards, the narrow end downwards. It occupies the centre of the front row of the tarsus between the middle cuneiform internally, the cuboid externally, the scaphoid behind, and the third metatarsal in front. It is distinguished from the internal cuneiform bone by its more regular wedge-like shape, and by the absence of the kidney-shaped articular surface: from the middle cuneiform, by the absence of the bent, or angular, facet, and by the two articular facets which mark both its inner and outer surfaces. It has six surfaces for examination.

The *anterior surface*, triangular in form, articulates with the third metatarsal bone. The *posterior surface* articulates with the most external facet of the scaphoid, and is rough below for the attachment of ligamentous fibres. The *internal surface* presents two articular facets separated by a rough depression; the anterior one, situated at the superior angle of the bone, articulates with the outer side of the base of the second metatarsal bone; the posterior one skirts the posterior border, and articulates with the middle cuneiform; the rough depression between the two gives attachment to an interosseous ligament. The *external surface* also presents two articular facets, separated by a rough non-articular surface; the anterior facet, situated at the superior angle of the bone, is small, and articulates with the inner side of the base of the fourth metatarsal; the posterior, and larger one, articulates with the cuboid; the rough non-articular surface serves for the attachment of an interosseous ligament. The three facets for articulation with the three metatarsal bones are continuous with one another, and covered by a prolongation of the same cartilage; the facets for articulation with the middle cuneiform and scaphoid are also continuous, but that for articulation with the cuboid is usually separate. The *superior* or *dorsal surface*, of an oblong form, is rough for the attachment of ligaments. The *inferior* or *plantar surface* is an obtuse rounded margin, and serves for the attachment of part of the tendon of the Tibialis posticus, part of the Flexor brevis pollicis, and ligaments.

To ascertain to which side it belongs, hold the bone with the broad dorsal surface upwards, the prolonged edge backwards; the separate articular facet for the cuboid will point to the proper side.

Articulations. With six bones: the scaphoid, middle cuneiform, cuboid, and second, third, and fourth metatarsal bones.

Attachment of Muscles. Part of Tibialis posticus, and Flexor brevis pollicis.

The Metatarsal Bones.

The Metatarsal bones are five in number; they are long bones, and subdivided into a shaft and two extremities.

Common characters. The *Shaft* is prismoid in form, tapers gradually from the tarsal to the phalangeal extremity, and is slightly curved longitudinally, so as to be concave below, slightly convex above. The *Posterior Extremity*, or *Base*, is wedge-shaped, articulating by its terminal surface with the tarsal bones, and by its lateral surfaces with the contiguous bones; its dorsal and plantar surfaces being rough for the attachment of ligaments. The *Anterior Extremity*, or *Head*, presents a terminal rounded articular surface, oblong from above downwards, and extending further backwards below than above. Its sides are flattened, and present a depression, surmounted by a tubercle, for ligamentous attachment. Its under surface is grooved in the middle line, for the passage of the Flexor tendon, and marked on each side by an articular eminence continuous with the terminal articular surface.

Peculiar characters. The *First* is remarkable for its great size, but is the shortest of all the metatarsal bones. The *shaft* is strong, and of well-marked prismoid form. The *posterior extremity* presents no lateral articular facets; its terminal articular surface is of large size, of semilunar form, and its circumference grooved for the tarso-metatarsal ligaments; its inferior angle presents a rough oval prominence for the insertion of the tendon of the Peroneus longus. The *head* is of large size; on its plantar surface are two grooved facets, over which glide sesamoid bones; the facets are separated by a smooth elevated ridge.

The *Second* is the longest and largest of the remaining metatarsal bones, being prolonged backwards into the recess formed between the three cuneiform bones. Its *tarsal extremity* is broad above, narrow and rough below. It presents four articular surfaces: one behind, of a triangular form, for articulation with the middle cuneiform; one at the upper part of its internal lateral surface, for articulation with the internal cuneiform; and two on its external lateral surface, a superior and an inferior, separated by a rough depression. Each of the latter articular surfaces is divided by a vertical ridge into two parts; the anterior segment of each facet articulates with the third metatarsal; the two posterior (sometimes continuous) with the external cuneiform.

The *Third* articulates behind, by means of a triangular smooth surface, with the external cuneiform; on its inner side, by two facets, with the second metatarsal; and on its outer side, by a single facet, with the third metatarsal. The latter facet is of circular form, and situated at the upper angle of the base.

The *Fourth* is smaller in size than the preceding; its *tarsal extremity* presents a terminal quadrilateral surface, for articulation with the cuboid; a smooth facet on the inner side, divided by a ridge into an anterior portion for articulation with the third metatarsal, and a posterior portion for articulation with the external cuneiform; on the outer side a single facet, for articulation with the fifth metatarsal.

The *Fifth* is recognised by the tubercular eminence on the outer side of its base. It articulates behind, by a triangular surface cut obliquely from without inwards, with the cuboid; and internally, with the fourth metatarsal.

Articulations. Each bone articulates with the tarsal bones by one extremity, and by the other with the first row of phalanges. The number of tarsal bones with which each metatarsal articulates, is one for the first, three for the second, one for the third, two for the fourth, and one for the fifth.

Attachment of Muscles. To the first metatarsal bone, three: part of the Tibialis anticus, the Peroneus longus, and First dorsal interosseous. To the second, three: the Adductor pollicis, and First and Second dorsal interosseous. To the third, four: the Adductor pollicis, Second and Third dorsal, and First plantar interosseous. To the fourth, four: the Adductor pollicis, Third and Fourth dorsal, and Second plantar interosseous. To the fifth, five: the Peroneus brevis, Peroneus tertius, Flexor brevis minimi digiti, Fourth dorsal, and Third plantar interosseous.

Phalanges.

The *Phalanges* of the foot, both in number and general arrangement, resemble those in the hand; there being two in the great toe, and three in each of the other toes.

The phalanges of the *first row* resemble closely those of the hand. The *shaft* is compressed from side to side, convex above, concave below. The *posterior extremity* is concave; and the *anterior extremity* presents a trochlear surface, for articulation with the second phalanges.

The phalanges of the *second row* are remarkably small and short, but rather broader than those of the first row.

The *unqual* phalanges, in form, resemble those of the fingers; but they are smaller, flattened from above downwards, presenting a broad base for articulation with the second row, and an expanded extremity for the support of the nail and end of the toe.

Articulations. The first row, with the metatarsal bones, and second phalanges;

the second of the great toe, with the first phalanx, and of the other toes, with the first and third phalanges; the third with the second row.

Attachment of Muscles. To the first phalanges, great toe: innermost tendon of Extensor brevis digitorum, Abductor pollicis, Adductor pollicis, Flexor brevis pollicis, Transversus pedis. Second toe: First and Second dorsal interosseous. Third toe: Third dorsal and First plantar interosseous. Fourth toe: Fourth dorsal and Second plantar interosseous. Fifth toe: Flexor brevis minimi digiti, Abductor minimi digiti, and Third plantar interosseous. Second phalanges, great toe: Extensor longus pollicis, Flexor longus pollicis. Other toes: Flexor brevis digitorum, one slip from the Extensor brevis digitorum (except in the little toe), and Extensor longus digitorum.—Third phalanges: two slips from the common tendon of the Extensor longus and Extensor brevis digitorum, and the Flexor longus digitorum.

Development of the Foot. (Fig. 115.)

The Tarsal bones are each developed by a single centre, excepting the os calcis, which has an epiphysis for its posterior extremity. The centres make their appearance in the following order: in the os calcis, at the sixth month of fœtal life; in

115.—Plan of the Development of the Foot.

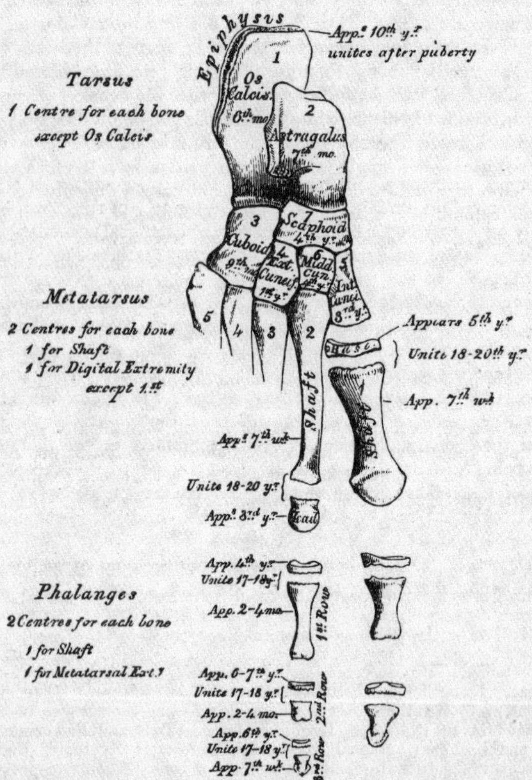

the astragalus, about the seventh month; in the cuboid, at the ninth month; external cuneiform, during the first year; internal cuneiform in the third year; middle cuneiform and scaphoid in the fourth year. The epiphysis for the posterior tuberosity of the os calcis appears at the tenth year, and unites with the rest of the bone soon after puberty.

The Metatarsal bones are each developed by *two* centres: one for the shaft, and one for the digital extremity, in the four outer metatarsal; one for the shaft, and one for the base, in the metatarsal bone of the great toe. Ossification commences in the centre of the shaft about the seventh week, and extends towards either extremity, and in the digital epiphyses about the third year; they become joined between the eighteenth and twentieth years.

The Phalanges are developed by *two* centres for each bone: one for the shaft, and one for the metatarsal extremity.

Sesamoid Bones.

These are small rounded masses, cartilaginous in early life, osseous in the adult, which are developed in those tendons which exert a great amount of pressure upon the parts over which they glide. It is said that they are more commonly found in the male than in the female, and in persons of an active muscular habit than in those who are weak and debilitated. They are invested throughout their whole surface by the fibrous tissue of the tendon in which they are found, excepting upon that side which lies in contact with the part over which they play, where they present a free articular facet. They may be divided into two kinds: those which glide over the articular surfaces of joints, and those which play over the cartilaginous facets found on the surfaces of certain bones.

The sesamoid bones of the joints are, in the lower extremity, the patella, which is developed in the tendon of the Quadriceps extensor; two small sesamoid bones, found in the tendons of the Flexor brevis pollicis, opposite the metatarso-phalangeal joint of the great toe, and occasionally one in the metatarso-phalangeal joint of the second toe, the little toe, and, still more rarely, the third and fourth toes.

In the upper extremity, there are two on the palmar surface of the metacarpo-phalangeal joint in the thumb, developed in the tendons of the Flexor brevis pollicis, occasionally one or two opposite the metacarpo-phalangeal articulations of the fore and little fingers, and, still more rarely, one opposite the same joints of the third and fourth fingers.

Those found in the tendons which glide over certain bones, occupy the following situations: one in the tendon of the Peroneus longus, where it glides through the groove in the cuboid bone; one appears late in life in the tendon of the Tibialis anticus, opposite the smooth facet on the internal cuneiform bone; one is found in the tendon of the Tibialis posticus, opposite the inner side of the astragalus; one in the outer head of the Gastrocnemius, behind the outer condyle of the femur; and one in the Psoas and Iliacus, where they glide over the body of the pubes. Sesamoid bones are found occasionally in the tendon of the Biceps, opposite the tuberosity of the radius; in the tendon of the Gluteus maximus, as it passes over the great trochanter; and in the tendons which wind round the inner and outer malleoli.

The author has to acknowledge valuable aid derived from the perusal of the works of Cloquet, Cruveilhier, Bourgery, and Boyer, especially of the latter. Reference has also been made to the following:—'Outlines of Human Osteology,' by F. O. Ward. 'A Treatise on the Human Skeleton, and Observations on the Limbs of Vertebrate Animals,' by G. M. Humphry. Holden's 'Human Osteology.' Henle's 'Handbuch der systematischen Anatomie des Menschen. Erster Band. Erste Abtheilung. Knochenlehre.' 'Osteological Memoirs (The Clavicle),' by Struthers. 'On the Archetype and Homologies of the Vertebrate Skeleton,' and 'On the Nature of Limbs,' by Owen.—Todd and Bowman's 'Physiological Anatomy,' and Kölliker's 'Manual of Human Microscopic Anatomy,' contain the most complete account of the structure and development of bone.—The development of the bones is minutely described in 'Quain's Anatomy,' edited by Sharpey and Ellis.—On the chemical analysis of bone, refer to 'Lehmann's Physiological Chemistry,' translated by Day; vol. iii. p. 12. 'Simon's Chemistry,' translated by Day; vol. ii. p. 396. A paper by Dr. Stark, 'On the Chemical Constitution of the Bones of the Vertebrated Animals' (Edinburgh Medical and Surgical Journal; vol. liii. p. 308); and Dr. Owen Rees' paper in the 21st vol. of the Medico-chirurgical Transactions.

The Articulations.

THE various bones of which the Skeleton consists are connected together at different parts of their surfaces, and such a connection is designated by the name of *Joint* or *Articulation.* If the joint is *immoveable*, as between the cranial and most of the facial bones, their adjacent margins are applied in almost close contact, a thin layer of fibrous membrane, the *sutural ligament*, and, at the base of the skull, in certain situations, a thin layer of cartilage being interposed. Where slight movement is required, combined with great strength, the osseous surfaces are united by tough and elastic fibro-cartilages, as in the joints of the spine, the sacro-iliac, and interpubic articulations; but in the *moveable* joints, the bones forming the articulation are generally expanded for greater convenience of mutual connection, covered by *cartilage*, held together by strong bands or capsules of fibrous tissue, called *ligaments*, and lined by a membrane, the *synovial membrane*, which secretes a fluid to lubricate the various parts of which the joint is formed: so that the structures which enter into the formation of a joint are bone, cartilage, fibro-cartilage, ligament, and synovial membrane.

Bone constitutes the fundamental element of all the joints. In the long bones, the extremities are the parts which form the articulations; they are generally somewhat enlarged, consisting of spongy cancellous tissue, with a thin coating of compact substance. In the flat bones, the articulations usually take place at the edges; and, in the short bones, at various parts of their surface. The layer of compact bone which forms the articular surface, and to which the cartilage is attached, is called the *articular lamella.* It is of a white colour, extremely dense, and varies in thickness. Its structure differs from ordinary bone tissue in this respect, that it contains no Haversian canals, and its lacunæ are much larger than in ordinary bone, and have no canaliculi. The vessels of the cancellous tissue, as they approach the articular lamella, turn back in loops, and do not perforate it; this layer is consequently more dense and firmer than ordinary bone, and is evidently designed to form a firm and unyielding support for the articular cartilage.

The articular will be found described along with the other kinds of cartilage in the Introduction.

Ligaments are found in nearly all the moveable articulations; they consist of bands of various forms, serving to connect together the articular extremities of bones, and composed mainly of bundles of *white fibrous tissue* placed parallel with, or closely interlaced with, one another, and presenting a white, shining silvery aspect. Ligament is pliant and flexible, so as to allow of the most perfect freedom of movement, but strong, tough, and inextensile, so as not readily to yield under the most severely applied force; it is consequently well adapted to serve as the connecting medium between the bones. Some ligaments consist entirely of *yellow elastic tissue*, as the ligamenta subflava, which connect together the adjacent arches of the vertebræ, and the ligamentum nuchæ. In these cases, it will be observed that the elasticity of the ligament is intended to act as a substitute for muscular power.

Synovial Membrane is a thin, delicate membrane, arranged in the form of a short wide tube, attached by its open ends to the margins of the articular extremities of the bones, and covering the inner surface of the various ligaments which connect the articulating surfaces. It resembles the serous membranes in structure, but differs in the nature of its secretion, which is thick, viscid, and glairy, like the white of egg; and hence termed *synovia.* The synovial membranes found in the body admit of subdivision into three kinds, articular, bursal, and vaginal.

The *articular synovial membranes* are found in all the freely moveable joints. In the fœtus, this membrane is said, by Toynbee, to be continued over the surface of the cartilages; but in the adult it is wanting, excepting at their circumference, upon which it encroaches for a short distance: it then invests the inner surface of the capsular or other ligaments enclosing the joint, and is reflected over the surface of any tendons passing through its cavity, as the tendon of the Popliteus in the knee, and the tendon of the Biceps in the shoulder. In most of the joints, the synovial membrane is thrown into folds, which project into the cavity. Some of these folds contain large masses of fat. These are especially distinct in the hip and the knee. Others are flattened folds, subdivided at their margins into fringe-like processes, the vessels of which have a convoluted arrangement. The latter generally project from the synovial membrane near the margin of the cartilage, and lie flat upon its surface. They consist of connective tissue, covered with epithelium, and contain fat cells in variable quantity, and, more rarely, isolated cartilage cells. They are found in most of the bursal and vaginal, as well as in the articular synovial membranes, and were described, by Clopton Havers, as mucilaginous glands, and as the source of the synovial secretion. Under certain diseased conditions, similar processes are found covering the entire surface of the synovial membrane, forming a mass of pedunculated fibro-fatty growths, which project into the joint.

The *bursæ* are found interposed between surfaces which move upon each other, producing friction, as in the gliding of a tendon, or of the integument over projecting bony surfaces. They admit of subdivision into two kinds, the *bursæ mucosæ*, and the *synovial bursæ*. The former are large, simple, or irregular cavities in the subcutaneous areolar tissue, enclosing a clear viscid fluid. They are found in various situations, as between the integument and front of the patella, over the olecranon, the malleoli, and other prominent parts. The *synovial bursæ* are found interposed between muscles or tendons as they play over projecting bony surfaces, as between the Glutei muscles and surface of the great trochanter. They consist of a thin wall of connective tissue, partially covered by epithelium and contain a viscid fluid. Where one of these exists in the neighbourhood of a joint, it usually communicates with its cavity, as is generally the case with the bursa between the tendon of the Psoas and Iliacus, and the capsular ligament of the hip, or the one interposed between the under surface of the Subscapularis and the neck of the scapula.

The *vaginal synovial membranes* (synovial sheaths) serve to facilitate the gliding of tendons in the osseo-fibrous canals through which they pass. The membrane is here arranged in the form of a sheath, one layer of which adheres to the wall of the canal, and the other is reflected upon the outer surface of the contained tendon; the space between the two free surfaces of the membrane being partially filled with synovia. These sheaths are chiefly found surrounding the tendons of the Flexor and Extensor muscles of the fingers and toes, as they pass through the osseo-fibrous canals in the hand or foot.

Synovia is a transparent, yellowish-white, or slightly reddish fluid, viscid like the white of egg, having an alkaline reaction, and slightly saline taste. It consists, according to Frerichs, in the ox, of 94·85 water, 0·56 mucus and epithelium, 0·07 fat, 3·51 albumen and extractive matter, and 0·99 salts.

The Articulations are divided into three classes: *Synarthrosis*, or immoveable; *Amphiarthrosis*, or mixed; and *Diarthrosis*, or moveable joints.

I. Synarthrosis. Immoveable Articulations.

Synarthrosis includes all those articulations in which the surfaces of the bones are in almost direct contact, not separated by an intervening synovial cavity, and immoveably connected with each other, as the joints between the bones of the cranium and face, excepting those of the lower jaw. The varieties of synarthrosis are three in number: Sutura, Schindylesis, and Gomphosis.

Sutura (a seam). Where the articulating surfaces are connected by a series of processes and indentations interlocked together, it is termed *sutura vera*; of which there are three varieties: sutura dentata, serrata and limbosa. The surfaces of the bones are not in direct contact, being separated by a layer of membrane continuous externally with the pericranium, internally with the dura mater. The *sutura dentata* (*dens*, a tooth) is so called from the tooth-like form of the projecting articular processes, as in the suture between the parietal bones. In the *sutura serrata* (*serra*, a saw), the edges of the two bones forming the articulation are serrated like the teeth of a fine saw, as between the two portions of the frontal bone. In the *sutura limbosa* (*limbus*, a selvage), besides the dentated processes, there is a certain degree of bevelling of the articular surfaces, so that the bones overlap one another, as in the suture between the parietal and frontal bones. When the articulation is formed by roughened surfaces placed in apposition with one another, it is termed the *false suture, sutura notha*, of which there are two kinds: the *sutura squamosa* (*squama*, a scale), formed by the overlapping of two contiguous bones by broad bevelled margins, as in the temporo-parietal (squamous) suture; and the *sutura harmonia* (*ἁρμονία, a joining together*,) where there is simple apposition of two contiguous rough bony surfaces, as in the articulation between the two superior maxillary bones, or of the horizontal plates of the palate.

Schindylesis (*σχινδύλησις, a fissure*) is that form of articulation in which a thin plate of bone is received into a cleft or fissure formed by the separation of two laminæ of another, as in the articulation of the rostrum of the sphenoid, and perpendicular plate of the ethmoid with the vomer, or in the reception of the latter in the fissure between the superior maxillary and palate bones.

Gomphosis (*γόμφος, a nail*) is an articulation formed by the insertion of a conical process into a socket, as a nail is driven into a board; this is not illustrated by any articulations between bones, properly so called, but is seen in the articulation of the teeth with the alveoli of the maxillary bones.

2. Amphiarthrosis. Mixed Articulations.

In this form of articulation, the contiguous osseous surfaces are connected together by broad flattened discs of fibro-cartilage, which adhere to the ends of both bones, as in the articulation between the bodies of the vertebræ, and first two pieces of the sternum; or the articulating surfaces are covered with fibro-cartilage, partially lined by synovial membrane, and connected together by external ligaments, as in the sacro-iliac and pubic symphyses; both these forms being capable of limited motion in every direction. The former resemble the synarthrodial joints in the continuity of their surfaces, and absence of synovial sac; the latter, the diarthrodial. These joints occasionally become obliterated in old age; as is frequently the case in the pubic articulation, and occasionally in the intervertebral and sacro-iliac.

3. Diarthrosis. Moveable Articulations.

This form of articulation includes the greater number of the joints in the body, mobility being their distinguishing character. They are formed by the approximation of two contiguous bony surfaces, covered with cartilage, connected by ligaments, and lined by synovial membrane. The varieties of joints in this class have been determined by the kind of motion permitted in each; they are four in number: Arthrodia, Enarthrosis, Ginglymus, Diarthrosis rotatorius.

Arthrodia is that form of joint which admits of a gliding movement; it is formed by the approximation of plane surfaces, or one slightly concave, the other slightly convex; the amount of motion between them being limited by the ligaments, or osseous processes, surrounding the articulation; as in the articular processes of the vertebræ, temporo-maxillary, sterno- and acromio-clavicular, inferior radio-ulnar, carpal, carpo-metacarpal, superior tibio-fibular, tarsal, and tarso-metatarsal articulations.

Enarthrosis is that form of joint which is capable of motion in all directions. It is formed by the reception of a globular head into a deep cup-like cavity (hence the name 'ball and socket'), the parts being kept in apposition by a capsular ligament strengthened by accessory ligamentous bands. Examples of this form of articulation are found in the hip and shoulder.

Ginglymus, Hinge-joint (γιγγλυμὸς, *a hinge*). In this form of joint, the articular surfaces are moulded to each other in such a manner, as to permit motion only in two directions, forwards and backwards, the extent of motion at the same time being considerable. The articular surfaces are connected together by strong lateral ligaments, which form their chief bond of union. The most perfect forms of ginglymus are the elbow and ankle; the knee is less perfect, as it allows a slight degree of rotation in certain positions of the limb: there are also the metatarso-phalangeal and phalangeal joints in the lower extremity, and the metacarpo-phalangeal and phalangeal joints in the upper extremity.

Diarthrosis rotatorius (Lateral Ginglymus). Where the movement is limited to rotation, the joint is formed by a pivot-like process turning within a ring, or the ring on the pivot, the ring being formed partly of bone, partly of ligament. In the articulation of the odontoid process of the axis with the atlas, the ring is formed in front by the anterior arch of the atlas; behind, by the transverse ligament; here the ring rotates round the odontoid process. In the superior radio-ulnar articulation, the ring is formed partly by the lesser sigmoid cavity of the ulna; in the rest of its extent, by the orbicular ligament; here, the head of the radius rotates within the ring.

Subjoined, in a tabular form, are the names, distinctive characters, and examples of the different kinds of articulations.

- *Synarthrosis*, or immoveable joint. Surfaces separated by fibrous membrane, without any intervening synovial cavity, and immoveably connected with each other. As in joints of cranium and face (except lower jaw).
  - *Sutura*. Articulation by processes and indentations interlocked together.
    - *Sutura vera* (true) articulate by indented borders.
      - *Dentata*, having tooth-like processes. As in interparietal suture.
      - *Serrata*, having serrated edges, like the teeth of a saw. As in interfrontal suture.
      - *Limbosa*, having bevelled margins, and dentated processes. As in fronto-parietal suture.
    - *Sutura notha* (false) articulate by rough surfaces.
      - *Squamosa*, formed by thin bevelled margins overlapping each other. As in Squamo-parietal suture.
      - *Harmonia*, formed by the apposition of contiguous rough surfaces. As in intermaxillary suture.
  - *Schindylesis*. Articulation formed by the reception of a thin plate of bone into a fissure of another. As in articulation of rostrum of sphenoid with vomer.
  - *Gomphosis*. Articulation formed by the insertion of a conical process into a socket. The teeth.

Amphiarthrosis, Mixed Articulation.	1. Surfaces connected by fibro-cartilage, not separated by synovial membrane, and having limited motion. As in joints between bodies of vertebræ. 2. Surfaces covered by fibro-cartilage; lined by a partial synovial membrane. As in sacro-iliac and pubic symphyses
Diarthrosis, Moveable Joint.	*Arthrodia.* Gliding joint; articulations by plane surfaces, which glide upon each other. As in sterno- and acromio-clavicular articulations. *Enarthrosis.* Ball and socket joint; capable of motion in all directions. Articulations by a globular head received into a cup-like cavity. As in hip and shoulder joints. *Ginglymus.* Hinge joint; motion limited to two directions, forwards and backwards. Articular surfaces fitted together so as to permit of movement in one plane. As in the elbow, ankle, and knee. *Diarthrosis rotatorius or Lateral Ginglymus.* Articulation by a pivot process turning within a ring, or ring around a pivot. As in superior radio-ulnar articulation, and atlo-axoid joint.

The Kinds of Movement admitted in Joints.

The movements admissible in joints may be divided into four kinds, gliding, angular movement, circumduction, and rotation.

Gliding movement is the most simple kind of motion that can take place in a joint, one surface gliding over another. It is common to all moveable joints; but in some, as in the articulations of the carpus and tarsus, it is the only motion permitted. This movement is not confined to plane surfaces, but may exist between any two contiguous surfaces, of whatever form, limited by the ligaments which enclose the articulation.

Angular movement occurs only between the long bones, and may take place in four directions, forwards and backwards, constituting flexion and extension, or inwards and outwards, constituting adduction and abduction. The strictly ginglymoid or hinge joints admit of flexion and extension only. Abduction and adduction, combined with flexion and extension, are met with in the more moveable joints; as in the hip, shoulder, and metacarpal joint of the thumb, and partially in the wrist and ankle.

Circumduction is that limited degree of motion which takes place between the head of a bone and its articular cavity, whilst the extremity and sides of the limb are made to circumscribe a conical space, the base of which corresponds with the inferior extremity of the limb, the apex with the articular cavity; this kind of motion is best seen in the shoulder and hip joints.

Rotation is the movement of a bone upon its own axis, the bone retaining the same relative situation with respect to the adjacent parts; as in the articulation between the atlas and axis, where the odontoid process serves as a pivot around which the atlas turns; or in the rotation of the radius upon the humerus, and also in the hip and shoulder.

The articulations may be arranged into those of the trunk, those of the upper extremity, and those of the lower extremity.

ARTICULATIONS OF THE TRUNK.

These may be divided into the following groups, viz. :—

I. Of the vertebral column.
II. Of the atlas with the axis.
III. Of the atlas with the occipital bone.
IV. Of the axis with the occipital bone.
V. Of the lower jaw.
VI. Of the ribs with the vertebræ.
VII. Of the cartilages of the ribs with the sternum, and with each other.
VIII. Of the sternum.
IX. Of the vertebral column with the pelvis.
X. Of the pelvis.

I. Articulations of the Vertebral Column.

The different segments of the spine are connected together by ligaments, which admit of the same arrangement as the vertebræ. They may be divided into five sets. 1. Those connecting the *bodies* of the vertebræ. 2. Those connecting the *laminæ*. 3. Those connecting the *articular processes*. 4. Those connecting the *spinous processes*. 5. Those of the *transverse processes*.

The articulations of the *bodies* of the vertebræ with each other form a series of amphiarthrodial joints: those between the *articular processes* form a series of arthrodial joints.

1. The Ligaments of the Bodies.

Anterior Common Ligament. Posterior Common Ligament.
Intervertebral Substance.

The *Anterior Common Ligament* (figs. 116, 117, 124, 127,) is a broad and strong band of ligamentous fibres, which extends along the front surface of the bodies of the vertebræ, from the axis to the sacrum. It is broader below than above, thicker in the dorsal than in the cervical or lumbar regions, and somewhat thicker opposite the front of the body of each vertebra, than opposite the intervertebral substance. It is attached, above, to the body of the axis by a pointed process, which is connected with the tendon of insertion of the Longus colli muscle; and extends down as far as the upper bone of the sacrum. It consists of dense longitudinal fibres, which are intimately adherent to the intervertebral substance, and the prominent margins of the vertebræ; but less closely to the middle of the bodies. In the latter situation the fibres are exceedingly thick, and serve to fill up the concavities on their front surface, and to make the anterior surface of the spine more even. This ligament is composed of several layers of fibres, which vary in length, but are closely interlaced with each other. The most superficial or longest fibres extend between four or five vertebræ. A second subjacent set extend between two or three vertebræ; whilst a third set, the shortest and deepest, extend from one vertebra to the next. At the side of the bodies, the ligament consists of a few short fibres, which pass from one vertebra to the next, separated from the median portion by large oval apertures, for the passage of vessels.

The *Posterior Common Ligament* (figs. 116, 120) is situated within the spinal canal, and extends along the posterior surface of the bodies of the vertebræ, from the body of the axis above, where it is continuous with the occipito-axoid ligament, to the sacrum below. It is broader at the upper than at the lower part of the spine, and thicker in the dorsal than in the cervical or lumbar regions. In the situation of the intervertebral substance and contiguous margins of the vertebræ, where the ligament is more intimately adherent, it is broad, and presents a series of dentations with intervening concave margins; but it is narrow and thick over the centre of the bodies, from which it is separated by the *venæ basis vertebræ*. This ligament is composed of smooth, shining, longitudinal fibres, denser and more compact than those of the anterior ligament, and composed of a superficial layer occupying the interval between three or four vertebræ, and of a deeper layer which extends between one vertebra and the next adjacent to it. It is separated from the dura mater of the spinal cord by some loose filamentous tissue, very liable to serous infiltration.

The *Intervertebral Substance* (fig. 116) is a lenticular disc of fibro-cartilage, interposed between the adjacent surfaces of the bodies of the vertebræ, from the axis to the sacrum, and forming the chief bond of connection between those bones. These discs vary in shape, size, and thickness, in different parts of the spine. In *shape* they accurately correspond with the surfaces of the bodies between which they are placed, being oval in the cervical and lumbar regions, and circular in the dorsal. Their *size* is greatest in the lumbar region. In *thickness* they vary not only in the different regions of the spine, but in different parts of the same region:

thus, they are uniformly thick in the lumbar region; thickest, in front, in the cervical and lumbar regions which are convex forwards; and behind, to a slight extent, in the dorsal region. They thus contribute, in a great measure, to the curvatures of the spine in the neck and loins; whilst the concavity of the dorsal region is chiefly due to the shape of the bodies of the vertebræ. The intervertebral discs form about one-fourth of the spinal column, exclusive of the first two vertebræ; they are not equally distributed, however, between the various bones; the dorsal portion of the spine having, in proportion to its length, a much smaller quantity than in the cervical and lumbar regions, which necessarily gives to the latter parts greater pliancy and freedom of movement. The intervertebral discs are adherent, by their surfaces, to the adjacent parts of the bodies of the vertebræ; and by their circumference are closely connected in front to the anterior, and behind to the posterior common ligament; whilst, in the dorsal region, they are connected laterally, by means of the interarticular ligament, to the heads of those ribs which articulate with two vertebræ; they, consequently, form part of the articular cavities in which the heads of these bones are received.

116.—Vertical Section of two Vertebræ and their Ligaments, from the Lumbar Region.

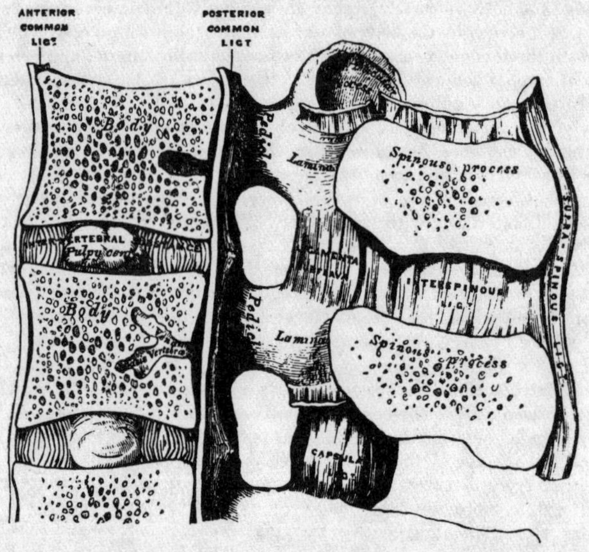

The intervertebral substance is composed, at its circumference, of laminæ of fibrous tissue and fibro-cartilage; and, at its centre, of a soft, elastic, pulpy matter. The laminæ are arranged concentrically one within the other, with their edges turned towards the corresponding surfaces of the vertebræ, and consist of alternate plates of fibrous tissue and fibro-cartilage. These plates are not quite vertical in their direction, those near the circumference being curved outwards and closely approximated; whilst those nearest the centre curve in the opposite direction, and are somewhat more widely separated. The fibres of which each plate is composed, are directed, for the most part, obliquely from above downwards; the fibres of an adjacent plate have an exactly opposite arrangement, varying in their direction in every layer; whilst in some few they are horizontal. This laminar arrangement belongs to about the outer half of each disc, the central part being occupied by a soft, pulpy, highly elastic substance, of a yellowish colour, which rises up considerably above the surrounding level, when the disc is divided horizontally. This substance presents no concentric arrangement, and consists of white fibrous tissue, with cells of variable shape and size interspersed. The pulpy matter, which is especially well developed in the lumbar region, is separated from immediate contact with the vertebræ by the interposition of thin plates of cartilage.

2. Ligaments connecting the Laminæ.

Ligamenta Subflava.

The *Ligamenta Subflava* (fig. 116) are interposed between the laminæ of the vertebræ, from the axis to the sacrum. They are most distinct when seen from the interior of the spinal canal; when viewed from the outer surface, they appear short, being overlapped by the laminæ. Each ligament consists of two lateral portions, which commence on each side at the root of either articular process, and pass backwards to the point where the laminæ converge to form the spinous process, where their margins are thickest, and separated by a slight interval, filled up with areolar tissue. These ligaments consist of yellow elastic tissue, the fibres of which, almost perpendicular in direction, are attached to the anterior surface of the margin of the lamina above, and to the posterior surface, as well as to the margin of the lamina below. In the cervical region, they are thin in texture, but very broad and long; they become thicker in the dorsal region: and in the lumbar acquire very considerable thickness. Their highly elastic property serves to preserve the upright posture, and to assist in resuming it, after the spine has been flexed. These ligaments do not exist between the occiput and atlas, or between the atlas and axis.

3. Ligaments connecting the Articular Processes.

Capsular.

The *Capsular Ligaments* (fig. 118) are thin and loose ligamentous sacs, attached to the contiguous margins of the articulating processes of each vertebræ, through the greater part of their circumference, and completed internally by the ligamenta subflava. They are longer and more loose in the cervical than in the dorsal or lumbar regions. The capsular ligaments are lined on their inner surface by synovial membrane.

4. Ligaments connecting the Spinous Processes.

Inter-spinous. Supra-spinous.

The *Inter-spinous Ligaments* (fig. 116), thin and membranous, are interposed between the spinous processes in the dorsal and lumbar regions. Each ligament extends from the root to near the summit of each spinous process, and connects together their adjacent margins. They are narrow and elongated in the dorsal region, broader, quadrilateral in form, and thicker in the lumbar region.

The *Supra-spinous Ligament* is a strong fibrous cord, which connects together the apices of the spinous processes from the seventh cervical to the spine of the sacrum. It is thicker and broader in the lumbar than in the dorsal region, and intimately blended, in both situations, with the neighbouring aponeuroses. The most superficial fibres of this ligament connect three or four vertebræ; those deeper seated pass between two or three vertebræ; whilst the deepest connect the contiguous extremities of neighbouring vertebræ.

5. Ligaments connecting the Transverse Processes.

Inter-transverse.

The *Inter-transverse Ligaments* consist of a few thin scattered fibres, interposed between the transverse processes. They are generally wanting in the cervical region; in the dorsal, they are rounded cords; in the lumbar region they are thin and membranous.

Actions. The movements permitted in the spinal column are, Flexion, Extension, Lateral movement, Circumduction, and Rotation.

In *Flexion*, or movement of the spine forwards, the anterior common ligament is relaxed, and the intervertebral substances are compressed in front; while the posterior common ligament, the ligamenta subflava, and the inter- and supraspinous ligaments, are stretched, as well as the posterior fibres of the intervertebral discs. The interspaces between the laminæ are widened, and the inferior articular processes of the vertebræ above glide upwards, upon the articular processes of the vertebræ below. Flexion is the most extensive of all the movements of the spine.

In *Extension*, or movement of the spine backwards, an exactly opposite disposition of the parts takes place. This movement is not extensive, being limited by the anterior common ligament, and by the approximation of the spinous processes.

Flexion and extension are most free in the lower part of the lumbar, and in the cervical regions; extension in the latter region being greater than flexion, the reverse of which is the case in the lumbar region. These movements are least free in the middle and upper part of the back.

In *Lateral Movement*, the sides of the intervertebral discs are compressed, the extent of motion being limited by the resistance offered by the surrounding ligaments, and by the approximation of the transverse processes. This movement may take place in any part of the spine, but is most free in the neck and loins.

Circumduction is very limited, and is produced merely by a succession of the preceding movements.

117.—Occipito-Atloid and Atlo-Axoid Ligaments. Front View.

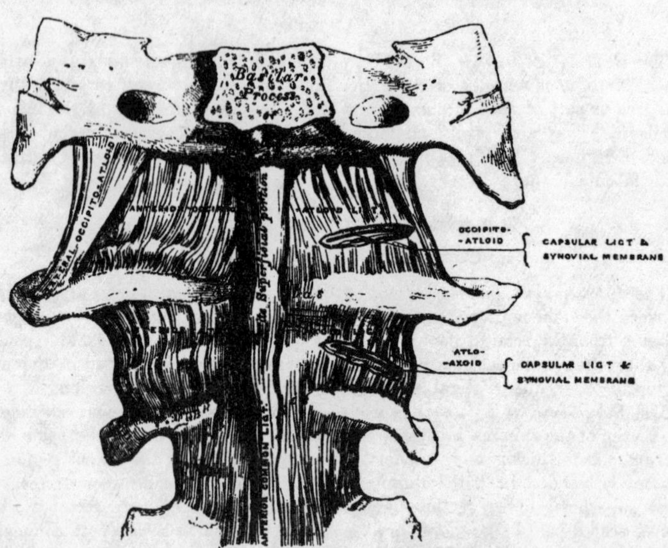

Rotation is produced by the twisting of the intervertebral substances; this, although only slight between any two vertebræ, produces a great extent of movement, when it takes place in the whole length of the spine, the front of the column being turned to one or the other side. This movement takes place only to a slight extent in the neck, but is more free in the lower part of the dorsal and lumbar regions.

It is thus seen, that the *cervical region* enjoys the greatest extent of each variety of movement, flexion and extension especially being very free. In the *dorsal region*, especially at its upper part, the movements are most limited; flexion, extension, and lateral motion taking place only to a slight extent.

II. Articulation of the Atlas with the Axis.

The articulation of the anterior arch of the atlas with the odontoid process forms a lateral ginglymoid joint, whilst that between the articulating processes of the two bones forms a double arthrodia. The ligaments which connect these bones are, the

Two Anterior Atlo-Axoid.	Transverse.
Posterior Atlo-Axoid.	Two Capsular.

Of the *Two Anterior Atlo-Axoid Ligaments* (fig. 117), the more superficial is a rounded cord, situated in the middle line; it is attached, above, to the tubercle on the anterior arch of the atlas; below, to the base of the odontoid process and body of the axis. The deeper ligament is a membranous layer, attached, above, to the lower border of the anterior arch of the atlas; below, to the base of the odontoid process, and body of the axis. These ligaments are in relation, in front, with the Recti antici majores.

The *Posterior Atlo-Axoid Ligament* (fig. 118) is a broad and thin membranous layer, attached, above, to the lower border of the posterior arch of the atlas;

118.—Occipito-Atloid and Atlo-Axoid Ligaments. Posterior View.

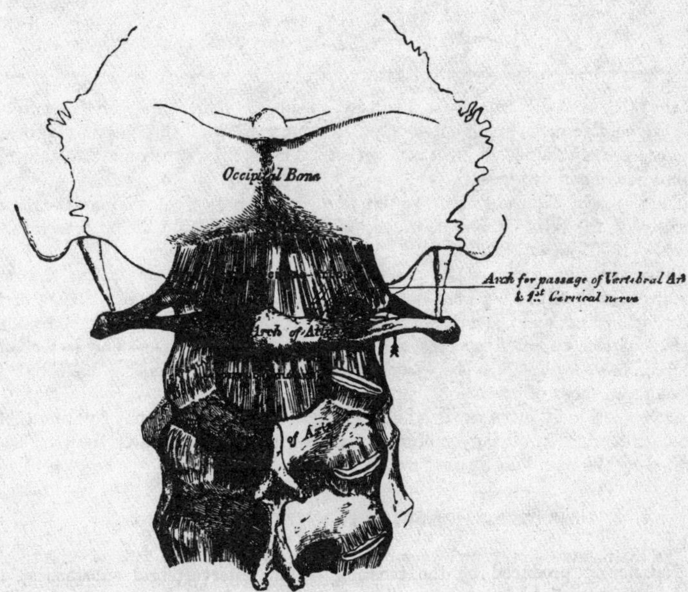

below, to the upper edge of the laminæ of the axis. This ligament supplies the place of the ligamenta subflava, and is in relation, behind, with the Inferior oblique muscles.

The *Transverse Ligament** (figs. 119, 120) is a thick and strong ligamentous band, which arches across the ring of the atlas, and serves to retain the odontoid process in firm connection with its anterior arch. This ligament is flattened from before backwards, broader and thicker in the middle than at either extremity, and firmly attached on each side of the atlas to a small tubercle on the inner surface of its lateral mass. As it crosses the odontoid process, a small fasciculus is derived from its upper and lower borders; the former passing upwards, to be inserted into the basilar process of the occipital bone; the latter, downwards, to be attached to the root of the odontoid process: hence, the whole ligament has received the name of *cruciform.* The transverse ligament divides the ring of the atlas into two unequal parts: of these, the posterior and larger serves for the transmission of the cord and its membranes: the anterior and smaller contains the odontoid process. Since the lower border of the space between the anterior arch

119.—Articulation between Odontoid Process and Atlas.

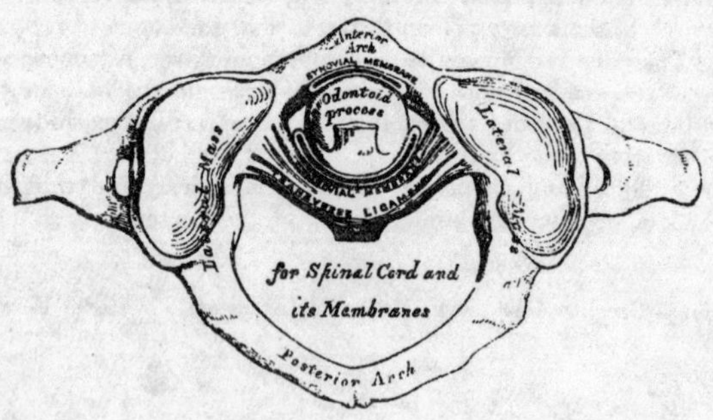

of the atlas and the transverse ligament is smaller than the upper (because the transverse ligament embraces firmly the narrow neck of the odontoid process), this process is retained in firm connection with the atlas when all the other ligaments have been divided.

The *Capsular Ligaments* are two thin and loose capsules, connecting the articular surfaces of the atlas and axis, the fibres being strongest on the anterior and external part of the articulation.

There are *four Synovial Membranes* in this articulation. One lining the inner surface of each of the capsular ligaments: one between the anterior surface of the odontoid process and the anterior arch of the atlas: and one between the posterior surface of the odontoid process and the transverse ligament. The latter often communicates with those between the condyles of the occipital bone and the articular surfaces of the atlas.

Actions. This joint is capable of great mobility, and allows the rotation of the atlas, and, with it, of the cranium upon the axis, the extent of rotation being limited by the odontoid ligaments.

Articulations of the Spine with the Cranium.

The ligaments connecting the spine with the cranium may be divided into two sets, those connecting the occipital bone with the atlas, and those connecting the occipital bone with the axis.

* It has been found necessary to describe the transverse ligament with those of the atlas and axis; but the student must remember that it is really a portion of the mechanism by which the movements of the head on the spine are regulated; so that the connections between the atlas and axis ought always to be studied together with those between the latter bones and the skull.

III. Articulation of the Atlas with the Occipital Bone.

This articulation is a double arthrodia. Its ligaments are the

Two Anterior Occipito-Atloid.
Posterior Occipito-Atloid.
Two Lateral Occipito-Atloid.
Two Capsular.

Of the *Two Anterior Ligaments* (fig. 117), the superficial is a strong, narrow, rounded cord, attached, above, to the basilar process of the occiput; below, to the tubercle on the anterior arch of the atlas: the deeper ligament is a broad and thin membranous layer which passes between the anterior margin of the foramen magnum above, and the whole length of the upper border of the anterior arch of the atlas below. This ligament is in relation, in front, with the Recti antici minores; behind, with the odontoid ligaments.

The *Posterior Occipito-Atloid Ligament* (fig. 118) is a very broad but thin membranous lamina, intimately blended with the dura mater. It is connected, above, to the posterior margin of the foramen magnum; below, to the upper border of the posterior arch of the atlas. This ligament is incomplete at each side, and forms, with the superior intervertebral notch, an opening for the passage of the vertebral artery and sub-occipital nerve. It is in relation, behind, with the Recti postici minores and Obliqui superiores; in front, with the dura mater of the spinal canal, to which it is intimately adherent.

120.—Occipito-Axoid and Atlo-Axoid Ligaments. Posterior View, obtained by removing the arches of the Vertebræ and the posterior part of the Skull.

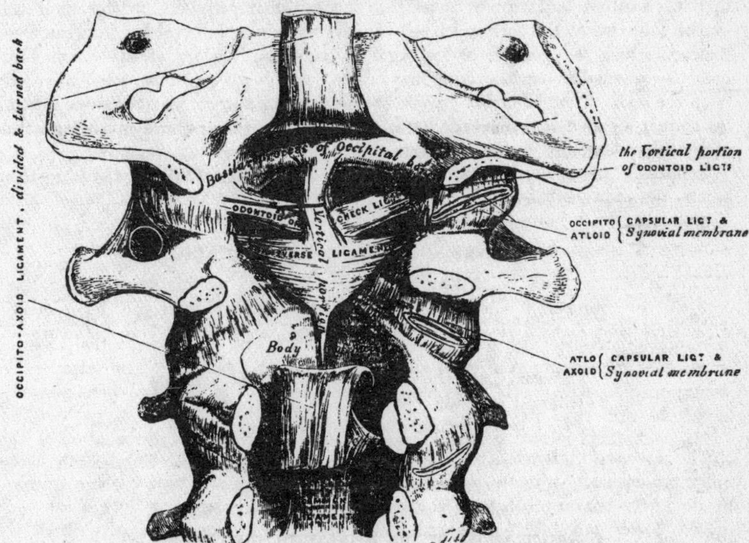

The *Lateral Ligaments* are strong fibrous bands, directed obliquely upwards and inwards, attached above to the jugular process of the occipital bone; below, to the base of the transverse process of the atlas.

The *Capsular Ligaments* surround the condyles of the occipital bone, and connect them with the articular surfaces of the atlas; they consist of thin and loose capsules, which enclose the synovial membrane of the articulation. The synovial membranes between the occipital bone and atlas communicate occasionally with that between the posterior surface of the odontoid process and transverse ligament.

Actions. The movements permitted in this joint are flexion and extension, which give rise to the ordinary forward or backward nodding of the head, besides slight lateral motion to one or the other side. When either of these actions is carried beyond a slight extent, the whole of the cervical portion of the spine assists in its production. According to Cruveilhier, there is a slight motion of rotation in this joint.

IV. Articulation of the Axis with the Occipital Bone.

Occipito-Axoid. Three Odontoid.

To expose these ligaments, the spinal canal should be laid open by removing the posterior arch of the atlas, the laminæ and spinous process of the axis, and the portion of the occipital bone behind the foramen magnum, as seen in fig. 120.

The *Occipito-Axoid Ligament* (Apparatus ligamentosus colli) is situated at the upper part of the front surface of the spinal canal. It is a broad and strong ligamentous band, which covers the odontoid process and its ligaments, and appears to be a prolongation upwards of the posterior common ligament of the spine. It is attached, below, to the posterior surface of the body of the axis, and becoming expanded as it ascends, is inserted into the basilar groove of the occipital bone, in front of the foramen magnum.

Relations. By its anterior surface, it is intimately connected with the transverse ligament, by its posterior surface with the dura mater. By cutting this ligament across, and turning its ends aside, the transverse and odontoid ligaments are exposed.

The *Odontoid* or *Check Ligaments* are strong, rounded, fibrous cords, which arise one on either side of the apex of the odontoid process, and passing obliquely upwards and outwards, are inserted into the rough depressions on the inner side of the condyles of the occipital bone. In the triangular interval left between these ligaments and the margin of the foramen magnum, a third strong ligamentous band (ligamentum suspensorium) may be seen, which passes almost perpendicularly from the apex of the odontoid process to the anterior margin of the foramen, being intimately blended with the anterior occipito-atloid ligament, and upper fasciculus of the transverse ligament of the atlas.

Actions. The odontoid ligaments serve to limit the extent to which rotation of the cranium may be carried; hence they have received the name of *check ligaments.*

V. Temporo-Maxillary Articulation.

This is an arthrodial joint; the parts entering into its formation are, on each side, the anterior part of the glenoid cavity of the temporal bone and the eminentia articularis above; with the condyle of the lower jaw below. The ligaments are the following:

External Lateral. Stylo-maxillary.
Internal Lateral. Capsular.
Interarticular Fibro-cartilage.

The *External Lateral Ligament* (fig. 121) is a short, thin, and narrow fasciculus attached above to the outer surface of the zygoma and to the rough tubercle on its lower border; below, to the outer surface and posterior border of the neck of the lower jaw. This ligament is broader above than below; its fibres are placed parallel with one another, and directed obliquely downwards and backwards. Externally, it is covered by the parotid gland, and by the integument. Internally, it is in relation with the interarticular fibro-cartilage and the synovial membranes.

The *Internal Lateral Ligament* (fig. 122) is a long, thin, and loose band, which is attached above to the spinous process of the sphenoid bone, and becoming broader as it descends, is inserted into the inner margin of the dental foramen. Its outer surface is in relation above with the External pterygoid muscle; lower down it is separated from the neck of the condyle by the internal maxillary artery; and still more inferiorly the inferior dental vessels and nerve separate it from the ramus of the jaw. Internally it is in relation with the Internal pterygoid.*

121.—Temporo-Maxillary Articulation. External View.

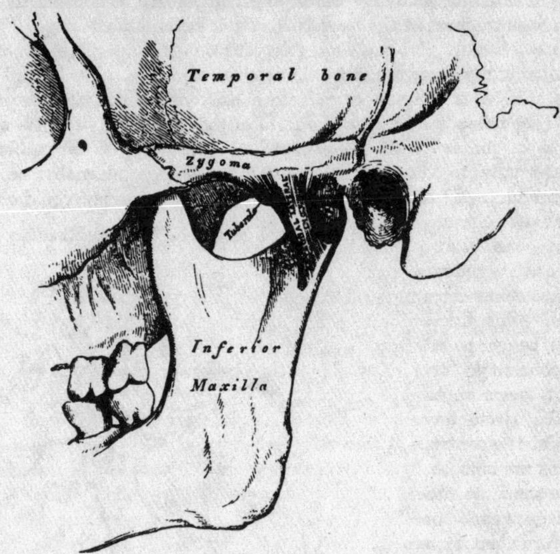

The *Stylo-Maxillary Ligament* is a thin aponeurotic cord, which extends from near the apex of the styloid process of the temporal bone, to the angle and posterior border of the ramus of the lower jaw, between the Masseter and Internal pterygoid muscles. This ligament separates the parotid from the sub-maxillary gland, and has attached to its inner side part of the fibres of origin of the Stylo-glossus muscle. Although usually classed among the ligaments of the jaw, it can only be considered as an accessory in the articulation.

122.—Temporo-Maxillary Articulation. Internal View.

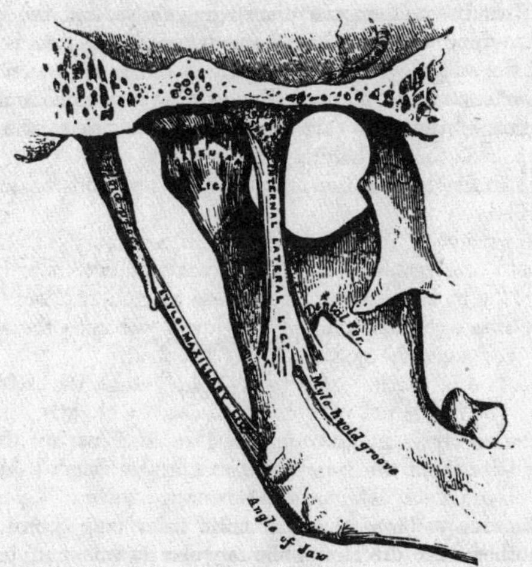

Along with the stylo-maxillary ligament, although in no way connected with the functions of the lower jaw, may be described the *stylo-hyoid ligament.* This is a fibrous cord, which continues the styloid process down to the hyoid bone, being attached to the tip of the former and the small cornu of the latter. It is often more or less ossified.

* Dr. Humphry describes the internal portion of the capsular ligament separately, as the short internal lateral ligament; and it certainly seems as deserving of a separate description as the external lateral ligament is.

The *Capsular Ligament* consists of a thin and loose ligamentous capsule, attached above to the circumference of the glenoid cavity and the articular surface immediately in front: below, to the neck of the condyle of the lower jaw. It consists of a few thin scattered fibres, and can hardly be considered as a distinct ligament; it is thickest at the back part of the articulation.

The *Interarticular fibro-cartilage* (fig. 123) is a thin plate of an oval form, placed horizontally between the condyle of the jaw and the glenoid cavity. Its upper surface is concave from before backwards, and a little convex transversely, to accommodate itself to the form of the glenoid cavity. Its under surface, where it is in contact with the condyle, is concave. Its circumference is connected externally to the external lateral ligament; internally, to the capsular ligament; and in front to the tendon of the External pterygoid muscle. It is thicker at its circumference, especially behind, than at its centre, where it is sometimes perforated. The fibres of which it is composed have a concentric arrangement, more apparent at the circumference than at the centre. Its surfaces are smooth, and divide the joint into two cavities, each of which is furnished with a separate synovial membrane. When the fibro-cartilage is perforated, the synovial membranes are continuous with one another.

123.—Vertical Section of Temporo-Maxillary Articulation.

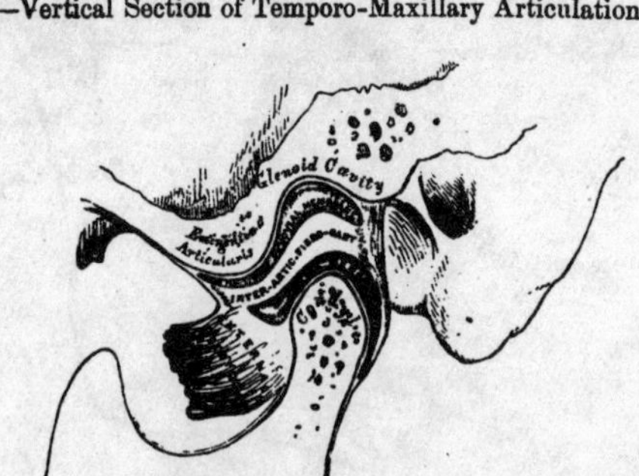

The *Synovial Membranes*, two in number, are placed one above, and the other below the fibro-cartilage. The upper one, the larger and looser of the two, is continued from the margin of the cartilage covering the glenoid cavity and eminentia articularis, over the upper surface of the fibro-cartilage. The lower one is interposed between the under surface of the fibro-cartilage and the condyle of the jaw, being prolonged downwards a little further behind than in front.

The *Nerves* of this joint are derived from the auriculo-temporal and masseteric branches of the inferior maxillary.

Actions. The movements permitted in this articulation are very extensive. Thus, the jaw may be depressed or elevated, or it may be carried forwards or backwards, or from side to side. It is by the alternation of these movements performed in succession, that a kind of rotatory movement of the lower jaw upon the upper takes place, which materially assists in the mastication of the food.

If the movement of depression is carried only to a slight extent, the condyles remain in the glenoid cavities, their anterior part descending only slightly; but if depression is considerable, the condyles glide from the glenoid fossæ on to the articular eminences, carrying with them the interarticular fibro-cartilages. When this movement is carried to too great an extent, as, for instance, during a convulsive yawn, dislocation of the condyle into the zygomatic fossa may occur; the interarticular cartilage being carried forwards, and the capsular ligament ruptured. When the jaw is elevated, after forced depression, the condyles and fibro-cartilages are carried backwards into their original position. When the jaw is carried horizontally forwards and backwards, or from side to side, a horizontal gliding movement of the fibro-cartilages and condyles upon the glenoid cavities takes place in the corresponding direction.

VI. Articulation of the Ribs with the Vertebræ.

The articulation of the ribs with the vertebral column, may be divided into two sets. 1. Those which connect the heads of the ribs with the bodies of the vertebræ; 2. Those which connect the neck and tubercle of the ribs with the transverse processes.

1. Articulation between the Heads of the Ribs and the Bodies of the Vertebræ. (Fig. 124.)

These constitute a series of ginglymoid joints, formed by the articulation of the heads of the ribs with the cavities on the contiguous margins of the bodies of the dorsal vertebræ, connected together by the following ligaments:—

Anterior Costo-vertebral or Stellate.
Capsular.
Interarticular.

The *Anterior Costo-vertebral* or *Stellate Ligament* connects the anterior part of the head of each rib with the sides of the bodies of two vertebræ, and the intervertebral disc between them. It consists of three flat bundles of ligamentous fibres, which radiate from the anterior part of the head of the rib. The superior fasciculus passes upwards to be connected with the body of the vertebra above; the inferior one descends to the body of the vertebra below; and the middle one, the smallest and least distinct, passes horizontally inwards to be attached to the intervertebral substance.

124.—Costo-vertebral and Costo-transverse Articulations. Anterior View.

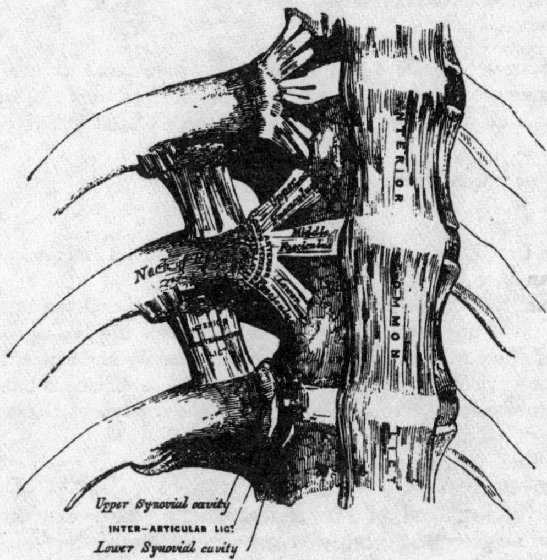

Relations. In front, with the thoracic ganglia of the sympathetic, the pleura, and, on the right side, with the vena azygos major; behind, with the interarticular ligament and synovial membranes.

In the first rib, which articulates with a single vertebra only, this ligament does not present a distinct division into three fasciculi; its superior fibres, however, pass to be attached to the body of the last cervical vertebra, as well as to the body of the vertebra with which the rib articulates. In the eleventh and twelfth ribs also, which likewise articulate with a single vertebra, the division does not exist; but the upper fibres of the ligament, in each case, are connected with the vertebra above, as well as that with which the ribs articulate.

The *Capsular Ligament* is a thin and loose ligamentous bag, which surrounds the joint between the head of the rib and the articular cavity formed by the junction of the vertebræ. It is very thin, firmly connected with the anterior ligament, and most distinct at the upper and lower parts of the articulation.

The *Interarticular Ligament* is situated in the interior of the joint. It consists of a short band of fibres, flattened from above downwards, attached by one extremity to the sharp crest on the head of the rib, and by the other to the intervertebral disc. It divides the joint into two cavities, which have no communication with one another, but are each lined by a separate synovial membrane. In the first, eleventh, and twelfth ribs, the interarticular ligament does not exist; consequently, there is but one synovial membrane.

Actions. The movements permitted in these articulations are limited to elevation, depression, and a slight amount of movement forwards and backwards. The mobility, however, of the different ribs varies very much. The first rib is almost immoveable, excepting in deep inspiration. The movement of the second rib is also not very extensive. In the other ribs, their mobility increases successively to the last two, which are very moveable. The ribs are generally more moveable in the female than in the male.

2. Articulation of the Neck and Tubercle of the Ribs with the Transverse Processes. (Fig. 125.)

The ligaments connecting these parts are—

Anterior Costo-Transverse.
Middle Costo-Transverse (Interosseous).
Posterior Costo-Transverse.
Capsular.

The *Anterior Costo-Transverse Ligament* is a broad and strong band of fibres, attached below to the sharp crest on the upper border of the neck of each rib, and passing obliquely upwards and outwards, to the lower border of the transverse

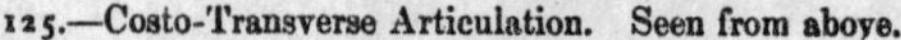

125.—Costo-Transverse Articulation. Seen from above.

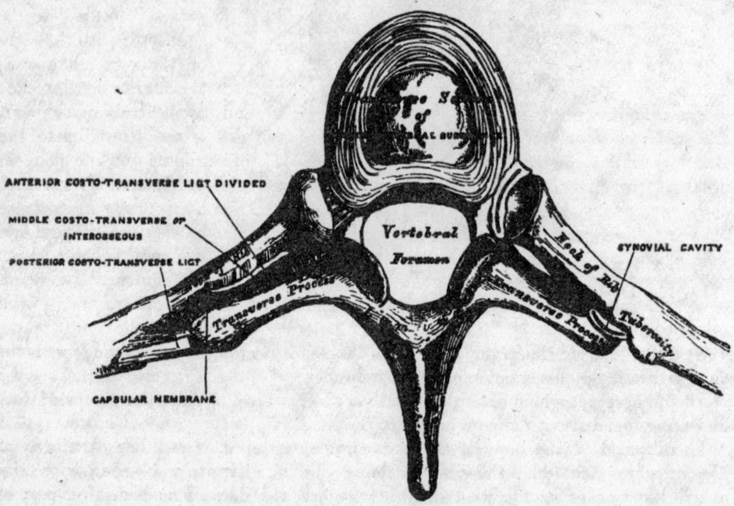

process immediately above. It is broader below than above, broader and thinner between the lower ribs than between the upper, and more distinct in front than behind. This ligament is in relation, in front, with the intercostal vessels and nerves; behind, with the Longissimus dorsi. Its *internal border* completes an aperture formed between it and the articular processes, through which pass the posterior branches of the intercostal vessels and nerves. Its *external border* is continuous with a thin aponeurosis, which covers the External intercostal muscle.

The *first* and *last ribs* have no anterior costo-transverse ligament.

The *Middle Costo-Transverse* or *Interosseous Ligament* consists of short, but strong, fibres, which pass between the rough surface on the posterior part of the neck of each rib, and the anterior surface of the adjacent transverse process. In order fully to expose this ligament, a horizontal section should be made across the transverse process and corresponding part of the rib; or the rib may be forcibly separated from the transverse process, and its fibres put on the stretch.

In the *eleventh* and *twelfth ribs*, this ligament is quite rudimentary.

The *Posterior Costo-Transverse Ligament* is a short, but thick and strong, fasciculus, which passes obliquely from the summit of the transverse process to the rough non-articular portion of the tubercle of the rib. This ligament is shorter and more oblique in the upper than in the lower ribs. Those corresponding to the superior ribs ascend, and those of the inferior one slightly descend.

In the *eleventh* and *twelfth ribs*, this ligament is wanting.

The articular portion of the tubercle of the rib, and adjacent transverse process, form an arthrodial joint, provided with a thin *capsular ligament* attached to the circumference of the articulating surfaces, and enclosing a small *synovial membrane*.

In the *eleventh* and *twelfth ribs*, this articulation is wanting.

Actions. The movement permitted in these joints, is limited to a slight gliding motion of the articular surfaces one upon the other.

VII. Articulation of the Cartilages of the Ribs with the Sternum, etc. (Fig. 126.)

The articulation to the cartilages of the true ribs with the sternum are arthrodial joints. The ligaments connecting them are—

Anterior Costo-Sternal.
Posterior Costo-Sternal.
Capsular.

The *Anterior Costo-Sternal Ligament* is a broad and thin membranous band that radiates from the inner extremity of the cartilages of the true ribs to the anterior surface of the sternum. It is composed of fasciculi, which pass in different directions. The *superior* fasciculi ascend obliquely, the *inferior* pass obliquely downwards, and the *middle* fasciculi horizontally. The superficial fibres of this ligament are the longest; they intermingle with the fibres of the ligaments above and below them, with those of the opposite side, and with the tendinous fibres of origin of the Pectoralis major; forming a thick fibrous membrane, which covers the surface of the sternum. This is more distinct at the lower than at the upper part.

The *Posterior Costo-Sternal Ligament*, less thick and distinct than the anterior, is composed of fibres which radiate from the posterior surface of the sternal end of the cartilages of the true ribs, to the posterior surface of the sternum, becoming blended with the periosteum.

The *Capsular Ligament* surrounds the joints formed between the cartilages of the true ribs and the sternum. It is very thin, intimately blended with the anterior and posterior ligaments, and strengthened at the upper and lower part of the articulation by a few fibres, which pass from the cartilage to the side of the sternum. These ligaments protect the synovial membranes.

Synovial Membranes. The cartilage of the *first rib* is directly continuous with the sternum, without any synovial membrane. The cartilage of the *second rib* is connected with the sternum by means of an interarticular ligament, attached by one extremity to the cartilage of the second rib, and by the other extremity to the cartilage which unites the first and second pieces of the sternum. This articulation is provided with two synovial membranes. That of the third rib has also two synovial membranes; and that of the fourth, fifth, sixth, and seventh, each a single synovial membrane. Thus there are *eight* synovial cavities in the articulations between the costal cartilages of the true ribs and the sternum. They may be demonstrated by removing a thin section from the anterior surface of the sternum and cartilages, as seen in the figure. After middle life, the articular surfaces lose their polish, become roughened, and the synovial membranes appear to be wanting. In old age, the articulations do not exist, the cartilages of most of the ribs becoming continuous with the sternum. The cartilage of the *seventh*

126.—Costo-Sternal, Costo-Xiphoid, and Intercostal Articulations. Anterior View.

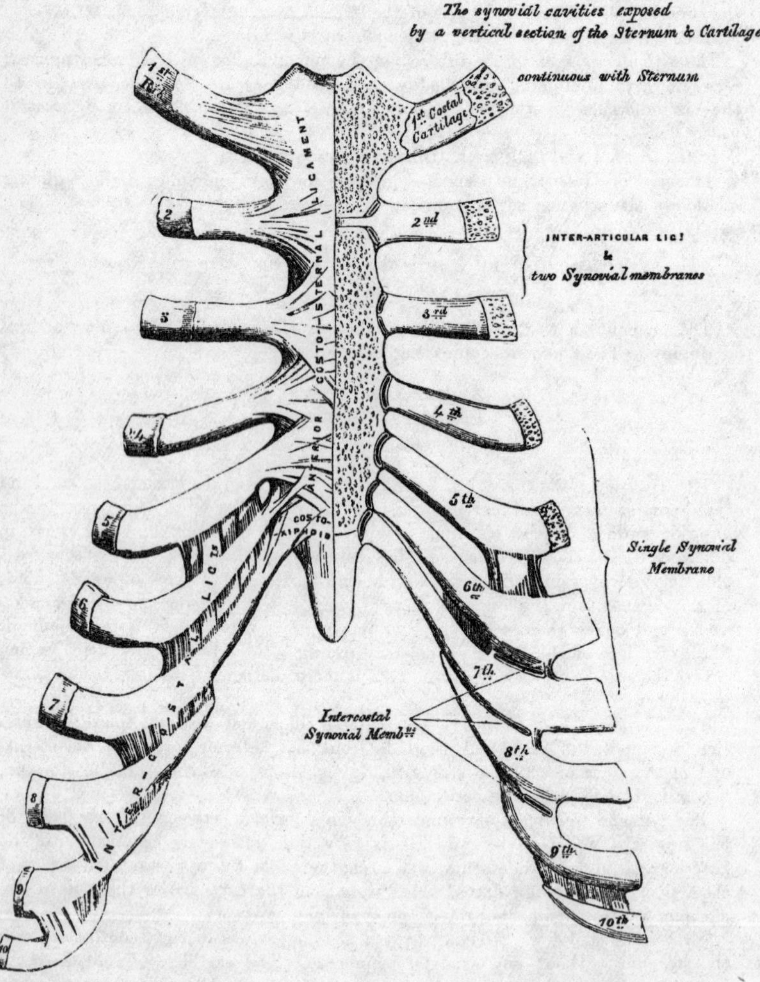

rib, and occasionally also that of the *sixth*, is connected to the anterior surface ot the ensiform appendix, by a band of ligamentous fibres, which varies in length and breadth in different subjects. It is called the *costo-xiphoid ligament.*

Actions. The movements which are permitted in the costo-sternal articulations, are limited to elevation and depression; and these only to a slight extent.

Articulation of the Cartilages of the Ribs with each other. (Fig. 126.)

The cartilages of the sixth, seventh, and eighth ribs articulate, by their lower borders with the corresponding margin of the adjoining cartilages, by means of a small, smooth, oblong-shaped facet. Each articulation is enclosed in a thin *capsular ligament*, lined by *synovial membrane*, and strengthened externally and internally by ligamentous fibres (intercostal ligaments), which pass from one cartilage to the other. Sometimes the cartilage of the fifth rib, more rarely that of the ninth, articulates, by its lower border, with the adjoining cartilage by a small oval facet; more frequently they are connected together by a few ligamentous fibres. Occasionally, the articular surfaces above mentioned are wanting.

Articulation of the Ribs with their Cartilages. (Fig. 126.)

The outer extremity of each costal cartilage is received into a depression in the sternal end of the ribs, and held together by the periosteum.

127.—Articulations of Pelvis and Hip. Anterior View.

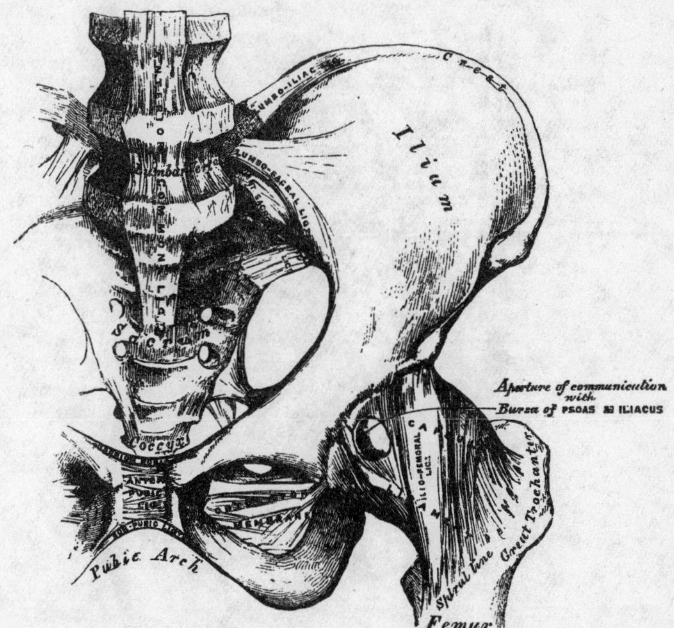

VIII. Ligaments of the Sternum.

The first and second pieces of the Sternum are united by a layer of cartilage which rarely ossifies, except at an advanced period of life. These two segments are connected by an anterior and posterior ligament.

The *anterior sternal ligament* consists of a layer of fibres, having a longitudinal direction; it blends with the fibres of the anterior costo-sternal ligaments on both sides, and with the aponeurosis of origin of the Pectoralis major. This ligament is rough, irregular, and much thicker at the lower than at the upper part of the bone.

The *posterior sternal ligament* is disposed in a somewhat similar manner on the posterior surface of the articulation.

IX. Articulation of the Pelvis with the Spine.

The ligaments connecting the last lumbar vertebra with the sacrum are similar to those which connect the segments of the spine with each other, viz.:—1. The continuation downwards of the anterior and posterior common ligaments. 2. The intervertebral substance connecting the flattened oval surfaces of the two bones, and forming an amphiarthrodial joint. 3. Ligamenta subflava, connecting the arch of the last lumbar vertebra with the posterior border of the sacral canal. 4. Capsular ligaments connecting the articulating processes and forming a double arthrodia. 5. Inter- and supra-spinous ligaments.

The two proper ligaments connecting the pelvis with the spine are the lumbo-sacral and lumbo-iliac.

The *Lumbo-sacral Ligament* (fig. 127) is a short, thick, triangular fasciculus, which is connected above to the lower and front part of the transverse process of

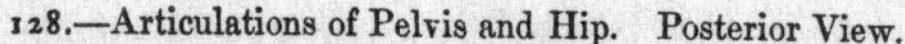

128.—Articulations of Pelvis and Hip. Posterior View.

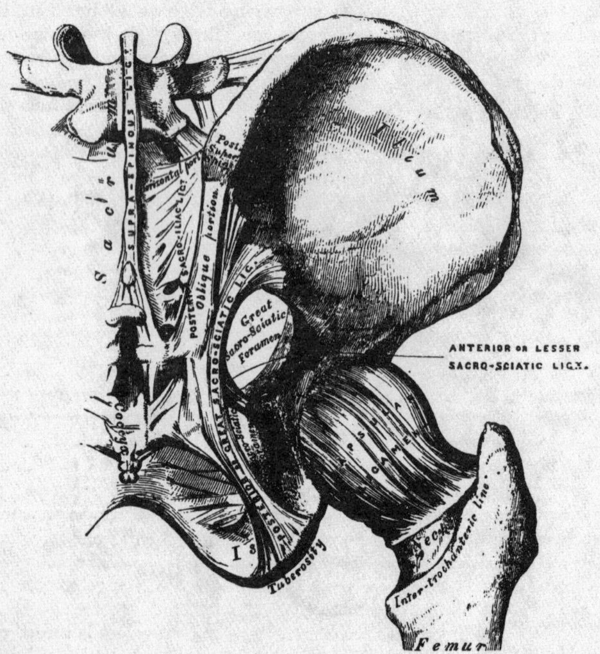

the last lumbar vertebra, passes obliquely outwards, and is attached below to the lateral surface of the base of the sacrum, becoming blended with the anterior sacro-iliac ligament. This ligament is in relation in front with the Psoas muscle.

The *Lumbo-iliac Ligament* (fig. 127) passes horizontally outwards from the apex of the transverse process of the last lumbar vertebra, to the crest of the ilium immediately in front of the sacro-iliac articulation. It is of a triangular form, thick and narrow internally, broad and thinner externally. It is in relation, in front, with the Psoas muscle; behind, with the muscles occupying the vertebral groove; above, with the Quadratus lumborum.

X. Articulations of the Pelvis.

The Ligaments connecting the bones of the pelvis with each other may be divided into four groups. 1. Those connecting the sacrum and ilium. 2. Those passing between the sacrum and ischium. 3. Those connecting the sacrum and coccyx. 4. Those between the two pubic bones.

1. Articulation of the Sacrum and Ilium.

The sacro-iliac articulation is an amphiarthrodial joint, formed between the lateral surfaces of the sacrum and ilium. The anterior or auricular portion of each articular surface is covered with a thin plate of cartilage, thicker on the sacrum than on the ilium. The surfaces of these cartilages in the adult are rough and irregular, and separated from one another by a soft yellow pulpy substance. At an early period of life, occasionally in the adult, and in the female during pregnancy, they are smooth, and lined by a delicate synovial membrane. The ligaments connecting these surfaces are the anterior and posterior sacro-iliac.

The *Anterior Sacro-iliac Ligament* (fig. 127) consists of numerous thin ligamentous bands, which connect the anterior surfaces of the sacrum and ilium.

The *Posterior Sacro-iliac* (fig. 128) is a strong interosseous ligament, situated in the deep depression between the sacrum and ilium behind, and forming the chief bond of connection between those bones. It consists of numerous strong fasciculi, which pass between the bones in various directions. Three of these are of large size; the *two superior*, nearly horizontal in direction, arise from the first and second transverse tubercles on the posterior surface of the sacrum, and are inserted into the rough uneven surface at the posterior part of the inner surface of the ilium. The third fasciculus, oblique in direction, is attached by one extremity to the third or fourth transverse tubercle on the posterior surface of the sacrum, and by the other to the posterior superior spine of the ilium; it is sometimes called *the oblique sacro-iliac ligament.*

2. Ligaments passing between the Sacrum and Ischium. (Fig. 128.)

The Great Sacro-Sciatic (Posterior).
The Lesser Sacro-Sciatic (Anterior).

The *Great* or *Posterior Sacro-Sciatic Ligament* is situated at the lower and back part of the pelvis. It is thin, flat, and triangular in form ; narrower in the middle than at the extremities; attached by its broad base to the posterior inferior spine of the ilium, to the third and fourth transverse tubercles on the sacrum, and to the lower part of the lateral margin of that bone and the coccyx; passing obliquely downwards, outwards, and forwards, it becomes narrow and thick; and at its insertion into the inner margin of the tuberosity of the ischium, it increases in breadth, and is prolonged forwards along the inner margin of the ramus, forming what is known as the falciform ligament. The free concave edge of this ligament has attached to it the obturator fascia, with which it forms a kind of groove, protecting the internal pudic vessels and nerve. One of its surfaces is turned towards the perinæum, the other towards the Obturator internus muscle.

The *posterior surface* of this ligament gives origin, by its whole extent, to fibres of the Gluteus maximus. Its *anterior surface* is united to the lesser sacro-sciatic ligament. Its *superior border* forms the lower boundary of the lesser sacro-sciatic foramen. Its *lower border* forms part of the boundary of the perinæum. It is pierced by the coccygeal branch of the sciatic artery.

The *Lesser* or *Anterior Sacro-Sciatic Ligament*, much shorter and smaller than the preceding, is thin, triangular in form, attached by its apex to the spine of the ischium, and internally, by its broad base, to the lateral margin of the sacrum and coccyx, anterior to the attachment of the great sacro-sciatic ligament, with which its fibres are intermingled.

It is in relation, *anteriorly*, with the Coccygeus muscle; *posteriorly*, it is covered by the posterior ligament, and crossed by the pudic vessels and nerve. Its *superior border* forms the lower boundary of the great sacro-sciatic foramen; its *inferior border*, part of the lesser sacro-sciatic foramen.

These two ligaments convert the sacro-sciatic notches into foramina. The *superior* or *great* sacro-sciatic foramen is bounded, in front and above, by the posterior border of the os innominatum; behind, by the great sacro-sciatic ligament; and below, by the lesser ligament. It is partially filled up, in the recent state, by the Pyriformis muscle. Above this muscle, the gluteal vessels and superior gluteal nerve emerge from the pelvis; and below it, the ischiatic vessels and nerves, the internal pudic vessels and nerve, and the nerve to the Obturator internus. The *inferior* or *lesser* sacro-sciatic foramen is bounded, in front, by the tuber ischii; above, by the spine and lesser ligament; behind, by the greater ligament. It transmits the tendon of the Obturator internus muscle, its nerve, and the pudic vessels and nerve.

3. Articulation of the Sacrum and Coccyx.

This articulation is an amphiarthrodial joint, formed between the oval surface, on the summit of the sacrum, and the base of the coccyx. It is analogous to the joints between the bodies of the vertebræ, and is connected by similar ligaments. They are the—

Anterior Sacro-Coccygeal.
Posterior Sacro-Coccygeal.
Interarticular Fibro-Cartilage.

The *Anterior Sacro-Coccygeal Ligament* consists of a few irregular fibres, which descend from the anterior surface of the sacrum to the front of the coccyx, becoming blended with the periosteum.

The *Posterior Sacro-Coccygeal Ligament* is a flat band of ligamentous fibres, of a pearly tint, which arises from the margin of the lower orifice of the sacral canal, and descends to be inserted into the posterior surface of the coccyx. This ligament completes the lower and back part of the sacral canal. Its superficial fibres are much longer than the deep-seated; the latter extend from the apex of the sacrum to the upper cornua of the coccyx. This ligament is in relation in front with the arachnoid membrane of the sacral canal, a portion of the sacrum and almost the whole of the posterior surface of the coccyx; behind, with the Gluteus maximus.

An *Interarticular Fibro-Cartilage* is interposed between the contiguous surfaces of the sacrum and coccyx; it differs from that interposed between the bodies of the vertebræ in being thinner, and its central part more firm in texture. It is somewhat thicker in front and behind than at the sides. Occasionally a synovial membrane is found where the coccyx is freely moveable, which is more especially the case during pregnancy.

The different segments of the coccyx are connected together by an extension downwards of the anterior and posterior sacro-coccygeal ligaments, a thin annular disc of fibro-cartilage being interposed between each of the bones. In the adult male, all the pieces become ossified; but in the female, this does not commonly occur until a later period of life. The separate segments of the coccyx are first united, and at a more advanced age the joint between the sacrum and the coccyx is obliterated.

Actions. The movements which take place between the sacrum and coccyx, and between the different pieces of the latter bone, are slightly forwards and backwards; they are very limited. Their extent increases during pregnancy.

4. Articulation of the Pubes. (Fig. 129.)

The articulation between the pubic bones is an amphiarthrodial joint, formed by the junction of the two oval articular surfaces of the ossa pubis. The articular surface has been described above under the name of *symphysis*, and the same name is given to the joint. The ligaments of this articulation are the

Anterior Pubic.	Posterior Pubic.
Superior Pubic.	Sub-Pubic.

Interarticular Fibro-Cartilage.

The *Anterior Pubic Ligament* consists of several superimposed layers, which pass across the front of the articulation. The superficial fibres pass obliquely from one bone to the other, decussating and forming an interlacement with the fibres of the aponeurosis of the External oblique muscle. The deep fibres pass transversely across the symphysis, and are blended with the interarticular fibro-cartilage.

The *Posterior Pubic Ligament* consists of a few thin, scattered fibres, which unite the two pubic bones posteriorly.

The *Superior Pubic Ligament* is a band of fibres, which connects together the two pubic bones superiorly.

The *Sub-Pubic Ligament* is a thick, triangular arch of ligamentous fibres, connecting together the two pubic bones below, and forming the upper boundary of the pubic arch. Above, it is blended with the interarticular fibro-cartilage; laterally, with the rami of the pubes. Its fibres are of a yellowish colour, closely connected, and have an arched direction.

129.—Vertical Section of the Symphysis Pubis. Made near its Posterior Surface.

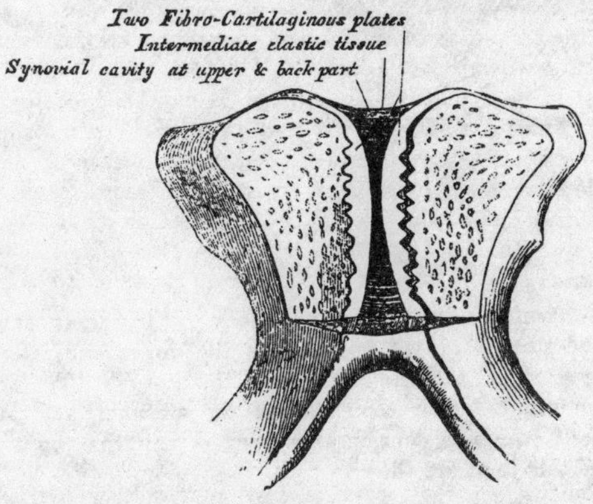

The *Interarticular Fibro-Cartilage* consists of two oval-shaped plates, one covering the surface of each symphysis pubis. They vary in thickness in different subjects, and project somewhat beyond the level of the bones, especially behind. The outer surface of each plate is firmly connected to the bone by a series of nipple-like processes, which accurately fit within corresponding depressions on the osseous surface. Their opposed surfaces are connected in the greater part of their extent, by an intermediate elastic fibrous tissue; and by their circumference to the various ligaments surrounding the joint. An interspace is left between the plates at the upper and back part of the articulation, where the fibrous tissue is deficient, and the surface of the fibro-cartilage is lined by epithelium. This space is found at all periods of life, both in the male and female; but it is larger in the latter, especially during pregnancy, and after parturition. It is most frequently limited to the upper and back part of the joint; but it occasionally reaches to the front, and may extend the entire length of the cartilages. This structure may be easily demonstrated, by making a vertical section of the symphysis pubis near its posterior surface.

The *Obturator Ligament* is more properly regarded as analogous to the muscular fasciæ, with which it will therefore be described.

ARTICULATIONS OF THE UPPER EXTREMITY.

The articulations of the Upper Extremity may be arranged in the following groups:—I. Sterno-clavicular articulation. II. Scapulo-clavicular articulation. III. Ligaments of the Scapula. IV. Shoulder-joint. V. Elbow-joint. VI. Radio-ulnar articulations. VII. Wrist-joint. VIII. Articulations of the Carpal bones. IX. Carpo-metacarpal articulations. X. Metacarpo-phalangeal articulations. XI. Articulations of the Phalanges.

130.—Sterno-Clavicular Articulation. Anterior View.

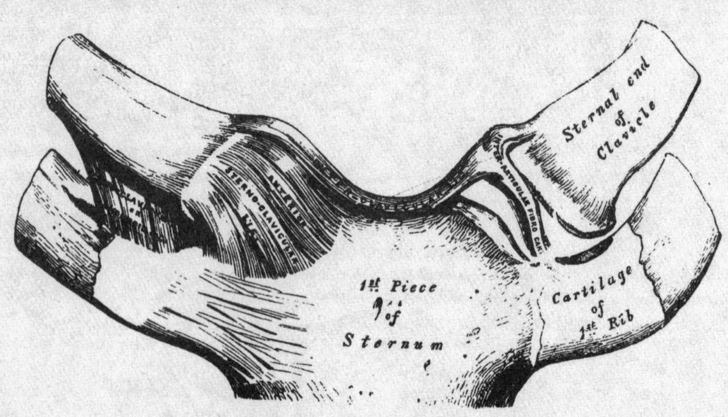

I. Sterno-Clavicular Articulation. (Fig. 130.)

The *Sterno-Clavicular* is an arthrodial joint. The parts entering into its formation are the sternal end of the clavicle, the upper and lateral part of the first piece of the sternum, and the cartilage of the first rib. The articular surface of the clavicle is much longer than that of the sternum, and invested with a layer of cartilage,* which is considerably thicker than that on the latter bone. The ligaments of this joint are the

Anterior Sterno-Clavicular.	Inter-Clavicular.
Posterior Sterno-Clavicular.	Costo-Clavicular (rhomboid).
Interarticular Fibro-Cartilage.	

The *Anterior Sterno-Clavicular Ligament* is a broad band of fibres, which covers the anterior surface of the articulation, being attached, above, to the upper and front part of the inner extremity of the clavicle; and, passing obliquely downwards and inwards, is attached, below, to the front and upper part of the first piece of the sternum. This ligament is covered in front by the sternal portion of the Sterno-cleido-mastoid and the integument; behind, it is in relation with the inter-articular fibro-cartilage and the two synovial membranes.

The *Posterior Sterno-Clavicular Ligament* is a similar band of fibres, which covers the posterior surface of the articulation, being attached, above, to the pos-

* According to Bruch, the sternal end of the clavicle is covered by a tissue, which is rather fibrous than cartilaginous in structure.

terior part of the inner extremity of the clavicle; and which, passing obliquely downwards and inwards, is connected, below, to the posterior and upper part of the sternum. It is in relation, in front, with the interarticular fibro-cartilage and synovial membranes; behind, with the Sterno-hyoid and Sterno-thyroid muscles.

The *Interclavicular Ligament* is a flattened band, which varies considerably in form and size in different individuals; it passes from the upper part of the inner extremity of one clavicle to the other, and is closely attached to the upper margin of the sternum. It is in relation, in front, with the integument; behind, with the Sterno-thyroid muscles.

The *Costo-Clavicular Ligament* (*rhomboid*) is short, flat, and strong: it is of a rhomboid form, attached, below, to the upper and inner part of the cartilage of the first rib: it ascends obliquely backwards and outwards, and is attached, above, to the rhomboid depression on the under surface of the clavicle. It is in relation, in front, with the tendon of origin of the Subclavius; behind, with the subclavian vein.

The *Interarticular Fibro-Cartilage* is a flat and nearly circular disc, interposed between the articulating surfaces of the sternum and clavicle. It is attached, above, to the upper and posterior border of the clavicle; below, to the cartilage of the first rib, at its junction with the sternum: and by its circumference to the anterior and posterior sterno-clavicular ligaments. It is thicker at the circumference, especially its upper and back part, than at its centre, or below. It divides the joint into two cavities, each of which is furnished with a separate synovial membrane; when the fibro-cartilage is perforated, which not unfrequently occurs, the synovial membranes communicate.

Of the *two Synovial Membranes* found in this articulation, one is reflected from the sternal end of the clavicle, over the adjacent surface of the fibro-cartilage, and cartilage of the first rib; the other is placed between the articular surface of the sternum and adjacent surface of the fibro-cartilage; the latter is the more loose of the two. They seldom contain much synovia.

Actions. This articulation is the centre of the movements of the shoulder, and admits of motion in nearly every direction—upwards, downwards, backwards, forwards, as well as circumduction; the sternal end of the clavicle and the interarticular cartilage gliding on the articular surface of the sternum.

II. Scapulo-Clavicular Articulation. (Fig. 131.)

The *Scapulo-Clavicular* is an arthrodial joint, formed between the outer extremity of the clavicle, and the upper edge of the acromion process of the scapula. Its ligaments are the

Superior Acromio-Clavicular.
Inferior Acromio-Clavicular.
Coraco-Clavicular { Trapezoid and Conoid.
Interarticular Fibro-Cartilage.

The *Superior Acromio-Clavicular Ligament* is a broad band, of a quadrilateral form, which covers the superior part of the articulation, extending between the upper part of the outer end of the clavicle, and the adjoining part of the acromion. It is composed of parallel fibres, which interlace with the aponeurosis of the Trapezius and Deltoid muscles; below, it is in contact with the interarticular fibro-cartilage and synovial membranes.

The *Inferior Acromio-Clavicular Ligament*, somewhat thinner than the preceding, covers the under part of the articulation, and is attached to the adjoining surfaces of the two bones. It is in relation, above, with the interarticular fibro-cartilage (when it exists) and the synovial membranes; below, with the tendon of the

Supraspinatus. These two ligaments are continuous with each other in front and behind, and form a complete capsule around the joint.

The *Coraco-Clavicular Ligament* serves to connect the clavicle with the coracoid process of the scapula. It consists of two fasciculi, called the trapezoid and conoid ligaments.

The *trapezoid ligament*, the anterior and external fasciculus, is broad, thin, and quadrilateral: it is placed obliquely between the coracoid process and the clavicle. It is attached, below, to the upper surface of the coracoid process; above, to the oblique line on the under surface of the clavicle. Its anterior border is free; its posterior border is joined with the conoid ligament, the two forming by their junction a projecting angle.

The *conoid ligament*, the posterior and internal fasciculus, is a dense band of fibres, conical in form, the base being turned upwards, the summit downwards. It is attached by its apex to a rough depression at the base of the coracoid process,

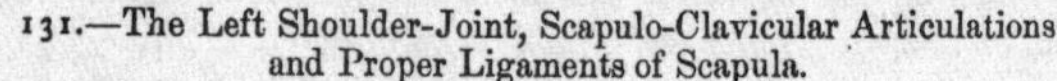
131.—The Left Shoulder-Joint, Scapulo-Clavicular Articulations and Proper Ligaments of Scapula.

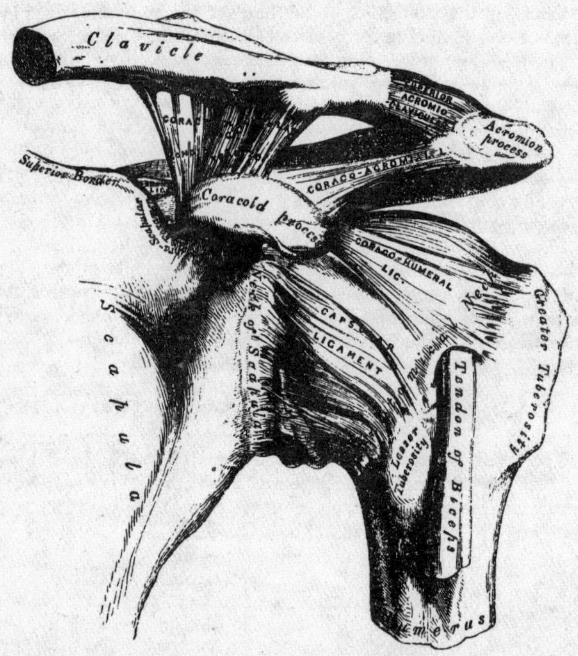

internal to the preceding; above, by its expanded base, to the conoid tubercle on the under surface of the clavicle, and to a line proceeding internally from it for half an inch. These ligaments are in relation, in front, with the Subclavius; behind, with the Trapezius: they serve to limit rotation of the scapula forwards and backwards.

The *Interarticular Fibro-Cartilage* is most frequently absent in this articulation. When it exists, it generally only partially separates the articular surfaces, and occupies the upper part of the articulation. More rarely, it completely separates the joint into two cavities.

There are *two Synovial Membranes* when a complete interarticular cartilage exists; more frequently there is only one synovial membrane.

Actions. The movements of this articulation are of two kinds. 1. A gliding motion of the articular end of the clavicle on the acromion. 2. Rotation of the scapula forwards and backwards upon the clavicle, the extent of this rotation being limited by the two portions of the coraco-clavicular ligament.

III. Proper Ligaments of the Scapula. (Fig. 131.)

The proper ligaments of the scapula are, the

Coraco-acromial. Transverse.

The *Coraco-acromial Ligament* is a broad, thin, flat band, of a triangular shape, extended transversely above the upper part of the shoulder-joint, between the coracoid and acromion processes. It is attached, by its apex, to the summit of the acromion just in front of the articular surface for the clavicle; and by its broad base, to the whole length of the outer border of the coracoid process. Its posterior fibres are directed obliquely backwards and outwards, its anterior fibres transversely. This ligament completes the vault formed by the coracoid and acromion processes for the protection of the head of the humerus. It is in relation, above, with the clavicle and under surface of the Deltoid; below, with the tendon of the Supraspinatus muscle, a bursa being interposed. Its anterior border is continuous with a dense cellular lamina that passes beneath the Deltoid upon the tendons of the Supra- and Infraspinatus muscles.

The *Transverse* or *Coracoid Ligament* converts the suprascapular notch into a foramen. It is a thin and flat fasciculus, narrower at the middle than at the extremities, attached, by one end, to the base of the coracoid process, and, by the other, to the inner extremity of the scapular notch. The suprascapular nerve passes through the foramen; the superscapular vessels above it.

IV. Shoulder-Joint. (Fig. 131.)

The Shoulder is an enarthrodial or ball-and-socket joint. The bones entering into its formation, are the large globular head of the humerus, received into the shallow glenoid cavity of the scapula, an arrangement which permits of very considerable movement, whilst the joint itself is protected against displacement by the strong ligaments and tendons which surround it, and above by an arched vault, formed by the under surface of the coracoid and acromion processes, and the coraco-acromial ligament. The articular surfaces are covered by a layer of cartilage: that on the head of the humerus is thicker at the centre than at the circumference, the reverse being the case in the glenoid cavity. The ligaments of the shoulder are, the

Capsular. Coraco-humeral.
Glenoid.*

The *Capsular Ligament* completely encircles the articulation; being attached, above, to the circumference of the glenoid cavity beyond the glenoid ligament; below, to the anatomical neck of the humerus, approaching nearer to the articular cartilage above than in the rest of its extent. It is thicker above than below, remarkably loose and lax, and much larger and longer than is necessary to keep the bones in contact, allowing them to be separated from each other more than an inch, an evident provision for that extreme freedom of movement which is peculiar to this articulation. Its external surface is strengthened, above, by the Supraspinatus; above and internally, by the coraco-humeral ligament; below, by the long head of the Triceps; externally, by the tendons of the Infraspinatus and

* The long tendon of origin of the Biceps muscle also acts as one of the ligaments of this joint.

Teres minor; and internally, by the tendon of the Subscapularis. The capsular ligament usually presents three openings: one at its inner side, below the coracoid process, partially filled up by the tendon of the Subscapularis; it establishes a communication between the synovial membrane of the joint and a bursa beneath the tendon of that muscle. The second, which is not constant, is at the outer part, where a communication sometimes exists between the joint and a bursal sac belonging to the Infraspinatus muscle. The third is seen in the lower border of the ligament, between the two tuberosities, for the passage of the long tendon of the Biceps muscle.

The *Coraco-humeral* or *Accessory Ligament* is a broad band which strengthens the upper and inner part of the capsular ligament. It arises from the outer border of the coracoid process, and passes obliquely downwards and outwards to the front of the great tuberosity of the humerus, being blended with the tendon of the Supraspinatus muscle. This ligament is intimately united to the capsular in the greater part of its extent.

The *Glenoid Ligament* is a firm fibrous band attached round the margin of the glenoid cavity. It is triangular on section, the thickest portion being fixed to the circumference of the cavity, the free edge being thin and sharp. It is continuous above with the long tendon of the Biceps muscle, which bifurcates at the upper part of the cavity into two fasciculi, encircling the margin of the glenoid cavity and uniting at its lower part. This ligament deepens the cavity for articulation, and protects the edges of the bone. It is lined by the synovial membrane.

The *Synovial Membrane* lines the margin of the glenoid cavity and the fibro-cartilaginous rim surrounding it; it is then reflected over the internal surface of the capsular ligament, covers the lower part and sides of the neck of the humerus, and is continued a short distance over the cartilage covering the head of the bone. The long tendon of the Biceps muscle which passes through the joint, is enclosed in a tubular sheath of synovial membrane, which is reflected upon it at the point where it perforates the capsule, and is continued around it as far as the summit of the glenoid cavity. The tendon of the Biceps is thus enabled to traverse the articulation, but is not contained in the interior of the synovial cavity. The synovial membrane communicates with a large bursal sac beneath the tendon of the Subscapularis, by an opening at the inner side of the capsular ligament; it also occasionally communicates with another bursal sac, beneath the tendon of the Infraspinatus, through an orifice at its outer part. A third bursal sac, which does not communicate with the joint, is placed between the under surface of the deltoid and the outer surface of the capsule.

The Muscles in relation with the joint are, above, the Supraspinatus; below, the long head of the Triceps; internally, the Subscapularis; externally, the Infraspinatus, and Teres minor; within, the long tendon of the Biceps. The Deltoid is placed most externally, and covers the articulation on its outer side, as well as in front and behind.

The Arteries supplying the joint, are articular branches of the anterior and posterior circumflex, and suprascapular.

The Nerves are derived from the circumflex and suprascapular.

Actions. The shoulder-joint is capable of movement in every direction, forwards, backwards, abduction, adduction, circumduction, and rotation.

V. Elbow-Joint.

The *Elbow* is a *ginglymus* or hinge joint. The bones entering into its formation are the trochlear surface of the humerus, which is received in the greater sigmoid cavity of the ulna, and admits of the movements peculiar to this joint, those of flexion and extension, whilst the cup-shaped depression on the head of the radius articulates with the radial tuberosity of the humerus, and the circumference of the head of the radius with the lesser sigmoid cavity of the ulna, allowing of the movement of rotation of the radius on the ulna, the chief action of the superior radio-ulnar articulation. The articular surfaces are covered with a thin layer of cartilage, and connected together by the following ligaments :—

Anterior.	Internal Lateral.
Posterior.	External Lateral.

The *Anterior Ligament* (fig. 132) is a broad and thin fibrous layer, which covers the anterior surface of the joint. It is attached to the front of the humerus immediately above the coronoid fossa; below, to the anterior surface of the coronoid process of the ulna and orbicular ligament, being continuous on each side with the lateral ligaments. Its superficial or oblique fibres pass from the inner condyle of the humerus outwards to the orbicular ligament. The middle fibres, vertical in direction, pass from the upper part of the coronoid depression, and become blended with the preceding. A third, or transverse set, intersect these at right angles. This ligament is in relation, in front, with the Brachialis anticus: behind, with the synovial membrane.

132.—Left Elbow-Joint, showing Anterior and Internal Ligaments.

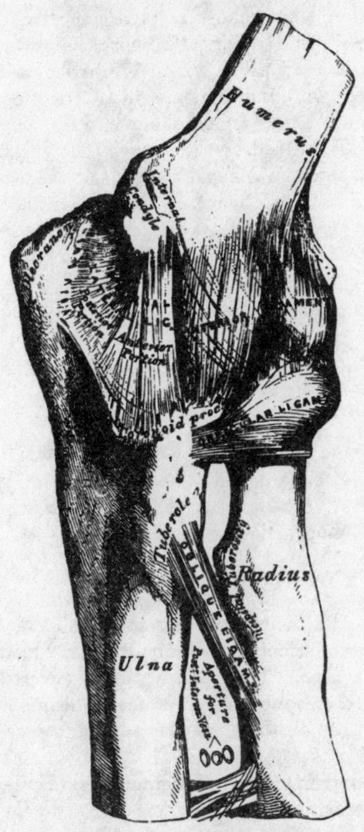

The *Posterior Ligament* (fig. 133) is a thin and loose membranous fold, attached, above, to the lower end of the humerus, immediately above the olecranon depression; below, to the margin of the olecranon. The superficial or transverse fibres pass between the adjacent margins of the olecranon fossa. The deeper portion consists of vertical fibres, which pass from the upper part of the olecranon fossa to the margin of the olecranon. This ligament is in relation, behind, with the tendon of the Triceps and the Anconeus; in front, with the synovial membrane.

The *Internal Lateral Ligament* (fig. 132) is a thick triangular band, consisting of two distinct portions, an anterior and posterior. The *anterior portion*, directed obliquely forwards, is attached, above, by its apex, to the front part of the internal condyle of the humerus; and, below, by its broad base, to the inner margin of the coronoid process. The *posterior portion*, also of triangular form, is attached, above, by its apex, to the lower and back part of the internal condyle; below, to the inner margin of the olecranon. This ligament is in relation, internally, with the Triceps and Flexor carpi ulnaris muscles, and the ulnar nerve.

The *External Lateral Ligament* (fig. 133) is a short and narrow fibrous fasciculus, less distinct than the internal, attached, above, to the external condyle of the humerus; below, to the orbicular ligament, some of its most posterior fibres passing over that ligament, to be inserted into the outer margin of the ulna. This ligament is intimately blended with the tendon of origin of the Supinator brevis muscle.

The *Synovial Membrane* is very extensive. It covers the margin of the articular surface of the humerus, and lines the coronoid and olecranon depressions on that bone; from these points, it is reflected over the anterior, posterior, and lateral ligaments; and forms a pouch between the lesser sigmoid cavity, the internal surface of the annular ligament, and the circumference of the head of the radius.

The *Muscles* in relation with the joint are, in front, the Brachialis anticus; behind, the Triceps and Anconeus; externally, the Supinator brevis, and the common tendon of origin of the Extensor muscles; internally, the common tendon of origin of the Flexor muscles, and the Flexor carpi ulnaris, with the ulnar nerve.

The *Arteries* supplying the joint are derived from the communicating branches between the superior profunda, inferior profunda, and anastomotic branches of the brachial, with the anterior, posterior, and interosseous recurrent branches of the ulnar, and the recurrent branch of the radial. These vessels form a complete chain of inosculation around the joint.

The *Nerves* are derived from the ulnar, as it passes between the internal condyle and the olecranon; and a few filaments from the musculo-cutaneous.

Actions. The elbow is a perfect hinge-joint; its movements are, consequently, limited to flexion and extension, the exact apposition of the articular surfaces preventing the least lateral motion.

133.—Left Elbow-Joint, showing Posterior and External Ligaments.

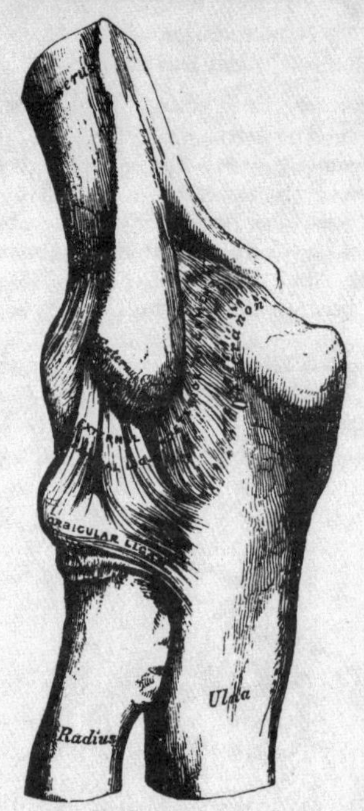

VI. Radio-Ulnar Articulations.

The articulation of the radius with the ulna is effected by ligaments, which connect together both extremities as well as the shafts of these bones. They may, consequently, be subdivided into three sets:—1. the superior radio-ulnar; 2. the middle radio-ulnar; and, 3. the inferior radio-ulnar articulations.

1. Superior Radio-Ulnar Articulation.

This articulation is a lateral ginglymus. The bones entering into its formation are the inner side of the circumference of the head of the radius, which rotates within the lesser sigmoid cavity of the ulna. These surfaces are covered with cartilage, and invested with a duplicature of synovial membrane, continuous with that which lines the elbow-joint. Its only ligament is the annular or orbicular.

The *Orbicular Ligament* (fig. 133) is a strong flat band of ligamentous fibres, which surrounds the head of the radius, and retains it in firm connection with the lesser sigmoid cavity of the ulna. It forms about three-fourths of a fibrous ring, attached by each end to the extremities of the sigmoid cavity, and is broader at the upper part of its circumference than below, by which means the head of the radius is more securely held in its position. Its *outer surface* is strengthened by the external lateral ligament of the elbow, and affords origin to part of the Supinator brevis muscle. Its *inner surface* is smooth, and lined by synovial membrane.

Actions. The movement which takes place in this articulation is limited to rotation of the head of the radius within the orbicular ligament, and upon the lesser sigmoid cavity of the ulna; rotation forwards being called *pronation*; rotation backward, *supination*.

2. Middle Radio-ulnar Articulation.

The interval between the shafts of the radius and ulna is occupied by two ligaments.

Oblique. Interosseous.

The *Oblique* or *Round Ligament* (fig. 132) is a small round fibrous cord, which extends obliquely downwards and outwards, from the tubercle of the ulna at the base of the coronoid process, to the radius a little below the bicipital tuberosity. Its fibres run in the opposite direction to those of the interosseous ligament; and it appears to be placed as a substitute for it in the upper part of the interosseous interval. This ligament is sometimes wanting.

The *Interosseous Membrane* is a broad and thin plane of aponeurotic fibres, descending obliquely downwards and inwards, from the interosseous ridge on the radius to that on the ulna. It is deficient above, commencing about an inch beneath the tubercle of the radius; is broader in the middle than at either extremity; and presents an oval aperture just above its lower margin for the passage of the anterior interosseous vessels to the back of the forearm. This ligament serves to connect the bones, and to increase the extent of surface for the attachment of the deep muscles. Between its upper border and the oblique ligament an interval exists, through which the posterior interosseous vessels pass. Two or three fibrous bands are occasionally found on the posterior surface of this membrane, which descends obliquely from the ulna towards the radius, and which have consequently a direction contrary to that of the other fibres. It is in relation, in *front*, by its upper three-fourths with the Flexor longus pollicis on the outer side, and with the Flexor profundus digitorum on the inner, lying upon the interval between which are the anterior interosseous vessels and nerve, by its lower fourth with the Pronator quadratus; *behind*, with the Supinator brevis, Extensor ossis metacarpi pollicis, Extensor primi internodii pollicis, Extensor secundi internodii pollicis, Extensor indicis; and, near the wrist, with the anterior interosseous artery and posterior interosseous nerve.

3. Inferior Radio-Ulnar Articulation.

This is a lateral ginglymus, formed by the head of the ulna received into the sigmoid cavity at the inner side of the lower end of the radius. The articular surfaces are covered by a thin layer of cartilage, and connected together by the following ligaments.

Anterior Radio-ulnar.
Posterior Radio-ulnar.
Triangular Interarticular Fibro-cartilage.

The *Anterior Radio-ulnar Ligament* (fig. 134) is a narrow band of fibres, extending from the anterior margin of the sigmoid cavity of the radius to the anterior surface of the head of the ulna.

The *Posterior Radio-ulnar Ligament* (fig. 135) extends between similar points on the posterior surface of the articulation.

The *Triangular Fibro-cartilage* (fig. 136) is placed transversely beneath the head of the ulna, binding the lower end of this bone and the radius firmly together. Its circumference is thicker than its centre, which is thin and occasionally perforated. It is attached by its apex to a depression which separates the styloid process of the ulna from the head of that bone; and, by its base, which is thin, to the prominent edge of the radius, which separates the sigmoid cavity from the carpal articulating surface. Its margins are united to the ligaments of the wrist joint. Its *upper surface*, smooth and concave, is contiguous with the head of the ulna; its *under surface*, also concave and smooth, with the cuneiform bone. Both surfaces are lined by a synovial membrane: the upper surface, by one peculiar to the radio-ulnar articulation; the under surface, by the synovial membrane of the wrist.

134.—Ligaments of Wrist and Hand. Anterior View.

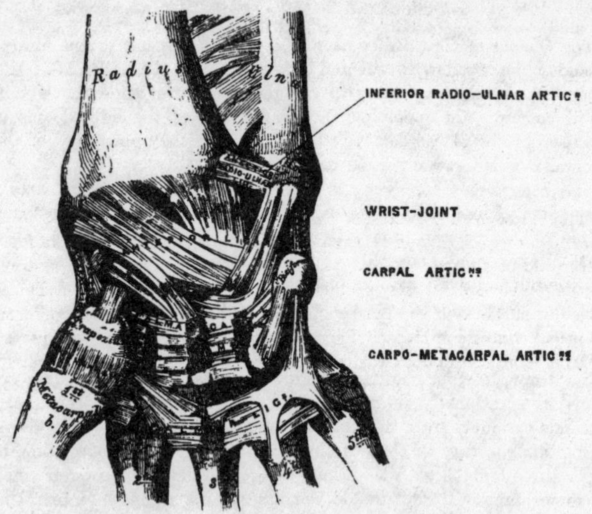

The *Synovial Membrane* (fig. 136) of this articulation has been called, from its extreme looseness, the *membrana sacciformis*; it covers the margin of the articular surface of the head of the ulna, and where reflected from this bone on to the radius,

135.—Ligaments of Wrist and Hand. Posterior View.

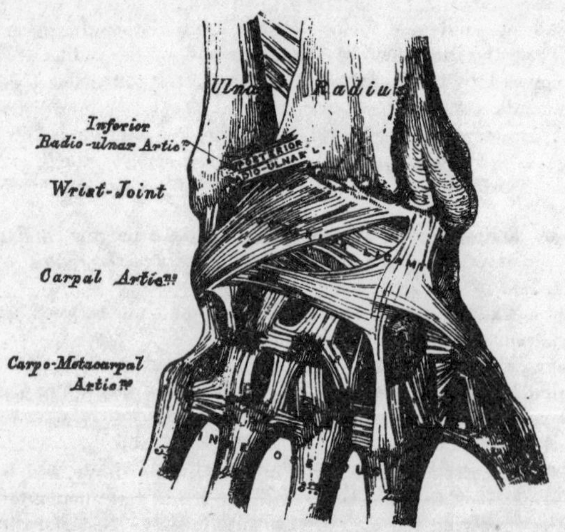

forms a very loose *cul-de-sac*; from the radius, it is continued over the upper surface of the fibro-cartilage. The quantity of synovia which it contains is usually considerable. When the fibro-cartilage is perforated, the synovial membrane is continuous with that which lines the wrist.

Actions. The movement which occurs in the inferior radio-ulnar articulation is just the reverse of that which takes place between the two bones above; it is limited to rotation of the radius round the head of the ulna; rotation forwards being termed *pronation*, rotation backwards *supination*. In pronation, the sigmoid cavity glides forward on the articular edge of the ulna; in supination, it rolls in the opposite direction, the extent of these movements being limited by the anterior and posterior ligaments.

VII. Wrist-Joint.

The *Wrist* presents some of the characters of an enarthrodial joint, but is more correctly regarded as an arthrodia. The parts entering into its formation are, the lower end of the radius, and under surface of the triangular interarticular fibro-cartilage above; and the scaphoid, semilunar, and cuneiform bones below. The articular surfaces of the radius and inter-articular fibro-cartilage form a transversely elliptical concave surface. The radius is subdivided into two parts by a line extending from before backwards; and these, together with the interarticular cartilage, form three facets, one for each carpal bone. The three carpal bones are connected together, and form a convex surface, which is received into the concavity above mentioned. All the bony surfaces of the articulation are covered with cartilage, and connected together by the following ligaments.

External Lateral.	Anterior.
Internal Lateral.	Posterior.

The *External Lateral Ligament* (*radio-carpal*) (fig. 134) extends from the summit of the styloid process of the radius to the outer side of the scaphoid, some of its fibres being prolonged to the trapezium and annular ligament.

The *Internal Lateral Ligament* (*ulno-carpal*) is a rounded cord, attached, above, to the extremity of the styloid process of the ulna; and dividing below into two fasciculi, which are attached, one to the inner side of the cuneiform bone, the other to the pisiform bone and annular ligament.

The *Anterior Ligament* is a broad membranous band, consisting of three fasciculi, attached, above, to the anterior margin of the lower end of the radius, its styloid process, and the ulna; its fibres pass downwards and inwards, to be inserted into the palmar surface of the scaphoid, semilunar, and cuneiform bones. This ligament is perforated by numerous apertures for the passage of vessels, and is in relation, in front, with the tendons of the Flexor profundus digitorum and Flexor longus pollicis; behind, with the synovial membrane of the wrist-joint.

The *Posterior Ligament* (fig. 135), less thick and strong than the anterior, is attached, above, to the posterior border of the lower end of the radius; its fibres pass obliquely downwards and inwards to be attached to the dorsal surface of the scaphoid, semilunar, and cuneiform bones, being continuous with those of the dorsal carpal ligaments. This ligament is in relation, behind, with the extensor tendons of the fingers; in front, with the synovial membrane of the wrist.

The *Synovial Membrane* (fig. 136) lines the under surface of the triangular interarticular fibro-cartilage above; and is reflected on the inner surface of the ligaments just described.

Relations. The wrist-joint is covered in front by the flexor, and behind by the extensor tendons; it is also in relation with the radial and ulnar arteries.

The Arteries supplying the joint are the anterior and posterior carpal branches of the radial and ulnar, the anterior and posterior interosseous, and some ascending branches from the deep palmar arch.

The Nerves are derived from the ulnar.

Actions. The movements permitted in this joint are flexion, extension, abduction, adduction, and circumduction. It is totally incapable of rotation, one of the characteristic movements in true enarthrodial joints.

VIII. Articulations of the Carpus.

These articulations may be subdivided into three sets.

1. The Articulations of the First Row of Carpal Bones.
2. The Articulations of the Second Row of Carpal Bones.
3. The Articulations of the Two Rows with each other.

1. Articulations of the First Row of Carpal Bones.

These are arthrodial joints. The articular surfaces are covered with cartilage, and connected together by the following ligaments:—

Two Dorsal. Two Palmar.
Two Interosseous.

The *Dorsal Ligaments* are placed transversely behind the bones of the first row; they connect the scaphoid and semilunar, and the semilunar and cuneiform.

The *Palmar Ligaments* connect the scaphoid and semilunar, and the semilunar and cuneiform bones; they are less strong than the dorsal, and placed very deep under the anterior ligament of the wrist.

The *Interosseous Ligaments* (fig. 136) are two narrow bundles of fibrous tissue, connecting the semilunar bone, on one side with the scaphoid, on the other with the cuneiform. They close the upper part of the interspaces between the scaphoid, semilunar, and cuneiform bones, their upper surfaces being smooth, and lined by the synovial membrane of the wrist-joint.

The articulation of the pisiform with the cuneiform is provided with a separate synovial membrane, protected by a thin capsular ligament. There are also two strong fibrous fasciculi, which connect this bone to the unciform, and base of the fifth metacarpal bone (fig. 134).

2. Articulations of the Second Row of Carpal Bones.

These are also arthrodial joints. The articular surfaces are covered with cartilage, and connected by the following ligaments:—

Three Dorsal. Three Palmar.
Two Interosseous.

The *three Dorsal Ligaments* extend transversely from one bone to another on the dorsal surface, connecting the trapezium with the trapezoid, the trapezoid with the os magnum, and the os magnum with the unciform.

The *three Palmar Ligaments* have a similar arrangement on the palmar surface.

The *two Interosseous Ligaments*, much thicker than those of the first row, are placed one on each side of the os magnum, connecting it with the trapezoid externally, and the unciform internally. The former is less distinct than the latter.

3. Articulations of the Two Rows of Carpal Bones with each other.

The articulations between the two rows of the carpus consist of a joint in the middle, formed by the reception of the head of the os magnum into a cavity formed by the scaphoid and semilunar bones, and of an arthrodial joint on each side, the outer one formed by the articulation of the scaphoid with the trapezium and trapezoid, the internal one by the articulation of the cuneiform and unciform. The articular surfaces are covered by a thin layer of cartilage, and connected by the following ligaments:—

Anterior or Palmar. External Lateral.
Posterior or Dorsal. Internal Lateral.

The *Anterior* or *Palmar Ligaments* consist of short fibres, which pass obliquely between the bones of the first and second row on the palmar surface.

The *Posterior* or *Dorsal Ligaments* have a similar arrangement on the dorsal surface of the carpus.

The *Lateral Ligaments* are very short; they are placed, one on the radial, the other on the ulnar side of the carpus; the former, the stronger and more distinct, connecting the scaphoid and trapezium bones, the latter the cuneiform and unciform; they are continuous with the lateral ligaments of the wrist-joint.

The *common Synovial Membrane of the Carpus* is very extensive; it lines the under surface of the scaphoid, semilunar, and cuneiform bones, sending upwards two prolongations between their contiguous surfaces; it is then reflected over the bones of the second row, and sends down three prolongations between them, which line their contiguous surfaces, and invest the carpal extremities of the four inner metacarpal bones. There is a separate synovial membrane between the pisiform and cuneiform bones.

Actions. The partial movement which takes place between the bones of each row is very inconsiderable; the movement between the two rows is more marked, but limited chiefly to flexion and extension.

IX. Carpo-Metacarpal Articulations.

1. Articulation of the Metacarpal Bone of the Thumb with the Trapezium.

This is an arthrodial joint, which enjoys great freedom of movement, on account of the shape of its articular surfaces, which are saddle-shaped, so that, on section, each bone appears to be received into a cavity in the other, according to the direction in which they are cut. Hence this joint is sometimes described as one 'by reciprocal reception.' Its ligaments are a capsular and synovial membrane.

The *capsular ligament* is a thick but loose capsule, which passes from the circumference of the upper extremity of the metacarpal bone to the rough edge bounding the articular surface of the trapezium; it is thickest externally and behind, and lined by a separate *synovial membrane.*

2. Articulation of the Metacarpal Bones of the Fingers with the Carpus.

The joints formed between the carpus and four inner metacarpal bones are connected together by dorsal, palmar, and interosseous ligaments.

The *Dorsal Ligaments*, the strongest and most distinct, connect the carpal and metacarpal bones on their dorsal surface. The second metacarpal bone receives two fasciculi, one from the trapezium, the other from the trapezoid; the third metacarpal receives one from the os magnum; the fourth two, one from the os magnum, and one from the unciform; the fifth receives a single fasciculus from the unciform bone.

The *Palmar Ligaments* have a somewhat similar arrangement on the palmar surface, with the exception of the third metacarpal, which has three ligaments, an external one from the trapezium, situated above the sheath of the tendon of the Flexor carpi radialis; a middle one, from the os magnum; and an internal one, from the unciform.

The *Interosseous Ligaments* consist of short thick fibres, which are limited to one part of the carpo-metacarpal articulation; they connect the contiguous inferior angles of the os magnum and unciform with the adjacent surfaces of the third and fourth metacarpal bones.

The *Synovial Membrane* is a continuation of that between the two rows of carpal bones. Occasionally, the articulation of the unciform with the fourth and fifth metacarpal bones has a separate synovial membrane.

The Synovial Membranes of the wrist (fig. 136) are thus seen to be five in number. The *first*, the membrana sacciformis, lines the lower end of the ulna, the sigmoid cavity of the radius, and upper surface of the triangular interarticular fibro-cartilage. The *second* lines the lower end of the radius and interarticular fibro-cartilage above, and the scaphoid, semilunar, and cuneiform bones below. The *third*, the most extensive, covers the contiguous surfaces of the two rows of carpal bones, and, passing between the bones of the second row, lines the carpal extremities of the four inner metacarpal bones. The *fourth* lines the adjacent surfaces of the trapezium and metacarpal bone of the thumb. The *fifth* lines the adjacent surfaces of the cuneiform and pisiform bones.

Actions. The movement permitted in the carpo-metacarpal articulations is limited to a slight gliding of the articular surfaces upon each other, the extent of which varies in the different joints. Thus the articulation of the metacarpal bone of the thumb with the trapezium is most moveable, then the fifth metacarpal, and then the fourth. The second and third are almost immoveable. In the articulation of the metacarpal bone of the thumb with the trapezium, the movements permitted are flexion, extension, adduction, abduction, and circumduction.

136.—Vertical Section through the Articulations at the Wrist, showing the Five Synovial Membranes.

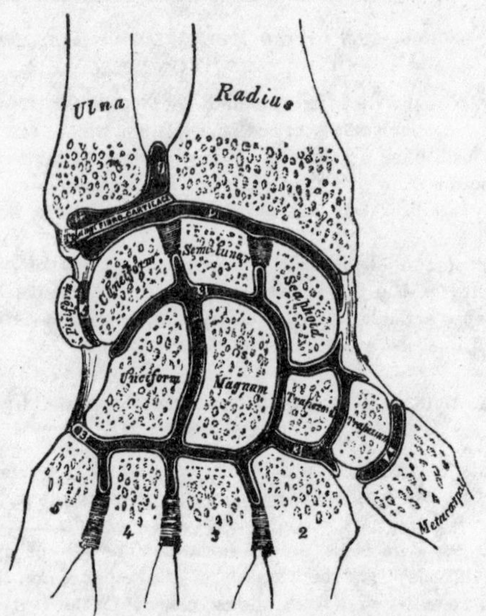

3. Articulations of the Metacarpal Bones with each other.

The carpal extremities of the metacarpal bones articulate with one another at each side by small surfaces covered with cartilage, and connected together by dorsal, palmar, and interosseous ligaments.

The *Dorsal* and *Palmar Ligaments* pass transversely from one bone to another on the dorsal and palmar surfaces. The *Interosseous Ligaments* pass between their contiguous surfaces, just beneath their lateral articular facets

The *Synovial Membrane* lining the lateral facets is a reflection from that between the two rows of carpal bones.

The digital extremities of the metacarpal bones are connected together by a narrow fibrous band, the transverse ligament (fig. 137) which passes transversely across their anterior surfaces, and is blended with the ligaments of the metacarpo-phalangeal articulations. Its *anterior surface* presents four grooves for the passage of the flexor tendons. Its *posterior surface* blends with the ligaments of the metacarpo-phalangeal articulation.

X. Metacarpo-phalangeal Articulations. (Fig. 137.)

These articulations are of the ginglymus kind, formed by the reception of the rounded head of the metacarpal bone, into a superficial cavity in the extremity of the first phalanx. They are connected by the following ligaments :—

Anterior. Two Lateral.

The *Anterior Ligaments* (*Glenoid ligaments* of Cruveilhier) are thick, dense, and fibro-cartilaginous in texture. Each is placed on the palmar surface of the joint, in the interval between the lateral ligaments, to which they are connected; they are loosely united to the metacarpal bone, but very firmly to the base of the first phalanges. Their palmar surface is intimately blended with the transverse ligament, forming a groove for the passage of the flexor tendons, the sheath surrounding which is connected to each side of the groove. By their internal surface, they form part of the articular surface for the head of the metacarpal bone, and are lined by a synovial membrane.

137.—Articulations of the Phalanges.

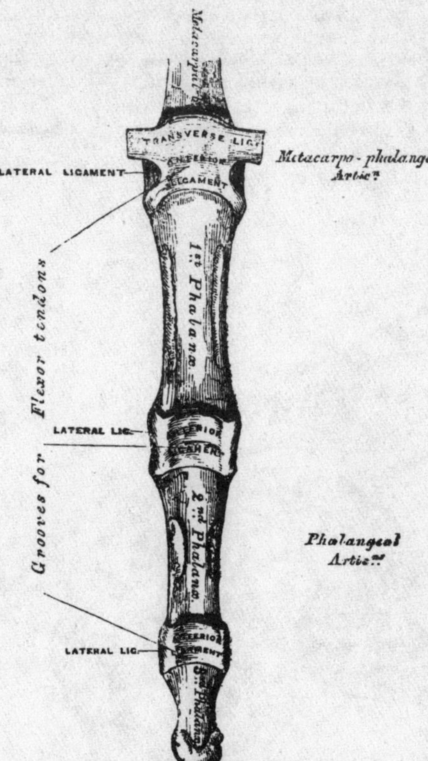

The *Lateral Ligaments* are strong rounded cords, placed one on each side of the joint, each being attached by one extremity to the tubercle on the side of the head of the metacarpal bone, and by the other to the contiguous extremity of the phalanx.

The *posterior ligament* is supplied by the extensor tendon placed over the back of each joint.

Actions. The movements which occur in these joints are flexion, extension, adduction, abduction, and circumduction; the lateral movements are very limited.

XI. Articulations of the Phalanges.

These are ginglymus joints, connected by the following ligaments :—

Anterior. Two Lateral.

The arrangement of these ligaments is similar to those in the metacarpo-phalangeal articulations; the extensor tendon supplies the place of a posterior ligament.

Actions. The only movements permitted in the phalangeal joints are flexion and extension; these movements are more extensive between the first and second phalanges than between the second and third. The movement of flexion is very considerable, but the extension is limited by the anterior and lateral ligaments.

ARTICULATIONS OF THE LOWER EXTREMITY.

The articulations of the Lower Extremity comprise the following groups. I. The hip-joint. II. The knee-joint. III. The articulations between the tibia and fibula. IV. The ankle-joint. V. The articulations of the tarsus. VI. The tarso-metatarsal articulations. VII. The metatarso-phalangeal articulations. VIII. The articulations of the phalanges.

I. Hip-Joint. (Fig. 138.)

This articulation is an enarthrodial, or ball-and-socket joint, formed by the reception of the head of the femur into the cup-shaped cavity of the acetabulum. The articulating surfaces are covered with cartilage, that on the head of the femur being thicker at the centre than at the circumference, and covering the entire

138.—Left Hip-Joint laid open.

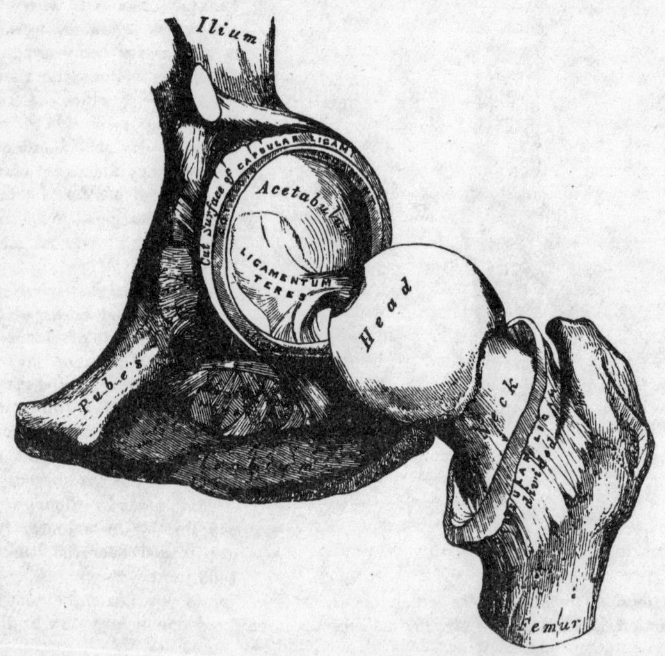

surface with the exception of a depression just below its centre for the ligamentum teres; that covering the acetabulum is much thinner at the centre than at the circumference, being deficient in the situation of the circular depression at the bottom of the cavity. The ligaments of the joint are the

Capsular.	Teres.
Ilio-femoral.	Cotyloid.
Transverse.	

The *Capsular Ligament* is a strong, dense, ligamentous capsule, embracing the margin of the acetabulum above, and surrounding the neck of the femur below. Its *upper circumference* is attached to the acetabulum two or three lines external to the cotyloid ligament; but opposite the notch where the margin of this cavity is deficient, it is connected with the transverse ligament, and by a few fibres to the edge of the obturator foramen. Its *lower circumference* surrounds the neck of the femur, being attached, in front, to the spiral or anterior intertrochanteric line; above, to the base of the neck; behind, to the middle of the neck of the bone, about half an inch above the posterior intertrochanteric line. It is much thicker at the upper and fore part of the joint where the greatest amount of resistance is required, than below, where it is thin, loose, and longer than in any other part. Its external surface (fig. 127) is rough, covered by numerous muscles, and separated in front from the Psoas and Iliacus by a synovial bursa, which not unfrequently communicates by a circular aperture with the cavity of the joint. It differs from the capsular ligament of the shoulder, in being much less loose and lax, and in not being perforated for the passage of a tendon.

The *Ilio-femoral Ligament* (fig. 127) is an accessory band of fibres, extending obliquely across the front of the joint: it is intimately connected with the capsular ligament, and serves to strengthen it in this situation. It is attached, above, to the anterior inferior spine of the ilium; below, to the anterior intertrochanteric line.

The *Ligamentum Teres* is a triangular band of fibres, implanted, by its apex, into the depression a little behind and below the centre of the head of the femur, and by its broad base, which consists of two bundles of fibres, into the margins of the notch at the bottom of the acetabulum, becoming blended with the transverse ligament. It is formed of a bundle of fibres, the thickness and strength of which is very variable, surrounded by a tubular sheath of synovial membrane. Sometimes, only the synovial fold exists, or the ligament may be altogether absent. The use of the round ligament is to check rotation outwards, as well as adduction in the flexed position: it thus assists in preventing dislocation of the head of the femur forwards and outwards, an accident likely to occur from the necessary mechanism of the joint, if not provided against by this ligament and the thick anterior part of the capsule.*

The *Cotyloid Ligament* is a fibro-cartilaginous rim attached to the margin of the acetabulum, the cavity of which it deepens; at the same time it protects the edges of the bone, and fills up the inequalities on its surface. It is prismoid in form, its base being attached to the margin of the acetabulum, and its opposite edge being free and sharp; whilst its two surfaces are invested by synovial membrane, the external one being in contact with the capsular ligament, the internal one being inclined inwards so as to narrow the acetabulum and embrace the cartilaginous surface of the head of the femur. It is much thicker above and behind than below and in front, and consists of close compact fibres, which arise from different points of the circumference of the acetabulum, and interlace with each other at very acute angles.

The *Transverse Ligament* is a strong flattened band of fibres, which crosses the notch at the lower part of the acetabulum, and converts it into a foramen. It is continuous at each side with the cotyloid ligament. An interval is left beneath the ligament for the passage of nutrient vessels to the joint.

The *Synovial Membrane* is very extensive. Commencing at the margin of the cartilaginous surface of the head of the femur, it covers all that portion of the neck which is contained within the joint; from the head it is reflected on the internal surface of the capsular ligament, covers both surfaces of the cotyloid ligament, and the mass of fat contained in the fossa at the bottom of the acetabulum, and is prolonged in the form of a tubular sheath around the ligamentum teres, as far as the head of the femur.

The Muscles in relation with the joint are, in front, the Psoas and Iliacus,

* See an interesting paper, 'On the Use of the Round Ligament of the Hip-Joint,' by Dr. J. Struthers. *Edinburgh Medical Journal*, 1858.

separated from the capsular ligament by a synovial bursa; above, the short head of the Rectus and Gluteus minimus, the latter being closely adherent to the capsule; internally, the Obturator externus and Pectineus; behind, the Pyriformis, Gemellus superior, Obturator internus, Gemellus inferior, Obturator externus, and Quadratus femoris.

The arteries supplying the joint are derived from the obturator, sciatic, internal circumflex, and gluteal.

The Nerves are articular branches from the sacral plexus, great sciatic, obturator, and accessory obturator nerves.

Actions. The movements of the hips, like all enarthrodial joints, are very extensive; they are, flexion, extension, adduction, abduction, circumduction, and rotation.

II. Knee-Joint.

The knee is a ginglymus, or hinge-joint; the bones entering into its formation are, the condyles of the femur above, the head of the tibia below, and the patella in front. The articular surfaces are covered with cartilage, and connected together by ligaments, some of which are placed on the exterior of the joint, whilst others occupy its interior.

External Ligaments.	*Internal Ligaments.*
Anterior, or Ligamentum Patellæ.	Anterior, or External Crucial.
Posterior, or Ligamentum Posticum Winslowii.	Posterior, or Internal Crucial.
Internal Lateral.	Two Semilunar Fibro-cartilages.
Two External Lateral.	Transverse.
Capsular.	Coronary.
	Ligamentum mucosum.
	Ligamenta alaria.

The *Anterior Ligament*, or *Ligamentum Patellæ* (fig. 139) is that portion of the common tendon of the extensor muscles of the thigh which is continued from the patella to the tubercle of the tibia, supplying the place of an anterior ligament. It is a strong, flat, ligamentous band, about three inches in length, attached, above, to the apex of the patella and the rough depression on its posterior surface; below, to the lower part of the tuberosity of the tibia; its superficial fibres being continuous across the front of the patella with those of the tendon of the Quadriceps extensor. Two synovial bursæ are connected with this ligament and the patella; one is interposed between the patella and the skin covering its anterior surface; the other of small size, between the ligamentum patellæ and the upper part of the tuberosity of the tibia. The posterior surface of this ligament is separated above from the knee-joint by a large mass of adipose tissue; its lateral margins are continuous with the aponeuroses derived from the Vasti muscles.

The *Posterior Ligament, Ligamentum Posticum Winslowii* (fig. 140), is a broad, flat, fibrous band, which covers over the whole of the back part of the joint. It consists of two lateral portions, formed chiefly of vertical fibres, which arise above from the condyles of the femur, and are connected below with the back part of the head of the tibia, being closely united with the tendons of the Gastrocnemius, Plantaris, and Popliteus muscles; the central portion is formed of fasciculi, obliquely directed and separated from one another by apertures for the passage of vessels. The strongest of these fasciculi is derived from the tendon of the Semi-membranosus, and passes from the back part of the inner tuberosity of the tibia, obliquely upwards and outwards to the back part of the outer condyle of the femur. The posterior ligament forms part of the floor of the popliteal space, and the popliteal artery rests upon it.

The *Internal Lateral Ligament* is a broad, flat, membranous band thicker behind than in front, and situated nearer to the back than the front of the joint. It is attached, above, to the inner tuberosity of the femur; below, to the inner tuberosity and inner surface of the shaft of the tibia, to the extent of about two inches. It is crossed, at its lower part, by the aponeurosis of the Sartorius, and the tendons of the Gracilis and Semitendinosus muscles, a synovial bursa being interposed. Its *deep surface* covers the anterior portion of the tendon of the Semimembranosus, the synovial membrane of the joint, and the inferior internal articular artery; it is intimately adherent to the internal semilunar fibro-cartilage.

139.—Right Knee-joint. Anterior View.

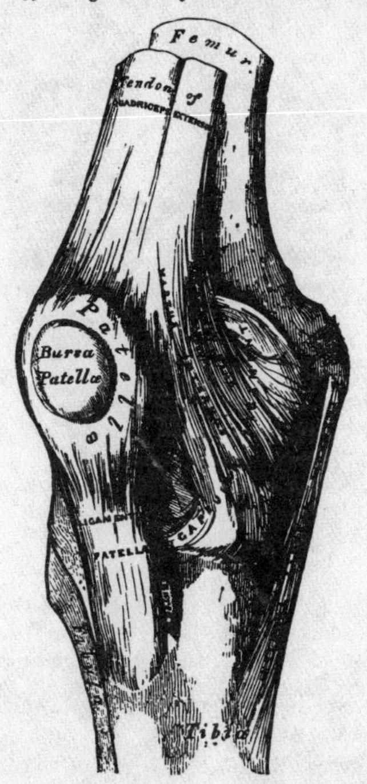

140.—Right Knee-joint. Posterior View.

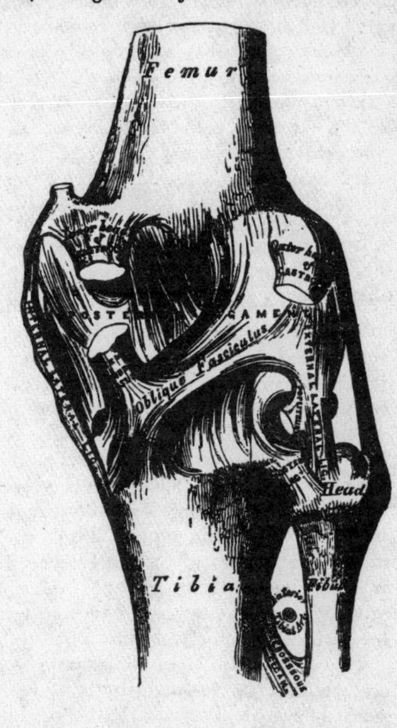

The *Long External Lateral Ligament* is a strong, rounded, fibrous cord, situated nearer to the back than the front of the joint. It is attached, above, to the outer condyle of the femur; below, to the outer part of the head of the fibula. Its *outer surface* is covered by the tendon of the Biceps, which divides at its insertion into two parts, separated by the ligament. The ligament has, passing beneath it, the tendon of the Popliteus muscle, and the inferior external articular vessels and nerve.

The *Short External Lateral Ligament* is an accessory bundle of fibres, placed behind and parallel with the preceding; attached, above, to the lower part of the outer condyle of the femur: below, to the summit of the styloid process of the fibula. This ligament is intimately connected with the capsular ligament, and has, passing beneath it, the tendon of the Popliteus muscle.

The *Capsular Ligament* consists of an exceedingly thin, but strong, fibrous membrane, which fills in the intervals left by the preceding ligaments. It is attached to the femur immediately above its articular surface; below, to the upper border and sides of the patella and the margins of the head of the tibia and inter-articular cartilages, and is continuous behind with the posterior ligament. This membrane is strengthened by fibrous expansions, derived from the fascia lata, from the Vasti and Crureus muscles, and from the Biceps, Sartorius, and tendon of the Semimembranosus.

The *Crucial* are two interosseous ligaments of considerable strength, situated in the interior of the joint, nearer its posterior than its anterior part. They are called *crucial*, because they cross each other, somewhat like the lines of the letter X; and have received the names *anterior* and *posterior*, from the position of their attachment to the tibia.

141.—Right Knee-Joint. Showing Internal Ligaments.

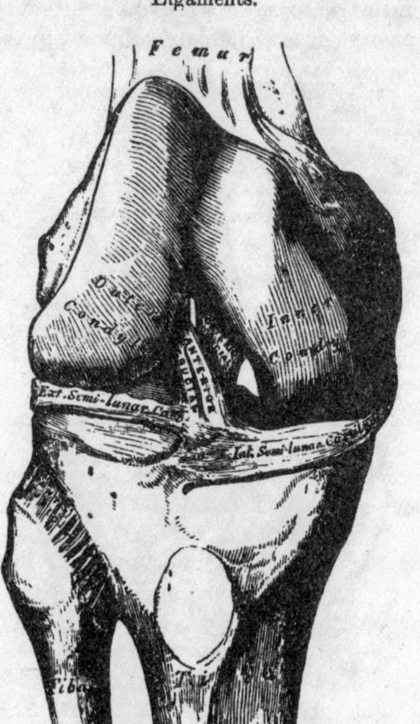

The *Anterior* or *External Crucial Ligament* (fig. 141), smaller than the posterior, is attached to the inner side of the depression in front of the spine of the tibia, being blended with the anterior extremity of the external semilunar fibro-cartilage, and passing obliquely upwards, backwards, and outwards, is inserted into the inner and back part of the outer condyle of the femur.

The *Posterior* or *Internal Crucial Ligament* is larger in size, but less oblique in its direction than the anterior. It is attached to the back part of the depression behind the spine of the tibia, and to the posterior extremity of the external semilunar fibro-cartilage; and passes upwards, forwards, and inwards, to be inserted into the outer and fore-part of the inner condyle of the femur. As it crosses the anterior crucial ligament, a fasciculus is given off from it, which blends with the posterior part of that ligament. It is in relation, in front, with the anterior crucial ligament; behind, with the ligamentum posticum Winslowii.

The *Semilunar Fibro-Cartilages* (fig. 142) are two crescentic lamellæ, which are attached to the margins of the head of the tibia, and serve to deepen its surface for articulation with the condyles of the femur. The circumference of each cartilage is thick and convex; the inner free border, thin and concave. Their upper surfaces are concave, and in relation with the condyles of the femur; their lower surfaces are flat, and rest upon the head of the tibia. Each cartilage covers nearly the outer two-thirds of the corresponding articular surface of the tibia, leaving the inner third uncovered; both surfaces are smooth, and invested by synovial membrane.

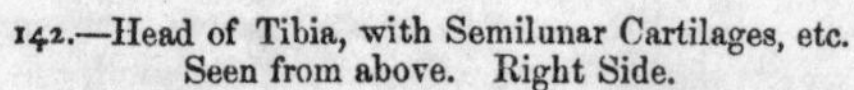
142.—Head of Tibia, with Semilunar Cartilages, etc. Seen from above. Right Side.

The *Internal Semilunar Fibro-Cartilage* is nearly semicircular in form, a little elongated from before backwards, and broader behind than in front; its convex border is united to the internal lateral ligament, and to the head of the tibia, by means of the coronary ligaments; its anterior extremity, thin and pointed, is firmly implanted into a depression in front of the inner articular surface of the tibia; its posterior extremity into the depression behind the spine, between the attachment of the external cartilage and posterior crucial ligament.

The *External Semilunar Fibro-Cartilage* forms nearly an entire circle, covering a larger portion of the articular surface than the internal one. It is grooved on its outer side, for the tendon of the Popliteus muscle. Its circumference is held in connection with the head of the tibia, by means of the coronary ligaments; and its two extremities are firmly implanted in the depressions in front and behind the spine of the tibia. These extremities, at their insertion, are interposed between the attachments of the internal cartilage. The external semilunar fibro-cartilage gives off from its *anterior border* a fasciculus, which forms the transverse ligament. By its *anterior extremity*, it is continuous with the anterior crucial ligament. Its *posterior extremity* divides into three slips; two of these pass upwards and forwards, and are inserted into the outer side of the inner tuberosity of the tibia, one in front, the other behind the posterior crucial ligament; the third fasciculus is inserted into the back part of the anterior crucial ligament.

The *Transverse Ligament* is a band of fibres, which passes transversely from the anterior convex margin of the external semilunar cartilage to the anterior extremity of the internal cartilage; its thickness varies considerably in different subjects.

The *Coronary Ligaments* consist of numerous short fibrous bands, which connect the convex border of the semilunar cartilages with the circumference of the head of the tibia, and with the other ligaments surrounding the joint.

The *Synovial Membrane* of the knee-joint is the largest and most extensive in the body. Commencing at the upper border of the patella, it forms a large *cul-de-sac* beneath the Extensor tendon of the thigh: this is sometimes replaced by a synovial bursa interposed between the tendon and the front of the femur, which in some subjects communicates with the synovial membrane of the knee-joint, by an orifice of variable size. On each side of the patella, the synovial membrane extends beneath the aponeurosis of the Vasti muscles, and more especially beneath that of the Vastus internus; and, below the patella, it is separated from the anterior ligament by a considerable quantity of adipose tissue. In this situation it sends off a triangular prolongation, containing a few ligamentous fibres, which extends from the anterior part of the joint below the patella, to the front of the inter-condyloid notch. This fold has been termed the *ligamentum mucosum.* The *ligamenta alaria* consist of two fringe-like folds, which extend from the sides of the ligamentum mucosum, upwards and outwards, to the sides of the patella. The synovial membrane invests the semilunar fibro-cartilages, and on the back part of the external one forms a *cul-de-sac* between the groove on its surface, and the tendon of the Popliteus; it is continued to the articular surface of the tibia: surrounds the crucial ligaments, and the inner surface of the ligaments which enclose the joint; lastly, it approaches the condyles of the femur, and from them is continued on to the lower part of the front of the shaft. The pouch of synovial membrane between the Extensor tendon and front of the femur is supported, during the movements of the knee, by a small muscle, the Subcruræus, which is inserted into it.

The Arteries supplying the joint are derived from the anastomotic branch of the femoral, articular branches of the popliteal, and recurrent branch of the anterior tibial.

The Nerves are derived from the obturator, anterior crural, and external and internal popliteal.

Actions.—The knee-joint allows of movements of flexion and extension, and of slight rotation inwards and outwards. The complicated mechanism of this joint renders it necessary to study each of these movements separately, pointing out incidentally the functions of each of the principal components of the joint. The tibia executes a rotatory movement during flexion around an imaginary axis drawn transversely through its upper end. This causes a change in the apposition of the tibia and femur. Thus, in extreme extension, it is the anterior portion of the tibia which is in contact with the femur; in the semiflexed position, its middle; in complete flexion, its posterior edge.* Also, during *flexion* the articular surface of the tibia, covered by the interarticular cartilages, glides backwards on the femur. The patella is attached by the inextensible ligamentum patellæ to the tubercle of the tibia, and as the tibia glides backwards, the patella falls more and more into the intercondyloid notch of the femur. The ligamentum patellæ is put on the stretch during flexion, as is also the posterior crucial ligament in extreme flexion. The other ligaments are all relaxed by flexion of the joint, though the relaxation of the anterior crucial ligament is very trifling. In partial flexion of the knee before the ligamentum patellæ comes upon the stretch, and while both crucial ligaments are somewhat relaxed, some rotation of the joint is permitted. Flexion is only checked during life by the contact of the leg with the thigh. In *extension*, the ligamentum patellæ becomes relaxed, and, in extreme extension, completely so, so as to allow free lateral movement to the patella, which then rests on the front of the condyles of the femur. The other ligaments are all on the stretch. When the limb has been brought into a straight line extension is checked, mainly by the tension of the posterior crucial ligament. The movements of *rotation* of which the knee is susceptible are permitted in the semiflexed condition by the partial relaxation of both crucial ligaments, as well as the lateral ligaments. Rotation inwards (or pronation of the leg) is checked by the anterior crucial ligament. The chief agent in effecting this movement is the Popliteus muscle. Rotation outwards (or supination) is checked by the posterior crucial ligament. It is effected mainly by the Biceps. The main function of the crucial ligaments is to act as a direct bond of union between the tibia and femur, preventing the former bone from being carried too far backwards or forwards. They also assist the lateral ligaments in resisting any lateral bending of the joint. The interarticular cartilages are intended, as it seems, to adapt the surface of the tibia to the shape of the femur to a certain extent, so as to fill up the intervals which would otherwise be left in the varying positions of the joint, and to interrupt the jars which would be so frequently transmitted up the limb in jumping or falls on the feet. The patella is a great defence to the knee-joint from any injury inflicted in front, and it distributes upon a large and tolerably even surface during kneeling the pressure which would otherwise fall upon the prominent ridges of the condyles: it also affords leverage to the Quadriceps extensor muscle to act upon the tibia, and Mr. Ward has pointed out † how this leverage varies in the various positions of the joint, so that the action of the muscle produces velocity at the expense of force in the commencement of extension, and, on the contrary, at the close of extension tends to diminish the velocity, and therefore the shock to the ligaments; whilst in the standing position it draws the tibia powerfully forwards, and thus maintains it in its place.

The folds of synovial membrane and the fatty processes contained in them act, as it seems, mainly as padding to fill up interspaces and obviate concussions.

The bursæ in connection with the synovial membrane will be found described in connection with the regional anatomy of the popliteal space.

III. Articulation between the Tibia and Fibula.

The articulations between the tibia and fibula are effected by ligaments which connect both extremities, as well as the shafts of the bone. They may, consequently, be subdivided into three sets. 1. The Superior Tibio-Fibular articulation. 2. The Middle Tibio-Fibular articulation. 3. The Inferior Tibio-Fibular articulation.

1. Superior Tibio-Fibular Articulation.

This articulation is anarthrodial joint. The contiguous surfaces of the bones present two flat oval surfaces covered with cartilage, and connected together by the following ligaments.

Anterior Superior Tibio-Fibular.
Posterior Superior Tibio-Fibular.

* See Plate XLVII. in Humphry, on the 'Skeleton.' † 'Human Osteology,' p. 405.

The *Anterior Superior Ligament* (fig. 141) consists of two or three broad and flat bands, which pass obliquely upwards and inwards, from the head of the fibula to the outer tuberosity of the tibia.

The *Posterior Superior Ligament* is a single thick and broad band, which passes from the back part of the head of the fibula to the back part of the outer tuberosity of the tibia. It is covered by the tendon of the Popliteus muscle.

A *Synovial Membrane* lines this articulation. It is occasionally continuous with that of the knee-joint at its upper and back part.

2. Middle Tibio-Fibular Articulation.

An interosseous membrane extends between the contiguous margins of the tibia and fibula, and separates the muscles on the front from those on the back of the leg. It consists of a thin aponeurotic lamina composed of oblique fibres, which pass between the interosseous ridges on the two bones. It is broader above than below, and presents at its upper part a large oval aperture for the passage of the anterior tibial artery forwards to the anterior aspect of the leg; and at its lower part an opening for the passage of the anterior peroneal vessels. It is continuous below with the inferior interosseous ligament; and is perforated in numerous parts for the passage of small vessels. It is in relation in front with the Tibialis anticus, Extensor longus digitorum, Extensor proprius pollicis, Peroneus tertius, and the anterior tibial vessels and nerve; behind, with the Tibialis posticus and Flexor longus pollicis.

3. Inferior Tibio-Fibular Articulation.

This articulation is formed by the rough convex surface of the inner side of the lower end of the fibula, connected with a similar rough surface on the outer side of the tibia. Below, to the extent of about two lines, these surfaces are smooth and covered with cartilage, which is continuous with that of the ankle-joint. The ligaments of this joint are—

Inferior Interosseous.
Anterior Inferior Tibio-fibular.
Posterior Inferior Tibio-fibular.
Transverse.

The *Inferior Interosseous Ligament* consists of numerous short, strong fibrous bands, which pass between the contiguous rough surfaces of the tibia and fibula, and constitute the chief bond of union between the bones. This ligament is continuous, above, with the interosseous membrane.

The *Anterior Inferior Ligament* (fig. 144) is a flat triangular band of fibres, broader below than above, which extends obliquely downwards and outwards between the adjacent margins of the tibia and fibula on the front aspect of the articulation. It is in relation, in front, with the Peroneus tertius, the aponeurosis of the leg, and the integument; behind, with the inferior interosseous ligament; and lies in contact with the cartilage covering the astragalus.

The *Posterior Inferior Ligament*, smaller than the preceding, is disposed in a similar manner on the posterior surface of the articulation.

The *Transverse Ligament* is a long narrow band, continuous with the preceding, passing transversely across the back of the joint, from the external malleolus to the tibia, a short distance from its malleolar process. This ligament projects below the margin of the bones, and forms part of the articulating surface for the astragalus.

The *Synovial Membrane* lining the articular surfaces is derived from that of the ankle-joint.

Actions. The movement permitted in these articulations is limited to a very slight gliding of the articular surfaces one upon another.

IV. Ankle-Joint.

The *Ankle* is a ginglymus or hinge joint. The bones entering into its formation are the lower extremity of the tibia and its malleolus, and the malleolus of the fibula. These bones are united above, and form an arch, to receive the upper convex surface of the astragalus and its two lateral facets. The bony surfaces are covered with cartilage, and connected together by the following ligaments—

Anterior. Internal Lateral.
External Lateral.

The *Anterior* or *Tibio-tarsal Ligament* (fig. 143) is a broad, thin, membranous layer, attached, above, to the margin of the articular surface of the tibia; below

143.—Ankle-Joint: Tarsal and Tarso-Metatarsal Articulations. Internal View. Right Side.

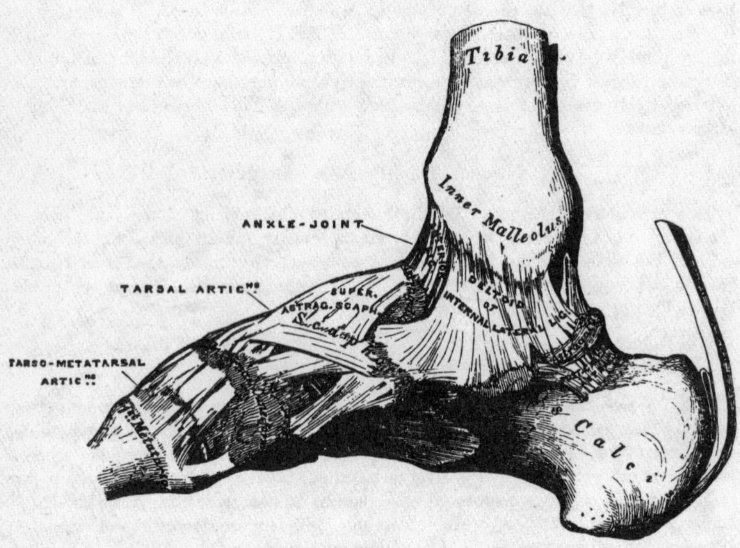

to the margin of the astragalus, in front of its articular surface. It is in relation, in front, with the extensor tendons of the toes, with the tendons of the Tibialis anticus and Peroneus tertius, and the anterior tibial vessels and nerve; behind, it lies in contact with the synovial membrane.

The *Internal Lateral* or *Deltoid Ligament* consists of two layers, superficial and deep. The *superficial layer* is a strong, flat, triangular band, attached, above, to the apex and anterior and posterior borders of the inner malleolus. The most anterior fibres pass forwards to be inserted into the scaphoid; the middle descend almost perpendicularly to be inserted into the os calcis; and the posterior fibres pass backwards and outwards to be attached to the inner side of the astragalus. The *deeper layer* consists of a short, thick, and strong fasciculus, which passes from the apex of the malleolus to the inner surface of the astragalus, below the articular surface. This ligament is covered by the tendons of the Tibialis posticus and Flexor longus digitorum muscles.

The *External Lateral Ligament* (fig. 144) consists of three fasciculi, taking different directions, and separated by distinct intervals; for which reason it is described by some anatomists as three distinct ligaments.* This would seem the preferable description, were it not that the old nomenclature has passed into general use.

The *anterior fasciculus*, the shortest of the three, passes from the anterior margin of the summit of the external malleolus, downwards and forwards, to the astragalus, in front of its external articular facet.

144.—Ankle-Joint: Tarsal and Tarso-Metatarsal Articulations. External View. Right Side.

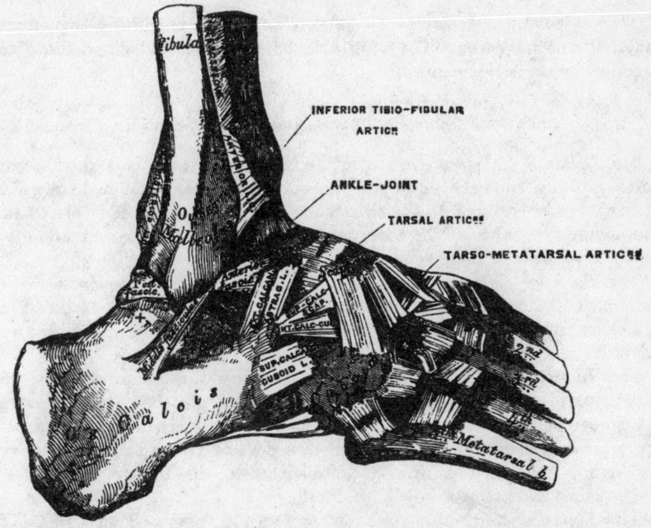

The *posterior fasciculus*, the most deeply seated, passes from the depression at the inner and back part of the external malleolus to the astragalus, behind its external malleolar facet. Its fibres are almost horizontal in direction.

The *middle fasciculus*, the longest of the three, is a narrow rounded cord, passing from the apex of the external malleolus downwards and slightly backwards to the middle of the outer side of the os calcis. It is covered by the tendons of the Peroneus longus and brevis.

There is no posterior ligament, its place being supplied by the transverse ligament of the inferior tibio-fibular articulation.

The *Synovial Membrane* invests the inner surface of the ligaments, and sends a duplicature upwards between the lower extremities of the tibia and fibula for a short distance.

Relations. The tendons, vessels, and nerves in connection with the joint are, in front, from within outwards, the Tibialis anticus, Extensor proprius pollicis, anterior tibial vessels, anterior tibial nerve, Extensor communis digitorum, and Peroneus tertius: behind, from within outwards, Tibialis posticus, Flexor longus digitorum, posterior tibial vessels, posterior tibial nerve, Flexor longus pollicis and, in the groove behind the external malleolus, the tendons of the Peroneus longus and brevis.

The *Arteries* supplying the joint are derived from the malleolar branches of the anterior tibial and peroneal.

* Humphry, on the 'Skeleton,' p. 559.

The *Nerves* are derived from the anterior tibial.

Actions. The movements of the joint are limited to flexion and extension. There is no lateral motion.

V. Articulations of the Tarsus.

These articulations may be subdivided into three sets: 1. The articulation of the first row of tarsal bones. 2. The articulation of the second row of tarsal bones. 3. The articulations of the two rows with each other.

1. Articulation of the First Row of Tarsal Bones.

The articulations between the astragalus and os calcis are two in number—anterior and posterior. They are arthrodial joints. The bones are connected together by three ligaments:—

External Calcaneo-Astragaloid. Posterior Calcaneo-Astragaloid.
Interosseous.

The *External Calcaneo-Astragaloid Ligament* (fig. 144) is a short strong fasciculus, passing from the outer surface of the astragalus, immediately beneath its external malleolar facet, to the outer edge of the os calcis. It is placed in front of the middle fasciculus of the external lateral ligament of the ankle-joint, with the fibres of which it is parallel.

The *Posterior Calcaneo-Astragaloid Ligament* (fig. 143) connects the posterior extremity of the astragalus with the upper contiguous surface of the os calcis; it is a short narrow band, the fibres of which are directed obliquely backwards and inwards.

The *Interosseous Ligament* forms the chief bond of union between the bones. It consists of numerous vertical and oblique fibres, attached, by one extremity, to the groove between the articulating surfaces of the astragalus; by the other, to a corresponding depression on the upper surface of the os calcis. It is very thick and strong, being at least an inch in breadth from side to side, and serves to unite the os calcis and astragalus solidly together.

The *Synovial Membranes* (fig. 146) are two in number: one for the posterior calcaneo-astragaloid articulation; a second for the anterior calcaneo-astragaloid joint. The latter synovial membrane is continued forwards between the contiguous surfaces of the astragalus and scaphoid bones.

2. Articulations of the Second Row of Tarsal Bones.

The articulations between the scaphoid, cuboid, and three cuneiform are effected by the following ligaments:—

Dorsal. Plantar.
Interosseous.

The *Dorsal Ligaments* are small bands of parallel fibres, which pass from each bone to the neighbouring bones with which it articulates.

The *Plantar Ligaments* have the same arrangement on the plantar surface.

The *Interosseous Ligaments* are four in number. They consist of strong transverse fibres, which pass between the rough non-articular surfaces of adjoining bones. There is one between the sides of the scaphoid and cuboid; a second between the internal and middle cuneiform bones; a third between the middle and external cuneiform; and a fourth between the external cuneiform and cuboid. The scaphoid and cuboid, when in contact, present each a small articulating facet, covered with cartilage, and lined either by a separate synovial membrane, or by an offset from the common tarsal synovial membrane.

3. Articulations of the Two Rows of the Tarsus with each other.

These may be conveniently divided into three sets. The joint between the os calcis and the cuboid. The ligaments connecting the os calcis with the scaphoid. The joint between the astragalus and the scaphoid.

The ligaments connecting the os calcis with the cuboid are four in number:—

Dorsal. { Superior Calcaneo-Cuboid.
Internal Calcaneo-Cuboid (Interosseous).

Plantar. { Long Calcaneo-Cuboid.
Short Calcaneo-Cuboid.

145.—Ligaments of Plantar Surface of the Foot.

The *Superior Calcaneo-Cuboid Ligament* (fig. 144) is a thin and narrow fasciculus which passes between the contiguous surfaces of the os calcis and cuboid, on the dorsal surface of the joint.

The *Internal Calcaneo-Cuboid (Interosseous) Ligament* (fig. 144) is a short, but thick and strong, band of fibres, arising from the Os calcis, in the deep groove which intervenes between it and the astragalus; and closely blended, at its origin, with the superior calcaneo-scaphoid ligament. It is inserted into the inner side of the cuboid bone. This ligament forms one of the chief bonds of union between the first and second row of the tarsus.

The *Long Calcaneo-Cuboid* (fig. 145), the more superficial of the two plantar ligaments, is the longest of all the ligaments of the tarsus; it is attached to the under surface of the os calcis, from near the tuberosities, as far forwards as the anterior tubercle; its fibres pass forwards to be attached to the ridge on the under surface of the cuboid bone, the more superficial fibres being continued onwards to the bases of the second, third, and fourth metatarsal bones. This ligament crosses the groove on the under surface of the cuboid bone, converting it into a canal for the passage of the tendon of the Peroneus longus.

The *Short Calcaneo-Cuboid Ligament* lies nearer to the bones than the preceding, from which it is separated by a little areolar adipose tissue. It is exceedingly broad, about an inch in length, and extends from the tubercle and the depression in front of it on the fore part of the under surface of the os calcis, to the inferior surface of the cuboid bone behind the peroneal groove. A separate synovial membrane is found in the calcaneo-cuboid articulation.

The ligaments connecting the os calcis with the scaphoid are two in number:—

Superior Calcaneo-Scaphoid. Inferior Calcaneo-Scaphoid.

The *Superior Calcaneo-Scaphoid* (fig. 144) arises, as already mentioned, with the internal calcaneo-cuboid in the deep groove between the astragalus and os calcis; it passes forward from the inner side of the anterior extremity of the os calcis to the outer side of the scaphoid bone. These two ligaments resemble the letter Y, being blended together behind, but separated in front.

The *Inferior Calcaneo-Scaphoid* (fig. 145) is by far the larger and stronger of the two ligaments of this articulation; it is a broad and thick band of fibres, which passes forwards and inwards from the anterior and inner extremity of the os calcis to the under surface of the scaphoid bone. This ligament not only serves to connect the os calcis and scaphoid, but supports the head of the astragalus, forming part of the articular cavity in which it is received. Its *upper surface* is lined by the synovial membrane continued from the anterior calcaneo-astragaloid articulation. Its *under surface* is in contact with the tendon of the Tibialis posticus muscle.*

The articulation between the astragalus and scaphoid is an arthrodial joint; the rounded head of the astragalus being received into the concavity formed by the posterior surface of the scaphoid, the anterior articulating surface of the calcaneum, and the upper surface of the calcaneo-scaphoid ligament, which fills up the triangular interval between those bones. The only ligament of this joint is the superior astragalo-scaphoid, a broad band, which passes obliquely forwards from the neck of the astragalus to the superior surface of the scaphoid bone. It is thin and weak in texture, and covered by the Extensor tendons. The inferior calcaneo-scaphoid supplies the place of an inferior ligament.

The *Synovial Membrane* which lines the joint is continued forwards from the anterior calcaneo-astragaloid articulation. This articulation permits of considerable mobility; but its feebleness is such as to allow occasionally of dislocation of the astragalus.

146.—Oblique Section of the Articulations of the Tarsus and Metatarsus. Showing the Six Synovial Membranes.

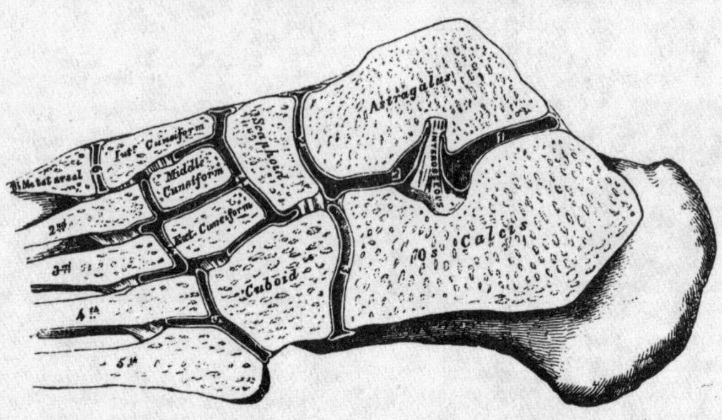

The *Synovial Membranes* (fig. 146) found in the articulations of the tarsus, are four in number; *one* for the posterior calcaneo-astragaloid articulation; a *second* for the anterior calcaneo-astragaloid and astragalo-scaphoid articulations; a *third* for the calcaneo-cuboid articulation; and a *fourth* for the articulations of the scaphoid with the three cuneiform, the three cuneiform with each other, the external cuneiform with the cuboid, and the middle and external cuneiform with the bases of the second and third metatarsal bones. The prolongation which lines the metatarsal bones passes forwards between the external and middle cuneiform bones. A small synovial membrane is sometimes found between the contiguous surfaces of the scaphoid and cuboid bones.

* Mr. Hancock describes an extension of this ligament upwards on the inner side of the foot, which completes the socket of the joint in that direction. *Lancet*, 1866, vol. i. p. 618.

Actions. The movements permitted between the bones of the first row, the astragalus and os calcis, are limited to a gliding upon each other from before backwards, and from side to side. The gliding movement which takes place between the bones of the second row is very slight, the articulation between the scaphoid and cuneiform bones being more moveable than those of the cuneiform with each other and with the cuboid. The movement which takes place between the two rows is more extensive, and consists in a sort of rotation, by means of which the sole of the foot may be slightly flexed, and extended, or carried inwards and outwards.

VI. Tarso-Metatarsal Articulations.

These are arthrodial joints. The bones entering into their formation are the internal, middle and external cuneiform, and the cuboid, which articulate with the metatarsal bones of the five toes. The metatarsal bone of the great toe articulates with the internal cuneiform; that of the second is deeply wedged in between the internal and external cuneiform, resting against the middle cuneiform, and being the most strongly articulated of all the metatarsal bones; the third metatarsal articulates with the extremity of the external cuneiform; the fourth with the cuboid and external cuneiform; and the fifth with the cuboid. The articular surfaces are covered with cartilage, lined by synovial membrane, and connected together by the following ligaments—

Dorsal. Plantar.
Interosseous.

The *Dorsal Ligaments* consist of strong, flat, fibrous bands, which connect the tarsal with the metatarsal bones. The first metatarsal is connected to the internal cuneiform by a single broad, thin, fibrous band; the second has three dorsal ligaments, one from each cuneiform bone; the third has one from the external cuneiform; and the fourth and fifth have one each from the cuboid.

The *Plantar Ligaments* consist of longitudinal and oblique fibrous bands connecting the tarsal and metatarsal bones, but disposed with less regularity than on the dorsal surface. Those for the first and second metatarsal are the most strongly marked; the second and third metatarsal receive strong fibrous bands, which pass obliquely across from the internal cuneiform; the plantar ligaments of the fourth and fifth metatarsal consist of a few scanty fibres derived from the cuboid.

The *Interosseous Ligaments* are three in number: internal, middle, and external. The *internal* one passes from the outer extremity of the internal cuneiform to the adjacent angle of the second metatarsal. The *middle* one, less strong than the preceding, connects the external cuneiform with the adjacent angle of the second metatarsal. The *external* interosseous ligament connects the outer angle of the external cuneiform with the adjacent side of the third metatarsal.

The *Synovial Membranes* of these articulations are three in number: one for the metatarsal bone of the great toe, with the internal cuneiform: one for the second and third metatarsal bones, with the middle and external cuneiform; this is a part of the great tarsal synovial membrane: and one for the fourth and fifth metatarsal bones with the cuboid. The synovial membranes of the tarsus and metatarsus are thus seen to be six in number (fig. 146).

Articulations of the Metatarsal Bones with each other.

The bases of the metatarsal bones, except the first, are connected together by dorsal, plantar, and interosseous ligaments. The *dorsal* and *plantar ligaments* pass from one metatarsal bone to another. The *interosseous ligaments* lie deeply between the rough non-articular portions of their lateral surfaces. The articular surfaces are covered with cartilage, and provided with synovial membrane, continued forwards from the tarso-metatarsal joints. The digital extremities of the metatarsal bones are united by the transverse metatarsal ligament. It connects the great toe with the rest of the metatarsal bones; in this respect it differs from the transverse ligament in the hand.

Actions. The movement permitted in the tarsal ends of the metatarsal bones is limited to a slight gliding of the articular surfaces upon one another; considerable motion, however, takes place in the digital extremities.

VII. Metatarso-Phalangeal Articulations.

The heads of the metatarsal bones are connected with the concave articular surfaces of the first phalanges by the following ligaments—

Plantar. Two Lateral.

They are arranged precisely similar to the corresponding parts in the hand. The expansion of the extensor tendon supplies the place of a dorsal ligament.

Actions. The movements permitted in the metatarso-phalangeal articulations are flexion, extension, abduction, and adduction.

VIII. Articulations of the Phalanges.

The ligaments of these articulations are similar to those found in the hand; each pair of phalanges being connected by an anterior or plantar and two lateral ligaments, and their articular surfaces lined by synovial membrane. Their actions are also similar.

For further information on this subject, the Student is referred to Cruveilhier's 'Anatomie Descriptive;' to Dr. Humphry's work on the 'Human Skeleton, including the Joints;' and to Arnold's 'Tabulæ Anatomicæ,' Fascic. 4. Pars 2. Icones Articulorum et Ligamentorum.

The Muscles and Fasciæ.*

THE Muscles are connected with the bones, cartilages, ligaments and skin, either directly or through the intervention of fibrous structures, called tendons or aponeuroses. Where a muscle is attached to bone or cartilage, the fibres terminate in blunt extremities upon the periosteum or perichondrium, and do not come into direct relation with the osseous or cartilaginous tissue. Where muscles are connected with the skin, they either lie as a flattened layer beneath it, or are connected with its areolar tissue by larger or smaller bundles of fibres, as in the muscles of the face.

The muscles vary extremely in their form. In the limbs, they are of considerable length, especially the more superficial ones, the deep ones being generally broad; they surround the bones, and form an important protection to the various joints. In the trunk, they are broad, flattened, and expanded, forming the parietes of the cavities which they enclose; hence, the reason of the terms, *long*, *broad*, *short*, etc., used in the description of a muscle.

There is considerable variation in the arrangement of the fibres of certain muscles with reference to the tendons to which they are attached. In some, the fibres are arranged longitudinally, and terminate at either end in a narrow tendon. If the fibres converge, like the plumes of a pen, to one side of a tendon, which runs the entire length of a muscle, the muscle is said to be *penniform*, as the Peronei; if they converge to both sides of the tendon, the muscle is called *bipenniform*, as the Rectus femoris; if they converge from a broad surface to a narrow tendinous point, the muscle is said to be *radiated*, as the Temporal and Glutei muscles.

They differ no less in size; the Gastrocnemius forms the chief bulk of the back of the leg, and the fibres of the Sartorius are nearly two feet in length, whilst the Stapedius, a small muscle of the internal ear, weighs about a grain, and its fibres are not more than two lines in length.

The names applied to the various muscles have been derived: 1, from their situation, as the Tibialis, Radialis, Ulnaris, Peroneus; 2, from their direction, as the Rectus abdominis, Obliqui capitis, Transversalis; 3, from their uses, as Flexors, Extensors, Abductors, etc.; 4, from their shape, as the Deltoid, Trapezius, Rhomboideus; 5, from the number of their divisions, as the Biceps, the Triceps; 6, from their points of attachment, as the Sterno-cleido-mastoid, Sterno-hyoid, Sterno-thyroid.

In the description of a muscle, the term *origin* is meant to imply its more fixed or central attachment; and the term *insertion* the moveable point upon which the force of the muscle is directed; but the origin is absolutely fixed in only a very small number of muscles, such as those of the face, which are attached by one extremity to the bone, and by the other to the moveable integument; in the greater number, the muscle can be made to act from either extremity.

* The Muscles and Fasciæ are described conjointly, in order that the student may consider the arrangement of the latter in his dissection of the former. It is rare for the student of anatomy in this country to have the opportunity of dissecting the fasciæ separately; and it is for this reason, as well as from the close connection that exists between the muscles and their investing aponeuroses, that they are considered together. Some general observations are first made on the anatomy of the muscles and fasciæ, the special description being given in connection with the different regions.

In the dissection of the muscles, the student should pay especial attention to the exact *origin*, *insertion*, and *actions* of each, and its more important *relations* with surrounding parts. An accurate knowledge of the points of attachment of the muscles is of great importance in the determination of their action. By a knowledge of the action of the muscles, the surgeon is able to explain the causes of displacement in various forms of fracture, and the causes which produce distortion in various deformities, and consequently, to adopt appropriate treatment in each case. The relations, also, of some of the muscles, especially those in immediate apposition with the larger blood-vessels, and the surface-markings they produce, should be especially remembered, as they form useful guides in the application of a ligature to those vessels.

Tendons are white, glistening, fibrous cords, varying in length and thickness, sometimes round, sometimes flattened, of considerable strength, and only slightly elastic. They consist almost entirely of white fibrous tissue, the fibrils of which have an undulating course parallel with each other, and are firmly united together. They are very sparingly supplied with blood-vessels, the smaller tendons presenting in their interior not a trace of them. Nerves also are not present in the smaller tendons; but the larger ones, as the tendo Achillis, receive nerves which accompany the nutrient vessels. The tendons consist principally of a substance which yields gelatine.

Aponeuroses are fibrous membranes, of a pearly-white colour, iridescent, glistening, and similar in structure to the tendons. They are destitute of nerves, and the thicker ones only sparingly supplied with blood-vessels.

The tendons and aponeuroses are connected, on the one hand, with the muscles; and, on the other hand, with the moveable structures, as the bones, cartilages, ligaments, fibrous membranes (for instance, the sclerotic), and the synovial membranes (subcruræus, subanconeus). Where the muscular fibres are in a direct line with those of the tendon or aponeurosis, the two are directly continuous, the muscular fibre being distinguishable from that of the tendon only by its striation. But where the muscular fibre joins the tendon or aponeurosis at an oblique angle, the former terminates, according to Kölliker, in rounded extremities, which are received into corresponding depressions on the surface of the latter, the connective tissue between the fibres being continuous with that of the tendon. The latter mode of attachment occurs in all the penniform and bipenniform muscles, and in those muscles the tendons of which commence in a membranous form, as the Gastrocnemius and Soleus.

The fasciæ (*fascia*, a bandage) are fibro-areolar or aponeurotic laminæ, of variable thickness and strength, found in all regions of the body, investing the softer and more delicate organs. The fasciæ have been subdivided, from the structure which they present, into two groups, fibro-areolar or superficial fasciæ, and aponeurotic or deep fasciæ.

The *fibro-areolar fascia* is found immediately beneath the integument over almost the entire surface of the body, and is generally known as the *superficial fascia*. It connects the skin with the deep or aponeurotic fascia, and consists of fibro-areolar tissue, containing in its meshes pellicles of fat in varying quantity. In the eyelids and scrotum, where adipose tissue is rarely deposited, this tissue is very liable to serous infiltration. The superficial fascia varies in thickness in different parts of the body: in the groin it is so thick as to be capable of being subdivided into several laminæ, but in the palm of the hand it is of extreme thinness, and intimately adherent to the integument. The superficial fascia is capable of separation into two or more layers, between which are found the superficial vessels and nerves, and superficial lymphatic glands; as the superficial epigastric vessels in the abdominal region, the radial and ulnar veins in the forearm, the saphenous veins in the leg and thigh; certain cutaneous muscles also are situated in the superficial fascia, as the Platysma myoides in the neck, and the Orbicularis palpebrarum around the eyelids. This fascia is most distinct at the lower part of the abdomen, the scrotum, perinæum, and extremities; is very thin in those regions where muscular fibres are inserted into the integument, as on the side of the neck, the face, and around the margin of the anus; and is almost entirely wanting in the palms of the hands and soles of the feet, where the integument is adherent to the subjacent aponeurosis. The superficial fascia connects the skin to the subjacent parts, facilitates the movement of the skin, serves as a soft nidus for the passage of vessels and nerves to the integument, and retains the warmth of the body, since the adipose tissue contained in its areolæ is a bad conductor of caloric.

The *aponeurotic* or *deep fascia* is a dense inelastic and unyielding fibrous membrane, forming sheaths for the muscles, and affording them broad surfaces for attachment. It consists of shining tendinous fibres, placed parallel with one another, and connected together by other fibres disposed in a reticular manner. It is usually exposed on the removal of the superficial fascia, forming a strong investment, which not only binds down collectively the muscles in each region, but gives a separate sheath to each, as well as to the vessels and nerves. The fasciæ are thick in unprotected situations, as on the outer side of a limb, and thinner on the inner side. Aponeurotic fasciæ are divided into two classes, aponeuroses of insertion, and aponeuroses of investment.

The *aponeuroses of insertion* serve for the insertion of muscles. Some of these are formed by the expansion of a tendon into an aponeurosis, as, for instance, the tendon of the Sartorius; others are connected directly to the muscle, as the aponeuroses of the abdominal muscles.

The *aponeuroses of investment* form a sheath for the entire limb, as well as for each individual muscle. Many aponeuroses, however, serve both for investment and insertion. Thus the aponeurosis given off from the tendon of the Biceps of the arm near its insertion is continuous with, and partly forms, the investing fascia of the forearm, and gives origin to the muscles in this region. The deep fasciæ assist the muscles in their action, by the degree of tension and pressure they make upon their surface; and, in certain situations, this is increased and regulated by muscular action, as, for instance, by the Tensor vaginæ femoris and Gluteus maximus in the thigh, by the Biceps in the leg, and Palmaris longus in the hand. In the limbs, the fasciæ not only invest the entire limb, but give off septa, which separate the various muscles, and are attached beneath to the periosteum; these prolongations of fasciæ are usually spoken of as intermuscular septa.

The Muscles and Fasciæ may be arranged, according to the general division of the body, into those of the head, face, and neck; those of the trunk; those of the upper extremity; and those of the lower extremity.

MUSCLES AND FASCIÆ OF THE HEAD AND FACE.

The Muscles of the Head and Face consist of ten groups, arranged according to the region in which they are situated.

1. Cranial Region.
2. Auricular Region.
3. Palpebral Region.
4. Orbital Region.
5. Nasal Region.
6. Superior Maxillary Region.
7. Inferior Maxillary Region.
8. Intermaxillary Region.
9. Temporo-Maxillary Region.
10. Pterygo-Maxillary Region.

The Muscles contained in each of these groups are the following:—

1. *Cranial Region.*

Occipito-frontalis.

2. *Auricular Region.*

Attollens aurem.
Attrahens aurem.
Retrahens aurem.

3. *Palpebral Region.*

Orbicularis palpebrarum.
Corrugator supercilii.
Tensor Tarsi.

4. *Orbital Region.*

Levator palpebræ.
Rectus superior.

Rectus inferior.
Rectus internus.
Rectus externus.
Obliquus superior.
Obliquus inferior.

5. *Nasal Region.*

Pyramidalis nasi.
Levator labii superioris alæque nasi.
Dilatator naris posterior.
Dilatator naris anterior.
Compressor nasi.
Compressor narium minor.
Depressor alæ nasi.

6. *Superior Maxillary Region.*

Levator labii superioris.
Levator anguli oris.
Zygomaticus major.
Zygomaticus minor.

7. *Inferior Maxillary Region.*

Levator labii inferioris.
Depressor labii inferioris.
Depressor anguli oris.

8. *Inter-Maxillary Region.*

Buccinator.
Risorius.
Orbicularis oris.

9. *Temporo-Maxillary Region.*

Masseter.
Temporal.

10. *Pterygo-Maxillary Region.*

Pterygoideus externus.
Pterygoideus internus.

1. Cranial Region—Occipito-Frontalis.

Dissection (fig. 147). The head being shaved, and a block placed beneath the back of the neck, make a vertical incision through the skin from before backwards, commencing at the root of the nose in front, and terminating behind at the occipital protuberance; make a second incision in a horizontal direction along the forehead and round the side of the head, from the anterior to the posterior extremity of the preceding. Raise the skin in front from the subjacent muscle from below upwards; this must be done with extreme care, on account of their intimate union. The tendon of the muscle is best avoided by removing the integument from the outer surface of the vessels and nerves which lie between the two.

147.—Dissection of the Head, Face, and Neck.

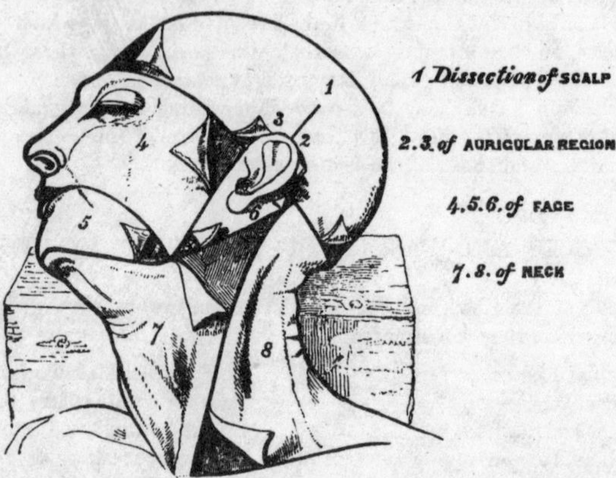

The *superficial fascia* in the cranial region is a firm, dense layer, intimately adherent to the integument, and to the Occipito-frontalis and its tendinous aponeurosis; it is continuous, behind, with the superficial fascia at the back part of the neck; and, laterally, is continued over the temporal aponeurosis: it contains between its layers the small muscles of the auricle, and the superficial temporal vessels and superficial nerves.

The *Occipito-frontalis* (fig. 148) is a broad musculo-fibrous layer, which covers the whole of one side of the vertex of the skull, from the occiput to the eyebrow.

148.—Muscles of the Head, Face, and Neck.

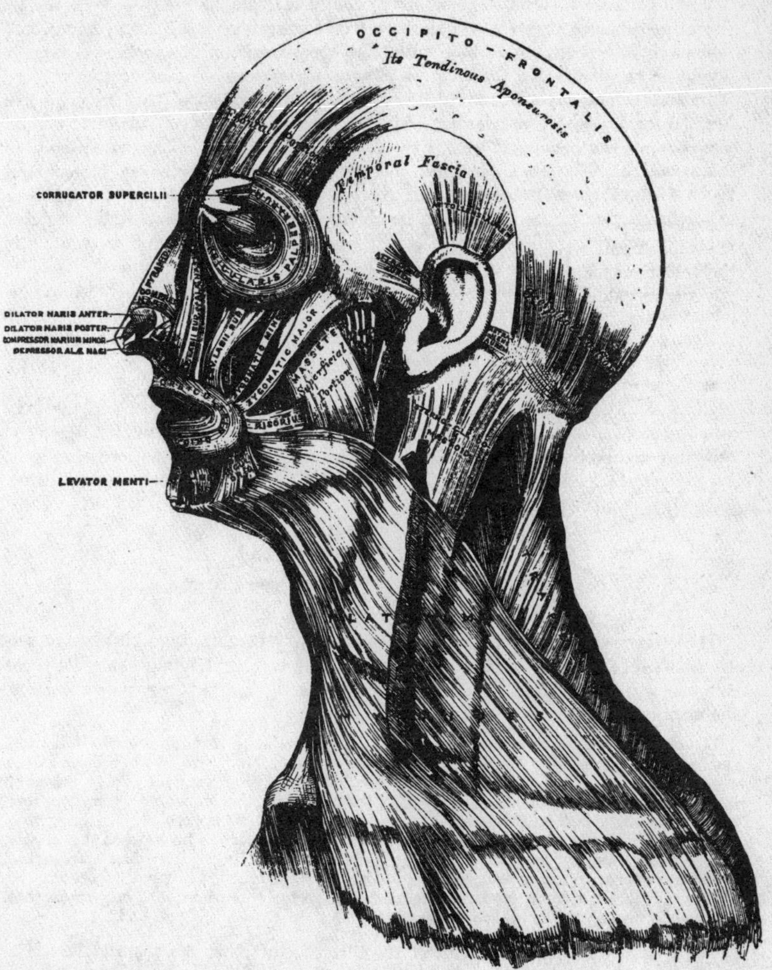

It consists of two muscular slips, separated by an intervening tendinous aponeurosis. The *occipital portion*, thin, quadrilateral in form, and about an inch and a half in length, arises from the outer two-thirds of the superior curved line of the occipital bone, and from the mastoid portion of the temporal. Its fibres of origin are tendinous, but they soon become muscular, and ascend in a parallel direction to terminate in the tendinous aponeurosis. The *frontal portion* is thin, of a quadrilateral form, and intimately adherent to the skin. It is broader, its fibres are longer, and their structure paler than the occipital portion. Its internal fibres are continuous with those of the Pyramidalis nasi. Its middle fibres become blended with the Corrugator supercilii and Orbicularis: and the outer fibres are also blended with the latter muscle over the external angular process. From this attachment, the fibres are directed upwards and join the aponeurosis below the coronal suture. The inner margins of the two frontal portions of the muscle are joined together for some distance above the root of the nose; but between the occipital portions there is a considerable though variable interval.

The *aponeurosis* covers the upper part of the vertex of the skull, being continuous across the middle line with the aponeurosis of the opposite muscle. Behind, it is attached, in the interval between the occipital origins, to the occipital protuberance and superior curved lines above the attachment of the Trapezius; in front, it forms a short angular prolongation between the frontal portions; and on each side, it has connected with it the Attollens and Attrahens aurem muscles; in this situation it loses its aponeurotic character, and is continued over the temporal fascia to the zygoma by a layer of laminated areolar tissue. This aponeurosis is closely connected to the integument by a dense fibro-cellular tissue, which contains much granular fat, and in which ramify the numerous vessels and nerves of the integument; it is loosely connected with the pericranium by a quantity of loose cellular tissue, which allows of a considerable degree of movement of the integument.

Nerves. The frontal portion of the Occipito-frontalis is supplied by the facial nerve; its occipital portion by the posterior auricular branch of the facial, and sometimes by the small occipital.

Actions. The frontal portion of the muscle raises the eyebrows and the skin over the root of the nose; at the same time throwing the integument of the forehead into transverse wrinkles, a predominant expression in the emotions of delight. By bringing alternately into action the occipital and frontal portions, the entire scalp may be moved from before backwards.

2. Auricular Region. (148.)

Attollens Aurem. Attrahens Aurem.
Retrahens Aurem.

These three small muscles are placed immediately beneath the skin around the external ear. In man, in whom the external ear is almost immoveable, they are rudimentary. They are the analogues of large and important muscles in some of the mammalia.

Dissection. This requires considerable care, and should be performed in the following manner:—To expose the Attollens aurem; draw the pinna or broad part of the ear downwards, when a tense band will be felt beneath the skin, passing from the side of the head to the upper part of the concha; by dividing the skin over the tendon, in a direction from below upwards, and then reflecting it on each side, the muscle is exposed. To bring into view the Attrahens aurem, draw the helix backwards by means of a hook, when the muscle will be made tense, and may be exposed in a similar manner to the preceding. To expose the Retrahens aurem, draw the pinna forwards, when the muscle being made tense may be felt beneath the skin, at its insertion into the back part of the concha, and may be exposed in the same manner as the other muscles.

The *Attollens Aurem*, the largest of the three, is thin, and fan-shaped; its fibres arise from the aponeurosis of the Occipito-frontalis, and converge to be inserted by a thin, flattened tendon into the upper part of the cranial surface of the pinna.

Relations. *Externally*, with the integument; *internally*, with the Temporal aponeurosis.

The *Attrahens Aurem*, the smallest of the three, is thin, fan-shaped, and its fibres pale and indistinct; they arise from the lateral edge of the aponeurosis of the Occipito-frontalis, and converge to be inserted into a projection on the front of the helix.

Relations. *Externally*, with the skin; *internally*, with the temporal fascia, which separates it from the temporal artery and vein.

The *Retrahens Aurem* consists of two or three fleshy fasciculi, which arise from the mastoid portion of the temporal bone by short aponeurotic fibres. They are inserted into the lower part of the cranial surface of the concha.

Relations. *Externally*, with the integument; *internally*, with the mastoid portion of the temporal bone.

Nerves. The Attollens aurem is supplied by the small occipital; the Attrahens aurem, by the facial; and the Retrahens aurem, by the posterior auricular branch of the facial.

Actions. In man, these muscles possess very little action; the Attollens aurem slightly raises the ear; the Attrahens aurem draws it forwards and upwards; and the Retrahens aurem draws it backwards.

3. Palpebral Region. (Fig. 148.)

Orbicularis Palpebrarum.	Levator Palpebræ.
Corrugator Supercilii.	Tensor Tarsi.

Dissection (fig. 147). In order to expose the muscles of the face, continue the longitudinal incision, made in the dissection of the Occipito-frontalis, down the median line of the face to the tip of the nose, and from this point onwards to the upper lip; and carry another incision along the margin of the lip to the angle of the mouth, and transversely across the face to the angle of the jaw. Then make an incision in front of the external ear, from the angle of the jaw upwards, to join the transverse incision made in exposing the Occipito-frontalis. These incisions include a square-shaped flap, which should be removed in the direction marked in the figure, with care, as the muscles at some points are intimately adherent to the integument.

The *Orbicularis Palpebrarum* is a sphincter muscle, which surrounds the circumference of the orbit and eyelids. It arises from the internal angular process of the frontal bone, from the nasal process of the superior maxillary in front of the lachrymal groove, and from the anterior surface and borders of a short tendon, the tendo palpebrarum, placed at the inner angle of the orbit. From this origin, the fibres are directed outwards, forming a broad, thin, and flat layer, which covers the eyelids, surrounds the circumference of the orbit, and spreads out over the temple, and downwards on the cheek, becoming blended with the Occipito-frontalis and Corrugator supercilii. The palpebral portion (ciliaris) of the Orbicularis is thin and pale; it arises from the bifurcation of the tendo palpebrarum, and forms a series of concentric curves, which are united on the outer side of the eyelids at an acute angle by a cellular raphé, some being inserted into the external tarsal ligament and malar bone. The orbicular portion (orbicularis latus) is thicker and of a reddish colour: its fibres are well developed, and form complete ellipses.

Relations. By its *superficial surface*, with the integument. By its *deep surface*, above, with the Occipito-frontalis and Corrugator supercilii, with which it is intimately blended, and with the supra-orbital vessels and nerve; below, it covers, the lachrymal sac, and the origin of the Levator labii superioris, and the Levator labii superioris alæque nasi muscles. *Internally*, it is occasionally blended with the Pyramidalis nasi. *Externally*, it lies on the temporal fascia. On the eyelids, it is separated from the conjunctiva by a fibrous membrane and the tarsal cartilages.

The *tendo palpebrarum* (tendo oculi) is a short tendon, about two lines in length and one in breadth, attached to the nasal process of the superior maxillary bone in front of the lachrymal groove. Crossing the lachrymal sac, it divides into two parts, each division being attached to the inner extremity of the corresponding tarsal cartilage. As the tendon crosses the lachrymal sac, a strong aponeurotic lamina is given off from the posterior surface, which expands over the sac, and is attached to the ridge on the lachrymal bone. This is the reflected aponeurosis of the tendo palpebrarum.

The *Corrugator Supercilii* is a small, narrow, pyramidal muscle, placed at the inner extremity of the eyebrow, beneath the Occipito-frontalis and Orbicularis palpebrarum muscles. It arises from the inner extremity of the superciliary ridge; from whence its fibres pass upwards and outwards, to be inserted into the under surface of the orbicularis, opposite the middle of the orbital arch.

Relations. By its *anterior surface*, with the Occipito-frontalis and Orbicularis palpebrarum muscles. By its *posterior surface*, with the frontal bone and supra-trochlear nerve.

The *Levator Palpebræ* will be described with the muscles of the orbital region.

The *Tensor Tarsi* is a small thin muscle, about three lines in breadth and six in length, situated at the inner side of the orbit, behind the tendo oculi. It arises from the crest and adjacent part of the orbital surface of the lachrymal bone, and passing across the lachrymal sac, divides into two slips, which cover the lachrymal canals, and are inserted into the tarsal cartilages near the puncta lachrymalia. Its fibres appear to be continuous with those of the palpebral portion of the Orbicularis; it is occasionally very indistinct.

Nerves. The Orbicularis palpebrarum Corrugator supercilii, and Tensor tarsi, are supplied by the facial nerve.

Actions. The Orbicularis palpebrarum is the sphincter muscle of the eyelids. The palpebral portion acts involuntarily in closing the lids, and independently of the orbicular portion, which is subject to the will. When the entire muscle is brought into action, the integument of the forehead, temple, and cheek is drawn inwards towards the inner angle of the eye, and the eyelids are firmly closed. The Levator palpebræ is the direct antagonist of this muscle; it raises the upper eyelid and exposes the globe. The Corrugator supercilii draws the eyebrow downwards and inwards, producing the vertical wrinkles of the forehead. This muscle may be regarded as the principal agent in the expression of grief. The Tensor tarsi draws the eyelids and the extremities of the lachrymal canals inwards, and compresses them against the surface of the globe of the eye; thus placing them in the most favourable situation for receiving the tears. It serves, also, to compress the lachrymal sac.

4. Orbital Region. (Fig. 149.)

Levator Palpebræ. Rectus Internus.
Rectus Superior. Rectus Externus.
Rectus Inferior. Obliquus Superior.
Obliquus Inferior.

Dissection. To open the cavity of the orbit, remove the skull-cap and brain; then saw through the frontal bone at the inner extremity of the supraorbital ridge, and externally at its junction with the malar. Break in pieces the thin roof of the orbit by a few slight blows of the hammer and take it away; drive forward the superciliary portion of the frontal bone by a smart stroke, but do not remove it, as that would destroy the pulley of the Obliquus superior. When the fragments are cleared away, the periosteum of the orbit will be exposed: this being removed, together with the fat which fills the cavity of the orbit, the several muscles of this region can be examined. The dissection will be facilitated by distending the globe of the eye. In order to effect this, puncture the optic nerve near the eyeball, with a curved needle, and push the needle onwards into the globe; insert the point of a blow-pipe through this aperture, and force a little air into the cavity of the eyeball; then apply a ligature round the nerve, so as to prevent the air escaping. The globe being now drawn forwards, the muscles will be put upon the stretch.

The *Levator Palpebræ Superioris* is thin, flat, and triangular in shape. It arises from the under surface of the lesser wing of the sphenoid, immediately in front of the optic foramen; and is inserted, by a broad aponeurosis, into the upper border of the superior tarsal cartilage. At its origin, it is narrow and tendinous; but soon becomes broad and fleshy, and finally terminates in a broad aponeurosis.

Relations. By its *upper surface*, with the frontal nerve and artery, the periosteum of the orbit; and, in front, with the inner surface of the broad tarsal ligament. By its *under surface*, with the Superior rectus; and in the lid, with the conjunctiva. A small branch of the third nerve enters its under surface.

The *Rectus superior*, the thinnest and narrowest of the four Recti, arises from the upper margin of the optic foramen, beneath the Levator Palpebræ and Superior oblique, and from the fibrous sheath of the optic nerve; and is inserted, by a ten-

149.—Muscles of the Right Orbit.

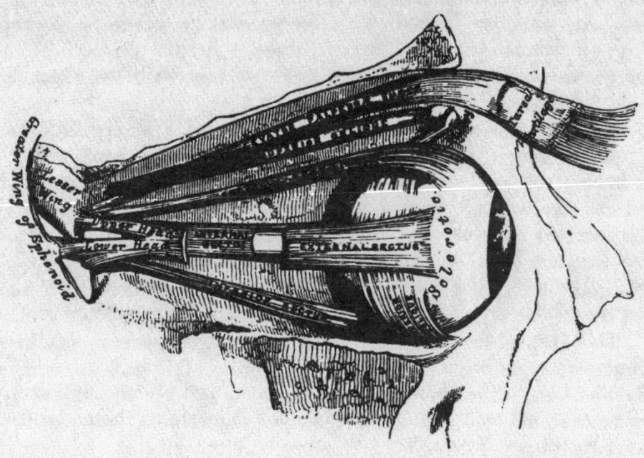

dinous expansion, into the sclerotic coat, about three or four lines from the margin of the cornea.

Relations. By its *upper surface*, with the Levator Palpebræ. By its *under surface*, with the optic nerve, the ophthalmic artery, the nasal nerve, and the branch of the third nerve, which supplies it; and, in front, with the tendon of the Superior oblique, and the globe of the eye.

The *Inferior* and *Internal Recti* arise by a common tendon (the ligament of Zinn), which is attached round the circumference of the optic foramen, except at its upper and outer part. The External rectus has two heads: the upper one arises from the outer margin of the optic foramen, immediately beneath the Superior rectus: the lower head, partly from the ligament of Zinn, and partly from a small pointed process of bone on the lower margin of the sphenoidal fissure. Each muscle passes forward in the position implied by its name, to be inserted, by a tendinous expansion (the *tunica albuginea*), into the sclerotic coat, about three or four lines from the margin of the cornea. Between the two heads of the External rectus is a narrow interval, through which pass the third, nasal branch of the fifth, and sixth nerves, and the ophthalmic vein. Although nearly all these muscles present a common origin, and are inserted in a similar manner in the sclerotic coat, there are certain differences to be observed in them, as regards their length and breadth. The Internal Rectus is the broadest; the External, the longest; and the Superior, the thinnest and narrowest.

150.—The Relative position and attachment of the Muscles of the Left Eyeball.

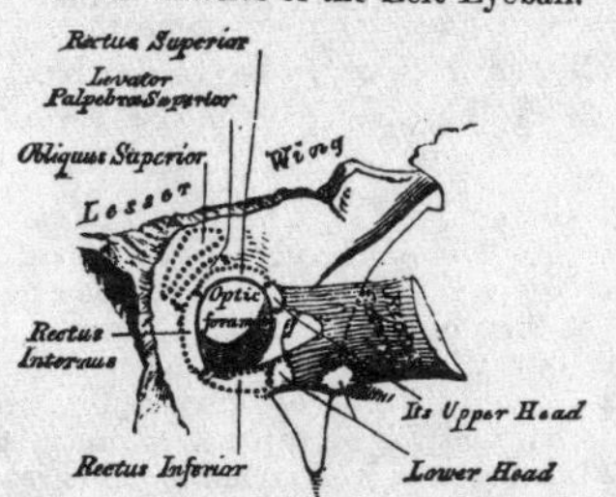

The *Superior Oblique* is a fusiform muscle, placed at the upper and inner side of the orbit, internal to the Levator palpebræ. It arises about a line above the inner margin of the optic foramen, and, passing forwards to the inner angle of the orbit, terminates in a rounded tendon, which plays in a ring or pulley, formed by fibro-cartilaginous tissue attached to a depression beneath the internal angular process of the frontal bone, the contiguous surfaces of the tendon and ring being lined by a delicate synovial membrane, and enclosed in a thin fibrous investment. The tendon is reflected backwards and outwards beneath the Superior rectus to the outer part of the globe of the eye, and is inserted into the sclerotic coat, midway between the cornea and entrance of the optic nerve, the insertion of the muscle lying between the Superior and External recti.

Relations. By its *upper surface*, with the periosteum covering the roof of the orbit, and the fourth nerve. The tendon, where it lies on the globe of the eye, is covered by the Superior rectus. By its *under surface*, with the nasal nerve, and the upper border of the Internal rectus.

The *Inferior Oblique* is a thin, narrow muscle, placed near the anterior margin of the orbit. It arises from a depression in the orbital plate of the superior maxillary bone, external to the lachrymal groove. Passing outwards and backwards beneath the Inferior rectus, and between the eyeball and the External rectus, it is inserted into the outer part of the sclerotic coat between the Superior and External rectus, near the tendon of insertion of the Superior oblique.

Relations. By its *upper surface*, with the globe of the eye, and with the Inferior rectus. By its *under surface*, with the periosteum covering the floor of the orbit, and with the External rectus. Its borders look forwards and backwards; the posterior one receives a branch of the third nerve.

Nerves. The Levator palpebræ, Inferior oblique, and all the recti excepting the External, are supplied by the third nerve; the Superior oblique, by the fourth; the External rectus, by the sixth.

Actions. The Levator palpebræ raises the upper eyelid, and is the direct antagonist of the Orbicularis palpebrarum. The four Recti muscles are attached in such a manner to the globe of the eye, that, acting singly, they will turn it either upwards, downwards, inwards, or outwards, as expressed by their names. If any two Recti act together, they carry the globe of the eye in the diagonal of these directions, viz. upwards and inwards, upwards and outwards, downwards and inwards, or downwards and outwards. The movement of circumduction, as in looking round a room, is performed by the alternate action of the four Recti. By some anatomists, these muscles have been considered the chief agents in adjusting the sight at different distances, by compressing the globe, and so lengthening its antero-posterior diameter. The Oblique muscles rotate the eyeball on its *antero-posterior axis*, this kind of movement being required for the correct viewing of an object, when the head is moved laterally, as from shoulder to shoulder, in order that the picture may fall in all respects on the same part of the retina of each eye.*

Surgical Anatomy. The position and exact point of insertion of the tendons of the Internal and External recti muscles into the globe, should be carefully examined from the front of the eyeball, as the surgeon is often required to divide one or the other muscle for the cure of strabismus. In convergent strabismus, which is the most common form of the disease, the eye is turned inwards, requiring the division of the Internal rectus. In the divergent form, which is more rare, the eye is turned outwards, the External rectus being especially implicated. The deformity produced in either case is to be remedied by division of one or the other muscle. The operation is thus effected: the lids are to be well separated; the eyeball being drawn outwards, the conjunctiva should be raised by a pair of forceps, and divided immediately beneath the lower border of the tendon of the Internal rectus, a little behind its insertion into the schlerotic; the submucous areolar tissue is then divided, and into the small aperture thus made, a blunt hook is passed upwards between the muscle and the globe, and the tendon of the muscle and conjunctiva covering it, divided by a pair of blunt-pointed scissors. Or the tendon may be divided by a subconjunctival incision, one blade of the scissors being passed upwards between the tendon and the conjunctiva, and the other between the tendon and the sclerotic. The student, when dissecting these muscles, should remove on one side of the subject the conjunctiva from the front of the eye, in order to see more accurately the position of the tendons, while on the opposite side the operation may be performed.

* '*On the Oblique Muscles of the Eye in Man and Vertebrate Animals*,' by John Struthers, M.D. '*Anatomical and Physiological Observations*.'

5. Nasal Region. (Fig. 148.)

Pyramidalis Nasi.
Levator Labii Superioris Alæque Nasi.
Dilatator Naris Posterior.
Dilatator Naris Anterior.
Compressor Nasi.
Compressor Narium Minor.
Depressor Alæ Nasi.

The *Pyramidalis Nasi* is a small pyramidal slip, prolonged downwards from the Occipito-frontalis upon the side of the nose, where it becomes tendinous and blends with the Compressor nasi. As the two muscles descend, they diverge, leaving an angular interval between them.

Relations. By its *upper surface*, with the skin. By its *under surface*, with the frontal and nasal bones.

The *Levator Labii Superioris Alæque Nasi* is a thin triangular muscle, placed by the side of the nose, and extending between the inner margin of the orbit and upper lip. It arises by a pointed extremity from the upper part of the nasal process of the superior maxillary bone, and passing obliquely downwards and outwards, divides into two slips, one of which is inserted into the cartilage of the ala of the nose; the other is prolonged into the upper lip, becoming blended with the Orbicularis and Levator labii proprius.

Relations. In front, with the integument; and with a small part of the Orbicularis palpebrarum above.

Lying upon the superior maxillary bone, beneath this muscle, is a longitudinal muscular fasciculus about an inch in length. It is attached by one end near the origin of the Compressor nasi, and by the other to the nasal process about an inch above it; it was described by Albinus as the 'Musculus anomalus,' and by Santorini, as the 'Rhomboideus.'

The *Dilatator Naris Posterior* is a small muscle, which is placed partly beneath the proper elevator of the nose and lip. It arises from the margin of the nasal notch of the superior maxilla, and from the sesamoid cartilages, and is inserted into the skin near the margin of the nostril.

The *Dilatator Naris Anterior* is a thin delicate fasciculus, passing from the cartilage of the ala of the nose to the integument near its margin. This muscle is situated in front of the preceding.

The *Compressor Nasi* is a small, thin, triangular muscle, arising by its apex from the superior maxillary bone, above and a little external to the incisive fossa; its fibres proceed upwards and inwards, expanding into a thin aponeurosis which is attached to the fibro-cartilage of the nose, and is continuous on the bridge of the nose with that of the muscle of the opposite side, and with the aponeurosis of the Pyramidalis nasi.

The *Compressor Narium Minor* is a small muscle, attached by one end to the alar cartilage, and by the other to the integument at the end of the nose.

The *Depressor Alæ Nasi* is a short, radiated muscle, arising from the incisive fossa of the superior maxilla; its fibres ascend to be inserted into the septum, and back part of the ala of the nose. This muscle lies between the mucous membrane and muscular structure of the lip.

Nerves. All the muscles of this group are supplied by the facial nerve.

Actions. The Pyramidalis nasi draws down the inner angle of the eyebrows; by some anatomists it is also considered as an elevator of the ala, and, consequently, a dilatator of the nose. The Levator labii superioris alæque nasi draws upwards the upper lip and ala of the nose; its most important action is upon the nose, which it dilates to a considerable extent. The action of this muscle produces a marked influence over the countenance, and it is the principal agent in the expression of contempt. The two Dilatatores nasi enlarge the aperture of the nose, and the Compressor nasi appears to press upon the nose so as to increase its breadth, and thus tends rather to open than to close the nostrils. The Depressor alæ nasi is a direct antagonist of the preceding muscles, drawing the ala of the nose downwards, and thereby constricting the aperture of the nares.

6. Superior Maxillary Region. (Fig. 148.)

Levator Labii Superioris.	Zygomaticus major.
Levator Anguli Oris.	Zygomaticus minor.

The *Levator Labii Superioris* is a thin muscle of a quadrilateral form. It arises from the lower margin of the orbit immediately above the infraorbital foramen, some of its fibres being attached to the superior maxilla, others to the malar bone; its fibres converge to be inserted into the muscular substance of the upper lip.

Relations. By its *superficial surface*, with the lower segment of the Orbicularis palpebrarum; below, it is subcutaneous. By its *deep surface*, it conceals the origin of the Compressor nasi and Levator anguli oris muscles, and the infraorbital vessels and nerves, as they escape from the infraorbital foramen.

The *Levator Anguli Oris* arises from the canine fossa, immediately below the infraorbital foramen; its fibres incline downwards and a little outwards, to be inserted into the angle of the mouth, intermingling with those of the Zygomatici, the Depressor anguli oris, and the Orbicularis.

Relations. By its *superficial surface*, with the Levator labii superioris and the infraorbital vessels and nerves. By its *deep surface*, with the superior maxilla, the Buccinator, and the mucous membrane.

The *Zygomaticus major* is a slender fasciculus, which arises from the malar bone, in front of the zygomatic suture, and, descending obliquely downwards and inwards, is inserted into the angle of the mouth, where it blends with the fibres of the Orbicularis and Depressor anguli oris.

Relations. By its *superficial surface*, with the subcutaneous adipose tissue. By its *deep surface*, with the malar bone, and the Masseter and Buccinator muscles.

The *Zygomaticus minor* arises from the malar bone, immediately behind the maxillary suture, and, passing downwards and inwards, is continuous with the outer margin of the Levator labii superioris. It lies in front of the preceding.

Relations. By its *superficial surface*, with the integument and the Orbicularis palpebrarum above. By its *deep surface*, with the Levator anguli oris.

Nerves. This group of muscles is supplied by the facial nerve.

Actions. The Levator labii superioris is the proper elevator of the upper lip, carrying it at the same time a little outwards. The Levator anguli oris raises the angle of the mouth and draws it inwards; whilst the Zygomatici raise the upper lip and draw it somewhat outwards, as in laughing.

7. Inferior Maxillary Region. (Fig. 148.)

Levator Labii Inferioris (Levator menti).
Depressor Labii Inferioris (Quadratus menti).
Depressor Anguli Oris (Triangularis menti).

Dissection. The Muscles in this region may be dissected by making a vertical incision through the integument from the margin of the lower lip to the chin: a second incision should then be carried along the margin of the lower jaw as far as the angle, and the integument carefully removed in the direction shown in fig. 147.

The *Levator Labii Inferioris* (*Levator menti*) is to be dissected by everting the lower lip and raising the mucous membrane. It is a small conical fasciculus, placed on the side of the frænum of the lower lip. It arises from the incisive fossa, external to the symphysis of the lower jaw: its fibres descend to be inserted into the integument of the chin.

Relations. On its *inner surface*, with the mucous membrane; in the *median line*, it is blended with the muscle of the opposite side; and on its *outer side*, with the Depressor labii inferioris.

The *Depressor Labii Inferioris* (*Quadratus menti*) is a small quadrilateral muscle, situated at the outer side of the preceding. It arises from the external oblique line of the lower jaw, between the symphysis and mental foramen, and passes obliquely upwards and inwards, to be inserted into the integument of the lower lip, its fibres blending with the Orbicularis, and with those of its fellow of the opposite side. It is continuous with the fibres of the Platysma at its origin. This muscle contains much yellow fat intermingled with its fibres.

Relations. By its *superficial surface*, with part of the Depressor anguli oris, and with the integument, to which it is closely connected. By its *deep surface*, with the mental vessels and nerves, the mucous membrane of the lower lip, the labial glands, and the Levator menti, with which it is intimately united.

The *Depressor Anguli Oris* is triangular in shape, arising, by its broad base, from the external oblique line of the lower jaw, from whence its fibres pass upwards, to be inserted, by a narrow fasciculus, into the angle of the mouth. It is continuous with the Platysma at its origin, and with the Orbicularis and Risorius at its insertion, and some of its fibres are directly continuous with those of the Levator anguli oris.

Relations. By its *superficial surface*, with the integument. By its *deep surface*, with the Depressor labii inferioris and Buccinator.

Nerves. This group of muscles is supplied by the facial nerve.

Actions. The Levator labii inferioris raises the lower lip, and protrudes it forwards, and at the same time wrinkles the integument of the chin. The Depressor labii inferioris draws the lower lip directly downwards and a little outwards. The Depressor anguli oris depresses the angle of the mouth, being the antagonist to the Levator anguli oris and Zygomaticus major: acting with these muscles, it will draw the angle of the mouth directly backwards.

8. Inter-Maxillary Region.

Orbicularis Oris. Buccinator. Risorius.

Dissection. The dissection of these muscles may be considerably facilitated by filling the cavity of the mouth with tow, so as to distend the cheeks and lips; the mouth should then be closed by a few stitches, and the integument carefully removed from the surface.

The *Orbicularis Oris* is a sphincter muscle, elliptic in form, composed of concentric fibres, which surround the orifice of the mouth. It consists of two thick semicircular planes of muscular fibre, which interlace on either side with those of the Buccinator and other muscles inserted into the lips. On the free margin of the lips the muscular fibres are continued uninterruptedly from one lip to the other, around the corner of the mouth, forming a roundish fasciculus of fine pale fibres closely approximated. To the outer part of each segment some special fibres are added, by which the lips are connected directly with the maxillary bones and septum of the nose. The additional fibres for the upper segment consist of four bands, two of which (Accessorii orbicularis superiores) arise from the alveolar border of the superior maxilla, opposite the incisor teeth, and arching outwards on each side, are continuous at the angles of the mouth with the other muscles inserted into this part. The two remaining muscular slips, called the Naso-labialis, connect the upper lip to the septum of the nose: as they descend from the septum, an interval is left between them, which corresponds to that left by the divergence of the accessory portions of the Orbicularis above described. It is this interval which forms the depression seen on the surface of the skin beneath the septum of the nose. The additional fibres for the lower segment (Accessorii orbicularis inferiores) arise from the inferior maxilla, externally to the Levator labii inferioris, and arch outwards to the angles of the mouth, to join the Buccinator and the other muscles attached to this part.

Relations. By its *superficial surface*, with the integument, to which it is closely connected. By its *deep surface*, with the buccal mucous membrane, the labial glands, and coronary vessels. By its *outer circumference*, it is blended with the numerous muscles which converge to the mouth from various parts of the face. Its *inner circumference* is free, and covered by the mucous membrane.

The *Buccinator* is a broad, thin muscle, quadrilateral in form, which occupies the interval between the jaws at the side of the face. It arises from the outer surface of the alveolar processes of the upper and lower jaws, corresponding to the three molar teeth; and, behind, from the anterior border of the pterygo-maxillary ligament. The fibres converge towards the angle of the mouth, where the central fibres intersect each other, those from below being continuous with the upper segment of the Orbicularis oris; and those from above with the inferior segment; the highest and lowest fibres continue forward uninterruptedly into the corresponding segment of the lip, without decussation.

Relations. By its *superficial surface*, behind, with a large mass of fat, which separates it from the ramus of the lower jaw, the Masseter, and a small portion of the Temporal muscle; anteriorly, with the Zygomatici, Risorius, Levator anguli oris, Depressor anguli oris, and Steno's duct, which pierces it opposite the second molar tooth of the upper jaw; the facial artery and vein cross it from below upwards; it is also crossed by the branches of the facial and buccal nerve. By its *internal surface*, with the buccal glands and mucous membrane of the mouth.

The *pterygo-maxillary ligament* separates the Buccinator muscle from the Superior constrictor of the pharynx. It is a tendinous band, attached by one extremity to the apex of the internal pterygoid plate, and by the other to the posterior extremity of the internal oblique line of the lower jaw. Its *inner surface* corresponds to the cavity of the mouth, and is lined by mucous membrane. Its *outer surface* is separated from the ramus of the jaw by a quantity of adipose tissue. Its *posterior border* gives attachment to the Superior constrictor of the pharynx; its *anterior border*, to the fibres of the Buccinator.

The *Risorius* (*Santorini*) consists of a narrow bundle of fibres, which arises in the fascia over the Masseter muscle, and passing horizontally forwards, is inserted into the angle of the mouth, joining with the fibres of the Depressor anguli oris. It is placed superficial to the Platysma, and is broadest at its outer extremity. This muscle varies much in its size and form.

Nerves. The Orbicularis oris is supplied by the facial, the Buccinator by the facial and by the buccal branch of the inferior maxillary nerve.

Actions. The Orbicularis oris is the direct antagonist of all those muscles which converge to the lips from the various parts of the face, its ordinary action producing the direct closure of the lips; and its forcible action throwing the integument into wrinkles, on account of the firm connection between the latter and the surface of the muscle. The Buccinators contract and compress the cheeks, so that, during the process of mastication, the food is kept under the immediate pressure of the teeth.

9. Temporo-Maxillary Region.

Masseter. Temporal.

The *Masseter* has been already exposed by the removal of the integument from the side of the face (fig. 148); it is a short thick muscle, somewhat quadrilateral in form, consisting of two portions, superficial and deep. The *superficial portion*, the larger, arises by a thick tendinous aponeurosis from the malar process of the superior maxilla, and from the anterior two-thirds of the lower border of the zygomatic arch: its fibres pass downwards and backwards, to be inserted into the angle and lower half of the ramus of the jaw. The *deep portion* is much smaller, and more muscular in texture; it arises from the posterior third of the lower border and the whole of the inner surface of the zygomatic arch; its fibres pass downwards and forwards, to be inserted into the upper half of the ramus and outer surface of the coronoid process of the jaw. The deep portion of the muscle is partly concealed, in front, by the superficial portion; behind, it is covered by the parotid gland. The fibres of the two portions are united at their insertion.

Relations. By its *superficial surface,* with the integument; above, with the Orbicularis palpebrarum and Zygomatici; and with Steno's duct, the branches of the facial nerve, and the transverse facial vessels, which cross it. By its *deep surface,* with the ramus of the jaw, and the Buccinator, from which it is separated by a mass of fat. Its *posterior margin* is overlapped by the parotid gland. Its *anterior margin* projects over the Buccinator muscle; and the facial artery lies on it below.

The *temporal fascia* is seen, at this stage of the dissection, covering in the Temporal muscle. It is a strong aponeurotic investment, affording attachment, by its inner surface, to the superficial fibres of the muscle. Above, it is a single layer, attached to the entire extent of the temporal ridge; but below, where it is attached to the zygoma, it consists of two layers, one of which is inserted into the outer, and the other into the inner border of the zygomatic arch. A small quantity of fat, the orbital branch of the temporal artery, and a filament from the orbital branch of the superior maxillary nerve, are contained between these two layers. It is covered, on its outer surface, by the aponeurosis of the Occipito-frontalis, the Orbicularis palpebrarum, and Attollens and Attrahens aurem muscles; the temporal vessels and nerves cross it from below upwards.

151.—The Temporal Muscle, the Zygoma and Masseter having been removed.

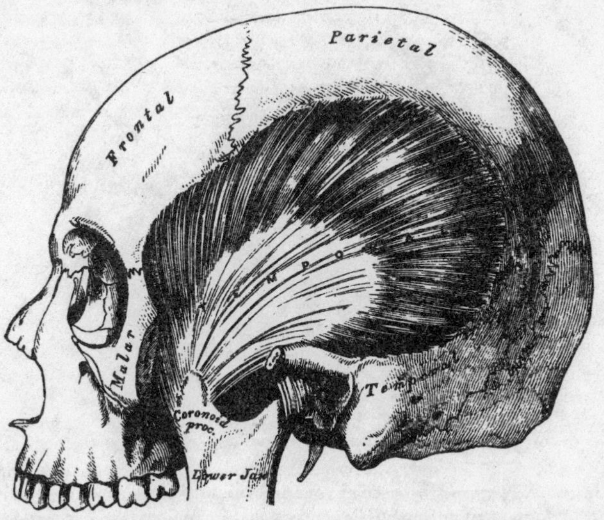

Dissection. In order to expose the Temporal muscle, remove the temporal fascia, which may be effected by separating it at its attachment along the upper border of the zygoma, and dissecting it upwards from the surface of the muscle. The zygomatic arch should then be divided, in front, at its junction with the malar bone; and, behind, near the external auditory meatus, and drawn downwards with the Masseter, which should be detached from its insertion into the ramus and angle of the jaw. The whole extent of the temporal muscle is then exposed.

The *Temporal* (fig. 151) is a broad radiating muscle, situated at the side of the head, and occupying the entire extent of the temporal fossa. It arises from the whole of the temporal fossa, which extends from the external angular process of the frontal in front, to the mastoid portion of the temporal behind; and from the curved line on the frontal and parietal bones above, to the pterygoid ridge on the great wing of the sphenoid below. It is also attached to the inner surface of the temporal fascia. Its fibres converge as they descend, and terminate in an aponeurosis, the fibres of which, radiated at its commencement, converge into a thick and flat tendon, which is inserted into the inner surface, apex, and anterior border of the coronoid process of the jaw, nearly as far forwards as the last molar tooth.

Relations. By its *superficial surface*, with the integument, the temporal fascia, the aponeurosis of the Occipito-frontalis, the Attollens and Attrahens aurem muscles, the temporal vessels and nerves, the zygoma and Masseter. By its *deep surface*, with the temporal fossa, the External pterygoid and part of the Buccinator muscles, the internal maxillary artery, its deep temporal branches, and the temporal nerves.

Nerves. Both muscles are supplied by the inferior maxillary nerve.

10. Pterygo-Maxillary Region. (Fig. 152.)

Internal Pterygoid. External Pterygoid.

Dissection. The Temporal muscle having been examined, saw through the base of the coronoid process, and draw it upwards, together with the Temporal muscle, which should be detached from the surface of the temporal fossa. Divide the ramus of the jaw just below the condyle, and also, by a transverse incision extending across the commencement of its lower third, just above the dental foramen; remove the fragment, and the Pterygoid muscles will be exposed.

152.—The Pterygoid Muscles; the Zygomatic Arch and a portion of the Ramus of the Jaw having been removed.

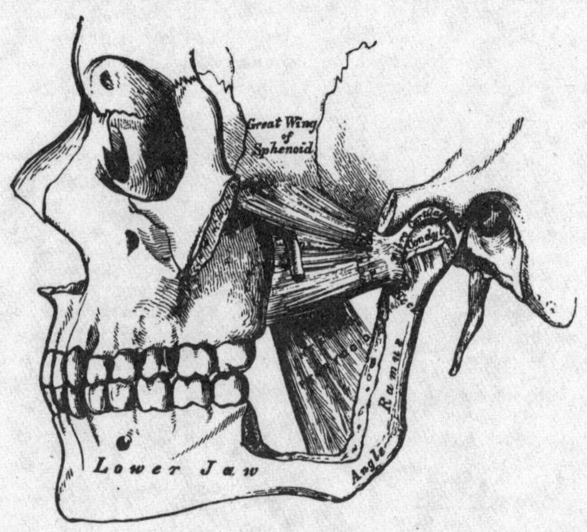

The *Internal Pterygoid* is a thick quadrilateral muscle, and resembles the Masseter, in form, structure, and the direction of its fibres. It arises from the pterygoid fossa, being attached to the inner surface of the external pterygoid plate, and to the grooved surface of the tuberosity of the palate bone; its fibres pass downwards, outwards, and backwards, to be inserted, by strong tendinous laminæ, into the lower and back part of the inner side of the ramus and angle of the lower jaw, as high as the dental foramen.

Relations. By its *external surface*, with the ramus of the lower jaw, from which it is separated, at its upper part, by the External pterygoid, the internal lateral ligament, the internal maxillary artery, and the dental vessels and nerves. By its *internal surface*, with the Tensor palati, being separated from the Superior constrictor of the pharynx by a cellular interval.

The *External Pterygoid* is a short thick muscle, somewhat conical in form, which extends almost horizontally between the zygomatic fossa and the condyle of the jaw. It arises from the pterygoid ridge on the great wing of the sphenoid, and the portion of bone included between it and the base of the pterygoid process; from the outer surface of the external pterygoid plate; and from the tuberosity of the palate and superior maxillary bones. Its fibres pass horizontally backwards and outwards, to be inserted into a depression in front of the neck of the condyle of the lower jaw, and into the corresponding part of the interarticular fibro-cartilage. This muscle, at its origin, appears to consist of two portions separated by a slight interval; hence the terms upper and lower head sometimes used in the description of the muscle.

Relations. By its *external surface*, with the ramus of the lower jaw, the internal maxillary artery, which crosses it,* the tendon of the Temporal muscle, and the Masseter. By its *internal surface*, it rests against the upper part of the Internal pterygoid, the internal lateral ligament, the middle meningeal artery, and inferior maxillary nerve; by its *upper border* it is in relation with the temporal and masseteric branches of the inferior maxillary nerve.

Nerves. These muscles are supplied by the inferior maxillary nerve.

Actions. The Temporal, Masseter, and Internal pterygoid raise the lower jaw against the upper with great force. The superficial portion of the Masseter, and the Internal pterygoid, assist the External pterygoid in drawing the lower jaw forwards upon the upper, the jaw being drawn back again by the deep fibres of the Masseter, and posterior fibres of the Temporal. The External pterygoid muscles are the direct agents in the trituration of the food, drawing the lower jaw directly forwards, so as to make the lower teeth project beyond the upper. If the muscle of one side acts, the corresponding side of the jaw is drawn forwards, and the other condyle remaining fixed, the symphysis deviates to the opposite side. The alternation of these movements on the two sides produces trituration.

MUSCLES AND FASCIÆ OF THE NECK.

The muscles of the Neck may be arranged into groups, corresponding with the region in which they are situated.

These groups are nine in number:—

1. Superficial Region.
2. Depressors of the Os Hyoides and Larynx.
3. Elevators of the Os Hyoides and Larynx.
4. Muscles of the Tongue.
5. Muscles of the Pharynx.
6. Muscles of the Soft Palate.
7. Muscles of the Anterior Vertebral Region.
8. Muscles of the Lateral Vertebral Region.
9. Muscles of the Larynx.

1. *Superficial Region.*

Platysma myoides.
Sterno-cleido-mastoid.

Infra-hyoid Region.

2. *Depressors of the Os Hyoides and Larynx.*

Sterno-hyoid.
Sterno-thyroid.
Thyro-hyoid.
Omo-hyoid.

Supra-hyoid Region.

3. *Elevators of the Os Hyoides and Larynx.*

Digastric.
Stylo-hyoid.
Mylo-hyoid.
Genio-hyoid.

Lingual Region.

4. *Muscles of the Tongue.*

Genio-hyo-glossus.
Hyo-glossus.
Lingualis.
Stylo-glossus.
Palato-glossus.

5. *Muscles of the Pharynx.*

Constrictor inferior.
Constrictor medius.
Constrictor superior.
Stylo-pharyngeus.
Palato-pharyngeus.

* This is the usual relation; but in many cases the artery will be found below the muscle.

6. *Muscles of the Soft Palate.*

Levator palati.
Tensor palati.
Azygos uvulæ.
Palato-glossus.
Palato-pharyngeus.

7. *Muscles of the Anterior Vertebral Region.*

Rectus capitis anticus major.
Rectus capitis anticus minor.
Rectus lateralis.
Longus colli.

8. *Muscles of the Lateral Vertebral Region.*

Scalenus anticus.
Scalenus medius.
Scalenus posticus.

9. *Muscles of the Larynx.*

Included in the description of the Larynx.

1. Superficial Cervical Region.

Platysma Myoides. Sterno-Cleido-Mastoid.

Dissection. A block having been placed at the back of the neck, and the face turned to the side opposite to that to be dissected, so as to place the parts upon the stretch, make two transverse incisions: one from the chin, along the margin of the lower jaw, to the mastoid process; and the other along the upper border of the clavicle. Connect these by an oblique incision made in the course of the Sterno-mastoid muscle, from the mastoid process to the sternum; the two flaps of integument having been removed in the direction shown in fig. 147, the superficial fascia will be exposed.

The *superficial cervical fascia* is exposed on the removal of the integument from the side of the neck; it is an extremely thin aponeurotic lamina, which is hardly demonstrable as a separate membrane. Beneath it is found the Platysma-myoides muscle, the external jugular vein, and some superficial branches of the cervical plexus of nerves.

The *Platysma Myoides* (fig. 148) is a broad thin plane of muscular fibres, placed immediately beneath the skin on each side of the neck. It arises from the clavicle and acromion, and from the fascia covering the upper part of the Pectoral, Deltoid, and Trapezius muscles; its fibres proceed obliquely upwards and inwards along the side of the neck, to be inserted into the lower jaw beneath the external oblique line, some passing forwards to the angle of the mouth, and others becoming lost in the cellular tissue of the face. The most anterior fibres interlace, in front of the jaw, with the fibres of the muscle of the opposite side; those next in order become blended with the Depressor labii inferioris and the Depressor anguli oris; others are prolonged upon the side of the cheek, and interlace, near the angle of the mouth, with the muscles in this situation, and may occasionally be traced to the Zygomatic muscles, or to the margin of the Orbicularis palpebrarum. Beneath the Platysma, the external jugular vein may be seen descending from the angle of the jaw to the clavicle. It is essential to remember the direction of the fibres of the Platysma, in connection with the operation of bleeding from this vessel; for if the point of the lancet is introduced in the direction of the muscular fibres, the orifice made will be filled up by the contraction of the muscle, and blood will not flow; but if the incision is made across the course of the fibres, they will retract, and expose the orifice in the vein, and so facilitate the flow of blood.

Relations. By its *external surface*, with the integument, to which it is united closely below, but m e loosely above. By its *internal surface*, with the Pectoralis major, Deltoid, and Trapezius, and with the clavicle. In the *neck*, with the external and anterior jugular veins, the deep cervical fascia, the superficial cervical plexus, the Sterno-mastoid, Sterno-hyoid, Omo-hyoid, and Digastric muscles. In front of the Sterno-mastoid, it covers the sheath of the carotid vessels; and behind it, the Scaleni muscles and the nerves of the brachial plexus. On the *face*, it is in relation with the parotid gland, the facial artery and vein, and the Masseter and Buccinator muscles.

The *deep cervical fascia* is exposed on the removal of the Platysma myoides.

It is a strong fibrous layer, which invests the muscles of the neck, and encloses the vessels and nerves. It commences, as an extremely thin layer, at the back part of the neck, where it is attached to the spinous processes of the cervical vertebræ, and to the ligamentum nuchæ; and, passing forwards to the posterior border of the Sterno-mastoid muscle, divides into two layers, one of which passes in front, and the other behind that muscle. These join again at the anterior border of the Sterno-mastoid; and, being continued forwards to the front of the neck, blend with the fascia of the opposite side. The superficial layer of the deep cervical fascia (that which passes in front of the Sterno-mastoid), if traced upwards, is found to pass across the parotid gland and Masseter muscle, forming the parotid and masseteric fascia, and is attached to the lower border of the zygoma, and more anteriorly to the lower border of the body of the jaw; if the same layer is traced downwards, it is seen to pass to the upper border of the clavicle and sternum, being pierced just above the former bone by the external jugular vein. In the middle line of the neck, the fascia is thin above, and connected to the hyoid bone; but it becomes thicker below, and divides, just below the thyroid gland, into two layers, the more superficial of which is attached to the upper border of the sternum and interclavicular ligament; the deeper and stronger layer is connected to the posterior border of that bone, covering in the Sterno-hyoid and Sterno-thyroid muscles. Between these two layers is a little areolar tissue and fat, and occasionally a small lymphatic gland. The deep layer of the cervical fascia (that which lies behind the posterior surface of the Sterno-mastoid) sends numerous prolongations, which invest the muscles and vessels of the neck; if traced upwards, a process of the fascia, of extreme density, is found passing behind and to the inner side of the parotid gland, to be attached to the base of the styloid process and angle of the lower jaw, termed the *stylo-maxillary ligament*; if traced downwards and outwards, the fascia will be found to enclose the posterior belly of the Omo-hyoid muscle, binding it down by a distinct process, which descends to be inserted into the clavicle and cartilage of the first rib. The deep layer of the cervical fascia also assists in forming the sheath which encloses the common carotid artery, internal jugular vein, and pneumogastric nerve. There are fibrous septa intervening between each of these parts, which, however, are included together in one common investment. More internally, a thin layer is continued across the trachea and thyroid gland, beneath the Sterno-thyroid muscles; and at the root of the neck this may be traced, over the large vessels, to be continuous with the fibrous layer of the pericardium.

The *Sterno-Cleido-Mastoid* (fig. 153) is a large thick muscle which passes obliquely across the side of the neck, being enclosed between the two layers of the deep cervical fascia. It is thick and narrow at its central part, but is broader and thinner at each extremity. It arises, by two heads, from the sternum and clavicle. The *sternal portion* is a rounded fasciculus, tendinous in front, fleshy behind, which arises from the upper and anterior part of the first piece of the sternum, and is directed upwards and backwards. The *clavicular portion* arises from the inner third of the superior border of the clavicle, being composed of fleshy and aponeurotic fibres; it is directed almost vertically upwards. These two portions are separated from one another, at their origin, by a triangular cellular interval; but become gradually blended, below the middle of the neck, into a thick rounded muscle, which is inserted, by a strong tendon, into the outer surface of the mastoid process, from its apex to its superior border, and by a thin aponeurosis into the outer two-thirds of the superior curved line of the occipital bone. The Sterno-mastoid varies much in its extent of attachment to the clavicle: in one case the clavicular may be as narrow as the sternal portion: in another, as much as three inches in breadth. When the clavicular origin is broad, it is occasionally subdivided into numerous slips, separated by narrow intervals. More rarely, the corresponding margins of the Sterno-mastoid and Trapezius have been found in contact. In the application of a ligature to the third part of the subclavian artery, it will be necessary, where the muscles come close together, to divide a portion of one or of both.

This muscle divides the quadrilateral space at the side of the neck into two triangles, an anterior and a posterior. The boundaries of the *anterior* triangle are in front, the median line of the neck; above, the lower border of the body of the jaw, and an imaginary line drawn from the angle of the jaw to the mastoid process; behind, the anterior border of the Sterno-mastoid muscle. The boundaries of the *posterior* triangle are, in front, the posterior border of the Sterno-mastoid; below, the upper border of the clavicle; behind, the anterior margin of the Trapezius.*

153.—Muscles of the Neck, and Boundaries of the Triangles.

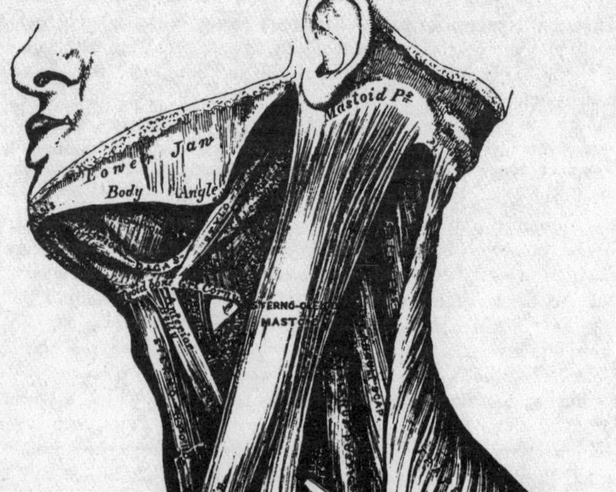

The anterior edge of the muscle forms a very prominent ridge beneath the skin, which it is important to notice, as it forms a guide to the surgeon in making the necessary incisions for ligature of the common carotid artery, and for œsophagotomy.

Relations. By its *superficial surface*, with the integument and Platysma, from which it is separated by the external jugular vein, the superficial branches of the cervical plexus, and the anterior layer of the deep cervical fascia. By its *deep surface*, it rests on the sterno-clavicular articulation, the deep layer of the cervical fascia, the Sterno-hyoid, Sterno-thyroid, Omo-hyoid, the posterior belly of the Digastric, Levator anguli scapulæ, the Splenius and Scaleni muscles. Below, it is in relation with the lower part of the common carotid artery, internal jugular vein, pneumogastric, descendens noni, and communicans noni nerves, and with the deep lymphatic glands; with the spinal accessory nerve, which pierces its upper third, the cervical plexus, the occipital artery, and a part of the parotid gland.

Nerves. The Platysma myoides is supplied by the facial and superficial cervical

* The anatomy of these triangles will be more exactly described with that of the vessels of the neck.

nerves; the sterno-cleido-mastoid by the spinal accessory and deep branches of the cervical plexus.

Actions. The Platysma myoides produces a slight wrinkling of the surface of the skin of the neck, in a vertical direction, when the entire muscle is brought into action. Its anterior portion, the thickest part of the muscle, depresses the lower jaw; it also serves to draw down the lower lip and angle of the mouth on each side, being one of the chief agents in the expression of melancholy. The Sterno-mastoid muscles, when both are brought into action, serve to depress the head upon the neck, and the neck upon the chest. Either muscle, acting singly, flexes the head, and (combined with the Splenius) draws it towards the shoulder of the same side, and rotates it so as to carry the face towards the opposite side.

Surgical Anatomy. The relations of the sternal and clavicular parts of the Sterno-mastoid should be carefully examined, as the surgeon is sometimes required to divide one or both portions of the muscle in *wry neck.* One variety of this distortion is produced by spasmodic contraction or rigidity of the Sterno-mastoid; the head being carried down towards the shoulder of the same side, and the face turned to the opposite side, and fixed in that position. When all other remedies for the relief of this disease have failed, subcutaneous division of the muscle is resorted to. This is performed by introducing a long narrow bistoury beneath it, about half an inch above its origin, and dividing it from behind forwards whilst the muscle is put well upon the stretch. There is seldom any difficulty in dividing the sternal portion. In dividing the clavicular portion care must be taken to avoid wounding the external jugular vein, which runs parallel with the posterior border of the muscle in this situation.

2.—Infra-Hyoid Region. (Figs. 153, 154.)

Depressors of the Os Hyoides and Larynx.

Sterno-Hyoid.	Thyro-Hyoid.
Sterno-Thyroid.	Omo-Hyoid.

Dissection. The muscles in this region may be exposed by removing the deep fascia from the front of the neck. In order to see the entire extent of the Omo-hyoid, it is necessary to divide the Sterno-mastoid at its centre, and turn its ends aside, and to detach the Trapezius from the clavicle and scapula. This, however, should not be done unless the Trapezius has been dissected.

The *Sterno-Hyoid* is a thin, narrow, riband-like muscle, which arises from the inner extremity of the clavicle, and the upper and posterior part of the first piece of the sternum; and, passing upwards and inwards, is inserted, by short tendinous fibres, into the lower border of the body of the os hyoides. This muscle is separated, below, from its fellow by a considerable interval; but they approach one another in the middle of their course, and again diverge as they ascend. It often presents, immediately above its origin, a transverse tendinous intersection, like those in the Rectus abdominis.

Variations. This muscle sometimes arises from the inner extremity of the clavicle, and the posterior sterno-clavicular ligament; or from the sternum and this ligament; from either bone alone, or from all these parts; and occasionally has a fasciculus connected with the cartilage of the first rib.

Relations. By its *superficial surface*, below, with the sternum, the sternal end of the clavicle, and the Sterno-mastoid; and above, with the Platysma and deep cervical fascia. By its *deep surface*, with the Sterno-thyroid, Crico-thyroid, and Thyro-hyoid muscles, the thyroid gland, the superior thyroid vessels, the crico-thyroid and thyro-hyoid membranes.

The *Sterno-Thyroid* is situated beneath the preceding muscle, but is shorter and wider than it. It arises from the posterior surface of the first bone of the sternum, below the origin of the Sterno-hyoid, and occasionally from the edge of the cartilage of the first rib; and is inserted into the oblique line on the side of the ala of the thyroid cartilage. This muscle is in close contact with its fellow at the lower part of the neck; and is frequently traversed by a transverse or oblique tendinous intersection, like those in the Rectus abdominis.

Variations. This muscle is sometimes continuous with the Thyro-hyoid and Inferior constrictor of the pharynx; and a lateral prolongation from it sometimes passes as far as the os hyoides.

Relations. By its *anterior surface*, with the Sterno-hyoid, Omo-hyoid, and Sterno-mastoid. By its *posterior surface*, from below upwards, with the trachea, vena innominata, common carotid (and on the right side the arteria innominata), the thyroid gland and its vessels, and the lower part of the larynx. The middle thyroid vein lies along its inner border, a relation which it is important to remember in the operation of tracheotomy.

154.—Muscles of the Neck. Anterior View.

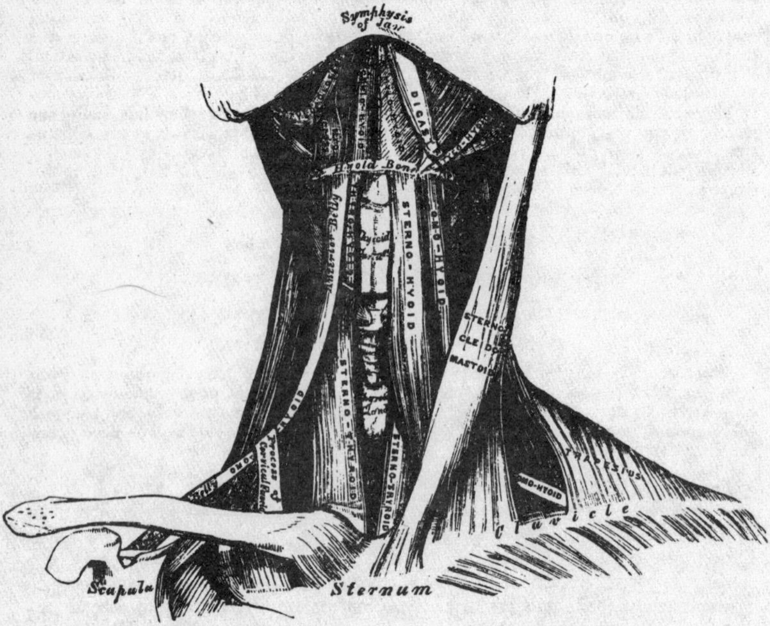

The *Thyro-Hyoid* is a small quadrilateral muscle appearing like a continuation of the Sterno-thyroid. It arises from the oblique line on the side of the thyroid cartilage, and passes vertically upwards to be inserted into the lower border of the body and greater cornu of the hyoid bone.

Relations. By its *external surface*, with the Sterno-hyoid and Omo-hyoid muscles. By its *internal surface*, with the thyroid cartilage, the Thyro-hyoid membrane, and the superior laryngeal vessels and nerve.

The *Omo-hyoid* passes across the side of the neck, from the scapula to the hyoid bone. It consists of two fleshy bellies, united by a central tendon. It arises from the upper border of the scapula, and occasionally from the transverse ligament which crosses the suprascapular notch; its extent of attachment to the scapula varying from a few lines to an inch. From this origin, the posterior belly forms a flat, narrow fasciculus, which inclines forwards across the lower part of the neck; behind the Sterno-mastoid muscle, where it becomes tendinous, it changes its direction, forming an obtuse angle, and ascends almost vertically upwards, close to the outer border of the Sterno-hyoid, to be inserted into the lower border of the body of the os hyoides, just external to the insertion of the Sterno-hyoid.

The tendon of this muscle, which varies much in its length and form in different subjects, is held in its position by a process of the deep cervical fascia, which includes it in a sheath, and is prolonged down to be attached to the cartilage of the first rib. It is by this means that the angular form of the muscle is maintained.

This muscle subdivides each of the two large triangles at the side of the neck into two smaller triangles; the two posterior ones being the *posterior superior* or *suboccipital*, and the *posterior inferior* or *subclavian*; the two anterior, the *anterior superior* or *superior carotid*, and the *anterior inferior* or *inferior carotid* triangle.

Relations. By its *superficial surface*, with the Trapezius, Subclavius, the clavicle, the Sterno-mastoid, deep cervical fascia, Platysma, and integument. By its *deep surface*, with the Scaleni, brachial plexus, sheath of the common carotid artery, and internal jugular vein, the descendens noni nerve, Sterno-thyroid and Thyro-hyoid muscles.

Nerves. The Thyro-hyoid is supplied by the hypo-glossal; the other muscles of this group by branches from the loop of communication between the descendens and communicans noni.

Actions. These muscles depress the larynx and hyoid bone, after they have been drawn up with the pharynx in the act of deglutition. The Omo-hyoid muscles not only depress the hyoid bone, but carry it backwards, and to one or the other side. They are also tensors of the cervical fascia. The Thyro-hyoid may act as an elevator of the thyroid cartilage, when the hyoid bone ascends, drawing upwards the thyroid cartilage behind the os hyoides.

3. Supra-Hyoid Region. (Figs. 153, 154.)

Elevators of the Os Hyoides—Depressors of the Lower Jaw.

Digastric.	Mylo-Hyoid.
Stylo-Hyoid.	Genio-Hyoid.

Dissection. To dissect these muscles, a block should be placed beneath the back of the neck, and the head drawn backwards, and retained in that position. On the removal of the deep fascia, the muscles are at once exposed.

The *Digastric* consists of two fleshy bellies united by an intermediate rounded tendon. It is a small muscle, situated below the side of the body of the lower jaw, and extending, in a curved form, from the side of the head to the symphysis of the jaw. The *posterior belly*, longer than the anterior, arises from the digastric groove on the inner side of the mastoid process of the temporal bone, and passes downwards, forwards, and inwards. The *anterior belly*, being reflected upwards and forwards, is inserted into a depression on the inner side of the lower border of the jaw, close to the symphysis. The tendon of the muscle perforates the Stylo-hyoid, and is held in connection with the side of the body of the hyoid bone by an aponeurotic loop, lined by a synovial membrane. A broad aponeurotic layer is given off from the tendon of the Digastric on each side, which is attached to the body and great cornu of the hyoid bone: this is termed the *supra-hyoid aponeurosis*. It forms a strong layer of fascia between the anterior portion of the two muscles, and a firm investment for the other muscles of the supra-hyoid region which lie beneath it.

The Digastric muscle divides the anterior superior triangle of the neck into two smaller triangles; the upper, or submaxillary, being bounded, above, by the lower jaw and mastoid process; below, by the two bellies of the Digastric muscle: the lower, or superior carotid triangle, being bounded, above, by the posterior belly of the Digastric: behind, by the Sterno-mastoid; below, by the Omo-hyoid.

Relations. By its *superficial surface*, with the Platysma, Sterno- and Trachelo-mastoid, part of the Stylo-hyoid muscle, and the parotid and submaxillary glands. By its *deep surface*, the anterior belly lies on the Mylo-hyoid; the posterior belly on the Stylo-glossus, Stylo-pharyngeus, and Hyo-glossus muscles, the external carotid artery and its lingual and facial branches, the internal carotid artery, internal jugular vein, and hypoglossal nerve.

The *Stylo-Hyoid* is a small, slender muscle, lying in front of, and above, the posterior belly of the Digastric. It arises from the middle of the outer surface of the styloid process; and, passing downwards and forwards, is inserted into the body of the hyoid bone, just at its junction with the greater cornu, and immediately above the Omo-hyoid. This muscle is perforated, near its insertion, by the tendon of the Digastric.

Relations. The same as the posterior belly of the Digastric.

The Digastric and Stylo-hyoid should be removed, in order to expose the next muscle.

The *Mylo-Hyoid* is a flat triangular muscle, situated immediately beneath the anterior belly of the Digastric, and forming, with its fellow of the opposite side, a muscular floor for the cavity of the mouth. It arises from the whole length of the mylo-hyoid ridge, from the symphysis in front to the last molar tooth behind. The posterior fibres pass obliquely forwards, to be inserted into the body of the os hyoides. The middle and anterior fibres are inserted into a median fibrous raphé, where they join at an angle with the fibres of the opposite muscle. This median raphé is sometimes wanting; the muscular fibres of the two sides are then directly continuous with one another.

Relations. By its *cutaneous surface*, with the Platysma, the anterior belly of the Digastric, the supra-hyoid fascia, the submaxillary gland, and submental vessels. By its *deep* or *superior surface*, with the Genio-hyoid, part of the Hyo-glossus, and Stylo-glossus muscles, the lingual and gustatory nerves, the sublingual gland, and the Buccal mucous membrane. Wharton's duct curves round its posterior border in its passage to the mouth.

Dissection. The Mylo-hyoid should now be removed, in order to expose the muscles which lie beneath; this is effected by detaching it from its attachments to the hyoid bone and jaw, and separating it by a vertical incision from its fellow of the opposite side.

The *Genio-Hyoid* is a narrow slender muscle, situated immediately beneath* the inner border of the preceding. It arises from the inferior genial tubercle on the inner side of the symphysis of the jaw, and passes downwards and backwards, to be inserted into the anterior surface of the body of the os hyoides. This muscle lies in close contact with its fellow of the opposite side, and increases slightly in breadth as it descends.

Relations. It is covered by the mylo-hyoid, and lies on the Genio-hyo-glossus.

Nerves. The Digastric is supplied, its anterior belly, by the mylo-hyoid branch of the inferior dental; its posterior belly, by the facial; the Stylo-hyoid, by the facial; the Mylo-hyoid, by the mylo-hyoid branch of the inferior dental; the Genio-hyoid, by the hypoglossal.

Actions. This group of muscles performs two very important actions. They raise the hyoid bone, and with it the base of the tongue, during the act of deglutition; or, when the hyoid bone is fixed by its depressors and those of the larynx, they depress the lower jaw. During the first act of deglutition, when the mass is being driven from the mouth into the pharynx, the hyoid bone, and with it the tongue, is carried upwards and forwards by the anterior belly of the Digastric, the Mylo-hyoid, and Genio-hyoid muscles. In the second act, when the mass is passing through the pharynx, the direct elevation of the hyoid bone takes place by the combined action of all the muscles; and after the food has passed, the hyoid bone is carried upwards and backwards by the posterior belly of the Digastric and Stylo-hyoid muscles, which assist in preventing the return of the morsel into the mouth.

4. Lingual Region.

Genio-Hyo-Glossus. Lingualis.
Hyo-Glossus. Stylo-Glossus.
Palato-Glossus.

* This refers to the depth of the muscles from the skin in dissecting. In the erect position of the body each of these muscles lies above the preceding.

Dissection. After completing the dissection of the preceding muscles, saw through the lower jaw just external to the symphysis. Then draw the tongue forwards, and attach it, by a stitch, to the nose; and its muscles, which are thus put on the stretch, may be examined.

The *Genio-Hyo-Glossus* has received its name from its triple attachment to the chin, hyoid bone, and tongue; it is a thin, flat, triangular muscle, placed vertically on either side of the middle line, its apex corresponding with its point of attachment to the lower jaw, its base with its insertion into the tongue and hyoid bone. It arises by a short tendon from the superior genial tubercle on the inner side of the symphysis of the chin, immediately above the Genio-hyoid; from this point, the muscle spreads out in a fan-like form, the inferior fibres passing downwards, to be inserted into the upper part of the body of the hyoid bone, a few being continued into the side of the pharynx; the middle fibres passing backwards, and the superior ones upwards and forwards, to be attached to the whole length of the under surface of the tongue, from the base to the apex.

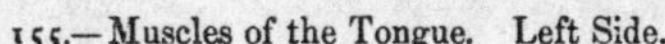
155.—Muscles of the Tongue. Left Side.

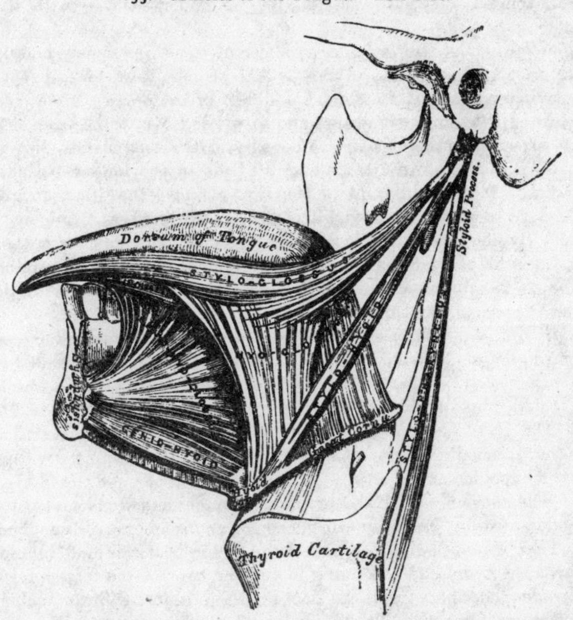

Relations. By its *internal surface*, it is in contact with its fellow of the opposite side, from which it is separated, at the back part of the tongue, by the fibrous septum, which extends through the middle of the organ. By its *external surface*, with the Lingualis, Hyo-glossus, and Stylo-glossus, the lingual artery and hypoglossal nerve, the gustatory nerve, and sublingual gland. By its *upper border*, with the mucous membrane of the floor of the mouth (frænum linguæ). By its *lower border*, with the Genio-hyoid.

The *Hyo-Glossus* is a thin, flat, quadrilateral muscle, which arises from the side of the body, the lesser cornu, and whole length of the greater cornu of the hyoid bone, and passing almost vertically upwards, is inserted into the side of the tongue, between the Stylo-glossus and Lingualis. Those fibres of this muscle which arise from the body, are directed upwards and backwards, overlapping those from the greater cornu, which are directed obliquely forwards. Those from the lesser cornu extend forwards and outwards along the side of the tongue, under cover of the portion arising from the body.

The difference in the direction of the fibres of this muscle, and their separate origin from different parts of the hyoid bone, led Albinus and other anatomists to describe it as three muscles, under the names of the Basio-glossus, the Kerato-glossus, and the Chondro-glossus.

Relations. By its *external surface*, with the Digastric, the Stylo-hyoid, Stylo-glossus, and Mylo-hyoid muscles, the gustatory and hypoglossal nerves, Wharton's duct, and the sublingual gland. By its *deep surface*, with the Genio-hyo-glossus, Lingualis, and Middle constrictor, the lingual vessels, and the glosso-pharyngeal nerve.

The *Lingualis* is a longitudinal band of muscular fibres, situated on the under surface of the tongue, lying in the interval between the Hyo-glossus and the Genio-hyo-glossus, and extending from the base to the apex of the organ. Posteriorly, some of its fibres are lost in the base of the tongue, and others are occasionally attached to the hyoid bone. It blends with the fibres of the Stylo-glossus, in front of the Hyo-glossus, and is continued forwards as far as the apex of the tongue. It is in relation, by its under surface, with the ranine artery.

The *Stylo-Glossus*, the shortest and smallest of the three styloid muscles, arises from the anterior and outer side of the styloid process, near its centre, and from the stylo-maxillary ligament, to which its fibres, in most cases, are attached by a thin aponeurosis. Passing downwards and forwards, so as to become nearly horizontal in its direction, it divides upon the side of the tongue into two portions: one longitudinal, which is inserted along the side of the tongue, blending with the fibres of the Lingualis in front of the Hyo-glossus; the other oblique, which overlaps the Hyo-glossus muscle, and decussates with its fibres.

Relations. By its *external surface*, from above downwards, with the parotid gland, the Internal pterygoid muscle, the sublingual gland, the gustatory nerve, and the mucous membrane of the mouth. By its *internal surface*, with the tonsil, the Superior constrictor, and the Hyo-glossus muscle.

The *Palato-Glossus*, or *Constrictor Isthmi Faucium*, although it is one of the muscles of the tongue, serving to draw its base upwards during the act of deglutition, is more nearly associated with the soft palate, both in its situation and function; it will, consequently, be described with that group of muscles.

Nerves. The Palato-glossus is supplied by the palatine branches of Meckel's ganglion; the Lingualis, by the chorda tympani; the remaining muscles of this group, by the hypoglossal.

Actions. The movements of the tongue, although numerous and complicated, may be understood by carefully considering the direction of the fibres of its muscles. The *Genio-hyo-glossi*, by means of their posterior and inferior fibres, draw upwards the hyoid bone, bringing it and the base of the tongue forwards, so as to protrude the apex from the mouth. The anterior fibres will draw the tongue back into the mouth. The whole length of these two muscles acting along the middle line of the tongue will draw it downwards, so as to make it concave from side to side, forming a channel along which fluids may pass towards the pharynx, as in sucking. The *Hyo-glossi* muscles draw down the sides of the tongue, so as to render it convex from side to side. The *Linguales*, by drawing downwards the centre and apex of the tongue, render it convex from before backwards. The *Palato-glossi* draw the base of the tongue upwards, and the Stylo-glossi upwards and backwards.

5. Pharyngeal Region.

Constrictor Inferior.	Constrictor Superior.
Constrictor Medius.	Stylo-Pharyngeus.

Palato-Pharyngeus.

Dissection (fig. 156). In order to examine the muscles of the pharynx, cut through the trachea and œsophagus just above the sternum, and draw them upwards by dividing the loose areolar tissue connecting the pharynx with the front of the vertebral column. The parts being drawn well forwards, apply the edge of the saw immediately behind the styloid processes, and saw the base of the skull through from below upwards. The pharynx and mouth should then be stuffed with tow, in order to distend its cavity and render the muscles tense and easier of dissection.

156.—Muscles of the Pharynx. External View.

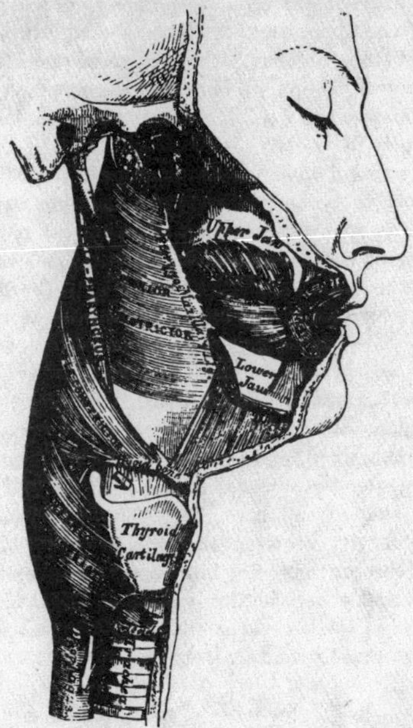

The *Inferior Constrictor*, the most superficial and thickest of the three constrictors, arises from the side of the cricoid and thyroid cartilages. To the cricoid cartilage it is attached in the interval between the crico-thyroid muscle, in front, and the articular facet for the thyroid cartilage behind. To the thyroid cartilage, it is attached to the oblique line on the side of the great ala, the cartilaginous surface behind it, nearly as far as its posterior border, and to the inferior cornu. From these attachments, the fibres spread backwards and inwards, to be inserted into the fibrous raphé in the posterior median line of the pharynx. The inferior fibres are horizontal, and continuous with the fibres of the œsophagus; the rest ascend, increasing in obliquity, and overlap the Middle constrictor. The superior laryngeal nerve passes near the upper border, and the inferior, or recurrent laryngeal, beneath the lower border of this muscle, previous to their entering the larynx.

Relations. It is covered by a dense cellular membrane which surrounds the entire pharynx. *Behind*, it is in relation with the vertebral column and the longus colli muscle; *laterally*, with the thyroid gland, the common carotid artery, and the Sterno-thyroid muscle; by its *internal surface*, with the Middle constrictor, the Stylo-pharyngeus, Palato-pharyngeus, and the mucous membrane of the pharynx.

The *Middle Constrictor* is a flattened, fan-shaped muscle, smaller than the preceding, and situated on a plane anterior to it. It arises from the whole length of the greater cornu of the hyoid bone, from the lesser cornu, and from the stylohyoid ligament. The fibres diverge from their origin: the lower ones descending beneath the Inferior constrictor, the middle fibres passing transversely, and the upper fibres ascending and overlapping the Superior constrictor. The muscle is inserted into the posterior median fibrous raphé, blending in the middle line with that of the opposite side.

Relations. This muscle is separated from the Superior constrictor by the glosso-pharyngeal nerve and the Stylo-pharyngeus muscle; and from the Inferior constrictor, by the superior laryngeal nerve. *Behind*, it lies on the vertebral column, the longus colli, and the Rectus anticus major. *On each side* it is in relation with the carotid vessels, the pharyngeal-plexus, and some lymphatic glands. Near its origin, it is covered by the Hyo-glossus, from which it is separated by the lingual artery. It lies upon the Superior constrictor, the Stylo-pharyngeus, the Palato-pharyngeus, and the mucous membrane.

The *Superior Constrictor* is a quadrilateral muscle, thinner and paler than the other constrictors, and situated at the upper part of the pharynx. It arises from the lower third of the margin of the internal pterygoid plate and its hamular process, from the contiguous portion of the palate bone and the reflected tendon of the Tensor palati muscle, from the pterygo-maxillary ligament, from the alveolar process above the posterior extremity of the mylo-hyoid ridge, and by a few fibres from the side of the tongue in connection with the Genio-hyo-glossus. From these points, the fibres curve backwards, to be inserted into the median raphé, being also prolonged by means of a fibrous aponeurosis to the pharyngeal spine on the basilar process of the occipital bone. The superior fibres arch beneath the Levator palati and the Eustachian tube, the interval between the upper border of the muscle and the basilar process being deficient in muscular fibres, and closed by fibrous membrane.

Relations. By its *outer surface*, with the vertebral column, the carotid vessels, the internal jugular vein, the three divisions of the eighth nerve and the ninth nerve, the Middle constrictor which overlaps it, and the Stylo-pharyngeus. It covers the Palato-pharyngeus and the tonsil, and is lined by mucous membrane.

The *Stylo-pharyngeus* is a long, slender muscle, round above, broad and thin below. It arises from the inner side of the base of the styloid process, passes downwards along the side of the pharynx between the Superior and Middle constrictors, and spreads out beneath the mucous membrane, where some of its fibres are lost in the constrictor muscles, and others joining with the Palato-pharyngeus, are inserted into the posterior border of the thyroid cartilage. The glosso-pharyngeal nerve runs on the outer side of this muscle, and crosses over it in passing forward to the tongue.

Relations. Externally, with the Stylo-glossus muscle, the external carotid artery, the parotid gland, and the Middle constrictor. *Internally*, with the internal carotid, the internal jugular vein, the Superior constrictor, Palato-pharyngeus and mucous membrane.

Nerves. The muscles of this group are supplied by branches from the pharyngeal plexus and glosso-pharyngeal nerve; and the Inferior constrictor, by an additional branch from the external laryngeal nerve.

Actions. When deglutition is about to be performed, the pharynx is drawn upwards and dilated in different directions, to receive the morsel propelled into it from the mouth. The Stylo-pharyngei, which are much farther removed from one another at their origin than at their insertion, draw the sides of the pharynx upwards and outwards, its breadth in the antero-posterior direction being increased by the larynx and tongue being carried forwards in their ascent. As soon as the morsel is received in the pharynx, the elevator muscles relax, the bag descends, and the Constrictors contract upon the morsel, and convey it gradually downwards into the œsophagus. Besides its action in deglutition, the pharynx also exerts an important influence in the modulation of the voice, especially in the production of the higher tones.

6. Palatal Region.

Levator Palati. Azygos Uvulæ.
Tensor Palati. Palato-Glossus.
Palato-Pharyngeus.

Dissection (fig. 157). Lay open the pharynx from behind, by a vertical incision extending from its upper to its lower part, and partially divide the occipital attachment by a transverse incision on each side of the vertical one; the posterior surface of the soft palate is then exposed. Having fixed the uvula so as to make it tense, the mucous membrane and glands should be carefully removed from the posterior surface of the soft palate, and the muscles of this part are at once exposed.

The *Levator Palati* is a long, thick, rounded muscle, placed on the outer side of the posterior nares. It arises from the under surface of the apex of the petrous portion of the temporal bone, and from the adjoining cartilaginous portion of the Eustachian tube; after passing into the pharynx, above the upper concave margin of the Superior constrictor, it descends obliquely downwards and inwards, its fibres spreading out in the posterior surface of the soft palate as far as the middle line, where they blend with those of the opposite side.

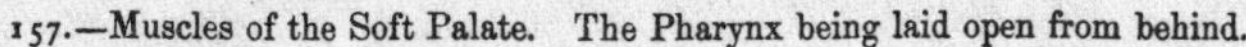

157.—Muscles of the Soft Palate. The Pharynx being laid open from behind.

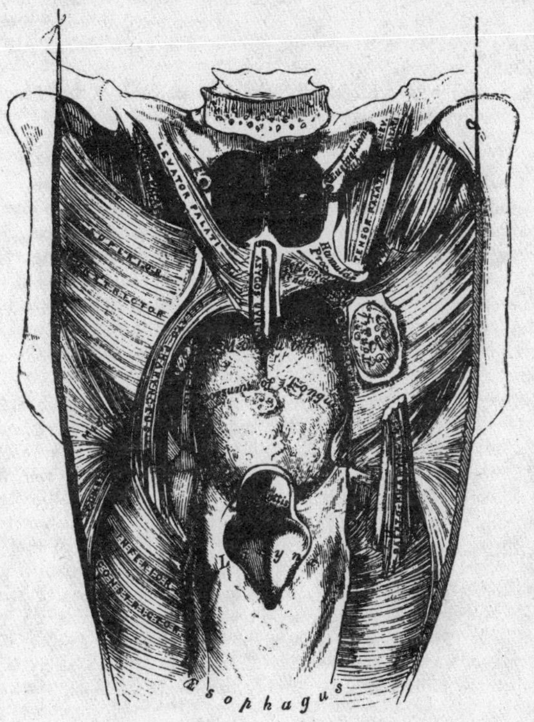

Relations. *Externally*, with the Tensor palati and Superior constrictor. *Internally*, with the mucous membrane of the pharynx. *Posteriorly*, with the mucous lining of the soft palate. This muscle must be removed and the pterygoid attachment of the Superior constrictor dissected away, in order to expose the next muscle.

The *Circumflexus* or *Tensor Palati* is a broad, thin, riband-like muscle, placed on the outer side of the preceding, and consisting of a vertical and a horizontal portion. The vertical portion arises by a broad, thin, and flat lamella from the scaphoid fossa at the base of the internal pterygoid plate, its origin extending as far back as the spine of the sphenoid; it also arises from the anterior aspect of the cartilaginous portion of the Eustachian tube; descending vertically between the internal pterygoid plate and the inner surface of the Internal pterygoid muscle, it terminates in a tendon which winds round the hamular process, being retained in this situation by some of the fibres of origin of the Internal pterygoid muscle, and lubricated by a bursa. The tendon or horizontal portion then passes horizontally inwards, and expands into a broad aponeurosis on the anterior surface of the soft palate, which unites in the median line with the aponeurosis of the opposite muscle, the fibres being attached in front to the transverse ridge on the posterior border of the horizontal portion of the palate bone.

Relations. Externally, with the Internal pterygoid. *Internally*, with the Levator palati, from which it is separated by the Superior constrictor, and the internal pterygoid plate. In the soft palate, its aponeurotic expansion is anterior to that of the Levator palati, being covered by mucous membrane.

The *Azygos Uvulæ* is not a single muscle as implied by its name, but a pair of narrow cylindrical fleshy fasciculi, placed side by side in the median line of the soft palate. Each muscle arises from the posterior nasal spine of the palate bone, and from the contiguous tendinous aponeurosis of the soft palate, and descends to be inserted into the uvula.

Relations. Anteriorly, with the tendinous expansion of the Levatores palati; *behind*, with the mucous membrane.

The two next muscles are exposed by removing the mucous membrane from the pillars of the soft palate throughout nearly their whole extent.

The *Palato-Glossus* (*Constrictor Isthmi Faucium*) is a small fleshy fasciculus, narrower in the middle than at either extremity, forming, with the mucous membrane covering its surface, the anterior pillar of the soft palate. It arises from the anterior surface of the soft palate on each side of the uvula, and passing forwards and outwards in front of the tonsil, is inserted into the side and dorsum of the tongue, where it blends with the fibres of the Stylo-glossus muscle. In the soft palate, the fibres of this muscle are continuous with those of the muscle of the opposite side.

The *Palato-Pharyngeus* is a long fleshy fasciculus, narrower in the middle than at either extremity, forming, with the mucous membrane covering its surface, the posterior pillar of the soft palate. It is separated from the preceding by an angular interval, in which the tonsil is lodged. It arises from the soft palate by an expanded fasciculus, which is divided into two parts by the Levator palati. The *anterior fasciculus*, the thickest, enters the soft palate between the Levator and Tensor, and joins in the middle line the corresponding part of the opposite muscle; the *posterior fasciculus* lies in contact with the mucous membrane, and also joins with the corresponding muscle in the middle line. Passing outwards and downwards behind the tonsil, the Palato-pharyngeus joins the Stylo-pharyngeus, and is inserted with that muscle into the posterior border of the thyroid cartilage, some of its fibres being lost on the side of the pharynx, and others passing across the middle line posteriorly, to decussate with the muscle of the opposite side.

Relations. In the soft palate, its *anterior* and *posterior surfaces* are covered by mucous membrane, from which it is separated by a layer of palatine glands. By its *superior border*, it is in relation with the Levator palati. Where it forms the posterior pillar of the fauces, it is covered by mucous membrane, excepting on its outer surface. In the *pharynx* it lies between the mucous membrane and the constrictor muscles.

Nerves. The Tensor palati is supplied by a branch from the otic ganglion; the Levator palati, and Azygos uvulæ, by the facial, through the connection of its trunk with the Vidian, by the petrosal nerves; the other muscles, by the palatine branches of Meckel's ganglion.

Actions. During the *first act* of deglutition, the morsel of food is driven back into the fauces by the pressure of the tongue against the hard palate; the base of the tongue being, at the same time, retracted, and the larynx raised with the pharynx, and carried forwards under it. During the *second* stage, the epiglottis is pressed over the superior aperture of the larynx, and the morsel glides past it; then the Palato-glossi muscles, the constrictors of the fauces, contract behind the food; the soft palate is slightly raised by the Levator palati, and made tense by the Tensor palati; and the Palato-pharyngei contract, and come nearly together, the uvula filling up the slight interval between them. By these means, the food is prevented passing into the upper part of the pharynx or the posterior nares; at the same time, the latter muscles form an inclined plane, directed obliquely downwards and backwards, along which the morsel descends into the lower part of the pharynx.

Surgical Anatomy. The muscles of the soft palate should be carefully dissected, the relations they bear to the surrounding parts especially examined, and their action attentively studied upon the dead subject, as the surgeon is required to divide one or more of these muscles in the operation of staphyloraphy. Sir W. Fergusson has shown, that in the congenital deficiency, called *cleft palate*, the edges of the fissure are forcibly separated by the action of the Levatores palati and Palato-pharyngei muscles, producing very considerable impediment to the healing process after the performance of the operation for uniting their margins by adhesion; he has, consequently, recommended the division of these muscles as one of the most important steps in the operation. This he effects by an incision made with a curved knife introduced behind the flap. The incision is to be half way between the hamular process and Eustachian tube, and perpendicular to a line drawn between them. This incision perfectly accomplishes the division of the Levator palati. The Palato-pharyngeus may be divided by cutting across the posterior pillar of the soft palate, just below the tonsil, with a pair of blunt-pointed curved scissors; and the anterior pillar may be divided also. To divide the Levator palati, the plan recommended by Mr. Pollock is to be greatly preferred. The flap being put upon the stretch, a double-edged knife is passed through the soft palate, just on the inner side of the hamular process, and above the line of the Levator palati. The handle being now alternately raised and depressed, a sweeping cut is made along the posterior surface of the soft palate, and the knife withdrawn, leaving only a small opening in the mucous membrane on the anterior surface. If this operation is performed on the dead body, and the parts afterwards dissected, the Levator palati will be found completely divided.

7. Vertebral Region (Anterior).

Rectus Capitis Anticus Major. Rectus Lateralis.
Rectus Capitis Anticus Minor. Longus Colli.

The *Rectus Capitis Anticus Major* (fig. 158), broad and thick above, narrow below, appears like a continuation upwards of the Scalenus anticus. It arises by four tendinous slips from the anterior tubercles of the transverse processes of the third, fourth, fifth, and sixth cervical vertebræ, and ascends, converging towards its fellow of the opposite side, to be inserted into the basilar process of the occipital bone.

Relations. By its *anterior surface*, with the pharynx, the sympathetic nerve, and the sheath enclosing the carotid artery, internal jugular vein, and pneumogastric nerve. By its *posterior surface*, with the Longus colli, the Rectus anticus minor, and the upper cervical vertebræ.

The *Rectus Capitis Anticus Minor* is a short flat muscle, situated immediately beneath the upper part of the preceding. It arises from the anterior surface of the lateral mass of the atlas, and from the root of its transverse process, and passing obliquely upwards and inwards, is inserted into the basilar process immediately behind the preceding muscle.

Relations. By its *anterior surface*, with the Rectus anticus major. By its *posterior surface*, with the front of the occipito-atlantal articulation. *Externally*, with the superior cervical ganglion of the sympathetic.

The *Rectus Lateralis* is a short, flat muscle, which arises from the upper surface of the transverse process of the atlas, and is inserted into the under surface of the jugular process of the occipital bone.

Relations. By its *anterior surface*, with the internal jugular vein. By its *posterior surface*, with the vertebral artery.

The *Longus Colli* is a long, flat muscle, situated on the anterior surface of the spine, between the atlas and the third dorsal vertebra. It is broad in the middle, narrow and pointed at each extremity, and consists of three portions, of a superior oblique, an inferior oblique, and a vertical portion. The *superior oblique portion* arises from the anterior tubercles of the transverse processes of the third, fourth, and fifth cervical vertebræ; and, ascending obliquely inwards, is inserted by a narrow tendon into the tubercle on the anterior arch of the atlas. The *inferior oblique portion*, the smallest part of the muscle, arises from the bodies of the first two or three dorsal vertebræ; and, passing obliquely outwards, is inserted into the transverse processes of the fifth and sixth cervical vertebræ.

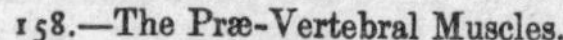
158.—The Præ-Vertebral Muscles.

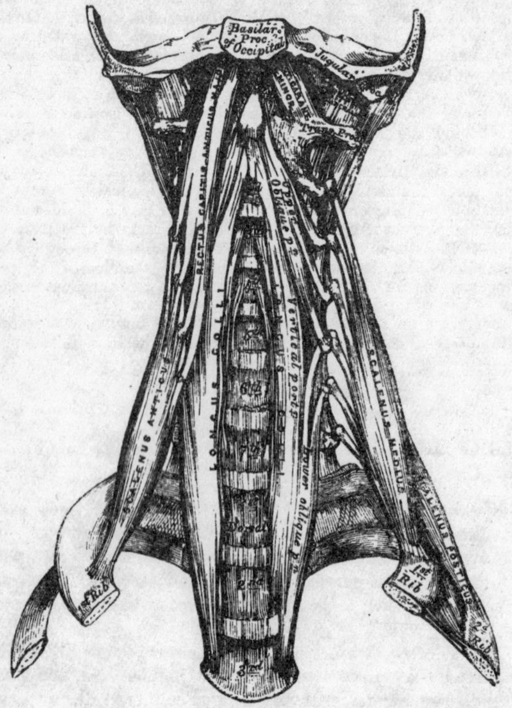

The *vertical portion* lies directly on the front of the spine, and is extended between the bodies of the lower three cervical and the upper three dorsal vertebræ below, and the bodies of the second, third, and fourth cervical vertebræ above.

Relations. By its *anterior surface*, with the pharynx, the œsophagus, sympathetic nerve, the sheath of the great vessels of the neck, the inferior thyroid artery, and recurrent laryngeal nerve. By its *posterior surface*, with the cervical and dorsal portions of the spine. Its *inner border* is separated from the opposite muscle by a considerable interval below; but they approach each other above.

8. Vertebral Region (Lateral).

Scalenus Anticus. Scalenus Medius.
Scalenus Posticus.

The *Scalenus Anticus* is a conical-shaped muscle, situated deeply at the side of the neck, behind the Sterno-mastoid. It arises by a narrow, flat tendon from the tubercle on the inner border and upper surface of the first rib; and, ascending almost vertically, is inserted into the anterior tubercles of the transverse processes of the third, fourth, fifth, and sixth cervical vertebræ. The lower part of this muscle separates the subclavian artery and vein: the latter being in front, and the former, with the brachial plexus, behind.

Relations. *In front*, with the clavicle, the Subclavius, Sterno-mastoid, and Omo-hyoid muscles, the transversalis colli, and ascending cervical arteries, the subclavian vein, and the phrenic nerve. By its *posterior surface*, with the pleura, the subclavian artery, and brachial plexus of nerves. It is separated from the Longus colli, on the inner side, by the vertebral artery.

The *Scalenus Medius*, the largest and longest of the three Scaleni, arises, by a broad origin, from the upper surface of the first rib, behind the groove for the subclavian artery, as far back as the tubercle; and, ascending along the side of the vertebral column, is inserted, by separate tendinous slips, into the posterior tubercles of the transverse processes of the lower six cervical vertebræ. It is separated from the Scalenus anticus by the subclavian artery below, and the cervical nerves above.

Relations. By its *anterior surface*, with the Sterno-mastoid; it is crossed by the clavicle, the Omo-hyoid muscle, and subclavian artery. To its *outer side* is the Levator anguli scapulæ, and the Scalenus posticus muscle.

The *Scalenus posticus*, the smallest of the three Scaleni, arises by a thin tendon from the outer surface of the second rib, behind the attachment of the Serratus magnus, and, enlarging as it ascends, is inserted, by two or three separate tendons, into the posterior tubercles of the transverse processes of the lower two or three cervical vertebræ. This is the most deeply placed of the three Scaleni, and is occasionally blended with the Scalenus medius.

Nerves. The Rectus capitis anticus major and minor are supplied by the suboccipital and deep branches of the cervical plexus; the Rectus lateralis, by the suboccipital; and the Longus colli and Scaleni, by branches from the lower cervical nerves.

Actions. The Rectus anticus major and minor are the direct antagonists of the muscles at the back of the neck, serving to restore the head to its natural position after it has been drawn backwards. These muscles also serve to flex the head, and, from their obliquity, rotate it, so as to turn the face to one or the other side. The Longus colli will flex and slightly rotate the cervical portion of the spine. The Scaleni muscles, taking their fixed point from below, draw down the transverse processes of the cervical vertebræ, bending the spinal column to one or the other side. If the muscles of both sides act, the spine will be kept erect. When they take their fixed point from above, they elevate the first and second ribs, and are, therefore, inspiratory muscles.

MUSCLES AND FASCIÆ OF THE TRUNK.

The Muscles of the Trunk may be arranged in four groups; the muscles of the Back, of the Abdomen, of the Thorax, and of the Perinæum.

Muscles of the Back.

The muscles of the Back are very numerous, and may be subdivided into five layers:—

First Layer.

Trapezius.
Latissimus dorsi.

Second Layer.

Levator anguli scapulæ.
Rhomboideus minor.
Rhomboideus major.

Third Layer.

Serratus posticus superior.
Serratus posticus inferior.
Splenius capitis.
Splenius colli.

Fourth Layer.

Sacral and lumbar regions.

Erector spinæ.

Dorsal region.

Sacro-lumbalis.
Musculus accessorius ad sacro-lumbalem.
Longissimus dorsi.
Spinalis dorsi.

Cervical region.

Cervicalis ascendens.
Transversalis colli.
Trachelo-mastoid.
Complexus.
Biventer cervicis.
Spinalis cervicis.

FIFTH LAYER.

Semispinalis dorsi.
Semispinalis colli.
Multifidus spinæ.
Rotatores spinæ.
Supraspinales.
Interspinales.
Extensor coccygis.
Intertransversales.
Rectus capitis posticus major.
Rectus capitis posticus minor.
Obliquus capitis superior.
Obliquus capitis inferior.

FIRST LAYER.

Trapezius. Latissimus dorsi.

Dissection (fig. 159). Place the body in the prone position, with the arms extended over the sides of the table, and the chest and abdomen supported by several blocks, so as to render the muscles tense. Then make an incision along the middle line of the back, from the occipital protuberance to the coccyx. Make a transverse incision from the upper end of this to the mastoid process; and a third incision from its lower end, along the crest of the ilium to about its middle. This large intervening space should, for convenience of dissection, be subdivided by a fourth incision, extending obliquely from the spinous process of the last dorsal vertebra, upwards and outwards, to the acromion process. This incision corresponds with the lower border of the Trapezius muscle. The flaps of integument are then to be removed in the direction shown in the figure.

The *Trapezius* is a broad, flat, triangular muscle, placed immediately beneath the skin, and covering the upper and back part of the neck and shoulders. It arises from the inner third of the superior curved line of the occipital bone; from the ligamentum nuchæ, the spinous process of the seventh cervical, and those of all the dorsal vertebræ; and from the corresponding portion of the supraspinous ligament. From this origin, the superior fibres proceed downwards and outwards, the inferior ones, upwards and outwards; and the middle fibres, horizontally; and are inserted, the superior ones, into the outer third of the posterior border of the clavicle; the middle fibres into the upper margin of the acromion process, and into the whole length of the upper border of the spine of the scapula; the inferior fibres converge near the scapula, and are attached to a triangular aponeurosis, which glides over a smooth surface at the inner extremity of the spine, and is inserted into a tubercle at the outer part of the surface. The Trapezius is fleshy in the greater part of its extent, but tendinous at its origin and insertion. At its occipital origin, it is connected to the bone by a thin fibrous lamina, firmly adherent to the skin, and wanting the lustrous, shining appearance of aponeurosis. At its origin from the spines of the vertebræ, it is connected to the bones by means of a broad semi-elliptical

159.—Dissection of the Muscles of the Back.

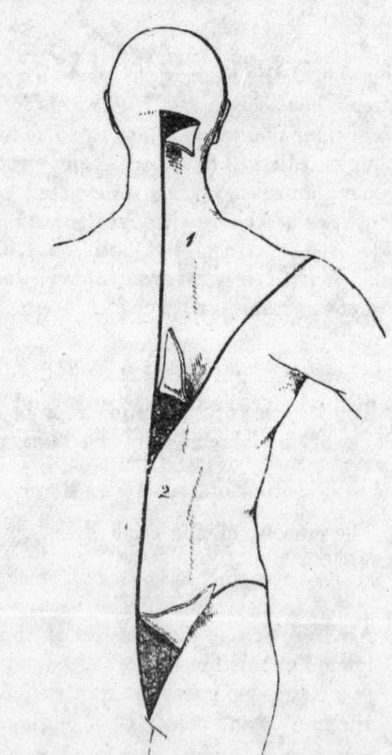

aponeurosis, which occupies the space between the sixth cervical and the third dorsal vertebræ, and forms, with the aponeurosis of the opposite muscle, a tendinous ellipse. The rest of the muscle arises by numerous short tendinous fibres. If the Trapezius is dissected on both sides, the two muscles resemble a trapezium, or diamond-shaped quadrangle; two angles corresponding to the shoulders; a third to the occipital protuberance; and the fourth to the spinous process of the last dorsal vertebra.

The clavicular insertion of this muscle varies as to the extent of its attachment; it sometimes advances as far as the middle of the clavicle, and may even become blended with the posterior edge of the Sterno-mastoid, or overlap it. This should be borne in mind in the operation for tying the third part of the subclavian artery.

Relations. By its *superficial surface*, with the integument, to which it is closely adherent above, but separated below by an aponeurotic lamina. By its *deep surface*, in the neck, with the Complexus, Splenius, Levator anguli scapulæ, and Rhomboideus minor; in the back, with the Rhomboideus major, Supraspinatus, Infraspinatus, a small portion of the Serratus posticus superior, the vertebral aponeurosis (which separates it from the Erector spinæ) and the Latissimus dorsi. The spinal accessory nerve passes beneath the anterior border of this muscle, near the clavicle. The outer margin of its cervical portion forms the posterior boundary of the posterior triangle of the neck, the other boundaries being the Sterno-mastoid in front, and the clavicle below.

The *ligamentum nuchæ* (fig. 160) is a thin band of condensed cellulo-fibrous membrane, placed in the line of union between the two Trapezii in the neck. It extends from the external occipital protuberance to the spinous process of the seventh cervical vertebra, where it is continuous with the supraspinous ligament. From its anterior surface a fibrous slip is given off to the spinous process of each of the cervical vertebræ, excepting the atlas, so as to form a septum between the muscles on each side of the neck. In man, it is merely the rudiment of an important elastic ligament, which, in some of the lower animals, serves to sustain the weight of the head.

The *Latissimus Dorsi* is a broad, flat muscle, which covers the lumbar and the lower half of the dorsal regions, and is gradually contracted into a narrow fasciculus at its insertion into the humerus. It arises by an aponeurosis from the spinous processes of the six inferior dorsal, from those of the lumbar and sacral vertebræ, and from the supraspinous ligament. Over the sacrum, the aponeurosis of this muscle blends with the tendon of the Erector spinæ. It also arises from the external lip of the crest of the ilium, behind the origin of the External oblique, and by fleshy digitations from the three or four lower ribs, which are interposed between similar processes of the External oblique muscle (fig. 163). From this extensive origin the fibres pass in different directions, the upper ones horizontally, the middle obliquely upwards, and the lower vertically upwards, so as to converge and form a thick fasciculus, which crosses the inferior angle of the scapula, and occasionally receives a few fibres from it. The muscle then curves around the lower border of the Teres major, and is twisted upon itself, so that the superior fibres become at first posterior and then inferior, and the vertical fibres at first anterior and then superior. It then terminates in a short quadrilateral tendon, about three inches in length, which, passing in front of the tendon of the Teres major, is inserted into the bottom of the bicipital groove of the humerus, above the insertion of the tendon of the Pectoralis major. The lower border of the tendon of this muscle is united with that of the Teres major, the surfaces of the two being separated by a bursa; another bursa is sometimes interposed between the muscle and the inferior angle of the scapula.

A muscular slip, varying from 3 to 4 inches in length, and from $\frac{1}{4}$ to $\frac{3}{4}$ of an inch in breadth, occasionally arises from the upper edge of the Latissimus dorsi, about the middle of the posterior fold of the axilla, and crosses the axilla in front of the axillary vessels and nerves, to join the under surface of the tendon of the Pectoralis major, the Coraco-brachialis, or

160.—Muscles of the Back. On the Left Side is exposed the First Layer; on the Right Side, the Second Layer and part of the Third.

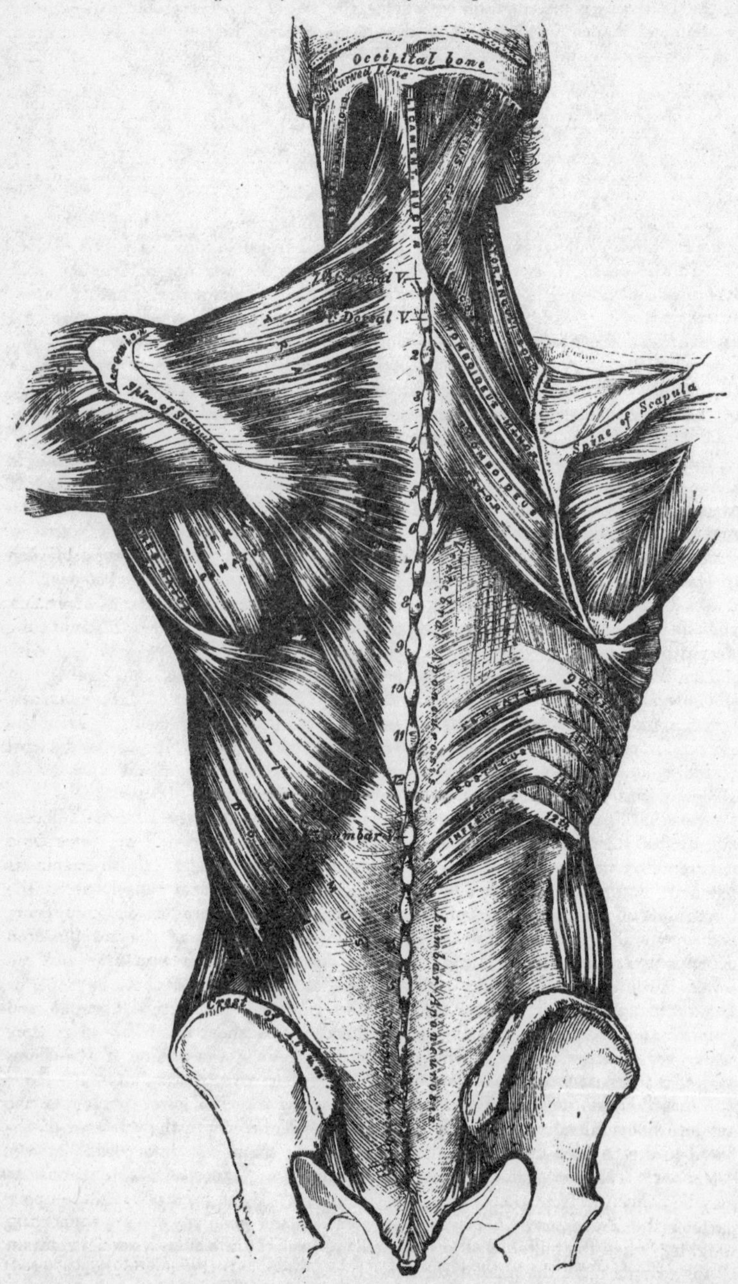

the fascia over the Biceps. The position of this abnormal slip is a point of interest in its relation to the axillary artery, as it crosses the vessel just above the spot usually selected for the application of a ligature, and may mislead the surgeon during the operation. It may be easily recognised by the transverse direction of its fibres. Dr. Struthers found it, in 8 out of 105 subjects, occurring seven times on both sides.

Relations. Its *superficial surface* is subcutaneous, excepting at its upper part, where it is covered by the Trapezius. By its *deep surface*, it is in relation with the Erector spinæ, the Serratus posticus inferior, the lower Intercostal muscles and ribs, the Serratus magnus, inferior angle of the scapula, Rhomboideus major, Infraspinatus, and Teres major. Its outer margin is separated below, from the External oblique, by a small triangular interval; and another triangular interval exists between its upper border and the margin of the Trapezius, in which the Intercostal and Rhomboideus major muscles are exposed.

Nerves. The Trapezius is supplied by the spinal accessory, and cervical plexus; the Latissimus dorsi by the subscapular nerves.

Second Layer.

Levator Anguli Scapulæ. Rhomboideus Minor.
Rhomboideus Major.

Dissection. The Trapezius must be removed in order to expose the next layer; to effect this, detach the muscle from its attachment to the clavicle and spine of the scapula, and turn it back towards the spine.

The *Levator Anguli Scapulæ* is situated at the back part and side of the neck. It arises by four tendinous slips from the posterior tubercles of the transverse processes of the three or four upper cervical vertebræ; these becoming fleshy are united so as to form a flat muscle, which, passing downwards and backwards, is inserted into the posterior border of the scapula, between the superior angle and the triangular smooth surface at the root of the spine.

Relations. By its *superficial* (*anterior*) *surface*, with the integument, Trapezius, and Sterno-mastoid. By its *deep* (*posterior*) *surface*, with the Splenius colli, Transversalis colli, Cervicalis ascendens, and Serratus posticus superior, and with the transverse cervical and posterior scapular arteries.

The *Rhomboideus Minor* arises from the ligamentum nuchæ, and spinous processes of the seventh cervical and first dorsal vertebræ. Passing downwards and outwards, it is inserted into the margin of the triangular smooth surface at the root of the spine of the scapula. This small muscle is usually separated from the Rhomboideus major by a slight cellular interval.

The *Rhomboideus Major* is situated immediately below the preceding, the adjacent margins of the two being occasionally united. It arises by tendinous fibres from the spinous processes of the four or five upper dorsal vertebræ and the supraspinous ligament, and is inserted into a narrow tendinous arch, attached above, to the triangular surface near the spine; below, to the inferior angle, the arch being connected to the border of the scapula by a thin membrane. When the arch extends, as it occasionally does, but a short distance, the muscular fibres are inserted into the scapula itself.

Relations. By their *superficial* (*posterior*) *surface*, with the integument and Trapezius; the Rhomboideus major, with the Latissimus dorsi. By their *deep* (*anterior*) *surface*, with the Serratus posticus superior, posterior scapular artery, part of the Erector spinæ, the Intercostal muscles and ribs.

Nerves. These muscles are supplied by branches from the fifth cervical nerve, and additional filaments from the deep branches of the cervical plexus are distributed to the Levator anguli scapulæ.

Actions. The movements effected by the preceding muscles are numerous, as may be conceived from their extensive attachment. If the head is fixed, the upper part of the Trapezius will elevate the point of the shoulder, as in supporting weights; when the middle and lower fibres are brought into action, partial rotation of the scapula upon the side of the chest is produced. If the shoulders are fixed, both Trapezii acting together will draw the head directly backwards, or if only one acts, the head is drawn to the corresponding side.

The *Latissimus dorsi*, when it acts upon the humerus, draws it backwards and downwards, and at the same time rotates it inwards. If the arm is fixed, the muscle may act in various ways upon the trunk; thus, it may raise the lower ribs and assist in forcible inspiration, or if both arms are fixed, the two muscles may assist the abdominal and great Pectoral muscles in drawing the whole trunk forwards, as in climbing or walking on crutches.

The *Levator anguli scapulæ* raises the superior angle of the scapula after it has been depressed by the lower fibres of the Trapezius, whilst the Rhomboid muscles carry the inferior angle backwards and upwards, thus producing a slight rotation of the scapula upon the side of the chest. If the shoulder be fixed, the Levator anguli scapulæ may incline the neck to the corresponding side. The Rhomboid muscles acting together with the middle and inferior fibres of the Trapezius, will draw the scapula directly backwards towards the spine.

Third Layer.

Serratus Posticus Superior. Serratus Posticus Inferior.

Splenius { Splenius Capitis.
Splenius Colli.

Dissection. To bring into view the third layer of muscles, remove the whole of the second, together with the Latissimus dorsi; by cutting through the Levator anguli scapulæ and Rhomboid muscles near their insertion, and reflecting them upwards, to expose the Serratus posticus superior, dividing the Latissimus dorsi in the middle by a vertical incision carried from its upper to its lower part, and reflecting the two halves of the muscle.

The *Serratus Posticus Superior* is a thin, flat, quadrilateral muscle, situated at the upper and back part of the thorax. It arises by a thin and broad aponeurosis, from the ligamentum nuchæ, and from the spinous processes of the last cervical and two or three upper dorsal vertebræ. Inclining downwards and outwards, it becomes muscular, and is inserted, by four fleshy digitations, into the upper borders of the second, third, fourth, and fifth ribs, a little beyond their angles.

Relations. By its *superficial surface*, with the Trapezius, Rhomboidei, and Serratus magnus. By its *deep surface*, with the Splenius, upper part of the Erector spinæ, Intercostal muscles and ribs.

The *Serratus Posticus Inferior* is situated at the junction of the dorsal and lumba regions: it is of an irregularly quadrilateral form, broader than the preceding, an separated from it by a considerable interval. It arises by a thin aponeurosis from the spinous processes of the last two dorsal and two or three upper lumbar vertebræ, and from the interspinous ligaments. Passing obliquely upwards and outwards, it becomes fleshy, and divides into four flat digitations, which are inserted into the lower borders of the four lower ribs, a little beyond their angles.

Relations. By its *superficial surface*, with the Latissimus dorsi, with the aponeurosis of which its own aponeurotic origin is inseparably blended. By its *deep surface*, with the lumbar fascia, the Erector spinæ, ribs, and Intercostal muscles. Its upper margin is continuous with the vertebral aponeurosis.

The *vertebral aponeurosis* is a thin aponeurotic lamina, extending along the whole length of the back part of the thoracic region, serving to bind down the Erector spinæ, and separating it from those muscles which connect the spine to the upper extremity. It consists of longitudinal and transverse fibres blended together, forming a thin lamella, which is attached in the median line to the spinous processes of the dorsal vertebræ; externally, to the angles of the ribs; and below, to the upper border of the Inferior serratus and tendon of the Latissimus dorsi; above, it passes beneath the Splenius, and blends with the deep fascia of the neck.

Now detach the Serratus posticus superior from its origin, and turn it outwards, when the Splenius muscle will be brought into view.

The *Splenius* is situated at the back of the neck and upper part of the dorsal region. At its origin, it is a single muscle, narrow and pointed in form; but it soon becomes broader, and divides into two portions, which have separate insertions. It arises, by tendinous fibres, from the lower half of the ligamentum nuchæ, from the spinous processes of the last cervical and of the six upper dorsal vertebræ, and from the supraspinous ligament. From this origin, the fleshy fibres proceed obliquely upwards and outwards, forming a broad flat muscle, which divides as it ascends into two portions, the Splenius capitis and Splenius colli.

The *splenius capitis* is inserted into the mastoid process of the temporal bone, and into the rough surface on the occipital bone beneath the superior curved line.

The *splenius colli* is inserted, by tendinous fasciculi, into the posterior tubercles of the transverse processes of the three or four upper cervical vertebræ.

The Splenius is separated from its fellow of the opposite side by a triangular interval, in which is seen the Complexus.

Relations. By its *superficial surface*, with the Trapezius, from which it is separated below by the Rhomboidei and the Serratus posticus superior. It is covered at its insertion by the Sterno-mastoid. By its *deep surface*, with the Spinalis dorsi, Longissimus dorsi, Semispinalis colli, Complexus, Trachelo-mastoid, and Transversalis colli.

Nerves. The Splenius and Superior serratus are supplied from the external posterior branches of the cervical nerves; the Inferior serratus, from the external branches of the dorsal nerves.

Actions. The Serrati are respiratory muscles acting in antagonism to each other. The Serratus posticus superior elevates the ribs; it is, therefore, an inspiratory muscle; while the Serratus inferior draws the lower ribs downwards, and is a muscle of expiration. This muscle is also probably a tensor of the vertebral aponeurosis. The Splenii muscles of the two sides, acting together, draw the head directly backwards, assisting the Trapezius and Complexus; acting separately, they draw the head to one or the other side, and slightly rotate it, turning the face to the same side. They also assist in supporting the head in the erect position.

Fourth Layer.

Sacral and Lumbar Regions.	*Cervical Region.*
Erector Spinæ.	Cervicalis Ascendens.
Dorsal Regions.	Transversalis Colli.
Sacro-Lumbalis.	Trachelo-Mastoid.
Musculus Accessorius ad Sacro-Lumbalem.	Complexus.
Longissimus Dorsi.	Biventer Cervicis.
Spinalis Dorsi.	Spinalis Colli.

Dissection. To expose the muscles of the fourth layer, remove entirely the Serrati and vertebral aponeurosis. Then detach the Splenius by separating its attachments to the spinous processes, and reflecting it outwards.

The *Erector Spinæ* (fig. 161), and its prolongations in the dorsal and cervical regions, fill up the vertebral groove on each side of the spine. It is covered in the lumbar region by the lumbar aponeurosis; in the dorsal region by the Serrati muscles and the vertebral aponeurosis; and in the cervical region by a layer of cervical fascia continued beneath the Trapezius. This large muscular and tendinous mass varies in size and structure at different parts of the spine. In the sacral region, the Erector spinæ is narrow and pointed, and its origin chiefly tendinous in structure. In the lumbar region, the muscle becomes enlarged, and forms a large fleshy mass. In the dorsal region, it subdivides into two parts, which gradually diminish in size as they ascend to be inserted into the vertebræ and ribs, and are gradually lost in the cervical region, where a number of special muscles are superadded, which are continued upwards to the head, and support it upon the spine.

The Erector spinæ arises from the sacro-iliac groove, and from the anterior

161.—Muscles of the Back. Deep Layers.

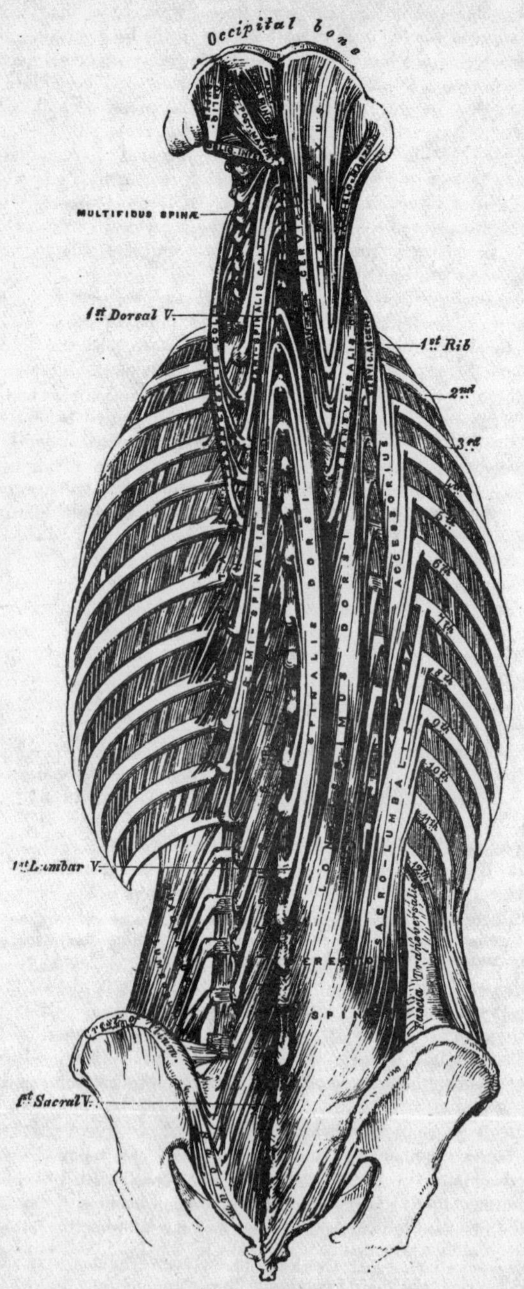

surface of a very broad and thick tendon, which is attached, internally, to the spines of the sacrum, to the spinous processes of the lumbar and three lower dorsal vertebræ, and the supraspinous ligament; externally, to the back part of the inner lip of the crest of the ilium, and to the series of eminences on the posterior part of the sacrum, which represent the transverse processes, where it blends with the great sacro-sciatic ligament. The muscular fibres form a single large fleshy mass, bounded in front by the transverse processes of the lumbar vertebræ, and by the middle lamella of the aponeurosis of origin of the Transversalis muscle. Opposite the last rib, it divides into two parts, the Sacro-lumbalis, and the Longissimus dorsi.

The *Sacro-Lumbalis* (Ilio-Costalis), the external and smaller portion of the Erector spinæ, is inserted, by six or seven flattened tendons, into the angles of the six lower ribs. If this muscle is reflected outwards, it will be seen to be reinforced by a series of muscular slips, which arise from the angles of the ribs; by means of these the Sacro-lumbalis is continued upwards to the upper ribs, and the cervical portion of the spine. The accessory portions form two additional muscles, the Musculus accessorius and the Cervicalis ascendens.

The *musculus accessorius ad sacro-lumbalem* arises by separate flattened tendons, from the angles of the six lower ribs; these become muscular, and are finally inserted, by separate tendons, into the angles of the six upper ribs.

The *cervicalis ascendens** is the continuation of the Accessorius upwards into the neck: it is situated on the inner side of the tendons of the Accessorius, arising from the angles of the four or five upper ribs, and is inserted by a series of slender tendons into the posterior tubercles of the transverse processes of the fourth, fifth, and sixth cervical vertebræ.

The *Longissimus Dorsi*, the inner and larger portion of the Erector spinæ, arises, with the Sacro-lumbalis, from the common origin above described. In the lumbar region, where it is as yet blended with the Sacro-lumbalis, some of the fibres are attached to the whole length of the posterior surface of the transverse processes of the lumbar vertebræ, to the tubercles at the back of the articular processes, and to the layer of lumbar fascia connected with the apices of the transverse processes. In the dorsal region, the Longissimus dorsi is inserted, by long thin tendons, into the tips of the transverse processes of all the dorsal vertebræ, and into from seven to eleven ribs between their tubercles and angles. This muscle is continued upwards, to the cranium and cervical portion of the spine, by means of two additional muscles, the Transversalis colli, and Trachelo-mastoid.

The *transversalis colli*, placed on the inner side of the Longissimus dorsi, arises by long thin tendons from the summit of the transverse processes of the third, fourth, fifth, and sixth dorsal vertebræ, and is inserted by similar tendons into the posterior tubercles of the transverse processes of the five lower cervical.

The *trachelo-mastoid* lies on the inner side of the preceding, between it and the Complexus muscle. It arises by four tendons, from the transverse processes of the third, fourth, fifth, and sixth dorsal vertebræ, and by additional separate tendons from the articular processes of the three or four lower cervical. The fibres form a small muscle, which ascends to be inserted into the posterior margin of the mastoid process, beneath the Splenius and Sterno-mastoid muscles. This small muscle is almost always crossed by a tendinous intersection near its insertion into the mastoid process.

The *Spinalis Dorsi* connects the spinous processes of the upper lumbar and the dorsal vertebræ together by a series of muscular and tendinous slips, which are intimately blended with the Longissimus dorsi. It is situated at the inner side of the Longissimus dorsi, arising, by three or four tendons, from the spinous processes of the first two lumbar and the last two dorsal vertebræ: these uniting,

* This muscle is sometimes called 'Cervicalis descendens.' The student should remember that these long muscles take their fixed point from above or from below, according to circumstances.

form a small muscle, which is inserted, by separate tendons, into the spinous processes of the dorsal vertebræ, the number varying from four to eight. It is intimately united with the Semispinalis dorsi, which lies beneath it.

The *Spinalis Colli* is a small muscle, connecting together the spinous processes of the cervical vertebræ, and analogous to the Spinalis dorsi in the dorsal region. It varies considerably in its size, and in its extent of attachment to the vertebræ, not only in different bodies, but on the two sides of the same body. It usually arises by fleshy or tendinous slips, varying from two to four in number, from the spinous processes of the fifth and sixth cervical vertebræ, and occasionally from the first and second dorsal, and is inserted into the spinous process of the axis, and occasionally into the spinous process of the two vertebræ below it. This muscle was found absent in five cases out of twenty-four.

The *Complexus* is a broad thick muscle, situated at the upper and back part of the neck, beneath the Splenius, and internal to the Transversalis colli and Trachelo-mastoid. It arises, by a series of tendons, about seven in number, from the tips of the transverse processes of the upper three dorsal and seventh cervical, and from the articular processes of the three cervical above this. The tendons uniting form a broad muscle, which passes obliquely upwards and inwards, and is inserted into the innermost depression between the two curved lines of the occipital bone. This muscle, about its middle, is traversed by a transverse tendinous intersection.

The *Biventer Cervicis* is a small fasciculus, situated on the inner side of the preceding, and in the majority of cases blended with it; it has received its name from having a tendon intervening between two fleshy bellies. It is sometimes described as a separate muscle, arising, by from two to four tendinous slips, from the transverse processes of as many of the upper dorsal vertebræ, and inserted, on the inner side of the Complexus, into the superior curved line of the occipital bone.

Relations. The muscles of the fourth layer are bound down to the vertebræ and ribs in the dorsal and lumbar regions by the lumbar fascia and vertebral aponeurosis. Their inner part covers the muscles of the fifth layer. In the neck they are in relation, by their *superficial surface*, with the Trapezius and Splenius; by their *deep surface*, with the Semispinalis dorsi and colli and the Recti and Obliqui. The Biventer cervicis is separated from its fellow of the opposite side by the ligamentum nuchæ, and the Complexus from the Semispinalis colli by the profunda cervicis artery, the princeps cervicis artery, and by the posterior cervical plexus of nerves.

Nerves. The Erector spinæ and its subdivisions in the dorsal region are supplied by the external posterior branches of the lumbar and dorsal nerves; the Cervicalis ascendens, Transversalis colli, Trachelo-mastoid, and Spinalis cervicis, by the external posterior branches of the cervical nerves; the Complexus by the internal posterior branches of the cervical nerves, the suboccipital and great occipital.

Fifth Layer.

Semispinalis Dorsi.	Extensor Coccygis.
Semispinalis Colli.	Intertransversales.
Multifidus Spinæ.	Rectus Capitis Posticus Major.
Rotatores Spinæ.	Rectus Capitis Posticus Minor.
Supraspinales.	Obliquus Capitis Superior.
Interspinales.	Obliquus Capitis Inferior.

Dissection. Remove the muscles of the preceding layer by dividing and turning aside the Complexus; then detaching the Spinalis and Longissimus dorsi from their attachments, dividing the Erector spinæ at its connection below to the sacral and lumbar spines, and turning it outwards. The muscles filling up the interval between the spinous and transverse processes are then exposed.

The *Semispinalis muscles* (fig. 161) connect the transverse and articular processes to the spinous processes of the vertebræ, extending from the lower part of the dorsal region to the upper part of the cervical.

The *semispinalis dorsi* consists of thin, narrow, fleshy fasciculi, interposed between tendons of considerable length. It arises by a series of small tendons from the transverse processes of the lower dorsal vertebræ, from the tenth or eleventh to the fifth or sixth; and is inserted by five or six tendons, into the spinous processes of the upper four dorsal and lower two cervical vertebræ.

The *semispinalis colli*, thicker than the preceding, arises by a series of tendinous and fleshy points from the transverse processes of the upper four dorsal vertebræ, and from the articular processes of the lower four cervical vertebræ; and is inserted into the spinous processes of four cervical vertebræ, from the axis to the fifth cervical. The fasciculus connected with the axis is the largest, and chiefly muscular in structure.

Relations. By their *superficial surface*, from below upwards, with the Longissimus dorsi, Spinalis dorsi, Splenius, Complexus, the profunda cervicis artery, the princeps cervicis artery, and the posterior cervical plexus of nerves. By their *deep surface*, with the Multifidus spinæ.

The *Multifidus Spinæ* consists of a number of fleshy and tendinous fasciculi, which fill up the groove on either side of the spinous processes of the vertebræ, from the sacrum to the axis. In the sacral region, these fasciculi arise from the back of the sacrum, as low as the fourth sacral foramen, and from the aponeurosis of origin of the Erector spinæ; in the iliac region, from the inner surface of the posterior superior spine, and posterior sacro-iliac ligaments; in the lumbar and cervical regions, from the articular processes; and in the dorsal region, from the transverse processes. Each fasciculus, ascending obliquely upwards and inwards, is inserted into the lamina and whole length of the spinous process of one of the vertebræ above. These fasciculi vary in length: the most superficial, the longest, pass from one vertebra to the third or fourth above; those next in order pass from one vertebra to the second or third above; whilst the deepest connect two contiguous vertebræ.

Relations. By its *superficial surface*, with the Longissimus dorsi, Spinalis dorsi, Semispinalis dorsi, and Semispinalis colli. By its *deep surface*, with the laminæ, and spinous processes of the vertebræ, and with the Rotatores spinæ in the dorsal region.

The *Rotatores Spinæ* are found only in the dorsal region of the spine, beneath the Multifidus spinæ; they are eleven in number on each side. Each muscle is small and somewhat quadrilateral in form; it arises from the upper and back part of the transverse process, and is inserted into the lower border and outer surface of the lamina of the vertebra above, the fibres extending as far inwards as the root of the spinous process. The first is found between the first and second dorsal; the last, between the eleventh and twelfth. Sometimes the number of these muscles is diminished by the absence of one or more from the upper or lower end.

The *Supraspinales* consist of a series of fleshy bands, which lie on the spinous processes in the cervical region of the spine.

The *Interspinales* are short muscular fasciculi, placed in pairs between the spinous processes of the contiguous vertebræ. In the *cervical region*, they are most distinct, and consist of six pairs, the first being situated between the axis and third vertebra, and the last between the last cervical and the first dorsal. They are small narrow bundles, attached, above and below, to the apices of the spinous processes. In the *dorsal region*, they are found between the first and second vertebræ, and occasionally between the second and third; and below, between the eleventh and twelfth. In the *lumbar region*, there are four pairs of these muscles in the intervals between the five lumbar vertebræ. There is also occasionally one in the interspinous space, between the last dorsal and first lumbar, and between the fifth lumbar and the sacrum.

The *Extensor Coccygis* is a slender muscular fasciculus, occasionally present, which extends over the lower part of the posterior surface of the sacrum and coccyx. It arises by tendinous fibres from the last bone of the sacrum, or first piece of the coccyx, and passes downwards to be inserted into the lower part of the coccyx. It is a rudiment of the Extensor muscle of the caudal vertebræ which exists in some animals.

The *Intertransversales* are small muscles placed between the transverse processes of the vertebræ. In the *cervical region*, they are most developed, consisting of two rounded muscular and tendinous fasciculi, which pass between the anterior and posterior tubercles of the transverse processes of two contiguous vertebræ, separated from one another by the anterior branch of a cervical nerve, which lies in the groove between them, and by the vertebral artery and vein. In this region there are seven pairs of these muscles, the first being between the atlas and axis, and the last between the seventh cervical and first dorsal vertebræ. In the *dorsal region*, they are least developed, consisting chiefly of rounded tendinous cords in the intertransverse spaces of the upper dorsal vertebræ; but between the transverse processes of the lower three dorsal vertebræ, and the first lumbar, they are muscular in structure. In the *lumbar region*, they are four in number, and consist of a single muscular layer, which occupies the entire interspace between the transverse processes of the lower lumbar vertebræ, whilst those between the transverse processes of the upper lumbar are not attached to more than half the breadth of the process.

The *Rectus Capitis Posticus Major* arises by a pointed tendinous origin from the spinous process of the axis, and, becoming broader as it ascends, is inserted into the inferior curved line of the occipital bone and the surface of bone immediately below it. As the muscles of the two sides ascend upwards and outwards, they leave between them a triangular space, in which are seen the Recti capitis postici minores muscles.

Relations. By its *superficial surface*, with the Complexus, and, at its insertion, with the superior oblique. By its *deep surface*, with the posterior arch of the atlas, the posterior occipito-atloid ligament, and part of the occipital bone.

The *Rectus Capitis Posticus Minor*, the smallest of the four muscles in this region, is of a triangular shape; it arises by a narrow pointed tendon from the tubercle on the posterior arch of the atlas, and, becoming broader as it ascends, is inserted into the rough surface beneath the inferior curved line, nearly as far as the foramen magnum, nearer to the middle line than the preceding.

Relations. By its *superficial surface*, with the Complexus. By its *deep surface* with the posterior occipito-atloid ligament.

The *Obliquus Inferior*, the larger of the two oblique muscles, arises from the apex of the spinous process of the axis, and passes almost horizontally outwards, to be inserted into the apex of the transverse process of the atlas.

Relations. By its *superficial surface*, with the Complexus, and with the posterior branch of the second cervical nerve which crosses it. By its *deep surface*, with the vertebral artery, and posterior atlo-axoid ligament.

The *Obliquus Superior*, narrow below, wide and expanded above, arises by tendinous fibres from the upper part of the transverse process of the atlas, joining with the insertion of the preceding, and, passing obliquely upwards and inwards, is inserted into the occipital bone, between the two curved lines, external to the Complexus. Between the two oblique muscles and the Rectus posticus major a triangular interval exists, in which is seen the vertebral artery, and the posterior branch of the suboccipital nerve.

Relations. By its *superficial surface*, with the Complexus and trachelo-mastoid. By its *deep surface*, with the posterior occipito-atloid ligament.

Nerves. The Semispinalis dorsi and Rotatores spinæ are supplied by the internal posterior branches of the dorsal nerves; the Semispinalis colli, Supraspinales, and Interspinales, by the internal posterior branches of the cervical nerves; the Intertransversales, by the internal posterior branches of the cervical, dorsal, and lumbar nerves; the Multifidus spinæ, by the same, with the addition of the internal posterior branches of the sacral nerves. The Recti and Obliqui muscles are all supplied by the suboccipital and great occipital nerves.

Actions. The Erector spinæ, comprising the Sacro-lumbalis, with its accessory muscles, the Longissimus dorsi and Spinalis dorsi, serves, as its name implies, to maintain the spine in the erect posture; it also serves to bend the trunk backwards when it is required to counterbalance the influence of any weight at the front of the body, as, for instance, when a heavy weight is suspended from the neck, or when there is any great abdominal development, as in pregnancy or dropsy; the peculiar gait under such circumstances depends upon the spine being drawn backwards, by the counterbalancing action of the Erector spinæ muscles. The muscles which form the continuation of the Erector spinæ upwards steady the head and neck, and fix them in the upright position. If the Sacro-lumbalis and Longissimus dorsi of one side act, they serve to draw down the chest and spine to the corresponding side. The Cervicalis ascendens, taking its fixed point from the cervical vertebræ, elevates those ribs to which it is attached. The Multifidus spinæ acts successively upon the different parts of the spine; thus, the sacrum furnishes a fixed point from which the fasciculi of this muscle act upon the lumbar region; these then become the fixed points for the fasciculi moving the dorsal region, and so on throughout the entire length of the spine; it is by the successive contraction and relaxation of the separate fasciculi of this and other muscles, that the spine preserves the erect posture without the fatigue that would necessarily have been produced, had this position been maintained by the action of a single muscle. The Multifidus spinæ, besides preserving the erect position of the spine, serves to rotate it, so that the front of the trunk is turned to the side opposite to that from which the muscle acts, this muscle being assisted in its action by the Obliquus externus abdominis. The Complexi, the analogues of the Multifidus spinæ in the neck, draw the head directly backward; if one muscle acts, it draws the head to one side, and rotates it so that the face is turned to the opposite side. The Rectus capitis posticus minor and the Superior oblique draw the head backwards; and the latter, from the obliquity in the direction of its fibres, may turn the face to the opposite side. The Rectus capitis posticus major and the Obliquus inferior rotate the atlas, and, with it, the cranium round the odontoid process, and turn the face to the same side.

Muscles of the Abdomen.

The muscles in this region are, the

Obliquus Externus.	Rectus.
Obliquus Internus.	Pyramidalis.
Transversalis.	Quadratus Lumborum.

Dissection (fig. 162). To dissect the abdominal muscles, make a vertical incision from the ensiform cartilage to the pubes, a second incision from the umbilicus obliquely upwards and outwards to the outer surface of the chest, as high as the lower border of the fifth or sixth rib, and a third, commencing midway between the umbilicus and pubes, transversely outwards to the anterior superior iliac spine, and along the crest of the ilium as far as its posterior third. Then reflect the three flaps included between these incisions from within outwards, in the line of direction of the muscular fibres. If necessary, the abdominal muscles may be made tense by inflating the peritoneal cavity through the umbilicus.

The *External* or *descendiug Oblique Muscle* (fig. 163) is situated on the side and fore part of the abdomen; being the largest and the most superficial of the three flat muscles in this region. It is broad, thin, and, irregularly quadrilateral, its muscular portion occupying the side, its aponeurosis the anterior wall of the abdomen. It arises, by eight fleshy digitations, from the external surface and lower borders of the eight inferior ribs; these digitations are arranged in an oblique line running downwards and backwards; the upper ones being attached close to the cartilages of the corresponding ribs; the lowest, to the apex of the cartilage of the last rib; the intermediate ones, to the ribs at some distance from their cartilages. The five superior serrations increase in size from above downwards, and are received between corresponding processes of the Serratus magnus; the three lower ones diminish in size from above downwards, receiving between them corresponding processes from the Latissimus dorsi. From these attachments, the fleshy fibres proceed in various directions. Those from the lowest ribs pass nearly vertically downwards, to be inserted into the anterior half of the outer lip of the crest of the ilium; the middle and upper fibres, directed downwards and forwards, terminate in tendinous fibres, which spread out into a broad aponeurosis. This aponeurosis, joined with that of the opposite muscle along the median line, covers the whole of the front of the abdomen: above, it is connected with the lower border of the Pectoralis major; below, its fibres are closely aggregated together, and extend obliquely across from the anterior superior spine of the ilium to the spine of the os pubis and the pectineal line. In the median line, it interlaces with the aponeurosis of the opposite muscle, forming the linea alba, and extends from the ensiform cartilage to the symphysis pubis.

162.—Dissection of Abdomen.

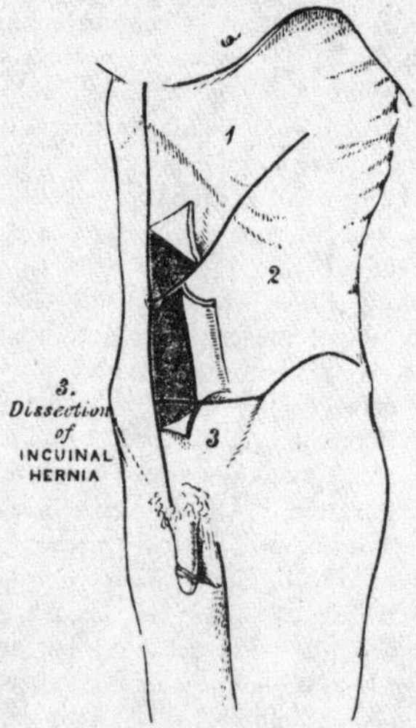

That portion of the aponeurosis which extends between the anterior superior spine of the ilium and the spine of the os pubis, is a broad band, folded inwards, and continuous below with the fascia lata; it is called *Poupart's ligament.* The portion which is reflected from Poupart's ligament into the pectineal line is called *Gimbernat's ligament.** From the point of attachment of the latter to the pectineal line, a few fibres pass upwards and inwards, beneath the inner pillar of the ring, to the linea alba. They diverge as they ascend, and form a thin, triangular, fibrous band, which is called the *triangular ligament.*

In the aponeurosis of the External oblique, immediately above the crest of the os pubis, is a triangular opening, the *external abdominal ring*, formed by a separation of the fibres of the aponeurosis in this situation: it serves for the transmission of the spermatic cord in the male, and the round ligament in the female. This opening is directed obliquely upwards and outwards, and corresponds with the course of the fibres of the aponeurosis. It is bounded, below, by the crest of the os pubis; above, by some curved fibres, which pass across the aponeurosis at the upper angle of the ring, so as to increase its strength; and, on either side, by the margins of the aponeurosis, which are called the *pillars of the ring.* Of these, the external, which is, at the same time, inferior, from the obliquity of its direction, is inserted into the spine of the os pubis. The internal, or superior pillar is attached to the front of the symphysis pubis, and interlaces with the corresponding fibres of the opposite muscle, the fibres of the right muscle being superficial. To the margins of the pillars of the external abdominal ring is attached an exceedingly thin and delicate fascia, which is prolonged down over the outer surface of the cord and testis. This has received the name of *intercolumnar fascia*, from its attachment to the pillars of the ring. It is also called the *external spermatic fascia*, from being the most external of the fasciæ which cover the spermatic cord.

Relations. By its *external surface*, with the superficial fascia, superficial epigastric and circumflex iliac vessels, and some cutaneous nerves. By its *internal*

* All these parts will be found more particularly described below, with the Surgical Anatomy of Hernia.

surface, with the Internal oblique, the lower part of the eight inferior ribs, and Intercostal muscles, the cremaster, the spermatic cord in the male, and round ligament in the female. Its *posterior border* is occasionally overlapped by the Latissimus dorsi; sometimes an interval exists between the two muscles, in which is seen a portion of the Internal oblique.

163.—The External Oblique Muscle.

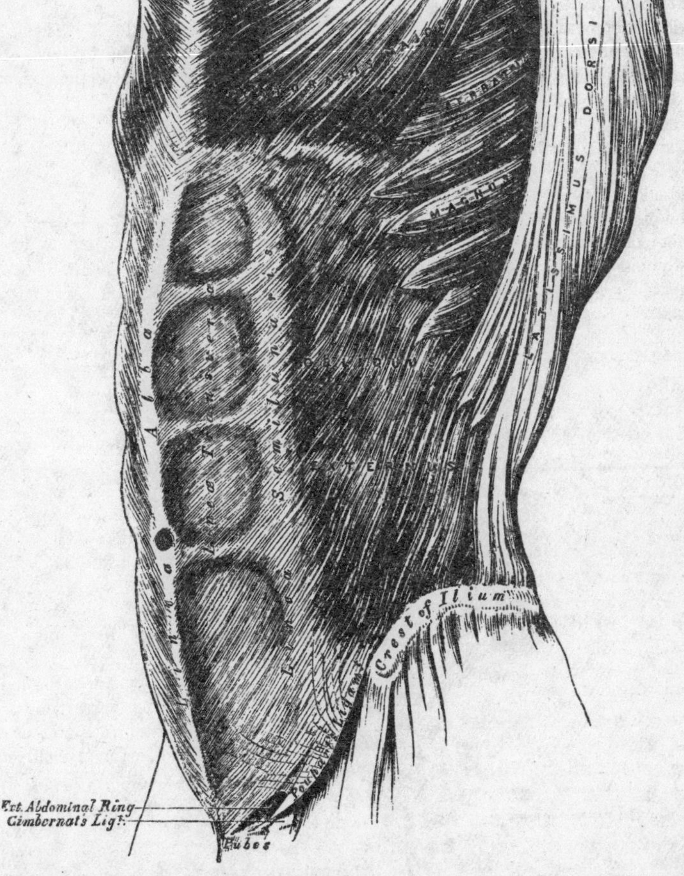

Dissection. Detach the External oblique by dividing it across, just in front of its attachment to the ribs, as far as its posterior border, and separating it below from the crest of ilium as far as the spine; then separate the muscle carefully from the Internal oblique, which lies beneath, and turn it towards the opposite side.

The *Internal* or *Ascending Oblique Muscle* (fig. 164), thinner and smaller than the preceding, beneath which it lies, is of an irregularly quadrilateral form, and situated at the side and fore part of the abdomen. It arises, by fleshy fibres, from the outer half of Poupart's ligament, being attached to the groove on its upper surface; from the anterior two-thirds of the middle lip of the crest of the ilium, and from the lumbar fascia. From this origin, the fibres diverge: those from Poupart's ligament, few in number and paler in colour than the rest, arch downwards and inwards across the spermatic cord, to be inserted, conjointly with those of the Transversalis, into the crest of the os pubis and pectineal line, to the extent of half an inch, forming what is known as the conjoined tendon of the Internal oblique and Transversalis; those from the anterior superior iliac spine are horizontal in their direction; whilst those which arise from the fore part of the crest of the ilium pass obliquely upwards and inwards, and terminate in an aponeurosis, which is continued forwards to the linea alba; the most posterior fibres ascend almost vertically upwards, to be inserted into the lower borders of the cartilages of the four lower ribs, being continuous with the Internal Intercostal muscles.

The conjoined tendon of the Internal oblique and Transversalis is inserted into the crest of the os pubis and pectineal line, immediately behind the external abdominal ring, serving to protect what would otherwise be a weak point in the abdomen.

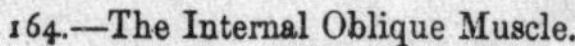

164.—The Internal Oblique Muscle.

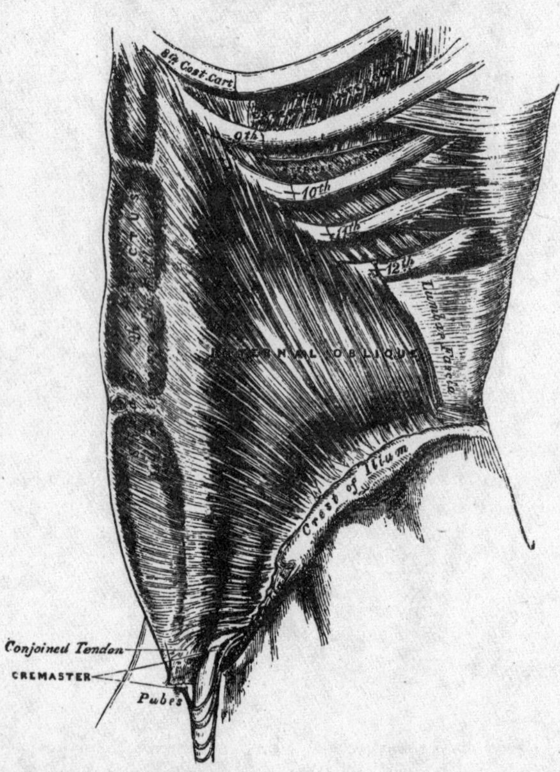

Sometimes this tendon is insufficient to resist the pressure from within, and is carried forward in front of the protrusion through the external ring, forming one of the coverings of direct inguinal hernia.

The aponeurosis of the Internal oblique is continued forward to the middle line of the abdomen, where it joins with the aponeurosis of the opposite muscle at the linea alba, and extends from the margin of the thorax to the pubes. At the outer margin of the Rectus muscles, this aponeurosis, for the upper three-fourths of its extent, divides into two lamellæ, which pass, one in front and the other behind the muscle, enclosing it in a kind of sheath, and reuniting on its inner border at the linea alba: the anterior layer is blended with the aponeurosis of the External oblique muscle; the posterior layer with that of the Transversalis. Along the lower fourth, the aponeurosis passes altogether in front of the Rectus without any separation.

Relations. By its *external surface*, with the External oblique, Latissimus dorsi, spermatic cord, and external ring. By its *internal surface*, with the Transversalis muscle, fascia transversalis, internal ring and spermatic cord. Its lower border forms the upper boundary of the spermatic canal.

165.—The Transversalis, Rectus, and Pyramidalis Muscles.

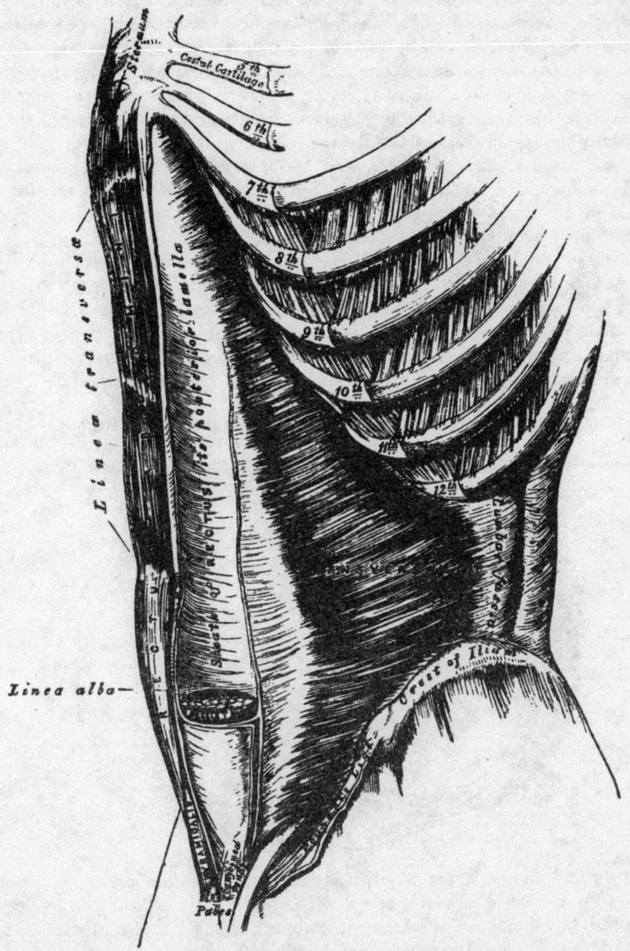

Dissection. Detach the Internal oblique in order to expose the Transversalis beneath. This may be effected by dividing the muscle, above, at its attachment to the ribs; below, at its connection with Poupart's ligament and the crest of the ilium; and behind, by a vertical incision extending from the last rib to the crest of the ilium. The muscle should previously be made tense by drawing upon it with the fingers of the left hand, and if its division is carefully effected, the cellular interval between it and the Transversalis, as well as the direction of the fibres of the latter muscle, will afford a clear guide to their separation; along the crest of the ilium the circumflex iliac vessels are interposed between them, and

form an important guide in separating them. The muscle should then be thrown forwards towards the linea alba.

The *Transversalis muscle* (fig. 165), so called from the direction of its fibres, is the most internal flat muscle of the abdomen, being placed immediately beneath the Internal oblique. It arises by fleshy fibres from the outer third of Poupart's ligament, from the inner lip of the crest of the ilium, its anterior three-fourths, from the inner surface of the cartilages of the six lower ribs, interdigitating with the Diaphragm, and by a broad aponeurosis from the spinous and transverse processes of the lumbar vertebræ. The lower fibres curve downwards, and are inserted, together with those of the Internal oblique, into the crest of the os pubis and pectineal line, forming what was described above as the conjoined tendon of these muscles. Throughout the rest of its extent the fibres pass horizontally inwards, and near the outer margin of the Rectus, terminate in an aponeurosis, which is inserted into the linea alba; its upper three-fourths passing behind the Rectus muscle, blending with the posterior lamella of the Internal oblique; its lower fourth passing in front of the Rectus.

Relations. By its *external surface*, with the Internal oblique, the inner surface of the lower ribs, and Internal intercostal muscles. Its *inner surface* is lined by the fascia transversalis, which separates it from the peritoneum. Its lower border forms the upper boundary of the spermatic canal.

Lumbar Fascia (fig. 166). The vertebral aponeurosis of the Transversalis divides into three layers; an anterior, very thin, which is attached to the front part of the apices of the transverse processes of the lumbar vertebræ, and, above, to the lower margin of the last rib, where it forms the ligamentum arcuatum externum; a middle layer, much stronger, which is attached to the apices of the transverse processes; and a posterior layer, attached to the apices of the spinous processes. Between

166.—A Transverse Section of the Abdomen in the Lumbar Region.

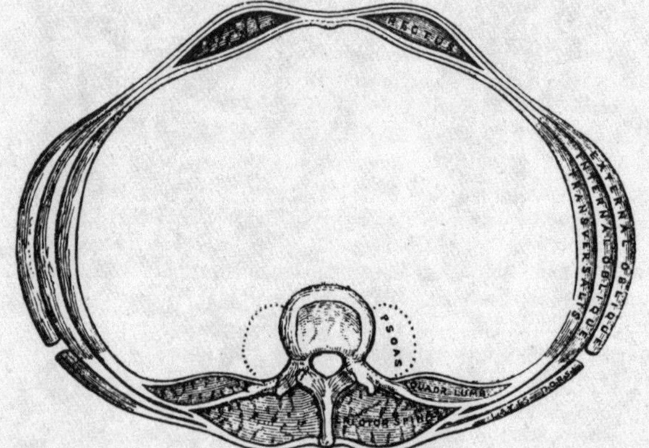

the anterior and middle layers is situated the Quadratus lumborum; between the middle and posterior, the Erector spinæ. The posterior lamella of this aponeurosis receives the attachment of the Internal oblique; it is also blended with the aponeurosis of the Serratus posticus inferior and with that of the Latissimus dorsi, forming the lumbar fascia.

Dissection. To expose the Rectus muscle, open its sheath by a vertical incision extending from the margin of the thorax to the pubes, and then reflect the two portions from the surface of the muscle, which is easily done, excepting at the lineæ transversæ, where so close an adhesion exists, that the greatest care is requisite in separating them. Now raise the outer edge of the muscle, in order to examine the posterior layer of the sheath. By dividing the muscle in the centre, and turning its lower part downwards, the point where the posterior wall of the sheath terminates in a thin curved margin will be seen.

The *Rectus Abdominis* is a long flat muscle, which extends along the whole length of the front of the abdomen, being separated from its fellow of the opposite side by the linea alba. It is much broader above than below, and arises by two tendons, the external or larger being attached to the crest of the os pubis; the internal, smaller portion, interlacing with its fellow of the opposite side, and being connected with the ligaments covering the symphysis pubis. The fibres ascend vertically, and the muscle, becoming broader and thinner at its upper part, is inserted by three portions of unequal size into the cartilages of the fifth, sixth, and seventh ribs. Some fibres are occasionally connected with the costo-xiphoid ligaments, and side of the ensiform cartilage.

The Rectus muscle is traversed by a series of tendinous intersections, which vary from two to five in number, and have received the name Lineæ transversæ. One of these is usually situated opposite the umbilicus, and two above that point; of the latter, one corresponds to the ensiform cartilage, and the other, to the interval between the ensiform cartilage and the umbilicus; there is occasionally one below the umbilicus. These intersections pass transversely or obliquely across the muscle in a zigzag course; they rarely extend completely through its substance, sometimes pass only halfway across it, and are intimately adherent to the sheath in which the muscle is enclosed.

The Rectus is enclosed in a sheath (fig. 166) formed by the aponeuroses of the Oblique and Transversalis muscles, which are arranged in the following manner. When the aponeurosis of the Internal oblique arrives at the margin of the Rectus, it divides into two lamellæ, one of which passes in front of the Rectus, blending with the aponeurosis of the External oblique; the other, behind it, blending with the aponeurosis of the Transversalis; and these, joining again at its inner border, are inserted into the linea alba. This arrangement of the fasciæ exists along the upper three-fourths of the muscle; at the commencement of the lower fourth, the posterior wall of the sheath terminates in a thin curved margin, or *falciform edge*, the concavity of which looks downwards towards the pubes; the aponeuroses of all three muscles passing in front of the Rectus without any separation. The Rectus muscle, in the situation where its sheath is deficient, is separated from the peritoneum by the transversalis fascia.

The *Pyramidalis* is a small muscle, triangular in shape, placed at the lower part of the abdomen, in front of the Rectus, and contained in the same sheath with that muscle. It arises by tendinous fibres from the front of the os pubis and the anterior pubic ligament; the fleshy portion of the muscle passes upwards, diminishing in size as it ascends, and terminates by a pointed extremity, which is inserted into the linea alba, midway between the umbilicus and the os pubis. This muscle is sometimes found wanting on one or both sides; the lower end of the Rectus then becomes proportionately increased in size. Occasionally it has been found double on one side, or the muscles of the two sides are of unequal size. Sometimes its length exceeds what is stated above.

The *Quadratus Lumborum* (fig. 161, p. 236) is situated in the lumbar region; it is irregularly quadrilateral in shape, broader below than above, and consists of two portions. One portion arises by aponeurotic fibres from the ilio-lumbar ligament, and the adjacent portion of the crest of the ilium for about two inches, and is inserted into the lower border of the last rib, about half its length, and by four small tendons, into the apices of the transverse processes of the third, fourth, and fifth lumbar vertebræ. The other portion of the muscle, situated in front of the preceding, arises from the upper borders of the transverse processes of the third, fourth, and fifth lumbar vertebræ, and is inserted into the lower margin of the last rib. The Quadratus lumborum is contained in a sheath formed by the anterior and middle lamellæ of the aponeurosis of origin of the Transversalis.

Nerves. The abdominal muscles are supplied by the lower intercostal, ilio-hypogastric, and ilio-inguinal nerves. The Quadratus lumborum receives filaments from the anterior branches of the lumbar nerves.

In the description of the abdominal muscles, mention has frequently been made of the linea alba, lineæ semilunares, lineæ transversæ; when the dissection of the muscles is completed, these structures should be examined.

The *linea alba* is a tendinous raphé or cord seen along the middle line of the abdomen, extending from the ensiform cartilage to the pubes. It is placed between the inner borders of the Recti muscles, and formed by the blending of the aponeuroses of the Oblique and Transversalis muscles. It is narrow below, corresponding to the narrow interval existing between the Recti, but broader above, as these muscles diverge from one another in their ascent, becoming of considerable breadth after great distension of the abdomen from pregnancy or ascites. It presents numerous apertures for the passage of vessels and nerves; the largest of these is the umbilicus, which in the fœtus transmits the umbilical vessels, but in the adult is obliterated, the cicatrix being stronger than the neighbouring parts; hence umbilical hernia occurs in the adult above the umbilicus, whilst in the fœtus it occurs at the umbilicus. The linea alba is in relation, in front, with the integument, to which it is adherent, especially at the umbilicus; behind, it is separated from the peritoneum by the transversalis fascia; and below, by the urachus, and the bladder, when that organ is distended.

The *lineæ semilunares* are two curved tendinous lines, placed one on each side of the linea alba. Each corresponds with the outer border of the Rectus muscle, extends from the cartilage of the eighth rib to the pubes, and is formed by the aponeurosis of the Internal oblique at its point of division to enclose the Rectus, where it is reinforced in front and behind by the External oblique and Transversalis.

The *lineæ transversæ* are three or four narrow transverse lines which intersect the Rectus muscle as already mentioned: they connect the lineæ semilunares with the linea alba.

Actions. The abdominal muscles perform a threefold action.

When the pelvis and thorax are fixed, they compress the abdominal viscera, by constricting the cavity of the abdomen, in which action they are materially assisted by the descent of the diaphragm. By these means the fœtus is expelled from the uterus, the fæces from the rectum, the urine from the bladder, and its contents from the stomach in vomiting.

If the spine is fixed, these muscles compress the lower part of the thorax, materially assisting in expiration. If the spine is not fixed, the thorax is bent directly forward, when the muscles of both sides act, or to either side when those of the two sides act alternately, rotation of the trunk at the same time taking place to the opposite side.

If the thorax is fixed, these muscles, acting together, draw the pelvis upwards, as in climbing; or, acting singly, they draw the pelvis upwards, and rotate the vertebral column to one side or the other. The Recti muscles, acting from below, depress the thorax, and consequently flex the vertebral column; when acting from above, they flex the pelvis upon the vertebral column. The Pyramidales are tensors of the linea alba.

The Quadratus lumborum, by the portion inserted into the last rib, draws down and fixes that bone, acting thereby as a muscle of forced expiration: by the portion inserted into the lumbar vertebræ, it draws the spine towards the ilium, and thus inclines the trunk towards its own side: or if the thorax and spine be fixed it may act upon the pelvis—raising it towards its own side when only one muscle is put in action, and when both muscles act together, either from below or above, they flex the trunk.

Muscles and Fasciæ of the Thorax.

The muscles exclusively connected with the bones in this region are few in number. They are the

Intercostales Externi.	Infracostales.
Intercostales Interni.	Triangularis Sterni.
Levatores Costarum.	

Intercostal Fasciæ. A thin but firm layer of fascia covers the outer surface of the External intercostal and the inner surface of the Internal intercostal muscles; and a third layer, more delicate, is interposed between the two planes of muscular fibres. These are the intercostal fasciæ; they are best marked in those situations where the muscular fibres are deficient, as between the External intercostal muscles and sternum, in front; and between the Internal intercostals and spine, behind.

The *Intercostal Muscles* (fig. 170) are two thin planes of muscular and tendinous structure, placed one over the other, filling up the intercostal spaces, and being directed obliquely between the margins of the adjacent ribs. They have received the name 'external' and 'internal,' from the position they bear to one another.

The *External Intercostals* are eleven in number on each side, being attached to the adjacent margins of each pair of ribs, and extending from the tubercles of the ribs, behind, to the commencement of the cartilages of the ribs, in front, where they terminate in a thin membranous aponeurosis, which is continued forwards to the sternum. They arise from the outer lip of the groove on the lower border of each rib, and are inserted into the upper border of the rib below. In the two lowest spaces they extend to the end of the ribs. Their fibres are directed obliquely downwards and forwards, in a similar direction with those of the External oblique muscle. They are thicker than the Internal intercostals.

Relations. By their *outer surface*, with the muscles which immediately invest the chest, viz. the Pectoralis major and minor, Serratus magnus, Rhomboideus major, Serratus posticus superior and inferior, Scalenus posticus, Sacro-lumbalis, Longissimus dorsi, Cervicalis ascendens, Transversalis colli, Levatores costarum, and the Obliquus externus abdominis. By their *internal surface*, with a thin layer of fascia, which separates them from the intercostal vessels and nerve, and the Internal intercostal muscles, and, behind, from the pleura.

The *Internal Intercostals*, also eleven in number on each side, are placed on the inner surface of the preceding, commencing anteriorly at the sternum, in the interspaces between the cartilages of the true ribs, and from the anterior extremities of the cartilages of the false ribs; and extend backwards as far as the angles of the ribs, where they are continued to the vertebral column by a thin aponeurosis. They arise from the inner lip of the groove on the lower border of each rib, as well as from the corresponding costal cartilage, and are inserted into the upper border of the rib below. Their fibres are directed obliquely downwards and backwards, decussating with the fibres of the preceding.

Relations. By their *external surface*, with the External intercostals, and the intercostal vessels and nerves. By their *internal surface*, with the Pleura costalis, Triangularis sterni, and Diaphragm.

The Intercostal muscles consist of muscular and tendinous fibres, the latter being longer and more numerous than the former; hence the walls of the intercostal spaces possess very considerable strength, to which the crossing of the muscular fibres materially contributes.

The *Infracostales* consist of muscular and aponeurotic fasciculi, which vary in number and length; they arise from the inner surface of one rib, and are inserted into the inner surface of the first, second, or third rib below. Their direction is most usually oblique, like the Internal intercostals. They are most frequent between the lower ribs.

The *Triangularis Sterni* is a thin plane of muscular and tendinous fibres, situated upon the inner wall of the front of the chest. It arises from the lower part of the side of the sternum, from the inner surface of the ensiform cartilage, and from the sternal ends of the costal cartilages of the three or four lower true ribs. Its fibres diverge upwards and outwards, to be inserted by fleshy digitations into the lower border and inner surfaces of the costal cartilages of the second, third, fourth, and fifth ribs. The lowest fibres of this muscle are horizontal in their direction, and are continuous with those of the Transversalis; those which succeed are oblique, whilst the superior fibres are almost vertical. This muscle varies much in its attachment, not only in different bodies, but on opposite sides of the same body.

Relations. *In front* with the sternum, ensiform cartilage, costal cartilages, Internal intercostal muscles, and internal mammary vessels. *Behind*, with the pleura, pericardium, and anterior mediastinum.

The *Levatores Costarum* (fig. 161), twelve in number on each side, are small tendinous and fleshy bundles, which arise from the extremities of the transverse processes of the dorsal vertebræ, and passing obliquely downwards and outwards, are inserted into the upper rough surface of the rib below them, between the tubercle and the angle. That for the first rib arises from the transverse process of the last cervical vertebra, and that for the last from the eleventh dorsal. The Inferior levatores divide into two fasciculi, one of which is inserted as above described; the other fasciculus passes down to the second rib below its origin; thus, each of the lower ribs receives fibres from the transverse processes of two vertebræ.

Nerves. The muscles of this group are supplied by the intercostal nerves.

Actions. The Intercostals are the chief agents in the movement of the ribs in ordinary respiration. The External intercostals raise the ribs, especially their fore part, and so increase the capacity of the chest from before backwards; at the same time they evert their lower borders, and so enlarge the thoracic cavity transversely. The Internal intercostals, at the side of the thorax, depress the ribs, and invert their lower borders, and so diminish the thoracic cavity; but at the fore part of the chest these muscles assist the External intercostals in raising the cartilages. The Levatores Costarum assist the External intercostals in raising the ribs. The Triangularis sterni draws down the costal cartilages; it is therefore an expiratory muscle.

Muscles of inspiration and expiration. The muscles which assist the action of the Diaphragm in inspiration are the Intercostals and the Levatores costarum as above stated, the Scaleni, the Serratus posticus superior, and to a slight extent the Subclavius. When the need for more forcible action exists, the shoulders and the base of the scapula are fixed, and then the powerful muscles of forced inspiration come into play; the chief of these are the Serratus magnus, Latissimus dorsi, and the Pectorales, particularly the Pectoralis minor.

The ordinary action of expiration is merely passive, the resilience of the ribs and the elasticity of the lungs producing a tendency to a vacuum. This causes the ascent of the abdominal viscera covered by the Diaphragm. Forced expiratory actions are performed mainly by the flat muscles (Obliqui and Transversalis) of the abdomen, assisted also by the Rectus. Other muscles of forced expiration are the Internal Intercostals and Triangularis Sterni (as above mentioned), the Serratus posticus inferior, the Quadratus lumborum and the Sacro-lumbalis.

Diaphragmatic Region.

Diaphragm.

The *Diaphragm* (*Διάφραγμα, a partition wall*) (fig. 167) is a thin musculo-fibrous septum, placed obliquely at the junction of the upper with the middle third of the trunk, and separating the thorax from the abdomen, forming the floor of the former cavity and the roof of the latter. It is elliptical, its longest diameter being from side to side, somewhat fan-shaped, the broad elliptical portion being horizontal, the narrow part, which represents the handle of the fan, vertical, and joined at right angles to the former. It is from this circumstance that some anatomists describe it as consisting of two portions, the upper or great muscle of the Diaphragm, and the lower or lesser muscle. It arises from the whole of the internal circumference of the thorax, being attached, in front, by fleshy fibres to the ensiform cartilage; on either side, to the inner surface of the cartilages and bony portions of the six or seven inferior ribs, interdigitating with the Transversalis; and behind, to two aponeurotic arches, named the ligamentum arcuatum externum and internum, and to the lumbar vertebræ. The fibres from these sources vary in length; those arising from the ensiform appendix are very short and occasionally aponeurotic; those from the ligamenta arcuata, and more especially those from the ribs at the side of the chest, are longer, describe well-marked curves as they ascend, and finally converge to be inserted into the circumference of the central tendon. Between the sides of the muscular slip from the ensiform appendix and the cartilages of the adjoining ribs, the fibres of the Diaphragm are deficient, the interval being filled by areolar tissue, covered on the thoracic side by the pleuræ; on the abdominal, by the peritoneum. This is, consequently, a weak point, and a portion of the contents of the abdomen may protrude into the chest, forming phrenic or diaphragmatic hernia, or a collection of pus in the mediastinum may descend through it, so as to point at the epigastrium.

The *ligamentum arcuatum internum* is a tendinous arch, thrown across the upper part of the Psoas magnus muscle, on each side of the spine. It is connected, by one end, to the outer side of the body of the first, and occasionally the second lumbar vertebra, being continuous with the outer side of the tendon of the

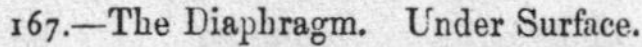

167.—The Diaphragm. Under Surface.

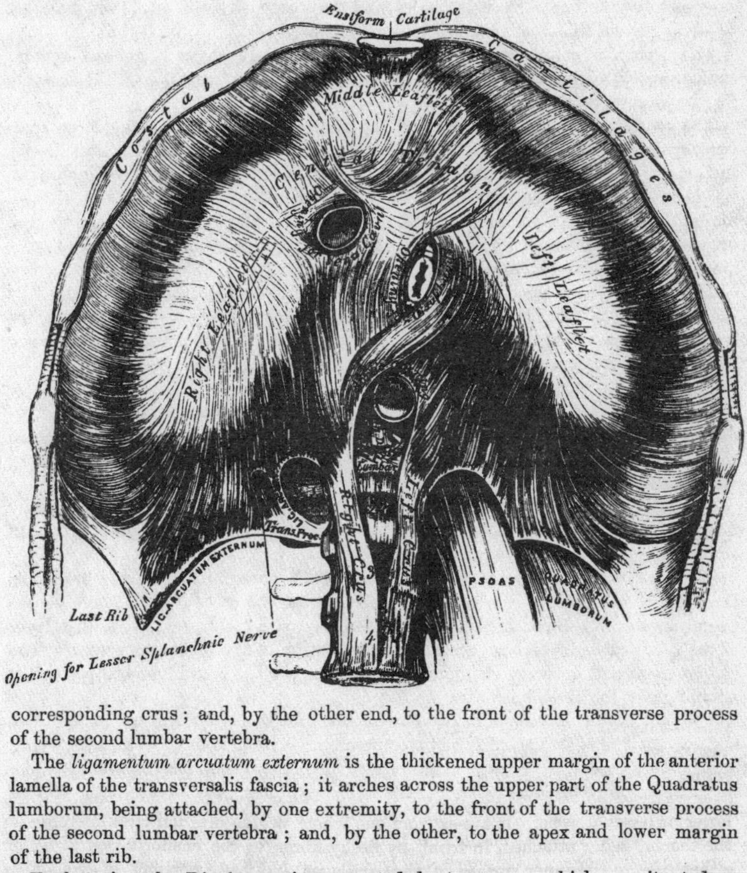

corresponding crus; and, by the other end, to the front of the transverse process of the second lumbar vertebra.

The *ligamentum arcuatum externum* is the thickened upper margin of the anterior lamella of the transversalis fascia; it arches across the upper part of the Quadratus lumborum, being attached, by one extremity, to the front of the transverse process of the second lumbar vertebra; and, by the other, to the apex and lower margin of the last rib.

To the spine, the Diaphragm is connected by two crura, which are situated on the bodies of the lumbar vertebræ, one on each side of the aorta. The crura, at their origin, are tendinous in structure; the right crus, larger and longer than the left, arising from the anterior surface of the bodies and intervertebral substances of the second, third, and fourth lumbar vertebræ; the left, from the second and third; both blending with the anterior common ligament of the spine. A tendinous arch is thrown across the front of the vertebral column, from the tendon of one crus to that of the other, beneath which passes the aorta, vena azygos major, and thoracic duct. The tendons terminate in two large fleshy bellies, which, with the tendinous portions above alluded to, are called the *crura*, or *pillars of the diaphragm.* The outer fasciculi of the two crura are directed upwards and outwards to the central tendon; but the inner fasciculi decussate in front of the aorta, and then diverge, so as to surround the œsophagus before ending in the central tendon. The anterior and larger of these fasciculi is formed by the right crus.

The *Central* or *Cordiform Tendon* of the Diaphragm is a thin tendinous aponeurosis, situated at the centre of the vault formed by the muscle, immediately beneath the pericardium, with which its circumference is blended. It is shaped somewhat like a trefoil leaf, consisting of three divisions, or leaflets, separated from one another by slight indentations. The right leaflet is the largest; the middle one, directed towards the ensiform cartilage, the next in size; and the left, the smallest. In structure, the tendon is composed of several planes of fibres, which intersect one another at various angles, and unite into straight or curved bundles, an arrangement which affords it additional strength.

The *Openings* connected with the Diaphragm, are three large and several smaller apertures. The former are the aortic, the œsophageal, and the opening for the vena cava.

The *aortic opening* is the lowest and the most posterior of the three large apertures connected with this muscle. It is situated in the middle line, immediately in front of the bodies of the vertebræ; and is, therefore, *behind* the Diaphragm, not in it. It is an osseo-aponeurotic aperture, formed by a tendinous arch thrown across the front of the bodies of the vertebræ, from the crus on one side to that on the other, and transmits the aorta, vena azygos major, thoracic duct, and occasionally the left sympathetic nerve.

The *œsophageal opening*, elliptical in form, muscular in structure, and formed by the two crura, is placed above, and, at the same time, anterior, and a little to the left of the preceding. It transmits the œsophagus and pneumogastric nerves. The anterior margin of this aperture is occasionally tendinous, being formed by the margin of the central tendon.

The *opening for the vena cava* is the highest; it is quadrilateral in form, tendinous in structure, and placed at the junction of the right and middle leaflets of the central tendon, its margins being bounded by four bundles of tendinous fibres, which meet at right angles.

The *right crus* transmits the sympathetic and the greater and lesser splanchnic nerves of the right side; the *left crus*, the greater and lesser splanchnic nerves of the left side, and the vena azygos minor.

The *Serous Membranes* in relation with the Diaphragm, are four in number; three lining its upper or thoracic surface; one its abdominal. The three serous membranes on its upper surface are the pleura on either side, and the serous layer of the pericardium, which covers the middle portion of the tendinous centre. The serous membrane covering its under surface, is a portion of the general peritoneal membrane of the abdominal cavity.

The Diaphragm is arched, being convex towards the chest, and concave to the abdomen. The *right portion* forms a complete arch from before backwards, being accurately moulded over the convex surface of the liver, and having resting upon it the concave base of the right lung. The *left portion* is arched from before backwards in a similar manner; but the arch is narrower in front, being encroached upon by the pericardium, and lower than the right, at its summit, by about three quarters of an inch. It supports the base of the left lung, and covers the great end of the stomach, the spleen and left kidney. The *central portion*, which supports the heart, is higher, in front at the sternum, and behind at the vertebræ, than the lateral portions; the reverse is the case in the parts further removed from the surface of the body.

The height of the Diaphragm is constantly varying during respiration, the muscle being carried upwards or downwards from the average level; its height also varies according to the degree of distension of the stomach and intestines, and the size of the liver. After a forced expiration, the right arch is on a level, in front, with the fourth costal cartilage; at the side, with the fifth, sixth, and seventh ribs: and behind, with the eighth rib; the left arch being usually from one to two ribs breadth below the level of the right one. In a forced inspiration, it descends from one to two inches; its slope would then be represented by a line drawn from the ensiform cartilage towards the tenth rib.

Nerves. The Diaphragm is supplied by the phrenic nerves.

Actions. The action of the Diaphragm modifies considerably the size of the chest, and the position of the thoracic and abdominal viscera. *During a forced inspiration*, the cavity of the thorax is enlarged in the vertical direction from two to three inches, partly by the ascent of the walls of the chest, partly by the descent of the Diaphragm. The chest, consequently, encroaches upon the abdomen: the lungs are expanded, and lowered, in relation with the ribs, nearly two inches; the heart being drawn down about an inch and a half; the descent of the latter organ taking place indirectly through the medium of its connection with the lungs, as well as directly by means of the central tendon to which the pericardium is attached. The abdominal viscera are also pushed down (the liver, to the extent of nearly three inches), so that these organs are no longer protected by the ribs. *During expiration*, when the Diaphragm is passive, it is pushed up by the action of the abdominal muscles; the cavity of the abdomen (with the organs contained in it), encroaches upon the chest, by which the lungs and heart are compressed upwards, and the vertical diameter of the thoracic cavity diminished. The Diaphragm is passive when raised or lowered, by the abdominal organs, independently of respiration, in proportion as they are large or small, full or empty; hence the oppression felt in the chest after a full meal, or from flatulent distension of the stomach and intestines.

In all expulsive acts the Diaphragm is called into action, to give additional power to each expulsive effort. Thus, before sneezing, coughing, laughing, and crying; before vomiting; previous to the expulsion of the urine and fæces, or of the fœtus from the womb, a deep inspiration takes place.*

MUSCLES AND FASCLÆ OF THE UPPER EXTREMITY.

The Muscles of the Upper Extremity are divisible into groups, corresponding with the different regions of the limb.

Anterior Thoracic Region.

Pectoralis major.
Pectoralis minor.
Subclavius.

Lateral Thoracic Region.

Serratus magnus.

Acromial Region.

Deltoid.

Anterior Scapular Region.

Subscapularis.

Posterior Scapular Region.

Supraspinatus.
Infraspinatus.
Teres minor.
Teres major.

Anterior Humeral Region.

Coraco-brachialis.
Biceps.
Brachialis anticus.

Posterior Humeral Region.

Triceps.
Subanconeus.

Anterior Brachial Region.

Superficial Layer.
- Pronator radii teres.
- Flexor carpi radialis.
- Palmaris longus.
- Flexor carpi ulnaris.
- Flexor sublimis digitorum.

Deep Layer.
- Flexor profundus digitorum.
- Flexor longus pollicis.
- Pronator quadratus.

Radial Region.

Supinator longus.
Extensor carpi radialis longior.
Extensor carpi radialis brevior.

* For a detailed description of the general relations of the Diaphragm, and its action, refer to Dr. Sibson's '*Medical Anatomy*.'

Posterior Brachial Region.

Superficial Layer.
- Extensor communis digitorum.
- Extensor minimi digiti.
- Extensor carpi ulnaris.
- Anconeus.

Deep Layer.
- Supinator brevis.
- Extensor ossis metacarpi pollicis.
- Extensor primi internodii pollicis.
- Extensor secundi internodii pollicis.
- Extensor indicis.

MUSCLES OF THE HAND.

Radial Region.

Abductor pollicis.
Flexor ossis metacarpi pollicis (opponens).
Flexor brevis pollicis.
Adductor pollicis.

Ulnar Region.

Palmaris brevis.
Abductor minimi digiti.
Flexor brevis minimi digiti.
Flexor ossis metacarpi minimi digiti (opponens).

Palmar Region.

Lumbricales.
Interossei palmares.
Interossei dorsales.

Dissection of Pectoral Region and Axilla (fig. 168). The arm being drawn away from the side nearly at right angles with the trunk, and rotated outwards, make a vertical incision through the integument in the median line of the chest, from the upper to the lower part of the sternum; a second incision along the lower border of the Pectoral muscle, from the ensiform cartilage to the other side of the axilla; a third, from the sternum along the clavicle, as far as its centre; and a fourth, from the middle of the clavicle obliquely downwards, along the interspace between the Pectoral and Deltoid muscles, as low as the fold of the arm-pit. The flap of integument is then to be dissected off in the direction indicated in the figure, but not entirely removed, as it should be replaced on completing the dissection. If a transverse incision is now made from the lower end of the sternum to the side of the chest, as far as the posterior fold of the arm-pit, and the integument reflected outwards, the axillary space will be more completely exposed.

FASCIÆ OF THE THORAX.

The *superficial fascia* of the thoracic region is a loose cellulo-fibrous layer, continuous with the superficial fascia of the neck and upper extremity above, and of the abdomen below; opposite the mamma, it subdivides into two layers, one of which passes in front, the other behind that gland; and from both of these layers numerous septa pass into its substance, supporting its various lobes: from the anterior layer, fibrous processes pass forward to the integument and nipple, enclosing in their areolæ masses of fat. These processes were called by Sir A. Cooper the *ligamenta suspensoria*, from the support they afford to the gland in this situation. On removing the superficial fascia, the *deep fascia* of the thoracic region is exposed: it is a thin aponeurotic lamina, covering the surface of the great Pectoral muscle, and sending numerous prolongations between its fasciculi: it is attached, in the middle line, to the front of the sternum; and, above, to the clavicle: it is very thin over the upper part of the muscle, somewhat thicker in the interval between the Pectoralis major and Latissimus dorsi, where it closes in the axillary space, and divides at the margin of the latter muscle into two layers, one of which passes in front, and the other

168.—Dissection of Upper Extremity.

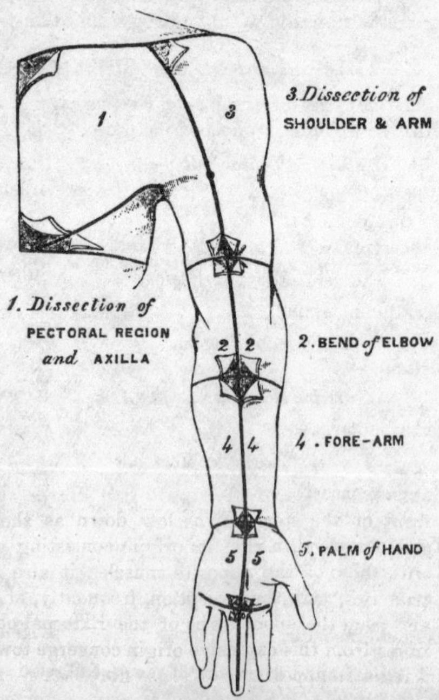

behind it; these proceed as far as the spinous processes of the dorsal vertebræ, to which they are attached. At the lower part of the thoracic region, this fascia is well developed, and is continuous with the fibrous sheath of the Recti muscles.

Anterior Thoracic Region.

Pectoralis Major. Pectoralis Minor.
Subclavius.

The *Pectoralis Major* (fig. 169) is a broad, thick, triangular muscle, situated at the upper and fore part of the chest, in front of the axilla. It arises from the

169.—Muscles of the Chest and Front of the Arm. Superficial View.

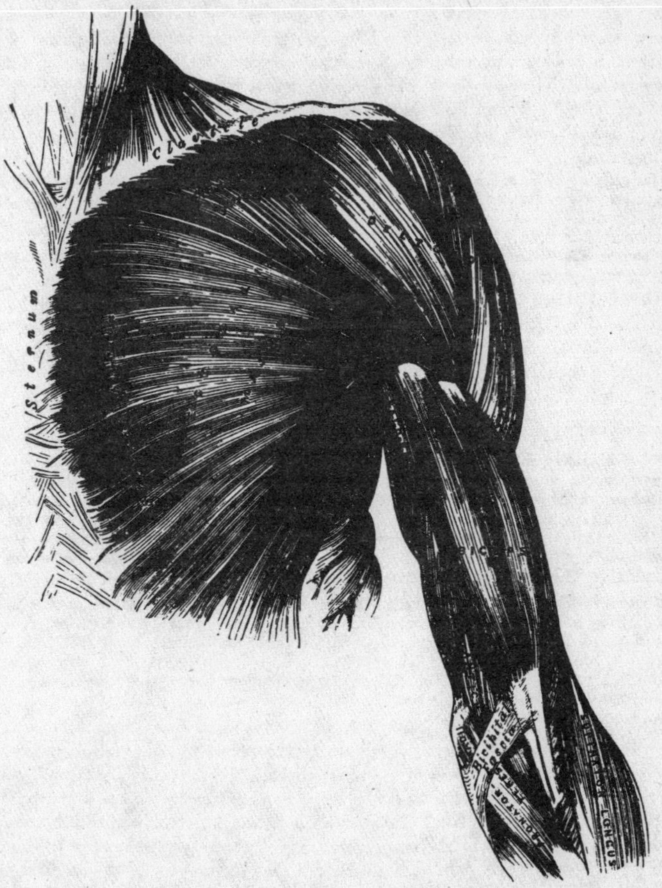

anterior surface of the sternal half of the clavicle; from half the breadth of the front of the sternum, as low down as the attachment of the cartilage of the sixth or seventh rib; its origin consisting of aponeurotic fibres, which intersect with those of the opposite muscle; it also arises from the cartilages of all the true ribs, with the exception, frequently, of the first, or of the seventh, or both; and from the aponeurosis of the External oblique muscle of the abdomen. The fibres from this extensive origin converge towards its insertion, giving to the muscle a radiated appearance. Those fibres which arise from the clavicle, pass obliquely outwards and downwards, and are usually separated from the rest by a cellular interval: those from the lower part of the sternum and the cartilages of the lower true ribs, pass upwards and outwards; whilst the middle fibres pass horizontally. As these three sets of fibres converge, they are so disposed that the upper overlap the middle, and the middle the lower portion, the fibres of the lower portion being folded backwards upon themselves; so that those fibres which are lowest in front, become highest at their point of insertion. They all terminate in a flat tendon, about two inches broad, which is inserted into the anterior bicipital ridge of the humerus. This tendon consists of two laminæ, placed one in front of the other, and usually blended together below. The anterior, the thicker, receives the clavicular and upper half of the sternal portion of the muscle; the posterior lamina receiving the attachment of the lower half of the sternal portion. From this arrangement it results, that the fibres of the upper and middle portions of the muscle are inserted into the lower part of the bicipital ridge; those of the lower portion, into the upper part. The tendon, at its insertion, is connected with that of the Deltoid; it sends up an expansion over the bicipital groove to the head of the humerus; another backwards, which lines the groove; and a third to the fascia of the arm.

Relations. By its *anterior surface*, with the Platysma, the mammary gland, the superficial fascia, and integument. By its *posterior surface*—its *thoracic portion*, with the sternum, the ribs and costal cartilages, the Subclavius, Pectoralis minor, Serratus magnus, and the Intercostals; its *axillary portion* forms the anterior wall of the axillary space, and covers the axillary vessels and nerves. Its *upper border* lies parallel with the Deltoid, from which it is separated by the cephalic vein and descending branch of the thoracico-acromialis artery. Its *lower border* forms the anterior margin of the axilla, being at first separated from the Latissimus dorsi by a considerable interval; but both muscles gradually converge towards the outer part of the space.

Peculiarities. In muscular subjects, the sternal origins of the two Pectoral muscles are separated only by a narrow interval; but this interval is enlarged where these muscles are ill developed. Very rarely, the whole of the sternal portion is deficient. Occasionally, one or two additional muscular slips arise from the aponeurosis of the External oblique, and become united to the lower margin of the Pectoralis major. A slender muscular slip is occasionally found lying parallel with the outer margin of the sternum, overlapping the origin of the pectoral muscle. It is attached, by one end, to the upper part of the sternum, near the origin of the sterno-mastoid; and, by the other, to the anterior wall of the sheath of the rectus abdominis. It has received the name 'rectus sternalis.'

Dissection. Detach the Pectoralis major by dividing the muscle along its attachment to the clavicle, and by making a vertical incision through its substance a little external to its line of attachment to the sternum and costal cartilages. The muscle should then be reflected outwards, and its tendon carefully examined. The Pectoralis minor is now exposed, and immediately above it, in the interval between its upper border and the clavicle, a strong fascia, the costo-coracoid membrane.

The *costo-coracoid membrane* protects the axillary vessels and nerves; it is very thick and dense externally, where it is attached to the coracoid process, and is continuous with the fascia of the arm; more internally, it is connected with the lower border of the clavicle, as far as the sternal extremity of the first rib: traced downwards, it passes behind the Pectoralis minor, surrounding, in a more or less complete sheath, the axillary vessels and nerves; and above, it sends a prolongation behind the Subclavius, which is attached to the lower border of the clavicle, and so encloses the muscle in a kind of sheath. The costo-coracoid membrane is pierced by the cephalic vein, the thoracico-acromialis artery and vein, superior thoracic artery, and anterior thoracic nerve.

The *Pectoralis Minor* (fig. 170) is a thin, flat, triangular muscle, situated at the upper part of the thorax, beneath the Pectoralis major. It arises, by three tendinous digitations, from the upper margin and outer surface of the third, fourth, and fifth ribs, near their cartilages, and from the aponeurosis covering the Intercostal muscles; the fibres pass upwards and outwards, and converge to form a flat tendon, which is inserted into the anterior border of the coracoid process of the scapula.

Relations. By its *anterior surface*, with the Pectoralis major, and the superior thoracic vessels and nerves. By its *posterior surface*, with the ribs, Intercostal muscles, Serratus magnus, the axillary space, and the axillary vessels and nerves. Its upper border is separated from the clavicle by a triangular interval, broad internally, narrow externally, bounded in front by the costo-coracoid membrane, and internally by the ribs. In this space are seen the axillary vessels and nerves.

The costo-coracoid membrane should now be removed, when the Subclavius muscle will be seen.

The *Subclavius* is a long, thin, spindle-shaped muscle, placed in the interval between the clavicle and the first rib. It arises by a short, thick tendon from the cartilage of the first rib, in front of the rhomboid ligament; the fleshy fibres proceed obliquely outwards, to be inserted into a deep groove on the under surface of the middle third of the clavicle.

170.—Muscles of the Chest and Front of the Arm, with the Boundaries of the Axilla.

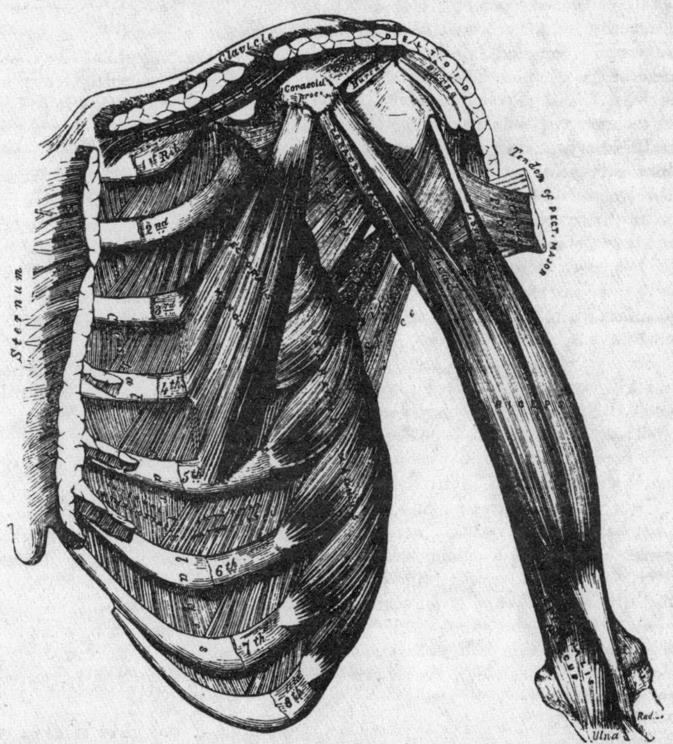

Relations. By its *upper surface*, with the clavicle. By its *under surface*, it is separated from the first rib by the axillary vessels and nerves. Its *anterior surface* is separated from the Pectoralis major by a strong aponeurosis, which, with the clavicle, forms an osteo-fibrous sheath in which the muscle is enclosed.

If the costal attachment of the Pectoralis Minor is divided across, and the muscle reflected outwards, the axillary vessels and nerves are brought fully into view, and should be examined.

Nerves. The Pectoral muscles are supplied by the anterior thoracic nerves; the Subclavius, by a filament from the cord formed by the union of the fifth and sixth cervical nerves.

Actions. If the arm has been raised by the Deltoid, the Pectoralis major will, conjointly with the Latissimus dorsi and Teres major, depress it to the side of the chest; and, if acting singly, it will draw the arm across the front of the chest. The Pectoralis minor depresses the point of the shoulder, drawing the scapula downwards and inwards to the thorax. The Subclavius depresses the shoulder, drawing the clavicle downwards and forwards. When the arms are fixed, all three muscles act upon the ribs, drawing them upwards and expanding the chest, and thus becoming very important agents in forced inspiration. Asthmatic patients always assume this attitude, fixing the shoulders, so that all these muscles may be brought into action to assist in dilating the cavity of the chest.

Lateral Thoracic Region.

Serratus Magnus.

The *Serratus Magnus* (fig. 170) is a broad, thin, and irregularly quadrilateral muscle, situated at the upper part and side of the chest. It arises by nine fleshy digitations from the outer surface and upper border of the eight upper ribs (the second rib having two), and from the aponeurosis covering the upper intercostal spaces, and is inserted into the whole length of the inner margin of the posterior border of the scapula. This muscle has been divided into three portions, a superior, middle, and inferior, on account of the difference in the direction, and in the extent of attachment of each part. The upper portion, separated from the rest by a cellular interval, is a narrow, but thick fasciculus, which arises by two digitations from the first and second ribs, and from the aponeurotic arch between them; its fibres proceed upwards, outwards, and backwards, to be inserted into the triangular smooth surface on the inner side of the superior angle of the scapula. The middle portion of the muscle arises by three digitations from the second, third and fourth ribs, it forms a thin and broad muscular layer, which proceeds horizontally backwards to be inserted into the posterior border of the scapula, between the superior and inferior angles. The lower portion arises from the fifth, sixth, seventh and eighth ribs, by four digitations, in the intervals between which are received corresponding processes of the External oblique; the fibres pass upwards, outwards, and backwards, to be inserted into the inner surface of the inferior angle of the scapula, by an attachment partly muscular, partly tendinous.

Relations. This muscle is covered, in front, by the Pectoral muscles; behind, by the Subscapularis; above, by the axillary vessels and nerves. Its *deep surface* rests upon the ribs and Intercostal muscles.

Nerves. The Serratus magnus is supplied by the posterior thoracic nerve.

Actions. The Serratus magnus is the most important external inspiratory muscle. When the shoulders are fixed, it elevates the ribs, and so dilates the cavity of the chest, assisting the Pectoral and Subclavius muscles. This muscle, especially its middle and lower segments, draws the base and inferior angle of the scapula forwards, and so raises the point of the shoulder by causing a rotation of the bone on the side of the chest; assisting the Trapezius muscle in supporting weights upon the shoulder, the thorax being at the same time fixed by preventing the escape of the included air.

Dissection. After completing the dissection of the axilla, if the muscles of the back have been dissected, the upper extremity should be separated from the trunk. Saw through the clavicle at its centre, and then cut through the muscles which connect the scapula and arm with the trunk, viz. the Pectoralis minor, in front, Serratus magnus, at the side, and the Levator anguli scapulæ, the Rhomboids, Trapezius, and Latissimus dorsi behind. These muscles should be cleaned and traced to their respective insertions. Then make an incision through the integument, commencing at the outer third of the clavicle, and extending along the margin of that bone, the acromion process, and spine of the scapula; the integument should be dissected from above downwards and outwards, when the fascia covering the Deltoid is exposed (fig. 168, No. 3).

The *superficial fascia* of the upper extremity, is a thin cellulo-fibrous lamina, containing between its layers the superficial veins and lymphatics, and the cutaneous nerves. It is most distinct in front of the elbow, and contains very large superficial veins and nerves; in the hand it is hardly demonstrable, the integument being closely adherent to the deep fascia by dense fibrous bands. Small subcutaneous bursæ are found in this fascia, over the acromion, the olecranon, and the knuckles. The deep fascia of the upper extremity comprises the aponeurosis of the shoulder, arm, and forearm, the anterior and posterior annular ligaments of the carpus, and the palmar fascia. These will be considered in the description of the muscles of the several regions.

Acromial Region.

Deltoid.

The *deep fascia* covering the Deltoid (deltoid aponeurosis), is a thick and strong fibrous layer, which encloses the outer surface of the muscle, and sends down numerous prolongations between its fasciculi; it is continuous, internally, with the fascia covering the great Pectoral muscle; behind, with that covering the Infraspinatus and back of the arm: above, it is attached to the clavicle, the acromion, and spine of the scapula.

The *Deltoid* (fig. 169) is a large, thick, triangular muscle, which forms the convexity of the shoulder, and has received its name from its resemblance to the Greek letter Δ reversed. It surrounds the shoulder joint in the greater part of its extent, covering it on its outer side, and in front and behind. It arises from the outer third of the anterior border and upper surface of the clavicle; from the outer margin and upper surface of the acromion process; and from the whole length of the lower border of the spine of the scapula. From this extensive origin, the fibres converge towards their insertion, the middle passing vertically, the anterior obliquely backwards, the posterior obliquely forwards; they unite to form a thick tendon, which is inserted into a rough prominence on the middle of the outer side of the shaft of the humerus. This muscle is remarkably coarse in texture, and intersected by three or four tendinous laminæ; these are attached, at intervals, to the clavicle and acromion, extend into the substance of the muscle, and give origin to a number of fleshy fibres. The largest of these laminæ extends from the summit of the acromion.

Relations. By its *superficial surface*, with the Platysma, supra-acromial nerves, the superficial fascia, and integument. Its *deep surface* is separated from the head of the humerus by a large sacculated synovial bursa, and covers the coracoid process, coraco-acromial ligament, Pectoralis minor, Coraco-brachialis, both heads of the Biceps, tendon of the Pectoralis major, Teres minor, Triceps (its scapular and external heads), the circumflex vessels and nerve, and the humerus. Its *anterior border* is separated from the Pectoralis major by a cellular interspace, which lodges the cephalic vein and descending branch of the thoracico-acromialis artery. Its *posterior border* rests on the Infraspinatus and Triceps muscles.

Nerves. The Deltoid is supplied by the circumflex nerve.

Actions. The Deltoid raises the arm directly from the side, so as to bring it at right angles with the trunk. Its anterior fibres, assisted by the Pectoralis major, draw the arm forwards; and its posterior fibres, aided by the Teres major and Latissimus dorsi, draw it backwards.

Dissection. Divide the Deltoid across, near its upper part, by an incision carried along the margin of the clavicle, the acromion process, and spine of the scapula, and reflect it downwards; the bursa will be seen on its under surface, as well as the circumflex vessels and nerve. The insertion of the muscle should be carefully examined.

Anterior Scapular Region.

Subscapularis.

The *subscapular aponeurosis* is a thin membrane, attached to the entire circumference of the subscapular fossa, and affording attachment by its inner surface to some of the fibres of the Subscapularis muscle: when this is removed, the Subscapularis muscle is exposed.

The *Subscapularis* (fig. 170) is a large triangular muscle, which fills up the subscapular fossa, arising from its internal two-thirds, with the exception of a narrow margin along the posterior border, and the inner side of the superior and inferior angles, which afford attachment to the Serratus magnus. Some fibres arise from tendinous laminæ, which intersect the muscle, and are attached to ridges on the bone; and others from an aponeurosis, which separates the muscle from the Teres major and the long head of the Triceps. The fibres pass outwards, and, gradually converging, terminate in a tendon, which is inserted into the lesser tuberosity of the humerus. Those fibres which arise from the axillary border of the scapula, are inserted into the neck of the humerus to the extent of an inch below the tuberosity. The tendon of the muscle is in close contact with the capsular ligament of the shoulder-joint, and glides over a large bursa, which separates it from the base of the coracoid process. This bursa communicates with the cavity of the joint by an aperture in the capsular ligament.

Relations. By its *anterior surface*, with the Serratus magnus, Coraco-brachialis, and Biceps, and the axillary vessels and nerves. By its *posterior surface*, with the scapula, the subscapular vessels and nerves, and the capsular ligament of the shoulder-joint. Its *lower border* is contiguous with the Teres major and Latissimus dorsi.

Nerves. It is supplied by the subscapular nerves.

Actions. The Subscapularis rotates the head of the humerus inwards; when the arm is raised, it draws the humerus downwards. It is a powerful defence to the front of the shoulder-joint, preventing displacement of the head of the bone forwards.

Posterior Scapular Region. (Fig. 171.)

Supraspinatus.	Teres Minor.
Infraspinatus.	Teres Major.

Dissection. To expose these muscles, and to examine their mode of insertion into the humerus, detach the Deltoid and Trapezius from their attachment to the spine of the scapula and acromion process. Remove the clavicle by dividing the ligaments connecting it with the coracoid process, and separate it at its articulation with its scapula: divide the acromion process near its root with a saw. The fragments being removed, the tendons of the posterior scapular muscles will be fully exposed, and can be examined. A block should be placed beneath the shoulder-joint, so as to make the muscles tense.

The *supraspinous aponeurosis* is a thick and dense membranous layer, which completes the osseo-fibrous case in which the Supraspinatus muscle is contained; affording attachment, by its inner surface, to some of the fibres of the muscle. It is thick internally, but thinner externally under the coraco-acromial ligament. When this fascia is removed, the Supraspinatus muscle is exposed.

The *Supraspinatus* muscle occupies the whole of the supraspinous fossa, arising from its internal two-thirds, and from the strong fascia which covers its surface. The muscular fibres converge to a tendon, which passes across the capsular ligament of the shoulder-joint, to which it is intimately adherent, and is inserted into the highest of the three facets on the great tuberosity of the humerus.

Relations. By its *upper surface*, with the Trapezius, the clavicle, the acromion, the coraco-acromial ligament, and the Deltoid. By its *under surface*, with the scapula, the suprascapular vessels and nerve, and upper part of the shoulder-joint.

The *infraspinous aponeurosis* is a dense fibrous membrane, covering in the Infraspinatus muscle, and attached to the circumference of the infraspinous fossa: it affords attachment, by its inner surface, to some fibres of that muscle, is continuous externally with the fascia of the arm, and gives off from its under surface intermuscular septa, which separate the Infraspinatus from the Teres minor, and the latter from the Teres major.

The *Infraspinatus* is a thick triangular muscle, which occupies the chief part of the infraspinous fossa, arising by fleshy fibres, from its internal two-thirds;

and by tendinous fibres, from the ridges on its surface: it also arises from a strong fascia which covers it externally, and separates it from the Teres major and minor. The fibres converge to a tendon, which glides over the concave border of the spine of the scapula, and, passing across the capsular ligament of the shoulder-joint, is inserted into the middle facet on the great tuberosity of the humerus. The tendon of this muscle is occasionally separated from the spine of the scapula by a synovial bursa, which communicates with the synovial membrane of the shoulder-joint.

Relations. By its *posterior surface*, with the Deltoid, the Trapezius, Latissimus dorsi, and the integument. By its *anterior surface*, with the scapula, from which it is separated by the suprascapular and dorsalis scapulæ vessels, and with the capsular ligament of the shoulder-joint. Its *lower border* is in contact with the Teres minor, and occasionally united with it, and with the Teres major.

The *Teres Minor* is a narrow, elongated muscle, which lies along the inferior border of the scapula. It arises from the dorsal surface of the axillary border of the scapula for the upper two-thirds of its extent, and from two aponeurotic laminæ, one of which separates this muscle from the Infraspinatus, the other from

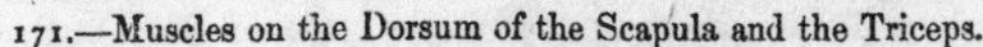
171.—Muscles on the Dorsum of the Scapula and the Triceps.

the Teres major; its fibres pass obliquely upwards and outwards, and terminate in a tendon, which is inserted into the lowest of the three facets on the great tuberosity of the humerus, and, by fleshy fibres, into the humerus immediately below it. The tendon of this muscle passes across the capsular ligament of the shoulder-joint.

Relations. By its *posterior surface*, with the Deltoid, Latissimus dorsi, and integument. By its *anterior surface*, with the scapula, the dorsal branch of the subscapular artery, the long head of the Triceps, and the shoulder-joint. By its *upper border*, with the Infraspinatus. By its *lower border*, with the Teres major, from which it is separated anteriorly by the long head of the Triceps.

The *Teres Major* is a broad and somewhat flattened muscle, which arises from the dorsal aspect of the inferior angle of the scapula, and from the fibrous septa interposed between it and the Teres minor and Infraspinatus; the fibres are directed upwards and outwards, and terminate in a flat tendon, about two inches in length, which is inserted into the posterior bicipital ridge of the humerus. The tendon of this muscle, at its insertion into the humerus, lies behind that of the Latissimus dorsi, from which it is separated by a synovial bursa.

Relations. By its *posterior surface*, with the integument, from which it is separated, internally, by the Latissimus dorsi; and externally, by the long head of the Triceps. By its *anterior surface*, with the Subscapularis, Latissimus dorsi, Coraco-brachialis, short head of the Biceps, the axillary vessels, and brachial plexus of nerves. Its *upper border* is at first in relation with the Teres minor, from which it is afterwards separated by the long head of the Triceps. Its *lower border* forms, in conjunction with the Latissimus dorsi, part of the posterior boundary of the axilla.

Nerves. The Supra and Infraspinatus muscles are supplied by the suprascapular nerve; the Teres minor, by the circumflex, and the Teres major, by the subscapular.

Actions. The Supraspinatus assists the Deltoid in raising the arm from the side; its action must, therefore, be very feeble, from the very disadvantageous manner in which the force is applied. The Infraspinatus and Teres minor rotate the head of the humerus outwards; when the arm is raised, they assist in retaining it in that position, and carrying it backwards. One of the most important uses of these three muscles is the great protection they afford to the shoulder-joint, the Supraspinatus supporting it above, and preventing displacement of the head of the humerus upwards, whilst the Infraspinatus and Teres minor protect it behind, and prevent dislocation backwards. The Teres major assists the Latissimus dorsi in drawing the humerus downwards and backwards when previously raised, and rotating it inwards; when the arm is fixed, it may assist the Pectoral and Latissimus dorsi muscles in drawing the trunk forwards.

Anterior Humeral Region. (Fig. 170.)

Coraco-Brachialis. Biceps. Brachialis Anticus.

Dissection.—The arm being placed on the table, with the front surface uppermost, make a vertical incision through the integument along the middle line, from the middle of the interval between the folds of the axilla, to about two inches below the elbow-joint, where it should be joined by a transverse incision, extending from the inner to the outer side of the forearm; the two flaps being reflected on either side, the fascia should be examined.

The *deep fascia* of the arm, continuous with that covering the shoulder and front of the great Pectoral muscle, is attached, above, to the clavicle, acromion, and spine of the scapula; it forms a thin, loose, membranous sheath investing the muscles of the arm, sending down septa between them, and composed of fibres disposed in a circular or spiral direction, and connected together by vertical fibres. It differs in thickness at different parts, being thin over the Biceps, but thicker where it covers the Triceps, and over the condyles of the humerus: it is strengthened by fibrous aponeuroses, derived from the Pectoralis major and Latissimus dorsi, on the inner side, and from the Deltoid externally. On either side it gives off a strong *intermuscular septum*, which is attached to the condyloid ridge and condyle of the humerus. These septa serve to separate the muscles of the anterior from those of the posterior brachial region. The external intermuscular septum extends from the lower part of the anterior bicipital ridge, along the external condyloid ridge, to the outer condyle; it is blended with the tendon of the Deltoid; gives attachment to the Triceps behind, to the Brachialis anticus, Supinator longus, and Extensor carpi radialis longior, in front; and is perforated by the musculo-spiral nerve, and superior profunda artery. The internal intermuscular septum, thicker than the preceding, extends from the lower part of the posterior lip of the bicipital groove below the Teres major, along the internal condyloid ridge to the inner condyle; it is blended with the tendon of the Coraco-brachialis, and affords attachment to the Triceps behind, and the Brachialis anticus in front. It is perforated by the ulnar nerve, and the inferior profunda and anastomotic arteries. At the elbow, the deep fascia is attached to all the prominent points round the joint, and is continuous with the fascia of the forearm. On the removal of this fascia, the muscles of the anterior humeral region are exposed.

The *Coraco-Brachialis*, the smallest of the three muscles in this region, is situated at the upper and inner part of the arm. It arises by fleshy fibres from the apex of the coracoid process, in common with the short head of the Biceps, and from the intermuscular septum between the two muscles; the fibres pass downwards, backwards, and a little outwards, to be inserted by means of a flat tendon into a rough ridge at the middle of the inner side of the shaft of the humerus. It is perforated by the musculo-cutaneous nerve. The inner border of the muscle forms a guide to the position of the vessel, in tying the brachial artery in the upper part of its course.

Relations. By its *anterior surface*, with the Deltoid and Pectoralis major above, and at its insertion with the brachial vessels and median nerve which cross it. By its *posterior surface*, with the tendons of the Subscapularis, Latissimus dorsi, and Teres major, the short head of the Triceps, the humerus, and the anterior circumflex vessels. By its *inner border*, with the brachial artery, and the median and musculo-cutaneous nerves. By its *outer border*, with the short head of the Biceps and Brachialis anticus.

The *Biceps* is a long fusiform muscle, occupying the whole of the anterior surface of the arm, and divided above into two portions or heads, from which circumstance it has received its name. The short head arises by a thick flattened tendon from the apex of the coracoid process, in common with the Coraco-brachialis. The long head arises from the upper margin of the glenoid cavity, by a long rounded tendon, which is continuous with the glenoid ligament. This tendon arches over the head of the humerus, being enclosed in a special sheath of the synovial membrane of the shoulder joint; it then pierces the capsular ligament at its attachment to the humerus, and descends in the bicipital groove, in which it is retained by a fibrous prolongation from the tendon of the Pectoralis major. The fibres from this tendon form a rounded belly, and, about the middle of the arm, join with the portion of the muscle derived from the short head. The belly of the muscle, narrow and somewhat flattened, terminates above the elbow in a flattened tendon, which is inserted into the back part of the tuberosity of the radius, a synovial bursa being interposed between the tendon and the front of the tuberosity. The tendon of the muscle is thin and broad; as it approaches the radius it becomes narrow and twisted upon itself, being applied by a flat surface to the back part of the tuberosity: opposite the bend of the elbow the tendon gives off, from its inner side, a broad aponeurosis, which passes obliquely downwards and inwards across the brachial artery, and is continuous with the fascia of the forearm (fig. 169). The inner border of this muscle forms a guide to the position of the vessel, in tying the brachial artery in the middle of the arm.*

Relations. Its *anterior surface* is overlapped above by the Pectoralis major and Deltoid; in the rest of its extent it is covered by the superficial and deep fasciæ and the integument. Its *posterior surface* rests on the shoulder-joint and humerus, from which it is separated by the Subscapularis, Teres major, Latissimus dorsi, Brachialis anticus, and the musculo-cutaneous nerve. Its *inner border* is in

* A third head to the Biceps is occasionally found (Theile says as often as once in eight or nine subjects), arising at the upper and inner part of the Brachialis anticus, with the fibres of which it is continuous, and inserted into the bicipital fascia and inner side of the tendon of the Biceps. In most cases, this additional slip passes behind the brachial artery in its course down the arm. Occasionally the third head consists of two slips, which pass down, one in front, the other behind the artery, concealing the vessel in the lower half of the arm.

relation with the Coraco-brachialis, the brachial vessels, and median nerve; its *outer border*, with the Deltoid and Supinator longus.

The *Brachialis Anticus* is a broad muscle, which covers the elbow-joint and the lower half of the front of the humerus. It is somewhat compressed from before backward, and is broader in the middle than at either extremity. It arises from the lower half of the outer and inner surfaces of the shaft of the humerus; and commences above at the insertion of the Deltoid, which it embraces by two angular processes. Its origin extends below, to within an inch of the margin of the articular surface, and is limited on each side by the external and internal borders of the shaft of the humerus. It also arises from the intermuscular septa on each side, but more extensively from the inner than the outer. Its fibres converge to a thick tendon, which is inserted into a rough depression on the under surface of the coronoid process of the ulna, being received into an interval between two fleshy slips of the Flexor digitorum profundus.

Relations. By its *anterior surface*, with the Biceps, the brachial vessels, musculo-cutaneous, and median nerves. By its *posterior surface*, with the humerus and front of the elbow-joint. By its *inner border*, with the Triceps, ulnar nerve, and Pronator radii teres, from which it is separated by the intermuscular septum. By its *outer border*, with the musculo-spiral nerve, radial recurrent artery, the Supinator longus, and Extensor carpi radialis longior.

Nerves. The muscles of this group are supplied by the musculo-cutaneous nerve. The Brachialis anticus usually receives an additional filament from the musculo-spiral.

Actions. The Coraco-brachialis draws the humerus forwards and inwards, and at the same time assists in elevating it towards the scapula. The Biceps and Brachialis anticus are flexors of the forearm; the former muscle is also a supinator, and serves to render tense the fascia of the forearm by means of the broad aponeurosis given off from its tendon. When the forearm is fixed, the Biceps and Brachialis anticus flex the arm upon the forearm, as is seen in efforts of climbing. The Brachialis anticus forms an important defence to the elbow-joint.

Posterior Humeral Region.

Triceps. Subanconeus.

The *Triceps* (fig. 171) is situated on the back of the arm, extending the entire length of the posterior surface of the humerus. It is of large size, and divided above into three parts; hence its name. These three portions have been named, (1) the middle, scapular, or long head, (2) the external, or long humeral, and (3) the internal, or short humeral head.

The *middle* or *scapular head* arises, by a flattened tendon, from a rough triangular depression, immediately below the glenoid cavity, being blended at its upper part with the capsular and glenoid ligaments; the muscular fibres pass downwards between the two other portions of the muscle, and join with them in the common tendon of insertion.

The *external head* arises from the posterior surface of the shaft of the humerus, between the insertion of the Teres minor and the upper part of the musculo-spiral groove, from the external border of the humerus and the external intermuscular septum: the fibres from this origin converge towards the common tendon of insertion.

The *internal head* arises from the posterior surface of the shaft of the humerus, below the groove for the musculo-spiral nerve, commencing above, narrow and pointed, below the insertion of the Teres major, and extending to within an inch of the trochlear surface: it also arises from the internal border of the humerus and internal intermuscular septum. The fibres of this portion of the muscle are directed, some downwards to the olecranon, whilst others converge to the common tendon of insertion.

The *common tendon* of the Triceps commences about the middle of the back part of the muscle: it consists of two aponeurotic laminæ, one of which is subcutaneous, and covers the posterior surface of the muscle for the lower half of its extent: the other is more deeply seated in the substance of the muscle; after receiving the attachment of the muscular fibres, they join together above the elbow, and are inserted into the back part of the upper surface of the olecranon process, a small bursa, occasionally multilocular, being interposed between the tendon and the front of this surface.

The long head of the Triceps descends between the Teres minor and Teres major, dividing the triangular space between these two muscles and the humerus into two smaller spaces, one triangular, the other quadrangular (fig. 171). The triangular space contains the dorsalis scapulæ vessels; it is bounded by the Teres minor above, the Teres major below, and the scapular head of the Triceps externally: the quadrangular space transmits the posterior circumflex vessels and nerve; it is bounded by the Teres minor above, the Teres major below, the scapular head of the Triceps internally, and the humerus externally.

Relations. By its *posterior surface* with the Deltoid above: in the rest of its extent it is subcutaneous. By its *anterior surface*, with the humerus, musculo-spiral nerve, superior profunda vessels, and back part of the elbow-joint. Its *middle* or *long head* is in relation, behind, with the Deltoid and Teres minor; in front, with the Subscapularis, Latissimus dorsi, and Teres major.

The *Subanconeus* is a small muscle, distinct from the Triceps, and analogous to the Subcrureus in the lower limb. It may be exposed by removing the Triceps from the lower part of the humerus. It consists of one or two slender fasciculi, which arise from the humerus, immediately above the olecranon fossa, and are inserted into the posterior ligament of the elbow-joint.

Nerves. The Triceps and Subanconeus are supplied by the musculo-spiral nerve.

Actions. The Triceps is the great Extensor muscle of the forearm; serving, when the forearm is flexed, to draw it into a right line with the arm. It is the direct antagonist of the Biceps and Brachialis anticus. When the arm is extended, the long head of the muscle may assist the Teres major and Latissimus dorsi in drawing the humerus backwards. The long head of the Triceps protects the under part of the shoulder-joint, and prevents displacement of the head of the humerus downwards and backwards.

Muscles of the Forearm.

Dissection. To dissect the forearm, place the limb in the position indicated in fig. 168; make a vertical incision along the middle line from the elbow to the wrist, and a transverse incision at each extremity of this; the flaps of integument being removed, the fascia of the forearm is exposed.

The *deep fascia* of the forearm, continuous above with that enclosing the arm, is a dense highly glistening aponeurotic investment, which forms a general sheath enclosing the muscles in this region; it is attached behind to the olecranon and posterior border of the ulna, and gives off from its inner surface numerous intermuscular septa, which enclose each muscle separately. It consists of circular and oblique fibres, connected together by numerous vertical fibres. It is much thicker on the dorsal than on the palmar surface, and at the lower than at the upper part of the forearm, and is strengthened by tendinous fibres, derived from the Brachialis anticus and Biceps in front, and from the Triceps behind. Its inner surface gives origin to muscular fibres, especially at the upper part of the inner and outer sides of the forearm, and forms the boundaries of a series of conical-shaped cavities, in which the muscles are contained. Besides the vertical septa separating each muscle, transverse septa are given off both on the anterior and posterior surfaces of the forearm, separating the deep from the superficial layer of muscles. Numerous apertures exist in the fascia for the passage of vessels and nerves; one of these, of large size, situated at the front of the elbow, serves for the passage of a communicating branch between the superficial and deep veins.

The muscles of the forearm may be subdivided into groups corresponding to the region they occupy. One group occupies the inner and anterior aspect of the forearm, and comprises the Flexor and Pronator muscles. Another group occupies its outer side; and a third, its posterior aspect. The two latter groups include all the Extensor and Supinator muscles.

Anterior Brachial Region.

Superficial Layer.

Pronator Radii Teres.
Flexor Carpi Radialis.
Palmaris Longus.
Flexor Carpi Ulnaris.
Flexor Sublimis Digitorum.

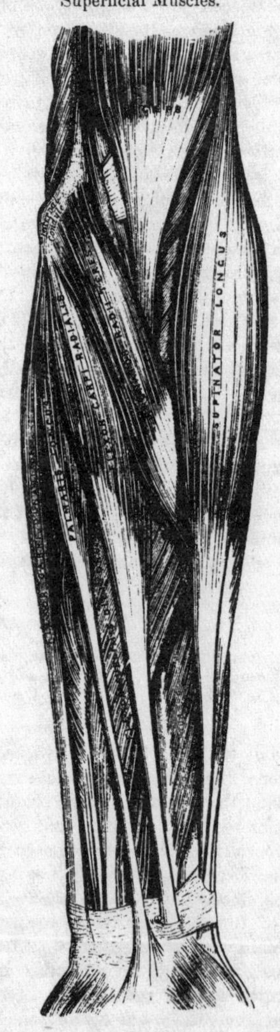

172.—Front of the Left Forearm. Superficial Muscles.

These muscles take origin from the internal condyle of the humerus by a common tendon.

The *Pronator Radii Teres* arises by two heads. One, the larger and more superficial, arises from the humerus, immediately above the internal condyle, and from the tendon common to the origin of the other muscles; also from the fascia of the forearm, and intermuscular septum between it and the Flexor carpi radialis. The other head is a thin fasciculus, which arises from the inner side of the coronoid process of the ulna, joining the preceding at an acute angle. Between the two heads passes the median nerve. The muscle passes obliquely across the forearm from the inner to the outer side, and terminates in a flat tendon, which turns over the outer margin of the radius, and is inserted into a rough ridge at the middle of the outer surface of the shaft of that bone.

Relations. By its *anterior surface,* with the deep fascia, the Supinator longus, and the radial vessels and nerve. By its *posterior surface,* with the Brachialis anticus, Flexor sublimis digitorum, the median nerve, and ulnar artery; the small, or deep, head being interposed between the two latter structures. Its *outer border* forms the inner boundary of a triangular space, in which is placed the brachial artery, median nerve, and tendon of the Biceps muscle. Its *inner border* is in contact with the Flexor carpi radialis.

The *Flexor Carpi Radialis* lies on the inner side of the preceding muscle. It arises from the internal condyle by the common tendon, from the fascia of the forearm, and from the intermuscular septa between it and the Pronator teres, on the outside; the Palmaris longus, internally; and the Flexor sublimis digitorum, beneath. Slender and aponeurotic in structure at its commencement, it increases in size, and terminates in a tendon which forms the lower two-thirds of its length. This tendon passes through a canal on the outer side of the annular ligament, runs through a groove in the os trapezium (which is converted into a canal by a fibrous sheath, and lined by a synovial membrane), and is inserted into the base of the metacarpal bone of the index finger. The radial artery lies between the tendon of this muscle and the Supinator longus, and may easily be tied in this situation.

Relations. By its *superficial surface*, with the deep fascia and the integument. By its *deep surface*, with the Flexor sublimis digitorum, Flexor longus pollicis and wrist joint. By its *outer border*, with the Pronator radii teres, and the radial vessels. By its *inner border*, with the Palmaris longus above, and the median nerve below.

The *Palmaris Longus* is a slender fusiform muscle, lying on the inner side of the preceding. It arises from the inner condyle of the humerus by the common tendon, from the deep fascia, and the intermuscular septa, between it and the adjacent muscles. It terminates in a slender flattened tendon, which is inserted into the annular ligament, expanding to end in the palmar fascia.

Variations. This muscle is often absent; when present, it presents many varieties. Its fleshy belly is sometimes very long, or may occupy the middle of the muscle, which is tendinous at either extremity; or the Palmaris may be muscular at its lower extremity, its upper part being tendinous. Occasionally there is a second Palmaris longus placed on the inner side of the preceding, terminating, below, partly in the annular ligament or fascia, and partly in the small muscles of the little finger.

Relations. By its *anterior surface*, with the deep fascia. By its *posterior surface*, with the Flexor digitorum sublimis. *Internally*, with the Flexor carpi ulnaris. *Externally*, with the Flexor carpi radialis. The median nerve lies close to the tendon, just above the wrist, on its inner and posterior side.

The *Flexor Carpi Ulnaris* lies along the ulnar side of the forearm. It arises by two heads, separated by a tendinous arch, beneath which passes the ulnar nerve, and posterior ulnar recurrent artery. One head arises from the inner condyle of the humerus by the common tendon; the other, from the inner margin of the olecranon, by an aponeurosis from the upper two-thirds of the posterior border of the ulna, and from the intermuscular septum between it and the Flexor sublimis digitorum. The fibres terminate in a tendon, which occupies the anterior part of the lower half of the muscle, and is inserted into the pisiform bone, some fibres being prolonged to the annular ligament and base of the metacarpal bone of the little finger. The ulnar artery lies on the outer side of the tendon of this muscle, in the lower two-thirds of the forearm; the tendon forming a guide in tying the vessel in this situation.

Relations. By its *anterior surface*, with the deep fascia, with which it is intimately connected for a considerable extent. By its *posterior surface*, with the Flexor sublimis, the Flexor profundus, the Pronator quadratus, and the ulnar vessels and nerve. By its *outer* or *radial border*, with the Palmaris longus above, and the ulnar vessels and nerve below.

The *Flexor Digitorum Sublimis* (*perforatus*) is placed beneath the preceding muscles, which therefore must be removed in order to bring its attachment into view. It is the largest of the muscles of the superficial layer, and arises by three heads. One head arises from the internal condyle of the humerus by the common tendon, from the internal lateral ligament of the elbow-joint, and from the intermuscular septum common to it and the preceding muscles. The second head arises from the inner side of the coronoid process of the ulna, above the ulnar origin of the Pronator radii teres (fig. 90). The third head arises from the oblique line of the radius, extending from the tubercle to the insertion of the Pronator radii teres. The fibres pass vertically downwards, forming a broad and thick muscle, which divides into four tendons about the middle of the forearm; as these tendons pass beneath the annular ligament into the palm of the hand, they are arranged in pairs, the anterior pair corresponding to the middle and ring fingers; the posterior pair to the index and little fingers. The tendons diverge from one another as they pass onwards, and are finally inserted into the lateral margins of the second phalanges, about their centre. Opposite the base of the first phalanges, each tendon divides, so as to leave a fissured interval, between which passes one of the tendons of the Flexor profundus, and the tendons of both the Flexors then enter an osseo-aponeurotic canal, formed by a strong fibrous band, which arches across them, and is attached on each side to the margins of the phalanges. The two portions into which the tendon of the Flexor sublimis divides, so as to admit of the passage of the deep flexor, expand somewhat, and form a grooved channel, into which the accompanying deep flexor tendon is received; the two divisions then unite, and finally subdivide a second time to be inserted into the fore part and sides of the second phalanges (fig. 177). The tendons, whilst contained in the fibro-osseous canals are connected to the phalanges by slender tendinous filaments, called *vincula accessoria tendinum*. A synovial sheath invests the tendons as they pass beneath the annular ligament; a prolongation from which surrounds each tendon as it passes along the phalanges.

Relations. In the forearm, by its *anterior surface*, with the deep fascia and all the preceding superficial muscles; by its *posterior surface*, with the Flexor profundus digitorum, Flexor longus pollicis, the ulnar vessels and nerve, and the median nerve. In the hand, its tendons are in relation, in front, with the palmar fascia, superficial palmar arch, and the branches of the median nerve; behind, with the tendons of the deep Flexor and the Lumbricales.

Anterior Brachial Region.

Deep Layer.

Flexor Profundus Digitorum. Flexor Longus Pollicis.
Pronator Quadratus.

Dissection. Divide each of the superficial muscles at its centre, and turn either end aside: the deep layer of muscles, together with the median nerve and ulnar vessels, will then be exposed.

The *Flexor Profundus Digitorum* (*perforans*) (fig. 173), is situated on the ulnar side of the forearm, immediately beneath the superficial Flexors. It arises from the upper two-thirds of the anterior and inner surfaces of the shaft of the ulna, embracing the insertion of the Brachialis anticus above, and extending, below, to within a short distance of the Pronator quadratus. It also arises from a depression on the inner side of the coronoid process, by an aponeurosis from the upper two-thirds of the posterior border of the ulna, and from the ulnar half of the interosseous membrane. The fibres form a fleshy belly of considerable size which divides into four tendons: these pass under the annular ligament beneath the tendons of the Flexor sublimis. Opposite the first phalanges, the tendons pass between the two slips of the tendons of the Flexor sublimis, and are finally inserted into the bases of the last phalanges. The tendon of the index finger is distinct; the rest are connected together by cellular tissue and tendinous slips, as far as the palm of the hand.

Four small muscles, the Lumbricales, are connected with the tendons of the Flexor profundus in the palm. They will be described with the muscles in that region.

Relations. By its *anterior surface*, in the forearm, with the Flexor sublimis digitorum, the Flexor carpi ulnaris, the ulnar vessels and nerve, and the median nerve; and in the hand, with the tendons of the superficial Flexor. By its *posterior surface*, in the forearm, with the ulnar, the interosseous membrane, the Pronator quadratus; and in the hand, with the Interossei, Adductor pollicis, and deep palmar arch. By its *ulnar border*, with the Flexor carpi ulnaris. By its *radial border*, with the Flexor longus pollicis, the anterior interosseous vessels and nerve being interposed.

173.—Front of the Left Forearm. Deep Muscles.

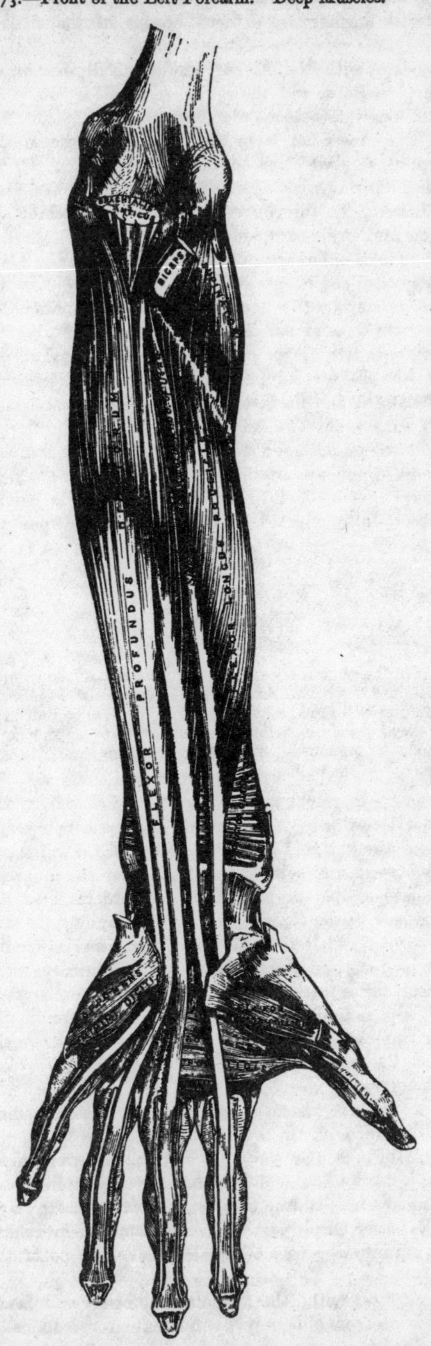

The *Flexor Longus Pollicis* is situated on the radial side of the forearm, lying on the same plane as the preceding. It arises from the upper two-thirds of the grooved anterior surface of the shaft of the radius; commencing, above, immediately below the tuberosity and oblique line, and extending, below, to within a short distance of the Pronator quadratus. It also arises from the adjacent part of the interosseous membrane, and occasionally by a fleshy slip from the inner side of the base of the coronoid process. The fibres pass downwards and terminate in a flattened tendon, which passes beneath the annular ligament, is then lodged in the interspace between the two heads of the Flexor brevis pollicis, and entering a tendino-osseous canal similar to those for the other flexor tendons, is inserted into the base of the last phalanx of the thumb.

Relations. By its *anterior surface*, with the Flexor sublimis digitorum, Flexor carpi radialis, Supinator longus, and radial vessels. By its *posterior surface*, with the radius, interosseous membrane, and Pronator quadratus. By its *ulnar border*, with the Flexor profundus digitorum, from which it is separated by the anterior interosseous vessels and nerve.

The *Pronator Quadratus* is a small, flat, quadrilateral muscle, extending transversely across the front of the radius and ulna, above their carpal extremities. It arises from the oblique line on the lower fourth of the anterior surface of the shaft of the ulna, and the surface of bone immediately below it; from the internal border of the ulna; and from a strong aponeurosis which covers the inner third of the muscle. The fibres pass horizontally outwards, to be inserted into the lower fourth of the anterior surface and external border of the shaft of the radius.

Relations. By its *anterior surface*, with the Flexor profundus digitorum, the Flexor longus pollicis, Flexor carpi radialis, and the radial vessels. By its *posterior surface*, with the radius, ulna, and interosseous membrane.

Nerves. All the muscles of the superficial layer are supplied by the median nerve, excepting the Flexor carpi ulnaris, which is supplied by the ulnar. Of the deep layer, the Flexor profundus digitorum is supplied conjointly by the ulnar and by the median, through its branch, the anterior interosseous nerve, which also supplies the Flexor longus pollicis and Pronator quadratus.

Actions. These muscles act upon the forearm, the wrist, and hand. Those acting on the forearm, are the Pronator radii teres and Pronator quadratus, which rotate the radius upon the ulna, rendering the hand prone; when pronation has been fully effected, the Pronator radii teres assists the other muscles in flexing the forearm. The flexors of the wrist are the Flexor carpi ulnaris and radialis; and the flexors of the phalanges are the Flexor sublimis and Profundus digitorum; the former flexing the second phalanges, and the latter the last. The Flexor longus pollicis flexes the last phalanx of the thumb. The three latter muscles, after flexing the phalanges, by continuing their action, act upon the wrist, assisting the ordinary flexors of this joint; and all those which are attached to the humerus assist in flexing the forearm upon the arm. The Palmaris longus is a tensor of the palmar fascia; when this action has been fully effected, it flexes the hand upon the forearm.

Radial Region. (Fig. 174.)

Supinator Longus. Extensor Carpi Radialis Longior.
Extensor Carpi Radialis Brevior.

Dissection. Divide the integument in the same manner as in the dissection of the anterior brachial region; and after having examined the cutaneous vessels and nerves and deep fascia, remove all those structures. The muscles will then be exposed. The removal of the fascia will be considerably facilitated by detaching it from below upwards. Great care should be taken to avoid cutting across the tendons of the muscles of the thumb, which cross obliquely the larger tendons running down the back of the radius.

The *Supinator Longus* is the most superficial muscle on the radial side of the forearm: it is fleshy for the upper two-thirds of its extent, tendinous below. It arises from the upper two-thirds of the external condyloid ridge of the humerus, and from the external intermuscular septum, being limited above by the musculo-spiral groove. The fibres terminate above the middle of the forearm in a flat tendon, which is inserted into the base of the styloid process of the radius.

Relations. By its *superficial surface*, with the integument and fascia for the greater part of its extent; near its insertion it is crossed by the Extensor ossis metacarpi pollicis and the Extensor primi internodii pollicis. By its *deep surface*, with the humerus, the Extensor carpi radialis longior and brevior, the insertion of the Pronator radii teres, and the Supinator brevis. By its *inner border*, above the elbow, with the Brachialis anticus, the musculo-spiral nerve, and radial recurrent artery; and in the forearm, with the radial vessels and nerve.

The *Extensor Carpi Radialis Longior* is placed partly beneath the preceding muscle. It arises from the lower third of the external condyloid ridge of the humerus, and from the external intermuscular septum. The fibres terminate at the upper third of the forearm in a flat tendon, which runs along the outer border of the radius, beneath the extensor tendons of the thumb; it then passes through a groove common to it and the Extensor carpi radialis brevior immediately behind the styloid process; and is inserted into the base of the metacarpal bone of the index finger, on its radial side.

Relations. By its *superficial surface*, with the Supinator longus, and fascia of the forearm. Its *outer side* is crossed obliquely by the extensor tendons of

174.—Posterior Surface of Forearm. Superficial Muscles.

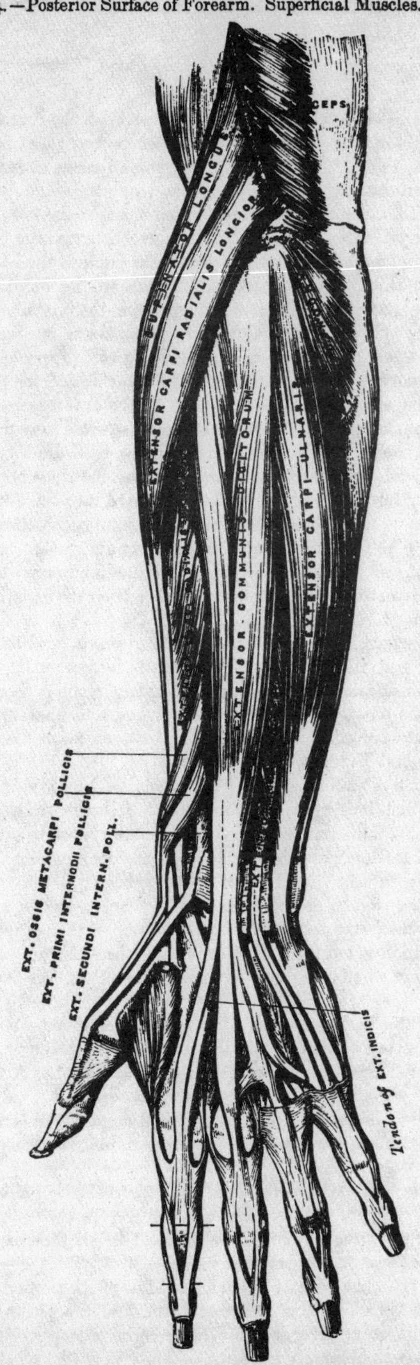

the thumb. By its *deep surface*, with the elbow-joint, the Extensor carpi radialis brevior, and back part of the wrist.

The *Extensor Carpi Radialis Brevior* is shorter, as its name implies, and thicker than the preceding muscle, beneath which it is placed. It arises from the external condyle of the humerus by a tendon common to it and the three following muscles; from the external lateral ligament of the elbow-joint; from a strong aponeurosis which covers its surface; and from the inter-muscular septa between it and the adjacent muscles. The fibres terminate about the middle of the forearm in a flat tendon, which is closely connected with that of the preceding muscle, accompanies it to the wrist, lying in the same groove on the posterior surface of the radius; passes beneath the annular ligament, and, diverging somewhat from its fellow, is inserted into the base of the metacarpal bone of the middle finger, on its radial side.

The tendons of the two preceding muscles pass through the same compartment of the annular ligament, and are lubricated by a single synovial membrane, but are separated from each other by a small vertical ridge of bone, as they lie in the groove at the back of the radius.

Relations. By its *superficial surface*, with the Extensor carpi radialis longior, and with the Extensor muscles of the thumb, which cross it. By its *deep surface*, with the Supinator brevis, tendon of the Pronator radii teres, radius, and wrist joint. By its *ulnar border*, with the Extensor communis digitorum.

Posterior Brachial Region. (Fig. 174.)

Superficial Layer.

Extensor Communis Digitorum.	Extensor Carpi Ulnaris.
Extensor Minimi Digiti.	Anconeus.

The *Extensor Communis Digitorum* is situated at the back part of the forearm. It arises from the external condyle of the humerus by the common tendon, from the deep fascia, and the intermuscular septa between it and the adjacent muscles. Just below the middle of the forearm it divides into three tendons, which pass, together with the Extensor indicis, through a separate compartment of the annular ligament, lubricated by a synovial membrane. The tendons then diverge, the innermost one dividing into two ; and all, after passing across the back of the hand, are inserted into the second and third phalanges of the fingers in the following manner: Each tendon becomes narrow and thickened opposite the metacarpo-phalangeal articulation, and gives off a thin fasciculus upon each side of the joint, which serves as the posterior ligament; after having passed the joint, it spreads out into a broad aponeurosis, which covers the whole of the dorsal surface of the first phalanx; being reinforced, in this situation, by the tendons of the Interossei and Lumbricales. Opposite the first phalangeal joint, this aponeurosis divides into three slips, a middle, and two lateral; the former is inserted into the base of the second phalanx; and the two lateral, which are continued onwards along the sides of the second phalanx, unite by their contiguous margins, and are inserted into the dorsal surface of the last phalanx. As the tendons cross the phalangeal joints, they furnish them with posterior ligaments. The tendons of the middle, ring, and little fingers are connected together, as they cross the hand, by small oblique tendinous slips. The tendons of the index and little fingers also receive, before their division, the special extensor tendons belonging to them.

Relations. By its *superficial surface*, with the fascia of the forearm and hand, the posterior annular ligament, and integument. By its *deep surface*, with the Supinator brevis, the Extensor muscles of the thumb and index finger, the posterior interosseous vessels and nerve, the wrist joint, carpus, metacarpus, and phalanges. By its *radial border*, with the Extensor carpi radialis brevior. By its *ulnar border*, with the Extensor minimi digiti, and Extensor carpi ulnaris.

The *Extensor Minimi Digiti* is a slender muscle, placed on the inner side of the Extensor communis, with which it is generally connected. It arises from the common tendon by a thin tendinous slip; and from the intermuscular septa between it and the adjacent muscles. Its tendon runs through a separate compartment in the annular ligament behind the inferior radio-ulnar joint, subdivides into two as it crosses the hand, and, at the metacarpo-phalangeal articulation, unites with the tendon derived from the common Extensor. The common tendon then spreads into a broad aponeurosis, which is inserted into the second and third phalanges of the little finger in a similar manner to the common extensor tendons of the other fingers.

The *Extensor Carpi Ulnaris* is the most superficial muscle on the ulnar side of the forearm. It arises from the external condyle of the humerus, by the common tendon; from the middle third of the posterior border of the ulna below the Anconeus, and from the fascia of the forearm. This muscle terminates in a tendon, which runs through a groove behind the styloid process of the ulna, passes through a separate compartment in the annular ligament, and is inserted into the base of the metacarpal bone of the little finger.

Relations. By its *superficial surface*, with the fascia of the forearm. By its *deep surface*, with the ulna, and the muscles of the deep layer.

The *Anconeus* is a small triangular muscle, placed behind and below the elbow-joint, and appears to be a continuation of the external portion of the Triceps. It arises by a separate tendon from the back part of the outer condyle of the humerus; and is inserted into the side of the olecranon, and upper third of the posterior surface of the shaft of the ulna; its fibres diverge fron their origin, the upper ones being directed transversely, the lower obliquely inwards.

Relations. By its *superficial surface* with a strong fascia derived from the Triceps. By its *deep surface*, with the elbow-joint, the orbicular ligament, the ulna, and a small portion of the Supinator brevis.

Posterior Brachial Region. (Fig. 175.)

Deep Layer.

Supinator Brevis.	Extensor Primi Internodii Pollicis.
Extensor Ossis Metacarpi Pollicis.	Extensor Secundi Internodii Pollicis.

Extensor Indicis.

The *Supinator Brevis* is a broad muscle, of a hollow cylindrical form, curved round the upper third of the radius. It arises from the external condyle of the humerus, from the external lateral ligament of the elbow-joint, and the orbicular ligament of the radius, from the ridge on the ulna, which runs obliquely downwards from the posterior extremity of the lesser sigmoid cavity, from the triangular depression in front of it, and from a tendinous expansion which covers the surface of the muscle. The muscle surrounds the upper part of the radius: the upper fibres forming a sling-like fasciculus, which encircles the neck of the radius above the tuberosity, and is attached to the back part of its inner surface; the middle fibres are attached to the outer edge of the bicipital tuberosity; the lower fibres to the oblique line of the radius, as low down as the insertion of the Pronator radii teres. This muscle is pierced by the posterior interosseous nerve.

Relations. By its *superficial surface*, with the superficial Extensor and Supinator muscles, and the radial vessels and nerve. By its *deep surface*, with the elbow-joint, the interosseous membrane, and the radius.

The *Extensor Ossis Metacarpi Pollicis* is the most external and the largest of the deep Extensor muscles: it lies immediately below the Supinator brevis, with which it is sometimes united. It arises from the posterior surface of the shaft of the ulna below the insertion of the Anconeus, from the interosseous ligament, and from the middle third of the posterior surface of the shaft of the radius. Passing obliquely downwards and outwards, it terminates in a tendon which runs through a groove on the outer side of the styloid process of the radius, accompanied by the tendon of the Extensor primi internodii pollicis, and is inserted into the base of the metacarpal bone of the thumb.

Relations. By its *superficial surface*, with the Extensor communis digitorum, Extensor minimi digiti, and fascia of the forearm; and with the branches of the posterior interosseous artery and nerve which cross it. By its *deep surface*, with the ulna, interosseous membrane, radius, the tendons of the Extensor carpi radialis longior and brevior, which it crosses obliquely; and, at the outer side of the wrist, with the radial vessels. By its *upper border*, with the Supinator brevis. By its *lower border*, with the Extensor primi internodii pollicis.

The *Extensor Primi Internodii Pollicis*, the smallest muscle of this group, lies on the inner side of the preceding. It arises from the posterior surface of the shaft of the radius, below the Extensor ossis metacarpi, and from the interosseous membrane. Its direction is similar to that of the Extensor ossis metacarpi, its tendon passing through the same groove on the outer side of the styloid process, to be inserted into the base of the first phalanx of the thumb.

Relations. The same as those of the Extensor ossis metacarpi pollicis.

The *Extensor Secundi Internodii Pollicis* is much larger than the preceding muscle, the origin of which it partly covers in. It arises from the posterior surface of the shaft of the ulna, below the origin of the Extensor ossis metacarpi pollicis, and from the interosseous membrane. It terminates in a tendon which passes through a separate compartment in the annular ligament, lying in a narrow oblique groove at the back part of the lower end of the radius. It then crosses obliquely the Extensor tendons of the carpus, being separated from the other Extensor tendons of the thumb by a triangular interval, in which the radial artery is found; and is finally inserted into the base of the last phalanx of the thumb.

Relations. By its *superficial surface*, with the same parts as the Extensor ossis metacarpi pollicis. By its *deep surface*, with the ulna, interosseous membrane, radius, the wrist, the radial vessels, and metacarpal bone of the thumb.

The *Extensor Indicis* is a narrow elongated muscle, placed on the inner side of, and parallel with, the preceding. It arises from the posterior surface of the shaft of the ulna, below the origin of the Extensor secundi internodii pollicis, and from the interosseous membrane. Its tendon passes with the Extensor communis digitorum through the same canal in the annular ligament, and subsequently joins that tendon of the Extensor communis which belongs to the index finger, opposite the lower end of the corresponding metacarpal bone. It is finally inserted into the second and third phalanges of the index finger, in the manner already described.

Relations. They are similar to those of the preceding muscles.

Nerves. The Supinator longus, Extensor carpi radialis longior, and Anconeus, are supplied by branches from the musculo-spiral nerve. The remaining muscles of the radial and posterior brachial regions, by the posterior interosseous nerve.

Actions. The muscles of the radial and posterior brachial regions, which comprise all the Extensor and Supinator muscles, act upon the forearm, wrist, and hand; they are the direct antagonists of the Pronator and Flexor muscles. The Anconeus assists the Triceps in extending the forearm. The Supinator longus and brevis are the supinators of the forearm and hand; the former muscle more especially acting as a supinator when the limb is pronated. When supination has been produced, the Supinator longus, if still continuing to act, flexes the forearm. The Extensor carpi radialis longior and brevior, and Extensor carpi ulnaris muscles, are the

175.—Posterior Surface of the Forearm. Deep Muscles.

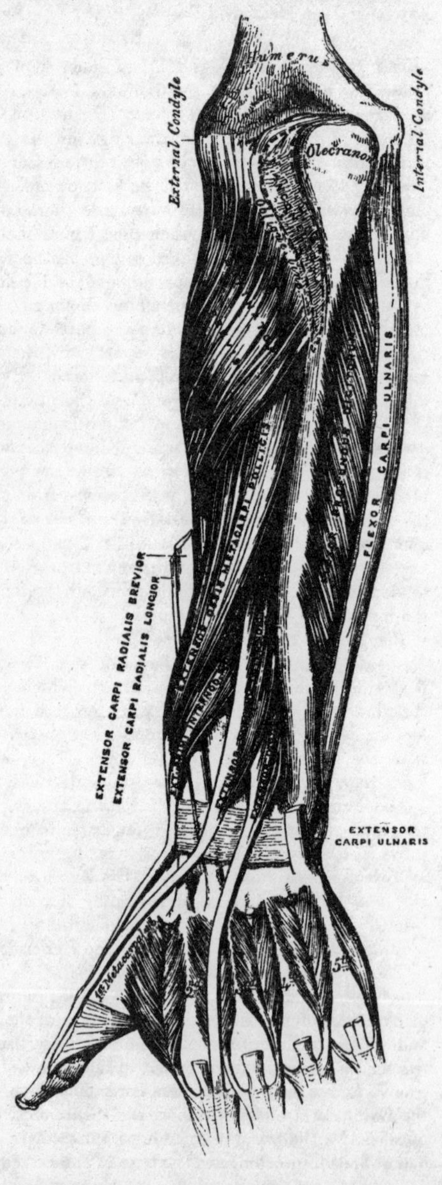

Extensors of the wrist; continuing their action, they serve to extend the forearm upon the arm; they are the direct antagonists of the Flexor carpi radialis and ulnaris. The common Extensor of the fingers, the Extensors of the thumb, and the Extensors of the index and little fingers, serve to extend the phalanges into which they are inserted; and are the direct antagonists of the Flexors. By continuing their action, they assist in extending the forearm. The Extensors of the thumb, in consequence of the oblique direction of their tendons, assist in supinating the forearm, when the thumb has been drawn inwards towards the palm.

Muscles and Fasciæ of the Hand.

Dissection (fig. 168). Make a transverse incision across the front of the wrist, and a second across the heads of the metacarpal bones: connect the two by a vertical incision in the middle line, and continue it through the centre of the middle finger. The anterior and posterior annular ligaments, and the palmar fascia, should first be dissected.

The *Anterior Annular Ligament* is a strong fibrous band, which arches over the carpus, converting the deep groove on the front of the carpal bones into a canal, beneath which pass the flexor tendons of the fingers. It is attached, internally, to the pisiform bone, and unciform process of the unciform; and externally, to the tuberosity of the scaphoid, and ridge on the trapezium. It is continuous, above, with the deep fascia of the forearm, and below, with the palmar fascia. It is crossed by the tendon of the Palmaris longus, by the ulnar vessels and nerve, and the cutaneous branches of the median and ulnar nerves. It has inserted into its upper and inner part the greater part of the tendon of the Flexor carpi ulnaris; and has, arising from it below, the small muscles of the thumb and little finger. It is pierced by the tendon of the Flexor carpi radialis; and, beneath it, pass the tendons of the Flexor sublimis and profundus digitorum, the Flexor longus pollicis, and the median nerve. There are two synovial membranes beneath this ligament; one of large size, enclosing the tendons of the Flexor sublimis and profundus; and a separate one for the tendon of the Flexor longus pollicis, which is also very extensive, reaching from above the wrist to the extremity of the last phalanx of the thumb.

176.—Transverse Section through the Wrist, showing the Annular Ligaments and the Canals for the Passage of the Tendons.

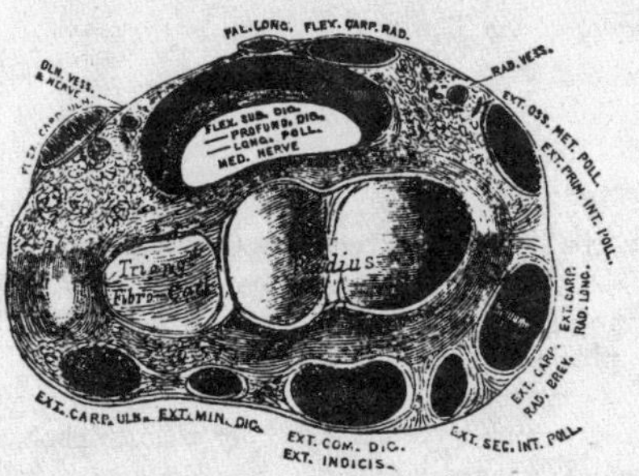

The *Posterior Annular Ligament* is a strong fibrous band, extending transversely across the back of the wrist, and continuous with the fascia of the forearm. It forms a sheath for the extensor tendons in their passage to the fingers, being attached, internally, to the ulna, the cuneiform and pisiform bones, and palmar fascia; externally, to the margin of the radius: and in its passage across the wrist, to the elevated ridges on the posterior surface of the radius. It presents six compartments for the passage of tendons, each of which is lined by a separate synovial membrane. These are, from without inwards—1. On the outer side of the styloid process for the tendons of the Extensor ossis metacarpi, and Extensor primi internodii pollicis. 2. Behind the styloid process, for the tendons of the Extensor carpi radialis longior and brevior. 3. Opposite the outer side of the posterior surface of the radius, for the tendon of the Extensor secundi internodii pollicis. 4. To the inner side of the latter, for the tendons of the Extensor communis digitorum, and Extensor indicis. 5. For the Extensor minimi digiti, opposite the interval between the radius and ulna. 6. For the tendon of the Extensor carpi ulnaris, grooving the back of the ulna. The synovial membranes lining these sheaths are usually very extensive, reaching from above the annular ligament, down upon the tendons almost to their insertion.

The *palmar fascia* forms a common sheath which invests the muscles of the hand. It consists of a central and two lateral portions.

The *central portion* occupies the middle of the palm, is triangular in shape, of great strength and thickness, and binds down the tendons in this situation. It is narrow above, being attached to the lower margin of the annular ligament, and receives the expanded tendon of the Palmaris longus muscle. Below, it is broad and expanded, and opposite the heads of the metacarpal bones divides into four slips, for the four fingers. Each slip subdivides into two processes, which enclose the tendons of the Flexor muscles, and are attached to the sides of the first phalanx, and to the glenoid ligament; by this arrangement, four arches are formed, under which the Flexor tendons pass. The intervals left in the fascia, between the four fibrous slips, transmit the digital vessels and nerves, and the tendons of the Lumbricales. At the point of division of the palmar fascia into the slips above mentioned, numerous strong transverse fibres bind the separate processes together. The palmar fascia is intimately adherent to the integument by numerous fibrous bands, and gives origin by its inner margin to the Palmaris brevis; it covers the superficial palmar arch, the tendons of the flexor muscles, and the branches of the median and ulnar nerves; and on each side it gives off a vertical septum, which is continuous with the interosseous aponeurosis, and separates the lateral from the middle palmar group of muscles.

The *lateral portions* of the palmar fascia are thin fibrous layers, which cover, on the radial side, the muscles of the ball of the thumb; and, on the ulnar side, the muscles of the little finger; they are continuous with the dorsal fascia, and in the palm with the middle portion of the palmar fascia.

Muscles of the Hand.

The muscles of the hand are subdivided into three groups.—1. Those of the thumb which occupy the radial side. 2. Those of the little finger which occupy the ulnar side. 3. Those in the middle of the palm and between the interosseous spaces.

Radial Region. (Fig. 177.)

Muscles of the Thumb.

Abductor Pollicis.
Opponens Pollicis (Flexor Ossis Metacarpi).
Flexor Brevis Pollicis.
Adductor Pollicis.

The *Abductor Pollicis* is a thin, flat muscle, placed immediately beneath the integument. It arises from the ridge of the os trapezium and annular ligament; and passing outwards and downwards, is inserted by a thin, flat tendon into the radial side of the base of the first phalanx of the thumb.

Relations. By its *superficial surface*, with the palmar fascia. By its *deep surface*, with the Opponens pollicis, from which it is separated by a thin aponeurosis. Its *inner border* is separated from the Flexor brevis pollicis by a narrow cellular interval.

The *Opponens Pollicis* is a small triangular muscle, placed beneath the preceding. It arises from the palmar surface of the trapezium and annular ligament, passes downwards and outwards, and is inserted into the whole length of the metacarpal bone of the thumb on its radial side.

Relations. By its *superficial surface*, with the Abductor pollicis. By its *deep surface*, with the trapezio-metacarpal articulation. By its *inner border*, with the Flexor brevis pollicis.

The *Flexor Brevis Pollicis* is much larger than either of the two preceding muscles, beneath which it is placed. It consists of two portions, in the interval between which lies the tendon of the Flexor longus pollicis. The anterior and

177.—Muscles of the Left Hand. Palmar Surface.

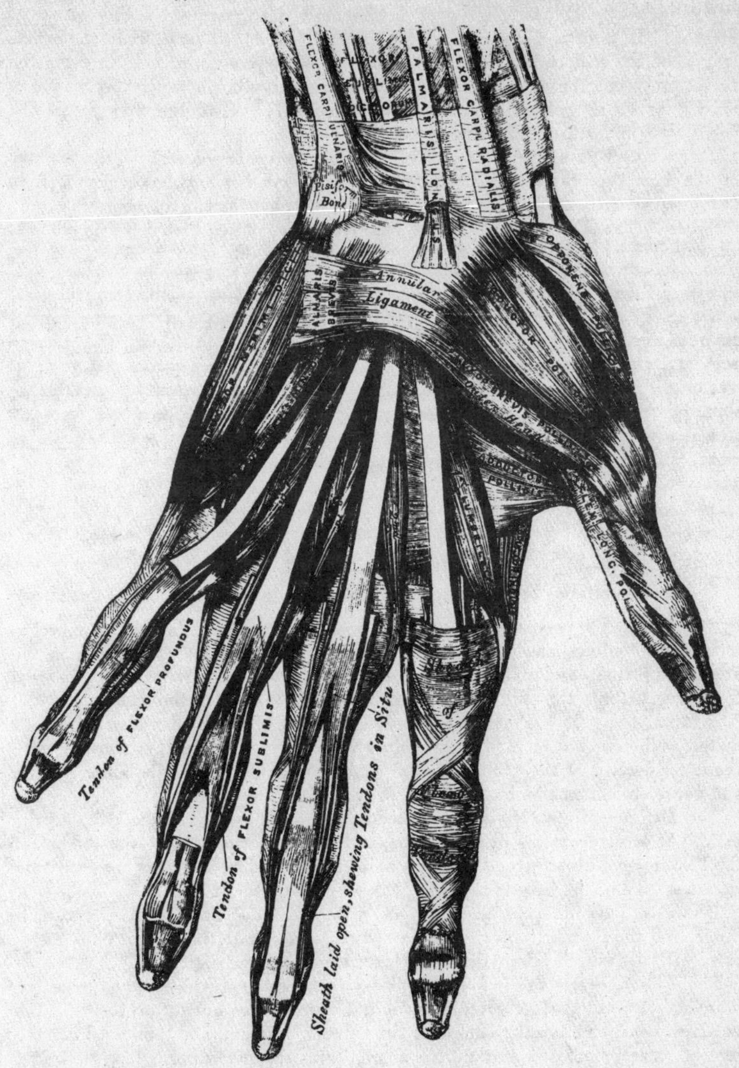

more superficial portion arises from the trapezium and outer two-thirds of the annular ligament; the deeper portion from the trapezoides, os magnum, base of the third metacarpal bone, and sheath of the tendon of the Flexor carpi radialis. The fleshy fibres unite to form a single muscle; this divides into two portions, which are inserted one on either side of the base of the first phalanx of the thumb, the outer portion being joined with the Abductor, and the inner with the Adductor. A sesamoid bone is developed in each tendon as it passes across the metacarpo-phalangeal joint.

Relations. By its *superficial surface*, with the palmar fascia. By its *deep surface*, with the Adductor pollicis, and tendon of the Flexor carpi radialis. By its *external surface*, with the Opponens pollicis. By its *internal surface*, with the tendon of the Flexor longus pollicis.

The *Adductor Pollicis* (fig. 173) is the most deeply-seated of this group of muscles. It is of a triangular form, arising, by its broad base, from the whole length of the metacarpal bone of the middle finger on its palmar surface; the fibres, proceeding outwards, converge, to be inserted with the innermost tendon of the Flexor brevis pollicis, into the ulnar side of the base of the first phalanx of the thumb, and into the internal sesamoid bone.

Relations. By its *superficial surface*, with the Flexor brevis pollicis, the tendons of the Flexor profundus and the Lumbricales. Its *deep surface* covers the first two interosseous spaces, from which it is separated by a strong aponeurosis.

Nerves. The Abductor, Opponens, and outer head of the Flexor brevis pollicis, are supplied by the median nerve; the inner head of the Flexor brevis, and the Adductor pollicis, by the ulnar nerve.

Actions. The actions of the muscles of the thumb are almost sufficiently indicated by their names. This segment of the hand is provided with three Extensors, an Extensor of the metacarpal bone, an Extensor of the first, and an Extensor of the second phalanx; these occupy the dorsal surface of the forearm and hand. There are, also, three Flexors on the palmar surface, a Flexor of the metacarpal bone, the Flexor ossis metacarpi (Opponens pollicis), the Flexor brevis pollicis, and the Flexor longus pollicis; there is also an Abductor and an Adductor. These muscles give to the thumb its extensive range of motion.

Ulnar Region. (Fig. 177.)

Muscles of the Little Finger.

Palmaris Brevis.	Flexor Brevis Minimi Digiti.
Abductor Minimi Digiti.	Opponens Minimi Digiti (Flexor Ossis Metacarpi).

The *Palmaris Brevis* is a thin quadrilateral muscle, placed beneath the integument on the ulnar side of the hand. It arises by tendinous fasciculi, from the annular ligament and palmar fascia; the fleshy fibres pass horizontally inwards, to be inserted into the skin on the inner border of the palm of the hand.

Relations. By its *superficial surface*, with the integument to which it is intimately adherent, especially by its inner extremity. By its *deep surface*, with the inner portion of the palmar fascia, which separates it from the ulnar vessels and nerve, and from the muscles of the ulnar side of the hand.

The *Abductor Minimi Digiti* is situated on the ulnar border of the palm of the hand. It arises from the pisiform bone, and from an expansion of the tendon of the Flexor carpi ulnaris; and terminates in a flat tendon, which is inserted into the ulnar side of the base of the first phalanx of the little finger.

Relations. By its *superficial surface*, with the inner portion of the palmar fascia, and the Palmaris brevis. By its *deep surface*, with the Flexor ossis metacarpi. By its *inner border*, with the Flexor brevis minimi digiti.

The *Flexor Brevis Minimi Digiti* lies on the same plane as the preceding muscle, on its radial side. It arises from the tip of the unciform process of the unciform bone, and anterior surface of the annular ligament, and is inserted into the base of the first phalanx of the little finger, with the preceding. It is separated from the Abductor at its origin, by the deep branches of the ulnar artery and nerve. This muscle is sometimes wanting: the Abductor is then, usually, of large size.

Relations. By its *superficial surface*, with the internal portion of the palmar fascia, and the Palmaris brevis. By its *deep surface*, with the Opponens.

The *Opponens Minimi Digiti* (fig. 173), is of a triangular form, and placed immediately beneath the preceding muscles. It arises from the unciform process of the unciform bone, and contiguous portion of the annular ligament; its fibres pass downwards and inwards, to be inserted into the whole length of the metacarpal bone of the little finger, along its ulnar margin.

Relations. By its *superficial surface*, with the Flexor brevis, and Abductor minimi digiti. By its *deep surface*, with the interossei muscles in the fourth metacarpal space, the metacarpal bone, and the Flexor tendons of the little finger.

Nerves. All the muscles of this group are supplied by the ulnar nerve.

Actions. The actions of the muscles of the little finger are expressed in their names. The Palmaris brevis corrugates the skin on the inner side of the palm of the hand.

Middle Palmar Region.

Lumbricales. Interossei Palmares.
Interossei Dorsales.

The *Lumbricales* (fig. 177) are four small fleshy fasciculi, accessories to the deep Flexor muscle. They arise by fleshy fibres from the tendons of the deep Flexor: the first and second, from the radial side and palmar surface of the tendons of the index and middle fingers; the third, from the contiguous sides of the tendons of the middle and ring fingers; and the fourth, from the contiguous sides of the tendons of the ring and little fingers. They pass forwards to the radial side of the corresponding fingers, and opposite the metacarpo-phalangeal articulation each tendon terminates in a broad aponeurosis, which is inserted into the tendinous expansion from the Extensor communis digitorum, covering the dorsal aspect of each finger.

178.—The Dorsal Interossei of Left Hand.

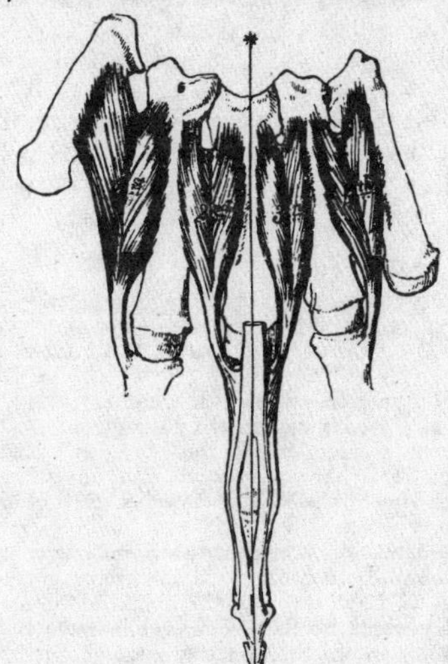

The *Interossei Muscles* are so named from occupying the intervals between the metacarpal bones. They are divided into two sets, a dorsal and palmar; the former are four in number, one in each metacarpal space; the latter, three in number, lie upon the metacarpal bones.

The *Dorsal Interossei* are four in number, larger than the palmar, and occupy the intervals between the metacarpal bones. They are bipenniform muscles, arising by two heads from the adjacent sides of the metacarpal bones, but more extensively from that side of the metacarpal bone which corresponds to the side of the finger in which the muscle is inserted. They are inserted into the base of the first phalanges and into the aponeurosis of the common Extensor tendon. Between the double origin of each of these muscles is a narrow triangular interval, through which passes a perforating branch from the deep palmar arch.

The *First Dorsal Interosseous* muscle, or Abductor indicis, is larger than the others. It is flat, triangular in form, and arises by two heads, separated by a fibrous arch, for the passage of the radial artery from the dorsum to the palm of the hand. The outer head arises from the upper half of the ulnar border of the first metacarpal bone; the inner head, from almost the entire length of the radial border of the second metacarpal bone; the tendon is inserted into the radial side of the index finger. The second and third dorsal interossei are inserted into the middle finger, the former into its radial, the latter into its ulnar side. The fourth is inserted into the ulnar side of the ring-finger.

The *Palmar Interossei*, three in number, are smaller than the Dorsal, and placed upon the palmar surface of the metacarpal bones, rather than between them. They arise from the entire length of the metacarpal bone of one finger, and are inserted into the side of the base of the first phalanx and aponeurotic expansion of the common Extensor tendon of the same finger.

The first arises from the ulnar side of the second metacarpal bone, and is inserted into the same side of the index finger. The second arises from the radial side of the fourth metacarpal bone, and is inserted into the same side of the ring-finger. The third arises from the radial side of the fifth metacarpal bone, and is inserted into the same side of the little finger. From this account it may be seen, that each finger is provided with two Interossei muscles, with the exception of the little finger, in which the Abductor muscle takes the place of one of the pair.

179.—The Palmar Interossei of Left Hand.

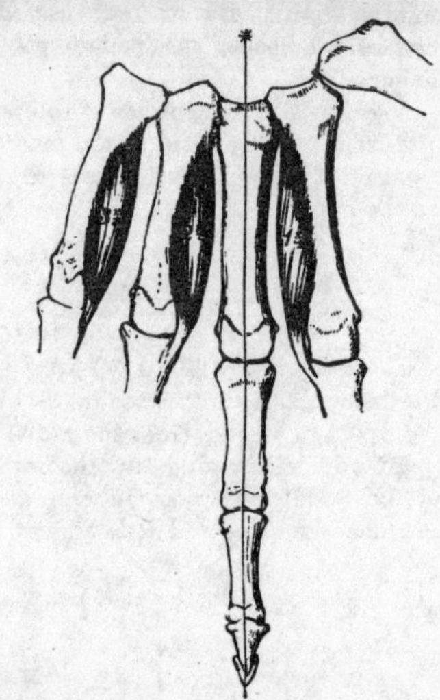

Nerves. The two outer Lumbricales are supplied by the median nerve; the rest of the muscles of this group, by the ulnar.

Actions. The Dorsal interossei muscles abduct the fingers from an imaginary line drawn longitudinally through the centre of the middle finger; and the Palmar interossei adduct the fingers towards that line. They usually assist the Extensor muscles; but when the fingers are slightly bent, they assist in flexing them.*

SURGICAL ANATOMY.

The Student, having completed the dissection of the muscles of the upper extremity, should consider the effects likely to be produced by the action of the various muscles in fracture of the bones.

In considering the actions of the various muscles upon fractures of the upper extremity, I have selected the most common forms of injury, both for illustration and description.

Fracture of the *clavicle* is an exceedingly common accident, and is usually caused by indirect violence, as a fall upon the shoulder; it occasionally, however, occurs from direct force. Its most usual situation is just external to the centre of the bone, but it may occur at the sternal or acromial end.

Fracture of the *middle of the clavicle* (fig. 180) is always attended with considerable displacement, the outer fragment being drawn downwards, forwards, and inwards; the inner fragment slightly upwards. The outer fragment is drawn down by the weight of the arm, and the action of the Deltoid, and forwards and inwards by the Pectoralis minor and Subclavius muscles: the inner fragment is slightly raised by the Sterno-cleido-mastoid, but only to a very limited extent, as the attachment of the costo-clavicular ligament and Pectoralis major below and in front would prevent any very great displacement upwards. The causes of displacement having been ascertained, it is easy to apply the appropriate treatment. The outer fragment is to be drawn outwards, and together with the scapula, raised upwards to a level with the inner fragment, and retained in that position.

In fracture of the *acromial end of the clavicle*, between the conoid and trapezoid ligaments,

* M. Duchenne gives a different account of the mechanism of the extension of the fingers and of the action of the interossei muscles from that usually accepted. According to him the extensor communis digitorum acts almost entirely on the first phalanges, extension of the second and third phalanges being effected by the interossei muscles, which also act to a certain extent as flexors of the first phalanges. This action of the interossei is additional to their action in abduction and adduction ('Physiologie des Mouvements,' pp. 261-298). M. Duchenness view of the action of these muscles certainly derives support from the phenomena observed in lead-palsy and from the results of galvanising the common extensor and the interossei, as Dr. W. Ogle has been kind enough to point out to me. Thus also in a case related by Mr. Hutchinson, in which the ulnar nerve had been divided below the part from which the extensor communis was supplied, (and therefore the interossei were paralysed while the extensor acted,) 'the first phalanges were bent backwards on the metacarpal bones' (extended) 'while the fingers were curved into the palm' (second and third phalanges flexed). *London Hospital Reports*, vol. iii. p. 307.

only slight displacement occurs, as these ligaments, from their oblique insertion, serve to hold both portions of the bone in apposition. Fracture, also of the *sternal end*, internal to the costo-clavicular ligament, is attended with only slight displacement, this ligament serving to retain the fragments in close apposition.

180.—Fracture of the Middle of the Clavicle.

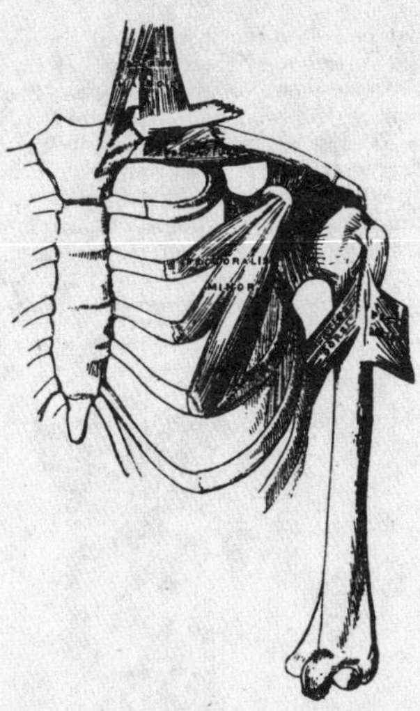

Fracture of the *acromion process* usually arises from violence applied to the upper and outer part of the shoulder; it is generally known by the rotundity of the shoulder being lost, from the Deltoid drawing the fractured portion downwards and forwards; and the displacement may easily be discovered by tracing the margin of the clavicle outwards, when the fragment will be found resting on the front and upper part of the head of the humerus. In order to relax the anterior and outer fibres of the Deltoid (the opposing muscle), the arm should be drawn forwards across the chest, and the elbow well raised, so that the head of the bone may press the acromion process upwards, and retain it in its position.

Fracture of the *coracoid process* is an extremely rare accident, and is usually caused by a sharp blow on the point of the shoulder. Displacement is here produced by the combined actions of the Pectoralis minor, short head of the Biceps, and Coraco-brachialis, the former muscle drawing the fragment inwards, and the latter directly downwards, the amount of displacement being limited by the connection of this process to the acromion by means of the coraco-acromial ligament. In order to relax these muscles and replace the fragments in close apposition, the forearm should be flexed so as to relax the Biceps, and the arm drawn forwards and inwards across the chest so as to relax the Coraco-brachialis; the humerus should then be pushed upwards against the coraco-acromial ligament, and the arm retained in that position.

Fracture of the *anatomical neck of the humerus* within the capsular ligament is a rare accident, attended with very slight displacement, an impaired condition of the motions of the joint, and crepitus.

Fracture of the *surgical neck* (fig. 181) is very common, is attended with considerable displacement, and its appearances correspond somewhat with those of dislocation of the head of the humerus into the axilla. The upper fragment is slightly elevated under the coraco-acromial ligament by the muscles attached to the greater and lesser tuberosities; the lower fragment is drawn inwards by the Pectoralis major, Latissimus dorsi, and Teres major; and the humerus is thrown obliquely outwards from the side by the Deltoid, and occasionally elevated so as to project beneath and in front of the coracoid process. The deformity is reduced by fixing the shoulder, and drawing the arm outwards and downwards. To counteract the opposing muscles, and to keep the fragments in position, the arm should be drawn from the side, and pasteboard splints applied on its four sides, a large conical-shaped pad should be placed in the axilla with the base turned upwards, and the elbow approximated to the side, and retained there by a broad roller passed round the chest; the forearm should then be flexed, and the hand supported in a sling, care being taken not to raise the elbow, otherwise the lower fragment may be displaced upwards.

181.—Fracture of the Surgical Neck of the Humerus.

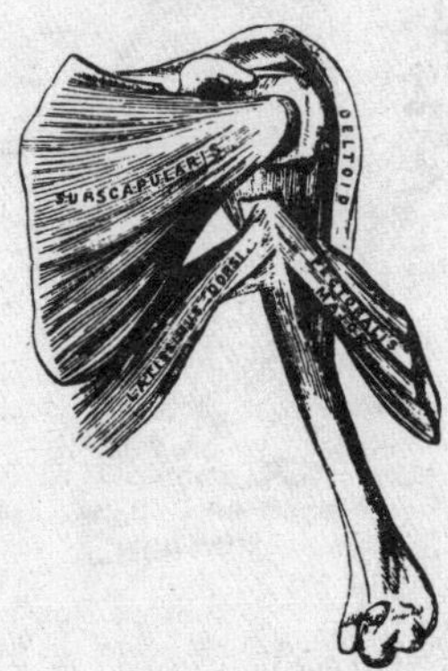

In fracture of the *shaft of the humerus* below the insertion of the Pectoralis major, Latissimus dorsi, and Teres major, and above the insertion of the Deltoid, there is also considerable deformity, the upper fragment being drawn inwards by the first-mentioned muscles, and the lower fragment upwards and outwards by the Deltoid, producing shortening of the limb, and a considerable prominence at the seat of fracture, from the fractured ends of the bone riding over one another, especially if the fracture takes place in an oblique direction. The fragments may be brought into apposition by extension from the elbow, and retained in that position by adopting the same means as in the preceding injury.

In fracture of the *shaft of the humerus* immediately below the insertion of the Deltoid, the amount of deformity depends greatly upon the direction of the fracture. If the fracture occurs in a transverse direction, only slight displacement occurs, the upper fragment being drawn a little forwards; but in oblique fracture, the combined actions of the Biceps and Brachialis anticus muscles in front, and the Triceps behind, draw upwards the lower fragment, causing it to glide over the upper fragment, either backwards or forwards, according to the direction of the fracture. Simple extension reduces the deformity, and the application of splints on the four sides of the arm will retain the fragments in apposition. Care should be taken not to raise the elbow; but the forearm and hand may be supported in a sling.

Fracture of the *humerus* (fig. 182) immediately above the condyles deserves very attentive consideration, as the general appearances correspond somewhat with those produced by separation of the epiphysis of the humerus, and with those of dislocation of the radius and ulna backwards. If the direction of the fracture is oblique from above, downwards, and forwards, the lower fragment is drawn upwards and backwards by the Brachialis anticus and Biceps in front, and the Triceps behind. This injury may be diagnosed from dislocation, by the increased mobility in fracture, the existence of crepitus, and the fact of the deformity being remedied by extension, on the discontinuance of which it is reproduced. The age of the patient is of importance in distinguishing this form of injury from separation of the epiphysis. If fracture occurs in the opposite direction to that shown in the accompanying figure, the lower fragment is drawn upwards and forwards, causing a considerable prominence in front; and the upper fragment projects backwards beneath the tendon of the Triceps muscle.

182.—Fracture of the Humerus above the Condyles.

Fracture of the *coronoid process of the ulna* is an accident of rare occurrence, and is usually caused by violent action of the Brachialis anticus muscle. The amount of displacement varies according to the extent of the fracture. If the tip of the process only is broken off, the fragment is drawn upwards by the Brachialis anticus on a level with the coronoid depression of the humerus, and the power of flexion is partially lost. If the process is broken off near its root, the fragment is still displaced by the same muscle; at the same time, on extending the forearm, partial dislocation backwards of the ulna occurs from the action of the Triceps muscle. The appropriate treatment would be to relax the Brachialis anticus by flexing the forearm, and to retain the fragments in apposition by keeping the arm in this position. Union is generally ligamentous.

183.—Fracture of the Olecranon.

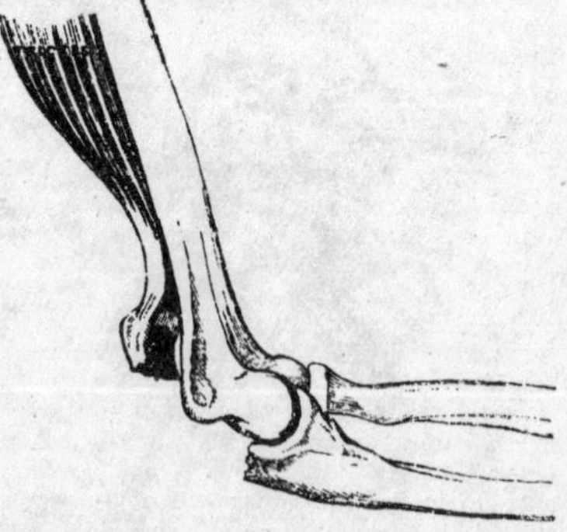

Fracture of the *olecranon process* (fig. 183) is a more frequent accident, and is caused either by violent action of the Triceps muscle, or by a fall or blow upon the point of the elbow. The detached fragment is displaced upwards, by the action of the Triceps muscle, from half an inch to two inches; the prominence of the elbow is consequently lost, and a deep hollow is felt at the back part of the joint, which is much increased on flexing the limb. The patient at the same time loses, more or less, the power of extending the forearm. The treatment consists in relaxing the Triceps by extending the limb, and retaining it in the extended position by means of a long straight splint applied to the front of the arm; the fragments are thus brought into close apposition, and may be further approximated by drawing down the upper fragment. Union is generally ligamentous.

Fracture of the *neck of the radius* is an exceedingly rare accident, and is generally caused by direct violence. Its diagnosis is somewhat obscure, on account of the slight deformity visible: the injured part being surrounded by a large number of muscles; but the movements of pronation and supination are entirely lost. The upper fragment is drawn outwards by the Supinator brevis, its extent of displacement being limited by the attachment of the orbicular ligament. The lower fragment is drawn forwards and slightly upwards by the Biceps, and inwards by the Pronator radii teres, its displacement forwards and upwards being counteracted in some degree by the Supinator brevis. The treatment essentially consists in relaxing the Biceps, Supinator brevis, and Pronator radii teres muscles, by flexing the forearm, and placing it in a position midway between pronation and supination, extension having been previously made so as to bring the parts in apposition.

184.—Fracture of the Shaft of the Radius.

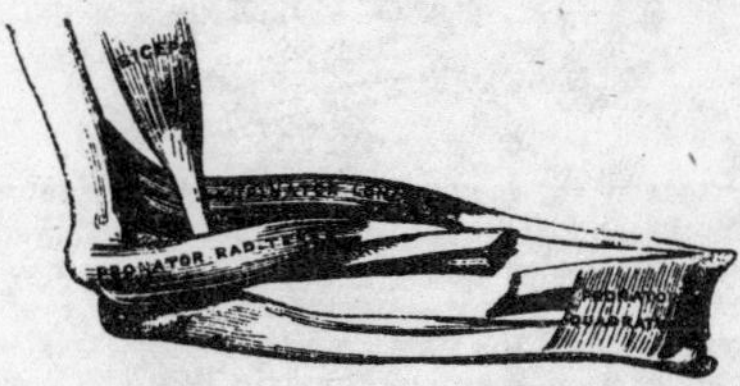

Fracture of the *radius* (fig. 184) is more common than fracture of the ulna, on account of the connection of the former bone with the wrist. Fracture of the shaft of the radius near its centre may occur from direct violence, but more frequently from a fall forwards, the weight of the body being received on the wrist and hand. The upper fragment is drawn upwards by the Biceps, and inwards by the Pronator radii teres, holding a position midway between pronation and supination, and a degree of fulness in the upper half of the forearm is thus produced; the lower fragment is drawn downwards and inwards towards the ulna by the Pronator quadratus, and thrown into a state of pronation by the same muscle; at the same time, the Supinator longus, by elevating the styloid process, into which it is inserted, will serve to depress the upper end of the lower fragment still more towards the ulna. In order to relax the opposing muscles the forearm should be bent, and the limb placed in a position midway between pronation and supination; the fracture is then easily reduced by extension from the wrist and elbow: well-padded splints should then be applied on both sides of the forearm from the elbow to the wrist; the hand being allowed to fall, will, by its own weight, counteract the action of the Pronator quadratus and Supinator longus, and elevate the lower fragment to the level of the upper one.

Fracture of the *shaft of the ulna* is not a common accident; it is usually caused by direct violence. The more protected position of the ulna on the inner side of the limb, the greater strength of its shaft, and its indirect connection with the wrist, render it less liable to injury than the radius. The fracture usually occurs a little below the middle, which is the weakest part of the bone. The upper fragment retains its usual position; but the lower fragment is drawn outwards towards the radius by the Pronator quadratus, producing a well-marked depression at the seat of fracture, and some fulness on the dorsal and palmar surfaces of the forearm. The fracture is easily reduced by extension from the wrist and forearm. The forearm should be flexed, and placed in a position midway between pronation and supination, and well-padded splints applied from the elbow to the ends of the fingers.

Fracture of the *shafts of the radius and ulna together* is not a very common accident; it may arise from a direct blow, or from indirect violence. The lower fragments are drawn upwards, sometimes forwards, sometimes backwards, according to the direction of the fracture, by the combined actions of the Flexor and Extensor muscles, producing a degree of fulness on the dorsal or palmar surface of the forearm; at the same time the two fragments are drawn into contact by the Pronator quadratus, the radius in a state of pronation: the upper fragment of the radius is drawn upwards and inwards by the Biceps and Pronator radii teres to a higher level than the ulna; the upper portion of the ulna is slightly elevated by the Brachialis anticus. The fracture may be reduced by extension from the wrist and elbow, and the forearm should be placed in the same position as in fracture of the ulna.

In the treatment of all cases of fracture of the bones of the forearm, the greatest care is requisite to prevent the ends of the bones from being drawn inwards towards the interosseous space: if this point is not carefully attended to, the radius and ulna may become anchylosed, and the movements of pronation and supination entirely lost. To obviate this, the splints applied to the limb should be well padded, so as to press the muscles down into their normal situation in the interosseous space, and thus prevent the approximation of the fragments.

Fracture of the *lower end of the radius* (fig. 185) is usually called *Colles's fracture*, from the name of the eminent Dublin surgeon who first accurately described it. It is generally produced by the patient falling upon the hand, which receives the entire weight of the body. This fracture usually takes place from half an inch to an inch above the articular surface if it occurs in the adult; but in the child, before the age of sixteen, it is more frequently a separation of the epiphysis from the apophysis. The displacement which is produced is very considerable, and bears some resemblance to dislocation of the carpus backwards, from which it should be carefully distinguished. The lower fragment is drawn

185.—Fracture of the Lower End of the Radius.

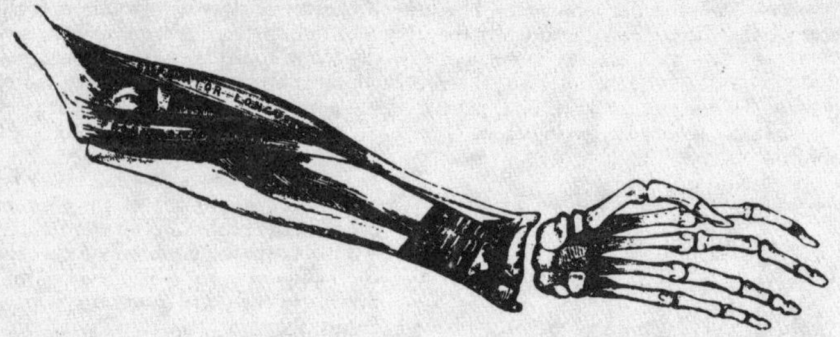

upwards and backwards behind the upper fragment by the combined actions of the Supinator longus and the flexors and the extensors of the thumb and carpus, producing a well-marked prominence on the back of the wrist, with a deep depression above it. The upper fragment projects forwards, often lacerating the substance of the Pronator quadratus, and is drawn by this muscle into close contact with the lower end of the ulna, causing a projection on the anterior surface of the forearm, immediately above the carpus, from the flexor tendons being thrust forwards. This fracture may be distinguished from dislocation by the deformity being removed on making sufficient extension, when crepitus may be occasionally detected; at the same time, on extension being discontinued, the parts immediately resume their deformed appearance. The age of the patient will also assist in determining whether the injury is fracture or separation of the epiphysis. The treatment consists in flexing the forearm, and making powerful extension from the wrist and elbow, depressing at the same time the radial side of the hand, and retaining the parts in that position by well-padded *pistol-shaped* splints.

MUSCLES AND FASCIÆ OF THE LOWER EXTREMITY.

The Muscles of the Lower Extremity are subdivided into groups, corresponding with the different regions of the limb.

Iliac region.

Psoas magnus.
Psoas parvus.
Iliacus.

THIGH.

Anterior femoral region.

Tensor vaginæ femoris.
Sartorius.
Rectus.
Vastus externus.
Vastus internus.
Crureus.
Subcrureus.

Internal femoral region.

Gracilis.
Pectineus.
Adductor longus.
Adductor brevis.
Adductor magnus.

HIP.

Gluteal region.

Gluteus maximus.
Gluteus medius.
Gluteus minimus.
Pyriformis.
Gemellus superior.
Obturator internus.
Gemellus inferior.
Obturator externus.
Quadratus femoris.

Posterior femoral region.

Biceps.
Semitendinosus.
Semimembranosus.

LEG.

Anterior tibio-fibular region.

Tibialis anticus.
Extensor longus digitorum.
Extensor proprius pollicis.
Peroneus tertius.

Posterior tibio-fibular region.

Superficial layer.

Gastrocnemius.
Plantaris.
Soleus.

Deep layer.

Popliteus.
Flexor longus pollicis.
Flexor longus digitorum.
Tibialis posticus.

Fibular region.

Peroneus longus.
Peroneus brevis.

FOOT.

Dorsal region.

Extensor brevis digitorum.

Plantar region.

First layer.

Abductor pollicis.
Flexor brevis digitorum.
Abductor minimi digiti.

Second layer.

Flexor accessorius.
Lumbricales.

Third layer.

Flexor brevis pollicis.
Adductor pollicis.
Flexor brevis minimi digiti.
Transversus pedis.

Fourth layer.

The Interossei.

ILIAC REGION.

Psoas Magnus. Psoas Parvus. Iliacus.

Dissection. No detailed description is required for the dissection of these muscles. On the removal of the viscera from the abdomen, they are exposed, covered by the peritoneum and a thin layer of fascia, the fascia iliaca.

The *iliac fascia* is the aponeurotic layer which lines the back part of the abdominal cavity, and encloses the Psoas and Iliacus muscles throughout their whole extent. It is thin above; and becomes gradually thicker below, as it approaches the femoral arch.

The *portion investing the Psoas*, is attached, above, to the ligamentum arcuatum internum; internally, to the sacrum; and by a series of arched processes to the intervertebral substances, and prominent margins of the bodies of the vertebræ; the intervals so left, opposite the constricted portions of the bodies, transmitting the lumbar arteries and filaments of the sympathetic nerve. Externally, this portion of the iliac fascia is continuous with the fascia lumborum.

The *portion investing the Iliacus* is connected, externally, to the whole length of the inner border of the crest of the ilium; and internally, to the brim of the true pelvis, where it is continuous with the periosteum, and receives the tendon of insertion of the Psoas parvus, when that muscle exists. External to the femoral vessels, this fascia is intimately connected with Poupart's ligament, and is continuous with the fascia transversalis; but, as the femoral vessels pass down into the thigh, it is prolonged down behind them, forming the posterior wall of the femoral sheath. Under the femoral sheath, the iliac fascia surrounds the Psoas and Iliacus muscles to their termination, and becomes continuous with the iliac portion of the fascia lata. Internal to the femoral vessels, the iliac fascia is connected to the iliopectineal line, and is continuous with the pubic portion of the fascia lata. The iliac vessels lie in front of the iliac fascia, but all the branches of the lumbar plexus behind it; it is separated from the peritoneum by a quantity of loose areolar tissue. In abscess accompanying caries of the lower part of the spine, the matter makes its way to the femoral arch, distending the sheath of the Psoas; and when it accumulates in considerable quantity, this muscle becomes absorbed, and the nervous cords contained in it are dissected out, and lie exposed in the cavity of the abscess; the femoral vessels, however, remain intact, and the peritoneum seldom becomes implicated.

Remove this fascia, and the muscles of the iliac region will be exposed.

The *Psoas Magnus* (fig. 187) is a long fusiform muscle, placed on the side of the lumbar region of the spine and margin of the pelvis. It arises from the sides of the bodies, from the corresponding intervertebral substances, and from the front of the bases of the transverse processes of the last dorsal and all the lumbar vertebræ. The muscle is connected to the bodies of the vertebræ by five slips; each slip is attached to the upper and lower margins of two vertebræ, and to the intervertebral substance between them; the slips themselves being connected by the tendinous arches which extend across the constricted part of the bodies, and beneath which pass the lumbar arteries and sympathetic nerves. These tendinous arches also give origin to muscular fibres and protect the blood-vessels and nerves from pressure during the action of the muscle. The first slip is attached to the contiguous margins of the last dorsal and first lumbar vertebræ; the last to the contiguous margins of the fourth and fifth lumbar, and to the intervertebral substance. From these points, the muscle passes down across the brim of the pelvis, and diminishing gradually in size, passes behind Poupart's ligament, and terminates in a tendon, which, after receiving the fibres of the Iliacus, is inserted into the lesser trochanter of the femur.

Relations. In the lumbar region. By its *anterior surface*, which is placed behind the peritoneum, with the ligamentum arcuatum internum, the kidney, Psoas parvus, renal vessels, ureter, spermatic vessels, genito-crural nerve, the colon, and along its pelvic border, with the common and external iliac artery and vein. By its *posterior surface*, with the transverse processes of the lumbar vertebræ and the Quadratus lumborum, from which it is separated by the anterior lamella of the aponeurosis of the Transversalis. The anterior crural nerve is at first situated in the substance of the muscle, and emerges from its outer border at the lower part. The lumbar plexus is situated in the posterior part of the substance of the muscle. By its *inner side*, the muscle is in relation with the bodies of the lumbar vertebræ, the lumbar arteries, the sympathetic ganglia, and their branches of communication with the spinal nerves; the lumbar glands; the vena cava on the right, and the aorta on the left side. In the thigh it is in relation, in front, with the fascia lata; behind, with the capsular ligament of the hip, from which it is separated by a synovial bursa, which sometimes communicates with the cavity of the joint through an opening of variable size; by its *inner border*, with the Pectineus and the femoral artery, which slightly overlaps it; by its *outer border*, with the anterior crural nerve and Iliacus muscle.

The *Psoas Parvus* is a long slender muscle, placed in front of the preceding. It arises from the sides of the bodies of the last dorsal and first lumbar vertebræ, and from the intervertebral substance between them. It forms a small flat muscular bundle, which terminates in a long flat tendon, inserted into the ilio-pectineal eminence, and continuous, by its outer border, with the iliac fascia. This muscle is present, according to M. Theile, in one out of every twenty subjects examined.

Relations. It is covered by the peritoneum, and at its origin by the ligamentum arcuatum internum; it rests on the Psoas magnus.

The *Iliacus* is a flat radiated muscle, which fills up the whole of the internal iliac fossa. It arises from the iliac fossa, and inner margin of the crest of the ilium; behind, from the ilio-lumbar ligament, and base of the sacrum; in front, from the anterior superior and anterior inferior spinous processes of the ilium, from the notch between them, and by a few fibres from the capsule of the hip-joint. The fibres converge to be inserted into the outer side of the tendon of the Psoas, some of them being prolonged into the oblique line which extends from the lesser trochanter to the linea aspera.

Relations. Within the pelvis: by its *anterior surface*, with the iliac fascia, which separates the muscle from the peritoneum, and with the external cutaneous nerve; on the right side, with the cæcum; on the left side, with the sigmoid flexure of the colon. By its *posterior surface*, with the iliac fossa. By its *inner border*, with the Psoas magnus, and anterior crural nerve. In the thigh, it is in relation, by its *anterior surface*, with the fascia lata, Rectus and Sartorius; behind, with the capsule of the hip-joint, a synovial bursa common to it and the Psoas magnus being interposed.

Nerves. The Psoas is supplied by the anterior branches of the lumbar nerves, the Iliacus by the anterior crural.

Actions. The Psoas and Iliacus muscles, acting from above, flex the thigh upon the pelvis, and, at the same time, rotate the femur outwards, from the obliquity of their insertion into the inner and back part of that bone. Acting from below, the femur being fixed, the muscles of both sides bend the lumbar portion of the spine and pelvis forwards. They also serve to maintain the erect position, by supporting the spine and pelvis upon the femur, and assist in raising the trunk when the body is in the recumbent posture.

The *Psoas parvus* is a tensor of the iliac fascia.

Anterior Femoral Region.

Tensor Vaginæ Femoris.	Vastus Externus.
Sartorius.	Vastus Internus.
Rectus.	Crureus.

Subcrureus.

Dissection. To expose the muscles and fasciæ in this region, make an incision along Poupart's ligament, from the spine of the ilium to the pubes, a vertical incision from the centre of this, along the middle line of the thigh to below the knee-joint, and a transverse incision from the inner to the outer side of the leg, at the lower end of the vertical incision. The flaps of integument having been removed, the superficial and deep fasciæ should be examined. The more advanced student should commence the study of this region by an examination of the anatomy of femoral hernia, and Scarpa's triangle, the incisions for the dissection of which are marked out in the accompanying figure.

186.—Dissection of Lower Extremity. Front View.

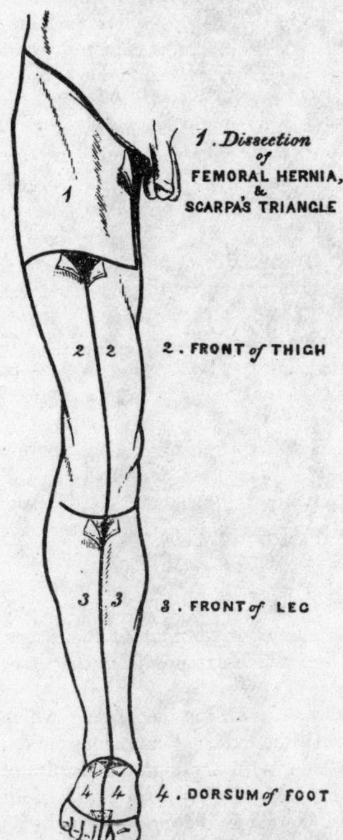

Fasciæ of the Thigh.

The *superficial fascia* forms a continuous layer over the whole of the lower extremity, consisting of areolar tissue, containing in its meshes much adipose matter, and capable of being separated into two or more layers, between which are found the superficial vessels and nerves. It varies in thickness, in different parts of the limb; in the sole of the foot it is so thin as to be scarcely demonstrable, the integument being closely adherent to the deep fascia beneath, but in the groin it is thicker, and the two layers are separated from one another by the superficial inguinal glands, the internal saphenous vein, and several smaller vessels. Of these two layers, the superficial is continuous above with the superficial fascia of the abdomen, the deep layer becoming blended with the fascia lata, a little below Poupart's ligament. The deep layer of superficial fascia is intimately adherent to the margins of the saphenous opening in the fascia lata, and pierced in this situation by numerous small blood and lymphatic vessels: hence the name *cribriform fascia,* which has been applied to it. Subcutaneous bursæ are found in the superficial fascia over the patella, point of the heel, and phalangeal articulations of the toes.

The *deep fascia* of the thigh is exposed on the removal of the superficial fascia, and is named, from its great extent, the fascia lata; it forms a uniform investment for the whole of this region of the limb, but varies in thickness in different parts; thus, it is thickest in the upper and outer part of the thigh, where it receives a fibrous expansion from the Gluteus maximus muscle, and the Tensor vaginæ femoris is inserted between its layers: it is very thin behind, and at the upper and inner part, where it covers the Adductor muscles, and again becomes stronger around the knee, receiving fibrous expansions from the tendon of the Biceps externally, and from the Sartorius, Gracilis, Semitendinosus, and Quadriceps extensor cruris in front. The fascia lata is attached, above, to Poupart's ligament, and the crest of the ilium; behind, to the margin of the sacrum and coccyx; internally, to the pubic arch, and pectineal line; and below, to all the prominent points around the knee-joint, the condyles of the femur, tuberosities of the tibia, and head of the fibula. That portion which invests the Gluteus medius (the Gluteal aponeurosis) is very thick and strong, and gives origin, by its inner surface, to some of the fibres of that muscle; at the upper border of the Gluteus maximus, it divides into two layers, the upper of which, very thin, covers the surface of the Gluteus maximus, and is continuous below with the fascia lata: the deep layer is thick above, where it blends with the great sacro-sciatic ligament, thin below, where it separates the Gluteus maximus from the deeper muscles. From the inner surface of the fascia lata are given off two strong intermuscular septa, which are attached to the whole length of the linea aspera; the external and stronger one, which extends from the insertion of the Gluteus maximus to the outer condyle, separates the Vastus externus in front from the short head of the Biceps behind, and gives partial origin to those muscles; the inner one, the thinner of the two, separates the Vastus internus from the Adductor muscles.

187.—Muscles of the Iliac and Anterior Femoral Regions.

Besides these, there are numerous smaller septa, separating the individual muscles, and enclosing each in a distinct sheath. At the upper and inner part of the thigh, a little below Poupart's ligament, a large oval-shaped aperture is observed after the superficial fascia has been cleared off: it transmits the internal saphenous vein, and other smaller vessels, and is termed the *saphenous opening*. In order more correctly to consider the mode of formation of this aperture, the fascia lata is described as consisting, in this part of the thigh, of two portions, an iliac portion, and a pubic portion.

The *iliac portion* is all that part of the fascia lata on the outer side of the saphenous opening. It is attached, externally, to the crest of the ilium, and its anterior superior spine, to the whole length of Poupart's ligament, as far internally as the spine of the pubes, and to the pectineal line in conjunction with Gimbernat's ligament. From the spine of the pubes, it is reflected downwards and outwards, forming an arched margin, the superior cornu, or outer boundary of the saphenous opening; this margin overlies, and is adherent to the anterior layer of the sheath of the femoral vessels; to its edge is attached the cribriform fascia, and, below, it is continuous with the pubic portion of the fascia lata.

The *pubic portion* is situated at the inner side of the saphenous opening; at the lower margin of this aperture it is continuous with the iliac portion; traced upwards, it is seen to cover the surface of the Pectineus muscle, and passing behind the sheath of the femoral vessels, to which it is closely united, is continuous with the sheath of the Psoas and Iliacus muscles, and is finally lost in the fibrous capsule of the hip-joint. This fascia is attached above to the pectineal line in front of the insertion of the aponeurosis of the External oblique, and internally to the margin of the pubic arch. From this description it may be observed, that the iliac portion of the fascia lata passes in front of the femoral vessels, and the pubic portion behind them, so that an apparent aperture exists between the two, through which the internal saphenous joins the femoral vein.*

The fascia should now be removed from the surface of the muscles. This may be effected by pinching it up between the forceps, dividing it, and separating it from each muscle in the course of its fibres.

The *Tensor Vaginæ Femoris* is a short flat muscle, situated at the upper and outer side of the thigh. It arises from the anterior part of the outer lip of the crest of the ilium, and from the outer surface of the anterior superior spinous process, between the Gluteus medius and Sartorius. The muscle passes obliquely downwards, and a little backwards, to be inserted into the fascia lata, about one-fourth down the outer side of the thigh.

Relations. By its *superficial surface*, with the fascia lata and the integument. By its *deep surface*, with the Gluteus medius, Rectus femoris, Vastus externus, and the ascending branches of the external circumflex artery. By its *anterior border*, with the Sartorius, from which it is separated below by a triangular space, in which is seen the Rectus femoris. By its *posterior border*, with the Gluteus medius.

The *Sartorius*, the longest muscle in the body, is flat, narrow, and riband-like: it arises by tendinous fibres from the anterior superior spinous process of the ilium and upper half of the notch below it, passes obliquely across the upper and anterior part of the thigh, from the outer to the inner side of the limb, then descends vertically, as far as the inner side of the knee, passing behind the inner condyle of the femur, and terminates in a tendon, which curving obliquely forwards, expands into a broad aponeurosis, inserted into the upper part of the inner surface of the shaft of the tibia, nearly as far forwards as the crest. This expansion covers the insertion of the tendons of the Gracilis and Semitendinosus, with which it is partially united, a synovial bursa being interposed between them. An offset is derived from this aponeurosis, which blends with the fibrous capsule

* These parts will be again more particularly described with the anatomy of Hernia.

of the knee-joint, and another, given off from its lower border, blends with the fascia on the inner side of the leg. The relations of this muscle to the femoral artery should be carefully examined, as its inner border forms the chief guide in tying the artery. In the upper third of the thigh, it forms, with the Adductor longus, the side of a triangular space, Scarpa's triangle, the base of which, turned upwards, is formed by Poupart's ligament; the femoral artery passes perpendicularly through the middle of this space from its base to its apex. In the middle third of the thigh, the femoral artery lies first along the inner border, and then behind the Sartorius.

Relations. By its *superficial surface*, with the fascia lata and integument. By its *deep surface*, with the Iliacus, Psoas, Rectus, Vastus internus, anterior crural nerve, sheath of the femoral vessels, Adductor longus, Adductor magnus, Gracilis, long saphenous nerve, and internal lateral ligament of the knee-joint.

The *Quadriceps extensor* includes the four remaining muscles on the front of the thigh. It is the great Extensor muscle of the leg, forming a large fleshy mass, which covers the front and sides of the femur, being united below into a single tendon, attached to the tibia, and above subdividing into separate portions, which have received distinct names. Of these, one occupying the middle of the thigh, connected above with the ilium, is called the *Rectus femoris*, from its straight course. The other divisions lie in immediate connection with the shaft of the femur, which they cover from the condyles to the trochanters. The portion on the outer side of the femur is termed the *Vastus externus*; that covering the inner side, the *Vastus internus*; and that covering the front of the femur, the *Crureus*. The two latter portions are, however, so intimately blended, as to form but one muscle.

The *Rectus femoris* is situated in the middle of the anterior region of the thigh; it is fusiform in shape, and its fibres are arranged in a bipenniform manner. It arises by two tendons; one, the straight tendon, from the anterior inferior spinous process of the ilium; the other is flattened, and curves outwards, to be attached to a groove above the brim of the acetabulum; this is the reflected tendon of the Rectus, it unites with the straight tendon at an acute angle, and then spreads into an aponeurosis, from which the muscular fibres arise. The muscle terminates in a broad and thick aponeurosis, which occupies the lower two-thirds of its posterior surface, and gradually becoming narrowed into a flattened tendon, is inserted into the patella in common with the Vasti and Crureus.

Relations. By its *superficial surface*, with the anterior fibres of the Gluteus medius, the Tensor vaginæ femoris, Sartorius, and the Psoas and Iliacus; by its lower three-fourths, with the fascia lata. By its *posterior surface*, with the hip-joint, the external circumflex vessels, and the Crureus and Vasti muscles.

The three remaining muscles have been described collectively by some anatomists, separate from the Rectus, under the name of the *Triceps extensor cruris*; in order to expose them, divide the Sartorius and Rectus across the middle, and turn them aside, when the muscles in question will be fully brought into view.

The *Vastus externus* is the largest part of the Quadriceps extensor. It arises by a broad aponeurosis, which is attached to the anterior border of the great trochanter, to a horizontal ridge on its outer surface, to a rough line leading from the trochanter major to the linea aspera, and to the whole length of the outer lip of the linea aspera: this aponeurosis covers the upper three-fourths of the muscle, and from its inner surface many fibres arise. A few additional fibres arise from the tendon of the Gluteus maximus, and from the external intermuscular septum between the Vastus externus, and short head of the Biceps. The fibres form a large fleshy mass, which is attached to a strong aponeurosis, placed on the under surface of the muscle at its lowest part; this becomes contracted and thickened into a flat tendon, which is inserted into the outer part of the upper border of the patella, blending with the great extensor tendon.

Relations. By its *superficial surface*, with the Rectus, the Tensor vaginæ femoris, the fascia lata, and the Gluteus maximus, from which it is separated by a synovial bursa. By its *deep surface*, with the Crureus, some large branches of the external circumflex artery and anterior crural nerve being interposed.

The *Vastus Internus* and *Crureus* are so inseparably connected together, as to form but one muscle, as which it will be accordingly described. It is the smallest portion of the Quadriceps extensor. The anterior portion of it, covered by the Rectus, is called the Crureus; the internal portion, which lies immediately beneath the fascia lata, the Vastus Internus. It arises by an aponeurosis, which is attached to the lower part of the line that extends from the inner side of the neck of the femur to the linea aspera, from the whole length of the inner lip of the linea aspera, and internal intermuscular septum. It also arises from nearly the whole of the internal, anterior, and external surfaces of the shaft of the femur, limited, above, by the line between the two trochanters, and extending, below, to within the lower fourth of the bone. From these different origins, the fibres converge to a broad aponeurosis, which covers the anterior surface of the middle portion of the muscle (the Crureus), and the deep surface of the inner division of the muscle (the Vastus internus), and which gradually narrows down to its insertion into the patella, where it blends with the other portions of the Quadriceps extensor.

Relations. By its *superficial surface*, with the Psoas and Iliacus, the Rectus, Sartorius, Pectineus, Adductors, and fascia lata, femoral vessels, and saphenous nerve. By its *deep surface*, with the femur, subcrureus, and synovial membrane of the knee-joint.

The student will observe the striking analogy that exists between the Quadriceps extensor and the Triceps muscle in the upper extremity. So close is this similarity, that M. Cruveilhier has described it under the name of the *Triceps femoralis.* Like the Triceps brachialis, it consists of three distinct divisions, or heads; a middle or long head, analogous to the long head of the Triceps, attached to the ilium, and two other portions which may be called the external and internal heads of the Triceps femoralis. These, it will be noticed, are strictly analogous to the outer and inner heads of the Triceps brachialis.

The *tendons* of the different portions of the Quadriceps extensor unite at the lower part of the thigh, so as to form a single strong tendon, which is inserted into the upper part of the patella. More properly, the patella may be regarded as a sesamoid bone, developed in the tendon of the Quadriceps; and the ligamentum patellæ, which is continued from the lower part of the patella to the tuberosity of the tibia, as the proper tendon of insertion of the muscle. A synovial bursa is interposed between the tendon and the upper part of the tuberosity of the tibia. From the tendons corresponding to the Vasti, a fibrous prolongation is derived, which is attached below to the upper extremities of the tibia and fibula, and which serves to protect the knee-joint, being strengthened on its outer side by the fascia lata.

The *Subcrureus* is a small muscle, usually distinct from the Crureus, but occasionally blended with it, which arises from the anterior surface of the lower part of the shaft of the femur, and is inserted into the upper part of the synovial pouch that extends upwards from the knee-joint behind the patella. It sometimes consists of two separate muscular bundles.

Nerves. The Tensor vaginæ femoris is supplied by the superior gluteal nerve; the other muscles of this region, by branches from the anterior crural.

Actions. The Tensor vaginæ femoris is a tensor of the fascia lata; continuing its action, the oblique direction of its fibres enables it to rotate the thigh inwards. In the erect posture, acting from below, it will serve to steady the pelvis upon the head of the femur. The Sartorius flexes the leg upon the thigh, and, continuing to act, flexes the thigh upon the pelvis, at the same time drawing the limb inwards, so as to cross one leg over the other. Taking its fixed point from the leg, it flexes the pelvis upon the thigh, and, if one muscle acts, assists in rotating the pelvis. The Quadriceps extensor extends the leg upon the thigh. Taking its fixed point from the leg, as in standing, this muscle will act upon the femur, supporting it perpendicularly upon the head of the tibia, and thus maintaining the entire weight of the body. The Rectus muscle assists the Psoas and Iliacus, in supporting the pelvis and trunk upon the femur, or in bending it forwards.

Internal Femoral Region.

Gracilis.
Pectineus.
Adductor Longus.
Adductor Brevis.
Adductor Magnus.

Dissection. These muscles are at once exposed by removing the fascia from the fore-part and inner side of the thigh. The limb should be abducted, so as to render the muscles tense, and easier of dissection.

188.—Deep Muscles of the Internal Femoral Region.

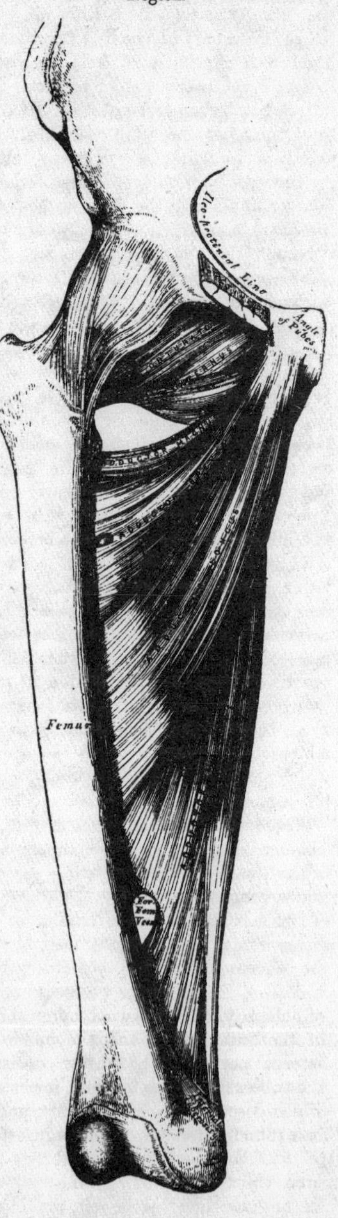

The *Gracilis* (figs. 187, 190) is the most superficial muscle on the inner side of the thigh. It is thin and flattened, broad above, narrow and tapering below. It arises by a thin aponeurosis between two and three inches in breadth, from the inner margin of the ramus of the pubes and ischium. The fibres pass vertically downwards, and terminate in a rounded tendon which passes behind the internal condyle of the femur, and curving round the inner tuberosity of the tibia, becomes flattened, and is inserted into the upper part of the inner surface of the shaft of the tibia, below the tuberosity. The tendon of this muscle is situated immediately above that of the Semitendinosus, and beneath the aponeurosis of the Sartorius, with which it is in part blended. As it passes across the internal lateral ligament of the knee-joint, it is separated from it by a synovial bursa common to it and the Semitendinosus muscle.

Relations. By its *superficial surface,* with the fascia lata and the Sartorius below; the internal saphenous vein crosses it obliquely near its lower part, lying superficial to the fascia lata. By its *deep surface,* with the three Adductors, and the internal lateral ligament of the knee-joint.

The *Pectineus* (fig. 187) is a flat quadrangular muscle, situated at the anterior part of the upper and inner aspect of the thigh. It arises from the linea ilio-pectinea, from the surface of bone in front of it, between the pectineal eminence and spine of the pubes, and from a tendinous prolongation of Gimbernat's ligament, which is attached to the crest of the pubes, and is continuous with the fascia covering the outer surface of the muscle; the fibres pass downwards, backwards, and outwards, to be inserted into a rough line leading from the trochanter minor to the linea aspera.

Relations. By its *anterior surface*, with the pubic portion of the fascia lata, which separates it from the femoral vessels and internal saphenous vein. By its *posterior surface*, with the hip-joint, the Adductor brevis and Obturator externus muscles, the obturator vessels and nerve being interposed. By its *outer border*, with the Psoas, a cellular interval separating them, upon which lies the femoral artery. By its *inner border*, with the margin of the Adductor longus.

The *Adductor Longus*, the most superficial of the three Adductors, is a flat triangular muscle, lying on the same plane as the Pectineus, with which it is often blended above. It arises, by a flat narrow tendon, from the front of the pubes, at the angle of junction of the crest with the symphysis; and soon expands into a broad fleshy belly, which, passing downwards, backwards, and outwards, is inserted, by an aponeurosis, into the middle third of the linea aspera, between the Vastus internus and the Adductor magnus.

Relations. By its *anterior surface*, with the fascia lata, and, near its insertion, with the femoral artery and vein. By its *posterior surface*, with the Adductor brevis and magnus, the anterior branches of the obturator vessels and nerve, and with the profunda artery and vein near its insertion. By its *outer border*, with the Pectineus. By its *inner border*, with the Gracilis.

The Pectineus and Adductor longus should now be divided near their origin, and turned downwards, when the Adductor brevis and Obturator externus will be exposed.

The *Adductor Brevis* is situated immediately behind the two preceding muscles. It is somewhat triangular in form, and arises by a narrow origin from the outer surface of the descending ramus of the pubes, between the Gracilis and Obturator externus. Its fibres, passing backwards, outwards, and downwards, are inserted, by an aponeurosis, into the upper part of the linea aspera, immediately behind the Pectineus and upper part of the Adductor longus.

Relations. By its *anterior surface*, with the Pectineus, Adductor longus, and anterior branches of the obturator vessels and nerve. By its *posterior surface*, with the Adductor magnus, and posterior branches of the obturator vessels and nerve. By its *outer border*, with the Obturator externus, and conjoined tendon of the Psoas and Iliacus. By its *inner border*, with the Gracilis and Adductor magnus. This muscle is pierced, near its insertion, by the middle perforating branch of the profunda artery.

The Adductor brevis should now be cut away near its origin, and turned outwards, when the entire extent of the Adductor magnus will be exposed.

The *Adductor Magnus* is a large triangular muscle, forming a septum between the muscles on the inner, and those on the back of the thigh. It arises from a small part of the descending ramus of the pubes, from the ascending ramus of the ischium, and from the outer margin and under surface of the tuberosity of the ischium. Those fibres which arise from the ramus of the pubes are very short, horizontal in direction, and are inserted into the rough line leading from the great trochanter to the linea aspera, internal to the Gluteus maximus; those from the ramus of the ischium are directed downwards and outwards with different degrees of obliquity, to be inserted, by means of a broad aponeurosis, into the whole length of the linea aspera and the upper part of its internal bifurcation below. The internal portion of the muscle, consisting principally of those fibres which arise from the tuberosity of the ischium, forms a thick fleshy mass consisting of coarse bundles which descend almost vertically, and terminate about the lower third of the thigh in a rounded tendon, which is inserted into the tubercle above the inner condyle of the femur, being connected by a fibrous expansion to the line leading upwards from the tubercle to the linea aspera. Between the two portions of the muscle an angular interval is left, tendinous in front, fleshy behind, for the passage of the femoral vessel into the popliteal space. The external portion of the muscle is pierced by four apertures: the three superior, for the three superior perforating arteries; the fourth, for the passage of the profunda. This muscle gives off an aponeurosis, which passes in front of the femoral vessels, and joins with the Vastus internus.

Relations. By its *anterior surface*, with the Pectineus, Adductor brevis, Adductor longus and the femoral vessels. By its *posterior surface*, with the great sciatic nerve, the Gluteus maximus, Biceps, Semitendinosus, and Semimembranosus. By its *superior* or *shortest border*, it lies parallel with the Quadratus femoris. By its *internal* or *longest border*, with the Gracilis, Sartorius, and fascia lata. By its *external* or *attached border*, it is inserted into the femur behind the Adductor brevis and Adductor longus, which separate it from the Vastus internus; and in front of the Gluteus Maximus and short head of the Biceps, which separate it from the Vastus externus.

Nerves. All the muscles of this group are supplied by the obturator nerve. The Pectineus receives additional branches from the accessory obturator and anterior crural; and the Adductor magnus an additional branch from the great sciatic.

Actions. The Pectineus and three Adductors adduct the thigh powerfully; they are especially used in horse exercise, the flanks of the horse being grasped between the knees by the action of these muscles. In consequence of the obliquity of their insertion into the linea aspera, they rotate the thigh outwards, assisting the external Rotators, and when the limb has been adducted, they draw it inwards, carrying the thigh across that of the opposite side. The Pectineus and Adductor brevis and longus assist the Psoas and Iliacus in flexing the thigh upon the pelvis. In progression, also, all these muscles assist in drawing forwards the hinder limb. The Gracilis assists the Sartorius in flexing the leg and drawing it inwards; it is also an Adductor of the thigh. If the lower extremities are fixed, these muscles may take their fixed point from below and act upon the pelvis, serving to maintain the body in the erect posture; or, if their action is continued, to flex the pelvis forwards upon the femur.

Gluteal Region.

Gluteus Maximus.
Gluteus Medius.
Gluteus Minimus.
Pyriformis.
Gemellus Superior.
Obturator Internus.
Gemellus Inferior.
Obturator Externus.
Quadratus Femoris.

Dissection (fig. 189). The subject should be turned on its face, a block placed beneath the pelvis to make the buttocks tense, and the limbs allowed to hang over the end of the table, with the foot inverted, and the thigh abducted. Make an incision through the integument along the back part of the crest of the ilium and margin of the sacrum to the tip of the coccyx, and carry a second incision from that point obliquely downwards and outwards to the outer side of the thigh, four inches below the great trochanter. The portion of integument included between these incisions, together with the superficial fascia, is to be removed in the direction shown in the figure, when the Gluteus maximus and the dense fascia covering the Gluteus medius will be exposed.

The *Gluteus Maximus* (fig. 190), the most superficial muscle in the gluteal region, is a very broad and thick fleshy mass, of a quadrilateral shape, which forms the prominence of the nates. Its large size is one of the most characteristic points in the muscular system in man, connected as it is with the power he has of maintaining the trunk in the erect posture. In structure the muscle is remarkably coarse, being made up of muscular fasciculi lying parallel with one another, and collected together into large bundles, separated by deep cellular intervals. It arises from the superior curved line of the ilium, and the portion of bone including the crest, immediately behind it; from the posterior surface of the last piece of the sacrum, the side of the coccyx, and posterior surface of the great sacro-sciatic and posterior sacro-iliac ligaments. The fibres are directed obliquely downwards and outwards; those forming the upper and larger portion of the muscle (after converging somewhat) terminate in a thick tendinous lamina, which passes across the great trochanter, and is inserted into the fascia lata covering the outer side of the thigh, the lower portion of the muscle being inserted into the rough line leading from the great trochanter to the linea aspera between the Vastus externus and Adductor magnus.

189.—Dissection of Lower Extremity. Posterior View.

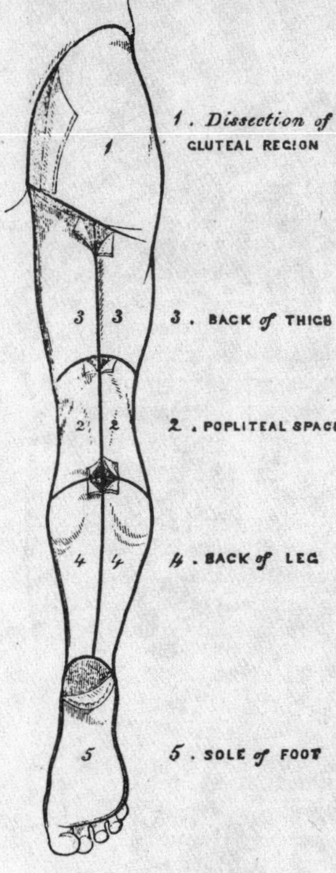

Three *synovial bursæ* are usually found separating the under surface of this muscle from the eminences which it covers. One of these, of large size, and generally multilocular, separates it from the great trochanter. A second, often wanting, is situated on the tuberosity of the ischium. A third is found between the tendon of this muscle and the Vastus externus.

Relations. By its *superficial surface*, with a thin fascia, which separates it from the subcutaneous tissue. By its *deep surface*, from above downwards, with the ilium, sacrum, coccyx, and great sacro-sciatic ligament, part of the Gluteus medius, Pyriformis, Gemelli, Obturator internus, Quadratus femoris, the tuberosity of the ischium, great trochanter, the origin of the Biceps, Semitendinosus, Semimembranosus, and Adductor magnus muscles. The gluteal vessels and superior gluteal nerve are seen issuing from the pelvis above the Pyriformis muscle, the ischiatic and internal pudic vessels and nerves, and the nerve to the Obturator internus muscle below it. Its *upper border* is thin, and connected with the Gluteus medius by the fascia lata. Its *lower border*, free and prominent, forms the fold of the nates, and is directed towards the perineum.

Dissection. Now divide the Gluteus maximus near its origin, by a vertical incision carried from its upper to its lower border: a cellular interval will be exposed, separating it from the Gluteus medius and External rotator muscles beneath. The upper portion of the muscle is to be altogether detached, and the lower portion turned outwards; the loose areolar tissue filling up the interspace between the trochanter major and tuberosity of the ischium being removed, the parts already enumerated as exposed by the removal of this muscle will be seen.

The *Gluteus Medius* is a broad, thick, radiated muscle, situated on the outer surface of the pelvis. Its posterior third is covered by the Gluteus maximus; its anterior two-thirds by the fascia lata, which separates it from the integument. It arises from the outer surface of the ilium, between the superior and middle curved lines, and from the outer lip of that portion of the crest which is between them; it also arises from the dense fascia covering its anterior part. The fibres converge to a strong flattened tendon, which is inserted into the oblique line which traverses the outer surface of the great trochanter. A synovial bursa separates the tendon of the muscle from the surface of the trochanter in front of its insertion.

190.—Muscles of the Hip and Thigh.

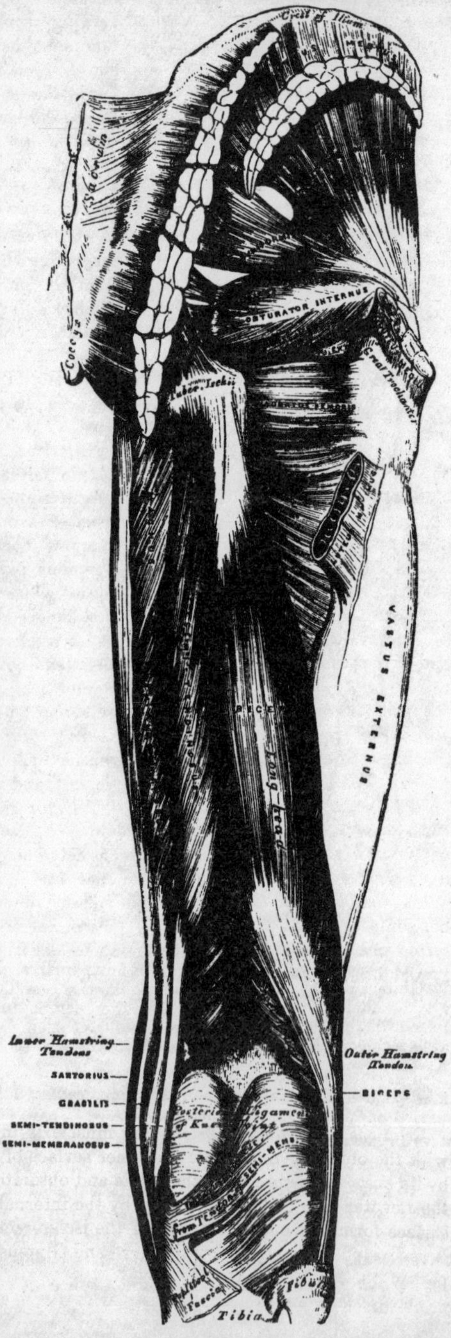

Relations. By its *superficial surface*, with the Gluteus maximus behind, the Tensor vaginæ femoris, and deep fascia in front. By its *deep surface*, with the Gluteus minimus and the gluteal vessels and superior gluteal nerve. Its *anterior border* is blended with the Gluteus minimus. Its *posterior border* lies parallel with the Pyriformis, the gluteal vessels intervening.

This muscle should now be divided near its insertion and turned upwards, when the Gluteus minimus will be exposed.

The *Gluteus Minimus*, the smallest of the three glutei, is placed immediately beneath the preceding. It is fan-shaped, arising from the outer surface of the ilium, between the middle and inferior curved lines, and behind, from the margin of the great sacro-sciatic notch; the fibres converge to the deep surface of a radiated aponeurosis, which, terminating in a tendon, is inserted into an impression on the anterior border of the great trochanter. A synovial bursa is interposed between the tendon and the great trochanter.

Relations. By its *superficial surface*, with the Gluteus medius, and the gluteal vessels and superior gluteal nerve. By its *deep surface*, with the ilium, the reflected tendon of the Rectus femoris, and capsular ligament of the hip-joint. Its *anterior margin* is blended with the Gluteus medius. Its *posterior margin* is often joined with the tendon of the Pyriformis.

The *Pyriformis*, is a flat muscle, pyramidal in shape, lying almost parallel with the lower margin of the Gluteus minimus. It is situated partly within the pelvis at its posterior part, and partly at the back of the hip joint. It arises from the front of the sacrum by three fleshy digitations, attached to the portions of bone between the first, second, third, and fourth anterior sacral foramina, and also from the grooves leading from the foramina: a few fibres also arise from the margin of the great sacro-sciatic foramen, and from the anterior surface of the great sacro-sciatic ligament. The muscle passes out of the pelvis through the great sacro-sciatic foramen, the upper part of which it fills, and is inserted by a rounded tendon into the upper border of the great trochanter, being generally blended with the tendon of the Obturator internus.

Relations. By its *anterior surface, within the pelvis*, with the Rectum (especially on the left side), the sacral plexus of nerves, and the internal iliac vessels; *external to the pelvis*, with the os innominatum and capsular ligament of the hip joint. By its *posterior surface, within the pelvis*, with the sacrum; and *external to it*, with the Gluteus maximus. By its *upper border*, with the Gluteus medius, from which it is separated by the gluteal vessels and superior gluteal nerve. By its *lower border*, with the Gemellus superior and Coccygeus; the sciatic vessels and nerves, the internal pudic vessels and nerve, and the nerve to the Obturator internus, passing from the pelvis in the interval between the two muscles.

Dissection. The next muscle, as well as the origin of the Pyriformis, can only be seen when the pelvis is divided, and the viscera removed.

The *Obturator membrane* is a dense layer of interlacing fibres, which completely closes the obturator foramen, except at its upper and outer part, where a small oval canal is left for the obturator vessels and nerve. Each obturator muscle is connected with this membrane.

The *Obturator Internus*, like the preceding muscle, is situated partly within the cavity of the pelvis, partly at the back of the hip joint. It arises from the inner surface of the anterior and external wall of the pelvis, around the inner side of the obturator foramen, being attached to the descending ramus of the pubes, and the ascending ramus of the ischium, and at the side to the inner surface of the body of the ischium, between the margin of the obturator foramen in front, the great sacro-sciatic notch behind, and the brim of the true pelvis above. It also arises from the inner surface of the obturator membrane and from the tendinous arch which completes the canal for the passage of the obturator vessels and nerve. The fibres are directed backwards and downwards, and terminate in four or five tendinous bands, which are found on its deep surface; these bands are reflected at a right angle over the inner surface of the tuberosity of the ischium, which is grooved for their reception: the groove is covered with cartilage, and lined with a synovial bursa. The muscle leaves the pelvis by the lesser sacro-sciatic notch; and the tendinous bands unite into a single flattened tendon, which passes horizontally outwards, and, after receiving the attachment of the Gemelli, is inserted into the upper border of the great trochanter in front of the Pyriformis. A synovial bursa, narrow and elongated in form, is usually found between the tendon of this muscle and the capsular ligament of the hip: it occasionally communicates with that between the tendon and the tuberosity of the ischium, the two forming a single sac.

In order to display the peculiar appearances presented by the tendon of this muscle, it must be divided near its insertion and reflected outwards.

Relations. Within the pelvis, this muscle is in relation, by its *anterior surface*, with the obturator membrane and inner surface of the anterior wall of the pelvis; by its *posterior surface*, with the pelvic and obturator fasciæ, which separate it from the Levator ani; and it is crossed by the internal pudic vessels and nerve. This surface forms the outer boundary of the ischio-rectal fossa. *External to the pelvis*, it is covered by the great sciatic nerve and Gluteus maximus, and rests on the back part of the hip joint.

The *Gemelli* are two small muscular fasciculi, accessories to the tendon of the

Obturator internus, which is received into a groove between them. They are called *superior* and *inferior*.

The *Gemellus Superior*, the smaller of the two, arises from the outer surface of the spine of the ischium, and passing horizontally outwards becomes blended with the upper part of the tendon of the Obturator internus, and is inserted with it into the upper border of the great trochanter. This muscle is sometimes wanting.

Relations. By its *superficial surface*, with the Gluteus maximus and the sciatic vessels and nerves. By its *deep surface*, with the capsule of the hip joint. By its *upper border*, with the lower margin of the Pyriformis. By its *lower border*, with the tendon of the Obturator internus.

The *Gemellus Inferior* arises from the upper part of the outer border of the tuberosity of the ischium, and passing horizontally outwards, is blended with the lower part of the tendon of the Obturator internus, and inserted with it into the upper border of the great trochanter.

Relations. By its *superficial surface*, with the Gluteus maximus, and the sciatic vessels and nerves. By its *deep surface*, with the capsular ligament of the hip joint. By its *upper border*, with the tendon of the Obturator internus. By its *lower border*, with the tendon of the Obturator externus and Quadratus femoris.

The *Quadratus Femoris* is a short, flat muscle, quadrilateral in shape (hence its name), situated between the Gemellus inferior and the upper margin of the Adductor magnus. It arises from the outer border of the tuberosity of the ischium, and proceeding horizontally outwards, is inserted into the upper part of the linea quadrati, on the posterior surface of the trochanter major. A synovial bursa is often found between the under surface of this muscle and the lesser trochanter, which it covers.

Relations. By its *posterior surface*, with the Gluteus maximus and the sciatic vessels and nerves. By its *anterior surface*, with the tendon of the Obturator externus and trochanter minor, and with the capsule of the hip joint. By its *upper border*, with the Gemellus inferior. Its *lower border* is separated from the Adductor magnus by the terminal branches of the internal circumflex vessels.

Dissection. In order to expose the next muscle (the Obturator externus), it is necessary to remove the Psoas, Iliacus, Pectineus, and Adductor brevis and longus muscles from the front and inner side of the thigh; and the Gluteus maximus and Quadratus femoris from the back part. Its dissection should consequently be postponed until the muscles of the anterior and internal femoral regions have been examined.

The *Obturator Externus* (fig. 188) is a flat triangular muscle, which covers the outer surface of the anterior wall of the pelvis. It arises from the margin of bone immediately around the inner side of the obturator foramen, viz., from the body and ramus of the pubes, and the ramus of the ischium; it also arises from the inner two-thirds of the outer surface of the obturator membrane, and from the tendinous arch which completes the canal for the passage of the obturator vessels and nerves. The fibres converging pass outwards and backwards, and terminate in a tendon which runs across the back part of the hip joint, and is inserted into the digital fossa of the femur.

Relations. By its *anterior surface*, with the Psoas, Iliacus, Pectineus, Adductor longus, Adductor brevis, and Gracilis; and more externally, with the neck of the femur and capsule of the hip joint. By its *posterior surface*, with the obturator membrane and Quadratus femoris.

Nerves. The Gluteus maximus is supplied by the inferior gluteal nerve and a branch from the sacral plexus; the Gluteus medius and minimus, by the superior gluteal; the Pyriformis, Gemelli, Obturator internus, and Quadratus femoris, by branches from the sacral plexus, and the Obturator externus, by the obturator nerve.

Actions. The Glutei muscles, when they take their fixed point from the pelvis, are all abductors of the thigh. The Gluteus maximus and the posterior fibres of the Gluteus medius, rotate the thigh outwards; the anterior fibres of the Gluteus medius and the Gluteus minimus rotate it inwards. The Gluteus maximus serves to extend the femur, and the Gluteus medius and minimus draw it forwards. The Gluteus maximus is also a tensor of the fascia lata. Taking their fixed point from the femur, the Glutei muscles act upon the pelvis, supporting it and the whole trunk upon the head of the femur, which is especially obvious in standing on one leg. In order to gain the erect posture after the effort of stooping, these muscles draw the pelvis backwards, assisted by the Biceps, Semitendinosus, and Semimembranosus muscles. The remaining muscles are powerful rotators of the thigh outwards. In the sitting posture, when the thigh is flexed upon the pelvis, their action as rotators ceases, and they become abductors, with the exception of the Obturator externus, which still rotates the femur outwards. When the femur is fixed, the Pyriformis and Obturator muscles serve to draw the pelvis forwards if it has been inclined backwards, and assist in steadying it upon the head of the femur.

Posterior Femoral Region.

Biceps. Semitendinosus. Semimembranosus.

Dissection (fig. 189). Make a vertical incision along the middle of the thigh, from the lower fold of the nates to about three inches below the back of the knee joint, and there connect it with a transverse incision, carried from the inner to the outer side of the leg. Make a third incision transversely at the junction of the middle with the lower third of the thigh. The integument having been removed from the back of the knee, and the boundaries of the popliteal space examined, the removal of the integument from the remaining part of the thigh should be continued, when the fascia and muscles of this region will be exposed.

The *Biceps* (fig. 190) is a large muscle, of considerable length, situated on the posterior and outer aspect of the thigh. It arises by two heads. One, the long head, arises from an impression at the upper and back part of the tuberosity of the ischium, by a tendon common to it and the Semitendinosus. The femoral, or short head, arises from the outer lip of the linea aspera, between the Adductor magnus and Vastus externus, extending from two inches below the insertion of the Gluteus maximus, to within two inches of the outer condyle; it also arises from the external intermuscular septum. The fibres of the long head form a fusiform belly, which passing obliquely downwards and a little outwards, terminate in an aponeurosis which covers the posterior surface of the muscle, and receives the fibres of the short head; this aponeurosis becomes gradually contracted into a tendon, which is inserted into the outer side of the head of the fibula. At its insertion, the tendon divides into two portions, which embrace the external lateral ligament of the knee joint, a strong prolongation being sent forwards to the outer tuberosity of the tibia, which gives off an expansion to the fascia of the leg. The tendon of this muscle forms the outer ham-string.

Relations. By its *superficial surface* with the Gluteus maximus above, the fascia lata and integument in the rest of its extent. By its *deep surface*, with the Semimembranosus, Adductor magnus, and Vastus externus, the great sciatic nerve, popliteal artery and vein, and near its insertion, with the external head of the Gastrocnemius, Plantaris, the superior external articular artery, and the external Popliteal nerve.

The *Semitendinosus*, remarkable for the great length of its tendon, is situated at the posterior and inner aspect of the thigh. It arises from the tuberosity of the ischium by a tendon common to it and the long head of the Biceps; it also arises from an aponeurosis which connects the adjacent surfaces of the two muscles to the extent of about three inches after their origin. It forms a fusiform muscle, which passing downwards and inwards, terminates a little below the middle of the thigh in a long round tendon which lies along the inner side of the popliteal space, then curves around the inner tuberosity of the tibia, and is inserted into the upper part of the inner surface of the shaft of that bone, nearly as far forwards as its anterior border. This tendon lies beneath the expansion of the Sartorius, and below that of the Gracilis, to which it is united. A tendinous intersection is usually observed about the middle of the muscle.

Relations. By its *superficial surface*, with the Gluteus maximus and fascia lata. By its *deep surface*, with the Semimembranosus, Adductor magnus, inner head of the Gastrocnemius, and internal lateral ligament of the knee joint.

The *Semimembranosus*, so called from the membranous expansion on its anterior and posterior surfaces, is situated at the back part and inner side of the thigh. It arises by a thick tendon from the upper and outer part of the tuberosity of the ischium, above and to the outer side of the Biceps and Semitendinosus, and is inserted into the inner and back part of the inner tuberosity of the tibia, beneath the internal lateral ligament. The tendon of the muscle at its origin expands into an aponeurosis, which covers the upper part of its anterior surface: from this aponeurosis, muscular fibres arise, and converge to another aponeurosis, which covers the lower part of its posterior surface, and this contracts into the tendon of insertion. The tendon of the muscle at its insertion divides into three portions; the middle portion is the fasciculus of insertion into the back part of the inner tuberosity; it sends down an expansion to cover the Popliteus muscle. The internal portion is horizontal, passing forwards beneath the internal lateral ligament, to be inserted into a groove along the inner side of the internal tuberosity. The posterior division passes upwards and backwards, to be inserted into the back part of the outer condyle of the femur, forming the chief part of the posterior ligament of the knee joint.

The tendons of the two preceding muscles, with those of the Gracilis and Sartorius, form the inner ham-string.

Relations. By its *superficial surface*, with the Semitendinosus, Biceps, and fascia lata. By its *deep surface*, with the popliteal vessels, Adductor magnus, and inner head of the Gastrocnemius, from which it is separated by a synovial bursa. By its *inner border*, with the Gracilis. By its *outer border*, with the great sciatic nerve, and its internal popliteal branch.

Nerves. The muscles of this region are supplied by the great sciatic nerve.

Actions. The three ham-string muscles flex the leg upon the thigh. When the knee is semi-fixed, the Biceps, in consequence of its oblique direction downwards and outwards, rotates the leg slightly outwards; and the Semimembranosus, in consequence of its oblique direction, rotates the leg inwards, assisting the Popliteus. Taking their fixed point from below, these muscles serve to support the pelvis upon the head of the femur, and to draw the trunk directly backwards, as in feats of strength, when the body is thrown backwards in the form of an arch.

Surgical Anatomy. The tendons of these muscles occasionally require subcutaneous division in some forms of spurious anchylosis of the knee joint, dependent upon permanent contraction and rigidity of the Flexor muscles, or from stiffening of the ligamentous and other tissues surrounding the joint, the result of disease. This is effected by putting the tendon upon the stretch, and inserting a narrow sharp-pointed knife between it and the skin; the cutting edge being then turned towards the tendon, it should be divided, taking care that the wound in the skin is not at the same time enlarged.

Muscles and Fasciæ of the Leg.

Dissection (fig. 186). The knee should be bent, a block placed beneath it, and the foot kept in an extended position; then make an incision through the integument in the middle line of the leg to the ankle, and continue it along the dorsum of the foot to the toes. Make a second incision transversely across the ankle, and a third in the same direction across the bases of the toes; remove the flaps of integument included between these incisions, in order to examine the deep fascia of the leg.

The *fascia of the leg* forms a complete investment to the whole of this region of the limb, excepting to the inner surface of the tibia. It is continuous above with the fascia lata, receiving an expansion from the tendon of the Biceps on the outer side, and from the tendons of the Sartorius, Gracilis, and Semitendinosus on the inner side; in front it blends with the periosteum covering the tibia and fibula; below, it is continuous with the annular ligaments of the ankle. It is thick and dense in the upper and anterior part of the leg, and gives attach-

191.—Muscles of the Front of the Leg.

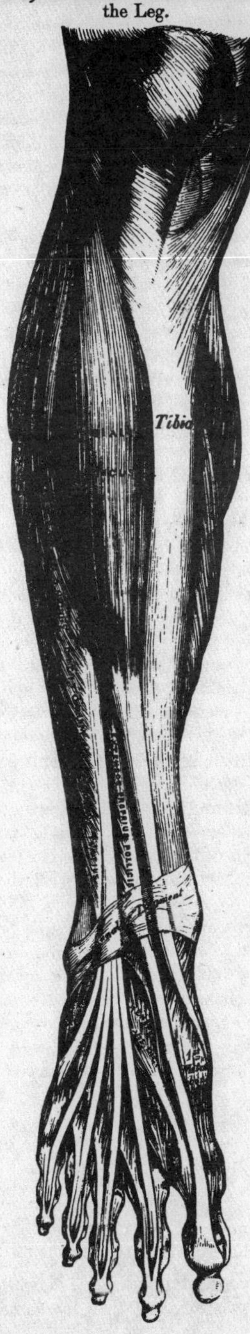

ment, by its inner surface, to the Tibialis anticus and Extensor longus digitorum muscles; but thinner behind, where it covers the Gastrocnemius and Soleus muscles. Its inner surface gives off, on the outer side of the leg, two strong intermuscular septa, which enclose the Peronei muscles, and separate them from the muscles on the anterior and posterior tibial regions, and several smaller and more slender processes, which enclose the individual muscles in each region; at the same time a broad transverse intermuscular septum intervenes between the superficial and deep muscles in the posterior tibio-fibular region.

Now remove the fascia by dividing it in the same direction as the integument, excepting opposite the ankle, where it should be left entire. Commence the removal of the fascia from below, opposite the tendons, and detach it in the line of direction of the muscular fibres.

Muscles of the Leg.

These may be subdivided into three groups: those on the anterior, those on the posterior, and those on the outer side.

Anterior Tibio-Fibular Region.

Tibialis Anticus.
Extensor Proprius Pollicis.
Extensor Longus Digitorum.
Peroneus Tertius.

The *Tibialis Anticus* is situated on the outer side of the tibia; it is thick and fleshy at its upper part, tendinous below. It arises from the outer tuberosity and upper two-thirds of the external surface of the shaft of the tibia; from the adjoining part of the interosseous membrane; from the deep fascia of the leg; and from the intermuscular septum between it and the Extensor longus digitorum: the fibres pass vertically downwards, and terminate in a tendon, which is apparent on the anterior surface of the muscle at the lower third of the leg. After passing through the innermost compartment of the anterior annular ligament, it is inserted into the inner and under surface of the internal cuneiform bone, and base of the metatarsal bone of the great toe.

Relations. By its *anterior surface*, with the deep fascia, and with the annular ligament. By its *posterior surface*, with the interosseous membrane, tibia, ankle joint, and inner side of the tarsus: this surface also overlaps the anterior tibial vessels and nerve in the upper part of the leg. By its *inner surface*, with the tibia. By its *outer surface*, with the Extensor longus digitorum, and Extensor proprius pollicis, and the anterior tibial vessels and nerve.

The *Extensor Proprius Pollicis* is a thin, elongated, and flattened muscle, situated between the Tibialis anticus and Extensor longus digitorum. It arises from the anterior surface of the fibula for about the middle two-fourths of its extent, its origin being internal to that of the Extensor longus digitorum; it also arises from the interosseous membrane to a similar extent. The fibres pass downwards, and terminate in a tendon, which occupies the anterior border of the muscle, passes through a distinct compartment in the annular ligament, crosses the anterior Tibial vessels near the bend of the ankle, and is inserted into the base of the last phalanx of the great toe. Opposite the metatarso-phalangeal articulation, the tendon gives off a thin prolongation on each side, which covers the surface of the joint.

Relations. By its *anterior border*, with the deep fascia, and the anterior annular ligament. By its *posterior border*, with the interosseous membrane, fibula, tibia, ankle joint, and Extensor brevis digitorum. By its *outer side*, with the Extensor longus digitorum above, the dorsalis pedis vessels and anterior tibial nerve below. By its *inner side*, with the Tibialis anticus and the anterior tibial vessels above.

The *Extensor Longus Digitorum* is an elongated, flattened, semipenniform muscle, situated the most externally of all the muscles on the forepart of the leg. It arises from the outer tuberosity of the tibia; from the upper three-fourths of the anterior surface of the shaft of the fibula; from the interosseous membrane, and deep fascia; and from the intermuscular septa between it and the Tibialis anticus on the inner, and the Peronei on the outer side. The muscle terminates in three tendons, which pass through a canal in the annular ligament, with the Peroneus tertius, run across the dorsum of the foot, and are inserted into the second and third phalanges of the four lesser toes, the innermost tendon having subdivided into two. The mode in which the tendons are inserted is the following: Each tendon opposite the metatarso-phalangeal articulation is joined, on its outer side, by the tendon of the Extensor brevis digitorum (except the fourth), and receives a fibrous expansion from the Interossei and Lumbricales; it then spreads into a broad aponeurosis, which covers the dorsal surface of the first phalanx: this aponeurosis, at the articulation of the first with the second phalanx, divides into three slips, a middle one, which is inserted into the base of the second phalanx; and two lateral slips, which, after uniting on the dorsal surface of the second phalanx, are continued onwards, to be inserted into the base of the third.

Relations. By its *anterior surface*, with the deep fascia and the annular ligament. By its *posterior surface*, with the fibula, interosseous membrane, ankle joint, and Extensor brevis digitorum. By its *inner side*, with the Tibialis anticus, Extensor proprius pollicis, and anterior tibial vessels and nerve. By its *outer side*, with the Peroneus longus and brevis.

The *Peroneus Tertius* is a part of the Extensor longus digitorum, and might be described as its fifth tendon. The fibres belonging to this tendon arise from the lower fourth of the anterior surface of the fibula, on its outer side; from the lower part of the interosseous membrane; and from an intermuscular septum between the Peroneus tertius and the Peroneus brevis. The tendon, after passing through the same canal in the annular ligament as the Extensor longus digitorum, is inserted into the base of the metatarsal bone of the little toe, on its dorsal surface. This muscle is sometimes wanting.

Nerves. These muscles are supplied by the anterior tibial nerve.

Actions. The Tibialis anticus and Peroneus tertius are the direct flexors of the tarsus upon the leg; the former muscle, from the obliquity in the direction of its tendon, raises the inner border of the foot; and the latter, acting with the Peroneus brevis and longus, will draw the outer border of the foot upwards, and the sole outwards. The Extensor longus digitorum and Extensor proprius pollicis extend the phalanges of the toes, and, continuing their action, flex the tarsus upon the leg. Taking their fixed point from below, in the erect posture, all these muscles serve to fix the bones of the leg in the perpendicular position, and give increased strength to the ankle joint.

Posterior Tibio-Fibular Region.

Dissection (fig. 189). **Make a vertical incision along the middle line of the back of the leg, from the lower part of the popliteal space to the heel, connecting it below by a transverse incision extending between the two malleoli; the flaps of integument being removed, the fascia and muscles should be examined.**

The muscles in this region of the leg are subdivided into two layers, superficial and deep. The superficial layer constitutes a powerful muscular mass, forming the calf of the leg. Their large size is one of the most characteristic features of the muscular apparatus in man, and bears a direct connection with his ordinary attitude and mode of progression.

192.—Muscles of the Back of the Leg. Superficial Layer.

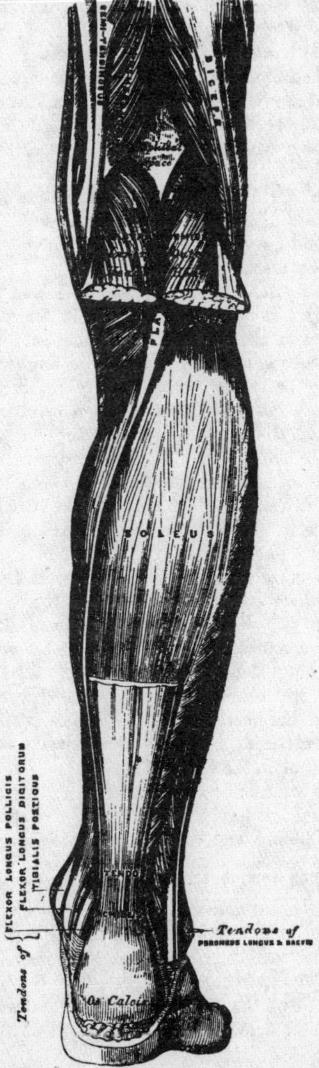

Superficial Layer.

Gastrocnemius. Soleus.
Plantaris.

The *Gastrocnemius* is the most superficial muscle, and forms the greater part of the calf. It arises by two heads, which are connected to the condyles of the femur by two strong flat tendons. The inner head, the larger, and a little the more posterior, arises from a depression at the upper and back part of the inner condyle. The outer head arises from the upper and back part of the external condyle, immediately above the origin of the Popliteus. Both heads, also, arise by a few tendinous and fleshy fibres from the ridges which are continued upwards from the condyles to the linea aspera. Each tendon spreads out into an aponeurosis, which covers the posterior surface of that portion of the muscle to which it belongs; that covering the inner head being longer and thicker than the outer. From the anterior surface of these tendinous expansions, muscular fibres are given off. The fibres in the median line, which correspond to the accessory portions of the muscle derived from the bifurcations of the linea aspera, unite at an angle upon a median tendinous raphé below; the remaining fibres converge to the posterior surface of an aponeurosis which covers the front of the muscle, and this, gradually contracting, unites with the tendon of the Soleus, and forms with it the Tendo Achillis.

Relations. By its *superficial surface*, with the fascia of the leg, which separates it from the external saphenous vein and nerve. By its *deep surface*, with the posterior ligament of the knee joint, the Popliteus, Soleus, Plantaris, popliteal vessels, and internal popliteal nerve. The tendon of the inner head corresponds with the back part of the inner condyle, from which it is separated by a synovial bursa, which, in some cases, communicates with the cavity of the knee joint. The tendon of the outer head contains a sesamoid fibro-cartilage (rarely osseous), where it plays over the corresponding outer condyle; and one is occasionally found in the tendon of the inner head.

The Gastrocnemius should be divided across, just below its origin, and turned downwards, in order to expose the next muscles.

The *Soleus* is a broad flat muscle, situated immediately beneath the preceding. It has received its name from its resemblance in shape to a sole-fish. It arises by tendinous fibres from the back part of the head of the fibula, and from the upper half of the posterior surface of its shaft; from the oblique line of the tibia, and from the middle third of its internal border; some fibres also arise from a tendinous arch placed between the tibial and fibular origins of the muscle, and beneath which the posterior tibial vessels and nerve pass. The fibres pass backwards to an aponeurosis which covers the posterior surface of the muscle, and this, gradually becoming thicker and narrower, joins with the tendon of the Gastrocnemius, and forms with it the Tendo Achillis.

Relations. By its *superficial surface*, with the Gastrocnemius and Plantaris. By its *deep surface*, with the Flexor longus digitorum, Flexor longus pollicis, Tibialis posticus, and posterior tibial vessels and nerve, from which it is separated by the transverse intermuscular septum or deep fascia of the leg.

The *Tendo Achillis*, the common tendon of the Gastrocnemius and Soleus, is the thickest and strongest tendon in the body. It is about six inches in length, and formed by the junction of the aponeuroses of the two preceding muscles. It commences about the middle of the leg, but receives fleshy fibres on its anterior surface, nearly to its lower end. Gradually becoming contracted below, it is inserted into the lower part of the posterior tuberosity of the os calcis, a synovial bursa being interposed between the tendon and the upper part of the tuberosity. The tendon spreads out somewhat at its lower end, so that its narrowest part is usually about an inch and a half above its insertion. The tendon is covered by the fascia and the integument, and is separated from the deep muscles and vessels by a considerable interval filled up with areolar and adipose tissue. Along its outer side, but superficial to it, is the external saphenous vein.

The *Plantaris* is an extremely diminutive muscle, placed between the Gastrocnemius and Soleus, and remarkable for its long and delicate tendon. It arises from the lower part of the outer bifurcation of the linea aspera, and from the posterior ligament of the knee joint. It forms a small fusiform belly, about two inches in length, which terminates in a long slender tendon which crosses obliquely between the two muscles of the calf, and running along the inner border of the Tendo Achillis, is inserted with it into the posterior part of the os calcis. This muscle is occasionally double, and is sometimes wanting. Occasionally, its tendon is lost in the internal annular ligament, or in the fascia of the leg.

Nerves. These muscles are supplied by the internal popliteal nerve.

Actions. The muscles of the calf possess considerable power, and are constantly called into use in standing, walking, dancing, and leaping; hence the large size they usually present. In walking, these muscles draw powerfully upon the os calcis, raising the heel, and, with it, the entire body, from the ground; the body being thus supported on the raised foot, the opposite limb can be carried forwards. In standing, the Soleus, taking its fixed point from below, steadies the leg upon the foot, and prevents the body from falling forwards, to which there is a constant tendency from the superincumbent weight. The Gastrocnemius, acting from below, serves to flex the femur upon the tibia, assisted by the Popliteus. The Plantaris is the rudiment of a large muscle which exists in some of the lower animals, and serves as a tensor of the plantar fascia.

Deep Layer.

Popliteus.	Flexor Longus Digitorum.
Flexor Longus Pollicis.	Tibialis Posticus.

Dissection. Detach the Soleus from its attachment to the fibula and tibia, and turn it downwards when the deep layer of muscles is exposed, covered by the deep fascia of the leg.

193.—Muscles of the Back of the Leg. Deep Layers.

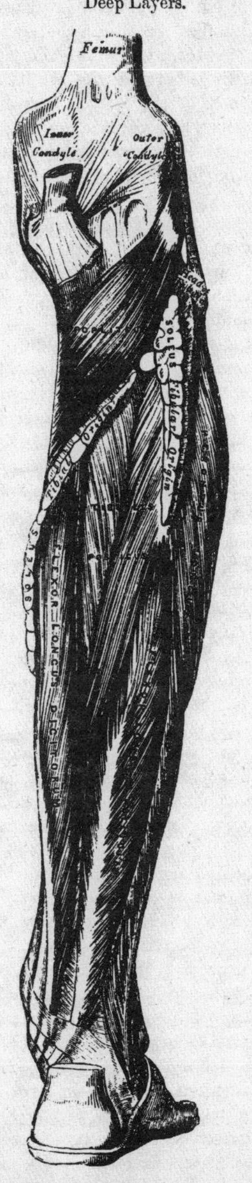

The *deep fascia* of the leg is a broad, transverse, intermuscular septum, interposed between the superficial and deep muscles in the posterior tibio-fibular region. On each side it is connected to the margins of the tibia and fibula. Above, where it covers the Popliteus, it is thick and dense, and receives an expansion from the tendon of the Semimembranosus; it is thinner in the middle of the leg; but below, where it covers the tendons passing behind the malleoli, it is thickened. It is continued onwards in the interval between the ankle and the heel, where it covers the vessels, and is blended with the internal annular ligament.

This fascia should now be removed, commencing from below opposite the tendons, and detaching it from the muscles in the direction of their fibres.

The *Popliteus* is a thin, flat, triangular muscle, which forms part of the floor of the popliteal space, and is covered by a tendinous expansion, derived from the Semimembranosus muscle. It arises by a strong flat tendon, about an inch in length, from a deep depression on the outer side of the external condyle of the femur, and from the posterior ligament of the knee joint; and is inserted into the inner two-thirds of the triangular surface above the oblique line on the posterior surface of the shaft of the tibia, and into the tendinous expansion covering the surface of the muscle. The tendon of the muscle is covered by that of the Biceps and the external lateral ligament of the knee joint; it grooves the outer surface of the external semilunar cartilage, and is invested by the synovial membrane of the knee joint.

Relations. By its *superficial surface*, with the fascia above mentioned, which separates it from the Gastrocnemius, Plantaris, popliteal vessels, and internal popliteal nerve. By its *deep surface*, with the superior tibio-fibular articulation, and back of the tibia.

The *Flexor Longus Pollicis* is situated on the fibular side of the leg, and is the most superficial and largest of the three next muscles. It arises from the lower two-thirds of the internal surface of the shaft of the fibula, with the exception of an inch at its lowest part; from the lower part of the interosseous membrane; from an intermuscular septum between it and the Peronei, externally; and from the fascia covering the Tibialis posticus. The fibres pass obliquely downwards and backwards, and terminate round a tendon which occupies nearly the whole length of the posterior surface of the muscle. This tendon passes through a groove on the posterior surface of the tibia, external to that for the Tibialis posticus and Flexor longus digitorum; it then passes through another groove on the posterior extremity of the astragalus, and along a third groove, beneath the tubercle of the os calcis, into the sole of the foot, where it runs forwards between the two heads of the Flexor brevis pollicis, and is inserted into the base of the last phalanx of the great toe. The grooves in the astragalus and os calcis which contain the tendon of the muscle, are converted by tendinous fibres into distinct canals, lined by synovial membrane; and as the tendon crosses the sole of the foot, it is connected to the common flexor by a tendinous slip.

Relations. By its *superficial surface*, with the Soleus and Tendo Achillis, from which it is separated by the deep fascia. By its *deep surface*, with the fibula, Tibialis posticus, the peroneal vessels, the lower part of the interosseous membrane, and the ankle joint. By its *outer border*, with the Peronei. By its *inner border*, with the Tibialis posticus, and Flexor longus digitorum.

The *Flexor Longus Digitorum* (*perforans*), is situated on the tibial side of the leg. At its origin, it is thin and pointed, but gradually increases in size as it descends. It arises from the posterior surface of the shaft of the tibia, immediately below the oblique line, to within three inches of its extremity, internal to the tibial origin of the Tibialis posticus; some fibres also arise from the intermuscular septum between it and the Tibialis posticus. The fibres terminate in a tendon, which runs nearly the whole length of the posterior surface of the muscle. This tendon passes, behind the inner malleolus, in a groove, common to it and the Tibialis posticus, but separated from the latter by a fibrous septum; each tendon being contained in a special sheath lined by a separate synovial membrane. It then passes obliquely forwards and outwards, beneath the arch of the os calcis, into the sole of the foot (fig. 195), where, crossing beneath the tendon of the Flexor longus pollicis, to which it is connected by a strong tendinous slip, it becomes expanded, is joined by the Flexor accessorius, and finally divides into four tendons which are inserted into the bases of the last phalanges of the four lesser toes, each tendon passing through a fissure in the tendon of the Flexor brevis digitorum, opposite the middle of the first phalanges.

Relations. In the leg: by its *superficial surface*, with the Soleus, and the posterior tibial vessels and nerve, from which it is separated by the deep fascia; by its *deep surface*, with the tibia and Tibialis posticus. *In the foot*, it is covered by the Abductor Pollicis, and Flexor brevis digitorum, and crosses beneath the Flexor longus pollicis.

The *Tibialis Posticus* lies between the two preceding muscles, and is the most deeply seated of all the muscles in the leg. It commences above, by two pointed processes, separated by an angular interval, through which the anterior tibial vessels pass forwards to the front of the leg. It arises from the whole of the posterior surface of the interosseous membrane, excepting its lowest part, from the posterior surface of the shaft of the tibia, external to the Flexor longus digitorum, between the commencement of the oblique line above, and the middle of the external border of the bone below, and from the upper two-thirds of the inner surface of the shaft of the fibula; some fibres also arise from the deep fascia, and from the intermuscular septa, separating it from the adjacent muscles on each side. This muscle, in the lower fourth of the leg, passes in front of the Flexor longus digitorum, terminates in a tendon, which passes through a groove behind the inner malleolus, with the tendon of that muscle, but enclosed in a separate sheath; it then passes through another sheath, over the internal lateral ligament, and beneath the calcaneo-scaphoid articulation, and is inserted into the tuberosity of the scaphoid, and internal cuneiform bones. The tendon of this muscle contains a sesamoid bone, near its insertion, and gives off fibrous expansions, one of which passes backwards to the os calcis, others outwards to the middle and external cuneiform, and some forwards to the bases of the third and fourth metatarsal bones (fig. 196).

Relations. By its *superficial surface*, with the Soleus, and Flexor longus digitorum, the posterior tibial vessels and nerve, and the peroneal vessels, from which it is separated by the deep fascia. By its *deep surface*, with the interosseous ligament, the tibia, fibula, and ankle joint.

Nerves. The Popliteus is supplied by the internal popliteal nerve, the remaining muscles of this group by the posterior tibial nerve.

Actions. The Popliteus assists in flexing the leg upon the thigh; when the leg is flexed, it will rotate the tibia inwards. The Tibialis posticus is a direct extensor of the tarsus upon the leg; acting in conjunction with the Tibialis anticus, it turns the sole of the foot inwards, antagonizing the Peroneus longus, which turns it outwards. The Flexor longus digitorum and Flexor longus pollicis are the direct Flexors of the phalanges, and, continuing their action, extend the foot upon the leg; they assist the Gastrocnemius and Soleus in extending the foot, as in the act of walking, or in standing on tiptoe. In consequence of the oblique direction of the tendon of the long extensor, the toes would be drawn inwards, were it not for the Flexor accessorius muscle, which is inserted into the outer side of that tendon, and draws it to the middle line of the foot during its action. Taking their fixed point from the foot, these muscles serve to maintain the upright posture, by steadying the tibia and fibula, perpendicularly, upon the ankle joint. They also serve to raise these bones from the oblique position they assume in the stooping posture.

Fibular Region.

Peroneus Longus. Peroneus Brevis.

Dissection. These muscles are readily exposed, by removing the fascia covering their surface, from below upwards, in the line of direction of their fibres.

The *Peroneus Longus* is situated at the upper part of the outer side of the leg, and is the more superficial of the two muscles. It arises from the head, and upper two-thirds of the outer surface of the shaft of the fibula, from the deep fascia, and from the intermuscular septa, between it and the muscles on the front, and those on the back of the leg. It terminates in a long tendon, which passes behind the outer malleolus, in a groove, common to it and the Peroneus brevis, the groove being converted into a canal by a fibrous band, and the tendons invested by a common synovial membrane; it is then reflected, obliquely forwards, across the outer side of the os calcis, being contained in a separate fibrous sheath, lined by a prolongation of the synovial membrane from that which lines the groove behind the malleolus. Having reached the outer side of the cuboid bone, it runs in a groove on the under surface of that bone, which is converted into a canal by the long calcaneo-cuboid ligament, and is lined by a synovial membrane: the tendon then crosses, obliquely, the sole of the foot, and is inserted into the outer side of the base of the metatarsal bone of the great toe. The tendon changes its direction at two points, first, behind the external malleolus; secondly, on the outer side of the cuboid bone; in both of these situations, the tendon is thickened, and, in the latter, a sesamoid bone is usually developed in its substance.

Relations. By its *superficial surface*, with the fascia and integument; by its *deep surface*, with the fibula, the Peroneus brevis, os calcis and cuboid bone; by its *anterior border*, with an intermuscular septum, which intervenes between it and the Extensor longus digitorum; by its *posterior border*, with an intermuscular septum, which separates it from the Soleus above, and the Flexor longus pollicis below.

The *Peroneus Brevis* lies beneath the Peroneus longus, and is shorter and smaller than it. It arises from the middle third of the external surface of the shaft of the fibula, internal to the Peroneus longus; from the anterior and posterior borders of the bone; and from the intermuscular septa separating it from the adjacent muscles on the front and back part of the leg. The fibres pass vertically downwards, and terminate in a tendon, which runs in front of that of the preceding muscle through the same groove, behind the external malleolus, being contained in the same fibrous sheath, and lubricated by the same synovial membrane; it then passes through a separate sheath on the outer side of the os calcis, above that for the tendon of the Peroneus longus, and is finally inserted into the base of the metatarsal bone of the little toe, on its dorsal surface.

Relations. By its *superficial surface*, with the Peroneus longus and the fascia of the leg and foot. By its *deep surface*, with the fibula and outer side of the os calcis.

Nerves. The Peroneus longus and brevis are supplied by the musculo-cutaneous branch of the external popliteal nerve.

Actions. The Peroneus longus and brevis extend the foot upon the leg, in conjunction with the Tibialis posticus, antagonizing the Tibialis anticus and Peroneus tertius, which are flexors of the foot. The Peroneus longus also everts the sole of the foot; hence the extreme eversion occasionally observed in fracture of the lower end of the fibula, where that bone offers no resistance to the action of this muscle. Taking their fixed point below, the Peronei serve to steady the leg upon the foot. This is especially the case in standing upon one leg, when the tendency of the superincumbent weight is to throw the leg inwards: the Peroneus longus overcomes this tendency, by drawing on the outer side of the leg, and thus maintains the perpendicular direction of the limb.

Surgical Anatomy. The student should now consider the position of the tendons of the various muscles of the leg, their relation with the ankle joint and surrounding bloodvessels, and especially their action upon the foot, as their rigidity and contraction give rise to one or the other forms of deformity known as *club-foot.* The most simple and common deformity, and one that is rarely if ever congenital, is the *talipes equinus,* the heel being raised by rigidity and contraction of the Gastrocnemius muscle, and the patient walking upon the ball of the foot. In the *talipes varus,* which is the more common congenital form, the heel is raised by the tendo Achillis, the inner border of the foot drawn upwards by the Tibialis anticus, and the anterior two-thirds of the foot twisted inwards by the Tibialis posticus and Flexor longus digitorum, the patient walking upon the outer edge of the foot, and in severe cases upon the dorsum and outer ankle. In the *talipes valgus,* the outer edge of the foot is raised by the Peronei muscles, and the patient walks on the inner ankle. In the *talipes calcaneus* the toes are raised by the extensor muscles, the heel is depressed, and the patient walks upon it. Other varieties of deformity are met with, as the *talipes equino-varus, equino-valgus,* and *calcaneo-valgus,* whose names sufficiently indicate their nature. Each of these deformities may be successfully relieved (after other remedies fail) by division of the opposing tendons and fascia; by this means, the foot regains its proper position, and the tendons heal by the organization of lymph thrown out between the divided ends. The operation is easily performed by putting the contracted tendon upon the stretch, and dividing it by means of a narrow sharp-pointed knife inserted between it and the skin.

Muscles and Fasciæ of the Foot.

The fibrous bands which bind down the tendons in front of and behind the ankle in their passage to the foot, should now be examined; they are termed the *annular ligaments,* and are three in number, anterior, internal, and external.

The *Anterior Annular Ligament* consists of a superior or vertical portion, which binds down the extensor tendons as they descend on the front of the tibia and fibula; and an inferior or horizontal portion, which retains them in connection with the tarsus, the two portions being connected by a thin intervening layer of fascia. The vertical portion is attached externally to the lower end of the fibula, internally to the tibia, and above is continuous with the fascia of the leg; it contains two separate sheaths, one internally, for the tendon of the Tibialis anticus; one externally, for the tendons of the Extensor longus digitorum and Peroneus tertius, the tendon of the Extensor proprius pollicis, and the anterior tibial vessels and nerve pass beneath it, but without any distinct sheath. The horizontal portion is attached externally to the upper surface of the os calcis, in front of the depression for the interosseous ligament, and internally to the inner malleolus and plantar fascia: it contains three sheaths; the most internal for the tendon of the Tibialis anticus, the next in order for the tendon of the Extensor proprius pollicis, and the most external for the Extensor longus digitorum and Peroneus tertius: the anterior tibial vessels and nerve lie altogether beneath it. These sheaths are lined by separate synovial membranes.

The *Internal Annular Ligament* is a strong fibrous band, which extends from the inner malleolus above, to the internal margin of the os calcis below, converting a series of bony grooves in this situation into osseo-fibrous canals, for the passage of the tendons of the flexor muscles and vessels into the sole of the foot. It is continuous above with the deep fascia of the leg, below with the plantar fascia and the fibres of origin of the Abductor pollicis muscle. The three canals which it forms, transmit from within outwards, first, the tendon of the Tibialis posticus; second, the tendon of the Flexor longus digitorum, then the posterior tibial vessels and nerve, which run through a broad space beneath the ligament; lastly, in a canal formed partly by the astragalus, the tendon of the Flexor longus pollicis. Each of these canals is lined by a separate synovial membrane.

The *External Annular Ligament* extends from the extremity of the outer malleolus to the outer surface of the os calcis: it binds down the tendons of the Peronei muscles in their passage beneath the outer ankle. The two tendons are enclosed in one synovial sac.

Dissection of the Sole of the Foot. The foot should be placed on a high block with the sole uppermost, and firmly secured in that position. Carry an incision round the heel and along the inner and outer borders of the foot to the great and little toes. This incision should divide the integument and thick layer of granular fat beneath, until the fascia is visible; the skin and fat should then be removed from the fascia in a direction from behind forwards, as seen in fig. 189.

The *Plantar Fascia*, the densest of all the fibrous membranes, is of great strength, and consists of dense pearly-white glistening fibres, disposed, for the most part, longitudinally: it is divided into a central and two lateral portions.

The *central portion*, the thickest, is narrow behind and attached to the inner tuberosity of the os calcis, behind the origin of the Flexor brevis digitorum, and becoming broader and thinner in front, divides opposite the middle of the metatarsal bones into five processes, one for each of the toes. Each of these processes divides opposite the metatarso-phalangeal articulation into two slips, which embrace the sides of the flexor tendons of the toes, and are inserted into the sides of the metatarsal bones, and into the transverse metatarsal ligament, thus forming a series of arches through which the tendons of the short and long flexors pass to the toes. The intervals left between the five processes allow the digital vessels and nerves, and the tendons of the Lumbricales and Interossei muscles to become superficial. At the point of division of the fascia into processes and slips, numerous transverse fibres are superadded, which serve to increase the strength of the fascia at this part, by binding the processes together, and connecting them with the integument. The central portion of the plantar fascia is continuous with the lateral portions at each side, and sends upwards into the foot, at their point of junction, two strong vertical intermuscular septa, broader in front than behind, which separate the middle from the external and internal plantar group of muscles; from these again thinner transverse septa are derived, which separate the various layers of muscles in this region. The upper surface of this fascia gives attachment behind to the Flexor brevis digitorum muscle.

The *lateral portions* of the plantar fascia are thinner than the central piece and cover the sides of the foot.

The *outer portion* covers the under surface of the Abductor minimi digiti; it is thick behind, thin in front, and extends from the os calcis forwards to the base of the fifth metatarsal bone, into the outer side of which it is attached; it is continuous internally with the middle portion of the plantar fascia, and externally with the dorsal fascia.

The *inner portion* is very thin, and covers the Abductor pollicis muscle; it is attached behind to the internal annular ligament, and is continuous around the side of the foot with the dorsal fascia, and externally with the middle portion of the plantar fascia.

Muscles of the Foot.

These are found in two regions: 1. On the dorsum; 2. On the plantar surface.

1. Dorsal Region.

Extensor Brevis Digitorum.

The *Fascia* on the dorsum of the foot is a thin membranous layer, continuous above with the anterior margin of the annular ligament; it becomes gradually lost opposite the heads of the metatarsal bones, and on each side blends with the lateral portions of the plantar fascia; it forms a sheath for the tendons placed on the dorsum of the foot. On the removal of this fascia, the muscles and tendons of the dorsal region of the foot are exposed.

The *Extensor Brevis Digitorum* (fig. 191) is a broad thin muscle, which arises from the outer side of the os calcis, in front of the groove for the Peroneus brevis; from the astragalo-calcanean ligament; and from the horizontal portion of the anterior annular ligament. It passes obliquely across the dorsum of the foot, and terminates in four tendons. The innermost, which is the largest, is inserted into the first phalanx of the great toe; the other three, into the outer sides of the long extensor tendons of the second, third, and fourth toes.

Relations. By its *superficial surface*, with the fascia of the foot, the tendons of the Extensor longus digitorum, and Extensor proprius pollicis. By its *deep surface*, with the tarsal and metatarsal bones, and the Dorsal interossei muscles.

Nerves. It is supplied by the anterior tibial nerve.

Actions. The Extensor brevis digitorum is an accessory to the long Extensor, extending the phalanges of the four inner toes, but acting only on the first phalanx of the great toe. The obliquity of its direction counteracts the oblique movement given to the toes by the long Extensor, so that, both muscles acting together, the toes are evenly extended.

2. Plantar Region.

The muscles in the plantar region of the foot may be divided into three groups, in a similar manner to those in the hand. Those of the internal plantar region are connected with the great toe, and correspond with those of the thumb; those of the external plantar region, are connected with the little toe, and correspond with those of the little finger; and those of the middle plantar region, are connected with the tendons intervening between the two former groups; but in order to facilitate the dissection of these muscles, it will be found more convenient to divide them into four layers, as they present themselves, in the order in which they are successively exposed.

First Layer.

Abductor Pollicis. Flexor Brevis Digitorum.
Abductor Minimi Digiti.

Dissection. Remove the fascia on the inner and outer sides of the foot, commencing in front over the tendons, and proceeding backwards. The central portion should be divided transversely in the middle of the foot, and the two flaps dissected forwards and backwards.

The *Abductor Pollicis* lies along the inner border of the foot. It arises from the inner tuberosity on the under surface of the os calcis; from the internal annular ligament; from the plantar fascia; and from the intermuscular septum between it and the Flexor brevis digitorum. The fibres terminate in a tendon, which is inserted, together with the innermost tendon of the Flexor brevis pollicis, into the inner side of the base of the first phalanx of the great toe. It is supplied by the internal plantar nerve.

Relations. By its *superficial surface*, with the plantar fascia. By its *deep surface*, with the Flexor brevis pollicis, the Flexor accessorius, and the tendons of the Flexor longus digitorum and Flexor longus pollicis, the Tibialis anticus and posticus, the plantar vessels and nerves, and the articulations of the tarsus.

194.—Muscles of the Sole of the Foot. First Layer.

The *Flexor Brevis Digitorum* (*perforatus*) lies in the middle of the sole of the foot, immediately beneath the plantar fascia, with which it is firmly united. It arises by a narrow tendinous process, from the inner tubercle of the os calcis; from the central part of the plantar fascia; and from the intermuscular septa between it and the adjacent muscles. It passes forwards, and divides into four tendons. Opposite the middle of the first phalanges each tendon presents a longitudinal slit, to allow of the passage of the corresponding tendon of the Flexor longus digitorum; the two portions form a groove for the reception of that tendon. The tendon of the short Flexor then reunites and immediately divides a second time into two processes, which are inserted into the sides of the second phalanges. The mode of division of the tendons of the Flexor brevis digitorum, and their insertion into the phalanges, is analogous to the Flexor sublimis in the hand. It is supplied by the internal plantar nerve.

Relations. By its *superficial surface*, with the plantar fascia. By its *deep surface*, with the Flexor accessorius, the Lumbricales, the tendons of the Flexor longus digitorum, and the external plantar vessels and nerve, from which it is separated by a thin layer of fascia. The *outer* and *inner borders* are separated from the adjacent muscles by means of vertical prolongations of the plantar fascia.

The *Abductor Minimi Digiti* lies along the outer border of the foot. It arises, by a very broad origin, from the outer tuberosity of the os calcis, from the under surface of the os calcis in front of the tubercles, from the plantar fascia, and the intermuscular septum between it and the Flexor brevis digitorum. Its tendon, after gliding over a smooth facet on the under surface of the base of the fifth metatarsal bone, is inserted with the short flexor of the little toe into the outer side of the base of the first phalanx of the little toe. It is supplied by the external plantar nerve.

Relations. By its *superficial surface*, with the plantar fascia. By its *deep surface*, with the Flexor accessorius, the Flexor brevis minimi digiti, the long plantar ligament, and Peroneus longus. On its *inner side* are the external plantar vessels and nerve, and it is separated from the Flexor brevis digitorum by a vertical septum of fascia.

Dissection. The muscles of the superficial layer should be divided at their origin, by inserting the knife beneath each, and cutting obliquely backwards, so as to detach them from the bone; they should then be drawn forwards, in order to expose the second layer, but not cut away at their insertion. The two layers are separated by a thin membrane, the deep plantar fascia, on the removal of which is seen the tendon of the Flexor longus digitorum, the Flexor accessorius, the Flexor longus pollicis, and the Lumbricales. The long flexor tendons cross each other at an acute angle, the Flexor longus pollicis running along the inner side of the foot, on a plane superior to that of the Flexor longus digitorum, the direction of which is obliquely outwards.

195.—Muscles of the Sole of the Foot. Second Layer.

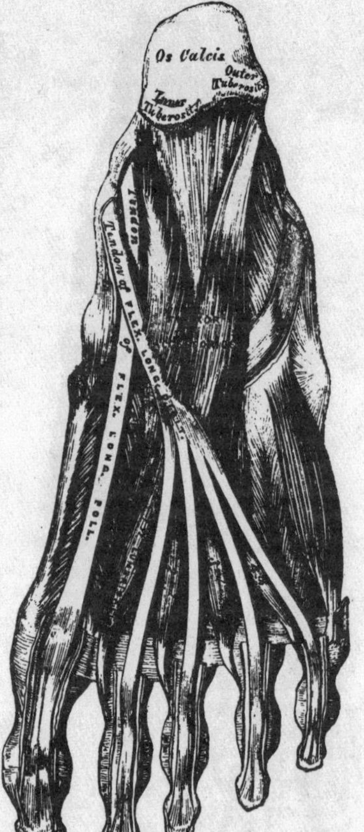

Second Layer.

Flexor Accessorius.
Lumbricales.

The *Flexor Accessorius* arises by two heads: the inner or larger, which is muscular, being attached to the inner concave surface of the os calcis, and to the calcaneo-scaphoid ligament; the outer head, flat and tendinous, to the under surface of the os calcis, in front of its outer tubercle, and to the long plantar ligament: the two portions join at an acute angle, and are inserted into the outer margin and upper and under surfaces of the tendon of the Flexor longus digitorum, forming a kind of groove, in which the tendon is lodged. It is supplied by the external plantar nerve.

Relations. By its *superficial surface*, with the muscles of the superficial layer, from which it is separated by the external plantar vessels and nerves. By its *deep surface*, with the os calcis and long calcaneo-cuboid ligament.

The *Lumbricales* are four small muscles, accessory to the tendons of the Flexor longus digitorum: they arise from the tendons of the long flexor, as far back as their angle of division, each arising from two tendons, except the internal one. Each muscle terminates in a tendon, which passes forwards on the inner side of each of the lesser toes, and is inserted into the expansion of the long extensor and base of the second phalanx of the corresponding toe. The two internal Lumbricales muscles are supplied by the internal, and the two external by the external plantar nerve.

Dissection. The flexor tendons should be divided at the back part of the foot, and the Flexor accessorius at its origin, and drawn forwards, in order to expose the third layer.

Third Layer.

Flexor Brevis Pollicis.	Flexor Brevis Minimi Digiti.
Adductor Pollicis.	Transversus Pedis.

The *Flexor Brevis Pollicis* arises, by a pointed tendinous process, from the inner border of the cuboid bone, from the contiguous portion of the external cuneiform, and from the prolongation of the tendon of the Tibialis posticus, which is attached to that bone. The muscle divides, in front, into two portions, which are inserted into the inner and outer sides of the base of the first phalanx of the great toe, a sesamoid bone being developed in each tendon at its insertion. The inner head of this muscle is blended with the Abductor pollicis previous to its insertion; the outer head, with the Adductor pollicis; and the tendon of the Flexor longus pollicis lies in a groove between them.

196.—Muscles of the Sole of the Foot. Third Layer.

Relations. By its *superficial surface*, with the Abductor pollicis, the tendon of the Flexor longus pollicis and plantar fascia. By its *deep surface*, with the tendon of the Peroneus longus, and metatarsal bone of the great toe. By its *inner border*, with the Abductor pollicis. By its *outer border*, with the Adductor pollicis.

The *Adductor Pollicis* is a large, thick, fleshy mass, passing obliquely across the foot, and occupying the hollow space between the four outer metatarsal bones. It arises from the tarsal extremities of the second, third, and fourth metatarsal bones, and from the sheath of the tendon of the Peroneus longus, and is inserted, together with the outer head of the Flexor brevis pollicis, into the outer side of the base of the first phalanx of the great toe.

The *Flexor Brevis Minimi Digiti* lies on the metatarsal bone of the little toe, and much resembles one of the interossei. It arises from the base of the metatarsal bone of the little toe, and from the sheath of the Peroneus longus; its tendon is inserted into the base of the first phalanx of the little toe, on its outer side.

Relations. By its *superficial surface*, with the plantar fascia and tendon of the Abductor minimi digiti. By its *deep surface*, with the fifth metatarsal bone.

The *Transversus Pedis* is a narrow, flat, muscular fasciculus, stretched transversely across the heads of the metatarsal bones, between them and the flexor tendons. It arises from the under surface of the head of the fifth metatarsal bone, and from the transverse ligament of the metatarsus; and is inserted into the outer side of the first phalanx of the great toe; its fibres being blended with the tendon of insertion of the Adductor pollicis.

Nervous Supply. The Flexor brevis pollicis is supplied by the internal plantar nerve, and sometimes (according to Meckel) receives a branch from the external plantar. The other three muscles of this layer are supplied by the external plantar nerve.

Relations. By its *under surface,* with the tendons of the long and short Flexors and Lumbricales. By its *upper surface,* with the Interossei.

Fourth Layer.

The Interossei.

The Interossei muscles in the foot are similar to those in the hand, with this exception, that they are grouped around the middle line of the second toe, instead of the middle line of the whole member, as in the hand. They are seven in number, and consist of two groups, dorsal and plantar.

The *Dorsal Interossei,* four in number, are situated between the metatarsal bones. They are bipenniform muscles, arising by two heads from the adjacent sides of the metatarsal bones between which they are placed; their tendons are inserted into the bases of the first phalanges, and into the aponeurosis of the common extensor tendon. In the angular interval left between each muscle at its posterior extremity, the perforating arteries pass to the dorsum of the foot; except in the first Interosseous muscle, where the interval allows the passage of the communicating branch of the dorsalis pedis artery. The first Dorsal interosseous muscle is inserted into the inner side of the second toe; the other three are inserted into the outer sides of the second, third, and fourth toes. They are all abductors from the middle line of the second toe.

The *Plantar Interossei,* three in number, lie beneath, rather than between, the metatarsal bones. They are single muscles, and are each connected with but one metatarsal bone. They arise from the base and inner sides of the shaft of the third, fourth, and fifth metatarsal bones, and are inserted into the inner sides of the bases of the first phalanges of the same toes, and into the aponeurosis of the common extensor tendon. These muscles are all adductors towards the middle line of the second toe.

All the Interossei muscles are supplied by the external plantar nerve.

197.—The Dorsal Interossei. Left Foot.

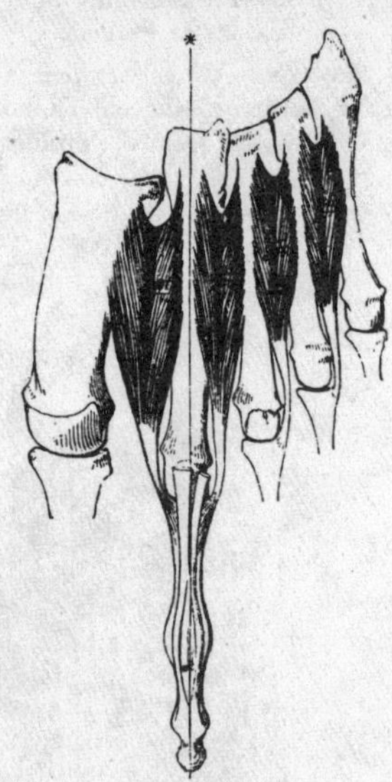

198.—The Plantar Interossei. Left Foot.

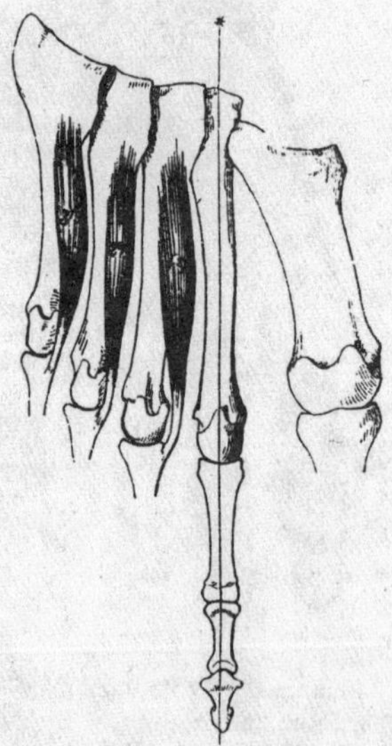

SURGICAL ANATOMY.

The student should now consider the effects produced by the action of the various muscles in fractures of the bones of the lower extremity. The more common forms of fracture are selected for illustration and description.

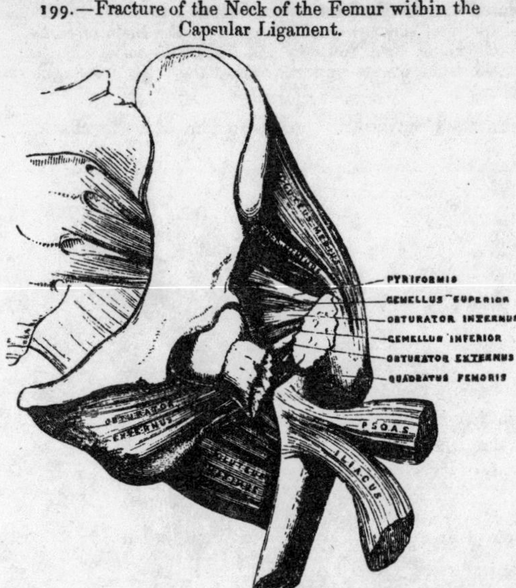

199.—Fracture of the Neck of the Femur within the Capsular Ligament.

Fracture of the *neck of the femur internal to the capsular ligament* (fig. 199) is a very common accident, and is most frequently caused by indirect violence, such as slipping off the edge of the kerbstone, the impetus and weight of the body falling upon the neck of the bone. It usually occurs in females, and seldom under fifty years of age. At this period of life, the cancellous tissue of the neck of the bone not unfrequently is atrophied, becoming soft and infiltrated with fatty matter; the compact tissue is partially absorbed: hence the bone is more brittle, and more liable to fracture. The characteristic marks of this accident are slight shortening of the limb, and eversion of the foot, neither of which symptoms occur, however, in some cases until some time after the injury. The eversion is caused by the combined action of the external rotator muscles, as well as by the Psoas and Iliacus, Pectineus, Adductors, and Glutei muscles. The shortening is produced by the action of the Glutei, and by the Rectus femoris in front, and the Biceps, Semitendinosus, and Semimembranosus behind.

200.—Fracture of the Femur below the Trochanters.

Fracture of the *femur just below the trochanters* (fig. 200) is an accident of not unfrequent occurrence, and is attended with great displacement, producing considerable deformity. The upper fragment, the portion chiefly displaced, is tilted forwards almost at right angles with the pelvis, by the combined action of the Psoas and Iliacus; and, at the same time, everted and drawn outwards by the external rotator and Glutei muscles, causing a marked prominence at the upper and outer side of the thigh, and much pain from the bruising and laceration of the muscles. The limb is shortened, in consequence of the lower fragment being drawn upwards by the Rectus in front, and the Biceps, Semimembranosus, and Semitendinosus behind; and, at the same time, everted, and the upper end thrown outwards, the lower inwards, by the Pectineus and Adductor muscles. This fracture may be reduced in two different methods: either by direct relaxation of all the opposing muscles, to effect which the limb should be placed on a double inclined plane; or by overcoming the contraction of the muscles, by continued extension, which may be effected by means of the long splint.

Oblique fracture of the femur *immediately above the condyles* (fig. 201) is a formidable injury, and attended with considerable displacement. On examination of the limb, the lower fragment may be felt deep in the popliteal space, being drawn backwards by the Gastrocnemius, Soleus, and Plantaris muscles; and upwards by the posterior femoral and Rectus muscles. The pointed end of the upper fragment is drawn inwards by the Pectineus and Adductor muscles, and tilted forwards by the Psoas and Iliacus, piercing the Rectus muscle, and, occasionally, the integument. Relaxation of these muscles, and direct approximation of the broken fragments is effected by placing the limb on a double inclined plane. The greatest care is requisite in keeping the pointed extremity of the upper fragment in proper position; otherwise, after union of the fracture, the power of extension of the limb is partially destroyed, from the Rectus muscle being held down by the fractured end of the bone, and from the patella, when elevated, being drawn upwards against the projecting fragment.

201.—Fracture of the Femur above the Condyles.

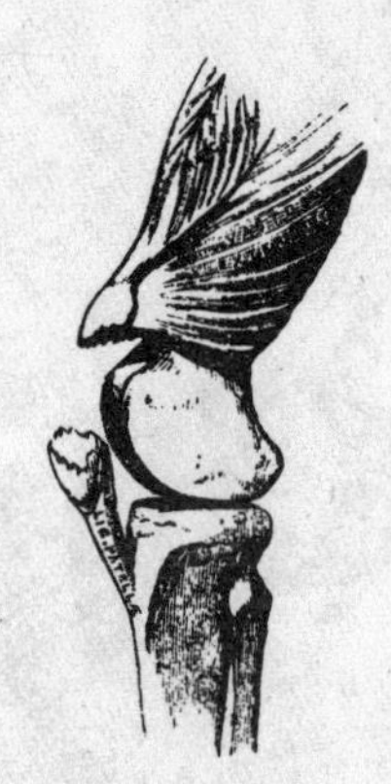

202.—Fracture of the Patella.

Fracture of the *patella* (fig. 202) may be produced by muscular action, or by direct violence. When produced by muscular action, it occurs thus: a person in danger of falling forwards, attempts to recover himself by throwing the body backwards, and the violent action of the Quadriceps extensor upon the patella snaps that bone transversely across. The upper fragment is drawn up the thigh by the Quadriceps extensor, the lower fragment being retained in its position by the ligamentum patellæ; the extent of separation of the two fragments depending upon the degree of laceration of the ligamentous structures around the bone. The patient is totally unable to straighten the limb; the prominence of the patella is lost; and a marked but varying interval can be felt between the fragments. The treatment consists in relaxing the opposing muscles, which may be effected by raising the trunk, and slightly elevating the limb, which should be kept in a straight position. Union is usually ligamentous. In fracture from direct violence, the bone is generally comminuted, or fractured obliquely or perpendicularly.

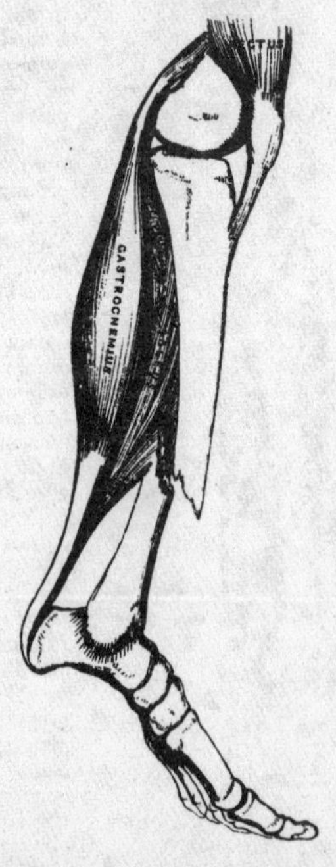

203.—Oblique Fracture of the Shaft of the Tibia.

Oblique fracture of the *shaft of the tibia* (fig. 203) usually occurs at the lower fourth of the bone, this being the narrowest and weakest part, and is usually accompanied with fracture of the fibula. If the fracture has taken place obliquely from above, downwards, and forwards, the fragments ride over one another, the lower fragments being drawn backwards and upwards by the powerful action of the muscles of the calf; the pointed extremity of the upper fragment projects forwards immediately beneath the integument, often protruding through it, and rendering the fracture a compound one. If the direction of the fracture is the reverse of that shown in the figure, the pointed extremity of the lower fragment projects forwards, riding upon the lower end of the upper one. By bending the knee, which relaxes the opposing muscles, and making extension from the knee and ankle, the fragments may be brought into apposition. It is often necessary, however, in compound fracture, to remove a portion of the projecting bone with the saw before complete adaptation can be effected.

204.—Fracture of the Fibula, with Displacement of the Tibia.—'Pott's Fracture.'

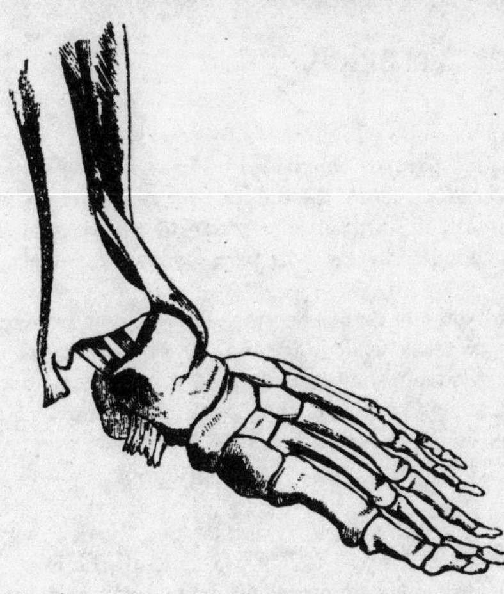

Fracture of the *fibula, with displacement of the tibia* (fig. 204), commonly known as 'Pott's Fracture,' is one of the most frequent injuries of the ankle joint. The end of the tibia is displaced from the corresponding surface of the astragalus; the internal lateral ligament is ruptured; and the inner malleolus projects inwards beneath the integument, which is tightly stretched over it, and in danger of bursting. The fibula is broken, usually from two to three inches above the ankle, and occasionally that portion of the tibia with which it is more directly connected below; the foot is everted by the action of the Peroneus longus, its inner border resting upon the ground, and, at the same time, the heel is drawn up by the muscles of the calf. This injury may be at once reduced by flexing the leg at right angles with the thigh, which relaxes all the opposing muscles, and by making slight extension from the knee and ankle.

On the Descriptive Anatomy of the Muscles, refer to Cruveilhier's 'Anatomie Descriptive;' 'Traité de Myologie et d'Angeiologie,' by F. G. Theile, Encyclopédie Anatomique, Paris, 1843; and Henle's 'Handbuch der Systematischen Anatomie,' before referred to.

Of the Arteries.

THE Arteries are cylindrical tubular vessels, which serve to convey blood from both ventricles of the heart to every part of the body. These vessels were named arteries (ἀὴρ, *air*; τηρεῖν, *to contain*), from the belief entertained by the ancients that they contained air. To Galen is due the honour of refuting this opinion; he showed that these vessels, though for the most part empty after death, contain blood in the living body.

The pulmonary artery, which arises from the right ventricle of the heart, carries venous blood directly into the lungs, whence it is returned by the pulmonary veins into the left auricle. This constitutes the lesser or pulmonic circulation. The great artery which arises from the left ventricle, the aorta, conveys arterial blood to the body generally; whence it is brought back to the right side of the heart by means of the veins. This constitutes the greater or systemic circulation.

The distribution of the systemic arteries is like a highly ramified tree, the common trunk of which, formed by the aorta, commences at the left ventricle of the heart, the smallest ramifications corresponding to the circumference of the body and the contained organs. The arteries are found in nearly every part of the body, with the exception of the hairs, nails, epidermis, cartilages, and cornea; and the larger trunks usually occupy the most protected situations, running, in the limbs, along the flexor side, where they are less exposed to injury.

There is considerable variation in the mode of division of the arteries; occasionally a short trunk subdivides into several branches at the same point, as we observe in the cœliac and thyroid axes; or the vessel may give off several branches in succession, and still continue as the main trunk, as is seen in the arteries of the limbs; but the usual division is dichotomous, as, for instance, the aorta dividing into the two common iliacs; and the common carotid, into the external and internal.

The branches of arteries arise at very variable angles; some, as the superior intercostal arteries from the aorta, arise at an obtuse angle; others, as the lumbar arteries, at a right angle; or, as the spermatic, at an acute angle. An artery from which a branch is given off, is smaller in size, but retains a uniform diameter until a second branch is derived from it. A branch of an artery is smaller than the trunk from which it arises; but if an artery divides into two branches, the combined area of the two vessels is, in nearly every instance, somewhat greater than that of the trunk; and the combined area of all the arterial branches greatly exceeds the area of the aorta; so that the arteries collectively may be regarded as a cone, the apex of which corresponds to the aorta; the base, to the capillary system.

The arteries, in their distribution, communicate freely with one another, forming what is called an *anastomosis* (ἀνὰ, *between*; στόμα, *mouth*), or inosculation; and this communication is very free between the large, as well as between the smaller branches. The anastomoses between trunks of equal size is found where great freedom and activity of the circulation is requisite, as in the brain; here the two vertebral arteries unite to form the basilar, and the two internal carotid arteries are connected by a short communicating trunk; it is also found in the abdomen, the intestinal arteries having very free anastomoses between their larger branches. In the limbs, the anastomoses are most frequent and of largest size around the joints; the branches of an artery above freely inosculating with branches from the vessels below; these anastomoses are of considerable interest to the surgeon, as it is by their enlargement that a *collateral circulation* is established after the application of a ligature to an artery for the cure of aneurism. The smaller branches of arteries anastomose more frequently than the larger; and between the smallest twigs, these inosculations become so numerous as to constitute a close network that pervades nearly every tissue of the body.

Throughout the body generally, the larger arterial branches pursue a perfectly straight course; but in certain situations they are tortuous: thus, the facial artery in its course over the face, and the arteries of the lips, are extremely tortuous in their course, to accommodate themselves to the movements of the parts. The uterine arteries are also tortuous, to accommodate themselves to the increase of size which the organ undergoes during pregnancy. Again, the internal carotid and vertebral arteries, previous to their entering the cavity of the skull, describe a series of curves, which are evidently intended to diminish the velocity of the current of blood, by increasing the extent of surface over which it moves, and adding to the amount of impediment which is produced by friction.

The arteries are dense in structure, of considerable strength, highly elastic, and, when divided, they preserve, although empty, their cylindrical form.

The minute structure of these vessels is described in the Introduction.

In the description of the arteries, we shall first consider the efferent trunk of the systemic circulation, the aorta, and its branches; and then the efferent trunk of the pulmonic circulation, the pulmonary artery.

The Aorta.

The *aorta* (ἀορτὴ; *arteria magna*) is the main trunk of a series of vessels, which, arising from the heart, convey the red oxygenated blood to every part of the body for its nutrition. This vessel commences at the upper part of the left ventricle, and, after ascending for a short distance, arches backwards to the left side, over the root of the left lung, descends within the thorax on the left side of the vertebral column, passes through the aortic opening in the Diaphragm, and entering the abdominal cavity, terminates, considerably diminished in size, opposite the fourth lumbar vertebra, where it divides into the right and left common iliac arteries. Hence its subdivision into the arch of the aorta, the thoracic aorta and the abdominal aorta, from the direction or position of its parts.

Arch of the Aorta.

Dissection. In order to examine the arch of the aorta, open the thorax, by dividing the cartilages of the ribs on each side of the sternum, raising this bone from below upwards, and then sawing through the sternum on a level with its articulation with the clavicle. By this means, the relations of the large vessels to the upper border of the sternum and root of the neck are kept in view.

The arch of the aorta extends from the origin of the vessel at the upper part of the left ventricle to the lower border of the body of the fourth dorsal vertebra. At its commencement, it ascends behind the sternum, obliquely upwards and forwards towards the right side, and opposite the upper border of the second costal cartilage of the right side, passes transversely from right to left, and from before backwards, to the left side of the third dorsal vertebra; it then descends upon the left side of the body of the fourth dorsal vertebra, at the lower border of which it takes the name of thoracic aorta. The arch of the aorta describes a curve, the convexity of which is directed upwards and to the right side; and it is subdivided, at the points where it changes its direction, so as to be described in three portions, the ascending, transverse, and descending portions of the arch of the aorta.

Ascending Part of the Arch.

The ascending portion of the arch of the aorta is about two inches in length. It commences at the upper part of the left ventricle, in front of the left auriculo-ventricular orifice, and opposite the middle of the sternum on a line with its junction to the third costal cartilage; it passes obliquely upwards in the direction of the heart's axis, to the right side, as high as the upper border of the second costal cartilage, describing a slight curve in its course, and being situated, when distended, about a quarter of an inch behind the posterior surface of the sternum.

205.—The Arch of the Aorta and its Branches.

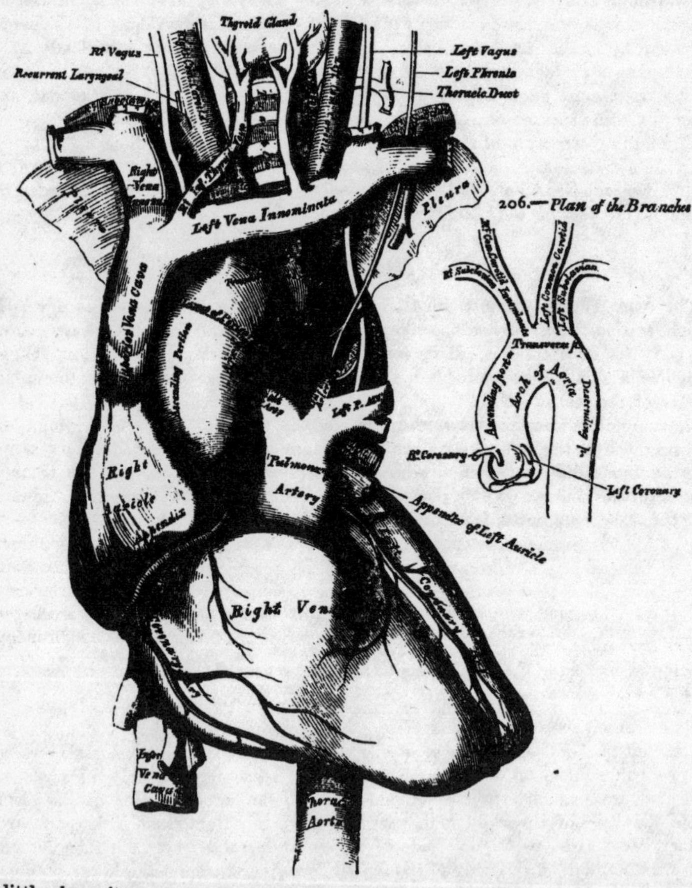

206.—Plan of the Branches

A little above its commencement, it is somewhat enlarged, and presents three small dilatations, called the *sinuses of the aorta* (sinuses of Valsalva), opposite to which are attached the three semilunar valves, which serve the purpose of preventing any regurgitation of blood into the cavity of the ventricle. A section of the aorta opposite this part has a somewhat triangular figure; but below the attachment of the valves it is circular. This portion of the arch is contained in the cavity of the pericardium, and, together with the pulmonary artery, is invested in a tube of serous membrane, continued on to them from the surface of the heart.

Relations. The ascending part of the arch is covered at its commencement by the trunk of the pulmonary artery and the right auricular appendix, and, higher up, is separated from the sternum by the pericardium, some loose areolar tissue, and the remains of the thymus gland; *behind*, it rests upon the right pulmonary vessels and root of the right lung. On the *right side*, it is in relation with the superior vena cava and right auricle; on the *left side*, with the pulmonary artery.

PLAN OF THE RELATIONS OE THE ASCENDING PART OF THE ARCH.

In Front.
Pulmonary artery.
Right auricular appendix.
Pericardium.
Remains of thymus gland.

Right side.
Superior cava.
Right auricle.

Arch of Aorta. Ascending Portion.

Left side.
Pulmonary artery.

Behind.
Right pulmonary vessels.
Root of right lung.

TRANSVERSE PART OF THE ARCH.

The second or transverse portion of the arch commences at the upper border of the second costo-sternal articulation of the right side in front, and passes from right to left, and from before backwards, to the left side of the second dorsal vertebra behind. Its upper border is usually about an inch below the upper margin of the sternum.

Relations. Its *anterior surface* is covered by the left pleura and lung, and crossed towards the left side by the left pneumogastric and phrenic nerves, and cardiac branches of the sympathetic. Its *posterior surface* lies on the trachea, just above its bifurcation, on the great cardiac plexus, the œsophagus, thoracic duct, and left recurrent laryngeal nerve. Its *upper border* is in relation with the left innominate vein; and from its upper part are given off the innominate, left carotid, and left subclavian arteries. Its *lower border* is in relation with the bifurcation of the pulmonary artery, and the remains of the ductus arteriosus, which is connected with the left division of that vessel; the left recurrent laryngeal nerve winds round it from before backwards, whilst the left bronchus passes below it.

PLAN OF THE RELATIONS OF THE TRANSVERSE PART OF THE ARCH.

Above.
Left innominate vein.
Arteria innominata.
Left carotid.
Left subclavian.

In front.
Left pleura and lung.
Left pneumogastric nerve.
Left phrenic nerve.
Cardiac nerves.

Arch of Aorta. Transverse Portion.

Behind.
Trachea.
Cardiac plexus.
Œsophagus.
Thoracic duct.
Left recurrent nerve.

Below.
Bifurcation of pulmonary artery.
Remains of ductus arteriosus.
Left recurrent nerve.
Left bronchus.

Descending Part of the Arch.

The descending portion of the arch has a straight direction, inclining downwards on the left side of the body of the fourth dorsal vertebra, at the lower border of which it takes the name of thoracic aorta.

Relations. Its *anterior surface* is covered by the pleura and root of the left lung; *behind*, it lies on the left side of the body of the fourth dorsal vertebra. On its *right side* are the œsophagus and thoracic duct; on its *left side* it is covered by the pleura.

Plan of the Relations of the Descending Part of the Arch.

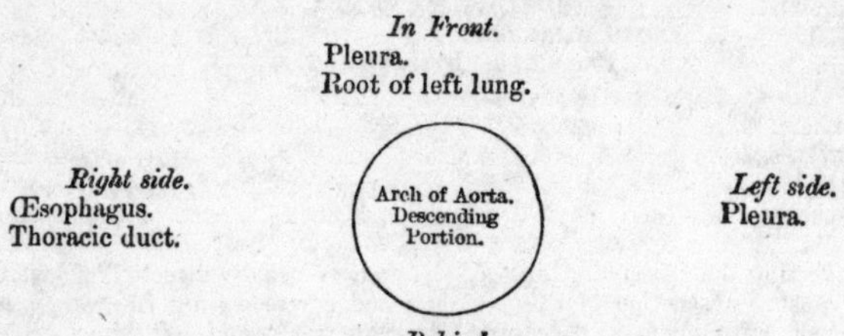

The ascending, transverse, and descending portions of the arch vary in position according to the movements of respiration, being lowered, together with the trachea, bronchi, and pulmonary vessels, during inspiration by the descent of the Diaphragm, and elevated during expiration, when the Diaphragm ascends. These movements are greater in the ascending than the transverse, and in the latter than the descending part.

Peculiarities. The height to which the aorta rises in the chest is usually about an inch below the upper border of the sternum; but it may ascend nearly to the top of that bone. Occasionally it is found an inch and a half, more rarely, three inches below this point.

In Direction. Sometimes the aorta arches over the root of the right instead of the left lung, as in birds, and passes down on the right side of the spine. In such cases all the viscera of the thoracic and abdominal cavities are transposed. Less frequently, the aorta, after arching over the root of the right lung, is directed to its usual position on the left side of the spine, this peculiarity not being accompanied by any transposition of the viscera.

In Conformation. The aorta occasionally divides, as in some quadrupeds, into an ascending and a descending trunk, the former of which is directed vertically upwards, and subdivides into three branches, to supply the head and upper extremities. Sometimes the aorta subdivides soon after its origin into two branches, which soon reunite. In one of these cases, the œsophagus and trachea were found to pass through the interval left by their division; this is the normal condition of the vessel in the reptilia.

Surgical Anatomy. Of all the vessels of the arterial system, the aorta, and more especially its arch, is most frequently the seat of disease; hence it is important to consider some of the consequences that may ensue from aneurism of this part.

It will be remembered, that the ascending part of the arch is contained in the pericardium, just behind the sternum, being crossed at its commencement by the pulmonary artery and right auricular appendix, and having the root of the right lung behind, the vena cava on the right side, and the pulmonary artery and left auricle on the left side.

Aneurism of the ascending aorta, in the situation of the aortic sinuses, in the great majority of cases, affects the right coronary sinus; this is mainly owing to the fact that the regurgitation of blood upon the sinuses takes place chiefly on the right anterior aspect of the vessel. As the aneurismal sac enlarges, it may compress any or all of the structures in immediate proximity with it, but chiefly projects towards the right anterior side; and, consequently, interferes mainly with those structures that have a corresponding relation with the vessel. In the majority of cases, it bursts in the cavity of the pericardium, the patient suddenly drops down dead, and, upon a post-mortem examination, the pericardial sac is found full of blood: or it may compress the right auricle, or the pulmonary artery, and adjoining part of the right ventricle, and open into one or the other of these parts, or may press upon the superior cava.

Aneurism of the ascending aorta, originating above the sinuses, most frequently implicates the right anterior wall of the vessel; this is probably mainly owing to the blood being impelled against this part. The direction of the aneurism is also chiefly towards the right of the median line. If it attains a large size and projects forwards, it may absorb the sternum and the cartilages of the ribs, usually on the right side, and appear as a pulsating tumour on the front of the chest, just below the manubrium; or it may burst into the pericardium, or may compress, or open into, the right lung, the trachea, bronchi, or œsophagus.

Regarding the transverse part of the arch, the student is reminded that the vessel lies on the trachea, the œsophagus, and thoracic duct; that the recurrent laryngeal nerve winds around it; and that from its upper part are given off three large trunks, which supply the head, neck, and upper extremities. Now an aneurismal tumour taking origin from the posterior part or right aspect of the vessel, its most usual site, may press upon the trachea, impede the breathing, or produce cough, hæmoptysis, or stridulous breathing, or it may ultimately burst into that tube, producing fatal hæmorrhage. Again, its pressure on the laryngeal nerves may give rise to symptoms which so accurately resemble those of laryngitis, that the operation of tracheotomy has in some cases been resorted to, from the supposition that disease existed in the larynx; or it may press upon the thoracic duct, and destroy life by inanition; or it may involve the œsophagus, producing dysphagia; or may burst into the œsophagus, when fatal hæmorrhage will occur. Again, the innominate artery, or the left carotid, or subclavian, may be so obstructed by clots, as to produce a weakness, or even a disappearance, of the pulse in one or the other wrist; or the tumour may present itself at or above the manubrium, generally either in the median line, or to the right of the sternum, and may simulate an aneurism of one of the arteries of the neck.

Aneurism affecting the descending part of the arch is usually directed backwards and to the left side, causing absorption of the vertebræ and corresponding ribs; or it may press upon the trachea, left bronchus, œsophagus, and the right and left lungs, generally the latter. When rupture of the sac occurs, it usually takes place into the left pleural cavity; less frequently into the left bronchus, the right pleura, or into the substance of the lungs or trachea. In this form of aneurism, pain is almost a constant and characteristic symptom, referred to either the back or chest, and usually radiating from the spine around the left side. This symptom depends upon the aneurismal sac compressing the intercostal nerves against the bone.

Branches of the Arch of the Aorta. (Figs. 205, 206.)

The branches given off from the arch of the aorta are five in number: two of small size from the ascending portion, the right and left coronary; and three of large size from the transverse portion, the innominate artery, the left carotid, and the left subclavian.

Peculiarities. Position of the Branches. The branches, instead of arising from the highest part of the arch (their usual position), may be moved more to the right, arising from the commencement of the transverse or upper part of the ascending portion; or the distance from one another at their origin may be increased or diminished, the most frequent change in this respect being the approximation of the left carotid towards the innominate artery.

The Number of the primary branches may be reduced to two: the left carotid arising from the innominate artery; or (more rarely), the carotid and subclavian arteries of the left side arising from a left innominate artery. But the number may be increased to four, from the right carotid and subclavian arteries arising directly from the aorta, the innominate being absent. In most of these latter cases, the right subclavian has been found to arise from the left end of the arch; in other cases, it was the second or third branch given off instead of the first. Lastly, the number of trunks from the arch may be increased to five or six; in these instances, the external and internal carotids arose separately from the arch, the common carotid being absent on one or both sides.

Number usual, Arrangement different. When the aorta arches over to the right side, the three branches have an arrangement the reverse of what is usual, the innominate supplying the left side; and the carotid and subclavian (which arises separately) the right side. In other cases, where the aorta takes its usual course, the two carotids may be joined in a common trunk, and the subclavians arise separately from the arch, the right subclavian generally arising from the left end of the arch.

Secondary Branches sometimes arise from the arch; most commonly such a secondary branch is the left vertebral, which usually takes origin between the left carotid and left subclavian, or beyond them. Sometimes, a thyroid branch is derived from the arch, or the right internal mammary, or left vertebral, or, more rarely, both vertebrals.

The Coronary Arteries.

The coronary arteries supply the heart; they are two in number, right and left, arising near the commencement of the aorta immediately above the free margin of the semilunar valves.

The *Right Coronary Artery*, about the size of a crow's quill, arises from the aorta immediately above the free margin of the right semilunar valve, between the pulmonary artery and the appendix of the right auricle. It passes forwards to the right side, in the groove between the right auricle and ventricle, and curving around the right border of the heart, runs along its posterior surface as far as the posterior interventricular groove, where it divides into two branches, one of which continues onwards in the groove between the left auricle and ventricle, and anastomoses with the left coronary; the other descends along the posterior interventricular furrow, supplying branches to both ventricles and to the septum, and anastomosing at the apex of the heart with the descending branch of the left coronary.

This vessel sends a large branch along the thin margin of the right ventricle to the apex, and numerous small branches to the right auricle and ventricle, and the commencement of the pu'monary artery.

The *Left Coronary*, smaller than the former, arises immediately above the free edge of the left semilunar valve, a little higher than the right; it passes forwards between the pulmonary artery and the left appendix auriculæ, and descends obliquely towards the anterior interventricular groove, where it divides into two branches. Of these, one passes transversely outwards in the left auriculo-ventricular groove, and winds around the left border of the heart to its posterior surface, where it anastomoses with the superior branch of the right coronary; the other descends along the anterior interventricular groove to the apex of the heart, where it anastomoses with the descending branch of the right coronary. The left coronary supplies the left auricle and its appendix, both ventricles, and numerous small branches to the pulmonary artery, and commencement of the aorta.

Peculiarities. These vessels occasionally arise by a common trunk, or their number may be increased to three; the additional branch being of small size. More rarely, there are two additional branches.

Arteria Innominata.

The innominate artery is the largest branch given off from the arch of the aorta. It arises from the commencement of the transverse portion in front of the left carotid, and, ascending obliquely to the upper border of the right sterno-clavicular articulation, divides into the right carotid and subclavian arteries. This vessel varies from an inch and a half to two inches in length.

Relations. In front, it is separated from the first bone of the sternum by the Sterno-hyoid and Sterno-thyroid muscles, the remains of the thymus gland, and by the left innominate and right inferior thyroid veins which cross its root. *Behind*, it lies upon the trachea, which it crosses obliquely. On the *right side* is the right vena innominata, right pneumogastric nerve, and the pleura; and on the *left side*, the remains of the thymus gland, and origin of the left carotid artery.

Plan of the Relations of the Innominate Artery.

In front.

Sternum.
Sterno-hyoid and Sterno-thyroid muscles.
Remains of thymus gland.
Left innominate and right inferior thyroid veins.
Inferior cervical cardiac branch from right pneumogastric nerve.

Right side.
Right vena innominata.
Right pneumogastric nerve.
Pleura.

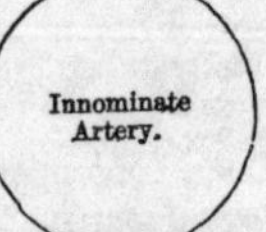

Left side.
Remains of thymus.
Left carotid.

Behind.
Trachea.

Peculiarities in point of division. When the bifurcation of the innominate artery varies from the point above mentioned, it sometimes ascends a considerable distance above the sternal end of the clavicle; less frequently it divides below it. In the former class of cases, its length may exceed two inches; and, in the latter, be reduced to an inch or less. These are points of considerable interest for the surgeon to remember in connection with the operation of tying this vessel.

Branches. The arteria innominata occasionally supplies a thyroid branch (middle thyroid artery), which ascends along the front of the trachea to the thyroid gland; and sometimes, a thymic or bronchial branch. The left carotid is frequently joined with the innominate artery at its origin. Sometimes, there is no innominate artery, the right subclavian arising directly from the arch of the aorta.

Position. When the aorta arches over to the right side, the innominate is directed to the left side of the neck instead of the right.

Collateral circulation. Allan Burns demonstrated, on the dead subject, the possibility of the establishment of the collateral circulation after ligature of the innominate artery, by tying and dividing that artery, after which, he says, "Even coarse injection impelled into the aorta, passed freely by the anastomosing branches into the arteries of the right arm, filling them and all the vessels of the head completely." (*Surgical Anatomy of the Head and Neck*, p. 62.) The branches by which this circulation would be carried on are very numerous; thus, all the communications across the middle line between the branches of the carotid arteries of opposite sides would be available for the supply of blood to the right side of the head and neck; while the anastomosis between the superior intercostal of the subclavian and the first aortic intercostal (see *infra* on the collateral circulation after obliteration of the thoracic aorta), would bring the blood, by a free and direct course, into the right subclavian: the numerous connections, also, between the lower intercostal arteries, and the branches of the axillary and internal mammary arteries would, doubtless, assist in the supply of blood to the right arm, while the epigastric, from the external iliac, would by means of its anastomosis with the internal mammary, compensate for any deficiency in the vascularity of the wall of the chest.

Surgical Anatomy. Although the operation of tying the innominate artery has been performed by several surgeons, for aneurism of the right subclavian extending inwards as far as the Scalenus, in only one instance has it been attended with success.* Mott's patient, however, on whom the operation was first performed, lived nearly four weeks, and Graefe's more than two months. The main obstacles to the operation are, as the student will perceive from his dissection of this vessel, the deep situation of the artery behind and beneath the sternum, and the number of important structures which surround it in every part.

In order to apply a ligature to this vessel, the patient is to be placed upon his back, with the shoulders raised, and the head bent a little backwards, so as to draw out the artery from behind the sternum into the neck. An incision two inches long is then made along the anterior border of the Sterno-mastoid muscle, terminating at the sternal end of the clavicle. From this point, a second incision is carried about the same length along the upper border of the clavicle. The skin is then dissected back, and the Platysma divided on a director: the sternal end of the Sterno-mastoid is now brought into view, and a director being passed beneath it, and close to its under surface, so as to avoid any small vessels, the muscle is to be divided transversely throughout the greater part of its attachment. By pressing aside any loose cellular tissue or vessels that may now appear, the Sterno-hyoid and Sterno-thyroid muscles will be exposed, and must be divided, a director being previously passed beneath them. The inferior thyroid veins now come into view, and must be carefully drawn either upwards or downwards, by means of a blunt hook. On no account should these vessels be divided, as it would add much to the difficulty of the operation, and endanger its ultimate success. After tearing through a strong fibro-cellular lamina, the right carotid is brought into view, and being traced downwards, the arteria innominata is arrived at. The left vena innominata should now be depressed, the right vena innominata, the internal jugular vein, and pneumogastric nerve drawn to the right side; and a curved aneurism needle may then be passed around the vessel, close to its surface, and in a direction from below upwards and inwards; care being taken to avoid the right pleural sac, the trachea, and cardiac nerves. The ligature should be applied to the artery as high as possible, in order to allow room between it and the aorta for the formation of a coagulum. The importance of avoiding the thyroid plexus of veins during the primary steps of the operation, and the pleural sac whilst including the vessel in the ligature, should be most carefully borne in mind, since secondary hæmorrhage or pleurisy have been the cause of death in all the cases hitherto operated on.

Common Carotid Arteries.

The common carotid arteries, although occupying a nearly similar position in the neck, differ in position, and, consequently, in their relations at their origin. The right carotid arises from the arteria innominata, behind the right sterno-

* The operation was performed by Dr. Smyth of New Orleans: see the New Sydenham Society's 'Biennial Retrospect,' for 1865–6, p. 346.

clavicular articulation; the left from the highest part of the arch of the aorta. The left carotid is, consequently, longer and placed more deeply in the thorax. It will, therefore, be more convenient to describe first the course and relations of that portion of the left carotid which intervenes between the arch of the aorta and the left sterno-clavicular articulations (see fig. 205).

The left carotid within the thorax ascends obliquely outwards from the arch of the aorta to the root of the neck. In *front*, it is separated from the first piece of the sternum by the Sterno-hyoid and Sterno-thyroid muscles, the left innominate vein, and the remains of the thymus gland; *behind*, it lies on the trachea, œsophagus, and thoracic duct. *Internally*, it is in relation with the arteria innominata; *externally*, with the left pneumogastric nerve, and left subclavian artery.

Plan of the Relations of the Left Common Carotid. Thoracic Portion.

In Front.

Sternum.
Sterno-hyoid and Sterno-thyroid muscles.
Left innominate vein.
Remains of thymus gland.

Internally.
Arteria innominata.

Left Common Carotid. Thoracic Portion.

Externally.
Left pneumogastric nerve.
Left subclavian artery.

Behind.
Trachea.
Œsophagus.
Thoracic duct.

In the neck, the two common carotids resemble each other so closely, that one description will apply to both. Each vessel passes obliquely upwards, from behind the sterno-clavicular articulation, to a level with the upper border of the thyroid cartilage, where it divides into the external and internal carotid; these names being derived from the distribution of the arteries to the external parts of the head and face, and to the internal parts of the cranium respectively. The course of the common carotid is indicated by a line drawn from the sternal end of the clavicle below, to a point midway between the angle of the jaw and the mastoid process above.

At the lower part of the neck the two common carotid arteries are separated from each other by a very small interval, which contains the trachea; but at the upper part, the thyroid body, the larynx and pharynx project forwards between the two vessels, and give the appearance of their being placed further back in that situation. The common carotid artery is contained in a sheath, derived from the deep cervical fascia, which also encloses the internal jugular vein and pneumogastric nerve, the vein lying on the outer side of the artery, and the nerve between the artery and vein, on a plane posterior to both. On opening the sheath, these three structures are seen to be separated from one another, each being enclosed in a separate fibrous investment.

Relations. At the lower part of the neck the common carotid artery is very deeply seated, being covered by the superficial fascia, Platysma, and deep fascia, the Sterno-mastoid, Sterno-hyoid, and Sterno-thyroid muscles, and by the Omo-hyoid opposite the cricoid cartilage; but in the upper part of its course, near its termination, it is more superficial, being covered merely by the integument, the superficial fascia, Platysma, and deep fascia, and inner margin of the Sterno-mastoid, and is contained in a triangular space, bounded behind by the Sterno-mastoid, above by the posterior belly of the Digastric, and below by the anterior belly of the Omo-hyoid. This part of the artery is crossed obliquely from within outwards by the sterno-mastoid artery; it is also crossed by the facial, lingual, and superior thyroid veins, which terminate in the internal jugular, and descending on its sheath in front, is seen the descendens noni nerve, this filament being joined with branches from the cervical nerves, which cross the vessel from without

207.—Surgical Anatomy of the Arteries of the Neck. Right Side.

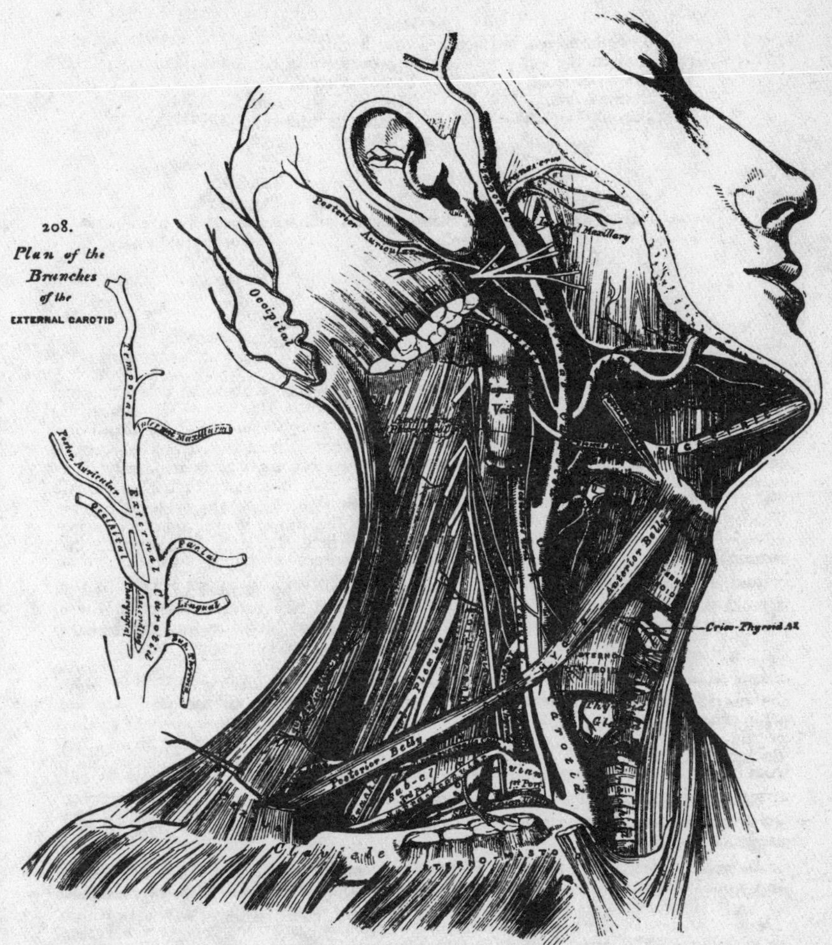

208. *Plan of the Branches of the* EXTERNAL CAROTID

inwards. Sometimes the descendens noni is contained within the sheath. The middle thyroid vein crosses the artery about its middle, and the anterior jugular vein below. *Behind*, the artery lies in front of the cervical portion of the spine, resting first on the Longus colli muscle, then on the Rectus anticus major, from which it is separated by the sympathetic nerve. The recurrent laryngeal nerve and inferior thyroid artery cross behind the vessel at its lower part. *Internally*, it is in relation with the trachea and thyroid gland, the inferior thyroid artery and recurrent laryngeal nerve being interposed; higher up, with the larynx and pharynx. On its *outer side* are placed the internal jugular vein and pneumogastric nerve.

At the lower part of the neck, the internal jugular vein on the right side diverges from the artery, but on the left side it approaches it, and often crosses its lower part. This is an important fact to bear in mind during the performance of any operation on the lower part of the left common carotid artery.

Plan of the Relations of the Common Carotid Artery.

In front.

Integument and fascia.
Platysma.
Sterno-mastoid.
Sterno-hyoid.
Sterno-thyroid.
Omo-hyoid.
Descendens noni nerve.
Sterno-mastoid artery.
Thyroid, lingual, and facial veins.
Anterior jugular vein.

Externally.

Internal jugular vein.
Pneumogastric nerve.

Common Carotid.

Internally.

Trachea.
Thyroid gland.
Recurrent laryngeal nerve.
Inferior thyroid artery.
Larynx.
Pharynx.

Behind.

Longus colli.
Rectus anticus major.
Sympathetic nerve.
Inferior thyroid artery.
Recurrent laryngeal nerve.

Peculiarities as to Origin. The *right common carotid* may arise above or below its usual point, the upper border of the sterno-clavicular articulation. This variation occurs in one out of about eight cases and a half, and the origin is more frequently above than below the usual point; or the artery may arise as a separate branch from the arch of the aorta, or in conjunction with the left carotid. The *left common carotid* varies more frequently in its origin than the right. In the majority of abnormal cases it arises with the innominate artery, or if the innominate artery is absent, the two carotids arise usually by a single trunk. The left carotid has a tendency towards the right side of the arch of the aorta, being occasionally the first branch given off from the transverse portion. It rarely joins with the left subclavian, except in cases of transposition of the arch.

Peculiarities as to Point of Division. The most important peculiarities of this vessel, in a surgical point of view, relate to its place of division in the neck. In the majority of abnormal cases, this occurs higher than usual, the artery dividing into two branches opposite the hyoid bone, or even higher; more rarely, it occurs below its usual place opposite the middle of the larynx, or the lower border of the cricoid cartilage; and one case is related by Morgagni, where the common carotid, only an inch and a half in length, divided at the root of the neck. Very rarely, the common carotid ascends in the neck without any subdivision, the internal carotid being wanting; and in two cases, the common carotid has been found to be absent, the external and internal carotids arising directly from the arch of the aorta. This peculiarity existed on both sides in one subject, on one side in the other.

Occasional Branches. The common carotid usually gives off no branches; but it occasionally gives origin to the superior thyroid, or a laryngeal branch, the inferior thyroid, or, more rarely, the vertebral artery.

Surgical Anatomy. The operation of tying the common carotid artery may be necessary in a wound of that vessel or its branches, in an aneurism, or in a case of pulsating tumour of the orbit or skull. If the wound involves the trunk of the common carotid, it will be necessary to tie the artery above and below the wounded part. But in cases of aneurism, or where one of the branches of the common carotid is wounded in an inaccessible situation, it may be judged necessary to tie the trunk. In such cases, the whole of the artery is accessible, and any part may be tied, except close to either end. When the case is such as to allow of a choice being made, the lower part of the carotid should never be selected as the spot upon which to place a ligature, for not only is the artery in this situation placed very deeply in the neck, but it is covered by three layers of muscles, and on the left side the jugular vein, in the great majority of cases, passes obliquely in front of it. Neither should the upper end be selected, for here the superior thyroid, lingual, and facial veins would give rise to very considerable difficulty in the application of a ligature. The point most favourable for the operation is opposite the lower part of the larynx, and here a ligature may be applied on the vessel, either above or below the point where it is crossed by the Omo-hyoid muscle. In the former situation the artery is most accessible, and it may be tied there in cases of wounds, or aneurism of any of the large branches of the carotid; whilst in cases of aneurism of the upper part of the carotid, that part of the vessel may be selected which is below the Omo-hyoid. It occasionally happens that the carotid artery bifurcates below its usual position: if the artery be exposed at its point of bifurcation, both divisions of the vessel should be tied near their origin, in preference to tying the trunk of the artery near its termination; and if, in consequence of the entire absence of the common carotid, or from its early division, two arteries, the external and internal carotids, are met with, the ligature should be placed on that vessel which is found on compression to be connected with the disease.

In this operation, the direction of the vessel and the inner margin of the Sterno-mastoid are the chief guides to its performance.

To tie the Common Carotid, above the Omo-hyoid. The patient should be placed on his back with the head thrown back; an incision is to be made, three inches long, in the direction of the anterior border of the Sterno-mastoid, from a little below the angle of the jaw to a level with the cricoid cartilage: after dividing the integument, superficial fascia, and Platysma, the deep fascia must be cut through on a director, so as to avoid wounding numerous small veins that are usually found beneath. The head may now be brought forwards so as to relax the parts somewhat, and the margins of the wound held asunder by copper spatulæ. The descendens noni nerve is now exposed, and must be avoided, and the sheath of the vessel having been raised by forceps, is to be opened over the artery to a small extent at its inner side. The internal jugular vein may now present itself alternately distended and relaxed; this should be compressed both above and below, and drawn outwards, in order to facilitate the operation. The aneurism needle is now passed from the outside, care being taken to keep the needle in close contact with the artery, and thus avoid the risk of injuring the jugular vein, or including the vagus nerve. Before the ligature is tied, it should be ascertained that nothing but the artery is included in it.

To tie the Common Carotid below the Omo-hyoid. The patient should be placed in the same position as above mentioned. An incision about three inches in length is to be made, parallel with the inner edge of the Sterno-mastoid, commencing on a level with the cricoid cartilage. The inner border of the Sterno-mastoid having been exposed, the sterno-mastoid artery and a large vein, the middle thyroid will be seen, and must be carefully avoided; the Sterno-mastoid is to be drawn outwards, and the Sterno-hyoid and thyroid muscles inwards. The deep fascia must now be divided below the Omo-hyoid muscle, and the sheath having been exposed, must be opened, care being taken to avoid the descendens noni, which here runs on the inner or tracheal side. The jugular vein and vagus nerve being then pressed to the outer side, the needle must be passed round the artery from without inwards, great care being taken to avoid the inferior thyroid artery, the recurrent laryngeal, and sympathetic nerves which lie behind it.

Collateral Circulation. After ligature of the common carotid, the collateral circulation can be perfectly established, by the free communication which exists between the carotid arteries of opposite sides, both without and within the cranium—and by enlargement of the branches of the subclavian artery on the side corresponding to that on which the vessel has been tied, the chief communication outside the skull taking place between the superior and inferior thyroid arteries, and the profunda cervicis, and arteria princeps cervicis of the occipital; the vertebral taking the place of the internal carotid within the cranium.

Sir A. Cooper had an opportunity of dissecting, thirteen years after the operation, the case in which he first successfully tied the common carotid (the second case in which the operation had been performed). *Guy's Hospital Reports*, i. 56. The injection, however, does not seem to have been a successful one. It showed merely that the arteries at the base of the brain (circle of Willis) were much enlarged on the side of the tied artery, the basilar artery on that side having been one of the chief means of restoring the circulation, and that the anastomosis between the branches of the external carotid on the affected side and those of the same artery on the sound side was free, so that the external carotid was pervious throughout.

External Carotid Artery.

The external carotid artery (fig. 207), arises opposite the upper border of the thyroid cartilage, and taking a slightly curved course, ascends upwards and forwards, and then inclines backwards, to the space between the neck of the condyle of the lower jaw, and the external meatus, where it divides into the temporal and internal maxillary arteries. It rapidly diminishes in size in its course up the neck, owing to the number and large size of the branches given off from it. In the child, it is somewhat smaller than the internal carotid; but in the adult, the two vessels are of nearly equal size. At its commencement, this artery is more superficial, and placed nearer the middle line than the internal carotid, and is contained in the triangular space bounded by the Sterno-mastoid behind, the Omo-hyoid below, and the posterior belly of the Digastric and Stylo-hyoid above; it is covered by the skin, Platysma, deep fascia, and anterior margin of the Sterno-mastoid, crossed by the hypoglossal nerve, and by the lingual and facial veins; it is afterwards crossed by the Digastric and Stylo-hyoid muscles, and higher up passes deeply into the substance of the parotid gland, where it lies beneath the facial nerve and the junction of the temporal and internal maxillary veins.

Internally is the hyoid bone, the wall of the pharynx, and the ramus of the jaw, from which it is separated by a portion of the parotid gland.

Behind it, near its origin, is the superior laryngeal nerve; and higher up, it is separated from the internal carotid by the Stylo-glossus and Stylo-pharyngeus muscles, the glosso-pharyngeal nerve, and part of the parotid gland.

Plan of the Relations of the External Carotid.

In front.
Integument, superficial fascia.
Platysma and deep fascia.
Hypoglossal nerve.
Lingual and facial veins.
Digastric and Stylo-hyoid muscles.
Parotid gland with facial nerve and temporo-maxillary vein in its substance.

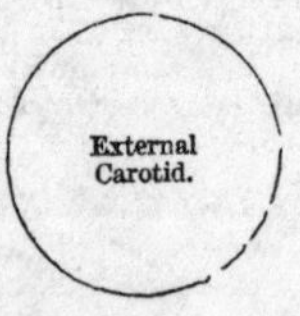

Behind.
Superior Laryngeal nerve.
Stylo-glossus.
Stylo-pharyngeus.
Glosso-pharyngeal nerve.
Parotid gland.

Internally.
Hyoid bone.
Pharynx.
Parotid gland.
Ramus of jaw.

Surgical Anatomy. The application of a ligature to the external carotid may be required in cases of wounds of this vessel, or of its branches when these cannot be tied, and in some cases of pulsating tumour of the scalp or face; the operation, however, is very rarely performed, ligature of the common carotid being preferable, on account of the number of branches given off from the external. To tie this vessel near its origin, below the point where it is crossed by the Digastric, an incision about three inches in length should be made along the margin of the Sterno-mastoid, from the angle of the jaw to the cricoid cartilage, as in the operation for tying the common carotid. To tie the vessel above the Digastric, between it and the parotid gland, an incision should be made from the lobe of the ear to the great cornu of the os hyoides, dividing successively the skin, Platysma, and fascia. By separating the posterior belly of the Digastric and Stylo-hyoid muscles which are seen at the lower part of the wound, from the parotid gland, the vessel will be exposed, and a ligature may be applied to it. The circulation is at once re-established by the free communication between most of the large branches of the artery (facial, lingual, superior thyroid, occipital) and the corresponding arteries of the opposite side, and by the free anastomosis of the facial with branches from the internal carotid, of the occipital with branches of the subclavian, etc.

Branches. The external carotid artery gives off eight branches, which, for convenience of description, may be divided into four sets. (See fig. 208, Plan of the Branches.)

Anterior.	*Posterior.*	*Ascending.*	*Terminal.*
Superior thyroid.	Occipital.	Ascending pharyngeal.	Temporal.
Lingual.	Posterior auricular.		Internal maxillary.
Facial.			

The student is here reminded that many variations are met with in the number, origin, and course of these branches in different subjects; but the above arrangement is that which is found in the great majority of cases.

The SUPERIOR THYROID ARTERY (figs. 207 and 212), is the first branch given off from the external carotid, being derived from that vessel just below the great cornu of the hyoid bone. At its commencement, it is quite superficial, being covered by the integument, fascia, and Platysma, and is contained in the triangular space bounded by the Sterno-mastoid, Digastric, and Omo-hyoid muscles. After running upwards and inwards for a short distance, it curves downwards and forwards in an arched and tortuous manner to the upper part of the thyroid gland, passing beneath the Omo-hyoid, Sterno-hyoid, and Sterno-thyroid muscles; and distributes numerous branches to the anterior surface of the gland, anastomosing with its fellow of the opposite side, and with the inferior thyroid arteries. Besides the arteries distributed to the muscles and the substance of the gland, the branches of the superior thyroid are the following:

Hyoid.	Superior Laryngeal.
Superficial descending branch (Sterno-mastoid).	Crico-Thyroid.

The *hyoid* is a small branch which runs along the lower border of the os hyoides, beneath the Thyro-hyoid muscle; after supplying the muscles connected to that bone, it forms an arch, by anastomosing with the vessel of the opposite side.

The *superficial descending branch* runs downwards and outwards across the sheath of the common carotid artery, and supplies the Sterno-mastoid and neighbouring muscles and integument. It is of importance that the situation of this vessel be remembered, in the operation for tying the common carotid artery. There is often a distinct branch from the external carotid distributed to the Sterno-mastoid muscle.

The *superior laryngeal*, larger than either of the preceding, accompanies the superior laryngeal nerve, beneath the Thyro-hyoid muscle; it pierces the thyro-hyoid membrane, and supplies the muscles, mucous membrane, and glands of the larynx and epiglottis, anastomosing with the branch from the opposite side.

The *crico-thyroid* (inferior laryngeal) is a small branch which runs transversely across the crico-thyroid membrane, communicating with the artery of the opposite side. The position of this vessel should be remembered, as it may prove the source of troublesome hæmorrhage during the operation of laryngotomy.

Surgical Anatomy. The superior thyroid, or some of its branches, are often divided in cases of cut throat, giving rise to considerable hæmorrhage. In such cases, the artery should be secured, the wound being enlarged for that purpose, if necessary. The operation may be easily performed, the position of the artery being very superficial, and the only structures of importance covering it being a few small veins. The operation of tying the superior thyroid artery, in bronchocele, has been performed in numerous instances with partial or temporary success. When, however, the collateral circulation between this vessel and the artery of the opposite side, and the inferior thyroid, is completely re-established, the tumour usually regains its former size.

The LINGUAL ARTERY (fig. 212) arises from the external carotid between the superior thyroid and facial; it runs obliquely upwards and inwards to the great cornu of the hyoid bone, then passes horizontally forwards parallel with the great cornu, and, ascending perpendicularly to the under surface of the tongue, turns forwards on its under surface as far as the tip of that organ, under the name of the ranine artery.

Relations. Its first, or oblique portion, is superficial, being contained in the triangular space already described, resting upon the Middle constrictor of the pharynx, and covered by the Platysma and fascia of the neck. Its second, or horizontal portion, also lies upon the Middle constrictor, being covered at first by the tendon of the digastric and the Stylo-hyoid muscle, and afterwards by the Hyo-glossus, the latter muscle separating it from the hypoglossal nerve. Its third, or ascending portion, lies between the Hyo-glossus and Genio-hyo-glossus muscles. The fourth, or terminal part, under the name of the ranine, runs along the under surface of the tongue to its tip: it is very superficial, being covered only by the mucous membrane, and rests on the Lingualis on the outer side of the Genio-hyo-glossus. The hypoglossal nerve lies nearly parallel with the lingual artery, separated from it, in the second part of its course, by the Hyo-glossus muscle.

The branches of the lingual artery are, the

Hyoid.	Sublingual.
Dorsalis Linguæ.	Ranine.

The *hyoid* branch runs along the upper border of the hyoid bone, supplying the muscles attached to it and anastomosing with its fellow of the opposite side.

The *dorsalis linguæ* (fig. 212) arises from the lingual artery beneath the Hyo-glossus muscle (which, in the figure, has been partly cut away, to show the vessel); ascending to the dorsum of the tongue, it supplies the mucous membrane, the tonsil, soft palate, and epiglottis; anastomosing with its fellow from the opposite side.

The *sublingual*, which may be described as a branch of bifurcation of the lingual artery, arises at the anterior margin of the Hyo-glossus muscle, and, running forwards and outwards beneath the Mylo-hyoid to the sublingual gland, supplies its substance, giving branches to the Mylo-hyoid and neighbouring muscles, the mucous membrane of the mouth and gums.

209.—The Arteries of the Face and Scalp.*

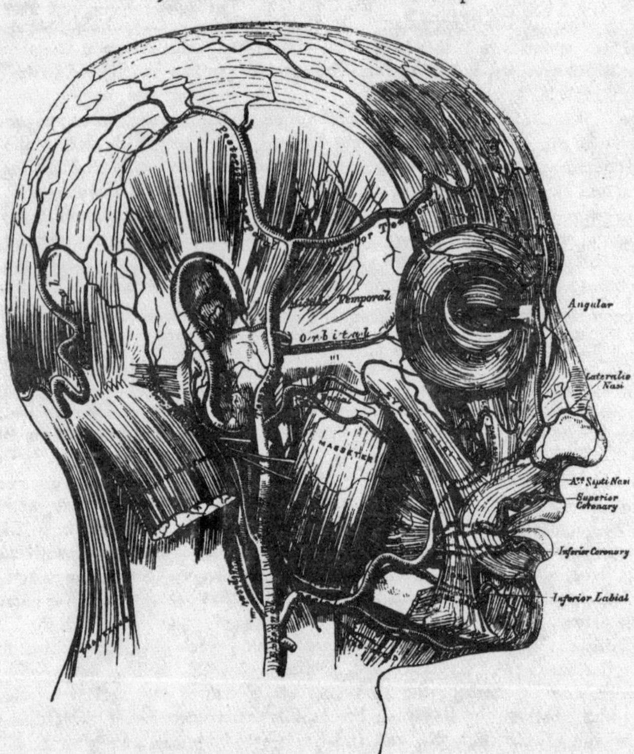

The *ranine* may be regarded as the other branch of bifurcation, or, as is more usual, as the continuation of the lingual artery; it runs along the under surface

* The muscular tissue of the lips must be supposed to have been cut away, in order to show the course of the coronary arteries.

of the tongue, resting on the Lingualis, and covered by the mucous membrane of the mouth; it lies on the outer side of the Genio-hyo-glossus, and is covered by the Hyo-glossus and Stylo-glossus, accompanied by the gustatory nerve. On arriving at the tip of the tongue, it anastomoses with the artery of the opposite side. These vessels in the mouth are placed one on each side of the frænum.

Surgical Anatomy. The lingual artery may be divided near its origin in cases of cut throat, a complication that not unfrequently happens in this class of wounds, or severe hæmorrhage which cannot be restrained by ordinary means, may ensue from a wound, or deep ulcer, of the tongue. In the former case, the primary wound may be enlarged if necessary, and the bleeding vessel secured. In the latter case, it has been suggested that the lingual artery should be tied near its origin. Ligature of the lingual artery is also occasionally practised, as a palliative measure, in cases of tumour of the tongue, in order to check the progress of the disease. The operation is a difficult one, on account of the depth of the artery, the number of important parts by which it is surrounded, the loose and yielding nature of the parts upon which it is supported, and its occasional irregularity of origin. An incision is to be made, about two and a half inches in length, running obliquely downwards and backwards, and having its centre opposite the point of the great cornu of the hyoid bone, which is the guide to the artery. The parts being gradually dissected, the hypoglossal nerve will first come into view, and then the artery must be carefully sought for among the loose tissue at the bottom of the wound, care being taken not to open the pharynx. Large veins, the internal jugular or some of its branches, may be met with, and prove a source of embarrassment.

Troublesome hæmorrhage may occur in the division of the frænum in children, if the ranine artery, which lies on each side of it, is cut through. The student should remember that the operation is always to be performed with a pair of blunt-pointed scissors, and the mucous membrane only is to be divided by a very superficial cut, which cannot endanger any vessel. The scissors, also, should be directed away from the tongue. Any further liberation of the tongue, which may be necessary, can be effected by tearing.

The Facial Artery (fig. 209), arises a little above the lingual, and ascends obliquely forwards and upwards, beneath the body of the lower jaw, to the submaxillary gland, in which it is imbedded; this may be called the cervical part of the artery. It then curves upwards over the body of the jaw at the anterior inferior angle of the Masseter muscle, ascends forwards and upwards across the cheek to the angle of the mouth, passes up along the side of the nose, and terminates at the inner canthus of the eye, under the name of the angular artery. This vessel, both in the neck, and on the face, is remarkably tortuous; in the former situation, to accommodate itself to the movements of the pharynx in deglutition; and in the latter, to the movements of the jaw, and the lips and cheeks.

Relations. In the neck, its origin is superficial, being covered by the integument, Platysma, and fascia; it then passes beneath the Digastric and Stylo-hyoid muscles, and the submaxillary gland. *On the face*, where it passes over the body of the lower jaw, it is comparatively superficial, lying immediately beneath the Platysma. In this situation, its pulsation may be distinctly felt, and compression of the vessel effectually made against the bone. In its course over the face, it is covered by the integument, the fat of the cheek, and, near the angle of the mouth, by the Platysma and Zygomatic muscles. It rests on the Buccinator, the Levator anguli oris, and the Levator labii superioris alæque nasi. It is accompanied by the facial vein throughout its entire course; the vein is not tortuous like the artery, and, on the face, is separated from that vessel by a considerable interval. The branches of the facial nerve cross the artery, and the infra-orbital nerve lies beneath it.

The branches of this vessel may be divided into two sets, those given off below the jaw (cervical), and those on the face (facial).

Cervical Branches.	*Facial Branches.*
Inferior or Ascending Palatine.	Muscular.
Tonsillar.	Inferior Labial.
Submaxillary.	Inferior Coronary.
Submental.	Superior Coronary.
	Lateralis Nasi.
	Angular.

The *inferior* or *ascending palatine* (fig. 212) passes up between the Stylo-glossus and Stylo-pharyngeus to the outer side of the pharynx. After supplying these muscles, the tonsil, and Eustachian tube, it divides, near the Levator palati, into two branches; one follows the course of the Tensor palati, and supplies the soft palate and the palatine glands; the other passes to the tonsil, which it supplies, anastomosing with the tonsillar artery. These vessels inosculate with the posterior palatine branch of the internal maxillary artery.

The *tonsillar* branch (fig. 212) passes up along the side of the pharynx, and, perforating the Superior constrictor, ramifies in the substance of the tonsil and root of the tongue.

The *submaxillary* consists of three or four large branches, which supply the submaxillary gland, some being prolonged to the neighbouring muscles, lymphatic glands, and integument.

The *submental*, the largest of the cervical branches, is given off from the facial artery, just as that vessel quits the submaxillary gland; it runs forwards upon the Mylo-hyoid muscle, just below the body of the jaw, and beneath the Digastric; after supplying the muscles attached to the jaw, and anastomosing with the sublingual artery, it arrives at the symphysis of the chin, where it divides into a superficial and a deep branch; the former turns round the chin, and, passing between the integument and Depressor labii inferioris, supplies both, and anastomoses with the inferior labial. The deep branch passes beneath the latter muscle and the bone, supplies the lip, and anastomoses with the inferior labial and mental arteries.

The *muscular* branches are distributed to the internal Pterygoid, Masseter, and Buccinator.

The *inferior labial* passes beneath the Depressor anguli oris, to supply the muscles and integument of the lower lip, anastomosing with the inferior coronary and submental branches of the facial, and with the mental branch of the inferior dental artery.

The *inferior coronary* is derived from the facial artery near the angle of the mouth; it passes upwards and inwards beneath the Depressor anguli oris, and, penetrating the Orbicularis muscle, runs in a tortuous course along the edge of the lower lip between this muscle and the mucous membrane, inosculating with the artery of the opposite side. This artery supplies the labial glands, the mucous membrane, and muscles of the lower lip; and anastomoses with the inferior labial, and mental branch of the inferior dental artery.

The *superior coronary* is larger, and more tortuous in its course than the preceding. It follows the same course along the edge of the upper lip, lying between the mucous membrane and the Orbicularis, and anastomoses with the artery of the opposite side. It supplies the textures of the upper lip, and gives off in its course two or three vessels which ascend to the nose. One, named the artery of the septum, ramifies on the septum of the nares as far as the point of the nose; another supplies the ala of the nose.

The *lateralis nasi* is derived from the facial, as that vessel is ascending along the side of the nose; it supplies the ala and dorsum of the nose, anastomosing with its fellow, the nasal branch of the ophthalmic, the artery of the septum, and the infra-orbital.

The *angular artery* is the termination of the trunk of the facial; it ascends to the inner angle of the orbit, accompanied by a large vein, the angular; it distributes some branches on the cheek which anastomose with the infraorbital, and, after supplying the lachrymal sac, and Orbicularis muscle, terminates by anastomosing with the nasal branch of the ophthalmic artery.

The anastomoses of the facial artery are very numerous, not only with the vessel of the opposite side, but with other vessels from different sources; viz., with the sublingual branch of the lingual, with the mental branch of the inferior dental as it emerges from the dental foramen, with the ascending pharyngeal and posterior palatine, and with the ophthalmic, a branch of the internal carotid; it also inosculates with the transverse facial, and with the infraorbital.

Peculiarities. The facial artery not unfrequently arises by a common trunk with the lingual. This vessel also is subject to some variations in its size, and in the extent to which it supplies the face. It occasionally terminates as the submental, and not unfrequently supplies the face only as high as the angle of the mouth or nose. The deficiency is then supplied by enlargement of one of the neighbouring arteries.

Surgical Anatomy. The passage of the facial artery over the body of the jaw would appear to afford a favourable position for the application of pressure in cases of hæmorrhage from the lips, the result either of an accidental wound, or from an operation; but its application is useless, except for a very short time, on account of the free communication of this vessel with its fellow, and with numerous branches from different sources. In a wound involving the lip, it is better to seize the part between the fingers, and evert it, when the bleeding vessel may be at once secured with a tenaculum. In order to prevent hæmorrhage in cases of excision, or in the removal of diseased growths from the part, the lip should be compressed on each side between the finger and thumb, whilst the surgeon excises the diseased part. In order to stop hæmorrhage where the lip has been divided in an operation, it is necessary in uniting the edges of the wound, to pass the sutures through the cut edges, almost as deep as its mucous surface; by these means, not only are the cut surfaces more neatly and securely adapted to each other, but the possibility of hæmorrhage is prevented by including in the suture the divided artery. If the suture is, on the contrary, passed through merely the cutaneous portion of the wound, hæmorrhage occurs into the cavity of the mouth. The student should, lastly, observe the relation of the angular artery to the lachrymal sac, and it will be seen that, as the vessel passes up along the inner margin of the orbit it ascends on its nasal side. In operating for fistula lachrymalis, the sac should always be opened on its outer side, in order that this vessel may be avoided.

The Occipital Artery (fig. 209) arises from the posterior part of the external carotid, opposite the facial, near the lower margin of the Digastric muscle. At its origin, it is covered by the posterior belly of the Digastric and Stylo-hyoid muscles, and part of the parotid gland, the hypoglossal nerve winding around it from behind forwards; higher up, it passes across the internal carotid artery, the internal jugular vein, and the pneumogastric and spinal accessory nerves; it then ascends to the interval between the transverse process of the atlas, and the mastoid process of the temporal bone, and passes horizontally backwards, grooving the surface of the latter bone, being covered by the Sterno-mastoid, Splenius, Digastric, and Trachelo-mastoid muscles, and resting upon the Complexus, Superior oblique, and Rectus posticus major muscles; it then ascends vertically upwards, piercing the cranial attachment of the Trapezius, and passes in a tortuous course over the occiput, as high as the vertex, where it divides into numerous branches.

The branches given off from this vessel are,

Muscular.	Inferior Meningeal.
Auricular.	Arteria Princeps Cervicis.

The *muscular branches* supply the Digastric, Stylo-hyoid, Sterno-mastoid, Splenius, and Trachelo-mastoid muscles. The branch distributed to the Sterno-mastoid is of large size.

The *auricular branch* supplies the back part of the concha.

The *meningeal branch* ascends with the internal jugular vein, and enters the skull through the foramen lacerum posterius, to supply the dura mater in the posterior fossa.

The *arteria princeps cervicis* (fig. 212) is a large branch which descends along the back part of the neck, and divides into a superficial and deep branch. The former runs beneath the Splenius, giving off branches which perforate that muscle to supply the Trapezius, anastomosing with the superficial cervical artery: the latter passes beneath the Complexus, between it and the Semispinalis colli, and anastomoses with the vertebral, and deep cervical branch of the superior intercostal. The anastomosis between these vessels serves mainly to establish the collateral circulation after ligature of the carotid or subclavian artery.

The cranial branches of the occipital artery are distributed upon the occiput; they are very tortuous, and lie between the integument and Occipito-frontalis, anastomosing with the artery of the opposite side, the posterior auricular, and temporal arteries. They supply the back part of the Occipito-frontalis muscle, the integument and pericranium, and one or two branches occasionally pass through the parietal or mastoid foramina, to supply the dura mater.

The Posterior Auricular Artery (fig. 209) is a small vessel which arises from the external carotid, above the Digastric and Stylo-hyoid muscles, opposite the apex of the styloid process. It ascends, under cover of the parotid gland, to the groove between the cartilage of the ear and the mastoid process, immediately above which it divides into two branches, an anterior, passing forwards to anastomose with the posterior division of the temporal; and a posterior, communicating with the occipital. Just before arriving at the mastoid process, this artery is crossed by the portio dura, and has beneath it the spinal accessory nerve.

Besides several small branches to the Digastric, Stylo-hyoid, and Sterno-mastoid muscles, and to the parotid gland, this vessel gives off two branches,

Stylo-Mastoid. Auricular.

The *stylo-mastoid branch* enters the stylo-mastoid foramen, and supplies the tympanum, mastoid cells, and semicircular canals. In the young subject, a branch from this vessel forms, with the tympanic branch from the internal maxillary, a vascular circle, which surrounds the auditory meatus, and from which delicate vessels ramify on the membrana tympani.

The *auricular branch* is distributed to the back part of the cartilage of the ear, upon which it ramifies minutely, some branches curving round the margin of the fibro-cartilage, others perforating it, to supply its anterior surface.

The Ascending Pharyngeal Artery (fig. 212), the smallest branch of the external carotid, is a long slender vessel, deeply seated in the neck, beneath the other branches of the external carotid and the Stylo-pharyngeus muscle. It arises from the back part of the external carotid, near the commencement of that vessel, and ascends vertically between the internal carotid and the side of the pharynx, to the under surface of the base of the skull, lying on the Rectus capitis anticus major. Its branches may be subdivided into three sets: 1. Those directed outwards to supply muscles and nerves. 2. Those directed inwards to the pharynx. 3. Meningeal branches.

The *external branches* are numerous small vessels, which supply the Recti antici muscles, the sympathetic, hypoglossal and pneumogastric nerves, and the lymphatic glands of the neck, anastomosing with the ascending cervical artery.

The *pharyngeal branches* are three or four in number. Two of these descend to supply the middle and inferior Constrictors and the Stylo-pharyngeus, ramifying in their substance and in the mucous membrane lining them. The largest of the pharyngeal branches passes inwards, running upon the Superior constrictor, and sends ramifications to the soft palate, Eustachian tube, and tonsil, which take the place of the ascending palatine branch of the facial artery, when that vessel is of small size.

The *meningeal branches* consist of several small vessels, which pass through foramina in the base of the skull, to supply the dura mater. One, the posterior meningeal, enters the cranium through the foramen lacerum posterius with the internal jugular vein. A second passes through the foramen lacerum medium; and occasionally a third through the anterior condyloid foramen. They are all distributed to the dura mater.

The Temporal Artery (fig. 209), the smaller of the two terminal branches of the external carotid, appears, from its direction, to be the continuation of that vessel. It commences in the substance of the parotid gland, in the interspace between the neck of the condyle of the lower jaw and the external meatus, crosses over the root of the zygoma, immediately beneath the integument, and divides about two inches above the zygomatic arch into two branches, an anterior and a posterior.

The *anterior temporal* inclines forwards over the forehead, supplying the muscles, integument, and pericranium in this region, and anastomoses with the supraorbital and frontal arteries, its branches being directed from before backwards.

The *posterior temporal*, larger than the anterior, curves upwards and backwards along the side of the head, lying above the temporal fascia, and inosculates with its fellow of the opposite side, and with the posterior auricular and occipital arteries.

The temporal artery, as it crosses the zygoma, is covered by the Attrahens aurem muscle, and by a dense fascia given off from the parotid gland; it is also usually crossed by one or two veins, and accompanied by branches of the facial and auriculo-temporal nerves. Besides some twigs to the parotid gland, the articulation of the jaw, and the Masseter muscle, its branches are the

Transverse Facial. Middle Temporal.
Anterior Auricular.

The *transverse facial* is given off from the temporal before that vessel quits the parotid gland; running forwards through its substance, it passes transversely across the face, between Steno's duct and the lower border of the zygoma, and divides on the side of the face into numerous branches, which supply the parotid gland, the Masseter muscle, and the integument, anastomosing with the facial and infraorbital arteries. This vessel rests on the Masseter, and is accompanied by one or two branches of the facial nerve. It is sometimes a branch of the external carotid.

The *middle temporal artery* arises immediately above the zygomatic arch, and perforating the temporal fascia, supplies the Temporal muscle, anastomosing with the deep temporal branches of the internal maxillary. It occasionally gives off an orbital branch, which runs along the upper border of the zygoma, between the two layers of the temporal fascia, to the outer angle of the orbit. This branch supplies the Orbicularis, and anastomoses with the lachrymal and palpebral branches of the ophthalmic artery.

The *anterior auricular branches* are distributed to the anterior portion of the pinna, the lobule, and part of the external meatus, anastomosing with branches of the posterior auricular.

Surgical Anatomy. It occasionally happens that the surgeon is called upon to perform the operation of arteriotomy upon this vessel in cases of inflammation of the eye or brain. Under these circumstances, the anterior branch is the one usually selected. If the student will consider the relations of the trunk of this vessel, as it crosses the zygomatic arch, with the surrounding structures, he will observe that it is covered by a thick and dense fascia, crossed by one or two veins, and accompanied by branches of the facial and temporo-auricular nerves. Bleeding should not be performed in this situation, as much difficulty may arise from the dense fascia over the vessel preventing a free flow of blood, and considerable pressure is requisite afterwards to repress the hæmorrhage. Again, a varicose aneurism may be formed by the accidental opening of one of the veins in front of the artery; or severe neuralgic pain may arise from the operation implicating one of the nervous filaments in the neighbourhood.

The anterior branch is, on the contrary, subcutaneous, is a large vessel, and as readily compressed as any other portion of the artery; it should consequently always be selected for the operation.

The Internal Maxillary (fig. 210), the larger of the two terminal branches of the external carotid, passes inwards, at right angles from that vessel, to the inner side of the neck of the condyle of the lower jaw, to supply the deep structures of the face. At its origin, it is imbedded in the substance of the parotid gland, being on a level with the lower extremity of the lobe of the ear.

In the first part of its course (maxillary portion), the artery passes horizontally forwards and inwards, between the ramus of the jaw and the internal lateral ligament. The artery here lies parallel with the auriculo-temporal nerve; it crosses the inferior dental nerve, and lies beneath the narrow portion of the External pterygoid muscle.

In the second part of its course (pterygoid portion), it runs obliquely forwards and upwards upon the outer surface of the External pterygoid muscle, being covered by the ramus of the lower jaw, and lower part of the Temporal muscle.

In the third part of its course (spheno-maxillary portion), it approaches the superior maxillary bone, and enters the spheno-maxillary fossa, in the interval between the processes of origin of the External pterygoid, where it lies in relation with Meckel's ganglion, and gives off its terminal branches.

210.—The Internal Maxillary Artery, and its Branches.

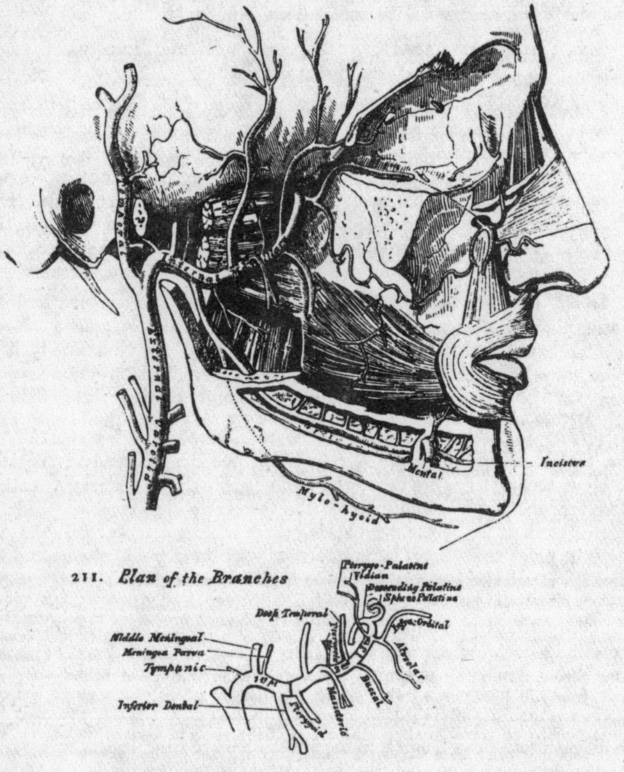

211. Plan of the Branches

Peculiarities. Occasionally, this artery passes between the two Pterygoid muscles. The vessel in this case passes forwards to the interval between the processes of origin of the External pterygoid, in order to reach the maxillary bone. Sometimes the vessel escapes from beneath the External pterygoid by perforating the middle of that muscle.

The branches of this vessel may be divided into three groups, corresponding with its three divisions.

Branches from the Maxillary Portion.

Tympanic.	Small Meningeal.
Middle Meningeal.	Inferior Dental.

The *tympanic branch* passes upwards behind the articulation of the lower jaw, enters the tympanum through the fissure of Glaser, supplies the Laxator tympani, and ramifies upon the membrana tympani, anastomosing with the stylo-mastoid and Vidian arteries.

The *middle meningeal* is the largest of the branches which supply the dura mater. It arises from the internal maxillary between the internal lateral ligament and the neck of the jaw, and passes vertically upwards to the foramen spinosum of the sphenoid bone. On entering the cranium, it divides into two branches, anterior and posterior. The anterior branch, the larger, crosses the great ala of the sphenoid, and reaches the groove, or canal, in the anterior inferior angle of the parietal bone; it then divides into branches, which spread out between the dura mater and internal surface of the cranium, some passing upwards over the parietal bone as far as the vertex, and others backwards to the occipital bone. The posterior branch crosses the squamous portion of the temporal, and on the inner surface of the parietal bone divides into branches which supply the posterior part of the dura mater and cranium. The branches of this vessel are distributed partly to the dura mater, but chiefly to the bones; they anastomose with the arteries of the opposite side, and with the anterior and posterior meningeal.

The middle meningeal, on entering the cranium, gives off the following collateral branches:—1. Numerous small vessels to the ganglion of the fifth nerve, and to the dura mater in this situation. 2. A branch to the facial nerve, which enters the hiatus Fallopii, supplies the facial nerve, and anastomoses with the stylo-mastoid branch of the posterior auricular artery. 3. Orbital branches, which pass through the sphenoidal fissure, or through separate canals in the great wing of the sphenoid to anastomose with the lachrymal or other branches of the ophthalmic artery. 4. Temporal branches, which pass through foramina in the great wing of the sphenoid, and anastomose in the temporal fossa with the deep temporal arteries.

The *small meningeal* is sometimes derived from the preceding. It enters the skull through the foramen ovale, and supplies the Casserian ganglion and dura mater. Before entering the cranium, it gives off a branch to the nasal fossa and soft palate.

The *inferior dental* descends with the dental nerve, to the foramen on the inner side of the ramus of the jaw. It runs along the dental canal in the substance of the bone, accompanied by the nerve, and opposite the first bicuspid tooth divides into two branches, incisor and mental: the former is continued forwards beneath the incisor teeth as far as the symphysis, where it anastomoses with the artery of the opposite side; the mental branch escapes with the nerve at the mental foramen, supplies the structures composing the chin, and anastomoses with the submental, inferior labial, and inferior coronary arteries. As the dental artery enters the foramen, it gives off a mylo-hyoid branch, which runs in the Mylo-hyoid groove, and ramifies on the under surface of the Mylo-hyoid muscle. The dental and incisor arteries during their course through the substance of the bone, give off a few twigs which are lost in the cancellous tissue, and a series of branches which correspond in number to the roots of the teeth: these enter the minute apertures at the extremities of the fangs, and supply the pulp of the teeth.

Branches of the Second, or Pterygoid Portion.

Deep Temporal.	Masseteric.
Pterygoid.	Buccal.

These branches are distributed, as their names imply, to the muscles in the maxillary region.

The *deep temporal branches*, two in number, anterior and posterior, each occupy that part of the temporal fossa indicated by its name. Ascending between the temporal muscle and pericranium, they supply that muscle, and anastomose with the other temporal arteries; the anterior branch communicating with the lachrymal through small branches which perforate the malar bone.

The *pterygoid branches*, irregular in their number and origin, supply the Pterygoid muscles.

The *masseteric* is a small branch which passes outwards above the sigmoid notch of the lower jaw, to the deep surface of the Masseter. It supplies that muscle, and anastomoses with the masseteric branches of the facial and with the transverse facial artery.

The *buccal* is a small branch which runs obliquely forwards between the Internal pterygoid and the ramus of the jaw, to the outer surface of the Buccinator, to which it is distributed, anastomosing with branches of the facial artery.

BRANCHES OF THE THIRD, OR SPHENO-MAXILLARY PORTION.

Alveolar.	Vidian.
Infraorbital.	Pterygo-Palatine.
Posterior or Descending Palatine.	Nasal or Spheno-Palatine.

The *alveolar* is given off from the internal maxillary by a common branch with the infraorbital, and just as the trunk of the vessel is passing into the spheno-maxillary fossa. Descending upon the tuberosity of the superior maxillary bone, it divides into numerous branches; one, the superior dental, larger than the rest, supplies the molar and bicuspid teeth, its branches entering the foramina in the alveolar process; some branches pierce the bone to supply the lining of the antrum, and others are continued forwards on the alveolar process to supply the gums.

The *infraorbital* appears, from its direction, to be the continuation of the trunk of the internal maxillary. It arises from that vessel by a common trunk with the preceding branch, and runs along the infraorbital canal with the superior maxillary nerve, emerging upon the face at the infraorbital foramen, beneath the Levator labii superioris. Whilst contained in the canal, it gives off branches which ascend into the orbit, and supply the Inferior rectus and Inferior oblique muscles, and the lachrymal gland. Other branches descend through canals in the bone, to supply the mucous membrane of the antrum, and the front teeth of the upper jaw. On the face, it supplies the lachrymal sac, and inner angle of the orbit, anastomosing with the facial artery and nasal branch of the ophthalmic; and other branches descend beneath the elevator of the upper lip, and anastomose with the transverse facial and buccal branches.

The four remaining branches arise from that portion of the internal maxillary which is contained in the spheno-maxillary fossa.

The *descending palatine* passes down along the posterior palatine canal with the posterior palatine branches of Meckel's ganglion, and emerging from the posterior palatine foramen, runs forwards in a groove on the inner side of the alveolar border of the hard palate to be distributed to the gums, the mucous membrane of the hard palate, and palatine glands. Whilst it is contained in the palatine canal, it gives off branches, which descend in the accessory palatine canals to supply the soft palate, anastomosing with the ascending palatine artery; and anteriorly it terminates in a small vessel, which ascends in the anterior palatine canal, and anastomoses with the artery of the septum, a branch of the spheno-palatine.

The *Vidian branch* passes backwards along the Vidian canal with the Vidian nerve. It is distributed to the upper part of the pharynx and Eustachian tube, sending a small branch into the tympanum.

The *pterygo-palatine* is also a very small branch, which passes backwards through the pterygo-palatine canal with the pharyngeal nerve, and is distributed to the upper part of the pharynx and Eustachian tube.

The *nasal* or *spheno-palatine* passes through the spheno-palatine foramen into the cavity of the nose, at the back part of the superior meatus, and divides into two branches; one internal, the artery of the septum, passes obliquely downwards and forwards along the septum nasi, supplies the mucous membrane, and anastomoses in front with the ascending branch of the descending palatine. The external branches, two or three in number, supply the mucous membrane covering the lateral wall of the nose, the antrum, and the ethmoid and sphenoid cells.

Surgical Anatomy of the Triangles of the Neck.

The student having considered the relative anatomy of the large arteries of the neck and their branches, and the relations they bear to the veins and nerves, should now examine these structures collectively, as they present themselves in certain regions of the neck, in each of which important operations are being constantly performed.

For this purpose, the Sterno-mastoid, or any other muscles that have been divided in the dissection of the vessels, should be replaced in their normal position; the head should be supported by placing a block at the back of the neck, and the face turned to the side opposite to that which is being examined.

The side of the neck presents a somewhat quadrilateral outline, limited, above, by the lower border of the body of the jaw, and an imaginary line extending from the angle of the jaw to the mastoid process; below, by the prominent upper border of the clavicle; in front, by the median line of the neck; behind, by the anterior margin of the Trapezius muscle. This space is subdivided into two large triangles by the Sterno-mastoid muscle, which passes obliquely across the neck, from the sternum and clavicle, below, to the mastoid process, above. The triangular space in front of this muscle, is called the *anterior triangle*; and that behind it, the *posterior triangle*.

Anterior Triangular Space.

The anterior triangle is limited, in front, by a line extending from the chin to the sternum; behind, by the anterior margin of the Sterno-mastoid; its base, directed upwards, is formed by the lower border of the body of the jaw, and a line extending from the angle of the jaw to the mastoid process; its apex is below, at the sternum. This space is covered by the integument, superficial fascia, Platysma, and deep fascia; it is crossed by branches of the facial and superficial cervical nerves, and is subdivided into three smaller triangles by the Digastric muscle, above, and the anterior belly of the Omo-hyoid, below. These smaller triangles are named from below upwards, the inferior carotid, the superior carotid, and the submaxillary triangle.

The *Inferior Carotid Triangle* is limited, in front, by the median line of the neck; behind, by the anterior margin of the Sterno-mastoid; above, by the anterior belly of the Omo-hyoid; and is covered by the integument, superficial fascia, Platysma, and deep fascia; ramifying between which is seen the descending branch of the superficial cervical nerve. Beneath these superficial structures, are the Sterno-hyoid and Sterno-thyroid muscles, which, together with the anterior margin of the Sterno-mastoid, conceal the lower part of the common carotid artery.*

This vessel is enclosed within its sheath, together with the internal jugular vein, and pneumogastric nerve; the vein lying on the outer side of the artery on the right side of the neck, but overlapping it, or passing directly across it on the left side: the nerve lying between the artery and vein, on a plane posterior to both. In front of the sheath are a few filaments descending from the loop of communication between the descendens and communicans noni; behind the sheath are seen the inferior thyroid artery, the recurrent laryngeal nerve, and the sympathetic nerve; and on its inner side, the trachea, the thyroid gland, much more prominent in the female than in the male, and the lower part of the larynx. By cutting into the upper part of this space, and slightly displacing the Sterno-mastoid muscle, the common carotid artery may be tied below the Omo-hyoid muscle.

* Therefore the carotid artery and jugular vein are not, strictly speaking, contained in this triangle, since they are covered by the Sterno-mastoid muscle, that is to say, lie behind the anterior border of that muscle, which forms the posterior border of the triangle. But as they lie very close to the structures which are really contained in the triangle, and whose position it is essential to remember in operating on this part of the artery, it has seemed expedient to study the relations of all these parts together.

The *Superior Carotid Triangle* is bounded, behind, by the Sterno-mastoid; below, by the anterior belly of the Omo-hyoid; and above, by the posterior belly of the Digastric muscle. Its floor is formed by parts of the Thyro-hyoid, Hyo-glossus, and the inferior and middle Constrictor muscles of the pharynx; and it is covered by the integument, superficial fascia, Platysma, and deep fascia; ramifying between which, are branches of the facial and superficial cervical nerves. This space contains the upper part of the common carotid artery, which bifurcates opposite the upper border of the thyroid cartilage into the external and internal carotid. These vessels are occasionally somewhat concealed from view by the anterior margin of the Sterno-mastoid muscle, which overlaps them. The external and internal carotids lie side by side, the external being the most anterior of the two. The following branches of the external carotid are also met with in this space: the superior thyroid, running forwards and downwards; the lingual directly forwards; the facial, forwards and upwards; the occipital, backwards; and the ascending pharyngeal directly upwards on the inner side of the internal carotid. The veins met with are: the internal jugular, which lies on the outer side of the common and internal carotid arteries; and veins corresponding to the above-mentioned branches of the external carotid, viz., the superior thyroid, the lingual, facial, ascending pharyngeal, and sometimes the occipital; all of which accompany their corresponding arteries, and terminate in the internal jugular. The nerves in this space are the following:—In front of the sheath of the common carotid is the descendens noni. The hypoglossal nerve crosses both carotids above, curving round the occipital artery at its origin. Within the sheath, between the artery and vein, and behind both, is the pneumogastric nerve; behind the sheath, the sympathetic. On the outer side of the vessels, the spinal accessory nerve runs for a short distance before it pierces the Sterno-mastoid muscle; and on the inner side of the internal carotid, just below the hyoid bone, may be seen the superior laryngeal nerve; and still more inferiorly, the external laryngeal nerve. The upper part of the larynx and lower part of the pharynx are also found in the front part of this space.

The *Submaxillary Triangle* corresponds to the part of the neck immediately beneath the body of the jaw. It is bounded, above, by the lower border of the body of the jaw, the parotid gland, and mastoid process; behind, by the posterior belly of the Digastric and Stylo-hyoid muscles; in front, by the middle line of the neck The floor of this space is formed by the anterior belly of the Digastric, the Mylo-hyoid, and Hyo-glossus muscles; and it is covered by the integument, superficial fascia, Platysma, and deep fascia; ramifying between which are branches of the facial and ascending filaments of the superficial cervical nerve. This space contains, in front, the submaxillary gland, imbedded in the substance of which are the facial artery and vein, and their glandular branches; beneath this gland, on the surface of the Mylo-hyoid muscle, is the submental artery, and the mylo-hyoid artery and nerve. The back part of this space is separated from the front part by the stylo-maxillary ligament; it contains the external carotid artery, ascending deeply in the substance of the parotid gland; this vessel here lies in front of, and superficial to, the internal carotid, being crossed by the facial nerve, and gives off in its course the posterior auricular, temporal, and internal maxillary branches; more deeply is the internal carotid, the internal jugular vein, and the pneumogastric nerve, separated from the external carotid by the Stylo-glossus and Stylo-pharyngeus muscles, and the glosso-pharyngeal nerve.*

Posterior Triangular Space.

The posterior triangular space is bounded, in front by the Sterno-mastoid

* The same remark will apply to this triangle as was made about the inferior carotid triangle. The structures enumerated, as contained in the back part of the space, lie, strictly speaking, beneath the muscles which form the posterior boundary of the triangle; but as it is very important to bear in mind their close relation to the parotid gland and its boundaries (on account of the frequency of surgical operations on this gland) all these parts are spoken of together.

muscle; behind, by the anterior margin of the Trapezius; its base corresponds to the upper border of the clavicle; its apex, to the occiput. This space is crossed, about an inch above the clavicle, by the posterior belly of the Omo-hyoid, which divides it unequally into two, an upper or occipital, and a lower or subclavian triangle.

The *Occipital*, the larger of the two posterior triangles, is bounded, in front, by the Sterno-mastoid; behind, by the Trapezius; below, by the Omo-hyoid. Its floor is formed from above downwards by the Splenius, Levator anguli scapulæ, and the middle and posterior Scaleni muscles. It is covered by the integument, the Platysma below, the superficial and deep fasciæ; and crossed, above, by the ascending branches of the cervical plexus; the spinal accessory nerve is directed obliquely across the space from the Sterno-mastoid, which it pierces, to the under surface of the Trapezius; below, the descending branches of the cervical plexus and the transversalis colli artery and vein cross the space. A chain of lymphatic glands is also found running along the posterior border of the Sterno-mastoid, from the mastoid process to the root of the neck.

The *Subclavian*, the smaller of the two posterior triangles, is bounded, above, by the posterior belly of the Omo-hyoid; below, by the clavicle; its base, directed forwards, being formed by the Sterno-mastoid. The size of this space varies according to the extent of attachment of the clavicular portion of the Sterno-mastoid and Trapezius muscles, and also according to the height at which the Omo-hyoid crosses the neck above the clavicle. The height also of this space varies much, according to the position of the arm, being much diminished by raising the limb, on account of the ascent of the clavicle, and increased by drawing the arm downwards, when that bone is depressed. This space is covered by the integument, superficial and deep fascia: and crossed by the descending branches of the cervical plexus. Just above the level of the clavicle, the third portion of the subclavian artery curves outwards and downwards from the outer margin of the Scalenus anticus, across the first rib to the axilla. Sometimes this vessel rises as high as an inch and a half above the clavicle, or to any point intermediate between this and its usual level. Occasionally, it passes in front of the Scalenus anticus, or pierces the fibres of that muscle. The subclavian vein lies behind the clavicle, and is usually not seen in this space; but it occasionally rises as high up as the artery, and has even been seen to pass with that vessel behind the Scalenus anticus. The brachial plexus of nerves lies above the artery, and in close contact with it. Passing transversely across the clavicular margin of the space, are the suprascapular vessels; and traversing its upper angle in the same direction, the transverse cervical vessels. The external jugular vein runs vertically downwards behind the posterior border of the Sterno-mastoid, to terminate in the Subclavian vein; it receives the transverse cervical and suprascapular veins, which occasionally form a plexus in front of the artery, and a small vein which crosses the clavicle from the cephalic. The small nerve to the Subclavius also crosses this triangle about its middle. A lymphatic gland is also found in the space.

Internal Carotid Artery.

The internal carotid artery commences at the bifurcation of the common carotid, opposite the upper border of the thyroid cartilage, and runs perpendicularly upwards, in front of the transverse processes of the three upper cervical vertebræ, to the carotid foramen in the petrous portion of the temporal bone. After ascending in it for a short distance, it passes forwards and inwards through the carotid canal, and, again ascending a little by the side of the sella Turcica, curves upwards by the anterior clinoid process, where it pierces the dura mater, and divides into its terminal branches.

This vessel supplies the anterior part of the brain, the eye, and its appendages. Its size, in the adult, is equal to that of the external carotid. In the child, it is larger than that vessel. It is remarkable for the number of curvatures that it presents in different parts of its course. In its cervical portion it occasionally presents one or two flexures near the base of the skull, whilst through the rest of its extent it describes a double curvature which resembles the italic letter ∫ placed horizontally ~. These curvatures most probably diminish the velocity of the current of blood, by increasing the extent of surface over which it moves, and adding to the amount of impediment produced from friction. In considering the

212.—The Internal Carotid and Vertebral Arteries. Right Side.

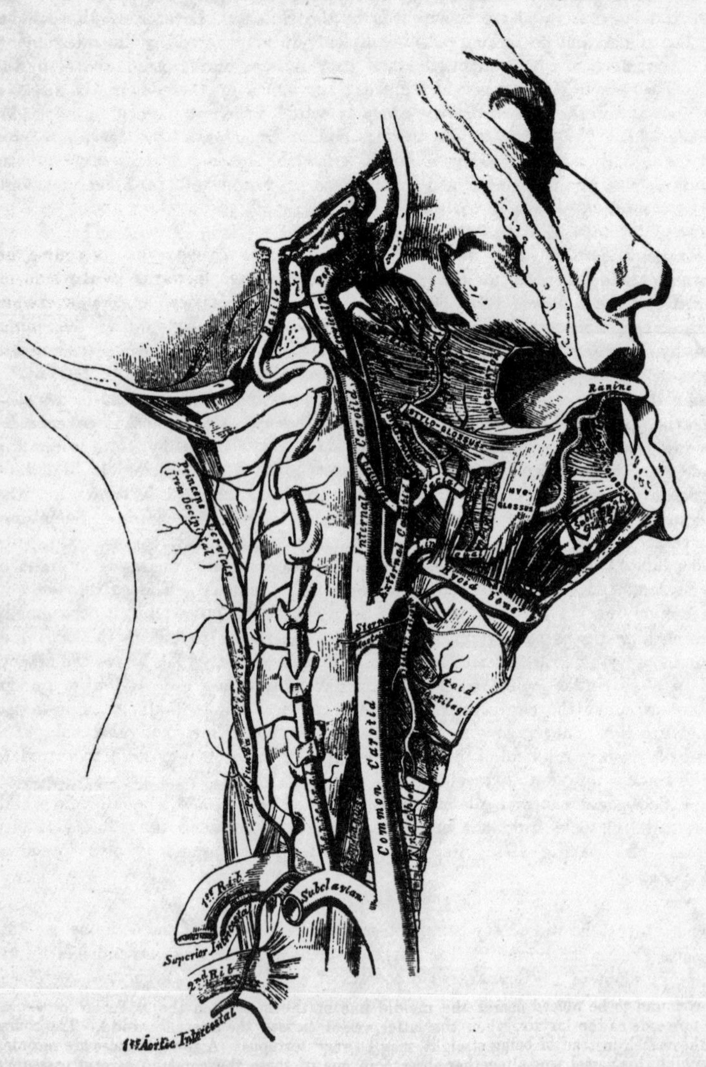

course and relations of this vessel, it may be conveniently divided into four portions, a cervical, petrous, cavernous, and cerebral.

Cervical Portion. This portion of the internal carotid is superficial at its commencement, being contained in the superior carotid triangle, and lying on the same level as the external carotid, but behind that artery, overlapped by the

Sterno-mastoid, and covered by the Platysma, deep fascia, and integument: it then passes beneath the parotid gland, being crossed by the hypoglossal nerve, the Digastric and Stylo-hyoid muscles, and the external carotid and occipital arteries. Higher up, it is separated from the external carotid by the Stylo-glossus and Stylo-pharyngeus muscles, the glosso-pharyngeal nerve, and pharyngeal branch of the vagus. It is in relation, *behind*, with the Rectus anticus major, the superior cervical ganglion of the sympathetic, and superior laryngeal nerve; *externally*, with the internal jugular vein, and pneumogastric nerve; *internally*, with the pharynx, tonsil, and ascending pharyngeal artery.

Petrous Portion. When the internal carotid artery enters the canal in the petrous portion of the temporal bone, it first ascends a short distance, then curves forwards and inwards, and again ascends as it leaves the canal to enter the cavity of the skull. In this canal, the artery lies at first anterior to the tympanum, from which it is separated by a thin bony lamella, which is cribriform in the young subject, and often absorbed in old age. It is separated from the bony wall of the carotid canal by a prolongation of dura mater, and is surrounded by filaments of the carotid plexus.

Cavernous Portion. The internal carotid artery, in this part of its course, at first ascends to the posterior clinoid process, then passes forwards by the side of the body of the sphenoid bone, being situated on the inner wall of the cavernous sinus, in relation, externally, with the sixth nerve, and covered by the lining membrane of the sinus. The third, fourth, and ophthalmic nerves are placed on the outer wall of the sinus, being separated from its cavity by the lining membrane.

Cerebral Portion. On the inner side of the anterior clinoid process the internal carotid curves upwards, perforates the dura mater bounding the sinus, and is received into a sheath of the arachnoid. This portion of the artery is on the outer side of the optic nerve; it lies at the inner extremity of the fissure of Sylvius, having the third nerve externally.

Plan of the Relations of the Internal Carotid Artery in the Neck.

In front.
Skin, superficial and deep fasciæ.
Parotid gland.
Stylo-glossus and Stylo-pharyngeus muscles.
Glosso-pharyngeal nerve.

Externally.
Internal jugular vein.
Pneumogastric nerve.

Internal Carotid Artery.

Internally.
Pharynx.
Ascending pharyngeal artery.
Tonsil.

Behind.
Rectus anticus major.
Sympathetic.
Superior laryngeal nerve.

Peculiarities. The length of the internal carotid varies according to the length of the neck, and also according to the point of bifurcation of the common carotid. Its origin sometimes takes place from the arch of the aorta; in such rare instances, this vessel has been found to be placed nearer the middle line of the neck than the external carotid, as far upwards as the larynx, when the latter vessel crossed the internal carotid. The course of the vessel, instead of being straight, may be very tortuous. A few instances are recorded in which this vessel was altogether absent: in one of these the common carotid passed up the neck, and gave off the usual branches of the external carotid: the cranial portion of the internal carotid being replaced by two branches of the internal maxillary, which entered the skull through the foramen rotundum and ovale, and joined to form a single vessel.

Surgical Anatomy The cervical part of the internal carotid is sometimes wounded by a stab or gun-shot wound in the neck, or even occasionally by a stab from within the mouth, as when a person receives a thrust from the end of a parasol, or falls down with a tobacco-pipe in his mouth. In such cases a ligature should be applied to the common carotid. The relation of the internal carotid with the tonsil should be especially remembered, as instances have occurred in which the artery has been wounded during the operation of scarifying the tonsil, and fatal hæmorrhage has supervened.

The branches given off from the internal carotid are:

From the Petrous Portion .	Tympanic.
From the Cavernous Portion	Arteriæ receptaculi. Anterior meningeal. Ophthalmic.
From the Cerebral Portion	Anterior cerebral. Middle cerebral. Posterior communicating. Anterior choroid.

The cervical portion of the internal carotid gives off no branches.

The *tympanic* is a small branch which enters the cavity of the tympanum, through a minute foramen in the carotid canal, and anastomoses with the tympanic branch of the internal maxillary, and with the stylo-mastoid artery.

The *arteriæ receptaculi* are numerous small vessels, derived from the internal carotid in the cavernous sinus; they supply the pituitary body, the Casserian ganglion, and the walls of the cavernous and inferior petrosal sinuses. One of these branches, distributed to the dura mater, is called the *anterior meningeal*; it anastomoses with the middle meningeal.

The Ophthalmic Artery arises from the internal carotid, just as that vessel

213.—The Ophthalmic Artery and its Branches, the Roof of the Orbit having been removed.

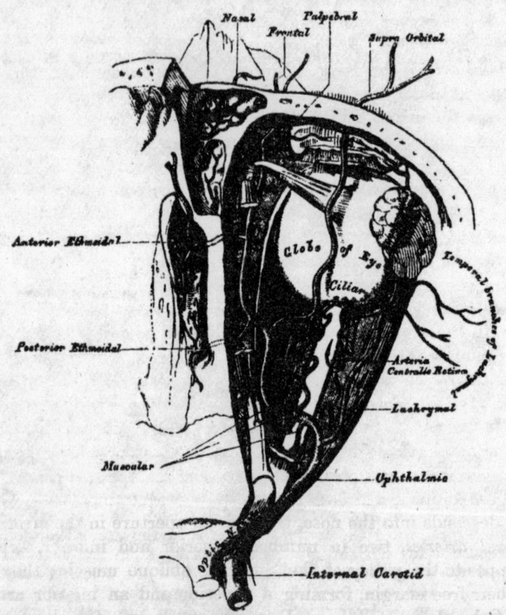

is emerging from the cavernous sinus, on the inner side of the anterior clinoid process, and enters the orbit through the optic foramen, below and on the outer side of the optic nerve. It then passes across the nerve, to the inner wall of the orbit, and thence horizontally forwards, beneath the lower border of the Superior oblique muscle to the inner angle of the eye, where it divides into two terminal branches, the frontal, and nasal.

Branches. The branches of this vessel may be divided into an orbital group, which are distributed to the orbit and surrounding parts; and an ocular group, which supply the muscles and globe of the eye.

Orbital Group.	*Ocular Group.*
Lachrymal.	Muscular.
Supraorbital.	Anterior ciliary.
Posterior ethmoidal.	Short ciliary.
Anterior ethmoidal.	Long ciliary.
Palpebral.	Arteria centralis retinæ.
Frontal.	
Nasal.	

The *lachrymal* is the first, and one of the largest branches, derived from the ophthalmic, arising close to the optic foramen: not unfrequently it is given off from the artery before it enters the orbit. It accompanies the lachrymal nerve along the upper border of the External rectus muscle, and is distributed to the lachrymal gland. Its terminal branches, escaping from the gland, are distributed to the upper eyelid and conjunctiva, anastomosing with the palpebral arteries. The lachrymal artery gives off one or two malar branches; one of which passes through a foramen in the malar bone to reach the temporal fossa, and anastomoses with the deep temporal arteries. The other appears on the cheek, and anastomoses with the transverse facial. A branch is also sent backwards, through the sphenoidal fissure, to the dura mater, which anastomoses with a branch of the middle meningeal artery.

Peculiarities. The lachrymal artery is sometimes derived from one of the anterior branches of the middle meningeal artery.

The *supraorbital artery*, the largest branch of the ophthalmic, arises from that vessel above the optic nerve. Ascending so as to rise above all the muscles of the orbit, it passes forwards, with the frontal nerve, between the periosteum and Levator palpebræ; and passing through the supraorbital foramen, divides into a superficial and deep branch, which supply the muscles and integument of the forehead and pericranium, anastomosing with the temporal, angular branch of the facial, and the artery of the opposite side. This artery in the orbit supplies the Superior rectus and the Levator palpebræ, sends a branch inwards, across the pulley of the Superior oblique muscle, to supply the parts at the inner canthus; and at the supraorbital foramen, frequently transmits a branch to the diploë.

The *ethmoidal branches* are two in number; posterior and anterior. The former, which is the smaller, passes through the posterior ethmoidal foramen, supplies the posterior ethmoidal cells, and entering the cranium, gives off a meningeal branch, which supplies the adjacent dura mater, and nasal branches, which descend into the nose through apertures in the cribriform plate, anastomosing with branches of the spheno-palatine. The anterior ethmoidal artery accompanies the nasal nerve through the anterior ethmoidal foramen, supplies the anterior ethmoidal cells and frontal sinuses, and entering the cranium, divides into a meningeal branch, which supplies the adjacent dura mater, and a nasal branch which descends into the nose, through an aperture in the cribriform plate.

The *palpebral arteries*, two in number, superior and inferior, arise from the ophthalmic, opposite the pulley of the Superior oblique muscle; they encircle the eyelids near their free margin, forming a superior and an inferior arch, which lie between the Orbicularis muscle and tarsal cartilages; the superior palpebral, inosculating at the outer angle of the orbit with the orbital branch of the temporal artery, the inferior palpebral with the orbital branch of the infra-orbital artery, at the inner side of the lid. From this anastomosis, a branch passes to the nasal duct, ramifying in its mucous membrane, as far as the inferior meatus.

214.—The Arteries of the Base of the Brain. The Right Half of the Cerebellum and Pons have been removed.

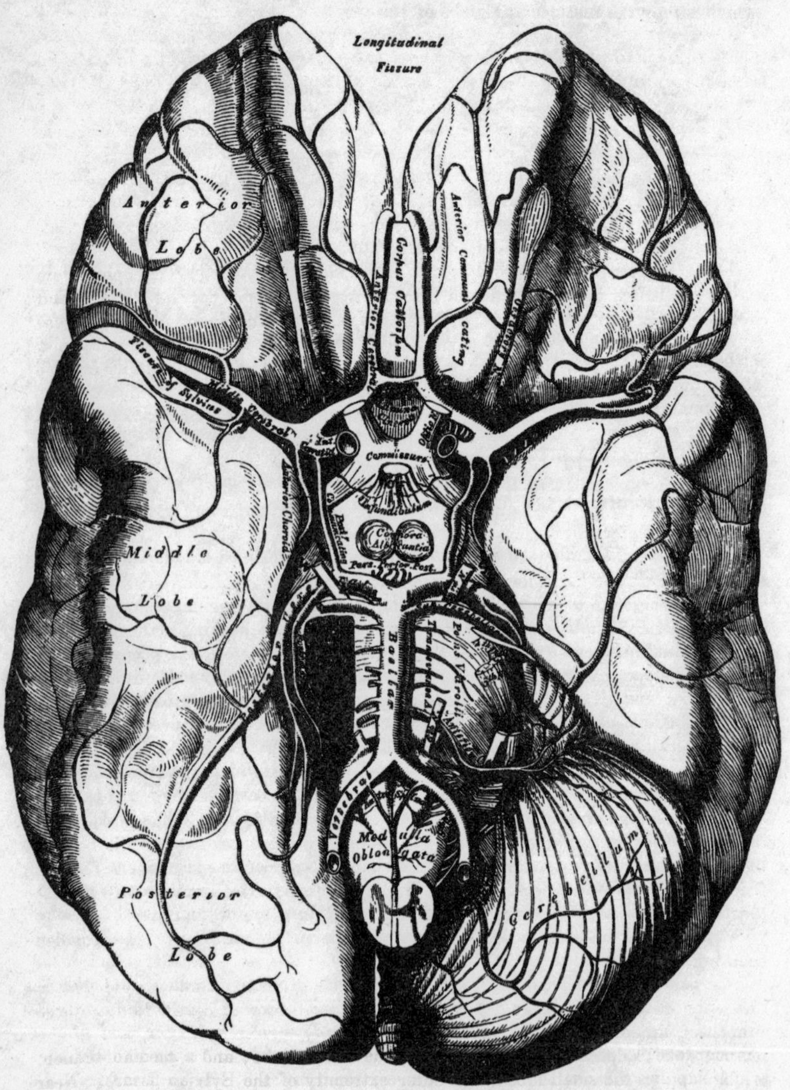

The *frontal artery*, one of the terminal branches of the ophthalmic, passes from the orbit at its inner angle, and ascending on the forehead, supplies the muscles, integument, and pericranium, anastomosing with the supraorbital artery.

The *nasal artery*, the other terminal branch of the ophthalmic, emerges from the orbit above the tendo oculi, and, after giving a branch to the lachrymal sac, divides into two, one of which anastomoses with the angular artery, the other branch, the dorsalis nasi, runs along the dorsum of the nose, supplies its entire surface, and anastomoses with the artery of the opposite side.

The *ciliary arteries* are divisible into three groups, the short, long, and anterior. The *short ciliary arteries*, from twelve to fifteen in number, arise from the ophthalmic, or some of its branches; they surround the optic nerve as they pass forwards to the posterior part of the eyeball, pierce the sclerotic coat around the entrance of the nerve, and supply the choroid coat and ciliary processes. The *long ciliary arteries*, two in number, also pierce the posterior part of the sclerotic, and run forwards, along each side of the eyeball, between the sclerotic and choroid, to the ciliary ligament, where they divide into two branches; these form an arterial circle around the circumference of the iris, from which numerous radiating branches pass forwards, in its substance, to its free margin, where they form a second arterial circle around its pupillary margin. The *anterior ciliary arteries* are derived from the muscular branches; they pierce the sclerotic a short distance from the cornea, and terminate in the great arterial circle of the iris.

The *arteria centralis retinæ* is one of the smallest branches of the ophthalmic artery. It arises near the optic foramen, pierces the optic nerve obliquely, and runs forwards, in the centre of its substance, to the retina, in which its branches are distributed as far forwards as the ciliary processes. In the human fœtus, a small vessel passes forwards, through the vitreous humour, to the posterior surface of the capsule of the lens.

The *muscular branches*, two in number, superior and inferior, supply the muscles of the eyeball. The superior, the smaller, often wanting, supplies the Levator palpebræ, Superior rectus, and Superior oblique. The inferior, more constant in its existence, passes forwards, between the optic nerve and Inferior rectus, and is distributed to the External and Inferior recti, and Inferior oblique. This vessel gives off most of the anterior ciliary arteries.

The Cerebral *branches* of the internal carotid are, the anterior cerebral, the middle cerebral, the posterior communicating, and the anterior choroid.

The *anterior cerebral* arises from the internal carotid, at the inner extremity of the fissure of Sylvius. It passes forwards in the great longitudinal fissure between the two anterior lobes of the brain, being connected, soon after its origin, with the vessel of the opposite side by a short anastomosing trunk, about two lines in length, the *anterior communicating*. The two anterior cerebral arteries, lying side by side, curve round the anterior border of the corpus callosum, and run along its upper surface to its posterior part, where they terminate by anastomosing with the posterior cerebral arteries. They supply the olfactory and optic nerves, the under surface of the anterior lobes, the third ventricle, the anterior perforated space, the corpus callosum, and the inferior surface of the hemispheres.

The *anterior communicating artery* is a short branch, about two lines in length, but of moderate size, connecting together the two anterior cerebral arteries across the longitudinal fissure. Sometimes this vessel is wanting, the two arteries joining together to form a single trunk, which afterwards subdivides. Or the vessel may be wholly or partially subdivided into two; frequently, it is longer and smaller than usual.

The *middle cerebral artery*, the largest branch of the internal carotid, passes obliquely outwards along the fissure of Sylvius, within which it divides into three branches: an anterior, which supplies the pia mater, investing the surface of the anterior lobe; a posterior, which supplies the middle lobe; and a median branch, which supplies the small lobe at the outer extremity of the Sylvian fissure. Near its origin, this vessel gives off numerous small branches, which enter the substantia perforata, to be distributed to the corpus striatum.

The *posterior communicating artery* arises from the back part of the internal carotid, runs directly backwards, and anastomoses with the posterior cerebral, a branch of the basilar. This artery varies considerably in size, being sometimes small, and occasionally so large that the posterior cerebral may be considered as arising from the internal carotid rather than from the basilar. It is frequently larger on one side than on the other side.

The *anterior choroid* is a small but constant branch which arises from the back part of the internal carotid, near the posterior communicating artery. Passing backwards and outwards, it enters the descending horn of the lateral ventricle, beneath the edge of the middle lobe of the brain. It is distributed to the hippocampus major, corpus fimbriatum, and choroid plexus.

ARTERIES OF THE UPPER EXTREMITY.

The artery which supplies the upper extremity, continues as a single trunk from its commencement, so far as the elbow; but different portions of it have received different names, according to the region through which it passes. That part of the vessel which extends from its origin, as far as the outer border of the first rib, is termed the subclavian; beyond this point to the lower border of the axilla, it is termed the axillary; and from the lower margin of the axillary space to the bend of the elbow, it is termed brachial; here, the single trunk terminates by dividing into two branches, the radial and ulnar, an arrangement precisely similar to what occurs in the lower limb.

Subclavian Arteries.

The subclavian artery on the right side arises from the arteria innominata, opposite the right sterno-clavicular articulation; on the left side it arises from the arch of the aorta. It follows, therefore, that these two vessels must, in the first part of their course, differ in their length, their direction, and in their relation with neighbouring parts.

In order to facilitate the description of these vessels, more especially in a surgical point of view, each subclavian artery has been divided into three parts. The first portion, on the right side, ascends obliquely outwards, from the origin of the vessel to the inner border of the Scalenus anticus. On the left side it ascends perpendicularly, to gain the inner border of that muscle. The second part passes outwards, behind the Scalenus anticus; and the third part passes from the outer margin of that muscle, beneath the clavicle, to the lower border of the first rib, where it becomes the axillary artery. The first portions of these two vessels differ so much in their course, and in their relation with neighbouring parts, that they will be described separately. The second and third parts are precisely alike, on both sides.

First Part of the Right Subclavian Artery. (Figs. 205, 207.)

The right subclavian artery arises from the arteria innominata, opposite the right sterno-clavicular articulation, passes upwards and outwards across the root of the neck, and terminates at the inner margin of the Scalenus anticus muscle. In this part of its course, it ascends a little above the clavicle, the extent to which it does so varying in different cases. It is covered, *in front*, by the integument, superficial fascia, Platysma, deep fascia, the clavicular origin of the Sterno-mastoid, the Sterno-hyoid, and Sterno-thyroid muscles, and another layer of the deep fascia. It is crossed by the internal jugular and vertebral veins, and by the pneumogastric, the cardiac branches of the sympathetic, and phrenic nerves. *Beneath*, the artery is invested by the pleura, and *behind*, it is separated by a cellular interval from the Longus colli, the transverse process of the seventh cervical vertebra, and the sympathetic; the recurrent laryngeal nerve winding around the lower and back part of the vessel. The subclavian vein lies below the subclavian artery, immediately behind the clavicle.

Plan of Relations of First Portion of Right Subclavian Artery.

In front.
Clavicular origin of Sterno-mastoid.
Sterno-hyoid and Sterno-thyroid.
Internal jugular and vertebral veins.
Pneumogastric, cardiac and phrenic nerves.

Right Subclavian Artery. First portion.

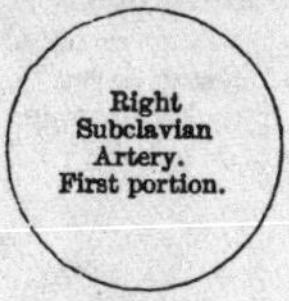

Beneath.
Pleura.

Behind.
Recurrent laryngeal nerve.
Sympathetic.
Longus colli.
Transverse process of seventh cervical vertebra.

First Part of the Left Subclavian Artery. (Fig. 205.)

The left subclavian artery arises from the end of the transverse portion of the arch of the aorta, opposite the second dorsal vertebra, and ascends to the inner margin of the first rib, behind the insertion of the Scalenus anticus muscle. This vessel is, therefore, longer than the right, situated more deeply in the cavity of the chest, and directed almost vertically upwards, instead of arching outwards like the vessel of the opposite side.

It is in relation, *in front*, with the pleura, the left lung, the pneumogastric, phrenic, and cardiac nerves, which lie parallel with it, the left carotid artery, left internal jugular and innominate veins, and is covered by the Sterno-thyroid, Sterno-hyoid, and Sterno-mastoid muscles; *behind*, it is in relation with the œsophagus, thoracic duct, inferior cervical ganglion of the sympathetic, Longus colli, and vertebral column. To its *inner side* are the œsophagus, trachea and thoracic duct; to its *outer side*, the pleura.

Plan of Relations of First Portion of Left Subclavian Artery.

In front.
Pleura and left lung.
Pneumogastric, cardiac, and phrenic nerves.
Left carotid artery.
Left internal jugular and innominate veins.
Sterno-thyroid, Sterno-hyoid, and Sterno-mastoid muscles.

Inner side.
Œsophagus.
Trachea.
Thoracic duct.

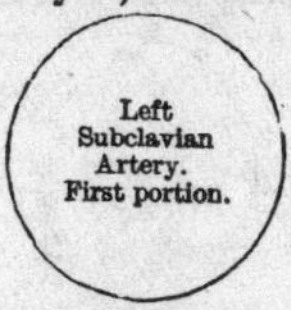

Left Subclavian Artery. First portion.

Outer side.
Pleura.

Behind.
Œsophagus and thoracic duct.
Inferior cervical ganglion of sympathetic.
Longus colli and vertebral column.

Second and Third Parts of the Subclavian Artery. (Fig. 207.)

The *Second Portion of the Subclavian Artery* lies behind the Scalenus anticus muscle; it is very short, and forms the highest part of the arch described by that vessel.

Relations. It is covered, *in front*, by the integument, Platysma, Sterno-mastoid, cervical fascia, and by the phrenic nerve, which is separated from the artery by the Scalenus anticus muscle. *Behind*, it is in relation with the Middle scalenus. *Above*, with the brachial plexus of nerves. *Below*, with the pleura. The subclavian vein lies below and in front of the artery, separated from it by the Scalenus anticus.

PLAN OF RELATIONS OF SECOND PORTION OF SUBCLAVIAN ARTERY.

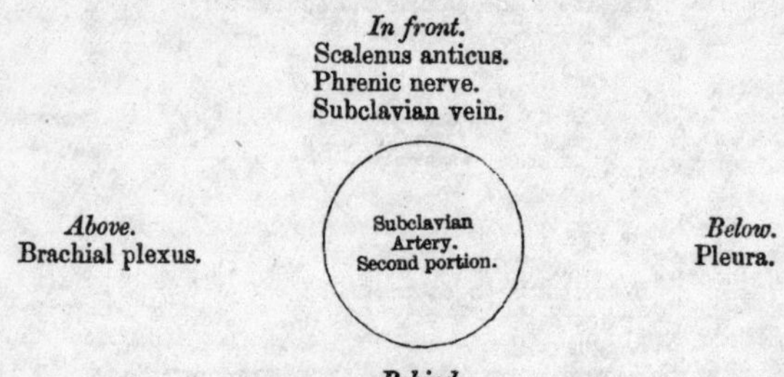

Behind.
Pleura and Middle Scalenus.

The *Third Portion of the Subclavian Artery* passes downwards and outwards from the outer margin of the Scalenus anticus to the lower border of the first rib, where it becomes the axillary artery. This portion of the vessel is the most superficial, and is contained in a triangular space, the base of which is formed in front by the Anterior scalenus, and the two sides by the Omo-hyoid above and the clavicle below.

Relations. It is covered, *in front*, by the integument, the superficial fascia, the Platysma, deep fascia; and by the clavicle, the Subclavius muscle, and the suprascapular artery and vein; the clavicular descending branches of the cervical plexus and the nerve to the Subclavius pass vertically downwards in front of the artery. The external jugular vein crosses it at its inner side, and receives the suprascapular and transverse cervical veins, which occasionally form a plexus in front of it. The subclavian vein is below the artery, lying close behind the clavicle. *Behind*, it lies on the Middle scalenus muscle. *Above* it, and to its outer side, is the brachial plexus, and Omo-hyoid muscle. *Below*, it rests on the outer surface of the first rib.

PLAN OF RELATIONS OF THIRD PORTION OF SUBCLAVIAN ARTERY.

In front.
Cervical fascia.
External jugular, suprascapular, and transverse cervical veins.
Descending branches of cervical plexus.
Subclavius muscle, suprascapular artery, and clavicle.

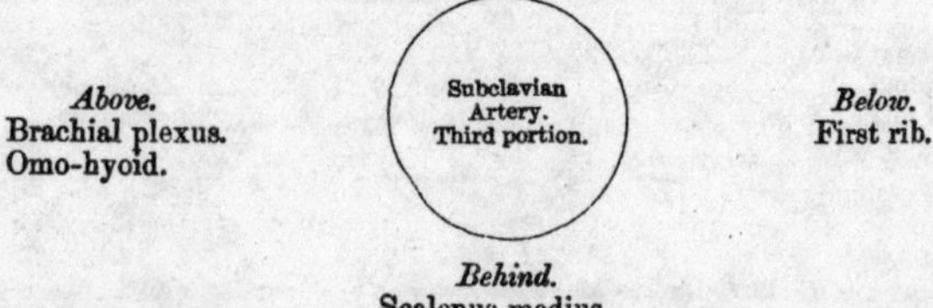

Behind.
Scalenus medius.

Peculiarities. The Subclavian arteries vary in their origin, their course, and the height to which they rise in the neck.

The origin of the right subclavian from the innominate takes place, in some cases, above the sterno-clavicular articulation; and occasionally, but less frequently, in the cavity of the thorax, below that joint. Or the artery may arise as a separate trunk from the arch of the aorta. In such cases it may be either the first, second, third, or even the last branch derived from that vessel: in the majority of cases, it is the first or last, rarely the second or third. When it is the first branch, it occupies the ordinary position of the innominate artery; when the second or third, it gains its usual position by passing behind the right carotid; and when the last branch, it arises from the left extremity of the arch, at its upper or back part, and passes obliquely towards the right side, usually behind the œsophagus and right carotid, sometimes between the œsophagus and trachea, to the upper border of the first rib, whence it follows its ordinary course. In very rare instances, this vessel arises from the thoracic aorta, as low down as the fourth dorsal vertebra. Occasionally it perforates the Anterior scalenus; more rarely it passes in front of that muscle: sometimes the subclavian vein passes with the artery behind the Scalenus. The artery sometimes ascends as high as an inch and a half above the clavicle, or to any intermediate point between this and the upper border of the bone, the right subclavian usually ascending higher than the left.

The left subclavian is occasionally joined at its origin with the left carotid.

Surgical Anatomy. The relations of the subclavian arteries of the two sides having been examined, the student should direct his attention to consider the best position in which compression of the vessel may be effected, or in what situation a ligature may be best applied in cases of aneurism or wounds.

Compression of the subclavian artery is required in cases of operations about the shoulder, in the axilla, or at the upper part of the arm; and the student will observe that there is only one situation in which it can be effectually applied, viz., where the artery passes across the outer surface of the first rib. In order to compress the vessel in this situation, the shoulder should be depressed, and the surgeon, grasping the side of the neck, may press with his thumb in the hollow behind the clavicle downwards against the rib; if from any cause the shoulder cannot be sufficiently depressed, pressure may be made from before backwards, so as to compress the artery against the Middle scalenus and transverse process of the seventh cervical vertebra. In appropriate cases, a preliminary incision may be made through the cervical fascia, and the finger may be pressed down directly upon the artery.

Ligature of the subclavian artery may be required in cases of wounds, or of aneurism in the axilla; and the third part of the artery is that which is most favourable for an operation, on account of its being comparatively superficial, and most remote from the origin of the large branches. In those cases where the clavicle is not displaced, this operation may be performed with comparative facility: but where the clavicle is pushed up by a large aneurismal tumour in the axilla, the artery is placed at a great depth from the surface, which materially increases the difficulty of the operation. Under these circumstances, it becomes a matter of importance to consider the height to which this vessel reaches above the bone. In ordinary cases, its arch is about half an inch above the clavicle, occasionally as high as an inch and a half, and sometimes so low as to be on a level with its upper border. If the clavicle is displaced, these variations will necessarily make the operation more or less difficult, according as the vessel is more or less accessible.

The chief points in the operation of tying the third portion of the subclavian artery are as follows: the patient being placed on a table in the horizontal position, and the shoulder depressed as much as possible, the integument should be drawn downwards upon the clavicle, and an incision made through it upon that bone from the anterior border of the Trapezius to the posterior border of the Sterno-mastoid, to which may be added a short vertical incision meeting the centre of the preceding; the Platysma and cervical fascia should be divided upon a director, and if the interval between the Trapezius and Sterno-mastoid muscles be insufficient for the performance of the operation, a portion of one or both may be divided. The external jugular vein will now be seen towards the inner side of the wound: this and the suprascapular and transverse cervical veins which terminate in it should be held aside, and if divided, both ends should be tied: the suprascapular artery should be avoided, and the Omo-hyoid muscle must now be looked for, and held aside if necessary. In the space beneath this muscle, careful search must be made for the vessel; the deep fascia having been divided with the finger-nail or silver scalpel, the outer margin of the Scalenus muscle must be felt for, and the finger being guided by it to the first rib, the pulsation of the subclavian artery will be felt as it passes over the rib. The aneurism needle may then be passed around the vessel from before backwards, by which means the vein will be avoided, care being taken not to include a branch of the brachial plexus instead of the artery in the ligature. If the clavicle is so raised by the tumour that the application of the ligature cannot be effected in this situation, the artery may be tied above the first rib, or even behind the Scalenus muscle: the difficulties of the operation in such a case will be materially increased, on account of the greater depth of the artery, and the alteration in position of the surrounding parts.

The second part of the subclavian artery, from being that portion which rises highest in the neck, has been considered favourable for the application of the ligature, when it is difficult to tie the artery in the third part of its course. There are, however, many objections to the operation in this situation. It is necessary to divide the Scalenus anticus muscle, upon which lies the phrenic nerve, and at the inner side of which is situated the internal jugular vein; and a wound of either of these structures might lead to the most dangerous consequences. Again, the artery is in contact, below, with the pleura, which must also be avoided; and, lastly, the proximity of so many of its larger branches arising internal to this point, must be a still further objection to the operation. If, however, it has been determined to perform the operation in this situation, it should be remembered that it occasionally happens, that the artery passes in front of the Scalenus anticus, or through the fibres of that muscle; and that the vein sometimes passes with the artery behind the Scalenus anticus.

In those cases of aneurism of the axillary or subclavian artery which encroach upon the outer portion of the Scalenus muscle to such an extent that a ligature cannot be applied in that situation, it may be deemed advisable, as a last resource, to tie the first portion of the subclavian artery. On the left side, this operation is almost impracticable; the great depth of the artery from the surface, its intimate relation with the pleura, and its close proximity to so many important veins and nerves, present a series of difficulties which it is next to impossible to overcome.* On the right side, the operation is practicable, and has been performed, though not with success. The main objection to the operation in this situation is the smallness of the interval which usually exists between the commencement of the vessel, and the origin of the nearest branch. This operation may be performed in the following manner:—The patient being placed on the table in the horizontal position, with the neck extended, an incision should be made parallel with the inner part of the clavicle, and a second along the inner border of the Sterno-mastoid, meeting the former at an angle. The sternal attachment of the Sterno-mastoid may now be divided on a director, and turned outwards; a few small arteries and veins, and occasionally the anterior jugular, must be avoided, and the Sterno-hyoid and Sterno-thyroid muscles divided in the same manner as the preceding muscle. After tearing through the deep fascia with the finger-nail, the internal jugular vein will be seen crossing the subclavian artery; this should be pressed aside, and the artery secured by passing the needle from below upwards, by which the pleura is more effectually avoided. The exact position of the vagus nerve, the recurrent laryngeal, the phrenic and sympathetic nerves should be remembered, and the ligature should be applied near the origin of the vertebral, in order to afford as much room as possible for the formation of a coagulum between the ligature and the origin of the vessel. It should be remembered, that the right subclavian artery is occasionally deeply placed in the first part of its course, when it arises from the left side of the aortic arch, and passes in such cases behind the œsophagus, or between it and the trachea.

Collateral Circulation. After ligature of the third part of the subclavian artery, the collateral circulation is mainly established by three sets of vessels, thus described in a dissection:—

'1. A posterior set, consisting of the suprascapular and posterior scapular branches of the subclavian, which anastomosed with the infrascapular from the axillary.

'2. An internal set produced by the connection of the internal mammary on the one hand, with the short and long thoracic arteries, and the infrascapular on the other.

'3. A middle or axillary set, which consisted of a number of small vessels derived from branches of the subclavian, above; and passing through the axilla, to terminate either in the main trunk, or some of the branches of the axillary, below. This last set presented most conspicuously the peculiar character of newly-formed, or, rather, dilated arteries,' being excessively tortuous, and forming a complete plexus.

'The chief agent in the restoration of the axillary artery below the tumour, was the infrascapular artery, which communicated most freely with the internal mammary, suprascapular, and posterior scapular branches of the subclavian, from all of which it received so great an influx of blood as to dilate it to three times its natural size.'†

Branches of the Subclavian Artery.

These are four in number. Three arising from the first portion of the vessel, the vertebral, the internal mammary, and the thyroid axis; and one from the second portion, the superior intercostal. The vertebral arises from the upper and back part of the first portion of the artery; the thyroid axis from the front, and the internal mammary from the under part of this vessel. The superior intercostal is given off from the upper and back part of the second portion of the artery. On the left side, the second portion usually gives off no branch, the superior intercostal arising at the inner side of the Scalenus anticus. On both sides of the

215.—Plan of the Branches of the Right Subclavian Artery.

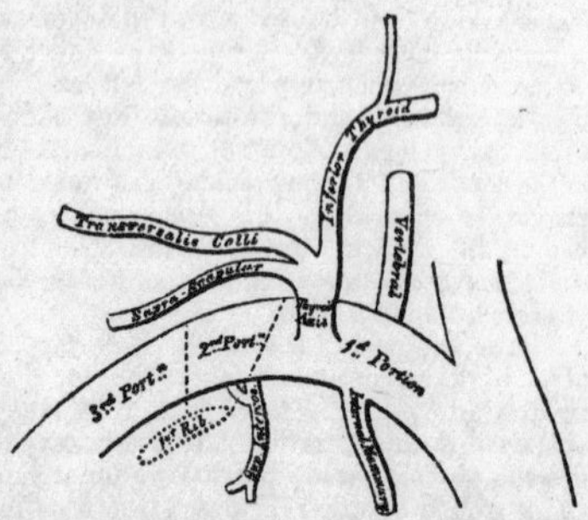

* The operation was, however, performed in New York, by Dr. J. K. Rodgers, and the case is very briefly mentioned in Mott's Translation of Velpeau, vol. ii. p. 365.

† *Guy's Hospital Reports*, vol. i. 1836. Case of axillary aneurism, in which Mr. Aston Key had tied the subclavian artery on the outer edge of the Scalenus muscle, twelve years previously.

body, the first three branches arise close together at the inner margin of the Scalenus anticus; in the majority of cases, a free interval of half an inch to an inch existing between the commencement of the artery and the origin of the nearest branch; in a smaller number of cases, an interval of more than an inch exists, never exceeding an inch and three-quarters. In a very few instances, the interval has been found less than half an inch.

The VERTEBRAL ARTERY (fig. 212) is generally the first and largest branch of the subclavian; it arises from the upper and back part of the first portion of the vessel, and, passing upwards, enters the foramen in the transverse process of the sixth cervical vertebra, and ascends through the foramina in the transverse processes of all the vertebræ above this. Above the upper border of the axis, it inclines outwards and upwards to the foramen in the transverse process of the atlas, through which it passes; it then winds backwards behind its articular process, runs in a deep groove on the surface of the posterior arch of this bone, and, piercing the posterior occipito-atloid ligament and dura mater, enters the skull through the foramen magnum. It then passes in front of the medulla oblongata, and unites with the vessel of the opposite side at the lower border of the pons Varolii, to form the basilar artery.

At its origin, it is situated behind the internal jugular vein, and inferior thyroid artery; and, near the spine, lies between the Longus colli and Scalenus anticus muscles, having the thoracic duct in front of it on the left side. Within the foramina formed by the transverse processes of the vertebræ, it is accompanied by a plexus of nerves from the sympathetic, and lies between the vertebral vein, which is in front, and the cervical nerves, which issue from the intervertebral foramina behind it. Whilst winding round the articular process of the atlas, it is contained in a triangular space formed by the Rectus posticus major, the Superior and the Inferior oblique muscles; and is covered by the Rectus posticus major and Complexus. Within the skull, as it winds round the medulla oblongata, it is placed between the hypoglossal nerve and the anterior root of the suboccipital nerve.

Branches. These may be divided into two sets, those given off in the neck, and those within the cranium.

Cervical Branches.	*Cranial Branches.*
Lateral spinal.	Posterior meningeal.
Muscular.	Anterior spinal.
	Posterior spinal.
	Inferior cerebellar.

The *lateral spinal branches* enter the spinal canal through the intervertebral foramina, each dividing into two branches. Of these, one passes along the roots of the nerves to supply the spinal cord and its membranes, anastomosing with the other spinal arteries; the other is distributed to the posterior surface of the bodies of the vertebræ.

Muscular branches are given off to the deep muscles of the neck, where the vertebral artery curves round the articular process of the atlas. They anastomose with the occipital and deep cervical arteries.

The *posterior meningeal* are one or two small branches given off from the vertebral opposite the foramen magnum. They ramify between the bone and dura mater in the cerebellar fossæ, and supply the falx cerebelli.

The *anterior spinal* is a small branch, larger than the posterior spinal, which arises near the termination of the vertebral, and unites with its fellow of the opposite side in front of the medulla oblongata. The single trunk, thus formed, descends a short distance on the front of the spinal cord, and joins with a succession of small branches which enter the spinal canal through some of the intervertebral foramina; these branches are derived from the vertebral and ascending cervical in the neck; from the intercostal, in the dorsal region; and from the lumbar, ilio-lumbar, and lateral sacral arteries in the lower part of the spine. They unite, by means of ascending and descending branches, to form a single anterior median artery, which extends as far as the lower part of the spinal cord. This vessel is placed beneath the pia mater along the anterior median fissure; it supplies that membrane and the substance of the cord, and sends off branches at its lower part to be distributed to the cauda equina.

The *posterior spinal* arises from the vertebral, at the side of the medulla oblongata; passing backwards to the posterior aspect of the spinal cord, it descends on either side, lying behind the posterior roots of the spinal nerves; and is reinforced by a succession of small branches, which enter the spinal canal through the intervertebral foramina, and by which it is continued to the lower part of the cord, and to the cauda equina. Branches from these vessels form a free anastomosis round the posterior roots of the spinal nerves, and communicate, by means of very tortuous transverse branches, with the vessel of the opposite side. At its commencement, it gives off an ascending branch, which terminates on the side of the fourth ventricle.

The *inferior cerebellar artery* (fig. 214), the largest branch of the vertebral, winds backwards round the upper part of the medulla oblongata, passing between the origin of the spinal accessory and pneumogastric nerves, over the restiform body, to the under surface of the cerebellum, where it divides into two branches; an internal one, which is continued backwards to the notch between the two hemispheres of the cerebellum; and an external one, which supplies the under surface of the cerebellum, as far as its outer border, where it anastomoses with the superior cerebellar. Branches from this artery supply the choroid plexus of the fourth ventricle.

The *Basilar artery*, so named from its position at the base of the skull, is a single trunk, formed by the junction of the two vertebral arteries; it extends from the posterior to the anterior border of the pons Varolii, where it divides into two terminal branches, the posterior cerebral arteries. Its branches are, on each side, the following:

Transverse.	Superior cerebellar.
Anterior cerebellar.	Posterior cerebral.

The *transverse* branches supply the pons Varolii and adjacent parts of the brain; one accompanies the auditory nerve into the internal auditory meatus; and another, of larger size, passes along the crus cerebelli, to be distributed to the anterior border of the under surface of the cerebellum. It is called the *anterior (inferior) cerebellar artery*.

The *superior cerebellar arteries* arise near the termination of the basilar. They wind round the crus cerebri, close to the fourth nerve, and arriving at the upper surface of the cerebellum, divide into branches which ramify in the pia mater and anastomose with the inferior cerebellar. Several branches are given to the pineal gland, and also to the velum interpositum.

The *posterior cerebral arteries*, the two terminal branches of the basilar, are larger than the preceding, from which they are separated near their origin by the third nerves. Winding round the crus cerebri, they pass to the under surface of the posterior lobes of the cerebrum, which they supply, anastomosing with the anterior and middle cerebral arteries. Near their origin, they receive the posterior communicating arteries from the internal carotid, and give off numerous branches, which enter the posterior perforated space. They also give off a branch, the posterior choroid, which supplies the velum interpositum and choroid plexus, entering the interior of the brain beneath the posterior border of the corpus callosum.

Circle of Willis. The remarkable anastomosis which exists between the branches of the internal carotid and vertebral arteries at the base of the brain, constitutes the circle of Willis. It is formed, in front, by the anterior cerebral and anterior communicating arteries; on each side, by the trunk of the internal carotid, and the posterior communicating; behind, by the posterior cerebral, and point of the basilar. It is by this anastomosis that the cerebral circulation is equalized, and provision made for effectually carrying it on if one or more of the branches are obliterated. The parts of the brain included within this arterial circle are, the lamina cinerea, the commissure of the optic nerves, the infundibulum, the tuber cinereum, the corpora albicantia, and the pars perforata postica.

The Thyroid Axis (fig. 207) is a short thick trunk, which arises from the fore part of the first portion of the subclavian artery, close to the inner side of the Scalenus anticus muscle, and divides, almost immediately after its origin, into three branches, the inferior thyroid, suprascapular, and transversalis colli.

The Inferior Thyroid Artery passes upwards, in a serpentine course, behind the sheath of the common carotid vessel and sympathetic nerve (the middle cervical ganglion resting upon it), and is distributed to the under surface of the thyroid gland, anastomosing with the superior thyroid, and with the corresponding artery of the opposite side. Its branches are the

Laryngeal.	Œsophageal.
Tracheal.	Ascending cervical.

The *laryngeal* branch ascends upon the trachea to the back part of the larynx, and supplies the muscles and the mucous membrane of this part.

The *tracheal* branches are distributed upon the trachea, anastomosing below with the bronchial arteries.

The *œsophageal* branches are distributed to the œsophagus.

The *ascending cervical* is a small branch which arises from the inferior thyroid, just where that vessel is passing behind the common carotid artery, and runs up the neck in the interval between the Scalenus anticus and Rectus anticus major. It gives branches to the muscles of the neck, which communicate with those sent out from the vertebral, and sends one or two through the intervertebral foramina, along the cervical nerves, to supply the bodies of the vertebræ, the spinal cord, and its membranes.

The Suprascapular Artery, smaller than the transversalis colli, passes obliquely from within outwards, across the root of the neck. It at first lies on the lower part of the Scalenus anticus, being covered by the Sterno-mastoid; it then crosses the subclavian artery, and runs outwards behind and parallel with the clavicle and Subclavius muscle, and beneath the posterior belly of the Omohyoid, to the superior border of the scapula, where it passes over the transverse ligament of the scapula to the supraspinous fossa. In this situation it lies close to the bone, and ramifies between it and the Supraspinatus muscle, to which it is mainly distributed, giving off a communicating branch, which crosses the neck of the scapula, to reach the infraspinous fossa, where it anastomoses with the dorsal branch of the subscapular artery. Besides distributing branches to the Sternomastoid, and neighbouring muscles, it gives off a supra-acromial branch, which, piercing the Trapezius muscle, supplies the cutaneous surface of the acromion, anastomosing with the acromial thoracic artery. As the artery passes across the suprascapular notch, a branch descends into the subscapular fossa, ramifies beneath that muscle, and anastomoses with the posterior and subscapular arteries. It also supplies the shoulder joint.

The Transversalis Colli passes transversely outwards, across the upper part of the subclavian triangle, to the anterior margin of the Trapezius muscle, beneath which it divides into two branches, the superficial cervical, and the posterior scapular. In its passage across the neck, it crosses in front of the Scaleni muscles and the brachial plexus, between the divisions of which it sometimes passes, and is covered by the Platysma, Sterno-mastoid, Omo-hyoid, and Trapezius muscles.

The *superficial cervical* ascends beneath the anterior margin of the Trapezius, distributing branches to it, and to the neighbouring muscles and glands in the neck.

The *posterior scapular*, the continuation of the transversalis colli, passes beneath the Levator anguli scapulæ to the superior angle of the scapula, and descends along the posterior border of that bone as far as the inferior angle, where it anastomoses with the subscapular branch of the axillary. In its course it is

216.—The Scapular and Circumflex Arteries.

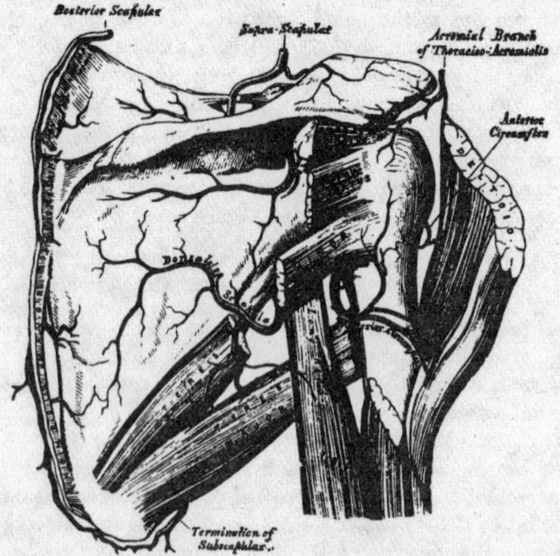

covered by the Rhomboid muscles, supplying these, the Latissimus dorsi and Trapezius, and anastomosing with the suprascapular and subscapular arteries, and with the posterior branches of some of the intercostal arteries.

Peculiarities. The *superficial cervical* frequently arises as a separate branch from the thyroid axis; and the posterior scapular, from the third, more rarely from the second, part of the subclavian.

The Internal Mammary arises from the under surface of the first portion of the subclavian artery, opposite the thyroid axis. It descends behind the clavicle, to the inner surface of the anterior wall of the chest, resting upon the costal cartilages, a short distance from the margin of the sternum; and, at the interval between the sixth and seventh cartilages, divides into two branches, the musculo-phrenic, and superior epigastric.

At its origin, it is covered by the internal jugular and subclavian veins, and crossed by the phrenic nerve. In the upper part of the thorax, it lies upon the costal cartilages, and internal Intercostal muscles in front, and is covered by the pleura behind. At the lower part of the thorax, the Triangularis sterni separates the artery from the pleura. It is accompanied by two veins, which join at the upper part of the thorax into a single trunk.

The branches of the internal mammary are,

Comes nervi phrenici (Superior phrenic).	Anterior Intercostal.
Mediastinal.	Perforating.
Pericardiac.	Musculo-phrenic.
Sternal.	Superior epigastric.

The *comes nervi phrenici* (*superior phrenic*), is a long slender branch, which accompanies the phrenic nerve, between the pleura and pericardium, to the Diaphragm, to which it is distributed; anastomosing with the other phrenic arteries from the internal mammary, and abdominal aorta.

The *mediastinal branches* are small vessels, which are distributed to the areolar tissue in the anterior mediastinum, and the remains of the thymus gland.

The *pericardiac branches* supply the upper part of the pericardium, the lower part receiving branches from the musculo-phrenic artery. Some *sternal* branches are distributed to the Triangularis sterni, and both surfaces of the sternum.

The *anterior intercostal arteries* supply the five or six upper intercostal spaces. The branch corresponding to each space passes outwards, and soon divides into two, which run along the opposite borders of the ribs, and inosculate with the intercostal arteries from the aorta. They are at first situated between the pleura and the internal Intercostal muscles, and then between the two layers of those muscles. They supply the Intercostal and Pectoral muscles, and the mammary gland.

The *perforating arteries* correspond to the five or six upper intercostal spaces. They arise from the internal mammary, pass forwards through the intercostal spaces, and curving outwards, supply the Pectoralis major, and the integument. Those which correspond to the first three spaces, are distributed to the mammary gland. In females, during lactation, these branches are of large size.

The *musculo-phrenic artery* is directed obliquely downwards and outwards, behind the cartilages of the false ribs, perforating the Diaphragm at the eighth or ninth rib, and terminating, considerably reduced in size, opposite the last intercostal space. It gives off anterior intercostal arteries to each of the intercostal spaces across which it passes; these diminish in size as the spaces decrease in length, and are distributed in a manner precisely similar to the anterior intercostals from the internal mammary. The musculo-phrenic also gives branches backwards to the Diaphragm, and downwards to the abdominal muscles.

The *superior epigastric* continues in the original direction of the internal mammary, descends behind the Rectus muscle, and perforating its sheath, divides into branches which supply the Rectus, anastomosing with the epigastric artery from the external iliac. Some vessels perforate the sheath of the Rectus, and supply the muscles of the abdomen and the integument, and a small branch which passes inwards upon the side of the ensiform appendix, anastomoses in front of that cartilage with the artery of the opposite side.

The Superior Intercostal (fig. 212) arises from the upper and back part of the subclavian artery, behind the anterior scalenus on the right side, and to the inner side of the muscle on the left side. Passing backwards, it gives off the *deep cervical* branch, and then descends behind the pleura in front of the necks of the first two ribs, and inosculates with the first aortic intercostal. In the first intercostal space, it gives off a branch which is distributed in a manner similar to the distribution of the aortic intercostals. The branch for the second intercostal space usually joins with one from the first aortic intercostal. Each intercostal gives off a branch to the posterior spinal muscles, and a small one, which passes through the corresponding intervertebral foramen to the spinal cord and its membranes.

The *deep cervical branch* (*profunda cervicis*) arises, in most cases, from the superior intercostal, and is analogous to the posterior branch of an aortic intercostal artery. Passing backwards, between the transverse process of the seventh cervical vertebra and the first rib, it runs up the back part of the neck, between the Complexus and Semispinalis colli muscles, as high as the axis, supplying these and adjacent muscles, and anastomosing with the arteria princeps cervicis of the occipital, and with branches which pass outwards from the vertebral.

Surgical Anatomy of the Axilla.

The *Axilla* is a pyramidal space, situated between the upper and lateral part of the chest, and the inner side of the arm.

Boundaries. Its *apex*, which is directed upwards towards the root of the neck, corresponds to the interval between the two Scaleni on the first rib. The *base*, directed downwards, is formed by the integument, and a thick layer of fascia, extending between the lower border of the Pectoralis major in front, and the lower border of the Latissimus dorsi behind; it is broad internally, at the chest, but narrow and pointed externally, at the arm. The *anterior boundary* is formed by the Pectoralis major and minor muscles, the former covering the whole of the anterior wall of the axilla, the latter covering only its central part. The *posterior boundary*, which extends somewhat lower than the anterior, is formed by the Subscapularis above, the Teres major and Latissimus dorsi below. On the *inner side* are the first four ribs with their corresponding Intercostal muscles, and part of the

217.—The Axillary Artery, and its Branches.

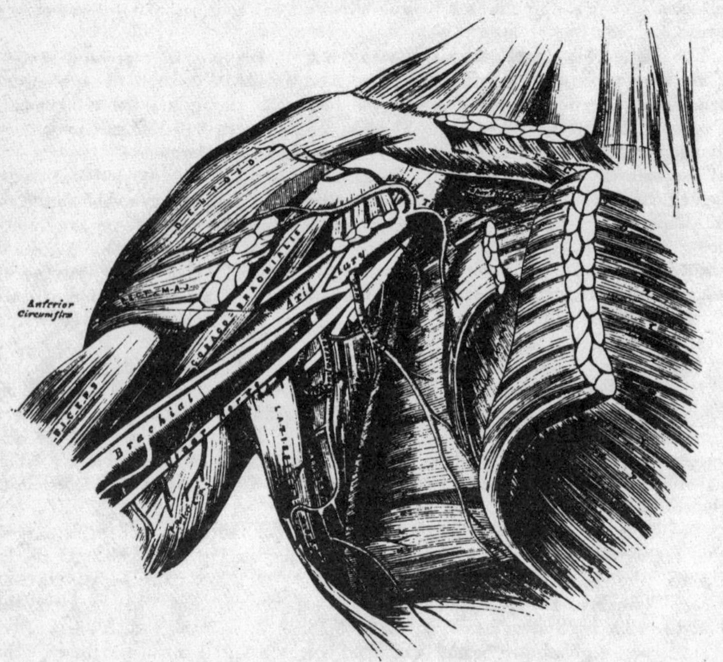

Serratus magnus. On the *outer side*, where the anterior and posterior boundaries converge, the space is narrow, and bounded by the humerus, the Coraco-brachialis and Biceps muscles.

Contents. This space contains the axillary vessels, and brachial plexus of nerves, with their branches, some branches of the intercostal nerves, and a large number of lymphatic glands, all connected together by a quantity of fat and loose areolar tissue.

Their Position. The axillary artery and vein, with the brachial plexus of nerves, extend obliquely along the outer boundary of the axillary space, from its apex to its base, and are placed much nearer the anterior than the posterior wall, the vein lying to the inner or thoracic side of the artery, and altogether concealing it. At the fore part of the axillary space, in contact with the Pectoral muscles, are the thoracic branches of the axillary artery, and along the anterior margin of the axilla the long thoracic artery extends to the side of the chest. At the back part, in contact with the lower margin of the Subscapularis muscle, are the subscapular vessels and nerves; winding around the lower border of this muscle, is the dorsalis scapulæ artery and veins; and towards the outer extremity of the muscle, the posterior circumflex vessels and the circumflex nerve are seen curving backwards to the shoulder.

Along the inner or thoracic side no vessel of any importance exists, the upper part of the space being crossed merely by a few small branches from the superior thoracic artery. There are some important nerves, however, in this situation; viz., the posterior thoracic or external respiratory nerve, descending on the surface of the Serratus magnus, to which it is distributed; and perforating the upper and anterior part of this wall, the intercosto-humeral nerve or nerves, passing across the axilla to the inner side of the arm.

The cavity of the axilla is filled by a quantity of loose areolar tissue, a large number of small arteries and veins, all of which are, however, of inconsiderable size, and numerous lymphatic glands; these are from ten to twelve in number, and situated chiefly on the thoracic side, and lower and back part of this space.

The student should attentively consider the relation of the vessels and nerves in the several parts of the axilla; for it not unfrequently happens, that the surgeon is called upon to extirpate diseased glands, or to remove a tumour from this situation. In performing such an operation, it will be necessary to proceed with much caution in the direction of the outer wall and apex of the space, as here the axillary vessels will be in danger of being wounded. Towards the posterior wall, it will be necessary to avoid the subscapular, dorsalis scapulæ, and posterior circumflex vessels, and, along the anterior wall, the thoracic branches. It is only along the inner or thoracic wall, and in the centre of the axillary cavity, that there are no vessels of any importance; a fortunate circumstance, for it is in this situation more especially that tumours requiring removal are usually situated.

The Axillary Artery.

The axillary artery, the continuation of the subclavian, commences at the lower border of the first rib, and terminates at the lower border of the tendons of the Latissimus dorsi and Teres major muscles, where it takes the name of brachial. Its direction varies with the position of the limb: when the arm lies by the side of the chest, the vessel forms a gentle curve, the convexity being upwards and outwards; when it is directed at right angles with the trunk, the vessel is nearly straight; and when it is elevated still higher, the artery describes a curve, the concavity of which is directed upwards. At its commencement the artery is very deeply situated, but near its termination is superficial, being covered only by the skin and fascia. The description of the relations of this vessel is facilitated by its division into three portions, the first portion being that above the Pectoralis minor; the second portion, behind; and the third, below that muscle.

The *first portion* of the axillary artery is in relation, *in front*, with the clavicular portion of the Pectoralis major, the costo-coracoid membrane, the Subclavius, and the cephalic vein; *behind*, with the first intercostal space, the corresponding Intercostal muscle, the first serration of the Serratus magnus, and the posterior thoracic nerve; on its *outer side* with the brachial plexus, from which it is separated by a little cellular interval; on its *inner*, or thoracic side, with the axillary vein.

Relations of First Portion of the Axillary Artery.

In front.
Pectoralis major.
Costo-coracoid membrane.
Subclavius.
Cephalic vein.

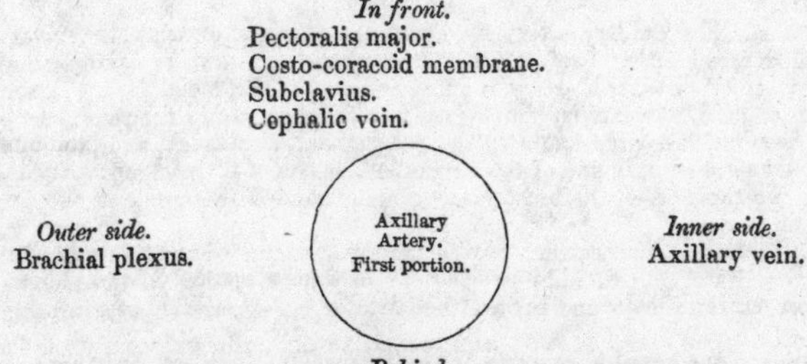

Behind.
First Intercostal space, and Intercostal muscle.
First serration of Serratus magnus.
Posterior thoracic nerve.

The *second portion* of the axillary artery lies behind the Pectoralis minor. It is covered, *in front*, by the Pectoralis major and minor muscles; *behind*, it is separated from the Subscapularis by a cellular interval; on the *inner side* is the axillary vein. The brachial plexus of nerves surrounds the artery, and separates it from direct contact with the vein and adjacent muscles.

RELATIONS OF THE SECOND PORTION OF THE AXILLARY ARTERY.

In front.
Pectoralis major and minor.

Outer side.
Outer cord of plexus.

Axillary Artery. Second portion.

Inner side.
Axillary vein.
Inner cord of plexus.

Behind.
Subscapularis.
Posterior cord of plexus.

The *third portion* of the axillary artery lies below the Pectoralis minor. It is in relation, *in front*, with the lower part of the Pectoralis major above, being covered only by the integument and fascia below; *behind*, with the lower part of the Subscapularis, and the tendons of the Latissimus dorsi and Teres major; on its *outer side*, with the Coraco-brachialis; on its *inner*, or thoracic side, with the axillary vein. The nerves of the brachial plexus bear the following relation to the artery in this part of its course: on the *outer side* is the median nerve, and the musculo-cutaneous for a short distance; on the *inner side*, the ulnar, the internal, and lesser internal cutaneous nerves; and *behind*, the musculo-spiral, and circumflex, the latter extending only to the lower border of the Subscapularis muscle.

RELATIONS OF THE THIRD PORTION OF THE AXILLARY ARTERY.

In front.
Integument and fascia.
Pectoralis major.

Outer side.
Coraco-brachialis.
Median nerve.
Musculo-cutaneous nerve.

Axillary Artery. Third portion.

Inner side.
Ulnar nerve.
Internal cutaneous nerves.
Axillary vein.

Behind.
Subscapularis.
Tendons of Latissimus dorsi, and Teres major.
Musculo-spiral, and circumflex nerves.

Peculiarities. The axillary artery, in about one case out of every ten, gives off a large branch, which forms either one of the arteries of the fore-arm, or a large muscular trunk. In the first set of cases, this artery is most frequently the radial (1 in 33), sometimes the ulnar (1 in 72), and, very rarely, the interosseous (1 in 506). In the second set of cases, the trunk has been found to give origin to the subscapular, circumflex, and profunda arteries of the arm. Sometimes, only one of the circumflex, or one of the profunda arteries, arose from the trunk. In these cases, the brachial plexus surrounded the trunk of the branches, and not the main vessel.

Surgical Anatomy. The student having carefully examined the relations of the axillary artery in its various parts, should now consider in what situation compression of this vessel may be most easily effected, and the best position for the application of a ligature to it when necessary.

Compression of the vessel is required in the removal of tumours, or in amputation of the upper part of the arm; and the only situation in which this can be effectually made, is in the lower part of its course; by pressing on it in this situation from within outwards against the humerus, the circulation may be effectually suspended.

The *application of a ligature to the axillary artery* may be required in cases of aneurism of the upper part of the brachial; and there are only two situations in which it can be secured, viz., in the first and in the third parts of its course; for the axillary artery at its central part is so deeply seated, and, at the same time, so closely surrounded with large nervous trunks, that the application of a ligature to it in that situation would be almost impracticable.

In the *third part* of its course, the operation is most simple, and may be performed in the following manner:—The patient being placed on a bed, and the arm separated from the side, with the hand supinated, the head of the humerus is felt for, and an incision made through the integument over it, about two inches in length, a little nearer to the anterior than the posterior fold of the axilla. After carefully dissecting through the areolar tissue and fascia, the median nerve and axillary vein are exposed; the former having been displaced to the outer, and the latter to the inner side of the arm, the elbow being at the same time bent, so as to relax the structures, and facilitate their separation, the ligature may be passed round the artery from the ulnar to the radial side. This portion of the artery is occasionally crossed by a muscular slip derived from the Latissimus dorsi, which may mislead the surgeon during an operation. The occasional existence of this muscular fasciculus was spoken of in the description of the muscles. It may easily be recognised by the transverse direction of its fibres.

The *first portion* of the axillary artery may be tied, in cases of aneurism encroaching so far upwards that a ligature cannot be applied in the lower part of its course. Notwithstanding that this operation has been performed in some few cases, and with success, its performance is attended with much difficulty and danger. The student will remark that in this situation, it would be necessary to divide a thick muscle, and after separating the costo-coracoid membrane, the artery would be exposed at the bottom of a more or less deep space, with the cephalic and axillary veins in such relation with it as must render the application of a ligature to this part of the vessel particularly hazardous. Under such circumstances it is an easier, and, at the same time, more advisable operation, to tie the subclavian artery in the third part of its course.

In a case of wound of the vessel, the general practice of cutting down upon, and tying it above and below the wounded point, should be adopted in all cases.

Collateral circulation after ligature of the axillary artery. If the artery be tied above the origin of the acromial thoracic, the collateral circulation will be carried on by the same branches as after the ligature of the subclavian; if at a lower point, between the acromial thoracic and subscapular arteries, the latter vessel, by its free anastomoses with the other scapular arteries, branches of the subclavian, will become the chief agent in carrying on the circulation, to which the long thoracic, if it be below the ligature, will materially contribute, by its anastomoses with the intercostal and internal mammary arteries. If the point included in the ligature be below the origin of the subscapular artery, the anastomoses are less free. The chief agents in restoring the circulation, will be the posterior circumflex, by its anastomoses with the suprascapular and acromial thoracic, and the communications between the subscapular and superior profunda, which will be afterwards referred to as performing the same office after ligature of the brachial. The cases in which the operation has been performed are few in number, and no published account of dissection of the collateral circulation appears to exist.

The branches of the axillary artery are,—

From 1st Part { Superior thoracic. / Acromial thoracic.

From 2nd Part { Thoracica longa. / Thoracica alaris.

From 3rd Part { Subscapular. / Anterior circumflex. / Posterior circumflex.

The *superior thoracic* is a small artery, which arises from the axillary separately, or by a common trunk with the acromial thoracic. Running forwards and inwards along the upper border of the Pectoralis minor, it passes between it and the Pectoralis major to the side of the chest. It supplies these muscles, and the parietes of the thorax, anastomosing with the internal mammary and intercostal arteries.

The *acromial thoracic* is a short trunk, which arises from the fore part of the axillary artery. Projecting forwards to the upper border of the Pectoralis minor, it divides into three sets of branches, thoracic, acromial, and descending. The thoracic branches, two or three in number, are distributed to the Serratus magnus, and Pectoral muscles, anastomosing with the intercostal branches of the internal mammary. The acromial branches are directed outwards towards the acromion, supplying the Deltoid muscle, and anastomosing, on the surface of the acromion, with the suprascapular and posterior circumflex arteries. The descending branch passes in the interspace between the Pectoralis major and Deltoid, accompanying the cephalic vein, and supplying both muscles.

The *long thoracic* passes downwards and inwards along the lower border of the Pectoralis minor to the side of the chest, supplying the Serratus magnus, the Pectoral muscles, and mammary gland, and sending branches across the axilla to the axillary glands and Subscapularis, which anastomose with the internal mammary and intercostal arteries.

The *thoracica alaris* is a small branch, which supplies the glands and areolar tissue of the axilla. Its place is frequently supplied by branches from some of the other thoracic arteries.

The *subscapular*, the largest branch of the axillary artery, arises opposite the lower border of the Subscapularis muscle, and passes downwards and backwards along its lower margin to the inferior angle of the scapula, where it anastomoses with the posterior scapula, a branch of the subclavian. It distributes branches to the Subscapularis, Serratus magnus, Teres major, and Latissimus dorsi muscles, and gives off, about an inch and a-half from its origin, a large branch, the dorsalis scapulæ, which curves round the inferior border of the scapula, leaving the axilla in the interspace between the Teres minor above, the Teres major below, and the long head of the Triceps in front. Three branches, or sets of branches, arise from the dorsalis scapulæ; the first enters the subscapular fossa, beneath the Subscapularis, which it supplies, anastomosing with the subscapular and suprascapular arteries; the second, the trunk of the artery (*dorsalis scapulæ*) turns round the axillary border of the scapula, and enters the infraspinous fossa, where it anastomoses with the suprascapular and posterior scapular arteries; and a third, or median branch, is continued along the axillary border of the scapula, between the Teres major and minor, and, at the dorsal surface of the inferior angle of the bone, anastomoses with the posterior scapular.

The *circumflex arteries* wind round the neck of the humerus. The *posterior circumflex* (fig. 216), the larger of the two, arises from the back part of the axillary, opposite the lower border of the Subscapularis muscle, and, passing backwards with the circumflex veins and nerve, through the quadrangular space bounded by the Teres major and minor, the scapular head of the Triceps and the humerus, winds round the neck of that bone, and is distributed to the Deltoid muscle and shoulder-joint, anastomosing with the anterior circumflex, suprascapular and acromial thoracic arteries. The *anterior circumflex* (figs. 216, 217), considerably smaller than the preceding, arises just below that vessel, from the outer side of the axillary artery. It passes horizontally outwards, beneath the Coraco-brachialis and short head of the Biceps, lying upon the fore part of the neck of the humerus, and, on reaching the bicipital groove, gives off an ascending branch, which passes upwards along the groove, to supply the head of the bone and the shoulder-joint. The trunk of the vessel is then continued outwards beneath the Deltoid, which it supplies, and anastomoses with the posterior circumflex, and acromial thoracic arteries.

Brachial Artery. (Fig. 218.)

The brachial artery commences at the lower margin of the tendon of the Teres major, and passing down the inner and anterior aspect of the arm, terminates about half an inch below the bend of the elbow, where it divides into the radial and ulnar arteries.

The direction of this vessel is marked by a line drawn from the outer side of the axillary space between the folds of the axilla, to a point midway between the condyles of the humerus, which corresponds to the depression along

218.—The Surgical Anatomy of the Brachial Artery.

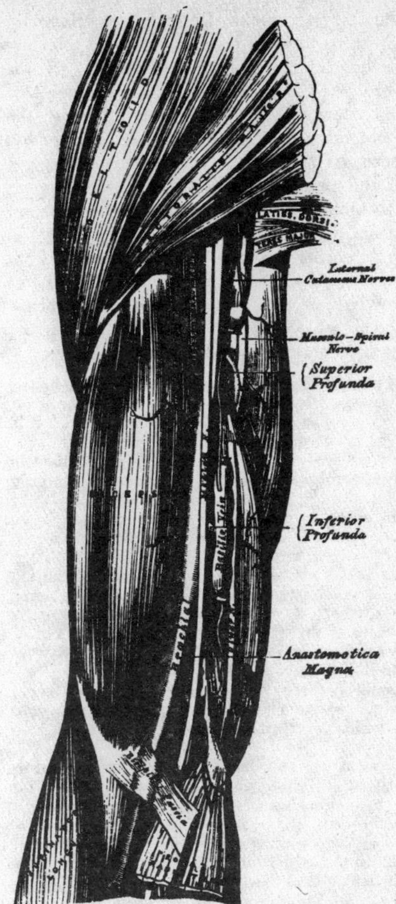

the inner border of the Coraco-brachialis and Biceps muscles. In the upper part of its course, the artery lies internal to the humerus; but below, it is in front of that bone.

Relations. This artery is superficial throughout its entire extent, being covered, *in front*, by the integument, the superficial and deep fascia; the bicipital fascia separates it opposite the elbow from the median basilic vein; the median nerve crosses it at its middle; and the basilic vein lies in the line of the artery, but separated from it by the fascia, in the lower half of its course. *Behind*, it is separated from the inner side of the humerus above, by the long and inner heads of the Triceps, the musculo-spiral nerve and superior profunda artery intervening; and from the front of the bone below, by the insertion of the Coraco-brachialis muscle, and by the Brachialis anticus. By its *outer side*, it is in relation with the commencement of the median nerve, and the Coraco-brachialis and Biceps muscles, which slightly overlap the artery. By its *inner side*, its upper half is in relation with the internal cutaneous and ulnar nerves, its lower half with the median nerve. It is accompanied by two venæ comites, which lie in close contact with the artery, being connected together at intervals by short transverse communicating branches.

Plan of the Relations of the Brachial Artery.

In front.
Integument and fasciæ.
Bicipital fascia, median basilic vein.
Median nerve.

Outer side.
Median nerve.
Coraco-brachialis.
Biceps.

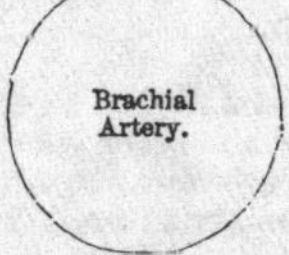

Inner side.
Internal cutaneous and ulnar nerve.
Median nerve.

Behind.
Triceps.
Musculo-spiral nerve.
Superior profunda artery.
Coraco-brachialis.
Brachialis anticus.

Bend of the Elbow.

At the bend of the elbow, the brachial artery sinks deeply into a triangular interval, the base of which is directed upwards towards the humerus, and the sides of which are bounded, externally, by the Supinator longus, internally, by the Pronator radii teres; its floor is formed by the Brachialis anticus, and Supinator brevis. This space contains the brachial artery, with its accompanying veins; the radial and ulnar arteries; the median and musculo-spiral nerves; and the tendon of the Biceps. The brachial artery occupies the middle line of this space, and divides opposite the coronoid process of the ulna into the radial and ulnar arteries; it is covered, *in front*, by the integument, the superficial fascia, and the median basilic vein, the vein being separated from direct contact with the artery by the bicipital fascia. *Behind*, it lies on the Brachialis anticus, which separates it from the elbow-joint. The median nerve lies on the inner side of the artery, but separated from it below by an interval of half an inch. The tendon of the Biceps lies to the outer side of the space, and the musculo-spiral nerve still more externally, lying upon the Supinator brevis, and partly concealed by the Supinator longus.

Peculiarities of the Artery as regards its Course. The brachial artery accompanied by the median nerve, may leave the inner border of the Biceps, and descend towards the inner condyle of the humerus, where it usually curves round a prominence of bone, to which it is connected by a fibrous band; it then inclines outwards, beneath or through the substance of the Pronator teres muscle, to the bend of the elbow. This variation bears considerable analogy with the normal condition of the artery in some of the carnivora: it has been referred to, above, in the description of the humerus.

As regards its Division. Occasionally the artery is divided for a short distance at its upper part into two trunks, which are united above and below. A similar peculiarity occurs in the main vessel of the lower limb.

The point of bifurcation may be above or below the usual point, the former condition being by far the most frequent. Out of 481 examinations recorded by Mr. Quain, some made on the right, and some on the left side of the body, in 386 the artery bifurcated in its normal position. In one case only was the place of division lower than usual, being two or three inches below the elbow-joint. 'In 94 cases out of 481, or about one in $5\frac{1}{9}$, there were two arteries instead of one in some part, or in the whole of the arm.'

There appears, however, to be no correspondence between the arteries of the two arms, with respect to their irregular division; for in sixty-one bodies it occurred on one side only in forty-three; on both sides, in different positions, in thirteen; on both sides, in the same position, in five.

The point of bifurcation takes place at different parts of the arm, being most frequent in the upper part, less so in the lower part, and least so in the middle, the most usual point for the application of a ligature; under any of these circumstances, two large arteries would be found in the arm instead of one. The most frequent (in three out of four) of these peculiarities is the high division of the radial. That artery often arises from the inner side of the brachial, and runs parallel with the main trunk to the elbow, where it crosses it, lying beneath the fascia; or it may perforate the fascia, and pass over the artery immediately beneath the integument.

The ulnar sometimes arises from the brachial high up, and then occasionally leaves that vessel at the lower part of the arm, and descends towards the inner condyle. In the forearm, it generally lies beneath the deep fascia, superficial to the flexor muscles; occasionally between the integument and deep fascia, and very rarely beneath the flexor muscles.

The interosseous artery sometimes arises from the upper part of the brachial or axillary: as it passes down the arm, it lies behind the main trunk, and, at the bend of the elbow, regains its usual position.

In some cases of high division of the radial, the remaining trunk (ulnar-interosseous) occasionally passes, together with the median nerve, along the inner margin of the arm to the inner condyle, and then passing from within outwards, beneath or through the Pronator teres, regains its usual position at the bend of the elbow.

Occasionally, the two arteries representing the brachial are connected at the bend of the elbow by a short transverse branch, and are even sometimes reunited.

Sometimes, long slender vessels, *vasa aberrantia*, connect the brachial or axillary arteries with one of the arteries of the fore-arm, or a branch from them. These vessels usually join the radial.

*Varieties in Muscular Relations.** The brachial artery is occasionally concealed, in some part of its course, by muscular or tendinous slips derived from various sources. In the upper third of the arm, the brachial vessels and median nerve have been seen concealed to the extent of three inches, by a muscular layer of considerable thickness, derived from the Coraco-

* See Struthers's *Anatomical and Physiological Observations.*

brachialis, which passed round to the inner side of the vessel, and joined the internal head of the Triceps. In the lower half of the arm the artery is occasionally concealed by a broad thin head to the Biceps muscle. A narrow fleshy slip from the Biceps has been seen to cross the artery, concealing it for an inch and a half, its tendon ending in the aponeurosis covering the Pronator teres. A muscular and tendinous slip has been seen to arise from the external bicipital ridge by a long tendon, cross obliquely behind the long tendon of the Biceps, and end in a fleshy belly, which appears on the inner side of the arm between the Biceps and Coraco-brachialis, passes down along the inner edge of the former, and crosses the artery very obliquely, so as to lie in front of it for three inches, and, finally, gives rise to a narrow flattened tendon, which is inserted into the aponeurosis over the Pronator teres. A tendinous slip, arising from the deep part of the tendon of the Pectoralis major, has been seen to cross the artery obliquely at or below the Coraco-brachialis, and join the intermuscular septum above the inner condyle. The Brachialis anticus not unfrequently projects at the outer side of the artery, and occasionally overlaps it, sending inwards, across the artery, an aponeurosis which binds the vessel down upon the Brachialis anticus. Sometimes, a fleshy slip from the muscle covers the vessel, in one case to the extent of three inches. In some cases of high origin of the Pronator teres, an aponeurosis extends from it to join the Brachialis anticus external to the artery; a kind of arch being thus formed under which the principal artery and median nerve pass, so as to be concealed for half an inch above the transverse level of the condyle.

Surgical Anatomy. Compression of the brachial artery is required in cases of amputation and some other operations in the arm and fore-arm; and it will be observed, that it may be effected in almost any part of the course of the artery. If pressure is made in the upper part of the limb, it should be directed from within outwards, and if in the lower part, from before backwards, as the artery lies on the inner side of the humerus above, and in front of it below. The most favourable situation is near the insertion of the Coraco-brachialis.

The application of a ligature to the brachial artery may be required in cases of wounds of the vessel, and in some cases of wound of the palmar arch. It is also sometimes necessary in cases of aneurism of the brachial, the radial, ulnar, or interosseous arteries. The artery may be secured in any part of its course. The chief guides in determining its position are the surface-markings produced by the inner margin of the Coraco-brachialis and Biceps, the known course of the vessel, and its pulsation, which should be carefully felt for before any operation is performed, as the vessel occasionally deviates from its usual position in the arm. In whatever situation the operation is performed, great care is necessary, on account of the extreme thinness of the parts covering the artery, and the intimate connection which the vessel has throughout its whole course with important nerves and veins. Sometimes a thin layer of muscular fibre is met with concealing the artery; if such is the case, it must be cut across, in order to expose the vessel.

In the upper third of the arm the artery may be exposed in the following manner:—The patient being placed horizontally upon a table, the affected limb should be raised from the side, and the hand supinated. An incision about two inches in length should be made on the ulnar side of the Coraco-brachialis muscle, and the subjacent fascia cautiously divided, so as to avoid wounding the internal cutaneous nerve or basilic vein, which sometimes run on the surface of the artery as high as the axilla. The fascia having been divided, it should be remembered, that the ulnar and internal cutaneous nerves lie on the inner side of the artery, the median on the outer side, the latter nerve being occasionally superficial to the artery in this situation, and that the venæ comites are also in relation with the vessel, one on either side. These being carefully separated, the aneurism needle should be passed round the artery from the ulnar to the radial side.

If two arteries are present in the arm, in consequence of a high division, they are usually placed side by side; and if they are exposed in an operation, the surgeon should endeavour to ascertain, by alternately pressing on each vessel, which of the two communicates with the wound or aneurism, when a ligature may be applied accordingly; or if pulsation or hæmorrhage ceases only when both vessels are compressed, both vessels may be tied, as it may be concluded that the two communicate above the seat of disease, or are reunited.

It should also be remembered, that two arteries may be present in the arm in a case of high division, and that one of these may be found along the inner intermuscular septum, in a line towards the inner condyle of the humerus, or in the usual position of the brachial, but deeply placed beneath the common trunk: a knowledge of these facts will suggest the precautions necessary in every case, and indicate the measures to be adopted when anomalies are met with.

In the middle of the arm the brachial artery may be exposed by making an incision along the inner margin of the Biceps muscle. The forearm being bent so as to relax the muscle, it should be drawn slightly aside, and the fascia being carefully divided, the median nerve will be exposed lying upon the artery (sometimes beneath); this being drawn inwards and the muscle outwards, the artery should be separated from its accompanying veins and secured. In this situation the inferior profunda may be mistaken for the main trunk, especially if enlarged, from the collateral circulation having become established; this may be avoided by directing the incision externally towards the Biceps rather than inwards or backwards towards the Triceps.

The lower part of the brachial artery is of extreme interest in a surgical point of view, on account of the relation which it bears to the veins most commonly opened in venesection. Of these vessels, the median basilic is the largest and most prominent, and, consequently, the one usually selected for the operation. It should be remembered, that this vein runs parallel with the brachial artery, from which it is separated by the bicipital fascia, and that in no case should this vessel be selected for venesection, except in a part which is not in contact with the artery.

Collateral Circulation. After the application of a ligature to the brachial artery in the upper third of the arm, the circulation is carried on by branches from the circumflex and subscapular arteries, anastomosing with ascending branches from the superior profunda. If the brachial is tied *below* the origin of the profunda arteries the circulation is maintained by the branches of the profundæ, anastomosing with the recurrent radial, ulnar, and interosseous arteries. In two cases described by Mr. South,* in which the brachial artery had been tied some time previously, in one 'a long portion of the artery had been obliterated, and sets of vessels are descending on either side from above the obliteration, to be received into others which ascend in a similar manner from below it. In the other, the obliteration is less extensive, and a single curved artery about as big as a crow-quill passes from the upper to the lower open part of the artery.'

The branches of the brachial artery are the

Superior profunda.	Inferior profunda.
Nutrient artery.	Anastomotica magna.

Muscular.

The *superior profunda* arises from the inner and back part of the brachial, opposite the lower border of the Teres major, and passes backwards to the interval between the outer and inner heads of the Triceps muscle, accompanied by the musculo-spiral nerve; it winds round the back part of the shaft of the humerus in the spiral groove, between the Triceps and the bone, and descends on the outer side of the arm to the space between the Brachialis anticus and Supinator longus, as far as the elbow, where it anastomoses with the recurrent branch of the radial artery. It supplies the Deltoid, Coraco-brachialis, and Triceps muscles, and whilst in the groove between the Triceps and the bone, it gives off the posterior articular artery, which descends perpendicularly between the Triceps and the bone, to the back part of the elbow-joint, where it anastomoses with the interosseous recurrent branch, and, on the inner side of the arm, with the posterior ulnar recurrent, and with the anastomotica magna or inferior profunda (fig. 221).

The *nutrient artery* of the shaft of the humerus arises from the brachial, about the middle of the arm. Passing downwards, it enters the nutrient canal of that bone, near the insertion of the Coraco-brachialis muscle.

The *inferior profunda*, of small size, arises from the brachial, a little below the middle of the arm; piercing the internal intermuscular septum, it descends on the surface of the inner head of the Triceps muscle, to the space between the inner condyle and olecranon, accompanied by the ulnar nerve, and terminates by anastomosing with the posterior ulnar recurrent, and anastomotica magna.

The *anastomotica magna* arises from the brachial, about two inches above the elbow-joint. It passes transversely inwards upon the Brachialis anticus, and piercing the internal intermuscular septum, winds round the back part of the humerus between the Triceps and the bone, forming an arch above the olecranon fossa, by its junction with the posterior articular branch of the superior profunda. As this vessel lies on the Brachialis anticus, an offset passes between the internal condyle and olecranon, which anastomoses with the inferior profunda and posterior ulnar recurrent arteries. Other branches ascend to join the inferior profunda; and some descend in front of the inner condyle, to anastomose with the anterior ulnar recurrent.

The *muscular* are three or four large branches, which are distributed to the

* CHELIUS' *Surgery*, vol. ii. p. 254. See also White's engraving referred to by Mr. South, of the anastomosing branches after ligature of the brachial, in WHITE'S *Cases in Surgery*. Porta also gives a case (with drawings) of the circulation after ligature of both brachial and radial—*Alterazioni Patologiche delle Arterie.*

muscles in the course of the artery. They supply the Coraco-brachialis, Biceps, and Brachialis anticus muscles.

219.—The Surgical Anatomy of the Radial and Ulnar Arteries.

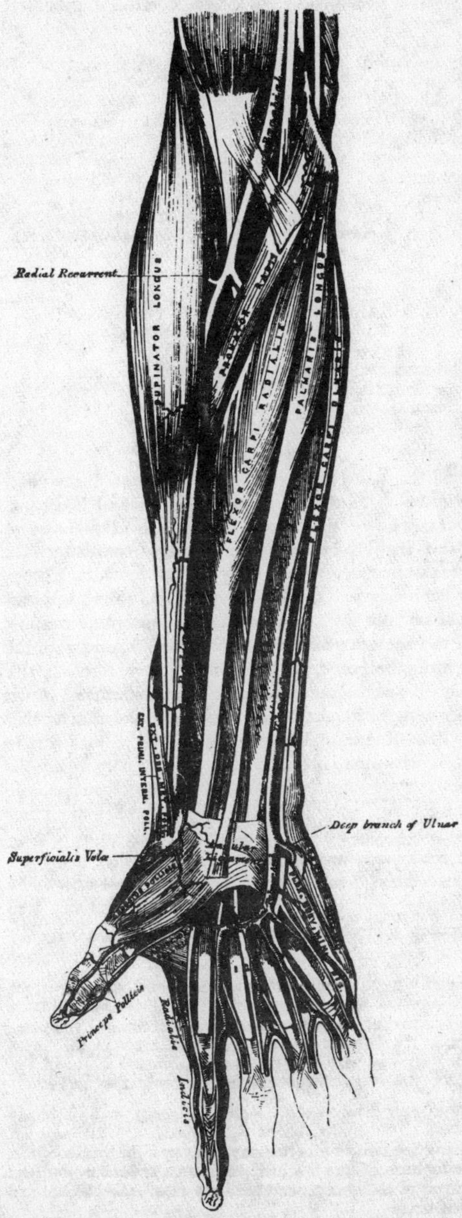

Radial Artery.

The Radial artery appears, from its direction, to be the continuation of the brachial, but, in size, it is smaller than the ulnar. It commences at the bifurcation of the brachial, just below the bend of the elbow, and passes along the radial side of the forearm to the wrist; it then winds backwards, round the outer side of the carpus, beneath the extensor tendons of the thumb, and, finally, passes forwards between the two heads of the first Dorsal interosseous muscle, into the palm of the hand, where it crosses the metacarpal bones to the ulnar border of the hand, to form the deep palmar arch. At its termination, it inosculates with the deep branch of the ulnar artery. The relations of this vessel may thus be conveniently divided into three parts, viz., in front of the forearm, at the back of the wrist, and in the hand.

Relations. In the forearm, this vessel extends from opposite the neck of the radius, to the fore-part of the styloid process, being placed to the inner side of the shaft of the bone, above, and in front of it below. It is superficial throughout its entire extent, being covered by the integument, the superficial and deep fascia, and slightly overlapped above by the Supinator longus. In its course downwards, it lies upon the tendon of the Biceps, the Supinator brevis, the Pronator radii teres, the radial origin of the Flexor sublimis digitorum, the Flexor longus pollicis, the Pronator quadratus, and the lower extremity of the radius. In the upper third of its course, it lies between the Supinator longus and the Pronator radii teres; in its lower two-thirds, between the tendons of the Supinator longus and the Flexor carpi radialis. The radial nerve lies along the outer side of the artery, in the middle third of its course; and some filaments of the musculo-cutaneous nerve, after piercing the deep fascia, run along the lower part of the artery as it winds round the wrist. The vessel is accompanied by venæ comites throughout its whole course.

Plan of the Relations of the Radial Artery in the Forearm.

In front.
Integument—superficial and deep fasciæ.
Supinator longus.

Inner side.
Pronator radii teres.
Flexor carpi radialis.

Radial Artery in Forearm.

Outer side.
Supinator longus.
Radial nerve (middle third).

Behind.
Tendon of Biceps.
Supinator brevis.
Pronator radii teres.
Flexor sublimis digitorum.
Flexor longus pollicis.
Pronator quadratus.
Radius.

At the wrist, as it winds round the outer side of the carpus, from the styloid process to the first interosseous space, it lies upon the external lateral ligament, being covered by the extensor tendons of the thumb, subcutaneous veins, some filaments of the radial nerve, and the integument. It is accompanied by two veins, and a filament of the musculo-cutaneous nerve.

In the hand, it passes from the upper end of the first interosseous space, between the heads of the Abductor indicis or first Dorsal interosseous muscle transversely across the palm, to the base of the metacarpal bone of the little finger, where it inosculates with the communicating branch from the ulnar artery, forming the deep palmar arch. It lies upon the carpal extremities of the metacarpal bones and the Interossei muscles, being covered by the flexor tendons of the fingers, the Lumbricales, the muscles of the little finger, and the Flexor brevis pollicis, and is accompanied by the deep branch of the ulnar nerve.

Peculiarities. The origin of the radial artery varies in the proportion nearly of one in eight cases. In one case the origin was lower than usual. In the other cases, the upper part of the brachial was a more frequent source of origin than the axillary. The variations in the position of this vessel in the arm, and at the bend of the elbow, have been already mentioned. In the forearm it deviates less frequently from its position than the ulnar. It has been found lying over the fascia, instead of beneath it. It has also been observed on the surface of the Supinator longus, instead of along its inner border: and in turning round the wrist, it has been seen lying over, instead of beneath, the Extensor tendons.

Surgical Anatomy. The operation of tying the radial artery is required in cases of wounds either of its trunk, or of some of its branches, or for aneurism: and it will be observed, that the vessel may be easily exposed in any part of its course through the forearm. The operation in the middle or inferior third of the forearm is easily performed; but in the upper third, near the elbow, it is attended with some difficulty, from the greater depth of the vessel, and from its being overlapped by the Supinator longus and Pronator teres muscles.

To tie the artery in the upper third, an incision three inches in length should be made through the integument, from the bend of the elbow obliquely downwards and outwards, on the radial side of the forearm, avoiding the branches of the median vein; the fascia of the arm being divided, and the Supinator longus drawn a little outwards, the artery will be exposed. The venæ comites should be carefully separated from the vessel, and the ligature passed from the radial to the ulnar side.

In the middle third of the forearm the artery may be exposed by making an incision of similar length on the inner margin of the Supinator longus. In this situation, the radial nerve lies in close relation with the outer side of the artery, and should, as well as the veins, be carefully avoided.

In the lower third, the artery is easily secured by dividing the integument and fasciæ in the interval between the tendons of the Supinator longus and Flexor carpi radialis muscles.

The branches of the radial artery may be divided into three groups, corresponding with the three regions in which the vessel is situated.

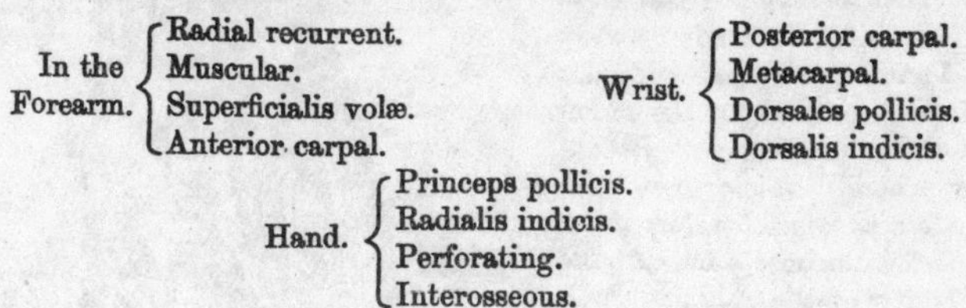

In the Forearm.	Radial recurrent. Muscular. Superficialis volæ. Anterior carpal.
Wrist.	Posterior carpal. Metacarpal. Dorsales pollicis. Dorsalis indicis.
Hand.	Princeps pollicis. Radialis indicis. Perforating. Interosseous.

The *radial recurrent* is given off immediately below the elbow. It ascends between the branches of the musculo-spiral nerve, lying on the Supinator brevis, and then between the Supinator longus and Brachialis anticus, supplying these muscles and the elbow-joint, and anastomosing with the terminal branches of the superior profunda.

The *muscular branches* are distributed to the muscles on the radial side of the forearm.

The *superficialis volæ* arises from the radial artery, just where this vessel is about to wind round the wrist. Running forwards, it passes between the muscles of the thumb, which it supplies, and sometimes anastomoses with the termination of the ulnar artery, completing the superficial palmar arch. This vessel varies considerably in size, usually it is very small, and terminates in the muscles of the thumb; sometimes it is as large as the continuation of the radial.

The *carpal branches* supply the joints of the wrist. The *anterior carpal* is a small vessel which arises from the radial artery near the lower border of the Pronator quadratus, and running inwards in front of the radius, anastomoses with the anterior carpal branch of the ulnar artery. From the arch thus formed, branches descend to supply the articulations of the wrist.

The *posterior carpal* is a small vessel which arises from the radial artery beneath the extensor tendons of the thumb; crossing the carpus transversely to the inner border of the hand, it anastomoses with the posterior carpal branch of the ulnar. It sends branches upwards, which anastomose with the termination of the anterior interosseous artery; other branches descend to the metacarpal spaces; they are the dorsal interosseous arteries for the third and fourth interosseous spaces; they anastomose with the posterior perforating branches from the deep palmar arch.

The *metacarpal* (*first dorsal interosseous branch*) arises beneath the extensor tendons of the thumb, sometimes with the posterior carpal artery; running forwards on the second dorsal interosseous muscle, it communicates, behind, with the corresponding perforating branch of the deep palmar arch; and, in front, inosculates with the digital branch of the superficial palmar arch, and supplies the adjoining sides of the index and middle fingers.

The *dorsales pollicis* are two small vessels which run along the sides of the dorsal aspect of the thumb. They arise separately, or occasionally by a common trunk, near the base of the first metacarpal bone.

The *dorsalis indicis*, also a small branch, runs along the radial side of the back of the index finger, sending a few branches to the Abductor indicis.

The *princeps pollicis* arises from the radial just as it turns inwards to the deep part of the hand; it descends between the Abductor indicis and Adductor pollicis, along the ulnar side of the metacarpal bone of the thumb, to the base of the first phalanx, where it divides into two branches, which run along the sides of the palmar aspect of the thumb, and form an arch on the under surface of the last phalanx, from which branches are distributed to the integument and cellular membrane of the thumb.

220.—Ulnar and Radial Arteries. Deep View.

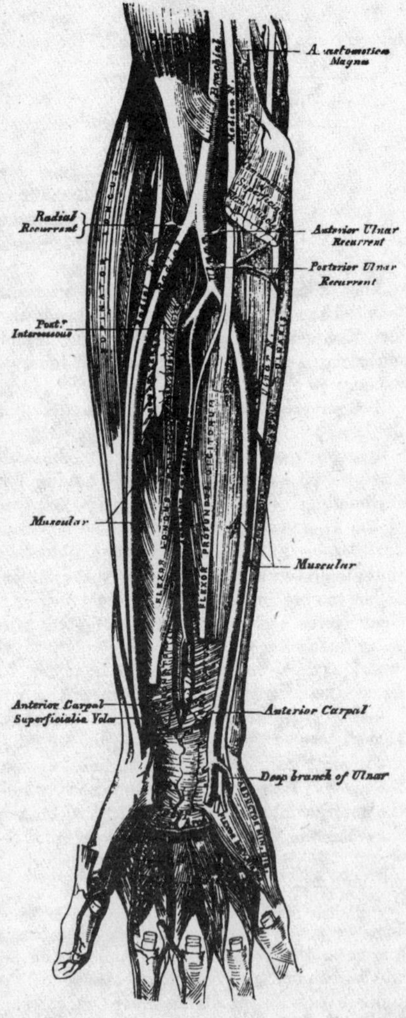

The *radialis indicis* arises close to the preceding, descends between the Abductor indicis and Adductor pollicis, and runs along the radial side of the index finger to its extremity, where it anastomoses with the collateral digital artery from the superficial palmar arch. At the lower border of the Adductor pollicis, this vessel anastomoses with the princeps pollicis, and gives a communicating branch to the superficial palmar arch.

The *perforating* arteries, three in number, pass backwards between the heads of the last three Dorsal interossei muscles, to inosculate with the dorsal interosseous arteries.

The *palmar interosseous*, three or four in number, are branches of the deep palmar arch; they run forwards upon the Interossei muscles, and anastomose at the clefts of the fingers with the digital branches of the superficial arch.

Ulnar Artery.

The *Ulnar Artery*, the larger of the two subdivisions of the brachial, commences a little below the bend of the elbow, and crosses the inner side of the forearm obliquely inwards, to the commencement of its lower half; it then runs along its ulnar border to the wrist, crosses the annular ligament on the radial side of the pisiform bone, and passes across the palm of the hand, forming the superficial palmar arch, which sometimes terminates by inosculating with the superficialis volæ.

Relations in the Forearm. In its *upper half*, it is deeply seated, being covered by all the superficial flexor muscles, excepting the Flexor carpi ulnaris; it is crossed by the median nerve, which lies to its inner side for about an inch, and it lies upon the Brachialis anticus and Flexor profundus digitorum muscles. In the *lower half* of the forearm, it lies upon the Flexor profundus, being covered by the integument, the superficial and deep fascia, and is placed between the Flexor carpi ulnaris and Flexor sublimis digitorum muscles. It is accompanied by two venæ comites; the ulnar nerve lies on its inner side for the lower two-thirds of its extent, and a small branch from the nerve descends on the lower part of the vessel to the palm of the hand.

Plan of Relations of the Ulnar Artery in the Forearm.

In front.

Superficial flexor muscles. } Upper half.
Median nerve. }
Superficial and deep fasciæ. Lower half.

Inner side.
Flexor carpi ulnaris.
Ulnar nerve (lower two-thirds).

Ulnar Artery in Forearm.

Outer side.
Flexor sublimis digitorum.

Behind.
Brachialis anticus.
Flexor profundus digitorum.

At the wrist (fig. 219), the ulnar artery is covered by the integument and fascia, and lies upon the anterior annular ligament. On its inner side is the pisiform bone. The ulnar nerve lies at the inner side, and somewhat behind the artery.

In the palm of the hand, the continuation of the ulnar artery is called the superficial palmar arch; it passes obliquely outwards to the interspace between the ball of the thumb and the index finger, where it occasionally anastomoses with the superficialis volæ, and a branch from the radialis indicis, thus completing the arch. The convexity of this arch is directed towards the fingers, its concavity towards the muscles of the thumb. If the thumb be put at right angles to the hand, the position of the superficial palmar arch will be roughly indicated by a line drawn along the lower margin of the thumb across the palm of the hand: the deep palmar arch is situated about a finger's breadth nearer to the carpus.

The superficial palmar arch is covered by the Palmaris brevis, the palmar fascia, and integument; and lies upon the annular ligament, the muscles of the little finger, the tendons of the superficial flexor, and the divisions of the median and ulnar nerves, the latter accompanying the artery a short part of its course.

Relations of the Superficial Palmar Arch.

In front.
Integument.
Palmaris brevis.
Palmar fascia.

Ulnar Artery in Hand.

Behind.
Annular ligament.
Origin of muscles of little finger.
Superficial flexor tendons.
Divisions of median and ulnar nerves.

Peculiarities. The ulnar artery has been found to vary in its origin nearly in the proportion of one in thirteen cases, in one case arising lower than usual, about two or three inches below the elbow, and in all the other cases much higher, the brachial being a more frequent source of origin than the axillary.

Variations in the position of this vessel are more frequent than in the radial. When its origin is normal, the course of the vessel is rarely changed. When it arises high up, it is almost invariably superficial to the flexor muscles in the forearm, lying commonly beneath the fascia, more rarely between the fascia and integument. In a few cases, its position was subcutaneous in the upper part of the forearm, subaponeurotic in the lower part.

Surgical Anatomy. The application of a ligature to this vessel is required in cases of wound of the artery, or of its branches, or in consequence of aneurism. In the upper half of the forearm, the artery is deeply seated beneath the superficial flexor muscles, and their division would be requisite in a case of recent wound of the artery in this situation, in order to secure it, but under no other circumstances. In the middle and lower third of the forearm, this vessel may be easily secured by making an incision on the radial side of the tendon of the Flexor carpi ulnaris; the deep fascia being divided, and the Flexor carpi ulnaris and its companion muscle, the Flexor sublimis, being separated from each other, the vessel will be exposed, accompanied by its venæ comites, the ulnar nerve lying on its inner side. The veins being separated from the artery, the ligature should be passed from the ulnar to the radial side, taking care to avoid the ulnar nerve.

The branches of the ulnar artery may be arranged into the following groups,

- *Forearm.*
  - Anterior ulnar recurrent.
  - Posterior ulnar recurrent.
  - Interosseous
    - Anterior interosseous.
    - Posterior interosseous.
  - Muscular.
- *Wrist.*
  - Anterior carpal.
  - Posterior carpal.
- *Hand.*
  - Deep or communicating branch.
  - Digital.

221.—Arteries of the Back of the Forearm and Hand.

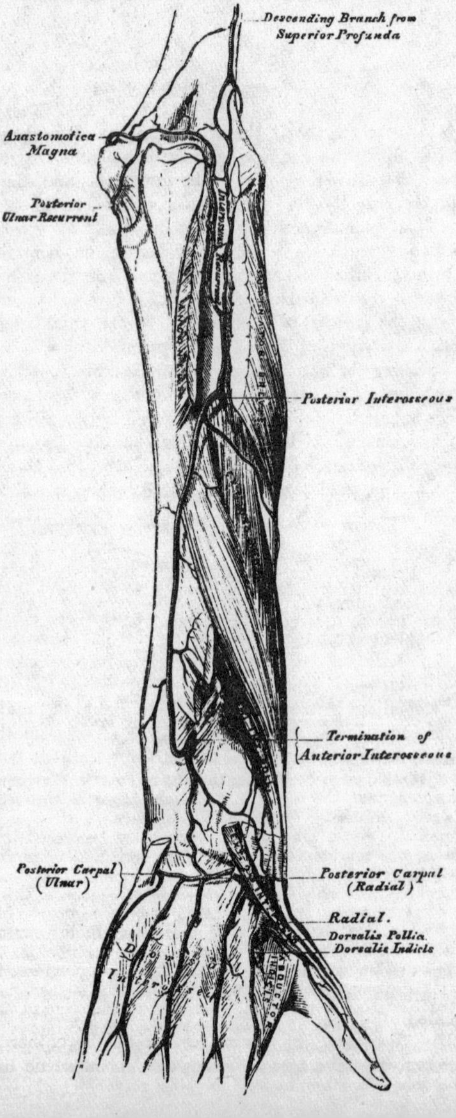

The *anterior ulnar recurrent* (fig. 220) arises immediately below the elbow-joint, passes upwards and inwards between the Brachialis anticus and Pronator radii teres, supplies those muscles, and, in front of the inner condyle, anastomoses with the anastomotica magna and inferior profunda.

The *posterior ulnar recurrent* is much larger, and arises somewhat lower than the preceding. It passes backwards and inwards, beneath the Flexor sublimis, and ascends behind the inner condyle of the humerus. In the interval between this process and the olecranon, it lies beneath the Flexor carpi ulnaris, ascending between the heads of that muscle, beneath the ulnar nerve; it supplies the neighbouring muscles and joint, and anastomoses with the inferior profunda, anastomotica magna, and interosseous recurrent arteries (fig. 221).

The *interosseous artery* (fig. 220) is a short trunk, about an inch in length, and of considerable size, which arises immediately below the tuberosity of the radius, and, passing backwards to the upper border of the interosseous membrane, divides into two branches, the anterior and posterior interosseous.

The *anterior interosseous* passes down the forearm

on the anterior surface of the interosseous membrane, to which it is connected by a thin aponeurotic arch. It is accompanied by the interosseous branch of the median nerve, and overlapped by the contiguous margins of the Flexor profundus digitorum and Flexor longus pollicis muscles, giving off in this situation muscular branches, and the nutrient arteries of the radius and ulna. At the upper border of the Pronator quadratus, a branch descends in front of that muscle, to anastomose in front of the carpus with branches from the anterior carpal and deep palmar arch. The continuation of the artery passes behind the Pronator quadratus (fig. 221), and, piercing the interosseous membrane, descends to the back of the wrist, where it anastomoses with the posterior interosseous and the posterior carpal branches of the radial and ulnar arteries. The anterior interosseous gives off a long, slender branch, which accompanies the median nerve, and gives offsets to its substance. This, the *median artery*, is sometimes much enlarged.

The *posterior interosseous artery* passes backwards through the interval between the oblique ligament and the upper border of the interosseous membrane, and runs down the back part of the forearm, between the superficial and deep layer of muscles, to both of which it distributes branches. Descending to the back of the wrist, it anastomoses with the termination of the anterior interosseous, and with the posterior carpal branches of the radial and ulnar arteries. This artery gives off, near its origin, the *interosseous recurrent* branch, a large vessel, which ascends to the interval between the external condyle and olecranon, beneath the Anconeus and Supinator brevis, anastomosing with a branch from the superior profunda, and with the posterior ulnar recurrent, and anastomotica magna.

The *muscular branches* are distributed to the muscles along the ulnar side of the forearm.

The *carpal branches* are intended for the supply of the wrist joint.

The *anterior carpal* is a small vessel which crosses the front of the carpus beneath the tendons of the Flexor profundus, and inosculates with a corresponding branch of the radial artery.

The *posterior carpal* arises immediately above the pisiform bone, winding backwards beneath the tendon of the Flexor carpi ulnaris; it gives off a branch which passes across the dorsal surface of the carpus beneath the extensor tendons, anastomosing with a corresponding branch of the radial artery, and forming the posterior carpal arch; it is then continued along the metacarpal bone of the little finger, forming its dorsal branch.

The *deep* or *communicating branch* (fig. 220), arises at the commencement of the palmar arch, and passes deeply inwards between the Abductor minimi digiti and Flexor brevis minimi digiti, near their origins; it anastomoses with the termination of the radial artery, completing the deep palmar arch.

The *digital branches* (fig. 219), four in number, are given off from the convexity of the superficial palmar arch. They supply the ulnar side of the little finger, and the adjoining sides of the little, ring, middle, and index-fingers; the radial side of the index-finger and thumb being supplied from the radial artery. The digital arteries at first lie superficial to the flexor tendons, but as they pass forwards with the digital nerves to the clefts between the fingers, they lie between them, and are there joined by the interosseous branches from the deep palmar arch. The digital arteries on the sides of the fingers lie beneath the digital nerves; and, about the middle of the last phalanx, the two branches for each finger form an arch, from the convexity of which branches pass to supply the matrix of the nail.

The Descending Aorta.

The descending aorta is divided into two portions, the thoracic, and abdominal, in correspondence with the two great cavities of the trunk in which it is situated.

The *Thoracic Aorta* commences at the lower border of the fourth dorsal vertebra, on the left side, and terminates at the aortic opening in the Diaphragm, in front of the last dorsal vertebra. At its commencement, it is situated on the left side of the spine; it approaches the median line as it descends; and, at its termination, lies directly in front of the column. The direction of this vessel being influenced by the spine, upon which it rests, it describes a curve which is concave forwards in the dorsal region. As the branches given off from it are small, the diminution in the size of the vessel is inconsiderable. It is contained in the back part of the posterior mediastinum, being in relation, *in front*, from above downwards, with the left pulmonary artery, the left bronchus, the pericardium, and the œsophagus; *behind*, with the vertebral column, and the vena azygos minor; on the *right side*, with the vena azygos major, and thoracic duct; on the *left side*, with the left pleura and lung. The œsophagus, with its accompanying nerves, lies on the right side of the aorta *above*: in front of the artery, in the *middle* of its course; whilst, at its *lower part*, it is on the left side, on a plane anterior to it.

Plan of the Relations of the Thoracic Aorta.

In front.
Left pulmonary artery.
Left bronchus.
Pericardium.
Œsophagus.

Right side.
Œsophagus (above).
Vena azygos major.
Thoracic duct.

Thoracic Aorta.

Left side.
Pleura.
Left lung.
Œsophagus (below).

Behind.
Vertebral column.
Vena azygos minor.

Surgical Anatomy. The student should now consider the effects likely to be produced by aneurism of the thoracic aorta, a disease of common occurrence. When we consider the great depth of the vessel from the surface, and the number of important structures which surround it on every side, it may be easily conceived what a variety of obscure symptoms may arise from disease of this part of the arterial system, and how they may be liable to be mistaken for those of other affections. Aneurism of the thoracic aorta most usually extends backwards, along the left side of the spine, producing absorption of the bodies of the vertebræ, with curvature of the spine; whilst the irritation or pressure on the cord will give rise to pain, either in the chest, back, or loins, with radiating pain in the left upper intercostal spaces, from pressure on the intercostal nerves; at the same time, the tumour may project backwards on each side of the spine, beneath the integument, as a pulsating swelling, simulating abscess connected with diseased bone; or it may displace the œsophagus, and compress the lung on one or the other side. If the tumour extend forward, it may press upon and displace the heart, giving rise to palpitation and other symptoms of disease of that organ; or it may displace, or even compress, the œsophagus, causing pain and difficulty of swallowing, as in stricture of that tube, and ultimately even open into it by ulceration, producing fatal hæmorrhage. If the disease make way to either side, it may press upon the thoracic duct; or it may burst into the pleural cavity, or into the trachea or lung; and lastly, it may open into the posterior mediastinum.

The aorta is, comparatively often, found to be obliterated at a particular spot, viz., at the junction of the arch with the thoracic aorta, just below the ductus arteriosus. Whether this is the result of disease, or of congenital malformation, is immaterial to our present purpose; it affords an interesting opportunity of observing the resources of the collateral circulation. The course of the anastomosing vessels, by which the blood is brought from the upper to the lower part of the artery, will be found well described in an account of two cases in the *Pathological Transactions*, vols. viii. and x. In the former (p. 162), Mr. Sydney Jones thus sums up the detailed description of the anastomosing vessels. 'The principal communications by which the circulation was carried on, were—Firstly, the internal mammary, anastomosing with the intercostal arteries, with the phrenic of the abdominal aorta by means of the musculo-phrenic and comes nervi phrenici, and largely with the deep epigastric. Secondly, the superior intercostal, anastomosing anteriorly by means of a large branch with the first aortic intercostal, and posteriorly, with the posterior branch of the same artery. Thirdly, the inferior thyroid, by means of a branch about the size of an ordinary radial, ormed a communication with the first aortic intercostal. Fourthly, the transversalis colli, by means of very large communications with the posterior branches of the intercostals. Fifthly, the branches (of the subclavian and axillary) going to the side of the chest were large, and anastomosed freely with the lateral branches of the intercostals.' In the second case also (vol. x. p. 97), Mr. Wood describes the anastomoses in a somewhat similar manner, adding the remark, that 'the blood which was brought into the aorta through the anastomoses of the intercostal arteries, appeared to be expended principally in supplying the abdomen and pelvis; while the supply to the lower extremities had passed through the internal mammary and epigastrics.'

Branches of the Thoracic Aorta.

Pericardiac. Œsophageal.
Bronchial. Posterior mediastinal.
Intercostal.

The *pericardiac* are a few small vessels, irregular in their origin, distributed to the pericardium.

The *bronchial* arteries are the nutrient vessels of the lungs, and vary in number, size, and origin. That of the right side arises from the first aortic intercostal, or by a common trunk with the left bronchial, from the front of the thoracic aorta. Those of the left side, usually two in number, arise from the thoracic aorta, one a little lower than the other. Each vessel is directed to the back part of the corresponding bronchus, along which they run, dividing and subdividing, upon the bronchial tubes, supplying them, the cellular tissue of the lungs, the bronchial glands, and the œsophagus.

The *œsophageal arteries*, usually four or five in number, arise from the front of the aorta, and pass obliquely downwards to the œsophagus, forming a chain of anastomoses along that tube, anastomosing with the œsophageal branches of the inferior thyroid arteries above, and with ascending branches from the phrenic and gastric arteries below.

The *posterior mediastinal arteries* are numerous small vessels which supply the glands and loose areolar tissue in the mediastinum.

The *Intercostal arteries* arise from the back part of the aorta. They are usually ten in number on each side, the superior intercostal space (and occasionally the second one) being supplied by the superior intercostal, a branch of the subclavian. The right intercostals are longer than the left, on account of the position of the aorta to the left side of the spine; they pass outwards, across the bodies of the vertebræ, to the intercostal spaces, being covered by the pleura, the œsophagus, thoracic duct, sympathetic nerve, and the vena azygos major; the left passing beneath the superior intercostal vein, the vena azygos minor, and sympathetic. In the intercostal spaces, each artery divides into two branches, an anterior, or proper intercostal branch; and a posterior, or dorsal branch.

The *anterior branch* passes outwards, at first lying upon the External intercostal muscle, covered in front by the pleura, and a thin fascia. It then passes between the two layers of Intercostal muscles, and, having ascended obliquely to the lower border of the rib above, divides, near the angle of that bone, into two branches; of these, the larger runs in the groove, on the lower border of the rib above; the smaller branch along the upper border of the rib below; passing forward, they supply the Intercostal muscles, and anastomose with the anterior intercostal branches of the internal mammary, and with the thoracic branches of the axillary artery. The first aortic intercostal anastomoses with the superior intercostal, and the last three pass between the abdominal muscles, inosculating with the epigastric in front, and with the phrenic, and lumbar arteries. Each intercostal artery is accompanied by a vein and nerve, the former being above, and the latter below, except in the upper intercostal spaces, where the nerve is at first above the artery. The arteries are protected from pressure during the action of the Intercostal muscles, by fibrous arches thrown across, and attached by each extremity to the bone.

The *posterior*, or *dorsal branch*, of each intercostal artery, passes backwards to the inner side of the anterior costo-transverse ligament, and divides into a spinal branch, which supplies the vertebræ, the spinal cord and its membranes, and a muscular branch, which is distributed to the muscles and integument of the back.

The Abdominal Aorta. (Fig. 222.)

The *Abdominal Aorta* commences at the aortic opening of the Diaphragm, in front of the body of the last dorsal vertebra, and, descending a little to the left

222.—The Abdominal Aorta and its Branches.

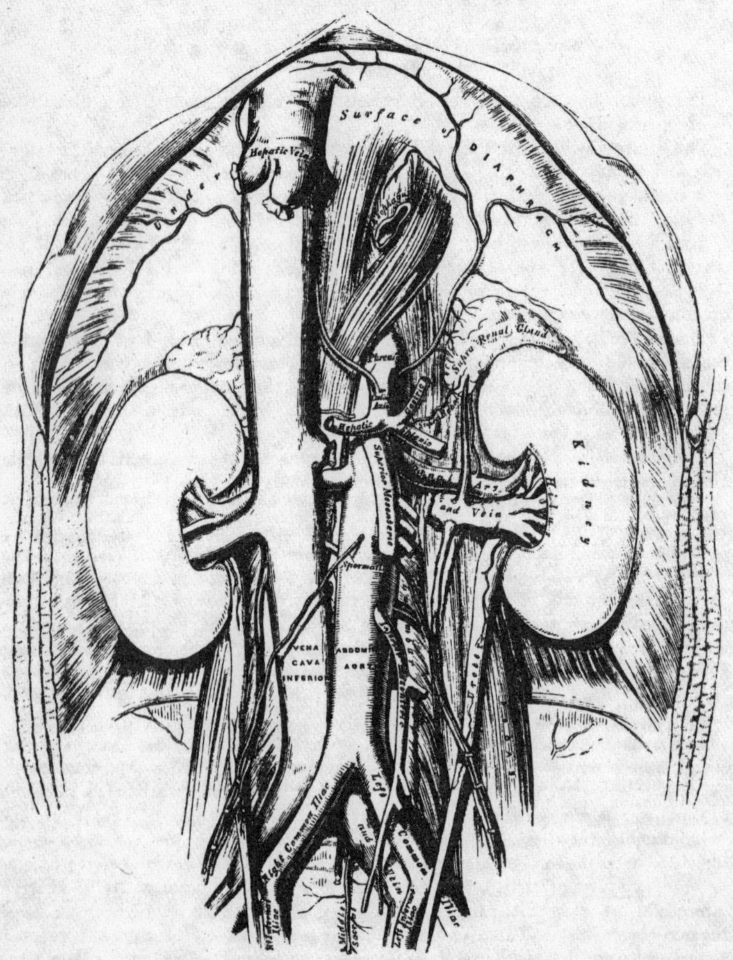

side of the vertebral column, terminates on the left side of the body of the fourth lumbar vertebra, where it divides into the two common iliac arteries. It diminishes rapidly in size, in consequence of the many large branches which it gives off. As it lies upon the bodies of the vertebræ, the curve which it describes is convex forwards, the greatest convexity corresponding to the third lumbar vertebra, which is a little above and to the left side of the umbilicus.

Relations. It is covered, *in front*, by the lesser omentum and stomach, behind which are the branches of the cœliac axis, and the solar plexus; below these, by the splenic vein, the pancreas, the left renal vein, the transverse portion of the duodenum, the mesentery, and aortic plexus. *Behind*, it is separated from the lumbar vertebræ by the left lumbar veins, the receptaculum chyli, and thoracic duct. On the *right side*, it is in relation with the inferior vena cava (the right crus of the Diaphragm being interposed above), the vena azygos, thoracic duct, and right semilunar ganglion; on the *left side*, with the sympathetic nerve, and left semilunar ganglion.

Plan of the Relations of the Abdominal Aorta.

In front.
Lesser omentum and stomach.
Branches of cœliac axis and solar plexus.
Splenic vein.
Pancreas.
Left renal vein.
Transverse duodenum.
Mesentery.
Aortic plexus.

Right side.
Right crus of Diaphragm.
Inferior vena cava.
Vena azygos.
Thoracic duct.
Right semilunar ganglion.

Abdominal Aorta.

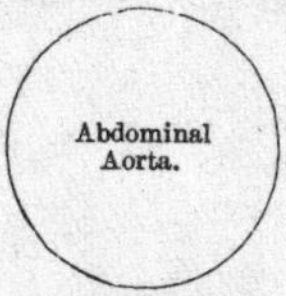

Left side.
Sympathetic nerve.
Left semilunar ganglion.

Behind.
Left lumbar veins.
Receptaculum chyli.
Thoracic duct.
Vertebral column.

Surgical Anatomy. Aneurisms of the abdominal aorta near the cœliac axis communicate in nearly equal proportion with the anterior and posterior parts of the artery.

When an aneurismal sac is connected with the back part of the abdominal aorta, it usually produces absorption of the bodies of the vertebræ, and forms a pulsating tumour, that presents itself in the left hypochondriac or epigastric regions, accompanied by symptoms of disturbance of the alimentary canal. Pain is invariably present, and is usually of two kinds, a fixed and constant pain in the back, caused by the tumour pressing on or displacing the branches of the solar plexus and splanchnic nerves, and a sharp lancinating pain, radiating along those branches of the lumbar nerves which are pressed on by the tumour; hence the pain in the loins, the testes, the hypogastrium, and in the lower limb (usually of the left side). This form of aneurism usually bursts into the peritoneal cavity, or behind the peritoneum, in the left hypochondriac region; or it may form a large aneurismal sac, extending down as low as Poupart's ligament; hæmorrhage in these cases being generally very extensive, but slowly produced, and not rapidly fatal.

When an aneurismal sac is connected with the front of the aorta near the cœliac axis, it forms a pulsating tumour in the left hypochondriac or epigastric regions, usually attended with symptoms of disturbance of the alimentary canal, as sickness, dyspepsia, or constipation, and accompanied by pain, which is constant, but nearly always fixed in the loins, epigastrium, or some part of the abdomen; the radiating pain being rare, as the lumbar nerves are seldom implicated. This form of aneurism may burst into the peritoneal cavity, or behind the peritoneum, between the layers of the mesentery, or, more rarely, into the duodenum; it rarely extends backwards so as to affect the spine.

The abdominal aorta has been tied several times, and although none of the patients permanently recovered, still, as one of them lived as long as ten days, the possibility of the re-establishment of the circulation may be considered to be proved. In the lower animals this artery is often successfully tied. The vessel may be reached in several ways. In the original operation, performed by Sir A. Cooper, an incision was made in the linea alba, the peritoneum opened in front, the finger carried down amongst the intestines towards the spine, the peritoneum again opened behind, by scratching through the mesentery, and the vessel thus reached. Or either of the operations, described below, for securing the common iliac artery, may, by extending the dissection a sufficient distance upwards, be made use of to expose the aorta. The chief difficulty in the dead subject consists in isolating the artery, in consequence of its great depth; but in the living subject, the embarrassment resulting from the proximity of the aneurismal tumour, and the great probability of disease in the vessel itself, add to the dangers and difficulties of this formidable operation so greatly, that it is very doubtful whether it ought ever to be performed.

The collateral circulation would be carried on by the anastomosis between the internal mammary and the epigastric; by the free communication between the superior and inferior mesenterics, if the ligature were placed above the latter vessel; or by the anastomosis between the inferior mesenteric and the internal pudic, when (as is more common) the point of ligature is below the origin of the inferior mesenteric; and possibly by the anastomoses of the lumbar arteries with the branches of the internal iliac.

The circulation through the abdominal aorta may be commanded, in thin persons, by firm pressure with the fingers. Mr. Lister has invented a tourniquet for this purpose, which is of the greatest use in amputation at the hip-joint, and some other operations.

Branches of the Abdominal Aorta.

Phrenic.

Cœliac axis { Gastric. Hepatic. Splenic.

Superior mesenteric.

Suprarenal.

Renal.

Spermatic.

Inferior mesenteric.

Lumbar.

Sacra media.

The branches may be divided into two sets: 1. Those supplying the viscera. 2. Those distributed to the walls of the abdomen.

Visceral Branches.

Cœliac axis { Gastric. Hepatic. Splenic.

Superior mesenteric.

Inferior mesenteric.

Suprarenal. Renal. Spermatic.

Parietal Branches.

Phrenic.

Lumbar.

Sacra media.

Cœliac Axis. (Fig. 223.)

To expose this artery, raise the liver, draw down the stomach, and then tear through the layers of the lesser omentum.

The Cœliac axis is a short thick trunk, about half an inch in length, which arises from the aorta, opposite the margin of the Diaphragm, and passing nearly horizontally forwards (in the erect posture), divides into three large branches, the gastric, hepatic, and splenic, occasionally giving off one of the phrenic arteries.

Relations. It is covered by the lesser omentum. On the *right side,* it is in relation with the right semilunar ganglion, and the lobus Spigelii: on the *left side,* with the left semilunar ganglion and cardiac end of the stomach. *Below,* it rests upon the upper border of the pancreas.

The Gastric Artery (*Coronaria ventriculi*), the smallest of the three branches of the cœliac axis, passes upwards and to the left side, to the cardiac orifice of the stomach, distributing branches to the œsophagus, which anastomose with the aortic œsophageal arteries; others supply the cardiac end of the stomach, inosculating with branches of the splenic artery: it then passes from left to right, along the lesser curvature of the stomach to the pylorus, lying in its course between the layers of the lesser omentum, and giving branches to both surfaces of the organ; at its termination it anastomoses with the pyloric branch of the hepatic.

The Hepatic Artery in the adult is intermediate in size between the gastric and splenic; in the fœtus, it is the largest of the three branches of the cœliac axis. It passes upwards to the right side, between the layers of the lesser omentum, and in front of the foramen of Winslow, to the transverse fissure of the liver, where it divides into two branches, right and left, which supply the corresponding lobes of that organ, accompanying the ramifications of the vena portæ and hepatic duct. The hepatic artery, in its course along the right border of the lesser omentum, is in relation with the ductus communis choledocus and portal vein, the duct lying to the right of the artery, and the vena portæ behind.

Its branches are the

Pyloric.
Gastro-duodenalis { Gastro-epiploica dextra. / Pancreatico-duodenalis.
Cystic.

223.—The Cœliac Axis and its Branches, the Liver having been raised, and the Lesser Omentum removed.

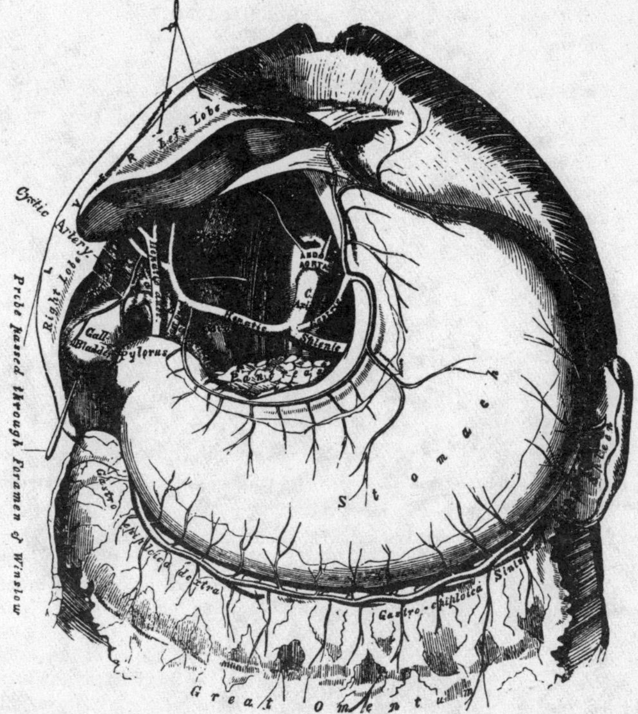

The *pyloric branch* arises from the hepatic, above the pylorus, descends to the pyloric end of the stomach, and passes from right to left along its lesser curvature, supplying it with branches, and inosculating with the gastric artery.

The *gastro-duodenalis* (fig. 224) is a short but large branch, which descends behind the duodenum, near the pylorus, and divides at the lower border of the stomach into two branches, the gastro-epiploica dextra and the pancreatico-duodenalis. Previous to its division, it gives off two or three small inferior pyloric branches to the pyloric end of the stomach and pancreas.

The *gastro-epiploica dextra* runs from right to left along the greater curvature of the stomach, between the layers of the great omentum, anastomosing about the middle of the lower border of the stomach with the gastro-epiploica sinistra from the splenic artery. This vessel gives off numerous branches, some of which ascend to supply both surfaces of the stomach, whilst others descend to supply the great omentum.

The *pancreatico-duodenalis* descends along the contiguous margins of the duodenum and pancreas. It supplies both these organs, and anastomoses with the inferior pancreatico-duodenal branch of the superior mesenteric artery.

In ulceration of the duodenum, which frequently occurs in connexion with severe burns, this artery may be involved, and death may occur from hæmorrhage into the intestinal canal.

The *cystic artery* (fig. 223), usually a branch of the right hepatic, passes upwards and forwards along the neck of the gall bladder, and divides into two branches, one of which ramifies on its free surface, the other, between it and the substance of the liver.

224.—The Cœliac Axis and its Branches, the Stomach having been raised, and the Transverse Meso-Colon removed.

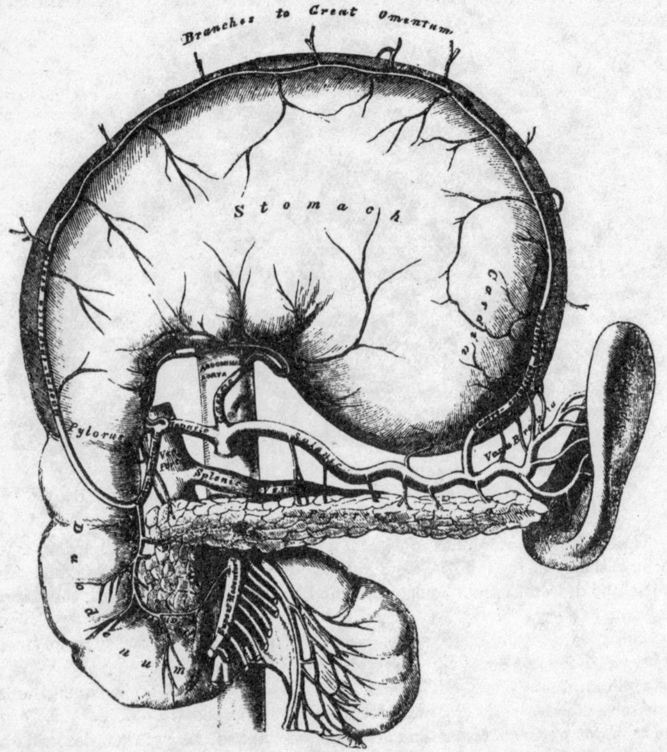

The Splenic Artery, in the adult, is the largest of the three branches of the cœliac axis, and is remarkable for the extreme tortuosity of its course. It passes horizontally to the left side behind the upper border of the pancreas, accompanied by the splenic vein, which lies below it; and on arriving near the spleen, divides into branches, some of which enter the hilum of that organ to be distributed to its structure, whilst others are distributed to the great end of the stomach.

The branches of this vessel are:—

Pancreaticæ parvæ.	Gastric (Vasa brevia).
Pancreatica magna.	Gastro-epiploica sinistra.

The *pancreatic* are numerous small branches derived from the splenic as it runs behind the upper border of the pancreas, supplying its middle and left parts. One of these, larger than the rest, is given off from the splenic near the left extremity of the pancreas; it runs from left to right near the posterior surface of the gland, following the course of the pancreatic duct, and is called the *pancreatica magna*. These vessels anastomose with the pancreatic branches of the pancreatico-duodenal arteries.

The *gastric* (*vasa brevia*) consist of from five to seven small branches, which arise either from the termination of the splenic artery, or from its terminal branches; and passing from left to right, between the layers of the gastro-splenic omentum, are distributed to the great curvature of the stomach; anastomosing with branches of the gastric and gastro-epiploica sinistra arteries.

The *gastro-epiploica sinistra*, the largest branch of the splenic, runs from left to right along the great curvature of the stomach, between the layers of the great omentum; and anastomoses with the gastro-epiploica dextra. In its course it distributes several branches to the stomach, which ascend upon both surfaces; others descend to supply the omentum.

Superior Mesenteric Artery. (Fig. 225.)

In order to expose this vessel, raise the great omentum and transverse colon, draw down the small intestines, and cut through the peritoneum where the transverse meso-colon and mesentery join: the artery will then be exposed, just as it issues from beneath the lower border of the pancreas.

The Superior Mesenteric Artery supplies the whole length of the small intestine, except the first part of the duodenum; it also supplies the cæcum, ascending and transverse colon; it is a vessel of large size, arising from the fore part of the aorta, about a quarter of an inch below the cœliac axis; being covered, at its origin, by the splenic vein and pancreas. It passes forwards, between the pancreas and transverse portion of the duodenum, crosses in front of this portion of the intestine, and descends between the layers of the mesentery to the right iliac fossa, where it terminates, considerably diminished in size. In its course it forms an arch, the convexity of which is directed forwards and downwards to the left side, the concavity backwards and upwards to the right. It is accompanied by the superior mesenteric vein, and is surrounded by the superior mesenteric plexus of nerves. Its branches are the

Inferior pancreatico-duodenal. Ileo-colic.
Vasa intestini tenuis. Colica dextra.
Colica media.

The *inferior pancreatico duonenal* is given off from the superior mesenteric behind the pancreas, and is distributed to the head of the pancreas, and the transverse and descending portions of the duodenum; anastomosing with the pancreatico-duodenal artery.

The *vasa intestini tenuis* arise from the convex side of the superior mesenteric artery. They are usually from twelve to fifteen in number, and are distributed to the jejunum and ileum. They run parallel with one another between the layers of the mesentery; each vessel dividing into two branches, which unite with a similar branch on each side, forming a series of arches, the convexities of which are directed towards the intestine. From this first set of arches branches arise, which again unite with similar branches from either side, and thus a second series of arches is formed; and from these latter, a third, and a fourth, or even fifth series of arches are constituted, diminishing in size the nearer they approach the intestine. From the terminal arches numerous small straight vessels arise which encircle the intestine, upon which they are distributed, ramifying thickly between its coats.

The *ileo-colic artery* is the lowest branch given off from the concavity of the superior mesenteric artery. It descends between the layers of the mesentery to the right iliac fossa, where it divides into two branches. Of these, the inferior one inosculates with the lowest branches of the vasa intestini tenuis, from the convexity of which branches proceed to supply the termination of the ileum, the cœcum and appendix cœci, and the ileo-cœcal valve. The superior division inosculates with the colica dextra, and supplies the commencement of the colon.

The *colica dextra* arises from about the middle of the concavity of the superior mesenteric artery, and passing beneath the peritoneum to the middle of the ascending colon, divides into two branches; a descending branch, which inosculates with the ileo-colic; and an ascending branch, which anastomoses with the colica media. These branches form arches, from the convexity of which vessels are distributed to the ascending colon. The branches of this vessel are covered with peritoneum only on their anterior aspect.

The *colica media* arises from the upper part of the concavity of the superior mesenteric, and, passing forwards between the layers of the transverse meso-colon, divides into two branches; the one on the right side inosculating with the colica dextra; that on the left side, with the colica sinistra, a branch of the inferior mesenteric. From the arches formed by their inosculation, branches are distributed to the transverse colon. The branches of this vessel lie between two layers of peritoneum.

225.—The Superior Mesenteric Artery and its Branches

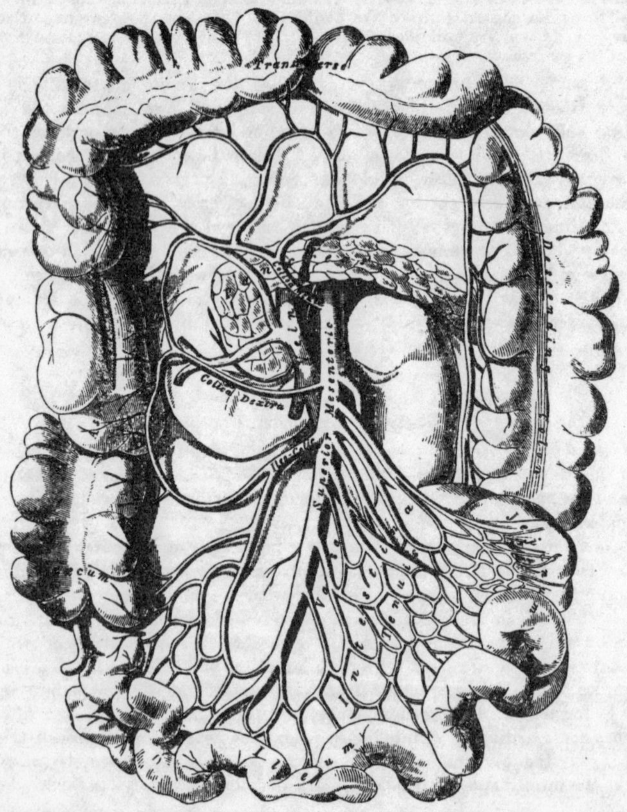

Inferior Mesenteric Artery. (Fig. 226.)

In order to expose this vessel, draw the small intestines and mesentery over to the right side of the abdomen, raise the transverse colon towards the thorax, and divide the peritoneum covering the left side of the aorta.

The Inferior Mesenteric Artery supplies the descending and sigmoid flexure of the colon, and the greater part of the rectum. It is smaller than the superior mesenteric; and arises from the left side of the aorta, between one and two inches above its division into the common iliacs. It passes downwards to the left iliac fossa, and then descends, between the layers of the meso-rectum, into the pelvis, under the name of the *superior hæmorrhoidal artery*. It lies at first in close relation with the left side of the aorta, and then passes in front of the left common iliac artery. Its branches are the

Colica sinistra. Sigmoid.
Superior hæmorrhoidal.

226.—The Inferior Mesenteric Artery and its Branches.

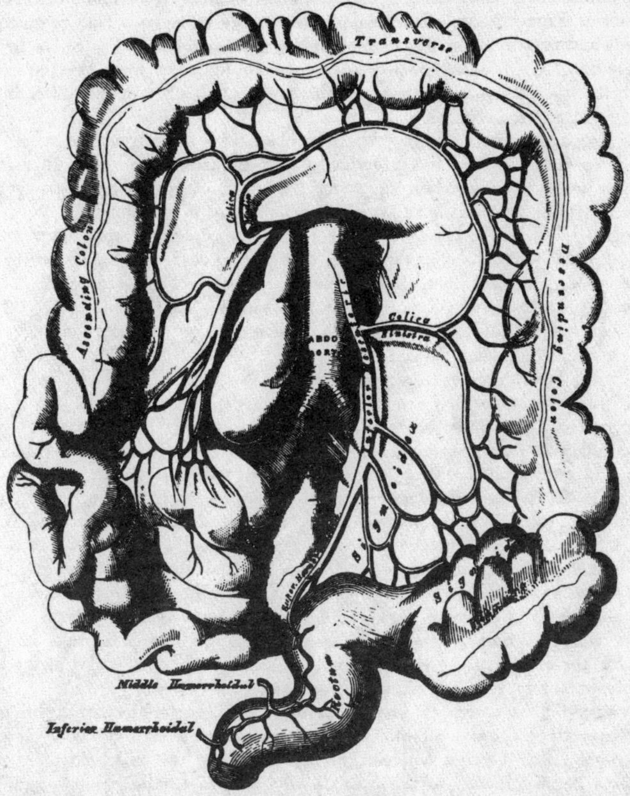

The *colica sinistra* passes behind the peritoneum, in front of the left kidney, to reach the descending colon, and divides into two branches; an ascending branch, which inosculates with the colica media; and a descending branch, which anastomoses with the sigmoid artery. From the arches formed by these inosculations, branches are distributed to the descending colon.

The *sigmoid artery* runs obliquely downwards across the Psoas muscle to the sigmoid flexure of the colon, and divides into branches which supply that part of the intestine; anastomosing above, with the colica sinistra; and below, with the superior hæmorrhoidal artery. This vessel is sometimes replaced by three or four small branches.

The *superior hæmorrhoidal artery*, the continuation of the inferior mesenteric, descends into the pelvis between the layers of the meso-rectum, crossing, in its course, the ureter, and left common iliac vessels. Opposite the middle of the sacrum it divides into two branches, which descend one on each side of the rectum, where they divide into several small branches, which are distributed between the mucous and muscular coats of that tube, nearly as far as its lower end; anastomosing with each other, with the middle hæmorrhoidal arteries, branches of the internal iliac, and with the inferior hæmorrhoidal, branches of the internal pudic.

The student should especially remark, that the trunk of the vessel descends along the back part of the rectum as far as the middle of the sacrum before it divides; this is about a finger's length or four inches from the anus. In disease of this tube, the rectum should never be divided beyond this point in that direction, for fear of involving this artery.

The Suprarenal Arteries (fig. 222), are two small vessels which arise, one on each side of the aorta, opposite the superior mesenteric artery. They pass obliquely upwards and outwards, to the under surface of the suprarenal capsules, to which they are distributed, anastomosing with capsular branches from the phrenic and renal arteries. In the adult these arteries are of small size; in the fœtus they are as large as the renal arteries.

The Renal Arteries are two large trunks, which arise from the sides of the aorta, immediately below the superior mesenteric artery. Each is directed outwards, so as to form nearly a right angle with the aorta. The right is longer than the left, on account of the position of the aorta; it passes behind the inferior vena cava. The left is somewhat higher than the right. Previously to entering the kidney, each artery divides into four or five branches, which are distributed to its substance. At the hilum, these branches lie between the renal vein and ureter, the vein being usually in front, the ureter behind. Each vessel gives off some small branches to the suprarenal capsules, the ureter, and the surrounding cellular membrane and muscles.

The Spermatic Arteries are distributed to the testes in the male, and to the ovaria in the female. They are two slender vessels, of considerable length, which arise from the front of the aorta, a little below the renal arteries. Each artery passes obliquely outwards and downwards, behind the peritoneum, crossing the ureter, and resting on the Psoas muscle, the right spermatic lying in front of the inferior vena cava, the left behind the sigmoid flexure of the colon. On reaching the margin of the pelvis, each vessel passes in front of the corresponding external iliac artery, and takes a different course in the two sexes.

In the male, it is directed outwards, to the internal abdominal ring, and accompanies the other constituents of the spermatic cord along the spermatic canal to the testis, where it becomes tortuous, and divides into several branches, two or three of which accompany the vas deferens, and supply the epididymis, anastomosing with the artery of the vas deferens; others pierce the back part of the tunica albuginea, and supply the substance of the testis.

In the female, the spermatic arteries (ovarian) are shorter than in the male, and do not pass out of the abdominal cavity. On arriving at the margins of the pelvis, each artery passes inwards, between the two laminæ of the broad ligament of the uterus, to be distributed to the ovary. One or two small branches supply the Fallopian tube; another passes on to the side of the uterus, and anastomoses with the uterine arteries. Other offsets are continued along the round ligament, through the inguinal canal, to the integument of the labium and groin.

At an early period of fœtal life, when the testes lie by the side of the spine, below the kidneys, the spermatic arteries are short; but as these organs descend from the abdomen into the scrotum, the arteries become gradually lengthened.

The Phrenic Arteries are two small vessels, which present much variety in their origin. They may arise separately from the front of the aorta, immediately below the cœliac axis, or by a common trunk, which may spring either from the aorta, or from the cœliac axis. Sometimes one is derived from the aorta, and the other from one of the renal arteries. In only one out of thirty-six cases examined, did these arteries arise as two separate vessels from the aorta. They diverge from one another across the crura of the Diaphragm, and then pass obliquely upwards and outwards upon its under surface. The left phrenic passes behind the œsophagus, and runs forwards on the left side of the œsophageal opening. The right phrenic passes behind the liver and inferior vena cava, and ascends along the right side of the aperture for transmitting that vein. Near the back part of the central tendon, each vessel divides into two branches. The internal branch runs forwards to the front of the thorax, supplying the Diaphragm, and anastomosing with its fellow of the opposite side, and with the musculo-phrenic, a branch of the internal mammary. The external branch passes towards the side of the thorax, and inosculates with the intercostal arteries. The internal branch of the right phrenic gives off a few vessels to the inferior vena cava; and the left one some branches to the œsophagus. Each vessel also sends capsular branches to the suprarenal capsule of its own side. The spleen on the left side, and the liver on the right, also receives a few branches from these vessels.

The LUMBAR ARTERIES are analogous to the intercostal. They are usually four in number on each side, and arise from the back part of the aorta, nearly at right angles with that vessel. They pass outwards and backwards, around the sides of the body of the corresponding lumbar vertebra, behind the sympathetic nerve and the Psoas muscle; those on the right side being covered by the inferior vena cava, and the two upper ones on each side by the crura of the Diaphragm. In the interval between the transverse processes of the vertebræ each artery divides into a dorsal and an abdominal branch.

The *dorsal branch* gives off, immediately after its origin, a spinal branch, which enters the spinal canal; it then continues its course backwards, between the transverse processes, and is distributed to the muscles and integument of the back, anastomosing with each other, and with the posterior branches of the intercostal arteries.

The *spinal branch*, besides supplying offsets which run along the nerves to the dura mater and cauda equina, anastomosing with the other spinal arteries, divides into two branches, one of which ascends on the posterior surface of the body of the vertebra above, and the other descends on the posterior surface of the body of the vertebra below, both vessels anastomosing with similar branches from neighbouring spinal arteries. The inosculations of these vessels on each side, throughout the whole length of the spine, form a series of arterial arches behind the bodies of the vertebræ, which are connected with each other, and with a median longitudinal vessel, extending along the middle of the posterior surface of the bodies of the vertebræ, by transverse branches. From these vessels offsets are distributed to the periosteum and bones.

The *abdominal branches* pass outwards, behind the Quadratus lumborum, the lowest branch occasionally in front of that muscle, and, being continued between the abdominal muscles, anastomose with branches of the epigastric and internal mammary *in front*, the intercostals *above*, and those of the ilio-lumbar and circumflex iliac, *below*.

The MIDDLE SACRAL ARTERY is a small vessel, about the size of a crow-quill, which arises from the back part of the aorta, just at its bifurcation. It descends upon the last lumbar vertebra, and along the middle line of the front of the sacrum, to the upper part of the coccyx, where it anastomoses with the lateral sacral arteries, and terminates in a minute branch, which runs down to the situation of the body presently to be described as 'Luschka's gland.' From it, branches arise which run through the meso-rectum, to supply the posterior surface of the rectum. Other branches are given off on each side, which anastomose with the lateral sacral arteries, and send off small offsets which enter the anterior sacral foramina.

Coccygeal Gland, or Luschka's Gland.—Lying near the tip of the coccyx, in a small tendinous interval formed by the union of the Levator ani muscles of either side, and just above the coccygeal attachment of the Sphincter ani, is a small conglobate body, about as large as a lentil or a pea, first described by Luschka,* and named by him *the coccygeal gland*, but the real nature and uses of which are doubtful, nor does it seem at present certain

* *Der Hirnanhang und die Steissdrüse des Menschen*, Berlin, 1860; *Anatomie des Menschen*, Tubingen, 1864, vol. ii. pt. 2. p. 187.

that it always exists. Its most obvious connections are with the arteries of the part. It receives comparatively large branches from the middle and lateral sacral arteries, and its structure, according to Julius Arnold,* consists in great measure of dilated arterial vessels. On this account Arnold proposes to call it not a gland, but 'glomerulus arterio-coccygeus.' It is sometimes single, sometimes formed of several lobes, surrounded by a very definite capsule, into which the sympathetic filaments from the ganglion impar are to be traced, and in which they are said by some observers to terminate. The structure of the body is composed of a number of cavities, which Luschka believes to be glandular follicles, but which are regarded by Arnold as fusiform dilatations of the terminal branches from the middle sacral arteries. Nerves pass into this little body both from the sympathetic and from the fifth sacral, and in the interstices of the lobules nerve-cells are described.

This body has been variously regarded as an appendage to the nervous or to the arterial system. The former seems to be Luschka's view, the latter is Arnold's. Arnold's view is supported by the observation of Dr. Macalister,† that he has found in several birds the middle sacral arteries terminating in a bunch of interlacing and anastomosing capillaries, but without any capsule, and it is rendered in the highest degree probable, if Arnold's observation be correct, that several small saccular bodies, of a somewhat similar kind, may be found connected with the middle sacral artery.

For a more detailed description of this body, we would refer to the elaborate account in 'Luschka's Anatomie,' and to the authorities quoted in Dr. Macalister's paper, as well as to a monograph by Dr. W. Mitchell Banks, reprinted in 1867 from the 'Glasgow Medical Journal.'

Common Iliac Arteries.

The abdominal aorta divides into the two common iliac arteries. The bifurcation usually takes place on the left side of the body of the fourth lumbar vertebra. This point corresponds to the left side of the umbilicus, and is on a level with a line drawn from the highest point of one iliac crest to the other. The common iliac arteries are about two inches in length; diverging from the termination of the aorta, they pass downwards and outwards to the margin of the pelvis, and divide opposite the intervertebral substance, between the last lumbar vertebra and the sacrum, into two branches, the external and internal iliac arteries; the former supplying the lower extremity; the latter, the viscera, and parietes of the pelvis.

The *right common iliac* is somewhat larger than the left, and passes more obliquely across the body of the last lumbar vertebra. In front of it are the peritoneum, the ileum, branches of the sympathetic nerve, and, at its point of division, the ureter. *Behind*, it is separated from the last lumbar vertebra by the two common iliac veins. On its *outer side*, it is in relation with the inferior vena cava, and right common iliac vein, above; and the Psoas magnus muscle, below.

The *left common iliac* is in relation, in front, with the peritoneum, branches of the sympathetic nerve, the rectum and superior hæmorrhoidal artery; and is crossed, at its point of bifurcation, by the ureter. The left common iliac vein lies partly on the inner side, and part beneath the artery; on its outer side, the artery is in relation with the Psoas magnus.

Branches. The common iliac arteries give off small branches to the peritoneum, Psoas muscles, ureters, and the surrounding cellular membrane, and occasionally give origin to the ilio-lumbar, or renal arteries.

* Virchow, *Arch.*, 1864, 5, 6; see also Krause and Meyer in Henle and Pfeiffer's *Zeitsch. f. rat. Medizin.*

† *British Medical Journal*, Jan. 11, 1868.

Plan of the Relations of the Common Iliac Arteries.

Right Common Iliac.

In front.
Peritoneum.
Small intestines.
Sympathetic nerves.
Ureter.

Outer side.
Vena cava.
Right common iliac vein.
Psoas muscle.

Behind.
Right and Left common iliac veins.

Left Common Iliac.

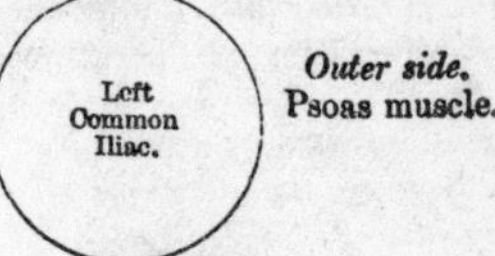

In front.
Peritoneum.
Sympathetic nerves.
Rectum.
Superior hæmorrhoidal artery.
Ureter.

Inner side.
Left common iliac vein.

Outer side.
Psoas muscle.

Behind.
Left common iliac vein.

Peculiarities. The *point of origin* varies according to the bifurcation of the aorta. In three-fourths of a large number of cases, the aorta bifurcated either upon the fourth lumbar vertebra, or upon the intervertebral disc between it and the fifth; the bifurcation being, in one case out of nine below, and in one out of eleven above this point. In ten out of every thirteen cases, the vessel bifurcated within half an inch above or below the level of the crest of the ilium; more frequently below than above.

The *point of division* is subject to great variety. In two thirds of a large number of cases, it was between the last lumbar vertebra and the upper border of the sacrum; being above that point in one case out of eight, and below it in one case out of six. The left common iliac artery divides lower down more frequently than the right.

The *relative length*, also, of the two common iliac arteries varies. The right common iliac was longest in sixty-three cases; the left, in fifty-two; whilst they were both equal in fifty-three. The length of the arteries varied in five-sevenths of the cases examined, from an inch and a-half to three inches; in about half of the remaining cases, the artery was longer; and in the other half, shorter; the minimum length being less than half an inch, the maximum four and a half inches. In one instance, the right common iliac was found wanting, the external and internal iliacs arising directly from the aorta.

Surgical Anatomy. The application of a ligature to the common iliac artery may be required on account of aneurism or hæmorrhage, implicating the external or internal iliacs, or on account of secondary hæmorrhage after amputation of the thigh high up. It has been seen that the origin of this vessel corresponds to the left side of the umbilicus on a level with a line drawn from the highest point of one iliac crest to the opposite one, and its course to a line extending from the left side of the umbilicus downwards towards the middle of Poupart's ligament. The line of incision required in the first steps of an operation for securing this vessel, would materially depend upon the nature of the disease. If the surgeon select the iliac region, a curved incision, about five inches in length, may be made, commencing on the left side of the umbilicus, carried outwards towards the anterior superior iliac spine, and then along the upper border of Poupart's ligament, as far as its middle. But if the aneurismal tumour should extend high up in the abdomen, along the external iliac, it is better to select the side of the abdomen, approaching the artery from above, by making an incision from four to five inches in length, from about two inches above and to the left of the umbilicus, carried outwards in a curved direction towards the lumbar region, and terminating a little below the anterior superior iliac spine. The abdominal muscles (in either case) having been cautiously divided in succession, the transversalis fascia must be carefully cut through, and the peritoneum, together with the ureter, separated from the artery, and pushed aside; the sacro-iliac articulation must then be felt for, and upon it the vessel will be felt pulsating, and may be fully exposed in close connection with its accompanying vein. On the right side, both common iliac veins, as well as the inferior vena cava, are in close connection with the artery, and must be carefully avoided. On the left side, the vein usually lies on the inner side, and behind the artery; but it occasionally happens that the two common iliac veins are joined on the left instead of the right side, which would add much to the difficulty of an operation in such a case. The common iliac artery may be so short that danger may be apprehended from secondary hæmorrhage if a ligature is applied to it. It would be preferable, in such a case, to tie both the external and internal iliacs near their origin.

Collateral Circulation. The principal agents in carrying on the collateral circulation after the application of a ligature to the common iliac, are, the anastomoses of the hæmorrhoidal branches of the internal iliac, with the superior hæmorrhoidal from the inferior mesenteric; the anastomoses of the uterine and ovarian arteries, and of the vesical arteries of opposite sides; of the lateral sacral, with the middle sacral artery; of the epigastric, with the internal mammary, inferior intercostal, and lumbar arteries; of the ilio-lumbar, with the last lumbar artery; of the obturator artery, by means of its pubic branch, with the vessel of the opposite side, and with the internal epigastric; and of the gluteal, with the posterior branches of the sacral arteries.

Internal Iliac Artery. (Fig. 227.)

The internal iliac artery supplies the walls and viscera of the pelvis, the generative organs, and inner side of the thigh. It is a short, thick vessel, smaller than the external iliac, and about an inch and a half in length, which arises at the point of bifurcation of the common iliac; and, passing downwards to the upper margin

227.—Arteries of the Pelvis.

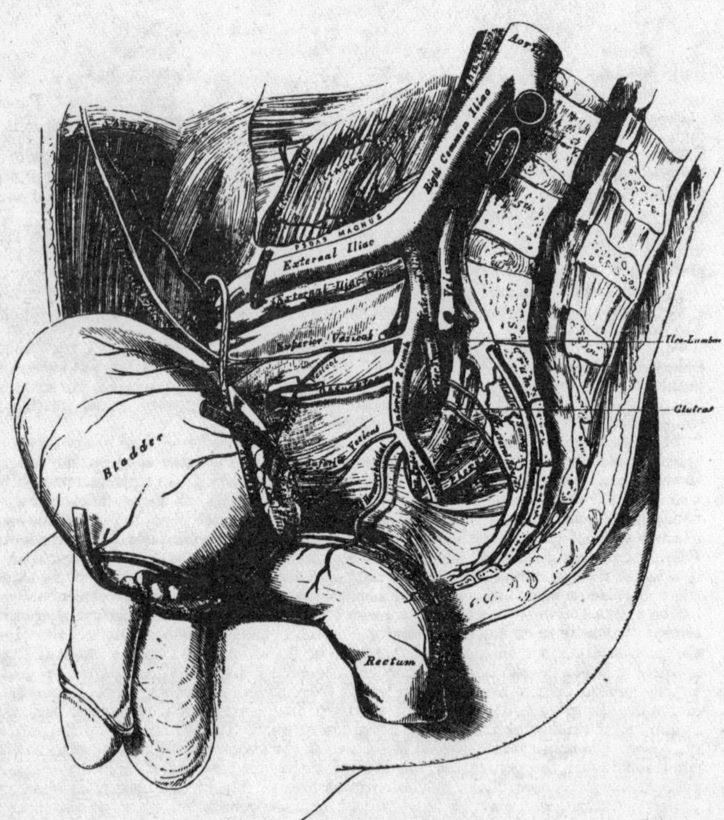

of the great sacro-sciatic foramen, divides into two large trunks, an anterior and posterior; a partially obliterated cord, the hypogastric artery, extending from the extremity of the vessel forwards to the bladder.

Relations. In *front*, with the ureter, which separates it from the peritoneum. *Behind*, with the internal iliac vein, the lumbo-sacral nerve, and Pyriformis muscle. By its *outer side*, near its origin with the Psoas muscle.

Plan of the Relations of the Internal Iliac Artery.

In front.
Peritoneum.
Ureter.

Outer side.
Psoas magnus.

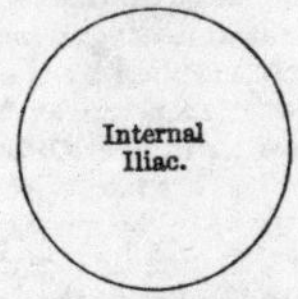

Internal Iliac.

Behind.
Internal iliac vein.
Lumbo-sacral nerve.
Pyriformis muscle.

In the fœtus, the internal iliac artery (hypogastric), is twice as large as the external iliac, and appears to be the continuation of the common iliac. Passing forwards to the bladder, it ascends along the side of that viscus to its summit, to which it gives branches; it then passes upwards along the back part of the anterior wall of the abdomen to the umbilicus, converging towards its fellow of the opposite side. Having passed through the umbilical opening, the two arteries twine round the umbilical vein, forming with it the umbilical cord; and, ultimately, ramify in the placenta. The portion of the vessel within the abdomen, is called the hypogastric artery; and that external to that cavity, the umbilical artery.

At birth, when the placental circulation ceases, that portion of the hypogastric artery which extends from the umbilicus to the summit of the bladder, contracts, and ultimately dwindles to a solid fibrous cord; the portion of the same vessel extending from the summit of the bladder to within an inch and a half of its origin is not totally impervious, though it becomes considerably reduced in size; and serves to convey blood to the bladder, under the name of the superior vesical artery.

Peculiarities, as regards length. In two-thirds of a large number of cases, the length of the internal iliac varied between an inch and an inch and a-half; in the remaining third, it was more frequently longer than shorter, the maximum length being three inches, the minimum half an inch.

The lengths of the common and internal iliac arteries bear an inverse proportion to each other, the internal iliac artery being long when the common iliac is short, and *vice versâ.*

As regards its place of division. The place of division of the internal iliac varies between the upper margin of the sacrum, and the upper border of the sacro-sciatic foramen.

The arteries of the two sides in a series of cases often differed in length, but neither seemed constantly to exceed the other.

Surgical Anatomy. The application of a ligature to the internal iliac artery may be required in cases of aneurism or hæmorrhage affecting one of its branches. The vessel may be secured by making an incision through the abdominal parietes in the iliac region, in a direction and to an extent similar to that for securing the common iliac; the transversalis fascia having been cautiously divided, and the peritoneum pushed inwards from the iliac fossa towards the pelvis, the finger may feel the pulsation of the external iliac at the bottom of the wound; and, by tracing this vessel upwards, the internal iliac is arrived at, opposite the sacro-iliac articulation. It should be remembered that the vein lies behind, and on the right side, a little external to the artery, and in close contact with it; the ureter and peritoneum, which lie in front, must also be avoided. The degree of facility in applying a ligature to this vessel, will mainly depend upon its length. It has been seen that, in the great majority of the cases examined, the artery was short, varying from an inch to an inch and a-half; in these cases, the artery is deeply seated in the pelvis; when, on the contrary, the vessel is longer, it is found partly above that cavity. If the artery is very short, as occasionally happens, it would be preferable to apply a ligature to the common iliac, or upon the external and internal iliacs at their origin.

Collateral Circulation. In Mr. Owen's dissection of a case in which the internal iliac artery had been tied by Stevens ten years before death, for aneurism of the sciatic artery, the internal iliac was found impervious for about an inch above the point where the ligature had been applied; but the obliteration did not extend to the origin of the external iliac, as the ilio-lumbar artery arose just above this point. Below the point of obliteration, the artery resumed its natural diameter, and continued so for half an inch; the obturator, lateral sacral, and gluteal, arising in succession from the latter portion. The obturator artery was entirely obliterated. The lateral sacral artery was as large as a crow's quill, and had a very free anastomosis with the artery of the opposite side, and with the middle sacral artery. The sciatic artery was entirely obliterated as far as its point of connection with the aneurismal tumour; but, on the distal side of the sac, it was continued down along the back of the thigh nearly as large in size as the femoral, being pervious about an inch below the sac by receiving an anastomosing vessel from the profunda.* In addition to the above, the circulation in the parts supplied by the internal iliac would be carried on by the anastomoses of the uterine and ovarian arteries; of the opposite vesical arteries; of the hæmorrhoidal branches of the internal iliac, with those from the inferior mesenteric; of the obturator artery, by means of its pubic branch, with the vessel of the opposite side, and with the epigastric and internal circumflex; by the anastomoses of the circumflex and perforating branches of the femoral, with the sciatic; of the gluteal, with the posterior branches of the sacral arteries; of the ilio-lumbar, with the last lumbar; of the lateral sacral, with the middle sacral; and by the anastomoses of the circumflex iliac, with the ilio-lumbar and gluteal.

Branches of the Internal Iliac.

From the Anterior Trunk.	*From the Posterior Trunk.*
Superior vesical.	Gluteal.
Middle vesical.	Ilio-lumbar.
Inferior vesical.	Lateral sacral.
Middle hæmorrhoidal.	
Obturator.	
Internal pudic.	
Sciatic.	
In female. { Uterine.	
Vaginal.	

The *superior vesical* is that part of the fœtal hypogastric artery which remains pervious after birth. It extends to the side of the bladder, distributing numerous branches to the body and fundus of the organ. From one of these a slender vessel is derived, which accompanies the vas deferens in its course to the testis, where it anastomoses with the spermatic artery. This is the *artery of the vas deferens.* Other branches supply the ureter.

The *middle vesical*, usually a branch of the superior, is distributed to the base of the bladder, and under surface of the vesiculæ seminales.

The *inferior vesical* arises from the anterior division of the internal iliac, in common with the middle hæmorrhoidal, and is distributed to the base of the bladder, the prostate gland, and vesiculæ seminales. The branches distributed to the prostate communicate with the corresponding vessel of the opposite side.

The *middle hæmorrhoidal artery* usually arises together with the preceding vessel. It supplies the rectum, anastomosing with the other hæmorrhoidal arteries.

The *uterine artery* passes downwards from the anterior trunk of the internal iliac to the neck of the uterus. Ascending, in a tortuous course, on the side of this viscus, between the layers of the broad ligament, it distributes branches to its substance, anastomosing, near its termination, with a branch from the ovarian artery. Branches from this vessel are also distributed to the bladder and ureter.

The *vaginal artery* is analogous to the inferior vesical in the male; it descends upon the vagina, supplying its mucous membrane, and sending branches to the neck of the bladder and contiguous part of the rectum.

The Obturator Artery usually arises from the anterior trunk of the internal iliac, frequently from the posterior. It passes forwards below the brim of the pelvis, to the canal in the upper border of the obturator foramen, and escaping from the pelvic cavity through this aperture, divides into an internal and an external branch. In the pelvic cavity this vessel lies upon the pelvic fascia, beneath the peritoneum, and a little below the obturator nerve; and, whilst passing through the obturator foramen, is contained in an oblique canal, formed by the horizontal branch of the pubes, above; and the arched border of the obturator membrane, below.

* *Medico-Chirurgical Trans.*, vol. xvi.

Branches. Within the pelvis, the obturator artery gives off an *iliac branch* to the iliac fossa, which supplies the bone and the Iliacus muscle, and anastomoses with the ilio-lumbar artery; a *vesical branch*, which runs backwards to supply the bladder; and a *pubic branch*, which is given off from the vessel just before it leaves the pelvic cavity. This branch ascends upon the back of the pubes, communicating with offsets from the epigastric artery, and with the corresponding vessel of the opposite side. This branch is placed on the inner side of the femoral ring. *External to the pelvis*, the obturator artery divides into an external and an internal branch, which are deeply situated beneath the Obturator externus muscle; skirting the circumference of the obturator foramen, they anastomose at the lower part of this aperture with each other, and with branches of the internal circumflex artery.

The *internal branch* curves inwards along the inner margin of the obturator foramen, distributing branches to the Obturator muscles, Pectineus, Adductors, and Gracilis, and anastomoses with the external branch, and with the internal circumflex artery.

The *external branch* curves round the outer margin of the foramen, to the space between the Gemellus inferior and Quadratus femoris, where it anastomoses with the sciatic artery. It supplies the Obturator muscles, anastomoses, as it passes backwards, with the internal circumflex, and sends a branch to the hip-joint through the cotyloid notch, which ramifies on the round ligament as far as the head of the femur.

Peculiarities. In two out of every three cases the obturator arises from the internal iliac; in one case in 3½ from the epigastric; and in about one in seventy-two cases by two roots from both vessels. It arises in about the same proportion from the external iliac artery. The origin of the obturator from the epigastric is not commonly found on both sides of the same body.

When the obturator artery arises at the front of the pelvis from the epigastric, it descends almost vertically to the upper part of the obturator foramen. The artery in this course usually lies in contact with the external iliac vein, and on the outer side of the femoral ring (228, fig. 1); in such cases it would not be endangered in the operation for femoral hernia. Occasionally, however, it curves inwards along the free margin of Gimbernat's ligament (228, fig. 2), and under such circumstances would almost completely encircle the neck of a hernial sac (supposing a hernia to exist in such a case), and would be in great danger of being wounded if an operation was performed.

228.—Variations in Origin and Course of Obturator Artery.

fig. 1.

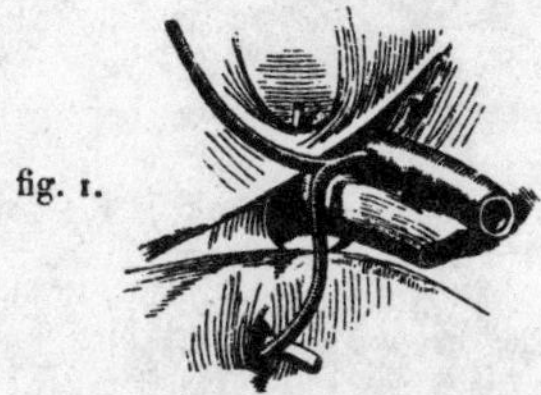

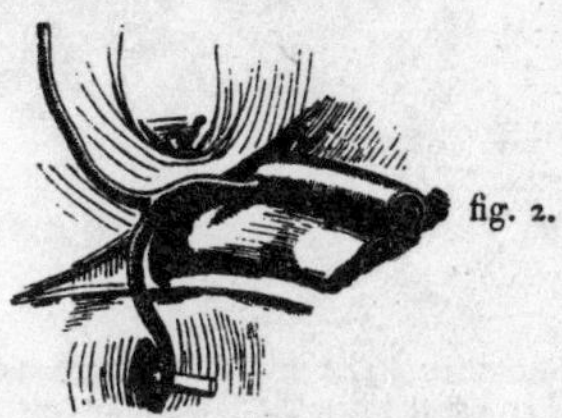

fig. 2.

The *Internal Pudic* is the smaller of the two terminal branches of the anterior trunk of the internal iliac, and supplies the external organs of generation. It passes downwards and outwards to the lower border of the great sacro-sciatic foramen, and emerges from the pelvis between the Pyriformis and Coccygeus muscles; it then crosses the spine of the ischium, and re-enters the pelvis through the lesser sacro-sciatic foramen. The artery now crosses the Obturator internus muscle, to the ramus of the ischium, being covered by the obturator fascia, and situated about an inch and a half from the margin of the tuberosity; it then ascends forwards and upwards along the ramus of the ischium, pierces the posterior layer of the deep perinæal fascia, and runs forwards along the inner margin of the ramus of the pubes; finally, it perforates the anterior layer of the deep perinæal fascia, and divides into its two terminal branches, the dorsal artery of the penis, and the artery of the corpus cavernosum.

Relations. In the first part of its course, within the pelvis, it lies in front of the Pyriformis muscle and sacral plexus of nerves, and on the outer side of the rectum (on the left side). As it crosses the spine of the ischium, it is covered by the Gluteus maximus, and great sacro-sciatic ligament. In the pelvis, it lies on the outer side of the ischio-rectal fossa, upon the surface of the Obturator internus muscle, contained in a fibrous canal formed by the obturator fascia and the falciform process of the great sacro-sciatic ligament. It is accompanied by the pudic veins, and the internal pudic nerve.

229. The Internal Pudic Artery and its Branches.

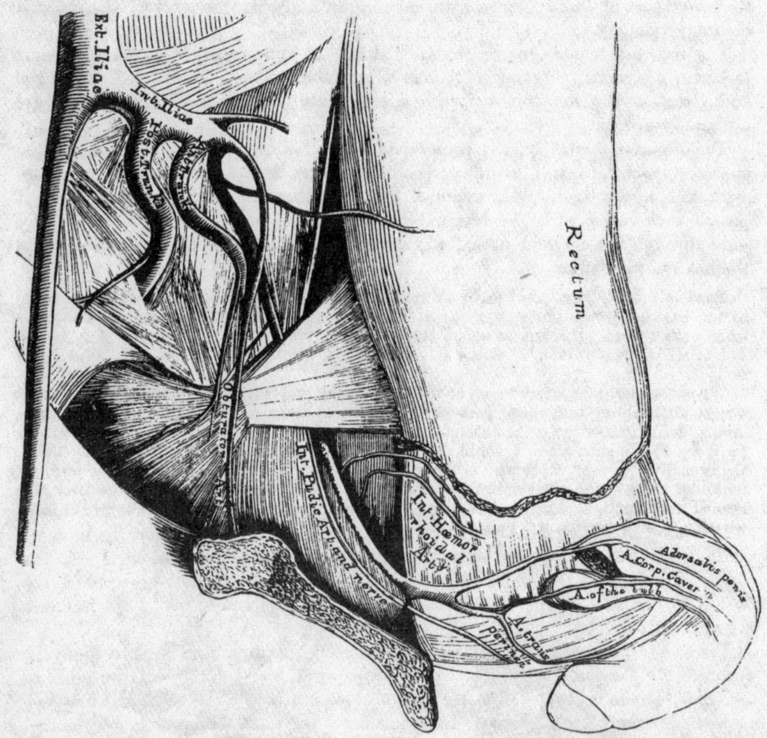

Peculiarities. The internal pudic is sometimes smaller than usual, or fails to give off one or two of its usual branches; in such cases, the deficiency is supplied by branches derived from an additional vessel, the *accessory pudic,* which generally arises from the pudic artery before its exit from the great sacro-sciatic foramen, and passes forwards near the base of the bladder, on the upper part of the prostate gland, to the perinæum, where it gives off the branches, usually derived from the pudic artery. The deficiency most frequently met with, is that in which the internal pudic ends as the artery of the bulb; the artery of the corpus cavernosum and arteria dorsalis penis being derived from the accessory pudic. Or the pudic may terminate as the superficial perinæal, the artery of the bulb being derived, with the other two branches, from the accessory vessel.

The relation of the accessory pudic to the prostate gland and urethra, is of the greatest interest in a surgical point of view, as this vessel is in danger of being wounded in the lateral operation of lithotomy.

Branches. Within the pelvis, the internal pudic gives off several small branches, which supply the muscles, sacral nerves, and pelvic viscera. *In the perinæum* the following branches are given off.

Inferior or external hæmorrhoidal.	Artery of the bulb.
Superficial perinæal.	Artery of the corpus cavernosum.
Transverse perinæal.	Dorsal artery of the penis.

The *external hæmorrhoidal* are two or three small arteries, which arise from the internal pudic as it passes above the tuberosity of the ischium. Crossing the ischio-rectal fossa, they are distributed to the muscles and integument of the anal region.

The *superficial perinæal artery* supplies the scrotum, and muscles and integument of the perinæum. It arises from the internal pudic, in front of the preceding branches, and piercing the lower border of the deep perinæal fascia, runs across the Transversus perinæi, and through the triangular space between the Accelerator urinæ and Erector penis, both of which it supplies, and is finally distributed to the skin of the scrotum and dartos. In its passage through the perinæum it lies beneath the superficial perinæal fascia.

The *transverse perinæal* is a small branch which arises either from the internal pudic, or from the superficial perinæal artery as it crosses the Transversus perinæi muscle. Piercing the lower border of the deep perinæal fascia, it runs transversely inwards along the cutaneous surface of the Transversus perinæi muscle, which it supplies, as well as the structures between the anus and bulb of the urethra.

The *artery of the bulb* is a large but very short vessel which arises from the internal pudic between the two layers of the deep perinæal fascia, and passing nearly transversely inwards, pierces the bulb of the urethra, in which it ramifies. It gives off a small branch which descends to supply Cowper's gland. This artery is of considerable importance in a surgical point of view, as it is in danger of being wounded in the lateral operation of lithotomy, an accident usually attended in the adult with alarming hæmorrhage. The vessel is sometimes very small, occasionally wanting, or even double. It sometimes arises from the internal pudic earlier than usual, and crosses the perinæum to reach the back part of the bulb. In such a case the vessel could hardly fail to be wounded in the performance of the lateral operation for lithotomy. If, on the contrary, it should arise from an accessory pudic, it lies more forward than usual, and is out of danger in the operation.

The *artery of the corpus cavernosum*, one of the terminal branches of the internal pudic, arises from that vessel while it is situated between the crus penis and the ramus of the pubes; piercing the crus penis obliquely, it runs forwards in the corpus cavernosum by the side of the septum pectiniforme, to which its branches are distributed.

The *dorsal artery of the penis* ascends between the crus and pubic symphysis, and piercing the suspensory ligament, runs forward on the dorsum of the penis to the glans, where it divides into two branches, which supply the glans and prepuce. On the dorsum of the penis, it lies immediately beneath the integument, parallel with the dorsal vein, and the corresponding artery of the opposite side. It supplies the integument and fibrous sheath of the corpus cavernosum.

The *internal pudic artery in the female* is smaller than in the male. Its origin and course are similar, and there is considerable analogy in the distribution of its branches. The superficial artery supplies the labia pudenda ; the artery of the bulb supplies the erectile tissue of the bulb of the vagina, whilst the two terminal branches supply the clitoris; the artery of the corpus cavernosum, the cavernous body of the clitoris ; and the arteria dorsalis clitoridis, the dorsum of that organ.

The *Sciatic Artery* (fig. 230), the larger of the two terminal branches of the anterior trunk of the internal iliac, is distributed to the muscles on the back of the pelvis. It passes down to the lower part of the great sacro-sciatic foramen, behind the internal pudic, resting on the sacral plexus of nerves and Pyriformis muscle, and escapes from the pelvis between the Pyriformis and Coccygeus. It then descends in the interval between the trochanter major and tuberosity of the ischium, accompanied by the sciatic nerves, and covered by the Gluteus maximus and divides into branches, which supply the deep muscles at the back of the hip.

Within the pelvis it distributes branches to the Pyriformis Coccygeus, and Levator ani muscles; some hæmorrhoidal branches, which supply the rectum, and occasionally take the place of the middle hæmorrhoidal artery; and vesical branches to the base and neck of the bladder, vesiculæ seminales, and prostate gland. *External to the pelvis*, it gives off the coccygeal, inferior gluteal, comes nervi ischiadici, muscular, and articular branches.

The *coccygeal branch* runs inwards, pierces the great sacro-sciatic ligament, and supplies the Gluteus maximus, the integument, and other structures on the back of the coccyx.

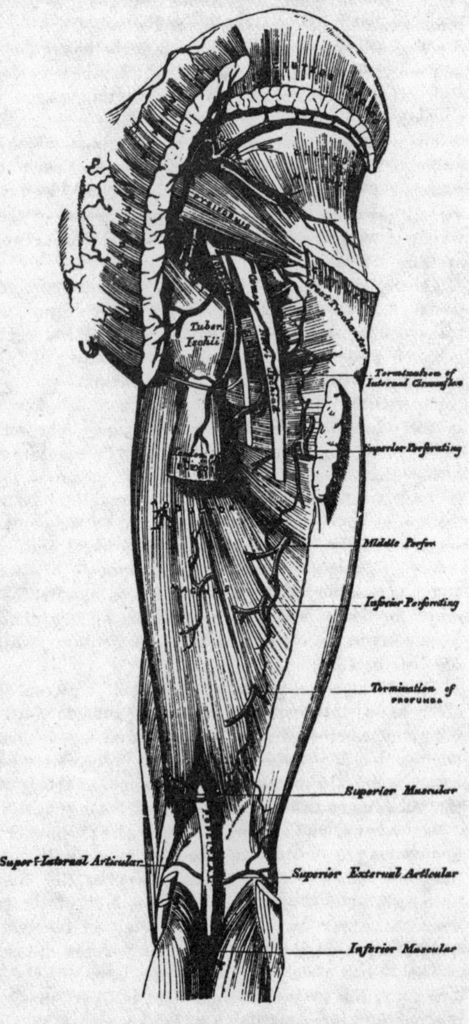

230.—The Arteries of the Gluteal and Posterior Femoral Regions.

The *inferior gluteal branches*, three or four in number, supply the Gluteus maximus muscle.

The *comes nervi ischiadici* is a long slender vessel, which accompanies the great sciatic nerve for a short distance; it then penetrates it, and runs in its substance to the lower part of the thigh.

The *muscular branches* supply the muscles on the back part of the hip, anastomosing with the gluteal, internal and external circumflex, and superior perforating arteries.

Some *articular branches* are distributed to the capsule of the hip-joint.

The *Gluteal Artery* is the largest branch of the internal iliac, and appears to be the continuation of the posterior division of that vessel. It is a short, thick trunk, which passes out of the pelvis above the upper border of the Pyriformis muscle, and immediately divides into a superficial and deep branch. Within the pelvis, it gives off a few muscular branches to the Iliacus, Pyriformis, and Obturator internus, and just previous to quitting that cavity a nutrient artery, which enters the ilium.

The *superficial branch* passes beneath the Gluteus maximus, and divides into numerous branches, some of which supply that muscle, whilst others perforate its tendinous origin, and supply the integument covering the posterior surface of the sacrum, anastomosing with the posterior branches of the sacral arteries.

The *deep branch* runs between the Gluteus medius and minimus, and subdivides into two. Of these, the *superior division*, continuing the original course of the vessel, passes along the upper border of the Gluteus minimus to the anterior superior spine of the ilium, anastomosing with the circumflex iliac and ascending branches of the external circumflex artery. The *inferior division* crosses the Gluteus minimus obliquely to the trochanter major, distributing branches to the Glutei muscles, and inosculates with the external circumflex artery. Some branches pierce the Gluteus minimus to supply the hip-joint.

The *Ilio-Lumbar Artery* ascends beneath the Psoas muscle and external iliac vessels, to the upper part of the iliac fossa, where it divides into a lumbar and an iliac branch.

The *lumbar branch* supplies the Psoas and Quadratus lumborum muscles, anastomosing with the last lumbar artery, and sends a small spinal branch through the intervertebral foramen, between the last lumbar vertebra and the sacrum, into the spinal canal, to supply the spinal cord and its membranes.

The *iliac branch* descends to supply the Iliacus internus, some offsets running between the muscle and the bone, one of which enters an oblique canal to supply the diploë, whilst others run along the crest of the ilium, distributing branches to the Gluteal and Abdominal muscles, and anastomosing in their course with the gluteal, circumflex iliac, external circumflex, and epigastric arteries.

The *Lateral Sacral Arteries* (fig. 227) are usually two in number on each side, superior and inferior.

The *superior*, which is of large size, passes inwards, and, after anastomosing with branches from the middle sacral, enters the first or second sacral foramen, is distributed to the contents of the sacral canal, and escaping by the corresponding posterior sacral foramen, supplies the skin and muscles on the dorsum of the sacrum.

The *inferior branch* passes obliquely across the front of the Pyriformis muscle and sacral nerves to the inner side of the anterior sacral foramina, descends on the front of the sacrum, and anastomoses over the coccyx with the sacra media and opposite lateral sacral arteries. In its course, it gives off branches, which enter the anterior sacral foramina: these, after supplying the bones and membranes of the interior of the spinal canal, escape by the posterior sacral foramina, and are distributed to the muscles and skin on the dorsal surface of the sacrum.

External Iliac Artery. (Fig. 227.)

The external iliac artery is the chief vessel which supplies the lower limb. It is larger in the adult than the internal iliac, and passes obliquely downwards and outwards along the inner border of the Psoas muscle, from the bifurcation of the common iliac to the femoral arch, where it enters the thigh, and becomes the femoral artery. The course of this vessel would be indicated by a line drawn from the left side of the umbilicus to a point midway between the anterior superior spinous process of the ilium and the symphysis pubis.

Relations. In front, with the peritoneum, subperitoneal areolar tissue, the intestines, and a thin layer of fascia, derived from the iliac fascia, which surrounds the artery and vein. At its origin it is occasionally crossed by the ureter. The spermatic vessels descend for some distance upon it near its termination, and it is crossed in this situation by a branch of the genito-crural nerve and the circumflex iliac vein; the vas deferens curves down along its inner side. *Behind*, it is in relation with the external iliac vein, which, at the femoral arch, lies at its inner side; on the left side the vein is altogether internal to the artery. *Externally*, it rests against the Psoas muscle, from which it is separated by the iliac fascia. The artery rests upon this muscle near Poupart's ligament. Numerous lymphatic vessels and glands are found lying on the front and inner side of the vessel.

Plan of the Relations of the External Iliac Artery.

In front.

Peritoneum, intestines, and iliac fascia.

Near Poupart's Ligament. { Spermatic vessels. Genito-crural nerve. Circumflex iliac vein. Lymphatic vessels and glands.

Outer side.
Psoas magnus.
Iliac fascia.

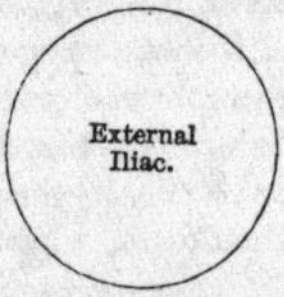

External Iliac.

Inner side.
External iliac vein and vas deferens at femoral arch.

Behind.
External iliac vein.
Psoas magnus.

Surgical Anatomy. The application of a ligature to the external iliac may be required in cases of aneurism of the femoral artery, or in cases of secondary hæmorrhage, after the latter vessel has been tied for popliteal aneurism. This vessel may be secured in any part of its course, excepting near its upper end, which is to be avoided on account of the proximity of the great stream of blood in the internal iliac, and near its lower end, which should also be avoided, on account of the proximity of the epigastric and circumflex iliac vessels. One of the chief points in the performance of the operation is to secure the vessel without injury to the peritoneum. The patient having been placed in the recumbent position, an incision should be made, commencing below at a point about three quarters of an inch above Poupart's ligament, and a little external to its centre, and running upwards and outwards, parallel to Poupart's ligament, to a point above the anterior superior spine of the ilium. When the artery is deeply seated, more room will be required, and may be obtained by curving the incision from the point last named inwards towards the umbilicus for a short distance; or, if the lower part of the artery is to be reached, the surgeon may commence the incision nearer the inner end of Poupart's ligament, taking care to avoid the epigastric artery. Abernethy, who first tied this artery, made his incision in the course of the vessel. The precise line of incision selected is of less moment, provided an easy access to the deeper parts is secured. The abdominal muscles and transversalis fascia having been cautiously divided, the peritoneum should be separated from the iliac fossa and pushed towards the pelvis; and on introducing the finger to the bottom of the wound the artery may be felt pulsating along the inner border of the Psoas muscle. The external iliac vein is generally found on the inner side of the artery, and must be cautiously separated from it by the finger-nail, or handle of the knife, and the aneurism needle should be introduced on the inner side, between the artery and vein.

Collateral Circulation. The principal anastomoses in carrying on the collateral circulation, after the application of a ligature to the external iliac, are—the ilio-lumbar with the circumflex iliac; the gluteal with the external circumflex; the obturator with the internal circumflex; the sciatic with the superior perforating and circumflex branches of the profunda artery; the internal pudic with the external pudic, and with the internal circumflex. When the obturator arises from the epigastric, it is supplied with blood by branches, either from the internal iliac, the lateral sacral, or the internal pudic. The epigastric receives its supply from the internal mammary and inferior intercostal arteries, and from the internal iliac, by the anastomoses of its branches with the obturator.

In the dissection of a limb, eighteen years after the successful ligature of the external iliac artery, by Sir A. Cooper, which is to be found in Guy's Hospital Reports, vol. i. p. 50, the anastomosing branches are described in three sets. *An anterior set.* 1. A very large branch from the ilio-lumbar artery to the circumflex iliac; 2. Another branch from the ilio-lumbar, joined by one from the obturator, and breaking up into numerous tortuous branches to anastomose with the external circumflex; 3. Two other branches from the obturator, which passed over the brim of the pelvis, communicated with the epigastric, and then broke up into a plexus to anastomose with the internal circumflex. *An internal set.* Branches given off from the obturator, after quitting the pelvis, which ramified among the adductor muscles on the inner side of the hip joint, and joined most freely with branches of the internal circumflex. *A posterior set.* 1. Three large branches from the gluteal to the external circumflex; 2. Several branches from the sciatic around the great sciatic notch to the internal and external circumflex, and the perforating branches of the profunda.

Branches. Besides several small branches to the Psoas muscles and the neighbouring lymphatic glands, the external iliac gives off two branches of considerable size, the

Epigastric. Circumflex iliac.

The *epigastric artery* arises from the external iliac, a few lines above Poupart's ligament. It at first descends to reach this ligament, and then ascends obliquely inwards between the peritoneum and transversalis fascia, to the margin of the sheath of the Rectus muscle. Having perforated the sheath near its lower third, it runs vertically upwards behind the Rectus, to which it is distributed, dividing into numerous branches, which anastomose above the umbilicus with the terminal branches of the internal mammary and inferior intercostal arteries. It is accompanied by two veins, which usually unite into a single trunk before their termination in the external iliac vein. As this artery ascends from Poupart's ligament to the Rectus, it lies behind the inguinal canal, to the inner side of the internal abdominal ring, and immediately above the femoral ring. The vas deferens in the male, and the round ligament in the female, cross behind the artery in descending into the pelvis.

Branches. The branches of this vessel are the following: the *cremasteric*, which accompanies the spermatic cord, and supplies the Cremaster muscle, anastomosing with the spermatic artery; a *pubic branch*, which runs across Poupart's ligament, and then descends behind the pubes to the inner side of the femoral ring, and anastomoses with offsets from the obturator artery; *muscular branches*, some of which are distributed to the abdominal muscles and peritoneum, anastomosing with the lumbar and circumflex iliac arteries; others perforate the tendon of the External oblique and supply the integument, anastomosing with branches of the external epigastric.

Peculiarities. The origin of the epigastric may take place from any part of the external iliac between Poupart's ligament and two inches and a half above it; or it may arise below this ligament, from the femoral, or from the deep femoral.

Union with Branches. It frequently arises from the external iliac, by a common trunk with the obturator. Sometimes the epigastric arises from the obturator, the latter vessel being furnished by the internal iliac, or the epigastric may be formed of two branches, one derived from the external iliac, the others from the internal iliac.

The *circumflex iliac artery* arises from the outer side of the external iliac, nearly opposite the epigastric artery. It ascends obliquely outwards behind Poupart's ligament, and runs along the inner surface of the crest of the ilium to about its middle, where it pierces the Transversalis, and runs backwards between that muscle and the Internal oblique, to anastomose with the ilio-lumbar and gluteal arteries. Opposite the anterior superior spine of the ilium, it gives off a large branch, which ascends between the Internal oblique and Transversalis muscles, supplying them and anastomosing with the lumbar and epigastric arteries. The circumflex iliac artery is accompanied by two veins. These unite into a single trunk which crosses the external iliac artery just above Poupart's ligament, and enters the external iliac vein.

Femoral Artery. (Fig. 231.)

The femoral artery is the continuation of the external iliac. It commences immediately beneath Poupart's ligament, midway between the anterior superior spine of the ilium and the symphysis pubis, and passing down the fore part and inner side of the thigh, terminates at the opening in the Adductor magnus, at the junction of the middle with the lower third of the thigh, where it becomes the popliteal artery. A line drawn from a point midway between the anterior superior spine of the ilium and the symphysis pubis to the inner side of the inner condyle of the femur, will be nearly parallel with the course of the artery. This vessel, at the upper part of the thigh, lies a little internal to the head of the femur; in the lower part of its course, on the inner side of the shaft of the bone, and between these two parts, the vessel is separated from the bone by a considerable interval.

In the upper third of the thigh the femoral artery is very superficial, being covered by the integument, inguinal glands, and the superficial and deep fasciæ, and is contained in a triangular space, called 'Scarpa's triangle.'

Scarpa's Triangle. Scarpa's triangle corresponds to the depression seen immediately below the fold of the groin. It is a triangular space, the apex of which is directed downwards, and the sides of which are formed externally by the

Sartorius, internally by the Adductor longus, and above by Poupart's ligament. The floor of this space is formed from without inwards by the Iliacus, Psoas, Pectineus, Adductor longus, and a small part of the Adductor brevis muscles; and it is divided into two nearly equal parts by the femoral vessels, which extend from the middle of its base to its apex; the artery giving off in this situation its cutaneous and profunda branches, the vein receiving the deep femoral and internal saphenous veins. In this space the femoral artery rests on the inner margin of the Psoas muscle, which separates it from the capsular ligament of the hip-joint. The artery in this situation has in front of it filaments from the crural branch of the genito-crural nerve, and branches from the anterior crural, one of which is usually of considerable size; behind the artery is the branch to the Pectineus from the anterior crural. The femoral vein lies at its inner side, between the margins of the Pectineus and Psoas muscles. The anterior crural nerve lies about half an inch to the outer side of the femoral artery, deeply imbedded between the Iliacus and Psoas muscles. The femoral artery and vein are enclosed in a strong fibrous sheath, formed by fibrous and cellular tissue, and by a process of fascia sent inwards, from the fascia lata; the vessels are separated, however, from one another by thin fibrous partitions.

231.—Surgical Anatomy of the Femoral Artery.

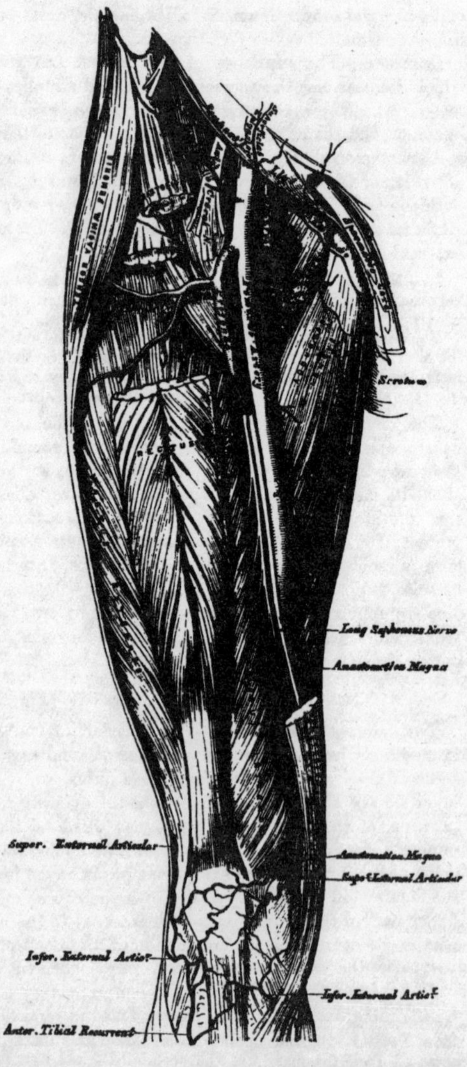

In the middle third of the thigh, the femoral artery is more deeply seated, being covered by the integument, the superficial and deep fascia, and the Sartorius, and is contained in an aponeurotic canal, formed by a dense fibrous band, which extends transversely from the Vastus internus to the tendons of the Adductor longus and magnus muscles. In this part of its course it lies in a depression, bounded externally by the Vastus internus, internally by the Adductor longus and Adductor magnus. The femoral vein lies on the outer side of the artery, in close apposition with it, and, still more externally, is the internal (long) saphenous nerve.

Relations. From above downwards, the femoral artery rests upon the Psoas muscle, which separates it from the margin of the pelvis and capsular ligament of the hip; it is next separated from the Pectineus by the profunda vessels and femoral vein; it then lies upon the Adductor longus; and lastly, upon the tendon of the Adductor magnus, the femoral vein being interposed. To its *inner side*, it is in relation above, with the femoral vein, and, lower down, with the Adductor longus, and Sartorius. To its *outer side*, the Vastus internus separates it from the femur, in the lower part of its course.

The *femoral vein*, at Poupart's ligament, lies close to the inner side of the artery, separated from it by a thin fibrous partition, but, as it descends, gets behind it, and then to its outer side.

The *internal saphenous nerve* is situated on the outer side of the artery, in the middle third of the thigh, beneath the aponeurotic covering, but not usually within the sheath of the vessels. Small cutaneous nerves cross the front of the sheath.

Plan of the Relations of the Femoral Artery.

In front.

Fascia lata.
Branch of anterior crural nerve.
Sartorius.
Long saphenous nerve.
Aponeurotic covering of Hunter's canal.

Inner side.

Femoral vein (at upper part.)
Adductor longus.
Sartorius.

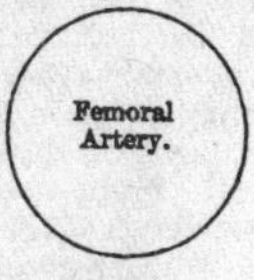

Femoral Artery.

Outer side.

Vastus internus.
Femoral vein (at lower part.)

Behind.

Psoas muscle.
Profunda vein.
Pectineus muscle.
Adductor longus.
Femoral vein.
Adductor magnus.

Peculiarities. Double femoral re-united. Four cases are at present recorded, in which the femoral artery divided into two trunks below the origin of the profunda, and became re-united near the opening in the Adductor magnus, so as to form a single popliteal artery. One of them occurred in a patient operated upon for popliteal aneurism.

Change of Position. A similar number of cases have been recorded, in which the femoral artery was situated at the back of the thigh, the vessel being continuous above with the internal iliac, escaping from the pelvis through the great sacro-sciatic foramen, and accompanying the great sciatic nerve to the popliteal space, where its division occurred in the usual manner.

Position of the Vein. The femoral vein is occasionally placed along the inner side of the artery, throughout the entire extent of Scarpa's triangle; or it may be slit, so that a large vein is placed on each side of the artery for a greater or less extent.

Origin of the Profunda. This vessel occasionally arises from the inner side, and more rarely, from the back of the common trunk; but the more important peculiarity, in a surgical point of view, is that which relates to the height at which the vessel arises from the femoral. In three-fourths of a large number of cases, it arose between one or two inches below Poupart's ligament; in a few cases, the distance was less than an inch; more rarely, opposite the ligament; and in one case, above Poupart's ligament, from the external iliac. Occasionally, the distance between the origin of the vessel and Poupart's ligament exceeds two inches, and in one case it was found to be as much as four inches.

Surgical Anatomy. Compression of the femoral artery, which is constantly requisite in amputations and other operations on the lower limb, is most effectually made immediately below Poupart's ligament. In this situation the artery is very superficial, and is merely separated from the margin of the acetabulum and front of the head of the femur, by the Psoas muscle; so that the surgeon, by means of his thumb, or a compressor, may effectually control the circulation through it. This vessel may also be compressed in the middle third of the thigh, by placing a compress over the artery, beneath the tourniquet, and directing the pressure from within outwards, so as to compress the vessel on the inner side of the shaft of the femur.

The *application of a ligature* to the femoral artery may be required in cases of wound or aneurism of the arteries of the leg, of the popliteal or femoral; and the vessel may be exposed and tied in any part of its course. The great depth of this vessel at its lower part, its close connection with important structures, and the density of its sheath, render the operation in this situation one of much greater difficulty than the application of a ligature at its upper part, where it is more superficial.

Ligature of the femoral artery, within two inches of its origin, is usually considered unsafe, on account of the connection of large branches with it, the epigastric and circumflex iliac arising just above its origin; the profunda, from one to two inches below, occasionally, also, one of the circumflex arteries arises from the vessel in the interspace between these. The profunda sometimes arises higher than the point above mentioned, and rarely between two or three inches (in one case four), below Poupart's ligament. It would appear, then, that the most favourable situation for the application of a ligature to the femoral, is between four and five inches from its point of origin. In order to expose the artery in this situation, an incision, between two and three inches long, should be made in the course of the vessel, the patient lying in the recumbent position, with the limb slightly flexed and abducted. A large vein is frequently met with, passing in the course of the artery to join the saphena; this must be avoided, and the fascia lata having been cautiously divided, and the Sartorius exposed, that muscle must be drawn outwards, in order to fully expose the sheath of the vessels. The finger being introduced into the wound, and the pulsation of the artery felt, the sheath should be divided over the artery to a sufficient extent to allow of the introduction of the ligature, but no further; otherwise the nutrition of the coats of the vessel may be interfered with, or muscular branches which arise from the vessel at irregular intervals may be divided. In this part of the operation, a small nerve which crosses the sheath should be avoided. The aneurism needle must be carefully introduced and kept close to the artery, to avoid the femoral vein, which lies behind the vessel in this part of its course.

To expose the artery in the middle of the thigh, an incision should be made through the integument, between three and four inches in length, over the inner margin of the Sartorius, taking care to avoid the internal saphenous vein, the situation of which may be previously known by compressing it higher up in the thigh. The fascia lata having been divided, and the Sartorius muscle exposed, it should be drawn outwards, when the strong fascia which is stretched across from the Adductors to the Vastus internus, will be exposed, and must be freely divided; the sheath of the vessels is now seen, and must be opened, and the artery secured by passing the aneurism needle between the vein and artery, in the direction from within outwards. The femoral vein in this situation lies on the outer side of the artery, the long saphenous nerve on its anterior and outer side.

It has been seen that the femoral artery occasionally divides into two trunks, below the origin of the profunda. If, in the operation for tying the femoral, two vessels are met with, the surgeon should alternately compress each, in order to ascertain which vessel is connected with the aneurismal tumour, or with the bleeding from the wound, and that one only should be tied which controls the pulsation or hæmorrhage. If, however, it is necessary to compress both vessels before the circulation in the tumour is controlled, both should be tied, as it would be probable that they became re-united, as in the four instances referred to above.

Collateral Circulation. When the common femoral is tied, the main channels for carrying on the circulation are the anastomoses of the gluteal and circumflex iliac arteries above with the external circumflex below; of the obturator and sciatic above with the internal circumflex below; of the iliolumbar with the external circumflex, and of the comes nervi ischiadici with the arteries in the ham.

The principal agents in carrying on the collateral circulation after ligature of the superficial femoral artery are, according to Sir A. Cooper, as follows.*

'The arteria profunda formed the new channel for the blood.' 'The first artery sent off passed down close to the back of the thigh bone, and entered the two superior articular branches of the popliteal artery.'

'The second new large vessel arising from the profunda at the same part with the former, passed down by the inner side of the Biceps muscle, to an artery of the popliteal which was distributed to the Gastrocnemius muscle; whilst a third artery dividing into several branches passed down with the sciatic nerve behind the knee-joint, and some of its branches united themselves with the inferior articular arteries of the popliteal, with some recurrent branches of those arteries, with arteries passing to the Gastrocnemii, and, lastly, with the origin of the anterior and posterior tibial arteries.'

'It appears then that it is those branches of the profunda which accompany the sciatic nerve, that are the principal supporters of the new circulation.'

In Porta's work† (Tab. xii. xiii.) is a good representation of the collateral circulation after the ligature of the femoral artery. The patient had survived the operation three years. The lower part of the artery is, at least, as large as the upper; about two inches of the vessel appear to have been obliterated. The external and internal circumflex arteries are seen anastomosing by a great number of branches with the lower branches of the femoral (mus-

* *Med. Chir. Trans.*, vol. ii. 1811.

† *Alterazioni patologiche delle Arterie.*

cular and anostomotica magna), and with the articular branches of the popliteal. The branches from the external circumflex are extremely large and numerous, one very distinct anastomosis can be traced between this artery on the outside, and the anastomotica magna on the inside, through the intervention of the superior external articular artery with which they both anastomose, and blood reaches even the anterior tibial recurrent from the external circumflex by means of an anastomosis with the same external articular artery. The perforating branches of the profunda are also seen bringing blood round the obliterated portion of the artery into long branches (muscular) which have been given off just below that portion. The termination of the profunda itself anastomoses most freely with the superior external articular. A long branch of anastomosis is also traced down from the internal iliac by means of the comes nervi ischiadici of the sciatic which anastomoses on the popliteal nerves with branches from the popliteal and posterior tibial arteries. In this case the anastomosis had been too free, since the pulsation and growth of the aneurism recurred, and the patient died after ligature of the external iliac.

There is an interesting preparation in the Museum of the Royal College of Surgeons, of a limb on which John Hunter had tied the femoral artery fifty years before the patient's death. The whole of the superficial femoral and popliteal artery seems to have been obliterated. The anastomosis by means of the comes nervi ischiadici, which is shown in Porta's plate, is distinctly seen; the external circumflex, and the termination of the profunda artery, seem to have been the chief channels of anastomosis; but the injection has not been a very successful one.

Branches. The branches of the femoral artery are the

Superficial epigastric.
Superficial circumflex iliac.
Superficial external pudic.
Deep external pudic.
Profunda. { External circumflex. Internal circumflex. Three perforating.
Muscular.
Anastomotica magna.

The *superficial epigastric* arises from the femoral, about half an inch below Poupart's ligament, and, passing through the saphenous opening in the fascia lata, ascends on to the abdomen, in the superficial fascia covering the external oblique muscle, nearly as high as the umbilicus. It distributes branches to the inguinal glands, the superficial fascia and the integument, anastomosing with branches of the deep epigastric and internal mammary arteries.

The *superficial circumflex iliac*, the smallest of the cutaneous branches, arises close to the preceding, and, piercing the fascia lata, runs outwards, parallel with Poupart's ligament, as far as the crest of the ilium, dividing into branches which supply the integument of the groin, the superficial fascia, and inguinal glands, anastomosing with the circumflex iliac, and with the gluteal and external circumflex arteries.

The *superficial external pudic* (superior), arises from the inner side of the femoral artery, close to the preceding vessels, and, after piercing the fascia lata at the saphenous opening, passes inwards, across the spermatic cord, to be distributed to the integument on the lower part of the abdomen, and of the penis and scrotum in the male, and to the labia in the female, anastomosing with branches of the internal pudic.

The *deep external pudic* (inferior), more deeply seated than the preceding, passes inwards on the Pectineus muscle, covered by the fascia lata, which it pierces opposite the ramus of the pubes, its branches being distributed, in the male, to the integument of the scrotum and perinæum, and in the female, to the labium, anastomosing with branches of the superficial perinæal artery.

The *Profunda Femoris* (deep femoral artery) nearly equals the size of the superficial femoral. It arises from the outer and back part of the femoral artery, from one to two inches below Poupart's ligament. It at first lies on the outer side of the superficial femoral, and then passes beneath it and the femoral vein to the inner side of the femur, and terminates at the lower third of the thigh in a small branch, which pierces the Adductor magnus, to be distributed to the flexor muscles on the back of the thigh, anastomosing with branches of the popliteal and inferior perforating arteries.

Relations. *Behind,* it lies first upon the Iliacus, and then on the Adductor brevis and Adductor magnus muscles. *In front,* it is separated from the femoral artery, above by the femoral and profunda veins, and below by the Adductor longus. On its *outer side,* the origin of the Vastus internus separates it from the femur.

PLAN OF THE RELATIONS OF THE PROFUNDA ARTERY.

In front.
Femoral and profunda veins.
Adductor longus.

Outer side.
Vastus internus.

Profunda.

Behind.
Iliacus.
Adductor brevis.
Adductor magnus.

The *External Circumflex Artery* supplies the muscles on the front of the thigh. It arises from the outer side of the profunda, passes horizontally outwards, between the divisions of the anterior crural nerve, and beneath the Sartorius and Rectus muscles, and divides into three sets of branches, ascending, transverse, and descending.

The *ascending branches* pass upwards, beneath the Tensor vaginæ femoris muscle, to the outer side of the hip, anastomosing with the terminal branches of the gluteal and circumflex iliac arteries.

The *descending branches,* three or four in number, pass downwards, beneath the Rectus, upon the Vasti muscles, to which they are distributed, one or two passing beneath the Vastus externus as far as the knee, anastomosing with the superior articular branches of the popliteal artery.

The *transverse branches,* the smallest and least numerous, pass outwards over the Crureus, pierce the Vastus externus, and wind round the femur to its back part, just below the great trochanter, anastomosing at the back of the thigh with the internal circumflex, sciatic, and superior perforating arteries.

The *Internal Circumflex Artery,* smaller than the external, arises from the inner and back part of the profunda, and winds round the inner side of the femur, between the Pectineus and Psoas muscles. On reaching the tendon of the Obturator externus, it gives off two branches, one of which ascends to be distributed to the Adductor muscles, the Gracilis, and Obturator externus, anastomosing with the obturator artery; the other descends, and passes beneath the Adductor brevis, to supply it and the great Adductor; while the continuation of the vessel passes backwards, between the Quadratus femoris and upper border of the Adductor magnus, anastomosing with the sciatic, external circumflex, and superior perforating arteries. Opposite the hip-joint, this branch gives off an articular vessel, which enters the joint beneath the transverse ligament; and, after supplying the adipose tissue, passes along the round ligament to the head of the bone.

The *Perforating Arteries* (fig. 230), usually three in number, are so called from their perforating the tendons of the Adductor brevis and magnus muscles to reach the back of the thigh. The first is given off above the Adductor brevis, the second in front of that muscle, and the third immediately below it.

The *first* or *superior perforating artery* passes backwards between the Pectineus and Adductor brevis (sometimes perforates the latter); it then pierces the Adductor magnus close to the linea aspera, and divides into branches which supply both Adductors, the Biceps, and Gluteus maximus muscle; anastomosing with the sciatic, internal circumflex, and middle perforating arteries.

The *second* or *middle perforating artery*, larger than the first, pierces the tendons of the Adductor brevis and Adductor magnus muscles, and divides into ascending and descending branches, which supply the flexor muscles of the thigh, anastomosing with the superior and inferior perforating. The nutrient artery of the femur is usually given off from this branch.

The *third* or *inferior perforating artery* is given off below the Adductor brevis; it pierces the Adductor magnus, and divides into branches which supply the flexor muscles of the thigh; anastomosing with the perforating arteries above, and with the terminal branches of the profunda below.

Muscular Branches are given off from the superficial femoral throughout its entire course. They vary from two to seven in number, and supply chiefly the Sartorius and Vastus internus.

The *Anastomotica Magna* arises from the femoral artery just before it passes through the tendinous opening in the Adductor magnus muscle, and divides into a superficial and deep branch.

The *superficial branch* accompanies the long saphenous nerve, beneath the Sartorius, and, piercing the fascia lata, is distributed to the integument.

The *deep branch* descends in the substance of the Vastus internus, lying in front of the tendon of the Adductor magnus, to the inner side of the knee, where it anastomoses with the superior internal articular artery and recurrent branch of the anterior tibial. A branch from this vessel crosses outwards above the articular surface of the femur, forming an anastomotic arch with the superior external articular artery, and supplies branches to the knee-joint.

Popliteal Artery.

The popliteal artery commences at the termination of the femoral, at the opening in the Adductor magnus, and, passing obliquely downwards and outwards behind the knee-joint to the lower border of the Popliteus muscle, divides into the anterior and posterior tibial arteries. Through the whole of this extent the artery lies in the popliteal space.

The Popliteal Space. (Fig. 232).

Dissection. A vertical incision about eight inches in length should be made along the back part of the knee-joint, connected above and below by a transverse incision from the inner to the outer side of the limb. The flaps of integument included between these incisions should be reflected in the direction shown in fig. 189, p. 295.

On removing the integument, the superficial fascia is exposed, and ramifying in it along the middle line are found some filaments of the small sciatic nerve, and towards the inner part some offsets from the internal cutaneous nerve.

The superficial fascia having been removed, the fascia lata is brought into view. In this region it is strong and dense, being strengthened by transverse fibres, and firmly attached to the tendons on the inner and outer sides of the space. It is perforated below by the external saphenous vein. This fascia having been reflected back in the same direction as the integument, the small sciatic nerve and external saphenous vein are seen immediately beneath it, in the middle line. If the loose adipose tissue is now removed, the boundaries and contents of the space may be examined.

Boundaries. The popliteal space, or the ham, occupies the lower third of the thigh and the upper fifth of the leg; extending from the aperture in the Adductor magnus, to the lower border of the Popliteus muscle. It is a lozenge-shaped space, being widest at the back part of the knee-joint, and deepest above the articular end of the femur. It is bounded, externally, above the joint, by the Biceps, and below the joint by the Plantaris and external head of the Gastrocnemius. Internally, above the joint, by the Semimembranosus, Semitendinosus, Gracilis, and Sartorius; below the joint, by the inner head of the Gastrocnemius.

Above, it is limited by the apposition of the inner and outer hamstring muscles; below by the junction of the two heads of the Gastrocnemius. The floor is formed by the lower part of the posterior surface of the shaft of the femur, the posterior ligament of the knee-joint, the upper end of the tibia, and the fascia covering the Popliteus muscle, and the space is covered in by the fascia lata.

Contents. It contains the Popliteal vessels and their branches, together with the termination of the external saphenous vein, the internal and external popliteal nerves and their branches, the small sciatic nerve, the articular branch from the obturator nerve, a few small lymphatic glands, and a considerable quantity of loose adipose tissue.

Position of contained parts. The internal popliteal nerve descends in the middle line of the space, lying superficial and a little external to the vein and artery. The external popliteal nerve descends on the outer side of the space, lying close to the tendon of the Biceps muscle. More deeply at the bottom of the space are the popliteal vessels, the vein lying superficial and a little external to the artery, to which it is closely united by dense areolar tissue; sometimes the vein is placed on the inner instead of the outer side of the artery; or the vein may be double, the artery lying between the two venæ comites, which are usually connected by short transverse branches. More deeply, and close to the surface of the bone, is the popliteal artery, and passing off from it at right angles are its articular branches. The articular branch from the obturator nerve descends upon the popliteal artery to supply the knee; and occasionally there is found deep in the space an articular filament from the great sciatic nerve. The popliteal lymphatic glands, four or five in number, are found surrounding the artery; one usually lies superficial to the vessel, another is situated between it and the bone, and the rest are placed on either side of it. The bursæ usually found in this space are: 1. On the outer side, one beneath the outer head of the Gastrocnemius (which sometimes communicates with the joint) and one beneath the tendon of the Popliteus, which is almost always an extension of the synovial membrane. Sometimes also there is a bursa above the tendon of the Popliteus, between it and the external lateral ligament. 2. On the inner side of the joint there is a large bursa between the inner head of the Gastrocnemius and the femur, which sends a prolongation between the tendons of the Gastrocnemius and Semimembranosus, and lies in contact with the ligament of Winslow. This bursa often communicates with the joint. There is a second bursa between the tendon of the Semimembranosus and the head of the tibia; and sometimes a bursa between the tendons of the Semitendinosus and Semimembranosus.

The Popliteal Artery, in its course downwards from the aperture in the Adductor magnus to the lower border of the Popliteus muscle, rests first on the inner, and then on the posterior surface of the femur; in the middle of its course, on the posterior ligament of the knee-joint; and below, on the fascia covering the Popliteus muscle. *Superficially*, it is covered, above, by the Semimembranosus; in the middle of its course, by a quantity of fat, which separates it from the deep fascia and integument; and below, it is overlapped by the Gastrocnemius, Plantaris, and Soleus muscles, the popliteal vein, and the internal popliteal nerve. The popliteal vein, which is intimately attached to the artery, lies superficial and external to it, until near its termination, when it crosses it and lies to its inner side. The popliteal nerve is still more superficial and external, crossing, however, the artery below the joint, and lying on its inner side. *Laterally*, the artery is bounded by the muscles which form the boundaries of the popliteal space.

Plan of Relations of Popliteal Artery.

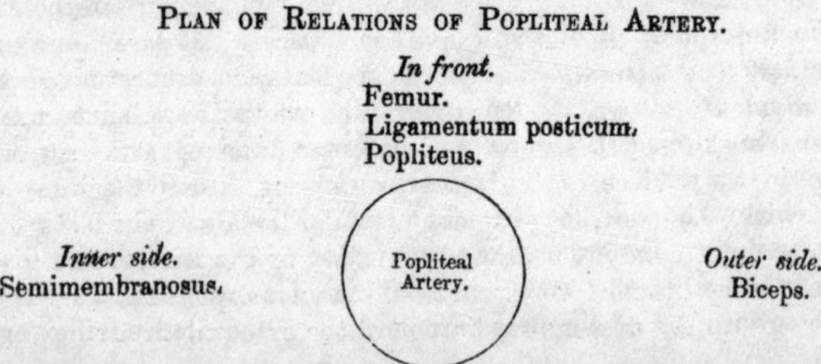

Behind.
Popliteal vein.
Internal popliteal nerve.
Fascia.

Peculiarities in point of division. Occasionally the popliteal artery divides prematurely into its terminal branches; this division occurs most frequently opposite the knee-joint.

Unusual branches. The artery sometimes divides into the anterior tibial and peroneal, the posterior tibial being wanting, or very small. In a single case, the popliteal was found to divide into three branches, the anterior and posterior tibial, and peroneal.

Surgical Anatomy. Ligature of the popliteal artery is required in cases of wound of that vessel, but for aneurism of the posterior tibial it is preferable to tie the superficial femoral. The popliteal may be tied in the upper or lower part of its course; but in the middle of the ham the operation is attended with considerable difficulty, from the great depth of the artery, and from the extreme degree of tension of the lateral boundaries of the space.

In order to expose the vessel in the upper part of its course, the patient should be placed in the prone position, with the limb extended. An incision about three inches in length should then be made through the integument, along the posterior margin of the Semimembranosus, and the fascia lata having been divided, this muscle must be drawn inwards, when the pulsation of the vessel will be detected with the finger; the nerve lies on the outer or fibular side of the artery, the vein, superficial and also to its outer side; the vein having been cautiously separated from the artery, the aneurism needle should be passed around the latter vessel from without inwards.

To expose the vessel in the lower part of its course, where the artery lies between the two heads of the Gastrocnemius, the patient should be placed in the same position as in the preceding operation. An incision should then be made through the integument in the middle line, commencing opposite the bend of the knee-joint, care being taken to avoid the external saphenous vein and nerve. After dividing the deep fascia, and separating some dense cellular membrane, the artery, vein, and nerve will be exposed, descending between the two heads of the Gastrocnemius. Some muscular branches of the popliteal should be avoided if possible, or if divided, tied immediately. The leg being now flexed, in order the more effectually to separate the two heads of the Gastrocnemius, the nerve should be drawn inwards and the vein outwards, and the aneurism needle passed between the artery and vein from without inwards.

The branches of the popliteal artery are, the

Muscular { Superior. / Inferior or Sural.
Cutaneous.
Superior external articular.
Superior internal articular.
Azygos articular.
Inferior external articular.
Inferior internal articular.

The *superior muscular branches*, two or three in number, arise from the upper part of the popliteal artery, and are distributed to the Vastus externus and flexor muscles of the thigh; anastomosing with the inferior perforating, and terminal branches of the profunda.

The *inferior muscular* (*Sural*) are two large branches, which are distributed to the two heads of the Gastrocnemius and to the Plantaris muscle. They arise from the popliteal artery opposite the knee-joint.

Cutaneous branches descend on each side and in the middle of the limb, between the Gastrocnemius and integument; they arise separately from the popliteal artery, or from some of its branches, and supply the integument of the calf.

The *superior articular arteries*, two in number, arise one on either side of the popliteal, and wind round the femur immediately above its condyles to the front of the knee-joint. The *internal branch* passes beneath the tendon of the Adductor magnus, and divides into two, one of which supplies the Vastus internus, inosculating with the anastomotica magna and inferior internal articular; the other ramifies close to the surface of the femur, supplying it and the knee-joint, and anastomosing with the superior external articular artery. The *external branch* passes above the outer condyle, beneath the tendon of the Biceps, and divides into a superficial and deep branch: the superficial branch supplies the Vastus externus, and anastomoses with the descending branch of the external circumflex artery; the deep branch supplies the lower part of the femur and knee-joint, and forms an anastomotic arch across the bone with the anastomotica magna artery.

The *azygos articular* is a small branch, arising from the popliteal artery opposite the bend of the knee-joint. It pierces the posterior ligament, and supplies the ligaments and synovial membrane in the interior of the articulation.

The *inferior articular arteries*, two in number, arise from the popliteal, beneath the Gastrocnemius, and wind round the head of the tibia, below the joint. The *internal* one passes below the inner tuberosity, beneath the internal lateral ligament, at the anterior border of which it ascends to the front and inner side of the joint, to supply the head of the tibia and the articulation of the knee. The *external* one passes outwards above the head of the fibula, to the front of the knee-joint, lying in its course beneath the outer head of the Gastrocnemius, the external lateral ligament, and the tendon of the Biceps muscle, and divides into branches, which anastomose with the inferior internal articular artery, the superior articular arteries, and the recurrent branch of the anterior tibial.

Anterior Tibial Artery. (Fig. 233.)

The anterior tibial artery commences at the bifurcation of the popliteal, at the lower border of the Popliteus muscle, passes forwards between the two heads of the Tibialis posticus, and through the aperture left between the bones at the upper part of the interosseous membrane, to the deep part of the front of the leg; it then descends on the anterior surface of the interosseous ligament, and of the tibia, to the front of the ankle-joint, where it lies more superficially, and becomes the dorsalis pedis. A line drawn from the inner side of the head of the fibula to midway between the two malleoli will mark the course of the artery.

Relations. In the upper two-thirds of its extent, it rests upon the interosseous ligament, to which it is connected by delicate fibrous arches thrown across it. In the lower third, upon the front of the tibia, and the anterior ligament of the ankle-joint. In the upper third of its course, it lies between the Tibialis anticus and Extensor longus digitorum; in the middle third, between the Tibialis anticus and Extensor proprius pollicis. In the lower third, it is crossed by the tendon of the Extensor proprius pollicis, and lies between it and the innermost tendon of the Extensor longus digitorum. It is covered, in the upper two-thirds of its course, by the muscles which lie on either side of it, and by the deep fascia; in the lower third, by the integument, annular ligament, and fascia.

The anterior tibial artery is accompanied by two veins (venæ comites), which lie one on either side of the artery; the anterior tibial nerve lies at first to its outer side, and about the middle of the leg is placed superficial to it; at the lower part of the artery the nerve is generally again on the outer side.

Plan of the Relations of the Anterior Tibial Artery.

In front.

Integument, superficial and deep fasciæ.
Tibialis anticus (overlaps it in upper part of leg).
Extensor longus digitorum } (overlap it slightly).
Extensor proprius pollicis } (overlap it slightly).
Anterior tibial nerve.

Inner side.

Tibialis anticus.
Extensor proprius pollicis (crosses it at its lower part).

Outer side.

Anterior tibial nerve.
Extensor longus digitorum.
Extensor proprius pollicis.

Behind.

Interosseous membrane.
Tibia.
Anterior ligament of ankle-joint.

Peculiarities in Size. This vessel may be diminished in size, may be deficient to a greater or less extent, or may be entirely wanting, its place being supplied by perforating branches from the posterior tibial, or by the anterior division of the peroneal artery.

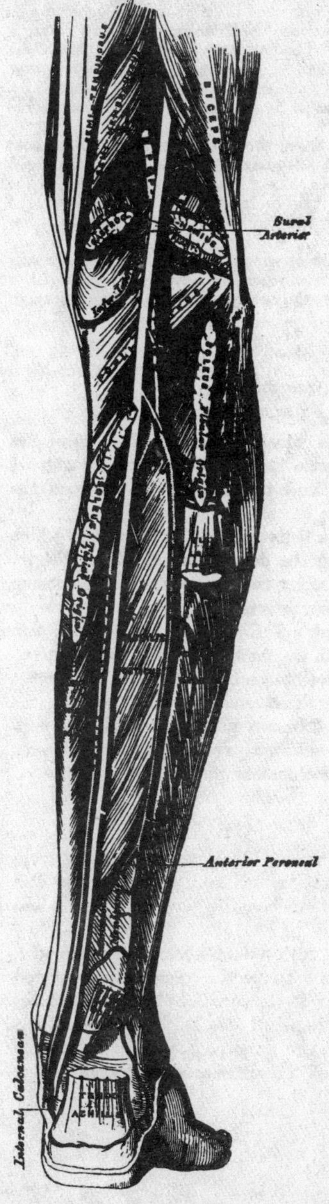

232.—The Popliteal, Posterior Tibial, and Peroneal Arteries.

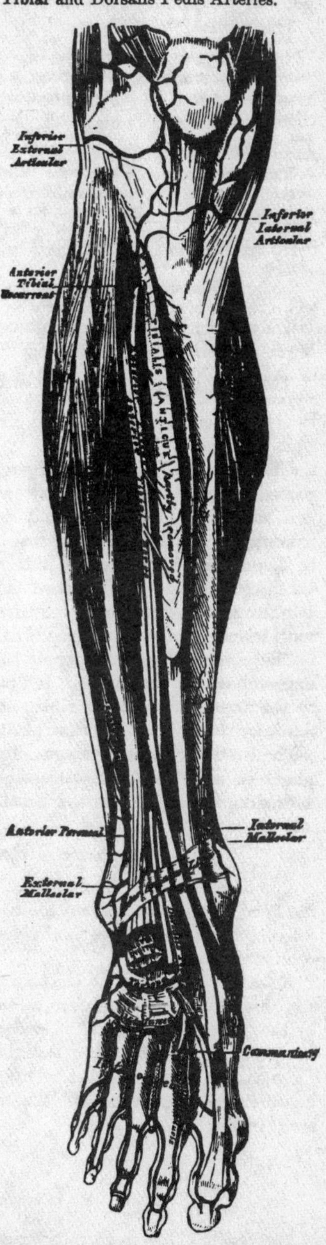

233.—Surgical Anatomy of the Anterior Tibial and Dorsalis Pedis Arteries.

Course. The artery occasionally deviates in its course towards the fibular side of the leg, regaining its usual position beneath the annular ligament at the front of the ankle. In two instances, the vessel has been found to approach the surface in the middle of the leg, being covered merely by the integument and fascia below that point.

Surgical Anatomy. The anterior tibial artery may be tied in the upper or lower part of the leg. In the upper part, the operation is attended with great difficulty, on account of the depth of the vessel from the surface. An incision, about four inches in length, should be made through the integument, midway between the spine and the tibia and the outer margin of the fibula, the fascia and intermuscular septum between the Tibialis anticus and Extensor longus digitorum being divided to the same extent. The foot must be flexed to relax these muscles, and they must be separated from each other by the finger. The artery is then exposed, deeply seated, lying upon the interosseous membrane, the nerve lying externally, and one of the venæ comites on either side; these must be separated from the artery before the aneurism needle is passed round it.

To tie the vessel in the lower third of the leg above the ankle-joint, an incision about three inches in length should be made through the integument between the tendons of the Tibialis anticus and Extensor proprius pollicis muscles, the deep fascia being divided to the same extent; the tendon on either side should be held aside, when the vessel will be seen lying upon the tibia, with the nerve superficial to it, and one of the venæ comites on either side.

In order to secure the artery over the instep, an incision should be made on the fibular side of the tendon of the Extensor proprius pollicis, between it and the innermost tendon of the long Extensor; the deep fascia having been divided, the artery will be exposed, the nerve lying either superficial to it, or to its outer side.

The branches of the anterior tibial artery are, the

Recurrent tibial.	Internal malleolar.
Muscular.	External malleolar.

The *recurrent branch* arises from the anterior tibial, as soon as that vessel has passed through the interosseous space; it ascends in the Tibialis anticus muscle, and ramifies on the front and sides of the knee-joint, anastomosing with the articular branches of the popliteal.

The *muscular branches* are numerous; they are distributed to the muscles which lie on either side of the vessel, some piercing the deep fascia to supply the integument, others passing through the interosseous membrane, and anastomosing with branches of the posterior tibial and peroneal arteries.

The *malleolar arteries* supply the ankle-joint. The *internal* arises about two inches above the articulation, and passes beneath the tendon of the Tibialis anticus to the inner ankle, upon which it ramifies, anastomosing with branches of the posterior tibial and internal plantar arteries. The *external* passes beneath the tendons of the Extensor longus digitorum and Extensor proprius pollicis, and supplies the outer ankle, anastomosing with the anterior peroneal artery, and with ascending branches from the tarsal branch of the dorsalis pedis.

Dorsalis Pedis Artery. (Fig. 233.)

The dorsalis pedis, the continuation of the anterior tibial, passes forwards from the bend of the ankle along the tibial side of the foot to the back part of the first interosseous space, where it divides into two branches, the dorsalis hallucis and communicating.

Relations. This vessel, in its course forwards, rests upon the astragalus, scaphoid, and internal cuneiform bones and the ligaments connecting them, being covered by the integument and fascia, and crossed near its termination by the innermost tendon of the Extensor brevis digitorum. On its *tibial side* is the tendon of the Extensor proprius pollicis; on its *fibular side*, the innermost tendon of the Extensor longus digitorum, and the termination of the anterior tibial nerve. It is accompanied by two veins.

Plan of the Relations of the Dorsalis Pedis Artery.

In front.
Integument and fascia.
Innermost tendon of Extensor brevis digitorum.

Tibial side.
Extensor proprius pollicis.

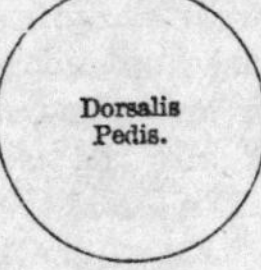

Fibular side.
Extensor longus digitorum.
Anterior tibial nerve.

Behind.
Astragalus,
Scaphoid,
Internal cuneiform,
and their ligaments.

Peculiarities in Size. The dorsal artery of the foot may be larger than usual, to compensate for a deficient plantar artery; or it may be deficient in its terminal branches to the toes, which are then derived from the internal plantar; or its place may be supplied altogether by a large anterior peroneal artery.

Position. This artery frequently curves outwards, lying external to the line between the middle of the ankle and the back part of the first interosseous space.

Surgical Anatomy. This artery may be tied, by making an incision through the integument, between two and three inches in length, on the fibular side of the tendon of the Extensor proprius pollicis, in the interval between it and the inner border of the short Extensor muscle. The incision should not extend further forwards than the back part of the first interosseous space, as the artery divides in that situation. The deep fascia being divided to the same extent, the artery will be exposed, the nerve lying upon its outer side.

Branches. The branches of the dorsalis pedis are, the

Tarsal.	Dorsalis pollicis, or hallucis.
Metatarsal.	Communicating.
Interosseous.	

The *tarsal artery* arises from the dorsalis pedis, as that vessel crosses the scaphoid bone; it passes in an arched direction outwards, lying upon the tarsal bones, and covered by the Extensor brevis digitorum; it supplies that muscle and the articulations of the tarsus, and anastomoses with branches from the metatarsal, external malleolar, peroneal, and external plantar arteries.

The *metatarsal* arises a little anterior to the preceding; it passes outwards to the outer part of the foot, over the bases of the metatarsal bones, beneath the tendons of the short Extensor, its direction being influenced by its point of origin; and it anastomoses with the tarsal and external plantar arteries. This vessel gives off three branches, the interosseous, which pass forwards upon the three outer Dorsal interossei muscles, and, in the clefts between the toes, divide into two dorsal collateral branches for the adjoining toes. At the back part of each interosseous space these vessels receive the posterior perforating branches from the plantar arch; and at the fore part of each interosseous space, they are joined by the anterior perforating branches, from the digital arteries. The outermost interosseous artery gives off a branch which supplies the outer side of the little toe.

The *dorsalis hallucis* runs forwards along the outer border of the first metatarsal bone, and, at the cleft between the first and second toes, divides into two branches, one of which passes inwards, beneath the tendon of the Extensor proprius pollicis, and is distributed to the inner border of the great toe; the other branch bifurcating to supply the adjoining sides of the great and second toes.

The *communicating artery* dips down into the sole of the foot, between the two heads of the first Dorsal interosseous muscle, and inosculates with the termination of the external plantar artery, to complete the plantar arch. It here gives off two digital branches; one runs along the inner side of the great toe, on its plantar surface; the other passes forwards along the first metatarsal space, and bifurcates for the supply of the adjacent sides of the great and second toes.

Posterior Tibial Artery. (Fig. 232.)

The posterior tibial is an artery of large size, which extends obliquely downwards from the lower border of the Popliteus muscle, along the tibial side of the leg, to the fossa between the inner ankle and the heel, where it divides beneath the origin of the Abductor pollicis, into the internal and external plantar arteries. At its origin it lies opposite the interval between the tibia and fibula; as it descends, it approaches the inner side of the leg, lying behind the tibia, and, in the lower part of its course, is situated midway between the inner malleolus and the tuberosity of the os calcis.

Relations. It lies successively upon the Tibialis posticus, the Flexor longus digitorum, the tibia and the back part of the ankle-joint. It is *covered* by the intermuscular fascia, which separates it above from the Gastrocnemius and Soleus muscles. In the lower third, where it is more superficial, it is covered only by the integument and fascia, and runs parallel with the inner border of the tendo Achillis. It is accompanied by two veins, and by the posterior tibial nerve, which lies at first to the inner side of the artery, but soon crosses it, and is, in the greater part of its course, on its outer side.

Plan of the Relations of the Posterior Tibial Artery.

In front.
Tibialis posticus.
Flexor longus digitorum.
Tibia.
Ankle joint.

Inner side.
Posterior tibial nerve,
upper third.

Posterior Tibial.

Outer side.
Posterior tibial nerve,
lower two-thirds.

Behind.
Gastrocnemius.
Soleus.
Deep fascia and integument.

Behind the Inner Ankle, the tendons and blood-vessels are arranged in the following order, from within outwards:—First, the tendons of the Tibialis posticus and Flexor longus digitorum, lying in the same groove, behind the inner malleolus, the former being the most internal. External to these is the posterior tibial artery, having a vein on either side; and, still more externally, the posterior tibial nerve. About half an inch nearer the heel is the tendon of the Flexor longus pollicis.

Peculiarities in Size. The posterior tibial is not unfrequently smaller than usual, or absent, its place being supplied by a large peroneal artery, which passes inwards at the lower end of the tibia, and either joins the small tibial artery, or continues alone to the sole of the foot.

Surgical Anatomy. The *application of a ligature* to the posterior tibial may be required in cases of wound of the sole of the foot, attended with great hæmorrhage, when the vessel should be tied at the inner ankle. In cases of wound of the posterior tibial, it will be necessary to enlarge the wound so as to expose the vessel at the wounded point, excepting where the vessel is injured by a punctured wound from the front of the leg. In cases of aneurism from wound of the artery low down, the vessel should be tied in the middle of the leg. But in aneurism of the posterior tibial high up, it would be better to tie the femoral artery.

To tie the posterior tibial artery at the ankle, a semilunar incision should be made through the integument, about two inches and a half in length, midway between the heel and inner ankle, or a little nearer the latter. The subcutaneous cellular membrane having been divided, a strong and dense fascia, the internal annular ligament, is exposed. This ligament is continuous above with the deep fascia of the leg, covers the vessels and nerves, and is intimately adherent to the sheaths of the tendons. This having been cautiously divided upon a director, the sheath of the vessels is exposed, and being opened, the artery is seen with one of the venæ comites on each side. The aneurism needle should be passed round the vessel from the heel towards the ankle, in order to avoid the posterior tibial nerve, care being at the same time taken not to include the venæ comites.

The vessel may also be tied in the lower third of the leg by making an incision about three inches in length, parallel with the inner margin of the tendo Achillis. The internal saphenous vein being carefully avoided, the two layers of fascia must be divided upon a director, when the artery is exposed along the outer margin of the Flexor longus digitorum, with one of its venæ comites on either side, and the nerve lying external to it.

To tie the posterior tibial in the middle of the leg is a very difficult operation, on account of the great depth of the vessel from the surface, and its being covered by the Gastrocnemius and Soleus muscles. The patient being placed in the recumbent position, the injured limb should rest on its outer side, the knee being partially bent, and the foot extended, so as to relax the muscles of the calf. An incision about four inches in length should then be made through the integument, along the inner margin of the tibia, taking care to avoid the internal saphenous vein. The deep fascia having been divided, the margin of the Gastrocnemius is exposed, and must be drawn aside, and the tibial attachment of the Soleus divided, a director being previously passed beneath it. The artery may now be felt pulsating beneath the deep fascia, about an inch from the margin of the tibia. The fascia having been divided, and the limb placed in such a position as to relax the muscles of the calf as much as possible, the veins should be separated from the artery, and the aneurism needle passed round the vessel from without inwards, so as to avoid wounding the posterior tibial nerve.

The branches of the posterior tibial artery are, the

Peroneal.	Nutrient.
Anterior peroneal.	Communicating.
Muscular.	Internal calcanean.

The *Peroneal Artery* lies, deeply seated, along the back part of the fibular side of the leg. It arises from the posterior tibial, about an inch below the lower border of the Popliteus muscle, passes obliquely outwards to the fibula, and then descends along the inner border of that bone to the lower third of the leg, where it gives off the anterior peroneal. It then passes across the articulation between the tibia and fibula, to the outer side of the os calcis, supplying the neighbouring muscles and back of the ankle, and anastomosing with the external malleolar, tarsal, and external plantar arteries.

Relations. This vessel rests at first upon the Tibialis posticus, and, in the greater part of its course, in the fibres of the Flexor longus pollicis, in a groove between the interosseous ligament and the bone. It is *covered* in the upper part of its course by the Soleus and deep fascia; *below*, by the Flexor longus pollicis.

PLAN OF THE RELATIONS OF THE PERONEAL ARTERY.

In front.
Tibialis posticus.
Flexor longus pollicis.

Outer side.
Fibula.

Peroneal Artery.

Behind.
Soleus.
Deep fascia.
Flexor longus pollicis.

Peculiarities in Origin. The peroneal artery may arise three inches below the Popliteus, or from the posterior tibial high up, or even from the popliteal.

Its Size is more frequently increased than diminished; and then it either reinforces the posterior tibial by its junction with it, or altogether takes the place of the posterior tibial in the lower part of the leg and foot, the latter vessel only existing as a short muscular branch. In those rare cases where the peroneal artery is smaller than usual, a branch from the posterior tibial supplies its place, and a branch from the anterior tibial compensates for the diminished anterior peroneal artery. In one case, the peroneal artery has been found entirely wanting.

The anterior peroneal is sometimes enlarged, and takes the place of the dorsal artery of the foot.

The peroneal artery, in its course, gives off branches to the Soleus, Tibialis posticus, Flexor longus pollicis, and Peronei muscles, and a nutrient branch to the fibula.

The *Anterior Peroneal*, the only named branch of the Peroneal artery, pierces the interosseous membrane, about two inches above the outer malleolus, to reach the fore part of the leg, and, passing down beneath the Peroneus tertius to the outer ankle, ramifies on the front and outer side of the tarsus, anastomosing with the external malleolar and tarsal arteries.

The *nutrient artery* of the tibia arises from the posterior tibial near its origin, and after supplying a few muscular branches, enters the nutrient canal of that bone, which it traverses obliquely from above downwards. This is the largest nutrient artery of bone in the body.

The *muscular branches* of the posterior tibial are distributed to the Soleus and deep muscles along the back of the leg.

The *communicating branch* to the peroneal runs transversely across the back of the tibia, about two inches above its lower end, passing beneath the Flexor longus pollicis.

The *internal calcanean* consist of several large branches, which arise from the posterior tibial just before its division; they are distributed to the fat and integument behind the tendo Achillis and about the heel, and to the muscles on the inner side of the sole, anastomosing with the peroneal and internal malleolar arteries.

The *Internal Plantar Artery* (figs. 234, 235), much smaller than the external, passes forwards along the inner side of the foot. It is at first situated above the Abductor pollicis, and then between it and the Flexor brevis digitorum, both of which it supplies. At the base of the first metatarsal bone, where it has become much diminished in size, it passes along the inner border of the great toe, inosculating with its digital branches.

The *External Plantar Artery*, much larger than the internal, passes obliquely outwards and forwards to the base of the fifth metatarsal bone. It then turns obliquely inwards to the interval between the bases of the first and second metatarsal bones, where it anastomoses with the communicating branch from the dorsalis pedis artery, thus completing the plantar arch. As this artery passes outwards, it is at first placed between the os calcis and Abductor pollicis, and then between the Flexor brevis digitorum and Flexor accessorius; and as it passes forwards to the base of the little toe, it lies more superficially between the Flexor brevis digitorum and Abductor minimi digiti, covered by the deep fascia and integument. The remaining portion of the vessel is deeply situated: it extends from the base of the metatarsal bone of the little toe to the back part of the first interosseous space, and forms the plantar arch; it is convex forwards, lies upon the Interossei muscles, opposite the tarsal ends of the metatarsal bones, and is covered by the Adductor pollicis, the flexor tendons of the toes, and the Lumbricales.

Branches. The plantar arch, besides distributing numerous branches to the muscles, integument, and fasciæ in the sole, gives off the following branches:

Posterior perforating. Digital—Anterior perforating.

The *Posterior Perforating* are three small branches, which ascend through the back part of the three outer interosseous spaces, between the heads of the Dorsal interossei muscles, and anastomose with the interosseous branches from the metatarsal artery.

The *Digital Branches* are four in number, and supply the three outer toes and half the second toe. The *first* passes outwards from the outer side of the plantar arch, and is distributed to the outer side of the little toe, passing in its course beneath the Abductor and short Flexor muscles. The *second*, *third*, and *fourth* run forwards along the metatarsal spaces, and on arriving at the clefts between the toes, divide into collateral branches, which supply the adjacent sides of the three outer toes and the outer side of the second. At the bifurcation of the toes, each digital artery sends upwards, through the fore part of the corresponding metatarsal space, a small branch, which inosculates with the interosseous branches of the metatarsal artery. These are the anterior perforating branches.

From the arrangement already described of the distribution of the vessels to the toes, it will be seen that both sides of the three outer toes, and the outer side

234.—The Plantar Arteries. Superficial View.

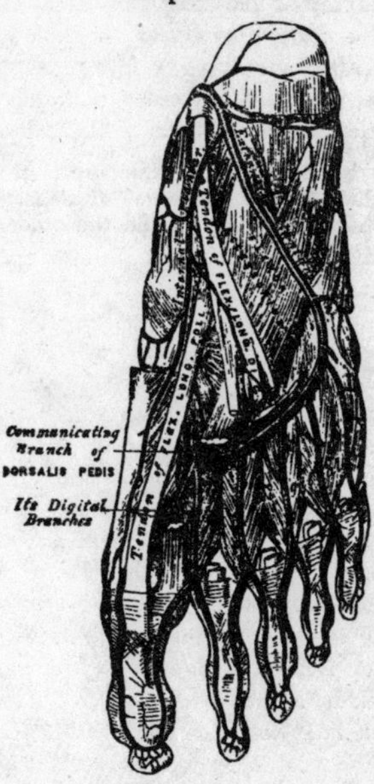

235.—The Plantar Arteries. Deep View.

of the second toe, are supplied by branches from the plantar arch; both sides of the great toe, and the inner side of the second, being supplied by the dorsal artery of the foot.

Pulmonary Artery.

The pulmonary artery conveys the venous blood from the right side of the heart to the lungs. It is a short wide vessel, about two inches in length, arising from the left side of the base of the right ventricle, in front of the aorta. It ascends obliquely upwards, backwards, and to the left side, as far as the under surface of the arch of the aorta, where it divides into two branches of nearly equal size, the right and left pulmonary arteries.

Relations. The greater part of this vessel is contained, together with the ascending part of the arch of the aorta, in the pericardium, being enclosed with it in a tube of serous membrane, continued upwards from the base of the heart, and has attached to it, above, the fibrous layer of the membrane. Behind, it rests at first upon the ascending aorta, and higher up lies in front of the left auricle. On either side of its origin is the appendix of the corresponding auricle, and a coronary artery; and higher up it passes to the left side of the ascending aorta.

A little to the left of its point of bifurcation, it is connected to the under surface of the arch of the aorta by a short fibrous cord, the remains of a vessel peculiar to fœtal life, the ductus arteriosus.

The *right pulmonary artery*, longer and larger than the left, runs horizontally outwards, behind the ascending aorta and superior vena cava, to the root of the right lung, where it divides into two branches, of which the lower, which is the larger, supplies the lower lobe; the upper giving a branch to the middle lobe.

The *left pulmonary artery*, shorter but somewhat smaller than the right, passes horizontally in front of the descending aorta and left bronchus to the root of the left lung, where it divides into two branches for the two lobes.

The terminal branches of the pulmonary artery will be described with the anatomy of the lung.

The author has to acknowledge valuable aid derived from the following works:—Harrison's 'Surgical Anatomy of the Arteries of the Human Body.' Dublin, 1824.—Richard Quain's 'Anatomy of the Arteries of the Human Body. London, 1844.—Sibson's 'Medical Anatomy;' and the other works on General and Microscopic Anatomy before referred to.

Of the Veins.

THE Veins are the vessels which serve to return the blood from the capillaries of the different parts of the body to the heart. They consist of two distinct sets of vessels, the pulmonary and systemic.

The *Pulmonary Veins*, unlike other vessels of this kind, contain arterial blood, which they return from the lungs to the left auricle of the heart.

The *Systemic Veins* return the venous blood from the body generally to the right auricle of the heart.

The *Portal Vein*, an appendage to the systemic venous system, is confined to the abdominal cavity, returning the venous blood from the viscera of digestion, and carrying it to the liver by a single trunk of large size, the vena portæ. From this organ, the same blood is conveyed to the inferior vena cava by means of the hepatic veins.

The veins, like the arteries, are found in nearly every tissue of the body. They commence by minute plexuses, which communicate with the capillaries. The branches which have their commencement in these plexuses unite together into trunks, and these, in their passage towards the heart, constantly increase in size as they receive branches, and join other veins similar in size to themselves. The veins are larger and altogether more numerous than the arteries; hence, the entire capacity of the venous system is much greater than that of the arterial; the pulmonary veins excepted, which do not exceed in capacity the pulmonary arteries. From the combined area of the smaller venous branches being greater than the main trunks, it results, that the venous system represents a cone, the summit of which corresponds to the heart; its base to the circumference of the body. In form, the veins are not perfectly cylindrical like the arteries, their walls being collapsed when empty, and the uniformity of their surface being interrupted at intervals by slight contractions, which indicate the existence of valves in their interior. They usually retain, however, the same calibre as long as they receive no branches.

The veins communicate very freely with one another, especially in certain regions of the body; and this communication exists between the larger trunks as well as between the smaller branches. Thus, in the cavity of the cranium, and between the veins of the neck, where obstruction would be attended with imminent danger to the cerebral venous system, we find that the sinuses and larger veins have large and very frequent anastomoses. The same free communication exists between the veins throughout the whole extent of the spinal canal, and between the veins composing the various venous plexuses in the abdomen and pelvis, as the spermatic, uterine, vesical, prostatic, etc.

The veins are subdivided into three sets: superficial, deep, and sinuses.

The *Superficial* or *Cutaneous Veins* are found between the layers of superficial fascia, immediately beneath the integument; they return the blood from these structures, and communicate with the deep veins by perforating the deep fascia.

The *Deep Veins* accompany the arteries, and are usually enclosed in the same sheath with those vessels. In the smaller arteries, as the radial, ulnar, brachial, tibial, peroneal, they exist generally in pairs, one lying on each side of the vessel, and are called *venæ comites*. The larger arteries, as the axillary, subclavian, popliteal, and femoral, have usually only one accompanying vein. In certain organs of the body, however, the deep veins do not accompany the arteries; for instance, the veins in the skull and spinal canal, the hepatic veins in the liver, and the larger veins returning blood from the osseous tissue.

Sinuses are venous channels, which, in their structure and mode of distribution, differ altogether from the veins. They are found only in the interior of the skull, and are formed by a separation of the layers of the dura mater; their outer coat consisting of fibrous tissue, their inner of a serous membrane continuous with the serous membrane of the veins.

Veins have thinner walls than arteries, the difference in thickness being due to the small amount of elastic and muscular tissues which the veins contain. The superficial veins usually have thicker coats than the deep veins, and the veins of the lower limb are thicker than those of the upper.

The minute structure of these vessels is described in the Introduction.

The veins may be arranged into three groups:—1. Those of the head and neck, upper extremity, and thorax, which terminate in the superior vena cava 2. Those of the lower limb, pelvis, and abdomen, which terminate in the inferior vena cava. 3. The cardiac veins, which open directly into the right auricle of the heart.

Veins of the Head and Neck.

The veins of the head and neck may be subdivided into three groups:—1. The veins of the exterior of the head. 2. The veins of the neck. 3. The veins of the diploë and interior of the cranium.

The veins of the exterior of the head are, the

Facial.	Temporo-maxillary.
Temporal.	Posterior auricular.
Internal maxillary.	Occipital.

The Facial Vein passes obliquely across the side of the face, extending from the inner angle of the orbit, downwards and outwards, to the anterior margin of the Masseter muscle. It lies to the outer side of the facial artery, and is not so tortuous as that vessel. It commences in the frontal region, where it is called the *frontal vein*; at the inner angle of the eye it has received the name of the *angular vein*; and from this point to its termination is called the *facial vein*.

The *frontal vein* commences on the anterior part of the skull, by a venous plexus, which communicates with the anterior branches of the temporal vein; the veins converge to form a single trunk, which runs downwards near the middle line of the forehead parallel with the vein of the opposite side, and unites with it at the root of the nose, by a transverse trunk, called the *nasal arch*. Occasionally the frontal veins join to form a single trunk, which bifurcates at the root of the nose into the two angular veins. At the nasal arch the branches diverge, and run along the side of the root of the nose. The frontal vein as it descends upon the forehead receives the supraorbital vein; the dorsal veins of the nose terminate in the nasal arch; and the angular vein receives the veins of the ala nasi on its inner side, and the superior palpebral veins on its outer side; it moreover communicates with the ophthalmic vein, which establishes an important anastomosis between this vessel and the cavernous sinus.

The *facial vein* commences at the inner angle of the orbit, being a continuation of the angular vein. It passes obliquely downwards and outwards, beneath the great zygomatic muscle, descends along the anterior border of the Masseter, crosses over the body of the lower jaw, with the facial artery, and, passing obliquely outwards and backwards, beneath the Platysma and cervical fascia, unites with a branch of communication from the temporo-maxillary vein, to form a trunk of large size which enters the internal jugular.

Branches. The facial vein receives, near the angle of the mouth, communicating branches from the pterygoid plexus. It is also joined by the inferior palpebral, the superior and inferior labial veins, the buccal veins from the cheek, and the masseteric veins. Below the jaw, it receives the submental, the inferior palatine, which returns the blood from the plexus around the tonsil and soft palate; the submaxillary vein, which commences in the submaxillary gland; and lastly, the ranine vein.

236.—Veins of the Head and Neck.

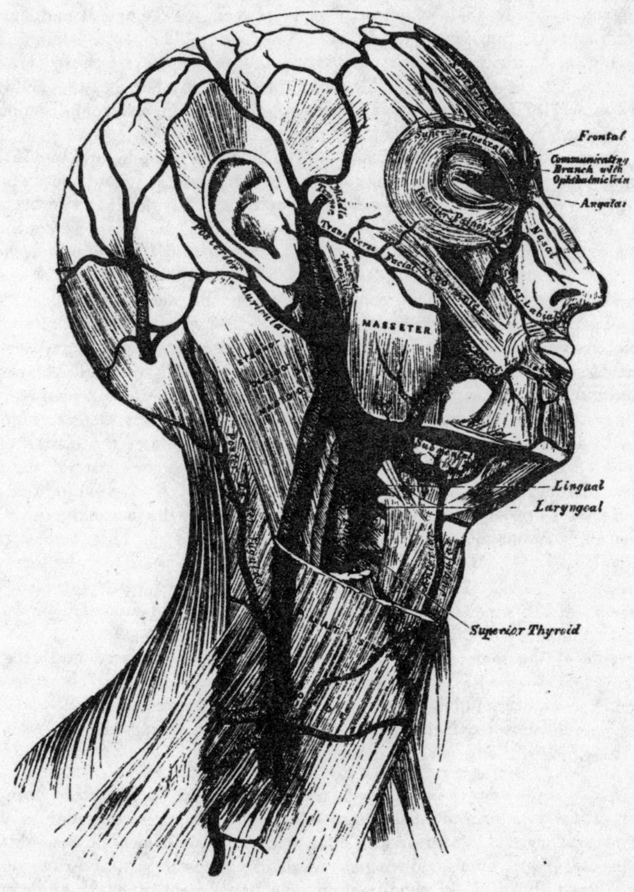

The *Temporal Vein* commences by a minute plexus on the side and vertex of the skull, which communicates with the frontal vein in front, the corresponding vein of the opposite side, and the posterior auricular and occipital veins behind. From this network, anterior and posterior branches are formed which unite above the zygoma, forming the trunk of the vein. This trunk is joined in this situation by a large vein, the middle temporal, which receives the blood from the substance of the Temporal muscle and pierces the fascia at the upper border of the zygoma. The temporal vein then descends between the external auditory meatus and the condyle of the jaw, enters the substance of the parotid gland, and unites with the internal maxillary vein, to form the temporo-maxillary.

Branches. The temporal vein receives in its course some parotid veins, an articular branch from the articulation of the jaw, anterior auricular veins from the external ear, and a vein of large size, the transverse facial, from the side of the face.

The *Internal Maxillary Vein* is a vessel of considerable size, receiving branches which correspond with those of the internal maxillary artery. Thus it receives the middle meningeal veins, the deep temporal, the pterygoid, masseteric, and buccal, some palatine veins, and the inferior dental. These branches form a large plexus, the pterygoid, which is placed between the Temporal and External pterygoid, and partly between the Pterygoid muscles. This plexus communicates very freely with the facial vein, and with the cavernous sinus, by branches through the base of the skull. The trunk of the vein then passes backwards, behind the neck of the lower jaw, and unites with the temporal vein, forming the temporo-maxillary.

The *Temporo-Maxillary Vein*, formed by the union of the temporal and internal maxillary vein, descends in the substance of the parotid gland, between the ramus of the jaw and the Sterno-mastoid muscle, and divides into two branches, one of which passes inwards to join the facial vein, the other is continuous with the external jugular. It receives near its termination the posterior auricular vein.

The *Posterior Auricular Vein* commences upon the side of the head, by a plexus which communicates with the branches of the temporal and occipital veins. The vein descends behind the external ear and joins the temporo-maxillary, just before that vessel terminates in the external jugular. This vessel receives the stylo-mastoid vein, and some branches from the back part of the external ear.

The *Occipital Vein* commences at the back part of the vertex of the skull, by a plexus in a similar manner with the other veins. It follows the course of the occipital artery, passing deeply beneath the muscles of the back part of the neck, and terminates in the internal jugular, occasionally in the external jugular. As this vein passes opposite the mastoid process, it receives the mastoid vein, which establishes a communication with the lateral sinus.

Veins of the Neck.

The veins of the neck, which return the blood from the head and face, are the

External jugular. Anterior jugular.
Posterior external jugular. Internal jugular.
Vertebral.

The *External Jugular Vein* receives the greater part of the blood from the exterior of the cranium and deep parts of the face, being a continuation of the temporo-maxillary and posterior auricular veins. It commences in the substance of the parotid gland, on a level with the angle of the lower jaw, and runs perpendicularly down the neck, in the direction of a line drawn from the angle of the jaw to the middle of the clavicle. In its course it crosses the Sterno-mastoid muscle, and runs parallel with its posterior border as far as its attachment to the clavicle, where it perforates the deep fascia, and terminates in the subclavian vein, on the outer side of the internal jugular. In the neck, it is separated from the Sterno-mastoid by the anterior layer of the deep cervical fascia, and is covered by the Platysma, the superficial fascia, and the integument. This vein is crossed about its centre by the superficial cervical nerve, and its upper half is accompanied by the auricularis magnus nerve. The external jugular vein varies in size, bearing an inverse proportion to that of the other veins of the neck: it is occasionally double. It is provided with two pairs of valves, the lower pair being placed at its entrance into the subclavian vein, the upper pair in most cases about an inch and a half above the clavicle. These valves do not prevent the regurgitation of the blood, or the passage of injection from below upwards.*

Branches. This vein receives the occipital occasionally, the posterior external jugular, and, near its termination, the suprascapular and transverse cervical veins. It communicates with the anterior jugular, and, in the substance of the parotid, receives a large branch of communication from the internal jugular.

The *Posterior External Jugular Vein* returns the blood from the integument and superficial muscles in the upper and back part of the neck, lying between the Splenius and Trapezius muscles. It runs down the back part of the neck, and opens into the external jugular just below the middle of its course.

The *Anterior Jugular Vein* collects the blood from the integument and muscles in the middle of the anterior region of the neck. It passes down between the median line and the anterior border of the Sterno-mastoid, and, at the lower part of the neck, passes beneath that muscle to open into the subclavian vein, near the termination of the external jugular. This vein varies considerably in size, bearing almost always an inverse proportion to the external jugular. Most frequently there are two anterior jugulars, a right and left; but occasionally only one. This vein receives some laryngeal branches, and occasionally an inferior thyroid vein. Just above the sternum, the two anterior jugular veins communicate by a transverse trunk, which receives branches from the inferior thyroid veins. It also communicates with the external and with the internal jugular. There are no valves in this vein.

The *Internal Jugular Vein* collects the blood from the interior of the cranium, from the superficial parts of the face, and from the neck. It commences at the jugular foramen, in the base of the skull, being formed by the coalescence of the lateral and inferior petrosal sinuses. At its origin it is somewhat dilated, and this dilatation is called the sinus, or gulf of the internal jugular vein. It runs down the side of the neck in a vertical direction, lying at first on the outer side of the internal carotid, and then on the outer side of the common carotid, and at the root of the neck unites with the subclavian vein, to form the vena innominata. The internal jugular vein, at its commencement, lies upon the Rectus lateralis, behind, and at the outer side of the internal carotid, and the eighth and ninth pairs of nerves; lower down, the vein and artery lie upon the same plane, the glosso-pharyngeal and hypoglossal nerves passing forwards between them; the pneumogastric descends between and behind them, in the same sheath; and the spinal accessory passes obliquely outwards, behind the vein. At the root of the neck the vein of the right side is placed at a little distance from the artery; on the left side, it usually crosses it at its lower part. The right internal jugular vein crosses the first part of the subclavian artery. This vein is of considerable size, but varies in different individuals, the left one being usually the smaller. It is provided with a pair of valves, which are placed at its point of termination, or from half to three quarters of an inch above it.

Branches. This vein receives in its course the facial, lingual, pharyngeal, superior and middle thyroid veins, and sometimes the occipital. At its point of junction with the branch common to the temporal and facial veins, it becomes greatly increased in size.

The *lingual veins* commence on the dorsum, sides, and under surface of the tongue, and passing backwards, following the course of the lingual artery and its branches, terminate in the internal jugular.

The *pharyngeal vein* commences in a minute plexus, the pharyngeal, at the back part and sides of the pharynx, and after receiving meningeal branches, and the Vidian and spheno-palatine veins, terminates in the internal jugular. It occasionally opens into the facial, lingual, or superior thyroid vein.

* The student may refer to an interesting paper by Dr. Struthers, 'On Jugular Venesection in Asphyxia, Anatomically and Experimentally Considered, including the Demonstration of Valves in the Veins of the Neck,' in the *Edinburgh Medical Journal*, for November, 1856.

The *superior thyroid vein* commences in the substance and on the surface of the thyroid gland, by branches corresponding with those of the superior thyroid artery, and terminates in the upper part of the internal jugular vein.

The *middle thyroid vein* collects the blood from the lower part of the lateral lobe of the thyroid gland, and, being joined by some branches from the larynx and trachea, terminates in the lower part of the internal jugular vein.

The *occipital vein* has been described above.

The *Vertebral Vein* commences in the occipital region, by numerous small branches, from the deep muscles at the upper and back part of the neck, passes outwards, and enters the foramen in the transverse process of the atlas, and descends by the side of the vertebral artery, in the canal formed by the transverse processes of the cervical vertebræ. Emerging from the foramen in the transverse process of the sixth cervical, it terminates at the root of the neck in the back part of the innominate vein near its origin, its mouth being guarded by a pair of valves. On the right side, it crosses the first part of the subclavian artery. This vein, in the lower part of its course, occasionally divides into two branches, one of which emerges with the artery at the sixth cervical vertebra; the other escapes through the foramen in the seventh cervical.

Branches. The vertebral vein receives in its course the posterior condyloid vein, muscular branches from the muscles in the prevertebral region; dorsi-spinal veins, from the back part of the cervical portion of the spine; meningo-rachidian veins, from the interior of the spinal canal; and lastly, the ascending and deep cervical veins.

Veins of the Diploë.

The diploë of the cranial bones is channelled, in the adult, by a number of tortuous canals, which are lined by a more or less complete layer of compact

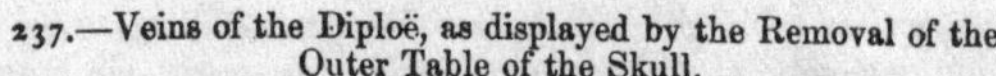

237.—Veins of the Diploë, as displayed by the Removal of the Outer Table of the Skull.

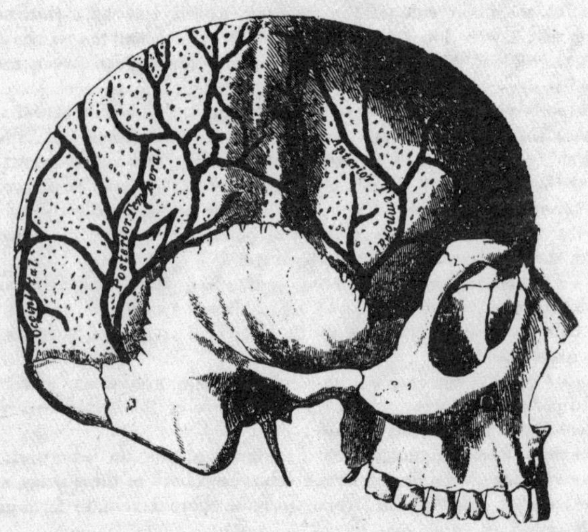

tissue. The veins they contain are large and capacious, their walls being thin, and formed only of epithelium, resting upon a layer of elastic tissue, and they present, at irregular intervals, pouch-like dilatations, or *culs de sac*, which serve as reservoirs for the blood. These are the veins of the diploë: they can only be displayed by removing the outer table of the skull.

In adult life, as long as the cranial bones are distinct and separable, these veins are confined to the particular bones; but in old age, when the sutures are united, they communicate with each other, and increase in size. These vessels communicate, in the interior of the cranium, with the meningeal veins, and with the sinuses of the dura mater; and on the exterior of the skull, with the veins of the pericranium. They are divided into the *frontal*, which opens into the supraorbital vein, by an aperture at the supraorbital notch; the *anterior temporal*, which is confined chiefly to the frontal bone, and opens into one of the deep temporal veins, after escaping by an aperture in the great wing of the sphenoid; the *posterior temporal*, which is confined to the parietal bone, and terminates in the lateral sinus by an aperture at the posterior inferior angle of the parietal bone; and the *occipital*, which is confined to the occipital bone and opens either into the occipital vein, or the occipital sinus.

Cerebral Veins.

The *Cerebral Veins* are remarkable for the extreme thinness of their coats, in consequence of the muscular tissue in them being wanting, and for the absence of valves. They may be divided into two sets, the superficial, which are placed on the surface, and the deep veins, which occupy the interior of the organ.

The *Superficial Cerebral Veins* ramify upon the surface of the brain, being lodged in the sulci, between the convolutions, a few running across the convolutions. They receive branches from the substance of the brain, and terminate in the sinuses. They are named, from the position they occupy, superior, inferior, internal, and external.

The *Superior Cerebral Veins*, seven or eight in number on each side, pass forwards and inwards towards the great longitudinal fissure, where they receive the internal cerebral veins, which return the blood from the convolutions of the flat surface of the corresponding hemisphere; near their termination, they become invested with a tubular sheath of the arachnoid membrane, and open into the superior longitudinal sinus, in the opposite direction to the course of the blood.

The *Inferior Anterior Cerebral Veins* commence on the under surface of the anterior lobes of the brain, and terminate in the cavernous sinuses.

The *Inferior Lateral Cerebral Veins* commence on the lateral parts of the hemispheres, and at the base of the brain: they unite to form from three to five veins, which open into the lateral sinus from before backwards.

The *Inferior Median Cerebral Veins*, which are very large, commence at the fore part of the under surface of the cerebrum, and from the convolutions of the posterior lobe, and terminate in the straight sinus behind the venæ Galeni.

The *Deep Cerebral*, or *Ventricular Veins* (venæ Galeni), are two in number, one from the right, the other from the left, ventricle. They are each formed by two veins, the vena corporis striati, and the choroid vein. They run backwards, parallel with one another, enclosed within the velum interpositum, and pass out of the brain at the great transverse fissure, between the under surface of the corpus callosum and the tubercula quadrigemina, to enter the straight sinus.

The *vena corporis striati* commences in the groove between the corpus striatum and thalamus opticus, receives numerous veins from both of these parts, and unites behind the anterior pillar of the fornix with the choroid vein, to form one of the venæ Galeni.

The *choroid vein* runs along the whole length of the outer border of the choroid plexus, receiving veins from the hippocampus major, the fornix and corpus callosum, and unites, at the anterior extremity of the choroid plexus, with the vein of the corpus striatum.

The *Cerebellar Veins* occupy the surface of the cerebellum, and are disposed in three sets, superior, inferior, and lateral. The superior pass forwards and inwards, across the superior vermiform process, and terminate in the straight sinus: some open into the venæ Galeni. The inferior cerebellar veins, of large size, run transversely outwards, and terminate by two or three trunks in the lateral sinuses. The lateral anterior cerebellar veins terminate in the superior petrosal sinuses.

Sinuses of the Dura Mater.

The sinuses of the dura mater are venous channels, analogous to the veins, their outer coat being formed by the dura mater; their inner, by a continuation of the serous membrane of the veins. They are fifteen in number, and are divided into two sets. 1. Those situated at the upper and back part of the skull. 2. Those at the base of the skull. The former are the

Superior longitudinal. Straight sinus.
Inferior longitudinal. Lateral sinuses.
Occipital sinuses.

The *Superior Longitudinal Sinus* occupies the attached margin of the falx cerebri. Commencing at the crista Galli, it runs from before backwards, groov-

238.—Vertical Section of the Skull, showing the Sinuses of the Dura Mater.

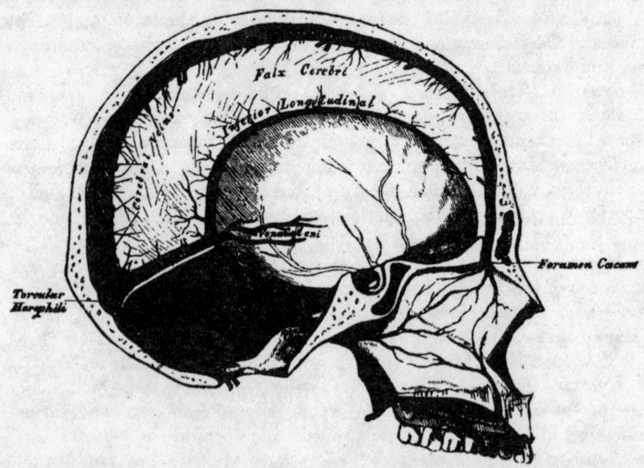

ing the inner surface of the frontal, the adjacent margins of the two parietal, and the superior division of the crucial ridge of the occipital bone, and terminates by dividing into the two lateral sinuses. This sinus is triangular in form, narrow in front, and gradually increasing in size as it passes backwards. On examining its inner surface, it presents the internal openings of the cerebral veins, which run, for the most part, from behind forwards, and open chiefly at the back part of the sinus, their orifices being concealed by fibrous areolæ; numerous fibrous bands (*chordæ Willisii*) are also seen, which extend transversely across the inferior angle of the sinus; and lastly, some small, white, projecting bodies, the glandulæ Pacchioni. This sinus receives the superior cerebral veins, numerous veins from the diploë and dura mater, and, at the posterior extremity of the sagittal suture, the parietal veins from the pericranium.

The point where the superior longitudinal and lateral sinuses are continuous is called the *confluence of the sinuses*, or the *torcular Herophili*. It presents a considerable dilatation, of very irregular form, and is the point of meeting of six sinuses, the superior longitudinal, the two lateral, the two occipital, and the straight.

The *Inferior Longitudinal Sinus*, more correctly described as the *inferior longitudinal vein*, is contained in the posterior part of the free margin of the falx cerebri. It is of a circular form, increases in size as it passes backwards, and terminates in the straight sinus. It receives several veins from the falx cerebri, and occasionally a few from the flat surface of the hemispheres.

The *Straight Sinus* is situated at the line of junction of the falx cerebri with the tentorium. It is triangular in form, increases in size as it proceeds backwards, and runs obliquely downwards and backwards from the termination of the inferior longitudinal sinus to the torcular Herophili. Besides the inferior longitudinal sinus, it receives the venæ Galeni, the inferior median cerebral veins, and the superior cerebellar. A few transverse bands cross its interior.

The *Lateral Sinuses* are of large size, and are situated in the attached margin of the tentorium cerebelli. They commence at the torcular Herophili, and passing horizontally outwards to the base of the petrous portion of the temporal bone, curve downwards and inwards on each side to reach the jugular foramen, where they terminate in the internal jugular vein. Each sinus rests, in its course, upon the inner surface of the occipital, the posterior inferior angle of the parietal, the mastoid portion of the temporal, and on the occipital again just before its termination. These sinuses are frequently of unequal size, and they increase in size as they proceed from behind forwards. The horizontal portion is of a triangular form, the curved portion semi-cylindrical; their inner surface is smooth, and not crossed by the fibrous bands found in the inner sinuses. These sinuses receive blood from the superior longitudinal, the straight, and the occipital sinuses; and in front they communicate with the superior and inferior petrosal. They communicate with the veins of the pericranium by means of the mastoid and posterior condyloid veins, and they receive the inferior cerebral and inferior cerebellar veins, and some veins from the diploë.

The *Occipital* are the smallest of the cranial sinuses. They are usually two in number, and situated in the attached margin of the falx cerebelli. They commence by several small veins around the posterior margin of the foramen magnum, which communicate with the posterior spinal veins, and terminate by separate openings (sometimes by a single aperture) in the torcular Herophili.

The sinuses at the base of the skull are the

Cavernous. Inferior petrosal.
Circular. Superior petrosal.
Transverse.

The *Cavernous Sinuses* are named from their presenting a reticulated structure. They are two in number, of large size, and placed one on each side of the sella Turcica, extending from the sphenoidal fissure to the apex of the petrous portion of the temporal bone: they receive anteriorly the ophthalmic vein through the sphenoidal fissure, and communicate behind with the petrosal sinuses, and with each other by the circular and transverse sinuses. On the inner wall of each sinus is found the internal carotid artery, accompanied by filaments of the carotid plexus and by the sixth nerve; and on its outer wall, the third, fourth, and ophthalmic nerves. These parts are separated from the blood flowing along the sinus by the lining membrane, which is continuous with the inner coat of the veins. The cavity of the sinus, which is larger behind than in front, is intersected by filaments of fibrous tissue and small vessels. The cavernous sinuses receive the inferior anterior cerebral veins; they communicate with the lateral sinuses by means of the superior and inferior petrosal, and with the facial vein through the ophthalmic.

The *ophthalmic* is a large vein, which connects the frontal vein at the inner angle of the orbit with the cavernous sinus; it pursues the same course as the ophthalmic artery, and receives branches corresponding to those derived from that vessel. Forming a short single trunk, it passes through the inner extremity of the sphenoidal fissure, and terminates in the cavernous sinus.

239.—The Sinuses at the Base of the Skull.

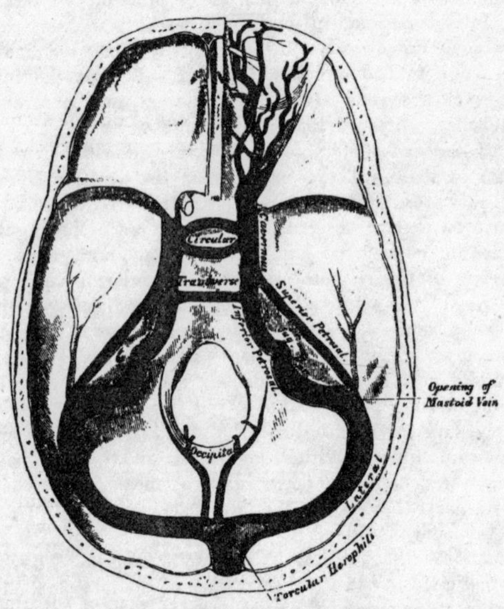

The *Circular Sinus* completely surrounds the pituitary body, and communicates on each side with the cavernous sinuses. Its posterior half is larger than the anterior; and in old age it is more capacious than at an early period of life. It receives veins from the pituitary body, and from the adjacent bone and dura mater.

The *Inferior Petrosal Sinus* is situated in the groove formed by the junction of the inferior border of the petrous portion of the temporal with the basilar process of the occipital. It commences in front at the termination of the cavernous sinus, and opens behind, into the jugular foramen, forming with the lateral sinus the commencement of the internal jugular vein. These sinuses are semi-cylindrical in form.

The *Transverse Sinus* is placed transversely across the fore part of the basilar process of the occipital bone serving to connect the two inferior petrosal and cavernous sinuses. A second is occasionally found opposite the foramen magnum.

The *Superior Petrosal Sinus* is situated along the upper border of the petrous portion of the temporal bone, in the front part of the attached margin of the tentorium. It is small and narrow, and connects together the cavernous and lateral sinuses at each side. It receives a cerebral vein (inferior lateral cerebral) from the under part of the middle lobe, and a cerebellar vein (anterior lateral cerebellar) from the anterior border of the cerebellum.

240.—The Superficial Veins of the Upper Extremity.

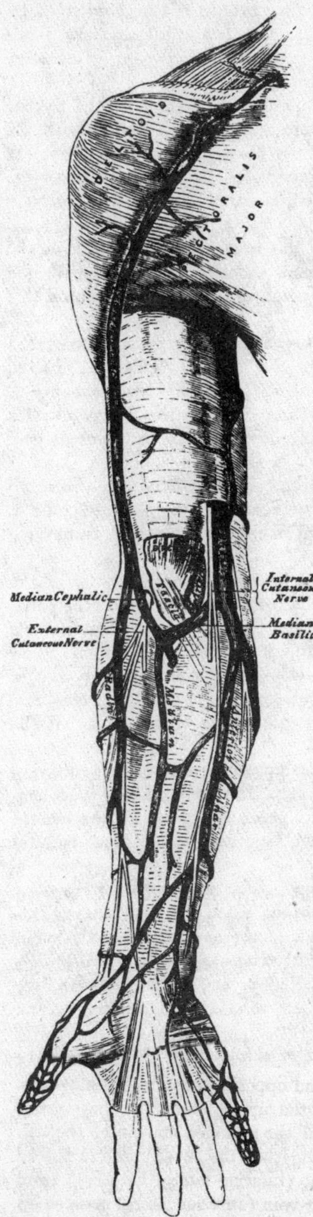

VEINS OF THE UPPER EXTREMITY.

The veins of the upper extremity are divided into two sets, superficial and deep.

The *Superficial Veins* are placed immediately beneath the integument between the two layers of superficial fascia; they commence in the hand chiefly on its dorsal aspect, where they form a more or less complete arch.

The *Deep Veins* accompany the arteries, and constitute the venæ comites of those vessels.

Both sets of vessels are provided with valves, which are more numerous in the deep than in the superficial.

The superficial veins of the upper extremity are the

Anterior ulnar.	Cephalic.
Posterior ulnar.	Median.
Basilic.	Median basilic.
Radial.	Median cephalic.

The *Anterior Ulnar Vein* commences on the anterior surface of the wrist and ulnar side of the hand, and ascends along the inner side of the fore-arm to the bend of the elbow, where it joins with the posterior ulnar vein to form the basilic. It communicates with branches of the median vein in front, and with the posterior ulnar behind.

The *Posterior Ulnar Vein* commences on the posterior surface of the ulnar side of the hand, and from the vein of the little finger (vena salvatella), situated over the fourth metacarpal space. It ascends on the posterior surface of the ulnar side of the fore-arm, and just below the elbow unites with the anterior ulnar vein to form the basilic.

The *Basilic* is a vein of considerable size, formed by the coalescence of the anterior and posterior ulnar veins; ascending along the inner side of the elbow, it receives the median basilic vein, and passing upwards along the inner side of the arm, pierces the deep fascia, and ascends in the course of the brachial artery, terminating either in one of the venæ comites of that vessel or in the axillary vein.

The *Radial Vein* commences from the dorsal surface of the thumb, index finger, and radial side of the hand, by branches communicating with the vena salvatella, and forming by their union a large vessel, which ascends along the radial side of the fore-arm, and receives numerous branches from both its surfaces. At the bend of the elbow it receives the median cephalic, when it becomes the cephalic vein.

The *Cephalic Vein* ascends along the outer border of the Biceps muscle, to the upper third of the arm ; it then passes in the interval between the Pectoralis major and Deltoid muscles, accompanied by the descending branch of the thoracica acromialis artery, and terminates in the axillary vein just below the clavicle. This vein is occasionally connected with the external jugular or subclavian, by a branch which passes from it upwards in front of the clavicle.

The *Median Vein* collects the blood from the superficial structures on the palmar surface of the hand and middle line of the fore-arm, communicating with the anterior ulnar and radial veins. At the bend of the elbow, it receives a branch of communication from the deep veins, accompanying the brachial artery, and divides into two branches, the median cephalic and median basilic, which diverge from each other as they ascend.

The *Median Cephalic*, usually the smaller of the two, passes outwards in the groove between the Supinator longus and Biceps muscles, and joins with the cephalic vein. The branches of the external cutaneous nerve pass behind this vessel.

The *Median Basilic* vein passes obliquely inwards, in the groove between the Biceps and Pronator radii teres, and joins with the basilic. This vein passes in front of the brachial artery, from which it is separated by a fibrous expansion, given off from the tendon of the Biceps to the fascia covering the Flexor muscles of the fore-arm. Filaments of the internal cutaneous nerve pass in front as well as behind this vessel.*

The *Deep Veins of the Upper Extremity* follow the course of the arteries, forming their venæ comites. They are generally two in number, one lying on each side of the corresponding artery, and they are connected at intervals by short transverse branches.

There are two digital veins, accompanying each artery along the sides of the fingers ; these, uniting at their base, pass along the interosseous spaces in the palm, and terminate in the two superficial palmar veins. Branches from these vessels on the radial side of the hand accompany the superficialis volæ, and on the ulnar side terminate in the deep ulnar veins. . The deep ulnar veins, as they pass in front of the wrist, communicate with the interosseous and superficial veins, and unite, at the elbow, with the deep radial veins, to form the venæ comites of the brachial artery.

The *Interosseous Veins* accompany the anterior and posterior interosseous arteries. The anterior interosseous veins commence in front of the wrist, where they communicate with the deep radial and ulnar veins; at the upper part of the fore-arm they receive the posterior interosseous veins, and terminate in the venæ comites of the ulnar artery.

The *Deep Palmar Veins* accompany the deep palmar arch, being formed by branches which accompany the ramifications of that vessel. They communicate with the superficial palmar veins at the inner side of the hand ; and on the outer side terminate in the venæ comites of the radial artery. At the wrist, they receive a dorsal and a palmar branch from the thumb, and unite with the deep radial veins. Accompanying the radial artery, these vessels terminate in the venæ comites of the brachial artery.

The *Brachial Veins* are placed one on each side of the brachial artery, receiving branches corresponding with those given off from that vessel ; at the lower margin of the axilla they unite with the basilic to form the axillary vein.

* Cruveilhier says: 'Numerous varieties are observed in the disposition of the veins of the elbow ; sometimes the common median vein is wanting ; but in those cases, its two branches of bifurcation are furnished by the radial vein, and the cephalic is almost always in a rudimentary condition. In other cases, only two veins are found at the bend of the elbow, the radial and ulnar, which are continuous, without any demarcation, with the cephalic and basilic.'

The deep veins have numerous anastomoses, not only with each other, but also with the superficial veins.

The Axillary Vein is of large size and formed by the continuation upwards of the basilic vein. It commences at the lower part of the axillary space, increases in size as it ascends, by receiving branches corresponding with those of the axillary artery, and terminates immediately beneath the clavicle at the outer margin of the first rib, where it becomes the subclavian vein. This vessel is covered in front by the Pectoral muscles and costo-coracoid membrane, and lies on the thoracic side of the axillary artery. Opposite the Subscapularis, it is joined by a large vein, formed by the junction of the venæ comites of the brachial; and near its termination it receives the cephalic vein. This vein is provided with a pair of valves, opposite the lower border of the Subscapularis muscle; valves are also found at the termination of the cephalic and subscapular veins.

The Subclavian Vein, the continuation of the axillary, extends from the outer margin of the first rib to the inner end of the sterno-clavicular articulation, where it unites with the internal jugular, to form the vena innominata. It is in relation, in front, with the clavicle and Subclavius muscle; behind, with the subclavian artery, from which it is separated internally by the Scalenus anticus and phrenic nerve. Below, it rests in a depression on the first rib and upon the pleura. Above, it is covered by the cervical fascia and integument.

The subclavian vein occasionally rises in the neck to a level with the third part of the subclavian artery, and in two instances has been seen passing with this vessel behind the Scalenus anticus. This vessel is provided with valves about an inch from its termination in the innominate, just external to the entrance of the external jugular vein.

Branches. It receives the external and anterior jugular veins and a small branch from the cephalic, outside the Scalenus; and on the inner side of that muscle, the internal jugular vein.

The Venæ Innominatæ (fig. 241) are two large trunks, placed one on each side of the root of the neck, and formed by the union of the internal jugular and subclavian veins of the corresponding side.

The *Right Vena Innominata* is a short vessel, about an inch and a half in length, which commences at the inner end of the clavicle, and, passing almost vertically downwards, joins with the left vena innominata just below the cartilage of the first rib, to form the superior vena cava. It lies superficial and external to the arteria innominata; on its right side the pleura is interposed between it and the apex of the lung. This vein, at its angle of junction with the subclavian, receives the right vertebral vein, and right lymphatic duct; and, lower down, the right internal mammary, right inferior thyroid, and right superior intercostal veins.

The *Left Vena Innominata*, about three inches in length, and larger than the right, passes obliquely from left to right across the upper and front part of the chest, to unite with its fellow of the opposite side, forming the superior vena cava. It is in relation, in front, with the sternal end of the clavicle, the sterno-clavicular articulation, and the first piece of the sternum, from which it is separated by the Sterno-hyoid and Sterno-thyroid muscles, the thymus gland or its remains, and some loose areolar tissue. Behind, it lies across the roots of the three large arteries arising from the arch of the aorta. This vessel is joined by the left vertebral, left inferior thyroid, left internal mammary, and the left superior intercostal veins, and occasionally some thymic and pericardiac veins. There are no valves in the venæ innominatæ.

Peculiarities. Sometimes the innominate veins open separately into the right auricle; in such cases the right vein takes the ordinary course of the superior vena cava, but the left vein, after communicating by a small branch with the right one, passes in front of the root of the left lung, and turning to the back of the heart, receives the cardiac veins, and terminates in the back of the right auricle. This occasional condition of the veins in the adult, is a regular one in the fœtus at an early period, and the two vessels are persistent in birds and some mammalia. The subsequent changes which take place in these vessels are the following: The communicating branch between the two trunks enlarges and forms the future left innominate vein; the remaining part of the left trunk is obliterated as far as the heart, where it remains pervious, and forms the coronary sinus; a remnant of the obliterated vessel is seen in adult life as a fibrous band passing along the back of the left auricle and in front of the root of the left lung, called by Mr. Marshall the vestigial fold of the pericardium.

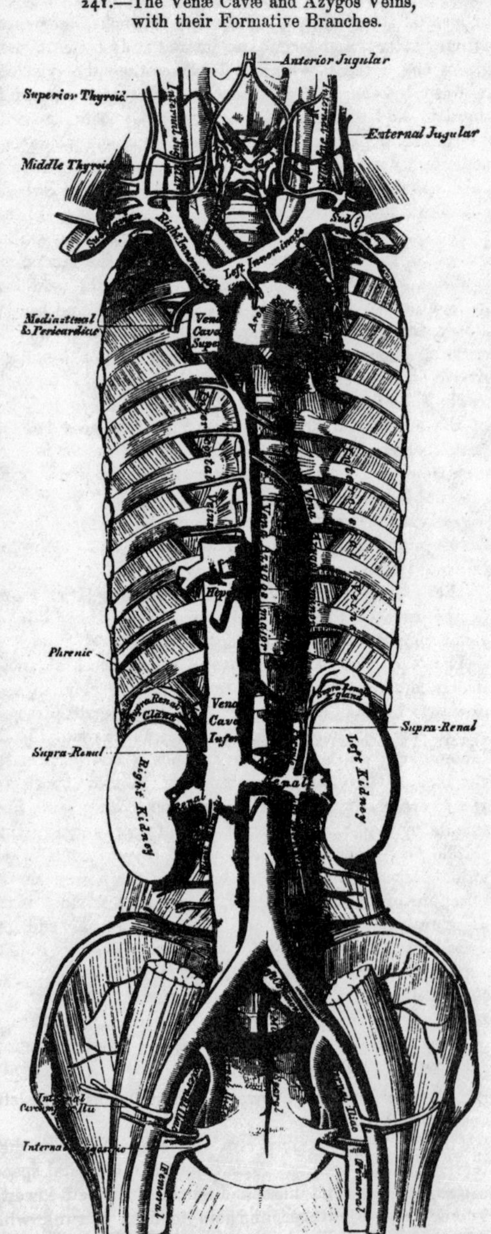

241.—The Venæ Cavæ and Azygos Veins, with their Formative Branches.

The *internal mammary veins*, two in number to each artery, follow the course of that vessel, and receive branches corresponding with those derived from it. The two veins unite into a single trunk, which terminates in the innominate vein.

The *inferior thyroid veins*, two, frequently three or four in number, arise in the venous plexus, on the thyroid body, communicating with the middle and superior thyroid veins. The left one descends in front of the trachea, behind the Sterno-thyroid muscles, communicating with its fellow by transverse branches, and terminates in the left vena innominata. The right one, which is placed a little to the right of the median line, opens into the right vena innominata, just at its junction with the superior cava. These veins receive tracheal and inferior laryngeal branches, and are provided with valves at their termination in the innominate veins.

The Superior Intercostal Veins return the blood from the upper intercostal spaces.

The *right superior intercostal*, much smaller than the left, closely corresponds with the superior intercostal artery, receiving the blood from the first, or first and second intercostal spaces, and terminates in the right vena innominata. Sometimes it passes down, and opens into the vena azygos major.

The *left superior intercostal* is always larger than the right, but varies in size in different subjects, being small when the left upper azygos vein is large, and *vice versâ*. It is usually formed by branches from the two or three upper intercostal spaces, and, passing across the arch of the aorta, terminates in the left vena innominata. The left bronchial vein opens into it.

The SUPERIOR VENA CAVA receives the blood which is conveyed to the heart from the whole of the upper half of the body. It is a short trunk, varying from two inches and a half to three inches in length, formed by the junction of the two venæ innominatæ. It commences immediately below the cartilage of the first rib on the right side, and, descending vertically, enters the pericardium about an inch and a half above the heart, and terminates in the upper part of the right auricle. In its course, it describes a slight curve, the convexity of which is turned to the right side.

Relations. In front, with the thoracic fascia, which separates it from the thymus gland, and from the sternum; *behind*, with the root of the right lung. On its *right side*, with the phrenic nerve and right pleura; on its *left side*, with the ascending part of the aorta. The portion contained within the pericardium is covered by the serous layer of that membrane, in its anterior three-fourths. It receives the vena azygos major, just before it enters the pericardium, and several small veins from the pericardium and parts in the mediastinum. The superior vena cava has no valves.

The AZYGOS VEINS connect together the superior and inferior venæ cavæ, supplying the place of those vessels in the part of the chest which is occupied by the heart.

The larger, or *right azygos vein*, commences opposite the first or second lumbar vertebra, by a branch from the right lumbar veins; sometimes by a branch from the renal vein, or from the inferior vena cava. It enters the thorax through the aortic opening in the Diaphragm, and passes along the right side of the vertebral column to the third dorsal vertebra, where it arches forward, over the root of the right lung, and terminates in the superior vena cava, just before that vessel enters the pericardium. Whilst passing through the aortic opening of the Diaphragm, it lies with the thoracic duct on the right side of the aorta; and in the thorax, it lies upon the intercostal arteries, on the right side of the aorta and thoracic duct, covered by the pleura.

Branches. It receives nine or ten lower intercostal veins of the right side, the vena azygos minor, several œsophageal, mediastinal, and vertebral veins; near its termination, the right bronchial vein; and is occasionally connected with the right superior intercostal vein. A few imperfect valves are found in this vein; but its branches are provided with complete valves.

The intercostal veins on the left side, below the two or three upper intercostal spaces, usually form two trunks, named the left lower, and the left upper, azygos veins.

The *left lower*, or *smaller azygos vein*, commences in the lumbar region, by a branch from one of the lumbar veins, or from the left renal. It passes into the thorax, through the left crus of the Diaphragm, and ascending on the left side of the spine, as high as the sixth or seventh dorsal vertebra, passes across the column, behind the aorta and thoracic duct, to terminate in the right azygos vein. It receives the four or five lower intercostal veins of the left side, and some œsophageal and mediastinal veins.

The *left upper azygos vein* varies according to the size of the left superior intercostal. It receives veins from the intercostal spaces between the left superior intercostal vein, and highest branch of the left lower azygos. They are usually two or three in number, and join to form a trunk which ends in the right azygos vein, or in the left lower azygos. When this vein is small, or altogether wanting, the left superior intercostal vein will extend as low as the fifth or sixth intercostal space.

The *bronchial veins* return the blood from the substance of the lungs; that of the right side opens into the vena azygos major, near its termination; that of the left side, into the left superior intercostal vein.

The Spinal Veins.

The numerous venous plexuses placed upon and within the spine may be arranged into four sets.

1. Those placed on the exterior of the spinal column (the dorsi-spinal veins).
2. Those situated in the interior of the spinal canal, between the vertebræ and the theca vertebralis (meningo-rachidian veins).
3. The veins of the bodies of the vertebræ.
4. The veins of the spinal cord (medulli-spinal).

1. The *Dorsi-Spinal Veins* commence by small branches, which receive their blood from the integument of the back of the spine, and from the muscles in the vertebral grooves. They form a complicated network, which surrounds the spinous processes, the laminæ, and the transverse and articular processes of all the vertebræ. At the bases of the transverse processes, they communicate, by means of ascending and descending branches, with the veins surrounding the contiguous vertebræ, and they join with the veins in the spinal canal by branches which perforate the ligamenta subflava; they terminate in the intervals between the arches of the vertebræ, by joining the vertebral veins in the neck, the intercostal veins in the thorax, and the lumbar and sacral veins in the loins and pelvis.

2. The principal veins contained in the spinal canal are situated between the theca vertebralis and the vertebræ. They consist of two longitudinal plexuses, one of which runs along the posterior surface of the bodies of the vertebræ throughout the entire length of the spinal canal (anterior longitudinal spinal veins), receiving the veins belonging to the bodies of the vertebræ (venæ basis vertebrarum). The other plexus (posterior longitudinal spinal veins) is placed on the inner, or anterior surface of the laminæ of the vertebræ, and extends also along the entire length of the spinal canal.

The *Anterior Longitudinal Spinal Veins* consist of two large, tortuous, venous canals, which extend along the whole length of the vertebral column, from the foramen magnum to the base of the coccyx, being placed one on each side of the posterior surface of the bodies of the vertebræ, external to the posterior common ligament. These veins communicate together opposite each vertebra, by transverse trunks, which pass beneath the ligament, and receive the large venæ basis vertebrarum, from the interior of the body of each vertebra. The anterior longitudinal spinal veins are least developed in the cervical and sacral regions. They are not of uniform size throughout, being alternately enlarged and constricted. At the intervertebral foramina, they communicate with the dorsi-spinal veins, and with the vertebral veins in the neck, with the intercostal veins in the dorsal region, and with the lumbar and sacral veins in the corresponding regions.

The *Posterior Longitudinal Spinal Veins*, smaller than the anterior, are situated one on either side, between the inner surface of the laminæ and the theca vertebralis. They communicate (like the anterior), opposite each vertebra, by transverse trunks; and with the anterior longitudinal veins, by lateral transverse branches, which pass from behind forwards. These veins, at the intervertebral foramina, join with the dorsi-spinal veins.

3. The *Veins of the Bodies of the Vertebræ* (venæ basis vertebrarum) emerge from the foramina on their posterior surface, and join the transverse trunk connecting the anterior longitudinal spinal veins. They are contained in large tortuous channels, in the substance of the bones, similar in every respect to those found in the diploë of the cranial bones. These canals lie parallel to the upper and lower surface of the bones, arise from the entire circumference of the vertebra, communicate with veins which enter through the foramina, on the anterior surface

of the bodies, and converge to the principal canal, which is sometimes double towards its posterior part. They become greatly developed in advanced age.

4. The *Veins of the Spinal Cord* (medulli-spinal) consist of a minute tortuous venous plexus, which covers the entire surface of the cord, being situated between

242.—Transverse Section of a Dorsal Vertebra, showing the Spinal Veins.

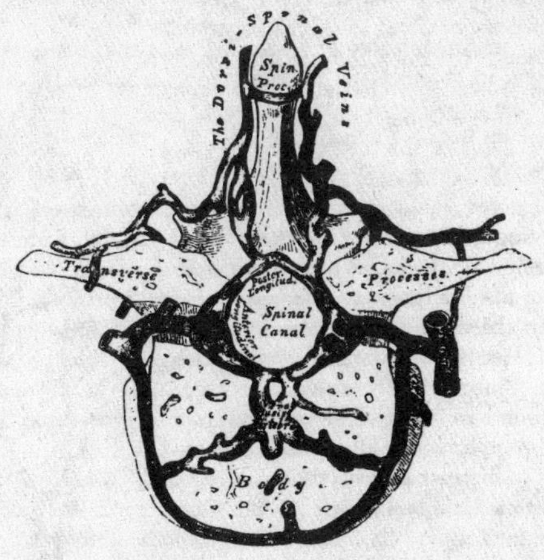

the pia mater and arachnoid. These vessels emerge chiefly from the posterior median furrow, and are largest in the lumbar region. Near the base of the skull they unite, and form two or three small trunks, which communicate with the ver-

243.—Vertical Section of two Dorsal Vertebræ, showing the Spinal Veins.

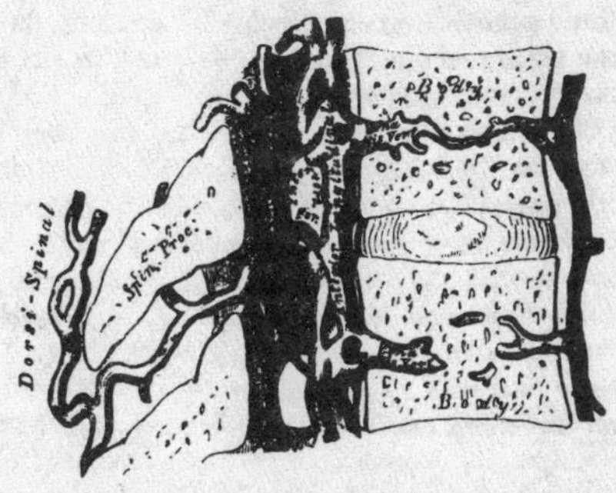

tebral veins, and then terminate in the inferior cerebellar veins, or in the petrosal sinuses. Each of the spinal nerves is accompanied by a branch as far as the intervertebral foramina, where they join the other veins from the spinal canal.

There are no valves in the spinal veins.

VEINS OF THE LOWER EXTREMITY.

The veins of the lower extremity are subdivided, like those of the upper, into two sets, superficial and deep; the superficial veins being placed beneath the integument, between the two layers of superficial fascia; the deep veins accompanying the arteries, and forming the venæ comites of those vessels. Both sets of veins are provided with valves, which are more numerous in the deep than in the superficial set. These valves are also more numerous in the lower than in the upper limb.

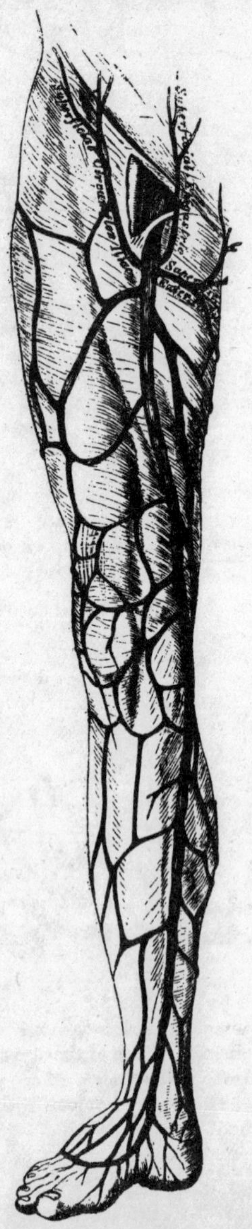

244.—The internal or Long Saphenous Vein and its Branches.

The *Superficial Veins* of the lower extremity are the internal or long saphenous, and the external or short saphenous.

The *internal saphenous vein* (fig. 244) commences from a minute plexus, which covers the dorsum and inner side of the foot; it ascends in front of the inner ankle, and along the inner side of the leg, behind the inner margin of the tibia, accompanied by the internal saphenous nerve. At the knee, it passes backwards behind the inner condyle of the femur, ascends along the inside of the thigh, and, passing through the saphenous opening in the fascia lata, terminates in the femoral vein about an inch and a half below Poupart's ligament. This vein receives in its course cutaneous branches from the leg and thigh, and at the saphenous opening, the superficial epigastric, superficial circumflex iliac, and external pudic veins. The veins from the inner and back part of the thigh frequently unite to form a large vessel, which enters the main trunk near the saphenous opening; and sometimes those on the outer side of the thigh join to form a large branch; so that occasionally three large veins are seen converging from different parts of the thigh towards the saphenous opening. The internal saphenous vein communicates in the foot with the internal plantar vein; in the leg, with the posterior tibial veins, by branches which perforate the tibial origin of the Soleus muscle, and also with the anterior tibial veins; at the knee, with the articular veins; in the thigh, with the femoral vein by one or more branches. The valves in this vein vary from two to six in number; they are more numerous in the thigh than in the leg.

The *external* or *short saphenous vein* (fig. 245) is formed by branches which collect the blood from the dorsum and outer side of the foot; it ascends behind the outer ankle, and along the outer border of the tendo Achillis, across which it passes at an acute angle to reach the middle line of the posterior aspect of the leg. Passing directly upwards, it perforates the deep fascia in the lower part of the popliteal space, and terminates in the popliteal vein, between the heads of the Gastrocnemius muscle. It is accompanied by the external saphenous nerve. It receives numerous large branches from the back part of the leg, and communicates with the deep veins on the dorsum of the foot, and behind the outer malleolus. This vein has only two valves, one of which is always found near its termination in the popliteal vein.

245.—External or Short Saphenous Vein.

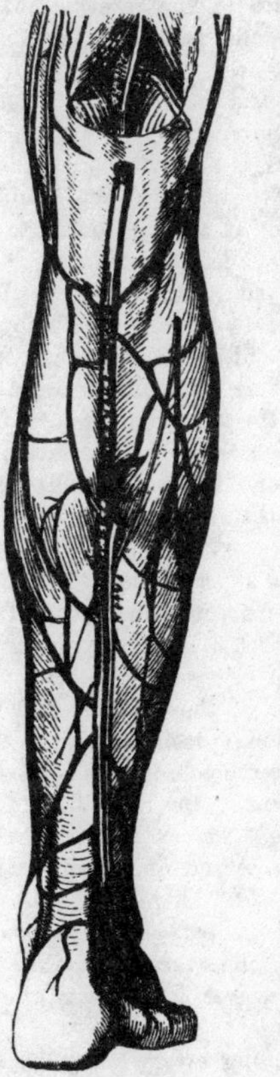

The *Deep Veins* of the lower extremity accompany the arteries and their branches, and are called the *venæ comites* of those vessels.

The external and internal plantar veins unite to form the posterior tibial. They accompany the posterior tibial artery, and are joined by the peroneal veins.

The *anterior tibial veins* are formed by a continuation upwards of the venæ comites of the dorsalis pedis artery. They perforate the interosseous membrane at the upper part of the leg, and form, by their junction with the posterior tibial, the popliteal vein.

The valves in the deep veins are very numerous.

The Popliteal Vein is formed by the junction of the venæ comites of the anterior and posterior tibial vessels; it ascends through the popliteal space to the tendinous aperture in the Adductor magnus, where it becomes the femoral vein. In the lower part of its course, it is placed internal to the artery; between the heads of the Gastrocnemius, it is superficial to that vessel; but above the knee-joint, it is close to its outer side. It receives the sural veins from the Gastrocnemius muscle, the articular veins, and the external saphenous. The valves in this vein are usually four in number.

The Femoral Vein accompanies the femoral artery through the upper two-thirds of the thigh. In the lower part of its course, it lies external to the artery; higher up, it is behind it; and beneath Poupart's ligament, it lies to its inner side, and on the same plane. It receives numerous muscular branches; the profunda femoris joins it about an inch and a half below Poupart's ligament, and near its termination the internal saphenous vein. The valves in this vein are four or five in number.

The External Iliac Vein commences at the termination of the femoral, beneath the crural arch, and passing upwards along the brim of the pelvis, terminates opposite the sacro-iliac symphysis, by uniting with the internal iliac to form the common iliac vein. On the right side, it lies at first along the inner side of the external iliac artery; but as it passes upwards, gradually inclines behind it. On the left side, it lies altogether on the inner side of the artery. It receives, immediately above Poupart's ligament, the epigastric and circumflex iliac veins. It has no valves.

The Internal Iliac Vein is formed by the venæ comites of the branches of the internal iliac artery, the umbilical arteries excepted. It receives the blood from the exterior of the pelvis by the gluteal, sciatic, internal pudic, and obturator veins; and from the organs in the cavity of the pelvis by the hæmorrhoidal and vesico-prostatic plexuses in the male, and the uterine and vaginal plexuses in the female. The vessels forming these plexuses are remarkable for their large size, their frequent anastomoses, and the number of valves which they contain. The internal iliac vein lies at first on the inner side and then behind the internal iliac artery, and terminates opposite the sacro-iliac articulation, by uniting with the external iliac, to form the common iliac vein. This vessel has no valves.

The *hæmorrhoidal plexus* surrounds the lower end of the rectum, being formed by the superior hæmorrhoidal veins, branches of the inferior mesenteric, and the middle and inferior hæmorrhoidal, which terminate in the internal iliac. The portal and general venous systems have a free communication by means of the branches composing this plexus.

The *vesico-prostatic plexus* surrounds the neck and base of the bladder and prostate gland. It communicates with the hæmorrhoidal plexus behind, and receives the dorsal vein of the penis, which enters the pelvis beneath the sub-pubic ligament. This plexus is supported upon the sides of the bladder by a reflection of the pelvic fascia. The veins composing it are very liable to become varicose, and often contain hard earthy concretions, called *phlebolithes*.

The *dorsal vein of the penis* is a vessel of large size, which returns the blood from the body of that organ. At first it consists of two branches, which are contained in the groove on the dorsum of the penis, and it receives veins from the glans, the corpus spongiosum, and numerous superficial veins; these unite near the root of the penis into a single trunk, which pierces the triangular ligament beneath the pubic arch, and divides into two branches, which enter the prostatic plexus.

The *vaginal plexus* surrounds the mucous membrane, being especially developed at the orifice of the vagina; it communicates with the vesical plexus in front, and with the hæmorrhoidal plexus behind.

The *uterine plexus* is situated along the sides and superior angles of the uterus, receiving large venous canals (the uterine sinuses) from its substance. The veins composing this plexus anastomose frequently with each other and with the ovarian veins. They are not tortuous like the arteries.

The Common Iliac Veins are formed by the union of the external and internal iliac veins in front of the sacro-vertebral articulation; passing obliquely upwards towards the right side, they terminate upon the intervertebral substance between the fourth and fifth lumbar vertebræ, where the veins of the two sides unite at an acute angle to form the inferior vena cava. The *right common iliac* is shorter than the left, nearly vertical in its direction, and ascends behind and then to the outer side of its corresponding artery. The *left common iliac*, longer and more oblique in its course, is at first situated on the inner side of the corresponding artery, and then behind the right common iliac. Each common iliac receives the ilio-lumbar, and sometimes the lateral sacral veins. The left receives, in addition, the middle sacral vein. No valves are found in these veins.

The *middle sacral vein* accompanies its corresponding artery along the front of the sacrum, and terminates in the left common iliac vein; occasionally in the commencement of the inferior vena cava.

Peculiarities. The left common iliac vein, instead of joining with the right in its usual position, occasionally ascends on the left side of the aorta as high as the kidney, where, after receiving the left renal vein, it crosses over the aorta, and then joins with the right vein to form the vena cava. In these cases, the two common iliacs are connected by a small communicating branch at the spot where they are usually united.

The Inferior Vena Cava returns to the heart the blood from all the parts below the Diaphragm. It is formed by the junction of the two common iliac veins on the right side of the intervertebral substance between the fourth and fifth lumbar vertebræ. It passes upwards along the front of the spine, on the right side of the aorta, and having reached the under surface of the liver, is contained in a groove in its posterior border. It then perforates the tendinous centre of the Diaphragm, enters the pericardium, where it is covered by its serous layer, and terminates in the lower and back part of the right auricle. At its termination in the auricle, it is provided with a valve, the Eustachian, which is of large size during fœtal life.

Relations. In front, from below upwards, with the mesentery, transverse portion of the duodenum, the pancreas, portal vein, and the posterior border of the liver, which partly and occasionally completely surrounds it; *behind*, with the vertebral column, the right crus of the Diaphragm, the right renal and lumbar arteries; on the *left side*, with the aorta. It receives in its course the following branches:

Lumbar.	Suprarenal.
Right spermatic.	Phrenic.
Renal.	Hepatic.

Peculiarities. In Position. This vessel is sometimes placed on the left side of the aorta, as high as the left renal vein, after receiving which, it crosses over to its usual position on the right side; or it may be placed altogether on the left side of the aorta, as far upwards as its termination in the heart: in such cases, the abdominal and thoracic viscera, together with the great vessels, are all transposed.

Point of Termination. Occasionally the inferior vena cava joins the right azygos vein, which is then of large size. In such cases, the superior cava receives the whole of the blood from the body before transmitting it to the right auricle, except the blood from the hepatic veins, which terminate directly in the right auricle.

The *lumbar veins*, three or four in number on each side, collect the blood by dorsal branches from the muscles and integument of the loins, and by abdominal branches from the walls of the abdomen, where they communicate with the epigastric veins. At the spine, they receive branches from the spinal plexuses, and then pass forwards round the sides of the bodies of the vertebræ beneath the Psoas magnus, and terminate at the back part of the inferior cava. The left lumbar veins are longer than the right, and pass behind the aorta. The lumbar veins communicate with each other by branches which pass in front of the transverse processes. Occasionally two or more of these veins unite to form a single trunk, the ascending lumbar, which serves to connect the common iliac, ilio-lumbar, lumbar, and azygos veins of the corresponding side of the body.

The *spermatic veins* emerge from the back of the testis, and receive branches from the epididymis; they form a branched and convoluted plexus, called the *spermatic plexus* (plexus pampiniformis), below the abdominal ring: the vessels composing this plexus are very numerous, and ascend along the cord in front of the vas deferens; having entered the abdomen, they coalesce to form two branches, which ascend on the Psoas muscle, behind the peritoneum, lying one on each side of the spermatic artery, and unite to form a single vessel, which opens on the right side in the inferior vena cava, at an acute angle, on the left side in the left renal vein, at a right angle. The spermatic veins are provided with valves. The left spermatic vein passes behind the sigmoid flexus of the colon, a part of the intestine in which fœcal accumulation is common; this circumstance, as well as the indirect communication of the vessel with the inferior vena cava, may serve to explain the more frequent occurrence of varicocele on the left side.

The *ovarian veins* are analogous to the spermatic in the male; they form a plexus near the ovary, and in the broad ligament and Fallopian tube, communicating with the uterine plexus. They terminate as in the male. Valves are occasionally found in these veins. These vessels, like the uterine veins, become much enlarged during pregnancy.

The *renal veins* are of large size, and placed in front of the renal arteries.* The left is longer than the right, and passes in front of the aorta, just below the origin of the superior mesenteric artery. It receives the left spermatic and left inferior phrenic veins. It usually opens into the vena cava, a little higher than the right.

The *suprarenal vein* terminates, on the right side, in the vena cava; on the left side, in the left renal or phrenic vein.

The *phrenic veins* follow the course of the phrenic arteries. The *two superior*,

* The student may observe that all veins above the diaphragm, which do not lie on the same plane as the arteries which they accompany, lie in front of them; and that all veins below the diaphragm, which do not lie on the same plane as the arteries which they accompany, lie behind them, except the renal and profunda femoris vein.

of small size, accompany the phrenic nerve and comes nervi phrenici artery; the right terminating opposite the junction of the two venæ innominatæ, the left in the left superior intercostal or left internal mammary. The *two inferior phrenic veins* follow the course of the phrenic arteries, and terminate, the right in the inferior vena cava, the left in the left renal vein.

The *hepatic veins* commence in the substance of the liver, in the capillary terminations of the vena portæ: these branches, gradually uniting, form three large veins, which converge towards the posterior border of the liver, and open into the inferior vena cava, whilst that vessel is situated in the groove at the back part of this organ. Of these three veins, one from the right, and another from the left lobe, open obliquely into the vena cava; that from the middle of the organ and lobulus Spigelii having a straight course. The hepatic veins run singly, and are in direct contact with the hepatic tissue. They are destitute of valves.

Portal System of Veins.

The portal venous system is composed of four large veins, which collect the venous blood from the viscera of digestion. The trunk formed by their union (vena portæ) enters the liver and ramifies throughout its substance; and its branches again emerging from that organ as the hepatic veins, terminate in the inferior vena cava. The branches of this vein are in all cases single, and destitute of valves.

The veins forming the portal system are, the

Inferior mesenteric.	Splenic.
Superior mesenteric.	Gastric.

The *inferior mesenteric vein* returns the blood from the rectum, sigmoid flexure, and descending colon, corresponding with the ramifications of the branches of the inferior mesenteric artery. Ascending beneath the peritoneum in the lumbar region, it passes behind the transverse portion of the duodenum and pancreas, and terminates in the splenic vein. Its hæmorrhoidal branches inosculate with those of the internal iliac, and thus establish a communication between the portal and the general venous system.*

The *superior mesenteric vein* returns the blood from the small intestines, and from the cæcum and ascending and transverse portions of the colon, corresponding with the distribution of the branches of the superior mesenteric artery. The large trunk formed by the union of these branches ascends along the right side and in front of the corresponding artery, passes in front of the transverse portion of the duodenum, and unites behind the upper border of the pancreas with the splenic vein, to form the vena portæ.

The *splenic vein* commences by five or six large branches, which return the blood from the substance of the spleen. These uniting form a single vessel, which passes from left to right behind the upper border of the pancreas, and terminates at its greater end by uniting at a right angle with the superior mesenteric to form the vena portæ. The splenic vein is of large size, and not tortuous like the artery. It receives the vasa brevia from the left extremity of the stomach, the left gastro-epiploic vein, pancreatic branches from the pancreas, the pancreatico-duodenal vein, and the inferior mesenteric vein.

The *gastric* is a vein of small size, which accompanies the gastric artery from left to right along the lesser curvature of the stomach, and terminates in the vena portæ.

The *Portal Vein* is formed by the junction of the superior mesenteric and

* Besides this anastomosis between the portal vein and the branches of the vena cava, other anastomoses between the portal and systemic veins are formed by the communication between the left renal vein and the veins of the intestines, especially of the colon and duodenum, and between superficial branches of the portal veins of the liver and the phrenic veins, as pointed out by Mr. Kiernan (Todd and Bowman).

splenic veins, their union taking place in front of the vena cava, and behind the upper border of the great end of the pancreas. Passing upwards through the right border of the lesser omentum to the under surface of the liver, it enters the transverse fissure, where it is somewhat enlarged, forming the sinus of the portal vein, and divides into two branches, which accompany the ramifications of the hepatic artery and hepatic duct throughout the substance of the liver. Of these two branches the right is the larger but the shorter of the two. The portal vein is about four inches in length, and, whilst contained in the lesser omentum, lies behind and between the hepatic duct and artery, the former being to the right, the latter to the left. These structures are accompanied by filaments of the hepatic plexus of nerves, and numerous lymphatics, surrounded by a quantity of loose

246.—Portal Vein and its Branches.

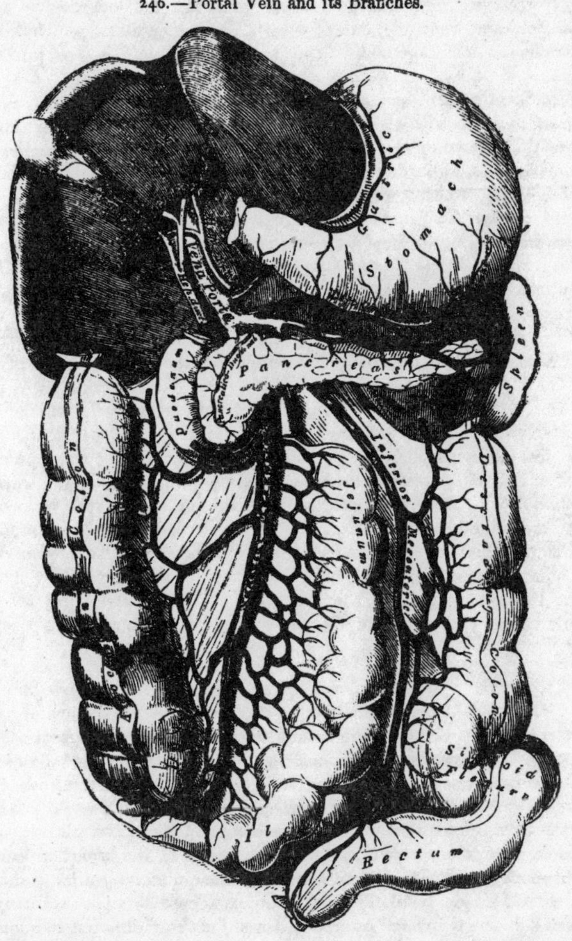

areolar tissue (capsule of Glisson), and placed between the layers of the lesser omentum. The vena portæ receives the gastric and cystic veins; the latter vein sometimes terminates in the right branch of the vena portæ. Within the liver the portal vein receives the blood from the branches of the hepatic artery.

Cardiac Veins.

The veins which return the blood from the substance of the heart are, the

Great cardiac vein.
Posterior cardiac vein.
Anterior cardiac veins.
Venæ Thebesii.

The *Great Cardiac Vein* is a vessel of considerable size, which commences at the apex of the heart, and ascends along the anterior interventricular groove to the base of the ventricles. It then curves to the left side, around the auriculo-ventricular groove, between the left auricle and ventricle, to the back part of the heart, and opens into the coronary sinus, its aperture being guarded by two valves. It receives the posterior cardiac vein, and the left cardiac veins from the left auricle and ventricle, one of which, ascending along the left margin of the ventricle, is of large size. The branches joining it are provided with valves.

The *Posterior Cardiac Vein* commences, by small branches, at the apex of the heart, communicating with those of the preceding. It ascends along the posterior interventricular groove to the base of the heart, and terminates in the coronary sinus, its orifice being guarded by a valve. It receives the veins from the posterior surface of both ventricles.

The *Anterior Cardiac Veins* are three or four small branches, which collect the blood from the anterior surface of the right ventricle. One of these (the vein of Galen), larger than the rest, runs along the right border of the heart. They open separately into the lower part of the right auricle.

The *Venæ Thebesii* are numerous minute veins, which return the blood directly from the muscular substance, without entering the venous current. They open, by minute orifices (*foramina Thebesii*), on the inner surface of the right auricle.

The Coronary Sinus is that portion of the great cardiac vein which is situated in the posterior part of the left auriculo-ventricular groove. It is about an inch in length, presents a considerable dilatation, and is covered by the muscular fibres of the left auricle. It receives the great cardiac vein, the posterior cardiac vein, and an oblique vein from the back part of the left auricle, the remnant of the obliterated left innominate trunk of the fœtus, described by Mr. Marshall. The coronary sinus terminates in the right auricle, between the inferior vena cava and the auriculo-ventricular aperture, its orifice being guarded by a semilunar fold of the lining membrane of the heart, the coronary valve. All the branches joining this vessel, excepting the oblique vein above mentioned, are provided with valves.

The Pulmonary Veins.

The *Pulmonary Veins* return the arterial blood from the lungs to the left auricle of the heart. They are four in number, two for each lung. The pulmonary differ from other veins in several respects. 1. They carry arterial, instead of venous blood. 2. They are destitute of valves. 3. They are only slightly larger than the arteries they accompany. 4. They accompany those vessels singly. They commence in a capillary network, upon the parietes of the bronchial cells, where they are continuous with the ramifications of the pulmonary artery, and, uniting together, form a single trunk for each lobule. These branches, uniting successively, form a single trunk for each lobe, three for the right, and two for the left lung. The vein from the middle lobe of the right lung unites with that from the upper lobe, in most cases, forming two trunks on each side, which open separately into the left auricle. Occasionally they remain separate; there are then three veins on the right side. Not unfrequently, the two left pulmonary veins terminate by a common opening.

Within the lung, the branches of the pulmonary artery are *in front*, the veins *behind*, and the bronchi between the two.

At the root of the lung, the veins are *in front*, the artery *in the middle*, and the bronchus *behind.*

Within the pericardium, their anterior surface is invested by the serous layer of this membrane. The right pulmonary veins pass behind the right auricle and ascending aorta; the left pass in front of the thoracic aorta, with the left pulmonary artery.

Of the Lymphatics.

THE Lymphatics have derived their name from the appearance of the fluid contained in their interior (*lympha*, water). They are also called *absorbents*, from the property they possess of absorbing certain materials from the tissues, and conveying them into the circulation.

The lymphatic system includes not only the lymphatic vessels and the glands through which they pass, but also the *lacteal*, or *chyliferous* vessels. The lacteals are the lymphatic vessels of the small intestine, and differ in no respect from the lymphatics generally, excepting that they contain a milk-white fluid, the chyle, during the process of digestion, and convey it into the blood through the thoracic duct.

The lymphatics are exceedingly delicate vessels, the coats of which are so transparent, that the fluid they contain is readily seen through them. They retain a nearly uniform size, being interrupted at intervals by constrictions, which give them a knotted or beaded appearance. These constrictions are due to the presence of valves in their interior. Lymphatics have been found in nearly every texture and organ of the body, with the exception of the substance of the brain and spinal cord, the eyeball, cartilage, tendon, the membranes of the ovum, the placenta, and umbilical cord, the nails, cuticle, and hair. Their existence in the substance of bone is doubtful.

The lymphatics are arranged into a superficial and deep set. The superficial lymphatics, on the surface of the body, are placed immediately beneath the integument, accompanying the superficial veins; they join the deep lymphatics in certain situations by perforating the deep fascia. In the interior of the body, they lie in the submucous areolar tissue, throughout the whole length of the gastro-pulmonary and genito-urinary tracts; and in the subserous areolar tissue in the cranial, thoracic, and abdominal cavities. These vessels probably arise in the form of a dense plexiform network interspersed among the proper elements and blood-vessels of the several tissues; the vessels composing which, as well as the meshes between them, are much larger than those of the capillary plexus. From these networks small vessels emerge, which pass, either to a neighbouring gland, or to join some larger lymphatic trunk. The deep lymphatics, fewer in number, and larger than the superficia l,accompany the deep blood-vessels. Their mode of origin is not known; it is, however, probably, similar to that of the superficial vessels. The lymphatics of any part or organ exceed, in number, the veins; but in size they are much smaller. Their anastomoses also, especially those of the large trunks, are more frequent, and are effected by vessels equal in diameter to those which they connect, the continuous trunks retaining the same diameter.

The lymphatic or absorbent glands, named also *conglobate glands*, are small solid glandular bodies, situated in the course of the lymphatic and lacteal vessels. They are found in the neck and on the external parts of the head; in the upper extremity, in the axilla and front of the elbow; in the lower extremity, in the groin and popliteal space. In the abdomen, they are found in large numbers in the mesentery, and along the side of the aorta, vena cava, and iliac vessels; and in the thorax, in the anterior and posterior mediastina. They are somewhat flattened, and of a round or oval form. In size, they vary from a hemp-seed to an almond, and their colour, on section, is of a pinkish grey tint, excepting the bronchial glands, which in the adult are mottled with black. Each gland has a layer, or capsule, of cellular tissue investing it, from which prolongations dip into its substance forming partitions. The lymphatic and lacteal vessels pass through these bodies in their passage to the thoracic and lymphatic ducts. A lymphatic or lacteal vessel, previous to entering a gland, divides into several small branches, which are named *afferent vessels*. As they enter, their external coat becomes continuous with the capsule of the gland, and the vessels, much thinned, and consisting only of their internal coat and epithelium, pass into the gland, where they subdivide and pursue a tortuous course; and they finally anastomose, so as to form a plexus. The vessels composing this plexus unite to form two or more *efferent* vessels, which, on emerging from the gland, are again invested with their external coat. Further details on the minute anatomy of the lymphatic vessels and glands will be found in the Introduction.

Thoracic Duct.

The thoracic duct (fig. 247) conveys the great mass of the lymph and chyle into the blood. It is the common trunk of all the lymphatic vessels of the body, excepting those of the right side of the head, neck, and thorax, and right upper extremity, the right lung, right side of the heart, and the convex surface of the liver. It varies from eighteen to twenty inches in length in the adult, and extends from the second lumbar vertebra to the root of the neck. It commences in the abdomen by a triangular dilatation, the receptaculum chyli (reservoir or cistern of Pecquet), which is situated upon the front of the body of the second lumbar vertebra, to the right side and behind the aorta, by the side of the right crus of the Diaphragm. It ascends into the thorax through the aortic opening in the Diaphragm, and is placed in the posterior mediastinum in front of the vertebral column, lying between the aorta and vena azygos. Opposite the fourth dorsal vertebra, it inclines towards the left side and ascends behind the arch of the aorta, on the left side of the œsophagus, and behind the first portion of the left subclavian artery, to the upper orifice of the thorax. Opposite the upper border of the seventh cervical vertebra, it curves downwards above the subclavian artery, and in front of the Scalenus muscle, so as to form an arch; and terminates near the angle of junction of the left internal jugular and subclavian veins. The thoracic duct, at its commencement, is about equal in size to the diameter of a goose-

247.—The Thoracic and Right Lymphatic Duct.

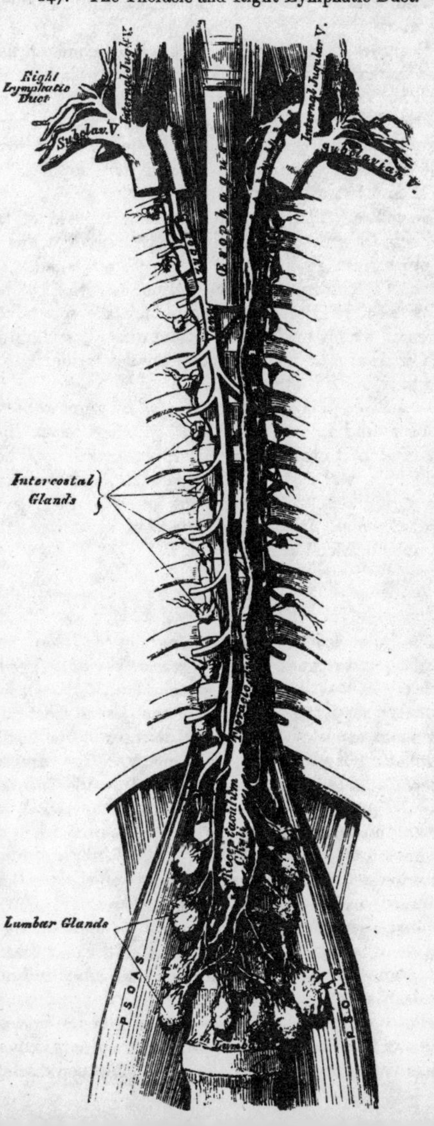

quill, diminishes considerably in its calibre in the middle of the thorax, and is again dilated just before its termination. It is generally flexuous in its course, and constricted at intervals so as to present a varicose appearance. The thoracic duct not unfrequently divides in the middle of its course into two branches of unequal size, which soon re-unite, or into several branches which form a plexiform interlacement. It occasionally bifurcates, at its upper part, into two branches, of which the one on the left side terminates in the usual manner, while that on the right opens into the right subclavian vein, in connection with the right lymphatic duct. The thoracic duct has numerous valves throughout its whole course, but they are more numerous in the upper than in the lower part; at its termination it is provided with a pair of valves, the free borders of which are turned towards the vein, so as to prevent the regurgitation of venous blood into the duct.

Branches. The thoracic duct, at its commencement, receives four or five large trunks from the abdominal lymphatic glands, and also the trunk of the lacteal vessels. Within the thorax, it is joined by the lymphatic vessels from the left half of the wall of the thoracic cavity, the lymphatics from the sternal and intercostal glands, those of the left lung, left side of the heart, trachea, and œsophagus; and just before its termination, receives the lymphatics of the left side of the head and neck, and left upper extremity.

Structure. The thoracic duct is composed of three coats, which differ in some respects from those of the lymphatic vessels. The *internal coat* consists of a layer of epithelium, resting upon some striped lamellæ, and an elastic fibrous coat, the fibres of which run in a longitudinal direction. The *middle coat* consists of a layer of connective tissue, beneath which are several laminæ of muscular tissue, the fibres of which are disposed transversely, and intermixed with the elastic fibres. The *external coat* is composed of areolar tissue, with elastic fibres and isolated fasciculi of muscular fibres.

The *Right Lymphatic Duct* is a short trunk, about an inch in length, and a line or a line and a half in diameter, which receives the lymph from the right side of the head and neck, the right upper extremity, the right side of the thorax, the right lung and right side of the heart, and from the convex surface of the liver, and terminates at the angle of union of the right subclavian and right internal jugular veins. Its orifice is guarded by two semilunar valves, which prevent the entrance of blood from the veins.

Lymphatics of the Head, Face, and Neck.

The *Superficial Lymphatic Glands of the Head* (fig. 248) are of small size, few in number, and confined to its posterior region. They are the *occipital*, placed at the back of the head along the attachment of the Occipito-frontalis; and the *posterior auricular*, near the upper end of the Sterno-mastoid. These glands are affected in cutaneous eruptions and other diseases of the scalp. In the face, the superficial lymphatic glands are more numerous: they are the *parotid*, some of which are superficial and others deeply placed in the substance of the parotid gland; the *zygomatic*, situated under the zygoma; the *buccal*, on the surface of the Buccinator muscle; and the *submaxillary*, the largest, beneath the body of the lower jaw.

The *superficial lymphatics of the head* are divided into an anterior and a posterior set, which follow the course of the temporal and occipital vessels. The temporal set accompany the temporal artery in front of the ear, to the parotid lymphatic glands, from which they proceed to the lymphatic glands of the neck. The occipital set follow the course of the occipital artery, descend to the occipital and posterior auricular lymphatic glands, and from thence join the cervical glands.

The *superficial lymphatics of the face* are more numerous than those of the head, and commence over its entire surface. Those from the frontal region accompany the frontal vessels; they then pass obliquely across the face, running with the facial vein, pass through the buccal glands on the surface of the Buccinator muscle, and join the submaxillary lymphatic glands. The latter receive the lymphatic vessels from the lips, and are often found enlarged in cases of malignant disease of those parts.

The *deep lymphatics of the face* are derived from the pituitary membrane of the nose, the mucous membrane of the mouth and pharynx, and the contents of the temporal and orbital fossæ; they accompany the branches of the internal maxillary artery, and terminate in the deep parotid and cervical lymphatic glands.

The *deep lymphatics of the cranium* consist of two sets, the meningeal and cerebral. The meningeal lymphatics accompany the meningeal vessels, escape

248.—The Superficial Lymphatics and Glands of the Head, Face, and Neck.

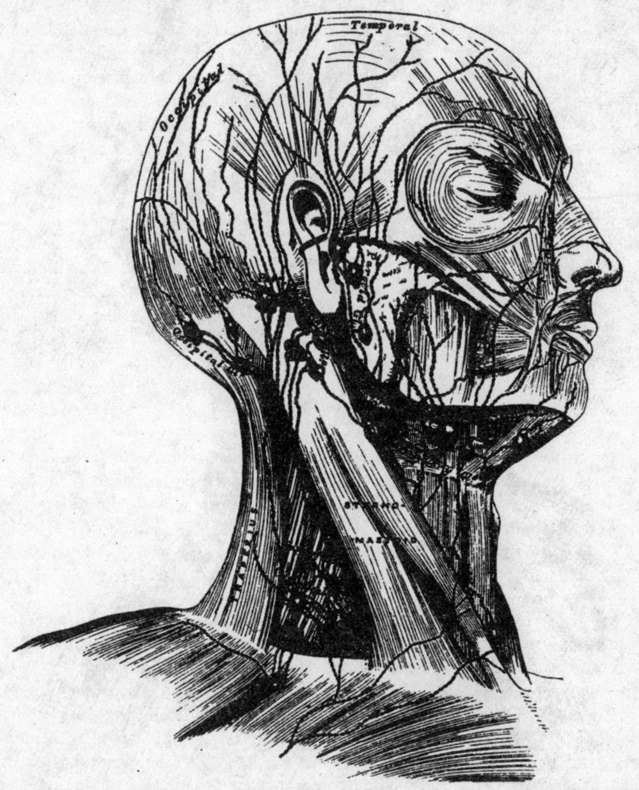

through foramina at the base of the skull, and join the deep cervical lymphatic glands. The cerebral lymphatics are described by Fohmann as being situated between the arachnoid and pia mater, as well as in the choroid plexuses of the lateral ventricles; they accompany the trunks of the carotid and vertebral arteries, and probably pass through foramina at the base of the skull, to terminate in the deep cervical glands. They have not at present been demonstrated in the dura mater, or in the substance of the brain.

The *Lymphatic Glands of the Neck* are divided into two sets, superficial and deep.

The *superficial cervical glands* are placed in the course of the external jugular vein, between the Platysma and Sterno-mastoid. They are most numerous at the root of the neck, in the triangular interval between the clavicle, the Sterno-mastoid, and the Trapezius, where they are continuous with the axillary glands. A few small glands are also found on the front and sides of the larynx.

The *deep cervical glands* (fig. 249) are numerous and of large size; they form an uninterrupted chain along the sheath of the carotid artery and internal jugular vein, lying by the side of the pharynx, œsophagus, and trachea, and extending from the base of the skull to the thorax, where they communicate with the lymphatic glands in that cavity.

249.—The Deep Lymphatics and Glands of the Neck and Thorax.

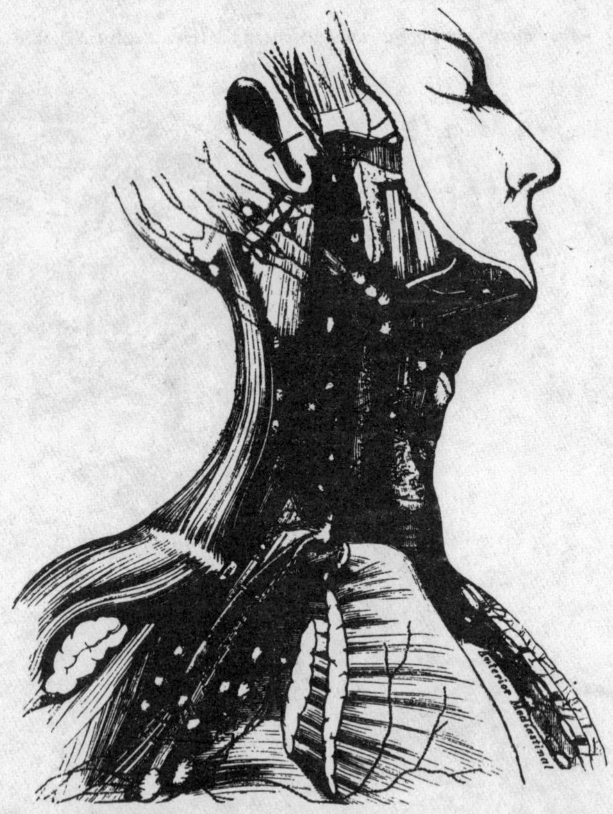

The *superficial and deep cervical lymphatics* are a continuation of those already described on the cranium and face. After traversing the glands in those regions, they pass through the chain of glands which lie along the sheath of the carotid vessels, being joined by the lymphatics from the pharynx, œsophagus, larynx, trachea, and thyroid gland. At the lower part of the neck, after receiving some lymphatics from the thorax, they unite into a single trunk, which terminates on the left side, in the thoracic duct; on the right side, in the right lymphatic duct.

Lymphatics of the Upper Extremity.

The *Lymphatic Glands* of the upper extremity (fig. 250) may be subdivided into two sets, superficial and deep.

The *superficial lymphatic glands* are few, and of small size. There are occasionally two or three in front of the elbow, and one or two above the internal condyle of the humerus, near the basilic vein.

The *deep lymphatic glands* are also few in number. In the forearm a few small ones are occasionally found in the course of the radial and ulnar vessels; and in the arm, there is a chain of small glands along the inner side of the brachial artery.

250.—The Superficial Lymphatics and Glands of the Upper Extremity.

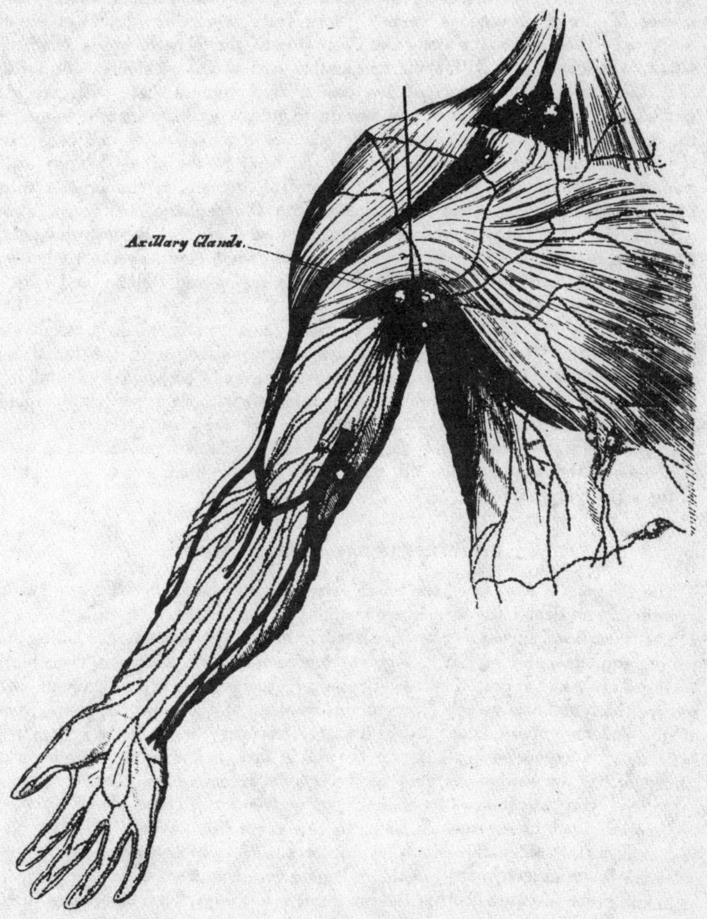

The *axillary glands* are of large size, and usually ten or twelve in number. A chain of these glands surrounds the axillary vessels imbedded in a quantity of loose areolar tissue; they receive the lymphatic vessels from the arm: others are dispersed in the areolar tissue of the axilla: the remainder are arranged in two series, a small chain running along the lower border of the Pectoralis major, as far as the mammary gland, receiving the lymphatics from the front of the chest and mamma; and others are placed along the lower margin of the posterior wall of the axilla, which receive the lymphatics from the integument of the back. Two or three subclavian lymphatic glands are placed immediately beneath the clavicle; it is through these that the axillary and deep cervical glands communicate with each other. One is figured by Mascagni near the umbilicus. In malignant diseases, tumours, or other affections implicating the upper part of the back and shoulder, the front of the chest and mamma, the upper part of the front and side of the abdomen, or the hand, forearm, and arm, the axillary glands are liable to be found enlarged.

The *superficial lymphatics* of the upper extremity arise from the skin of the hand, and run along the sides of the fingers chiefly on the dorsal surface of the hand; they then pass up the forearm, and subdivide into two sets, which take the course of the subcutaneous veins. Those from the inner border of the hand accompany the ulnar veins along the inner side of the forearm to the bend of the elbow, where they join with some lymphatics from the outer side of the forearm; they then follow the course of the basilic vein, communicate with the glands immediately above the elbow, and terminate in the axillary glands, joining with the deep lymphatics. The superficial lymphatics from the outer and back part of the hand accompany the radial veins to the bend of the elbow. They are less numerous than the preceding. At the bend of the elbow, the greater number join the basilic group; the rest ascend with the cephalic vein on the outer side of the arm, some crossing the upper part of the Biceps obliquely, to terminate in the axillary glands, whilst one or two accompany the cephalic vein in the cellular interval between the Pectoralis major and Deltoid, and enter the subclavian lymphatic glands.

The *deep lymphatics* of the upper extremity accompany the deep blood-vessels. In the forearm, they consist of three sets, corresponding with the radial, ulnar, and interosseous arteries; they pass through the glands occasionally found in the course of those vessels, and communicate at intervals with the superficial lymphatics. In their course upward, some of them pass through the glands which lie upon the brachial artery; they then enter the axillary and subclavian glands, and at the root of the neck terminate, on the left side, in the thoracic duct, and on the right side in the right lymphatic duct.

Lymphatics of the Lower Extremity.

The *Lymphatic Glands* of the lower extremity may be subdivided into two sets, superficial and deep; the former are confined to the inguinal region.

The *superficial inguinal glands*, placed immediately beneath the integument, are of large size, and vary from eight to ten in number. They are divisible into two groups; an upper, disposed irregularly along Poupart's ligament, which receive the lymphatic vessels from the integument of the scrotum, penis, parietes of the abdomen, perinæum, and gluteal regions; and an inferior group, which surround the saphenous opening in the fascia lata, a few being sometimes continued along the saphenous vein to a variable extent. The latter receive the superficial lymphatic vessels from the lower extremity. These glands frequently become enlarged in diseases implicating the parts from which their lymphatics originate. Thus, in malignant or syphilitic affections of the prepuce and penis, or of the labia majora in the female, in cancer scroti, in abscess in the perinæum, or in any other disease affecting the integument and superficial structures in those parts, or the sub-umbilical part of the abdomen or gluteal region, the upper chain of glands is almost invariably enlarged, the lower chain being implicated in diseases affecting the lower limb.

The *deep lymphatic glands* are, the anterior tibial, popliteal, deep inguinal, gluteal, and ischiatic.

The *anterior tibial gland* is not constant in its existence. It is generally found by the side of the anterior tibial artery, upon the interosseous membrane at the upper part of the leg. Occasionally two glands are found in this situation.

The *deep popliteal glands*, four or five in number, are of small size; they surround

251.—The Superficial Lymphatics and Glands of the Lower Extremity.

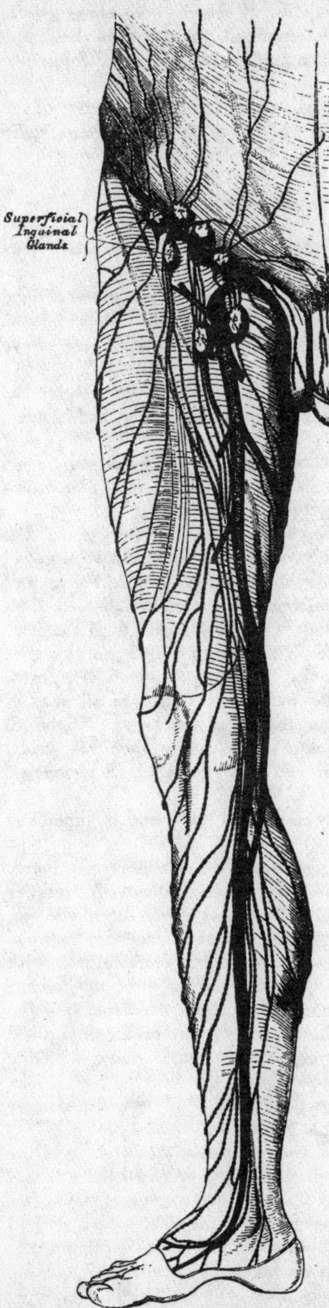

the popliteal vessels, imbedded in the cellular tissue and fat of the popliteal space.

The *deep inguinal glands* are placed beneath the deep fascia around the femoral artery and vein. They are of small size, and communicate with the superficial inguinal glands through the saphenous opening.

The *gluteal* and *ischiatic glands* are placed, the former above, the latter below the Pyriformis muscle, resting on their corresponding vessels as they pass through the great sacro-sciatic foramen.

The *Lymphatics* of the lower extremity, like the veins, may be divided into two sets, superficial and deep.

The *superficial lymphatics* are placed between the integument and superficial fascia, and are divisible into two groups, an internal group, which follow the course of the internal saphenous vein; and an external group, which accompany the external saphenous. The *internal group*, the larger, commence on the inner side and dorsum of the foot; they pass, some in front, and some behind the inner ankle, run up the leg with the internal saphenous vein, pass with it behind the inner condyle of the femur, and accompany it to the groin, where they terminate in the group of inguinal glands which surround the saphenous opening. Some of the efferent vessels from these glands pierce the cribriform fascia and sheath of the femoral vessels, and terminate in a lymphatic gland contained in the femoral canal, thus establishing a communication between the lymphatics of the lower extremity and those of the trunk; others pierce the fascia lata, and join the deep inguinal glands. The *external group* arise from the outer side of the foot, ascend in front of the leg, and, just below the knee, cross the tibia from without inwards, to join the lymphatics on the inner side of the thigh. Others commence on the outer side of the foot, pass behind the outer malleolus, and accompany the external saphenous vein along the back of the leg, where they enter the popliteal glands.

The *deep lymphatics* of the lower extremity are few in number, and accompany the deep blood-vessels. In the leg, they consist of three sets, the anterior tibial, peroneal, and posterior tibial, which accompany the corresponding blood-vessels, two or three to each artery; they ascend with the blood-vessels, and enter the lymphatic glands in the popliteal space; the efferent vessels from these glands accompany the femoral vein, and join the deep inguinal glands; from these, the vessels pass beneath Poupart's ligament, and communicate with the chain of glands surrounding the external iliac vessels.

The deep lymphatics of the gluteal and ischiatic regions follow the course of the blood-vessels, and join the gluteal and ischiatic glands at the great sacro-sciatic foramen.

Lymphatics of the Pelvis and Abdomen.

The *deep lymphatic glands in the pelvis* are, the external iliac, the internal iliac, and the sacral. Those of the abdomen are the lumbar glands.

The *external iliac glands* form an uninterrupted chain round the external iliac vessels, three being placed round the commencement of the vessel just behind the crural arch. They communicate below with the femoral lymphatics, and above with the lumbar glands.

The *internal iliac glands* surround the internal iliac vessels; they receive the lymphatics corresponding to the branches of the internal iliac artery, and communicate with the lumbar glands.

The *sacral glands* occupy the sides of the anterior surface of the sacrum, some being situated in the meso-rectal fold. These and the internal iliac glands are affected in malignant disease of the bladder, rectum, or uterus.

The *lumbar glands* are very numerous; they are situated on the front of the lumbar vertebræ, surrounding the common iliac vessels, the aorta, and vena cava; they receive the lymphatic vessels from the lower extremities and pelvis, as well as from the testes and some of the abdominal viscera; the efferent vessels from these glands unite into a few large trunks, which, with the lacteals, form the commencement of the thoracic duct. In some cases of malignant disease, these glands become enormously enlarged, completely surrounding the aorta and vena cava, and occasionally greatly contracting the calibre of those vessels. In all cases of malignant disease of the testis, and in malignant disease of the lower limb, before any operation is attempted, careful examination of the abdomen should be made, in order to ascertain if any enlargement exists; and if any should be detected, all operative measures should be avoided, as fruitless.

The *lymphatics of the pelvis and abdomen* may be divided into two sets, superficial and deep.

The *superficial lymphatics of the walls of the abdomen and pelvis* follow the course of the superficial blood-vessels. Those derived from the integument of the lower part of the abdomen below the umbilicus, follow the course of the superficial epigastric vessels, and converge to the superior group of the superficial inguinal glands; the deep set accompany the deep epigastric vessels, and communicate with the external iliac glands. The superficial lymphatics from the sides and lumbar part of the abdominal wall wind round the crest of the ilium, accompanying the superficial circumflex iliac vessels, to join the superior group of the superficial inguinal glands; the greater number, however, run backwards along with the ilio-lumbar and lumbar vessels, to join the lumbar glands.

The *superficial lymphatics of the gluteal region* turn horizontally round the outer side of the nates, and join the superficial inguinal glands.

The *superficial lymphatics of the scrotum and perinæum* follow the course of the external pudic vessels, and terminate in the superficial inguinal glands.

The *superficial lymphatics of the penis* occupy the sides and dorsum of the organ, the latter receiving the lymphatics from the skin covering the glans penis; they all converge to the upper chain of the superficial inguinal glands. The deep lymphatic vessels of the penis follow the course of the internal pudic vessels, and join the internal iliac glands.

In the female, the lymphatic vessels of the mucous membrane of the labia, nymphæ, and clitoris, terminate in the upper chain of the inguinal glands.

The *deep lymphatics of the pelvis and abdomen* take the course of the principal blood-vessels. Those of the parietes of the pelvis, which accompany the gluteal, ischiatic, and obturator vessels, follow the course of the internal iliac artery, and ultimately join the lumbar lymphatics.

252.—The Deep Lymphatic Vessels and Glands of the Abdomen and Pelvis.

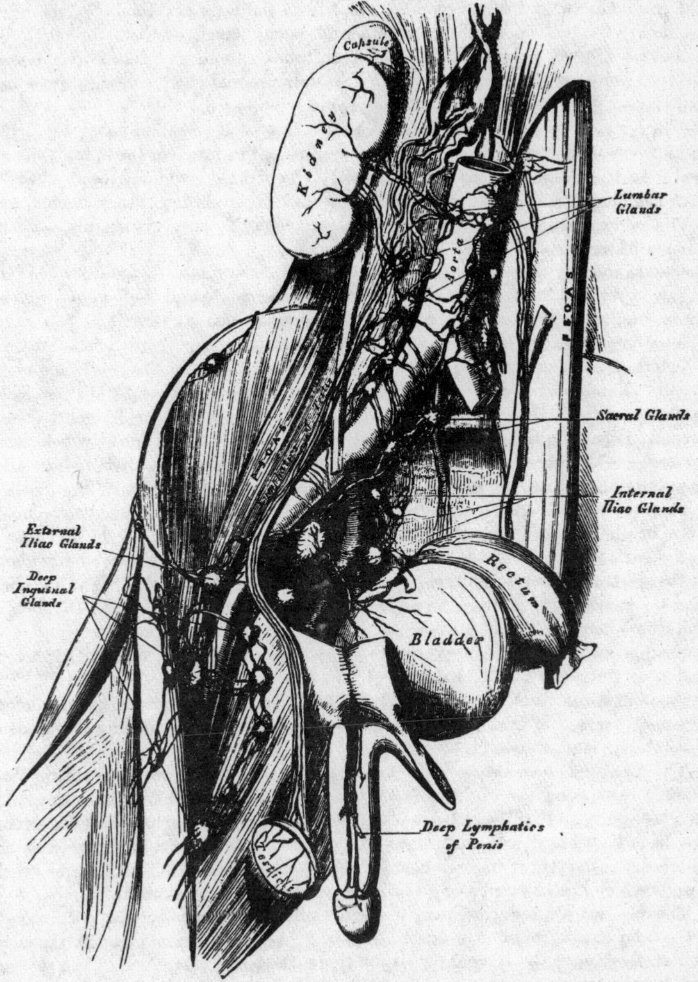

The efferent vessels from the inguinal glands enter the pelvis beneath Poupart's ligament, where they lie in close relation with the femoral vein; they then pass through the chain of glands surrounding the external iliac vessels, and finally terminate in the lumbar glands. They receive the deep epigastric, circumflex iliac, and ilio-lumbar lymphatics.

The *lymphatics of the bladder* arise from the entire surface of the organ; the greater number run beneath the peritoneum on its posterior surface, and, after passing through the lymphatic glands in that situation, join with the lymphatics from the prostate and vesiculæ seminales, and enter the internal iliac glands.

The *lymphatics of the rectum* are of large size; after passing through some small glands that lie upon its outer wall and in the meso-rectum, they pass to the sacral or lumbar glands.

The *lymphatics of the uterus* consist of two sets, superficial and deep; the former being placed beneath the peritoneum, the latter in the substance of the organ. The lymphatics of the cervix uteri, together with those from the vagina, enter the internal iliac and sacral glands; those from the body and fundus of the uterus pass outwards in the broad ligaments, and, being joined by the lymphatics from the ovaries, broad ligaments, and Fallopian tubes, ascend with the ovarian vessels to open into the lumbar glands. In the unimpregnated uterus, they are small; but during gestation, they become very greatly enlarged.

The *lymphatics of the testicle* consist of two sets, superficial and deep; the former commence on the surface of the tunica vaginalis, the latter in the epididymis and body of the testis. They form several large trunks, which ascend with the spermatic cord, and accompanying the spermatic vessels into the abdomen, open into the lumbar glands; hence the enlargement of these glands in malignant disease of the testis.

The *lymphatics of the kidney* arise on the surface, and also in the interior of the organ; they join at the hilum, and, after receiving the lymphatic vessels from the ureter and suprarenal capsule, open into the lumbar glands.

The *lymphatics of the liver* are divisible into two sets, superficial and deep. The former arise in the sub-peritoneal areolar tissue over the entire surface of the organ. Those on the convex surface may be divided into four groups: 1. Those which pass from behind forwards, consisting of three or four branches, which ascend in the longitudinal ligament, and unite to form a single trunk, which passes up between the fibres of the Diaphragm, behind the ensiform cartilage, to enter the anterior mediastinal glands, and finally ascends to the root of the neck, to terminate in the right lymphatic duct. 2. Another group, which also incline from behind forwards, are reflected over the anterior margin of the liver to its under surface, and from thence pass along the longitudinal fissure to the glands in the gastro-hepatic omentum. 3. A third group incline outwards to the right lateral ligament, and uniting into one or two large trunks, pierce the Diaphragm, and run along its upper surface to enter the anterior mediastinal glands; or, instead of entering the thorax, turn inwards across the crus of the Diaphragm, and open into the commencement of the thoracic duct. 4. The fourth group incline outwards from the surface of the left lobe of the liver to the left lateral ligament, pierce the Diaphragm, and passing forwards, terminate in the glands in the anterior mediastinum.

The *superficial lymphatics on the under surface of the liver* are divided into three sets: 1. Those on the right side of the gall-bladder enter the lumbar glands. 2. Those surrounding the gall-bladder form a remarkable plexus: they accompany the hepatic vessels, and open into the glands in the gastro-hepatic omentum. 3. Those on the left of the gall-bladder pass to the œsophageal glands, and to the glands which are situated along the lesser curvature of the stomach.

The *deep lymphatics* accompany the branches of the portal vein and the hepatic artery and duct through the substance of the liver; passing out at the transverse fissure, they enter the lymphatic glands along the lesser curvature of the stomach and behind the pancreas, or join with one of the lacteal vessels previous to its termination in the thoracic duct.

The *lymphatic glands of the stomach* are of small size: they are placed along the lesser and greater curvatures, some within the gastro-splenic omentum, whilst others surround the cardiac and pyloric orifices.

The *lymphatics of the stomach* consist of two sets, superficial and deep; the former originating in the subserous, and the latter in the submucous coat. They follow the course of the blood-vessels, and may, consequently, be arranged into three groups. The *first group* accompany the coronary vessels along the lesser curvature, receiving branches from both surfaces of the organ, and pass to the glands around the pylorus. The *second group* pass from the great end of the stomach, accompany the vasa brevia, and enter the splenic lymphatic glands. The *third group* run along the greater curvature with the right gastro-epiploic vessels, and terminate at the root of the mesentery in one of the principal lacteal vessels.

The *lymphatic glands of the spleen* occupy the hilum. Its *lymphatic vessels* consist of two sets, superficial and deep; the former are placed beneath its peritoneal covering, the latter in the substance of the organ: they accompany the blood-vessels, passing through a series of small glands, and after receiving the lymphatics from the pancreas, ultimately pass into the thoracic duct.

The Lymphatic System of the Intestines.

The *lymphatic glands of the small intestine* are placed between the layers of the mesentery, occupying the meshes formed by the superior mesenteric vessels, and hence called *mesenteric glands*. They vary in number from a hundred to a hundred and fifty; and in size, from that of a pea to that of a small almond. These glands are most numerous, and largest, above near the duodenum, and below opposite the termination of the ileum in the colon. This latter group becomes enlarged and infiltrated with deposit in cases of fever accompanied with ulceration of the intestines.

The *lymphatic glands of the large intestine* are much less numerous than the mesenteric glands: they are situated along the vascular arches formed by the arteries previous to their distribution, and even sometimes upon the intestine itself. They are fewest in number along the transverse colon, where they form an uninterrupted chain with the mesenteric glands.

The *lymphatics of the small intestine* are called *lacteals*, from the milk-white fluid they usually contain; they consist of two sets, superficial and deep; the former lie beneath the peritoneal coat, taking a longitudinal course along the outer side of the intestine; the latter occupy the submucous tissue, and course transversely round the intestine, accompanied by the branches of the mesenteric vessels: they pass between the layers of the mesentery, enter the mesenteric glands, and finally unite to form two or three large trunks, which terminate in the thoracic duct.

The *lymphatics of the large intestine* consist of two sets: those of the cæcum, ascending and transverse colon, which, after passing through their proper glands, enter the mesenteric glands; and those of the descending colon and rectum, which pass to the lumbar glands.

The Lymphatics of the Thorax.

The *deep lymphatic glands of the thorax* are the intercostal, internal mammary, anterior mediastinal, and posterior mediastinal.

The *intercostal glands* are small, irregular in number, and situated on each side of the spine, near the costo-vertebral articulations, some being placed between the two planes of intercostal muscles.

The *internal mammary glands* are placed at the anterior extremity of each intercostal space, by the side of the internal mammary vessels.

The *anterior mediastinal glands* are placed in the loose areolar tissue of the anterior mediastinum, some lying upon the Diaphragm in front of the pericardium, and others round the great vessels at the base of the heart.

The *posterior mediastinal glands* are situated in the areolar tissue in the posterior mediastinum, forming a continuous chain by the side of the aorta and œsophagus; they communicate on each side with the intercostal, below with the lumbar glands, and above with the deep cervical.

The *superficial lymphatics of the front of the thorax* run across the great Pectoral muscle, and those on the back part of this cavity lie upon the Trapezius and Latissimus dorsi; they all converge to the axillary glands. The lymphatics from the mamma run along the lower border of the Pectoralis major, through a chain of small lymphatic glands, and communicate with the axillary glands.

The *deep lymphatics of the thorax* are the intercostal, internal mammary, and diaphragmatic.

The *intercostal lymphatics* follow the course of the intercostal vessels, receiving lymphatics from the intercostal muscles and pleura; they pass backwards to the spine, and unite with lymphatics from the back part of the thorax and spinal canal. After traversing the intercostal glands, they incline down the spine, and terminate in the thoracic duct.

The *internal mammary lymphatics* follow the course of the internal mammary vessels; they commence in the muscles of the abdomen above the umbilicus, communicating with the epigastric lymphatics, ascend between the fibres of the Diaphragm at its attachment to the ensiform appendix, and in their course behind the costal cartilages are joined by the intercostal lymphatics, terminating on the right side in the right lymphatic duct, on the left side in the thoracic duct.

The *lymphatics of the Diaphragm* follow the course of their corresponding vessels, and terminate, some in front, in the anterior mediastinal and internal mammary glands, some behind in the intercostal and hepatic lymphatics.

The *bronchial glands* are situated round the bifurcation of the trachea and roots of the lungs. They are ten or twelve in number, the largest being placed opposite the bifurcation of the trachea, the smallest round the bronchi and their primary divisions for some little distance within the substance of the lungs. In infancy, they present the same appearance as lymphatic glands in other situations, in the adult they assume a brownish tinge, and in old age a deep black colour. Occasionally they become sufficiently enlarged to compress and narrow the canal of the bronchi; and they are often the seat of tubercle or deposits of phosphate of lime.

The *lymphatics of the lung* consist of two sets, superficial and deep: the former are placed beneath the pleura, forming a minute plexus, which covers the outer surface of the lung; the latter accompany the blood-vessels, and run along the bronchi: they both terminate at the root of the lungs in the bronchial glands. The efferent vessels from these glands, two or three in number, ascend upon the trachea to the root of the neck, traverse the tracheal and œsophageal glands, and terminate on the left side in the thoracic duct, and on the right side in the right lymphatic duct.

The *cardiac lymphatics* consist of two sets, superficial and deep; the former arise in the subserous areolar tissue of the surface, and the latter beneath the internal lining membrane of the heart. They follow the course of the coronary vessels; those of the right side unite into a trunk at the root of the aorta, which, ascending across the arch of that vessel, passes backwards to the trachea, upon which it ascends, to terminate at the root of the neck in the right lymphatic duct. Those of the left side unite into a single vessel at the base of the heart, which, passing along the pulmonary artery, and traversing some glands at the root of the aorta, ascends on the trachea to terminate in the thoracic duct.

The *thymic lymphatics* arise from the spinal surface of the thymus gland, and terminate on each side in the internal jugular veins.

The *thyroid lymphatics* arise from either lateral lobe of this organ; they converge to form a short trunk, which terminates on the right side in the right lymphatic duct, on the left side in the thoracic duct.

The *lymphatics of the œsophagus* form a plexus round that tube, traverse the glands in the posterior mediastinum, and, after communicating with the pulmonary lymphatic vessels near the roots of the lungs, terminate in the thoracic duct.

Nervous System.

THE Nervous System is composed: 1. of a series of connected central organs, called, collectively, the *cerebro-spinal centre*, or *axis*; 2. of the *ganglia*; and, 3. of the *nerves*.

The Cerebro-Spinal Centre consists of two parts, the spinal cord and the encephalon: the latter may be subdivided into the cerebrum, the cerebellum, the pons Varolii, and the medulla oblongata.

The Spinal Cord and its Membranes.

Dissection. To dissect the cord and its membranes, it will be necessary to lay open the whole length of the spinal canal. For this purpose, the muscles must be separated from the vertebral grooves, so as to expose the spinous processes and laminæ of the vertebræ; and the latter must be sawn through on each side, close to the roots of the transverse processes, from the third or fourth cervical vertebra, above, to the sacrum below. The vertebral arches having been displaced, by means of a chisel, and the separate fragments removed, the dura mater will be exposed, covered by a plexus of veins and a quantity of loose areolar tissue, often infiltrated with a serous fluid. The arches of the upper vertebræ are best divided by means of a strong pair of cutting bone-forceps.

Membranes of the Cord.

The membranes which envelope the spinal cord are three in number. The most external is the dura mater, a strong fibrous membrane, which forms a loose sheath around the cord. The most internal is the pia mater, a cellulo-vascular membrane, which closely invests the entire surface of the cord. Between the two is the arachnoid membrane, an intermediate serous sac, which envelopes the cord, and is then reflected on the inner surface of the dura mater.

The Dura Mater of the cord, continuous with that which invests the brain, is a loose sheath which surrounds the cord, and is separated from the bony walls of the spinal canal by a quantity of loose areolar adipose tissue, and a plexus of veins. It is attached to the circumference of the foramen magnum, and to the posterior common ligament, throughout the whole length of the spinal canal, by fibrous tissue, and extends, below, as far as the top of the sacrum; but, beyond this point, it is impervious, being continued, in the form of a slender cord, to the back of the coccyx, where it blends with the periosteum. This sheath is much larger than is necessary for its contents, and its size is greator in the cervical and lumbar regions than in the dorsal. Its inner surface is smooth, covered by a layer of polygonal cells; and on each side may be seen the double openings which transmit the two roots of the corresponding spinal nerve, the fibrous layer of the dura mater being continued in the form of a tubular prolongation on them as they issue from these apertures. These prolongations of the dura mater are short in the upper part of the spine, but become gradually longer below, forming a number of tubes of fibrous membrane, which enclose the sacral nerves, and are contained in the spinal canal.

The chief peculiarities of the dura mater of the cord, as compared with that investing the brain, are the following:

The dura mater of the cord is not adherent to the bones of the spinal canal, which have an independent periosteum.

It does not send partitions into the fissures of the cord, as in the brain.

Its fibrous laminæ do not separate, to form venous sinuses, as in the brain.

Structure. The dura mater consists of white fibrous tissue, arranged in bands which intersect one another. It is sparingly supplied with vessels; and no nerves have as yet been traced into it.

253.—The Spinal Cord and its Membranes.

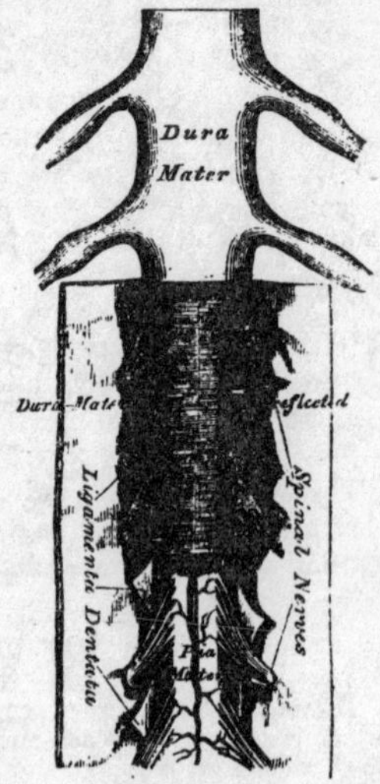

The *Arachnoid* is exposed by slitting up the dura mater, and reflecting that membrane to either side (fig. 253). It is a thin, delicate, serous membrane, which invests the outer surface of the cord, and is then reflected upon the inner surface of the dura mater, to which it is intimately adherent. Above, it is continuous with the cerebral arachnoid; below, it is reflected on the various nerves, so that its parietal and visceral layers become continuous with each other. The *visceral layer* is the portion which surrounds the cord, and that which lines the inner surface of the dura mater is called *the parietal layer*;* the interval between the two, is called the *cavity of the arachnoid.* The visceral layer forms a loose sheath around the cord, so as to leave a considerable interval between the two, which is called the *sub-arachnoïdean space.* This space is largest at the lower part of the spinal canal, and encloses the mass of nerves which form the cauda equina. It contains an abundant serous secretion, the cerebro-spinal fluid, and usually communicates with the general ventricular cavity of the brain, by means of an opening in the fibrous layer of the inferior boundary of the fourth ventricle. This secretion is sufficient in amount to expand the arachnoid membrane, so as to completely fill up the whole of the space included in the dura mater. The sub-arachnoidean space is crossed, at the back part of the cord, by numerous fibrous bands, which stretch from the arachnoid to the pia mater, especially in the cervical region, and is partially subdivided by a longitudinal membranous partition, which serves to connect the arachnoid with the pia mater, opposite the posterior median fissure. This partition is incomplete, and cribriform in structure, consisting of bundles of white fibrous tissue, interlacing with each other. The visceral layer of the arachnoid surrounds the spinal nerves where they arise from the cord, and encloses them in a tubular sheath as far as their point of exit from the dura mater, where it becomes continuous with the parietal layer.

254.—Transverse Section of the Spinal Cord and its Membranes.

The arachnoid is destitute of vessels. No nerves have as yet been traced into this membrane.

The *Pia Mater* of the cord is exposed on the removal of the arachnoid (fig. 253). It is less vascular in structure than the pia mater of the brain, with which it is continuous, being thicker, more dense in structure, and composed of fibrous tissue, arranged in longitudinal bundles. It covers the entire surface of the cord to

* Kölliker denies that the inner surface of the dura mater is covered by an outer layer of the arachnoid, and states that nothing is found here except an epithelial layer, no trace of a special membrane existing.

which it is very intimately adherent, forming its neurilemma, and sends a process downwards into its anterior fissure, and another, extremely delicate, into the posterior fissure. It also forms a sheath for each of the filaments of the spinal nerves, and invests the nerves themselves. A longitudinal fibrous band extends along the middle line on its anterior surface, called by Haller the linea splendens; and a somewhat similar band, the ligamentum denticulatum, is situated on each side. At the point where the cord terminates, the pia mater becomes contracted, and is continued down as a long, slender filament (*filum terminale*), which descends through the centre of the mass of nerves forming the cauda equina, and is blended with the impervious sheath of dura mater, on a level with the top of the sacral canal. It assists in maintaining the cord in its position during the movements of the trunk, and is, from this circumstance, called the central ligament of the spinal cord. It contains a little nervous substance, which may be traced for some distance into its upper part, and is accompanied by a small artery and vein.

Structure. The pia mater of the cord, though less vascular than that which invests the brain, contains a network of delicate vessels in its substance. It is also supplied with nerves, which are derived from the sympathetic, and from the posterior roots of the spinal nerves. At the upper part of the cord, the pia mater presents a greyish, mottled tint, which is owing to yellowish or brown pigment cells being scattered within its tissue.

The *Ligamentum Denticulatum* (fig. 253) is a narrow, fibrous band, situated on each side of the spinal cord, throughout its entire length, and separating the anterior from the posterior roots of the spinal nerves, having received its name from the serrated appearance which it presents. Its inner border is continuous with the pia mater, at the side of the cord. Its outer border presents a series of triangular, dentated serrations, the points of which are fixed, at intervals, to the dura mater, serving to unite together the two layers of the arachnoid membrane. These serrations are about twenty in number, on each side, the first being attached to the dura mater, opposite the margin of the foramen magnum, between the vertebral artery and the hypoglossal nerve; and the last near the lower end of the cord. Its use is to support the cord in the fluid by which it is surrounded.

The Spinal Cord.

The spinal cord (*medulla spinalis*) is the cylindrical elongated part of the cerebro-spinal axis, which is contained in the spinal canal. Its length is usually about sixteen or seventeen inches, and its weight, when divested of its membranes and nerves, about one ounce and a half, its proportion to the encephalon being about 1 to 33. It does not nearly fill the canal in which it is contained, its investing membranes being separated from the surrounding walls by areolar tissue and a plexus of veins. It occupies, in the adult, the upper two-thirds of the spinal canal, extending from the foramen magnum to the lower border of the body of the first lumbar vertebra, where it terminates in a slender filament of grey substance, which is continued for some distance into the *filum terminale*. In the fœtus, before the third month, it extends to the bottom of the sacral canal; but, after this period, it gradually recedes from below, as the growth of the bones composing the canal is more rapid in proportion than that of the cord; so that, in the child at birth, the cord extends as far as the third lumbar vertebra. Its position varies also according to the degree of curvature of the spinal column, being raised somewhat in flexion of the spine. On examining its surface, it presents a difference in its diameter in different parts, being marked by two enlargements, an upper or cervical, and a lower or lumbar. The cervical enlargement, which is the larger, extends from the third cervical to the first dorsal vertebra: its greatest diameter is in the transverse direction, and it corresponds with the origin of the nerves which supply the upper extremities. The lower, or lumbar enlargement, is situated opposite the last dorsal vertebra, its greatest diameter being from before backwards. It corresponds with the origin of the nerves which supply the lower extremities. In form, the spinal cord is a flattened cylinder. It presents, on its anterior surface, along the middle line, a longitudinal fissure, the anterior median fissure; and, on its posterior surface, another fissure, which also extends along the entire length of the cord, the posterior median fissure. These fissures serve to divide the cord into two symmetrical halves, which are united in the middle line, throughout their entire length, by a transverse band of nervous substance, the commissure.

The *Anterior median fissure* is wider, but of less depth than the posterior, extending into the cord for about one-third of its thickness, and is deepest at the lower part of the cord. It contains a prolongation from the pia mater; and its floor is formed by the anterior white commissure, which is perforated by numerous blood-vessels, passing to the centre of the cord.

255.—Spinal Cord. Side View. Plan of the Fissures and Columns.

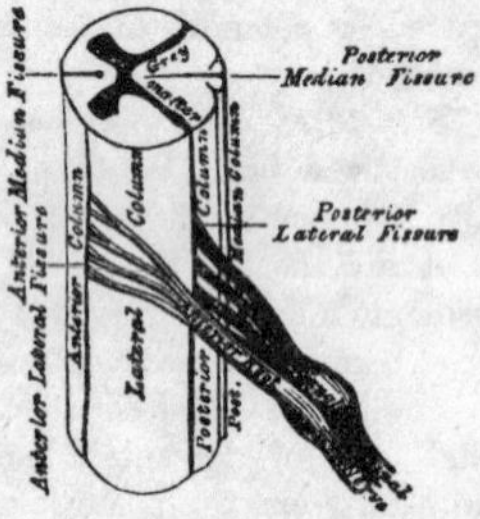

The *Posterior median fissure* is much more delicate than the anterior, and most distinct at the upper and lower parts of the cord. It extends into the cord to about one-half of its depth. It contains a very slender process of the pia mater, and numerous blood-vessels, and its floor is formed by a thin layer of white substance, the posterior white commissure. Some anatomists state, that the bottom of this fissure corresponds to the grey matter, except in the cervical region, and at a point corresponding to the enlargement in the lumbar region.

Lateral Fissures. On either side of the anterior median fissure, a linear series of foramina may be observed, indicating the points where the anterior roots of the spinal nerves emerge from the cord. This is called, by some anatomists, the anterior lateral fissure of the cord, although no actual fissure exists in this situation. And on either side of the posterior median fissure, along the line of attachment of the posterior roots of the nerves, a delicate fissure may be seen, leading down to the grey matter which approaches the surface in this situation: this is called the posterior lateral fissure of the spinal cord. On the posterior surface of the spinal cord, on either side of the posterior median fissure, is a slight longitudinal furrow, marking off two slender tracts, the posterior median columns. These are most distinct in the cervical region, but are stated by Foville to exist throughout the whole length of the cord.

Columns of the Cord. The fissures divide each half of the spinal cord into four columns, an anterior column, a lateral column, a posterior column, and a posterior median column.

The *anterior column* includes all the portion of the cord between the anterior median fissure and the anterior lateral fissure, from which the anterior roots of the nerves arise. It is continuous with the anterior pyramid of the medulla oblongata.

The *lateral column*, the largest segment of the cord, includes all the portion between the anterior and posterior lateral fissures. It is continuous with the lateral column of the medulla. By some anatomists, the anterior and lateral columns are included together, under the name of the antero-lateral column, which forms rather more than two-thirds of the entire circumference of the cord.

The *posterior column* is situated between the posterior median and posterior lateral fissures. It is continuous with the restiform body of the medulla.

The *posterior median column* is that narrow segment of the cord which is seen on each side of the posterior median fissure, usually included with the preceding, as the posterior column.

Structure of the Cord. If a transverse section of the spinal cord be made, it will be seen to consist of white and grey nervous substance. The white matter is situated externally, and constitutes the greater part. The grey substance occupies the centre, and is so arranged as to present on the surface of the section two crescentic masses placed one in each lateral half of the cord, united together by a transverse band of grey matter, the grey commissure. Each crescentic mass has an anterior and posterior horn. The posterior horn is long and narrow, and approaches the surface of the posterior lateral fissure, near which it presents a slight enlargement. The anterior horn is short and thick, and does not quite reach the surface, but extends towards the point of attachment of the anterior roots of the nerves. Its margin presents a dentate or stellate appearance. Owing to the projections towards the surface of the anterior and posterior horns of the grey matter, each half of the cord is divided, more or less completely, into three columns, anterior, middle, and posterior; the anterior and middle being joined to form the antero-lateral column, as the anterior horn does not quite reach the surface.

The grey commissure, which connects the two crescentic masses of grey matter, is separated from the bottom of the anterior median fissure by the anterior white commissure; and from the bottom of the posterior fissure by the posterior white commissure. The grey commissure consists of a transverse band of grey matter, and of white fibres, derived from the opposite half of the cord and the posterior roots of the nerves. The anterior commissure is formed of fibres, partly from the anterior column, and partly from the fibrils of the anterior roots of the spinal nerves, which decussate as they pass across from one to the other side.

256.—Transverse Sections of the Cord.

Opposite Middle of Cervical regn.

Opposite Middle of Dorsal regn.

Opposite Lumbar region.

The mode of arrangement of the grey matter, and its amount in proportion to the white, vary in different parts of the cord. Thus, the posterior horns are long and narrow, in the cervical region; short and narrower, in the dorsal; short, but wider, in the lumbar region. In the cervical region, the crescentic portions are small, the white matter more abundant than in any other region of the cord. In the dorsal region, the grey matter is least developed, the white matter being also small in quantity. In the lumbar region, the grey matter is more abundant than in any other region of the cord. Towards the lower end of the cord, the white matter gradually ceases. The crescentic portions of the grey matter soon blend into a single mass, which forms the only constituent of the extreme point of the cord. The minute anatomy of the cord is described in the Introduction.

The Brain and its Membranes.

Dissection. To examine the brain with its membranes, the skull-cap must be removed. In order to effect this, saw through the external table, the section commencing, in front, about an inch above the margin of the orbit, and extending, behind, to a level with the occipital protuberance. Then break the internal table with the chisel and hammer, to avoid injuring the investing membranes or brain; loosen, and forcibly detach the skull, when the dura mater will be exposed. The adhesion between the bone and the dura mater is very intimate, and much more so in the young subject than in the adult.

The membranes of the brain are, the dura mater, arachnoid membrane, and pia mater.

Dura Mater.

The dura mater is a thick and dense inelastic fibrous membrane, which lines the interior of the skull. Its outer surface is rough and fibrillated, and adheres closely to the inner surface of the bones, forming their internal periosteum, this adhesion being more intimate opposite the sutures and at the base of the skull; at the margin of the foramen magnum, it becomes continuous with the dura mater lining the spinal canal. Its inner surface is smooth and epitheliated, being lined by the parietal layer of the arachnoid. The dura mater is, therefore, a fibro-serous membrane, composed of an external fibrous lamella, and an internal serous layer. It sends numerous processes inwards, into the cavity of the skull, for the support and protection of the different parts of the brain; and is prolonged to the outer surface of the skull, through the various foramina which exist at the base, and thus becomes continuous with the pericranium; its fibrous layer forms sheaths for the nerves which pass through these apertures. At the base of the skull, it sends a fibrous prolongation into the foramen cæcum; it lines the olfactory groove, and sends a series of tubular prolongations round the filaments of the olfactory nerves as they pass through the cribriform plate; a prolongation is also continued through the sphenoidal fissure into the orbit, and another is continued into the same cavity through the optic foramen, forming a sheath for the optic nerve, which is continued as far as the eyeball. In certain situations in the skull already mentioned, the fibrous layers of this membrane separate, to form sinuses for the passage of venous blood. Upon the upper surface of the dura mater, in the situation of the longitudinal sinus, may be seen numerous small whitish bodies, the glandulæ Pacchioni.

Structure. The dura mater consists of white fibrous and elastic tissues, arranged in flattened laminæ, which intersect one another in every direction.

Its *arteries* are very numerous, but are chiefly distributed to the bones. Those found in the anterior fossa, are the anterior meningeal, branches of the anterior and posterior ethmoidal, and internal carotid. In the middle fossa are the middle and small meningeal, branches of the internal maxillary, and a third branch from the ascending pharyngeal, which enters the skull through the foramen lacerum basis cranii. In the posterior fossa, are the meningeal branch of the occipital, which enters the skull through the jugular foramen; the posterior meningeal, from the vertebral; and occasionally meningeal branches from the ascending pharyngeal, which enter the skull, one at the jugular foramen, the other at the anterior condyloid foramen.

The *veins*, which return the blood from the dura mater, and partly from the bones, anastomose with the diploic veins. These vessels terminate in the various sinuses, with the exception of two which accompany the middle meningeal artery, and pass out of the skull at the foramen spinosum.

The *nerves* of the dura mater are, the recurrent branch of the fourth, and filaments from the Casserian ganglion, from the ophthalmic nerve, and from the sympathetic.

The so-called glandulæ Pacchioni are numerous small whitish granulations, usually collected into clusters of variable size, which are found in the following situations: 1. Upon the outer surface of the dura mater, in the vicinity of the superior longitudinal sinus, being received into little depressions on the inner surface of the calvarium. 2. On the inner surface of the dura mater. 3. In the superior longitudinal sinus. 4. On the pia mater near the margin of the hemispheres.

These bodies are not glandular in structure, but consist of a fibro-cellular matrix originally developed from the pia mater; by their growth, they produce absorption or separation of the fibres of the dura mater; in a similar manner they make their way into the superior longitudinal sinus, where they are covered by the lining membrane. The cerebral layer of the arachnoid in the situation of these growths is usually thickened and opaque, and adherent to the parietal portion.

These bodies are not found in infancy, and very rarely until the third year. They are usually found after the seventh year; and from this period they increase in number as age advances. Occasionally they are wanting.

Processes of the Dura Mater. The processes of the dura mater, sent inwards into the cavity of the skull, are three in number, the falx cerebri, the tentorium cerebelli, and the falx cerebelli.

The *falx cerebri*, so named from its sickle-like form, is a strong arched process of the dura mater, which descends vertically in the longitudinal fissure between the two hemispheres of the brain. It is narrow in front, where it is attached to the crista galli process of the ethmoid bone; and broad behind, where it is connected with the upper surface of the tentorium. Its upper margin is convex, and attached to the inner surface of the skull as far back as the internal occipital protuberance. In this situation it is broad, and contains the superior longitudinal sinus. Its lower margin is free, concave, and presents a sharp curved edge, which contains the inferior longitudinal sinus.

The *tentorium cerebelli* is an arched lamina of dura mater, elevated in the middle, and inclining downwards towards the circumference. It covers the upper surface of the cerebellum, supporting the posterior lobes of the brain, and preventing their pressure upon it. It is attached, behind, by its convex border, to the transverse ridges upon the inner surface of the occipital bone, and there encloses the lateral sinuses; in front, to the superior margin of the petrous portion of the temporal bone, enclosing the superior petrosal sinuses, and from the apex of this bone, on each side, is continued to the anterior and posterior clinoid processes. Along the middle line of its upper surface, the posterior border of the falx cerebri is attached, the straight sinus being placed at their point of junction. Its anterior border is free and concave, and presents a large oval opening for the transmission of the crura cerebri.

The *falx cerebelli* is a small triangular process of dura mater, received into the indentation between the two lateral lobes of the cerebellum behind. Its base is attached, above, to the under and back part of the tentorium; its posterior margin, to the lower division of the vertical crest on the inner surface of the occipital bone. As it descends, it sometimes divides into two smaller folds, which are lost on the sides of the foramen magnum.

Arachnoid Membrane.

The arachnoid (ἀράχνη, εἶδος, *like a spider's web*), so named from its extreme thinness, is the serous membrane which envelopes the brain, and is then reflected on the inner surface of the dura mater. Like other serous membranes, it is a shut sac, and consists of a parietal and a visceral layer.

The *parietal layer* covers the inner surface of the dura mater,* and gives that membrane its smooth and polished surface; it is also reflected over those processes which separate the hemispheres of the brain and cerebellum.

The *visceral layer* invests the brain more loosely, being separated from direct contact with the cerebral substance by the pia mater, and a quantity of loose areolar tissue, the subarachnoidean. On the upper surface of the cerebrum, the arachnoid is thin and transparent, and may be easily demonstrated by injecting a stream of air beneath it by means of a blowpipe; it passes over the convolutions without dipping down into the sulci between them. At the base of the brain the arachnoid is thicker, and slightly opaque towards the central part; it covers the anterior lobes, and is extended across between the two middle lobes, so as to leave a considerable interval between it and the brain, the *anterior subarachnoidean space*; it is closely adherent to the pons and under surface of the cerebellum; but between

* Kölliker denies this; and states, that the inner surface of the dura mater is covered with pavement epithelium, but has no other investment which can be regarded as a parietal layer of the arachnoid.

the hemispheres of the cerebellum and the medulla oblongata another considerable interval is left between it and the brain, called the *posterior subarachnoidean space.* These two spaces communicate together across the crura cerebri. The arachnoid membrane surrounds the nerves which arise from the brain, and encloses them in loose sheaths as far as their point of exit from the skull, where it becomes continuous with the parietal layer.

The *subarachnoid space* is the interval between the arachnoid and pia mater: this space is narrow on the surface of the hemispheres; but at the base of the brain a wide interval is left between the two middle lobes, and behind, between the hemispheres of the cerebellum and the medulla oblongata. This space is the seat of an abundant serous secretion, the cerebro-spinal fluid, which fills up the interval between the arachnoid and pia mater. The subarachnoid space usually communicates with the general ventricular cavity of the brain by means of an opening in the inferior boundary of the fourth ventricle.

The sac of the arachnoid also contains serous fluid; this is, however, small in quantity compared with the cerebro-spinal fluid.

Structure. The arachnoid consists of bundles of white fibrous and elastic tissues intimately blended together. Its outer surface is covered with a layer of scaly epithelium. It is destitute of vessels, and the existence of nerves in it has not been satisfactorily demonstrated.

The *cerebro-spinal fluid* fills up the subarachnoid space, keeping the opposed surfaces of the arachnoid membrane in contact. It is a clear, limpid fluid, having a saltish taste, and a slightly alkaline reaction. According to Lassaigne, it consists of 98·5 parts of water, the remaining 1·5 per cent. being solid matters, animal and saline. It varies in quantity, being most abundant in old persons, and is quickly reproduced. Its chief use is probably to afford mechanical protection to the nervous centres, and to prevent the effects of concussions communicated from without.

Pia Mater.

The pia mater is a vascular membrane, and derives its blood from the internal carotid and vertebral arteries. It consists of a minute plexus of blood-vessels, held together by an extremely fine areolar tissue. It invests the entire surface of the brain, dipping down between the convolutions and laminæ, and is prolonged into the interior, forming the velum interpositum and choroid plexuses of the fourth ventricle. Upon the surfaces of the hemispheres, where it covers the grey matter of the convolutions, it is very vascular, and gives off from its inner surface a multitude of minute vessels, which extend perpendicularly for some distance into the cerebral substance. At the base of the brain, in the situation of the substantia perforata and locus perforatus, a number of long straight vessels are given off, which pass through the white matter to reach the grey substance in the interior. On the cerebellum the membrane is more delicate, and the vessels from its inner surface are shorter. Upon the crura cerebri and pons Varolii its characters are altogether changed; it here presents a dense fibrous structure, marked only by slight traces of vascularity.

According to Fohmann and Arnold, this membrane contains numerous lymphatic vessels. Its nerves are derived from the sympathetic, and also from the third, sixth, seventh, eighth, and spinal accessory. They accompany the branches of the arteries.

The Brain.

The brain (*encephalon*) is that portion of the cerebro-spinal axis that is contained in the cranial cavity. It is divided into four principal parts: viz., the cerebrum, the cerebellum, the pons Varolii, and medulla oblongata.

The *cerebrum* forms the largest portion of the encephalon, and occupies a considerable part of the cavity of the cranium, resting in the anterior and middle fossæ of the base of the skull, and separated posteriorly from the cerebellum by the tentorium cerebelli. About the middle of its under surface is a narrow constricted portion, part of which, the crura cerebri, is continued onwards into the pons Varolii below, and through it to the medulla oblongata and spinal cord; whilst another portion, the crura cerebelli, passes down into the cerebellum.

The *cerebellum* (little brain or after brain) is situated in the inferior occipital fossæ, being separated from the under surface of the posterior lobes of the cerebrum by the tentorium cerebelli. It is connected to the rest of the encephalon by means of connecting bands, called *crura*: of these, two ascend to the cerebrum, two descend to the medulla oblongata, and two blend together in front, forming the pons Varolii.

The *pons Varolii* is that portion of the encephalon which rests upon the upper part of the basilar process. It constitutes the bond of union of the various segments above named, receiving, above, the crura from the cerebrum; at the sides, the crura from the cerebellum; and below, the medulla oblongata.

The *medulla oblongata* extends from the lower border of the pons Varolii to the upper part of the spinal cord. It lies beneath the cerebellum, resting on the lower part of the basilar groove of the occipital bone.

Weight of the encephalon. The average weight of the brain, in the adult male, is 49½ oz., or little more than 3 lb. avoirdupois; that of the female, 44 oz.; the average difference between the two being from 5 to 6 oz. The prevailing weight of the brain, in the male, ranges between 46 oz. and 53 oz.; and, in the female, between 41 oz. and 47 oz. In the male, the maximum weight out of 278 cases was 65 oz., and the minimum weight 34 oz. The maximum weight of the adult female brain, out of 191 cases, was 56 oz., and the minimum weight 31 oz. It appears that the weight of the brain increases rapidly up to the seventh year, more slowly to between sixteen and twenty, and still more slowly to between thirty and forty, when it reaches its maximum. Beyond this period, as age advances and the mental faculties decline, the brain diminishes slowly in weight, about an ounce for each subsequent decennial period. These results apply alike to both sexes.

The size of the brain appears to bear a general relation to the intellectual capacity of the individual. Cuvier's brain weighed rather more than 64 oz., that of the late Dr. Abercrombie 63 oz., and that of Dupuytren 62½ oz. On the other hand, the brain of an idiot seldom weighs more than 23 oz.

The human brain is heavier than that of all the lower animals excepting the elephant and whale. The brain of the former weighs from 8 lb. to 10 lb.; and that of the whale, in a specimen seventy-five feet long, weighed rather more than 5 lb.

Medulla Oblongata.

The medulla oblongata is the upper enlarged part of the spinal cord, and extends from the upper border of the atlas to the lower border of the pons Varolii. It is directed obliquely downwards and backwards; its anterior surface rests on the basilar groove of the occipital bone, its posterior surface is received into the fossa between the hemispheres of the cerebellum, forming the floor of the fourth ventricle. It is pyramidal in form, its broad extremity directed upwards, its lower end being narrow at its point of connection with the cord. It measures an inch and a quarter in length, three-quarters of an inch in breadth at its widest part, and half an inch in thickness. Its surface is marked, in the median line, in front and behind, by an anterior and posterior median fissure, which are continuous with those of the spinal cord. The anterior fissure contains a fold of pia mater, and terminates just below the pons in a *cul-de-sac*, the foramen cæcum. The posterior is a deep but narrow fissure, continued upwards along the floor of the fourth ventricle, where it is finally lost. These two fissures divide the medulla into two symmetrical halves, each lateral half being subdivided by minor grooves into four columns, which, from before backwards, are named the *anterior pyramid, lateral tract and olivary body*, the *restiform body*, the *posterior pyramid*.

The *anterior pyramids*, or *corpora pyramidalia*, are two pyramidal-shaped bundles of white matter, placed one on either side of the anterior median fissure, and separated from the olivary body, which is external to them, by a slight depression. At the lower border of the pons they are somewhat constricted; they then become enlarged, and taper slightly as they descend, being continuous below with the anterior columns of the cord. On separating the pyramids below, it will be observed that their innermost fibres form from four to five bundles on each side, which decussate with one another; this decussation, however, is not formed entirely of fibres from the pyramids, but mainly from the deep portion of the lateral columns of the cord which pass forwards to the surface between the diverging anterior columns. The outermost fibres do not decussate; they are derived from the anterior columns of the cord, and are continued directly upwards through the pons Varolii.

257.—Medulla Oblongata and Pons Varolii. Anterior Surface.

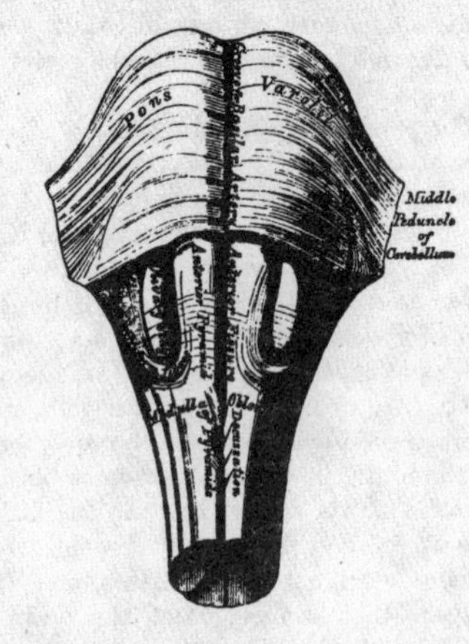

Lateral tract and *olivary body*. The lateral tract is continuous with the lateral column of the cord. Below, it is broad, and includes that part of the medulla between the anterior pyramid and restiform body; but, above, it is pushed a little backwards, and narrowed by the projection forwards of the olivary body.

The *olivary bodies* are two prominent, oval masses, situated behind the anterior pyramids, from which they are separated by slight grooves. They equal, in breadth, the anterior pyramids, are a little broader above than below, and are about half an inch in length, being separated, above, from the pons Varolii by a slight depression. Numerous white fibres (*fibræ arciformes*) are seen winding round the lower end of each body; sometimes crossing their surface.

The *restiform bodies* (fig. 258) are the largest columns of the medulla, and continuous, below, with the posterior columns of the cord. They are two rounded, cord-like eminences, placed between the lateral tracts, in front, and the posterior pyramids, behind; from both of which they are separated by slight grooves. As they ascend, they diverge from each other, assist in forming the lateral boundaries of the fourth ventricle, and then enter the corresponding hemisphere of the cerebellum, forming its inferior peduncle; it is probable that some fibres are continued from the restiform bodies into the cerebrum.

The *posterior pyramids* (*fasciculi graciles*) are two narrow, white cords, placed one on each side of the posterior median fissure, and separated from the restiform bodies by a narrow groove. They consist entirely of white fibres, and are continuous with the posterior median columns of the spinal cord. These bodies lie, at first, in close contact. Opposite the apex of the fourth ventricle they form an enlargement (*processus clavatus*), and then, diverging, are lost in the corresponding restiform body. The upper part of the posterior pyramids form the lateral boundaries of the calamus scriptorius.

The *posterior surface of the medulla oblongata* forms part of the floor of the fourth ventricle. It is of a triangular form, bounded on each side by the diverging posterior pyramids, and is that part of the ventricle which, from its resemblance to the point of a pen, is called the *calamus scriptorius*. The divergence of the posterior pyramids and restiform bodies, opens to view the grey matter of the medulla, which is continuous, below, with the grey commissure of the cord. In the middle line is seen a longitudinal furrow, continuous with the posterior median fissure of the cord, terminating, below, at the point of the ventricle, in a *cul-de-sac*, the ventricle of Arantius, which descends into the medulla for a slight extent. It is the remains of a canal, which, in the fœtus, extends throughout the entire length of the cord.

258.—Posterior Surface of Medulla Oblongata.

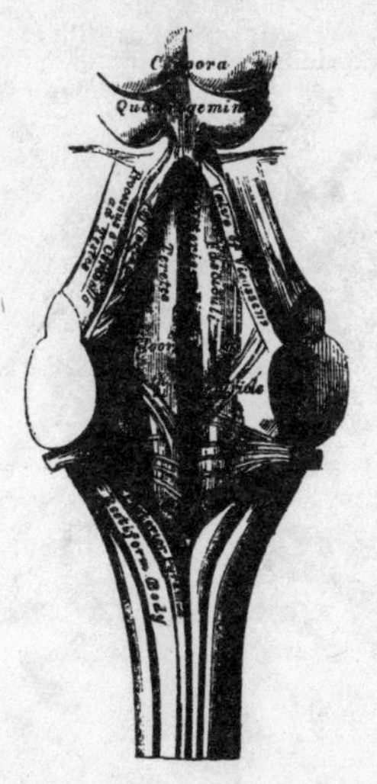

Structure. The columns of the cord are directly continuous with those of the medulla oblongata, below; but, higher up, both the white and grey constituents are rearranged before they are continued upwards to the cerebrum and cerebellum.

The *anterior pyramid* is composed of fibres derived from the anterior column of the cord of its own side, and from the lateral column of the opposite half of the cord, and is continued upwards into the cerebrum and cerebellum. The cerebellar fibres form a superficial and deep layer, which pass beneath the olivary body to the restiform body, and spread out into the structure of the cerebellum. A deeper fasciculus encloses the olivary body, and, receiving fibres from it, enters the pons as the olivary fasciculus or fillet; but the chief mass of fibres from the pyramid, the cerebral fibres, enter the pons in their passage upwards to the cerebrum. The anterior pyramids contain no grey matter.

The *lateral tract* is continuous, below, with the lateral column of the cord. Its fibres pass in three different directions. The most external join the restiform body, and pass to the cerebellum. The internal, more numerous, pass forwards, pushing aside the fibres of the anterior column, and form part of the opposite anterior pyramid. The middle fibres ascend, beneath the olivary body, to the cerebrum, passing along the back of the pons, and form, together with fibres from the restiform body, the *fasciculi teretes*, in the floor of the fourth ventricle.

259.—Transverse Section of Medulla Oblongata.

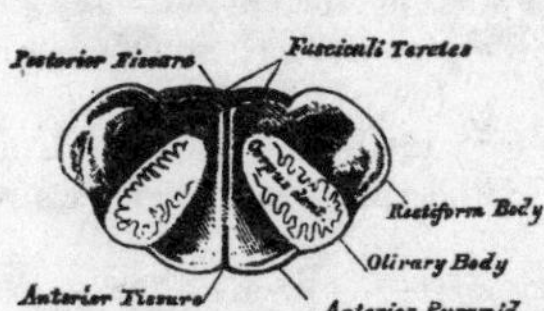

Olivary body. If a transverse section is made through either olivary body, it will be found to consist of a small ganglionic mass, deeply imbedded in the medulla, partly appearing on the surface as a smooth, olive-shaped eminence (fig. 259). It consists, externally, of white substance; and, internally, of a grey nucleus, the corpus dentatum. The grey matter is arranged in the form of a hollow capsule, open at its upper and inner part, and presenting a zigzag, or dentated outline. White fibres pass into, or from the interior of this body, by the aperture in the posterior part of the capsule. They join with those fibres of the anterior column which ascend on the outer side, and beneath the olivary body, to form the olivary fasciculus, which ascends to the cerebrum.

The *restiform body* is formed chiefly of fibres from the posterior column of the cord; but it receives some from the lateral column, and a fasciculus from the anterior, and is continued, upwards, to the cerebrum and cerebellum. On entering the pons, it divides into two fasciculi, above the point of the fourth ventricle. The external fasciculus enters the cerebellum: the inner fasciculus joins the posterior pyramid, is continued up along the fourth ventricle, and is traced up to the cerebrum with the fasciculi teretes.

260.—The Columns of the Medulla Oblongata, and their connection with the Cerebrum and Cerebellum.

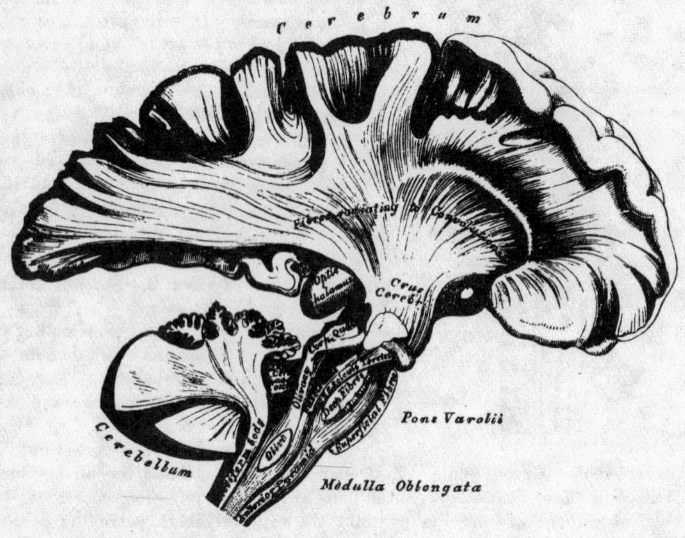

Septum of the medulla oblongata. Above the decussation of the anterior pyramids, numerous white fibres extend, from behind forwards, in the median line, forming a septum, which subdivides the medulla into two lateral halves. Some of these fibres emerge at the anterior median fissure, and form a band which curves round the lower border of the olivary body, or passes transversely across it, and round the sides of the medulla, forming the *arciform fibres* of Rolando. Others appear in the floor of the fourth ventricle, issuing from the posterior median fissure, and form the white striæ in that situation.

Grey matter of the medulla oblongata. The grey matter of the medulla is a continuation of that contained in the interior of the spinal cord, besides a series of special deposits or nuclei.

In the lower part of the medulla, the grey matter is arranged as in the cord, but at the upper part it becomes more abundant, and is disposed with less apparent regularity, becoming blended with all the white fibres, except the anterior pyramids. The part corresponding to the transverse grey commissure of the cord is exposed to view in the floor of the medulla oblongata, by the divergence of the restiform bodies, and posterior pyramids, becoming blended with the ascending fibres of the lateral column, and thus forming the fasciculi teretes. The lateral crescentic portions, but especially the posterior horns, become enlarged, blend with the fibres of the restiform bodies, and form the *tuberculo cinereo* of Rolando.

Special deposits of grey matter are found both in the anterior and posterior parts of the medulla; forming, in the former situation, the corpus dentatum within the olivary body, and, in the latter, a series of special masses, or nuclei, connected with the roots of origin of the spinal accessory, vagus, glosso-pharyngeal, and hypoglossal nerves. It thus appears that the closest analogy in structure, and also probably in general endowments, exists between the medulla oblongata and the spinal cord. The larger size and peculiar form of the medulla depends on the enlargement, divergence, and decussation of the various columns; and also on the addition of special deposits of grey matter in the olivary bodies and other parts, evidently in adaptation to the more extended range of function which this part of the cerebro-spinal axis possesses.

Pons Varolii.

The pons Varolii (*mesocephale*, Chaussier) is the bond of union of the various segments of the encephalon, connecting the cerebrum above, the medulla oblongata below, and the cerebellum behind. It is situated above the medulla oblongata, below the crura cerebri, and between the hemispheres of the cerebellum.

Its *under surface* presents a broad transverse band of white fibres, which arches like a bridge across the upper part of the medulla, extending between the two hemispheres of the cerebellum. This surface projects considerably beyond the level of these parts, is of a quadrangular form, rests upon the basilar groove of the occipital bone, and is limited before and behind by very prominent margins. It presents along the middle line a longitudinal groove, wider in front than behind, which lodges the basilar artery; numerous transverse striæ are also observed on each side, which indicate the course of its superficial fibres.

Its *upper surface* forms part of the floor of the fourth ventricle, and at each side it becomes contracted into a thick rounded cord, the crus cerebelli, which enters the substance of the cerebellum, constituting its middle peduncle.

Structure. The pons Varolii consists of alternate layers of transverse and longitudinal fibres intermixed with grey matter (fig. 260).

The *transverse fibres* connect together the two lateral hemispheres of the cerebellum, and constitute its great transverse commissure. They consist of a superficial and a deep layer. The superficial layer passes uninterruptedly across the surface of the pons, forming a uniform layer, which consists of fibres derived from the crus cerebelli on each side, meeting in the median line. The deep layer of transverse fibres decussates with the longitudinal fibres continued up from the medulla, and contains much grey matter between its fibres.

The *longitudinal fibres* are continued up through the pons. 1. From the anterior pyramid. 2. From the olivary body. 3. From the lateral and posterior columns of the cord, receiving special fibres from the grey matter of the pons itself.

1. The fibres from the anterior pyramid ascend through the pons, embedded between two layers of transverse fibres, being subdivided in their course into smaller bundles; at the upper border of the pons they enter the crus cerebri, forming its fasciculated portion.

2. The olivary fasciculus divides in the pons into two bundles, one of which ascends to the corpora quadrigemina; the other is continued to the cerebrum with the fibres of the lateral column.

3. The fibres from the lateral and posterior columns of the cord, with a bundle from the olivary fasciculus, are intermixed with much grey matter, and appear in the floor of the fourth ventricle as the fasciculi teretes: they ascend to the deep or cerebral part of the crus cerebri.

Foville believes that a few fibres from each of the longitudinal tracts of the medulla turn forwards, and are continuous with the transverse fibres of the pons.

Septum. The pons is subdivided into two lateral halves by a median septum, which extends through its posterior half. The septum consists of antero-posterior and transverse fibres. The former are derived from the floor of the fourth ventricle and from the transverse fibres of the pons, which bend backwards before passing across to the opposite side. The latter are derived from the floor of the fourth ventricle; they pierce the longitudinal fibres, and are then continued across from one to the other side of the medulla, piercing the antero-posterior fibres.

The two halves of the pons, in front, are connected together by transverse commissural fibres.

Cerebrum. Upper Surface. (Fig. 261.)

The cerebrum, in man, constitutes the largest portion of the encephalon. Its upper surface is of an ovoidal form, broader behind than in front, convex in its general outline, and divided into two lateral halves or hemispheres, right and left, by the great longitudinal fissure, which extends throughout the entire length of the cerebrum in the middle line, reaching down to the base of the brain in front and behind, but interrupted in the middle by a broad transverse commissure of white matter, the corpus callosum, which connects the two hemispheres together. This fissure lodges the falx cerebri, and indicates the original development of the brain by two lateral halves.

Each hemisphere presents an outer surface, which is convex, to correspond with the vault of the cranium; an inner surface, flattened, and in contact with the opposite hemisphere (the two inner surfaces forming the sides of the longitudinal fissure); and an under surface or base, of more irregular form, which rests, in front, on the anterior and middle fossæ at the base of the skull, and behind, upon the tentorium.

Convolutions. If the pia mater is removed with the forceps, the entire surface of each hemisphere will be seen to present a number of convoluted eminences, *the convolutions*, separated from each other by depressions (*sulci*) of various depths. The outer surface of each convolution, as well as the sides and bottom of the sulci between them, are composed of grey matter, which is here called the *cortical substance.* The interior of each convolution is composed of white matter; and white fibres also blend with the grey matter at the sides and bottom of the sulci. By this arrangement the convolutions are adapted to increase the amount of grey matter without occupying much additional space, while they also afford a greater extent of surface for the termination of the white fibres in grey matter. On closer examination, however, the cortical substance is found subdivided into four layers, two of which are composed of grey and two of white matter. The most external is an outer white stratum, not equally thick over all parts of the brain, being most marked on the convolutions in the longitudinal fissure and on the under part of the brain, especially on the middle lobe, near the descending horn of the lateral ventricle. Beneath this is a thick reddish-grey lamina, and then another thin white stratum; lastly, a thin stratum of grey matter, which lies in close contact with the white fibres of the hemispheres: consequently white and grey laminæ alternate with one another in the convolutions. In certain convolutions, however, the cortical substance consists of no less than six layers, three grey and three white, an additional white stratum dividing the most superficial grey one into two; this is especially marked in those convolutions which are situated near the corpus callosum.

There is no accurate resemblance between the convolutions in different brains, nor are they symmetrical on the two sides of the same brain. Occasionally the free borders or the sides of a deep convolution present a fissured or notched appearance.

The *sulci* are generally an inch in depth; they also vary in different brains, and in different parts of the same brain; they are usually deepest on the outer convex surface of the hemispheres; the deepest is situated on the inner surface of the hemisphere, on a level with the corpus callosum, and corresponds to the projection in the posterior horn of the lateral ventricle, the hippocampus minor.

The number and extent of the convolutions, as well as their depth, appear to bear a close relation to the intellectual power of the individual, as is shown in their increasing complexity of arrangement as we ascend from the lowest mammalia up to man. Thus they are absent in some of the lower orders of mammalia, and they increase in number and extent through the higher orders. In man they present the most complex arrangement. Again, in the child at birth before the intellectual faculties are exercised, the convolutions have a very simple arrangement, presenting few undulations; and the sulci between them are less deep than in the adult. In old age, when the mental faculties have diminished in activity, the convolutions become much less prominently marked.

Those convolutions which are the largest and most constantly present, are the convolution of the corpus callosum, the convolution of the longitudinal fissure, the supraorbital convolution, and the convolutions of the outer surface of the hemisphere.

261.—Upper Surface of the Brain, the Pia Mater having been removed.

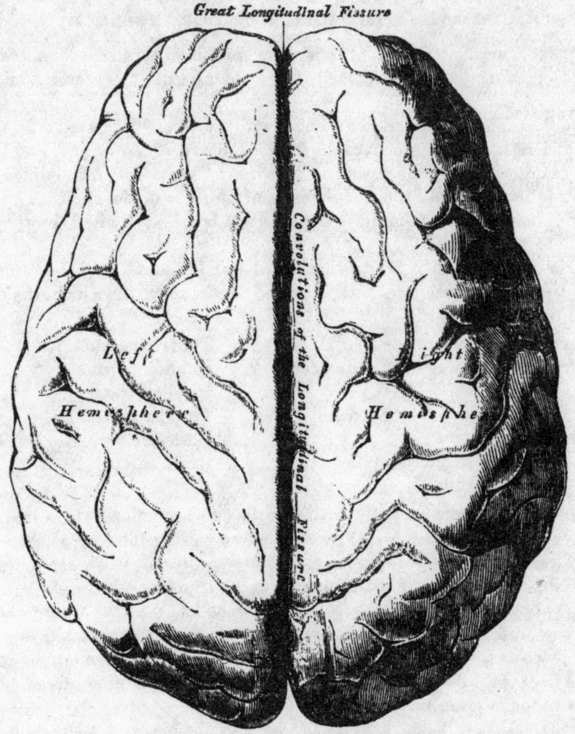

The *convolution of the corpus callosum* (*gyrus fornicatus*) is always well marked. It lies parallel with the free surface of the corpus callosum, commencing, on the under surface of the brain, in front of the anterior perforated space; it winds round the curved border of the corpus callosum, and passes along its upper surface as far as its posterior extremity, where it is connected with the convolutions of the posterior lobe; it then curves downwards and forwards, embracing the cerebral peduncle, passes into the middle lobe, forming the hippocampus major, and terminates just behind the point from whence it arose.

The *supraorbital convolution* on the under surface of the anterior lobe is well marked.

The *convolution of the longitudinal fissure* bounds the margin of the fissure on the upper surface of the hemisphere. It commences on the under surface of the brain, at the anterior perforated space, passes forwards along the inner margin of the anterior lobe, being here divided by a deep sulcus, in which the olfactory nerve is received; it then curves over the anterior and upper surface of the hemisphere, along the margin of the longitudinal fissure, to its posterior extremity, where it curves forwards along the under surface of the hemisphere as far as the middle lobe.

The convolutions on the outer convex surface of the hemisphere, the general direction of which is more or less oblique, are the largest and the most complicated convolutions of the brain, frequently becoming branched like the letter Y in their course upwards and backwards towards the longitudinal fissure: these convolutions attain their greatest development in man, and are especially characteristic of the human brain. They are seldom symmetrical on the two sides.

Cerebrum. Under Surface or Base. (Fig. 262.)

The under surface of each hemisphere presents a subdivision, as already mentioned, into three lobes, named from their position, anterior, middle, and posterior.

The *anterior lobe*, of a triangular form, with its apex backwards, is somewhat concave, and rests upon the convex surface of the roof of the orbit, being separated from the middle lobe by the fissure of Sylvius. The *middle lobe*, which is more prominent, is received into the middle fossa of the base of the skull. The *posterior lobe* rests upon the tentorium, its extent forwards being limited by the anterior margin of the cerebellum.

The various objects exposed to view on the under surface of the cerebrum, in and near the middle line, are here arranged in the order in which they are met with from before backwards.

- Longitudinal fissure.
- Corpus callosum and its peduncles.
- Lamina cinerea.
- Olfactory nerve.
- Fissure of Sylvius.
- Anterior perforated space.
- Optic commissure.
- Tuber cinereum.
- Infundibulum.
- Pituitary body.
- Corpora albicantia.
- Posterior perforated space.
- Crura cerebri.

The *longitudinal fissure* partially separates the two hemispheres from one another; it divides the two anterior lobes in front: and on raising the cerebellum and pons, it will be seen completely separating the two posterior lobes, the intermediate portion of the fissure being filled up by the great transverse band of white matter, the corpus callosum. Of these two portions of the longitudinal fissure, that which separates the posterior lobes is the longest. In the fissure between the two anterior lobes the anterior cerebral arteries may be seen ascending to the corpus callosum; and at the back part of this portion of the fissure, the anterior curved portion of the corpus callosum descends to the base of the brain.

The *corpus callosum* terminates at the base of the brain by a concave margin, which is connected with the tuber cinereum through the intervention of a thin layer of grey substance, the lamina cinerea. This may be exposed by gently raising and drawing back the optic commissure. A broad white band may be observed on each side, passing from the under surface of the corpus callosum backwards and outwards, to the commencement of the fissure of Sylvius; these bands are called the *peduncles of the corpus callosum.* Laterally, the corpus callosum extends into the anterior lobe.

The *lamina cinerea* is a thin layer of grey substance, extending backwards above the optic commissure from the termination of the corpus callosum to the tuber cinereum; it is continuous on either side with the grey matter of the anterior perforated space, and forms the anterior part of the inferior boundary of the third ventricle.

The *olfactory nerve*, with its bulb, is seen on either side of the longitudinal fissure, upon the under surface of each anterior lobe.

The *fissure of Sylvius* separates the anterior and middle lobes, and lodges the middle cerebral artery. At its commencement is seen a point of medullary substance, corresponding to a subjacent band of white fibres, connecting the anterior and middle lobes, and called the *fasciculus unciformis*; on following this fissure outwards, it divides into two branches, which enclose a triangular-shaped prominent cluster of isolated convolutions, the *island of Reil*. These convolutions, from being covered in by the sides of the fissure, are called the *gyri operti*.

262.—Base of the Brain.

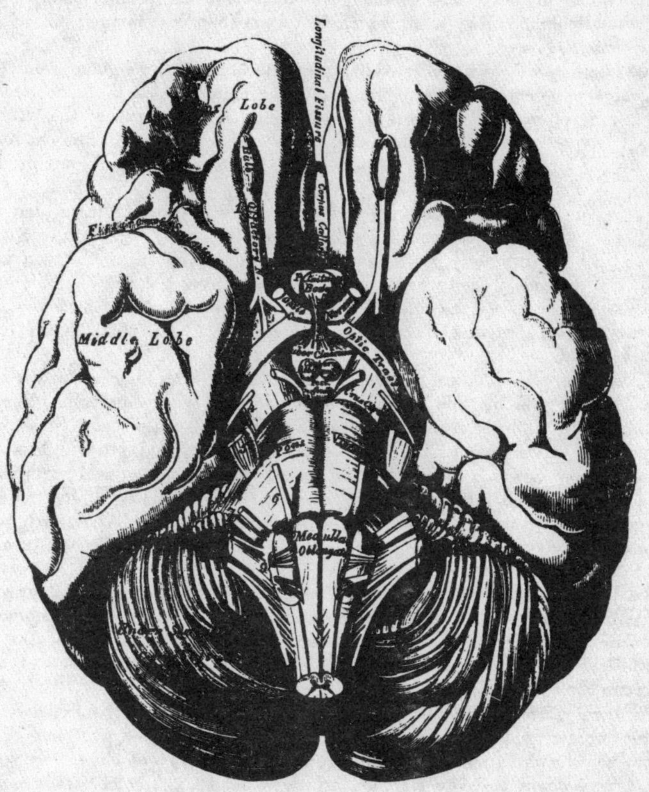

The *anterior perforated space* is situated at the inner side of the fissure of Sylvius. It is of a triangular shape, bounded in front by the convolution of the anterior lobe and the roots of the olfactory nerve; behind, by the optic tract; externally, by the middle lobe and commencement of the fissure of Sylvius; internally, it is continuous with the lamina cinerea, and crossed by the peduncle of the corpus callosum. It is of a greyish colour, and corresponds to the under surface of the corpus striatum, a large mass of grey matter, situated in the interior of the brain; it has received its name from being perforated by numerous minute apertures for the transmission of small straight vessels into the substance of the corpus striatum.

The *optic commissure* is situated in the middle line, immediately behind the lamina cinerea. It is the point of junction between the two optic nerves.

Immediately behind the diverging optic tracts, and between them and the peduncles of the cerebrum (*crura cerebri*), is a lozenge-shaped interval, the interpeduncular space, in which are found the following parts, arranged in the following order from before backwards: the tuber cinereum, infundibulum, pituitary body, corpora albicantia, and the posterior perforated space.

The *tuber cinereum* is an eminence of grey matter, situated between the optic tracts and the corpora albicantia; it is connected with the surrounding parts of the cerebrum, forms part of the floor of the third ventricle, and is continuous with the grey substance in that cavity. From the middle of its under surface a conical tubular process of grey matter, about two lines in length, is continued downwards and forwards to be attached to the posterior lobe of the pituitary body; this is the infundibulum. Its canal, which is funnel-shaped, communicates with the third ventricle.

The *pituitary body* is a small reddish-grey vascular mass, weighing from five to ten grains, and of an oval form, situated in the sella Turcica, in connection with which it is retained by the dura mater forming the inner wall of the cavernous sinus. It is very vascular, and consists of two lobes, separated from one another by a fibrous lamina. Of these, the anterior is the larger, of an oblong form, and somewhat concave behind, where it receives the posterior lobe, which is round. The anterior lobe consists externally of firm yellowish-grey substance, and internally of a soft pulpy substance of a yellowish-white colour. The posterior lobe is darker than the anterior. In the fœtus it is larger proportionately than in the adult, and contains a cavity which communicates through the infundibulum with the third ventricle. In the adult it is firmer and more solid, and seldom contains any cavity. Its structure, especially the anterior lobe, is similar to that of the ductless glands.

The *corpora albicantia* are two small round white masses, each about the size of a pea, placed side by side immediately behind the tuber cinereum. They are formed by the anterior crura of the fornix, hence called the *bulbs of the fornix*, which, after descending to the base of the brain, are folded upon themselves, before passing upwards to the thalami optici. They are composed externally of white substance, and internally of grey matter; the grey matter of the two being connected by a transverse commissure of the same material. At an early period of fœtal life they are blended together into one large mass, but become separated about the seventh month.

The *posterior perforated space* (*pons Tarini*) corresponds to a whitish-grey substance, placed between the corpora albicantia in front, the pons Varolii behind, and the crura cerebri on either side. It forms the back part of the floor of the third ventricle, and is perforated by numerous small orifices for the passage of blood-vessels to the thalami optici.

The *crura cerebri* (*peduncles of the Cerebrum*) are two thick cylindrical bundles of white matter, which emerge from the anterior border of the pons, and diverge as they pass forwards and outwards to enter the under part of either hemisphere. Each crus is about three-quarters of an inch in length, and somewhat broader in front than behind. They are marked upon their surface with longitudinal striæ, and each is crossed, just before entering the hemisphere, by a flattened white band, the optic tract, which is adherent by its upper border to the peduncle. In the interior of the crura is contained a mass of dark grey matter, called *locus niger*. The third nerves may be seen emerging from the inner side of either crus; and the fourth nerve winding around its outer side from above.

Each crus consists of a superficial and deep layer of longitudinal white fibres, continued upwards from the pons: these layers are separated from each other by the locus niger.

The *superficial longitudinal fibres* are continued upwards, from the anterior pyramids to the cerebrum. They consist of coarse fasciculi, which form the free part of the crus, and have received the name of the fasciculated portion of the peduncle, or crust.

The *deep layer of longitudinal fibres* are continued upwards, to the cerebrum, from the lateral and posterior columns of the medulla, and from the olivary fasciculus, these fibres consisting of some derived from the same, and others from the opposite lateral tract of the medulla. More deeply, are a layer of finer fibres, mixed with grey matter, derived from the cerebellum, blended with the former. The cerebral surface of the crus cerebri is formed of these fibres, and is named the tegmentum.

The *locus niger* is a mass of grey matter, situated between the superficial and deep layer of fibres above described. It is placed nearer the inner than the outer side of the crus.

The posterior lobes of the cerebrum are concealed from view by the upper surface of the cerebellum, and pons Varolii. When these parts are removed, the two hemispheres are seen to be separated by the great longitudinal fissure, this fissure being interrupted, in front, by the posterior rounded border of the corpus callosum.

General Arrangement of the Parts composing the Cerebrum.

As the peduncles of the cerebrum enter the hemispheres, they diverge from one another, so as to leave an interval between them, the interpeduncular space. As they ascend, the component fibres of each pass through two large masses of grey matter, the ganglia of the brain, called the *thalamus opticus* and *corpus striatum*, which project as rounded eminences from the upper and inner side of each peduncle. The hemispheres are connected together, above these masses, by the great transverse commissure, the corpus callosum, and the interval left between its under surface, the upper surface of the ganglia, and the parts closing the interpeduncular space, forms the general ventricular cavity. The upper part of this cavity is subdivided into two, by a vertical septum, the septum lucidum; and thus the two lateral ventricles are formed. The lower part of the cavity forms the third ventricle, which communicates with the lateral ventricles, above, and with the fourth ventricle, behind. The fifth ventricle is the interval left between the two layers composing the septum lucidum.

Interior of the Cerebrum.

If the upper part of either hemisphere is removed with a scalpel, about half an inch above the level of the corpus callosum, its internal white matter will be exposed. It is an oval-shaped centre, of white substance, surrounded on all sides by a narrow, convoluted margin of grey matter which presents an equal thickness in nearly every part. This white, central mass, has been called the *centrum ovale minus*. Its surface is studded with numerous minute red dots (*puncta vasculosa*), produced by the escape of blood from divided blood-vessels. In inflammation, or great congestion of the brain, these are very numerous, and of a dark colour. If the remaining portion of one hemisphere is slightly separated from the other, a broad band of white substance will be observed connecting them, at the bottom of the longitudinal fissure: this is the corpus callosum. The margins of the hemispheres, which overlap this portion of the brain, are called the *labia cerebri*. Each labium is part of the convolution of the corpus callosum (*gyrus fornicatus*), already described; and the space between it and the upper surface of the corpus callosum, has been termed the *ventricle of the corpus callosum*.

The hemispheres should now be sliced off, to a level with the corpus callosum, when the white substance of that structure will be seen connecting together both hemispheres. The large expanse of medullary matter now exposed, surrounded by the convoluted margin of grey substance, is called the *centrum ovale majus* of Vieussens.

The *corpus callosum* is a thick stratum of transverse fibres, exposed at the bottom of the longitudinal fissure. It connects the two hemispheres of the brain, forming their great transverse commissure; and forms the roof of a space in the interior of each hemisphere, the lateral ventricle. It is about four inches in length, extending to within an inch and a half of the anterior, and to within two inches and a half of the posterior, part of the brain. It is somewhat broader behind than in front, and is thicker at either end than in its central part, being thickest behind. It presents a somewhat arched form, from before backwards, terminating anteriorly in a rounded border, which curves downwards and backwards, between the anterior lobes to the base of the brain. In its course, it forms a distinct bend, named the *knee*, or *genu*, and the reflected portion, named the *beak* (*rostrum*), becoming gradually narrower, is attached to the anterior cerebral lobe, and is

263.—Section of the Brain. Made on a level with the Corpus Callosum.

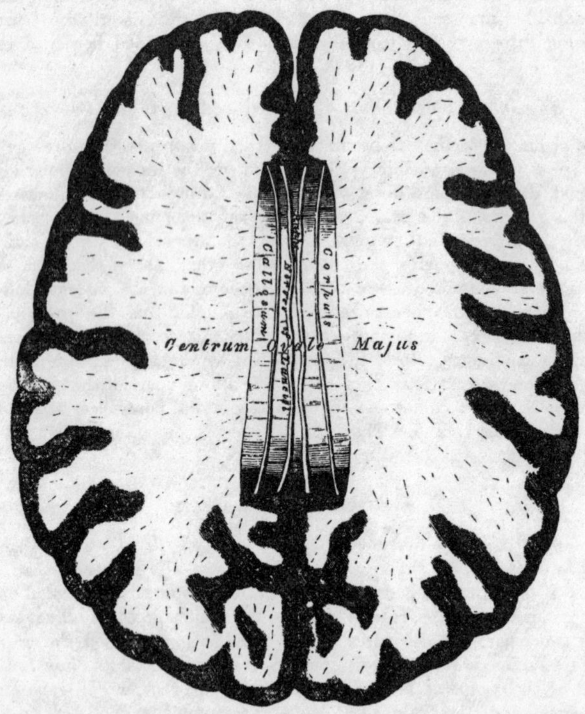

connected through the lamina cinerea with the optic commissure. The reflected portion of the corpus callosum gives off, near its termination, two bundles of white substance, which, diverging from one another, pass backwards, across the anterior perforated space, to the entrance of the fissure of Sylvius. They are called the *peduncles* of the corpus callosum. Posteriorly, the corpus callosum forms a thick, rounded fold, which is free for a little distance, as it curves forwards, and is then continuous with the fornix. On its upper surface, its fibrous structure is very apparent to the naked eye, being collected into coarse, transverse bundles. Along the middle line is a linear depression, the raphe, bounded laterally by two or more slightly elevated longitudinal bands, called the *striæ longitudinales*, or *nerves of Lancisi*; and, still more externally, other longitudinal striæ are seen, beneath the convolutions which rest on the corpus callosum. These are the *striæ longitudinales laterales*. The under surface of the corpus callosum is continuous behind with the fornix, being separated from it in front by the septum lucidum, which forms a vertical partition between the two ventricles. On either side, the fibres of the corpus callosum penetrate into the substance of the hemispheres, and connect together the anterior, middle, and part of the posterior lobes. It is the large number of fibres derived from the anterior and posterior lobes which explains the great thickness of the two extremities of this commissure.

An incision should now be made through the corpus callosum, on either side of the raphe, when two large irregular-shaped cavities will be exposed, which extend through a great part of the length of each hemisphere. These are the lateral ventricles.

The *lateral ventricles* are serous cavities, formed by the upper part of the general ventricular space in the interior of the brain. They are lined by a thin diaphanous lining membrane, covered with ciliated epithelium, and moistened by a serous

264.—The Lateral Ventricles of the Brain.

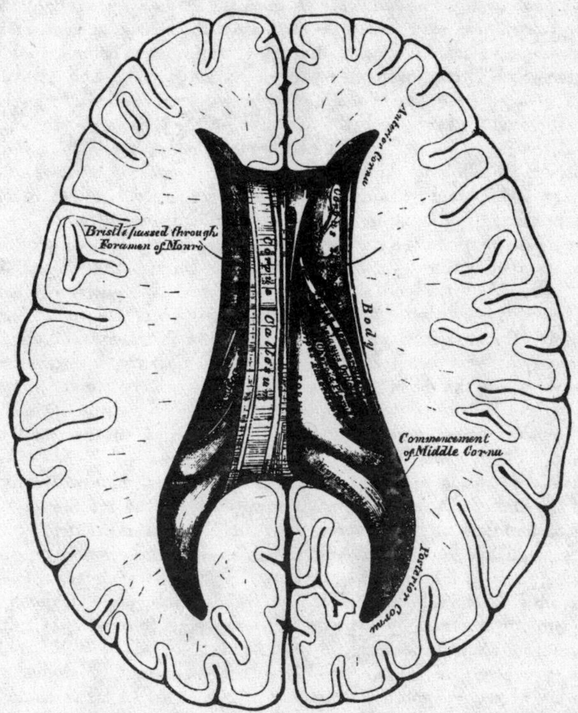

fluid, which is sometimes, even in health, secreted in considerable quantity. These cavities are two in number, one in each hemisphere, and they are separated from each other by a vertical septum, the *septum lucidum*.

Each lateral ventricle consists of a central cavity, or body, and three smaller cavities or cornua, which extend from it in different directions. The anterior cornu curves forwards and outwards, into the substance of the anterior lobe. The posterior cornu, called the *digital cavity*, curves backwards into the posterior lobe. The middle cornu descends into the middle lobe.

The *central cavity*, or body of the lateral ventricle, is triangular in form. It is bounded, above, by the under surface of the corpus callosum, which forms the roof of the cavity. Internally, is a vertical partition, the septum lucidum, which separates it from the opposite ventricle, and connects the under surface of the corpus callosum with the fornix. Its floor is formed by the following parts, enumerated in their order of position, from before backwards: the corpus striatum, tænia semicircularis, thalamus opticus, choroid plexus, corpus fimbriatum, and fornix.

The *anterior cornu* is triangular in form, passing outwards into the anterior lobe, and curving round the anterior extremity of the corpus striatum. It is bounded above and in front, by the corpus callosum; behind, by the corpus striatum.

The *posterior cornu*, or digital cavity, curves backwards into the substance of the posterior lobe, its direction being backwards and outwards, and then inwards. On its floor is seen a longitudinal eminence, which corresponds with a deep sulcus between two convolutions: this is called the *hippocampus minor*. Between the middle and posterior horns a smooth eminence is observed, which varies considerably in size in different subjects. It is called the *eminentia collateralis*.

The *corpus striatum* has received its name from the striped appearance which its section presents, in consequence of diverging white fibres being mixed with the grey matter which forms the greater part of its substance. The intraventricular portion is a large pear-shaped mass, of a grey colour externally; its broad extremity is directed forwards, into the fore-part of the body, and anterior cornu of the lateral ventricle: its narrow end is directed outwards and backwards, being separated from its fellow by the thalami optici: it is covered by the serous lining of the cavity, and crossed by some veins of considerable size. The extraventricular portion is imbedded in the white substance of the hemisphere.

The *tænia semicircularis* is a narrow, whitish, semi-transparent band of medullary substance, situated in the depression between the corpus striatum and thalamus opticus. Anteriorly, it descends in connection with the anterior pillar of the fornix; behind, it is continued into the descending horn of the ventricle, where it becomes lost. Its surface, especially at its fore-part, is transparent, and dense in structure, and this was called by Tarinus the *horny band*. It consists of longitudinal white fibres, the deepest of which run between the corpus striatum and thalamus opticus. Beneath it is a large vein (*vena corporis striati*), which receives numerous smaller veins from the surface of the corpus striatum and thalamus opticus, and joins the venæ Galeni.

The *choroid plexus* is a highly vascular, fringe-like membrane, occupying the margin of the fold of pia mater (*velum interpositum*), in the interior of the brain. It extends, in a curved direction, across the floor of the lateral ventricle. In front, where it is small and tapering, it communicates with the choroid plexus of the opposite side, through a large oval aperture, the *foramen of Monro*. Posteriorly, it descends into the middle horn of the lateral ventricle, where it joins with the pia mater through the transverse fissure. In structure, it consists of minute and highly vascular villous processes, the villi being covered by a single layer of epithelium, composed of large, round corpuscles, containing, besides a central nucleus, a bright yellow spot. The arteries of the choroid plexus enter the ventricle at the descending cornu, and, after ramifying through its substance, send branches into the substance of the brain. The veins of the choroid plexus terminate in the venæ Galeni.

The *corpus fimbriatum* (*tænia hippocampi*) is a narrow, white, tape-like band, situated immediately behind the choroid plexus. It is the lateral edge of the posterior pillar of the fornix, and is attached along the inner border of the hippocampus major as it descends into the middle horn of the lateral ventricle. It may be traced as far as the pes hippocampi.

The *thalami optici* and *fornix* will be described when more completely exposed, in a later stage of the dissection of the brain.

The middle cornu should now be exposed, throughout its entire extent, by introducing the little finger gently into it, and cutting outwards, along the finger through the substance of the hemisphere, which should be removed, to an extent sufficient to expose the entire cavity.

The *middle*, or *descending cornu*, the largest of the three, traverses the middle lobe of the brain, forming in its course a remarkable curve round the back of the optic thalamus. It passes, at first, backwards, outwards, and downwards, and then curves round the crus cerebri, forwards and inwards, nearly to the point of the middle lobe, close to the fissure of Sylvius. Its upper boundary is formed by the medullary substance of the middle lobe, and the under surface of the thalamus opticus. Its lower boundary, or floor, presents for examination the following parts: the hippocampus major, pes hippocampi, pes accessorius, corpus fimbriatum, choroid plexus, fascia dentata, transverse fissure.

The *hippocampus major*, or *cornu Ammonis*, so called from its resemblance to a ram's horn, is a white eminence, of a curved elongate form, extending along the

265.—The Fornix, Velum Interpositum, and Middle or Descending Cornu of the Lateral Ventricle.

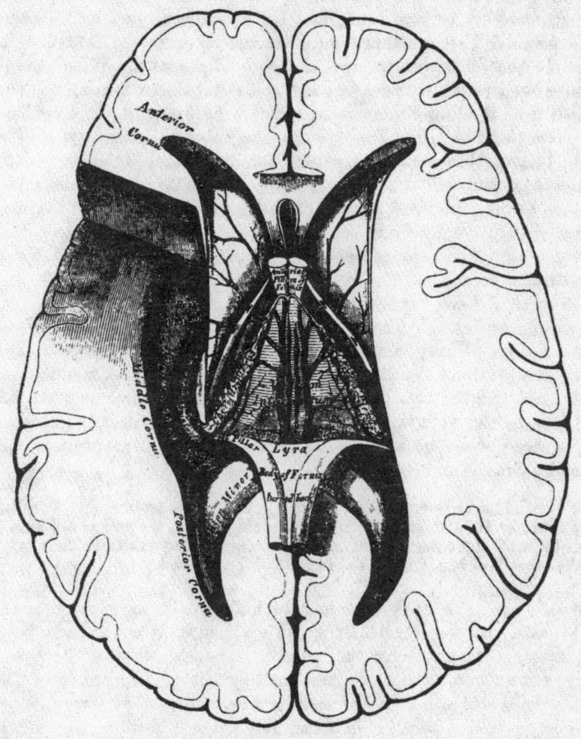

entire length of the floor of the middle horn of the lateral ventricle. At its lower extremity it becomes enlarged, and presents a number of rounded elevations with intervening depressions, which, from presenting some resemblance to the paw of an animal, is called the *pes hippocampi*. If a transverse section is made through the hippocampus major, it will be seen that this eminence is the inner surface of the convolution of the corpus callosum, doubled upon itself like a horn, the white convex portion projecting into the cavity of the ventricle; the grey portion being on the surface of the cerebrum, the edge of which, slightly indented, forms the fascia dentata. The white matter of the hippocampus major is continuous through the corpus fimbriatum, with the fornix and corpus callosum.

The *pes accessorius*, or *eminentia collateralis*, has been already mentioned, as a white eminence, varying in size, placed between the hippocampus major and minor, at the junction of the posterior with the descending cornu. Like the hippocampi, it is formed of white matter corresponding to one of the sulci, between two convolutions protruding into the cavity of the ventricle.

The *corpus fimbriatum* is a continuation of the posterior pillar of the fornix, prolonged, as already mentioned, from the central cavity of the lateral ventricle.

Fascia dentata. On separating the inner border of the corpus fimbriatum from the choroid plexus, and raising the edge of the former, a serrated band of grey substance, the edge of the grey substance of the middle lobe, will be seen beneath it: this is the fascia dentata. Correctly speaking, it is placed external to the cavity of the descending cornu.

The *transverse fissure* is seen on separating the corpus fimbriatum from the thalamus opticus. It is situated beneath the fornix, extending from the middle line behind, downwards on either side, to the end of the descending cornu, being bounded on one side by the fornix and the hemisphere, and on the other by the thalamus opticus. Through this fissure the pia mater passes from the exterior of the brain into the ventricles, to form the choroid plexuses. Where the pia mater projects into the lateral ventricle, beneath the edge of the fornix, it is covered by a prolongation of the lining membrane, which excludes it from the cavity.

The *septum lucidum* (fig. 264) forms the internal boundary of the lateral ventricle. It is a thin, semi-transparent septum, attached, above, to the under surface of the corpus callosum; below, to the anterior part of the fornix; and, in front of this, to the prolonged portion of the corpus callosum. It is triangular in form, broad in front, and narrow behind, its surfaces looking towards the cavities of the ventricles. The septum consists of two laminæ, separated by a narrow interval, the fifth ventricle.

Fifth Ventricle. Each lamina of the septum lucidum consists of an internal layer of white substance, covered by the lining membrane of the fifth ventricle; and an outer layer of grey matter, covered by the lining membrane of the lateral ventricle. The cavity of the fifth ventricle is lined by a serous membrane, covered with epithelium, and contains fluid. In the fœtus, and in some animals, this cavity communicates, below, with the third ventricle; but in the adult, it forms a separate cavity. In cases of serous effusion into the ventricles, the septum is often found softened and partially broken down.

The fifth ventricle may be exposed by cutting through the septum, and attached portion of the corpus callosum, with scissors; after examining which, the corpus callosum should be cut across, towards its anterior part, and the two portions carefully dissected, the one forwards, the other backwards, when the fornix will be exposed.

The *fornix* (figs. 264, 265) is a longitudinal lamella of white fibrous matter, situated beneath the corpus callosum, with which it is continuous behind, but separated from it in front by the septum lucidum. It may be described as consisting of two symmetrical halves, one for either hemisphere. These two portions are joined together in the middle line, where they form the body, but are separated from one another in front and behind; forming the anterior and posterior crura.

The *body* of the fornix is triangular; narrow in front, broad behind. Its upper surface is connected, in the median line, to the septum lucidum in front, and the corpus callosum behind. Its under surface rests upon the velum interpositum, which separates it from the third ventricle, and the inner portion of the optic thalami. Its lateral edges form, on each side, part of the floor of the lateral ventricles, and are in contact with the choroid plexuses.

The *anterior crura* arch downwards towards the base of the brain, separated from each other by a narrow interval. They are composed of white fibres, which descend through a quantity of grey matter in the lateral walls of the third ventricle, and are placed immediately behind the anterior commissure. At the base of the brain, the white fibres of each crus form a sudden curve upon themselves, spread out and form the outer part of the corresponding corpus albicans, from which point they may be traced upwards into the substance of the corresponding thalamus opticus. The anterior crura of the fornix are connected in their course with the optic commissure, the white fibres covering the optic thalamus, the peduncle of the pineal gland, and the superficial fibres of the tænia semicircularis.

The *posterior crura*, at their commencement, are intimately connected by their upper surfaces with the corpus callosum; diverging from one another, they pass downwards into the descending horn of the lateral ventricle, being continuous with the concave border of the hippocampus major. The lateral thin edges of the posterior crura have received the name *corpus fimbriatum*, already described. On the under surface of the fornix, towards its posterior part, between the diverging posterior crura, may be seen some transverse lines, and others longitudinal or oblique. This appearance has been termed the *lyra*, from the fancied resemblance it bears to the strings of a harp.

Between the anterior pillars of the fornix and the anterior extremities of the thalami optici, an oval aperture is seen on each side, the *foramen of Monro*. The two openings descend towards the middle line, and joining together, lead into the upper part of the third ventricle. These openings communicate with the lateral ventricles on each side, and below with the third ventricle.

Divide the fornix across anteriorly, and reflect the two portions, theone forwards, the other backwards, when the velum interpositum will be exposed.

The *velum interpositum* (fig. 265) is a vascular membrane, reflected from the pia mater into the interior of the brain through the transverse fissure, passing beneath the posterior rounded border of the corpus callosum and fornix, and above the corpora quadrigemina, pineal gland, and optic thalami. It is of a triangular form, and separates the under surface of the body of the fornix from the cavity of the third ventricle. Its posterior border forms an almost complete investment for the pineal gland. Its anterior extremity, or apex, is bifid; each bifurcation being continued into the corresponding lateral ventricle, behind the anterior crura of the fornix, forming the anterior extremity of the choroid plexus. On its under surface are two vascular fringes, which diverge from each other behind, and project into the cavity of the third ventricle. These are the choroid plexuses of the third ventricle. To its lateral margins are connected the choroid plexuses of the lateral ventricles. The arteries of the velum interpositum enter from behind, beneath the corpus callosum. Its veins, the venæ Galeni, two in number, run along its under surface; they are formed by the venæ corporis striati and the veins of the choroid plexuses: the venæ Galeni unite posteriorly into a single trunk, which terminates in the straight sinus.

The velum interpositum should now be removed. This must be effected carefully, especially at its posterior part, where it invests the pineal gland; the thalami optici will then be exposed with the cavity of the third ventricle between them (fig. 266).

The *thalami optici* are two large oblong masses, placed between the diverging portions of the corpora striata; they are of a white colour superficially; internally, they are composed of white fibres intermixed with grey matter. Each thalamus rests upon its corresponding crus cerebri, which it embraces. *Externally*, it is bounded by the corpus striatum, and tænia semicircularis; and is continuous with the hemisphere. *Internally*, it forms the lateral boundary of the third ventricle; and running along its upper border is seen the peduncle of the pineal gland. Its *upper surface* is free, being partly seen in the lateral ventricle; it is partly covered by the fornix, and marked in front by an eminence, the anterior tubercle. Its *under surface* forms the roof of the descending cornu of the lateral ventricle; into it the crus cerebri passes. Its *posterior and inferior part*, which projects into the descending horn of the lateral ventricle, presents two small round eminences, the *internal and external geniculate bodies.* Its *anterior extremity*, which is narrow, forms the posterior boundary of the foramen of Monro.

The *third ventricle* is the narrow oblong fissure placed between the thalami optici, and extending to the base of the brain. It is bounded, above, by the under surface of the velum interpositum, from which are suspended the choroid plexuses of the third ventricle; and, laterally, by two white tracts, one on either side, the peduncles of the pineal gland. Its floor, somewhat oblique in its direction, is formed, from before backwards, by the parts which close the interpeduncular space,

266.—The Third and Fourth Ventricles.

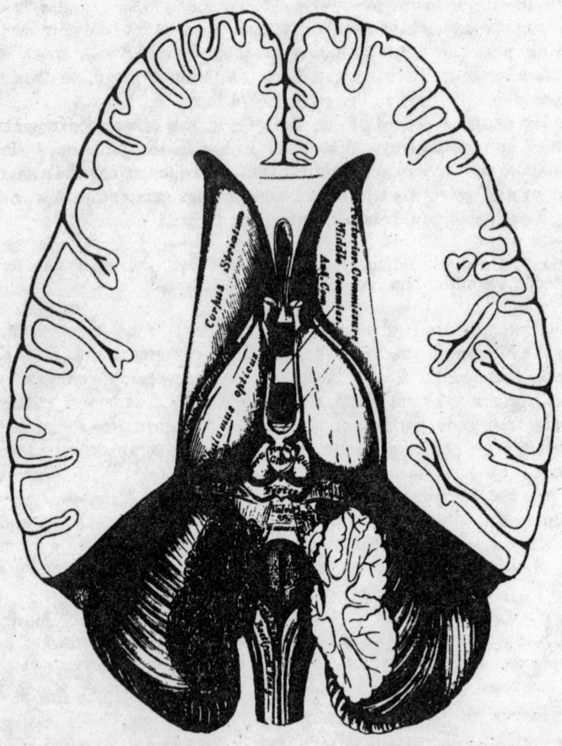

viz., the lamina cinerea, the tuber cinereum and infundibulum, the corpora albicantia, and the locus perforatus posticus; its sides, by the optic thalami; it is bounded, in front, by the anterior crura of the fornix, and part of the anterior commissure; behind, by the posterior commissure, and the *iter a tertio ad quartum ventriculum.*

The cavity of the third ventricle is crossed by three commissures, named, from their position, *anterior*, *middle*, and *posterior*.

The *anterior commissure* is a rounded cord of white fibres, placed in front of the anterior crura of the fornix. It perforates the corpus striatum on either side, and spreads out into the substance of the hemispheres, over the roof of the descending horn of each lateral ventricle.

The *middle* or *soft commissure* consists almost entirely of grey matter. It connects together the thalami optici, and is continuous with the grey matter lining the anterior part of the third ventricle. It is frequently broken in examining the brain, and might then be supposed to have been wanting.

The *posterior commissure*, smaller than the anterior, is a flattened white band of fibres, connecting together the two thalami optici posteriorly. It bounds the third ventricle posteriorly, and is placed in front of and beneath the pineal gland, above the opening leading to the fourth ventricle.

The third ventricle has four openings connected with it. In front are the two oval apertures of the foramen of Monro, one on either side, through which the third communicates with the lateral ventricles. Behind is a third opening leading into the fourth ventricle by a canal, the aquæduct of Sylvius, or *iter a tertio ad quartum ventriculum*. The fourth, situated in the anterior part of the floor of the ventricle, is a deep pit, which leads downwards to the funnel-shaped cavity of the infundibulum (*iter ad infundibulum*).

The lining membrane of the lateral ventricles is continued through the foramen of Monro into the third ventricle, and extends along the iter a tertio into the fourth ventricle; at the bottom of the iter ad infundibulum, it ends in a cul-de-sac.

Grey matter of the third ventricle. A layer of grey matter covers the greater part of the surface of the third ventricle. In the floor of this cavity it exists in great abundance, and is prolonged upwards on the sides of the thalami, extending across the cavity as the soft commissure; below, it enters into the corpora albicantia, and surrounds in part the anterior pillars of the fornix.

Behind the third ventricle, and in front of the cerebellum, are the corpora quadrigemina; and resting upon these, the pineal gland.

The *pineal gland* (*conarium*), so named from its peculiar shape (*pinus*, a fir-cone), is a small reddish-grey body, conical in form, placed immediately behind the posterior commissure, and between the nates, upon which it rests. It is retained in its position by a duplicature of pia mater, derived from the under surface of the velum interpositum, which almost completely invests it. The pineal gland is about four lines in length, and from two to three in width at its base, and is said to be larger in the child than in the adult, and in the female than in the male. Its base is connected with the cerebrum by some transverse commissural fibres, derived from the posterior commissure; and by four slender peduncles, formed of medullary fibres. Of these, the two superior pass forwards upon the upper and inner margin of the optic thalami to the anterior crura of the fornix, with which they become blended. The inferior peduncles pass vertically downwards from the base of the pineal gland, along the back part of the inner surface of the thalami, and are only seen on a vertical section through the gland. The pineal gland is very vascular, and consists chiefly of grey matter, with a few medullary fibres. In its base is a small cavity, said by some to communicate with that of the third ventricle. It contains a transparent viscid fluid, and occasionally a quantity of sabulous matter, named *acervulus cerebri*, composed of phosphate and carbonate of lime, phosphate of magnesia and ammonia, with a little animal matter. These concretions are almost constant in their existence, and are found at all periods of life. When this body is solid, the sabulous matter is found upon its surface, and occasionally upon its peduncles.

On the removal of the pineal gland and adjacent portion of pia mater, the corpora quadrigemina are exposed.

The *corpora* or *tubercula quadrigemina* (*optic lobes*) are four rounded eminences placed in pairs, two in front, and two behind, and separated from one another by a crucial depression. They are situated immediately behind the third ventricle and posterior commissure, beneath the posterior border of the corpus callosum, and above the *iter a tertio ad quartum ventriculum*. The anterior pair, the nates, are the larger, oblong from before backwards, and of a grey colour. The posterior pair, the testes, are hemispherical in form, and lighter in colour than the preceding. They are connected on each side with the thalamus opticus, and commencement of the optic tracts, by means of two white prominent bands, termed *brachia*. Those connecting the nates with the thalamus (*brachia anteriora*), are the larger, and pass obliquely outwards. Those connecting the testes with the thalamus, are called the *brachia posteriora.* Both pairs, in the adult, are quite solid, being composed of white matter externally, and grey matter within. These bodies are larger in the lower animals than in man. In fishes, reptiles, and birds, they are only two in number, are called the *optic lobes*, from their connection with the optic nerves, and are hollow in their interior; but in mammalia, they are four in number, as in man, and quite solid. In the human fœtus, they are developed at a very early period, and form a large proportion of the cerebral mass; at first, they are only two in number, as in the lower mammalia, and hollow in their interior.

These bodies receive, from below, white fibres from the olivary fasciculus or fillet; they are also connected with the cerebellum, by means of a large white cord on each side, the *processus ad testes*, or superior peduncles of the cerebellum, which pass up to the thalami from the tubercula quadrigemina.

The *valve of Vieussens* is a thin translucent lamina of medullary substance, stretched between the two *processus e cerebello ad testes*; it covers in the canal leading from the third to the fourth ventricle, forming part of the roof of the latter cavity. It is narrow in front, where it is connected with the testes; and broader behind, at its connection with the vermiform process of the cerebellum. A slight elevated ridge, the *frenulum*, descends upon the upper part of the valve from the corpora quadrigemina, and on either side of it may be seen the fibres of origin of the fourth nerve. Its lower half is covered by a thin transversely grooved lobule of grey matter prolonged from the anterior border of the cerebellum; this is called by the Italian anatomists the *linguetta laminosa.*

The *corpora geniculata* are two small flattened, oblong masses, placed on the outer side of the corpora quadrigemina, and on the under and back part of each optic thalamus, and named, from their position, *corpus geniculatum externum* and *internum.* They are placed one on the outer and one on the inner side of each optic tract. In this situation, the optic tract may be seen dividing into two bands, one of which is connected with the external geniculate body and nates, the other being connected with the internal geniculate body and testis.

Structure of the cerebrum. The white matter of each hemisphere consists of three kinds of fibres. 1. Diverging or peduncular fibres, which connect the hemisphere with the cord and medulla oblongata. 2. Transverse commissural fibres, which connect together the two hemispheres. 3. Longitudinal commissural fibres, which connect distant parts of the same hemisphere.

The *diverging* or *peduncular fibres* consist of a main body, and of certain accessory fibres. The main body originate in the columns of the cord and medulla oblongata, and enter the cerebrum through the crus cerebri, where they are arranged in two bundles, separated by the locus niger. Those fibres which form the inferior or fasciculated portion of the crus, are derived from the anterior pyramid, and, ascending, pass mainly through the centre of the striated body; those on the opposite surface of the crus, which form the tegmentum, are derived from the posterior pyramid and fasciculi teretes; as they ascend, they pass, some through the under part of the thalamus, and others through both thalamus and corpus striatum, decussating in these bodies with each other, and with the fibres of the corpus callosum. The optic thalami also receive accessory fibres from the *processus ad testes*, the olivary fasciculus, the corpora quadrigemina, and corpora geniculata. Some of the diverging fibres end in the cerebral ganglia, whilst others pass through and receive additional fibres from them, and as they emerge, radiate into the anterior, middle, and posterior lobes of the hemisphere, decussating again with the fibres of the corpus callosum, before passing to the convolutions. These fibres have received the name of *corona radiata.*

The *transverse commissural fibres* connect together the two hemispheres across the middle line. They are formed by the corpus callosum, and the anterior and posterior commissures.

The *longitudinal commissural fibres* connect together distant parts of the same hemisphere, the fibres being disposed in a longitudinal direction. They form the fornix, the tænia semicircularis, and peduncles of the pineal gland, the striæ longitudinales, the fibres of the gyrus fornicatus, and the fasciculus unciformis.

The Cerebellum.

The Cerebellum, or little brain, is that portion of the encephalon which is contained in the inferior occipital fossæ. It is situated beneath the posterior lobes of the cerebrum, from which it is separated by the tentorium. Its average weight in the male is 5 oz. 4 drs. It attains its maximum weight between the twenty-fifth and fortieth year; its increase in weight after the fourteenth year being relatively greater in the female than in the male. The proportion between the cerebellum and cerebrum is, in the male, as 1 to $8\frac{4}{7}$; and in the female, as 1 to $8\frac{1}{4}$. In the infant, the cerebellum is proportionally much smaller than in the adult, the relation between it and the cerebrum being, according to Chaussier, between 1 to 13, and 1 to 26; by Cruveilhier the proportion was found to be 1 to 20. In form, the cerebellum is oblong, and flattened from above downwards, its greatest diameter being from side to side. It measures from three and a-half to four inches transversely, and from two to two and a-half inches from before backwards, being about two inches thick in the centre, and about six lines at the circumference, which is the thinnest part. It consists of grey and white matter: the former, darker than that of the cerebrum, occupies the surface; the latter the interior. The surface of the cerebellum is not convoluted like the cerebrum, but traversed by numerous curved furrows or sulci, which vary in depth at different parts, and separate the laminæ of which its exterior is composed.

Its *upper surface* (fig. 267) is somewhat elevated in the median line, and depressed towards its circumference; it consists of two lateral hemispheres, connected together by an elevated median portion or lobe, the superior vermiform process.

267.—Upper Surface of the Cerebellum.

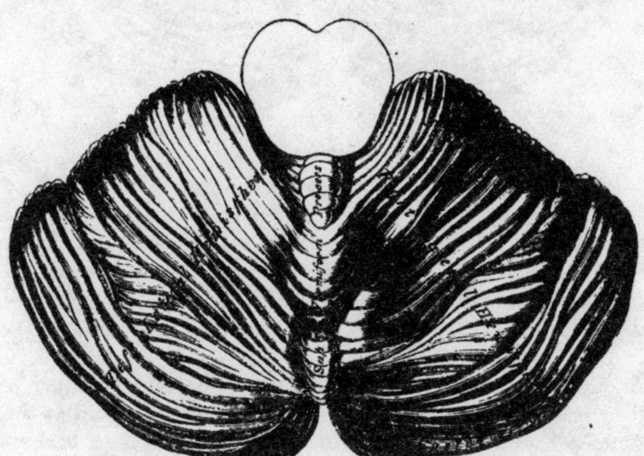

The median lobe is the fundamental part, and in some animals, as fishes and reptiles, the only part which exists; the hemispheres being additions, and attaining their maximum size in man. The hemispheres are separated, in front, by a deep notch, the *incisura cerebelli anterior*, which encircles the corpora quadrigemina behind; they are also separated by a similar notch behind, the *incisura cerebelli posterior*, in which is received the upper part of the falx cerebelli. The superior vermiform process (upper part of the median lobe of the cerebellum), extends from the notch on the anterior to that on the posterior border. It is divided into three lobes: the *lobulus centralis*, a small lobe, situated in the incisura anterior; the *monticulus cerebelli*, the central projecting part of the process; and the *commisura simplex*, a small lobe near the incisura posterior.

The *under surface* of the cerebellum (fig. 268) is subdivided into two lateral hemispheres by a depression, the valley, which extends from before backwards in the middle line. The lateral hemispheres are lodged in the inferior occipital fossæ; the median depression, or valley, receives the back part of the medulla oblongata, is broader in the centre than at either extremity, and has, projecting from its floor, part of the median lobe of the cerebellum, called the *inferior vermiform process*. The parts entering into the composition of this body are, from behind forwards, the *commissura brevis*, situated in the incisura posterior; in front of this, a laminated conical projection, the *pyramid*; more anteriorly, a larger eminence, the *uvula*, which is placed between the two rounded lobes which occupy the sides of the valley, the *amygdalæ* or *tonsils*, and is connected with them by a commissure of grey matter, indented on the surface, called the *furrowed band*. In front of the uvula is the *nodule*; it is the anterior pointed termination of the nferior vermiform process, and projects into the cavity of the fourth ventricle; it has been named by Malacarne the *laminated tubercle*. On each side of the nodule is a thin layer of white substance, attached externally to the flocculus, and

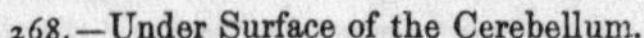

268.—Under Surface of the Cerebellum.

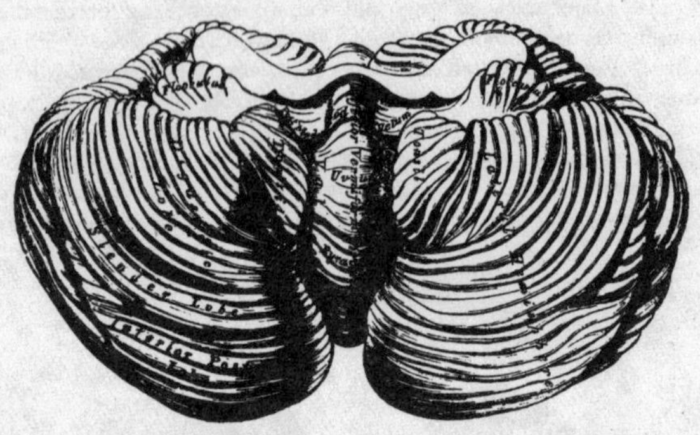

internally to the nodule; these form together the *posterior medullary velum*, or commissure of the flocculus. It is usually covered in and concealed by the amygdalæ, and cannot be seen until they are drawn aside. This band is of a semilunar form on each side, its anterior margin being free and concave, its posterior attached just in front of the furrowed band. Between it and the nodulus and uvula behind, is a deep fossa, called the *swallow's nest* (*nidus hirundinis*).

Lobes of the cerebellum. Each hemisphere is divided into an upper and a lower portion by the great horizontal fissure, which commences in front at the pons, and passes horizontally round the free margin of either hemisphere, backwards to the middle line. From this primary fissure numerous secondary fissures proceed, which separate the cerebellum into lobes.

Upon the upper surface of either hemisphere there are two lobes, separated from each other by a fissure. These are the anterior or square lobe, which extends as far back as the posterior edge of the vermiform process, and the posterior or semilunar lobe, which passes from the termination of the preceding to the great horizontal fissure.

Upon the under surface of either hemisphere there are five lobes, separated by sulci; these are from before backwards; the *flocculus* or *sub-peduncular lobe*, a prominent tuft, situated behind and below the middle peduncle of the cerebellum; its surface is composed of grey matter, subdivided into a few small laminæ: it is sometimes called the *pneumogastric lobule*, from being situated behind the pneumogastric nerve. The *amygdala* or *tonsil* is situated on either side of the great median fissure or valley, and projects into the fourth ventricle. The *digastric lobe* is situated on the outside of the tonsil, being connected in part with the pyramid. Behind the digastric is the *slender lobe*, which is connected with the back part of the pyramid and the commissura brevis: and more posteriorly is the *inferior posterior lobe*, which also joins the commissura brevis in the valley.

Structure. If a vertical section is made through either hemisphere of the cerebellum, midway between its centre and the superior vermiform process, the interior will be found to consist of a central stem of white matter, which contains in its interior a *dentate body*. From the surface of this central stem, a series of plates of medullary matter are detached, which, covered with grey matter, form the laminæ; and from the anterior part of each hemisphere arise three large processes or peduncles, superior, middle, and inferior, by which the cerebellum is connected with the rest of the encephalon.

The *laminæ* are about ten or twelve in number, including those on both surfaces of the cerebellum, those in front being detached at a right angle, and those behind at an acute angle; as each lamina proceeds outwards, other secondary laminæ are detached from it, and, from these, tertiary laminæ. The arrangement thus described gives to the cut surface of the organ a foliated appearance, to which the name *arbor vitæ* has been given. Each lamina consists of white matter, covered externally by a layer of grey substance. The white matter of each lamina is derived partly from the central stem: in addition to which white fibres pass from one lamina to another. The grey matter resembles somewhat the cortical substance of the convolutions. It consists of two layers: the external one, soft and of a greyish colour; the internal one, firmer and of a rust colour.

269.—Vertical Section of the Cerebellum.

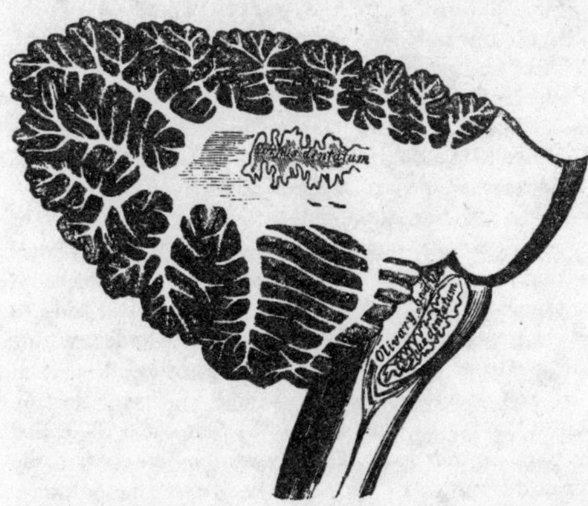

The *corpus dentatum*, or *ganglion of the cerebellum*, is situated a little to the inner side of the centre of the stem of white matter. It consists of an open bag or capsule of grey matter, the section of which presents a grey dentated outline, open at its anterior part. It is surrounded by white fibres; white fibres are also contained in its interior, which issue from it to join the superior peduncles.

The *peduncles* of the cerebellum, superior, middle, and inferior, serve to connect it with the rest of the encephalon.

The *superior peduncles* (*processus e cerebello ad testes*) connect the cerebellum with the cerebrum; they pass forwards and upwards to the testes, beneath which they ascend to the crura cerebri and optic thalami, forming part of the diverging cerebral fibres: each peduncle forms part of the lateral boundary of the fourth ventricle, and is connected with its fellow of the opposite side by the valve of Vieussens. The peduncles are continuous behind with the folia of the inferior vermiform process, and with the white fibres in the interior of the corpus dentatum. Beneath the corpora quadrigemina, the innermost fibres of each peduncle decussate with each other, so that some fibres from the right half of the cerebellum are continued to the left half of the cerebrum.

The *inferior peduncles* (*processus ad medullam*) connect the cerebellum with the medulla oblongata. They pass downwards, to the back part of the medulla, and form part of the restiform bodies. Above, the fibres of each process are connected chiefly with the laminæ, on the upper surface of the cerebellum; and below, they are connected with all three tracts of one half of the medulla, and, through these, with the corresponding half of the cord, excepting the posterior median columns.

The *middle peduncles* (*processus ad pontem*), the largest of the three, connect together the two hemispheres of the cerebellum, forming their great transverse commissure. They consist of a mass of curved fibres, which arise in the lateral parts of the cerebellum, and pass across to the same points on the opposite side. They form the transverse fibres of the pons Varolii.

Fourth Ventricle. (Fig. 266.)

The fourth ventricle, or ventricle of the cerebellum, is the space between the posterior surface of the medulla oblongata and pons in front, and the cerebellum behind. It is lozenge-shaped, being contracted above and below, and broadest across its central part. It is bounded laterally by the *processus e cerebello ad testes* above, and by the diverging posterior pyramids and restiform bodies below.

The *roof* is arched; it is formed by the valve of Vieussens and the under surface of the cerebellum, which presents, in this situation, four small eminences or lobules, the nodulus, uvula, and amygdalæ.

The anterior boundary, or *floor*, is formed by the posterior surface of the medulla oblongata and pons. In the median line is seen the posterior median fissure, which becomes gradually obliterated above, and terminates below in the point of the calamus scriptorius, formed by the convergence of the posterior pyramids. At this point is the orifice of a short canal terminating in a *cul-de-sac*, the remains of the canal which extends in fœtal life through the centre of the cord. On each side of the median fissure are two slightly convex longitudinal eminences, the fasciculi teretes: they extend the entire length of the floor, being indistinct below and of a greyish colour, but well marked and whitish above. Each eminence consists of fibres derived from the lateral tract and restiform body, which ascend to the cerebrum. Opposite the crus cerebelli, on the outer side of the fasciculi teretes, is a small eminence of dark grey substance, which presents a bluish tint through the thin stratum covering it; this is called the *locus cœruleus*; and a thin streak of the same colour continued up from this on either side of the fasciculi teretes, as far as the top of the ventricle, is called the *tænia violacea*. The lower part of the floor of the ventricle is crossed by several white transverse lines, *lineæ transversæ*; they emerge from the posterior median fissure; some enter the crus cerebelli, others enter the roots of origin of the auditory nerve, whilst some pass upwards and outwards on the floor of the ventricle.

The *lining membrane* of the fourth ventricle is continuous with that of the third, through the aquæduct of Sylvius, and its cavity communicates below with the sub-arachnoid space of the brain and cord through an aperture in the layer of pia mater extending between the cerebellum and medulla oblongata. Laterally, this membrane is reflected outwards a short distance between the cerebellum and medulla.

The *choroid plexuses* of the fourth ventricle are two in number; they are delicate vascular fringes, which project into the ventricle on each side, passing from the point of the inferior vermiform process to the outer margin of the restiform bodies.

The *grey matter* in the floor of the ventricle consists of a tolerably thick stratum, continuous below with the grey commissure of the cord, and extending up as high as the aquæduct of Sylvius, besides some special deposits connected with the roots of origin of certain nerves. In the upper half of the ventricle is a projection situated over the nucleus, from which the sixth and facial nerves take a common origin. In the lower half are three eminences on each side for the roots of origin of the eighth and ninth nerves.

For further information on the Descriptive Anatomy of the Nervous Centres, consult:— Cruveilhier's 'Anatomie Descriptive;' Todd's 'Descriptive Anatomy of the Brain, Spinal Cord, and Ganglions;' Herbert Mayo's 'Plates of the Brain and Spinal Cord;' and Arnold's 'Tabulæ Anatomicæ Fascic. 1. Icones Cerebri et Medullæ Spinalis.'

Cranial Nerves.

THE Cranial Nerves, nine in number on each side, arise from some part of the cerebro-spinal centre, and are transmitted through foramina in the base of the cranium. They have been named numerically, according to the order in which they pass out of the cranial cavity. Other names are also given to them derived from the parts to which they are distributed, or from their functions. Taken in their order, from before backwards, they are as follows:

1st.	Olfactory.	7th.	Facial (Portio dura).
2nd.	Optic.		Auditory (Portio mollis).
3rd.	Motor oculi.	8th.	Glosso-pharyngeal.
4th.	Pathetic.		Pneumogastric (Par vagum).
5th.	Trifacial (Trigeminus).		Spinal accessory.
6th.	Abducens.	9th.	Hypoglossal.

If, however, the 7th pair be considered as two, and the 8th pair as three distinct nerves, then their number will be increased to twelve, which is the arrangement adopted by Sömmering.

The cranial nerves may be subdivided into four groups, according to the peculiar function possessed by each, viz., nerves of special sense; nerves of common sensation; nerves of motion; and mixed nerves. These groups may be thus arranged:

Nerves of Special Sense.

Olfactory.
Optic.
Auditory.
Part of glosso-pharyngeal.
Lingual or gustatory branch of fifth.

Nerves of Motion.

Motor oculi.
Pathetic.
Part of third division of fifth.
Abducens.
Facial.
Hypoglossal.

Nerves of Common Sensation.

Fifth (greater portion).
Part of glosso-pharyngeal.

Mixed Nerves.

Pneumogastric.
Spinal accessory.

All the cranial nerves are connected to some part of the surface of the brain. This is termed their *superficial*, or *apparent origin*. But their fibres may, in all cases, be traced deeply into the substance of the organ. This part is called their *deep*, or *real origin*.

Olfactory Nerve. (Fig. 262.)

The First, or Olfactory Nerve, the special nerve of the sense of smell, may be regarded as a lobe, or portion of the cerebral substance, pushed forward in direct relation with the organ to which it is distributed. It arises by three roots.

The *external*, or *long root*, is a narrow, white, medullary band, which passes outwards across the fissure of Sylvius, into the substance of the middle lobe of the cerebrum. Its deep origin has been traced to the corpus striatum,* the superficial fibres of the optic thalamus,† the anterior commissure,‡ and the convolutions of the island of Reil.

* Vieussens, Winslow, Monro, Mayo. † Valentin. ‡ Cruveilhier.

The *middle*, or *grey root*, arises from a papilla of grey matter (caruncula mammillaris), imbedded in the anterior lobe. This root is prolonged into the nerve from the adjacent part of the brain, and contains white fibres in its interior, which are connected with the corpus striatum.

The *internal*, or *short root*, is composed of white fibres, which arise from the inner and back part of the anterior lobe, being connected, according to Foville, with the longitudinal fibres of the gyrus fornicatus.

These three roots unite, and form a flat band, narrower in the middle than at either extremity, and of a somewhat prismoid form on section. It is soft in texture, and contains a considerable amount of grey matter in its substance. As it passes forwards, it is contained in a deep sulcus, between two convolutions, lying on the under surface of the anterior lobe, on either side of the longitudinal fissure, and is retained in position by the arachnoid membrane which covers it. On reaching the cribriform plate of the ethmoid bone, it expands into an oblong mass of greyish-white substance, the olfactory bulb. From the under part of this bulb are given off numerous filaments, about twenty in number, which pass through the cribriform foramina, and are distributed to the mucous membrane of the nose. Each filament is surrounded by a tubular prolongation from the dura mater, and pia mater; the former being lost on the periosteum lining the nose; the latter, in the neurilemma of the nerve. The filaments, as they enter the nares, are divisible into three groups; an inner group, larger than those on the outer wall, spread out over the upper third of the septum; a middle set, confined to the roof of the nose; and an outer set, which are distributed over the superior and middle turbinated bones, and the surface of the ethmoid in front of them. As the filaments descend, they unite in a plexiform network, and become gradually lost in the lining membrane. Their mode of termination is unknown.

The olfactory differs in structure from other nerves, in containing grey matter in its interior, and being soft and pulpy in structure. Its filaments are deficient in the white substance of Schwann, are not divisible into fibrillæ, and resemble the gelatinous fibres, in being nucleated, and of a finely-granular texture.

Optic Nerve.

The Second, or Optic Nerve, the special nerve of the sense of sight, is distributed exclusively to the eyeball. The nerves of opposite sides are connected together at the commissure; and from the back of the commissure, they may be traced to the brain, under the name of the optic tracts.

270.—The Optic Nerves and Optic Tract.

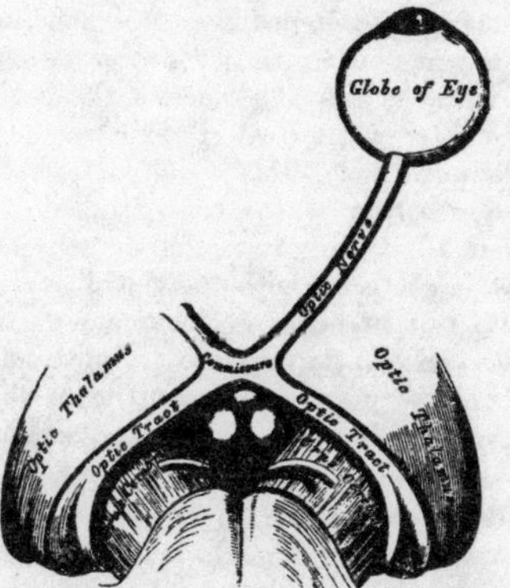

The *optic tract*, at its connection with the brain, is divided into two bands which arise from the optic thalami, the corpora geniculata, and the corpora quadrigemina. The fibres of origin from the thalamus may be traced partly from its surface, and partly from its interior. From this origin, the tract winds obliquely across the under surface of the crus cerebri, in the form of a flattened band, destitute of neurilemma, and is attached to the crus by its anterior margin. It now assumes a cylindrical form, and, as it passes forwards, is connected with the tuber cinereum, and lamina cinerea, from both of which it receives fibres. According to Foville, it is also connected with the tænia semicircularis, and the anterior termination of the gyrus fornicatus. It finally joins with the nerve of the opposite side, to form the optic commissure.

The *commissure*, or *chiasma*, somewhat quadrilateral in form, rests upon the olivary process of the sphenoid bone, being bounded, in front, by the lamina cinerea; behind, by the tuber cinereum; on either side by the anterior perforated space. Within the commissure, the optic nerves of the two sides undergo a partial decussation. The fibres which form the inner margin of each tract, are continued across from one to the other side of the brain, and have no connection with the optic nerves. These may be regarded as commissural fibres (inter-cerebral) between the thalami of opposite sides. Some fibres are continued across the anterior border of the chiasma, and connect the optic nerves of the two sides, having no relation with the optic tracts. They may be regarded as commissural fibres between the two retinæ (inter-retinal fibres). The outer fibres of each tract are continued into the optic nerve of the same side. The central fibres of each tract are continued into the optic nerve of the opposite side, decussating in the commissure with similar fibres of the opposite tract.*

271.—Course of the Fibres in the Optic Commissure.

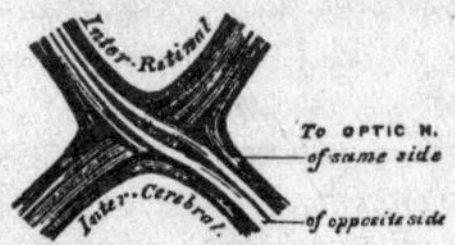

The *optic nerves* arise from the fore part of the commissure, and, diverging from one another, become rounded in form and firm in texture, and are enclosed in a sheath derived from the arachnoid. As each nerve passes through the corresponding optic foramen, it receives a sheath from the dura mater; and as it enters the orbit, this sheath subdivides into two layers, one of which becomes continuous with the periosteum of the orbit; the other forms a sheath for the nerve, and surrounds it as far as the sclerotic. The nerve passes through the cavity of the orbit, pierces the sclerotic and choroid coats at the back part of the eyeball, a little to the nasal side of its centre, and expands into the retina. A small artery, the arteria centralis retinæ, perforates the optic nerve a little behind the globe, and runs along its interior in a tubular canal of fibrous tissue. It supplies the inner surface of the retina, and is accompanied by corresponding veins.

Auditory Nerve.

The Auditory Nerve (portio mollis of the seventh pair) is the special nerve of the sense of hearing, being distributed exclusively to the internal ear. The portio dura of the seventh pair, or facial nerve, is the motor nerve of the muscles of the face. It will be described with the cranial motor nerves.

The auditory nerve arises by numerous white striæ, the *lineæ transversæ*, which emerge from the posterior median fissure in the anterior wall, or floor, of the fourth ventricle. It is also connected with the grey matter of the medulla, corresponding to the locus cæruleus. According to Foville, the roots of this nerve are connected, on the under surface of the middle peduncle, with the grey substance of the cerebellum, with the flocculus, and with the grey matter at the borders of the calamus scriptorius. The nerve winds round the restiform body, from which it receives fibres, and passes forwards across the posterior border of the crus cerebelli, in company with the facial nerve, from which it is partially separated by a small artery. It then enters the meatus auditorius, in company with the facial nerve, and, at the bottom of the meatus, divides into two branches, cochlear and vestibular, which are distributed, the former to the cochlea, the latter to the vestibule and semicircular canals. The auditory nerve is very soft in texture (hence the name, *portio mollis*), destitute of neurilemma, and, within the meatus, receives one or two filaments from the facial.

The other nerves of special sense must be described with the glosso-pharyngeal and fifth nerves, of which they are parts.

* A specimen of congenital absence of the optic commissure is to be found in the Museum of the Westminster Hospital.

THIRD NERVE. (Figs. 272, 273.)

The THIRD NERVE (*motor oculi*) supplies all the muscles of the eyeball except the Superior oblique and External rectus; it also sends motor filaments to the iris. It is a rather large nerve, of rounded form and firm texture, having its apparent origin from the inner surface of the crus cerebri, immediately in front of the pons Varolii.

The *deep origin* may be traced into the substance of the crus, where some of its fibres are connected with the locus niger; others run downwards, among the longitudinal fibres of the pons; whilst others ascend, to be connected with the tubercula quadrigemina, and valve of Vieussens. According to Stilling, the fibres of the nerve pierce the peduncle and locus niger, and arise from a grey nucleus in the floor of the aquæduct of Sylvius. On emerging from the brain, the nerve is invested with a sheath of pia mater, and enclosed in a prolongation from the arachnoid. It then pierces the dura mater on the outer side of the anterior clinoid process, where its serous covering is reflected from it, and it passes along the outer wall of the cavernous sinus, above the other orbital nerves, receiving in its course one or two filaments from the cavernous plexus of the sympathetic. It then divides into two branches, which enter the orbit through the sphenoidal fissure, between the two heads of the External rectus muscle. On passing through the fissure, the nerve is placed below the fourth, and the frontal and lachrymal branches of the ophthalmic nerve.

272.—Nerves of the Orbit. Seen from above.

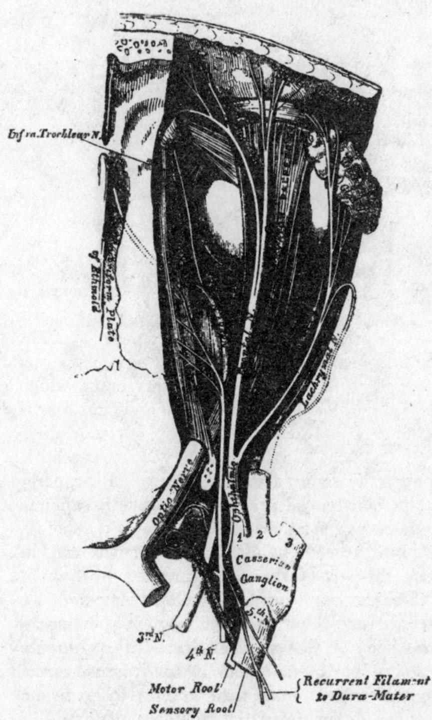

The *superior division*, the smaller, passes inwards across the optic nerve, and supplies the Superior rectus and Levator palpebræ.

The *inferior division*, the larger, divides into three branches. One passes beneath the optic nerve to the Internal rectus; another to the Inferior rectus; and the third, the largest of the three, passes forwards between the Inferior and External recti, to the Inferior oblique. From the latter, a short thick branch is given off to the lower part of the lenticular ganglion, forming its inferior root, as well as two filaments to the Inferior rectus. All these branches enter the muscles on their ocular surface.

FOURTH NERVE. (Fig. 272.)

The FOURTH, or trochlear nerve, the smallest of the cranial nerves, supplies the Superior oblique muscle. It arises from the upper part of the valve of Vieussens, immediately behind the testis, and divides beneath the corpora quadrigemina, into two fasciculi; the anterior one arising from a nucleus of grey matter, close to the middle line of the floor of the Sylvian aquæduct; the posterior one from a grey nucleus, at the upper part of the floor of the fourth ventricle, close to the origin of the fifth nerve. The two nerves are connected together at their origin, by a transverse band of white fibres, which crosses the surface of the valve of Vieussens. The nerve winds round the outer side of the crus cerebri, immediately above the pons Varolii, pierces the dura mater in the free border of the tentorium cerebelli, near the posterior clinoid process, above the oval opening for the fifth nerve, and passes forwards through the outer wall of the cavernous sinus, below the third: but, as it enters the orbit, through the sphenoidal fissure, it becomes the highest of all the nerves. In the orbit, it passes inwards, above the origin of the Levator palpebræ, and finally enters the orbital surface of the Superior oblique muscle.

In the outer wall of the cavernous sinus, this nerve receives some filaments from the carotid plexus of the sympathetic. It is not unfrequently blended with the ophthalmic division of the fifth; and occasionally gives off a branch to assist in the formation of the lachrymal nerve. It also gives off a recurrent branch, which passes backwards between the layers of the tentorium, dividing into two or three filaments, which may be traced as far back as the wall of the lateral sinus.

273.—Nerves of the Orbit and Ophthalmic Ganglion. Side view.

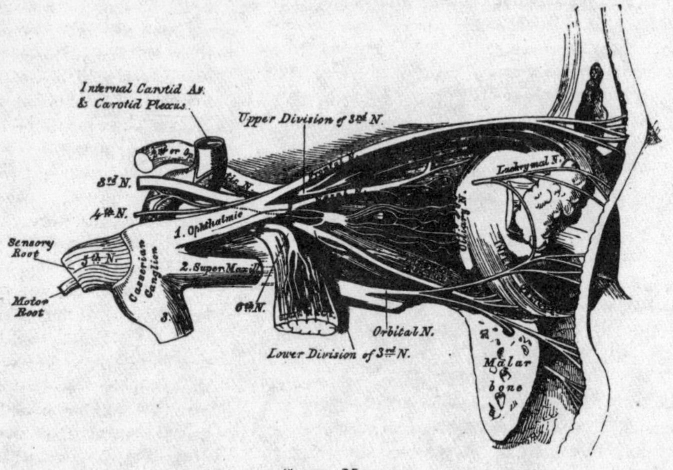

Sixth Nerve.

The Sixth Nerve (*Abducens*) supplies the External rectus muscle. Its apparent origin is by several filaments from the constricted part of the corpus pyramidale, close to the pons, or from the lower border of the pons itself.

The *deep origin* of this nerve has been traced, by Mayo, between the fasciculi of the corpus pyramidale, to the posterior part of the medulla, where Stilling has shown its connection with a grey nucleus in the floor of the fourth ventricle. The nerve pierces the dura mater, immediately below the posterior clinoid process, lying in a groove by the side of the body of the sphenoid bone. It passes forwards through the cavernous sinus, lying on the outer side of the internal carotid artery, where it is joined by several filaments from the carotid plexus, by one from Meckel's ganglion (Böck), and another from the ophthalmic nerve. It enters the orbit through the sphenoidal fissure, and lies above the ophthalmic vein, from which it is separated by a lamina of dura mater. It then passes between the two heads of the External rectus, and is distributed to that muscle on its ocular surface.

The above-mentioned nerves, as well as the ophthalmic division of the fifth, as they pass to the orbit, bear a certain relation to each other in the cavernous sinus, at the sphenoidal fissure, and in the cavity of the orbit, which will be now described.

In the *cavernous sinus*, the third, fourth, and ophthalmic division of the fifth, are placed in the dura mater of the outer wall of the sinus, in their numerical order, both from above downwards, and from within outwards. The sixth nerve lies at the outer side of the internal carotid artery. As these nerves pass forwards to the sphenoidal fissure, the third and fifth nerves become divided into branches, and the sixth approaches the rest; so that their relative position becomes considerably changed.

In the *sphenoidal fissure*, the fourth, and the frontal and lachrymal divisions of the ophthalmic, lie upon the same plane, the former being most internal, the latter external; and they enter the cavity of the orbit above the muscles. The remaining nerves enter the orbit between the two heads of the External rectus. The superior division of the third is the highest of these; beneath this lies the nasal branch of the fifth; then the inferior division of the third; and the sixth lowest of all.

In the *orbit*, the fourth, and the frontal and lachrymal divisions of the ophthalmic, lie on the same plane immediately beneath the periosteum, the fourth nerve being internal and resting on the Superior oblique, the frontal resting on the Levator palpebræ, and the lachrymal on the External rectus. Next in order comes the superior division of the third nerve lying immediately beneath the Superior rectus, and then the nasal division of the fifth crossing the optic nerve from the outer to the inner side of the orbit. Beneath these is found the optic nerve, surrounded in front by the ciliary nerves, and having the lenticular ganglion on its outer side, between it and the External rectus. Below the optic is the inferior division of the third, and the sixth, which lies on the outer side of the orbit.

Facial Nerve.

The Facial Nerve, or *portio dura* of the seventh pair, is the motor nerve of all the muscles of expression in the face, and of the Platysma and Buccinator. It supplies also the muscles of the external ear, the posterior belly of the Digastric, and the Stylo-hyoid. Through the chorda tympani it supplies the Lingualis; by its tympanic branch, the Stapedius and Laxator tympani; through the otic ganglion, the Tensor tympani; and through the connection of its trunk with the Vidian nerve, by the petrosal nerves, it probably supplies the Levator palati and Azygos uvulæ. It arises from the lateral tract of the medulla oblongata, in the groove between the olivary and restiform bodies. Its deep origin may be traced to the floor of the fourth ventricle, where it is connected with the same nucleus as the sixth nerve. This nerve is situated a little nearer to the middle line than the portio mollis, close to the lower border of the pons Varolii, from which some of its fibres are derived.

Connected with this nerve, and lying between it and the portio mollis, is a small fasciculus (*portio inter duram et mollem* of Wrisberg, or *portio intermedia*). This accessory portion arises from the lateral column of the cord.

274.—The Course and Connection of the Facial Nerve, in the Temporal Bone.

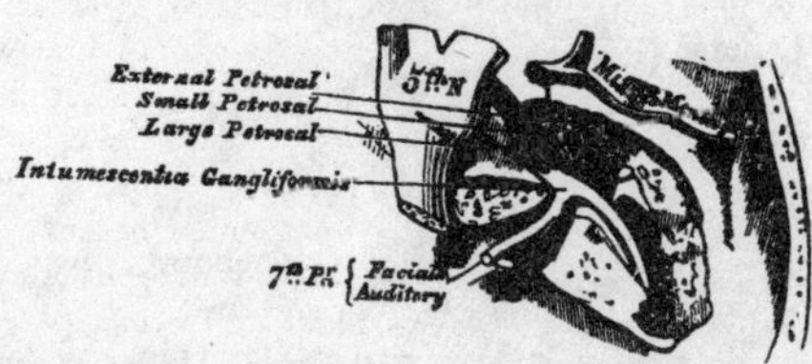

The nerve passes forwards and outwards upon the crus cerebelli, and enters the internal auditory meatus with the auditory nerve. Within the meatus, the facial nerve lies first to the inner side of the auditory, and then in a groove upon that nerve, and is connected to it by one or two filaments.

At the bottom of the meatus, it enters the aqueductus Fallopii, and follows the serpentine course of that canal through the petrous portion of the temporal bone, from its commencement at the internal meatus to its termination at the stylo-mastoid foramen. It is at first directed outwards towards the hiatus Fallopii, where it forms a reddish gangliform swelling (intumescentia ganglioformis), and is joined by several nerves; then bending suddenly backwards, it runs in the internal wall of the tympanum, above the fenestra ovalis, and at the back of that cavity passes vertically downwards to the stylo-mastoid foramen.

On emerging from this aperture, it runs forwards in the substance of the parotid gland, crosses the external carotid artery, and divides behind the ramus of the lower jaw into two primary branches, temporo-facial and cervico-facial, from which numerous offsets are distributed over the side of the head, face, and upper part of the neck, supplying the superficial muscles in these regions. As the primary branches and their offsets diverge from each other, they present somewhat the appearance of a bird's claws; hence the name of *pes anserinus* is given to the divisions of the facial nerve in and near the parotid gland.

The communications of the facial nerve may be thus arranged:—

In the internal auditory meatus .	With the auditory nerve.
In the aquæductus Fallopii . .	With Meckel's ganglion by the large petrosal nerve. With the otic ganglion by the small petrosal nerve. With the sympathetic on the middle meningeal by the external petrosal nerve.
At its exit from the stylo-mastoid foramen	With the pneumogastric. " glosso-pharyngeal. " carotid plexus. " auricularis magnus. " auriculo-temporal.
On the face	With the three divisions of the fifth.

In the internal auditory meatus, some minute filaments pass between the facial and auditory nerves.

Opposite the hiatus Fallopii, the gangliform enlargement on the facial nerve communicates by means of the large petrosal nerve, with Meckel's ganglion, forming its motor root; by a filament from the small petrosal, with the otic ganglion; and by the external petrosal, with the sympathetic filaments accompanying the middle meningeal artery (Bidder). From the gangliform enlargement, according to Arnold, a twig is sent back to the auditory nerve.

At its exit from the stylo-mastoid foramen, it sends a twig to the pneumogastric, another to the glosso-pharyngeal nerve, and communicates with the carotid plexus of the sympathetic, with the great auricular branch of the cervical plexus, with the auriculo-temporal branch of the inferior maxillary nerve in the parotid gland, and on the face with the terminal branches of the three divisions of the fifth.

Branches of Distribution.

Within aquæductus Fallopii	Tympanic. Chorda tympani.	
At exit from stylo-mastoid foramen	Posterior auricular. Digastric. Stylo-hyoid.	
On the face . . .	Temporo-facial	Temporal. Malar. Infraorbital.
	Cervico-facial	Buccal. Supramaxillary. Inframaxillary.

The *Tympanic branch* arises from the nerve opposite the pyramid; it is a small filament, which supplies the Stapedius and Laxator tympani muscles.

The *Chorda tympani* is given off from the facial as it passes vertically downwards at the back of the tympanum, about a quarter of an inch before its exit from the stylo-mastoid foramen. It ascends from below upwards in a distinct canal, parallel with the aquæductus Fallopii, and enters the cavity of the tympanum through an opening between the base of the pyramid and the attachment of the membrana tympani, and becomes invested with mucous membrane. It

275.—The Nerves of the Scalp, Face, and Side of the Neck.

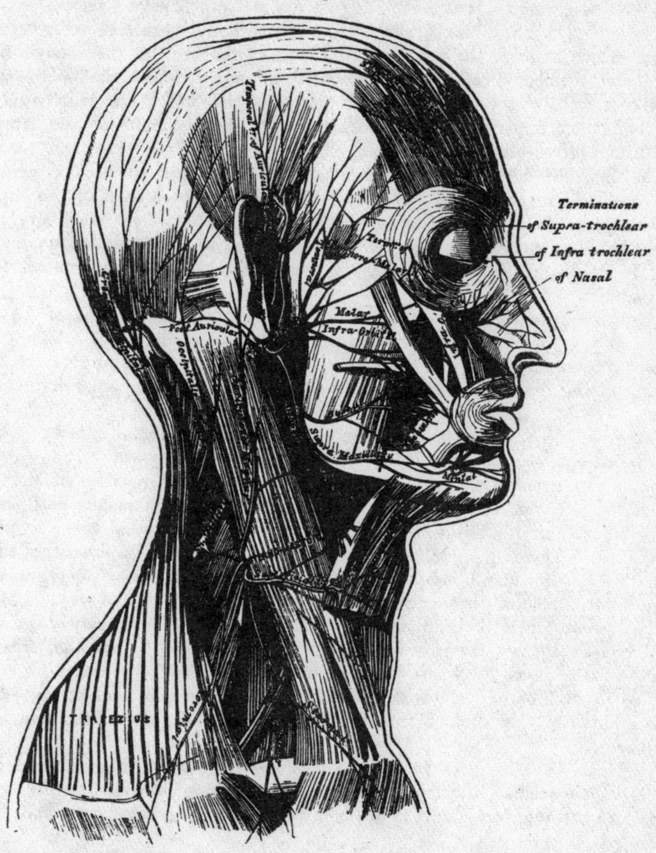

passes forwards through the cavity of the tympanum, between the handle of the malleus and vertical ramus of the incus, to its anterior inferior angle, and emerges from that cavity through a foramen (the canal of Huguier) at the inner side of the Glaserian fissure. It then descends between the two Pterygoid muscles, and meets the gustatory nerve at an acute angle, after communicating with which, it accompanies it to the submaxillary gland; it then joins the submaxillary ganglion, and terminates in the Lingualis muscle.

The *Posterior auricular nerve* arises close to the stylo-mastoid foramen, and passes upwards in front of the mastoïd process, where it is joined by a filament from the auricular branch of the pneumogastric, and communicates with the deep branch of the auricularis magnus; as it ascends between the meatus and mastoid process it divides into two branches. The *auricular branch* supplies the Retrahens aurem. The *occipital branch*, the larger, passes backwards along the superior curved line of the occipital bone, and supplies the occipital portion of the Occipito-frontalis.

The *Stylo-hyoid* is a long slender branch, which passes inwards, entering the Stylo-hyoid muscle about its middle; it communicates with the sympathetic filaments on the external carotid artery.

The *Digastric branch* usually arises by a common trunk with the preceding; it divides into several filaments, which supply the posterior belly of the Digastric; one of these perforates that muscle to join the glosso-pharyngeal nerve.

The *Temporo-facial*, the larger of the two terminal branches, passes upwards and forwards through the parotid gland, crosses the neck of the condyle of the jaw, being connected in this situation with the auriculo-temporal branch of the inferior maxillary nerve, and divides into branches, which are distributed over the temple and upper part of the face; these are divided into three sets, temporal, malar, and infraorbital.

The *temporal branches* cross the zygoma to the temporal region, supplying the Attrahens aurem muscle, and join with the temporal branch of the superior maxillary, and with the auriculo-temporal branch of the inferior maxillary. The more anterior branches supply the frontal portion of the Occipito-frontalis, and the Orbicularis palpebrarum muscle, joining with the supraorbital branch of the ophthalmic.

The *malar branches* pass across the malar bone to the outer angle of the orbit, where they supply the Orbicularis and Corrugator supercilii muscles, joining with filaments from the lachrymal and supraorbital nerves: others supply the lower eyelid, joining with filaments of the malar branches of the superior maxillary nerve.

The *infraorbital*, of larger size than the rest, pass horizontally forwards to be distributed between the lower margin of the orbit and the mouth. The *superficial branches* run beneath the skin and above the superficial muscles of the face, which they supply; some supply the lower eyelid and Pyramidalis nasi, joining at the inner angle of the orbit, with the infratrochlear and nasal branches of the ophthalmic. The *deep branches* pass beneath the Levator labii superioris, supply it and the Levator anguli oris, and form a plexus (infraorbital) by joining with the infraorbital branch of the superior maxillary nerve.

The *Cervico-facial* division of the facial nerve passes obliquely downwards and forwards through the parotid gland, where it is joined by branches from the great auricular nerve; opposite the angle of the lower jaw it divides into branches which are distributed on the lower half of the face and upper part of the neck. These may be divided into three sets: buccal, supramaxillary, and inframaxillary.

The *buccal branches* cross the Masseter muscle, join the infraorbital branches of the temporo-facial division of the nerve, and with filaments of the buccal branch of the inferior maxillary nerve. They supply the Buccinator and Orbicularis oris.

The *supramaxillary branches* pass forwards beneath the Platysma and Depressor anguli oris, supplying the muscles of the lip and chin, and anastomosing with the mental branch of the inferior dental nerve.

The *inframaxillary branches* run forward beneath the Platysma, and form a series of arches across the side of the neck over the supra-hyoid region. One of these branches descends vertically to join with the superficial cervical nerve from the cervical plexus; others supply the Platysma.

Ninth, or Hypoglossal Nerve.

The Ninth Nerve (*hypoglossal*) is the motor nerve of the tongue. It arises by several filaments, from ten to fifteen in number, from the groove between the pyramidal and olivary bodies, in a continuous line with the anterior roots of the spinal nerves. According to Stilling, these roots may be traced to a grey nucleus in the floor of the medulla oblongata, between the posterior median furrow and the nuclei of the glosso-pharyngeal and vagus nerves. The filaments of this nerve are collected into two bundles, which perforate the dura mater separately,

276.—Hypoglossal Nerve, Cervical Plexus, and their Branches.

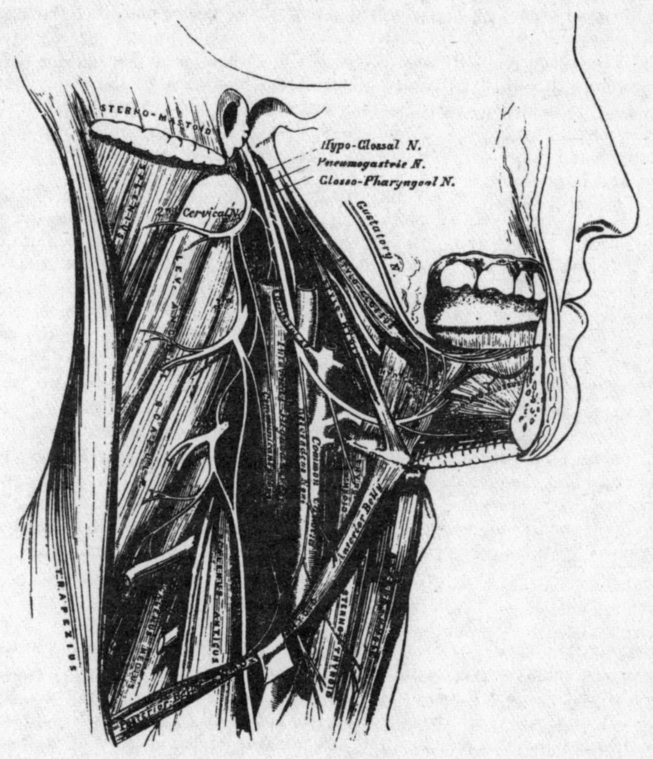

opposite the anterior condyloid foramen, and unite together after their passage through it. In those cases in which the anterior condyloid foramen in the occipital bone is double, these two portions of the nerve are separated by a small piece of bone, which divides the foramen. The nerve descends almost vertically to a point corresponding with the angle of the jaw. It is at first deeply seated beneath the internal carotid and internal jugular vein, and intimately connected with the pneumogastric nerve; it then passes forwards between the vein and artery, and at a lower part of the neck becomes superficial below the Digastric muscle. The nerve then loops round the occipital artery, and crosses the external carotid below the tendon of the Digastric muscle. It passes beneath the Mylohyoid muscle, lying between it and the Hyo-glossus, and is connected at the anterior border of the latter muscle with the gustatory nerve; it is then continued forwards into the Genio-hyo-glossus muscle as far as the tip of the tongue, distributing branches to its substance.

Branches of this nerve communicate with the

Pneumogastric. First and second cervical nerves.
Sympathetic. Gustatory.

The communication with the pneumogastric takes place close to the exit of the nerve from the skull, numerous filaments passing between the hypoglossal and second ganglion of the pneumogastric, or both being united so as to form one mass.

The communication with the sympathetic takes place opposite the atlas, by branches derived from the superior cervical ganglion, and in the same situation the ninth is joined by a filament derived from the loop connecting the first two cervical nerves.

The communication with the gustatory takes place near the anterior border of the Hyo-glossus muscle by numerous filaments which ascend upon it.

The branches of distribution are the

Descendens noni. Thyro-hyoid.
Muscular.

The *descendens noni* is a long slender branch, which quits the hypoglossal where it turns round the occipital artery. It descends obliquely across the sheath of the carotid vessels, and joins just below the middle of the neck, to form a loop with the communicating branches from the second and third cervical nerves. From the convexity of this loop branches pass forwards to supply the Sterno-hyoid, Sterno-thyroid, and both bellies of the Omo-hyoid. According to Arnold, another filament descends in front of the vessels into the chest, which joins the cardiac and phrenic nerves. The descendens noni is occasionally contained in the sheath of the carotid vessels, being sometimes placed over and sometimes beneath the internal jugular vein.

The *thyro-hyoid* is a small branch, arising from the hypoglossal near the posterior border of the Hyo-glossus; it passes obliquely across the great cornu of the hyoid bone, and supplies the Thyro-hyoid muscle.

The *muscular branches* are distributed to the Stylo-glossus, Hyo-glossus, Genio-hyoid, and Genio-hyo-glossus muscles. At the under surface of the tongue, numerous slender branches pass upwards into the substance of the organ.

Fifth Nerve.

The Fifth Nerve (*trifacial*, *trigeminus*) is the largest cranial nerve, and resembles a spinal nerve, in its origin by two roots, and in the existence of a ganglion on its posterior root. The functions of this nerve are various. It is a nerve of special sense, of common sensation, and of motion. It is the great sensitive nerve of the head and face, the motor nerve of the muscles of mastication, and its lingual branch is one of the nerves of the special sense of taste. It arises by two roots, a posterior larger or sensory, and an anterior smaller or motor root. Its *superficial origin* is from the side of the pons Varolii, a little nearer to the upper than the lower border. The smaller root consists of three or four bundles; in the larger, the bundles are more numerous, varying in number from seventy to a hundred: the two roots are separated from one another by a few of the transverse fibres of the pons. The *deep origin* of the larger or sensory root may be traced between the transverse fibres of the pons Varolii to the lateral tract of the medulla oblongata, immediately behind the olivary body. According to some anatomists, it is connected with the grey nucleus at the back part of the medulla, between the fasciculi teretes and restiform columns. By others, it is said to be continuous with the fasciculi teretes and lateral column of the cord; and, according to Foville, some of its fibres are connected with the transverse fibres of the pons; whilst others enter the cerebellum, spreading out on the surface of its middle peduncle. The motor root has been traced by Bell and Retzius to be connected with the pyramidal body. The two roots of the nerve pass forwards through an oval opening in the dura mater, at the apex of the petrous portion of the temporal bone: here the fibres of the larger root enter a large semilunar ganglion (Casserian), while the smaller root passes beneath the ganglion without having any connection with it, and joins outside the cranium with one of the trunks derived from it.

The Casserian, or Semilunar Ganglion, is lodged in a depression near the apex of the petrous portion of the temporal bone. It is of a somewhat crescentic form, with its convexity turned forwards. Its upper surface is intimately adherent to the dura mater.

Branches. This ganglion receives, on its *inner side*, filaments from the carotid plexus of the sympathetic; and from it some minute branches are given off to the tentorium cerebelli, and the dura mater, in the middle fossa of the cranium. From its *anterior border*, which is directed forwards and outwards, three large branches proceed: the ophthalmic, superior maxillary, and inferior maxillary. The ophthalmic and superior maxillary consist exclusively of fibres derived from the larger root and ganglion, and are solely nerves of common sensation. The third division, or inferior maxillary, is composed of fibres from both roots. This, therefore, strictly speaking, is the only portion of the fifth nerve which can be said to resemble a spinal nerve.

Ophthalmic Nerve. (Fig. 272.)

The Ophthalmic, or first division of the fifth, is a sensory nerve. It supplies the eyeball, the lachrymal gland, the mucous lining of the eye and nose, and the integument and muscles of the eyebrow and forehead. It is the smallest of the three divisions of the fifth, arising from the upper part of the Casserian ganglion. It is a short, flattened band, about an inch in length, which passes forwards along the outer wall of the cavernous sinus, below the other nerves, and just before entering the orbit, through the sphenoidal fissure, divides into three branches, lachrymal, frontal, and nasal. The ophthalmic nerve is joined by filaments from the cavernous plexus of the sympathetic, and gives off recurrent filaments which pass between the layers of the tentorium, with a branch from the fourth nerve.

Its branches are, the

Lachrymal. Frontal. Nasal.

The *Lachrymal* is the smallest of the three branches of the ophthalmic. Not unfrequently it arises by two filaments, one from the ophthalmic, the other from the fourth, and this Swan considers to be its usual condition. It passes forwards in a separate tube of dura mater, and enters the orbit through the narrowest part of the sphenoidal fissure. In the orbit, it runs along the upper border of the External rectus muscle, with the lachrymal artery, and is connected with the orbital branch of the superior maxillary nerve. Within the lachrymal gland it gives off several filaments, which supply the gland and the conjunctiva. Finally it pierces the palpebral ligament, and terminates in the integument of the upper eyelid, joining with filaments of the facial nerve.

The *Frontal* is the largest division of the ophthalmic, and may be regarded, both from its size and direction, as the continuation of this nerve. It enters the orbit above the muscles, through the highest and broadest part of the sphenoidal fissure, and runs forwards along the middle line, between the Levator palpebræ and the periosteum. Midway between the apex and base of the orbit it divides into two branches, supratrochlear, and supraorbital.

The *supratrochlear branch*, the smaller of the two, passes inwards, above the pulley of the Superior oblique muscle, and gives off a descending filament, which joins with the infratrochlear branch of the nasal nerve. It then escapes from the orbit between the pulley of the Superior oblique and the supraorbital foramen, curves up on to the forehead close to the bone, and ascends behind the Corrugator supercilii, and Occipito-frontalis muscles, to both of which it is distributed; finally, it is lost in the integument of the forehead.

The *supraorbital branch* passes forwards through the supraorbital foramen, and gives off, in this situation, palpebral filaments to the upper eyelid. It then ascends upon the forehead, and terminates in muscular, cutaneous, and pericranial branches. The *muscular branches* supply the Corrugator supercilii, Occipito-frontalis, and Orbicularis palpebrarum, furnishing these muscles with common sensation, and joining in the substance of the latter muscle with the facial nerve. The *cutaneous branches*, two in number, an inner and an outer, supply the integument of the cranium as far back as the occiput. They are at first situated beneath the Occipito-frontalis, the inner branch perforating the frontal portion of the muscle, the outer branch its tendinous aponeurosis. The *pericranial branches* are distributed to the pericranium over the frontal and parietal bones. They are derived from the cutaneous branches whilst beneath the muscle.

The *Nasal nerve* is intermediate in size between the frontal and lachrymal, and more deeply placed than the other branches of the ophthalmic. It enters the orbit between the two heads of the External rectus, and passes obliquely inwards across the optic nerve, beneath the Levator palpebræ and Superior rectus muscles, to the inner wall of the orbit, where it enters the anterior ethmoidal foramen, immediately below the Superior oblique. It then enters the cavity of the cranium, traverses a shallow groove on the front of the cribriform plate of the ethmoid bone, and passes down, through the slit by the side of the crista galli, into the nose, where it divides into two branches, an internal and an external. The *internal branch* supplies the mucous membrane near the fore part of the septum of the nose. The *external branch* descends in a groove on the inner surface of the nasal bone, and supplies a few filaments to the mucous membrane covering the fore part of the outer wall of the nares as far as the inferior spongy bone; it then leaves the cavity of the nose, between the lower border of the nasal bone and the upper lateral cartilage of the nose, and, passing down beneath the Compressor nasi, supplies the integument of the ala and tip of the nose, joining with the facial nerve.

The branches of the nasal nerve are, the ganglionic, ciliary, and infratrochlear.

The *ganglionic* is a long, slender branch, about half an inch in length, which usually arises from the nasal, between the two heads of the external rectus. It passes forwards on the outer side of the optic nerve, and enters the superior and posterior angle of the ciliary ganglion, forming its superior, or long root. It is sometimes joined by a filament from the cavernous plexus of the sympathetic, or from the superior division of the third nerve.

The *long ciliary nerves*, two or three in number, are given off from the nasal as it crosses the optic nerve. They join the short ciliary nerves from the ciliary ganglion, pierce the posterior part of the sclerotic, and, running forwards between it and the choroid, are distributed to the ciliary muscle and iris.

The *infratrochlear branch* is given off just as the nasal nerve passes through the anterior ethmoidal foramen. It runs forwards along the upper border of the Internal rectus, and is joined, beneath the pulley of the Superior oblique, by a filament from the supratrochlear nerve. It then passes to the inner angle of the eye, and supplies the Orbicularis palpebrarum, the integument of the eyelids, and side of the nose, the conjunctiva, lachrymal sac, and caruncula lachrymalis.

Ophthalmic Ganglion. (Fig. 273.)

Connected with the three divisions of the fifth nerve are four small ganglia, which form the whole of the cephalic portion of the sympathetic. With the first division is connected the ophthalmic ganglion; with the second division, the spheno-palatine, or Meckel's ganglion; and with the third, the otic and submaxillary ganglia. All the four receive sensitive filaments from the fifth, and motor and sympathetic filaments from various sources; these filaments are called the *roots of the ganglia*. The ganglia are also connected with each other, and with the cervical portion of the sympathetic.

The *Ophthalmic, Lenticular,* or *Ciliary Ganglion,* is a small, quadrangular, flattened ganglion, of a reddish-grey colour, and about the size of a pin's head, situated at the back part of the orbit between the optic nerve and the External rectus muscle, lying generally on the outer side of the ophthalmic artery. It is enclosed in a quantity of loose fat, which makes its dissection somewhat difficult.

Its *branches of communication,* or *roots,* are three, all of which enter its posterior border. One, the long root, is derived from the nasal branch of the ophthalmic, and joins its superior angle. Another branch, the short root, is a short thick nerve, occasionally divided into two parts, which is derived from the branch of the third nerve for the Inferior oblique muscle, and is connected with the inferior angle of the ganglion. A third branch, the sympathetic root, is a slender filament from the cavernous plexus of the sympathetic. This is occasionally blended with the long root, and sometimes passes to the ganglion separately. According to Tiedemann, this ganglion receives a filament of communication from the spheno-palatine ganglion.

Its *branches of distribution* are the short ciliary nerves. These are delicate filaments from ten to twelve in number, which arise from the fore part of the ganglion in two bundles, connected with its superior and inferior angles; the upper bundle consisting of four filaments, and the lower of six or seven. They run forwards with the ciliary arteries in a wavy course, one set above and the other below the optic nerve, pierce the sclerotic at the back part of the globe, pass forwards in delicate grooves on its inner surface, and are distributed to the ciliary muscle and iris. A small filament is described by Tiedemann, penetrating the optic nerve with the arteria centralis retinæ.

Superior Maxillary Nerve. (Fig. 277.)

The superior maxillary, or second division of the fifth, is a sensory nerve. It is intermediate, both in position and size, between the ophthalmic and inferior maxillary. It commences at the middle of the Casserian ganglion as a flattened plexiform band, and passes forwards through the foramen rotundum, where it becomes more cylindrical in form, and firmer in texture. It then crosses the spheno-maxillary fossa, traverses the infraorbital canal in the floor of the orbit, and appears upon the face at the infraorbital foramen. At its termination, the nerve lies beneath the Levator labii superioris muscle, and divides into a leash of branches, which spread out upon the side of the nose, the lower eyelid, and upper lip, joining with filaments of the facial nerve.

The branches of this nerve may be divided into three groups:—1. Those given off in the spheno-maxillary fossa. 2. Those in the infraorbital canal. 3. Those on the face.

Spheno-maxillary fossa	Orbital. Spheno-palatine. Posterior dental.
Infraorbital canal . .	Anterior dental.
On the face	Palpebral. Nasal. Labial.

The *Orbital branch* arises in the spheno-maxillary fossa, enters the orbit by the spheno-maxillary fissure, and divides at the back of that cavity into two branches, temporal and malar.

The *temporal branch* runs in a groove along the outer wall of the orbit (in the malar bone), receives a branch of communication from the lachrymal, and, passing through a foramen in the malar bone, enters the temporal fossa. It ascends between the bone and substance of the Temporal muscle, pierces this muscle and the temporal fascia about an inch above the zygoma, and is distributed to the integument covering the temple and side of the forehead, communicating with the facial and auriculo-temporal branch of the inferior maxillary nerve.

The *malar branch* passes along the external inferior angle of the orbit, emerges upon the face through a foramen in the malar bone, and perforating the Orbicularis palpebrarum muscle on the prominence of the cheek, joins with the facial.

The *Spheno-palatine branches*, two in number, descend to the spheno-palatine ganglion.

The *posterior dental branches* arise from the trunk of the nerve just as it is about to enter the infraorbital canal; they are two in number, posterior and anterior.

277.—Distribution of the Second and Third Divisions of the Fifth Nerve and Sub-maxillary Ganglion.

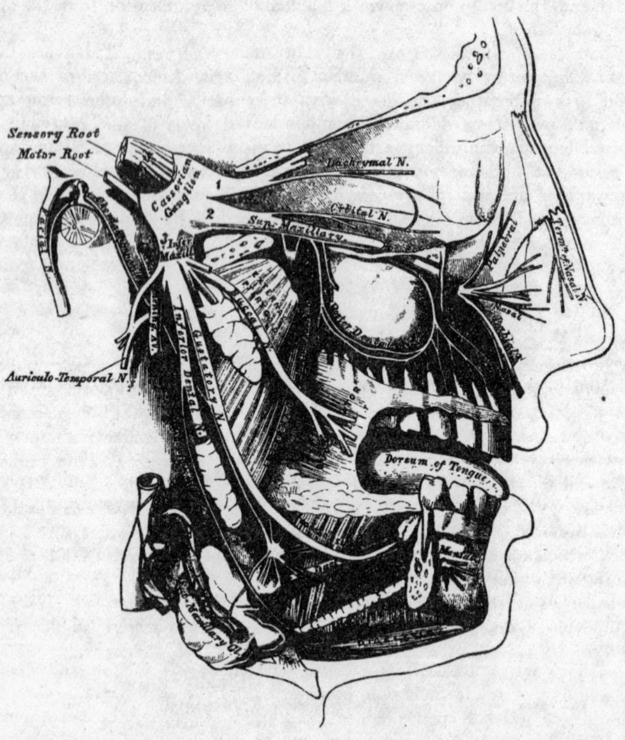

The *posterior branch* passes from behind forwards in the substance of the superior maxillary bone, and joins opposite the canine fossa with the anterior dental. Numerous filaments are given off from the lower border of this nerve, which form a minute plexus in the outer wall of the superior maxillary bone, immediately above the alveolus. From this plexus filaments are distributed to the pulps of the molar and second bicuspid teeth, the lining membrane of the antrum, and corresponding portion of the gums.

The *anterior branch* is distributed to the gums and Buccinator muscle.

The *anterior dental*, of large size, is given off from the superior maxillary nerve just before its exit from the infraorbital foramen; it enters a special canal in the anterior wall of the antrum, and anastomoses with the posterior dental. From this nerve some filaments are distributed to the incisor, canine, and first bicuspid teeth; others are lost upon the lining membrane covering the fore part of the inferior meatus.

The *palpebral branches* pass upwards beneath the Orbicularis palpebrarum. They supply this muscle, the integument, and conjunctiva of the lower eyelid with sensation, joining at the outer angle of the orbit with the facial nerve and malar branch of the orbital.

The *nasal branches* pass inwards; they supply the muscles and integument of the side of the nose, and join with the nasal branch of the ophthalmic.

The *labial branches*, the largest and most numerous, descend beneath the Levator labii superioris, and are distributed to the integument and muscles of the upper lip, the mucous membrane of the mouth, and labial glands.

All these branches are joined, immediately beneath the orbit, by filaments from the facial nerve, forming an intricate plexus, the infraorbital.

Spheno-Palatine Ganglion.

The *spheno-palatine ganglion* (Meckel's) (fig. 278), the largest of the cranial ganglia, is deeply placed in the spheno-maxillary fossa, close to the spheno-palatine foramen. It is triangular, or heart-shaped in form, of a reddish-grey colour, and placed mainly behind the palatine branches of the superior maxillary nerve, at the point where the sympathetic root joins the ganglion. It consequently does not involve those nerves which pass to the palate and nose. Like the other ganglia of the fifth nerve, it possesses a motor, a sensory, and a sympathetic root. Its motor root is derived from the facial, through the Vidian; its sensory root from the fifth; and its sympathetic root from the carotid plexus, through the Vidian. Its branches are divisible into four groups: ascending, which pass to the orbit; descending, to the palate; internal, to the nose; and posterior branches, to the pharynx.

The *ascending branches* are two or three delicate filaments, which enter the orbit by the spheno-maxillary fissure, and supply the periosteum. Arnold describes and delineates these branches as ascending to the optic nerve; one, to the sixth nerve (Böck); and one, to the ophthalmic ganglion (Tiedemann).

The *descending* or *palatine branches* are distributed to the roof of the mouth, the soft palate, tonsil, and lining membrane of the nose. They are almost a direct continuation of the spheno-palatine branches of the superior maxillary nerve, and are three in number: anterior, middle, and posterior.

The anterior, or large palatine nerve, descends through the posterior palatine canal, emerges upon the hard palate, at the posterior palatine foramen, and passes forwards through a groove in the hard palate, nearly as far as the incisor teeth. It supplies the gums, the mucous membrane and glands of the hard palate, and communicates in front with the termination of the naso-palatine nerve. While in the posterior palatine canal, it gives off inferior nasal branches, which enter the nose through openings in the palate bone, and ramify over the middle meatus, and the middle and inferior spongy bones; and at its exit from the canal, a palatine branch is distributed to both surfaces of the soft palate.

The middle, or external palatine nerve, descends in the same canal as the preceding, to the posterior palatine foramen, distributing branches to the uvula, tonsil, and soft palate. It is occasionally wanting.

The posterior, or small palatine nerve, descends with a small artery through the small posterior palatine canal, emerging by a separate opening behind the posterior palatine foramen. It supplies the Levator palati and Azygos uvulæ muscles, the soft palate, tonsil, and uvula.

The *internal branches* are distributed to the septum, and outer wall of the nasal fossæ. They are the superior nasal (anterior), and the naso-palatine.

The superior nasal branches (anterior), four or five in number, enter the back part of the nasal fossa by the spheno-palatine foramen. They supply the mucous membrane covering the superior and middle spongy bones, and that lining the posterior ethmoidal cells, a few being prolonged to the upper and back part of the septum.

The naso-palatine nerve (Cotunnius), enters the nasal fossa with the other nasal nerves, and passes inwards across the roof of the nose, below the orifice of the sphenoidal sinus, to reach the septum; it then runs obliquely downwards and forwards along the lower part of the septum, to the anterior palatine foramen, lying between the periosteum and mucous membrane. It descends to the roof of the mouth by a distinct canal, which opens below in the anterior palatine fossa; the right nerve, also in a separate canal, being posterior to the left one. In the mouth, they become united, supply the mucous membrane behind the incisor teeth, and join with the anterior palatine nerve. The naso-palatine nerve occasionally furnishes a few small filaments to the mucous membrane of the septum.

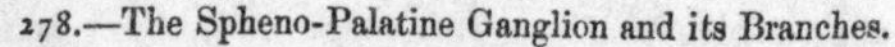
278.—The Spheno-Palatine Ganglion and its Branches.

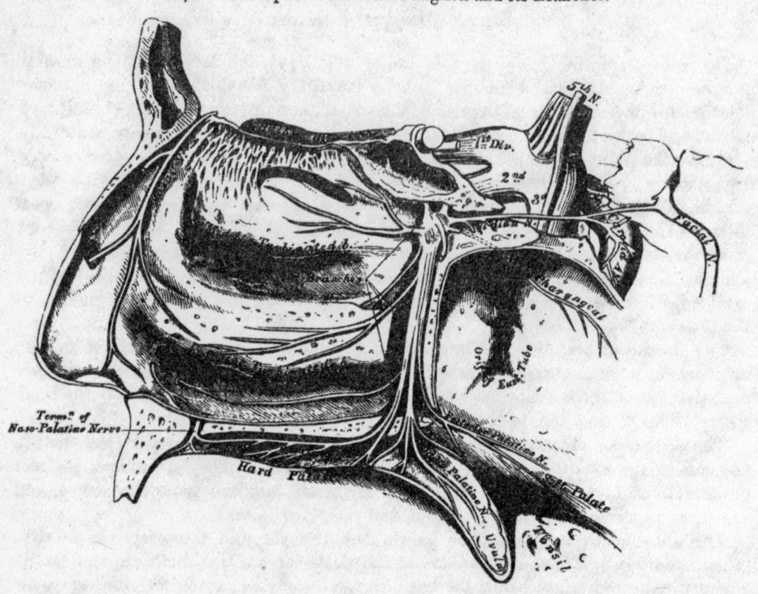

The *posterior branches* are the Vidian and pharyngeal (pterygo-palatine).

The Vidian nerve, if traced *from* Meckel's ganglion, may be said to arise from the back part of the spheno-palatine ganglion, and then passes through the Vidian canal, enters the cartilage filling in the foramen lacerum basis cranii, and divides into two branches, the large petrosal, and the carotid. In its course along the Vidian canal, it distributes a few filaments to the lining membrane at the back part of the roof of the nose and septum, and that covering the end of the Eustachian tube. These are upper posterior nasal branches.

The large petrosal branch (*nervus petrosus superficialis major*), enters the cranium through the foramen lacerum basis cranii, having pierced the cartilaginous substance which fills in this aperture. It runs beneath the Casserian ganglion and dura mater contained in a groove in the anterior surface of the petrous portion of the temporal bone, enters the hiatus Fallopii, and, being continued through it into the aquæductus Fallopii, joins the gangliform enlargement on the facial nerve. Properly speaking, this nerve passes from the facial to the spheno-palatine ganglion, forming its motor root.

The carotid branch is shorter, but larger than the petrosal, of a reddish-grey colour, and soft in texture. It crosses the foramen lacerum, surrounded by the cartilaginous substance which fills in that aperture, and enters the carotid canal on the outer side of the carotid artery, to join the carotid plexus.

This description of the Vidian nerve as a branch from the ganglion, is the more convenient anatomically, inasmuch as the nerve is generally dissected *from* the ganglion, as a single trunk *dividing* into two branches. But it is more correct, physiologically, to describe the Vidian as being formed by the *union* of the two branches (great petrosal and carotid) from the facial and the sympathetic, and as running *into* the ganglion. The filaments, which are described above as given off from the Vidian nerve, would then be regarded as branches from the ganglion which are merely enclosed in the same sheath as the Vidian.

The pharyngeal nerve (pterygo-palatine) is a small branch arising from the back part of the ganglion, occasionally together with the Vidian nerve. It passes through the pterygo-palatine canal with the pterygo-palatine artery, and is distributed to the lining membrane of the pharynx, behind the Eustachian tube.

Inferior Maxillary Nerve. (Fig. 277.)

The Inferior Maxillary Nerve distributes branches to the teeth and gums of the lower jaw, the integument of the temple and external ear, the lower part of the face and lower lip, and the muscles of mastication: it also supplies the tongue with one of its special nerves of the sense of taste. It is the largest of the three divisions of the fifth, and consists of two portions, the large or sensory root proceeding from the inferior angle of the Casserian ganglion; and the small or motor root, which passes beneath the ganglion, and unites with the inferior maxillary nerve, just after its exit through the foramen ovale. Immediately beneath the base of the skull this nerve divides into two trunks, anterior and posterior.

The anterior, and smaller division, which receives nearly the whole of the motor root, divides into five branches, which supply the muscles of mastication. They are the masseteric, deep temporal, buccal, and two pterygoid.

The *masseteric branch* passes outwards, above the External pterygoid muscle, in front of the temporo-maxillary articulation, and crosses the sigmoid notch, with the masseteric artery, to the Masseter muscle, in which it ramifies nearly as far as its anterior border. It occasionally gives a branch to the Temporal muscle, and a filament to the articulation of the jaw.

The *deep temporal branches*, two in number, anterior and posterior, supply the deep surface of the Temporal muscle. The *posterior branch*, of small size, is placed at the back of the temporal fossa. It is sometimes joined with the masseteric branch. The *anterior branch* is reflected upwards, at the pterygoid ridge of the sphenoid, to the front of the temporal fossa. It is occasionally joined with the buccal nerve.

The *buccal branch* pierces the External pterygoid, and passes downwards beneath the inner surface of the coronoid process of the lower jaw, or through the fibres of the Temporal muscle to reach the surface of the Buccinator, upon which it divides into a superior and an inferior branch. It gives a branch to the External pterygoid during its passage through that muscle, and a few ascending filaments to the Temporal muscle, one of which occasionally joins with the anterior branch of the deep temporal nerve. The *upper branch* supplies the integument and upper part of the Buccinator muscle, joining with the facial nerve round the facial vein. The *lower branch* passes forwards to the angle of the mouth; it supplies the integument and Buccinator muscle, as well as the mucous membrane lining the inner surface of that muscle, and joins the facial nerve.

The *pterygoid branches* are two in number, one for each pterygoid muscle. The *branch to the Internal pterygoid* is long and slender, and passes inwards to enter the deep surface of the muscle. This nerve is intimately connected at its origin with the otic ganglion. The *branch to the External pterygoid* is most frequently derived from the buccal, but it may be given off separately from the anterior trunk of the nerve.

The posterior and larger division of the inferior maxillary nerve also receives a few filaments from the motor root. It divides into three branches, auriculo-temporal, gustatory, and inferior dental.

The Auriculo-temporal Nerve generally arises by two roots, between which the middle meningeal artery passes. It runs backwards beneath the External pterygoid muscle to the inner side of the articulation of the lower jaw. It then turns upwards with the temporal artery, between the external ear and condyle of the jaw, under cover of the parotid gland, and escaping from beneath this structure, divides into two temporal branches. The *posterior temporal*, the smaller of the two, is distributed to the upper part of the pinna and the neighbouring tissues. The *anterior temporal* accompanies the temporal artery to the vertex of the skull, and supplies the integument of the temporal region, communicating with the facial nerve.

The auriculo-temporal nerve has *branches of communication* with the facial and otic ganglion. Those joining the facial nerve, usually two in number, pass forwards behind the neck of the condyle of the jaw, and join this nerve at the posterior border of the Masseter muscle. They form one of the principal branches of communication between the facial and the fifth nerve. The filaments of communication with the otic ganglion are derived from the commencement of the auriculo-temporal nerve.

The *auricular branches* are two in number, inferior and superior. The *inferior auricular* arises behind the articulation of the jaw, and is distributed to the ear below the external meatus; other filaments twine round the internal maxillary artery, and communicate with the sympathetic. The *superior auricular* arises in front of the external ear, and supplies the integument covering the tragus and pinna.

Branches to the meatus auditorius, two in number, arise from the point of communication between the auriculo-temporal and facial nerves, and are distributed to the meatus.

The *branch to the temporo-maxillary articulation* is usually derived from the auriculo-temporal nerve.

The *parotid branches* supply the parotid gland.

The Gustatory or Lingual Nerve, one of the special nerves of the sense of taste, supplies the papillæ and mucous membrane of the tongue. It is deeply placed throughout the whole of its course. It lies at first beneath the External pterygoid muscle, together with the inferior dental nerve, being placed to the inner side of the latter nerve, and is occasionally joined to it by a branch which crosses the internal maxillary artery. The chorda tympani also joins it at an acute angle in this situation. The nerve then passes between the Internal pterygoid muscle and the inner side of the ramus of the jaw, and crosses obliquely to the side of the tongue over the Superior constrictor muscle of the pharynx, and beneath the Stylo-glossus muscle and deep part of the submaxillary gland; the nerve lastly runs across Wharton's duct, and along the side of the tongue to its apex, being covered by the mucous membrane of the mouth.

Its *branches of communication* are with the submaxillary ganglion and hypoglossal nerve. The branches to the submaxillary ganglion are two or three in number; those connected with the hypoglossal nerve form a plexus at the anterior margin of the Hyo-glossus muscle.

Its *branches of distribution* are few in number. They supply the mucous membrane of the mouth, the gums, the sublingual gland, the conical and fungiform papillæ and mucous membrane of the tongue, the terminal filaments anastomosing at the tip of the tongue with the hypoglossal nerve.

The Inferior Dental is the largest of the three branches of the inferior maxillary nerve. It passes downwards with the inferior dental artery, at first beneath the External pterygoid muscle, and then between the internal lateral ligament and the ramus of the jaw to the dental foramen. It then passes forwards in the dental canal in the inferior maxillary bone, lying beneath the teeth, as far as the mental foramen, where it divides into two terminal branches, incisor and mental. The *incisor branch* is continued onwards within the bone to the middle line, and supplies the canine and incisor teeth. The *mental branch* emerges from the bone at the mental foramen, and divides beneath the Depressor anguli oris into an external branch, which supplies that muscle, the Orbicularis oris, and the integument, communicating with the facial nerve; and an inner branch, which ascends to the lower lip beneath the Quadratus menti; it supplies that muscle and the mucous membrane and integument of the lip, communicating with the facial nerve.

The branches of the inferior dental are, the mylo-hyoid, and dental.

The *mylo-hyoid* is derived from the inferior dental just as that nerve is about to enter the dental foramen. It descends in a groove on the inner surface of the ramus of the jaw, in which it is retained by a process of fibrous membrane. It supplies the cutaneous surface of the Mylo-hyoid muscle, and the anterior belly of the Digastric, occasionally sending one or two filaments to the submaxillary gland.

The *dental branches* supply the molar and bicuspid teeth. They correspond in number to the fangs of those teeth; each nerve entering the orifice at the point of the fang, and supplying the pulp of the tooth.

Two small ganglia are connected with the inferior maxillary nerve: the otic, with the trunk of the nerve; and the submaxillary with its lingual branch, the gustatory.

Otic Ganglion.

The *otic ganglion* (Arnold's) (fig. 279), is a small, oval-shaped, flattened ganglion, of a reddish-grey colour, situated immediately below the foramen ovale, on the

279.—The Otic Ganglion and its Branches.

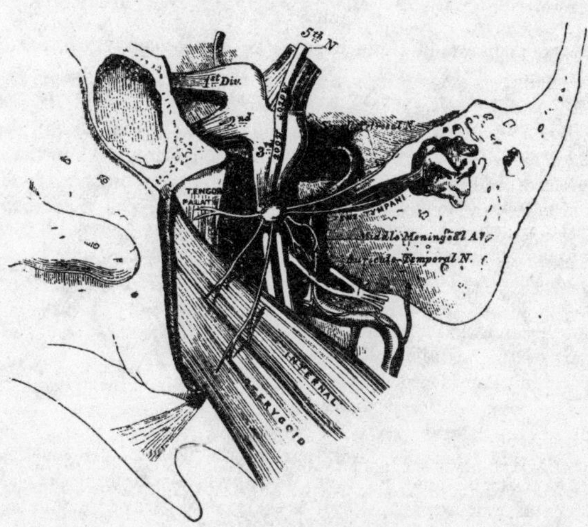

inner surface of the inferior maxillary nerve, and round the origin of the internal pterygoid nerve. It is in relation, *externally*, with the trunk of the inferior maxillary nerve, at the point where the motor root joins the sensory portion;

internally, with the cartilaginous part of the Eustachian tube, and the origin of the Tensor palati muscle; *behind* it, is the middle meningeal artery.

Branches of communication. This ganglion is connected with the inferior maxillary nerve, and its internal pterygoid branch, by two or three short, delicate filaments, and also with the auriculo-temporal nerve: from the former it obtains its motor; from the latter, its sensory root; its communication with the sympathetic being effected by a filament from the plexus surrounding the middle meningeal artery. This ganglion also communicates with the glosso-pharyngeal and facial nerves, through the small petrosal nerve continued from the tympanic plexus.

Its *branches of distribution* are a filament to the Tensor tympani, and one to the Tensor palati. The former passes backwards, on the outer side of the Eustachian tube; the latter arises from the ganglion, near the origin of the internal pterygoid nerve, and passes forwards.

Submaxillary Ganglion.

The *submaxillary ganglion* (fig. 277) is of small size, circular in form, and situated above the deep portion of the submaxillary gland, near the posterior border of the Mylo-hyoid muscle, being connected by filaments with the lower border of the gustatory nerve.

Branches of communication. This ganglion is connected with the gustatory nerve by a few filaments which join it separately, at its fore and back part. It also receives a branch from the chorda tympani, by which it communicates with the facial; and communicates with the sympathetic by filaments from the nervi molles—the sympathetic plexus around the facial artery.

Branches of distribution. These are five or six in number; they arise from the lower part of the ganglion, and supply the mucous membrane of the mouth and Wharton's duct, some being lost in the submaxillary gland. According to Meckel, a branch from this ganglion occasionally descends in front of the Hyo-glossus muscle, and, after joining with one from the hypoglossal, passes to the Genio-hyo-glossus muscle.

Eighth Pair.

The *eighth pair* consists of three nerves, the glosso-pharyngeal, pneumogastric, and spinal accessory.

The Glosso-Pharyngeal Nerve is distributed, as its name implies, to the tongue and pharynx, being the nerve of sensation to the mucous membrane of the pharynx, fauces, and tonsil; of motion to the Pharyngeal muscles; and a special nerve of taste in all the parts of the tongue to which it is distributed. It is the smallest of the three divisions of the eighth pair, and arises by three or four filaments, closely connected together, from the upper part of the medulla oblongata, immediately behind the olivary body.

280.—Nerves of the Eighth Pair, their Origin, Ganglia, and Communications.

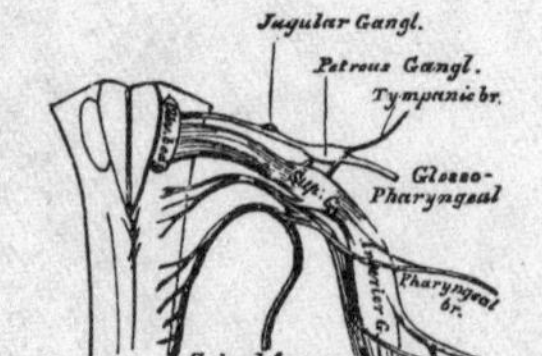

Its *deep origin* may be traced through the fasciculi of the lateral tract, to a nucleus of grey matter at the lower part of the floor of the fourth ventricle, external to the fasciculi teretes. From its superficial origin, it passes outwards across the flocculus, and leaves the skull at the central part of the jugular foramen, in a separate sheath of the dura mater and arachnoid, in front of the pneumogastric and spinal accessory nerves. In its passage through the jugular foramen, it grooves the lower border of the petrous portion of the temporal bone; and, at its exit from the skull, passes forwards between the jugular vein and internal carotid artery, and descends in front of the latter vessel, and beneath the styloid process and the muscles connected with it, to the lower border of the Stylo-pharyngeus. The nerve now curves inwards, forming an arch on the side of the neck, and lying upon the Stylo-pharyngeus and the Middle constrictor of the pharynx, above the superior laryngeal nerve. It then passes beneath the Hyo-glossus, and is finally distributed to the mucous membrane of the fauces, and base of the tongue, the mucous glands of the mouth and tonsil.

281.—Course and Distribution of the Eighth Pair of Nerves.

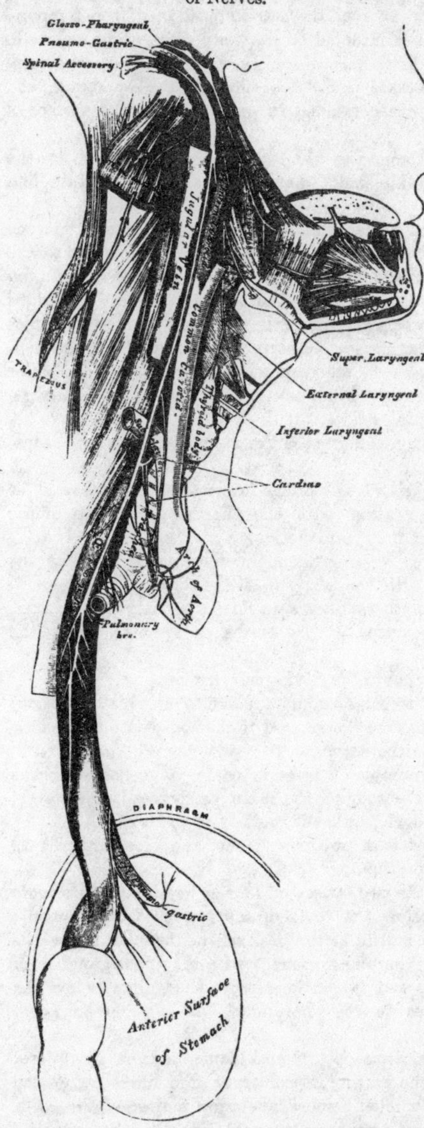

In passing through the jugular foramen, the nerve presents, in succession, two gangliform enlargements. The superior, the smaller, is called the *jugular ganglion*, the inferior and larger the *petrous ganglion*, or the ganglion of Andersch.

The *superior*, or *jugular ganglion*, is situated in the upper part of the groove in which the nerve is lodged during its passage through the jugular foramen. It is of very small size, and involves only the outer side of the trunk of the nerve, a small fasciculus passing beyond it, which is not connected directly with it.

The *inferior*, or *petrous ganglion*, is situated in a depression in the lower border of the petrous portion of the temporal bone; it is larger than the former, and involves the whole of the fibres of the nerve. From this ganglion arise those filaments which connect the glosso-pharyngeal with other nerves at the base of the skull.

Its branches of communication are with the pneumogastric, sympathetic, and facial, and the tympanic branch.

The branches to the pneumogastric are two filaments, one to its auricular branch, and one to the upper ganglion of the pneumogastric.

The branch to the sympathetic is connected with the superior cervical ganglion.

The branch of communication with the facial perforates the posterior belly of the Digastric. It arises from the trunk of the nerve below the petrous ganglion, and joins the facial just after its exit from the stylo-mastoid foramen.

The *tympanic branch* (Jacobson's nerve) arises from the petrous ganglion, and enters a small bony canal at the base of the petrous portion of the temporal bone; the lower opening of which is situated on the bony ridge which separates the carotid canal from the jugular fossa. Jacobson's nerve ascends to the tympanum, enters that cavity by an aperture in its floor close to the inner wall, and divides into three branches, which are contained in grooves upon the surface of the promontory.

Its branches of distribution are, one to the fenestra rotunda, one to the fenestra ovalis, and one to the lining membrane of the Eustachian tube and tympanum.

Its branches of communication are three, and occupy separate grooves on the surface of the promontory. One of these arches forwards and downwards to the carotid canal to join the carotid plexus. A second runs vertically upwards to join the greater superficial petrosal nerve, as it lies in the hiatus Fallopii. The third branch, the lesser petrosal, runs upwards and forwards towards the anterior surface of the petrous bone, and passes through a small aperture in the sphenoid and temporal bones, to the exterior of the skull, where it joins the otic ganglion. This nerve, in its course through the temporal bone, passes by the ganglionic enlargement of the facial, and has a connecting filament with it.

The branches of the glosso-pharyngeal nerve are the carotid, pharyngeal, muscular, tonsillar, and lingual.

The *carotid branches* descend along the trunk of the internal carotid artery as far as its point of bifurcation, communicating with the pharyngeal branch of the pneumogastric, and with branches of the sympathetic.

The *pharyngeal branches* are three or four filaments which unite opposite the Middle constrictor of the pharynx with the pharyngeal branches of the pneumogastric, superior laryngeal, and sympathetic nerves, to form the pharyngeal plexus, branches from which perforate the muscular coat of the pharynx to supply the mucous membrane.

The *muscular branches* are distributed to the Stylo-pharyngeus.

The *tonsillar branches* supply the tonsil, forming a plexus (circulus tonsillaris) around this body, from which branches are distributed to the soft palate and fauces, where they anastomose with the palatine nerves.

The *lingual branches* are two in number; one supplies the mucous membrane covering the surface of the base of the tongue, the other perforates its substance, and supplies the mucous membrane and papillæ of the side of the organ.

The Spinal Accessory Nerve consists of two parts; one, the accessory part to the vagus, and the other the spinal portion.

The *accessory part*, the smaller of the two, arises by four or five delicate filaments from the lateral tract of the cord below the roots of the vagus; these filaments may be traced to a nucleus of grey matter at the back of the medulla, below the origin of the vagus. It joins, in the jugular foramen, with the upper ganglion of the vagus by one or two filaments, and is continued into the vagus below the second ganglion. It gives branches to the pharyngeal and superior laryngeal branches of the vagus.

The *spinal portion*, firm in texture, arises by several filaments from the lateral tract of the cord, as low down as the sixth cervical nerve; the fibres pierce the tract, and are connected with the anterior horn of the grey matter of the cord. This portion of the nerve ascends between the ligamentum denticulatum and the posterior roots of the spinal nerves, enters the skull through the foramen magnum, and is then directed outwards to the jugular foramen, through which it passes, lying in the same sheath as the pneumogastric, separated from it by a fold of the arachnoid, and is here connected with the accessory portion. At its exit from the jugular foramen, it passes backwards behind the internal jugular vein, and descends obliquely behind the Digastric and Stylo-hyoid muscles to the upper part of the Sterno-mastoid. It pierces that muscle, and passes obliquely across the sub-occipital triangle, to terminate in the deep surface of the Trapezius. This nerve gives several branches to the Sterno-mastoid during its passage through it, and joins in its substance with branches from the third cervical. In the occipital triangle it joins with the second and third cervical nerves, and assists in the formation of the cervical plexus, and occasionally of the great auricular nerve. On the front of the Trapezius, it is reinforced by branches from the third, fourth, and fifth cervical nerves, joins with the posterior branches of the spinal nerves, and is distributed to the Trapezius, some filaments ascending and others descending in the substance of the muscle as far as its inferior angle.

The Pneumogastric Nerve (*nervus vagus*, or *par vagum*), one of the three divisions of the eighth pair, has a more extensive distribution than any of the other cranial nerves, passing through the neck and thorax to the upper part of the abdomen. It is composed of both motor and sensitive filaments. It supplies the organs of voice and respiration with motor and sensitive fibres ; and the pharynx, œsophagus, stomach, and heart with motor influence. Its *superficial origin* is by eight or ten filaments from the lateral tract immediately behind the olivary body and below the glosso-pharyngeal; its fibres may, however, be traced deeply through the fasciculi of the medulla, to terminate in a grey nucleus near the lower part of the floor of the fourth ventricle. The filaments become united, and form a flat cord, which passes outwards across the flocculus to the jugular foramen, through which it emerges from the cranium. In passing through this opening, the pneumogastric accompanies the spinal accessory, being contained in the same sheath of dura mater with it, a membranous septum separating it from the glosso-pharyngeal, which lies in front. The nerve in this situation presents a well-marked ganglionic enlargement, which is called *ganglion jugulare*, or the *ganglion of the root of the pneumogastric*: to it the accessory part of the spinal accessory nerve is connected. After the exit of the nerve from the jugular foramen, a second gangliform swelling is formed upon it, called the *ganglion inferius*, or the *ganglion of the trunk of the nerve*; below which it is again joined by filaments from the *accessory* nerve. The nerve passes vertically down the neck within the sheath of the carotid vessels, lying between the internal carotid artery and internal jugular vein as far as the thyroid cartilage, and then between the same vein and the common carotid to the root of the neck. Here the course of the nerve becomes different on the two sides of the body.

On the *right side*, the nerve passes across the subclavian artery between it and the subclavian vein, and descends by the side of the trachea to the back part of the root of the lung, where it spreads out in a plexiform network (posterior pulmonary), from the lower part of which two cords descend upon the œsophagus, on which they divide, forming, with branches from the opposite nerve, the œsophageal plexus; below, these branches are collected into a single cord, which runs along the back part of the œsophagus, enters the abdomen, and is distributed to the posterior surface of the stomach, joining the left side of the cœliac plexus, and the splenic plexus.

On the *left side*, the pneumogastric nerve enters the chest between the left carotid and subclavian arteries, behind the left innominate vein. It crosses the arch of the aorta, and descends behind the root of the left lung and along the anterior surface of the œsophagus to the stomach, distributing branches over its anterior surface, some extending over the great *cul-de-sac*, and others along the lesser curvature. Filaments from these latter branches enter the gastro-hepatic omentum, and join the left hepatic plexus.

The *ganglion of the root* is of a greyish colour, circular in form, about two lines in diameter, and resembles the ganglion on the large root of the fifth nerve.

Connecting branches. To this ganglion the accessory portion of the spinal accessory nerve is connected by several delicate filaments; it also has an anastomotic twig with the petrous ganglion of the glosso-pharyngeal, with the facial nerve by means of the auricular branch, and with the sympathetic by means of an ascending filament from the superior cervical ganglion.

The *ganglion of the trunk* (inferior) is a plexiform cord, cylindrical in form, of a reddish colour, and about an inch in length; it involves the whole of the fibres of the nerve, except the portion of the nerve derived from the spinal accessory, which blends with the nerve beyond the ganglion.

Connecting branches. This ganglion is connected with the hypoglossal, the superior cervical ganglion of the sympathetic, and the loop between the first and second cervical nerves.

The *branches of the pneumogastric* are—

In the jugular fossa . . .	Auricular.
In the neck	Pharyngeal. Superior laryngeal. Recurrent laryngeal. Cervical cardiac.
In the thorax	Thoracic cardiac. Anterior pulmonary. Posterior pulmonary. Œsophageal.
In the abdomen	Gastric.

The *auricular branch* (Arnold's) arises from the ganglion of the root, and is joined soon after its origin by a filament from the glosso-pharyngeal; it crosses the jugular fossa to an opening near the root of the styloid process. Traversing the substance of the temporal bone, it crosses the aquæductus Fallopii about two lines above its termination at the stylo-mastoid foramen; it here gives off an ascending branch, which joins the facial, and a descending branch, which anastomoses with the posterior auricular branch of the same nerve: the continuation of the nerve reaches the surface between the mastoid process and the external auditory meatus, and supplies the integument at the back part of the pinna.

The *pharyngeal branch*, the principal motor nerve of the pharynx and soft palate, arises from the upper part of the inferior ganglion of the pneumogastric, receiving a filament from the accessory portion of the spinal accessory; it passes across the internal carotid artery (in front or behind), to the upper border of the Middle constrictor, where it divides into numerous filaments, which anastomose with those from the glosso-pharyngeal, superior laryngeal, and sympathetic, to form the pharyngeal plexus, from which branches are distributed to the muscles and mucous membrane of the pharynx. As this nerve crosses the internal carotid, some filaments are distributed, together with those from the glosso-pharyngeal, upon the wall of this vessel.

The *superior laryngeal* is the nerve of sensation to the larynx. It is larger than the preceding, and arises from the middle of the inferior ganglion of the pneumogastric. It descends, by the side of the pharynx, behind the internal carotid, where it divides into two branches, the external and internal laryngeal.

The external laryngeal branch, the smaller, descends by the side of the larynx, beneath the Sterno-thyroid, to supply the Crico-thyroid muscle and the thyroid gland. It gives branches to the pharyngeal plexus, and the Inferior constrictor, and communicates with the superior cardiac nerve, behind the common carotid.

The internal laryngeal branch descends to the opening in the thyro-hyoid membrane, through which it passes with the superior laryngeal artery, and is distributed to the mucous membrane of the larynx, and the Arytenoid muscle, anastomosing with the recurrent laryngeal.

The branches to the mucous membrane are distributed, some in front, to the epiglottis, the base of the tongue and the epiglottidean glands; while others pass backwards, in the aryteno-epiglottidean fold, to supply the mucous membrane surrounding the superior orifice of the larynx, as well as the membrane which lines the cavity of the larynx as low down as the vocal chord.

The filament to the Arytenoid muscle is distributed partly to it, and partly to the mucous lining of the larynx.

The filament which joins with the recurrent laryngeal, descends beneath the mucous membrane on the posterior surface of the larynx, behind the lateral part of the thyroid cartilage, where the two nerves become united.

The *inferior* or *recurrent laryngeal*, so called from its reflected course, is the motor nerve of the larynx. It arises on the right side, in front of the subclavian artery: winds from before backwards round that vessel, and ascends obliquely to the side of the trachea, behind the common carotid and inferior thyroid arteries. On the left side, it arises in front of the arch of the aorta, and winds from before backwards round the aorta at the point where the obliterated remains of the ductus arteriosus are connected with it, and then ascends to the side of the trachea. The nerves on both sides ascend in the groove between the trachea and œsophagus, and, passing under the lower border of the Inferior constrictor muscle, enter the larynx behind the articulation of the inferior cornu of the thyroid cartilage with the cricoid, being distributed to all the muscles of the larynx, excepting the Crico-thyroid, and joining with the superior laryngeal.

The recurrent laryngeal, as it winds round the subclavian artery and aorta, gives off several cardiac filaments, which unite with cardiac branches from the pneumogastric and sympathetic. As it ascends in the neck, it gives off œsophageal branches, more numerous on the left than on the right side, which supply the mucous membrane and muscular coat of the œsophagus; tracheal branches to the mucous membrane and muscular fibres of the trachea; and some pharyngeal filaments to the Inferior constrictor of the pharynx.

The *cervical cardiac branches*, two or three in number, arise from the pneumogastric, at the upper and lower part of the neck.

The superior branches are small, and communicate with the cardiac branches of the sympathetic, and with the great cardiac plexus.

The inferior cardiac branches, one on each side, arise at the lower part of the neck, just above the first rib. On the right side, this branch passes in front of the arteria innominata, and anastomoses with the superior cardiac nerve. On the left side, it passes in front of the arch of the aorta, and anastomoses either with the superior cardiac nerve, or with the cardiac plexus.

The *thoracic cardiac branches*, on the right side, arise from the trunk of the pneumogastric, as it lies by the side of the trachea: passing inwards, they terminate in the deep cardiac plexus. On the left side, they arise from the left recurrent laryngeal nerve.

The *anterior pulmonary branches*, two or three in number, and of small size, are distributed on the anterior aspect of the root of the lungs. They join with filaments from the sympathetic, and form the anterior pulmonary plexus.

The *posterior pulmonary branches*, more numerous and larger than the anterior, are distributed on the posterior aspect of the root of the lung: they are joined by filaments from the third and fourth thoracic ganglia of the sympathetic, and form the posterior pulmonary plexus. Branches from both plexuses accompany the ramifications of the air-tubes through the substance of the lungs.

The *œsophageal branches* are given off from the pneumogastric both above and below the pulmonary branches. The latter are the more numerous and largest. They form, together with branches from the opposite nerve, the œsophageal plexus.

The *gastric branches* are the terminal filaments of the pneumogastric nerve. The nerve on the right side is distributed to the posterior surface of the stomach, and joins the left side of the cœliac plexus, and the splenic plexus. The nerve on the left side is distributed over the anterior surface of the stomach, some filaments passing across the great *cul-de-sac*, and others along the lesser curvature. They unite with branches of the right nerve and sympathetic, some filaments passing through the lesser omentum to the left hepatic plexus.

For the following brief account of the most recent views relating to the origin of the cranial nerves, the editor is indebted to his friend Mr. Lockhart Clarke.

The third cerebral nerve arises chiefly from two large masses of grey substance at the floor of the iter e tertio ad quartum ventriculum beneath the corpora quadrigemina.

The fourth arises from two nuclei at the floor of iter e tertio ad quartum ventriculum, and from the valve of Vieussens, where the opposite nerves decussate each other.

The large roots of the fifth or trigeminal arise chiefly from the grey tubercles of Rolando, or the upper expanded extremities of the posterior grey horns of the spinal cord; the small or motor roots arise from two masses of large, multipolar cells situated each on the inner side, and close to the grey tubercle, and intimately connected with it.

The sixth nerve arises in common with the facial from the grey substance of the *fasciculus teres* on the floor of the fourth ventricle.

The *facial* nerve has two origins: 1. From the grey substance of the *fasciculus teres* on the floor of the fourth ventricle. 2. From the nucleus of the motor root of the trigeminus; between these two origins it forms a loop along the floor of the ventricle.

The Auditory Nerve has three origins: 1. From the superior vermiform process of the cerebellum; 2 and 3. From the inner and outer auditory nuclei formed chiefly by the grey substance of the posterior pyramid and restiform body.

The vagus and glossopharyngeal nerves have each two origins: 1. From a special nucleus in the floor of the fourth ventricle. 2. From the anterior or motor part of the medulla.

The *Spinal-Accessory* nerve has three origins: 1. The lower roots from the anterior grey horn of the spinal cord in common with the motor roots of the cervical nerves. 2. From the grey nucleus of the hypoglossal nerve. 3. From a special nucleus behind the central canal of the medulla oblongata.

For further information on the origin of these nerves, and on the connection between their several nuclei, see Mr. Lockhart Clarke's memoir 'On the Intimate Structure of the Brain,' 1st and 2nd Ser. Phil. Trans. 1858 and 1868.

For fuller detail concerning the Cranial Nerves, the student may refer to F. Arnold's 'Icones Nervorum Capitis.'

The Spinal Nerves.

THE *spinal nerves* are so called, because they take their origin from the spinal cord, and are transmitted through the intervertebral foramina on either side of the spinal column. There are thirty-one pairs of spinal nerves, which are arranged into the following groups, corresponding to the region of the spine through which they pass:

Cervical . . .	8 pairs.
Dorsal . . .	12 ,,
Lumbar . . .	5 ,,
Sacral . . .	5 ,,
Coccygeal . . .	1 ,,

It will be observed, that each group of nerves corresponds in number with the vertebræ in that region, except the cervical and coccygeal.

Each spinal nerve arises by two roots, an anterior, or motor root, and a posterior, or sensory root.

Roots of the Spinal Nerves.

The *anterior roots* arise somewhat irregularly from a linear series of foramina, on the antero-lateral column of the spinal cord, gradually approaching towards the anterior median fissure as they descend.

The fibres of the anterior roots, according to the researches of Mr. Lockhart Clarke, are attached to the anterior part of the antero-lateral column; and, after penetrating horizontally through the longitudinal fibres of this tract, enter the grey substance, where their fibrils cross each other and diverge in all directions, like the expanded hairs of a brush, some of them running more or less longitudinally upwards and downwards, and others decussating with those of the opposite side through the anterior commissure in front of the central canal. Kölliker states that many fibres of the anterior root enter the lateral column of the same side, where, turning *upwards*, they pursue their course as longitudinal fibres. In other respects, the description of the origin of the anterior roots by these observers is very similar.

The *posterior roots* are all attached immediately to the posterior white columns only; but some of them pass through the grey substance into both the lateral and anterior white columns. Within the grey substance, they run, longitudinally, upwards and downwards, transversely, through the posterior commissure to the opposite side, and into the anterior column of their own side.

The posterior roots of the nerves are larger, but the individual filaments are finer and more delicate than those of the anterior. As their component fibrils pass outwards, towards the aperture in the dura mater, they coalesce into two bundles, receive a tubular sheath from that membrane, and enter the ganglion which is developed upon each root.

The posterior root of the first cervical nerve forms an exception to these characters. It is smaller than the anterior, has frequently no ganglion developed upon it, and, when the ganglion exists, it is often situated within the dura mater.

The anterior roots are the smaller of the two, devoid of any ganglionic enlargement, and their component fibrils are collected into two bundles, near the intervertebral foramina.

Ganglia of the Spinal Nerves.

A ganglion is developed upon the posterior root of each of the spinal nerves. These ganglia are of an oval form, and of a reddish colour; they bear a proportion in size to the nerves upon which they are formed, and are placed in the intervertebral foramina, external to the point where the nerves perforate the dura mater. Each ganglion is bifid internally, where it is joined by the two bundles of the posterior root, the two portions being united into a single mass externally. The ganglia upon the first and second cervical nerves form an exception to these characters, being placed on the arches of the vertebræ over which the nerves pass. The ganglia, also, of the sacral nerves are placed within the spinal canal; and that on the coccygeal nerve, also in the canal about the middle of its posterior root. Immediately beyond the ganglion, the two roots coalesce, their fibres intermingle, and the trunk thus formed passes out of the intervertebral foramen, and divides into an anterior branch, for the supply of the anterior part of the body; and a posterior branch for the posterior part, each branch containing fibres from both roots.

Anterior Branches of the Spinal Nerves.

The *anterior branches of the spinal nerves* supply the parts of the body in front of the spine, including the limbs. They are for the most part larger than the posterior branches; this increase of size being proportioned to the larger extent of structures they are required to supply. Each branch is connected by slender filaments with the sympathetic. In the dorsal region, the anterior branches of the spinal nerves are completely separate from each other, and are uniform in their distribution; but in the cervical, lumbar, and sacral regions, they form intricate plexuses previous to their distribution.

Posterior Branches of the Spinal Nerves.

The *posterior branches of the spinal nerves* are generally smaller than the anterior; they arise from the trunk, resulting from the union of the roots in the intervertebral foramina, and, passing backwards, divide into external and internal branches, which are distributed to the muscles and integument behind the spine. The first cervical and lower sacral nerves are exceptions to these characters.

Cervical Nerves.

The *roots of the cervical nerves* increase in size from the first to the fifth, and then maintain the same size to the eighth. The posterior roots bear a proportion to the anterior as 3 to 1, which is much greater than in any other region, the individual filaments being also much larger than those of the anterior roots. In direction, the roots of the cervical are less oblique than those of the other spinal nerves. The first cervical nerve is directed a little upwards and outwards; the second is horizontal; the others are directed obliquely downwards and outwards, the lowest being the most oblique, and consequently longer than the upper, the distance between their place of origin and their point of exit from the spinal canal never exceeding the depth of one vertebra.

The *trunk of the first cervical nerve* (*suboccipital*), leaves the spinal canal, between the occipital bone and the posterior arch of the atlas; the second between the posterior arch of the atlas and the lamina of the axis; and the eighth (the last), between the last cervical and first dorsal vertebræ.

Each nerve, at its exit from the intervertebral foramen, divides into an anterior and a posterior branch. The anterior branches of the four upper cervical nerves form the cervical plexus. The anterior branches of the four lower cervical nerves, together with the first dorsal, form the brachial plexus.

Anterior Branches of the Cervical Nerves.

The *anterior branch of the first*, or *suboccipital nerve*, is of small size. It escapes from the spinal canal, through a groove upon the posterior arch of the atlas. In this groove it lies beneath the vertebral artery, to the inner side of the Rectus lateralis. As it crosses the foramen in the transverse process of the atlas, it receives a filament from the sympathetic. It then descends, in front of this process, to communicate with an ascending branch from the second cervical nerve.

Communicating filaments from this nerve join the pneumogastric, the hypoglossal and sympathetic, and some branches are distributed to the Rectus lateralis, and the two Anterior recti. According to Valentin, the anterior branch of the suboccipital also distributes filaments to the occipito-atloid articulation, and mastoid process of the temporal bone.

The *anterior branch of the second cervical nerve* escapes from the spinal canal, between the posterior arch of the atlas and the lamina of the axis, and, passing forwards on the outer side of the vertebral artery, divides in front of the Inter-transverse muscle, into an ascending branch, which joins the first cervical; and two descending branches which join the third.

The *anterior branch of the third cervical nerve* is double the size of the preceding. At its exit from the intervertebral foramen, it passes downwards and outwards beneath the Sterno-mastoid, and divides into two branches. The ascending branch joins the anterior division of the second cervical, communicates with the sympathetic and spinal accessory nerves, and subdivides into the superficial cervical, and great auricular nerves. The descending branch passes down in front of the Scalenus anticus, anastomoses with the fourth cervical nerve, and becomes continuous with the clavicular nerves.

The *anterior branch of the fourth cervical* is of the same size as the preceding. It receives a branch from the third, sends a communicating branch to the fifth cervical, and, passing downwards and outwards, divides into numerous filaments, which cross the posterior triangle of the neck, towards the clavicle and acromion. It usually gives a branch to the phrenic nerve, whilst it is contained in the inter-transverse space.

The *anterior branches of the fifth, sixth, seventh, and eighth cervical nerves*, are remarkable for their large size. They are much larger than the preceding nerves, and are all of equal size. They assist in the formation of the brachial plexus.

Cervical Plexus.

The cervical plexus (fig. 276) is formed by the anterior branches of the four upper cervical nerves. It is situated in front of the four upper vertebræ, resting upon the Levator anguli scapulæ, and Scalenus medius muscles, and covered in by the Sterno-mastoid.

Its branches may be divided into two groups, superficial and deep, which may be thus arranged:—

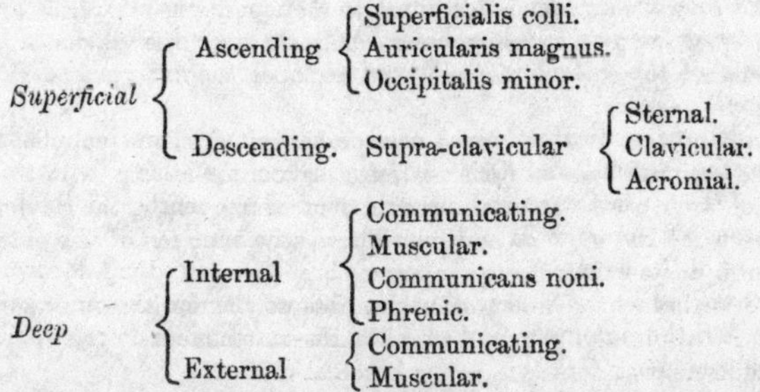
Superficial
Ascending: Superficialis colli. Auricularis magnus. Occipitalis minor.
Descending. Supra-clavicular: Sternal. Clavicular. Acromial.
Deep
Internal: Communicating. Muscular. Communicans noni. Phrenic.
External: Communicating. Muscular.

Superficial Branches of the Cervical Plexus.

The *Superficialis Colli* arises from the second and third cervical nerves, turns round the posterior border of the Sterno-mastoid about its middle, and passing obliquely forwards behind the external jugular vein to the anterior border of that muscle, perforates the deep cervical fascia, and divides beneath the Platysma into two branches, which are distributed to the anterior and lateral parts of the neck.

The *ascending branch* gives a filament, which accompanies the external jugular vein; it then passes upwards to the submaxillary region, and divides into branches, some of which form a plexus with the cervical branches of the facial nerve beneath the Platysma; others pierce that muscle, supply it, and are distributed to the integument of the upper half of the neck, at its fore part, as high as the chin.

The *descending branch* pierces the Platysma, and is distributed to the integument of the side and front of the neck, as low as the sternum.

This nerve is occasionally represented by two or more filaments.

The *Auricularis Magnus* is the largest of the ascending branches. It arises from the second and third cervical nerves, winds round the posterior border of the Sterno-mastoid, and after perforating the deep fascia, ascends upon that muscle beneath the Platysma to the parotid gland, where it divides into numerous branches.

The *facial branches* pass across the parotid, and are distributed to the integument of the face; others penetrate the substance of the gland, and communicate with the facial nerve.

The *posterior* or *auricular branches* ascend vertically to supply the integument of the back part of the pinna, communicating with the auricular branches of the facial and pneumogastric nerves.

The *mastoid branch* joins the posterior auricular branch of the facial, and crossing the mastoid process, is distributed to the integument behind the ear.

The *Occipitalis Minor* arises from the second cervical nerve; it curves round the posterior border of the Sterno-mastoid above the preceding, and ascends vertically along the posterior border of that muscle to the back part of the side of the head. Near the cranium it perforates the deep fascia, and is continued upwards along the side of the head behind the ear, supplying the integument and Occipito-frontalis muscle, and communicating with the occipitalis major, auricularis magnus, and posterior auricular branch of the facial.

This nerve gives off an *auricular branch*, which supplies the Attollens aurem and the integument of the upper and back part of the auricle. This branch is occasionally derived from the great occipital nerve. The occipitalis minor varies in size; it is occasionally double.

The *Descending* or *supra-clavicular branches* arise from the third and fourth cervical nerves; emerging beneath the posterior border of the Sterno-mastoid, they descend in the interval between that muscle and the Trapezius, and divide into branches, which are arranged, according to their position, into three groups.

The *inner* or *sternal branch* crosses obliquely over the clavicular and sternal attachments of the Sterno-mastoid, and supplies the integument as far as the median line.

The *middle* or *clavicular branch* crosses the clavicle, and supplies the integument over the Pectoral and Deltoid muscles, communicating with the cutaneous branches of the upper intercostal nerves. Not unfrequently, the clavicular branch passes through a foramen in the clavicle, at the junction of the outer with the middle third of the bone.

The *external* or *acromial branch* passes obliquely across the outer surface of the Trapezius and the acromion, and supplies the integument of the upper and back part of the shoulder.

Deep Branches of the Cervical Plexus. Internal Series.

The *communicating branches* consist of several filaments, which pass from the loop between the first and second cervical nerves in front of the atlas to the pneumogastric, hypoglossal, and sympathetic.

Muscular branches supply the Anterior recti and Rectus lateralis muscles; they proceed from the first cervical nerve, and from the loop formed between it and the second.

The *Communicans Noni* (fig. 276) consists usually of two filaments, one being derived from the second, and the other from the third cervical. These filaments pass vertically downwards on the outer side of the internal jugular vein, cross in front of the vein a little below the middle of the neck, and form a loop with the descendens noni in front of the sheath of the carotid vessels. Occasionally, the junction of these nerves takes place within the sheath.

The *Phrenic Nerve* (*internal respiratory of Bell*) arises from the third and fourth cervical nerves, and receives a communicating branch from the fifth. It descends to the root of the neck, lying obliquely across the front of the Scalenus anticus, passes over the first part of the subclavian artery, between it and the subclavian vein, and, as it enters the chest, crosses the internal mammary artery near its root. Within the chest, it descends nearly vertically in front of the root of the lung, and by the side of the pericardium, between it and the mediastinal portion of the pleura, to the Diaphragm, where it divides into branches, which separately pierce that muscle, and are distributed to its under surface.

The two phrenic nerves differ in their length, and also in their relations at the upper part of the thorax.

The *right nerve* is situated more deeply, and is shorter and more vertical in direction than the left; it lies on the outer side of the right vena innominata and superior vena cava.

The *left nerve* is rather longer than the right, from the inclination of the heart to the left side, and from the Diaphragm being lower on this than on the opposite side. At the upper part of the thorax, it crosses in front of the arch of the aorta to the root of the lung.

Each nerve supplies filaments to the pericardium and pleura, and near the chest is joined by a filament from the sympathetic, by another derived from the fifth and sixth cervical nerves, and, occasionally, by one from the union of the descendens noni with the spinal nerves: this filament is found, according to Swan, only on the left side.

From the *right nerve*, one or two filaments pass to join in a small ganglion with phrenic branches of the solar plexus; and branches from this ganglion are distributed to the hepatic plexus, the suprarenal capsule, and inferior vena cava. From the *left nerve*, filaments pass to join the phrenic plexus, but without any ganglionic enlargement.

Deep Branches of the Cervical Plexus. External Series.

Communicating branches. The cervical plexus communicates with the spinal accessory nerve, in the substance of the Sterno-mastoid muscle, in the occipital triangle, and beneath the Trapezius.

Muscular branches are distributed to the Sterno-mastoid, Levator anguli scapulæ, Scalenus medius and Trapezius.

The branch for the Sterno-mastoid is derived from the second cervical, the Levator anguli scapulæ receiving branches from the third, and the Trapezius branches from the third and fourth.

Posterior Branches of the Cervical Nerves.

The *posterior branches* of the cervical nerves, with the exception of those of the first two, pass backwards, and divide, behind the posterior Intertransverse muscles, into external and internal branches.

The *external branches* supply the muscles at the side of the neck, viz., the Cervicalis ascendens, Transversalis colli, and Trachelo-mastoid.

The external branch of the second cervical nerve is the largest; it is often joined with the third, and supplies the Complexus, Splenius, and Trachelo-mastoid muscles.

The *internal branches*, the larger, are distributed differently in the upper and lower part of the neck. Those derived from the third, fourth, and fifth nerves pass between the Semispinalis and Complexus muscles, and having reached the spinous processes, perforate the aponeurosis of the Splenius and Trapezius, and are continued outwards to the integument over the Trapezius; whilst those derived from the three lowest cervical nerves are the smallest, and are placed beneath the Semispinalis, which they supply, and do not furnish any cutaneous filaments. These internal branches supply the Complexus, Semispinalis colli, Interspinales, and Multifidus spinæ.

The posterior branches of the three first cervical nerves require a separate description.

The *posterior branch of the first cervical nerve* (*suboccipital*) is larger than the anterior, and escapes from the spinal canal between the occipital bone and the posterior arch of the atlas, lying behind the vertebral artery. It enters the triangular space formed by the Rectus posticus major, the Obliquus superior, and Obliquus inferior, and supplies the Recti and Obliqui muscles, and the Complexus. From the branch which supplies the Inferior oblique a filament is given off, which joins the second cervical nerve. This nerve also occasionally gives off a cutaneous filament, which accompanies the occipital artery, and communicates with the occipitalis major and minor nerves.

The posterior division of the first cervical has no branch analogous to the external branch of the other posterior cervical nerves.

The *posterior branch of the second cervical nerve* is three or four times greater than the anterior branch, and the largest of all the posterior cervical nerves. It emerges from the spinal canal between the posterior arch of the atlas and lamina of the axis, below the Inferior oblique. It supplies this muscle, and receives a communicating filament from the first cervical. It then divides into an external and an internal branch.

The internal branch, called, from its size and distribution, the *occipitalis major*, ascends obliquely inwards between the Obliquus inferior and Complexus, and pierces the latter muscle and the Trapezius near their attachments to the cranium. It is now joined by a filament from the third cervical nerve, and ascending on the back part of the head with the occipital artery, divides into two branches, which supply the integument of the scalp as far forwards as the vertex, communicating with the occipitalis minor. It gives off an auricular branch to the back part of the ear, and muscular branches to the Complexus.

The *posterior branch of the third cervical* is smaller than the preceding, but larger than the fourth; it differs from the posterior branches of the other cervical nerves in its supplying an additional filament to the integument of the occiput. This occipital branch arises from the internal or cutaneous branch beneath the Trapezius; it pierces that muscle, and supplies the skin on the lower and back part of the head. It lies to the inner side of the occipitalis major, with which it is connected.

The internal branches of the posterior divisions of the first three cervical nerves are occasionally joined beneath the Complexus by communicating branches. This communication is described by Cruveilhier as the *posterior cervical plexus*.

The Brachial Plexus. (Fig. 282.)

The brachial plexus is formed by the union of the anterior branches of the four lower cervical and first dorsal nerves. It extends from the lower part of the side of the neck to the axilla. It is very broad, and presents little of a plexiform arrangement at its commencement, is narrow opposite the clavicle, becomes broad, and forms a more dense interlacement in the axilla, and divides opposite the coracoid process into numerous branches for the supply of the upper limb. The nerves which form the plexus are all similar in size, and their mode of communication is the following. The fifth and sixth nerves unite near their exit from the spine into a common trunk; the seventh nerve joins this trunk near the outer border of the Middle scalenus; and the three nerves thus form one large single cord. The eighth cervical and first dorsal nerves unite behind the Anterior scalenus into a common trunk. Thus two large trunks are formed, the upper one by the union of the fifth, sixth, and seventh cervical; and the lower one by the eighth cervical and first dorsal. These two trunks accompany the subclavian artery to the axilla, lying upon its outer side, the trunk formed by the union of the last cervical and first dorsal being nearest to the vessel. Opposite the clavicle, and sometimes in the axilla, each of these cords gives off a fasciculus, by the union of which a third trunk is formed, so that in the middle of the axilla three cords are found, one lying on the outer side of the axillary artery, one on its inner side, and one behind.* The brachial plexus communicates with the cervical plexus by a branch from the fourth to the fifth nerve, and with the phrenic nerve by a branch from the fifth cervical, which joins that nerve on the Anterior scalenus muscle: the cervical and first dorsal nerves are also joined by filaments from the middle and inferior cervical ganglia of the sympathetic, close to their exit from the intervertebral foramina.

282.—Plan of the Brachial Plexus.

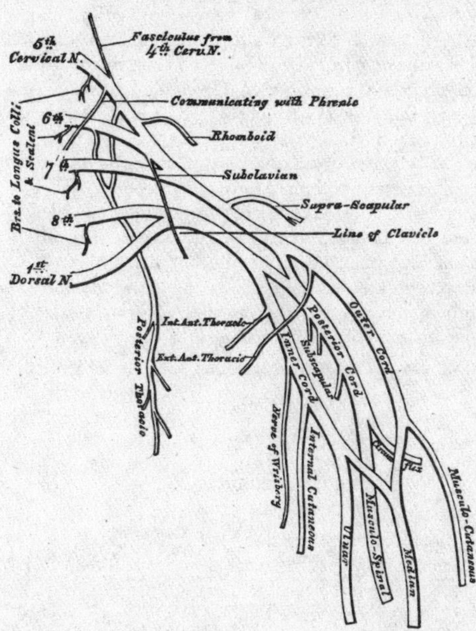

Relations. In the neck, the brachial plexus lies at first between the Anterior and Middle scaleni muscles, and then above and to the outer side of the subclavian artery; it then passes behind the clavicle and Subclavius muscle, lying upon the first serration of the Serratus magnus and the Subscapularis muscles. *In the axilla*, it is placed on the outer side of the first portion of the axillary artery; it surrounds the artery in the second part of its course, one cord lying upon the outer side of that vessel, one on the inner side, and one behind it; and at the lower part of the axillary space gives off its terminal branches to the upper extremity.

Branches. The branches of the brachial plexus are arranged into two groups, viz., those given off above the clavicle, and those below that bone.

Branches above the Clavicle.

Communicating.	Posterior thoracic.
Muscular.	Suprascapular.

* This is the most common mode of formation of the plexus; but it is also very common for the third, or posterior, cord to be formed by the seventh cervical nerve, running undivided, and receiving a branch from each of the other cords.

The *communicating branch* with the phrenic is derived from the fifth cervical nerve; it joins the phrenic on the Anterior scalenus muscle.

The *muscular branches* supply the Longus colli, Scaleni, Rhomboidei, and Subclavius muscles. Those for the Scaleni and Longus colli arise from the lower cervical nerves at their exit from the intervertebral foramina. The rhomboid branch arises from the fifth cervical, pierces the Scalenus medius, and passes beneath the Levator anguli scapulæ, which it occasionally supplies, to the Rhomboid muscles. The nerve to the Subclavius is a small filament, which arises from the trunk formed by the junction of the fifth and sixth cervical nerves; it descends in front of the subclavian artery to the Subclavius muscle, and is usually connected by a filament with the phrenic nerve.

The *posterior thoracic nerve* (long thoracic, external respiratory of Bell), (fig. 285), supplies the Serratus magnus, and is remarkable for the length of its course. It arises by two roots, from the fifth and sixth cervical nerves, immediately after their exit from the intervertebral foramina. These unite in the substance of the Middle scalenus muscle, and, after emerging from it, the nerve passes down behind the brachial plexus and the axillary vessels, resting on the outer surface of the Serratus magnus. It extends along the side of the chest to the lower border of that muscle, and supplies it with numerous filaments.

The *suprascapular nerve* (fig. 286) arises from the cord formed by the fifth, sixth, and seventh cervical nerves; passing obliquely outwards beneath the Trapezius, it enters the supraspinous fossa, through the notch in the upper border of the scapula; and, passing beneath the Supraspinatus muscle, curves in front of the spine of the scapula to the infraspinous fossa. In the supraspinous fossa, it gives off two branches to the Supraspinatus muscle, and an articular filament to the shoulder joint; and in the infraspinous fossa, it gives off two branches to the Infraspinatus muscle, besides some filaments to the shoulder joint and scapula.

BRANCHES BELOW THE CLAVICLE.

To chest	Anterior thoracic.
To shoulder	Subscapular. Circumflex.
To arm, fore-arm, and hand	Musculo-cutaneous. Internal cutaneous. Lesser internal cutaneous. Median. Ulnar. Musculo-spiral.

The branches given off below the clavicle, are derived from the three cords of the brachial plexus, in the following manner:

From the outer cord, arises the external of the two anterior thoracic nerves, the musculo-cutaneous nerve, and the outer head of the median.

From the inner cord, arises the internal of the two anterior thoracic nerves, the internal cutaneous, the lesser internal cutaneous (nerve of Wrisberg), the ulnar and inner head of the median.

From the posterior cord, arise the three subscapular nerves; and the cord then divides into the musculo-spiral and circumflex nerves.

The ANTERIOR THORACIC NERVES (fig. 285), two in number, supply the Pectoral muscles.

The *external*, or superficial branch, the larger of the two, arises from the outer cord of the brachial plexus, passes inwards, across the axillary artery and vein, and is distributed to the under surface of the Pectoralis major. It sends down a communicating filament to join the internal branch.

The *internal*, or deep branch, arises from the inner cord, and passes upwards between the axillary artery and vein (sometimes perforates the vein), and joins with the filament from the superficial branch. From the loop thus formed, branches are distributed to the under surface of the Pectoralis minor and major muscles.

The Subscapular Nerves, three in number, supply the Subscapularis, Teres major, and Latissimus dorsi muscles.

The *upper subscapular nerve*, the smallest, enters the upper part of the Subscapularis muscle.

283.—Cutaneous Nerves of Right Upper Extremity. Anterior View.

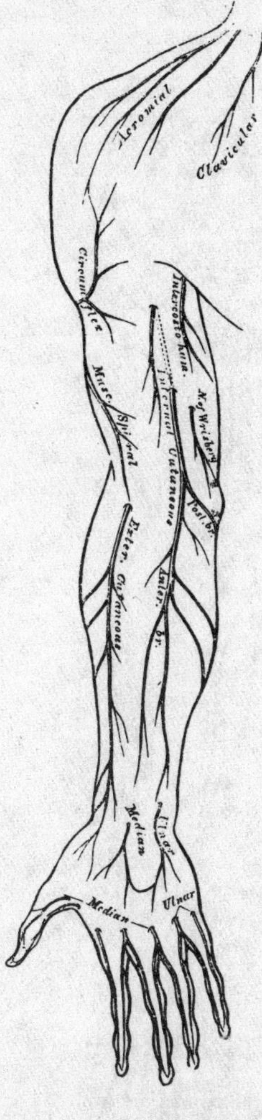

The *lower subscapular nerve* enters the axillary border of the Subscapularis, and terminates in the Teres major. The latter muscle is sometimes supplied by a separate branch.

The *long subscapular*, the largest of the three, descends along the lower border of the Subscapularis to the Latissimus dorsi, through which it may be traced as far as its lower border.

The Circumflex Nerve (fig. 286) supplies some of the muscles, and the integument of the shoulder, and the shoulder joint. It arises from the posterior cord of the brachial plexus, in common with the musculo-spiral nerve. It passes down behind the axillary artery, and in front of the Subscapularis; and, at the lower border of that muscle, passes backwards, and divides into two branches.

The *upper branch* winds round the neck of the humerus, beneath the Deltoid, with the posterior circumflex vessels, as far as the anterior border of the muscle, supplying it, and giving off cutaneous branches, which pierce it to ramify in the integument covering its lower part.

The *lower branch*, at its origin, distributes filaments to the Teres minor and back part of the Deltoid muscles. Upon the filament to the former muscle a gangliform enlargement usually exists. The nerve then pierces the deep fascia, and supplies the integument over the lower two-thirds of the posterior surface of the Deltoid, as well as that covering the long head of the Triceps.

The circumflex nerve, before its division, gives off an articular filament, which enters the shoulder joint below the Subscapularis.

The Musculo-Cutaneous Nerve (fig. 285) (external cutaneous, *perforans Casserii*), supplies some of the muscles of the arm, and the integument of the fore-arm. It arises from the outer cord of the brachial plexus, opposite the lower border of the Pectoralis minor. It then perforates the Coraco-brachialis muscle, and passes obliquely between the Biceps and Brachialis anticus, to the outer side of the arm, a little above the elbow, where it perforates the deep fascia and becomes cutaneous. This nerve, in its course through the arm, supplies the Coraco-brachialis, Biceps, and Brachialis anticus muscles, besides sending some filaments to the elbow joint and humerus.

The cutaneous portion of the nerve passes behind the median cephalic vein, and divides, opposite the elbow joint, into an anterior and a posterior branch.

The *anterior branch* descends along the radial border of the fore-arm to the wrist. It is here placed in front of the radial artery, and, piercing the deep fascia, accompanies that vessel to the back of the wrist. It communicates with a branch from the radial nerve, and distributes filaments to the integument of the ball of the thumb.

284.—Cutaneous Nerves of Right Upper Extremity. Posterior View.

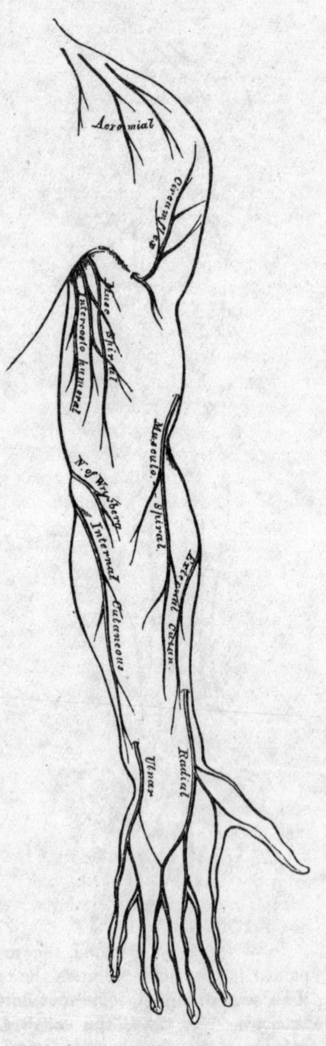

The *posterior branch* is given off about the middle of the fore-arm, and passes downwards, along the back part of its radial side to the wrist. It supplies the integument of the lower third of the fore-arm, communicating with the radial nerve, and the external cutaneous branch of the musculo-spiral.

The Internal Cutaneous Nerve (fig. 285) is one of the smallest branches of the brachial plexus. It arises from the inner cord, in common with the ulnar and internal head of the median, and, at its commencement, is placed on the inner side of the brachial artery. It passes down the inner side of the arm, pierces the deep fascia with the basilic vein, about the middle of the limb, and, becoming cutaneous, divides into two branches.

This nerve gives off, near the axilla, a cutaneous filament, which pierces the fascia, and supplies the integument covering the Biceps muscle, nearly as far as the elbow. This filament lies a little external to the common trunk from which it arises.

The *anterior branch*, the larger of the two, passes usually in front of, but occasionally behind, the median basilic vein. It then descends on the anterior surface of the ulnar side of the fore-arm, distributing filaments to the integument as far as the wrist, and communicating with a cutaneous branch of the ulnar nerve.

The *posterior branch* passes obliquely downwards on the inner side of the basilic vein, winds over the internal condyle of the humerus to the back of the fore-arm, and descends, on the posterior surface of its ulnar side, to a little below the middle, distributing filaments to the integument. It anastomoses above the elbow, with the lesser internal cutaneous, and above the wrist, with the dorsal branch of the ulnar nerve (Swan).

The Lesser Internal Cutaneous Nerve (nerve of Wrisberg) (fig. 285), is distributed to the integument on the inner side of the arm. It is the smallest of the branches of the brachial plexus, and usually arises from the inner cord, with the internal cutaneous and ulnar nerves. It passes through the axillary space, at first lying behind, and then on the inner side of the axillary vein, and communicates with the intercosto-humeral nerve. It then descends along the inner side of the brachial artery, to the middle of the arm, where it pierces the deep fascia, and is distributed to the integument of the back part of the lower third of the arm, extending as far as the elbow, where some filaments are lost in the integument in front of the inner condyle, and others over the olecranon. It communicates with the inner branch of the internal cutaneous nerve.

In some cases the nerve of Wrisberg and intercosto-humeral are connected by two or three filaments, which form a plexus at the back part of the axilla. In other cases, the intercosto-humeral is of large size, and takes the place of the nerve of Wrisberg, receiving merely a filament of communication from the brachial plexus, which represents the latter nerve. In other cases, this filament is wanting, the place of the nerve of Wrisberg being supplied entirely from the intercosto-humeral.

The MEDIAN NERVE (fig. 285) has received its name from the course it takes along the middle of the arm and fore-arm to the hand, lying between the ulnar and the musculo-spiral and radial nerves. It arises by two roots, one from the outer, and one from the inner cord of the brachial plexus; these embrace the lower part of the axillary artery, uniting either in front or on the outer side of that vessel. As it descends through the arm, it lies at first on the outer side of the brachial artery, crosses that vessel in the middle of its course, usually in front, but occasionally behind it, and lies on its inner side to the bend of the elbow, where it is placed beneath the bicipital fascia, and is separated from the elbow joint by the Brachialis anticus. *In the fore-arm*, it passes between the two heads of the Pronator radii teres, and descends beneath the Flexor sublimis, to within two inches above the annular ligament, where it becomes more superficial, lying between the Flexor sublimis and Flexor carpi radialis, covered by the integument and fascia. It then passes beneath the annular ligament into the hand.

Branches. No branches are given off from the median nerve in the arm. *In the fore-arm* its branches are, muscular, anterior interosseous, and palmar cutaneous.

The *muscular branches* supply all the superficial muscles on the front of the fore-arm, except the Flexor carpi ulnaris. These branches are derived from the nerve near the elbow. The branch furnished to the Pronator radii teres often arises above the joint.

The *anterior interosseous* supplies the deep muscles on the front of the fore-arm, except the Flexor carpi ulnaris and inner half of the Flexor profundus digitorum. It accompanies the anterior interosseous artery along the interosseous membrane, in the interval between the Flexor longus pollicis and Flexor profundus digitorum muscles, both of which it supplies, and terminates below in the Pronator quadratus.

The *palmar cutaneous branch* arises from the median nerve at the lower part of the fore-arm. It pierces the fascia above the annular ligament, and divides into two branches; of which the *outer* supplies the skin over the ball of the thumb, and communicates with the external cutaneous nerve; and the *inner* supplies the integument of the palm of the hand, anastomosing with the cutaneous branch of the ulnar. Both nerves cross the annular ligament previous to their distribution.

In the palm of the hand, the median nerve is covered by the integument and palmar fascia, and rests upon the tendons of the Flexor muscles. In this situation it becomes enlarged, somewhat flattened, of a reddish colour, and divides into two branches. Of these, the *external* supplies a muscular branch to some of the muscles of the thumb, and digital branches to the thumb and index-finger; the *internal branch* supplying digital branches to the contiguous sides of the index and middle, and of the middle and ring fingers.

The *branch to the muscles of the thumb* is a short nerve, which subdivides to

285.—Nerves of the Left Upper Extremity. Front View.

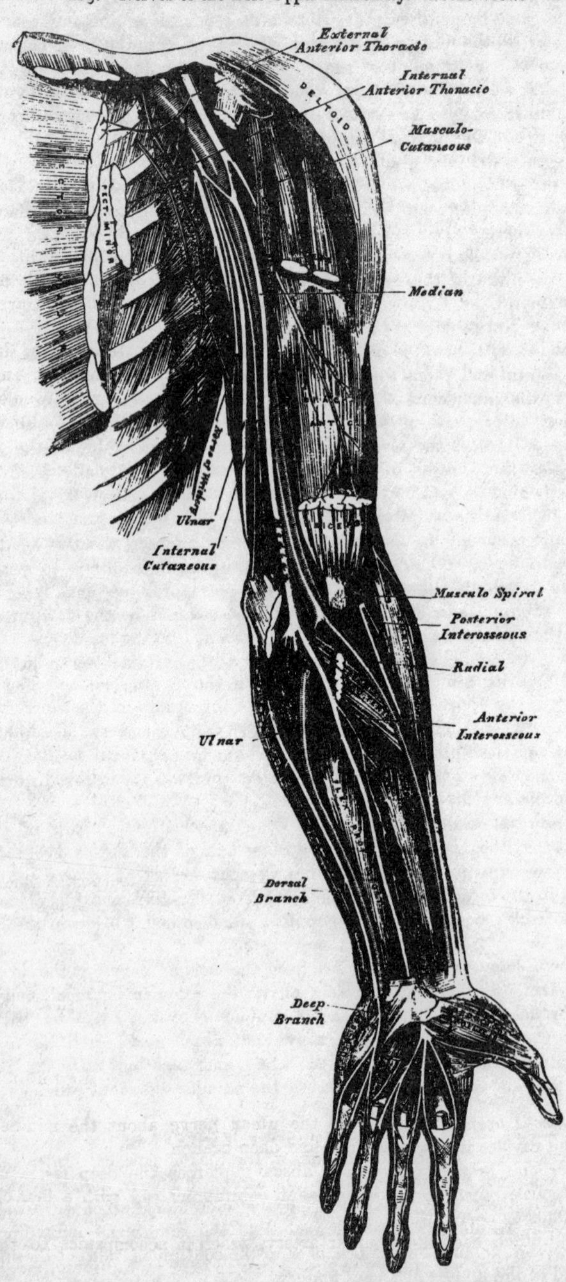

supply the Abductor, Opponens, and outer head of the Flexor brevis pollicis muscles; the remaining muscles of this group being supplied by the ulnar nerve.

The *digital branches* are five in number. The *first* and *second* pass along the borders of the thumb, the external branch communicating with branches of the radial nerve. The *third* passes along the radial side of the index-finger, and supplies the first Lumbrical muscle. The *fourth* subdivides to supply the adjacent sides of the index and middle fingers, and sends a branch to the second Lumbrical muscle. The *fifth* supplies the adjacent sides of the middle and ring fingers, and communicates with a branch from the ulnar nerve.

Each digital nerve, opposite the base of the first phalanx, gives off a dorsal branch, which joins the dorsal digital nerve, and runs along the side of the dorsum of the finger, ending in the integument over the last phalanx. At the end of the finger, the digital nerve divides into a palmar and a dorsal branch; the former of which supplies the extremity of the finger, and the latter ramifies round and beneath the nail. The digital nerves, as they run along the fingers, are placed superficial to the digital arteries.

The Ulnar Nerve (fig. 285) is placed along the inner or ulnar side of the upper limb, and is distributed to the muscles and integument of the fore-arm and hand. It is smaller than the median, behind which it is placed, diverging from it in its course down the arm. It arises from the inner cord of the brachial plexus, in common with the inner head of the median and the internal cutaneous nerve. At its commencement, it lies at the inner side of the axillary artery, and holds the same relation with the brachial artery to the middle of the arm. From this point, it runs obliquely across the internal head of the Triceps, pierces the internal intermuscular septum, and descends to the groove between the internal condyle and olecranon, accompanied by the inferior profunda artery. *At the elbow*, it rests upon the back of the inner condyle, and passes into the fore-arm between the two heads of the Flexor carpi ulnaris. *In the fore-arm*, it descends in a perfectly straight course along its ulnar side, lying upon the Flexor profundus digitorum, its upper half being covered by the Flexor carpi ulnaris, its lower half lying on the outer side of the muscle, covered by the integument and fascia. The ulnar artery, in the upper part of its course, is separated from the ulnar nerve by a considerable interval; but in the rest of its extent, the nerve lies to its inner side. *At the wrist*, the ulnar nerve crosses the annular ligament on the outer side of the pisiform bone, a little behind the ulnar artery, and immediately beyond this bone divides into two branches, superficial and deep palmar.

The branches of the ulnar nerve are:—

In fore-arm	Articular (elbow). Muscular. Cutaneous. Dorsal cutaneous. Articular (wrist).	In hand	Superficial palmar. Deep palmar.

The *articular branches* distributed to the elbow-joint consist of several small filaments. They arise from the nerve as it lies in the groove between the inner condyle and olecranon.

The *muscular branches* are two in number; one supplying the Flexor carpi ulnaris; the other, the inner half of the Flexor profundus digitorum. They arise from the trunk of the nerve near the elbow.

The *cutaneous branch* arises from the ulnar nerve about the middle of the fore-arm, and divides into a superficial and deep branch.

The superficial branch (frequently absent) pierces the deep fascia near the wrist, and is distributed to the integument, communicating with a branch of the internal cutaneous nerve.

The deep branch lies on the ulnar artery, which it accompanies to the hand, some filaments entwining round the vessel, which end in the integument of the palm, communicating with branches of the median nerve.

The *dorsal cutaneous branch* arises about two inches above the wrist; it passes backwards beneath the Flexor carpi ulnaris, perforates the deep fascia, and running along the ulnar side of the wrist and hand, supplies the inner side of the little finger, and the adjoining sides of the little and ring fingers; it also sends a communicating filament to that branch of the radial nerve which supplies the adjoining sides of the middle and ring fingers.

The *articular filaments to the wrist* are also supplied by the ulnar nerve.

The *superficial palmar branch* supplies the Palmaris brevis, and the integument on the inner side of the hand, and terminates in two digital branches, which are distributed, one to the ulnar side of the little finger, the other to the adjoining sides of the little and ring fingers, the latter communicating with a branch from the median.

The *deep palmar branch* passes between the Abductor and Flexor brevis minimi digiti muscles, and follows the course of the deep palmar arch beneath the flexor tendons. At its origin, it supplies the muscles of the little finger. As it crosses the deep part of the hand it sends two branches to each interosseous space, one for the Dorsal and one for the Palmar interosseous muscle, the branches to the second and third Palmar interossei supplying filaments to the two inner Lumbricales muscles. At its termination between the thumb and index finger, it supplies the Adductor pollicis and the inner head of the Flexor brevis pollicis.

The Musculo-Spiral Nerve (fig. 286), the largest branch of the brachial plexus, supplies the muscles of the back part of the arm and fore-arm, and the integument of the same parts, as well as that of the hand. It arises from the posterior cord of the brachial plexus by a common trunk with the circumflex nerve. At its commencement it is placed

286.—The Suprascapular, Circumflex, and Musculo-Spiral Nerves.

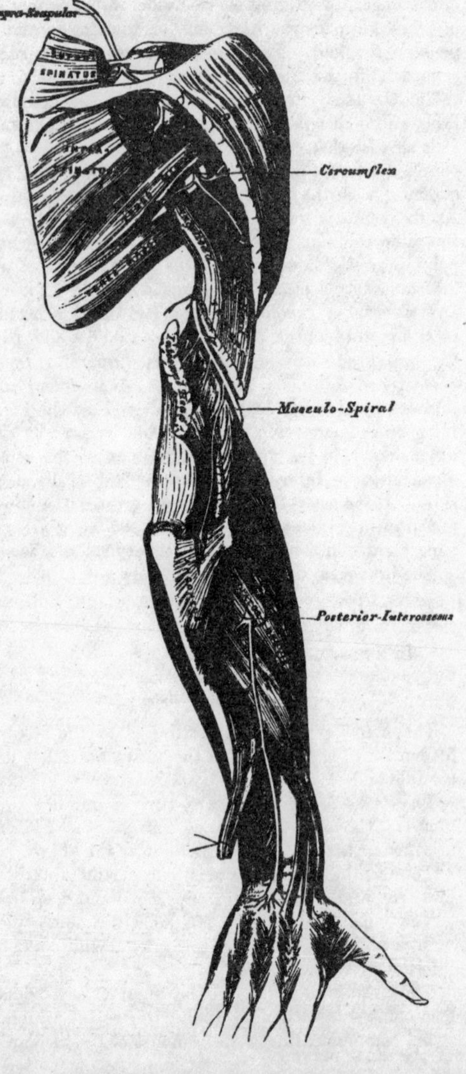

behind the axillary and upper part of the brachial arteries, passing down in front of the tendons of the Latissimus dorsi and Teres major. It winds round the humerus in the spiral groove with the superior profunda artery, passing from the inner to the outer side of the bone, beneath the Triceps muscle. At the outer side of the arm, it descends between the Brachialis anticus and Supinator longus to the front of the external condyle, where it divides into the radial and posterior interosseous nerves.

The branches of the musculo-spiral nerve are:

Muscular.	Radial.
Cutaneous.	Posterior interosseous.

The *muscular branches* supply the Triceps, Anconeus, Supinator longus, Extensor carpi radialis longior, and Brachialis anticus. These branches are derived from the nerve, at the inner side, back part, and outer side of the arm.

The internal muscular branches supply the inner and middle heads of the Triceps muscle. That to the inner head of the Triceps, is a long, slender filament, which lies close to the ulnar nerve, as far as the lower third of the arm.

The posterior muscular branch, of large size, arises from the nerve in the groove between the Triceps and the humerus. It divides into branches which supply the outer head of the Triceps and Anconeus muscles. The branch for the latter muscle is a long, slender filament, which descends in the substance of the Triceps to the Anconeus in the same course with the posterior articular branch from the superior profunda artery.

The external muscular branches supply the Supinator longus, Extensor carpi radialis longior, and, usually, the Brachialis anticus.

The *cutaneous branches* are three in number, one internal and two external.

The internal cutaneous branch arises in the axillary space, with the inner muscular branch. It is of small size, and passes across the axilla to the inner side of the arm, supplying the integument on its posterior aspect nearly as far as the olecranon.

The two external cutaneous branches perforate the outer head of the Triceps, at its attachment to the humerus. The upper and smaller one follows the course of the cephalic vein to the front of the elbow, supplying the integument of the lower half of the upper arm on its anterior aspect. The lower branch pierces the deep fascia below the insertion of the Deltoid, and passes down along the outer side of the arm and elbow, and along the back part of the radial side of the fore-arm to the wrist, supplying the integument in its course, and joining, near its termination, with a branch of the external cutaneous nerve.

The *radial nerve* passes along the front of the radial side of the fore-arm, to the commencement of its lower third. It lies at first a little to the outer side of the radial artery, concealed beneath the Supinator longus. In the middle third of the fore-arm, it lies beneath the same muscle, in close relation with the outer side of the artery. It quits the artery about three inches above the wrist, passes beneath the tendon of the Supinator longus, and, piercing the deep fascia at the outer border of the fore-arm, divides into two branches.

The external branch, the smaller of the two, supplies the integument of the radial side and ball of the thumb, joining with the posterior branch of the external cutaneous nerve.

The internal branch communicates, above the wrist, with a branch from the external cutaneous, and, on the back of the hand, forms an arch with the dorsal branch of the ulnar nerve. It then divides into four digital nerves, which are distributed as follows: The first supplies the ulnar side of the thumb; the second, the radial side of the index finger; the third, the adjoining sides of the index and middle fingers; and the fourth, the adjacent borders of the middle and ring fingers. The latter nerve communicates with a filament from the dorsal branch of the ulnar nerve.

The *posterior interosseous nerve* pierces the Supinator brevis, winds to the back of the fore-arm, in the substance of that muscle, and, emerging from its lower border, passes down between the superficial and deep layer of muscles, to the middle of the fore-arm. Considerably diminished in size, it descends on the interosseous membrane, beneath the Extensor secundi internodii pollicis, to the back of the carpus, where it presents a gangliform enlargement, from which filaments are distributed to the ligaments and articulations of the carpus. It supplies all the muscles of the radial and posterior brachial regions, excepting the Anconeus, Supinator longus, and Extensor carpi radialis longior.

Dorsal Nerves.

The *dorsal nerves* are twelve in number on each side. The first appears between the first and second dorsal vertebræ, and the last between the last dorsal and first lumbar.

The *roots of origin* of the dorsal nerves are few in number, of small size, and vary but slightly from the second to the last. Both roots are very slender; the posterior roots only slightly exceeding the anterior in thickness. These roots gradually increase in length from above downwards, and remain in contact with the spinal cord for a distance equal to the height of, at least, two vertebræ, in the lower part of the dorsal region. They then join in the intervertebral foramen, and, at their exit, divide into two branches, a posterior, or dorsal, and an anterior, or intercostal branch.

The first and last dorsal nerves are peculiar in several respects (see next page).

The *posterior branches of the dorsal nerves*, which are smaller than the intercostal, pass backwards between the transverse processes, and divide into external and internal branches.

The *external branches* increase in size from above downwards. They pass through the Longissimus dorsi, corresponding to the cellular interval between it and the Sacro-lumbalis, and supply those muscles, as well as their continuations upwards to the head, and the Levatores costarum; the five or six lower nerves also give off cutaneous filaments.

The *internal branches* of the six upper nerves pass inwards to the interval between the Multifidus spinæ, and Semispinalis dorsi muscles, which they supply; and then, piercing the origin of the Rhomboidei and Trapezius, become cutaneous by the side of the spinous processes. The internal branches of the six lower nerves are distributed to the Multifidus spinæ, without giving off any cutaneous filaments.

The *cutaneous branches* of the dorsal nerves are twelve in number, the six upper being derived from the internal branches, and the six lower from the external branches. The former pierce the Rhomboid and Trapezius muscles, close to the spinous processes, and ramify in the integument. They are frequently furnished with gangliform enlargements. The six lower cutaneous branches pierce the Serratus posticus inferior, and Latissimus dorsi, in a line with the angles of the ribs.

Intercostal Nerves.

The *intercostal nerves* (anterior branches of the dorsal nerves), are twelve in number on each side. They are distributed to the parietes of the thorax and abdomen, separately from each other, without being joined in a plexus; in which respect they differ from the other spinal nerves. Each nerve is connected with the adjoining ganglia of the sympathetic by one or two filaments. The intercostal nerves may be divided into two sets, from the difference they present in their distribution. The six upper, with the exception of the first, are limited in their distribution to the parietes of the chest. The six lower supply the parietes of the chest and abdomen.

Upper Intercostal Nerves.

The *upper intercostal nerves* pass forwards in the intercostal spaces with the intercostal vessels, being situated below them. At the back of the chest they lie between the pleura and the External intercostal muscle, but are soon placed between the two planes of Intercostal muscles as far as the costal cartilages, where they lie between the pleura and the Internal intercostal muscles. Near the sternum, they cross the internal mammary artery, and Triangularis sterni, pierce the Internal intercostal and Pectoralis major muscles, and supply the integument of the mamma and front of the chest, forming the anterior cutaneous nerves of the thorax; the branch from the second nerve becoming joined with the clavicular nerve.

Branches. Numerous slender muscular filaments supply the Intercostal and Triangularis sterni muscles. Some of these branches, at the front of the chest, cross the costal cartilages from one to another intercostal space.

Lateral cutaneous nerves. These are derived from the intercostal nerves, midway between the vertebræ and sternum; they pierce the External intercostal and Serratus magnus muscles, and divide into two branches, anterior and posterior.

The *anterior branches* are reflected forwards to the side and the fore part of the chest, supplying the integument of the chest and mamma, and the upper digitations of the External oblique.

The *posterior branches* are reflected backwards, to supply the integument over the scapula and over the Latissimus dorsi.

The first intercostal nerve has no lateral cutaneous branch. The lateral cutaneous branch of the second intercostal nerve is of large size, and named, from its origin and distribution, the *intercosto-humeral nerve* (fig. 285). It pierces the External intercostal muscle, crosses the axilla to the inner side of the arm, and joins with a filament from the nerve of Wrisberg. It then pierces the fascia, and supplies the skin of the upper half of the inner and back part of the arm, communicating with the internal cutaneous branch of the musculo-spiral nerve. The size of this nerve is in inverse proportion to the size of the other cutaneous nerves, especially the nerve of Wrisberg. A second intercosto-humeral nerve is frequently given off from the third intercostal. It supplies filaments to the arm-pit and inner side of the arm.

Lower Intercostal Nerves.

The *lower intercostal nerves* (excepting the last) have the same arrangement as the upper ones as far as the anterior extremities of the intercostal spaces, where they pass behind the costal cartilages, and between the Internal oblique and Transversalis muscles, to the sheath of the Rectus, which they perforate. They supply the Rectus muscle, and terminate in branches which become subcutaneous near the linea alba. These branches, which are named the anterior cutaneous nerves of the abdomen, supply the integument of the front of the belly; they are directed outwards as far as the lateral cutaneous nerves. The lower intercostal nerves supply the Intercostal and Abdominal muscles, and, about the middle of their course, give off lateral cutaneous branches, which pierce the External intercostal and External oblique muscles, and are distributed to the integument of the abdomen, the anterior branches passing nearly as far forwards as the margin of the Rectus; the posterior branches passing to supply the skin over the Latissimus dorsi, where they join the dorsal cutaneous nerves.

Peculiar Dorsal Nerves.

First dorsal nerve. Its *roots of origin* are similar to those of a cervical nerve. Its *posterior* or *dorsal branch* resembles, in its mode of distribution, the dorsal branches of the cervical nerves. Its *anterior branch* enters almost wholly into the formation of the brachial plexus, giving off, before it leaves the thorax, a small intercostal branch, which runs along the first intercostal space, and terminates on the front of the chest, by forming the first anterior cutaneous nerve of the thorax. The first intercostal nerve gives off no lateral cutaneous branch.

The *last dorsal* is larger than the other dorsal nerves. Its anterior branch runs along the lower border of the last rib in front of the Quadratus lumborum, perforates the aponeurosis of the Transversalis, and passes forwards between it and the Internal oblique, to be distributed in the same manner as the preceding nerves. It communicates with the ilio-hypogastric branch of the lumbar plexus, and is occasionally connected with the first lumbar nerve by a slender branch, the dorsi-lumbar nerve, which descends in the substance of the Quadratus lumborum.

The *lateral cutaneous branch* of the last dorsal is remarkable for its large size; it perforates the Internal and External oblique muscles, passes downwards over the crest of the ilium, and is distributed to the integument of the front of the hip, some of its filaments extending as low down as the trochanter major.

Lumbar Nerves.

The *lumbar nerves* are five in number on each side; the first appears between the first and second lumbar vertebræ, and the last between the last lumbar and the base of the sacrum.

The *roots* of the lumbar nerves are the largest, and their filaments the most numerous, of all the spinal nerves, and they are closely aggregated together upon the lower end of the cord. The anterior roots are the smaller: but there is not the same disproportion between them and the posterior roots as in the cervical nerves. The roots of these nerves have a vertical direction, and are of considerable length, more especially the lower ones, since the spinal cord does not extend beyond the first lumbar vertebra. The roots become joined in the intervertebral foramina; and the nerves, so formed, divide at their exit into two branches, anterior and posterior.

The *posterior branches* of the lumbar nerves diminish in size from above downwards; they pass backwards between the transverse processes, and divide into external and internal branches.

The external branches supply the Erector spinæ and Intertransverse muscles. From the three upper branches, cutaneous nerves are derived, which pierce the Sacro-lumbalis and Latissimus dorsi muscles, and descend over the back part of the crest of the ilium, to be distributed to the integument of the gluteal region, some of the filaments passing as far as the trochanter major.

The internal branches, the smaller, pass inwards close to the articular processes of the vertebræ, and supply the Multifidus spinæ and Interspinales muscles.

The *anterior branches* of the lumbar nerves increase in size from above downwards. At their origin, they communicate with the lumbar ganglia of the sympathetic by long slender filaments, which accompany the lumbar arteries round the sides of the bodies of the vertebræ, beneath the Psoas muscle. The nerves pass obliquely outwards behind the Psoas magnus, or between its fasciculi, distributing filaments to it and the Quadratus lumborum. The anterior branches of the four upper nerves are connected together in this situation by anastomotic loops, and form the *lumbar plexus.* The anterior branch of the fifth lumbar, joined with a branch from the fourth, descends across the base of the sacrum to join the anterior branch of the first sacral nerve, and assist in the formation of the sacral plexus. The cord resulting from the union of these two nerves, is called the *lumbo-sacral nerve.*

Lumbar Plexus.

The *lumbar plexus* is formed by the loops of communication between the anterior branches of the four upper lumbar nerves. The plexus is narrow above, and occasionally connected with the last dorsal by a slender branch, the dorsi-lumbar nerve; it is broad below, where it is joined to the sacral plexus by the lumbo-sacral cord. It is situated in the substance of the Psoas muscle near its posterior part, in front of the transverse processes of the lumbar vertebræ.

The mode in which the plexus is formed is the following:—The first lumbar nerve gives off the ilio-hypogastric and ilio-inguinal nerves, and a communicating branch to the second lumbar nerve. The second gives off the external cutaneous and genito-crural, and a communicating branch to the third nerve. The third nerve gives a descending filament to the fourth, and divides into two branches, which assist in forming the anterior crural and obturator nerves; sometimes, also, it furnishes a part of the accessory obturator. The fourth nerve completes the formation of the anterior crural, and the obturator, and gives off a communicating branch to the fifth lumbar; sometimes it also furnishes part of the accessory obturator.

The branches of the lumbar plexus are the

Ilio-hypogastric.	Obturator.
Ilio-inguinal.	Accessory obturator.
Genito-crural.	Anterior crural.
External cutaneous.	

These branches may be divided into two groups, according to their mode of distribution. One group, including the ilio-hypogastric, ilio-inguinal, and part of the genito-crural nerves, supplies the lower part of the parietes of the abdomen;

287.—The Lumbar Plexus and its Branches.

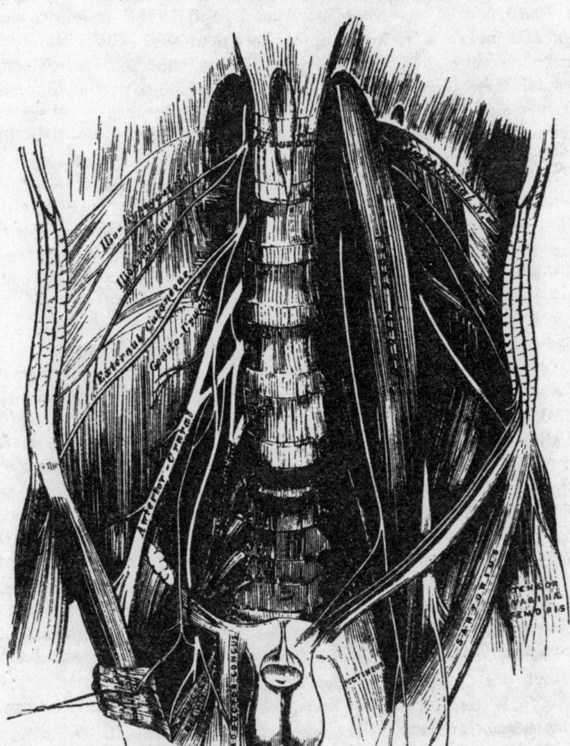

the other group, which includes the remaining nerves, supplies the fore part of the thigh and inner side of the leg.

The Ilio-hypogastric Nerve (*superior musculo-cutaneous*) arises from the first lumbar nerve. It pierces the outer border of the Psoas muscle at its upper part, and crosses obliquely in front of the Quadratus lumborum to the crest of the ilium. It then perforates the Transversalis muscle at its back part, and divides between it and the Internal oblique into two branches, iliac and hypogastric.

The *iliac branch* pierces the Internal and External oblique muscles immediately above the crest of the ilium, and is distributed to the integument of the gluteal region, behind the lateral cutaneous branch of the last dorsal nerve (fig. 290). The size of this nerve bears an inverse proportion to that of the cutaneous branch of the last dorsal nerve.

The *hypogastric branch* (fig. 288) continues onwards between the Internal oblique and Transversalis muscles. It first pierces the Internal oblique, and near the middle line perforates the External oblique above the external abdominal ring, and is distributed to the integument covering the hypogastric region.

The Ilio-inguinal Nerve (*inferior musculo-cutaneous*), smaller than the preceding, arises with it from the first lumbar nerve. It pierces the outer border of the Psoas just below the ilio-hypogastric, and, passing obliquely across the Quadratus lumborum and Iliacus muscles, perforates the Transversalis, near the fore part of the crest of the ilium, and communicates with the Ilio-hypogastric nerve between that muscle and the Internal oblique. The nerve then pierces the Internal oblique, distributing filaments to it, and, accompanying the spermatic cord, escapes at the external abdominal ring, and is distributed to the integument of the scrotum and upper and inner part of the thigh in the male, and to the labium in the female. The size of this nerve is in inverse proportion to that of the ilio-hypogastric. Occasionally it is very small, and ends by joining the ilio-hypogastric; in such cases, a branch from the ilio-hypogastric takes the place of the ilio-inguinal, or the latter nerve may be altogether absent.

The Genito-crural Nerve arises from the second lumbar, and by a few fibres from the cord of communication between it and the first. It passes obliquely through the substance of the Psoas, descends on its surface to near Poupart's ligament, and divides into a genital and a crural branch.

The *genital branch* descends on the external iliac artery, sending a few filaments round that vessel; it then pierces the fascia transversalis, and, passing through the internal abdominal ring, descends along the back part of the spermatic cord to the scrotum, and supplies, in the male, the Cremaster muscle. In the female, it accompanies the round ligament, and is lost upon it.

The *crural branch* passes along the inner margin of the Psoas muscle, beneath Poupart's ligament, into the thigh, where it pierces the fascia lata, and is distributed to the integument of the upper and anterior aspect of the thigh, communicating with the middle cutaneous nerve.

A few filaments from this nerve may be traced on to the femoral artery; they are derived from the nerve as it passes beneath Poupart's ligament.

The External Cutaneous Nerve arises from the second lumbar, or from the loop between it and the third. It perforates the outer border of the Psoas muscle about its middle, and crosses the Iliacus muscle obliquely, to the notch immediately beneath the anterior superior spine of the ilium, where it passes beneath Poupart's ligament into the thigh, and divides into two branches of nearly equal size.

The *anterior branch* descends in an aponeurotic canal formed in the fascia lata, becomes superficial about four inches below Poupart's ligament, and divides into branches, which are distributed to the integument along the anterior and outer part of the thigh, as far down as the knee. This nerve occasionally communicates with the long saphenous nerve.

The *posterior branch* pierces the fascia lata, and subdivides into branches which pass across the outer and posterior surface of the thigh, supplying the integument as far as the middle of the thigh.

The Obturator Nerve supplies the Obturator externus and Adductor muscles of the thigh, the articulations of the hip and knee, and occasionally the integument

288.—Cutaneous Nerves of Lower Extremity. Front View.

289.—Nerves of the Lower Extremity. Front View.

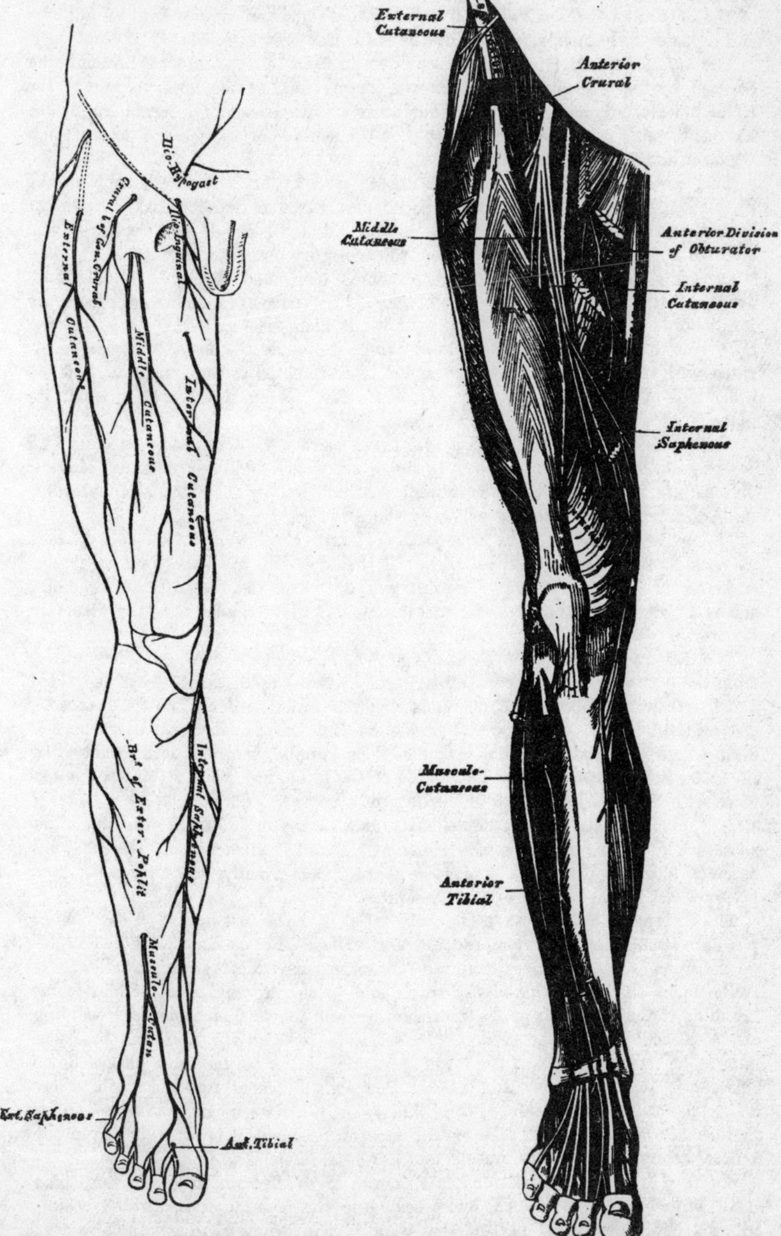

of the thigh and leg. It arises by two branches: one from the third, the other from the fourth lumbar nerve. It descends through the inner fibres of the Psoas muscle, and emerges from its inner border near the brim of the pelvis; it then runs along the lateral wall of the pelvis, above the obturator vessels, to the upper part of the obturator foramen, where it enters the thigh, and divides into an anterior and a posterior branch, separated by the Adductor brevis muscle.

The *anterior branch* (fig. 289) passes down in front of the Adductor brevis, being covered by the Pectineus and Adductor longus; and at the lower border of the latter muscle, communicates with the internal cutaneous and internal saphenous nerves, forming a kind of plexus. It then descends upon the femoral artery, upon which it is finally distributed.

This nerve, near the obturator foramen, gives off an articular branch to the hip-joint. Behind the Pectineus, it distributes muscular branches to the Adductor longus and Gracilis, and occasionally to the Adductor brevis and Pectineus, and receives a communicating branch from the accessory obturator nerve.

Occasionally this communicating branch is continued down, as a cutaneous branch, to the thigh and leg. This occasional cutaneous branch emerges from the lower border of the Adductor longus, descends along the posterior margin of the Sartorius to the inner side of the knee, where it pierces the deep fascia, communicates with the long saphenous nerve, and is distributed to the integument of the inner side of the leg, as low down as its middle. When this branch is small, its place is supplied by the internal cutaneous nerve.

The *posterior branch* of the obturator nerve pierces the Obturator externus, and passes behind the Adductor brevis to the front of the Adductor magnus, where it divides into numerous muscular branches, which supply the Obturator externus, the Adductor magnus, and occasionally the Adductor brevis.

The *articular branch* for the knee-joint perforates the lower part of the Adductor magnus, and enters the popliteal space; it then descends upon the popliteal artery, as far as the back part of the knee-joint, where it perforates the posterior ligament, and is distributed to the synovial membrane. It gives filaments to the artery in its course.

The Accessory Obturator Nerve (fig. 287) is of small size, and arises either from the obturator nerve near its origin, or by separate filaments from the third and fourth lumbar nerves. It descends along the inner border of the Psoas muscle, crosses the body of the pubes, and passes beneath the Pectineus muscle, where it divides into numerous branches. One of these supplies the Pectineus, penetrating its under surface; another is distributed to the hip-joint; while a third communicates with the anterior branch of the obturator nerve. This branch, when of large size, is prolonged (as already mentioned), as a cutaneous branch, to the leg. The accessory obturator nerve is not constantly found; when absent, the hip-joint receives branches from the obturator nerve. Occasionally it is very small, and becomes lost in the capsule of the hip-joint.

The Anterior Crural Nerve (figs. 287, 289) is the largest branch of the lumbar plexus. It supplies muscular branches to the Iliacus, Pectineus, and all the muscles on the front of the thigh, excepting the Tensor vaginæ femoris; cutaneous filaments to the front and inner side of the thigh, and to the leg and foot; and articular branches to the knee. It arises from the third and fourth lumbar nerves, receiving also a fasciculus from the second. It descends through the fibres of the Psoas muscle, emerging from it at the lower part of its outer border; and passes down between it and the Iliacus, and beneath Poupart's ligament, into the thigh, where it becomes somewhat flattened, and divides into an anterior or cutaneous, and a posterior or muscular part. Beneath Poupart's ligament, it is separated from the femoral artery by the Psoas muscle, and lies beneath the iliac fascia.

Within the pelvis, the anterior crural nerve gives off from its outer side some small branches to the Iliacus, and a branch to the femoral artery, which is distributed upon the upper part of that vessel. The origin of this branch varies; it occasionally arises higher than usual, or it may arise lower down in the thigh.

External to the pelvis, the following branches are given off:

From the Anterior Division.	*From the Posterior Division.*
Middle cutaneous.	Muscular.
Internal cutaneous.	Articular.
Long saphenous.	

The *middle cutaneous nerve* (fig. 288) pierces the fascia lata (occasionally the Sartorius also), about three inches below Poupart's ligament, and divides into two branches, which descend in immediate proximity along the fore part of the thigh, distributing numerous branches to the integument as low as the front of the knee, where the middle cutaneous communicates with a branch of the internal saphenous nerve. Its outer branch communicates, above, with the crural branch of the genito-crural nerve; and the inner branch with the internal cutaneous nerve below. The Sartorius muscle is supplied by this or the following nerve.

The *internal cutaneous nerve* passes obliquely across the upper part of the sheath of the femoral artery, and divides in front, or at the inner side, of that vessel, into two branches, anterior and internal.

The anterior branch perforates the fascia lata at the lower third of the thigh, and divides into two branches, one of which supplies the integument as low down as the inner side of the knee; the other crosses the patella to the outer side of the joint, communicating in its course with the long saphenous nerve. A cutaneous filament is occasionally given off from this nerve, which accompanies the long saphenous vein; and it sometimes communicates with the internal branch of the nerve.

The inner branch descends along the posterior border of the Sartorius muscle to the knee, where it pierces the fascia lata, communicates with the long saphenous nerve, and gives off several cutaneous branches. The nerve then passes down the inner side of the leg, to the integument of which it is distributed. This nerve, beneath the fascia lata, joins in a plexiform network, by uniting with branches of the long saphenous and obturator nerves (fig. 289). When the communicating branch from the latter nerve is large, and continued to the integument of the leg, the inner branch of the internal cutaneous is small, and terminates at the plexus, occasionally giving off a few cutaneous filaments.

This nerve, before subdividing, gives off a few filaments, which pierce the fascia lata, to supply the integument of the inner side of the thigh, accompanying the long saphenous vein. One of these filaments passes through the saphenous opening; a second becomes subcutaneous about the middle of the thigh; and a third pierces the fascia at its lower third.

The *long*, or *internal saphenous nerve*, is the largest of the cutaneous branches of the anterior crural. It approaches the femoral artery where this vessel passes beneath the Sartorius, and lies on its outer side, beneath the aponeurotic covering, as far as the opening in the lower part of the Adductor magnus. It then quits the artery, and descends vertically along the inner side of the knee, beneath the Sartorius, pierces the deep fascia between the tendons of the Sartorius and Gracilis, and becomes subcutaneous. The nerve then passes along the inner side of the leg, accompanied by the internal saphenous vein, descends behind the internal border of the tibia, and, at the lower third of the leg, divides into two branches: one continues its course along the margin of the tibia, terminating at the inner ankle; the other passes in front of the ankle, and is distributed to the integument along the inner side of the foot, as far as the great toe.

Branches. The long saphenous nerve, *about the middle of the thigh*, gives off a communicating branch, which joins the plexus formed by the obturator and internal cutaneous nerves.

At *the inner side of the knee*, it gives off a large branch (*n. cutaneus patellæ*), which pierces the Sartorius and fascia lata, and is distributed to the integument in front of the patella. This nerve communicates *above* the knee with the anterior branch of the internal cutaneous; *below* the knee, with other branches of the long saphenous; and, on the *outer side* of the joint, with branches of the middle and external cutaneous nerves, forming a plexiform network, the *plexus patellæ*. The cutaneous nerve of the patella is occasionally small, and terminates by joining the internal cutaneous, which supplies its place in front of the knee.

Below the knee, the branches of the long saphenous nerve are distributed to the integument of the front and inner side of the leg, communicating with the cutaneous branches from the internal cutaneous, or obturator nerve.

The DEEP GROUP of branches of the anterior crural nerve are muscular and articular.

The *muscular branches* supply the Pectineus, and all the muscles on the front of the thigh, except the Tensor vaginæ femoris, which is supplied from the superior gluteal nerve, and the Sartorius, which is supplied by filaments from the middle or internal cutaneous nerves.

The branches to the *Pectineus*, usually two in number, pass inwards behind the femoral vessels, and enter the muscle on its anterior surface.

The branch to the *Rectus* muscle enters its under surface high up.

The branch to the *Vastus externus*, of large size, follows the course of the descending branch of the external circumflex artery, to the lower part of the muscle. It gives off an articular filament.

The branches to the *Vastus internus* and Crureus enter the middle of those muscles.

The *articular branches*, two in number, supply the knee-joint. One, a long slender filament, is derived from the nerve to the Vastus externus. It penetrates the capsular ligament of the joint on its anterior aspect. The other is derived from the nerve to the Vastus internus. It descends along the internal intermuscular septum, accompanying the deep branch of the anastomotica magna artery, pierces the capsular ligament of the joint on its inner side, and supplies the synovial membrane.

THE SACRAL AND COCCYGEAL NERVES.

The *sacral nerves* are five in number on each side. The four upper ones pass from the sacral canal, through the sacral foramina; the fifth through the foramen between the sacrum and coccyx.

The *roots of origin* of the upper sacral (and lumbar) nerves are the largest of all the spinal nerves; whilst those of the lowest sacral and coccygeal nerve are the smallest.

The roots of these nerves are of very considerable length, being longer than those of any of the other spinal nerves, on account of the spinal cord not extending beyond the first lumbar vertebra. From their great length, and the appearance they present in connection with the spinal cord, the roots of origin of these nerves are called collectively the *cauda equina*. Each sacral and coccygeal nerve divides into two branches, anterior and posterior.

The *posterior sacral nerves* are small, diminish in size from above downwards, and emerge, except the last, from the sacral canal by the posterior sacral foramina.

The *three upper ones* are covered, at their exit from the sacral canal, by the Multifidus spinæ, and divide into external and internal branches.

The *internal branches* are small, and supply the Multifidus spinæ.

The *external branches* communicate with one another, and with the last lumbar and fourth sacral nerves, by means of anastomosing loops. These branches pass outwards, to the outer surface of the great sacro-sciatic ligament, where they form a second series of loops beneath the Gluteus maximus. Cutaneous branches from this second series of loops, usually three in number, pierce the Gluteus maximus, one near the posterior inferior spine of the ilium; another opposite the end of the sacrum; and the third, midway between the other two. They supply the integument over the posterior part of the gluteal region.

The *two lower posterior sacral nerves* are situated below the Multifidus spinæ. They are of small size, and join with each other, and with the coccygeal nerve, so as to form loops on the back of the sacrum, filaments from which supply the integument over the coccyx.

The *coccygeal nerve* divides into its anterior and posterior branch in the spinal canal. The *posterior* branch is the smaller. It receives, as already mentioned, a communicating branch from the last sacral, and is lost in the fibrous structure on the back of the coccyx.

The *anterior sacral nerves* diminish in size from above downwards. The four upper ones emerge from the anterior sacral foramina; the anterior branch of the fifth, together with the coccygeal nerve, between the sacrum and the coccyx. All the anterior sacral nerves communicate with the sacral ganglia of the sympathetic, at their exit from the sacral foramina. The *first* nerve, of large size, unites with the lumbo-sacral nerve. The *second* equals in size the preceding, with which it joins. The *third*, about one-fourth the size of the second, unites with the preceding nerves, to form the sacral plexus.

The *fourth anterior sacral nerve* sends a branch to join the sacral plexus. The remaining portion of the nerve divides into visceral and muscular branches: and a communicating filament descends to join the fifth sacral nerve. The *visceral branches* are distributed to the viscera of the pelvis, communicating with the sympathetic nerve. These branches ascend upon the rectum and bladder: in the female, upon the vagina and bladder, communicating with branches of the sympathetic to form the hypogastric plexus. The *muscular branches* are distributed to the Levator ani, Coccygeus, and Sphincter ani. Cutaneous filaments arise from the latter branch, which supply the integument between the anus and coccyx.

The *fifth anterior sacral nerve*, after passing from the lower end of the sacral canal, pierces the Coccygeus muscle, and descends upon its anterior surface to the tip of the coccyx, where it perforates that muscle, to be distributed to the integument over the back part and side of the coccyx. This nerve communicates above with the fourth sacral, and below with the coccygeal nerve, and supplies the Coccygeus muscle.

The *anterior branch* of the coccygeal nerve is a delicate filament which escapes at the termination of the sacral canal. It pierces the sacro-sciatic ligament and Coccygeus muscle, is joined by a branch from the fifth anterior sacral, and becomes lost in the integument at the back part and side of the coccyx.

Sacral Plexus.

The sacral plexus is formed by the lumbo-sacral, the anterior branches of the three upper sacral nerves, and part of that of the fourth. These nerves proceed in different directions; the upper ones obliquely outwards, the lower one nearly horizontally, and they all unite into a single, broad, flat cord. The sacral plexus is triangular in form, its base corresponding with the exit of the nerves from the sacrum, its apex with the lower part of the great sacro-sciatic foramen. It rests upon the anterior surface of the Pyriformis, and is covered in front by the pelvic fascia, which separates it from the sciatic and pudic branches of the internal iliac artery, and from the viscera of the pelvis.

The branches of the sacral plexus are:

Muscular. Pudic.
Superior gluteal. Small sciatic.
Great sciatic.

The *muscular branches* supply the Pyriformis, Obturator internus, the two Gemelli, and the Quadratus femoris. The branch to the Pyriformis arises either from the plexus, or from the upper sacral nerves: the branch to the Obturator internus arises at the junction of the lumbo-sacral and first sacral nerves; it crosses behind the spine of the ischium, and passes through the lesser sacro-sciatic foramen to the inner surface of the Obturator internus: the branch to the Gemellus superior arises from the lower part of the plexus, near the pudic nerve: the small branch to the Gemellus inferior and Quadratus femoris also arises from the lower part of the plexus; it passes beneath the Gemelli and tendon of the Obturator internus, and supplies an articular branch to the hip-joint. This branch is occasionally derived from the upper part of the great sciatic nerve.

The SUPERIOR GLUTEAL NERVE (fig. 291) arises from the back part of the lumbo-sacral; it passes from the pelvis through the great sacro-sciatic foramen above the Pyriformis muscle, accompanied by the gluteal vessels, and divides into a superior and an inferior branch.

The *superior branch* follows the line of origin of the Gluteus minimus, and supplies it and the Gluteus medius.

The *inferior branch* crosses obliquely between the Gluteus minimus and medius, distributing filaments to both these muscles, and terminates in the Tensor vaginæ femoris, extending nearly to its lower end.

The PUDIC NERVE arises from the lower part of the sacral plexus, and leaves the pelvis, through the great sacro-sciatic foramen, below the Pyriformis. It then crosses the spine of the ischium, and re-enters the pelvis through the lesser sacro-sciatic foramen. It accompanies the pudic vessels upwards and forwards along the outer wall of the ischio-rectal fossa, being covered by the obturator fascia, and divides into two terminal branches, the perineal nerve, and the dorsal nerve of the penis. Near its origin, it gives off the inferior hæmorrhoidal nerve.

The *inferior hæmorrhoidal nerve* is occasionally derived from the sacral plexus. It passes across the ischio-rectal fossa, with its accompanying vessels, towards the lower end of the rectum, and is distributed to the External sphincter and the integument round the anus. Branches of this nerve communicate with the inferior pudendal and superficial perineal nerves on the inner margin of the thigh.

The *perineal nerve*, the inferior and larger of the two terminal branches of the pudic, is situated below the pudic artery. It accompanies the superficial perineal artery in the perineum, dividing into cutaneous and muscular branches.

The cutaneous branches (superficial perineal) are two in number, posterior and anterior. The *posterior branch* passes to the back part of the ischio-rectal fossa, distributing filaments to the Sphincter ani and integument in front of the anus, which communicate with the inferior hæmorrhoidal nerve; it then passes forwards, with the anterior branch, to the back of the scrotum, communicating with the anterior branch and with the inferior pudendal. The *anterior branch* passes to the fore part of the ischio-rectal fossa, in front of the preceding, and accompanies it to the scrotum and under part of the penis. This branch gives one or two filaments to the Levator ani.

The muscular branches are distributed to the Transversus perinei, Accelerator urinæ, Erector penis, and Compressor urethræ. The nerve of the bulb supplies the corpus spongiosum; some of its filaments run for some distance on the surface, before penetrating to the interior.

The *dorsal nerve of the penis* is the superior division of the pudic nerve; it accompanies the pudic artery along the ramus of the ischium, and between the two layers of the deep perineal fascia; it then pierces the suspensory ligament of the penis, and accompanies the arteria dorsalis penis to the glans, to which it is distributed. On the penis, this nerve gives off a cutaneous branch, which runs along the side of the organ; it is joined with branches of the sympathetic, and supplies the integument of the upper surface and sides of the penis and prepuce, giving a large branch to the corpus cavernosum.

In the female, the pudic nerve is distributed to the parts analogous to those in the male; its superior division terminating in the clitoris, its inferior in the external labia and perineum.

The SMALL SCIATIC NERVE (fig. 291) supplies the integument of the perineum and back part of the thigh and leg, and one muscle, the Gluteus maximus. It is

290.—Cutaneous Nerves of Lower Extremity. Posterior View.

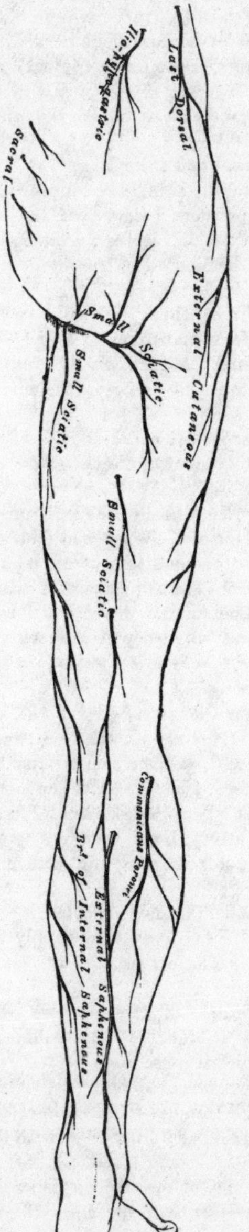

291.—Nerves of the Lower Extremity. Posterior View.

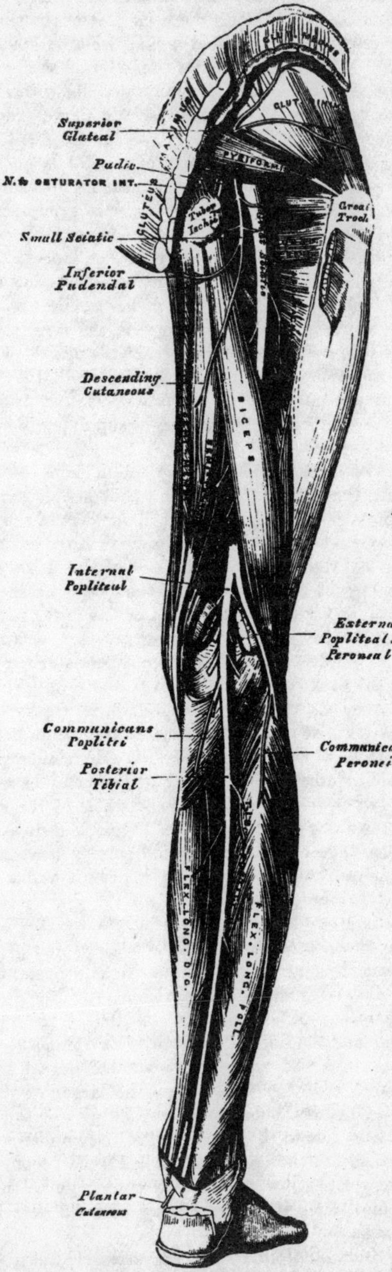

usually formed by the union of two branches, which arise from the lower part of the sacral plexus. It issues from the pelvis below the Pyriformis muscle, descends beneath the Gluteus maximus with the sciatic artery, and at the lower border of that muscle passes along the back part of the thigh, beneath the fascia lata, to the lower part of the popliteal region, where it pierces the fascia and becomes cutaneous. It then accompanies the external saphenous vein below the middle of the leg, its terminal filaments communicating with the external saphenous nerve.

The branches of the small sciatic nerve are muscular (inferior gluteal) and cutaneous.

The *inferior gluteal* consist of several large branches given off to the under surface of the Gluteus maximus, near its lower part.

The *cutaneous branches* consist of two groups, internal and ascending.

The *internal cutaneous branches* are distributed to the skin at the upper and inner side of the thigh, on its posterior aspect. One branch, longer than the rest, the *inferior pudendal*, curves forward below the tuber ischii, pierces the fascia lata on the outer side of the ramus of the ischium, and is distributed to the integument of the scrotum, communicating with the superficial perineal nerve.

The *ascending cutaneous branches* consist of two or three filaments, which turn upwards round the lower border of the Gluteus maximus, to supply the integument covering its surface. One or two filaments occasionally descend along the outer side of the thigh, supplying the integument as far as the middle of that region.

Two or three branches are given off from the lesser sciatic nerve as it descends beneath the fascia of the thigh; they supply the integument of the back part of the thigh, popliteal region, and upper part of the leg.

The Great Sciatic Nerve (fig. 291) supplies nearly the whole of the integument of the leg, the muscles of the back of the thigh, and those of the leg and foot. It is the largest nervous cord in the body, measuring three-quarters of an inch in breadth, and is the continuation of the lower part of the sacral plexus. It passes out of the pelvis through the great sacro-sciatic foramen, below the Pyriformis muscle. It descends between the trochanter major and tuberosity of the ischium, along the back part of the thigh to about its lower third, where it divides into two large branches, the internal and external popliteal nerves.

This division may take place at any point between the sacral plexus and the lower third of the thigh. When the division occurs at the plexus, the two nerves descend together, side by side; or they may be separated, at their commencement, by the interposition of part or the whole of the Pyriformis muscle. As the nerve descends along the back of the thigh, it rests at first upon the External rotator muscles, together with the small sciatic nerve and artery, being covered by the Gluteus maximus; lower down, it lies upon the Adductor magnus, and is covered by the long head of the Biceps.

The *branches* of the nerve, before its division, are articular and muscular.

The *articular branches* arise from the upper part of the nerve; they supply the hip-joint, perforating its fibrous capsule posteriorly. These branches are sometimes derived from the sacral plexus.

The *muscular branches* are distributed to the Flexors of the leg; viz., the Biceps, Semitendinosus, and Semimembranosus, and a branch to the Adductor magnus. These branches are given off beneath the Biceps muscle.

The *Internal Popliteal Nerve*, the larger of the two terminal branches of the great sciatic, descends along the back part of the thigh through the middle of the popliteal space, to the lower part of the Popliteus muscle, where it passes with the artery beneath the arch of the Soleus, and becomes the posterior tibial. It lies at first very superficial, and at the outer side of the popliteal vessels; opposite the knee-joint, it is in close relation with the vessels, and crosses the artery to its inner side.

The *branches* of this nerve are articular, muscular, and a cutaneous branch, the external or short saphenous nerve.

The *articular branches*, usually three in number, supply the knee joint; two of these branches accompany the superior and inferior internal articular arteries; and a third, the azygos.

The *muscular branches*, four or five in number, arise from the nerve as it lies between the two heads of the Gastrocnemius muscle; they supply that muscle, the Plantaris, Soleus, and Popliteus.

The *external* or *short saphenous nerve* (fig. 290) descends between the two heads of the Gastrocnemius muscle, and, about the middle of the back of the leg, pierces the deep fascia, and receives a communicating branch (*communicans peronei*) from the external popliteal nerve. The nerve then continues its course down the leg near the outer margin of the tendo Achillis, in company with the external saphenous vein, winds round the outer malleolus, and is distributed to the integument along the outer side of the foot and little toe, communicating on the dorsum of the foot with the musculo-cutaneous nerve.

The *posterior tibial nerve* (fig. 291) commences at the lower border of the Popliteus muscle, and passes along the back part of the leg with the posterior tibial vessels to the interval between the inner malleolus and the heel, where it divides into the external and internal plantar nerves. It lies upon the deep muscles of the leg, and is covered by the deep fascia, the superficial muscles, and integument. In the upper part of its course, it lies to the inner side of the posterior tibial artery; but it soon crosses that vessel, and lies to its outer side as far as the ankle. In the lower third of the leg, it is placed parallel with the inner margin of the tendo Achillis.

292.—The Plantar Nerves.

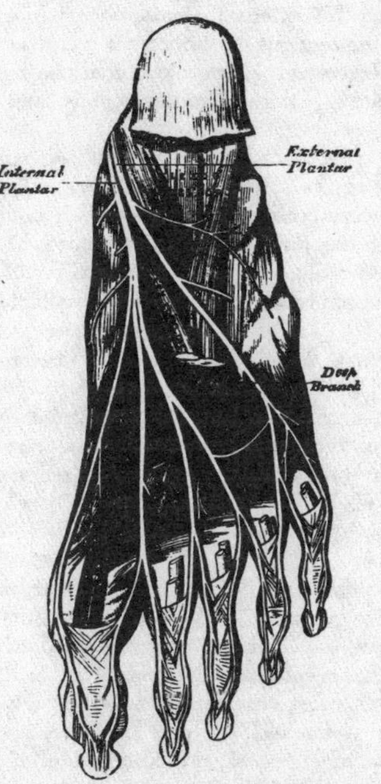

The *branches* of the posterior tibial nerve are muscular and plantar-cutaneous.

The *muscular branches* arise either separately or by a common trunk from the upper part of the nerve. They supply the Tibialis posticus, Flexor longus digitorum, and Flexor longus pollicis muscles; the branch to the latter muscle accompanying the peroneal artery.

The *plantar cutaneous branch* perforates the internal annular ligament, and supplies the integument of the heel and inner side of the sole of the foot.

The *internal plantar nerve* (fig. 292), the larger of the two terminal branches of the posterior tibial, accompanies the internal plantar artery along the inner side of the foot. From its origin at the inner ankle it passes forwards between the Abductor pollicis and Flexor brevis digitorum, divides opposite the bases of the metatarsal bones into four digital branches, and communicates with the external plantar nerve.

Branches. In its course, the internal plantar nerve gives off *cutaneous branches*, which pierce the plantar fascia, and supply the integument of the sole of the foot; *muscular branches*, which supply the Abductor pollicis and Flexor brevis digitorum; *articular branches* to the articulations of the tarsus and metatarsus; and *four digital branches*. These pierce the plantar fascia in the clefts between the toes, and are distributed in the following manner: The *first* supplies the inner border of the great toe, and sends a filament to the Flexor brevis pollicis muscle; the *second* bifurcates, to supply the adjacent sides of the great and second toes, sending a filament to the first Lumbrical muscle; the *third* digital branch supplies the adjacent sides of the second and third toes, and the second Lumbrical muscle; the *fourth* supplies the corresponding sides of the third and fourth toes, and receives a communicating branch from the external plantar nerve. It will be observed, that the distribution of these branches is precisely similar to that of the median nerve in the hand. Each digital nerve gives off cutaneous and articular filaments; and opposite the last phalanx sends a dorsal branch, which supplies the structures round the nail, the continuation of the nerve being distributed to the ball of the toe.

The *external plantar nerve*, the smaller of the two, completes the nervous supply to the structures of the foot, being distributed to the little toe and one half of the fourth, as well as to most of the deep muscles, its distribution being similar to that of the ulnar in the hand. It passes obliquely forwards with the external plantar artery to the outer side of the foot, lying between the Flexor brevis digitorum and Flexor accessorius; and, in the interval between the former muscle and Abductor minimi digiti, divides into a superficial and a deep branch. Before its division, it supplies the Flexor accessorius and Abductor minimi digiti.

The *superficial branch* separates into two digital nerves; one, the smaller of the two, supplies the outer side of the little toe, the Flexor brevis minimi digiti, and the two interosseous muscles of the fourth metatarsal space; the other, and larger digital branch, supplies the adjoining sides of the fourth and fifth toes, and communicates with the internal plantar nerve.

The *deep* or *muscular branch* accompanies the external plantar artery into the deep part of the sole of the foot, beneath the tendons of the Flexor muscles and Adductor pollicis, and supplies all the Interossei (except those in the fourth metatarsal space), the two outer Lumbricales, the Adductor pollicis, and the Transversus pedis.

The *External Popliteal* or *Peroneal Nerve* (fig. 291), about one half the size of the internal popliteal, descends obliquely along the outer side of the popliteal space to the fibula, close to the margin of the Biceps muscle. It is easily felt beneath the skin behind the head of the fibula, at the inner side of the tendon of the Biceps. About an inch below the head of the fibula it pierces the origin of the Peroneus longus, and divides beneath that muscle into the anterior tibial and musculo-cutaneous nerves.

The *branches* of the peroneal nerve, previous to its division, are articular and cutaneous.

The *articular branches*, two in number, accompany the superior and inferior external articular arteries to the outer side of the knee. The upper one occasionally arises from the great sciatic nerve before its bifurcation. A third (recurrent) articular nerve is given off at the point of division of the peroneal nerve; it ascends with the tibial recurrent artery through the Tibialis anticus muscle to the front of the knee, which it supplies.

The *cutaneous branches*, two or three in number, supply the integument along the back part and outer side of the leg, as far as its middle or lower part; one of these, larger than the rest, the *communicans peronei*, arises near the head of the fibula, crosses the external head of the Gastrocnemius to the middle of the leg, and joins with the external saphenous. This nerve occasionally exists as a separate branch, which is continued down as far as the heel.

The *Anterior Tibial Nerve* (fig. 289) commences at the bifurcation of the peroneal nerve, between the fibula and upper part of the Peroneus longus, passes obliquely forwards beneath the Extensor longus digitorum to the fore part of the interosseous membrane, and reaches the outer side of the anterior tibial artery above the middle of the leg; it then descends with the artery to the front of the ankle joint, where it divides into an external and an internal branch. This nerve lies at first on the outer side of the anterior tibial artery, then in front of it, and again at its outer side at the ankle joint.

The *branches* of the anterior tibial nerve, in its course through the leg, are the muscular nerves to the Tibialis anticus, Extensor longus digitorum, and Extensor proprius pollicis muscles.

The *external* or *tarsal branch* of the anterior tibial, passes outwards across the tarsus, beneath the Extensor brevis digitorum, and, having become ganglionic, like the posterior interosseous nerve at the wrist, supplies the Extensor brevis digitorum and the articulations of the tarsus and metatarsus.

The *internal branch*, the continuation of the nerve, accompanies the dorsalis pedis artery along the inner side of the dorsum of the foot, and, at the first interosseous space, divides into two branches, which supply the adjacent sides of the great and second toes, communicating with the internal division of the musculo-cutaneous nerve.

The *Musculo-Cutaneous Nerve* (fig. 289) supplies the muscles on the fibular side of the leg, and the integument of the dorsum of the foot. It passes forwards between the Peronei muscles and the Extensor longus digitorum, pierces the deep fascia at the lower third of the leg, on its front and outer side, and divides into two branches. This nerve, in its course between the muscles, gives off muscular branches to the Peroneus longus and brevis, and cutaneous filaments to the integument of the lower part of the leg.

The *internal branch* of the musculo-cutaneous nerve, passes in front of the ankle joint, and along the dorsum of the foot, supplying the inner side of the great toe, and the adjoining sides of the second and third toes. It also supplies the integument of the inner ankle and inner side of the foot, communicating with the internal saphenous nerve, and joins with the anterior tibial nerve, between the great and second toes.

The *external branch*, the larger, passes along the outer side of the dorsum of the foot, to be distributed to the adjoining sides of the third, fourth, and fifth toes. It also supplies the integument of the outer ankle and outer side of the foot, communicating with the short saphenous nerve.

The distribution of these branches of the musculo-cutaneous nerve will be found to vary; together, they supply all the toes excepting the outer side of the little toe, and the adjoining sides of the great and second toes.

The Sympathetic Nerve.

THE Sympathetic Nerve is so called from the opinion entertained that through it is produced a *sympathy* between the affections of distant organs. It consists of a series of ganglia, connected together by intervening cords, extending on each side of the vertebral column from the base of the skull to the coccyx. It may, moreover, be traced up into the head, where the ganglia (which are all in connection with the fifth cranial nerve) occupy spaces between the cranial and facial bones. These two gangliated cords lie parallel with one another as far as the sacrum, on which bone they converge, communicating together through a single ganglion (*ganglion impar*), placed in front of the coccyx. Some anatomists also state that the two cords are joined at their cephalic extremity, through a small ganglion (the ganglion of Ribes), situated upon the anterior communicating artery. Moreover, the chains of opposite sides communicate between these two extremities in several parts, by means of the nervous cords that arise from them.

The ganglia are somewhat less numerous than the vertebræ: thus there are only three in the cervical region, twelve in the dorsal, four in the lumbar, five in the sacral, and one in the coccygeal.

The sympathetic nerve, for convenience of description, may be divided into several parts, according to the position occupied by each; and the number of ganglia of which each part is composed, may be thus arranged:

Cephalic portion	. .	4 ganglia.
Cervical ,,	. .	3 ,,
Dorsal ,,	. .	12 ,,
Lumbar ,,	. .	4 ,,
Sacral ,,	. .	5 ,,
Coccygeal ,,	. .	1 ,,

Each ganglion may be regarded as a distinct centre, from or to which branches pass in various directions. These branches may be thus arranged:—1. Branches of communication between the ganglia. 2. Branches of communication with the cerebral or spinal nerves. 3. Primary branches passing to be distributed to the arteries in the vicinity of the ganglia, and to the viscera, or proceeding to other ganglia placed in the thorax, abdomen, or pelvis.

1. The branches of communication between the ganglia are composed of grey and white nerve-fibres, the latter being continuous with those fibres of the spinal nerves which pass to the ganglia.

2. The branches of communication between the ganglia and the cerebral or spinal nerves also consist of a white and a grey portion; the former proceeding from the spinal nerve *to* the ganglion, the latter passing *from* the ganglion to the spinal nerve.

3. The primary branches of distribution also consist of two kinds of nerve-fibres, the sympathetic and spinal. They have a remarkable tendency to form intricate plexuses, which encircle the blood-vessels, and are conducted by them to the viscera. The greater number, however, of these branches pass to a series of visceral ganglia: these are ganglionic masses, of variable size, situated in the large cavities of the trunk, the thorax, and abdomen; and are connected with the roots of the great arteries of the viscera. These ganglia are single and unsymmetrical, and are called the cardiac and semilunar. From these visceral ganglia numerous plexuses of nerves are derived, which entwine round the blood-vessels, and are conducted by them to the viscera.

293.—The Sympathetic Nerve.

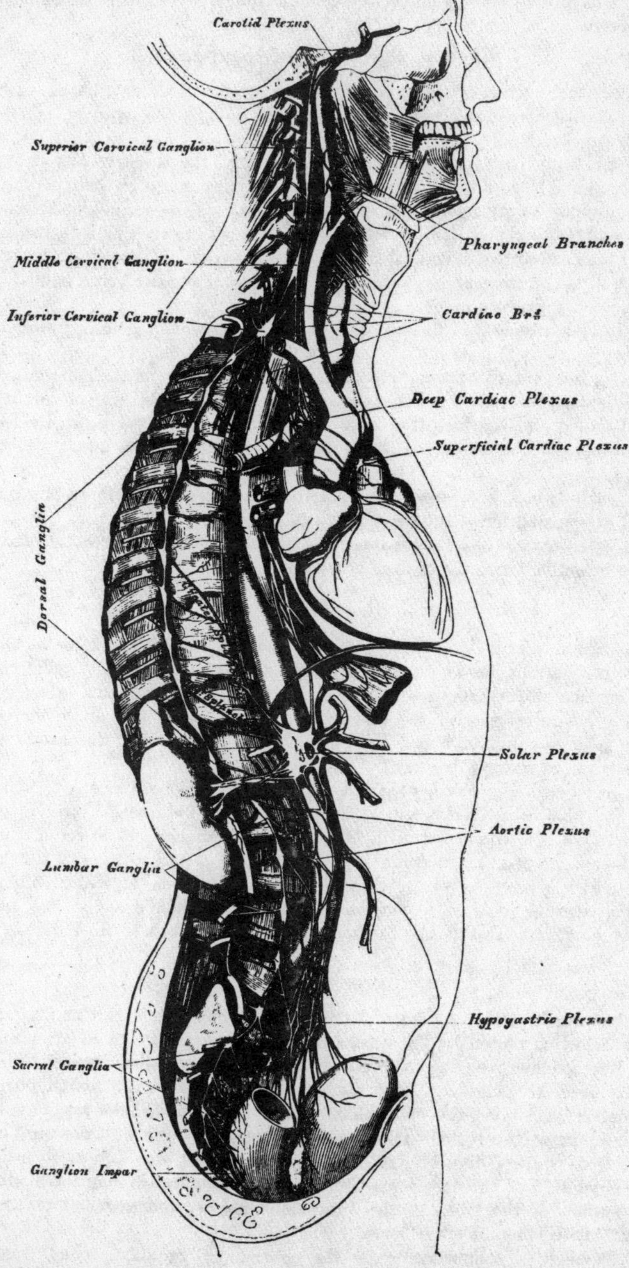

The *cephalic portion* of the sympathetic, consists of four ganglia. 1. The ophthalmic ganglion. 2. The spheno-palatine, or Meckel's ganglion. 3. The otic, or Arnold's ganglion. 4. The submaxillary ganglion.

These have been already described in connection with the three divisions of the fifth nerve.

Cervical Portion of the Sympathetic.

The cervical portion of the sympathetic consists of three ganglia on each side, which are distinguished according to their position, as the superior, middle, and inferior cervical.

The Superior Cervical Ganglion, the largest of the three, is placed opposite the second and third cervical vertebræ, and sometimes as low as the fourth or fifth. It is of a reddish-grey colour, and usually fusiform in shape: sometimes broad, and occasionally constricted at intervals, so as to give rise to the opinion, that it consists of the coalescence of several smaller ganglia. It is in relation, in front, with the sheath of the internal carotid artery, and internal jugular vein; behind, it lies on the Rectus capitis anticus major muscle.

Its branches may be divided into superior, inferior, external, internal, and anterior.

The *superior branch* appears to be a direct continuation of the ganglion. It is soft in texture, and of a reddish colour. It ascends by the side of the internal carotid artery, and, entering the carotid canal in the temporal bone, divides into two branches, which lie, one on the outer, and the other on the inner side, of that vessel.

The *outer branch*, the larger of the two, distributes filaments to the internal carotid artery, and forms the carotid plexus.

The *inner branch* also distributes filaments to the internal carotid, and, continuing onwards, forms the cavernous plexus.

Carotid Plexus.

The carotid plexus is situated on the outer side of the internal carotid. Filaments from this plexus occasionally form a small gangliform swelling on the under surface of the artery, which is called the carotid ganglion. The carotid plexus communicates with the Casserian ganglion, with the sixth nerve, and spheno-palatine ganglion, and distributes filaments to the wall of the carotid artery, and to the dura mater (Valentin).

The *communicating branches* with the sixth nerve consist of one or two filaments, which join that nerve as it lies upon the outer side of the internal carotid. Other filaments are also connected with the Casserian ganglion. The communication with the spheno-palatine ganglion is effected by the carotid portion of the Vidian nerve, which passes forwards, through the cartilaginous substance filling the foramen lacerum medium, along the pterygoid or Vidian canal, to the spheno-palatine ganglion. In this canal it joins the petrosal branch of the Vidian.

Cavernous Plexus.

The cavernous plexus is situated below, and internal to that part of the internal carotid, which is placed by the side of the sella Turcica, in the cavernous sinus, and is formed chiefly by the internal division of the ascending branch from the superior cervical ganglion. It communicates with the third, fourth, fifth, and sixth nerves, and with the ophthalmic ganglion, and distributes filaments to the wall of the internal carotid. The branch of communication with the third nerve joins it at its point of division; the branch to the fourth nerve joins it as it lies on the outer wall of the cavernous sinus; other filaments are connected with the under surface of the trunk of the ophthalmic nerve; and a second filament of communication joins the sixth nerve.

The filament of connection with the ophthalmic ganglion arises from the anterior part of the cavernous plexus; it accompanies the nasal nerve, or continues forwards as a separate branch.

The terminal filaments from the carotid and cavernous plexuses are prolonged along the internal carotid, forming plexuses which entwine round the cerebral and ophthalmic arteries; along the former vessel they may be traced on to the pia mater; along the latter, into the orbit, where they accompany each of the subdivisions of the vessel, a separate plexus passing with the arteria centralis retinæ into the interior of the eyeball.

The *inferior* or *descending branch* of the superior cervical ganglion communicates with the middle cervical ganglion.

The *external branches* are numerous, and communicate with the cranial nerves, and with the four upper spinal nerves. Sometimes, the branch to the fourth spinal nerve may come from the cord connecting the upper and middle cervical ganglia. The branches of communication with the cranial nerves consist of delicate filaments, which pass from the superior cervical ganglion to the ganglion of the trunk of the pneumogastric, and to the ninth nerve. A separate filament from the cervical ganglion subdivides and joins the petrosal ganglion of the glosso-pharyngeal, and the ganglion of the root of the pneumogastric in the jugular foramen.

The *internal branches* are three in number: pharyngeal, laryngeal, and the superior cardiac nerve. The pharyngeal branches pass inwards to the side of the pharynx, where they join with branches from the pneumogastric, glosso-pharyngeal, and external laryngeal nerves to form the *pharyngeal plexus*. The laryngeal branches unite with the superior laryngeal nerve and its branches.

The superior cardiac nerve will be described in connection with the other cardiac nerves.

The *anterior branches* ramify upon the external carotid artery and its branches, forming round each a delicate plexus, on the nerves composing which small ganglia are occasionally found. These ganglia have been named, according to their position, intercarotid (one placed at the angle of bifurcation of the common carotid), lingual, temporal, and pharyngeal. The plexuses accompanying some of these arteries have important communications with other nerves. That surrounding the external carotid, is connected with the digastric branch of the facial; that surrounding the facial, communicates with the submaxillary ganglion by one or two filaments; and that accompanying the middle meningeal artery, sends offsets which pass to the otic ganglion and to the intumescentia gangliofornis of the facial nerve.

The MIDDLE CERVICAL GANGLION (thyroid ganglion) is the smallest of the three cervical ganglia, and is occasionally altogether wanting. It is placed opposite the fifth cervical vertebra, usually upon, or close to, the inferior thyroid artery; hence the name 'thyroid ganglion,' assigned to it by Haller.

Its *superior branches* ascend to communicate with the superior cervical ganglion.

Its *inferior branches* descend to communicate with the inferior cervical ganglion.

Its *external branches* pass outwards to join the fifth and sixth spinal nerves. Those branches are not constantly found.

Its *internal branches* are, the thyroid, and the middle cardiac nerve.

The *thyroid branches* are small filaments, which accompany the inferior thyroid artery to the thyroid gland; they communicate, on the artery, with the superior cardiac nerve, and, in the gland, with branches from the recurrent and external laryngeal nerves.

The middle cardiac nerve is described with the other cardiac nerves.

The INFERIOR CERVICAL GANGLION is situated between the base of the transverse process of the last cervical vertebra and the neck of the first rib, on the inner side of the superior intercostal artery. Its form is irregular; it is larger in size than the preceding, and frequently joined with the first thoracic ganglion.

Its *superior branches* communicate with the middle cervical ganglion.

Its *inferior branches* descend, some in front of, others behind the subclavian artery, to join the first thoracic ganglion. The most important of these branches constitutes the inferior cardiac nerve, to be presently described.

The *external branches* consist of several filaments, some of which communicate with the seventh and eighth spinal nerves; others accompany the vertebral artery along the vertebral canal, forming a plexus round the vessel, supplying it with filaments, and communicating with the cervical spinal nerves as high as the fourth.

Cardiac Nerves.

The cardiac nerves are three in number on each side: superior, middle, and inferior, one being derived from each of the cervical ganglia.

The *superior cardiac nerve* (nervus superficialis cordis) arises by two or more branches from the superior cervical ganglion, and occasionally receives a filament from the cord of communication between the first and second cervical ganglia. It runs down the neck behind the common carotid artery, lying upon the Longus colli muscle; and crosses in front of the inferior thyroid artery, and the recurrent laryngeal nerve.

The *right superior cardiac nerve*, at the root of the neck, passes either in front of or behind the subclavian artery, and along the arteria innominata, to the back part of the arch of the aorta, where it joins the deep cardiac plexus. This nerve, in its course, is connected with other branches of the sympathetic; about the middle of the neck it receives filaments from the external laryngeal nerve; lower down, one or two twigs from the pneumogastric; and as it enters the thorax, it joins with the recurrent laryngeal. Filaments from this nerve accompany the inferior thyroid artery to the thyroid gland.

The *left superior cardiac nerve* runs by the side of the left carotid artery, and in front of the arch of the aorta, to the superficial cardiac plexus; but occasionally it passes behind the aorta, and terminates in the deep cardiac plexus.

The *middle cardiac nerve* (nervus cardiacus magnus), the largest of the three, arises from the middle cervical ganglion, or from the cord between the middle and inferior ganglia. On the right side, it descends behind the common carotid artery; and, at the root of the neck, passes either in front of or behind the subclavian artery; it then descends on the trachea, receives a few filaments from the recurrent laryngeal nerve, and joins the deep cardiac plexus. In the neck, it communicates with the superior cardiac and recurrent laryngeal nerves. On the left side, the middle cardiac nerve enters the chest between the left carotid and subclavian arteries, and joins the left side of the deep cardiac plexus.

The *inferior cardiac nerve* (nervus cardiacus minor) arises from the inferior cervical or first thoracic ganglion. It passes down behind the subclavian artery, and along the front of the trachea, to join the deep cardiac plexus. It communicates freely behind the subclavian artery with the recurrent laryngeal and middle cardiac nerves.

The *great* or *deep cardiac plexus* (*plexus magnus profundus*—Scarpa) is situated in front of the trachea at its bifurcation, above the point of division of the pulmonary artery, and behind the arch of the aorta. It is formed by the cardiac nerves derived from the cervical ganglia of the sympathetic, and the cardiac branches of the recurrent laryngeal and pneumogastric. The only cardiac nerves which do not enter into the formation of this plexus, are the left superior cardiac nerve, and the left inferior cardiac branch from the pneumogastric. The branches derived from the great cardiac plexus form the posterior coronary plexus, and part of the anterior coronary plexus; whilst a few filaments proceed to the pulmonary plexuses, and to the auricles of the heart.

The branches from the *right side* of this plexus pass, some in front of and others behind the right pulmonary artery; the former, the more numerous, transmit a few filaments to the anterior pulmonary plexus, and are continued along the trunk of the pulmonary artery, to form part of the anterior coronary plexus; those behind the pulmonary artery distribute a few filaments to the right auricle, and form part of the posterior coronary plexus.

The branches from the *left side* of the deep cardiac plexus distribute a few filaments to the left auricle of the heart and the anterior pulmonary plexus, and then pass on to form the greater part of the posterior coronary plexus, a few branches passing to the superficial cardiac plexus.

The *superficial (anterior) cardiac plexus* lies beneath the arch of the aorta, in front of the right pulmonary artery. It is formed by the left superior cardiac nerve, the left (and occasionally the right) inferior cardiac branches of the pneumogastric, and filaments from the deep cardiac plexus. A small ganglion (cardiac ganglion of Wrisberg) is occasionally found connected with these nerves at their point of junction. This ganglion, when present, is situated immediately beneath the arch of the aorta, on the right side of the ductus arteriosus. The superficial cardiac plexus forms the chief part of the anterior coronary plexus, and several filaments pass along the pulmonary artery to the left anterior pulmonary plexus.

The *posterior coronary plexus* is chiefly formed by filaments prolonged from the left side of the deep cardiac plexus, and by a few from the right side. It surrounds the branches of the coronary artery at the back of the heart, and its filaments are distributed with those vessels to the muscular substance of the ventricles.

The *anterior coronary plexus* is formed chiefly from the superficial cardiac plexus, but receives filaments from the deep cardiac plexus. Passing forwards between the aorta and pulmonary artery, it accompanies the right coronary artery on the anterior surface of the heart.

Valentin has described nervous filaments ramifying under the endocardium; and Remak has found, in several mammalia, numerous small ganglia on the cardiac nerves, both on the surface of the heart and in its muscular substance. The elaborate dissections lately completed by Dr. Robert Lee have demonstrated without any doubt the existence of a dense mesh of nerves distributed both to the surface, and in the substance of the heart, having numerous ganglia developed upon them.

Thoracic Part of the Sympathetic.

The thoracic portion of the sympathetic consists of a series of ganglia, which usually correspond in number to that of the vertebræ; but, from the occasional coalescence of two, their number is uncertain. These ganglia are placed on each side of the spine, resting against the heads of the ribs, and covered by the pleura costalis: the last two are, however, anterior to the rest, being placed on the side of the bodies of the vertebræ. The ganglia are small in size, and of a greyish colour. The first, larger than the rest, is of an elongated form, and usually blended with the last cervical. They are connected together by cord-like prolongations from their substance.

The *external branches* from each ganglion, usually two in number, communicate with each of the dorsal spinal nerves.

The *internal branches from the six upper ganglia* are very small: they supply filaments to the thoracic aorta and its branches, besides small branches to the bodies of the vertebræ and their ligaments. Branches from the third and fourth ganglia form part of the posterior pulmonary plexus.

The *internal branches from the six lower ganglia* are large and white in colour; they distribute filaments to the aorta, and unite to form the three splanchnic nerves. These are named, the *great*, the *lesser*, and the *smallest* or *renal splanchnic*.

The *great splanchnic nerve* is of a white colour, firm in texture, and bears a marked contrast to the ganglionic nerves. It is formed by branches from the thoracic ganglia between the sixth and tenth, receiving filaments (according to Dr. Beck) from all the thoracic ganglia above the sixth. These roots unite to form a large round cord of considerable size. It descends obliquely inwards in front of the bodies of the vertebræ along the posterior mediastinum, perforates the crus of the Diaphragm, and terminates in the semilunar ganglion, distributing filaments to the renal plexus and suprarenal capsule.

The *lesser splanchnic nerve* is formed by filaments from the tenth and eleventh ganglia, and from the cord between them. It pierces the Diaphragm with the preceding nerve, and joins the cœliac plexus. It communicates in the chest with the great splanchnic nerve, and occasionally sends filaments to the renal plexus.

The *smallest* or *renal splanchnic nerve* arises from the last ganglion, and piercing the Diaphragm, terminates in the renal plexus and lower part of the cœliac plexus. It occasionally communicates with the preceding nerve.

A striking analogy appears to exist between the splanchnic and the cardiac nerves. The cardiac nerves are three in number; they arise from the three cervical ganglia, and are distributed to a large and important organ in the thoracic cavity. The splanchnic nerves, also three in number, are connected probably with all the dorsal ganglia, and are distributed to important organs in the abdominal cavity.

The *epigastric* or *solar plexus* supplies all the viscera in the abdominal cavity. It consists of a dense network of nerves and ganglia, situated behind the stomach and in front of the aorta and crura of the Diaphragm. It surrounds the cœliac axis and root of the superior mesenteric artery, extending downwards as low as the pancreas, and outwards to the suprarenal capsules. This plexus, and the ganglia connected with it, receive the great splanchnic nerve of both sides, part of the lesser splanchnic nerves, and the termination of the right pneumogastric. It distributes filaments, which accompany, under the name of plexuses, all the branches from the front of the abdominal aorta.

The *semilunar ganglia*, of the solar plexus, two in number, one on each side, are the largest ganglia in the body. They are large irregular gangliform masses, formed by the aggregation of smaller ganglia, having interspaces between them. They are situated by the side of the cœliac axis and superior mesenteric artery, close to the suprarenal capsules: the one on the right side lies beneath the vena cava; the upper part of each ganglion is joined by the greater and lesser splanchnic nerves, and to the inner side of each the branches of the solar plexus are connected. From the solar plexus are derived the following:—

Phrenic or Diaphragmatic plexus.
Cœliac plexus.
Gastric plexus.
Hepatic plexus.
Splenic plexus.
Suprarenal plexus.
Renal plexus.
Superior mesenteric plexus.
Spermatic plexus.
Inferior mesenteric plexus.

The *phrenic plexus* accompanies the phrenic artery to the Diaphragm, which it supplies, some filaments passing to the suprarenal capsule. It arises from the upper part of the semilunar ganglion, and is larger on the right than on the left side. In connection with this plexus, on the right side, at its point of junction with the phrenic nerve, is a small ganglion (ganglion diaphragmaticum). This ganglion is placed on the under surface of the Diaphragm, near the suprarenal capsule. Its branches are distributed to the vena cava, suprarenal capsule, and the hepatic plexus. There is no ganglion on the left side.

The *suprarenal plexus* is formed by branches from the solar plexus, from the semilunar ganglion, and from the splanchnic and phrenic nerves, a ganglion being formed at the point of junction of the latter nerve. It supplies the suprarenal gland. The branches of this plexus are remarkable for their large size, in comparison with the size of the organ they supply.

The *renal plexus* is formed by filaments from the solar plexus, the outer part of the semilunar ganglion, and the aortic plexus. It is also joined by filaments from the lesser and smallest splanchnic nerves. The nerves from these sources, fifteen or twenty in number, have numerous ganglia developed upon them. They accompany the branches of the renal artery into the kidney; some filaments on the right side being distributed to the vena cava, and others to the spermatic plexus, on both sides.

The *spermatic plexus* is derived from the renal plexus, receiving branches from the aortic plexus. It accompanies the spermatic vessels to the testes.

In the female, the *ovarian plexus* is distributed to the ovaries and fundus of the uterus.

The *cœliac plexus*, of large size, is a direct continuation from the solar plexus: it surrounds the cœliac axis, and subdivides into the gastric, hepatic, and splenic plexuses. It receives branches from one or more of the splanchnic nerves, and, on the left side, a filament from the pneumogastric.

The *gastric plexus* accompanies the gastric artery along the lesser curvature of the stomach, and joins with branches from the left pneumogastric nerve. It is distributed to the stomach.

The *hepatic plexus*, the largest offset from the cœliac plexus, receives filaments from the left pneumogastric and right phrenic nerves. It accompanies the hepatic artery, ramifying in the substance of the liver, upon its branches, and upon those of the vena portæ.

Branches from this plexus accompany all the divisions of the hepatic artery. Thus there is a pyloric plexus accompanying the pyloric branch of the hepatic, which joins with the gastric plexus, and pneumogastric nerves. There is also a gastro-duodenal plexus, which subdivides into the pancreatico-duodenal plexus, which accompanies the pancreatico-duodenal artery, to supply the pancreas and duodenum, joining with branches from the mesenteric plexus; and a gastro-epiploic plexus, which accompanies the right gastro-epiploic artery along the greater curvature of the stomach, and anastomoses with branches from the splenic plexus. A cystic plexus, which supplies the gall-bladder, also arises from the hepatic plexus, near the liver.

The *splenic plexus* is formed by branches from the right and left semilunar ganglia, and from the right pneumogastric nerve. It accompanies the splenic artery and its branches to the substance of the spleen, giving off, in its course, filaments to the pancreas (pancreatic plexus), and the left gastro-epiploic plexus, which accompanies the gastro-epiploica sinistra artery along the convex border of the stomach.

The *superior mesenteric plexus* is a continuation of the lower part of the great solar plexus, receiving a branch from the junction of the right pneumogastric nerve with the cœliac plexus. It surrounds the superior mesenteric artery, which it accompanies into the mesentery, and divides into a number of secondary plexuses, which are distributed to all the parts supplied by the artery, viz., pancreatic branches to the pancreas; intestinal branches, which supply the whole of the small intestine; and ileo-colic, right colic, and middle colic branches, which supply the corresponding parts of the great intestine. The nerves composing this plexus are white in colour, and firm in texture, and have numerous ganglia developed upon them near their origin.

The *aortic plexus* is formed by branches derived, on each side, from the semilunar ganglia and renal plexuses, receiving filaments from some of the lumbar ganglia. It is situated upon the sides and front of the aorta, between the origins of the superior and inferior mesenteric arteries. From this plexus arise the inferior mesenteric, part of the spermatic, and the hypogastric plexuses; and it distributes filaments to the inferior vena cava.

The *inferior mesenteric plexus* is derived chiefly from the left side of the aortic plexus. It surrounds the inferior mesenteric artery, and divides into a number of secondary plexuses, which are distributed to all the parts supplied by the artery, viz., the left colic and sigmoid plexuses, which supply the descending and sigmoid flexure of the colon; and the superior hæmorrhoidal plexus, which supplies the upper part of the rectum, and joins in the pelvis with branches from the left hypogastric plexus.

The Lumbar Portion of the Sympathetic.

The lumbar portion of the sympathetic is situated in front of the vertebral column, along the inner margin of the Psoas muscle. It consists usually of four ganglia, connected together by interganglionic cords. The ganglia are of small size, of a greyish colour, shaped like a barley-corn, and placed much nearer the median line than the thoracic ganglia.

The *superior* and *inferior branches* of the lumbar ganglia, serve as communicating branches between the chain of ganglia in this region. They are usually single, and of a white colour.

The *external branches* communicate with the lumbar spinal nerves. From the situation of the lumbar ganglia, these branches are longer than in the other regions. They are usually two in number for each ganglion, and accompany the lumbar arteries around the sides of the bodies of the vertebræ, passing beneath the fibrous arches from which some of the fibres of the Psoas muscle arise.

The *internal branches* pass inwards, in front of the aorta, and form the lumbar aortic plexus, already described. Other branches descend in front of the common iliac arteries, and join, over the promontory of the sacrum, to form the hypogastric plexus. Numerous delicate filaments are also distributed to the bodies of the vertebræ, and the ligaments connecting them.

Pelvic Portion of the Sympathetic.

The pelvic portion of the sympathetic is situated in front of the sacrum, along the inner side of the anterior sacral foramina. It consists of four or five small ganglia on each side, connected together by interganglionic cords. Below, these cords converge and unite on the front of the coccyx, by means of a small ganglion (ganglion impar).

The *superior* and *inferior branches*, are the cords of communication between the ganglia above and below.

The *external branches*, exceedingly short, communicate with the sacral nerves. They are two in number to each ganglion. The coccygeal nerve communicates either with the last sacral, or coccygeal ganglion.

The *internal branches* communicate, on the front of the sacrum, with the corresponding branches from the opposite side; some, from the first two ganglia, pass to join the pelvic plexus, and others form a plexus, which accompanies the middle sacral artery.

The *hypogastric plexus* supplies the viscera of the pelvic cavity. It is situated in front of the promontory of the sacrum, between the two common iliac arteries, and is formed by the union of numerous filaments, which descend on each side from the aortic plexus, from the lumbar ganglia, and from the first two sacral ganglia. This plexus contains no ganglia, and bifurcates, below, into two lateral portions, which form the inferior hypogastric, or pelvic plexuses.

Inferior Hypogastric, or Pelvic Plexus.

The inferior hypogastric, or pelvic plexus, is situated at the side of the rectum and bladder in the male, and at the side of the rectum, vagina, and bladder, in the female. It is formed by a continuation of the hypogastric plexus, by branches from the second, third, and fourth sacral nerves, and by a few filaments from the sacral ganglia. At the point of junction of these nerves, small ganglia are found. From this plexus numerous branches are distributed to all the viscera of the pelvis. They accompany the branches of the internal iliac artery.

The *inferior hæmorrhoidal plexus* arises from the back part of the pelvic plexus. It supplies the rectum, joining with branches of the superior hæmorrhoidal plexus.

The *vesical plexus* arises from the fore part of the pelvic plexus. The nerves composing it are numerous, and contain a large proportion of spinal nerve-fibres.

294.—Ganglia and Nerves of the Gravid Uterus at the end of the Ninth Month. After Dr. R. Lee.

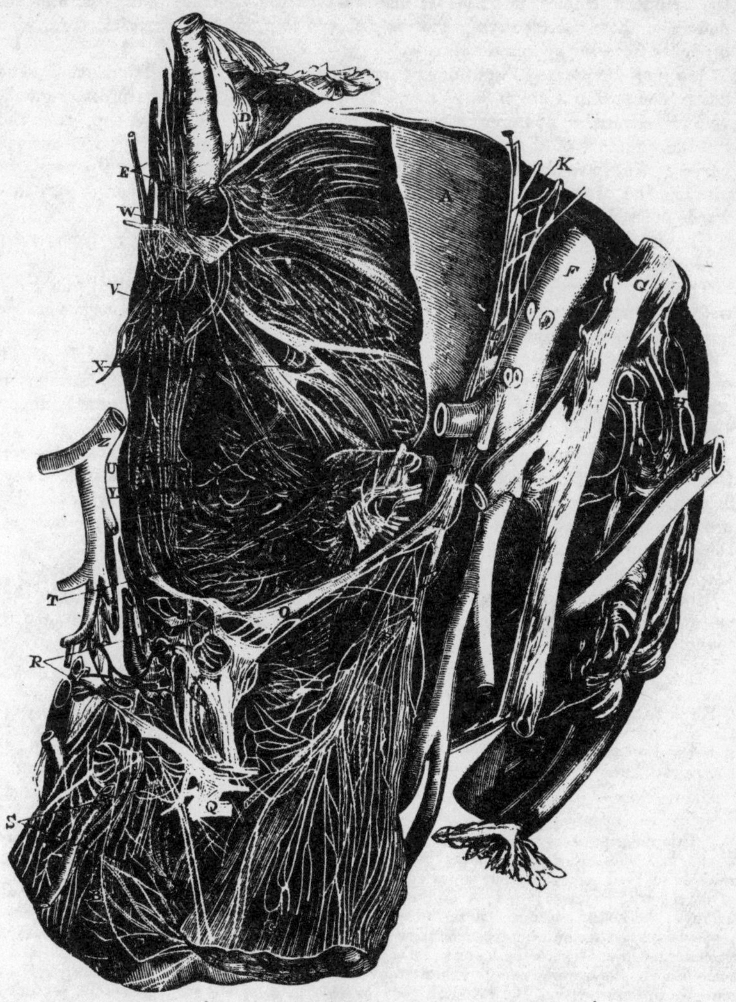

A. The fundus and body of the uterus, having the peritoneum dissected off from the left side. B. The vagina covered with nerves proceeding from the inferior border of the left hypogastric ganglion. C. The rectum. D. The left ovarium and Fallopian tube. E. The trunk of the left spermatic vein and artery surrounded by the left spermatic ganglion. F. The aorta divided a little above the origin of the right spermatic artery, and about three inches above its division into the two common iliac arteries. G. The vena cava. H. Trunk of the right spermatic vein entering the vena cava. I. Right ureter. K. The two cords of the great sympathetic nerve passing down along the front of the aorta. L. Trunk of the inferior mesenteric artery, passing off from the aorta, and covered with a great plexus of nerves sent off from the left and right cords of the great sympathetic. M. M. The two cords of the great sympathetic passing down below the bifurcation of the aorta to the point where they separate into the right and left hypogastric nerves. N. The right hypogastric nerve with its artery injected proceeding to the neck of the uterus, to terminate in the right hypogastric ganglion. O. The left hypogastric nerve where it is entering the left hypogastric ganglion, and giving off branches to the left subperitoneal ganglion. P. Hæmorrhoidal nerves accompanying the hæmorrhoidal artery and proceeding from the great plexus which surrounded the inferior mesenteric artery. Q. The sacral nerves entering the whole outer surface of the hypogastric ganglion. R. The left hypogastric ganglion with its arteries injected. S. The nerves of the vagina. T. Nerves with an injected artery proceeding from the upper part of the left hypogastric ganglion along the body of the uterus, and terminating in the left spermatic ganglion. U. Continuation of these nerves and the branches which they give off to the subperitoneal plexuses. V. The same nerves passing upward beneath the subperitoneal plexuses, and anastomosing freely with them. W. The left spermatic ganglion, in which the nerves and artery from the hypogastric ganglion, and the branches of the left subperitoneal plexuses terminate, and from which the nerves of the fundus uteri are supplied. X. The left subperitoneal plexuses covering the body of the uterus. Y. The left subperitoneal ganglion, with numerous branches of nerves extending between it and the left hypogastric nerve and ganglion. Z. The left common iliac artery cut across and turned aside, that the left hypogastric nerve and ganglion might be traced and exposed.

They accompany the vesical arteries, and are distributed to the side and base of the bladder. Numerous filaments also pass to the vesiculæ seminales, and vas deferens: those accompanying the vas deferens join, on the spermatic cord, with branches from the spermatic plexus.

The *prostatic plexus* is continued from the lower part of the pelvic plexus. The nerves composing it are of large size. They are distributed to the prostate gland, vesiculæ seminales, and erectile structure of the penis. The nerves supplying the erectile structure of the penis consist of two sets, the small and large cavernous nerves. They are slender filaments, which arise from the fore part of the prostatic plexus; and after joining with branches from the internal pudic nerve, pass forwards beneath the pubic arch.

The *small cavernous nerves* perforate the fibrous covering of the penis, near its roots.

The *large cavernous nerve* passes forwards along the dorsum of the penis, joins with the dorsal branch of the pudic nerve, and is distributed to the corpus cavernosum and spongiosum.

The *vaginal plexus* arises from the lower part of the pelvic plexus. It is lost on the walls of the vagina, being distributed to the erectile tissue at its anterior part, and to the mucous membrane. The nerves composing this plexus contain, like the vesical, a large proportion of spinal nerve-fibres.

The *uterine nerves* arise from the lower part of the hypogastric plexus, above the point where the branches from the sacral nerves join the pelvic plexus. They accompany the uterine arteries to the side of the organ between the layers of the broad ligament, and are distributed to the cervix and lower part of the body of the uterus, penetrating its substance.

Other filaments pass separately to the body of the uterus and Fallopian tube.

Branches from the hypogastric plexus accompany the uterine arteries into the substance of the uterus. Upon these filaments ganglionic enlargements are found.*

For a detailed account of the supply of nerves to the uterus, and for a description of the changes which these nerves and their ganglia undergo during pregnancy, the reader is referred to the papers on 'The Anatomy of the Nerves of the Uterus,' published by Dr. Robert Lee.

---

* This description of the sympathetic nerve accords with the works which are received as standard authorities in anatomy: but when the result of Dr. Robert Lee's dissections shall have been published, our knowledge of the distribution of the sympathetic system of nerves will be placed upon a different footing. Dr. Lee's beautiful dissections of the nerves of the heart, and those of the uterus, have been long known to anatomists. Other preparations, as accurate and beautiful as these, showing the communication between the pneumogastric and sympathetic, and the supply of nerves to the œsophagus, stomach, and some other viscera, were recently exhibited by Dr. Lee to the College of Physicians; and it may be anticipated that Dr. Lee will soon be able to complete and publish the account of the whole sympathetic system.

Organs of Sense.

THE organs of the senses are five in number, viz., those of touch, of taste, of smell, of hearing, and of sight. The skin, which is the principal seat of the sense of touch, has been described in the Introduction.

The Tongue.

The tongue is the organ of the special sense of taste. It is situated in the floor of the mouth, in the interval between the two lateral portions of the body of

295.—Upper Surface of the Tongue.

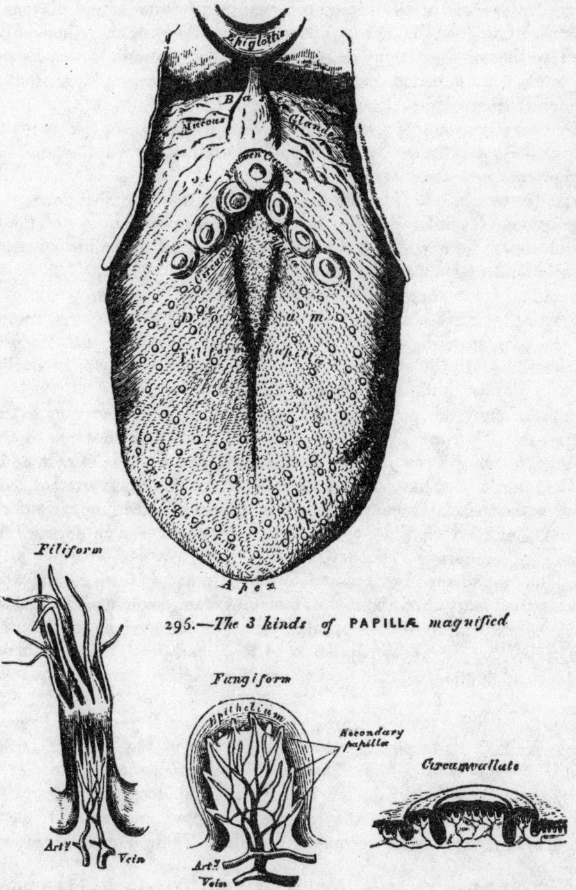

296.—The 3 kinds of PAPILLÆ magnified

the lower jaw. Its base, or root, is directed backwards, and connected with the os hyoides by numerous muscles, with the epiglottis by three folds of mucous membrane, which form the glosso-epiglottic ligaments, and with the soft palate and pharynx by means of the anterior and posterior pillars of the fauces. Its apex or tip, thin and narrow, is directed forwards against the inner surface of the lower incisor teeth. The under surface of the tongue is connected with the lower jaw by the Genio-hyo-glossi muscles; from its sides, the mucous membrane is reflected to the inner surface of the gums; and, in front, a distinct fold of that membrane, the *frænum linguæ*, is formed beneath its under surface.

The *tip* of the tongue, part of its under surface, its sides, and dorsum, are free.

The *dorsum* of the tongue is convex, marked along the middle line by a raphe, which divides it into two symmetrical halves; and this raphe terminates behind, about half an inch from the base of the organ, a little in front of a deep mucous follicle, the *foramen cæcum*. The anterior two-thirds of this surface are rough, and covered with papillæ; the posterior third is more smooth, and covered by the projecting orifices of numerous muciparous glands.

The *mucous membrane* invests the entire extent of the free surface of the tongue. On the under surface of the organ it is thin and smooth, and may be traced on either side of the frænum, through the ducts of the submaxillary glands; and between the sides of the tongue and the lower jaw, through the ducts of the sublingual glands. As it passes over the borders of the organ, it gradually assumes its papillary character.

The mucous membrane of the tongue consists of structures analogous to those of the skin, namely, a *cutis* or *corium*, supporting numerous *papillæ*, and covered, as well as the papillæ, with *epithelium*.

The cutis is tough, but thinner and less dense than in most parts of the skin, and is composed of similar tissue. It contains the ramifications of the numerous vessels and nerves from which the papillæ are supplied, and affords insertion to all the intrinsic muscular fibres of the organ.

The papillæ of the tongue are thickly distributed over the whole of its upper surface, giving to it its characteristic roughness. They are more prominent than those of the skin, standing out from the surface like the villi of the intestine. The principal varieties are the papillæ minimæ (circumvallatæ), papillæ mediæ (fungiformes), and papillæ minimæ (conicæ or filiformes).

The *papillæ maximæ* (circumvallatæ) are of large size, and vary from eight to ten in number. They are situated at the back part of the dorsum of the tongue, near its base, forming a row on each side, which, running backwards and inwards, meet in the middle line, like the two lines of the letter V inverted. Each papilla consists of a central flattened projection of mucous membrane, circular in form, from $\frac{1}{20}$ to $\frac{1}{12}$ of an inch wide, attached to the bottom of a cup-shaped depression of the mucous membrane; the exposed part being covered with numerous small papillæ. The cup-shaped depression forms a kind of fossa round the papilla, having a circular margin of about the same elevation, covered with smaller papillæ. The fissure corresponding to the papilla, which is situated at the junction of the two lines of the circumvallate papillæ, is so large and deep, that the name *foramen cæcum* has been applied to it. In the smaller papillæ, the fissure exists only on one side.

The *papillæ mediæ* (fungiformes), more numerous than the preceding, are scattered irregularly and sparingly over the dorsum of the tongue; but are found chiefly at its sides and apex. They are easily recognised, among the other papillæ, by their large size, rounded eminences, and deep red colour. They are narrow at their attachment to the tongue, but broad and rounded at their free extremities, and covered with secondary papillæ. Their epithelial investment is very thin.

The *papillæ minimæ* (conicæ—filiformes) cover the anterior two-thirds of the dorsum of the tongue. They are very minute, more or less conical or filiform in shape, and arranged in lines corresponding in direction with the two rows of the papillæ circumvallatæ; excepting at the apex of the organ, where their direction is transverse. The filiform papillæ are of a whitish tint, owing to the thickness and density of their epithelium; they are covered with numerous secondary papillæ, are firmer and more elastic than the papillæ of mucous membrane generally, and often enclose minute hairs.

Simple papillæ, similar to those of the skin, are dispersed very unequally among the compound forms, and exist sparingly on the surface of the tongue behind the circumvallate variety, buried under a layer of epithelium.

Structure of the papillæ. The papillæ apparently resemble in structure those of the cutis, consisting of a cone-shaped projection of homogeneous tissue, covered with a thick layer of squamous epithelium, and contain one or more capillary loops, amongst which nerves are distributed in great abundance. If the epithelium is removed, it will be found that they are not simple processes like the papillæ of the skin, for the surface of each is studded with minute conical processes of the mucous membrane, which form secondary papillæ (Todd and Bowman). In the papillæ circumvallatæ, the nerves are numerous and of large size; in the papillæ fungiformes they are also numerous, and terminate in a plexiform network, from which brush-like branches proceed; in the papillæ filiformes, their mode of termination is uncertain.

Besides the papillæ, the mucous membrane of the tongue is provided with numerous follicles and glands.

The *follicles* are found scattered over its entire surface, but are especially numerous between the papillæ circumvallatæ and the epiglottis.

The *mucous glands* (lingual), similar in structure to the labial and buccal, are found chiefly beneath the mucous membrane of the posterior third of the dorsum of the tongue. There is a small group of these glands beneath the tip of the tongue, a few along the borders of the organ, and some in front of the circumvallate papillæ projecting into the muscular substance. Their ducts open either upon the surface, or into the depressions round the large papillæ.

The *epithelium* is of the scaly variety like that of the epidermis. It covers the free surface of the tongue, as may be easily demonstrated by maceration, or boiling, when it can be detached entire: it is much thinner than in the skin: the intervals between the large papillæ are not filled up by it, but each papilla has a separate investment from root to summit. The deepest cells may sometimes be detached as a separate layer, corresponding to the rete mucosum, but they never contain colouring matter.

The tongue consists of two symmetrical halves, separated from each other, in the middle line, by a fibrous septum. Each half is composed of muscular fibres arranged in various directions, containing much interposed fat, and supplied by vessels and nerves: the entire organ is invested by mucous membrane, and a submucous fibrous stratum. The latter membrane invests the greater part of the surface of the tongue, and into it the muscular fibres are inserted that pass to the surface. It is thicker behind than in front, and is continuous with the sheaths of the muscles attached to it.

The *fibrous septum* consists of a vertical layer of fibrous tissue, extending throughout the entire length of the middle line of the tongue, from the base to the apex. It is thicker behind than in front, and occasionally contains a small fibro-cartilage, about a quarter of an inch in length. It is well displayed by making a vertical section across the organ. Another strong fibrous lamina, termed the *hyo-glossal membrane*, connects the under surface of the base of the tongue to the body of the hyoid bone. This membrane receives, in front, some of the fibres of the Genio-hyo-glossi.

Each half of the tongue consists of extrinsic and intrinsic muscles. The former have been already described; they are the Hyo-glossus, Genio-hyo-glossus, Stylo-glossus, Palato-glossus, and part of the Superior constrictor. The intrinsic muscles are the Superior longitudinal, Inferior longitudinal, and Transverse.

The *superior longitudinal fibres* (*lingualis superficialis*) form a superficial stratum of oblique and longitudinal fibres on the upper surface of the organ, beneath the mucous membrane, and extend from the apex backwards to the hyoid bone, the individual fibres being attached in their course to the submucous and glandular structures.

The *inferior longitudinal fibres* are formed by the Lingualis muscle, already described (p. 222).

The *transverse fibres* are placed between the two preceding layers; they are intermixed with a considerable quantity of adipose substance, and form the chief part of the substance of the organ. They are attached internally to the median fibrous septum; and, passing outwards, the posterior ones taking an arched course, are inserted into the dorsum and margins of the organ, intersecting the other muscular fibres.

The *arteries* of the tongue are derived from the lingual, the facial, and ascending pharyngeal.

The *nerves* of the tongue are three in number in each half: the gustatory branch of the fifth, which is distributed to the papillæ at the fore part and sides of the tongue; the lingual branch of the glosso-pharyngeal, which is distributed to the mucous membrane at the base and side of the tongue, and to the papillæ circumvallatæ; and the hypoglossal nerve, which is distributed to the muscular substance of the tongue. The two former are nerves of common sensation and of taste; the latter is the motor nerve of the tongue.

The Nose.

The Nose is the special organ of the sense of smell: by means of the peculiar properties of its nerves, it protects the lungs from the inhalation of deleterious gases, and assists the organ of taste in discriminating the properties of food.

The organ of smell consists of two parts, one external, the nose; the other internal, the nasal fossæ.

The *nose* is the more anterior and prominent part of the organ of smell. It is of a triangular form, directed vertically downwards, and projects from the centre of the face, immediately above the upper lip. Its summit, or root, is connected directly with the forehead. Its inferior part, the base of the nose, presents two

297, 298.—Cartilages of the Nose.

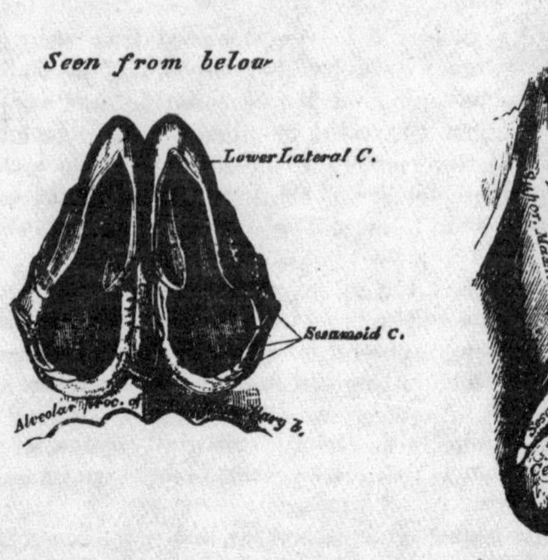

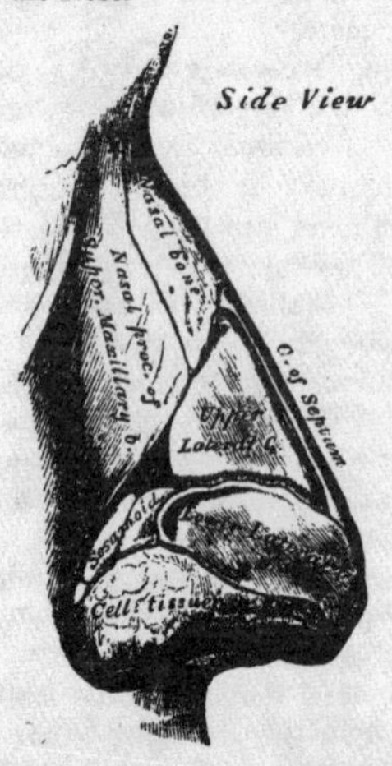

elliptical orifices, the nostrils, separated from each other by an antero-posterior septum, the *columna*. The margins of these orifices are provided with a number of stiff hairs, or *vibrissæ*, which arrest the passage of foreign substances carried with the current of air intended for respiration. The lateral surfaces of the nose form, by their union, the dorsum, the direction of which varies considerably in different individuals. The dorsum terminates below in a rounded eminence, the lobe of the nose.

The nose is composed of a framework of bones and cartilages, the latter being slightly acted upon by certain muscles. It is covered externally by the integument, internally by mucous membrane, and supplied with vessels and nerves.

The *bony framework* occupies the upper part of the organ: it consists of the nasal bones, and the nasal processes of the superior maxillary.

The *cartilaginous framework* consists of five pieces, the two upper, and the two lower lateral cartilages, and the cartilage of the septum.

The *upper lateral cartilages* are situated below the free margin of the nasal bones: each cartilage is flattened, and triangular in shape. Its anterior margin is thicker than the posterior, and connected with the fibro-cartilage of the septum. Its posterior margin is attached to the nasal process of the superior maxillary and nasal bones. Its inferior margin is connected by fibrous tissue with the lower lateral cartilage: one surface is turned outwards, the other inwards towards the nasal cavity.

The *lower lateral cartilages* are two thin, flexible plates, situated immediately below the preceding, and curved in such a manner as to form the inner and outer walls of each orifice of the nostril. The portion which forms the inner wall, thicker than the rest, is loosely connected with the same part of the opposite cartilage, and forms a small part of the columna. Its outer extremity, free, rounded, and projecting, forms, with the thickened integument and subjacent tissue, the lobe of the nose. The part which forms the outer wall is curved to correspond with the ala of the nose: it is oval and flattened, narrow behind, where it is connected with the nasal process of the superior maxilla by a tough fibrous membrane, in which are found three or four small cartilaginous plates (sesamoid cartilages), *cartilagines minores.* Above, it is connected to the upper lateral cartilage and front part of the cartilage of the septum; below, it is separated from the margin of the nostril by dense cellular tissue; and in front, it forms, with its fellow, the prominence of the tip of the nose.

299.—Bones and Cartilages of Septum of Nose. Right Side.

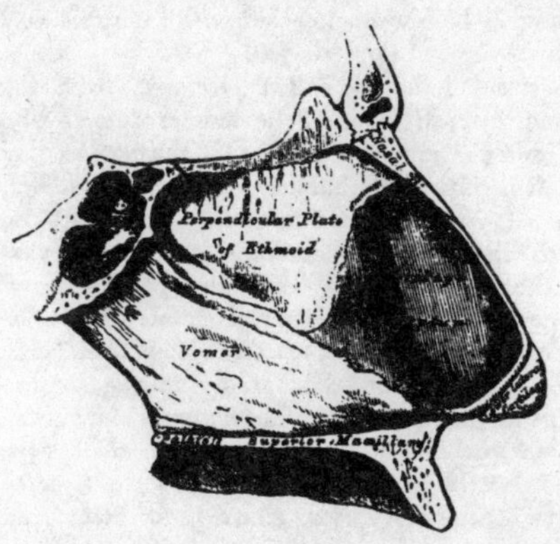

The *cartilage of the septum* is somewhat triangular in form, thicker at its margins than at its centre, and completes the separation between the nasal fossæ in front. Its anterior margin, thickest above, is connected from above downwards with the nasal bones, the front part of the two upper lateral cartilages, and the inner portion of the two lower lateral cartilages. Its posterior margin is connected with the perpendicular lamella of the ethmoid; its inferior margin with the vomer and the palate processes of the superior maxillary bones.

These various cartilages are connected to each other, and to the bones, by a tough fibrous membrane, the perichondrium, which allows the utmost facility of movement between them.

The *muscles of the nose* are situated immediately beneath the integument: they are (on each side) the Pyramidalis nasi, the Levator labii superioris alæque nasi, the Dilatator naris, anterior and posterior, the Compressor nasi, the Compressor narium minor, and the Depressor alæ nasi. They have been described above (p. 207).

The *integument* covering the dorsum and sides of the nose is thin, and loosely connected with the subjacent parts; but where it forms the tip, or lobe, and the alæ of the nose, it is thicker and more firmly adherent. It is furnished with a large number of sebaceous follicles, the orifices of which are usually very distinct.

The *mucous membrane*, lining the interior of the nose, is continuous with the skin externally, and with that which lines the nasal fossæ within.

The *arteries of the nose* are the lateralis nasi, from the facial, and the nasal artery of the septum, from the superior coronary, which supplies the alæ and septum; the sides and dorsum being supplied from the nasal branch of the ophthalmic and the infraorbital.

The *veins of the nose* terminate in the facial and ophthalmic.

The *nerves of the nose* are branches from the facial, infraorbital, and infratrochlear, and a filament from the nasal branch of the ophthalmic.

Nasal Fossæ.

The nasal fossæ are two irregular cavities, situated in the middle of the face and extending from before backwards. They open in front by the two anterior nares, and terminate in the pharynx, behind, by the posterior nares. The boundaries of these cavities, and the openings which are connected with them, as they exist in the skeleton, have been already described (pp. 72–74).

The *mucous membrane* lining the nasal fossæ is called *pituitary*, from the nature of its secretion; or Schneiderian, from Schneider, the first anatomist who showed that the secretion proceeded from the mucous membrane, and not, as was formerly imagined, from the brain. It is intimately adherent to the periosteum, or perichondrium, over which it lies. It is continuous externally with the skin, through the anterior nares, and with the mucous membrane of the pharynx, through the posterior nares. From the nasal fossæ its continuity may be traced with the conjunctiva, through the nasal duct and lachrymal canals; with the lining membrane of the tympanum and mastoid cells, through the Eustachian tube; and with the frontal, ethmoidal, and sphenoidal sinuses, and the antrum maxillare, through the several openings in the meatuses. The mucous membrane is thickest, and most vascular, over the turbinated bones. It is also thick over the septum; but, in the intervals between the spongy bones, and on the floor of the nasal fossæ, it is very thin. Where it lines the various sinuses and the antrum maxillare, it is thin and pale.

The surface of the membrane is covered with a layer of tesselated epithelium, at the upper part of the nasal fossæ, corresponding with the distribution of the olfactory nerve, but is ciliated throughout the rest of its extent, excepting near the aperture of the nares.

This membrane is also provided with a nearly continuous layer of branched mucous glands, the ducts of which open upon its surface. They are most numerous at the middle and back parts of the nasal fossæ, and largest at the lower and back part of the septum.

Owing to the great thickness of this membrane, the nasal fossæ are much narrower, and the turbinated bones, especially the lower ones, appear larger, and more prominent, than in the skeleton. From the same circumstance, also, the various apertures communicating with the meatuses are either narrowed or completely closed.

In the *superior meatus*, the aperture of communication with the posterior ethmoidal cells is considerably diminished in size, and the spheno-palatine foramen completely covered in.

In the *middle meatus*, the opening of the infundibulum is partially hidden by a projecting fold of mucous membrane, and the orifice of the antrum is contracted to a small circular aperture, much narrower than in the skeleton.

In the *inferior meatus*, the orifice of the nasal duct is partially hidden by either a single or double valvular mucous fold, and the anterior palatine canal either completely closed in, or a tubular *cul-de-sac* of mucous membrane is continued a short distance into it.

In *the roof*, the opening leading to the sphenoidal sinus is narrowed, and the apertures in the cribriform plate of the ethmoid completely closed in.

The *arteries of the nasal fossæ* are the anterior and posterior ethmoidal, from the ophthalmic, which supply the ethmoidal cells, frontal sinuses, and roof of the nose; the spheno-palatine, from the internal maxillary, which supplies the mucous membrane covering the spongy bones, the meatuses, and septum; and the alveolar branch of the internal maxillary, which supplies the lining membrane of the antrum. The ramifications of these vessels form a close, plexiform network, beneath and in the substance of the mucous membrane.

The *veins of the nasal fossæ* form a close network beneath the mucous membrane. They pass, some with the veins accompanying the spheno-palatine artery, through the spheno-palatine foramen; and others, through the alveolar branch, join the facial vein; some accompany the ethmoidal arteries, and terminate in the ophthalmic vein; and, lastly, a few communicate with the veins in the interior of the skull, through the foramina in the cribriform plate of the ethmoid bone, and the foramen cæcum.

300.—Nerves of Septum of Nose. Right Side.

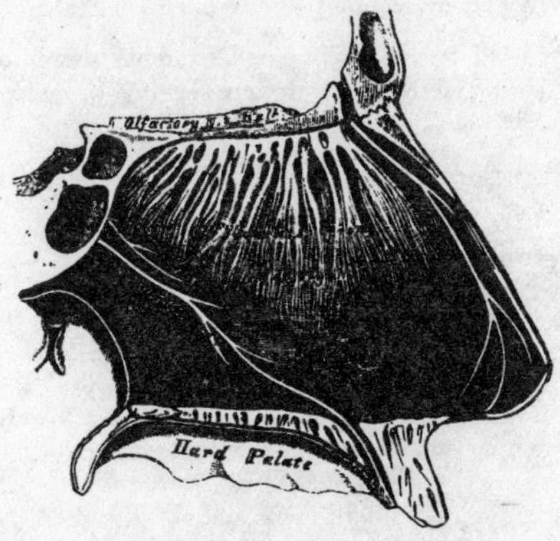

The *nerves* are, the olfactory, the nasal branch of the ophthalmic, filaments from the anterior dental branch of the superior maxillary, the Vidian, naso-palatine, descending anterior palatine, and spheno-palatine branches of Meckel's ganglion.

The *olfactory*, the special nerve of the sense of smell, is distributed over the upper third of the septum, and over the surface of the superior and middle spongy bones.

The *nasal branch of the ophthalmic* distributes filaments to the upper and anterior part of the septum, and outer wall of the nasal fossæ.

Filaments from the anterior dental branch of the superior maxillary supply the inferior meatus and inferior turbinated bone.

The *Vidian nerve* supplies the upper and back part of the septum, and superior spongy bone; and the upper anterior nasal branches from the spheno-palatine ganglion, have a similar distribution.

The *naso-palatine nerve* supplies the middle of the septum.

The *larger*, or *anterior palatine nerve*, supplies the middle and lower spongy bones.

The Eye.

The eyeball is contained in the cavity of the orbit. In this situation it is securely protected from injury, whilst its position is such as to ensure the most extensive range of sight. It is acted upon by numerous muscles, by which it is capable of being directed to any part, supplied by vessels and nerves, and is additionally protected in front by several appendages, such as the eyebrow, eyelids, &c.

The eyeball is spherical in form, having the segments of a smaller and more prominent sphere engrafted upon its anterior part. It is from this circumstance that the antero-posterior diameter of the eyeball, which measures about an inch, exceeds the transverse diameter by about a line. The segment of the larger sphere, which forms about five-sixths of the globe, is opaque, and formed by the sclerotic, the tunic of protection to the eyeball; the smaller sphere, which forms the remaining sixth, is transparent, and formed by the cornea. The axes of the eyeballs are nearly parallel, and do not correspond to the axes of the orbits, which are directed outwards. The optic nerves follow the direction of the axes of the orbits, and enter the eyeball a little to their inner or nasal side. The eyeball is composed of several investing tunics, and of fluid and solid refracting media, called *humours*.

The tunics are three in number:—

1. Sclerotic and Cornea.
2. Choroid, Iris, and Ciliary Processes.
3. Retina.

The refracting media, or humours, are also three:

Aqueous. Crystalline (lens) and Capsule. Vitreous.

The sclerotic and cornea form the external tunic of the eyeball; they are essentially fibrous in structure, the sclerotic being opaque, and forming the posterior five-sixths of the globe; the cornea, which forms the remaining sixth, being transparent.

The *Sclerotic* (σκληρός, *hard*) (fig. 301) has received its name from its extreme density and hardness; it is a firm, unyielding, fibrous membrane, serving to maintain the form of the globe. It is much thicker behind than in front. Its *external surface* is of a white colour, quite smooth, except at the points where the Recti and Obliqui muscles are inserted into it, and covered, for part of its extent, by the conjunctival membrane; hence the whiteness and brilliancy of the front of the eyeball. Its *inner surface* is stained of a brown colour, marked by grooves, in which are lodged the ciliary nerves, and connected by an exceedingly fine cellular tissue (*lamina fusca*) with the outer surface of the choroid. Behind, it is pierced by the optic nerve a little to its inner or nasal side, and is continuous with

301.—A Vertical Section of the Eyeball. (Enlarged.)

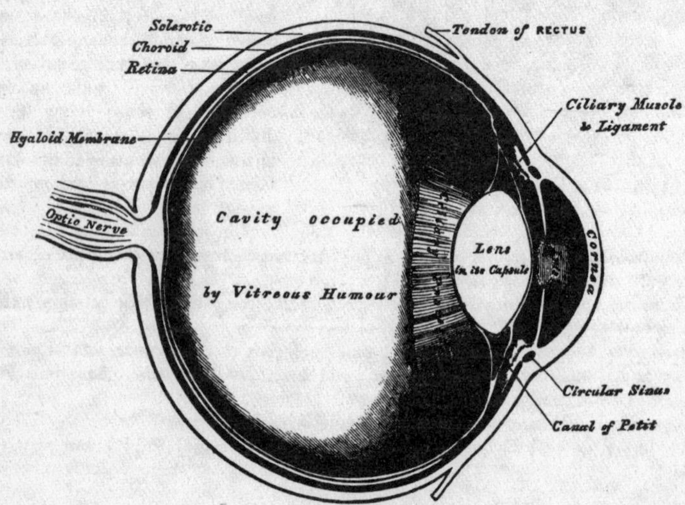

the fibrous sheath of the nerve, which is derived from the dura mater. At the point where the optic nerve passes through the sclerotic, this membrane forms a thin cribriform lamina (the *lamina cribrosa*); the minute orifices in this layer serve for the transmission of the nervous filaments, and the fibrous septa dividing them from one another are continuous with the membranous processes which separate the bundles of nerve-fibres. One of these openings, larger than the rest, occupies the centre of the lamella; it is called the *porus opticus*, and transmits the arteria centralis retinæ to the interior of the eyeball. Around the cribriform lamella are numerous smaller apertures for the transmission of the ciliary vessels and nerves. In front, the sclerotic is continuous with the cornea by direct continuity of tissue; but the opaque sclerotic overlaps the cornea rather more on its outer than upon its inner surface.

Structure. The sclerotic is formed of white fibrous tissue intermixed with fine elastic fibres, and fusiform nucleated cells. These are aggregated into bundles, which are arranged chiefly in a longitudinal direction. It yields gelatin on boiling. Its vessels are not numerous, the capillaries being of small size, uniting at long and wide intervals. The existence of nerves in it is doubtful.

The *Cornea* is the projecting transparent part of the external tunic of the eyeball, and forms the anterior sixth of the globe. It is not quite circular, being a little broader in the transverse than in the vertical direction, in consequence of the sclerotic overlapping the margin above and below. It is convex anteriorly, and projects forwards from the sclerotic in the same manner that a watch-glass does from its case. Its degree of curvature varies in different individuals, and in the same individual at different periods of life, it being more prominent in youth than in advanced life, when it becomes flattened. The cornea is dense and of uniform thickness throughout; its posterior surface is perfectly circular in outline, and exceeds the anterior surface slightly in extent, from the latter being overlapped by the sclerotic.

Structure. The cornea consists of five layers: namely, of a thick central fibrous structure, the cornea proper; in front of this is the anterior elastic lamina, covered by the conjunctiva; behind, the posterior elastic lamina, covered by the lining membrane of the anterior chamber of the eyeball.

The *proper substance* of the cornea is fibrous, tough, unyielding, perfectly transparent, and continuous with the sclerotic, with which it is in structure identical. The anastomosing fusiform cells of which it is composed are arranged in superimposed flattened laminæ, at least sixty in number, all of which have the same direction, the contiguous laminæ becoming united at frequent intervals. If the relative position of the component parts of this tissue is in any way altered, either by pressure or by an increase of its natural tension, it immediately presents an opaque milky appearance. The interstices between the laminæ are tubular, and usually contain a small amount of transparent fluid.

The *anterior* and *posterior elastic laminæ*, which cover the proper structure of the cornea behind and in front, present an analogous structure. They consist of a hard, elastic, and perfectly transparent homogeneous membrane, of extreme thinness, which is not rendered opaque by either water, alcohol, or acids. This membrane is intimately connected by means of a fine cellular web to the proper substance of the cornea both in front and behind. Its most remarkable property is its extreme elasticity, and the tendency which it presents to curl up, or roll upon itself, with the attached surface innermost, when separated from the proper substance of the cornea. Its use appears to be (as suggested by Dr. Jacob), 'to preserve the requisite permanent correct curvature of the flaccid cornea proper.'

The *conjunctival epithelium*, which covers the front of the anterior elastic lamina, consists of two or three layers of transparent nucleated cells, the deepest being of an oblong form and placed perpendicular to the surface, the superficial ones more flattened.

The *epithelial lining* of the aqueous chamber covers the posterior surface of the posterior elastic lamina. It consists of a single layer of polygonal transparent nucleated cells, similar to those found lining other serous cavities.

Arteries and Nerves. The cornea is a non-vascular structure, the capillary vessels terminating in loops at its circumference. Lymphatic vessels have not as yet been demonstrated in it. The nerves are numerous, twenty or thirty in number: they are derived from the ciliary nerves, and enter the laminated substance of the cornea. They ramify throughout its substance in a delicate network.

Dissection. In order to separate the sclerotic and cornea, so as to expose the second tunic, the eyeball should be immersed in a small vessel of water. A fold of the sclerotic near its anterior part having been pinched up, an operation not easily performed, from the extreme tension of the membrane, it should be divided with a pair of blunt-pointed scissors. As soon as the choroid is exposed, the end of a blowpipe should be introduced into the orifice, and a stream of air forced into it, so as to separate the slight cellular connection between the sclerotic and choroid. The sclerotic should now be divided around its entire circumference, and may be removed in separate portions. The front segment being then drawn forwards, the handle of the scalpel should be pressed gently against it at its connection with the iris, and these being separated, a quantity of perfectly transparent fluid will escape; this is the aqueous humour. In the course of the dissection, the ciliary nerves may be seen lying in the loose cellular tissue between the choroid and sclerotic, or contained in delicate grooves on the inner surface of the latter membrane.

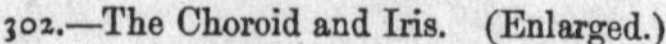

302.—The Choroid and Iris. (Enlarged.)

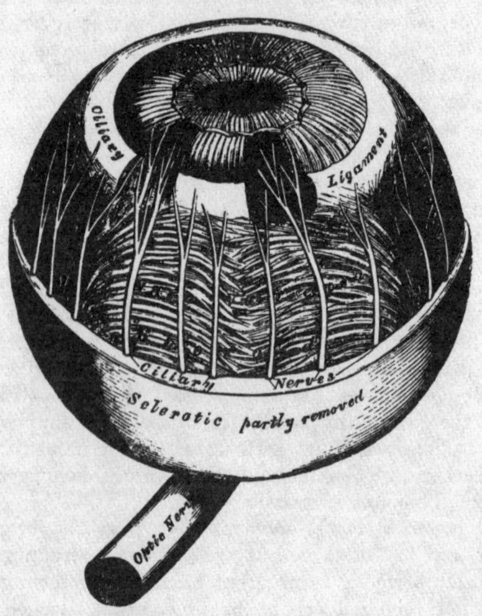

Second Tunic. This is formed by the choroid behind; the iris and ciliary processes in front; and by the ciliary ligament, and Ciliary muscle, at the point of junction of the sclerotic and cornea.

The choroid is the vascular and pigmentary tunic of the eyeball, investing the posterior five-sixths of the globe, and extending as far forwards as the cornea; the ciliary processes being appendages of the choroid developed from its inner surface in front. The iris is the circular muscular septum, which hangs vertically behind the cornea, presenting in its centre a large circular aperture, the pupil. The ciliary ligament and Ciliary muscle form the white ring observed at the point where the choroid and iris join with each other, and with the sclerotic and cornea.

The *Choroid* is a thin, highly vascular membrane, of a dark brown or chocolate colour, which invests the posterior five-sixths of the central part of the globe. It is pierced behind by the optic nerve, and terminates in front at the ciliary ligament, where it bends inwards, and forms on its inner surface a series of folds or plaitings, the ciliary processes. It is thicker behind than in front. Externally, it is connected by a fine cellular web (*membrana fusca*) with the inner surface of the sclerotic. Its inner surface is smooth, and lies in contact with the retina. The choroid is composed of three layers, external, middle, and internal.

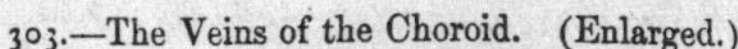

303.—The Veins of the Choroid. (Enlarged.)

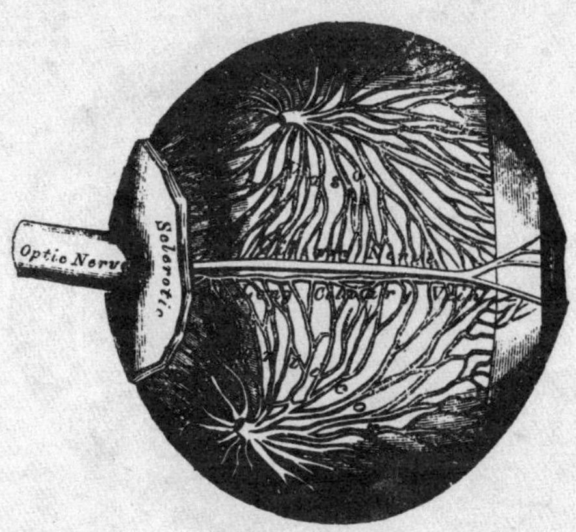

The *external layer* consists, in part, of the larger branches ot the short ciliary arteries, which run forwards between the veins before they bend downwards to terminate on the inner surface. This coat is formed, however, principally of veins, which are named, from their distribution, *venæ vorticosæ*. They converge to four or five equidistant trunks, which pierce the sclerotic midway between the margin of the cornea and the entrance of the optic nerve. Interspersed between the vessels are lodged dark star-shaped pigment-cells, the fibrous offsets from which, communicating with similar branchings from neighbouring cells, form a delicate network, which, towards the inner surface of the choroid, loses its pigmentary character.

The *middle layer* consists of an exceedingly fine capillary plexus, formed by the short ciliary vessels, and is known as the *tunica Ruychiana*. The network is close, and finer at the hinder part of the choroid than in front. About half an inch behind the cornea, its meshes become larger, and are continuous with those of the ciliary process.

The *internal*, or *pigmentary layer*, is a delicate membrane, consisting of a single layer of hexagonal nucleated cells, loaded with pigment-granules, and applied to each other, so as to resemble a tesselated pavement. Each cell contains a nucleus, and is filled with grains of pigment, which are in greater abundance at the circumference of the cell. In perfect albinos this epithelium contains no pigment, and none is present in the star-shaped cells found in the other layers of the choroid.

The ciliary processes should be next examined: they may be exposed, either by detaching the iris from its connection with the ciliary ligament, or by making a transverse section of the globe, and examining them from behind.

The *Ciliary processes* are formed by the plaiting or folding inwards of the middle and internal layers of the choroid, at its anterior margin, and are received between corresponding foldings of the suspensory ligament of the lens, thus establishing a communication between the choroid and inner tunic of the eye. They are arranged in a circle, behind the iris, round the margin of the lens. They vary between sixty and eighty in number, lie side by side, and may be divided into large and small; the latter, consisting of about one-third of the entire number, are situated in the spaces between the former, but without regular alternation. The larger processes are each about one-tenth of an inch in length, and

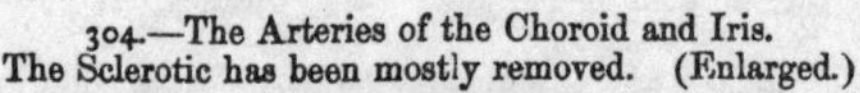

304.—The Arteries of the Choroid and Iris.
The Sclerotic has been mostly removed. (Enlarged.)

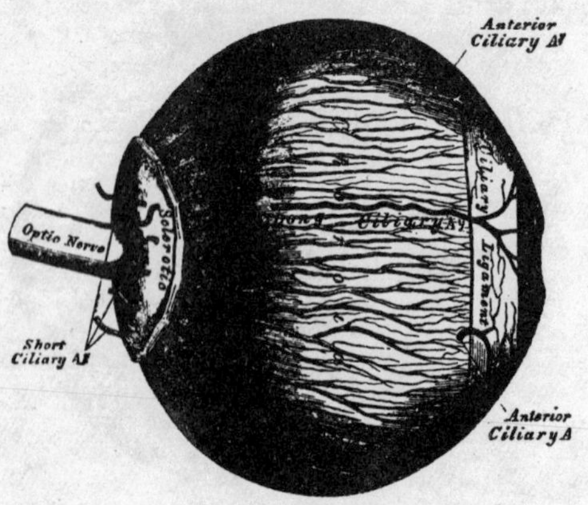

hemispherical in shape, their periphery being attached to the ciliary ligament, and continuous with the middle and inner layers of the choroid: the opposite margin is free, and rests upon the circumference of the lens. Their anterior surface is turned towards the back of the iris, with the circumference of which it is continuous. The posterior surface is closely connected with the suspensory ligament of the lens.

Structure. The ciliary processes are similar in structure to the choroid: the vessels are larger, having chiefly a longitudinal direction. Externally they are covered with several layers of pigment-cells; the component cells are small, rounded, and full of pigment-granules.

The *Iris* (*iris*, a rainbow) has received its name from its various colour in different individuals. It is a thin, circular-shaped, contractile curtain, suspended in the aqueous humour behind the cornea, and in front of the lens, being perforated at the nasal side of its centre by a circular aperture, the pupil, for the transmission of light. By its circumference it is intimately connected with the choroid; externally to this is the ciliary ligament, by which it is connected to the sclerotic and cornea; its inner edge forms the margin of the pupil; its surfaces are flattened, and look forwards and backwards, the anterior surface towards the cornea, the posterior towards the ciliary processes and lens. The anterior surface is variously coloured in different individuals, and marked by lines which converge towards the pupil. The posterior surface is of a deep purple tint, from being covered by dark pigment; it is hence named *uvea*, from its resemblance in colour to a ripe grape.

Structure. The iris is composed of a fibrous stroma, muscular fibres, and pigment-cells.

The *fibrous stroma* consists of fine, delicate bundles of fibrous tissue, which have a circular direction at the circumference; but the chief mass radiate towards the pupil. They form, by their interlacement, a delicate mesh, in which the pigment-cells, vessels, and nerves are contained.

The *muscular fibre* is involuntary, and consists of circular and radiating fibres. The *circular fibres* (sphincter of the pupil) surround the margin of the pupil on the posterior surface of the iris, like a sphincter, forming a narrow band, about one-thirtieth of an inch in width; those near the free margin being closely aggregated; those more external somewhat separated, and forming less complete circles. The *radiating fibres* (dilator of the pupil) converge from the circumference towards the centre, and blend with the circular fibres near the margin of the pupil.

The *pigment-cells* are found in the stroma of the iris, and also as a distinct layer on its anterior and posterior surfaces. In the stroma, the cells are ramified, and contain yellow or brown pigment, according to the colour of the eye. On the front of the iris, there is a single layer of oval or rounded cells, with branching offsets. On the back of the iris, there are several layers of small round cells, filled with dark pigment. This layer is continuous with the pigmentary covering of the ciliary processes.

The *arteries of the iris* are derived from the long and anterior ciliary, and from the vessels of the ciliary processes.

Membrana pupillaris. In *the fœtus*, the pupil is closed by a delicate, transparent, vascular membrane, the membrana pupillaris, which divides the space in which the iris is suspended into two distinct chambers. This membrane contains numerous minute vessels continued from the margin of the iris to those on the front part of the capsule of the lens. These vessels have a looped arrangement, converging towards each other without anastomosing. Between the seventh and eighth month, the membrane begins to disappear, by its gradual absorption from the centre towards the circumference, and at birth only a few fragments remain. It is said sometimes to remain permanent, and produce blindness.

The *Ciliary ligament* is a narrow ring of circular fibres, about one-fortieth of an inch thick, and of a whitish colour, which serves to connect the external and middle tunics of the eye. It is placed around the circumference of the iris, at its point of connection with the external layer of the choroid, the cornea, and sclerotic. Its component fibres are delicate, and resemble those of elastic tissue. At its point of connection with the sclerotic a minute canal is situated between the two, called the *sinus circularis iridis*.

The *Ciliary muscle* (Bowman) consists of unstriped fibres: it forms a greyish, semitransparent, circular band, about one-eighth of an inch broad, on the outer surface of the fore part of the choroid. It is thickest in front, and gradually becomes thinner behind. Its fibres are soft, of a yellowish-white colour, longitudinal in direction, and arise at the point of junction of the cornea and sclerotic. Passing backwards, they are attached to the choroid, in front of the retina, and correspond by their inner surface to the plicated part of the former membrane. Mr. Bowman supposes that this muscle is so placed as to advance the lens, by exercising compression on the vitreous body, and by drawing the ciliary processes towards the line of junction of the sclerotic and cornea, and by this means to adjust the eye to the vision of near objects.

The *Retina* may be exposed by carefully removing the choroid from its external surface. It is a delicate nervous membrane, upon the surface of which the images of external objects are received. Its outer surface is in contact with the pigmentary layer of the choroid; its inner surface, with the vitreous body. Behind, it is continuous with the optic nerve; it gradually diminishes in thickness from behind forwards; and, in front, extends nearly as far forwards as the ciliary ligament, where it terminates by a jagged margin, the *ora serrata*. It is soft, and semi-transparent, in the fresh state; but soon becomes clouded, opaque, and of a pinkish tint. Exactly in the centre of the posterior part of the retina, and at a point corresponding to the axis of the eye, in which the sense of vision is most perfect, is a round, elevated, yellowish spot, called, after its discoverer, the *yellow spot* or *limbus luteus*, of Sömmerring; having a central depression at its summit, the *fovea centralis*. The retina in the situation of the fovea centralis is exceedingly thin; so much so, that the dark colour of the choroid is distinctly seen through it; so that it presents more the appearance of a foramen, and hence the name 'foramen of Sömmerring' at first given to it. It exists only in man, the quadrumana, and some saurian reptiles. Its use is unknown. About $\frac{1}{10}$ of an inch to the inner side of the yellow spot, is the point of entrance of the optic nerve; the arteria centralis retinæ piercing its centre. This is the only part of the surface of the retina from which the power of vision is absent.

305.—The Arteria Centralis Retinæ, Yellow Spot, etc., the Anterior Half of the Eyeball being removed. (Enlarged.)

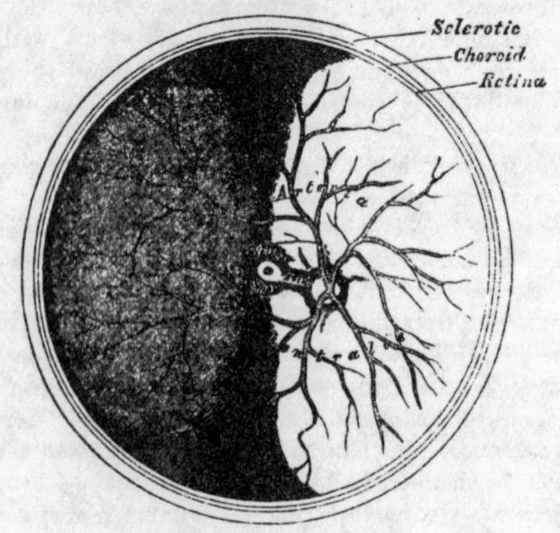

Structure. The retina is composed of three layers, together with blood-vessels:

External or columnar layer (Jacob's membrane).
Middle or granular layer.
Internal or nervous layer.

The blood-vessels do not form a distinct layer; they ramify in the substance of the internal layer.

The *external*, or *Jacob's membrane*, is exceedingly thin, and can be detached from the external surface of the retina by the handle of the scalpel, in the form of a flocculent film. It is thicker behind than in front, and consists of rod-like bodies of two kinds: 1. *Columnar rods*, solid, nearly of uniform size, and arranged perpendicularly to the surface. 2. *Bulbous particles*, or *cones*, which are interspersed at regular intervals among the former; these are conical or flask-shaped, their broad ends resting upon the granular layer, the narrow pointed extremity being turned towards the choroid; they are not solid, like the columnar rods, but consist of an external membrane with fluid contents. By their deep ends, both kinds are joined to the fibres of Müller.

The *middle* or *granular layer* forms about one-third of the entire thickness of the retina. It consists of two laminæ of rounded or oval nuclear particles, separated from each other by an intermediate layer, which is transparent, finely fibrillated, and contains no blood-vessels. The outermost layer is the thicker, and its constituent particles are globular. The innermost layer is the thinner; its component particles are flattened, looking like pieces of money seen edgeways; hence it has been called, by Bowman, the *nummular layer*.

The *internal* or *nervous layer* is a thin semitransparent membrane, consisting of an expansion of the terminal fibres of the optic nerve and nerve-cells. The nerve-fibres are collected into bundles, which radiate from the point at which the trunk of the optic nerve terminates. As they proceed in a tolerably straight course towards the anterior margin of the retina, the bundles interlace, forming a delicate net, with flattened elongated meshes. The nerve-fibres which form this layer, differ from the fibres of the optic nerve in this respect; they lose their dark outline, and their tendency to become varicose; and consist only of the central part, or axis, of the nerve-tubes. The mode of termination of the nerve-fibres is unknown. According to some observers, they terminate in loops; according to others, in free extremities. Recent observers have stated that some of the nerve-fibres are continuous with the caudate prolongations of the nerve-cells external to the fibrous layer. The nerve-cells are placed on both sides of the fibrous layer, but chiefly upon its inner surface, and embedded within the meshes formed by the interlacing nerve-fibres; they are round or pear-shaped transparent cells, nucleated, with granular contents, furnished with caudate prolongations, some of which join the fibres of the optic nerve, whilst others are directed externally towards the granular layer. It is probable that these cells are identical with the ganglion-corpuscles of vesicular nervous substance.

An extremely thin and delicate structureless membrane lines the inner surface of the retina, and separates it from the vitreous body; it is called the *membrana limitans.*

The *radiating fibres* of the retina, described by Heinrich Müller, consist of extremely fine fibrillated threads, which are connected externally with each of the rods of the columnar layer, of which they appear to be direct continuations, and, passing through the entire substance of the retina, are united to the outer surface of the membrana limitans. In their course through the retina, they become connected with the nuclear particles of the granular layer, and give off branching processes opposite its innermost lamina; as they approach the fibrous expansion of the optic nerve, they are collected into bundles, which pass through the areolæ between its fibres, and are finally attached to the inner surface of the membrana limitans, where each fibre terminates in a triangular enlargement.

The *arteria centralis retinæ* and its accompanying vein pierce the optic nerve, and enter the globe of the eye through the porus opticus. It immediately divides into four or five branches, which at first run between the hyaloid membrane and the nervous layer; but they soon enter the latter membrane, and form a close capillary network in its substance. At the ora serrata, they terminate in a single vessel which bounds the terminal margin of the retina.

The structure of the retina at the yellow spot, presents some modifications. Jacob's membrane is thinner, and of its constituents only the cones are present; but they are small, and more closely aggregated than in any other part. The granular layer is absent over the fovea centralis. Of the two elements of the nervous layer, the nerve-fibres extend only to the circumference of the spot; but the nerve-cells cover its entire surface. The radiating fibres are found at the circumference, and here only extend to the inner strata of the granular layer. Of the capillary vessels, the larger branches pass round the spot; but the smaller capillaries meander through it. The colour of the spot appears to imbue all the layers except Jacob's membrane; it is of a rich yellow, deepest towards the centre, and does not appear to consist of pigment-cells, but resembles more a staining of the constituent parts.

Humours of the Eye.

The *aqueous humour* completely fills the anterior and posterior chambers of the eyeball. It is small in quantity (scarcely exceeding, according to Petit, four or five grains in weight), has an alkaline reaction, in composition is little more than water, less than one-fiftieth of its weight being solid matter, chiefly chloride of sodium.

The *anterior chamber* is the space bounded in front by the cornea; behind, by the front of the iris and ciliary ligament.

The *posterior chamber*, smaller than the anterior, is bounded in front by the iris; behind, by the capsule of the lens and its suspensory ligament, and the ciliary processes.

In the adult, these two chambers communicate through the pupil; but in the fœtus before the seventh month, when the pupil is closed by the membrana pupillaris, the two chambers are quite separate.

It has been generally supposed that the two chambers are lined by a distinct membrane, the secreting membrane of the aqueous humour, analogous in structure to that of a serous sac. An epithelial covering can, however, only be found on the posterior surface of the cornea. That the two chambers do, however, secrete this fluid separately, is shown from its being found in both spaces before the removal of the membrana pupillaris. It is probable that the parts concerned in the secretion of the fluid, are the posterior surface of the cornea, both surfaces of the iris, and the ciliary processes.

Vitreous Body.

The *vitreous body* forms about four-fifths of the entire globe. It fills the concavity of the retina, and is hollowed in front for the reception of the lens and its capsule. It is perfectly transparent, of the consistence of thin jelly, and consists of an albuminous fluid enclosed in a delicate transparent membrane, the *hyaloid*. This membrane invests the outer surface of the vitreous body; it is intimately connected in front with the suspensory ligament of the lens; and is continued into the back part of the capsule of the lens. It has been supposed, by Hannover, that from its inner surface numerous thin lamellæ are prolonged inwards in a radiating manner, forming spaces in which the fluid is contained. In the adult, these lamellæ cannot be detected even after careful microscopic examination; but in the fœtus a peculiar fibrous texture pervades the mass, the fibres joining at numerous points, and presenting minute nuclear granules at their point of junction. The fluid from the vitreous body resembles nearly pure water; it contains, however, some salts, and a little albumen.

In the *fœtus*, the centre of the vitreous humour presents a tubular canal, through which a minute artery passes along the vitreous body to the capsule of the lens. In the *adult*, no vessels penetrate its substance; so that its nutrition must be carried on by the vessels of the retina and ciliary processes, situated upon its exterior.

Crystalline Lens and its Capsule.

The crystalline lens, enclosed in its capsule, is situated immediately behind the pupil, in front of the vitreous body, and surrounded by the ciliary processes, which slightly overlap its margin.

The *capsule of the lens* is a transparent, highly elastic, and brittle membrane, which closely surrounds the lens. It rests, behind, in a depression in the fore part of the vitreous body: in front, it forms part of the posterior chamber of the eye; and it is retained in its position chiefly by the suspensory ligament of the lens. The capsule is much thicker in front than behind, structureless in texture; and when ruptured, the edges roll up with the outer surface innermost, like the elastic laminæ of the cornea. The lens is connected to the inner surface of the capsule by a single layer of transparent, polygonal, nucleated cells. These, after death, absorb moisture from the fluids of the eye; and, breaking down, form the *liquor Morgagni.*

In the fœtus, a small branch from the arteria centralis retinæ runs forwards, as already mentioned, through the vitreous humour to the posterior part of the capsule of the lens, where its branches radiate and form a plexiform network, which covers its surface, and are continuous round the margin of the capsule with the vessels of the pupillary membrane, and with those of the iris. In the adult no vessels enter its substance.

The *lens* is a transparent, double-convex body, the convexity being greater on the posterior, than on the anterior, surface. It measures about a third of an inch in the transverse diameter, and about one-fourth in the antero-posterior. It consists of concentric layers, of which the external, in the fresh state, are soft and easily detached; those beneath are firmer, the central ones forming a hardened nucleus. These laminæ are best demonstrated by boiling, or immersion in alcohol. The same re-agents demonstrate that the lens consists of three triangular segments, the sharp edges of which are directed towards the centre, the bases towards the circumference. The laminæ consist of minute parallel fibres, which are united to each other by means of wavy margins, the convexities upon one fibre fitting accurately into the concavities of the adjoining fibre.

306.—The Crystalline Lens, hardened and divided. (Enlarged.)

The *changes produced in the lens by age*, are the following:—

In *the fœtus*, its form is nearly spherical, its colour of a slightly reddish tint, it is not perfectly transparent, and is so soft as to break down readily on the slightest pressure.

In *the adult*, the posterior surface is more convex than the anterior; it is colourless, transparent, and firm in texture.

In *old age*, it becomes flattened on both surfaces, slightly opaque, of an amber tint, and increases in density.

The *suspensory ligament of the lens* is a thin, transparent membranous structure, placed between the vitreous body and the ciliary processes of the choroid; it connects the anterior margin of the retina with the anterior surface of the lens near its circumference. It assists in retaining the lens in its position. Its outer surface presents a number of folds or plaitings, in which the corresponding folds of the ciliary processes are received. These plaitings are arranged round the lens in a radiating form, and are stained by the pigment of the ciliary processes. The suspensory ligament consists of two layers, which commence behind, at the ora serrata. The external, a tough, milky, granular membrane, covers the inner surface of the ciliary processes, and extends as far forwards as their anterior free extremities. The inner layer, an elastic, transparent, fibro-membranous structure, extends as far forwards as the anterior surface of the capsule of the lens, near its circumference. That portion of this membrane which intervenes between the ciliary processes and the capsule of the lens, forms part of the boundary of the posterior chamber of the eye. The posterior surface of this layer is turned towards the hyaloid membrane, being separated from it at the circumference of the lens by a space called the canal of Petit.

The *canal of Petit* is about one-tenth of an inch wide. It is bounded in front by the suspensory ligament; behind, by the hyaloid membrane, its base being formed by the capsule of the lens. When inflated with air, it is sacculated at intervals, owing to the foldings on its anterior surface.

The *Vessels of the globe of the eye* are the short, long, and anterior ciliary arteries, and the arteria centralis retinæ.

The *short ciliary arteries* pierce the back part of the sclerotic, round the entrance of the optic nerve, and divide into branches which run parallel with the axis of the eyeball: they are distributed to the middle layer of the choroid, and to the ciliary processes.

The *long ciliary arteries*, two in number, pierce the back part of the sclerotic, and run forward, between that membrane and the choroid, to the Ciliary muscle, where they each divide into an upper and lower branch; these anastomose, and form a vascular circle round the outer circumference of the iris; from this circle branches are given off, which unite, near the margin of the pupil, in a smaller vascular circle. These branches, in their course, supply the muscular structure.

The *anterior ciliary arteries*, five or six in number, are branches of the muscular and lachrymal branches of the ophthalmic. They pierce the eyeball, at the anterior part of the sclerotic, immediately behind the margin of the cornea, and are distributed to the ciliary processes, some branches joining the greater vascular circle of the iris.

The *arteria centralis retinæ* has been already described.

The *veins*, usually four in number, are formed mainly by branches from the surface of the choroid. They perforate the sclerotic, midway between the cornea and the optic nerve, and end in the ophthalmic vein.

The *nerves of the eyeball*, are the optic, the long ciliary nerves from the nasal branch of the ophthalmic, and the short ciliary nerves from the ciliary ganglion.

Appendages of the Eye.

The appendages of the eye (*tutamina oculi*), include the eyebrows, the eyelids, the conjunctiva, and the lachrymal apparatus, viz., the lachrymal gland, the lachrymal sac, and the nasal duct.

The *eyebrows* (*supercilia*) are two arched eminences of integument, which surmount the upper circumference of the orbit on each side, and support numerous short, thick hairs, directed obliquely on the surface. In structure, the eyebrows consist of thickened integument, connected beneath with the Orbicularis palpebrarum, Corrugator supercilii, and Occipito-frontalis muscles. These muscles serve, by their action on this part, to control to a certain extent the amount of light admitted into the eye.

The *eyelids* (*palpebræ*) are two thin, moveable folds, placed in front of the eye, protecting it from injury by their closure. The upper lid is the larger, and the more moveable of the two, and is furnished with a separate elevator muscle, the *Levator palpebræ superioris*. When the eyelids are opened, an elliptical space (*fissura palpebrarum*) is left between their margins, the angles of which correspond to the junction of the upper and lower lids, and are called *canthi*.

The *outer canthus* is more acute than the inner, and the lids here lie in close contact with the globe: but the *inner canthus* is prolonged for a short distance inwards, towards the nose, and the two lids are separated by a triangular space, the *lacus lachrymalis*. At the commencement of the lacus lachrymalis, on the margin of each eyelid, is a small conical elevation, the *lachrymal papilla*, or tubercle, the apex of which is pierced by a small orifice, the *punctum lachrymale*, the commencement of the lachrymal canal.

Structure of the eyelids. The eyelids are composed of the following structures, taken in their order from without inwards:

Integument, areolar tissue, fibres of the Orbicularis muscle, tarsal cartilage, fibrous membrane, Meibomian glands, and conjunctiva. The upper lid has, in addition, the aponeurosis of the Levator palpebræ.

The *integument* is extremely thin, and continuous at the margin of the lids with the conjunctiva.

The *subcutaneous areolar tissue* is very lax and delicate, seldom contains any fat, and is extremely liable to serous infiltration.

The *fibres of the Orbicularis muscle*, where they cover the palpebræ, are thin, pale in colour, and possess an involuntary action.

The *tarsal cartilages* are two thin, elongated plates of fibro-cartilage, about an inch in length. They are placed one in each lid, contributing to their form and support.

The *superior*, the larger, is of a semilunar form, about one-third of an inch in breadth at the centre, and becoming gradually narrowed at each extremity. Into the fore part of this cartilage the aponeurosis of the Levator palpebræ is attached.

The *inferior* tarsal cartilage, the smaller, is thinner, and of an elliptical form.

The *free*, or *ciliary margin* of the cartilages is thick, and presents a perfectly straight edge. The *attached*, or *orbital margin*, is connected to the circumference of the orbit by the fibrous membrane of the lids. The outer angle of each cartilage is attached to the malar bone by the external palpebral or tarsal ligament. The inner angles of the two cartilages terminate at the commencement of the lacus lachrymalis, being fixed to the margins of the orbit by the tendo oculi.

The *fibrous membrane of the lids*, or tarsal ligament, is a layer of fibrous membrane, beneath the Orbicularis, attached, externally, to the margin of the orbit, and internally to the orbital margin of the lids. It is thick and dense at the outer part of the orbit, but becomes thinner as it approaches the cartilages. This membrane serves to support the eyelids, and retains the tarsal cartilages in their position.

The *Meibomian glands* (fig. 307) are situated upon the inner surface of the eyelids, between the tarsal cartilages and conjunctiva, and may be distinctly seen through the mucous membrane on everting the eyelids, presenting the appearance of parallel strings of pearls. They are about thirty in number in the upper car-

307.—The Meibomian Glands, etc, seen from the Inner Surface of the Eyelids.

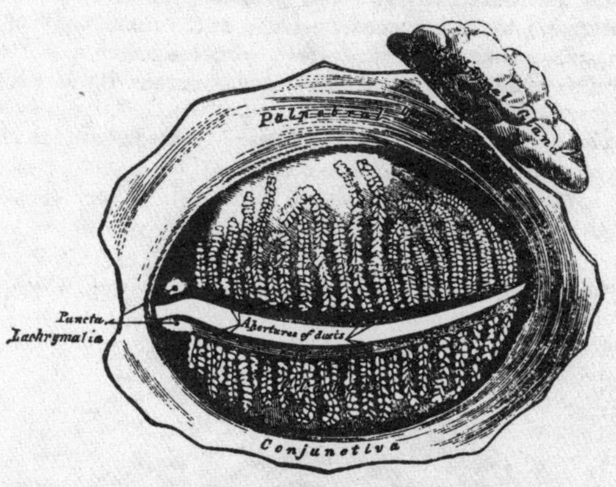

tilage, and somewhat fewer in the lower. They are embedded in grooves in the inner surface of the cartilages, and correspond in length with the breadth of each cartilage; they are, consequently, longer in the upper than in the lower eyelid. Their ducts open on the free margin of the lids by minute foramina, which correspond in number to the follicles. These glands are a variety of the cutaneous sebaceous glands, each consisting of a single straight tube or follicle, having a cæcal termination, into which open a number of small secondary follicles. The tubes consist of basement membrane, covered by a layer of scaly epithelium; the cells are charged with sebaceous matter, which constitutes the secretion. The peculiar parallel arrangement of these glands side by side, forms a smooth layer, adapted to the surface of the globe, over which they constantly glide. The use of their secretion is to prevent adhesion of the lids.

The *eyelashes* (*cilia*) are attached to the free edges of the eyelids; they are short, thick, curved hairs, arranged in a double or triple row at the margin of the lids: those of the upper lid, more numerous and longer than the lower, curve upwards; those of the lower lid curve downwards, so that they do not interlace in closing the lids.

The *conjunctiva* is the mucous membrane of the eye. It lines the inner surface of the eyelids, and is reflected over the fore part of the sclerotic and cornea. In each of these situations, its structure presents some peculiarities.

The *palpebral portion of the conjunctiva* is thick, opaque, highly vascular, and covered with numerous papillæ, which, in the disease called *granular lids*, become greatly hypertrophied. At the margin of the lids, it becomes continuous with the lining membrane of the ducts of the Meibomian glands, and, through the lachrymal canals, with the lining membrane of the lachrymal sac and nasal duct. At the outer angle of the upper lid, it may be traced along the lachrymal ducts into the lachrymal gland; and at the inner angle of the eye, it forms a semilunar fold, the *plica semilunaris*. The folds formed by the reflection of the conjunctiva from the lids on to the eye are called the *superior* and *inferior palpebral folds*, the former being the deeper of the two. Upon the *sclerotic*, the conjunctiva is loosely connected to the globe: it becomes thinner, loses its papillary structure, is transparent, and only slightly vascular in health. Upon the *cornea*, the conjunctiva is extremely thin and closely adherent, and no vessels can be traced into it in the adult in a healthy state. In the *fœtus*, fine capillary loops extend, for some little distance forwards, into this membrane; but in the *adult*, they pass only to the circumference of the cornea.

The *caruncula lachrymalis* is a small, reddish, conical-shaped body, situated at the inner canthus of the eye, and filling up the small triangular space in this situation, the *lacus lachrymalis*. It consists of a cluster of follicles similar in structure to the Meibomian, covered with mucous membrane, and is the source of the whitish secretion which constantly collects at the inner angle of the eye. A few slender hairs are attached to its surface. On the outer side of the caruncula is a slight semilunar fold of mucous membrane, the concavity of which is directed towards the cornea; it is called the *plica semilunaris*. Between its two layers is found a thin plate of cartilage. This structure is considered to be the rudiment of the third eyelid in birds, the *membrana nictitans*.

Lachrymal Apparatus. (Fig. 308.)

The lachrymal apparatus consists of the lachrymal gland, which secretes the

308.—The Lachrymal Apparatus. Right Side.

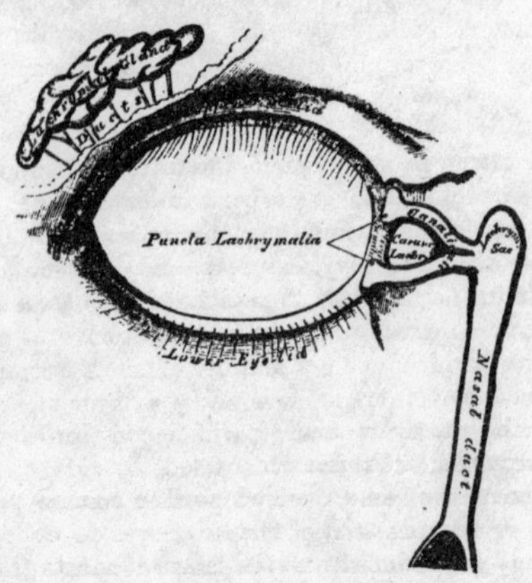

tears, and its excretory ducts, which convey the fluid to the surface of the eye. This fluid is carried away by the lachrymal canals into the lachrymal sac, and along the nasal duct into the cavity of the nose.

The *lachrymal gland* is lodged in a depression at the outer angle of the orbit, on the inner side of the external angular process of the frontal bone. It is of an oval form, about the size and shape of an almond. Its upper convex surface is in contact with the periosteum of the orbit, to which it is connected by a few fibrous bands. Its under concave surface rests upon the convexity of the eyeball, and upon the Superior and External recti muscles. Its vessels and nerves enter its posterior border, whilst its anterior margin is closely adherent to the back part of the upper eyelid, and is covered, on its inner surface, by a reflexion of the conjunctiva. This margin is separated from the rest of the gland by a slight depression, hence it is sometimes described as a separate lobe, called the *palpebral portion of the gland.* In structure and general appearance the lachrymal resembles the salivary glands. Its ducts, about seven in number, run obliquely beneath the mucous membrane for a short distance, and separating from each other, open by a series of minute orifices on the upper and outer half of the conjunctiva, near its reflexion on to the globe. These orifices are arranged in a row, so as to disperse the secretion over the surface of the membrane.

The *lachrymal canals* commence at the minute orifices, *puncta lachrymalia*, seen on the margin of the lids, at the outer extremity of the lacus lachrymalis. They commence on the summit of a slightly elevated papilla, the *papilla lachrymalis*, and lead into minute canals, the *canaliculi*, which proceed inwards to terminate in the lachrymal sac. The *superior canal*, the smaller and longer of the two, at first ascends, and then bends at an acute angle, and passes inwards and downwards to the lachrymal sac. The *inferior canal* at first descends, and then abruptly changing its course, passes almost horizontally inwards. They are dense and elastic in structure, and somewhat dilated at their angle.

The *lachrymal sac* is the upper dilated extremity of the nasal duct, and is lodged in a deep groove formed by the lachrymal bone and nasal process of the superior maxillary. It is oval in form, its upper extremity being closed in and rounded, whilst below it is continued into the nasal duct. It is covered by the Tensor tarsi muscle and by a fibrous expansion derived from the tendo oculi, which is attached to the ridge on the lachrymal bone. In structure, it consists of a fibrous elastic coat, lined internally by mucous membrane; the latter being continuous, through the canaliculi, with the mucous lining of the conjunctiva, and through the nasal duct with the pituitary membrane of the nose.

The *nasal duct* is a membranous canal, about three quarters of an inch in length, which extends from the lower part of the lachrymal sac to the inferior meatus of the nose, where it terminates by a somewhat expanded orifice, provided with an imperfect valve formed by the mucous membrane. It is contained in an osseous canal, formed by the superior maxillary, the lachrymal, and the inferior turbinated bones, is narrower in the middle than at each extremity, and takes a direction downwards, backwards, and a little outwards. It is lined by mucous membrane, which is continuous below with the pituitary lining of the nose. In the canaliculi, this membrane is provided with scaly epithelium; but in the lachrymal sac and nasal duct, the epithelium is ciliated as in the nose.

The Ear.

The organ of hearing has three parts; the external ear, the middle ear or tympanum, and the internal ear or labyrinth.

The *external ear* consists of an expanded portion named pinna, or auricle, and the auditory canal, or meatus. The former serves to collect the vibrations of the air by which sound is produced, and the latter conducts those vibrations to the tympanum.

The *pinna*, or *auricle* (fig. 309), is formed by a layer of cartilage, covered with integument, and connected to the commencement of the auditory canal; it is of an ovoid form, its surface uneven, with its larger end directed upwards. Its outer surface is irregularly concave, directed slightly forwards, and presents numerous eminences and depressions which result from the foldings of its fibro-cartilaginous element. To each of these names have been assigned. Thus, the external prominent rim of the auricle is called the *helix*. Another curved prominence parallel with, and in front of the helix, is called the *antihelix*; this bifurcates above, so as to enclose a triangular depression, the fossa of the antihelix. The narrow curved depression between the helix and antihelix is called the *fossa of the helix* (*fossa innominata, scaphoidea*). The antihelix describes a curve round a deep, capacious cavity, the *concha*, which is partially divided into two parts by the commencement of the helix. In front of the concha, and projecting backwards over the meatus, is a small pointed eminence, the *tragus*; so called from its being generally covered, on its under surface, with a tuft of hair, resembling a goat's beard. Opposite the tragus, and separated from it by a deep notch (*incisura intertragica*), is a small tubercle, the *antitragus*. Below this is the *lobule*, composed of tough areolar and adipose tissues, wanting the firmness and elasticity of the rest of the pinna.

309.—The Pinna, or Auricle. Outer Surface.

Structure of the pinna. The pinna is composed of a thin plate of yellow cartilage, covered with integument, and connected to the surrounding parts by ligaments, and a few muscular fibres.

The *integument* is thin, closely adherent to the cartilage, and furnished with sebaceous glands, which are most numerous in the concha and scaphoid fossa.

The *cartilage of the pinna* consists of one single piece; it gives form to this part of the ear, and upon its surface are found all the eminences and depressions above described. It does not enter into the construction of all parts of the auricle; thus it does not form a constituent part of the lobule; it is deficient, also, between the tragus and beginning of the helix, the notch between them being filled up by dense fibrous tissue. It presents several intervals or fissures in its substance, which partially separate the different parts. The fissure of the helix is a short, vertical slit, situated at the fore part of the pinna, immediately behind a small conical projection of cartilage, opposite the first curve of the helix (process of the helix). Another fissure, the fissure of the tragus, is seen upon the anterior surface of the tragus. The antihelix is divided below, by a deep fissure, into two parts; one part terminates by a pointed, tail-like extremity (*processus caudatus*); the other is continuous with the antitragus. The cartilage of the pinna is very pliable, elastic, of a yellowish colour, and similar in structure to the cartilages of the nose.

The *ligaments of the pinna* consist of two sets. 1. Those connecting it to the side of the head. 2. Those connecting the various parts of its cartilage together.

The former, the most important, are two in number, anterior and posterior. The *anterior ligament* extends from the process of the helix to the root of the zygoma. The *posterior ligament* passes from the posterior surface of the concha to the outer surface of the mastoid process of the temporal bone. A few fibres connect the tragus to the root of the zygoma.

The ligaments connecting the various parts of the cartilage together are also two in number. Of these, one is a strong fibrous band, stretching across from the tragus to the commencement of the helix, completing the meatus in front, and partly encircling the boundary of the concha; the other extends between the concha and the processus caudatus.

The *muscles of the pinna* (fig. 310), like the ligaments, consist of two sets; 1. Those which connect it with the side of the head, moving the pinna as a whole,

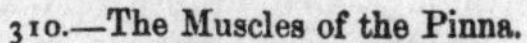

310.—The Muscles of the Pinna.

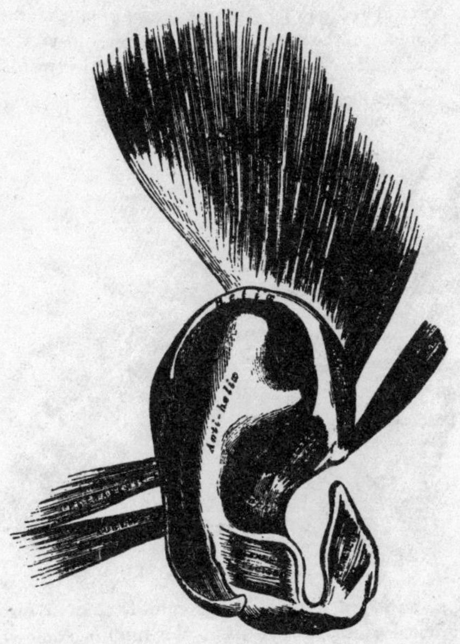

viz., the Attollens, Attrahens, and Retrahens aurem (p. 202); and the proper muscles of the pinna, which extend from one part of the auricle to another. These are, the

Helicis major.	Antitragicus.
Helicis minor.	Transversus auriculæ.
Tragicus.	Obliquus auris.

The *Helicis major* is a narrow, vertical band of muscular fibres, situated upon the anterior margin of the helix. It arises, below, from the tubercle of the helix, and is inserted into the anterior border of the helix, just where it is about to curve backwards. It is pretty constant in its existence.

The *Helicis minor* is an oblique fasciculus, attached to that part of the helix which commences from the bottom of the concha.

The *Tragicus* is a short, flattened band of muscular fibres situated upon the outer surface of the tragus, the direction of its fibres being vertical.

The *Antitragicus* arises from the outer part of the antitragus; its fibres are inserted into the processus caudatus of the helix. This muscle is usually very distinct.

The *Transversus auriculæ* is placed on the cranial surface of the pinna. It consists of radiating fibres, partly tendinous and partly muscular, extending from the convexity of the concha to the prominence corresponding with the groove of the helix.

The *Obliquus auris* (Todd) consists of a few fibres extending from the upper and back part of the concha to the convexity immediately above it.

The *arteries of the pinna* are, the posterior auricular, from the external carotid; the anterior auricular, from the temporal; and an auricular branch from the occi pital artery.

The *veins* accompany the corresponding arteries.

The *nerves* are, the auricularis magnus, from the cervical plexus; the posterio auricular, from the facial; the auricular branch of the pneumogastric; and the auriculo-temporal branch of the inferior maxillary nerve.

The AUDITORY CANAL (fig. 311), (*meatus auditorius externus*), extends from the bottom of the concha to the membrana tympani. It is about an inch and a quarter in length, its direction is obliquely forwards and inwards, and it is slightly

311.—A Front View of the Organ of Hearing. Right Side.

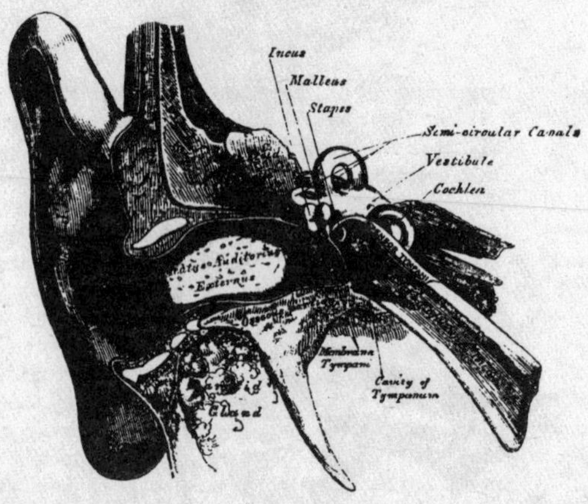

curved upon itself, so as to be higher in the middle than at either extremity. It forms an oval cylindrical canal, narrowest at the middle, somewhat flattened from before backwards, the greatest diameter being in the vertical direction at the external orifice; but, in the transverse direction, at the tympanic end. The membrana tympani, which occupies the termination of the meatus, is obliquely directed, in consequence of the floor of the canal being longer than the roof, and the anterior wall longer than the posterior. The auditory canal is formed partly by cartilage and membrane, and partly by bone.

The *cartilaginous portion* is about half an inch in length, being rather less than half the canal; it is formed by the cartilage of the concha and tragus, prolonged inwards, and firmly attached to the circumference of the auditory process. The cartilage is deficient at its upper and back part, its place being supplied by fibrous membrane. This part of the canal is rendered extremely moveable by two or three deep fissures (*incisuræ Santorini*), which extend through the cartilage in a vertical direction.

The *osseous portion* is about three-quarters of an inch in length, and narrower than the cartilaginous portion. It is directed inwards and a little forwards, forming a slight curve in its course, the convexity of which is upwards and backwards. Its inner end, which communicates with the cavity of the tympanum, is smaller than the outer, and sloped, the anterior wall projecting beyond the posterior about two lines; it is marked, except at its upper part, by a narrow groove for the insertion of the membrana tympani. Its outer end is dilated, and rough, in the greater part of its circumference, for the attachment of the cartilage of the pinna. Its vertical transverse section is oval, the greatest diameter being from above downwards. The front and lower parts of this canal are formed by a curved plate of bone, which, in the fœtus, exists as a separate ring (tympanic bone), incomplete at its upper part.

The *skin* lining the meatus is very thin, adheres closely to the cartilaginous and osseous portions of the tube, and covers the surface of the membrana tympani, forming its outer layer. After maceration, the thin pouch of epidermis, when withdrawn, preserves the form of the meatus. The skin near the orifice is furnished with hairs and sebaceous glands. In the thick subcutaneous tissue of the cartilaginous part of the meatus are numerous ceruminous glands, which secrete the ear-wax: their ducts open on the surface of the skin.

The *arteries* supplying the meatus are branches from the posterior auricular, internal maxillary, and temporal.

The *nerves* are chiefly derived from the auriculo-temporal branch of the inferior maxillary nerve.

Middle Ear, or Tympanum.

The middle ear, or tympanum, is an irregular cavity, compressed from without inwards, and situated within the petrous bone. It is placed above the jugular fossa, the carotid canal lying in front, the mastoid cells behind, the meatus auditorius externally, and the labyrinth internally. It is filled with air, and communicates with the pharynx by the Eustachian tube. The tympanum is traversed by a chain of moveable bones, which connect the membrana tympani with the labyrinth, and serve to convey the vibrations communicated to the membrana tympani across the cavity of the tympanum to the internal ear.

The *cavity of the tympanum* measures about five lines from before backwards, three lines in the vertical direction, and between two and three in the transverse, being a little broader behind and above than it is below and in front. It is bounded externally by the membrani tympani and meatus; internally, by the outer surface of the internal ear; and communicates, behind, with the mastoid cells; and, in front, with the Eustachian tube and canal for the Tensor tympani. Its roof and floor are formed by thin osseous laminæ, which connect the squamous and petrous portions of the temporal bone.

The *roof* is broad, flattened, and formed of a thin plate of bone, which separates the cranial and tympanic cavities.

The *floor* is narrow, and corresponds to the jugular fossa, which lies beneath.

The *outer wall* is formed by the membrana tympani, a small portion of bone being seen above and below this membrane. It presents three small apertures, the iter chordæ posterius, the Glaserian fissure, and the iter chordæ anterius.

The *aperture of the iter chordæ posterius* is behind the aperture for the membrana tympani, close to its margin, on a level with its centre; it leads into a minute canal, which descends in front of the aquæductus Fallopii, and terminates in that canal near the stylo-mastoid foramen. Through it the chorda tympani nerve enters the tympanum.

The *Glaserian fissure* opens just above and in front of the orifice of the membrana tympani; in this situation it is a mere slit, about a line in length. It gives passage to the long process of the malleus, the Laxator tympani muscle, and some tympanic vessels.

The *aperture of the iter chordæ anterius* is seen just above the preceding fissure; it leads into a canal (canal of Huguier), which runs parallel with the Glaserian fissure. Through it the chorda tympani nerve leaves the tympanum.

The *internal wall of the tympanum* (fig. 312) is vertical in direction, and looks directly outwards. It presents for examination the following parts:

Fenestra ovalis.	Ridge of the Aquæductus Fallopii.
Fenestra rotunda.	Pyramid.
Promontory.	Opening for the Stapedius.

The *fenestra ovalis* is a reniform opening, leading from the tympanum into the vestibule; its long diameter is directed horizontally, and its convex border is upwards. The opening in the recent state is closed by the lining membrane common to both cavities, and is occupied by the base of the stapes. This membrane is placed opposite the membrana tympani, and is connected with it by the ossicula auditûs.

The *fenestra rotunda* is an oval aperture, placed at the bottom of a funnel-shaped depression, leading into the cochlea. It is situated below and rather behind

312.—View of Inner Wall of Tympanum. (Enlarged.)

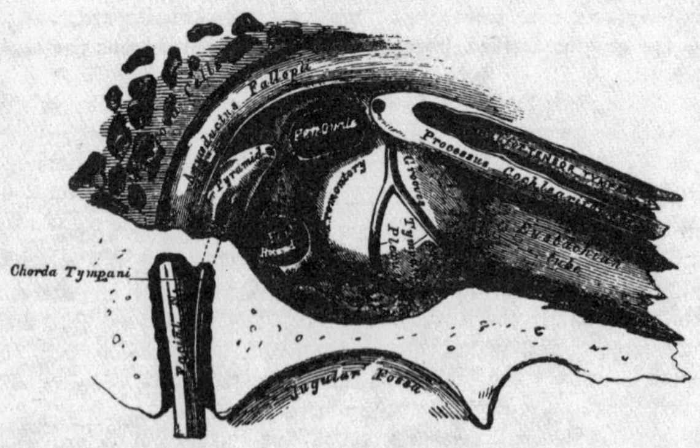

the fenestra ovalis, from which it is separated by a rounded elevation, the promontory; it is closed in the recent state by a membrane (*membrana tympani secundaria*, Scarpa). This membrane is concave towards the tympanum, convex towards the cochlea. It consists of three layers: the external, or mucous, derived from the mucous lining of the tympanum; the internal, or serous, from the lining membrane of the cochlea; and an intermediate, or fibrous layer.

The *promontory* is a rounded hollow prominence, formed by the projection outwards of the first turn of the cochlea; it is placed between the fenestræ, and is furrowed on its surface by three small grooves, which lodge branches of the tympanic plexus.

The *rounded eminence of the aquæductus Fallopii* is placed beneath the fenestra ovalis and roof of the tympanum; it is the prominence of the bony canal in which the portio dura is contained. It traverses the inner wall of the tympanum above the fenestra ovalis, and behind that opening curves nearly vertically downwards along the posterior wall.

The *pyramid* is a conical eminence, situated immediately behind the fenestra ovalis, and in front of the vertical portion of the eminence above described; it is hollow in the interior, and contains the Stapedius muscle; its summit projects forwards towards the fenestra ovalis, and presents a small aperture, which transmits the tendon of the muscle. The cavity in the pyramid is prolonged into a minute canal, which communicates with the aquæductus Fallopii, and transmits the nerve which supplies the Stapedius.

The *posterior wall of the tympanum* is wider above than below, and presents for examination the

Openings of the Mastoid Cells.

These consist of one large irregular aperture, and several smaller openings, situated at the upper part of the posterior wall; they lead into canals, which communicate with large irregular cavities contained in the interior of the mastoid process. These cavities vary considerably in number, size, and form; they are lined by mucous membrane, continuous with that covering the cavity of the tympanum.

The *anterior wall of the tympanum* is wider above than below; it corresponds with the carotid canal, from which it is separated by a thin plate of bone; it presents for examination the

Canal for the Tensor tympani. Orifice of the Eustachian Tube.
The Processus Cochleariformis.

The orifice of the canal for the Tensor tympani, and the orifice of the Eustachian tube, are situated at the upper part of the anterior wall, being separated from each other by a thin, delicate horizontal plate of bone, the processus cochleariformis. These canals run from the tympanum forward, inward, and a little downward, to the retiring angle between the squamous and petrous portions of the temporal bone.

The *canal for the Tensor tympani* is the superior and the smaller of the two; it is rounded, and lies beneath the upper surface of the petrous bone, close to the hiatus Fallopii. The tympanic end of this canal forms a conical eminence, which is prolonged backwards into the cavity of the tympanum, and is perforated at its summit by an aperture, which transmits the tendon of the muscle contained in it. This eminence is sometimes called the *anterior pyramid.* The canal contains the Tensor tympani muscle.

The *Eustachian tube* is the channel through which the tympanum communicates with the pharynx. Its length is from an inch and a half to two inches, and its direction downwards, forwards, and inwards. It is formed partly of bone, partly of cartilage and fibrous tissue.

The *osseous portion* is about half an inch in length. It commences in the lower part of the anterior wall of the tympanum, below the processus cochleariformis, and gradually narrowing, terminates in an oval dilated opening, at the angle of junction of the petrous and squamous portions, its extremity presenting a jagged margin, which serves for the attachment of the cartilaginous portion.

The *cartilaginous portion*, about an inch in length, is formed of a triangular plate of cartilage, curled upon itself, an interval being left below, between the margins of the cartilage, which is completed by fibrous tissue. Its canal is narrow behind, wide, expanded, and somewhat trumpet-shaped in front, terminating by an oval orifice, at the upper part and side of the pharynx, behind the back part of the inferior meatus. Through this canal the mucous membrane of the pharynx is continuous with that which lines the tympanum.

The *membrana tympani* separates the cavity of the tympanum from the bottom of the external meatus. It is a thin semi-transparent membrane, nearly oval in form, somewhat broader above than below, and directed very obliquely downwards and inwards. Its circumference is contained in a groove at the inner end of the meatus, which skirts the circumference of this part excepting above. The handle of the malleus descends vertically between the inner and middle layers of this membrane, as far down as its centre, where it is firmly attached, drawing the membrane inwards, so that its outer surface is concave, its inner convex.

Structure. This membrane is composed of three layers, an external (cuticular), a middle (fibrous), and an internal (mucous). The *cuticular lining* is derived from the integument lining the meatus. The *fibrous layer* consists of fibrous and elastic tissues; some of the fibres radiate from near the centre to the circumference; others are arranged, in the form of a dense circular ring, round the attached margin of the membrane. The *mucous lining* is derived from the mucous lining of the tympanum. The vessels pass to the membrana tympani along the handle of the malleus, and are distributed between its layers.

Ossicles of the Tympanum. (Fig. 313.)

The tympanum is traversed by a chain of moveable bones, three in number, the malleus, incus, and stapes. The former is attached to the membrana tympani, the latter to the fenestra ovalis, the incus being placed between the two, to both of which it is connected by delicate articulations.

The *Malleus*, so named from its fancied resemblance to a hammer, consists of a head, neck, and three processes: the handle or manubrium, the processus gracilis, and the processus brevis.

The *head* is the large upper extremity of the bone; it is oval in shape, and articulates posteriorly with the incus, being free in the rest of its extent.

The *neck* is the narrow contracted part just beneath the head; and below this is a prominence, to which the various processes are attached.

The *manubrium* is a vertical process of bone, which is connected by its outer margin with the membrana tympani. It decreases in size towards its extremity, where it is curved slightly forwards, and flattened from within outwards.

The *processus gracilis* is a long and very delicate process, which passes from the eminence below the neck forwards and outwards to the Glaserian fissure, to which it is connected by bone and ligamentous fibres. It gives attachment to the Laxator tympani.

The *processus brevis* is a slight conical projection, which springs from the root of the manubrium, and lies in contact with the membrana tympani. Its summit gives attachment to the Tensor tympani.

The *Incus* has received its name from its supposed resemblance to an anvil, but it is more like a bicuspid tooth, with two roots, which differ in length, and are widely separated from each other. It consists of a body and two processes.

313.—The Small Bones of the Ear, seen from the Outside. (Enlarged.)

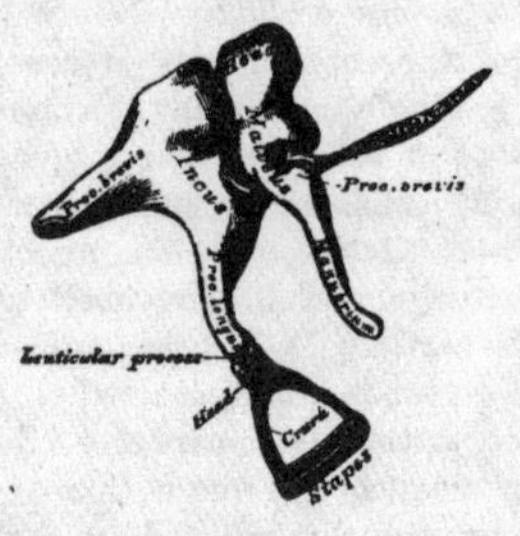

The *body* is somewhat quadrilateral, but compressed laterally. Its summit is deeply concave, and articulates with the malleus; in the fresh state, it is covered with cartilage and lined with synovial membrane.

The two processes diverge from one another nearly at right angles.

The *short process*, somewhat conical in shape, projects nearly horizontally backwards, and is attached to the margin of the opening leading into the mastoid cells, by ligamentous fibres.

The *long process*, longer and more slender than the preceding, descends nearly vertically behind the handle of the malleus, and bending inwards, terminates in a rounded globular projection, the *os orbiculare*, or lenticular process, which is tipped with cartilage, and articulates with the head of the stapes. In the fœtus the os orbiculare exists as a separate bone, but becomes united to the long process of the incus in the adult.

The *Stapes*, so called from its close resemblance to a stirrup, consists of a head, neck, two branches, and a base.

The *head* presents a depression, tipped with cartilage, which articulates with the os orbiculare.

The *neck*, the constricted part of the bone below the head, receives the insertion of the Stapedius muscle.

The *two branches* (*crura*) diverge from the neck, and are connected at their extremities by a flattened, oval-shaped plate (the *base*), which forms the foot of the stirrup, and is fixed to the margin of the fenestra ovalis by ligamentous fibres.

Ligaments of the Ossicula. These small bones are connected with each other, and with the walls of the tympanum, by ligaments, and moved by small muscles. The articular surfaces of the malleus and incus, the orbicular process of the incus and head of the stapes, are covered with cartilage, connected together by delicate capsular ligaments, and lined by synovial membrane. The ligaments connecting the ossicula with the walls of the tympanum are three in number, one for each bone.

The *suspensory ligament of the malleus* is a delicate, round bundle of fibres, which descends perpendicularly from the roof of the tympanum to the head of the malleus.

The *posterior ligament of the incus* is a short, thick, ligamentous band, which connects the extremity of the short process of the incus to the posterior wall of the tympanum, near the margin of the opening of the mastoid cells.

The *annular ligament of the stapes* connects the circumference of the base of this bone to the margin of the fenestra ovalis.

A suspensory ligament of the incus has been described by Arnold, descending from the roof of the tympanum to the upper part of the incus, near its articulation with the malleus.

The muscles of the tympanum are three:—

Tensor tympani. Laxator tympani. Stapedius.

The *Tensor tympani*, the largest, is contained in a bony canal, above the osseous portion of the Eustachian tube, from which it is separated by the processus cochleariformis. It arises from the under surface of the petrous bone, from the cartilaginous portion of the Eustachian tube, and from the osseous canal in which it is contained. Passing backwards, it terminates in a slender tendon, which is reflected outwards over the processus cochleariformis, and is inserted into the handle of the malleus, near its root. It is supplied by a branch from the otic ganglion.

The *Laxator tympani major* (Sömmerring) arises from the spinous process of the sphenoid bone, and from the cartilaginous portion of the Eustachian tube, and passing backwards through the Glaserian fissure, is inserted into the neck of the malleus, just above the processus gracilis. It is supplied by the tympanic branch of the facial.

The *Laxator tympani minor* (Sömmerring) arises from the upper and back part of the external meatus, passes forwards and inwards between the middle and inner layers of the membrana tympani, and is inserted into the handle of the malleus, and processus brevis. This is regarded as a ligament by some anatomists.

The *Stapedius* arises from the sides of a conical cavity hollowed out of the interior of the pyramid: its tendon emerges from the orifice at the apex of the pyramid, and passing forwards, is inserted into the neck of the stapes. Its surface is aponeurotic, its interior fleshy; and its tendon occasionally contains a slender bony spine, which is constant in some mammalia. It is supplied by a filament from the facial nerve.

Actions. The Tensor tympani draws the membrana tympani inwards, and thus heightens its tension. The Laxator tympani draws the malleus outwards, and thus the tympanic membrane, especially at its fore part, is relaxed. The Stapedius depresses the back part of the base of the stapes, and raises its fore part. It probably compresses the contents of the vestibule.

The *mucous membrane of the tympanum* is thin, vascular, and continuous with the mucous membrane of the pharynx, through the Eustachian tube. It invests the ossicula, and the muscles and nerves contained in the tympanic cavity; forms the internal layer of the membrana tympani; covers the foramen rotundum; and is reflected into the mastoid cells, which it lines throughout. In the tympanum and mastoid cells, this membrane is pale, thin, slightly vascular, and covered with ciliated epithelium. In the osseous portion of the Eustachian tube, the membrane is thin; but in the cartilaginous portion it is very thick, highly vascular, covered with laminar ciliated epithelium, and provided with numerous mucous glands.

The *arteries supplying the tympanum* are five in number. Two of them are larger than the rest, viz., the tympanic branch of the internal maxillary, which supplies the membrana tympani; and the stylo-mastoid branch of the posterior auricular, which supplies the back part of the tympanum and mastoid cells. The smaller branches are, the petrosal branch of the middle meningeal, and branches from the ascending pharyngeal and internal carotid.

The *veins of the tympanum* terminate in the middle meningeal and pharyngeal veins, and, through these, in the internal jugular.

The *nerves of the tympanum* may be divided into: 1. Those supplying the muscles; 2. Those distributed to the lining membrane; 3. Branches communicating with other nerves.

Nerves to muscles. The Tensor tympani is supplied by a branch from the otic ganglion; the Laxator tympani, and the Stapedius, by a filament from the facial (Sömmerring).

The *nerves distributed to the lining membrane* are derived from the tympanic plexus.

Communications between the following nerves take place in the tympanum: the tympanic branch from the petrous ganglion of the glosso-pharyngeal; a filament from the carotid plexus; a branch which joins the great superficial petrosal nerve from the Vidian; and a branch to the otic ganglion (small superficial petrosal nerve.)

The *tympanic branch of the glosso-pharyngeal* (Jacobson's nerve), enters the tympanum by an aperture in its floor, close to the inner wall, and ascends on to the promontory. It distributes filaments to the lining membrane of the tympanum, and divides into three branches, which are contained in grooves on the promontory, and serve to connect this with other nerves. One branch runs in a groove, forwards and downwards, to an aperture situated at the junction of the anterior and inner walls, just above the floor, and enters the carotid canal, to communicate with the carotid plexus of the sympathetic. The second branch is contained in a groove which runs vertically upwards to an aperture on the inner wall of the tympanum, just beneath the anterior pyramid, and in front of the fenestra ovalis. The canal leading from this opens into the hiatus Fallopii, where the nerve contained in it joins the great petrosal nerve. The third branch ascends towards the anterior surface of the petrous bone; it then passes through a small aperture in the sphenoid and temporal bones to the exterior of the skull, and joins the otic ganglion. As this nerve passes by the gangliform enlargement of the facial, it has a connecting filament with it.

The *chorda tympani* quits the facial near the stylo-mastoid foramen, enters the tympanum at the base of the pyramid, and arches forwards across its cavity between the handle of the malleus and long process of the incus, to an opening internal to the fissura Glaseri. It is invested by a reflection of the lining membrane of the tympanum.

Internal Ear, or Labyrinth.

The internal ear is the essential part of the organ, receiving the ultimate distribution of the auditory nerve. It is called the *labyrinth*, from the complexity of its shape, and consists of three parts, the vestibule, semicircular canals, and cochlea. It is formed by a series of cavities, channelled out of the substance of the petrous bone, communicating externally with the cavity of the tympanum, through the fenestra ovalis and rotunda; and internally, with the meatus auditorius internus, which contains the auditory nerve. Within the osseous labyrinth is contained the membranous labyrinth, upon which the ramifications of the auditory nerve are distributed.

The *Vestibule* (fig. 314) is the common central cavity of communication between the parts of the internal ear. It is situated on the inner side of the tympanum, behind the cochlea, and in front of the semicircular canals. It is somewhat ovoidal in shape from before backwards, flattened from side to side, and measures about one-fifth of an inch from before backwards, as well as from above downwards, being narrower from without inwards. On its *outer*, or *tympanic wall*, is the fenestra ovalis, closed, in the recent state, by the base of the stapes, and its annular ligament. On its *inner wall*, at the fore part, is a small circular depression, *fovea hemispherica*, which is perforated, at its anterior and inferior part, by several minute holes (*macula cribrosa*), for the passage of the filaments of the auditory nerve; and behind this depression is a vertical ridge, the pyramidal eminence. At the hinder part of the inner wall is the orifice of the *aquæductus vestibuli*, which extends to the posterior surface of the petrous portion of the temporal bone. It transmits a small vein, and, according to some, contains a tubular prolongation of the lining membrane of the vestibule, which ends in a *cul-de-sac*, between the layers of the dura mater within the cranial cavity. On the *upper wall* or *roof* is a transversely-oval depression, *fovea semi-elliptica*, separated from the fovea hemispherica by the pyramidal eminence, already mentioned. Behind, the semicircular canals open into the vestibule by five orifices. In front is a large oval opening which communicates with the scala vestibuli of the cochlea by a single orifice, *apertura scalæ vestibuli cochleæ.*

The *Semicircular canals* are three bony canals, situated above and behind the vestibule. They are of unequal length, compressed from side to side, and describe the greater part of a circle. They measure about one one-twentieth of an inch in diameter, and each presents a dilatation at one end, called the *ampulla*, which

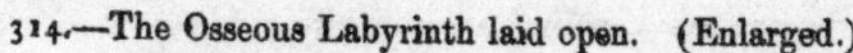
314.—The Osseous Labyrinth laid open. (Enlarged.)

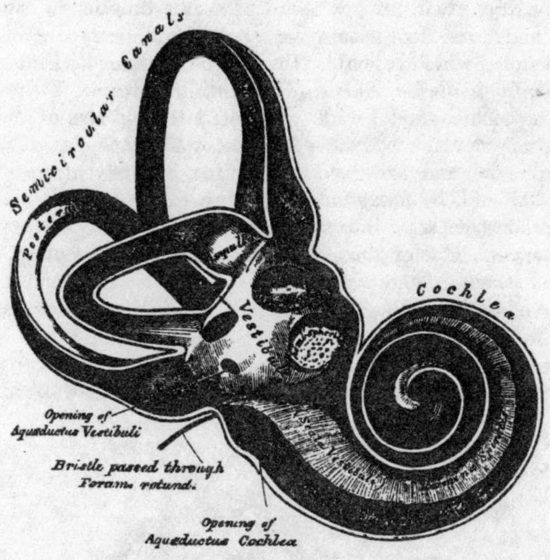

measures more than twice the diameter of the tube. These canals open into the vestibule by five orifices, one of the apertures being common to two of the canals.

The *superior semicircular canal* is vertical in direction, and stretches across the petrous portion of the temporal bone, at right angles to its posterior surface: its arch forms a round projection on the anterior surface of the petrous bone. It describes about two-thirds of a circle. Its outer extremity, which is ampullated, commences by a distinct orifice in the upper part of the vestibule; the opposite end of the canal, which is not dilated, joins with the corresponding part of the posterior canal, and opens by a common orifice with it in the back part of the vestibule.

The *posterior semicircular canal*, also vertical in direction, is directed backwards, nearly parallel to the posterior surface of the petrous bone: it is the longest of the three, its ampullated end commencing at the lower and back part of the vestibule, its opposite end joining to form the common canal already mentioned.

The *external*, or *horizontal canal*, is the shortest of the three, its arch being directed outwards and backwards; thus each semicircular canal stands at right angles to the other two. Its ampullated end corresponds to the upper and outer angle of the vestibule, just above the fenestra ovalis; its opposite end opens by a distinct orifice at the upper and back part of the vestibule.

The *Cochlea* bears some resemblance to a common snail-shell: it forms the anterior part of the labyrinth, is conical in form, and placed almost horizontally in front of the vestibule; its apex is directed forwards and outwards towards the upper and front part of the inner wall of the tympanum; its base corresponds with the anterior depression at the bottom of the internal auditory meatus; and is perforated by numerous apertures, for the passage of the cochlear branch of the auditory nerve. It measures about a quarter of an inch in length, and its breadth towards the base is about the same. It consists of a conical-shaped central axis, the modiolus or columella; of a canal wound spirally round the axis for two turns and a half, from the base to the apex; and of a delicate lamina (the *lamina spiralis*) contained within the canal, which follows its windings, and subdivides it into two.

The *central axis*, or *modiolus*, is conical in form, and extends from the base to the apex of the cochlea. Its base is broad, corresponds with the first turn of the cochlea, and is perforated by numerous orifices, which transmit filaments of the cochlear branch of the auditory nerve; the axis diminishes rapidly in size in the second coil, and terminates within the last half coil, or cupola, in an expanded, delicate, bony lamella, which resembles the half of a funnel, divided longitudinally, and called the *infundibulum*; the broad part of this funnel is directed towards the summit of the cochlea, and blends with the last half-turn of the spiral canal of the cochlea, the cupola. The outer surface of the modiolus is formed of the wall of the spiral canal, and is dense in structure; but its centre is channelled, as far as the last half-coil, by numerous branching canals, which transmit nervous filaments in regular succession into the canal of the cochlea, or on to the surface of the lamina spiralis. One of these, larger than the rest, occupies the centre of the modiolus, and is named the *tubulus centralis modioli*; it extends from the base to the extremity of the modiolus, and transmits a small nerve and artery (*arteria centralis modioli*).

The *spiral canal* (fig. 315) takes two turns and a half round the modiolus. It is about an inch and a half in length, measured along its outer wall; and diminishes

315.—The Cochlea laid open. (Enlarged.)

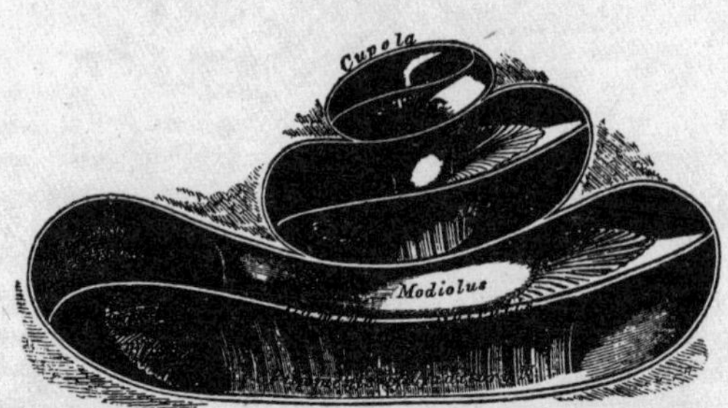

gradually in size from the base to the summit, where it terminates in a *cul-de-sac*, the cupola, which forms the apex of the cochlea. The commencement of this canal is about the tenth of an inch in diameter: it diverges from the modiolus towards the tympanum and vestibule, and presents three openings. One, the *fenestra rotunda*, communicates with the tympanum: in the recent state, this aperture is closed by a membrane, the *membrana tympani secundaria*. Another aperture, of an oval form, enters the vestibule. The third is the aperture of the *aquæductus cochleæ* leading to a minute funnel-shaped canal, which opens on the basilar surface of the petrous bone, and transmits a small vein.

The interior of the spiral canal is divided into two passages (*scalæ*) by a thin, osseous, and membranous lamina, which winds spirally round the modiolus. This is the *lamina spiralis*, the essential part of the cochlea upon which the nerve-tubules are distributed. The osseous part of the spiral lamina extends about half way across the diameter of the spiral canal; it is called the *osseous zone*. It commences in the vestibule between the tympanic and vestibular opening of the cochlea, and, gradually becoming narrower in its course, terminates in a projecting hook, the hamular process, just where the expansion of the infundibulum commences. The lamina spiralis consists of two thin lamellæ of bone, between which are numerous canals for the passage of nervous filaments, which open chiefly on the lower or tympanic surface. At the point where the osseous lamina is attached to the modiolus, and following its windings, is a small canal, called by Rosenthal the *canalis spiralis modioli*. In the recent state, the osseous zone is continued to the opposite wall of the canal by a membranous and muscular layer (membranous zone), so as to form a complete partition in the tube of the cochlea. Two passages, or scalæ, are thus formed, by the division of the canal of the cochlea into two. One, the *scala tympani*, is closed below by the membrane of the fenestra rotunda; the other, the *scala vestibuli*, communicates, by an oval aperture, with the vestibule. Near the termination of the scala vestibuli, close by the fenestra rotunda, is the orifice of the aquæductus cochleæ. The scalæ communicate, at the apex of the cochlea, by an opening common to both, the *helicotrema*, which exists in consequence of the deficiency of the lamina spiralis in the last half coil of the canal.

In *structure*, the membranous zone is a transparent glassy lamina, presenting near its centre a number of minute transverse lines, which radiate outwards, and give it a fibrous appearance; and at its circumference, where connected with the outer wall of the spiral canal, it is composed of a semi-transparent structure, which is described by Todd and Bowman as a muscle (the Cochlearis), and by Kölliker as connective tissue.

The vestibular surface of the osseous portion of the lamina spiralis is covered for about the outer fifth of its surface with a thin layer, resembling cartilage in texture. It is described as the *denticulate lamina* (Todd and Bowman), from its presenting a series of wedge-shaped teeth which form its free margin, and which project into the vestibular scala.

The *inner surface* of the osseous labyrinth is lined by an exceedingly thin fibro-serous membrane, analogous to a periosteum, from its close adhesion to the inner surface of these cavities, and performing the office of a serous membrane by its free surface. It lines the vestibule, and from this cavity is continued into the semicircular canals and the scala vestibuli of the cochlea, and through the helicotrema into the scala tympani. Two delicate tubular processes are prolonged along the aquæducts of the vestibule and cochlea, to the inner surface of the dura mater. This membrane is continued across the fenestra ovalis and rotunda, and consequently has no communication with the lining membrane of the tympanum. Its attached surface is rough and fibrous, and closely adherent to the bone; its free surface is smooth and pale, covered with a layer of epithelium, and secretes a thin, limpid fluid, the *aqua labyrinthi* (*perilymph* [Blainville], *liquor Cotunnii*). In the vestibule and semicircular canals, it separates the osseous from the membranous labyrinth; but in the cochlea it lines the two surfaces of the bony lamina spiralis; and being continued from its free margin across the canal to its outer wall, forms the lamina spiralis membranacea, serving to complete the separation between the two scalæ.

The Membranous Labyrinth.

The membranous labyrinth (fig. 316) is a closed membranous sac, containing fluid. The ramifications of the auditory nerve are distributed upon the wall of the sac. It has the same general form as the vestibule and semicircular canals, in which it is enclosed; but is considerably smaller, and separated from their lining membrane by the perilymph.

The *vestibular portion* consists of two sacs, the utricle and the saccule.

The *utricle* is the larger of the two, of an oblong form, compressed laterally, and occupies the upper and back part of the vestibule, lying in contact with the fovea

316.—The Membranous Labyrinth detached. (Enlarged.)

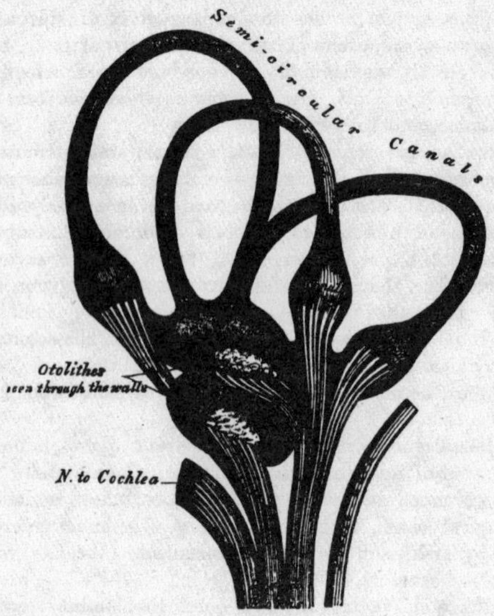

semi-elliptica. Numerous filaments of the auditory nerve are distributed on the wall of this sac; and its cavity communicates behind with the membranous semicircular canals by five orifices.

The *saccule* is the smaller of the two vestibular sacs: it is globular in form, lies in the fovea hemispherica, near the opening of the vestibular scala of the cochlea, and receives numerous nervous filaments, which enter from the bottom of the depression in which it is contained. Its cavity is apparently distinct from that of the utricle.

The *membranous semicircular canals* are about one-third the diameter of the osseous canals, but in number, shape, and general form they are precisely similar; they are hollow, and open by five orifices into the utricle, one opening being common to two canals. Their ampullæ are thicker than the rest of the tubes, and nearly fill the cavities in which they are contained.

The membranous labyrinth is held in its position by the numerous nervous filaments distributed to the utricle, to the saccule, and to the ampulla of each canal. These nerves enter the vestibule through the minute apertures on its inner wall.

Structure. The wall of the membranous labyrinth is semi-transparent, and consists of three layers. The *outer* layer is a loose and flocculent tissue, containing blood-vessels and numerous pigment-cells, analogous to those in the choroid. The *middle layer*, thicker and more transparent, bears some resemblance to the hyaloid membrane, but it presents in parts marks of longitudinal fibrillation and elongated nuclei on the addition of acetic acid. The *inner layer* is formed of polygonal nucleated epithelial cells, which secrete the endolymph.

The *endolymph* (*liquor Scarpæ*) is a limpid serous fluid, which fills the membranous labyrinth; in composition, it closely resembles the perilymph.

The *otoliths* are two small rounded bodies, consisting of a mass of minute crystalline grains of carbonate of lime, held together in a mesh of delicate fibrous tissue, and contained in the wall of the utricle and saccule, opposite the distribution of the nerves. A calcareous material is also, according to Bowman, sparingly scattered in the cells lining the ampulla of each semicircular canal.

The *Arteries of the labyrinth* are the internal auditory, from the basilar or superior cerebellar, the stylo-mastoid, from the posterior auricular, and, occasionally, branches from the occipital. The internal auditory divides at the bottom of the internal meatus into two branches, cochlear and vestibular.

The cochlear branch subdivides into from twelve to fourteen twigs, which traverse the canals in the modiolus, and are distributed, in the form of a capillary network, in the substance of the lamina spiralis.

The vestibular branches accompany the nerves, and are distributed, in the form of a minute capillary network, in the substance of the membranous labyrinth.

The *Veins* of the vestibule and semicircular canals accompany the arteries, and receiving those of the cochlea at the base of the modiolus terminate in the superior petrosal sinus.

The *Auditory nerve*, the special nerve of the sense of hearing, divides, at the bottom of the internal auditory meatus, into two branches, the cochlear and vestibular. The trunk of the nerve, as well as the branches, contains numerous ganglion-cells with caudate prolongations.

The *vestibular nerve*, the posterior of the two, divides into three branches, superior, middle, and inferior.

The superior vestibular branch, the largest, divides into numerous filaments, which pass through minute openings at the upper and back part of the *cul-de-sac* at the bottom of the meatus, and entering the vestibule, are distributed to the utricle, and to the ampulla of the external and superior semicircular canals.

The middle vestibular branch consists of numerous filaments, which enter the vestibule by a smaller cluster of foramina, placed below those above mentioned, and which correspond to the bottom of the fovea hemispherica; they are distributed to the saccule.

The inferior and smallest branch passes backwards in a canal behind the foramina for the nerves of the saccule, and is distributed to the ampulla of the posterior semicircular canal.

The nervous filaments enter the ampullary enlargement at a deep depression seen on their external surface, and a corresponding elevation is seen within, the nerve-fibres ending in loops, and in free extremities. In the utricle and saccule, the nerve-fibres spread out, some blending with calcareous matter, others radiating on the inner surface of the wall of each cavity, becoming blended with a layer of nucleated cells, and terminating in a thin fibrous film.

The *cochlear nerve* divides into numerous filaments at the base of the modiolus, which ascend along its canals, and, then, bending outwards at right angles, pass between the plates of the bony lamina spiralis, close to its tympanic surface. Between the plates of the spiral lamina, the nerves form a plexus, which contains ganglion-cells; and from the margin of the osseous zone, branches of this plexus are distributed to the membranous part of the septum, where they are arranged in small, conical-shaped bundles, parallel with one another. The filaments which supply the apical portion of the lamina spiralis, are conducted to this part through the tubulus centralis modioli.

Organs of Digestion.

THE Apparatus for the digestion of the food consists of the alimentary canal, and of certain accessory organs.

The *alimentary canal* is a musculo-membranous tube, about thirty feet in length, extending from the mouth to the anus, and lined throughout its entire extent by mucous membrane. It has received different names in the various parts of its course: at its commencement, the mouth, we find provision made for the mechanical division of the food (mastication), and for its admixture with a fluid secreted by the salivary glands (insalivation); beyond this are the organs of deglutition, the pharynx and the œsophagus, which convey the food into that part of the alimentary canal (the stomach) in which the principal chemical changes occur; in the stomach, the reduction and solution of the food takes place; in the small intestines, the nutritive principles of the food (the chyle), by its admixture with the bile and pancreatic fluid, are separated from that portion which passes into the large intestine, most of which is expelled from the system.

Alimentary Canal.

Mouth.	Small intestine .	Duodenum. Jejunum. Ileum.
Pharynx.		
Œsophagus.	Large intestine .	Cæcum. Colon. Rectum.
Stomach.		

Accessory Organs.

Teeth.		
Salivary glands	Parotid. Submaxillary. Sublingual.	Liver. Pancreas. Spleen.

The MOUTH (fig. 317) is placed at the commencement of the alimentary canal; it is a nearly oval-shaped cavity, in which the mastication of the food takes place. It is bounded, in front, by the lips; laterally, by the cheeks and the alveolar processes of the upper and lower jaw; above, by the hard palate and teeth of the upper jaw; below, by the tongue, and by the mucous membrane stretched between the under surface of that organ and the inner surface of the jaws, and by the teeth of the lower jaw; behind, by the soft palate and fauces.

The *mucous membrane* lining the mouth, is continuous with the integument at the free margin of the lips, and with the mucous lining of the fauces behind; it is of a rose-pink tinge during life, and very thick where it covers the hard parts bounding the cavity.

The LIPS are two fleshy folds, which surround the orifice of the mouth, formed externally of integument, and internally of mucous membrane, between which is found the Orbicularis oris muscle, the coronary vessels, some nerves, areolar tissue, and fat, and numerous small labial glands. The inner surface of each lip is connected in the middle line to the gum of the corresponding jaw by a fold of mucous membrane, the *frænum labii superioris* and *inferioris*, the former being the larger of the two.

The *labial glands* are situated between the mucous membrane and the Orbicularis oris, round the orifice of the mouth. They are rounded in form, about the size of small peas, their ducts opening by small orifices upon the mucous membrane. In structure, they resemble the other salivary glands.

The CHEEKS form the sides of the face, and are continuous in front with the lips. They are composed, externally, of integument; internally, of mucous membrane, and between the two, of a muscular stratum, besides a large quantity of fat, areolar tissue, vessels, nerves, and buccal glands.

The *mucous membrane* lining the cheek, is reflected above and below upon the gums, and is continuous behind with the lining membrane of the soft palate. Opposite the second molar tooth of the upper jaw is a papilla, the summit of

317.—Sectional View of the Nose, Mouth, Pharynx, etc.

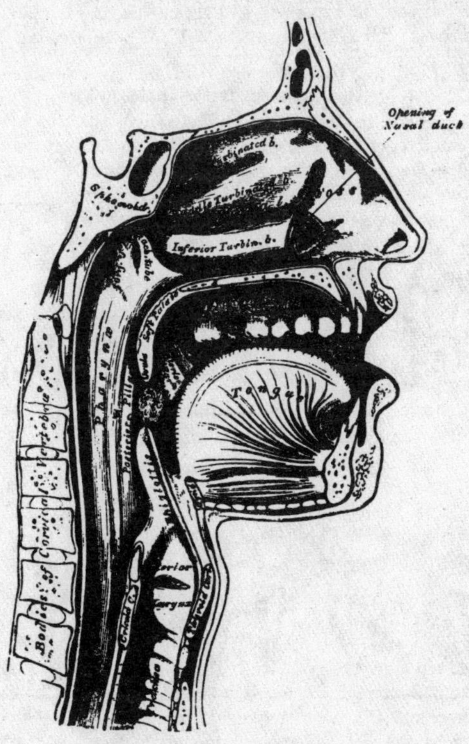

which presents the aperture of the duct of the parotid gland. The principal muscle of the cheek is the Buccinator; but numerous other muscles enter into its formation; viz., the Zygomatici, Masseter, and Platysma myoides.

The *buccal glands* are placed between the mucous membrane and Buccinator muscle: they are similar in structure to the labial glands, but smaller. Two or three, of larger size than the rest, are placed between the Masseter and Buccinator muscles; their ducts open into the mouth, opposite the last molar tooth. They are called *molar glands*.

The GUMS are composed of a dense fibrous tissue, closely connected to the periosteum of the alveolar processes, and surrounding the necks of the teeth. They are covered by smooth and vascular mucous membrane, which is remarkable for its limited sensibility. Around the necks of the teeth, this membrane presents numerous fine papillæ; and from this point it is reflected into the alveolus, where it is continuous with the periosteal membrane lining that cavity.

The Teeth.

The human subject is provided with two sets of teeth, which make their appearance at different periods of life. The first set appear in childhood, and are called the *temporary*, *deciduous*, or *milk teeth.* The second set, which also appear at an early period, continue until old age, and are named *permanent.*

The *temporary teeth* are twenty in number; four incisors, two canine, and four molars, in each jaw.

The *permanent teeth* are thirty-two in number; four incisors, two central and two lateral, two canine, four bicuspids, and six molars, in each jaw.

General characters. Each tooth consists of three portions; the crown or body,

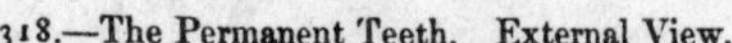
318.—The Permanent Teeth. External View.

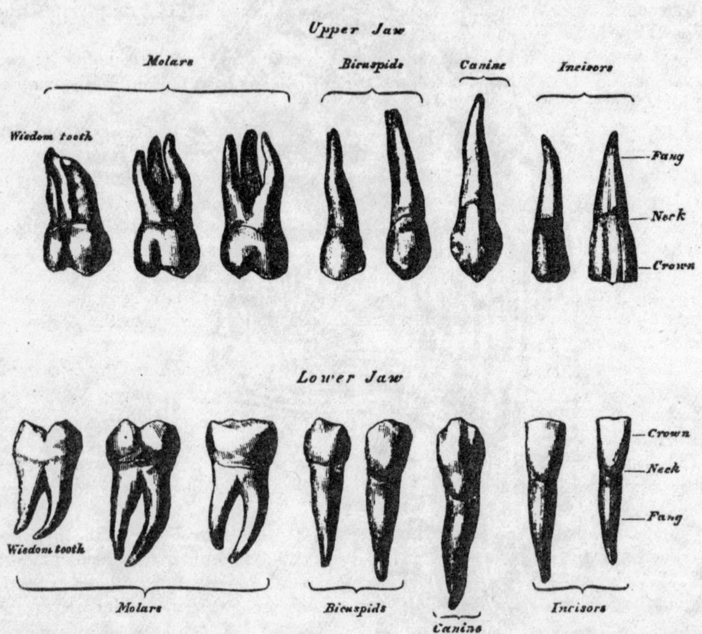

projecting above the gum; the root, or fang, entirely concealed within the alveolus; and the neck, the constricted portion between the other two.

The *roots of the teeth* are firmly implanted within the alveoli: these depressions are lined with periosteum, which is reflected on to the tooth at the point of the fang, and covers it as far as the neck. At the margin of the alveolus, the periosteum becomes continuous with the fibrous structure of the gums.

Permanent Teeth.

The Incisors, or cutting teeth, are so named from their presenting a sharp, cutting edge, adapted for cutting the food. They are eight in number, and form the four front teeth in each jaw.

The *crown* is directed vertically, is wedge-like in form, being bevelled at the expense of its posterior surface, so as to terminate in a sharp, horizontal cutting edge, which, before being subject to attrition, presents three small prominent points. It is convex, smooth, and highly polished in front; slightly concave behind, where it is frequently marked by slight longitudinal furrows.

The *neck* is constricted.

The *fang* is long, single, conical, transversely flattened, thicker before than behind, and slightly grooved on each side in the longitudinal direction.

The *incisors of the upper jaw* are altogether larger and stronger than those of the lower jaw. They are directed obliquely downwards and forwards. The two central ones are larger than the two lateral, and their free edges are sharp and chisel-like, being bevelled at the expense of their posterior edge: the root is more rounded.

The *incisors of the lower jaw* are smaller than the upper: the two central ones are smaller than the two lateral, and are the smallest of all the incisor teeth.

The Canine Teeth (*cuspidati*) are four in number, two in the upper, and two in the lower jaw; one being placed behind each lateral incisor. They are larger and stronger than the incisors, especially the root, which sinks deeply into the jaw, and causes a well-marked prominence upon its surface.

The *crown* is large and conical, very convex in front, a little hollowed and uneven posteriorly, and tapering to a blunted point, or cusp, which rises above the level of the other teeth.

The *root* is single, but longer and thicker than that of the incisors, conical in form, compressed laterally, and marked by a slight groove on each side.

The *upper canine teeth* (vulgarly called eye-teeth) are larger and longer than the two lower, and situated a little behind them.

The *lower canine teeth* are placed in front of the upper, so that their summits correspond to the interval between the upper canine tooth and the neighbouring incisors on each side.

The Bicuspid Teeth (small, or false molars), are eight in number, four in each jaw, two being placed immediately behind each of the canine teeth. They are smaller and shorter than the canine.

The *crown* is compressed from without inwards, and surmounted by two pyramidal eminences, or cusps, separated by a groove, hence their name, *bicuspidate.* The outer of these cusps is larger and more prominent than the inner.

The *neck* is oval.

The *root* is generally single, compressed, and presents a deep groove on each side, which indicates a tendency in the root to become double. The apex is generally bifid.

The *upper bicuspids* are larger, and present a greater tendency to the division of their roots than the lower: this is especially marked in the second upper bicuspid.

The Molar Teeth (*multicuspidati*, true, or large molars), are the largest of the permanent set, and are adapted, from the great breadth of their crowns, for grinding and pounding the food. They are twelve in number, six in each jaw, three being placed behind each of the posterior bicuspids.

The *crown* is nearly cubical in form, rounded on each of its lateral surfaces, flattened in front and behind; the upper surface being surmounted by four or five tubercles, or cusps (four in the upper, five in the lower molars), separated from each other by a crucial depression, hence their name, *multicuspidati.*

The *neck* is distinct, large, and rounded.

The *root* is subdivided into from two to five fangs, each of which presents an aperture at its summit.

The *first molar tooth* is the largest and broadest of all: its crown has usually five cusps, three outer and two inner. In the upper jaw, the root consists of three fangs, widely separated from one another, two being external, the other internal. The latter is the largest and the longest, slightly grooved, and sometimes bifid. In the lower jaw, the root consists of two fangs, one being placed in front, the other behind: they are both compressed from before backwards, and grooved on their contiguous faces, indicating a tendency to division.

The *second molar* is a little smaller than the first.

The crown has four cusps in the upper, and five in the lower jaw.

The root has three fangs in the upper jaw, and two in the lower, the characters of which are similar to the preceding tooth.

The *third molar tooth* is called the *wisdom tooth* (*dens sapientiæ*), from its late appearance through the gum. It is smaller than the others, and its axis is directed inwards.

The crown is small and rounded, and furnished with three tubercles.

The root is generally single, short, conical, slightly curved, and grooved so as to present traces of a subdivision into three fangs in the upper, and two in the lower jaw.

Temporary Teeth.

The temporary, or milk teeth, are smaller, but resemble in form those of the permanent set. The hinder of the two temporary molars is the largest of all the

319.—The Temporary, or Milk Teeth.
External View.

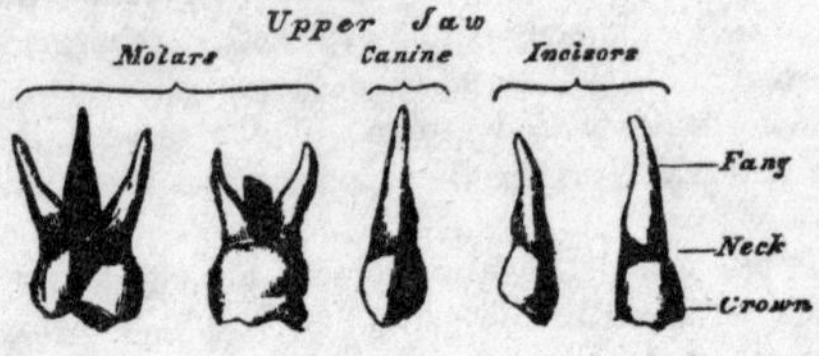

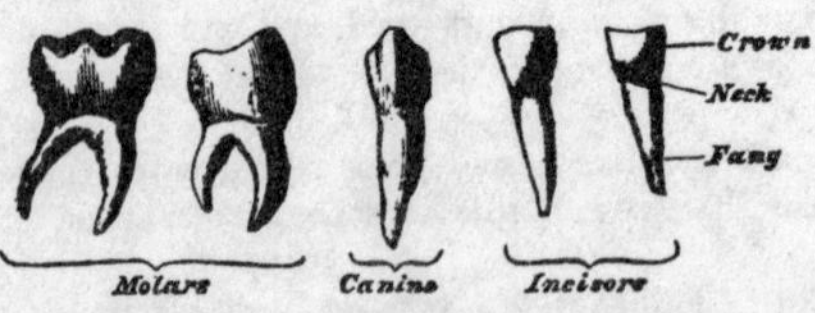

milk teeth, and is succeeded by the second permanent bicuspid. The first upper molar has only three cusps, two external, one internal: the second upper molar has four cusps. The first lower molar has four cusps: the second lower molar has five. The fangs of the temporary molar teeth are smaller, and more diverging than those of the permanent set; but, in other respects, bear a strong resemblance to them.

Structure. On making a vertical section of a tooth (fig. 320), a hollow cavity will be found in the interior. This cavity is situated at the base of the crown, and is continuous with a canal which traverses the centre of each fang, and opens by a minute orifice at its extremity. The shape of the cavity corresponds somewhat with that of the tooth: it forms what is called the *pulp cavity*, and contains a soft, highly vascular, and sensitive substance, the dental pulp. The pulp is richly supplied with vessels and nerves, which enter the cavity through the small aperture at the point of each fang.

320.—Vertical Section of a Molar Tooth.

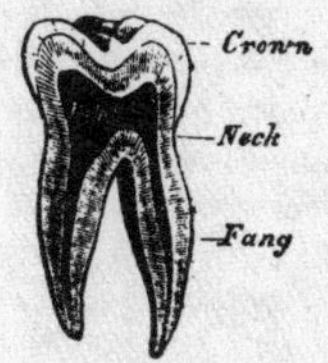

The solid portion of the tooth consists of three distinct structures, viz., ivory (tooth-bone, or dentine), which forms the larger portion of the tooth; enamel, which covers the exposed part, or crown; and the cortical substance, or cement (*crusta petrosa*), which is disposed as a thin layer on the surface of the fang.

The Ivory, or dentine (fig. 321), forms the principal mass of a tooth; in its central part is the cavity enclosing the pulp. It is a modification of the osseous tissue, from which it differs, however, in structure and chemical composition. On examination with the microscope, it is seen to consist of a number of minute wavy and branching tubes, having distinct parietes. They are called the *dental tubuli*, and are imbedded in a dense homogeneous substance, the *intertubular tissue*.

The *dental tubuli* are placed parallel with one another, and open at their inner ends into the pulp cavity. They pursue a wavy and undulating course towards the periphery. The direction of these tubes varies; they are vertical in the upper portion of the crown, oblique in the neck and upper part of the root, and towards the lower part of the root they are inclined downwards. The tubuli, at their commencement, are about $\frac{1}{4500}$ of an inch in diameter; in their course they divide and subdivide dichotomously, so as to give to the cut surface of the dentine a striated appearance. From the sides of the tubes, especially in the fang, ramifications of extreme minuteness are given off, which join together in loops in the intertubular substance, or terminate in small dilatations, from which branches are given off. Near the periphery of the dentine, the finer ramifications of the tubuli terminate in a somewhat similar manner. In the fang, these ramifications occasionally pass into the crusta petrosa. The dental tubuli have comparatively thick walls, and contain, according to Mr. Tomes, slender cylindrical prolongations of the pulp-tissue.

321.—Vertical Section of a Bicuspid Tooth. (Magnified.)

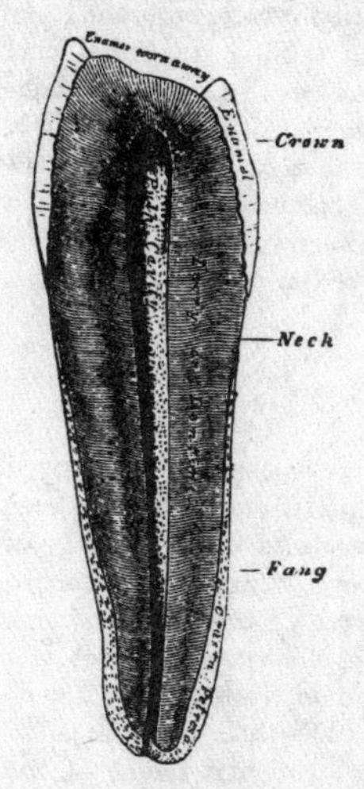

The *intertubular substance* is translucent, finely granular, and contains the chief part of the earthy matter of the dentine. After the earthy matter has been removed, by steeping a tooth in weak acid, the animal basis remaining is described by Henle as consisting of bundles of pale, granular, flattened fibres, running parallel with the tubes; but by Mr. Nasmyth as consisting of a mass of brick-shaped cells surrounding the tubules. By Czermak and Mr. Salter it is supposed to consist of laminæ which run parallel with the pulp cavity, across the direction of the tubes.

Chemical Composition. According to Berzolius and Bibra, dentine consists of 28 parts of animal, and 72 of earthy matter. The animal matter is resolvable by boiling into gelatin. The earthy matter consists of phosphate of lime, carbonate of lime, a trace of fluoride of calcium, phosphate of magnesia and other salts.

The Enamel is the hardest and most compact part of a tooth, and forms a thin crust over the exposed part of the crown, as far as the commencement of the fang. It is thickest on the grinding surface of the crown, until worn away by attrition, and becomes thinner towards the neck. It consists of a congeries of minute hexagonal rods. They lie parallel with one another, resting by one extremity upon the dentine, which presents a number of minute depressions for their reception; and forming the free surface of the crown by the other extremity. These fibres are directed vertically on the summit of the crown, horizontally at the sides; they are about the $\frac{1}{5500}$ of an inch in diameter, and pursue a more or less wavy course, which gives to the cut surface of the enamel the appearance of a series of concentric lines.

Numerous minute interstices intervene between the enamel-fibres near their dentinal surface, a provision calculated to allow of the permeation of fluids from the dentinal tubuli into the substance of the enamel. The enamel-rods consist of solid hexagonal or four-sided prisms, connected by their surfaces and ends, and filled with calcareous matter. If the latter is removed, by weak acid, from newly-formed or growing enamel, it will be found to present a network of delicate prismatic cells of animal matter.

Chemical Composition. According to Bibra, enamel consists of 96·5 per cent. of earthy matter, and 3·5 per cent. of animal matter. The earthy matter consists of phosphate of lime, with traces of fluoride of calcium, carbonate of lime, phosphate of magnesia, and other salts.

The CORTICAL SUBSTANCE, or cement (*crusta petrosa*), is disposed as a thin layer on the roots of the teeth, from the termination of the enamel, as far as the apex of the fang, where it is usually very thick. In structure and chemical composition, it resembles bone. It contains, sparingly, the lacunæ and canaliculi which characterise true bone: the lacunæ placed near the surface, have the canaliculi radiating from the side of the lacunæ towards the periodontal membrane; and those more deeply placed, join with the adjacent dental tubuli. In the thicker portions of the crusta petrosa, the lamellæ and Haversian canals peculiar to bone are also found. As age advances, the cement increases in thickness, and gives rise to those bony growths, or exostoses, so common in the teeth of the aged; the pulp cavity becomes also partially filled up by a hard substance, intermediate in structure between dentine and bone (*osteo-dentine*, Owen; *secondary dentine*, Tomes). It appears to be formed by a slow conversion of the dental pulp, which shrinks, or even disappears.

DEVELOPMENT OF THE TEETH. (Figs. 322 to 327.)

According to the observations of Arnold and Goodsir, the teeth are developed from the mucous membrane covering the edges of the maxillary arches. About the sixth week of fœtal life (fig. 322), the mucous membrane covering the edge of the upper jaw, presents a semicircular depression or groove: this is the primitive dental groove (Goodsir), from the floor of which the germs of the ten deciduous or milk-teeth are developed. The germ of each tooth is formed by a conical elevation or papilla of mucous membrane (fig. 323) which constitutes the rudimentary pulp of a milk-tooth. The germs of the milk-teeth make their appearance in the following order: at the seventh week, the germ of the first molar of the upper jaw appears; at the eighth week, that for the canine tooth is developed; the two incisor papillæ appear about the ninth week (the central preceding the lateral); lastly, the second molar papilla is seen at the tenth week, behind the anterior molar. The teeth of the lower jaw appear rather later, the first molar papilla being only just visible at the seventh week; and the tenth papilla not being developed before the eleventh week. This completes the first or papillary stage of their development.

The dental groove now becomes contracted, its margins thickened and prominent, and the groove is converted into follicles for the reception of the papillæ, by the growth of membranous septa, which pass across the groove between its borders (fig. 324). The follicles by this means become the alveoli, lined by periosteum, from the bottom of which the process of the mucous membrane of the gum rises, which is the germ of the future tooth. The follicle for the first molar is complete about the tenth week; the canine follows next,

Development of Teeth.

FIG. 322.

Primitive Dental Groove

FIG. 323.

Papilla, or Germ of Milk-tooth

Papillary Stage

FIG. 324.

Opercula

Follicular Stage

FIG. 325.

Saccular Stage

FIG. 326.

Germ of Permanent tooth in Cavity of Reserve

FIG. 327.

Eruption of Milk-tooth

Permanent tooth

Development of Permanent tooth

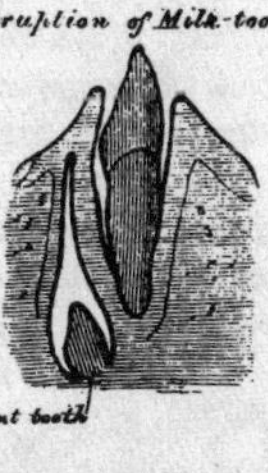

succeeded by the follicles for the incisors, which are completed about the eleventh or twelfth week; and, lastly, the follicle of the posterior molar is completed about the fourteenth week. These changes constitute the second or follicular stage.

About the thirteenth week, the papillæ begin to grow rapidly, project from the follicles, and assume a form corresponding with that of the future teeth; the follicles soon become deeper, and from their margins small membranous processes, or opercula, are developed, which, meeting, unite and form a lid to the now closed cavity (fig. 325). These processes correspond in shape to the form of the crown of the tooth, and in number to the tubercles on its surface. The follicles of the incisor teeth have two opercula, the canine three, and the molars four or five each. The follicles are thus converted into dental sacs, and the contained papillæ become pulps. The lips of the dental groove gradually advance over the follicles from behind forwards, and, uniting, gradually obliterate it. This completes the third or saccular stage, which takes place about the end of the fifteenth week.

The deep portion of the primitive dental groove is now closed in; but the more superficial portion, near the surface of the gum, still remains open; it is called, by Mr. Goodsir, the *secondary dental groove*; from it are developed the ten anterior permanent teeth. About the fourteenth week, certain lunated depressions are formed, one behind each of the sacs of the rudimentary milk-teeth. They are ten in number in each jaw, and are formed successively from before backwards; they are the rudimentary follicles of the four permanent incisors, the two canine, and the four bicuspids. As the secondary dental groove closes in, the follicles become closed cavities of reserve (fig. 325). The cavities soon elongate, and recede from the surface into the substance of the gum, behind the sacs of the deciduous teeth, and a papilla projects from the bottom of each, which is the germ of the permanent tooth; at the same time, one or more opercula are developed from the sides of the cavity; and these uniting, divide it into two portions; the lower portion containing the papilla of the permanent tooth, the upper narrower portion becoming gradually contracted in the same way that the primitive dental groove was obliterated over the sacs of the deciduous teeth (fig. 326).

The six posterior permanent teeth in each jaw, three on each side, arise from successive extensions backwards of the back part of the primitive dental groove. During the fourth month, that portion of the dental groove which lies behind the last temporary molar follicle, remains open, and from it is developed the papilla, the rudiment of the first permanent molar. The follicle in which it is contained, becomes closed by its operculum, and the upper part of the now-formed sac elongates backwards to form a cavity of reserve, in which the papilla of the second permanent molar appears at the seventh month after birth. After a considerable interval, during which the sacs of the first and second permanent molars have considerably increased in size, the remainder of the cavity of reserve presents for the last time a series of changes similar to the preceding, and gives rise to the sac and papilla of the wisdom-tooth, which appears at the sixth year.

Growth of the Teeth. As soon as the dental sacs are formed by the closing in of the follicles, they gradually enlarge, as well as their contained papillæ. Each sac consists of two layers: an internal, highly vascular layer, lined by epithelium; and an external or areolo-fibrous membrane, analogous to the corium of the mucous membrane.

The *dental pulps* soon become moulded to the form of the future teeth, and are adherent by their bases to the bottom of the dental sacs; in the case of the molars the base of the pulp is divided into two or more portions, which form the future fangs. During the fourth or fifth month of fœtal life, a thin lamina or cap of dentine is formed on the most prominent point of the pulp of all the milk-teeth. In the incisor and canine teeth, this newly-formed lamina has the form of a hollow cone; in the molar teeth, as many separate laminæ are found as there are eminences upon its crown. These laminæ grow at the expense of the pulp-substance, increasing in breadth by a growth round their margins, and in thickness by a similar formation in its substance; the separate cones (if a molar tooth) ultimately coalesce, and the crown is completely formed. The pulp now becomes constricted, so as to form the cervix; and the remaining portion becomes narrow and elongated, to form the fang. The growth of dentine takes place from the surface towards the interior, until nothing but the small pulp-cavity remains in the centre of the tooth, communicating by the aperture left at the point of each fang with the dental vessels and nerves.

As soon as the formation of the dentine has commenced, there is developed from the inner wall of the dental sac a soft pulpy mass, the *enamel organ*, which is intimately united to the surface of the dental pulp, or its cap of dentine. It consists of a mesh of fibres, elastic and spongy, containing within its reticulations fluid albumen; and at the point of junction of each fibre, a transparent nucleus is visible. The surface towards the dentinal pulp is covered by a layer of elongated nucleated cells, the *enamel membrane.* The deposition of the enamel takes place on the outer surface of the cap of dentine.

The *cementum* appears to be formed at a later period of life, by the periodontal membrane, extending from the margin of the enamel downwards.

Eruption. When the calcification of the different tissues of the tooth is sufficiently advanced to enable it to bear the pressure to which it will be afterwards subjected, its eruption takes place, the tooth making its way through the gum. The gum is absorbed by the pressure of the crown of the tooth against it, which is itself pressed up by the increasing size of the fang (fig. 327). At the same time, the septa between the dental sacs, at first fibrous in structure, ossify, and constitute the alveoli; these firmly embrace the necks of the teeth, and afford them a solid basis of support.

The eruption of the temporary teeth commences at the seventh month, and is complete about the end of the second year, those of the lower jaw preceding the upper.

The periods for the eruption of the temporary set are:

7th month, central incisors.
7th to 10th month, lateral incisors.
12th to 14th month, anterior molars.
14th to 20th month, canine.
18th to 36th month, posterior molars.

Calcification of the permanent teeth commences a little before birth, and proceeds in the following order in the upper jaw, in the lower jaw a little earlier. First molar, five or six months; the central incisor a little later; lateral incisors and canine, about the eighth or ninth month; the bicuspids at the second year; second molar, five or six years; wisdom-tooth, about twelve years.

Previous to the permanent teeth penetrating the gum, the bony partitions which separate their sacs from the deciduous teeth are absorbed, the fangs of the temporary teeth disappear, and the permanent teeth become placed under the loose crowns of the deciduous teeth; the latter finally become detached, and the permanent teeth take their place in the mouth.

The eruption of the permanent teeth takes place at the following periods, the teeth of the lower jaw preceding those of the upper by a short interval:

6½ years, first molars.	10th year, second bicuspid.
7th year, two middle incisors.	11th to 12th year, canine.
8th year, two lateral incisors.	12th to 13th year, second molars.
9th year, first bicuspid.	17th to 21st year, wisdom-teeth.

The Palate.

The Palate forms the roof of the mouth; it consists of two portions, the hard palate in front, the soft palate behind.

The *hard palate* is bounded in front and at the sides by the alveolar arches and gums; behind, it is continuous with the soft palate. It is covered by a dense structure, formed by the periosteum and mucous membrane of the mouth, which are intimately adherent together. Along the middle line is a linear ridge or raphe, which terminates anteriorly in a small papilla, corresponding with the inferior opening of the anterior palatine fossa. This papilla receives filaments from the naso-palatine and anterior palatine nerves. On either side and in front of the raphe, the mucous membrane is thick, pale in colour, and corrugated; behind, it is thin, smooth, and of a deeper colour: it is covered with squamous epithelium, and furnished with numerous glands (palatal glands), which lie between the mucous membrane and the surface of the bone.

The *soft palate* (*Velum pendulum palati*) is a moveable fold, suspended from the posterior border of the hard palate, and forming an incomplete septum between the mouth and pharynx. It consists of a fold of mucous membrane, enclosing muscular fibres, an aponeurosis, vessels, nerves, and mucous glands. When occupying its usual position (i.e. relaxed and pendent), its anterior surface is concave, continuous with the roof of the mouth, and marked by a median ridge or raphe, which indicates its original separation into two lateral halves. Its posterior surface is convex, and continuous with the mucous membrane covering the floor of the posterior nares. Its upper border is attached to the posterior margin of the hard palate, and its sides are blended with the pharynx. Its lower border is free.

Hanging from the middle of its lower border is a small conical-shaped pendulous process, the *uvula*; and arching outwards and downwards from the base of the uvula on each side, are two curved folds of mucous membrane, containing muscular fibres, called the *arches* or *pillars of the soft palate*.

The *anterior pillar* runs downwards and forwards to the side of the base of the tongue, and is formed by the projection of the Palato-glossus muscle, covered by mucous membrane.

The *posterior pillars* are nearer to each other and larger than the anterior; they run downwards and backwards to the sides of the pharynx, and are formed by the projection of the Palato-pharyngei muscles, covered by mucous membrane. The anterior and posterior pillars are separated below by a triangular interval, in which the tonsil is lodged.

The space left between the arches of the palate on the two sides is called the *isthmus of the fauces*. It is bounded above by the free margin of the palate;

below, by the tongue; and on each side, by the pillars of the soft palate and tonsils.

The *mucous membrane* of the soft palate is thin, and covered with squamous epithelium on both surfaces, excepting near the orifice of the Eustachian tube, where it is columnar and ciliated. The palatine glands form a continuous layer on its posterior surface and round the uvula.

The *aponeurosis* of the soft palate is a thin but firm fibrous layer, attached above to the hard palate, and becoming thinner towards the free margin of the velum. It is blended with the aponeurotic tendon of the Tensor palati muscle.

The *muscles* of the soft palate are five on each side: the Levator palati, Tensor palati, Palato-glossus, Palato-pharyngeus, and Azygos uvulæ (see p. 224).

The *tonsils* (*amygdalæ*) are two glandular organs, situated one on each side of the fauces, between the anterior and posterior pillars of the soft palate. They are of a rounded form, and vary considerably in size in different individuals. Externally, the tonsil is in relation with the inner surface of the Superior constrictor, and with the internal carotid and ascending pharyngeal arteries, and corresponds to the angle of the lower jaw. Its *inner surface* presents from twelve to fifteen orifices, leading into small recesses, from which numerous follicles branch out into the substance of the gland. These follicles are lined by a continuation of the mucous membrane of the pharynx, covered with epithelium, their walls being formed by a layer of closed capsules embedded in the submucous tissue. These capsules are analogous to those of Peyer's glands; they contain a thick greyish secretion.

The *arteries* supplying the tonsil are the dorsalis linguæ from the lingual, the ascending palatine and tonsillar from the facial, the ascending pharyngeal from the external carotid, and the descending palatine branch of the internal maxillary.

The *veins* terminate in the tonsillar plexus, on the outer side of the tonsil.

The *nerves* are derived from the fifth, and from the glosso-pharyngeal.

The Salivary Glands. (Fig. 328.)

The principal salivary glands communicating with the mouth, and pouring their secretion into its cavity, are the parotid, submaxillary, and sublingual.

The *Parotid gland*, so called from being placed near the ear (*παρά, near*; *οὖς, ὠτός, the ear*), is the largest of the three salivary glands, varying in weight from half an ounce to an ounce. It lies upon the side of the face, immediately below and in front of the external ear. It is limited above by the zygoma; below, by the angle of the jaw, and by an imaginary line drawn between it and the Sterno-mastoid muscle; anteriorly, it extends to a variable extent over the Masseter muscle; posteriorly, it is bounded by the external meatus, the mastoid process, and the Sterno-mastoid and Digastric muscles, slightly overlapping the former.

Its *anterior surface* is grooved to embrace the posterior margin of the ramus of the lower jaw, and advances forwards beneath the ramus, between the two pterygoid muscles. Its outer surface, slightly lobulated, is covered by the integument and fascia, and has one or two lymphatic glands resting on it. Its inner surface extends deeply into the neck, by means of two large processes, one of which dips behind the styloid process, and projects beneath the mastoid process and the Sterno-mastoid muscle; the other is situated in front of the styloid process, and passes into the back part of the glenoid fossa, behind the articulation of the lower jaw. Embedded in its substance is the external carotid artery, which ascends behind the ramus of the jaw; the posterior auricular artery emerges from the gland behind; the temporal artery above; the transverse facial in front; and the internal maxillary winds through it inwards, behind the neck of the jaw. Superficial to the external carotid is the trunk formed by the union of the temporal and internal maxillary veins; a branch, connecting this trunk with the internal jugular, also traverses the gland. It is also traversed, from before backwards, by the facial nerve and its branches, which emerge at its anterior border; the great auricular nerve pierces the gland to join the facial, and the temporal branch of the inferior maxillary nerve lies above the upper part of the gland. The internal carotid artery and internal jugular vein lie close to its deep surface.

The *duct* of the parotid gland (Steno's) is about two inches and a half in length. It opens upon the inner surface of the cheek by a small orifice, opposite the second molar tooth of the upper jaw; and from this orifice it may be traced obliquely for a short distance beneath the mucous membrane, and thence through the substance of the Buccinator muscle, and across the Masseter to the anterior border of the gland, in the substance of which it commences by numerous branches.

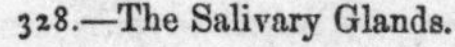

328.—The Salivary Glands.

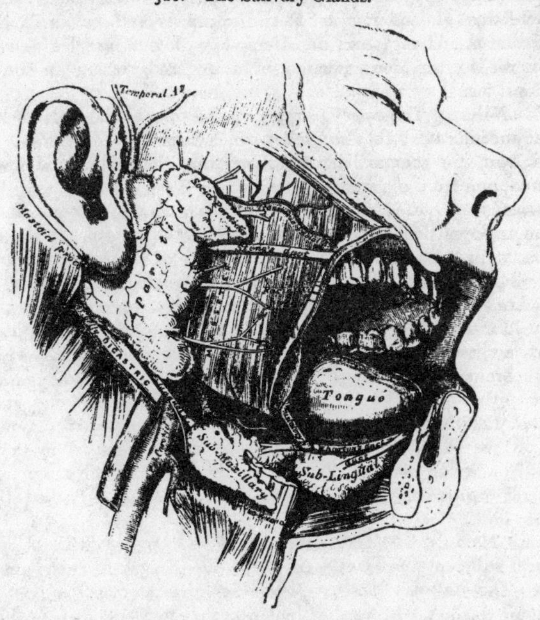

The direction of the duct corresponds to a line drawn across the face about a finger's breadth below the zygoma, from the lower part of the concha, to midway between the free margin of the upper lip and the ala of the nose. While crossing the Masseter, it receives the duct of a small detached portion of the gland, *socia parotidis*, which occasionally exists as a separate lobe, just beneath the zygomatic arch. The parotid duct is dense, of considerable thickness, and its canal about the size of a crow-quill; it consists of an external or fibrous coat, of considerable density, containing contractile fibres, and of an internal or mucous coat, lined with columnar epithelium.

Vessels and Nerves. The *arteries* supplying the parotid gland are derived from the external carotid, and from the branches of that vessel in or near its substance. The *veins* follow a similar course. The *lymphatics* terminate in the superficial and deep cervical glands, passing in their course through two or three lymphatic glands, placed on the surface and in the substance of the parotid. The *nerves* are derived from the carotid plexus of the sympathetic, the facial, the superficial temporal, the auriculo temporal, and great auricular nerves.

The *Submaxillary gland* is situated below the jaw, in the anterior part of the submaxillary triangle of the neck. It is irregular in form, and weighs about two drachms. It is covered by the integument, Platysma, deep cervical fascia, and the body of the lower jaw, corresponding to a depression on the inner surface of that bone; and lies upon the Mylo-hyoid, Hyo-glossus, and Stylo-glossus muscles, a portion of the gland passing beneath the posterior border of the Mylo-hyoid. In front of it is the anterior belly of the Digastric; behind, it is separated from the parotid gland by the stylo-maxillary ligament, and from the sublingual gland in front by the Mylo-hyoid muscle. The facial artery lies imbedded in a groove in its posterior and upper border.

The *duct* of the submaxillary gland (Wharton's) is about two inches in length, and its walls are much thinner than those of the parotid duct. It opens by a narrow orifice on the summit of a small papilla, at the side of the frænum linguæ. Traced from thence, it is found to pass between the sublingual gland and the Genio-hyo-glossus muscle, then backwards and outwards between the Mylo-hyoid, and the Hyo-glossus and Genio-hyo-glossus muscles, and beneath the gustatory nerve, to the deep portion of the gland, where it commences by numerous branches.

Vessels and Nerves. The *arteries* supplying the submaxillary gland are branches of the facial and lingual. Its *veins* follow the course of the arteries. The *nerves* are derived from the submaxillary ganglion, from the mylo-hyoid branch of the inferior dental, and from the sympathetic.

The *Sublingual gland* is the smallest of the salivary glands. It is situated beneath the mucous membrane of the floor of the mouth, at the side of the frænum linguæ, in contact with the inner surface of the lower jaw, close to the symphysis. It is narrow, flattened, in shape somewhat like an almond, and weighs about a drachm. It is in relation, *above*, with the mucous membrane; *below*, with the Mylo-hyoid muscle; *in front*, with the depression on the side of the symphysis of the lower jaw, and with its fellow of the opposite side; *behind*, with the deep part of the submaxillary gland; and *internally*, with the Genio-hyo-glossus, from which it is separated by the lingual nerve and Wharton's duct. Its excretory ducts (*ductus Riviniani*), from eight to twenty in number, open separately into the mouth, on the elevated crest of mucous membrane, caused by the projection of the gland, on either side of the frænum linguæ. One or more join to form a tube which opens into the Whartonian duct: this is called the *duct of Bartholine.*

Vessels and Nerves. The sublingual gland is supplied with blood from the sublingual and submental arteries. Its nerves are derived from the gustatory.

Structure. The salivary are conglomerate glands, consisting of numerous lobes, which are made up of smaller lobules, connected together by dense areolar tissue, vessels, and ducts. Each lobule consists of numerous closed vesicles, which open into a common duct: the wall of each vesicle is formed of a delicate basement membrane, lined by epithelium, and covered on its outer surface with a dense capillary network. In the submaxillary and sublingual glands, the lobes are larger and more loosely united than in the parotid.

The Pharynx.

The pharynx is that part of the alimentary canal which is placed behind the nose, mouth, and larynx. It is a musculo-membranous sac, somewhat conical in form, with the base upwards, and the apex downwards, extending from the under surface of the skull to the cricoid cartilage in front, and the fifth cervical vertebra behind.

The pharynx is about four inches and a half in length, and broader in the transverse than in the antero-posterior diameter. Its greatest breadth is opposite the cornua of the hyoid bone; its narrowest point at its termination in the œsophagus. It is limited, *above*, by the basilar process of the occipital bone; *below*, it is continuous with the œsophagus; *posteriorly*, it is connected by loose areolar tissue with the cervical portion of the vertebral column, and the Longi colli and Recti capitis antici muscles; *anteriorly*, it is incomplete, and is attached in succession to the internal pterygoid plate, the pterygo-maxillary ligament, the lower jaw, the tongue, hyoid bone, and larynx; *laterally*, it is connected to the styloid processes and their muscles, and is in contact with the common and internal carotid arteries, the internal jugular veins, and the eighth, ninth, and sympathetic nerves, and above, with a small part of the Internal pterygoid muscles.

It has seven openings communicating with it: the two posterior nares, the two Eustachian tubes, the mouth, larynx, and œsophagus.

The *posterior nares* are the two large apertures situated at the upper part of the anterior wall of the pharynx.

The *two Eustachian tubes* open one at each side of the upper part of the pharynx, at the back part of the inferior meatus. Below the nasal fossæ is the posterior surface of the soft palate and uvula, the large aperture of the mouth, the base of the tongue, the epiglottis, and the cordiform opening of the larynx.

The *œsophageal opening* is the lower contracted portion of the pharynx.

Structure. The pharynx is composed of three coats: a mucous coat, a muscular layer, and a fibrous coat.

The *fibrous coat* is situated between the mucous and muscular layers, and is called the *pharyngeal aponeurosis*. It is thick above, where the muscular fibres are wanting, and is firmly connected to the basilar process of the occipital and petrous portion of the temporal bones. As it descends, it diminishes in thickness, and is gradually lost.

The *mucous coat* is continuous with that lining the Eustachian tubes, the nares, the mouth, and the larynx. It is covered by columnar ciliated epithelium, as low down as on a level with the floor of the nares; below that point, it is of the squamous variety.

The *muscular coat* has been already described (p. 223).

The *pharyngeal glands* are of two kinds: the simple, or compound follicular which are found in considerable numbers beneath the mucous membrane, throughout the entire pharynx; and the racemose, which are especially numerous at the upper part of the pharynx, and form a thick layer, across the back of the fauces, between the two Eustachian tubes.

The Œsophagus.

The œsophagus is a membranous canal, about nine inches in length, extending from the pharynx to the stomach. It commences at the lower border of the cricoid cartilage, opposite the fifth cervical vertebra, descends along the front of the spine, through the posterior mediastinum, passes through the diaphragm, and, entering the abdomen, terminates at the cardiac orifice of the stomach, opposite the ninth dorsal vertebra. The general direction of the œsophagus is vertical; but it presents two or three slight curvatures in its course. At its commencement, it is placed in the median line; but it inclines to the left side as far as the root of the neck, gradually passes to the middle line again, and, finally, again deviates to the left, as it passes forwards to the œsophageal opening of the Diaphragm. The œsophagus also presents an antero-posterior flexure, corresponding to the curvature of the cervical and thoracic portions of the spine. It is the narrowest part of the alimentary canal, being most contracted at its commencement, and at the point where it passes through the Diaphragm.

Relations. In the *neck*, the œsophagus is in relation, *in front*, with the trachea; and, at the lower part of the neck, where it projects to the left side, with the thyroid gland and thoracic duct; *behind*, it rests upon the vertebral column and Longus colli muscle; on *each side*, it is in relation with the common carotid artery (especially the left, as it inclines to that side), and part of the lateral lobes of the thyroid gland; the recurrent laryngeal nerves ascend between it and the trachea.

In the *thorax*, it is at first situated a little to the left of the median line: it then passes across the left side of the transverse part of the aortic arch, and descends in the posterior mediastinum, along the right side of the aorta, nearly to the Diaphragm, where it passes in front and a little to the left of the artery, previous to entering the abdomen. It is in relation, *in front*, with the trachea, the arch of the aorta, the left bronchus, and the posterior surface of the pericardium; *behind*, it rests upon the vertebral column, the Longus colli, and the intercostal vessels; and below, near the Diaphragm, upon the front of the aorta; *laterally*, it is covered by the pleuræ; the vena azygos major lies on the right, and the descending aorta on the left side. The pneumogastric nerves descend in close contact with it, the right nerve passing down behind, and the left nerve in front of it.

Surgical Anatomy.—The relations of the œsophagus are of considerable practical interest to the surgeon, as he is frequently required, in cases of stricture of this tube, to dilate the canal by a bougie, when it becomes of importance that the direction of the œsophagus, and its relations to surrounding parts, should be remembered. In cases of malignant disease of the œsophagus, where its tissues have become softened from infiltration of the morbid deposit, the greatest care is requisite in directing the bougie through the strictured part, as a false passage may easily be made, and the instrument may pass into the mediastinum, or into one or the other pleural cavity, or even into the pericardium.

The student should also remember that contraction of the œsophagus, and consequent symptoms of stricture, are occasionally produced by an aneurism of some part of the aorta pressing upon this tube. In such a case, the passage of a bougie could only hasten the fatal issue.

It occasionally happens that a foreign body becomes impacted in the œsophagus, which can neither be brought upwards nor moved downwards. When all ordinary means for its removal have failed, excision is the only resource. This, of course, can only be performed when it is not very low down. If the foreign body is allowed to remain, extensive inflammation and ulceration of the œsophagus may ensue. In one case with which I am acquainted, the foreign body ultimately penetrated the intervertebral substance, and destroyed life by inflammation of the membranes and substance of the cord.

The operation of œsophagotomy is thus performed. The patient being placed upon his back, with the head and shoulders slightly elevated, an incision, about four inches in length, should be made on the left side of the trachea, from the thyroid cartilage downwards, dividing the skin and Platysma. The edges of the wound being separated, the Omo-hyoid muscle, and the fibres of the Sterno-hyoid and Sterno-thyroid muscles, must be drawn inwards; the sheath of the carotid vessels being exposed, should be drawn outwards, and retained in that position by retractors; the œsophagus will then be exposed, and should be divided over the foreign body, which should then be removed. Great care is necessary to avoid wounding the thyroid vessels, the thyroid gland, and the laryngeal nerves.

Structure. The œsophagus has three coats: an external, or muscular; a middle, or cellular; and an internal, or mucous coat.

The *muscular coat* is composed of two planes of fibres of considerable thickness, an external longitudinal, and an internal circular.

The *longitudinal fibres* are arranged at the commencement of the tube, in three fasciculi: one in front, which is attached to the vertical ridge on the posterior surface of the cricoid cartilage; and one at each side, which are continuous with the fibres of the Inferior constrictor: as they descend they blend together, and form a uniform layer, which covers the outer surface of the tube.

The *circular fibres* are continuous above with the Inferior constrictor: their direction is transverse at the upper and lower parts of the tube, but oblique in the central part.

The muscular fibres in the upper part of the œsophagus are of a red colour, and consist chiefly of the striped variety; but below, they consist entirely of the involuntary muscular fibre.

The *cellular coat* connects loosely the mucous and muscular coats.

The *mucous coat* is thick, of a reddish colour above, and pale below. It is disposed in longitudinal folds, which disappear on distension of the tube. Its surface is studded with minute papillæ, and it is covered throughout with a thick layer of squamous epithelium.

The *œsophageal glands* are numerous small compound glands, scattered throughout the tube; they are lodged in the submucous tissue, and open upon the surface by a long excretory duct. They are most numerous at the lower part of the tube, where they form a ring round the cardiac orifice.

The Abdomen.

The abdomen is the largest cavity in the body, and is separated, below, from the pelvic cavity by the brim of the pelvis. It is of an oval form, the extremities

329.—The Regions of the Abdomen and their Contents.
(Edge of Costal Cartilages in dotted outline.)

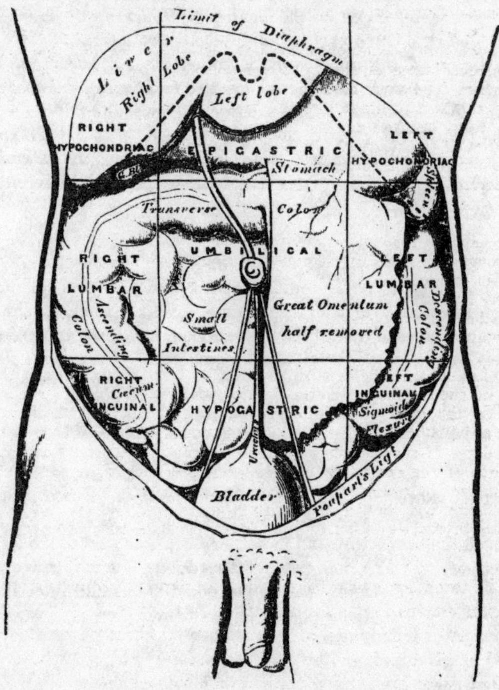

of the oval being directed upwards and downwards; it is wider above than below, and measures more in the vertical than in the transverse diameter.

Boundaries.—It is bounded, in *front* and at the *sides*, by the lower ribs, the Transversalis muscle, and venter ilii; *behind*, by the vertebral column, and the Psoas and Quadratus lumborum muscles; *above*, by the Diaphragm; *below*, by the brim of the pelvis. The muscles forming the boundaries of the cavities are lined upon their inner surface by a layer of fascia, differently named according to the part to which it is attached.

The abdomen contains the greater part of the alimentary canal; some of the accessory organs to digestion, viz., the liver, pancreas, and spleen; and the kidneys and suprarenal capsules. Most of these structures, as well as the wall of the cavity in which they are contained, are covered by an extensive and complicated serous membrane, the *peritoneum*.

The *apertures* found in the walls of the abdomen, for the transmission of structures to or from it, are the *umbilicus*, for the transmission (in the fœtus) of the umbilical vessels; the *caval opening* in the Diaphragm, for the transmission of the inferior vena cava; the *aortic opening*, for the passage of the aorta, vena azygos, and thoracic duct; and the *œsophageal opening*, for the œsophagus and pneumogastric nerves. *Below*, there are two apertures on each side: one for the passage of the femoral vessels, and the other for the transmission of the spermatic cord in the male, and the round ligament in the female.

Regions. For convenience of description of the viscera, as well as of reference to the morbid condition of the contained parts, the abdomen is artificially divided into nine regions. Thus, if two circular lines are drawn round the body, the one parallel with the cartilages of the ninth ribs, and the other with the highest point of the crests of the ilia, the abdominal cavity is divided into three zones, an upper, a middle, and a lower. If two parallel lines are drawn from the cartilage of the eighth rib on each side, down to the centre of Poupart's ligament, each of these zones is subdivided into three parts, a middle and two lateral.

The middle region of the upper zone is called the *epigastric* (ἐπί, *over*; γαστήρ, *the stomach*); and the two lateral regions, the *right* and *left hypochondriac* (ὑπό, *under*; χόνδροι, *the cartilages*). The central region of the middle zone is the *umbilical*; and the two lateral regions the *right* and *left lumbar*. The middle region of the lower zone is the *hypogastric* or *pubic region*; and the lateral regions are the *right* and *left inguinal*. The viscera contained in these different regions are the following (fig. 329):—

Right Hypochondriac.	*Epigastric Region.*	*Left Hypochondriac.*
The right lobe of the liver and the gall-bladder, the duodenum, pancreas, hepatic flexure of the colon, upper part of the right kidney, and the right suprarenal capsule.	The middle and pyloric end of the stomach, left lobe of the liver and lobulus Spigelii, and the pancreas.	The splenic end of the stomach, the spleen and extremity of the pancreas, the splenic flexure of the colon, upper half of the left kidney, and the left suprareual capsule.
Right Lumbar.	*Umbilical Region.*	*Left Lumbar.*
Ascending colon, lower part of the right kidney, and some convolutions of the small intestines.	The transverse colon, part of the great omentum and mesentery, transverse part of the duodenum, and some convolutions of the jejunum and ileum.	Descending colon, part of the omentum, lower part of the left kidney, and some convolutions of the small intestines.
Right Inguinal.	*Hypogastric Region.*	*Left Inguinal.*
The cæcum, appendix cæci, and ureter.	Convolutions of the small intestines, the bladder in children, and in adults if distended, and the uterus during pregnancy.	Sigmoid flexure of the colon, and ureter.

The Peritoneum.

The peritoneum (περιτείνειν, *to extend around*) is a serous membrane; and, like all membranes of this class, a shut sac. In the female, however, it is not completely closed, the Fallopian tubes communicating with it by their free extremities; and thus the serous membrane is continuous with their mucous lining.

The peritoneum partially invests all the viscera contained in the abdominal and pelvic cavities, forming the visceral layer of the membrane; it is then reflected upon the internal surface of the parietes of those cavities, forming the parietal layer. The *free* surface of the peritoneum is smooth, moist, and covered by a thin squamous epithelium; its *attached* surface is rough, being connected to the viscera and inner surface of the parietes by means of areolar tissue, called the *subperitoneal areolar tissue.* The parietal portion is loosely connected with the fascia lining the abdomen and pelvis; but more closely to the under surface of the Diaphragm, and in the middle line of the abdomen.

In order to trace the reflections of this membrane (fig. 330), the abdomen having been opened, the liver should be raised and supported in that position, and the stomach should be depressed, when a thin membranous layer is seen passing

330.—The Reflections of the Peritoneum, as seen in a vertical Section of the Abdomen.

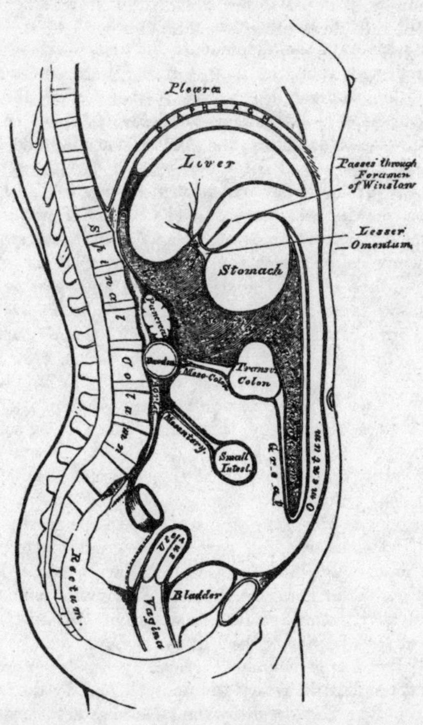

from the transverse fissure of the liver to the upper border of the stomach: this is the *lesser* or *gastro-hepatic omentum.* It consists of two delicate layers of peritoneum, an anterior and a posterior, between which are contained the hepatic vessels and nerves. Of these two layers, the anterior should first be traced, and then the posterior.

The *anterior layer* descends to the lesser curvature of the stomach, and covers its anterior surface as far as the great curvature; it descends for some distance in front of the small intestines, and, returning upon itself to the transverse colon, forms the *external layer of the great omentum*; it then covers the under surface of transverse colon, and, passing to the back part of the abdominal cavity, forms the *inferior layer of the transverse meso-colon.* It then descends in front of the duodenum, the aorta, and vena cava, as far as the superior mesenteric artery, along which it passes to invest the small intestines, and, returning to the vertebral column, form the *mesentery*; whilst, on either side, it covers the ascending and descending colon, and is thus continuous with the peritoneum lining the walls of the abdomen. From the root of the mesentery, it descends along the front of the spine into the pelvis, and surrounds the upper part of the rectum, which it holds in its position by means of a distinct fold, the *meso-rectum*. Its course in the male and female now differs.

In the *male*, it forms a fold between the rectum and bladder, the *recto-vesical fold*, and ascends over the posterior surface of the latter organ as far as its summit.

In the *female*, it descends into the pelvis in front of the rectum, covers a small part of the posterior wall of the vagina, and is then reflected on to the uterus, the fundus and body of which it covers. From the sides of the uterus, it is reflected on each side of the wall of the pelvis, forming the broad ligaments; and from the anterior surface of the uterus, it ascends upon the posterior wall of the bladder, as far as its summit. From this point it may be traced, as in the male, ascending upon the anterior parietes of the abdomen, to the under surface of the Diaphragm; from which it is reflected upon the liver, forming the upper layer of the coronary, and the lateral and longitudinal ligaments. It then covers the upper and under surfaces of the liver, and at the transverse fissure becomes continuous with the anterior layer of the lesser omentum, the point whence its reflection was originally traced.

The *posterior layer* of the lesser omentum descends to the lesser curvature of the stomach, and covers its posterior surface as far as the great curvature; it then descends for some distance in front of the small intestines, and, returning upon itself to the transverse colon, forms the internal layer of the great omentum; it covers the upper surface of the transverse colon, and, passing backwards to the spine, forms the upper layer of the transverse meso-colon. Ascending in front of the pancreas and crura of the Diaphragm, it lines the back part of the under surface of that muscle, from which it is reflected on to the posterior border of the liver, forming the inferior layer of the coronary ligament. From the under surface of the liver it may be traced to the transverse fissure, where it is continuous with the posterior layer of the lesser omentum, the point whence its reflection was originally traced.

The space included in the reflections of this layer of the peritoneum is called the *lesser cavity of the peritoneum*, or *cavity of the great omentum*. It is bounded, in *front*, by the lesser omentum, the stomach, and the descending part of the great omentum; *behind*, by the ascending part of the great omentum, the transverse colon, transverse meso-colon, and its ascending layer; *above*, by the liver; and *below*, by the folding of the great omentum. This space communicates with the general peritoneal cavity through the foramen of Winslow, which is situated behind the right, or free, border of the lesser omentum.

The *foramen of Winslow* is bounded *in front* by the lesser omentum, enclosing the vena portæ and the hepatic artery and duct; *behind*, by the inferior vena cava; *above*, by the lobulus Spigelii; *below*, by the hepatic artery curving forwards from the cæliac axis.

This foramen is nothing more than a constriction of the general peritoneal cavity at this point, caused by the hepatic and gastric arteries passing forwards from the cæliac axis to reach their respective viscera.

The viscera thus shown to be almost entirely invested by peritoneum are the liver, stomach, spleen, first portion of the duodenum, the jejunum, and ileum, the transverse colon, sigmoid flexure, upper end of the rectum, the uterus, and ovaries.

The viscera only partially invested by peritoneum are the descending and transverse portions of the duodenum, the cæcum, the ascending and descending colon, the middle portion of the rectum, and the upper part of the vagina and posterior wall of the bladder. The kidneys, suprarenal capsules, and pancreas are covered by the membrane without receiving any special investment from it.

The lower end of the rectum, the neck, base, and anterior surface of the bladder, the whole of the front, and the lower part of the posterior, wall of the vagina, have no peritoneal covering.

Numerous folds are formed by the peritoneum, extending between the various organs. These serve to hold them in position, and, at the same time, enclose the vessels and nerves proceeding to each part. Some of the folds are called *ligaments*, from their serving to support the organs in position. Others, which connect certain parts of the intestine with the abdominal wall, constitute the *mesenteries*; and lastly, those are called *omenta*, which proceed from the stomach to certain viscera in its neighbourhood.

The Ligaments, formed by folds of the peritoneum, include those of the liver, spleen, bladder, and uterus. They will be found described with their respective organs.

The Omenta are the lesser omentum, the great omentum, and the gastro-splenic omentum.

The *lesser omentum* (*gastro-hepatic*) is the duplicature which extends between the transverse fissure of the liver, and the lesser curvature of the stomach. It is extremely thin, and consists, as before said, of two layers of peritoneum. At the left border, its two layers pass on to the end of the œsophagus; but at the right border, where it is free, they are continuous, and form a free rounded margin, which contains between its layers the hepatic artery, the ductus communis choledochus, the portal vein, lymphatics, and the hepatic plexus of nerves—all these structures being enclosed in loose areolar tissue, called *Glisson's capsule*.

The *great omentum* (*gastro-colic*) is the largest peritoneal fold. It consists of four layers of peritoneum, two of which descend from the stomach, one from its anterior, the other from its posterior surface, and, uniting at its lower border, descend in front of the small intestines, as low down as the pelvis; they then turn upon themselves, and ascend again as far as the transverse colon, where they separate and enclose that part of the intestine. These separate layers may be easily demonstrated in the young subject; but, in the adult, they are more or less inseparably blended. The left border of the great omentum is continuous with the gastro-splenic omentum; its right border extends as far only as the duodenum. The great omentum is usually thin, presents a cribriform appearance, and always contains some adipose tissue, which, in fat subjects, accumulates in considerable quantity. Its use appears to be to protect the intestines from cold, and to facilitate their movement upon each other during their vermicular action.

The *gastro-splenic omentum* is the fold which connects the concave surface of the spleen to the *cul-de-sac* of the stomach, being continuous by its lower border with the great omentum. It contains the splenic vessels and vasa brevia.

The Mesenteries are, the mesentery proper, the meso-cæcum, the ascending, transverse, and descending meso-colon, the sigmoid meso-colon, and the meso-rectum.

The *mesentery* (*μέσον*, *ἔντερον*), so called from being connected to the middle of the cylinder of the small intestine, is the broad fold of peritoneum which connects the convolutions of the jejunum and ileum with the posterior wall of the abdomen. Its *root*, the part connected with the vertebral column, is narrow, about six inches in length, and directed obliquely from the left side of the second lumbar vertebra to the right sacro-iliac symphysis. Its intestinal border is much longer; and here its two layers separate, so as to enclose the intestine, and form its peritoneal coat. Its breadth, between its vertebral and intestinal border, is about four inches. Its *upper border* is continuous with the under surface of the transverse meso-colon; its *lower border*, with the peritoneum covering the cæcum and ascending colon. It serves to retain the small intestines in their position, and contains between its layers the mesenteric vessels and nerves, the lacteal vessels, and mesenteric glands.

The *meso-cæcum*, when it exists, serves to connect the back part of the cæcum with the right iliac fossa; more frequently, the peritoneum passes merely in front of this portion of the large intestine.

The *ascending meso-colon* is the fold which connects the back part of the ascending colon with the posterior wall of the abdomen; and the *descending meso-colon* retains the descending colon in connection with the posterior abdominal wall: more frequently, the peritoneum merely covers the anterior surface and sides of these two portions of the intestine.

The *transverse meso-colon* is a broad fold, which connects the transverse colon with the posterior wall of the abdomen. It is formed by the two ascending layers of the great omentum, which, after separating to surround the transverse colon, join behind it, and are continued backwards to the spine, where they diverge in front of the duodenum, as already mentioned. This fold contains between its layers the vessels which supply the transverse colon.

The *sigmoid meso-colon* is the fold of peritoneum which retains the sigmoid flexure in connection with the left iliac fossa.

The *meso-rectum* is the narrow fold which connects the upper part of the rectum with the front of the sacrum. It contains the hæmorrhoidal vessels.

The *appendices epiploicæ* are small pouches of the peritoneum filled with fat, and situated along the colon and upper part of the rectum. They are chiefly appended to the transverse colon.

The Stomach.

The stomach is the principal organ of digestion. It is the most dilated part of the alimentary canal, serving for the solution and reduction of the food, which constitutes the process of chymification. It is situated in the left hypochondriac, the epigastric, and part of the right hypochondriac regions. Its form is irregularly conical, curved upon itself, and presenting a rounded base, turned to the left side. It is placed immediately behind the anterior wall of the abdomen, above the transverse colon, below the liver and the Diaphragm. Its size varies considerably in different subjects, and also according to its state of distension. When moderately full, its transverse diameter is about twelve inches, its vertical diameter about four. Its weight, according to Clendenning, is about four ounces and a half. It presents for examination two extremities, two orifices, two borders, and two surfaces.

Its *left extremity* is called the *greater*, or *splenic end*. This is the largest part of the stomach, and extends two or three inches to the left of the point of entrance of the œsophagus. This expanded part is called the great *cul-de-sac*, or *fundus*. It lies beneath the ribs, in contact with the spleen, to which it is connected by the gastro-splenic omentum.

The *lesser*, or *pyloric end*, is much smaller than the fundus, and situated on a plane anterior and inferior to it. It lies in contact with the wall of the abdomen, the under surface of the liver, and the neck of the gall-bladder.

The *œsophageal* or *cardiac orifice* communicates with the œsophagus: it is the highest part of the stomach, and somewhat funnel-shaped.

The *pyloric orifice* communicates with the duodenum, the aperture being guarded by a kind of valve—the *pylorus*.

The *lesser curvature* extends between the œsophageal and pyloric orifices, along the upper border of the organ, and is connected to the under surface of the liver by the lesser omentum.

The *greater curvature* extends between the same points, along the lower border, and gives attachment to the great omentum. The surfaces of the organ are limited by these two curvatures.

The *anterior surface* is directed upwards and forwards, and is in relation with the Diaphragm, the under surface of the left lobe of the liver, and with the abdominal parietes, in the epigastric region.

The *posterior surface* is directed downwards and backwards, and is in relation with the pancreas and great vessels of the abdomen, the crura of the Diaphragm, and the solar plexus.

The stomach is held in position by the lesser omentum, which extends from the transverse fissure of the liver to its lesser curvature, and by a fold of peritoneum, which passes from the Diaphragm on to the œsophageal end of the stomach, the gastro-phrenic ligament; this constitutes the most fixed point of the stomach,

331.—The Mucous Membrane of the Stomach and Duodenum with the Bile Ducts.

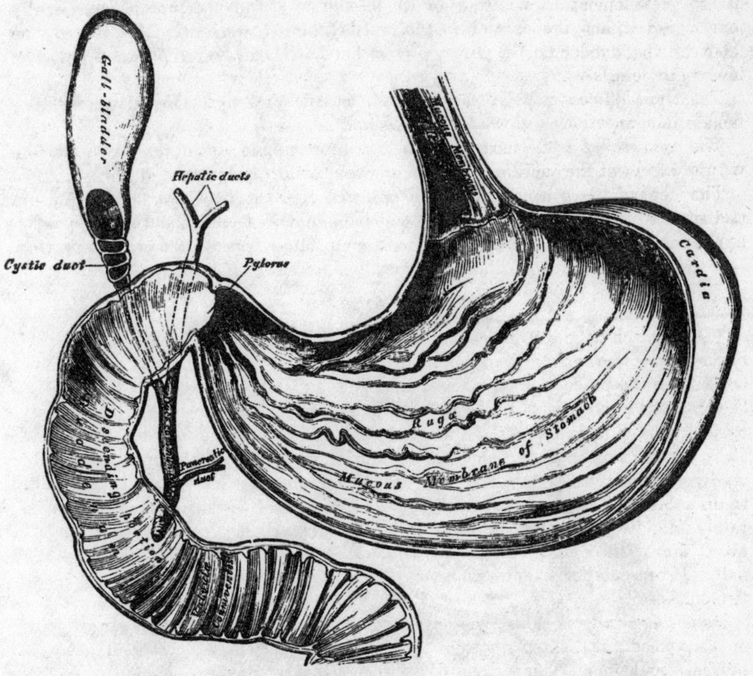

whilst the pyloric end and greater curvature are the most moveable parts: hence, when the stomach becomes greatly distended, the greater curvature is directed forwards, whilst the anterior and posterior surfaces are directed, the former upwards, and the latter downwards.

Alterations in Position.—There is no organ in the body the position and connections of which present such frequent alterations as the stomach. *During inspiration*, it is displaced downwards by the descent of the Diaphragm, and elevated by the pressure of the abdominal muscles during expiration. Its position in relation to the surrounding viscera is also changed, according to the empty or distended state of the organ. *When empty*, it occupies only a small part of the left hypochondriac region, the spleen lying behind it; the left lobe of the liver covers it in front, and the under surface of the heart rests upon it above, and in front; being separated from it by the left lobe of the liver, besides the diaphragm and pericardium. This close relation between the stomach and the heart explains the fact that, in gastralgia, the pain is generally referred to the heart, and is often accompanied by palpitation and intermission of the pulse. *When the stomach is distended* the Diaphragm is forced upwards, contracting the cavity of the chest; hence the dyspnœa complained of, from inspiration being impeded. The heart is also displaced upwards; hence the oppression in this region, and the palpitation experienced in extreme distension of the stomach. *Pressure from without*, as from tight lacing, pushes the stomach down towards the pelvis. In disease, also, the position and connections of the organ may be greatly changed, from the accumulation of fluid in the chest or abdomen, or from alteration in size of any of the surrounding viscera.

On looking into the pyloric end of the stomach, the mucous membrane is found projecting inwards in the form of a circular fold, the pylorus, leaving a narrow circular aperture, about half an inch in diameter, by which the stomach communicates with the duodenum.

The *pylorus* is formed by a reduplication of the mucous membrane of the stomach, containing numerous muscular fibres, which are aggregated into a thick circular ring, the longitudinal fibres and serous membrane being continued over the fold without assisting in its formation. The aperture is occasionally oval. Sometimes, the circular fold is replaced by two crescentic folds, placed, one above, and the other below, the pyloric orifice; and, more rarely, there is only one crescentic fold.

Structure. The stomach consists of four coats: a serous, a muscular, a cellular, and a mucous coat, together with vessels and nerves.

The *serous coat* is derived from the peritoneum, and covers the entire surface of the organ, excepting along the greater and lesser curvatures, at the points of attachment of the greater and lesser omenta; here the two layers of peritoneum leave a small triangular space, along which the nutrient vessels and nerves pass.

The *muscular coat* (fig. 332) is situated immediately beneath the serous covering. It consists of three sets of fibres: longitudinal, circular, and oblique.

332.—The Muscular Coat of the Stomach.

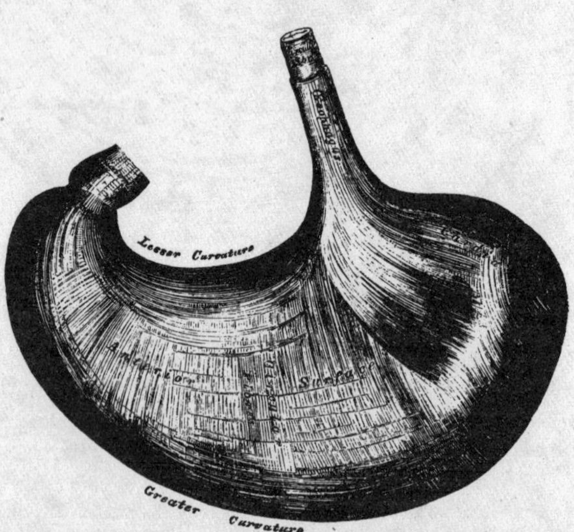

The *longitudinal fibres* are most superficial; they are continuous with the longitudinal fibres of the œsophagus, radiating in a stellate manner from the cardiac orifice. They are most distinct along the curvatures, especially the lesser; but are very thinly distributed over the surfaces. At the pyloric end, they are more thickly distributed, and continuous with the longitudinal fibres of the small intestine.

The *circular fibres* form a uniform layer over the whole extent of the stomach, beneath the longitudinal fibres. At the pylorus they are most abundant, and are aggregated into a circular ring, which projects into the cavity, and forms, with the fold of mucous membrane covering its surface, the pyloric valve.

The *oblique fibres* are limited chiefly to the cardiac end of the stomach, where they are disposed as a thick uniform layer covering both surfaces, some passing obliquely from left to right, others from right to left, round the cardiac orifice.

The *cellular coat* consists of a loose, filamentous, areolar tissue, connecting the mucous and muscular layers. It is sometimes called the *submucous coat.* It supports the blood-vessels previous to their distribution to the mucous membrane; hence it is sometimes called the *vascular coat.*

The *mucous membrane of the stomach* is thick; its surface smooth, soft, and velvety. During infancy, and immediately after death, it is of a pinkish tinge; but in adult life, and in old age, it becomes of a pale straw or ash-grey colour. It is thin at the cardiac extremity, but thicker towards the pylorus. During the contracted state of the organ, it is thrown into numerous plaits or rugæ, which, for the most part, have a longitudinal direction, and are most marked towards the lesser end of the stomach, and along the greater curvature (fig. 331). These folds are entirely obliterated when the organ becomes distended.

Structure of the mucous membrane (fig. 333). When examined with a lens, the inner surface of the mucous membrane presents a peculiar honeycomb appearance,

333.—Minute Anatomy of Mucous Membrane of Stomach.

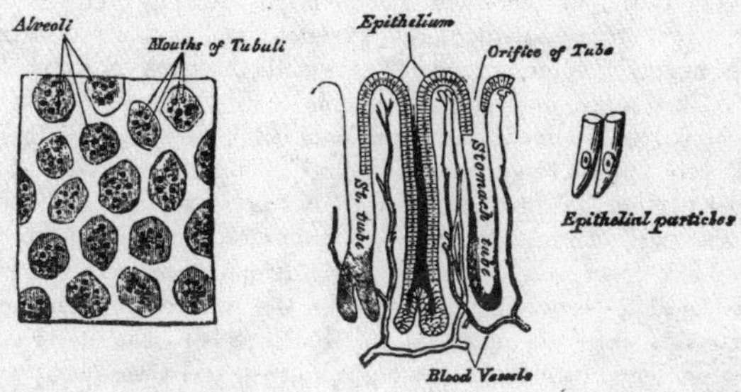

from being covered with small shallow depressions or alveoli, of a polygonal or hexagonal form, which vary from 1–100th to 1–350th of an inch in diameter, and are separated by slightly elevated ridges. In the bottom of the alveoli are seen the orifices of minute tubes, the *gastric follicles*, which are situated perpendicularly side by side, in the entire substance of the mucous membrane. They are short, and simply tubular in character towards the cardia; but at the pyloric end, they are longer, more thickly set, convoluted, and terminate in dilated saccular extremities, or are subdivided into from two to six tubular branches. The gastric follicles are composed of a homogeneous basement membrane, lined upon its free surface by a layer of cells, which differ in their character in different parts of the stomach. Towards the pylorus, these tubes are lined throughout with columnar epithelium; they are termed the *mucous glands*, and are supposed to secrete the gastric mucus. In other parts of the organ, the deep part of each tube is filled with nuclei, and a mass of granules; above these are a mass of nucleated cells, the upper fourth of the tube being lined by columnar epithelium. These are called the *peptic glands*, and are the supposed agents in the secretion of the gastric juice.

Simple follicles are found in greater or less number over the entire surface of the mucous membrane; they are most numerous near the pyloric end of the stomach, and are especially distinct in early life. The epithelium lining the mucous membrane of the stomach and its alveoli is of the columnar variety.

Vessels and Nerves. The *arteries* supplying the stomach are, the coronaria ventriculi, the pyloric and right gastro-epiploic branches of the hepatic, the left gastro-epiploic and vasa brevia from the splenic. They supply the muscular coat, ramify in the submucous coat, and are finally distributed to the mucous membrane. The *veins* accompany the arteries, and terminate in the splenic and portal veins. The *lymphatics* are numerous; they consist of a superficial and deep set, which pass through the lymphatic glands found along the two curvatures of the organ. The *nerves* are the terminal branches of the right and left pneumogastric, the former being distributed upon the back, and the latter upon the front part of the organ. A great number of branches from the sympathetic also supply the organ.

The Small Intestines.

The small intestine is that part of the alimentary canal in which the chyme is mixed with the bile, the pancreatic juice, and the secretions of the various glands embedded in the mucous membrane of the intestines, and where the separation of the nutritive principles of the food, the chyle, is effected: this constitutes chylification.

The small intestine is a convoluted tube, about twenty feet in length, which gradually diminishes in size from its commencement to its termination. It is contained in the central and lower parts of the abdominal and pelvic cavities, surrounded above and at the sides by the large intestine; in relation, in front, with the great omentum and abdominal parietes; and connected to the spine by a fold of peritoneum, the mesentery. The small intestine is divisible into three portions: the duodenum, jejunum, and ileum.

The *duodenum* has received its name from being about equal in length to the breadth of twelve fingers (eight or ten inches). It is the shortest, the widest, and the most fixed part of the small intestine; it has no mesentery, and is only partially covered by the peritoneum. Its course presents a remarkable curve, somewhat like a horseshoe in form; the convexity being directed towards the right, and the concavity to the left, embracing the head of the pancreas. Commencing at the pylorus, it ascends obliquely upwards and backwards to the under surface of the liver; it then descends in front of the right kidney, and passes nearly transversely across the front of the spine, terminating in the jejunum on the left side of the second lumbar vertebra. Hence the duodenum has been divided into three portions: ascending, descending, and transverse.

The first, or *ascending portion* (fig. 334), about two inches in length, is free, moveable, and nearly completely invested by the peritoneum. It is in relation, above and in front, with the liver and neck of the gall-bladder; behind, with the right border of the lesser omentum, the hepatic artery and duct, and vena portæ. This portion of the intestine is usually found, after death, stained with bile, especially on its anterior surface.

The second, or *descending portion*, about three inches in length, is firmly fixed by the peritoneum and pancreas. It passes from the neck of the gall-bladder vertically downwards, in front of the right kidney, as far as the third lumbar vertebra. It is covered by peritoneum only on its anterior surface. It is in relation, in front, with the right arch of the colon and the meso-colon; behind, with the front of the right kidney; at its inner side is the head of the pancreas, and the common choledic duct. The common bile duct and the pancreatic duct perforate the inner side of this portion of the intestine obliquely, a little below its middle.

The third, or *transverse portion*, the longest and narrowest part of the duodenum, passes across the front of the spine, ascending from the third to the second lumbar vertebra, and terminating in the jejunum on the left side of that bone. In front, it is covered by the descending layer of the transverse meso-colon, and crossed by the superior mesenteric vessels; behind, it rests upon the aorta, the vena cava, and the crura of the Diaphragm; above it, is the lower border of the pancreas, the superior mesenteric vessels passing forwards between the two.

Vessels and Nerves. The *arteries* supplying the duodenum are the pyloric and pancreatico-duodenal branches of the hepatic, and the inferior pancreatico-duodenal branch of the superior mesenteric. The *veins* terminate in the gastro-duodenal and superior mesenteric. The *nerves* are derived from the solar plexus.

334.—Relations of the Duodenum. (The Pancreas has been cut away, except its head.)

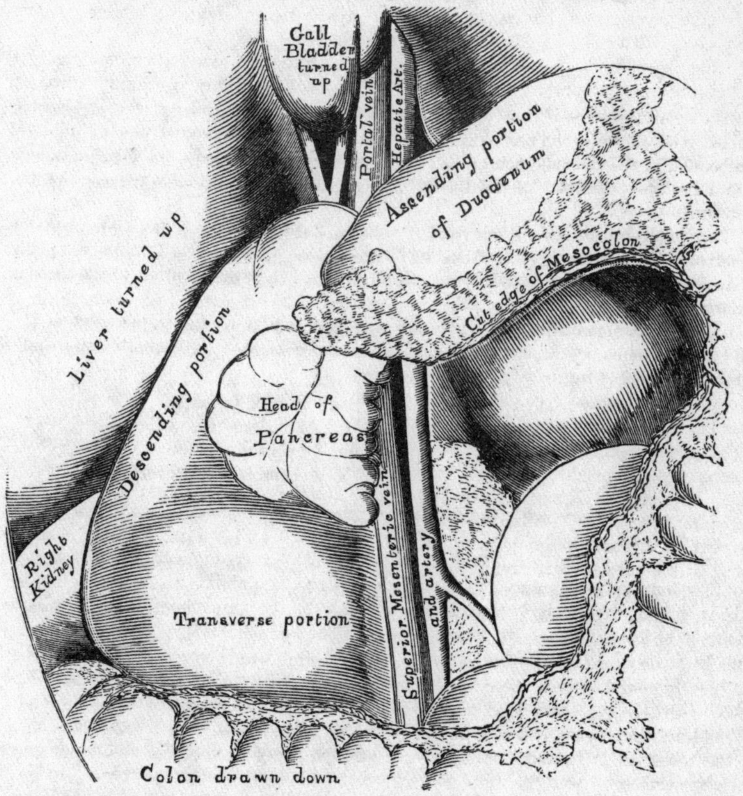

The *jejunum* (*jejunus*, empty), so called from being usually found empty after death, includes the upper two-fifths of the rest of the small intestine. It commences at the duodenum on the left side of the second lumbar vertebra, and terminates in the ileum; its convolutions being chiefly confined to the umbilical and left iliac regions. The jejunum is wider, its coats thicker, more vascular, and of a deeper colour than those of the ileum; but there is no characteristic mark to distinguish the termination of the one and the commencement of the other.

The *ileum* (*εἰλεῖν, to twist*), so called from its numerous coils or convolutions, includes the remaining three-fifths of the small intestine. It occupies chiefly the umbilical hypogastric, right iliac, and occasionally the pelvic regions, and terminates in the right iliac fossa by opening into the inner side of the commencement of the large intestine. The ileum is narrower, its coats thinner and less vascular than those of the jejunum; a given length of it weighing less than the same length of jejunum.

Structure. The wall of the small intestine is composed of four coats: serous, muscular, cellular, and mucous.

The *serous coat* is derived from the peritoneum. The first, or ascending portion of the duoderum, is almost completely surrounded by this membrane; the second, or descending portion, is covered by it only in front; and the third, or transverse, portion lies behind the descending layer of the transverse meso-colon, by which it is covered in front. The remaining portion of the small intestine is surrounded by the peritoneum, excepting along its attached or mesenteric border; here a space is left for the vessels and nerves to pass to the gut.

The *muscular coat* consists of two layers of fibres, an external or longitudinal, and an internal or circular layer. The *longitudinal* fibres are thinly scattered over the surface of the intestine, and are most distinct along its free border. The *circular* fibres form a thick, uniform layer; they surround the cylinder of the intestine in the greater part of its circumference, but do not form complete rings. The muscular coat is thicker at the upper than at the lower part of the small intestine.

The *cellular*, or *submucous coat*, connects together the mucous and muscular layers. It consists of a loose, filamentous, areolar tissue, which forms a nidus for the subdivision of the nutrient vessels, previous to their distribution to the mucous surface.

The *mucous membrane* is thick and highly vascular at the upper part of the small intestine, but somewhat paler and thinner below. It presents for examination the following constituents:

Epithelium.		Simple follicles.
Valvulæ conniventes.	Glands.	Duodenal glands.
Villi.		Solitary glands.
		Agminate or Peyer's glands.

The *epithelium*, covering the mucous membrane of the small intestine, is of the columnar variety.

The *valvulæ conniventes* (valves of Kerkring) are reduplications or foldings of the mucous membrane and submucous tissue, containing no muscular fibres. They extend transversely across the cylinder of the intestine for about three-fourths or five-sixths of its circumference. The larger folds are about two inches in length, and two-thirds of an inch in depth at their broadest part; but the greater number are of smaller size. The larger and smaller folds alternate with each other. They are not found at the commencement of the duodenum, but begin to appear about one or two inches beyond the pylorus. In the lower part of the descending portion, below the point where the common choledic and pancreatic ducts enter the intestine, they are very large and closely approximated. In the transverse portion of the duodenum and upper half of the jejunum, they are large and numerous; and from this point, down to the middle of the ileum, they diminish considerably in size. In the lower part of the ileum, they almost entirely disappear; hence the comparative thinness of this portion of the intestine, as compared with the duodenum and jejunum. The valvulæ conniventes retard the passage of the food along the intestines, and afford a more extensive surface for absorption.

The *villi* are minute, highly vascular processes, projecting from the mucous membrane of the small intestine throughout its whole extent, and giving to its surface a velvety appearance. In shape, some are triangular and laminated, others conical or cylindrical, with clubbed or filiform extremities. They are largest and most numerous in the duodenum and jejunum, and become fewer and smaller in the ileum. Krause estimates their number in the upper part of the small intestine at from fifty to ninety in a square line; and in the lower part from forty to seventy; the total number for the whole length of the intestine being about four millions.

In structure each villus consists of a network of capillary and lacteal vessels, with nuclear corpuscles and fat-globules in their interstices, enclosed in a thin prolongation of basement membrane covered by a single layer of columnar epithelium, the particles of which are arranged perpendicular to the surface. A layer of organic muscular fibre has been described forming a thin hollow cone round the central lacteal. It is possible that this assists in the propulsion of the chyle along the vessel. The mode of origin of the lacteals within the villi is unknown.

335.—Two Villi magnified.

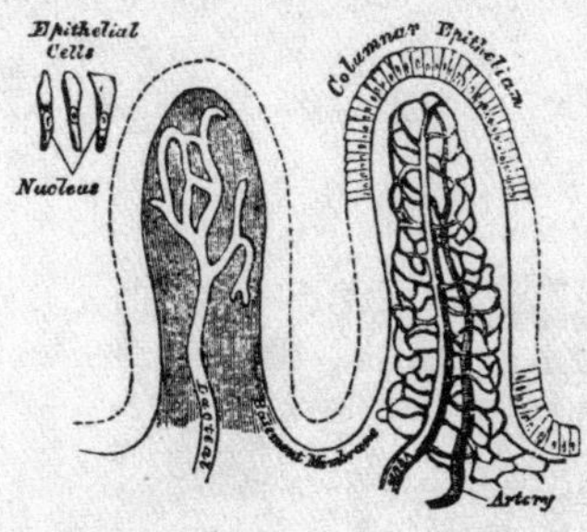

The *simple follicles*, or crypts of Lieberkühn, are found in considerable numbers over every part of the mucous membrane of the small intestine. They consist of minute tubular depressions of the mucous membrane, arranged perpendicularly to the surface, upon which they open by small circular apertures. They may be seen with the aid of a lens, their orifices appearing as minute dots, scattered between the villi. Their walls are thin, consisting of a layer of basement membrane, lined by cylindrical epithelium, and covered on their exterior by capillary vessels. Their contents vary, even in health, and the purpose served by their secretion is still very doubtful.

The *duodenal*, or Brunner's glands, are limited to the duodenum and commencement of the jejunum. They are small, flattened, granular bodies, imbedded in the submucous areolar tissue, and open upon the surface of the mucous membrane by minute excretory ducts. They are most numerous and largest near the pylorus. They may be compared to the elementary lobules of a salivary gland, spread out over a broad surface, instead of being collected in a mass. In structure they resemble the pancreas.

The *solitary glands* (*glandulæ solitariæ*), are found scattered throughout the mucous membrane of the small intestine, but are most numerous in the lower part of the ileum. They are small, round, whitish bodies, from half a line to a line in diameter, consisting of a closed saccular cavity, having no excretory duct, and containing an opaque white secretion. Their free surface is covered with villi, and each gland is surrounded by openings like those of the follicles of Lieberkühn. Their use is not known.

Peyer's glands may be regarded as aggregations of solitary glands, forming circular or oval patches from twenty to thirty in number, and varying in length from half an inch to four inches. They are largest and most numerous in the ileum. In the lower part of the jejunum they are small, of a circular form, and few in number. They are occasionally seen in the duodenum. They are placed lengthwise in the intestine, covering the portion of the tube most distant from the attachment of the mesentery. Each patch is formed of a group of small, round, whitish vesicles, covered with mucous membrane. Each vesicle consists of a moderately thick external capsule, having no excretory duct, and containing an opaque white secretion. Each is surrounded by a zone, or wreath of simple follicles, and the interspaces between them are covered with villi. These vesicles are usually closed; but it is supposed they open at intervals to discharge the secretion contained within them. The mucous and submucous coats of the intestine are intimately adherent, and highly vascular, opposite the Peyerian glands. Their use is not known. They are largest and most developed during the digestive process.

The Large Intestine.

The large intestine extends from the termination of the ileum to the anus. It is about five feet in length, being one-fifth of the whole extent of the intestinal

336.—Patch of Peyer's Glands. From the lower part of the Ileum.

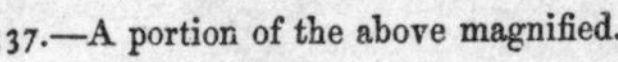

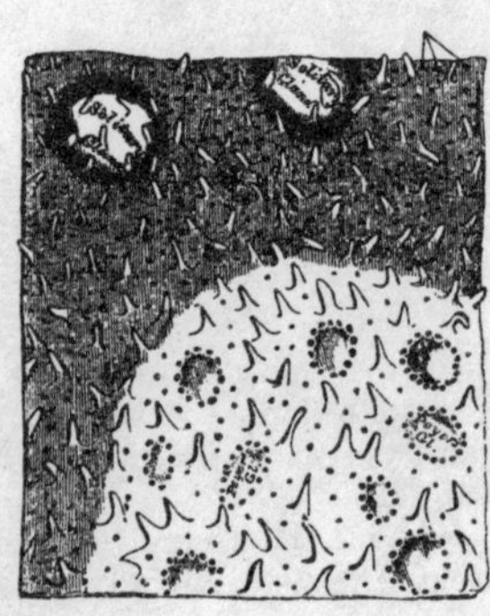

337.—A portion of the above magnified.

canal. It is largest at its commencement at the cæcum, and gradually diminishes as far as the rectum, where there is a dilatation of considerable size, just above the anus. It differs from the small intestine in its greater size, its more fixed position, and its sacculated form. The large intestine, in its course, describes an arch, which surrounds the convolutions of the small intestine. It commences in the right iliac fossa, in a dilated part, the cæcum. It ascends through the right lumbar and hypochondriac regions to the under surface of the liver; passes transversely across the abdomen, on the confines of the epigastric and umbilical regions, to the left hypochondriac region; descends through the left lumbar region to the left iliac fossa, where it becomes convoluted, and forms the sigmoid flexure; finally, it enters the pelvis, and descends along its posterior wall to the anus. The large intestine is divided into the cæcum, colon, and rectum.

The *Cæcum* (*cæcus*, blind) (fig. 338) is the large blind pouch, or *cul-de-sac*, in which the large intestine commences. It is the most dilated part of the tube, measuring about two and a half inches, both in its vertical and transverse diameters. It is situated in the right iliac fossa, immediately behind the anterior abdominal wall, being retained in its place by the peritoneum, which passes over its anterior surface and sides; its posterior surface being connected by loose areolar tissue with the iliac fascia. Occasionally, it is almost completely surrounded by peritoneum, which forms a distinct fold, the meso-cæcum, connecting its back part with the iliac fossa. When this fold exists the cæcum obtains considerable freedom of movement. Attached to its lower and back part is the appendix vermiformis, a long, narrow, warm-shaped tube, the rudiment of the lengthened cæcum found in all the mammalia, except the ourang-outang and wombat. The appendix varies from three to six inches in length, its average diameter being about equal to that of a goose-quill. It is usually directed upwards and inwards behind the cæcum, coiled upon itself, and terminates in a blunt point, being retained in its position by a fold of peritoneum, which sometimes forms a mesentery for it. Its canal is small, and communicates with the cæcum by an orifice, which is sometimes guarded with an incomplete valve. Its coats are thick, and its mucous lining furnished with a large number of solitary glands.

Ileo-Cæcal Valve. The lower end of the ileum terminates at the inner and back part of the large intestine, opposite the junction of the cæcum with the colon. At this point the mucous membrane forms two valvular folds, which project into the large intestine, and are separated from each other by a narrow elongated aperture. This is the ileo-cæcal valve (*valvula Bauhini*). Each fold is semilunar in form. The upper one, nearly horizontal in direction, is attached

338.—The Cæcum and Colon laid open to show the Ileo-Cæcal Valve.

by its convex border to the point of junction of the ileum with the colon; the lower segment, to the point of junction of the ileum with the cæcum. Their free concave margins project into the intestine, separated from one another by a narrow slit-like aperture, directed transversely. At each end of this aperture the two segments of the valve coalesce, and are continued, as a narrow membranous ridge, around the canal of the intestine for a short distance, forming the *fræna*, or *retinacula* of the valve. The left end of this aperture is rounded: the right end is narrow and pointed.

Each segment of the valve is formed by a reduplication of the mucous membrane, and of the circular muscular fibres of the intestine, the longitudinal fibres and peritoneum being continued uninterruptedly across from one intestine to the other. When these are divided or removed, the ileum may be drawn outwards, and all traces of the valve will be lost, the ileum appearing to open into the large intestine by a funnel shaped orifice of large size.

The surface of each segment of the valve directed towards the ileum is covered with villi, and presents the characteristic structure of the mucous membrane of the small intestine; whilst that turned towards the large intestine is destitute of villi, and marked with the orifices of the numerous tubuli peculiar to the membrane in the large intestine. These differences in structure continue as far as the free margin of the valve.

When the cæcum is distended, the margins of the opening are approximated, so as to prevent any reflux into the ileum.

The *colon* is divided into four parts, the ascending, transverse, descending, and the sigmoid flexure.

The *ascending colon* is smaller than the cæcum. It passes upwards, from the right iliac fossa to the under surface of the liver, on the right of the gall-bladder, where it bends abruptly inwards to the left, forming the *hepatic flexure*. It is retained in contact with the posterior wall of the abdomen by the peritoneum, which covers its anterior surface and sides, its posterior surface being connected by loose areolar tissue with the Quadratus lumborum and right kidney; sometimes the peritoneum almost completely invests it, and forms a distinct but narrow meso-colon. It is in relation, in front, with the convolutions of the ileum and the abdominal parietes; behind, it lies on the Quadratus lumborum muscle, and right kidney.

The *transverse colon*, the longest part of the large intestine, passes transversely from right to left across the abdomen, opposite the confines of the epigastric and umbilical zones, into the left hypochondriac region, where it curves downwards beneath the lower end of the spleen, forming the *splenic flexure*. In its course it describes an arch, the concavity of which is directed backwards towards the vertebral column; hence the name, *transverse arch of the colon*. This is the most moveable part of the colon, being almost completely invested by peritoneum, and connected to the spine behind by a large and wide duplicature of that membrane, the *transverse meso-colon*. It is in relation, by its upper surface, with the liver and gall-bladder, the great curvature of the stomach, and the lower end of the spleen; by its under surface, with the small intestines; by its anterior surface, with the anterior layers of the great omentum and the abdominal parietes; by its posterior surface, with the transverse meso-colon.

The *descending colon* passes almost vertically downwards through the left hypochondriac and lumbar regions to the upper part of the left iliac fossa, where it terminates in the sigmoid flexure. It is retained in position by the peritoneum, which covers its anterior surface and sides, its posterior surface being connected by areolar tissue with the left crus of the Diaphragm, the left kidney, and the Quadratus lumborum. It is smaller in calibre and more deeply placed than the ascending colon.

The *sigmoid flexure* is the narrowest part of the colon; it is situated in the left iliac fossa, commencing at the termination of the descending colon, at the margin of the crest of the ilium, and ending in the rectum, opposite the left sacro-iliac symphysis. It curves in the first place upwards, and then descends vertically, and to one or the other side, like the letter S—hence its name. It is retained in its place by a loose fold of peritoneum, the *sigmoid meso-colon*. It is in relation, in front, with the small intestines and abdominal parietes; behind, with the iliac fossa.

The *Rectum* is the terminal part of the large intestine, and extends from the sigmoid flexure to the anus; it varies in length from six to eight inches, and has received its name from being less flexuous than any other part of the intestinal canal. It commences opposite the left sacro-iliac symphysis, passes obliquely downwards from left to right to the middle of the sacrum, forming a gentle curve to the right side; then, regaining the middle line, it descends in front of the lower part of the sacrum and coccyx, and, near the extremity of the latter bone, inclines backwards to terminate at the anus, being curved both in the lateral and antero-posterior directions. The rectum is, therefore, not straight, the upper part being directed obliquely from the left side to the median line, the middle portion being curved in the direction of the hollow of the sacrum and coccyx, and the lower portion presenting a short curve in the opposite direction. The rectum is cylindrical, not sacculated like the rest of the large intestine; it is narrower at its upper part than the sigmoid flexure, gradually increases in size as it descends, and immediately above the anus presents a considerable dilatation, capable of acquiring an enormous size. The rectum is divided into three portions, upper, middle and lower.

The *upper portion*, which includes about half the length of the tube, extends obliquely from the left sacro-iliac symphysis to the middle of the third piece of the sacrum. It is almost completely surrounded by peritoneum, and connected to the sacrum behind by a duplicature of that membrane, the *meso-rectum*. It is in relation behind with the Pyriformis muscle, the sacral plexus of nerves, and the branches of the internal iliac artery of the left side, which separate it from the sacrum and sacro-iliac symphysis; in front it is separated, in the male, from the posterior surface of the bladder: in the female, from the posterior surface of the uterus, and its appendages, by some convolutions of the small intestine.

The *middle portion* of the rectum is about three inches in length, and extends as far as the tip of the coccyx. It is closely connected to the concavity of the sacrum, and covered by peritoneum only on the upper part of its anterior surface. It is in relation, in front, in the male, with the triangular portion of the base of the bladder, the vesiculæ seminales, and vasa deferentia; more anteriorly, with the under surface of the prostate. In the female, it is adherent to the posterior wall of the vagina.

The *lower portion* is about an inch in length; it curves backwards at the fore part of the prostate gland, and terminates at the anus. This portion of the intestine receives no peritoneal covering. It is invested by the Internal sphincter, supported by the Levatores ani muscles, and surrounded at its termination by the External sphincter. In the male, it is separated from the membranous portion and bulb of the urethra by a triangular space; and, in the female, a similar space intervenes between it and the vagina. This space forms by its base the perinæum.

Structure. The large intestine has four coats: serous, muscular, cellular, and mucous.

The *serous coat* is derived from the peritoneum, and invests the different portions of the large intestine to a variable extent. The cæcum is covered only on its anterior surface and sides; more rarely, it is almost completely invested, being held in its position by a duplicature, the meso-cæcum. The ascending and descending colon are usually covered only in front. The transverse colon is almost completely invested, the parts corresponding to the attachment of the great omentum and transverse meso-colon being alone excepted. The sigmoid flexure is nearly completely surrounded, the point corresponding to the attachment of the sigmoid meso-colon being excepted. The upper part of the rectum is almost completely invested by the peritoneum: the middle portion is covered only on its anterior surface; and the lower portion is entirely devoid of any serous covering. In the course of the colon, and upper part of the rectum, the peritoneal coat is thrown into a number of small pouches filled with fat, called *appendices epiploicæ*. They are chiefly appended to the transverse colon.

The *muscular coat* consists of an external longitudinal and an internal circular layer of muscular fibres.

The *longitudinal fibres* are found as a uniform layer over the whole surface of the large intestine. In the cæcum and colon, they are especially collected into three flat longitudinal bands, each being about half an inch in width. These bands commence at the attachment of the appendix vermiformis to the cæcum; one, the posterior, is placed along the attached border of the intestine; the anterior band, the largest, becomes inferior along the arch of the colon, where it corresponds to the attachment of the great omentum, but is in front in the ascending and descending colon and sigmoid flexure; the third, or lateral band, is found on the inner side of the ascending and descending colon, and on the under border of the transverse colon. These bands are nearly one-half shorter than the other parts of the intestine, and serve to produce the sacculi which are characteristic of the cæcum and colon; accordingly, when they are dissected off, the tube can be lengthened, and its sacculated character becomes lost. In the sigmoid flexure, the longitudinal fibres become more scattered, but upon its lower part, and round the rectum, they spread out, and form a thick uniform layer.

The *circular fibres* form a thin layer over the cæcum and colon, being especially accumulated in the intervals between the sacculi; in the rectum, they form a thick layer, especially at its lower end, where they become numerous, and form the Internal sphincter.

The *cellular coat* connects the muscular and mucous layers closely together.

The *mucous membrane*, in the cæcum and colon, is pale, and of a greyish or pale yellow colour. It is quite smooth, destitute of villi, and raised intc numerous crescentic folds, which correspond to the intervals between the sacculi. In the rectum, it is thicker, of a darker colour, more vascular, and connected loosely to the muscular coat as in the œsophagus. When the lower part of the rectum is contracted, its mucous membrane is thrown into a number of folds, some of which, near the anus, are longitudinal in direction, and are effaced by the distension of the gut. Besides these, there are certain permanent folds, of a semilunar shape, described by Mr. Houston.* They are usually three in number; sometimes a fourth is found, and, occasionally, only two are present. One is situated near the commencement of the rectum, on the right side; another extends inwards from the left side of the tube, opposite the middle of the sacrum; the largest and most constant one projects backwards from the fore part of the rectum, opposite the base of the bladder. When a fourth is present, it is situated about an inch above the anus on the back of the rectum. These folds are about half an inch in width, and contain some of the circular fibres of the gut. In the empty state of the intestine they overlap each other, as Mr. Houston remarks, so effectually as to require considerable manœuvring to conduct a bougie or the finger along the canal of the intestine. Their use seems to be, 'to support the weight of fæcal matter, and prevent its urging towards the anus, where its presence always excites a sensation demanding its discharge.' The mucous membrane of the large intestine presents for examination, epithelium, simple follicles, and solitary glands.

339.—Minute Structure of Large Intestine.

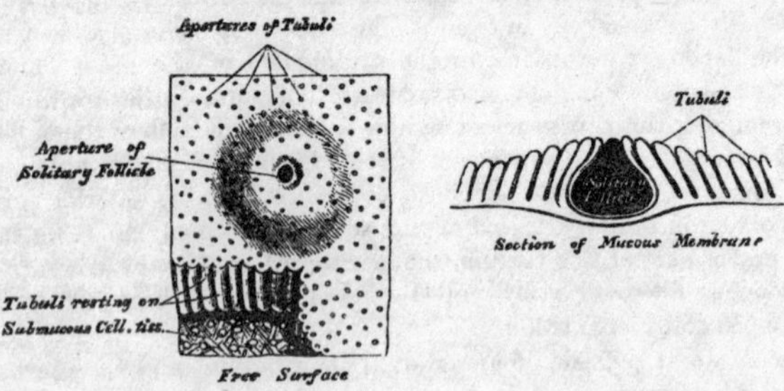

The *epithelium* is of the columnar kind.

The *simple follicles* are minute tubular prolongations of the mucous membrane, arranged perpendicularly, side by side, over its entire surface; they are longer, more numerous, and placed in much closer apposition than those of the small intestine; and they open by minute rounded orifices upon the surface, giving it a cribriform appearance.

The *solitary glands* in the large intestine, are most abundant in the cæcum and appendix vermiformis; but are irregularly scattered also over the rest of the intestine. They are small, prominent, flask-shaped bodies, of a whitish colour, perforated upon the central part of their free surface by a minute orifice, which in the majority, is permanent.

* *Dub. Hosp. Reports*, vol. v. p. 163.

The Liver.

The liver is a glandular organ of large size, intended mainly for the secretion of the bile, but effecting also important changes in certain constituents of the blood in their passage through the gland. It is situated in the right hypochondriac region, and extends across the epigastrium into the left hypochondrium. It is the largest gland in the body, weighing from three to four pounds (from fifty to sixty ounces avoirdupois). It measures, in its transverse diameter, from ten to twelve inches; from six to seven in its antero-posterior; and is about three inches thick at the back part of the right lobe, which is the thickest part.

Its *upper surface* is convex, directed upwards and forwards, smooth, covered by peritoneum. It is in relation with the under surface of the Diaphragm; and below, to a small extent, with the abdominal parietes. The surface is divided into two unequal lobes, the right and left, by a fold of peritoneum, the *suspensory* or *broad ligament.*

Its *under surface* is concave, directed downwards and backwards, and in relation with the stomach and duodenum, the hepatic flexure of the colon, and the right kidney and suprarenal capsule. The surface is divided by a longitudinal fissure into a right and left lobe.

The *posterior border* is rounded and broad, and connected to the Diaphragm by the coronary ligament; it is in relation with the aorta, the vena cava, and the crura of the Diaphragm.

The *anterior border* is thin and sharp, and marked, opposite the attachment of the broad ligament, by a deep notch. In adult males, this border usually corresponds with the margin of the ribs; but in women and children, it usually projects below the ribs.

The *right extremity* of the liver is thick and rounded; whilst the *left* is thin and flattened.

Changes of Position. The student should make himself acquainted with the different circumstances under which the liver changes its position, as they are of importance in determining the existence of enlargement, or other disease of the organ.

Its position varies according to the posture of the body; in the upright and sitting postures, its lower border may be felt below the edges of the ribs; in the recumbent posture, it usually recedes behind the ribs. Its position varies, also, with the ascent or descent of the Diaphragm. In a deep inspiration, the liver descends below the ribs; in expiration, it is raised to its ordinary level. Again, in emphysema, where the lungs are distended, and the Diaphragm descends very low, the liver is pushed down; in some other diseases, as phthisis, where the Diaphragm is much arched, the liver rises very high up. Pressure from without, as in tight lacing, by compressing the lower part of the chest, displaces the liver considerably, its anterior edge often extending as low as the crest of the ilium; and its convex surface is often, at the same time, deeply indented from pressure of the ribs. Again, its position varies greatly, according to the greater or less distension of the stomach and intestines. When the intestines are empty, the liver descends in the abdomen; but when they are distended, it is pushed upwards. Its relations to surrounding organs may also be changed by the growth of tumours, or by collections of fluid in the thoracic or abdominal cavities.

Ligaments. The ligaments of the liver (fig. 340) are five in number: four being formed of folds of peritoneum; the fifth, the *ligamentum teres*, is a round, fibrous cord, resulting from the obliteration of the umbilical vein. The ligaments are the longitudinal, two lateral, coronary, and round.

The *longitudinal ligament* (broad, falciform, or suspensory ligament) is a broad and thin antero-posterior peritoneal fold, falciform in shape, its base being directed forwards, its apex backwards. It is attached by one margin to the under surface of the Diaphragm, and the posterior surface of the sheath of the right Rectus muscle as low down as the umbilicus; by its hepatic margin, it extends from the notch on the anterior margin of the liver, as far back as its posterior border. It consists of two layers of peritoneum closely united together. Its anterior free edge contains the round ligament between its layers.

The *lateral ligaments*, two in number, right and left, are triangular in shape. They are formed of two layers of peritoneum united, and extend from the sides of the Diaphragm to the adjacent margins of the posterior border of the liver. The left is the longer of the two, and lies in front of the œsophageal opening in the Diaphragm.

The *coronary ligament* connects the posterior border of the liver to the Diaphragm. It is formed by the reflection of the peritoneum from the Diaphragm on to the upper and lower margins of the posterior border of the organ. The coronary ligament consists of two layers, which are continuous on each side with

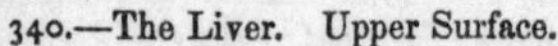
340.—The Liver. Upper Surface.

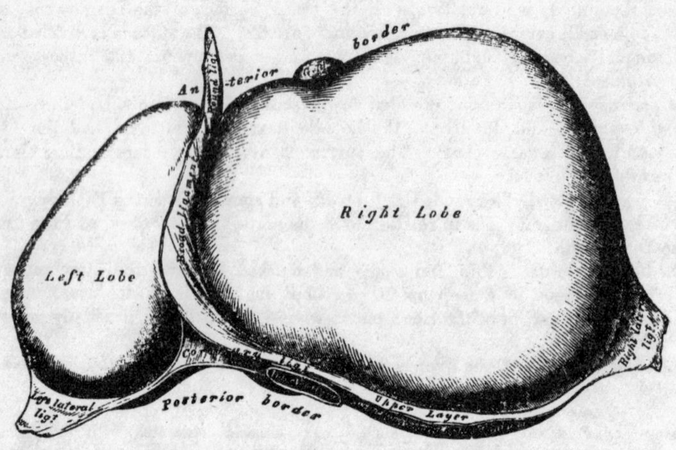

the lateral ligaments; and in front, with the longitudinal ligament. Between the layers, a large oval interspace is left uncovered by peritoneum, and connected to the Diaphragm by a firm areolar tissue. This space is subdivided, near its left extremity, into two parts by a deep notch (sometimes a canal), which lodges the inferior vena cava, and into which open the hepatic veins.

The *round ligament* (fig. 341) is a fibrous cord resulting from the obliteration of the umbilical vein. It ascends from the umbilicus, in the anterior free margin of the longitudinal ligament, to the notch in the anterior border of the liver, from which it may be traced along the longitudinal fissure on the under surface of the liver, as far back as the inferior vena cava.

Fissures (fig. 341). Five fissures are seen upon the under surface of the liver, which serve to divide it into five lobes. They are the longitudinal fissure, the fissure of the ductus venosus, the transverse fissure, the fissure for the gall-bladder, and the fissure for the vena cava.

The *longitudinal fissure* is a deep groove, which extends from the notch on the anterior margin of the liver, to the posterior border of the organ. It separates the right and left lobes; the transverse fissure joins it, at right angles, about one-third from its posterior extremity, and divides it into two parts. The anterior half is called the *umbilical fissure*: it is deeper than the posterior part, and lodges the umbilical vein in the fœtus, or its remains (the round ligament) in the adult. This fissure is often partially bridged over by a prolongation of the hepatic substance, the *pons hepatis*.

The *fissure of the ductus venosus* is the back part of the longitudinal fissure;

it is shorter and shallower than the anterior portion. It lodges in the fœtus the ductus venosus, and in the adult a slender fibrous cord, the obliterated remains of that vessel.

The *transverse* or *portal fissure*, is a short but deep fissure, about two inches in length, extending transversely across the under surface of the right lobe, nearer to its posterior than its anterior border. It joins, nearly at right angles, with the longitudinal fissure. By the older anatomists, this fissure was considered the gateway (*porta*) of the liver; hence the large vein which enters at this point was called the *portal vein*. Besides this vein, the fissure transmits the hepatic artery and nerves, and the hepatic duct and lymphatics. At their entrance into the fissure, the hepatic duct lies in front to the right, the hepatic artery to the left, and the portal vein behind (fig. 345).

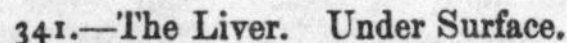
341.—The Liver. Under Surface.

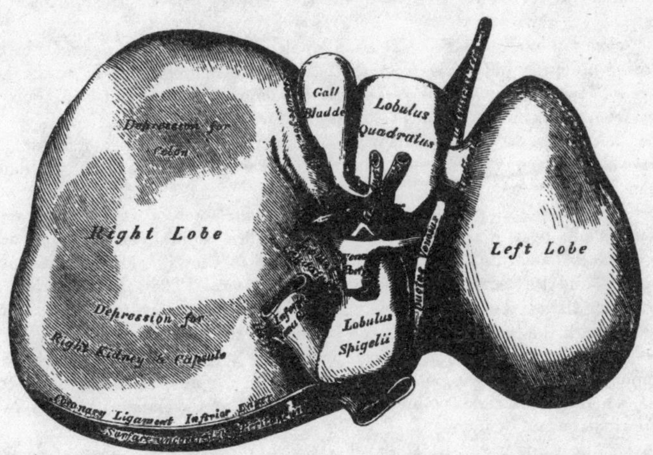

The *fissure for the gall-bladder* (*fossa cystis felleæ*) is a shallow, oblong fossa, placed on the under surface of the right lobe, parallel with the longitudinal fissure. It extends from the anterior free margin of the liver, which is occasionally notched for its reception, to near the right extremity of the transverse fissure.

The *fissure for the vena cava* is a short deep fissure, occasionally a complete canal, which extends obliquely upwards from a little behind the right extremity of the transverse fissure, to the posterior border of the organ, where it joins the fissure for the ductus venosus. On slitting open the inferior vena çava which is contained in it, a deep fossa is seen, at the bottom of which the hepatic veins communicate with this vessel. This fissure is separated from the transverse fissure by the lobus caudatus, and from the longitudinal fissure by the lobulus Spigelii.

Lobes. The lobes of the liver, like the ligaments and fissures, are five in number: the right lobe, the left lobe, the lobus quadratus, the lobulus Spigelii, and the lobus caudatus.

The *right lobe* is much larger than the left; the proportion between them being as six to one. It occupies the right hypochondrium, and is separated from the left lobe, on its upper surface, by the longitudinal ligament; on its under surface, by the longitudinal fissure; and in front, by a deep notch. It is of a quadrilateral form, its under surface being marked by three fissures: the transverse fissure, the fissure for the gall-bladder, and the fissure for the inferior vena cava; and by two shallow impressions, one in front (*impressio colica*), for the hepatic flexure of the colon; and one behind (*impressio renalis*), for the right kidney and suprarenal capsule.

The *left lobe* is smaller and more flattened than the right. It is situated in the epigastric and left hypochondriac regions, sometimes extending as far as the upper border of the spleen. Its upper surface is convex; its under concave surface rests upon the front of the stomach; and its posterior border is in relation with the cardiac orifice of the stomach.

The *lobus quadratus*, or square lobe, is situated on the under surface of the right lobe, bounded in front by the free margin of the liver; behind, by the transverse fissure; on the right, by the fissure for the gall-bladder; and, on the left, by the umbilical fissure.

The *lobulus Spigelii* projects from the back part of the under surface of the right lobe. It is bounded, in front, by the transverse fissure; on the right, by the fissure for the vena cava; and, on the left, by the fissure for the ductus venosus.

The *lobus caudatus*, or tailed lobe, is a small elevation of the hepatic substance, extending obliquely outwards, from the base of the lobulus Spigelii, to the under surface of the right lobe. It separates the right extremity of the transverse fissure from the commencement of the fissure for the inferior cava.

Vessels. The vessels connected with the liver are also five in number: they are the hepatic artery, the portal vein, the hepatic vein, the hepatic duct, and the lymphatics.

The *hepatic artery*, *portal vein*, and *hepatic duct*, accompanied by numerous lymphatics and nerves, ascend to the transverse fissure, between the layers of the gastro-hepatic omentum; the hepatic duct lying to the right, the hepatic artery to the left, and the portal vein behind the other two. They are enveloped in a loose areolar tissue, the capsule of Glisson, which accompanies the vessels in their course through the *portal canals*, in the interior of the organ.

The *hepatic veins* convey the blood from the liver. They commence at the circumference of the organ, and proceed towards the deep fossa in its posterior border, where they terminate by two large and several smaller branches, in the inferior vena cava.

The hepatic veins have no cellular investment; consequently their parietes are adherent to the walls of the canals through which they run; so that, on a section of the organ, these veins remain widely open and solitary, and may be easily distinguished from the branches of the portal vein, which are more or less collapsed, and always accompanied by an artery and duct.

The *lymphatics* are large and numerous, consisting of a deep and superficial set. They have been already described.

Nerves. The nerves of the liver are derived from the hepatic plexus of the sympathetic, from the pneumogastric nerves, especially the left, and from the right phrenic.

Structure. The substance of the liver is composed of lobules, held together by an extremely fine areolar tissue, and of the ramifications of the portal vein, hepatic duct, hepatic artery, hepatic veins, lymphatics, and nerves; the whole being invested by a fibrous and a serous coat.

The *serous coat* is derived from the peritoneum, and invests the entire surface of the organ, excepting at the attachment of its various ligaments, and at the bottom of the different fissures, where it is deficient. It is intimately adherent to the fibrous coat.

The *fibrous coat* lies beneath the serous investment, and covers the entire surface of the organ. It is difficult of demonstration, excepting where the serous coat is deficient. At the transverse fissure, it is continuous with the capsule of Glisson; and, on the surface of the organ, with the areolar tissue separating the lobules.

The *lobules* form the chief mass of the hepatic substance: they may be seen either on the surface of the organ, or by making a section through the gland.

They are small granular bodies, about the size of a millet-seed, measuring from one-twentieth to one-tenth of an inch in diameter. If divided longitudinally, they have a foliated, and, if transversely, a polygonal outline. The bases of the lobules are clustered round the smallest branches (sublobular) of the hepatic veins, to which each is connected by means of a small branch, which issues from the centre of the lobule (intralobular). The remaining part of the surface of each lobule is imperfectly isolated from the surrounding lobules, by a thin stratum of areolar tissue, and by the smaller vessels and ducts.

342.

H. Longitudinal section of an hepatic vein. *a.* Portion of the canal, from which the vein has been removed; *b.* orifices of ultimate twigs of the vein (sublobular), situated in the centre of the lobules. After Kiernan.

If one of the hepatic veins be laid open, the bases of the lobules may be seen through the thin wall of the vein, on which they rest, arranged in the form of a tesselated pavement, the centre of each polygonal space presenting a minute aperture, the mouth of a sublobular vein.

Each lobule is composed of a mass of cells; of a plexus of biliary ducts; of a venous plexus, formed by branches of the portal vein; of a branch of an hepatic vein (intralobular); of minute arteries; and, probably, of nerves and lymphatics.

The hepatic cells form the chief mass of the substance of a lobule, and lie in the interspaces of the capillary plexus, being probably contained in a tubular network, which forms the origin of the biliary ducts. The smallest branches of the vena portæ pass between the lobules, around which they form a plexus, the interlobular. Branches from this plexus enter the lobules, and form a network in their circumference. The radicles of the portal vein communicate with those of the hepatic vein, which occupy the centre of the lobule; and the latter converge to form the intralobular vein, which issues from the base of the lobule, and joins the hepatic vein. The portal vein carries the blood to the liver, from which the bile is secreted; the hepatic vein carries the superfluous blood from the liver, and the bile-duct carries the bile secreted by the hepatic cells.

343.

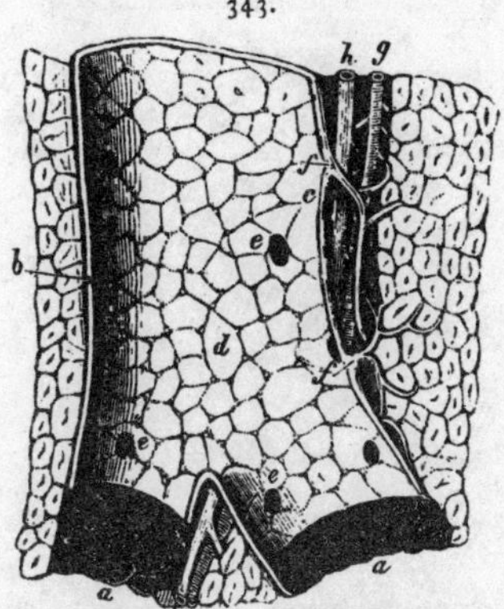

Longitudinal section of a small portal vein and canal, after Kiernan. *a.* Portions of the canal from which the vein has been removed; *b.* side of the portal vein in contact with the canal; *c.* the side of the vein which is separated from the canal by the hepatic artery and duct, with areolar tissue (Glisson's capsule); *d.* internal surface of the portal vein, through which are seen the outlines of the lobules and the openings (*e*) of the interlobular veins; *f.* vaginal veins of Kiernan; *g.* hepatic artery; *h.* hepatic duct.

The *hepatic cells* are of a more or less spheroidal form; but may be rounded, flattened, or many-sided, from mutual compression. They vary in size from the $\frac{1}{1000}$ to the $\frac{1}{2000}$ of an inch in diameter, and contain a distinct nucleus in the interior, or sometimes two. In the nucleus is a highly refracting nucleolus, with granules. The cell-contents are viscid, and contain yellow particles, the colouring matter of the bile, and oil-globules. The cells adhere together by their surfaces, so as to form rows, which radiate from the centre towards the circumference of the lobule. These cells are probably the chief agents in the secretion of the bile.

Biliary ducts. The precise mode of origin of the biliary ducts is uncertain. Mr. Kiernan's original view, which is supported by the researches of Dr. Beale, is that the ducts commence within the lobules, in a plexiform network (lobular biliary plexus), in which the hepatic cells lie. According to Henle, Handfield Jones, and Kölliker, the cells are packed in the interspaces of the capillary plexus, and, by means of temporary communications, transmit their contents into the minute bile-ducts which originate in the spaces *between* the lobules, never entering within them. The ducts form a plexus (interlobular) between the lobules; and the interlobular branches unite and form vaginal branches, which lie in the portal canals, with branches of the portal vein and hepatic duct. The ducts finally form two large trunks, which leave the liver at the transverse fissure, and the union of these is the hepatic duct.

The *Portal vein*, on entering the liver at the transverse fissure, divides into primary branches, which are contained in the portal canals, together with branches of the hepatic artery and duct, and the nerves and lymphatics. In the larger portal canals, the vessels are separated from the parietes, and joined to each other, by a loose cellular web, the capsule of Glisson. The veins, as they lie in the portal canals, give off vaginal branches, which form a plexus (vaginal plexus) in Glisson's capsule. From this plexus, and from the portal vein itself, small branches are given off, which pass between the lobules (interlobular veins); these cover the entire surface of the lobules, excepting their bases. The lobular branches are derived from the interlobular veins: they penetrate into the lobule, and form a capillary plexus within them. From this plexus the intralobular vein arises.

The *Hepatic artery* appears destined chiefly for the nutrition of the coats of the large vessels, the ducts, and the investing membranes of the liver. It enters the liver at the transverse fissure, with the portal vein and hepatic duct, and ramifies with these vessels through the portal canals. It gives off vaginal branches, which ramify in the capsule of Glisson; and other branches, which are distributed to the coats of the vena portæ and hepatic duct. From the vaginal plexus, interlobular branches are given off, which ramify through the interlobular fissures, a few branches being distributed to the lobules. Kiernan supposes that the branches of the hepatic artery terminate in a capillary plexus, which communicates with the branches of the vena portæ.

344.

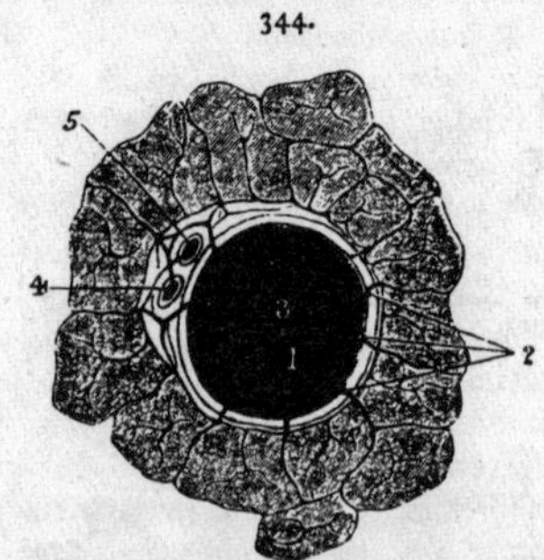

A transverse section of a small portal canal and its vessels, after Kiernan. 1. Portal vein; 2. interlobular branches; 3. branches of the vein, termed, by Mr. Kiernan, vaginal, also giving off interlobular branches; 4. hepatic duct; 5. hepatic artery.

The *Hepatic veins* commence in the interior of each lobule by a plexus, the branches of which converge to form the intralobular vein.

The *intralobular vein* passes through the centre of the lobule, and leaves it at its base to terminate in a sublobular vein.

The *sublobular veins* unite with neighbouring branches to form larger veins; and these join to form the large hepatic trunks, which terminate in the vena cava.

Gall-Bladder.

The gall-bladder is the reservoir for the bile; it is a conical or pear-shaped membranous sac, lodged in a fossa on the under surface of the right lobe of the liver, and extending from near the right extremity of the transverse fissure to the anterior free margin of the organ. It is about four inches in length, one inch in breadth at its widest part, and holds from eight to ten drachms. It is divided into a fundus, body, and neck. The *fundus*, or broad extremity, is directed downwards, forwards, and to the right, and occasionally projects from the anterior border of the liver: the *body* and *neck* are directed upwards and backwards to the left. The gall-bladder is held in its position by the peritoneum, which, in the majority of cases, passes over its under surface, but the serous membrane occasionally invests the gall-bladder, which then is connected to the liver by a kind of mesentery.

Relations. The *body* of the gall-bladder is in relation, by its upper surface, with the liver, to which it is connected by areolar tissue and vessels; by its under surface, with the first portion of the duodenum, occasionally the pyloric end of the stomach, and the hepatic flexure of the colon. The *fundus* is completely invested by peritoneum; it is in relation, in front, with the abdominal parietes, immediately below the tenth costal cartilage; behind, with the transverse arch of the colon. The *neck* is narrow, and curves upon itself like the italic letter *ſ*; at its point of connection with the body and with the cystic duct, it presents a well-marked constriction.

When the gall-bladder is distended with bile or calculi, the fundus may be felt through the abdominal parietes, especially in an emaciated subject; the relations of this sac will also serve to explain the occasional occurrence of abdominal biliary fistulæ, through which biliary calculi may pass out, and of the passage of calculi from the gall-bladder into the stomach, duodenum, or colon, which occasionally happens.

Structure. The gall-bladder consists of three coats; serous, fibrous and muscular, and mucous.

The *external* or *serous coat* is derived from the peritoneum; it completely invests the fundus, but covers the body and neck only on their under surface.

The *middle* or *fibrous coat* is a thin but strong fibrous layer, which forms the framework of the sac, consisting of dense fibres which interlace in all directions. Plain muscular fibres are also found in this coat, disposed chiefly in a longitudinal direction, a few running transversely.

The *internal* or *mucous coat* is loosely connected with the fibrous layer. It is generally tinged with a yellowish-brown colour, and is everywhere elevated into minute rugæ, by the union of which numerous meshes are formed; the depressed intervening spaces having a polygonal outline. The meshes are smaller at the fundus and neck, being most developed about the centre of the sac. Opposite the neck of the gall-bladder, the mucous membrane projects inwards so as to form a large valvular fold.

The mucous membrane is covered with columnar epithelium, and secretes an abundance of thick viscid mucus; it is continuous through the hepatic duct with the mucous membrane lining the ducts of the liver, and through the ductus communis choledochus with the mucous membrane of the alimentary canal.

The *Biliary Ducts* are, the hepatic, the cystic, and the ductus communis choledochus.

The *hepatic duct* is formed of two trunks of nearly equal size, which issue from the liver at the transverse fissure, one from the right, the other from the left lobe; these unite, and pass downwards and to the right for about an inch and a half, to join at an acute angle with the cystic duct, and so form the ductus communis choledochus.

The *cystic duct*, the smallest of the three biliary ducts, is about an inch in length. It passes obliquely downwards and to the left from the neck of the gall-bladder, and joins the hepatic duct to form the common duct. It lies in the gastro-hepatic omentum in front of the vena cava, the cystic artery lying to its left side. The mucous membrane lining its interior is thrown into a series of crescentic folds, from five to twelve in number, which project into the duct in regular succession, and are directed obliquely round the tube, presenting much the appearance of a continuous spiral valve. They exist only in the human subject. When the duct has been distended, the interspaces between the folds are dilated, so as to give to its exterior a sacculated appearance.

The *ductus communis choledochus*, the largest of the three, is the common excretory duct of the liver and gall-bladder. It is about three inches in length, of the diameter of a goose-quill, and formed by the junction of the cystic and hepatic ducts. It descends along the right border of the lesser omentum, behind the first

345.—The Parts in the Gastro-hepatic Omentum, its anterior Layer being removed.

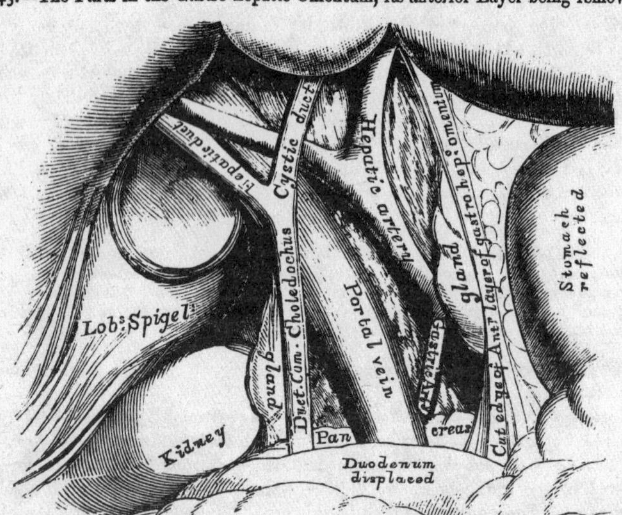

portion of the duodenum, in front of the vena portæ, and to the right of the hepatic artery; it then passes between the pancreas and descending portion of the duodenum, and, running for a short distance along the right side of the pancreatic duct, near its termination, passes with it obliquely between the mucous and muscular coats, the two opening by a common orifice upon the summit of a papilla, situated at the inner side of the descending portion of the duodenum, a little below its middle.

Structure. The coats of the biliary ducts are, an external or fibrous, and an internal or mucous. The fibrous coat is composed of a strong areolar fibrous tissue. The mucous coat is continuous with the lining membrane of the hepatic ducts and gall-bladder, and also with that of the duodenum. It is provided with numerous glands, the orifices of which are scattered irregularly in the larger ducts, but in the smaller hepatic ducts are disposed in two longitudinal rows, one on each side of the vessel. These glands are of two kinds. Some are ramified tubes, which occasionally anastomose, and from the sides of which saccular dilatations are given off; others are small clustered cellular glands, which open either separately into the hepatic duct, or into the ducts of the tubular glands.

The Pancreas.

Dissection. The pancreas may be exposed for dissection in three different ways: 1. By raising the liver, drawing down the stomach, and tearing through the gastro-hepatic omentum. 2. By raising the stomach, the arch of the colon, and great omentum, and then dividing the inferior layer of the transverse meso-colon. 3. By dividing the two layers of peritoneum, which descend from the great curvature of the stomach to form great omentum; turning the stomach upwards, and then cutting through the ascending layer of the transverse meso-colon.

The *Pancreas* (*παν-κρέας, all flesh*) is a conglomerate gland, analogous in its structure to the salivary glands. In shape it is transversely oblong, flattened from before backwards, and bears some resemblance to a dog's tongue, its right extremity being broad, and presenting a sort of angular bend from above downwards, called the *head*, whilst its left extremity gradually tapers to form the tail, the intermediate portion being called the *body*. It is situated transversely across the posterior wall of the abdomen, at the back of the epigastric and both hypochondriac regions. Its length varies from six to eight inches, its breadth is an inch and a half, and its thickness from half an inch to an inch, being greater at its right extremity and along its upper border. Its weight varies from two to three and a half ounces, but it may reach six ounces.

The *right extremity* or *head* of the pancreas (fig. 346) is curved upon itself from above downwards, and is embraced by the concavity of the duodenum. The

346.—The Pancreas and its Relations.

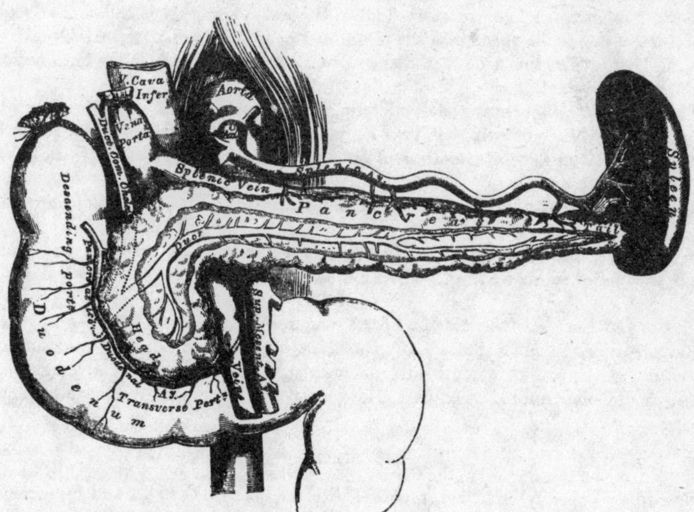

common bile-duct descends behind, between the duodenum and pancreas; and the pancreatico-duodenal artery descends in front between the same parts. On the posterior aspect of the pancreas is a lobular fold of the gland, which passes transversely to the left, behind the superior mesenteric vessels, forming the back part of the canal in which they are contained. It is sometimes detached from the rest of the gland, and is called the *lesser pancreas*.

The *lesser end* or *tail* of the pancreas is narrow; it extends to the left as far as the spleen, and is placed over the left kidney and suprarenal capsule.

The *body* of the pancreas is convex in front, and covered by the ascending layer of the transverse meso-colon and the posterior surface of the stomach.

The *posterior surface* is concave, and has the following structures interposed between it and the first lumbar vertebra: the superior mesenteric artery and vein, the commencement of the vena portæ, the vena cava, the aorta, the left kidney, the suprarenal capsule, and the corresponding renal vessels.

The *upper border* is thick, and has resting upon it, near its centre, the cœliac axis; the splenic artery and vein are lodged in a deep groove or canal in this border; and to the right, the first part of the duodenum and the hepatic artery are in relation with it.

The *lower border*, thinner than the upper, is separated from the transverse portion of the duodenum by the superior mesenteric artery and vein; to the left of these the inferior mesenteric vein ascends behind the pancreas to join the splenic vein.

The *pancreatic duct*, called the *canal of Wirsung* from its discoverer, extends transversely from left to right through the substance of the pancreas, nearer to its lower than its upper border, and lying nearer its anterior than its posterior surface. In order to expose it, the superficial portion of the gland must be removed. Traced backwards, it is found to commence by an orifice common to it and the ductus communis choledochus, upon the summit of an elevated papilla, situated at the inner side of the descending portion of the duodenum, a little below its middle: from this papilla it passes very obliquely through the mucous and muscular coats, separates itself from the ductus communis choledochus, and, ascending slightly, runs from right to left through the middle of the gland, giving off numerous branches, which commence in its lobules.

Sometimes the pancreatic duct and ductus communis choledochus open separately into the duodenum. The excretory duct of the lesser pancreas is called the *ductus pancreaticus minor*; it opens into the main duct near the duodenum, and sometimes separately into that intestine, at a distance of an inch or more from the termination of the principal duct.

The pancreatic duct, near the duodenum, is about the size of an ordinary quill; its walls are thin, consisting of two coats, an external fibrous and an internal mucous; the latter is thin, smooth, and furnished, near its termination, with a few scattered follicles.

Sometimes the pancreatic duct is double, up to its point of entrance into the duodenum.

In *structure*, the pancreas closely resembles the salivary glands; but it is looser and softer in its texture. The fluid secreted by it is almost identical with saliva.

Vessels and Nerves. The *arteries* of the pancreas are derived from the splenic, the pancreatico-duodenal branch of the hepatic, and the superior mesenteric. Its *veins* open into the splenic and superior mesenteric veins. Its *lymphatics* terminate in the lumbar glands. Its *nerves* are filaments from the splenic plexus.

The Spleen.

The spleen is usually classified, together with the thyroid, thymus, and suprarenal capsules, as one of the ductless, or blood-glands. It possesses no excretory duct. It is of an oblong flattened form, soft, of very brittle consistence, highly vascular, of a dark bluish-red colour, and situated in the left hypochondriac region, embracing the cardiac end of the stomach. It is invested by peritoneum, and connected with the stomach by the gastro-splenic omentum.

Relations. The *external surface* is convex, smooth, and in relation with the under surface of the Diaphragm, which separates it from the ninth, tenth, and eleventh ribs of the left side. The *internal surface* is slightly concave, and divided by a vertical fissure, the *hilum*, into an anterior or larger, and a posterior or smaller portion. The hilum is pierced by several large irregular apertures, for the entrance and exit of vessels and nerves. At the margins of the hilum, the two layers of peritoneum are reflected from the surface of the spleen on to the cardiac end of the stomach, forming the gastro-splenic omentum, which contains between its layers the splenic vessels and nerves, and the vasa brevia. The internal surface is in relation, in front, with the great end of the stomach; below, with the tail of the pancreas; and behind, with the left crus of the Diaphragm and corresponding suprarenal capsule. The *upper end*, thick and rounded, is in relation with the Diaphragm, to which it is connected by a fold of peritoneum, the suspensory ligament. The *lower end* is pointed; it is in relation with the left extremity of the transverse arch of the colon. The *anterior margin* is free, rounded, and often notched, especially below. The *posterior margin* is rounded, and lies in relation with the left kidney, to which it is connected by loose areolar tissue.

The spleen is held in its position by two folds of peritoneum: one, the *gastro-splenic omentum*, connects it with the stomach; and the other, the *suspensory ligament*, with the under surface of the Diaphragm.

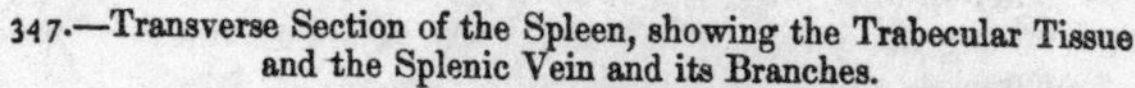
347.—Transverse Section of the Spleen, showing the Trabecular Tissue and the Splenic Vein and its Branches.

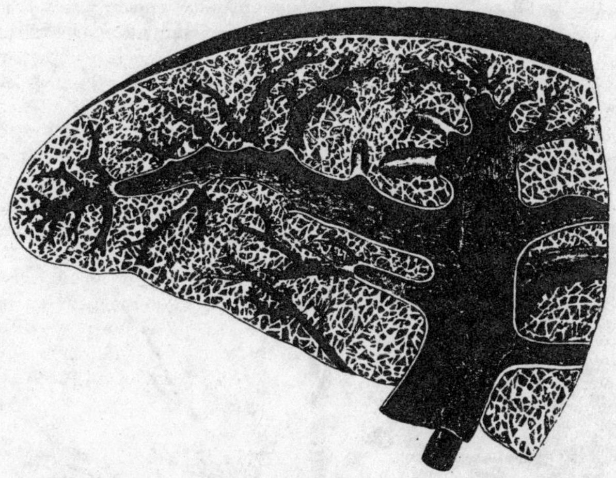

The size and weight of the spleen are liable to very extreme variations at different periods of life, in different individuals, and in the same individual under different conditions. In the *adult*, in whom it attains its greatest size, it is usually about five inches in length, three or four inches in breadth, and an inch or an inch and a half in thickness, and weighs about seven ounces. At *birth*, its weight, in proportion to the entire body, is almost equal to what is observed in the adult, being as 1 to 350; whilst in the adult it varies from 1 to 320 and 400. In *old age*, the organ not only decreases in weight, but decreases considerably in proportion to the entire body, being as 1 to 700. The size of the spleen is increased during and after digestion, and varies considerably, according to the state of nutrition of the body, being large in highly-fed, and small in starved animals. In intermittent and other fevers, it becomes much enlarged, weighing occasionally from 18 to 20 pounds.

Structure. The spleen is invested by two coats; an external serous, and an internal fibrous elastic coat.

The *external*, or *serous coat*, is derived from the peritoneum; it is thin, smooth, and in the human subject intimately adherent to the fibrous elastic coat. It invests almost the entire organ; being reflected from it, at the hilum, on to the great end of the stomach, and at the upper end of the organ on to the Diaphragm.

The *fibrous elastic coat* forms the framework of the spleen. It invests the exterior of the organ, and the hilum is reflected inwards upon the vessels in the form of vaginæ or sheaths. From these sheaths, as well as from the inner surface of the fibro-elastic coat, numerous small fibrous bands, *trabeculæ* (fig. 347), are given off in all directions; these uniting, constitute the areolar framework of the spleen. The proper coat, the sheaths of the vessels, and the trabeculæ, consist of a dense mesh of white and yellow elastic fibrous tissues, the latter considerably predominating. It is owing to the presence of this tissue, that the spleen possesses a considerable amount of elasticity, to allow of the very considerable variations in size that it presents under certain circumstances. In some of the mammalia, in addition to the usual constituents of this tunic, there are found numerous pale, flattened, spindle-shaped, nucleated fibres, like unstriped muscular fibres. It is probably owing to this structure, that the spleen possesses, when acted upon by the galvanic current, faint traces of contractility.

The *proper substance* of the spleen occupies the interspaces of the areolar framework of the organ; it is a soft, pulpy mass, of a dark reddish-brown colour, consisting of colourless and coloured elements.

The *colourless elements* consist of granular matter; nuclei, about the size of the red blood-discs, homogeneous or granular in structure; and nucleated vesicles

348.—The Malpighian Corpuscles, and their Relation with the Splenic Artery and its Branches.

in small numbers. These elements form, probably, one-half or two-thirds of the whole substance of the pulp, filling up the interspaces formed by the partitions of the spleen, and lying in close contact with the walls of the capillary vessels, so as to be readily acted upon by the nutrient fluid which permeates them. Thus in well-nourished animals, they form a large part of the entire bulk of the spleen, whilst they diminish in number, and occasionally are wanting in starved animals. The application of chemical tests shows that they are essentially a proteine compound.

The *coloured elements* of the pulp consist of red blood-globules and of coloured corpuscles, either free, or included in cells. Sometimes, unchanged blood-discs are seen included in a cell; but more frequently the included blood-discs are altered both in form and colour. Besides these, numerous deep-red, or reddish yellow, or black corpuscles and crystals, either single or aggregated in masses, are seen diffused throughout the pulp-substance: these, in chemical composition, are closely allied to the hæmatin of the blood.

Malpighian Corpuscles. On examining the cut surface of a healthy spleen, a number of small semi-opaque bodies, of gelatinous consistence, are seen disseminated throughout its substance; these are the splenic or Malpighian corpuscles (fig. 348). They may be seen at all periods of life; but they are more distinct in early than in adult life or old age; and they are much smaller in man than in most mammalia. They are of a spherical or ovoid form, vary considerably in size and number, and are of a semi-opaque whitish colour. They are appended to the sheaths of the smaller arteries and their branches, presenting a resemblance to the buds of the moss rose. Each consists of a membranous capsule, composed of fine pale fibres, which interlace in all directions. In man, the capsule is homogeneous in structure, and formed by a prolongation from the sheaths of the small arteries to which the corpuscles are attached. The blood-vessels ramifying on the surface of the corpuscles, consist of the larger ramifications of the arteries to which the sacculus is connected; and also of a delicate capillary plexus, similar

349.—One of the Splenic Corpuscles, showing its Relations with the Blood-vessels.

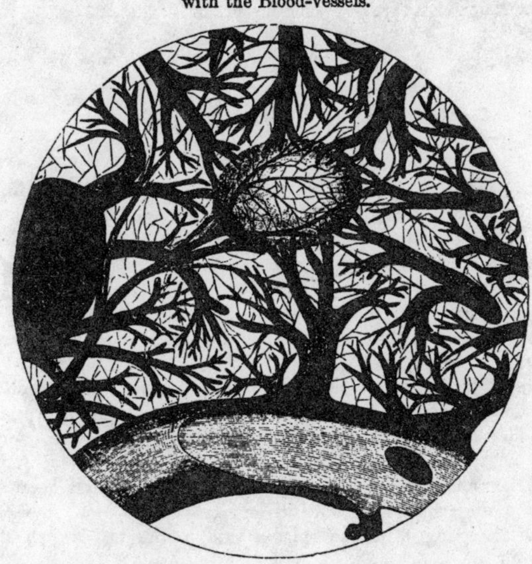

to that surrounding the vesicles of other glands. These vesicles have also a close relation with the veins (fig. 349). The latter vessels, which are of considerable size even at their origin, commence on the surface of each vesicle throughout the whole of its circumference, forming a dense mesh of veins, in which the Malpighian corpuscle is enclosed. It is probable, that from the blood contained in the capillary network, the material is separated which is occasionally stored up in the cavity of the corpuscle; the veins being so placed as to carry off, under certain conditions, those contents to be discharged again into the circulation. Each capsule contains a soft, white, semi-fluid substance, consisting of granular matter, nuclei similar to those found in the pulp, and a few nucleated cells, the composition of which is apparently albuminous. These bodies are very large after digestion is completed, in well-fed animals, and especially in those fed upon albuminous diet. In starved animals, they disappear altogether.

The *splenic artery* is remarkable for its large size, in proportion to the size of the organ; and also for its tortuous course. It divides into from four to six branches, which enter the hilum of the organ, and ramify throughout its substance (fig. 350), receiving sheaths from an involution of the external fibrous tunic, the same sheaths also investing the nerves and veins. Each branch runs in the transverse axis of the organ, from within outwards, diminishing in size during its transit, and giving off, in its passage, smaller branches, some of which pass to the

350.—Transverse Section of the Human Spleen, showing the Distribution of the Splenic Artery and its Branches.

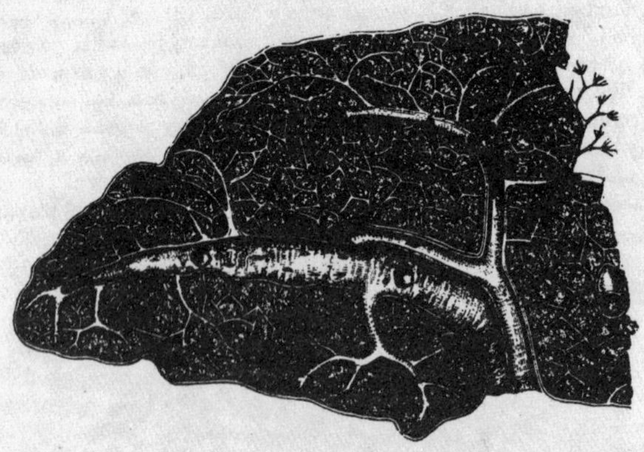

anterior, others to the posterior part; these ultimately terminate in the proper substance of the spleen, in small tufts or pencils of capillary vessels, which lie in direct contact with the pulp. Each of the larger branches of the arteries supplies chiefly that region of the organ in which the branch ramifies, having no anastomosis with the majority of the other branches.

The *capillaries*, supported by the minute trabeculæ, traverse the pulp in all directions, and terminate either directly in the veins, or open into lacunar spaces, from which the veins originate.

The *veins* are of large size, as compared with the size of the organ; and their distribution is limited, like that of the arteries, to the supply of a particular part of the gland; they are much larger and more numerous than the arteries. They originate, 1st, as continuations of the capillaries of the arteries; 2nd, by intercellular spaces communicating with each other; 3rd, by distinct cæcal pouches. By their junction they form from four to six branches, which emerge from the hilum; and these uniting, form the splenic vein, the largest branch of the vena portæ.

The *lymphatics* form a deep and superficial set; they pass through the lymphatic glands at the hilum, and terminate in the thoracic duct.

The *nerves* are derived from branches of the right and left semilunar ganglia, and right pneumogastric nerve.

The Thorax.

THE Thorax is a conical framework, formed partly of bones, and partly of the soft tissues by which they are connected together. It is supported and its back part is formed by the middle, or dorsal, region of the spine. It is narrow above, broad below, flattened before and behind, and somewhat cordiform on a transverse section.

Boundaries. The thorax is bounded in front by the sternum, the six upper costal cartilages, the ribs, and intercostal muscles; at the sides, by the ribs and intercostal muscles; and behind, by the same structures and the dorsal portion of the vertebral column.

The *superior opening* of the thorax is bounded on each side by the first rib; in front, by the upper border of the sternum; and behind, by the first dorsal vertebra. It is broader from side to side than from before backwards; and its direction is backwards and upwards.

The *lower opening*, or *base*, is bounded in front by the ensiform cartilage; behind, by the last dorsal vertebra; and on each side by the last rib, the Diaphragm filling in the intervening space. Its direction is obliquely downwards and backwards; so that the cavity of the thorax is much deeper on the posterior than on the anterior wall. It is wider transversely than from before backwards. Its outer surface is convex; but it is more flattened at the centre than at the sides. Its floor is higher on the right than on the left side, corresponding in the dead body to the upper border of the fifth costal cartilage on the right side; and to the corresponding part of the sixth cartilage on the left side.

The parts which pass through the upper opening of the thorax are, from before backwards, the Sterno-hyoid and Sterno-thyroid muscles, the remains of the thymus gland, the trachea, œsophagus, thoracic duct, and the Longus colli muscles of each side; on the sides, the arteria innominata, the left carotid and left subclavian arteries, the internal mammary and superior intercostal arteries, the right and left venæ innominatæ, and the inferior thyroid veins, the pneumogastric, sympathetic, phrenic, and cardiac nerves, the anterior branch of the first dorsal nerve, and the recurrent laryngeal nerve of the left side. The apex of each lung, covered by the pleura, also projects through this aperture, a little above the margin of the first rib.

The viscera contained in the thorax are, the heart, enclosed in its membranous bag, the pericardium; and the lungs, invested by the pleuræ.

The Pericardium.

The pericardium is a conical membranous sac, in which the heart and the commencement of the great vessels are contained. It is placed behind the sternum, and the cartilages of the third, fourth, fifth, sixth, and seventh ribs of the left side, in the interval between the pleuræ.

Its *apex* is directed upwards, and surrounds the great vessels about two inches above their origin from the base of the heart. Its *base* is attached to the central tendon of the Diaphragm, extending a little farther to the left than the right side. In *front*, it is separated from the sternum by the remains of the thymus gland above, and a little loose areolar tissue below; and is covered by the margins of the lungs, especially the left. *Behind*, it rests upon the bronchi, the œsophagus, and the descending aorta. *Laterally*, it is covered by the pleuræ; the phrenic vessels and nerve descending between the two membranes on either side.

The *pericardium* is a fibro-serous membrane, and consists, therefore, of two layers: an external fibrous, and an internal serous.

The *fibrous layer* is a strong, dense membrane. Above, it surrounds the great vessels arising from the base of the heart, on which it is continued in the form of tubular prolongations, which are gradually lost upon their external coats; the strongest being that which encloses the aorta. The pericardium may be traced, over these vessels, to become continuous with the deep layer of the cervical fascia. Below, it is attached to the central tendon of the Diaphragm; and, on the left side, to its muscular fibres.

The vessels receiving fibrous prolongations from this membrane are the aorta, the superior vena cava, and the pulmonary arteries and veins. As the inferior

351.—Front View of the Thorax. The Ribs and Sternum are represented in Relation to the Lungs, Heart, and other Internal Organs.

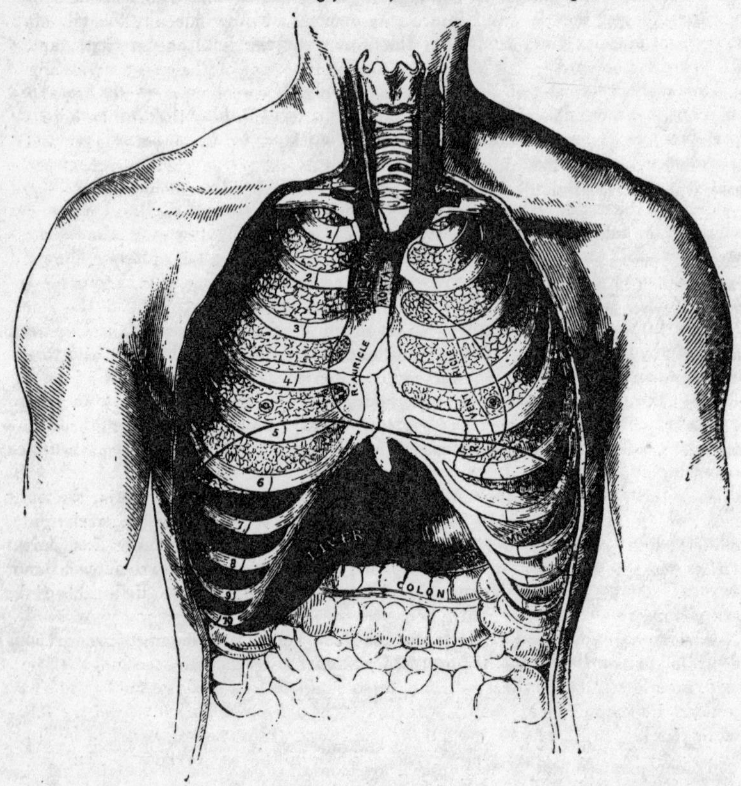

vena cava enters the pericardium, through the central tendon of the Diaphragm, it receives no covering from the fibrous layer.

The *serous layer* invests the heart, and is then reflected on the inner surface of the pericardium. It consists, therefore, of a visceral and a parietal portion. The former invests the surface of the heart, and the commencement of the great vessels, to the extent of two inches from their origin; from these, it is reflected upon the inner surface of the fibrous layer, lining, below, the upper surface of the central tendon of the Diaphragm. The serous membrane encloses the aorta and pulmonary artery in a single tube; but it only partially covers the superior and inferior vena cava and the four pulmonary veins. Its inner surface is smooth and glistening, and secretes a thin fluid, which serves to facilitate the movements of the heart.

The arteries of the pericardium are derived from the internal mammary, the bronchial, the œsophageal, and the phrenic.

The Heart.

The heart is a hollow muscular organ of a conical form, placed between the lungs, and enclosed in the cavity of the pericardium.

Position. The heart is placed obliquely in the chest: the broad attached end, or base, is directed upwards and backwards to the right, and corresponds to the interval between the fifth and eighth dorsal vertebræ: the apex is directed forwards and to the left, and corresponds to the interspace between the cartilage of the fifth and sixth ribs, one inch to the inner side, and two inches below the left nipple. The heart is placed behind the lower two-thirds of the sternum, and projects further into the left than into the right cavity of the chest, extending from the median line about three inches in the former direction, and only one and a half in the latter. Its upper border would correspond to a line drawn across the sternum, on a level with the upper border of the third costal cartilages; and its lower border, to a line drawn across the lower end of the gladiolus, from the costo-xiphoid articulation of the right side, to the point above mentioned, as the situation of the apex. Its anterior surface is rounded and convex, directed upwards and forwards, and formed chiefly by the right ventricle and part of the left. Its posterior surface is flattened, and rests upon the Diaphragm, and is formed chiefly by the left ventricle. The right border is long, thin, and sharp; the left border short, but thick and round.

Size. The heart, in the adult, measures about five inches in length, three inches and a half in breadth in its broadest part, and two inches and a half in thickness. The prevalent weight, in the male, varies from ten to twelve ounces; in the female, from eight to ten: its proportion to the body being as 1 to 169 in males; 1 to 149 in females. The heart continues increasing in weight, and also in length, breadth and thickness, up to an advanced period of life; this increase is more marked in men than in women.

Component parts. The heart is subdivided by a longitudinal muscular septum into two lateral halves, which are named respectively, from their position, right and left; and a transverse constriction divides each half of the organ into two cavities, the upper cavity on each side being called the *auricle*, the lower the *ventricle.* The right is the venous side of the heart, receiving into its auricle the dark venous blood from the entire body, by the superior and inferior vena cava, and coronary sinus. From the auricle, the blood passes into the right ventricle; and from the right ventricle, through the pulmonary artery, into the lungs. The blood, arterialized by its passage through the lungs, is returned to the left side of the heart by the pulmonary veins, which open into the left auricle; from the left auricle the blood passes into the left ventricle, and from the left ventricle is distributed, by the aorta and its subdivisions, through the entire body. This constitutes the circulation of the blood in the adult.

This division of the heart into four cavities, is indicated by grooves upon its surface. The great transverse groove separating the auricles from the ventricles, is called the *auriculo-ventricular* groove. It is deficient, in front, from being crossed by the root of the pulmonary artery, and contains the trunk of the nutrient vessels of the heart. The auricular portion occupies the base of the heart, and is subdivided into two cavities by a median septum. The two ventricles are also separated into a right and left, by two longitudinal furrows, which are situated one on the anterior, the other on the posterior surface: these extend from the base to the apex of the organ: the former being situated nearer to the left border of the heart, and the latter to the right. It follows, therefore, that the right ventricle forms the greater portion of the anterior surface of the heart, and the left ventricle more of its posterior surface.

Each of these cavities should now be separately examined.

The Right Auricle is a little larger than the left, its walls somewhat thinner, measuring about one line; and its cavity is capable of containing about two ounces. It consists of two parts, a principal cavity, or sinus, and an appendix auriculæ.

The *sinus* is the large quadrangular cavity, placed between the two venæ cavæ: its walls are extremely thin: it is connected below with the right ventricle, and internally with the left auricle, being free in the rest of its extent.

352.—The Right Auricle and Ventricle laid open, the Anterior Walls of both being removed.

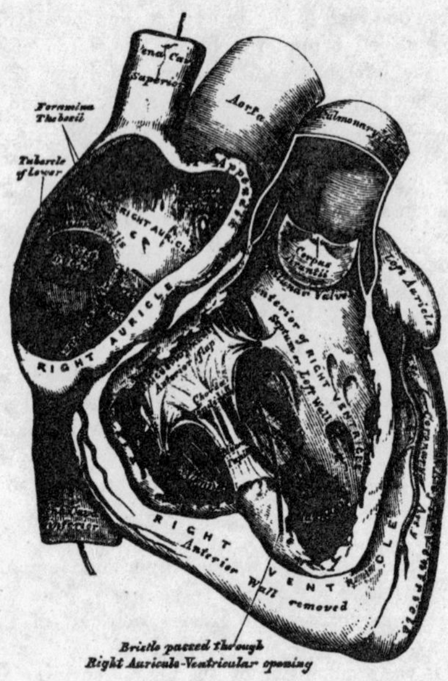

The *appendix auriculæ*, so called from its fancied resemblance to a dog's ear, is a small conical muscular pouch, the margins of which present a dentated edge. It projects from the sinus forwards and to the left side, overlapping the root of the pulmonary artery.

To examine the interior of the auricle, a transverse incision should be made along its ventricular margin, from its right border to the appendix; and, from the middle of this, a second incision should be carried upwards, along the inner side of the two venæ cavæ.

The following parts present themselves for examination:

Openings.
- Superior cava.
- Inferior cava.
- Coronary Sinus.
- Foramina Thebesii.
- Auriculo-ventricular.

Valves.
- Eustachian.
- Coronary.

Relics of fœtal structure.
- Annulus ovalis.
- Fossa ovalis.

Musculi pectinati.

Openings. The *superior vena cava* returns the blood from the upper half of the body, and opens into the upper and front part of the auricle, the direction of its orifice being downwards and forwards.

The *inferior vena cava*, larger than the superior, returns the blood from the lower half of the body, and opens into the lowest part of the auricle, near the septum, the direction of its orifice being upwards and inwards. The direction of a current of blood through the superior vena cava would consequently be towards the auriculo-ventricular orifice; whilst the direction of the blood through the inferior cava would be towards the auricular septum. This is the normal direction of the two currents in fœtal life.

The *tuberculum Loweri* is a small projection on the right wall of the auricle, between the two cavæ. This is most distinct in the hearts of quadrupeds: in man, it is scarcely visible. It was supposed by Lower to direct the blood from the superior cava towards the auriculo-ventricular opening.

The *coronary sinus* opens into the auricle, between the inferior vena cava and the auriculo-ventricular opening. It returns the blood from the substance of the heart, and is protected by a semicircular fold of the lining membrane of the auricle, the coronary valve. The sinus, before entering the auricle, is considerably dilated. Its wall is partly muscular, and, at its junction with the great coronary vein, is somewhat constricted, and furnished with a valve, consisting of two unequal segments.

The *foramina Thebesii* are numerous minute apertures, the mouths of small veins (*venæ cordis minimæ*), which open on various parts of the inner surface of the auricle. They return the blood directly from the muscular substance of the heart. Some of these foramina are minute depressions in the walls of the heart, presenting a closed extremity.

The *auriculo-ventricular opening* is the large oval aperture of communication between the auricle and ventricle, to be presently described.

Valves. The *Eustachian valve* is situated between the anterior margin of the inferior cava and the auriculo-ventricular orifice. It is semilunar in form, its convex margin being attached to the wall of the vein; its concave margin, which is free, terminating in two cornua, of which the left is attached to the anterior edge of the annulis ovalis; the right being lost on the wall of the auricle. The valve is formed by a duplicature of the lining membrane of the auricle, containing a few muscular fibres.

In the *fœtus*, this valve is of large size, and serves to direct the blood from the inferior cava, through the foramen ovale, into the left auricle.

In the *adult*, it is occasionally persistent, and may assist in preventing the reflux of blood into the inferior cava: more commonly, it is small, and its free margin presents a cribriform, or filamentous appearance; occasionally, it is altogether wanting.

The *coronary valve* is a semicircular fold of the lining membrane of the auricle, protecting the orifice of the coronary sinus. It prevents the regurgitation of blood into the sinus during the contraction of the auricle. This valve is occasionally double.

The *fossa ovalis* is an oval depression, corresponding to the situation of the foramen ovale in the fœtus. It is situated at the lower part of the septum auricularum, above the orifice of the inferior vena cava.

The *annulus ovalis* is the prominent oval margin of the foramen ovale. It is most distinct above, and at the sides; below it is deficient. A small slit-like valvular opening is occasionally found, at the upper margin of the fossa ovalis, which leads upwards, beneath the annulus, into the left auricle, and is the remains of the aperture between the two auricles in the fœtus.

The *musculi pectinati* are small, prominent muscular columns, which run across the inner surface of the appendix auriculæ, and adjoining portion of the wall of the sinus. They have received the name, *pectinati*, from the fancied resemblance they bear to the teeth of a comb.

The Right Ventricle is triangular in form, and extends from the right auricle to near the apex of the heart. Its anterior or upper surface is rounded and convex, and forms the larger part of the front of the heart. Its posterior or under surface is flattened, rests upon the Diaphragm, and forms only a small part of the back of the heart. Its inner wall is formed by the partition between the two ventricles, the *septum ventriculorum*, the surface of which is convex, and bulges into the cavity of the right ventricle. Superiorly, the ventricle forms a conical prolongation, the *infundibulum*, or *conus arteriosus*, from which the pulmonary artery arises. The walls of the right ventricle are thinner than those of the left, the proportion between them being as 1 to 2 (Bizot). The wall is thickest at the base, and gradually becomes thinner towards the apex. The cavity, which equals that of the left ventricle, is capable of containing about two fluid ounces.

To examine the interior, an incision should be made a little to the right of the anterior ventricular groove from the pulmonary artery to the apex of the heart, and should be carried up from thence along the right border of the ventricle, as far as the auriculo-ventricular opening.

The following parts present themselves for examination:

Openings . . . { Auriculo-ventricular.
Opening of the pulmonary artery.

Valves . . . { Tricuspid.
Semilunar.

And a muscular and tendinous apparatus connected with the tricuspid valve:

Columnæ carneæ. Chordæ tendineæ.

The *auriculo-ventricular orifice* is the large oval aperture of communication between the auricle and ventricle. It is situated at the base of the ventricle, near the right border of the heart, and corresponds to the centre of the sternum between the third costal cartilages. The opening is about an inch in diameter,* oval from side to side, surrounded by a fibrous ring, covered by the lining membrane of the heart, and rather larger than the corresponding aperture on the left side, being sufficiently large to admit the ends of three fingers. It is guarded by the tricuspid valve.

The *opening of the pulmonary artery* is circular in form, and situated at the summit of the conus arteriosus, close to the septum ventriculorum. It is placed on the left side of the auriculo-ventricular opening, upon the anterior aspect of the heart, and corresponds to the upper border of the third costal cartilage of the left side, close to the sternum. Its orifice is guarded by the pulmonary semilunar valves.

The *tricuspid valve* consists of three segments of a triangular or trapezoidal shape, formed by a duplicature of the lining membrane of the heart, strengthened by a layer of fibrous tissue, and containing, according to Kürschner and Senac, muscular fibres. These segments are connected by their bases to the auriculo-ventricular orifice, and by their sides with one another, so as to form a continuous annular membrane, which is attached round the margin of the auriculo-ventricular opening, their free margins and ventricular surfaces affording attachment to a number of delicate tendinous cords, the *chordæ tendineæ*. The largest and most

* In the *Pathological Transactions*, vol. vi. p. 119, Dr. Peacock has given some careful researches upon the weight and dimensions of the heart in health and disease. He states as the result of his investigations, that, in the healthy adult heart, the right auriculo-ventricular aperture has a mean circumference of 54·4 lines, or $4\frac{20}{24}$ inches; the left auriculo-ventricular aperture a mean circumference of 44·3 lines, or $3\frac{22}{24}$ inches; the pulmonic orifice of 40 lines, or $3\frac{13}{24}$ inches; and the aortic orifice of 35·5 lines, or $3\frac{4}{24}$ inches: but the dimensions of the orifices varied greatly in different cases, the right auriculo-ventricular aperture having a range of from 45 to 60 lines, and the others in the same proportion.

moveable segment is placed towards the left side of the auriculo-ventricular opening: interposed between that opening and the pulmonary artery. Another segment corresponds to the front of the ventricle; and a third to its posterior wall. The central part of each segment is thick and strong; the lateral margins are thin and indented. The chordæ tendineæ are connected with the adjacent margins of the principal segments of the valve, and are further attached to each segment in the following manner: 1. Three or four reach the attached margin of each segment, where they are continuous with the auriculo-ventricular tendinous ring. 2. Others, four to six in number, are attached to the central thickened part of each segment. 2. The most numerous and finest are connected with the marginal portion of each segment.

The *columnæ carneæ* are the rounded muscular columns which project from nearly the whole of the inner surface of the ventricle, excepting near the opening of the pulmonary artery. They may be classified, according to their mode of connection with the ventricle, into three sets. The first set merely form prominent ridges on the inner surface of the ventricle, being attached by their entire length on one side, as well as by their extremities. The second set are attached by their two extremities, but are free in the rest of their extent; whilst the third set (*columnæ papillares*), three or four in number, are attached by one extremity to the wall of the heart, the opposite extremity giving attachment to the *chordæ tendineæ*.

The *semilunar valves*, three in number, guard the orifice of the pulmonary artery. They consist of three semicircular folds, formed by a duplicature of the lining membrane, strengthened by fibrous tissue. They are attached, by their convex margins, to the wall of the artery, at its junction with the ventricle, the straight border being free, and directed upwards in the course of the vessel, against the sides of which the valve-flaps are pressed during the passage of the blood along the artery. The free margin of each is somewhat thicker than the rest of the valve, is strengthened by a bundle of tendinous fibres; and presents, at its middle, a small projecting fibro-cartilaginous nodule, called *corpus Arantii*. From this nodule, tendinous fibres radiate through the valve to its attached margin, and these fibres form a constituent part of its substance throughout its whole extent, excepting two narrow lunated portions, placed one on either side of the nodule, immediately behind the free margin; here, the valve is thin, and formed merely by the lining membrane. During the passage of the blood along the pulmonary artery, these valves are pressed against the sides of its cylinder, and the course of the blood along the tube is uninterrupted; but during the ventricular diastole, when the current of blood along the pulmonary artery is checked, and partly thrown back by its elastic walls, these valves become immediately expanded, and effectually close the entrance of the tube. When the valves are closed, the lunated portions of each are brought into contact with one another by their opposed surfaces, the three fibro-cartilaginous nodules filling up the small triangular space that would be otherwise left by the approximation of the three semilunar folds.

Between the semilunar valves and the commencement of the pulmonary artery are three pouches or dilatations, one behind each valve. These are the pulmonary sinuses (*sinuses of Valsalva*). Similar sinuses exist between the semilunar valves and the commencement of the aorta; they are larger than the pulmonary sinuses. The blood, in its regurgitation towards the heart, finds its way into these sinuses, and so shuts down the valve-flaps.

The Left Auricle is rather smaller but thicker than the right, measuring about one line and a half; it consists, like the right, of two parts, a principal cavity or sinus, and an appendix auriculæ.

The *sinus* is cuboidal in form, and concealed in front by the pulmonary artery and aorta; internally, it is separated from the right auricle by the septum auricularum; behind, it receives on each side the pulmonary veins, being free in the rest of its extent.

The *appendix auriculæ* is somewhat constricted at its junction with the auricle;

it is longer, narrower, and more curved than that of the right side, and its margins more deeply indented, presenting a kind of foliated appearance. Its direction is forwards and towards the right side, overlapping the root of the pulmonary artery.

In order to examine its interior, a horizontal incision should be made along the attached border of the auricle to the ventricle; and from the middle of this, a second incision should be carried upwards.

The following parts then present themselves for examination:

The openings of the four pulmonary veins.
Auriculo-ventricular opening.
Musculi pectinati.

The *pulmonary veins*, four in number, open, two into the right, and two into the left side of the auricle. The two left veins frequently terminate by a common opening. They are not provided with valves.

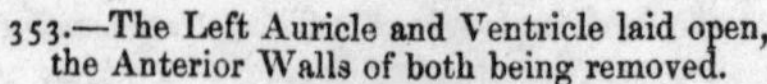
353.—The Left Auricle and Ventricle laid open, the Anterior Walls of both being removed.

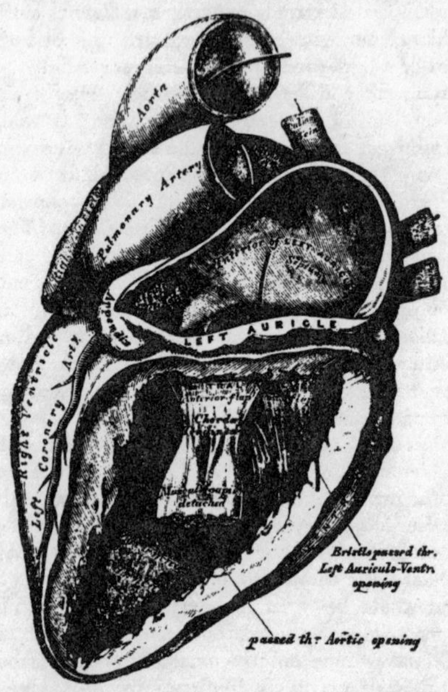

The *auriculo-ventricular opening* is the large oval aperture of communication between the auricle and ventricle. It is rather smaller than the corresponding opening on the opposite side.

The *musculi pectinati* are fewer in number and smaller than on the right side; they are confined to the inner surface of the appendix.

On the inner surface of the septum auricularum may be seen a lunated impression, bounded below by a crescentic ridge, the concavity of which is turned upwards. The depression is just above the fossa ovalis in the right auricle.

The Left Ventricle is longer and more conical in shape than the right ventricle. It forms a small part of the left side of the anterior surface of the heart, and a considerable part of its posterior surface. It also forms the apex of the heart by its projection beyond the right ventricle. Its walls are much thicker than those of the right ventricle, the proportion being as 2 to 1 (Bizot). They are also thickest in the broadest part of the ventricle, becoming gradually thinner towards the base, and also towards the apex, which is the thinnest part.

Its cavity should be opened, by making an incision through its anterior wall along the left side of the ventricular septum, and carrying it round the apex and along its posterior surface to the auriculo-ventricular opening.

The following parts present themselves for examination:

Openings.	Auriculo-ventricular. Aortic.	Valves.	Mitral. Semilunar.
	Chordæ tendineæ.	Columnæ carneæ.	

The *auriculo-ventricular opening* is placed to the left of the aortic orifice, beneath the right auriculo-ventricular opening, opposite the centre of the sternum. It is a little smaller than the corresponding aperture of the opposite side; and, like it, is broader in the transverse than in the antero-posterior diameter. It is surrounded by a dense fibrous ring, covered by the lining membrane of the heart, and guarded by the mitral valve.

The *aortic opening* is a small circular aperture, in front and to the right side of the auriculo-ventricular, from which it is separated by one of the segments of the mitral valve. Its orifice is guarded by the semilunar valves. Its position corresponds to the sternum, on a line with the lower border of the third left costal cartilage.

The *mitral valve* is attached to the circumference of the auriculo-ventricular orifice, in the same way that the tricuspid valve is on the opposite side. It is formed by a duplicature of the lining membrane, strengthened by fibrous tissue, and contains a few muscular fibres. It is larger in size, thicker, and altogether stronger than the tricuspid, and consists of two segments of unequal size. The larger segment is placed in front, between the auriculo-ventricular and aortic orifices, and is said to prevent the filling of the aorta during the distension of the ventricle. Two smaller segments are usually found at the angle of junction of the larger. The mitral valve-flaps are furnished with chordæ tendineæ: the mode of attachment of which is precisely similar to those on the right side; but they are thicker, stronger, and less numerous.

The *semilunar valves* surround the orifice of the aorta; they are similar in structure, and in their mode of attachment, to those of the pulmonary artery. They are, however, larger, thicker, and stronger than those of the right side; the lunulæ are more distinct, and the corpora Arantii larger and more prominent. Between each valve and the cylinder of the aorta is a deep depression, the *sinus aortici* (sinuses of Valsalva); they are larger than those at the root of the pulmonary artery.

The *columnæ carneæ* admit of a subdivision into three sets, like those upon the right side; but they are smaller, more numerous, and present a dense interlacement, especially at the apex, and upon the posterior wall. Those attached by one extremity only, the *musculi papillares*, are two in number, being connected one to the anterior, the other to the posterior wall; they are of large size, and terminate by free rounded extremities, from which the chordæ tendineæ arise.

The *Endocardium* is the serous membrane which lines the internal surface of the heart; it assists in forming the valves by its reduplications, and is continuous with the lining membrane of the great blood-vessels. It is a thin, smooth, transparent membrane, giving to the inner surface of the heart its glistening appearance. It is more opaque on the left than on the right side of the heart, thicker in the auricles than in the ventricles, and thickest in the left auricle. It is thin on the musculi pectinati, and on the columnæ carneæ; but thicker on the smooth part of the auricular and ventricular walls, and on the tips of the musculi papillares.

Structure. The heart consists of muscular fibres, and of fibrous rings which serve for their attachment.

The *fibrous rings* surround the auriculo-ventricular and arterial orifices: they are stronger upon the left than on the right side of the heart. The auriculo-ventricular rings serve for the attachment of the muscular fibres of the auricles and ventricles, and also for the mitral and tricuspid valves; the ring on the left side is closely connected, by its right margin, with the aortic arterial ring. Between these and the right auriculo-ventricular ring, is a fibro-cartilaginous mass; and in some of the larger animals, as the ox and elephant, a portion of bone.

The fibrous rings surrounding the arterial orifices, serve for the attachment of the great vessels and semilunar valves. Each ring receives, by its ventricular margin, the attachment of the muscular fibres of the ventricles; its opposite margin presents three deep semicircular notches, within which the middle coat of the rtery (which presents three convex semicircular segments) is firmly fixed; the attachment of the artery to its fibrous ring being strengthened by the thin cellular coat and serous membrane externally, and by the endocardium within. It is opposite the margins of these semicircular notches, in the arterial rings, that the endocardium, by its reduplication, forms the semilunar valves, the fibrous structure of the ring being continued into each of the segments of the valve at this part. The middle coat of the artery in this situation is thin, and the sides of the vessel dilated to form the sinuses of Valsalva.

The *muscular structure* of the heart consists of bands of fibres, which present an exceedingly intricate interlacement. They are of a deep red colour, and marked with transverse striæ.

The muscular fibres of the heart admit of a subdivision into two kinds: those of the auricles, and those of the ventricles; which are quite independent of one another.

Fibres of the auricles. These are disposed in two layers: a superficial layer common to both cavities, and a deep layer proper to each. The *superficial fibres* are more distinct on the anterior surface of the auricles, across the bases of which they run in a transverse direction, forming a thin, but incomplete, layer. Some of these fibres pass into the septum auricularum. The *internal* or *deep fibres* proper to each auricle consist of two sets, looped, and annular fibres. The *looped fibres* pass upwards over each auricle, being attached by both extremities to the corresponding auriculo-ventricular rings, in front and behind. The *annular fibres* surround the whole extent of the appendices auriculæ, and are continued upon the walls of the venæ cavæ and coronary sinus on the right side, and upon the pulmonary veins on the left side, at their connection with the heart. In the appendices, they interlace with the longitudinal fibres.

Fibres of the ventricles. These, as in the auricles, are disposed in layers, some of which are common to both ventricular cavities, whilst others belong exclusively to one ventricle, the latter being chiefly found towards the base of the heart. The greater majority of these fibres are connected by both ends with the auriculo-ventricular fibrous rings, either directly or indirectly through the chordæ tendineæ; some, however, are attached to the fibrous rings surrounding the arterial orifices.

The *superficial fibres* are either longitudinal, or more commonly oblique or spiral in their direction, and towards the apex are arranged in the form of twisted loops; the deeper fibres are circular.

The spiral fibres are disposed in layers of various degrees of thickness: the most superficial, on the front of the ventricles, run obliquely from right to left, and from above downwards. On the back of the ventricles they are directed more vertically, and pass from left to right.

The superficial fibres coil inwards at the apex of the heart, round which they are arranged in a whorl-like form, called the *vortex*, dipping beneath the edge of the deeper and shorter layers. If these fibres are carefully uncoiled, in a heart previously boiled, the cavity of the left, and then that of the right ventricle, will be exposed at this point. The layers of fibres successively met with have a similar arrangement; the more superficial and longer turning inwards, and including the deeper and shorter bands. All these fibres ascend and spread out upon the inner surface of the ventricles, forming the walls, the septum, and the musculi papillares, which project from these cavities; and they are finally inserted into the auriculo-ventricular fibrous rings directly, or, indirectly, through the chordæ tendineæ. Of these spiral fibres, some enter at the interventricular furrows, and surround either ventricle singly; others pass across the furrows and embrace both cavities. On tracing those which form the vortex back into the interventricular septum, they are found to be interlaced with similar fibres from the right ventricle, and ascend vertically upon the right side of the septum, as far as its base, in the form of a long and broad band.

Circular fibres. The circular fibres are situated deeply in the substance of the heart; towards the base they enter the anterior and posterior longitudinal furrows, so as to include each cavity singly, or, passing across them, surround both ventricles, more fibres passing across the posterior than the anterior furrow. They finally ascend in the substance of the ventricle, to be inserted into the fibrous rings at its base.

Vessels and Nerves. The *arteries* supplying the heart, are the left or anterior and right or posterior coronary (p. 323).

The *veins* accompany the arteries, and terminate in the right auricle. They are the great cardiac vein, the small, or anterior cardiac veins, and the venæ cordis minimæ (*venæ Thebesii*) (p. 440).

The *lymphatics* terminate in the thoracic and right lymphatic ducts.

The *nerves* are derived from the cardiac plexuses, which are formed partly from the cranial nerves, and partly from the sympathetic. They are freely distributed both on the surface, and in the substance of the heart; the separate filaments being furnished with small ganglia.*

Peculiarities in the Vascular System of the Fœtus.

The chief peculiarities in the heart of the fœtus, are the direct communication between the two auricles through the foramen ovale, and the large size of the Eustachian valve. There are also several minor peculiarities. Thus, the position of the heart is vertical until the fourth month, when it commences to assume an oblique direction. Its size is also very considerable, as compared with the body, the proportion at the second month being as 1 to 50: at birth it is as 1 to 120: whilst, in the adult, the average is about 1 to 160. At an early period of fœtal life, the auricular portion of the heart is larger than the ventricular, the right auricle being more capacious than the left; but, towards birth, the ventricular portion becomes the larger. The thickness of both ventricles is, at first, about equal; but, towards birth, the left becomes much the thicker of the two.

The *foramen ovale* is situated at the lower and back part of the septum auricularum, forming a communication between the auricles. It attains its greatest size at the sixth month.

The *Eustachian valve* is developed from the anterior border of the inferior vena cava, at its entrance into the auricle. It is directed upwards on the left side of the opening of this vein, and serves to direct the blood from the inferior vena cava through the foramen ovale into the left auricle.

The peculiarities in the arterial system of the fœtus are the communication between the pulmonary artery and descending part of the arch of the aorta, by means of the ductus arteriosus, and the communication between the internal iliac arteries and the placenta, by means of the umbilical arteries.

The *ductus arteriosus* is a short tube, about half an inch in length at birth,

* For full and accurate descriptions of the nerves and ganglia of the heart, the student is referred to Dr. R. Lee's papers on the subject.

and of the diameter of a goose-quill. In the early condition, it forms the continuation of the pulmonary artery, and opens into the arch of the aorta, just below the origin of the left subclavian artery; and so conducts the chief part of the blood from the right ventricle into the descending aorta. When the branches of the pulmonary artery have become larger relatively to the ductus arteriosus, the latter is chiefly connected to the left pulmonary artery; and the fibrous cord, which is all

354.—Plan of the Fœtal Circulation.

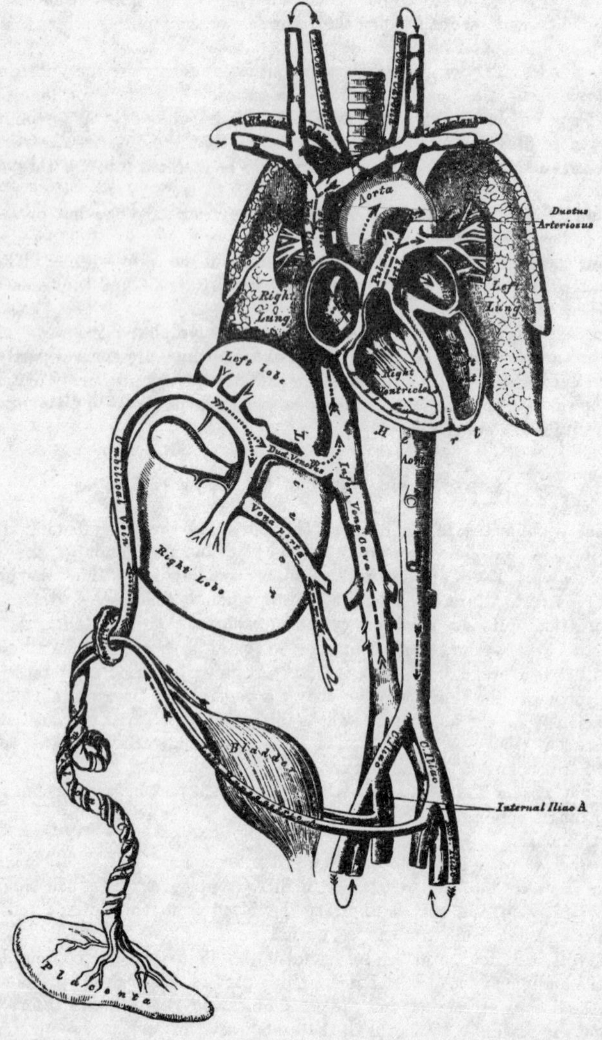

In this plan the figured arrows represent the kind of blood, as well as the direction which it takes in the vessels. Thus—arterial blood is figured ⋙-------⋗; venous blood, ⋙------⋗; mixed (arterial and venous blood), ⋙-·-·-⋗.

that remains of the ductus arteriosus in later life, will be found to be attached to the root of that vessel.

The *umbilical*, or *hypogastric arteries*, arise from the internal iliacs, in addition to the branches given off from those vessels in the adult. Ascending along the sides of the bladder to its fundus, they pass out of the abdomen at the umbilicus, and are continued along the umbilical cord to the placenta, coiling round the umbilical vein. They return to the placenta the blood which has circulated in the system of the fœtus.

The peculiarity in the venous system of the fœtus is the communication established between the placenta, and the liver and portal vein, through the umbilical vein, and with the inferior vena cava by the ductus venosus.

Fœtal Circulation.

The arterial blood destined for the nutrition of the fœtus, is carried from the placenta to the fœtus, along the umbilical cord, by the umbilical vein. The umbilical vein enters the abdomen at the umbilicus, and passes upwards along the free margin of the suspensory ligament of the liver, to the under surface of that organ, where it gives off two or three branches to the left lobe, one of which is of large size; and others to the lobus quadratus and lobulus Spigelii. At the transverse fissure it divides into two branches; of these, the larger is joined by the portal vein, and enters the right lobe: the smaller branch continues onwards, under the name of the ductus venosus, and joins the left hepatic vein at the point of junction of that vessel with the inferior vena cava. The blood, therefore, which traverses the umbilical vein, reaches the inferior cava in three different ways. The greater quantity circulates through the liver with the portal venous blood, before entering the vena cava by the hepatic veins: some enters the liver directly, and is also returned to the inferior cava by the hepatic veins: the smaller quantity passes directly into the vena cava, by the junction of the ductus venosus with the left hepatic vein.

In the inferior cava, the blood carried by the ductus venosus and hepatic veins, becomes mixed with that returning from the lower extremities and viscera of the abdomen. It enters the right auricle, and, guided by the Eustachian valve, passes through the foramen ovale into the left auricle, where it becomes mixed with a small quantity of blood returned from the lungs by the pulmonary veins. From the left auricle it passes into the left ventricle; and, from the left ventricle, into the aorta, from whence it is distributed almost entirely to the head and upper extremities, a small quantity being probably carried into the descending aorta. From the head and upper extremities, the blood is returned by the branches of the superior vena cava to the right auricle, where it becomes mixed with a small portion of the blood from the inferior cava. From the right auricle it descends over the Eustachian valve into the right ventricle; and, from the right ventricle, passes into the pulmonary artery. The lungs of the fœtus being solid, and almost impervious, only a small quantity of the blood of the pulmonary artery is distributed to them, by the right and left pulmonary arteries, which is returned by the pulmonary veins to the left auricle: the greater part passes through the ductus arteriosus into the commencement of the descending aorta, where it becomes mixed with a small quantity of blood transmitted by the left ventricle into the aorta. Along this vessel it descends to supply the lower extremities and viscera of the abdomen and pelvis, the chief portion being, however, conveyed by the umbilical arteries to the placenta.

From the preceding account of the circulation of the blood in the fœtus, it will be seen:

1. That the placenta serves the double purpose of a respiratory and nutritive organ, receiving the venous blood from the fœtus, and returning it again reoxygenated, and charged with additional nutritive material.

2. That nearly the whole of the blood of the umbilical vein traverses the liver

before entering the inferior cava; hence the large size of this organ, especially at an early period of fœtal life.

3. That the right auricle is the point of meeting of a double current, the blood in the inferior cava being guided by the Eustachian valve into the left auricle, whilst that in the superior cava descends into the right ventricle. At an early period of fœtal life, it is highly probable that the two streams are quite distinct; for the inferior cava opens almost directly into the left auricle, and the Eustachian valve would exclude the current along the vein from entering the right ventricle. At a later period, as the separation between the two auricles becomes more distinct, it seems probable that some mixture of the two streams must take place.

4. The blood carried from the placenta to the fœtus by the umbilical vein, mixed with the blood from the inferior cava, passes almost directly to the arch of the aorta, and is distributed by the branches of that vessel to the head and upper extremities: hence the large size and perfect development of those parts at birth.

5. The blood contained in the descending aorta, chiefly derived from that which has already circulated through the head and limbs, together with a small quantity from the left ventricle, is distributed to the lower extremities: hence the small size and imperfect development of these parts at birth.

Changes in the Vascular System at Birth.

At birth, when respiration is established, an increased amount of blood from the pulmonary artery passes through the lungs, which now perform their office as respiratory organs, and, at the same time, the placental circulation is cut off. The foramen ovale becomes gradually closed-in by about the tenth day after birth, a valvular fold rises up on the left side of its margin, and ultimately above its upper part; this valve becomes adherent to the margins of the foramen for the greater part of its circumference, but, above, a valvular opening is left between the two auricles, which sometimes remains persistent.

The *ductus arteriosus* begins to contract immediately after respiration is established, becomes completely closed from the fourth to the tenth day, and ultimately degenerates into an impervious cord, which serves to connect the left pulmonary artery to the concavity of the arch of the aorta.

Of the *umbilical* or *hypogastric arteries*, the portion continued on to the bladder from the trunk of the corresponding internal iliac remains pervious, as the superior vesical artery; and the part between the fundus of the bladder and the umbilicus becomes obliterated between the second and fifth days after birth, and forms the anterior true ligament of the bladder.

The *umbilical vein* and *ductus venosus* become completely obliterated between the second and fifth days after birth, and ultimately dwindle to fibrous cords; the former becoming the round ligament of the liver, the latter, the fibrous cord, which, in the adult, may be traced along the fissure of the ductus venosus.

Organs of Voice and Respiration.

The Larynx.

THE LARYNX is the organ of voice, placed at the upper part of the air-passage. It is situated between the trachea and base of the tongue, at the upper and fore part of the neck, where it forms a considerable projection in the middle line. On either side of it lie the great vessels of the neck; behind, it forms part of the boundary of the pharynx, and is covered by the mucous membrane lining that cavity.

The Larynx is narrow and cylindrical below, but broad above, where it presents the form of a triangular box, flattened behind and at the sides, whilst in front it is bounded by a prominent vertical ridge. It is composed of cartilages, which are connected together by ligaments and moved by numerous muscles: the interior is lined by mucous membrane, and supplied with vessels and nerves.

The Cartilages of the larynx are nine in number, three single, and three pairs:

Thyroid.	Two Arytenoid.
Cricoid.	Two Cornicula Laryngis.
Epiglottis.	Two Cuneiform.

The *Thyroid* (*θύρεος, a shield*) is the largest cartilage of the larynx. It consists of two lateral lamellæ or alæ, united at an acute angle in front, forming a vertical projection in the middle line, which is prominent above, and called the *pomum Adami*. This projection is subcutaneous, more distinct in the male than in the female, and occasionally separated from the integument by a bursa mucosa.

355.—Side View of the Thyroid and Cricoid Cartilages.

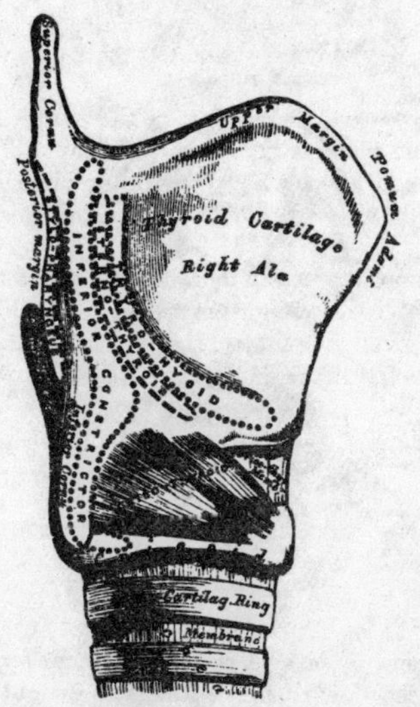

Each lamella is quadrilateral in form. Its *outer surface* presents an oblique ridge, which passes downwards and forwards from a tubercle, situated near the root of the superior cornu. This ridge gives attachment to the Sterno-thyroid and Thyro-hyoid muscles; the portion of cartilage included between it and the posterior border, to part of the Inferior constrictor muscle.

The *inner surface* of each ala is smooth, concave, and covered by mucous membrane above and behind; but in front, in the receding angle formed by their junction, is attached the epiglottis, the true and false chordæ vocales, the Thyro-arytenoid, and Thyro-epiglottidean muscles.

The *upper border* of the thyroid cartilage is deeply notched in the middle line, immediately above the pomum Adami, whilst on either side it is slightly concave. This border gives attachment throughout its whole extent to the thyro-hyoid membrane.

The *lower border* is connected to the cricoid cartilage, in the median line, by the crico-thyroid membrane; and, on each side, by the Crico-thyroid muscle.

The *posterior borders*, thick and rounded, terminate, above, in the superior cornua; and, below, in the inferior cornua. The two superior cornua are long and narrow, directed backwards, upwards, and inwards; and terminate in a conical extremity, which gives attachment to the thyro-hyoid ligament. The two inferior cornua are short and thick; they pass forwards and inwards, and present, on their inner surfaces, a small, oval, articular facet for articulation with the side of the cricoid cartilage. The posterior border receives the insertion of the Stylo-pharyngeus and Palato-pharyngeus muscles on each side.

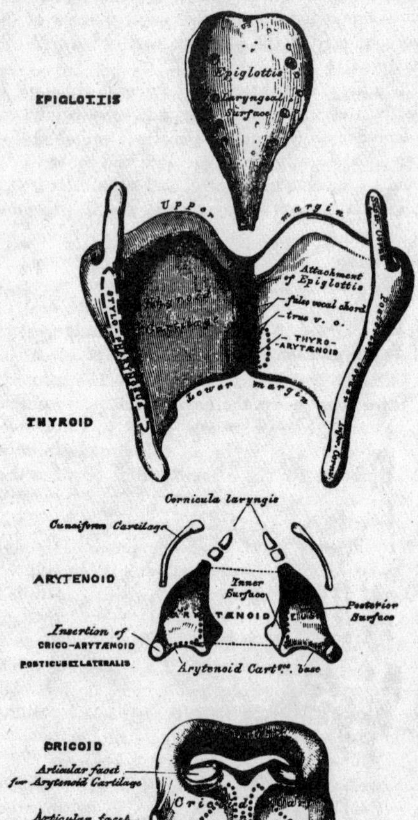

356.—The Cartilages of the Larynx. Posterior View.

The *Cricoid Cartilage* is so called from its resemblance to a signet ring (κρίκος, *a ring*). It is smaller but thicker and stronger than the thyroid cartilage, and forms the lower and back part of the cavity of the larynx.

Its *anterior half* is narrow, convex, affording attachment in front and at the sides to the Crico-thyroid muscles, and, behind those, to part of the Inferior constrictor.

Its *posterior half* is very broad, both from side to side and from above downwards; it presents in the middle line a vertical ridge for the attachment of the longitudinal fibres of the œsophagus; and on either side a broad depression for the Crico-arytænoideus posticus muscle.

At the point of junction of the two halves of the cartilage on either side, is a small round elevation, for articulation with the inferior cornu of the thyroid cartilage.

The *lower border* of the cricoid cartilage is horizontal, and connected to the upper ring of the trachea by fibrous membrane.

Its *upper border* is directed obliquely upwards and backwards, owing to the great depth of its posterior surface. It gives attachment, in front, to the crico-thyroid membrane; at the sides, to part of the same membrane and to the lateral Crico-arytenoid muscle; behind, the highest point of the upper border is surmounted on each side by a smooth oval surface, for articulation with the arytenoid cartilage. Between the articular surfaces is a slight notch, for the attachment of part of the Arytænoideus muscle.

The *inner surface* of the cricoid cartilage is smooth, and lined by mucous membrane.

The *Arytenoid Cartilages* are so called from the resemblance they bear, when approximated, to the mouth of a pitcher (*ἀρυταινα, a pitcher*). They are two in number, and situated at the upper border of the cricoid cartilage, at the back of the larynx. Each cartilage is pyramidal in form, and presents for examination three surfaces, a base, and an apex.

The *posterior surface* is triangular, smooth, concave, and lodges part of the Arytenoid muscle.

The *anterior surface*, somewhat convex and rough, gives attachment to the Thyro-arytenoid muscle, and to the false vocal cord.

The *internal surface* is narrow, smooth, and flattened, covered by mucous membrane, and lies almost in apposition with the cartilage of the opposite side.

The *base* of each cartilage is broad, and presents a concave smooth surface, for articulation with the cricoid cartilage. Of its three angles, the external is short, rounded, and prominent, receiving the insertion of the posterior and lateral Crico-arytenoid muscles. The anterior angle, also prominent, but more pointed, gives attachment to the true vocal cord.

The *apex* of each cartilage is pointed, curved backwards and inwards, and surmounted by a small conical-shaped, cartilaginous nodule, *corniculum laryngis* (cartilage of Santorini). This cartilage is sometimes united to the arytenoid, and serves to prolong it backwards and inwards. To it is attached the aryteno-epiglottidean fold.

The *cuneiform cartilages* (cartilages of Wrisberg) are two small, elongated, cartilaginous bodies, placed one on each side, in the fold of mucous membrane which extends from the apex of the arytenoid cartilage to the side of the epiglottis (*aryteno-epiglottidean fold*); they give rise to small whitish elevations on the inner surface of the mucous membrane, just in front of the arytenoid cartilages.

The *epiglottis* is a thin lamella of fibro-cartilage, of a yellowish colour, shaped like a leaf, and placed behind the tongue in front of the superior opening of the larynx. During respiration, its direction is vertically upwards, its free extremity curving forwards towards the base of the tongue; but when the larynx is drawn up beneath the base of the tongue during deglutition, it is carried downwards and backwards, so as to completely close the opening of the larynx. Its free extremity is broad and rounded; its attached end is long and narrow, and connected to the receding angle between the two alæ of the thyroid cartilage, just below the median notch, by a long, narrow, ligamentous band, the *thyro-epiglottic ligament*. It is also connected to the posterior surface of the body of the hyoid bone, by an elastic ligamentous band, the *hyo-epiglottic ligament.*

Its *anterior* or *lingual surface* is curved forwards towards the tongue, and covered by mucous membrane, which is reflected on to the sides and base of the organ, forming a median and two lateral folds, the *glosso-epiglottidean ligaments.*

Its *posterior* or *laryngeal surface* is smooth, concave from side to side, convex from above downwards, and covered by mucous membrane; when this is removed, the surface of the cartilage is seen to be studded with a number of small mucous glands, which are lodged in little pits upon its surface. To its sides the aryteno-epiglottidean folds are attached.

Structure. The epiglottis, cuneiform cartilages, and cornicula laryngis are composed of yellow cartilage, which shows little tendency to ossification; but the other cartilages resemble in structure the costal cartilages, becoming more or less ossified in old age.

Ligaments. The ligaments of the larynx are *extrinsic*, i.e. those connecting the thyroid cartilage with the os hyoides; and *intrinsic*, those which connect the several cartilaginous segments to each other.

The ligaments connecting the thyroid cartilage with the os hyoides are three in number; the thyro-hyoid membrane, and the two lateral thyro-hyoid ligaments.

The *thyro-hyoid membrane* is a broad, fibro-elastic, membranous layer, attached below to the upper border of the thyroid cartilage, and above to the upper border of the inner surface of the hyoid bone: being separated from the posterior surface of the hyoid bone by a synovial bursa. It is thicker in the middle line than at either side, in which situation it is pierced by the superior laryngeal vessels and nerve.

The *two lateral thyro-hyoid ligaments* are rounded, elastic cords, which pass between the superior cornua of the thyroid cartilage and the extremities of the greater cornua of the hyoid bone. A small cartilaginous nodule (*cartilago triticea*), sometimes bony, is found in each.

The ligaments connecting the thyroid cartilage to the cricoid are also three in number; the crico-thyroid membrane, and the capsular ligaments and synovial membrane.

The *crico-thyroid membrane* is composed mainly of yellow elastic tissue. It is of triangular shape; thick in front, where it connects together the contiguous margins of the thyroid and cricoid cartilages; thinner at each side, where it extends from the superior border of the cricoid cartilage, to the inferior margin of the true vocal cords, with which it is closely united in front.

The anterior portion of the crico-thyroid membrane is convex, concealed on each side by the Crico-thyroid muscle, subcutaneous in the middle line, and crossed horizontally by a small anastomotic arterial arch, formed by the junction of the two crico-thyroid arteries.

The lateral portions are lined internally by mucous membrane, and covered by the lateral Crico-arytenoid and Thyro-arytenoid muscles.

A *capsular ligament* encloses the articulation of the inferior cornu of the thyroid with the cricoid cartilage on each side. The articulation is lined by synovial membrane.

The ligaments connecting the arytenoid cartilages to the cricoid, are two thin and loose capsular ligaments connecting together the articulating surfaces lined internally by synovial membrane, and strengthened behind by a strong posterior crico-arytenoid ligament, which extends from the cricoid to the inner and back part of the base of the arytenoid cartilage.

The ligaments of the epiglottis are the hyo-epiglottic, the thyro-epiglottic, and the three glosso-epiglottic folds of mucous membrane which connect the epiglottis to the sides and base of the tongue. The latter have been already described.

The *hyo-epiglottic ligament* is an elastic fibrous band, which extends from the anterior surface of the epiglottis, near its apex, to the posterior surface of the body of the hyoid bone.

357.—The Larynx and adjacent parts, seen from above.

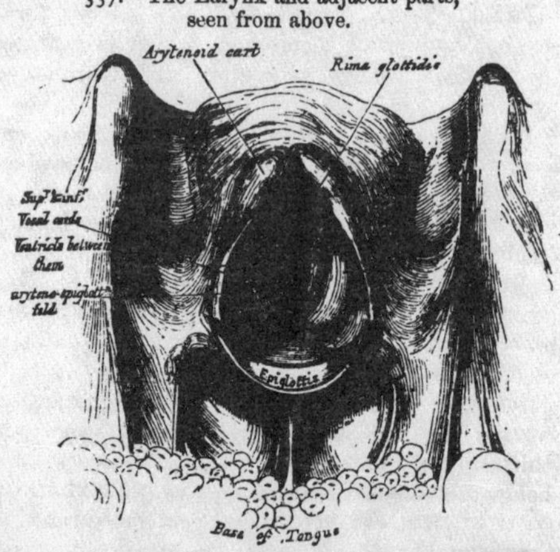

The *thyro-epiglottic ligament* is a long, slender, elastic cord, which connects the apex of the epiglottis with the receding angle of the thyroid cartilage, immediately beneath the median notch, above the attachment of the vocal cords.

Interior of the Larynx. The *superior aperture of the larynx* (fig. 357) is a triangular or cordiform opening, wide in front, narrow behind, and sloping obliquely downwards and backwards. It is bounded in front by the epiglottis; behind, by the apices of the arytenoid cartilages, and the cornicula laryngis; and laterally, by a fold of mucous membrane, enclosing ligamentous and muscular fibres, stretched between the sides of the epiglottis and the apex of the arytenoid cartilage: these are the aryteno-epiglottidean folds, on the margins of which the cuneiform cartilages form a more or less distinct whitish prominence.

The *cavity of the larynx* extends from the aperture behind the epiglottis to the lower border of the cricoid cartilage. It is divided into two parts by the projection inwards of the vocal cords, and Thyro-arytenoid muscles; between the two cords is a long and narrow triangular fissure or chink, the *glottis*, or *rima glottidis.* The portion of the cavity of the larynx above the glottis, is broad and triangular in shape above, and corresponds to the interval between the alæ of the thyroid cartilage; the portion below the glottis is at first of an elliptical, and lower down of a circular form.

The *glottis* (*rima glottidis*) is the interval between the inferior, or true, vocal cords. The two superior or false vocal cords are placed above the latter, and are formed almost entirely by a folding inwards of the mucous membrane; whilst the two inferior or true vocal cords are thick, strong, and formed partly by mucous membrane, and partly by ligamentous fibres. Between the true and false vocal cords, on each side, is an oval depression, the sinus, or *ventricle of the larynx,* which leads upwards, on the outer side of the superior vocal cord, into a cæcal pouch of variable size, the *sacculus laryngis.*

The *rima glottidis* is the narrow fissure or chink between the inferior or true vocal cords. It is the narrowest part of the cavity of the larynx, and corresponds to the lower level of the arytenoid cartilages. Its length, in the male, measures rather less than an inch, its breadth, when dilated, varying at its widest part from a third to half an inch. In the female, these measurements are less by two or three lines. The form of the glottis varies. In its half-closed condition, it is a narrow fissure, a little enlarged and rounded behind. In quiet breathing, it is widely open, somewhat triangular, the base of the triangle directed backwards, and corresponding to the space between the separated arytenoid cartilages. In forcible expiration, it is smaller than during inspiration. When sound is produced, it is more narrowed, the margins of the arytenoid cartilages being brought into contact, and the edges of the vocal cords approximated and made parallel, the degree of approximation and tension corresponding to the height of the note produced.*

358.—Vertical Section of the Larynx and Upper part of the Trachea.

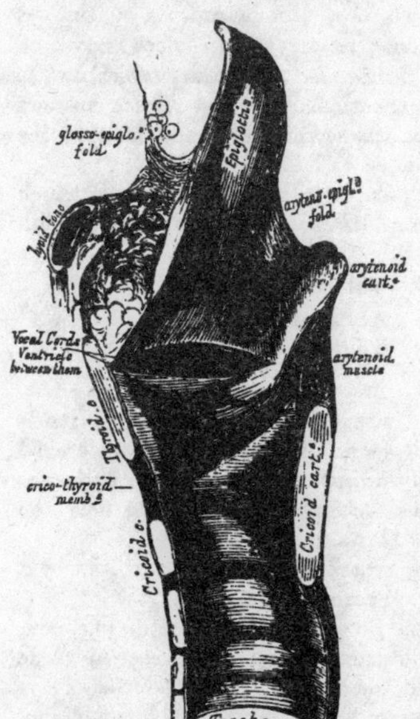

The *superior* or *false vocal cords,* so called because they are not directly concerned in the production of the voice, are two folds of mucous membrane, enclosing a delicate narrow fibrous band, the *superior thyro-arytenoid ligament.* This ligament consists of a thin band of elastic tissue, attached in front to the angle of the thyroid cartilage below the epiglottis, and behind to the anterior surface of the arytenoid cartilage.

* On the shape of the glottis in the various conditions of breathing and speaking, see 'Czermak on the Laryngoscope,' translated for the *New Sydenham Society.*

The lower border of this ligament, enclosed in mucous membrane, forms a free crescentic margin, which constitutes the upper boundary of the ventricle of the larynx.

The *inferior* or *true vocal cords*, so called from their being concerned in the production of sound, are two strong fibrous bands (*inferior thyro-arytenoid ligaments*), covered on their surface by a thin layer of mucous membrane. Each ligament consists of a band of yellow elastic tissue, attached in front to the depression between the two alæ of the thyroid cartilage, and behind to the anterior angle of the base of the arytenoid. Its lower border is continuous with the thin lateral part of the crico-thyroid membrane. Its upper border forms the lower boundary of the ventricle of the larynx. Externally, the Thyro-arytænoideus muscle lies parallel with it. It is covered internally by mucous membrane, which is extremely thin, and closely adherent to its surface.

The *ventricle of the larynx* is an oblong fossa, situated between the superior and inferior vocal cords on each side, and extending nearly their entire length. This fossa is bounded above by the free crescentic edge of the superior vocal cord; below, by the straight margin of the true vocal cord; externally, by the corresponding Thyro-arytænoideus muscle. The anterior part of the ventricle leads up by a narrow opening into a cæcal pouch of mucous membrane of variable size, called the *laryngeal pouch*.

The *sacculus laryngis*, or laryngeal pouch, is a membranous sac, placed between the superior vocal cord and the inner surface of the thyroid cartilage, occasionally extending as far as its upper border; it is conical in form, and curved slightly backwards, like a Phrygian cap. On the surface of its mucous membrane are the openings of sixty or seventy small follicular glands, which are lodged in the submucous areolar tissue. This sac is enclosed in a fibrous capsule, continuous below with the superior thyro-arytenoid ligament; its laryngeal surface is covered by the Arytæno-epiglottideus inferior muscle (*Compressor sacculi laryngis*, Hilton); whilst its exterior is covered by the Thyro-epiglottideus muscle. These muscles compress the sacculus laryngis, and discharge the secretion it contains upon the chordæ vocales, the surfaces of which it is intended to lubricate.

Muscles. The intrinsic muscles of the larynx are eight in number; five of which are the muscles of the chordæ vocales and rima glottidis; three are connected with the epiglottis.

The five muscles of the chordæ vocales and rima glottidis are the

Crico-thyroid.	Arytænoideus.
Crico-arytænoideus posticus.	Thyro-arytænoideus.
Crico-arytænoideus lateralis.	

The *Crico-thyroid* is triangular in form, and situated at the fore part and side of the cricoid cartilage. It arises from the front and lateral part of the cricoid cartilage; its fibres diverge, passing obliquely upwards and outwards, to be inserted into the lower and inner borders of the thyroid cartilage, from near the median line in front, as far back as the inferior cornu.

The inner borders of these two muscles are separated in the middle line by a triangular interval, occupied by the crico-thyroid membrane.

The *Crico-arytænoideus posticus* arises from the broad depression occupying each lateral half of the posterior surface of the cricoid cartilage; its fibres pass upwards and outwards, and converge to be inserted into the outer angle of the base of the arytenoid cartilage. The upper fibres are nearly horizontal, the middle oblique, and the lower almost vertical.*

* Dr. Merkel of Leipsic has lately described a muscular slip which occasionally extends between the outer border of the posterior surface of the cricoid cartilage, and the posterior margin of the inferior cornu of the thyroid; this he calls the 'Musculus kerato-cricoideus.' It is not found in every larynx, and when present exists usually only on one side, but is occasionally found on both sides. Mr. Turner (*Edinburgh Medical Journal*, Feb. 1860) states that it is found in about one case in five. Its action is to fix the lower horn of the thyroid cartilage backwards and downwards, opposing in some measure the part of the crico-thyroid muscle, which is connected to the anterior margin of the horn.

The *Crico-arytænoideus lateralis* is smaller than the preceding, and of an oblong form. It arises from the upper border of the side of the cricoid cartilage, and, passing obliquely upwards and backwards, is inserted into the outer angle of the base of the arytenoid cartilage, in front of the preceding muscle.

359.—Muscles of Larynx. Side View.
Right Ala of Thyroid Cartilage removed.

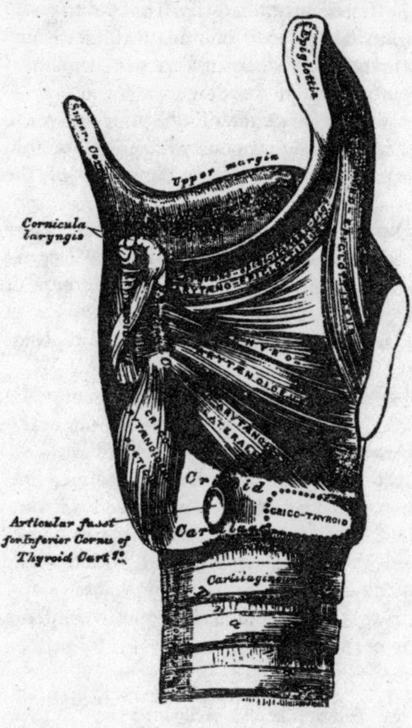

The *Thyro-arytænoideus* is a broad, flat muscle, which lies parallel with the outer side of the true vocal cord. It arises in front from the lower half of the receding angle of the thyroid cartilage, and from the crico-thyroid membrane. Its fibres pass horizontally backwards and outwards, to be inserted into the base and anterior surface of the arytenoid cartilage. This muscle consists of two fasciculi. The *inferior*, the thicker, is inserted into the anterior angle of the base of the arytenoid cartilage, and into the adjacent portion of its anterior surface; it lies parallel with the true vocal cord, to which it is occasionally adherent. The *superior* fasciculus, the thinner, is inserted into the anterior surface and outer border of the arytenoid cartilage above the preceding fibres; it lies on the outer side of the sacculus laryngis, immediately beneath its mucous lining.

360.—Interior of the Larynx, seen from above.
(Enlarged.)

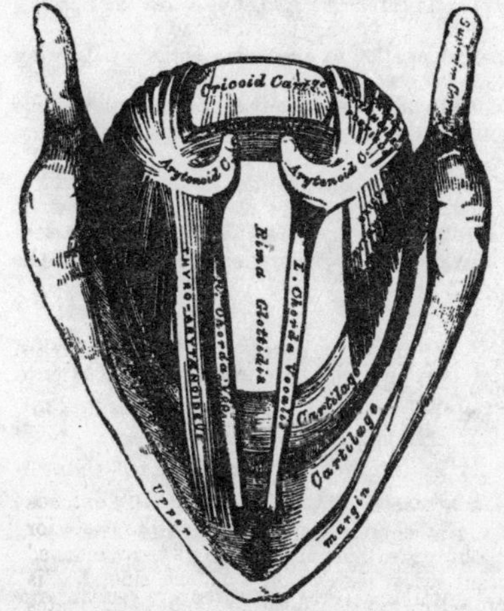

The *Arytænoideus* is a single muscle, filling up the posterior concave surface of the arytenoid cartilages. It arises from the posterior surface and outer border of one arytenoid cartilage, and is inserted into the corresponding parts of the opposite cartilage. It consists of three planes of fibres: two oblique, and one transverse. The *oblique fibres*, the most superficial, form two fasciculi, which pass from the base of one cartilage to the apex of the opposite one. The *transverse fibres*, the deepest and most numerous, pass transversely across between the two cartilages; hence the Arytænoideus was formerly considered as several muscles, under the names of *transversi* and *obliqui*. A few of the oblique fibres are occasionally continued round the outer margin of the cartilage, and blend with the Thyro-arytenoid or the Arytæno-epiglottideus muscle.

The muscles of the epiglottis are, the

Thyro-epiglottideus.
Arytæno-epiglottideus superior.
Arytæno-epiglottideus inferior.

The *Thyro-epiglottideus* is a delicate fasciculus, which arises from the inner surface of the thyroid cartilage, just external to the origin of the Thyro-arytenoid muscle, and spreading out upon the outer surface of the sacculus laryngis, some of its fibres are lost in the aryteno-epiglottidean fold, whilst others are continued forwards to the margin of the epiglottis (*Depressor epiglottidis*).

The *Arytæno-epiglottideus superior* consists of a few delicate muscular fasciculi, which arise from the apex of the arytenoid cartilage, and become lost in the fold of mucous membrane extending between the arytenoid cartilage and side of the epiglottis (*aryteno-epiglottidean folds*).

The *Arytæno-epiglottideus inferior* (*Compressor sacculi laryngis*, Hilton) arises from the arytenoid cartilage, just above the attachment of the superior vocal cord; passing forwards and upwards, it spreads out upon the inner and upper part of the sacculus laryngis, and is inserted, by a broad attachment, into the margin of the epiglottis. This muscle is separated from the preceding by an indistinct areolar interval.

Actions. In considering the action of the muscles of the larynx, they may be conveniently divided into two groups, viz.: 1. Those which open and close the glottis. 2. Those which regulate the degree of tension of the vocal cords.

1. The muscles which open the glottis are the Crico-arytænoidei postici; and those which close it, are the Arytænoideus, and the Crico-arytænoidei laterales. 2. The muscles which regulate the tension of the vocal cords are, the Crico-thyroidei, which tense and elongate them; and the Thyro-arytænoidei, which relax and shorten them. The Thyro-epiglottideus is a depressor of the epiglottis, and the Arytæno-epiglottidei constrict the superior aperture of the larynx, compress the sacculi laryngis, and empty them of their contents.

The *Crico-arytænoidei postici* separate the chordæ vocales, and, consequently, open the glottis, by rotating the base of the arytenoid cartilages outwards and backwards; so that their anterior angles, and the ligaments attached to them, become widely separated, the vocal cords, at the same time, being made tense.

The *Crico-arytænoidei laterales* close the glottis, by rotating the base of the arytenoid cartilages inwards, so as to approximate their anterior angles.

The *Arytænoideus muscle* approximates the arytenoid cartilages, and thus closes the opening of the glottis, especially at its back part.

The *Crico-thyroid muscles* produce tension and elongation of the vocal cords, by drawing down the thyroid cartilage over the cricoid.

The *Thyro-arytænoidei muscles* draw the arytenoid cartilages, together with the part of the cricoid to which they are connected, forwards towards the thyroid, and thus shorten and relax the vocal cords.

The *Thyro-epiglottidei* depress the epiglottis, and assist in compressing the sacculi laryngis. The Arytæno-epiglottideus superior constricts the superior aperture of the larynx, when it is drawn upwards, during deglutition, and the opening closed by the epiglottis. The Arytæno-epiglottideus inferior, together with some fibres of the Thyro-arytænoidei, compress the sacculus laryngis.

The *Mucous Membrane of the Larynx* is continuous, above, with that lining the mouth and pharynx, and is prolonged through the trachea and bronchi into the lungs. It lines both surfaces of the epiglottis, to which it is closely adherent, and forms the aryteno-epiglottidean folds, which encircle the superior aperture of the larynx. It lines the whole of the cavity of the larynx; forms, by its reduplication, the chief part of the superior, or false, vocal cord; and, from the ventricle, is continued into the sacculus laryngis. It is then reflected over the true vocal cords, where it is thin, and very intimately adherent; covers the inner surface of the crico-thyroid membrane, and cricoid cartilage; and is ultimately continuous with the lining membrane of the trachea. It is covered with columnar ciliated epithelium, below the superior vocal cord; but, above this point, the ciliæ are found only in front, as high as the middle of the epiglottis. In the rest of its extent, the epithelium is of the squamous variety.

Glands. The mucous membrane of the larynx is furnished with numerous muciparous glands, the orifices of which are found in nearly every part: they are very numerous upon the epiglottis, being lodged in little pits in its substance; they are also found in large numbers along the posterior margin of the aryteno-epiglottidean fold, in front of the arytenoid cartilages, where they are termed the *arytenoid glands.* They exist also in large numbers upon the inner surface of the sacculus laryngis. None are found on the vocal cords.

Vessels and Nerves. The *arteries* of the larynx are the laryngeal branches derived from the superior and inferior thyroid. The *veins* empty themselves into the superior, middle, and inferior thyroid veins. The *lymphatics* terminate in the deep cervical glands. The *nerves* are the superior laryngeal, and the inferior or recurrent laryngeal branches of the pneumogastric nerves, joined by filaments

361.—Front View of Cartilages of Larynx: the Trachea and Bronchi.

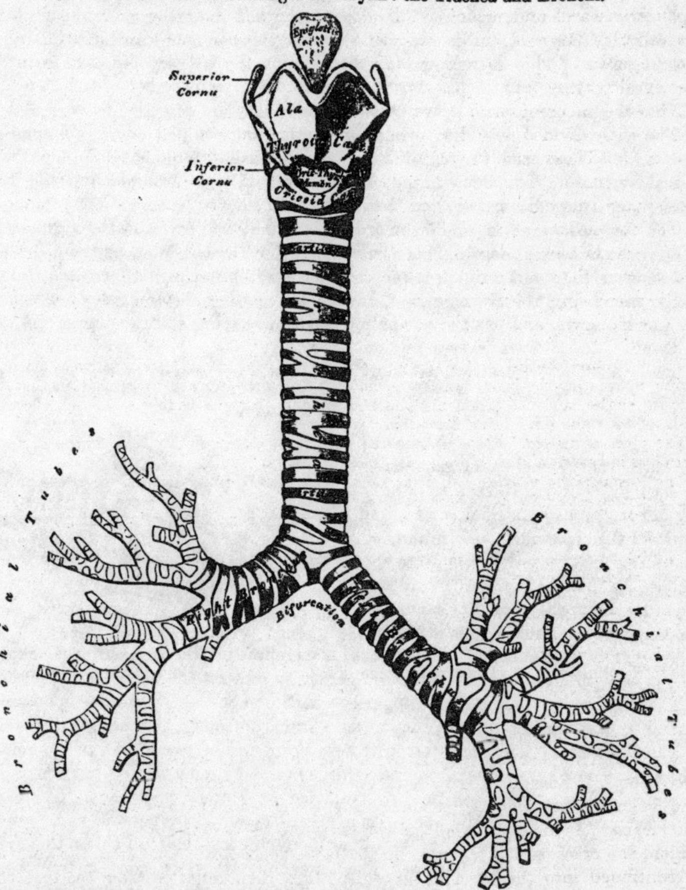

from the sympathetic. The superior laryngeal nerves supply the mucous membrane of the larynx, and the Crico-thyroid muscles. The inferior laryngeal nerves supply the remaining muscles. The Arytenoid muscle is supplied by both nerves.

The Trachea. (Fig. 361.)

The trachea, or air-tube, is a cartilaginous and membranous cylindrical tube, flattened posteriorly, which extends from the lower part of the larynx, on a level with the fifth cervical vertebra, to opposite the third dorsal, where it divides into the two bronchi, one for each lung. The trachea measures about four inches and a half in length; its diameter, from side to side, is from three-quarters of an inch to an inch, being always greater in the male than in the female.

Relations. The anterior surface of the trachea is convex, and covered, *in the neck*, from above downwards, by the isthmus of the thyroid gland, the inferior thyroid veins, the arteria thyroidea ima (when that vessel exists), the Sterno-hyoid and Sterno-thyroid muscles, the cervical fascia (in the interval between those muscles), and, more superficially, by the anastomosing branches between the anterior jugular veins: *in the thorax*, it is covered from before backwards by the first piece of the sternum, the remains of the thymus gland, the arch of the aorta, the innominate and left carotid arteries, and the deep cardiac plexus. It lies upon the œsophagus, which is directed to the left, near the arch of the aorta; laterally, *in the neck*, it is in relation with the common carotid arteries, the lateral lobes of the thyroid gland, the inferior thyroid arteries, and recurrent laryngeal nerves; and, *in the thorax*, it lies in the interspace between the pleuræ, having the pneumogastric nerve on each side of it.

The *Right Bronchus*, wider, shorter, and more horizontal in direction than the left, is about an inch in length, and enters the right lung, opposite the fourth dorsal vertebra. The vena azygos arches over it, from behind; and the right pulmonary artery lies below, and then in front of it.

The *Left Bronchus* is smaller, more oblique, and longer than the right, being nearly two inches in length. It enters the root of the left lung, opposite the fifth dorsal vertebra, about an inch lower than the right bronchus. It crosses in front of the œsophagus the thoracic duct, and the descending aorta; passes beneath the arch of the aorta, and has the left pulmonary artery lying at first above, and then in front of it. If a transverse section is made across the trachea, a short distance above its point of bifurcation, and a bird's eye view taken of its interior (fig. 362), the septum placed at the bottom of the trachea and separating the two bronchi will be seen to occupy the left of the median line, as was first shown by Mr. Goodall, of Dublin, so that any solid body dropping into the trachea, would naturally be directed towards the right bronchus, and this tendency is undoubtedly aided by the larger size of this tube, as compared with its fellow. This fact serves to explain why a foreign substance in the trachea generally falls into the right bronchus.

362.—Transverse Section of the Trachea, just above its Bifurcation, with a bird's eye view of the interior.

The trachea is composed of imperfect cartilaginous rings, fibrous membrane, muscular fibres, longitudinal yellow elastic fibres, mucous membrane, and glands.

The *Cartilages* vary from sixteen to twenty in number: each forms an imperfect ring, which surrounds about two-thirds of the cylinder of the trachea, being imperfect behind, where the tube is completed by fibrous membrane. The cartilages are placed horizontally above each other, separated by narrow membranous intervals. They measure about two lines in depth, and half a line in thickness. Their outer surfaces are flattened, but, internally, they are convex, from being thicker in the middle than at the margins. The cartilages are connected together at their margins, by an elastic fibrous membrane, which covers both their surfaces; and in the space between their extremities, behind, forms a distinct layer. The peculiar cartilages are the first and the last.

The *first cartilage* is broader than the rest, and sometimes divided at one end: it is connected by fibrous membrane with the lower border of the cricoid cartilage, with which, or with the succeeding cartilage, it is sometimes blended.

The *last cartilage* is thick and broad in the middle, in consequence of its lower border being prolonged downwards, and, at the same time, curved backwards, at the point of bifurcation of the trachea. It terminates on each side in an imperfect ring, which encloses the commencement of the bronchi. The cartilage above the last, is somewhat broader than the rest at its centre. Two or more of the cartilages often unite, partially or completely, and are sometimes bifurcated at their extremities. They are highly elastic, and seldom ossify, even in advanced life. In the right bronchus, the cartilages vary in number from six to eight; in the left, from nine to twelve. They are shorter and narrower than those of the trachea.

The *Muscular Fibres* are disposed in two layers, longitudinal and transverse. The longitudinal fibres are the most external, and arise by minute tendons from the termination of the tracheal cartilages, and from the fibrous membrane.

The *transverse fibres* (trachealis muscle, Todd and Bowman), the most internal, form a thin layer, which extends transversely between the ends of the cartilages, at the posterior part of the trachea. The muscular fibres are of the unstriped variety.

The *Elastic Fibres* are situated beneath the mucous membrane, enclosing the entire cylinder of the trachea; they are most abundant at its posterior part, where they are collected into longitudinal bundles.

The *Mucous Membrane* lining the tube is covered with columnar ciliated epithelium. It is continuous above with that of the larynx, and below with that of the lungs.

The *Tracheal Glands* are found in great abundance at the posterior part of the trachea. They are small, flattened, ovoid bodies, placed between the fibrous and muscular coats, each furnished with an excretory duct, which opens on the surface of the mucous membrane. Some glands of smaller size are also found at the sides of the trachea, between the layers of fibrous tissue connecting the rings, and others immediately beneath the mucous coat. The secretion from these glands serves to lubricate the inner surface of the trachea.

Vessels and Nerves. The trachea is supplied with blood by the inferior thyroid arteries. The *veins* terminate in the thyroid venous plexus. The *nerves* are derived from the pneumogastric and its recurrent branches, and from the sympathetic.

Surgical Anatomy. The air-passage may be opened in three different situations; through the crico-thyroid membrane (*laryngotomy*), through the cricoid cartilage and upper ring of the trachea (*laryngo-tracheotomy*), or through the trachea below the isthmus of the thyroid gland (*tracheotomy*). The student should, therefore, carefully consider the relative anatomy of the air-tube in each of these situations.

Beneath the integument of the laryngo-tracheal region, on either side of the median line, are the two anterior jugular veins. Their size and position vary; there is nearly always one, and frequently two: at the lower part of the neck they diverge, passing beneath the Sterno-mastoid muscles, and are frequently connected by a transverse communicating branch. These veins should, if possible, always be avoided in any operation on the larynx or trachea. If cut through, considerable hæmorrhage occurs.

Beneath the cervical fascia are the Sterno-hyoid and Sterno-thyroid muscles, the contiguous edges of the former being near the median line; and beneath these muscles the following parts are met with, from above downwards: the thyroid cartilage, the crico-thyroid membrane, the cricoid cartilage, the trachea, and the isthmus of the thyroid gland.

The crico-thyroid space is very superficial, and may be easily felt beneath the skin as a depression, about an inch below the pomum Adami; it is crossed transversely by a small artery, the crico-thyroid, the division of which is seldom accompanied by any troublesome hæmorrhage.

The isthmus of the thyroid gland usually crosses the second and third rings of the trachea; above it, is found a large transverse communicating branch between the superior thyroid veins, and the isthmus is covered by a venous plexus, formed between the thyroid veins, of opposite sides. On the sides of the thyroid gland, and below it, the veins converge to a single median vessel, or to two trunks which descend along the median line of the front of the trachea, to open into the innominate veins by valved orifices. In the infant, the thymus gland ascends a variable distance along the front of the trachea: and the innominate artery crosses the tube obliquely at the root of the neck, from left to right. The arterior thyroidea ima, when that vessel exists, passes from below upwards along the front of the trachea. The upper part of the trachea lies comparatively superficial; but the lower part passes obliquely downwards and backwards, so as to be deeply placed between the converging Sterno-mastoid muscles. In the child, the trachea is smaller, more deeply placed, and more moveable than in the adult. In fat, or short-necked people, or in those in whom the muscles of the neck are prominently developed, the trachea is more deeply placed than in the opposite conditions.

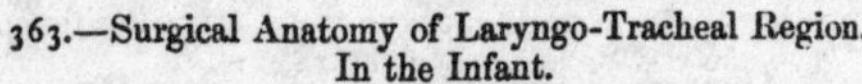

363.—Surgical Anatomy of Laryngo-Tracheal Region.
In the Infant.

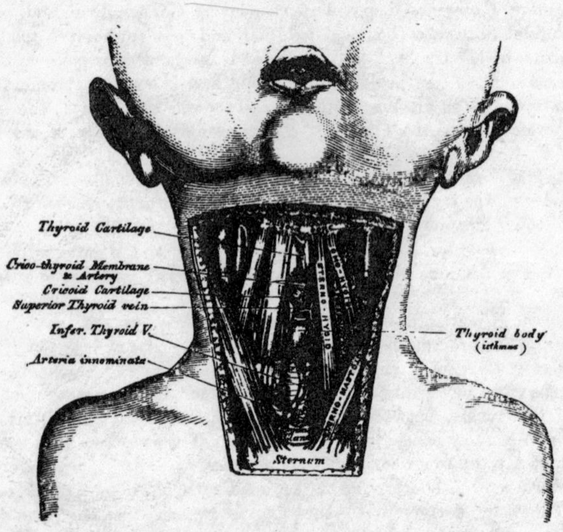

From these observations, it must be evident that laryngotomy is anatomically the most simple operation, can most readily be performed, and should always be preferred when particular circumstances do not render the operation of tracheotomy absolutely necessary. The operation is performed thus: The head being thrown back and steadied by an assistant, the finger is passed over the front of the neck, and the crico-thyroid depression felt for. A vertical incision is then made through the skin, in the middle line over this spot, and the crico-thyroid membrane is divided to a sufficient extent to allow of the introduction of a large curved tube. The crico-thyroid artery is the only vessel of importance crossing this space. If it should be of large size, its division might produce troublesome hæmorrhage.

Laryngo-tracheotomy, anatomically considered, is more dangerous than tracheotomy, on account of the small interspace between the cricoid cartilage and the isthmus of the thyroid gland: the communicating branches between the superior thyroid veins, which cover this spot, can hardly fail to be divided; and the greatest care will not, in some cases, prevent the division of part of the thyroid isthmus. If either of these structures is divided, the hæmorrhage may be considerable.

Tracheotomy below the isthmus of the thyroid gland is performed thus: The head being thrown back and steadied by an assistant, an incision, an inch and a half or two inches in length, is made through the skin, in the median line of the neck, from a little below the cricoid cartilage, to the top of the sternum. The anterior jugular veins should be avoided, by keeping exactly in the median line; the deep fascia should then be divided, and the contiguous borders of the Sterno-hyoid muscles separated from each other. A quantity of loose areolar tissue, containing the inferior thyroid veins, must then be separated from the front of the trachea, with the handle of the scalpel; and when the trachea is well exposed, it should be opened by inserting the knife into it, dividing two or three of its rings from below upwards. It is a matter of the greatest importance to restrain, if possible, all hæmorrhage before the tube is opened; otherwise, blood may pass into the trachea, and suffocate the patient.

The Pleuræ.

Each lung is invested, upon its external surface, by an exceedingly delicate serous membrane, the pleura, which encloses the organ as far as its root, and is then reflected upon the inner surface of the thorax. The portion of the serous membrane investing the surface of the lung is called the *pleura pulmonalis*

364.—A Transverse Section of the Thorax, showing the relative Position of the Viscera, and the Reflections of the Pleuræ.

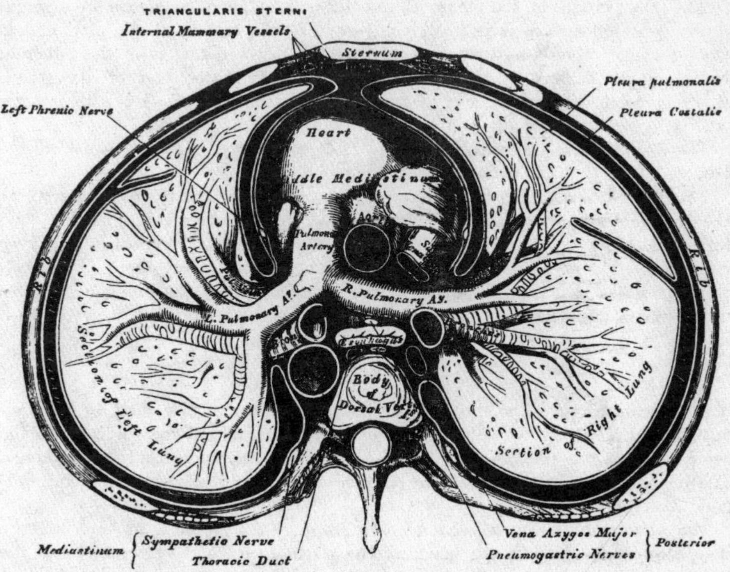

(visceral layer of pleura); whilst that which lines the inner surface of the chest is called the *pleura costalis* (parietal layer of pleura). The interspace or cavity between these two layers is called the *cavity of the pleura.* Each pleura is therefore a shut sac, one occupying the right, the other the left half of the thorax; and they are perfectly separate, not communicating with each other. The two pleuræ do not meet in the middle line of the chest, excepting at one point in front; an interspace being left between them, which contains all the viscera of the thorax, excepting the lungs: this is the *mediastinum.*

Reflections of the pleura (fig. 364). Commencing at the sternum, the pleura passes outwards, covers the costal cartilages, the inner surface of the ribs and Intercostal muscles, and at the back of the thorax passes over the thoracic ganglia and their branches, and is reflected upon the sides of the bodies of the vertebræ, where it is separated by a narrow interspace from the opposite pleura, the *posterior mediastinum.* From the vertebral column the pleura passes to the side of the pericardium, which it covers to a slight extent; it then covers the back part of the root of the lung, from the lower border of which a triangular fold descends vertically by the side of the posterior mediastinum to the Diaphragm. This fold is the broad ligament of the lung, the *ligamentum latum pulmonis,* and serves to retain the lower part of that organ in position. From the *root,* the pleura may be traced over the convex surface of the lung, the summit and base, and also over the sides of the fissures between the lobes. It covers its anterior surface, and the front part of its root, and is reflected upon the side of the pericardium to the inner surface of the sternum. *Below*, it covers the upper surface of the Diaphragm. *Above*, its apex projects, in the form of a *cul-de-sac*, through the superior opening of the thorax into the neck, extending about an inch above the margin of the first rib, and receives the summit of the corresponding lung: this sac is strengthened, according to Dr. Sibson, by a dome-like expansion of fascia, derived from the lower part of the Scaleni muscles.

A little above the middle of the sternum, the contiguous surfaces of the two pleuræ are sometimes in contact for a slight extent; but above and below this point, the interval left between them forms the anterior mediastinum.

The inner surface of the pleura is smooth, polished, and moistened by a serous fluid: its outer surface is intimately adherent to the surface of the lung, and to the pulmonary vessels as they emerge from the pericardium; it is also adherent to the upper surface of the Diaphragm: throughout the rest of its extent, it is somewhat thicker, and may be separated from the adjacent parts with extreme facility.

The right pleural sac is shorter, wider, and reaches higher in the neck than the left.

Vessels and Nerves. The *arteries* of the pleura are derived from the intercostal, the internal mammary, the phrenic, inferior thyroid, thymic, pericardiac, and bronchial. The *veins* correspond to the arteries. The *lymphatics* are very numerous. The *nerves* are derived from the phrenic and sympathetic (Luschka). Kölliker states that nerves accompany the ramifications of the bronchial arteries in the pleura pulmonalis.

Mediastinum.

The mediastinum is the space left in the median line of the chest by the non-approximation of the two pleuræ. It extends from the sternum in front to the spine behind, and contains all the viscera in the thorax, excepting the lungs. The mediastinum is subdivided, for convenience of description, into the anterior, middle, and posterior.

The *anterior mediastinum* is bounded in front by the sternum, on each side by the pleura, and behind by the pericardium. Owing to the oblique position of the heart towards the left side, this space is not parallel with the sternum, but directed obliquely from above downwards, and to the left of the median line; it is broad below, narrow above, very narrow opposite the second piece of the sternum, the contiguous surfaces of the two pleuræ being occasionally united over a small space. The anterior mediastinum contains the origins of the Sterno-hyoid and Sterno-thyroid muscles, the Triangularis sterni, the internal mammary vessels of the left side, the remains of the thymus gland, and a quantity of loose areolar tissue, in which some lymphatic vessels are found ascending from the convex surface of the liver.

The *middle mediastinum* is the broadest part of the interpleural space. It contains the heart enclosed in the pericardium, the ascending aorta, the superior vena cava, the bifurcation of the trachea, the pulmonary arteries and veins, and the phrenic nerves.

The *posterior mediastinum* is an irregular triangular space, running parallel with the vertebral column; it is bounded in front by the pericardium and roots of the lungs, behind by the vertebral column, and on either side by the pleura. It contains the descending aorta, the greater and lesser azygos veins and left superior intercostal vein, the pneumogastric and splanchnic nerves, the œsophagus, thoracic duct, and some lymphatic glands.

365.—The Posterior Mediastinum.

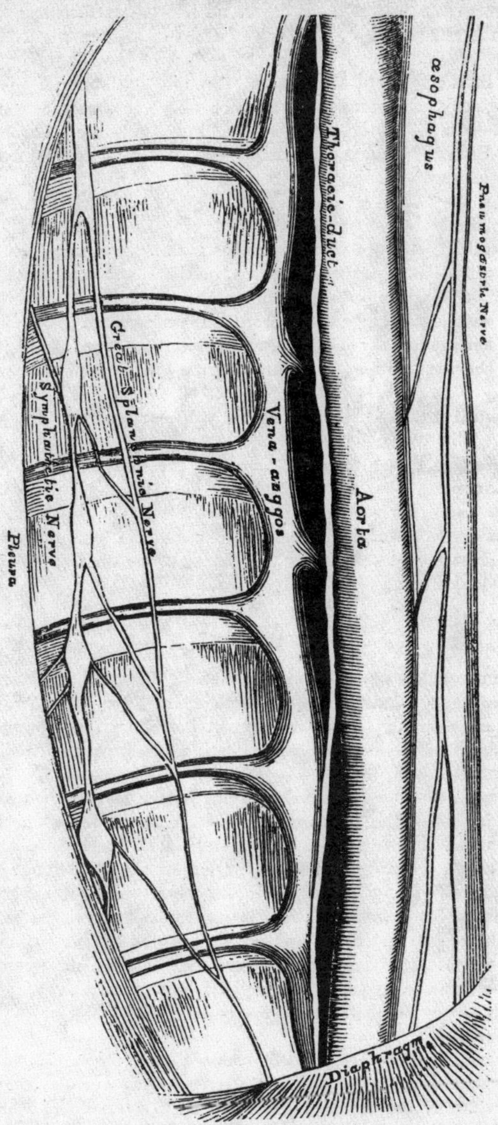

The Lungs.

The lungs are the essential organs of respiration; they are two in number; placed one in each of the lateral cavities of the chest, separated from each other by the heart and other contents of the mediastinum. Each lung is conical in shape, and presents for examination an apex, a base, two borders, and two surfaces (see fig. 351, p. 640).

The *apex* forms a tapering cone, which extends into the root of the neck, about an inch to an inch and a half above the level of the first rib.

The *base* is broad, concave, and rests upon the convex surface of the Diaphragm; its circumference is thin, and fits into the space between the lower ribs and the costal attachment of the Diaphragm, extending lower down externally and behind than in front.

The *external* or *thoracic surface* is smooth, convex, of considerable extent, and corresponds to the form of the cavity of the chest, being deeper behind than in front.

The *inner surface* is concave. It presents, in front, a depression corresponding to the convex surface of the pericardium, and behind, a deep fissure (the hilum pulmonis) which gives attachment to the root of the lung.

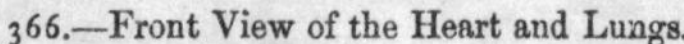

366.—Front View of the Heart and Lungs.

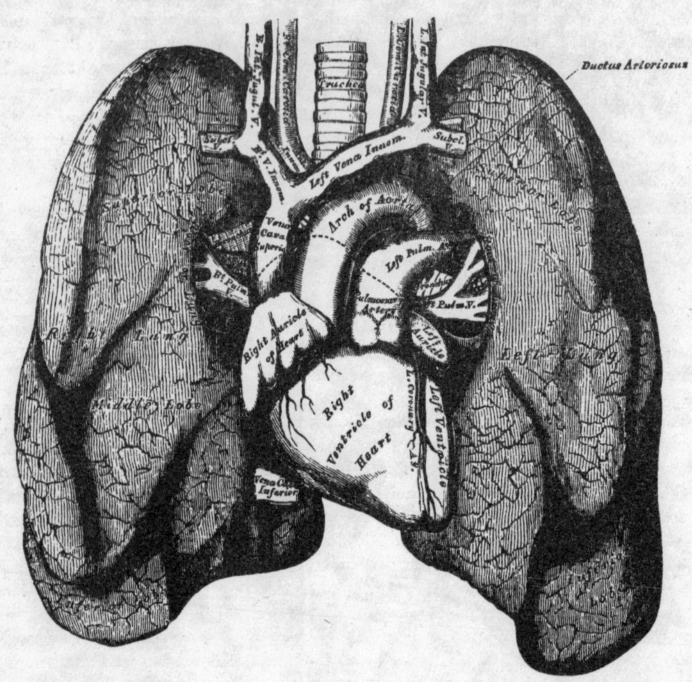

The *posterior border* is rounded and broad, and is received in the deep concavity on either side of the spinal column. It is much longer than the anterior border, and projects below between the ribs and the Diaphragm.

The *anterior border* is thin and sharp, and overlaps the front of the pericardium.

The anterior border of the right lung corresponds to the median line of the sternum, and is in contact with its fellow, the pleuræ being interposed, as low as the fourth costal cartilage; below this, the contiguous borders are separated by an irregularly shaped interval, formed at the expense of the anterior border of the left lung, and in which the pericardium is exposed.

Each lung is divided into two lobes, an upper and lower, by a long and deep fissure, which extends from the upper part of the posterior border of the organ, about three inches from its apex, downwards and forwards to the lower part of its anterior border. This fissure penetrates nearly to the root. In the right lung the upper lobe is partially divided by a second and shorter fissure, which extends from the middle of the preceding, forwards and upwards, to the anterior margin of the organ, marking off a small triangular portion, the middle lobe.

The *right lung* is the largest; it is broader than the left, owing to the inclination of the heart to the left side; it is also shorter by an inch, in consequence of the Diaphragm rising higher on the right side to accommodate the liver. The right lung has three lobes.

The *left lung* is smaller, narrower, and longer than the right, and has only two lobes.

A little above the middle of the inner surface of each lung, and nearer its posterior than its anterior border, is its root, by which the lung is connected to the heart and the trachea. The root is formed by the bronchial tube, the pulmonary artery, the pulmonary veins, the bronchial arteries and veins, the pulmonary plexus of nerves, lymphatics, bronchial glands, and areolar tissue, all of which are enclosed by a reflection of the pleura. The root of the right lung lies behind the superior cava and upper part of the right auricle, and below the vena azygos. That of the left lung passes beneath the arch of the aorta, and in front of the descending aorta; the phrenic nerve and the anterior pulmonary plexus lie in front of each, and the pneumogastric and posterior pulmonary plexus behind each.

The chief structures composing the root of each lung are arranged in a similar manner from before backwards on both sides, viz.: the pulmonary veins most anterior; the pulmonary artery in the middle; and the bronchus, together with the bronchial vessels, behind. From above downwards, on the two sides, their arrangement differs, thus:

On the right side, their position is, bronchus, pulmonary artery, pulmonary veins; but on the left side their position is, pulmonary artery, bronchus, pulmonary veins; which is accounted for by the bronchus being placed on a lower level on the left than on the right side.

The *weight* of both lungs together is about forty-two ounces, the right lung being two ounces heavier than the left; but much variation is met with according to the amount of blood or serous fluid they may contain. The lungs are heavier in the male than in the female, their proportion to the body being, in the former, as 1 to 37, in the latter as 1 to 43. The specific gravity of the lung-tissue varies from 345 to 746, water being 1,000.

The *colour* of the lungs at birth is a pinkish white; in adult life, a dark slate colour, mottled in patches; and, as age advances, this mottling assumes a black colour. The colouring matter consists of granules of a carbonaceous substance, deposited in the areolar tissue near the surface of the organ. It increases in quantity as age advances, and is more abundant in males than in females. The posterior border of the lung is usually darker than the anterior. The surface of the lung is smooth, shining, and marked out into numerous polyhedral spaces, indicating the lobules of the organ: the area of each of these spaces is crossed by numerous lighter lines.

The *substance* of the lung is of a light, porous, spongy texture; it floats in water, and crepitates when handled, owing to the presence of air in the tissue; it is also highly elastic; hence the collapsed state of these organs when they are removed from the closed cavity of the thorax.

Structure. The lungs are composed of an external serous coat, a subserous areolar tissue, and the pulmonary substance or parenchyma.

The *serous coat* is derived from the pleura; it is thin, transparent, and invests the entire organ as far as the root.

The *subserous areolar tissue* contains a large proportion of elastic fibres; it invests the entire surface of the lung, and extends inwards between the lobules.

The *parenchyma* is composed of lobules, which, although closely connected together by an interlobular areolar tissue, are quite distinct from one another, and are easily separable in the fœtus. The lobules vary in size; those on the surface are large, of a pyramidal form, the base turned towards the surface; those in the interior smaller, and of various forms. Each lobule is composed of one of the ramifications of the bronchial tube and its terminal air-cells, and of the ramifications of the pulmonary and bronchial vessels, lymphatics, and nerves: all of these structures being connected together by areolar fibrous tissue.

The *bronchus*, upon entering the substance of the lung, divides and subdivides dichotomously throughout the entire organ. Sometimes three branches arise together, and occasionally small lateral branches are given off from the sides of a main trunk. Each of the smaller subdivisions of the bronchi enters a pulmonary lobule (lobular bronchial tube), and, again subdividing, ultimately terminates in the intercellular passages and air-cells of which the lobule is composed. Within the lungs the bronchial tubes are circular, not flattened, and their constituent elements present the following peculiarities of structure.

The *cartilages* are not imperfect rings, but consist of thin laminæ, of varied form and size, scattered irregularly along the sides of the tube, being most distinct at the points of division of the bronchi. They may be traced into tubes, the diameter of which is only one-fourth of a line. Beyond this point, the tubes are wholly membranous. The fibrous coat, and longitudinal elastic fibres, are continued into the smallest ramifications of the bronchi. The muscular coat is disposed in the form of a continuous layer of annular fibres, which may be traced upon the smallest bronchial tubes: they consist of the unstriped variety of muscular fibre. The mucous membrane lines the bronchi and its ramifications throughout, and is covered with columnar ciliated epithelium.

According to the observations of Mr. Rainey,* the lobular bronchial tubes, on entering the substance of the lobules, divide and subdivide from four to nine times, according to the size of the lobule, continuing to diminish in size until they attain a diameter of $\frac{1}{50}$th to $\frac{1}{30}$th of an inch. They then become changed in structure, losing their cylindrical form, and are continued onwards as irregular passages (intercellular passages, Rainey—air-sacs, Waters), through the substance of the lobule, their sides and extremities being closely covered by numerous saccular dilatations, the air-cells. This arrangement resembles most closely the naked eye appearances observed in the reticulated structure of the lung of the tortoise, and other reptilia. Opinions have differed as to the existence of communications or anastomoses between the inter-cellular passages, or air-sacs. According to Dr. Waters,† these air-sacs, as he terms them, are arranged in groups, or 'lobulettes' of five or six, which spring from the terminal dilatation of a single bronchial tube, but have no other communication with each other, or with neighbouring lobulettes, than that which is afforded by their common connection with the bronchial tubes.

The *air-cells*, or *alveoli* (Waters), are small, polyhedral, alveolar recesses, separated from each other by thin septa, and communicating freely with the intercellular passages or air-sacs. They are well seen on the surface of the lung, and vary from $\frac{1}{200}$th to $\frac{1}{70}$th of an inch in diameter; being largest on the surface, at the thin borders, and at the apex; and smallest in the interior.

At the termination of the bronchial tubes, in the intercellular passages, their constituent elements become changed: their walls are formed by an interlacing of the longitudinal elastic bundles with fibrous tissue; the muscular fibres disappear, and the mucous membrane becomes thin and delicate, and lined with a layer of squamous epithelium. The latter membrane lines the air-cells, and forms by its reduplications the septa intervening between them.

* *Medico-Chirurgical Transactions*, vol. xxviii. 1845.

† '*The Anatomy of the Human Lung*,' 1860, pp. 136—150.

The *Pulmonary Artery* conveys the venous blood to the lungs: it divides into branches which accompany the bronchial tubes, and terminates in a dense capillary network upon the walls of the intercellular passages and air-cells. From this network, the radicles of the pulmonary veins arise; coalescing into large branches they accompany the arteries, and return the blood, purified by its passage through the capillaries, to the left auricle of the heart. In the lung, the branches of the pulmonary artery are usually above and in front of a bronchial tube, the vein below.

The *Pulmonary Capillaries* form plexuses which lie immediately beneath the mucous membrane, on the walls and septa of the air-cells, and upon the walls of the intercellular passages. In the septa between the cells, the capillary network forms a single layer. The capillaries form a very minute network, the meshes of which are smaller than the vessels themselves:* their walls are also exceedingly thin. The vessels of neighbouring lobules are distinct from each other, and do not anastomose; and, according to Dr. Waters, those of the separate groups of intercellular passages, or air-sacs (which groups he denominates lobulettes), are also independent; so that in the septa between two adjoining lobulettes, there would be a double layer of capillaries, one layer belonging to each of the adjacent air-sacs, or intercellular passages. If this is really the arrangement of the vessels, it would follow, that in the septa between the air-cells (or alveoli), the blood in the capillaries would be exposed on all sides to the action of the air, since it is circulating in a single layer of vessels, which is in contact with the membrane of the air-passages on both sides; but that, in the septa between the intercellular passages (or air-sacs) the blood in the double layer of capillaries will be in contact with the air on one side only.

The *Bronchial Arteries* supply blood for the nutrition of the lung: they are derived from the thoracic aorta, and, accompanying the bronchial tubes, are distributed to the bronchial glands, and upon the walls of the larger bronchial tubes and pulmonary vessels, and terminate in the deep bronchial veins. Others are distributed in the interlobular areolar tissue, and terminate partly in the deep, partly in the superficial, bronchial veins. Lastly, some ramify upon the walls of the smallest bronchial tubes, and terminate in the pulmonary veins.

The *Superficial and Deep Bronchial Veins* unite at the root of the lung, and terminate on the right side in the vena azygos; on the left side, in the superior intercostal vein.

According to Dr. Waters, the bronchial veins do not exist within the proper substance of the lung, but commence at or near the root of the lung, by branches which lie on the large bronchial tubes. He also denies that the bronchial arteries contribute to the formation of the pulmonary plexus, believing that the communication between the bronchial and pulmonary system of vessels takes place in the pulmonary veins. If this view be correct, almost the whole of the blood carried by the bronchial arteries must be returned to the heart by the pulmonary veins, and thus the great mass of pure, or arterial, blood which is carried by the pulmonary veins, would be adulterated by a small quantity of carbonized or venous blood, which has passed through the bronchial circulation.

The *Lymphatics* consist of a superficial and deep set: they terminate at the root of the lung, in the bronchial glands.

Nerves. The lungs are supplied from the anterior and posterior pulmonary plexuses, formed chiefly by branches from the sympathetic and pneumogastric. The filaments from these plexuses accompany the bronchial tubes, upon which they are lost. Small ganglia are found upon these nerves.

* The meshes are only 0·002''' to 0·008''' in width while the vessels are 0·003''' to 0·005'''. Kölliker, *Human Microscopic Anatomy*.

Thyroid Gland.

The thyroid gland bears much resemblance in structure to other glandular organs, and is usually classified together with the thymus, supra-renal capsules, and spleen, under the head of *ductless glands*, since it has no excretory duct. Its function is unknown, but, from its situation in connection with the trachea and larynx, the thyroid body is usually described with those organs, although it takes no part in the function of respiration. It is situated at the upper part of the trachea, and consists of two lateral lobes, placed one on each side of that tube, and connected together by a narrow transverse portion, the isthmus.

Its *anterior surface* is convex, and covered by the Sterno-hyoid, Sterno-thyroid, and Omo-hyoid muscles.

Its *lateral surfaces*, also convex, lie in contact with the sheath of the common carotid artery.

Its *posterior surface* is concave, and embraces the trachea and larynx. The posterior borders of the gland extend as far back as the lower part of the pharynx.

The thyroid is of a brownish red colour. Its weight varies from one to two ounces. It is larger in females than in males, and becomes slightly increased in size during menstruation. It occasionally becomes enormously hypertrophied, constituting the disease called bronchocele, or goître. Each lobe is somewhat conical in shape, about two inches in length, and three-quarters of an inch in breadth, the right lobe being rather the larger of the two.

The *isthmus* connects the lower third of the two lateral lobes; it measures about half an inch in breadth, and the same in depth, and usually covers the second and third rings of the trachea. Its situation presents, however, many variations, a point of importance in the operation of tracheotomy. Sometimes the isthmus is altogether wanting.

A third lobe, of conical shape, called the *pyramid*, occasionally arises from the left side of the upper part of the isthmus, or from the left lobe, and ascends as high as the hyoid bone. It is occasionally quite detached, or divided into two parts, or altogether wanting.

A few muscular bands are occasionally found attached, above, to the body of the hyoid bone, and below, to the isthmus of the gland, or its pyramidal process. These form a muscle, which was named by Sömmerring the *Levator glandulæ thyroideæ*.

Structure. The thyroid consists of numerous minute closed vesicles, composed of a homogeneous membrane, enclosed in a dense capillary plexus, and connected together into imperfect lobules, by areolar tissue. These vessels are spherical or oblong, perfectly distinct, and contain a yellowish fluid, in which are found floating numerous 'dotted corpuscles' and cells. The fluid coagulates by heat or alcohol, but preserves its transparency. In the fœtus, and in young subjects, the corpuscles lie in a single layer, in contact with the inner surface of these cavities, and become detached during the process of growth.

Vessels and Nerves. The *arteries* supplying the thyroid, are the superior and inferior thyroid, and sometimes an additional branch (thyroidea media, or ima) from the arteria innominata, or the arch of the aorta, which ascends upon the front of the trachea. The arteries are remarkable for their large size and frequent anastomoses. The *veins* form a plexus on the surface of the gland, and on the front of the trachea, from which arise the superior, middle, and inferior thyroid veins; the two former terminating in the internal jugular, the latter in the vena innominata. The *lymphatics* are numerous, of large size, and terminate in the thoracic and right lymphatic ducts. The *nerves* are derived from the pneumogastric, and from the middle and inferior cervical ganglia of the sympathetic.

Chemical Composition. The thyroid gland consists of albumen, traces of gelatine, stearine, oleine, extractive matter, alkaline, and earthy salts, and water. The salts are chloride of sodium, alkaline sulphate, phosphate of potash, lime, magnesia, and a trace of oxide of iron.

Thymus Gland.

The thymus gland presents much resemblance in structure to other glandular organs, and is another of the organs denominated *ductless glands.*

The thymus gland is a temporary organ, attaining its full size at the end of the second year, when it ceases to grow, aud gradually dwindles, until, at puberty, it has almost disappeared. If examined when its growth is most active, it will be found to consist of two lateral lobes, placed in close contact along the middle line, situated partly in the anterior mediastinum, partly in the neck, and extending from the fourth costal cartilage upwards, as high as the lower border of the thyroid gland. It is covered by the sternum, and by the origins of the Sterno-hyoid and Sterno-thyroid muscles. In the mediastinum, it rests upon the pericardium, being separated from the arch of the aorta and great vessels by the thoracic fascia. In the neck, it lies on the front and sides of the trachea, behind the Sterno-hyoid and Sterno-thyroid muscles. The two lobes generally differ in size; they are occasionally united, so as to form a single mass; and sometimes separated by an intermediate lobe. The thymus is of a pinkish-grey colour, soft, and lobulated on its surfaces. It is about two inches in length, one and a half in breadth, below, and about three or four lines in thickness. At birth, it weighs about half an ounce.

Structure. Each lateral lobe is composed of numerous lobules, held together by delicate areolar tissue; the entire gland being enclosed in an investing capsule of a similar, but denser structure. The primary lobules vary in size from a pin's head to a small pea. Each lobule contains, in its interior, a small cavity, which is surrounded with smaller or secondary lobules, also hollow. The cavities of the secondary and primary lobules communicate; those of the latter opening into the great central cavity, or *reservoir of the thymus,* which extends through the entire length of each lateral half of the gland. The central cavity is lined by a vascular membrane, which is prolonged into all the subordinate cavities, and contains a milk-white fluid resembling chyle.

If the investing capsule and vessels, as well as the areolar tissue connecting the lobules, are removed from the surface of either lateral lobe, it will be seen that the central cavity is folded upon itself, and admits of being drawn out into a lengthened tubular cord, around which the primary lobules are attached in a spiral manner, like knots upon a rope. Such is the condition of the organ at an early period of its development; for Mr. Simon has shown, that the primitive form of the thymus is a linear tube, from which, as its development proceeds, lateral diverticula lead outwards, the tube ultimately becoming obscure, from its surface being covered with numerous lobules.

According to Oesterlen and Mr. Simon, the cavities in the secondary lobules are surrounded by rounded saccular dilatations or vesicles, which open into it. These vesicles are formed of a homogeneous membrane, enclosed in a dense capillary plexus.

The whitish fluid contained in the vesicles and central cavity of the thymus, contains numerous dotted corpuscles, similar to those found in the chyle. The corpuscles are flattened circular discs, measuring about $\frac{1}{1000}$ of an inch in diameter.

Vessels and Nerves. The *arteries* supplying the thymus are derived from the internal mammary, and from the superior and inferior thyroid. The *veins* terminate in the left vena innominata, and in the thyroid veins. The *lymphatics* are of large size, arise in the substance of the gland, and are said to terminate in the internal jugular vein. Sir A. Cooper believed that these vessels carried into the blood the secretion formed in the substance of the thymus. The *nerves* are exceedingly minute; they are derived from the pneumogastric and sympathetic. Branches from the descendens noni and phrenic reach the investing capsule, but do not penetrate into the substance of the gland.

Chemical Composition. The solid animal constituents of the thymus are albumen and fibrine in large quantities, gelatine and other animal matters. The salts are alkaline and earthy phosphates, with chloride of potassium. It contains about 80 per cent. of water.

The Urinary Organs.

The Kidneys.

THE KIDNEYS are glandular organs, intended for the secretion of the urine. They are situated at the back part of the abdominal cavity, behind the peritoneum, one in each lumbar region, extending from the eleventh rib to near the crest of the ilium; the right being lower than the left, in consequence of the large size of the liver. They are usually surrounded by a considerable quantity of fat, and are retained in their position by the vessels which pass to and from them.

Relations. The *anterior surface* of the kidney is convex, partially covered by the peritoneum, and is in relation, on the right side, with the back part of the right lobe of the liver, the descending portion of the duodenum, and the ascending colon; and on the left side with the great end of the stomach, the lower end of the spleen, the tail of the pancreas, and the descending colon.

The *posterior surface* is flattened, and rests upon the corresponding crus of the Diaphragm, in front of the eleventh and twelfth ribs, on the anterior lamella of the aponeurosis of the transversalis, which separates the kidney from the Quadratus lumborum, and on the Psoas magnus.

The *superior extremity*, directed inwards, is thick and rounded, and embraced by the supra-renal capsule. It corresponds, on the left side, to the upper border and on the right side to the lower border of the eleventh rib.

The *inferior extremity*, small and flattened, extends nearly as low as the crest of the ilium.

The *external border* is convex, and directed outwards towards the parietes of the abdomen.

The *internal border* is concave, and presents a deep notch, the *hilum of the kidney*, more marked behind than in front. At the hilum, the vessels, excretory duct and nerves, pass into or from the organ; the branches of the renal vein lying in front, the artery and its branches next, the excretory duct or ureter behind and below. On the vessels the nerves and lymphatics ramify, and much cellular tissue and fat surrounds the whole. The hilum leads into a hollow space, the *sinus*, which occupies the interior of the gland.

Each kidney is about four inches in length, two inches in breadth, and about one inch in thickness; the left being somewhat longer and thinner than the right. The weight of the kidney in the adult male varies from 4½ oz. to 6 oz.; in the female, from 4 oz. to 5½ oz. The left kidney is nearly always heavier than the right, by about two drachms. Their weight in proportion to the body is about 1 to 240. The renal substance is dense, firm to the touch, but very fragile, and of a deep red colour.

The kidney is invested by a fibrous capsule, formed of dense fibro-areolar tissue. This capsule is thin, smooth, and easily removed from the surface of the kidney, to which it is connected by fine fibrous processes and vessels; and at the hilum is continued inwards, lining the sides of the sinus, and at the bottom of that cavity forms sheaths around the blood-vessels, and the subdivisions of the excretory duct.

On making a vertical section through the organ, from its convex to its concave border, it appears to consist of two different substances, viz., an external or cortical, and an internal or medullary, substance.

The *cortical substance* forms about three-fourths of the gland. It occupies the surface of the kidney, forming a layer about two lines in thickness, where it covers the pyramids, and sends numerous prolongations inwards, towards the sinus between the pyramids.

The cortical substance is soft, reddish, granular, easily lacerated, and contains numerous small, red bodies scattered through it in every part, excepting towards the free surface. These are the Malpighian bodies. The cortical substance is composed of a mass of convoluted tubuli uriniferi, blood-vessels, lymphatics, and nerves, connected together by a firm, transparent, granular substance, which contains small granular cells.

367.—Vertical Section of Kidney.

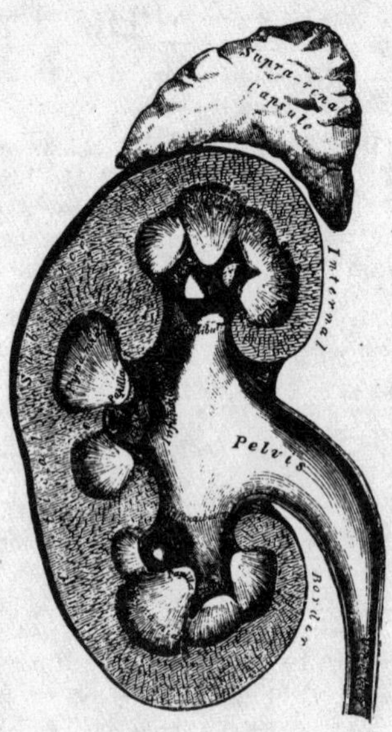

The *medullary substance* consists of pale, reddish-coloured, conical masses, the *pyramids of Malpighi*; they vary in number from eight to eighteen; their bases are directed towards the circumference of the organ; whilst their apices, which are free from the cortical substance, converge towards the sinus, and terminate in smooth, rounded extremities, called the *papillæ* (mammillæ) of the kidney. Sometimes, two of the masses are joined, and have between them only one papilla. The kidney is thus seen to consist of a number of conical-shaped masses, each enclosed, excepting at the apex, by an investment of the cortical substance: these represent the separate lobules of which the human kidney in the fœtus consists, a condition which is permanent in the kidneys of many of the lower animals. As the human kidney becomes developed, the adjacent lobules coalesce, so as to form a single gland, the surface of which, even in the adult, occasionally presents faint traces of a lobular subdivision.

The medullary substance is denser in structure than the cortical, darker in colour, and presents a striated appearance, from being composed of a number of minute converging tubes (*tubuli uriniferi*). If traced backwards the tubuli uriniferi are found to commence at the apices of the cones by small orifices, which vary from $\frac{1}{300}$ to $\frac{1}{200}$ of an inch; as they pass up in the medullary substance, towards the periphery, they pursue a diverging course, dividing and subdividing at very acute angles, until they reach the cortical substance, where they become convoluted, anastomose freely with each other, and retain the same diameter. The number of orifices on the entire surface of a single papilla is, according to Huschke, about a thousand; from four to five hundred large, and as many smaller ones. The tubuli uriniferi are formed of a transparent homogeneous basement membrane, lined by spheroidal epithelium, which occupies about two-thirds of the diameter of the tube. The tubes are separated from one another, in the medullary cones, by capillary vessels, which form oblong meshes parallel with the tubuli, and by an intermediate parenchymatous substance composed of cells.

368.—Minute Structure of Kidney.

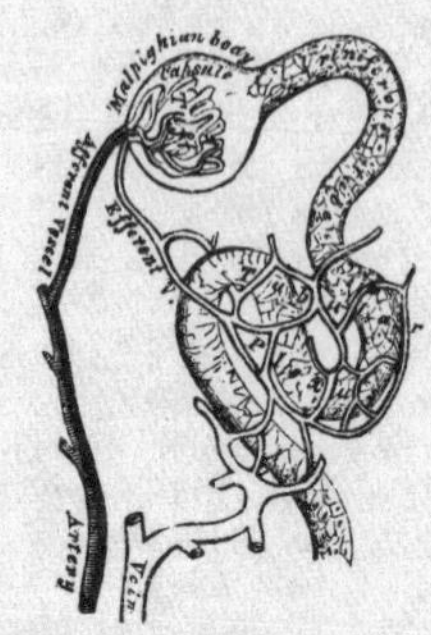

As soon as the tubuli uriniferi enter the cortical substance (fig. 368), they become convoluted, and anastomose freely with each other; they are sometimes called the *tubes of Ferrein.* At the bases of the pyramids, the straight tubes are described as being collected into small conical bundles, the tortuous tubuli corresponding to which are prolonged upwards into the cortical portion of the kidney as far as the surface, forming a number of small conical masses, which are named the *pyramids of Ferrein,* several of which correspond to each medullary cone and its corresponding portion of cortical substance. According to Mr. Bowman, the tubuli uriniferi commence in the cortical substance as small, dilated, membranous capsules, the capsules of the Malpighian bodies; they also form loops, either by the junction of adjacent tubes, or, according to Toynbee, by the union of two branches proceeding from the same tube; they have also been seen to arise by free closed extremities.

The *Malpighian bodies* are found only in the cortical substance of the kidney. They are small round bodies, of a deep red colour, and of the average diameter of the $\frac{1}{120}$th of an inch. Each body is composed of a vascular tuft enclosed in a thin membranous capsule, the dilated commencement of a uriniferous tubule. The vascular tuft consists of the ramifications of a minute artery, the *afferent vessel,* which, after piercing the capsule, divides, in a radiated manner, into several branches, which ultimately terminate in a finer set of capillary vessels. From these, a small vein, the *efferent vessel,* proceeds; this pierces the capsule near the artery, and forms a close venous plexus, with the efferent vessels from other Malpighian bodies, round the adjacent tubuli.

The capsular dilatation of the Malpighian body is not always placed at the commencement of the tube; it may occupy one side (Gerlach); hence their subdivision into lateral or terminal. The membrane composing it is thicker than that of the tubule; the epithelium lining its inner surface is thin, and, in the frog, provided with cilia at the neck of the dilated portion; but in the human subject cilia have not been detected. According to Mr. Bowman, the surface of the vascular tuft lies free and uncovered in the interior of its capsule; but, according to Gerlach, it is covered with a thick layer of nucleated cells, similar to those lining the inner surface of the capsule.

Ducts. The ureter, as it approaches the hilum, becomes dilated into a funnel-shaped membranous sac, the *pelvis.* It then enters the sinus, and subdivides usually into three prolongations, the *infundibula*; one placed at each extremity, and one in the middle of the organ; these subdivide into from seven to thirteen smaller tubes, the *calices,* each of which embraces the base of one of the papillæ. Sometimes, a *calix* encloses two or more papillæ. The ureter, the pelvis, and the calices consist of three coats: fibrous, muscular, and mucous.

The *external* or *fibro-elastic coat* is continuous, round the bases of the papillæ, with the tunica propria investing the surface of the organ.

The *muscular coat* is placed between the fibrous and mucous coats. It consists of an external or longitudinal, and an internal or circular stratum.

The *internal* or *mucous coat* invests the papillæ of the kidney, and is continued into the orifices upon their surfaces. It is lined by epithelium of the spheroidal kind.

Vessels and nerves. The *renal artery* is large in proportion to the size of the organ which it supplies. Each vessel divides into four or five branches, which enter the hilum, and are invested by sheaths derived from the fibrous capsule; they penetrate the substance of the organ between the papillæ, and enter the cortical substance in the intervals between the medullary cones; dividing and subdividing in their course towards the bases of the pyramids, where they form arches by their anastomoses; from these arches, numerous vessels are distributed to the cortical substance, some of which enter the Malpighian corpuscles; whilst others form a capillary network round the uriniferous tubes.

The *veins* of the kidney commence upon the surface of the organ, where they have a stellate arrangement; they pass inwards, and open into larger veins, which unite into arches round the bases of the medullary cones. After receiving the venous plexus from the tubular portion, they accompany the branches of the arteries of the sinus of the kidney, where they finally unite to form a single vein, which terminates in the inferior vena cava.

The *lymphatics* of the kidney consist of a superficial and deep set; they accompany the blood-vessels, and terminate in the lumbar glands.

The *nerves* are derived from the renal plexus, which is formed by filaments from the solar plexus and lesser splanchnic nerve; they accompany the branches of the arteries. From the renal plexus, some filaments pass to the spermatic plexus and ureter.

The Ureters.

The Ureter is the excretory duct of the kidney. It is a cylindrical membranous tube, from sixteen to eighteen inches in length, and of the diameter of a goose-quill. It is placed at the back part of the abdomen, behind the peritoneum: and extends obliquely downwards and inwards, from the lower part of the pelvis of the kidney, enters the cavity of the pelvis, and then passes downwards, forwards, and inwards, to the base of the bladder, into which it opens by a constricted orifice, after passing obliquely, for nearly an inch, between its muscular and mucous coats.

Relations. In its course from above downwards, it rests upon the Psoas muscle, being covered by the peritoneum, and crossed in front very obliquely by the spermatic vessels; the *right* ureter lying close to the outer side of the inferior vena cava. Opposite the sacrum, it crosses the common or the external iliac artery, lying behind the ileum on the right side, and the sigmoid flexure of the colon on the left. In the pelvis, it enters the posterior false ligament of the bladder, below the obliterated hypogastric artery; the vas deferens, in the male, passing between it and the bladder. In the female, the ureter passes along the sides of the cervix uteri and upper part of the vagina. At the base of the bladder, it is situated about two inches from its fellow; lying, in the male, about an inch and a half behind the base of the prostate, at the posterior angle of the trigone.

Structure. The ureter is composed of three coats: fibrous, muscular, and mucous.

The *fibrous coat* is continuous with that surrounding the pelvis of the kidney.

The *muscular coat* consists of two layers of longitudinal fibres, and an intermediate transverse layer.

The *mucous coat* is smooth, and presents a few longitudinal folds, which become effaced by distension. It is continuous with the mucous membrane of the bladder below; whilst, above, it is prolonged over the papillæ into the tubuli uriniferi. The epithelial cells lining it are spheroidal.

The *arteries* supplying the ureter are branches of the renal, spermatic, internal iliac, and inferior vesical.

The *nerves* are derived from the inferior mesenteric, spermatic, and hypogastric plexuses.

Suprarenal Capsules.

The suprarenal capsules are usually classified, together with the spleen, thymus, and thyroid, under the head of 'ductless glands,' as they have no excretory duct. They are two small flattened glandular bodies, of a yellowish colour, situated at the back part of the abdomen, behind the peritoneum, and immediately in front of the upper end of either kidney; hence their name. The right one is somewhat triangular in shape, bearing a resemblance to a cocked hat; the left is more semilunar, and usually larger and higher than the right. They vary in size in different individuals, being sometimes so small as to be scarcely detected; their usual size is from an inch and a quarter to nearly two inches in length, rather less in width, and from two to three lines in thickness. In weight, they vary from one to two drachms.

Relations. The *anterior surface* is in relation, on the right side, with the under surface of the liver; and on the left, with the pancreas and spleen. The *posterior surface* rests upon the crus of the Diaphragm, opposite the tenth dorsal vertebra. The *upper* thin convex border is directed upwards and inwards. The *lower* thick concave border rests upon the upper end of the kidney, to which it is connected by areolar tissue. The *inner border* is in relation with the great splanchnic nerve and semilunar ganglion, and lies in contact on the right side with the inferior vena cava, and on the left side with the aorta. The surface of the suprarenal gland is surrounded by areolar tissue containing much fat, and closely invested by a thin fibrous coat, which is difficult to remove, on account of the numerous fibrous processes and vessels which enter the organ through the furrows on its anterior surface and base.

Structure. On making a perpendicular section, the gland is seen to consist of two substances: external or cortical, and internal or medullary.

The *cortical substance* forms the chief part of the organ; it is of a deep yellow colour, and consists of narrow columnar masses placed perpendicularly to the surface.

The *medullary substance* is soft, pulpy, and of a dark brown or black colour; hence the name, *atrabiliary capsules*, formerly given to these organs. In the centre is often seen a space formed by the breaking down of the component parts of the tissue.

According to the researches of Oesterlen and Mr. Simon, the narrow columnar masses of which the cortical substance is composed measure about $\frac{1}{700}$th of an inch in diameter, and consist of small closed parallel tubes of limitary membrane containing dotted nuclei, together with much granular matter, oil globules, and nucleated cells. According to Ecker, the apparently tubular canals consist of rows of closed vesicles placed endwise, so as to resemble tubes; whilst Kölliker states, that these vesicles are merely loculi or spaces in the stroma of the organ, having no limitary membrane, which, from being situated endwise, present the appearance of linear tubes. Nucleated cells exist in large numbers in the suprarenal glands of ruminants, more sparingly in man and other animals, but the granular matter appears to form the chief constituent of the gland; the granules vary in size, and they present the singular peculiarity of undergoing no change when acted upon by most chemical re-agents. The columnar masses are surrounded by a close capillary network, which runs parallel with them.

The medullary substance consists of nuclei and granular matter, uniformly scattered throughout a plexus of minute veins.

The *arteries* supplying the suprarenal capsules are numerous and of large size, they are derived from the aorta, the phrenic, and the renal; they subdivide into numerous minute branches previous to entering the substance of the gland.

The *suprarenal vein* returns the blood from the medullary venous plexus, and receives several branches from the cortical substance; it opens on the right side into the inferior vena cava, on the left side into the renal vein.

The *lymphatics* terminate in the lumbar glands.

The *nerves* are exceedingly numerous; they are derived from the solar and renal plexuses, and, according to Bergmann, from the phrenic and pneumogastric nerves. They have numerous small ganglia developed upon them.

THE PELVIS.

The cavity of the pelvis is that part of the general abdominal cavity which is below the level of the linea ilio-pectinea and the promontory of the sacrum.

Boundaries. It is bounded, behind, by the sacrum, the coccyx, and the great sacro-sciatic ligaments; in front and at the sides, by the pubes and ischia, covered by the Obturator muscles; above, it communicates with the cavity of the abdomen;

and below, it is limited by the Levatores ani and Coccygei muscles, and the visceral layer of the pelvic fascia, which is reflected from the wall of the pelvis on to the viscera.

Contents. The viscera contained in this cavity are the urinary bladder, the rectum, and some of the generative organs peculiar to each sex: they are partially covered by the peritoneum, and supplied with blood and lymphatic vessels and nerves.

THE BLADDER.

The bladder is the reservoir for the urine. It is a musculo-membranous sac, situated in the pelvis, behind the pubes, and in front of the rectum in the male, the uterus and vagina intervening between it and that intestine in the female. The shape, position, and relations of the bladder are greatly influenced by age, sex, and the degree of distension of the organ. *During infancy*, it is conical in shape, and projects above the upper border of the pubes into the hypogastric region. *In the adult*, when quite empty and contracted, it is a small triangular sac, placed deeply in the pelvis, flattened from before backwards, its apex reaching

369.—Vertical Section of Bladder, Penis, and Urethra.

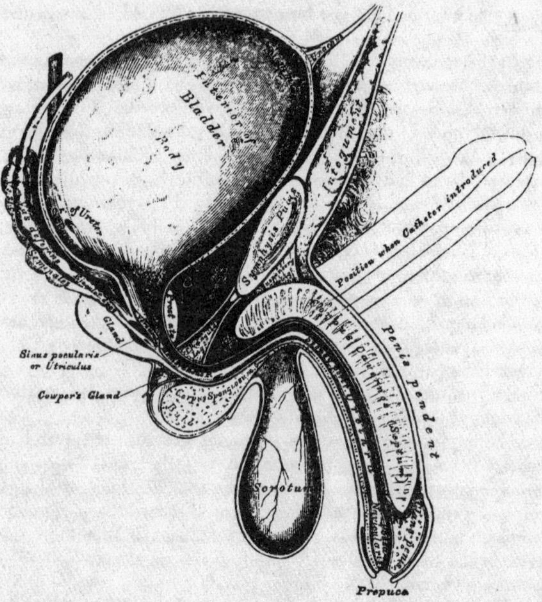

as high as the upper border of the symphysis pubis. When slightly distended, it has a rounded form, and partially fills the pelvic cavity; and when greatly distended, it is ovoid in shape, rising into the abdominal cavity, and often extending nearly as high as the umbilicus. It is larger in its vertical diameter than from side to side, and its long axis is directed from above obliquely downwards and backwards, in a line directed from some point between the pubes and umbilicus (according to its distension) to the end of the coccyx. The bladder, when distended, is slightly curved forwards towards the anterior wall of the abdomen, so as to be more convex behind than in front. In the female, it is larger in the transverse than in the vertical diameter, and its capacity is said to be greater than in the male. When moderately distended, it measures about five inches in length, and three inches across, and the ordinary amount which it contains is about a pint.

The bladder is divided into a summit, body, base, and neck.

The *summit*, or apex, of the bladder is rounded and directed forwards and upwards; it is connected to the umbilicus by a fibro-muscular cord, the urachus, and also by means of two rounded fibrous cords, the obliterated portions of the hypogastric arteries, which are placed one on each side of the urachus. The summit of the bladder behind the urachus is covered by peritoneum, whilst the portion in front of the urachus has no peritoneal covering, but rests upon the abdominal wall.

The urachus is the obliterated remains of a tubular canal which exists in the embryo, and connects the cavity of the bladder with a membranous sac placed external to the abdomen, opposite the umbilicus, called the *allantois*. In the infant, at birth, it is occasionally found pervious, so that the urine escapes at the umbilicus, and calculi have been found in its canal.

The *body* of the bladder in front is not covered by peritoneum, and is in relation with the triangular ligament of the urethra, the posterior surface of the symphysis pubis, the Internal obturator muscles, and, when distended, with the abdominal parietes.

The posterior surface is covered by peritoneum throughout. It corresponds, in the male, with the rectum; in the female, with the uterus, some convolutions of the small intestine being interposed.

The side of the bladder is crossed obliquely from below, upwards and forwards, by the obliterated hypogastric artery: above and behind this cord, the bladder is covered by peritoneum; but below and in front of it, the serous covering is wanting, and it is connected to the pelvic fascia. The vas deferens passes, in an arched direction, from before backwards, along the side of the bladder, towards its base, crossing in its course the obliterated hypogastric artery, and passing along the inner side of the ureter.

The *base* (*fundus*) of the bladder is directed downwards and backwards. It varies in extent according to the state of distension of the organ, being very broad when full, but much narrower when empty. *In the male*, it rests upon the second portion of the rectum, from which it is separated by a reflection of the recto-vesical fascia. It is covered posteriorly, for a slight extent, by the peritoneum, which is reflected from it upon the rectum, forming the recto-vesical fold. The portion of the bladder in relation with the rectum corresponds to a triangular space, bounded behind by the recto-vesical fold; on either side, by the vesicula seminalis and vas deferens; and touching the prostate gland in front. When the bladder is very full, the peritoneal fold is raised with it, and the distance between its reflection and the anus is about four inches; but this distance is much diminished when the bladder is empty and contracted. *In the female*, the base of the bladder lies in contact with the lower part of the cervix uteri, is adherent to the anterior wall of the vagina, and separated from the upper part of the anterior surface of the cervix uteri by a fold of the peritoneum.

The *neck* (*cervix*) of the bladder is the constricted portion continuous with the urethra. In the male, its direction is oblique in the erect posture, and it is surrounded by the prostate gland. In the female, its direction is obliquely downwards and forwards.

Ligaments. The bladder is retained in its place by ligaments, which are divided into true and false. The true ligaments are five in number, two anterior, and two lateral, formed by the recto-vesical fascia, and the urachus. The false ligaments, also five in number, are formed by folds of the peritoneum.

The *anterior ligaments* (*pubo-prostatic*) extend from the back of the pubes, one on each side of the symphysis, to the front of the neck of the bladder, and upper surface of the prostate gland. These ligaments contain a few muscular fibres, prolonged from the bladder.

The *lateral ligaments*, broader and thinner than the preceding, are attached to the lateral parts of the prostate, and to the sides of the base of the bladder.

The *urachus* is the fibro-muscular cord already mentioned, extending between the summit of the bladder and the umbilicus. It is broad below, at its attachment to the bladder, and becomes narrower as it ascends.

The *false ligaments* of the bladder are, two posterior, two lateral, and one superior.

The *two posterior* pass forwards, in the male, from the sides of the rectum; in the female, from the sides of the uretus, to the posterior and lateral aspect of the bladder: they form the lateral boundaries of the recto-vesical fold of peritoneum, and contain the obliterated hypogastric arteries, and the ureters, beside vessels and nerves.

The *two lateral* ligaments are reflections of the peritoneum, from the iliac fossæ to the sides of the bladder.

The *superior* ligament is the prominent fold of peritoneum extending from the summit of the bladder to the umbilicus. It covers the urachus, and the obliterated hypogastric arteries.

Structure. The bladder is composed of four coats: a serous, a muscular, a cellular, and a mucous coat.

The *serous coat* is partial, and derived from the peritoneum. It invests the posterior surface, from opposite the termination of the two ureters to its summit, and is reflected from this point and from the sides, on to the abdominal and pelvic walls.

The *muscular coat* consists of two layers of unstriped muscular fibre, an external layer, composed of longitudinal fibres, and an internal layer of circular fibres.

The longitudinal fibres are most distinct on the anterior and posterior surfaces of the organ. They arise in front, from the anterior ligaments of the bladder, from the neck of the bladder, and, in the male, from the adjacent portion of the prostate gland. They spread out, and form a plexiform mesh, on the anterior surface of the bladder, being continued over the posterior surface and base of the organ to the neck, where they are inserted into the prostate, in the male, and into the vagina in the female.

Other longitudinal fibres arise, in the male, from the sides of the prostate, and spread out upon the sides of the bladder, intersecting with one another.

The circular fibres are very thinly and irregularly scattered on the body of the organ; but, towards its lower part, round the cervix and commencement of the urethra, they are disposed as a thick circular layer, forming the sphincter vesicæ, which is continuous with the muscular fibres of the prostate gland.

Two bands of oblique fibres, originating behind the orifices of the ureters, converge to the back part of the prostate gland, and are inserted, by means of a fibrous process, into the middle lobe of that organ. They are the *muscles of the ureters*, described by Sir C. Bell, who supposed that, during the contraction of the bladder, they served to retain the oblique direction of the ureters, and so prevent the reflux of urine into them.

The *cellular coat* consists of a layer of areolar tissue, connecting together the muscular and mucous coats, and intimately united to the latter.

The *mucous coat* is thin, smooth, and of a pale rose colour. It is continuous through the ureters with the lining membrane of the uriniferous tubes, and below, with that of the urethra. It is connected loosely to the muscular coat, by a layer of areolar tissue, excepting at the trigone, where its adhesion is more close. It is provided with a few mucous follicles; and numerous small racemose glands, lined with columnar epithelium, exist near the neck of the organ. The epithelium covering it is intermediate in form between the columnar and squamous varieties.

Interior of the bladder. Upon the inner surface of the base of the bladder, immediately behind the urethral orifice, is a triangular, smooth surface, the apex of which is directed forwards; this is the *trigonum vesicæ*, or *trigone vesicale.* It is paler in colour than the rest of the mucous membrane, and never presents any rugæ, even in the collapsed condition of the organ, owing to its intimate adhesion to the subjacent tissues. It is bounded on each side by two slight ridges, which pass backwards and outwards to the orifices of the ureters, and correspond with the muscles of these tubes; and at each posterior angle, by the orifices of the ureters, which are placed nearly two inches from each other, and about an inch and a half behind the orifice of the urethra. The trigone corresponds with the interval at the base of the bladder, bounded by the prostate in front, and the vesiculæ and vasa deferentia on the sides. Projecting from the lower and anterior part of the bladder, into the orifice of the urethra, is a slight elevation of mucous membrane, called the *uvula vesicæ.* It is formed by a thickening of the prostate.

370.—The Bladder and Urethra laid open. Seen from above.

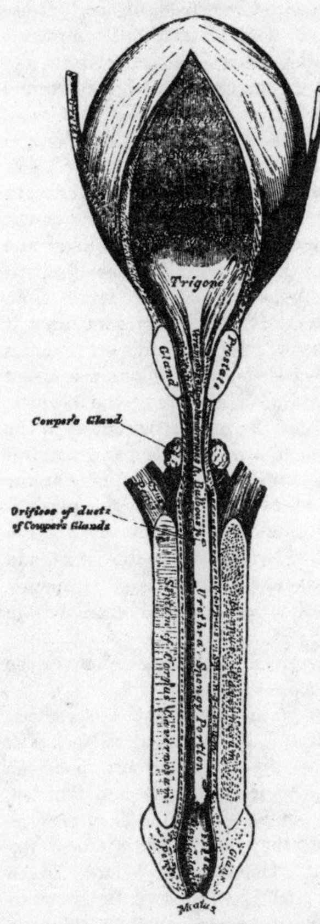

The *arteries* supplying the bladder are the superior, middle, and inferior vesical, in the male, with additional branches from the uterine, in the female. They are all derived from the anterior trunk of the internal iliac.

The *veins* form a complicated plexus round the neck, sides, and base of the bladder, and terminate in the internal iliac vein.

The *lymphatics* accompany the blood-vessels, passing through the glands surrounding them.

The *nerves* are derived from the hypogastric and sacral plexuses; the former supplying the upper part of the organ, the latter its base and neck.

Male Urethra.

The urethra extends from the neck of the bladder to the meatus urinarius. It presents a double curve in the flaccid state of the penis, but in the erect state it forms only a single curve, the concavity of which is directed upwards (fig. 369). Its length varies from eight to nine inches; and it is divided into three portions, the prostatic, membranous, and spongy, the structure and relations of which are essentially different.

The *Prostatic portion* is the widest and most dilatable part of the canal. It passes through the prostate gland, from its base to its apex, lying nearer its upper than its lower surface. It is about an inch and a quarter in length; the form of the canal is spindle-shaped, being wider in the middle than at either extremity, and narrowest in front, where it joins the membranous portion. A transverse section of the canal in this situation is triangular, the apex directed downwards.

Upon the floor of the canal is a narrow longitudinal ridge, the *verumontanum*, or *caput gallinaginis*, formed by an elevation of the mucous membrane and its subjacent tissue. It is eight or nine lines in length, and a line and a half in height; and contains, according to Kobelt, muscular and erectile tissues. When distended, it may serve to prevent the passage of the semen backwards into the bladder. On each side of the verumontanum is a slightly depressed fossa, the *prostatic sinus*, the floor of which is perforated by numerous apertures, the *orifices of the prostatic ducts*, the ducts of the middle lobe opening behind the crest. At the fore part of the verumontanum, in the middle line, is a depression, the *sinus pocularis* (*vesicula prostatica*); and upon or within its margins are the slit-like openings of the ejaculatory ducts. The sinus pocularis forms a *cul-de-sac* about a quarter of an inch in length, which runs upwards and backwards in the substance of the prostate, beneath the middle lobe; its prominent upper wall partly forms the verumontanum. Its walls are composed of fibrous tissue, muscular fibres, and mucous membrane; and numerous small glands open on its inner surface. It has been called by Weber, who discovered it, the *uterus masculinus*, from its supposed homology with the female organ.

The *Membranous portion* of the urethra extends between the apex of the prostate and the bulb of the corpus spongiosum. It is the narrowest part of the canal (excepting the orifice), and measures three-quarters of an inch along its upper and half an inch along its lower surface, in consequence of the bulb projecting backwards beneath it below. Its upper concave surface is placed about an inch beneath the pubic arch, from which it is separated by the dorsal vessels and nerves of the penis, and some muscular fibres. Its lower convex surface is separated from the rectum by a triangular space, which constitutes the perinæum. The membranous portion of the urethra perforates the deep perinæal fascia; and two layers from this membrane are prolonged round it, the one forwards, the other backwards; it is also surrounded by the Compressor urethræ muscle. Its coverings are mucous membrane, elastic fibrous tissue, a thin layer of erectile tissue, muscular fibres, and a prolongation from the deep perinæal fascia.

The *Spongy portion* is the longest part of the urethra, and is contained in the corpus spongiosum. It is about six inches in length, and extends from the termination of the membranous portion to the meatus urinarius. Commencing below the symphysis pubis, it ascends for a short distance, and then curves downwards. It is narrow, and of uniform size in the body of the penis, measuring about a quarter of an inch in diameter; being dilated behind, within the bulb; and again anteriorly within the glans penis, forming the fossa navicularis. A cross section of this canal in the body of the penis has its long diameter transverse; but in the glans, that diameter is directed vertically.

The *Bulbous portion* is a name given, in some descriptions of the urethra, to the posterior dilated part of the spongy portion contained within the bulb.

The *meatus urinarius* is the most contracted part of the urethra; it is a vertical slit, about three lines in length, bounded on each side by two small labia. The inner surface of the lining membrane of the urethra, especially on the floor of the spongy portion, presents the orifices of numerous mucous glands and follicles, situated in the submucous tissue, and named the *glands of Littre*. They vary in size, and their orifices are directed forwards, so that they may easily intercept the point of a catheter in its passage along the canal. One of these lacunæ, larger than the rest, is situated on the upper surface of the fossa navicularis, about an inch and a half from the orifice; it is called the *lacuna magna*. Into the bulbous portion are found opening the ducts of Cowper's glands.

Structure. The urethra is composed of three coats: a mucous, muscular, and erectile.

The *mucous coat* forms part of the genito-urinary mucous membrane. It is continuous with the mucous membrane of the bladder, ureters, and kidneys; externally, with the integument covering the glans penis; and is prolonged into the ducts of the glands which open into the urethra, viz., Cowper's glands, the prostate gland, and the vasa deferentia and vesiculæ seminales, through the ejaculatory ducts. In the spongy and membranous portions the mucous membrane is arranged in longitudinal folds when the organ is contracted. Small papillæ are found upon it, near the orifice; and its epithelial lining is of the columnar variety, excepting near the meatus, where it is laminated.

The *muscular coat* consists of two layers of plain muscular fibres, an external longitudinal layer, and an internal circular. The muscular tissue is most abundant in the prostatic portion of the canal.

A thin layer of *erectile tissue* is continued from the corpus spongiosum round the membranous and prostatic portions of the urethra to the neck of the bladder.

Male Generative Organs.

Prostate Gland.

THE Prostate Gland ($\pi\rho o\ddot{\iota}\sigma\tau\eta\mu\iota$, *to stand before*) is a pale, firm, glandular body, which surrounds the neck of the bladder and commencement of the urethra. It is placed in the pelvic cavity, behind and below the symphysis pubis, posterior to the deep perinæal fascia, and upon the rectum, through which it may be distinctly felt, especially when enlarged. In shape and size it resembles a horse-chestnut.

Its *base* is directed backwards towards the neck of the bladder.

The *apex* is directed forwards to the deep perinæal fascia, which it touches.

Its *under surface* is smooth, and rests on the rectum, to which it is connected by dense areolar fibrous tissue.

Its *upper surface* is flattened, marked by a slight longitudinal furrow, and placed about three-quarters of an inch below the pubic symphysis.

It measures about an inch and a half in its transverse diameter at the base, an inch in its antero-posterior diameter, and three-quarters of an inch in depth. Its weight is about six drachms. It is held in its position by the anterior ligaments of the bladder (*pubo-prostatic*); by the posterior layer of the deep perinæal fascia, which invests the commencement of the membranous portion of the urethra and prostate gland; and by the anterior portion of the Levator ani muscle (*levator prostatæ*), which passes down on each side from the symphysis pubis and anterior ligament of the bladder to the sides of the prostate.

The prostate consists of three lobes: two lateral and a middle lobe.

The *two lateral lobes* are of equal size, separated behind by a deep notch, and marked by a slight furrow upon their upper and lower surface, which indicates the bi-lobed condition of the organ in some animals.

The third, or *middle lobe*, is a small transverse band, occasionally a rounded or triangular prominence, placed between the two lateral lobes, at the under and posterior part of the organ. It lies immediately beneath the neck of the bladder, behind the commencement of the urethra, and above the ejaculatory ducts. Its existence is not constant; but it is occasionally found at an early period of life, as well as in adults, and in old age. In advanced life it often becomes considerably enlarged, and projects into the bladder, so as to impede the passage of the urine. According to Dr. Messer's researches, conducted at Greenwich Hospital,* it would seem that this obstruction exists in 20 per cent. of all prostates over sixty years of age.

The prostate gland is perforated by the urethra and common seminal ducts. The urethra usually lies about one-third nearer its upper than its lower surface; occasionally the prostate surrounds only the lower three-fourths of this tube, and more rarely the urethra runs through the lower instead of the upper part of the gland. The ejaculatory ducts pass forwards obliquely through a conical canal, situated in the lower part of the prostate, and open into the prostatic portion of the urethra.

Structure. The prostate is enclosed in a thin but firm fibrous capsule, distinct from that derived from the posterior layer of the deep perinæal fascia, and separated from it by a plexus of veins. Its substance is of a pale reddish-grey colour, very friable, but of great density. It consists of glandular substance and muscular tissue.

The *glandular substance* is composed of numerous follicular pouches, opening into elongated canals, which join to form from twelve to twenty small excretory ducts. The follicles are connected together by areolar tissue, supported by prolongations from the fibrous capsule, and enclosed in a delicate capillary plexus.

* *Med.-Chir. Trans.*, vol. xliii. p. 152.

The epithelium lining the canals is columnar, whilst that in the terminal vesicles is of the squamous variety.

The *muscular tissue* of the prostate is arranged in the form of circular bands round the urethra; it is continuous behind with the circular fibres of the sphincter vesicæ, and in front with the circular fibres of the urethra. The muscular fibres are of the involuntary kind. The prostatic ducts open into the floor of the prostatic portion of the urethra.

Vessels and Nerves. The *arteries* supplying the prostate are derived from the internal pudic, vesical, and hæmorrhoidal. Its *veins* form a plexus around the sides and base of the gland; they communicate in front with the dorsal vein of the penis, and terminate in the internal iliac vein. The *nerves* are derived from the hypogastric plexus.

The *Prostatic Secretion* is a milky fluid, having an acid reaction, and presenting, on microscopic examination, molecular matter, the squamous and columnar forms of epithelium, and granular nuclei. In old age, this gland is liable to be enlarged, and its ducts are often filled with innumerable small concretions, of a brownish-red colour, and of the size of a millet seed, composed of carbonate of lime and animal matter.

Cowper's Glands.

Cowper's Glands are two small rounded and somewhat lobulated bodies, of a yellowish colour, about the size of peas, placed beneath the fore part of the membranous portion of the urethra, between the two layers of the deep perinæal fascia. They lie close behind the bulb, and are enclosed by the transverse fibres of the Compressor urethræ muscle. Each gland consists of several lobules, held together by a fibrous investment. The excretory duct of each gland, nearly an inch in length, passes obliquely forwards beneath the mucous membrane, and opens by a minute orifice on the floor of the bulbous portion of the urethra. Their existence is said to be constant; they gradually diminish in size as age advances.

The Penis.

The penis is the organ of copulation, and contains in its interior the larger portion of the urethra. It consists of a root, body, and extremity or glans penis.

The *root* is broad, and firmly connected to the rami of the pubes by two strong tapering fibrous processes, the crura, and to the front of the symphysis pubis by a fibrous membrane, the suspensory ligament.

The *extremity*, or *glans penis*, presents the form of an obtuse cone, flattened from above downwards. At its summit is a vertical fissure, the orifice of the urethra (meatus urinarius): at the back part of this orifice a fold of mucous membrane passes backwards to the bottom of a depressed raphe, where it is continuous with the prepuce; this fold is termed the *frœnum preputii.* The base of the glans forms a rounded projecting border, the *corona glandis;* and behind the corona is a deep constriction, the *cervix.* Upon both of these parts numerous small lenticular sebaceous glands are found, the *glandulæ Tysonii, odoriferæ.* They secrete a sebaceous matter of very peculiar odour, which probably contains caseine, and becomes easily decomposed.

The *body* of the penis is the part between the root and the extremity. In the flaccid condition of the organ it is cylindrical, but when erect has a triangular prismatic form with rounded angles, the broadest side being turned upwards, and called the *dorsum.* It is covered by integument remarkable for its thinness, its dark colour, and its looseness of connection with the deeper parts of the organ, and containing no adipose tissue. At the root of the penis the integument is continuous with that upon the pubes and scrotum; and at the neck of the glans it leaves the surface, and becomes folded upon itself to form the prepuce.

The internal layer of the prepuce, which also becomes attached to the cervix, approaches in character to a mucous membrane; it is reflected over the glans penis, and at the meatus urinarius is continuous with the mucous lining of the urethra.

The mucous membrane covering the glans penis contains no sebaceous glands; but projecting from its free surface are a number of small, highly sensitive papillæ.

The penis is composed of a mass of erectile tissue, enclosed in three cylindrical fibrous compartments. Of these, two, the corpora cavernosa, are placed side by side along the upper part of the organ; the third, or corpus spongiosum, encloses the urethra, and is placed below.

The *Corpora Cavernosa* form the chief part of the body of the penis. They consist of two fibrous cylindrical tubes, placed side by side, and intimately connected along the median line for their anterior three-fourths, their posterior fourth being separated to form the two crura, by which the penis is connected to the rami of the pubes. Each crus commences by a thick-pointed process in front of the tuberosity of the ischium; and, near its junction with its fellow, presents a slight enlargement, named, by Kobelt, the *bulb of the corpus cavernosum.* Just beyond this point they become constricted, and retain an equal diameter to their anterior extremity, where they form a single rounded end, which is received into a fossa in the base of the glans penis. A median groove on the upper surface lodges the dorsal vein of the penis, and the groove on the under surface receives the corpus spongiosum. The root of the penis is connected to the symphysis pubis by the suspensory ligament.

Structure. Each corpus cavernosum consists of a strong fibrous envelope, enclosing a fibrous reticular structure, which contains erectile tissue in its meshes. It is separated from its fellow by an incomplete fibrous septum.

The *fibrous investment* is extremely dense, of considerable thickness, and highly elastic; it not only invests the surface of the organ, but sends off numerous fibrous bands (*trabeculæ*) from its inner surface, as well as from the surface of the septum, which cross its interior in all directions, subdividing it into a number of separate compartments, which present a spongy structure, in which the erectile tissue is contained.

The *trabecular* structure fills the interior of the corpora cavernosa. Its component fibres are larger and stronger round the circumference than at the centre of the corpora cavernosa; they are also thicker behind than in front. The interspaces, on the contrary, are larger at the centre than at the circumference, their long diameter being directed transversely; and they are largest anteriorly. They are lined by a layer of squamous epithelium.

The *fibrous septum* forms an imperfect partition between the two corpora cavernosa; it is thick and complete behind; but in front it is incomplete, and consists of a number of vertical bands of fibrous tissue, which are arranged like the teeth of a comb, hence the name, *septum pectiniforme*; these bands extend between the dorsal and urethral surface of the corpora cavernosa.

The fibrous investment and septum consist of longitudinal bands of white fibrous tissue, with numerous elastic and muscular fibres. The trabeculæ also consist of white fibrous tissue, elastic fibres, and plain muscular fibres, and enclose arteries and nerves.

The *Corpus Spongiosum* encloses the urethra, and is situated in the groove on the under surface of the corpora cavernosa. It commences posteriorly in front of the deep perinæal fascia, between the diverging crura of the corpora cavernosa, where it forms a rounded enlargement, the bulb; and terminates, anteriorly, in another expansion, the glans penis, which overlaps the anterior rounded extremity of the corpora cavernosa. The central portion, or body of the corpus spongiosum, is cylindrical, and tapers slightly from behind forwards.

The *bulb* varies in size in different subjects; it receives a fibrous investment from the anterior layer of the deep perinæal fascia, and is surrounded by the Accelerator urinæ muscle. The urethra enters the bulb nearer its upper than its lower surface, being surrounded by a layer of erectile tissue, a thin prolongation of which is continued backwards round the membranous and prostatic portions of the canal to the neck of the bladder, lying immediately beneath the mucous membrane. The portion of the bulb below the urethra presents a partial division into two lobes, being marked externally by a linear raphe, whilst internally there projects inwards, for a short distance, a thin fibrous septum, more distinct in early life.

Structure. The corpus spongiosum consists of a strong fibrous envelope, enclosing a trabecular structure, which contains in its meshes erectile tissue. The fibrous envelope is thinner, whiter in colour, and more elastic than that of the corpus cavernosum. The trabeculæ are delicate, uniform in size, and the meshes between them small; their long diameter, for the most part, corresponding with that of the penis. A thin layer of muscular fibres, continuous behind with those of the bladder, forms part of the outer coat of the corpus spongiosum.

Erectile tissue consists essentially of an intricate venous plexus, lodged in the interspaces between the trabeculæ. The veins forming this plexus are so numerous, and communicate so freely with one another, as to present a cellular appearance when examined by means of a section; their walls are extremely thin, and lined by squamous epithelium. The veins are smaller in the glans penis, corpus spongiosum, and circumference of the corpora cavernosa, than in the central part of the latter, where they are of large size, and much dilated. They return the blood by a series of vessels, some of which emerge in considerable numbers from the base of the glans penis, and converge on the dorsum of the organ to form the dorsal vein; others pass out on the upper surface of the corpora cavernosa, and join the dorsal vein; some emerge from the under surface of the corpora cavernosa, and, receiving branches from the corpus spongiosum, wind round the sides of the penis to terminate in the dorsal vein; but the greater number pass out at the root of the penis, and join the prostatic plexus and pudendal veins.

The *arteries of the penis* are derived from the internal pudic. Those supplying the corpora cavernosa are the arteries of the corpora cavernosa, and branches from the dorsal artery of the penis, which perforate the fibrous capsule near the fore part of the organ. Those to the corpus spongiosum are the arteries of the bulb. Additional branches are described, by Kobelt, as arising from the trunk of the internal pudic; they enter the bulbous enlargements on the corpora cavernosa and corpus spongiosum. The arteries, on entering the cavernous structure, divide into branches, which are supported and enclosed by the trabeculæ; according to Müller, some of these branches terminate in a capillary network, which communicates with the veins as in other parts; whilst others are more convoluted, and assume a tendril-like appearance; hence the name, *helicine arteries*, which is given to these vessels. The helicine arteries are more abundant in the back part of the corpora cavernosa and corpus spongiosum; they have not been seen in the glans penis. They are small twigs, given off in bunches from the sides of the arteries as they lie on the trabeculæ, and they terminate in dilated extremities hanging down into the cavity of a vein. Whether the extremities of the twigs are open or closed, appears uncertain. The existence of these vessels is denied by Valentin, who describes the smallest branches of the arteries as terminating by wide, funnel-shaped orifices, which open directly into the venous cavities.

The *lymphatics* of the penis consist of a superficial and deep set; the former terminate in the inguinal glands; the latter emerge from the corpora cavernosa and corpus spongiosum, and, passing beneath the pubic arch, join the deep lymphatics of the pelvis.

The *nerves* are derived from the internal pudic nerve and the hypogastric plexus. On the glans and bulb some filaments of the cutaneous nerves have Pacinian bodies connected with them.

The Testes and their Coverings.

The testes are two small glandular organs, which secrete the semen; they are situated in the scrotum, being suspended by the spermatic cords. At an early period of fœtal life, the testes are contained in the abdominal cavity, behind the peritoneum. Before birth, they descend to the inguinal canal, along which they pass with the spermatic cord, and, emerging at the external abdominal ring, they descend into the scrotum, becoming invested in their course by numerous coverings, derived from the serous, muscular, and fibrous layers of the abdominal parietes, as well as by the scrotum. The coverings of the testis are, the

Skin } Scrotum.
Dartos }
Intercolumnar, or External spermatic fascia.
Cremaster muscle.
Infundibuliform, or Fascia propria (Internal spermatic fascia).
Tunica vaginalis.

The SCROTUM is a cutaneous pouch, which contains the testes and part of the spermatic cords. It is divided into two lateral halves, by a median line, or raphe, which is continued forwards to the under surface of the penis, and backwards along the middle line of the perinæum to the anus. Of these two lateral portions, the left is longer than the right, and corresponds with the greater length of the spermatic cord on the left side. Its external aspect varies under different circumstances: thus, under the influence of warmth, and in old and debilitated persons, it becomes elongated and flaccid: but, under the influence of cold, and in the young and robust, it is short, corrugated, and closely applied to the testes.

The scrotum consists of two layers, the integument and the dartos.

The *integument* is very thin, of a brownish colour, and generally thrown into folds or rugæ. It is provided with sebaceous follicles, the secretion of which has a peculiar odour, and is beset with thinly scattered, crisp hairs, the roots of which are seen through the skin.

The *dartos* is a thin layer of loose reddish tissue, endowed with contractility; it forms the proper tunic of the scrotum, is continuous, around the base of the scrotum, with the superficial fascia of the groin, perinæum, and inner side of the thighs, and sends inwards a distinct septum, *septum scroti*, which divides it into two cavities for the two testes, the septum extending between the raphe and the under surface of the penis, as far as its root.

The dartos is closely united to the skin externally, but connected with the subjacent parts by delicate areolar tissue, upon which it glides with the greatest facility. The dartos is very vascular, and consists of a loose areolar tissue, containing unstriped muscular fibre. Its contractility is slow, and excited by cold and mechanical stimuli, but not by electricity.

The *intercolumnar fascia* is a thin membrane, derived from the margin of the pillars of the external abdominal ring, during the descent of the testis in the fœtus, being prolonged downwards around the surface of the cord and testis. It is separated from the dartos by loose areolar tissue, which allows of considerable movement of the latter upon it, but is intimately connected with the succeeding layer.

The *cremasteric fascia* consists of scattered bundles of muscular fibres (*Cremaster muscle*), derived from the lower border of the Internal oblique muscle, during the descent of the testis (p. 712).

The *fascia propria* is a thin membranous layer, which loosely invests the surface of the cord. It is a continuation downwards of the infundibuliform process of the fascia transversalis, and is also derived during the descent of the testis in the fœtus.

The *tunica vaginalis* is described with the proper coverings of the testis. A more detailed account of the other coverings of the testis will be found in the description of the surgical anatomy of inguinal hernia.

Vessels and Nerves. The *arteries* supplying the coverings of the testis are: the superficial and deep external pudic, from the femoral; the superficial perinæal branch of the internal pudic; and the cremasteric branch from the epigastric. The *veins* follow the course of the corresponding arteries. The *lymphatics* terminate in the inguinal glands. The *nerves* are: the ilio-inguinal and ilio-hypogastric branches of the lumbar plexus, the two superficial perinæal branches of the internal pudic nerve, the inferior pudendal branch of the small sciatic nerve, and the genital branch of the genito-crural nerve.

The Spermatic Cord extends from the internal abdominal ring, where the structures of which it is composed converge, to the back part of the testicle. It is composed of arteries, veins, lymphatics, nerves, and the excretory duct, of the testicle. These structures are connected together by areolar tissue, and invested by the fasciæ brought down by the testicle in its descent. In the abdominal wall the cord passes obliquely along the inguinal canal, lying at first beneath the Internal oblique, and upon the fascia transversalis; but nearer the pubes, it rests upon Poupart's ligament, having the aponeurosis of the External oblique in front of it, and the conjoined tendon behind it. It then escapes at the external ring, and descends nearly vertically into the scrotum. The left cord is rather longer than the right, consequently the left testis hangs somewhat lower than its fellow.

The *arteries of the cord* are: the spermatic, from the aorta; the artery of the vas deferens, from the superior vesical; and the cremasteric, from the epigastric artery.

The spermatic artery supplies the testicle. On approaching the gland, it gives off some branches which supply the epididymis, and others which perforate the tunica albuginea behind, and spread out on its inner surface, or pass through the fibrous septum in its interior, to be distributed on the membranous septa between the lobes.

The artery of the vas deferens is a long slender vessel, which accompanies the vas deferens, ramifying upon the coats of that duct, and anastomosing with the spermatic artery near the testis.

The cremasteric branch from the epigastric supplies the Cremaster muscle, and other coverings of the cord.

The *spermatic veins* leave the back part of the testis, and receiving branches from the epididymis, unite to form a plexus (*pampiniform plexus*), which forms the chief mass of the cord. They pass up in front of the vas deferens, and unite to form a single trunk, which terminates, on the right side, in the inferior vena cava, on the left side, in the left renal vein.

The *lymphatics* are of large size, accompany the blood-vessels, and terminate in the lumbar glands.

The *nerves* are the spermatic plexus from the sympathetic. This plexus is derived from the renal and aortic plexuses, joined by filaments from the hypogastric plexus, which accompany the artery of the vas deferens.

Testes.

The testes are two small glandular organs, suspended in the scrotum by the spermatic cords. Each gland is of an oval form, compressed laterally and behind, and having an oblique position in the scrotum; the upper extremity being directed forwards and a little outwards; the lower, backwards and a little inwards; the anterior convex border looks forwards and downwards, the posterior or straight border, to which the cord is attached, backwards and upwards.

The anterior and lateral surfaces, as well as both extremities of the organ, are convex, free, smooth, and invested by the tunica vaginalis. The posterior border, to which the cord is attached, receives only a partial investment from that membrane. Lying upon the outer edge of this border, is a long, narrow, flattened body, named, from its relation to the testis, the epididymis (δίδυμος, testis). It consists of a central portion, or body, an upper enlarged extremity, the globus major, or head; and a lower pointed extremity, the tail, or globus minor. The globus major is intimately connected with the upper end of the testicle by means of its efferent ducts; and the globus minor is connected with its lower end by cellular tissue, and a reflection of the tunica vaginalis. The outer surface and upper and lower ends of the epididymis are free and covered by serous membrane; the body is also completely invested by it, excepting along its posterior border, and connected to the back of the testis by a fold of the serous membrane. Attached to the upper end of the testis, or to the epididymis, is a small pedunculated body, the use of which is unknown.

371.—The Testis in Situ. The Tunica Vaginalis having been laid open.

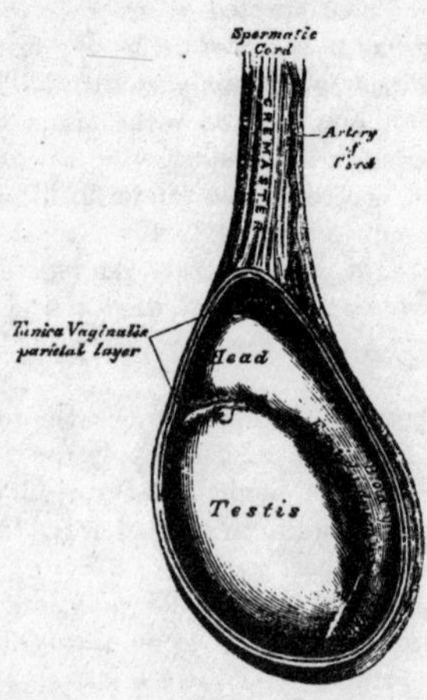

Size and Weight. The average dimensions of this gland are from one and a half to two inches in length, one inch in breadth, and an inch and a quarter in the antero-posterior diameter; and the weight varies from six to eight drachms, the left testicle being a little the larger.

The testis is invested by three tunics the tunica vaginalis, tunica albuginea, and tunica vasculosa.

The *Tunica Vaginalis* is the serous covering of the testis. It is a pouch of serous membrane, derived from the peritoneum during the descent of the testis in the fœtus, from the abdomen into the scrotum. After its descent, that portion of the pouch which extends from the internal ring to near the upper part of the gland becomes obliterated, the lower portion remaining as a shut sac, which invests the outer surface of the testis, and is reflected on the internal surface of the scrotum; hence it may be described as consisting of a visceral and parietal portion.

The *visceral portion* (*tunica vaginalis propria*), covers the outer surface of the testis, as well as the epididymis, connecting the latter to the testis by means of a distinct fold. From the posterior border of the gland, it is reflected on to the internal surface of the scrotum.

The *parietal portion* of the serous membrane (*tunica vaginalis reflexa*), is far more extensive than the visceral portion, extending upwards for some distance in front, and on the inner side of the cord, and reaching below the testis. The inner surface of the tunica vaginalis is free, smooth, and covered by a layer of squamous epithelium. The interval between the visceral and parietal layers of this membrane, constitutes the cavity of the tunica vaginalis.

The *Tunica Albuginea* is the fibrous covering of the testis. It is a dense fibrous membrane, of a bluish-white colour, composed of bundles of white fibrous tissue, which interlace in every direction. Its outer surface is covered by the tunica vaginalis, except along its posterior border, and at the points of attachment of the epididymis; hence the tunica albuginea is usually considered as a fibro-serous membrane, like the dura mater and pericardium. This membrane surrounds the glandular structure of the testicle, and, at its posterior and upper border, is reflected into the interior of the gland, forming an incomplete vertical septum, called the *mediastinum testis* (*corpus Highmorianum*).

The *mediastinum testis* extends from the upper, nearly to the lower border of the gland, and is wider above than below. From the front and sides of this septum, numerous slender fibrous cords (*trabeculæ*) are given off, which pass to be attached to the inner surface of the tunica albuginea: they serve to maintain the form of the testis, and join with similar cords given off from the inner surface of the tunica albuginea, to form spaces which enclose the separate lobules of the organ. The mediastinum supports the vessels and ducts of the testis in their passage to and from the substance of the gland.

The *Tunica Vasculosa* (*pia mater testis*), is the vascular layer of the testis, consisting of a plexus of blood-vessels, held together by a delicate areolar tissue. It covers the inner surface of the tunica albuginea, sending off numerous processes between the lobules, which are supported by the fibrous prolongations from the mediastinum testis.

Structure. The glandular structure of the testis consists of numerous lobules (*lobuli testis*). Their number, in a single testis, is estimated by Berres at 250, and by Krause at 400. They differ in size according to their position, those in the middle of the gland being larger and longer. The lobules are conical in shape, the base being directed towards the circumference of the organ, the apex towards the mediastinum. Each lobule is contained in one of the intervals between the fibrous cords and vascular processes, which extend between the mediastinum testis and the tunica albuginea, and consists of from one to three, or more, minute convoluted tubes, the tubuli seminiferi. The tubes may be separately unravelled, by careful dissection under water, and may be seen to commence either by free cæcal ends, or by anastomotic loops. The total number of tubes is considered by Monro to be about 300, and the length of each about sixteen feet: by Lauth, their number is estimated at 840, and their average length two feet and a quarter. Their diameter varies from $\frac{1}{200}$th to $\frac{1}{150}$th of an inch. The tubuli are pale in colour in early life, but, in old age, they acquire a deep yellow tinge, from containing much fatty matter. They consist of a basement membrane, lined by epithelium, consisting of nucleated granular corpuscles, and are enclosed in a delicate plexus of capillary vessels. In the apices of the lobules, the tubuli become less convoluted, assume a nearly straight course, and unite together to form from twenty to thirty larger ducts, of about $\frac{1}{50}$th of an inch in diameter, and these, from their straight course, are called *vasa recta.*

372.—Vertical Section of the Testicle, to show the arrangement of the Ducts.

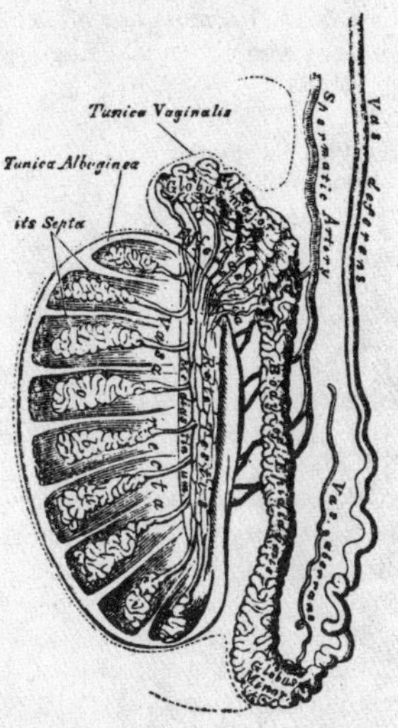

The *vasa recta* enter the fibrous tissue of the mediastinum, and pass upwards and backwards, forming, in their ascent, a close network of anastomosing tubes, with exceedingly thin parietes; this constitutes the *rete testis.* At the upper end of the mediastinum, the vessels of the rete testis terminate in from twelve to fifteen or twenty ducts, the *vasa efferentia*: they perforate the tunica albuginea, and carry the seminal fluid from the testis to the epididymis. Their course is at first straight; they then become enlarged, and exceedingly convoluted, and form a series of conical masses, the *coni vasculosi,* which, together, constitute the globus major of the epididymis. Each cone consists of a single convoluted duct, from six to eight inches in length, the diameter of which gradually decreases from the testis to the epididymis. Opposite the bases of the cones, the efferent vessels open at narrow intervals into a single duct, which constitutes, by its complex convolutions, the body and globus minor of the epididymis. When the convolutions of this tube are unravelled, it measures upwards of twenty feet in length, and increases in breadth and thickness as it approaches the vas deferens. The convolutions are held together by fine areolar tissue, and by bands of fibrous tissue. A long narrow tube, the *vasculum aberrans* of Haller, is occasionally found connected with the lower part of the canal of the epididymis, or with the commencement of the vas deferens, and extending up into the cord for about two or three inches, where it terminates by a blind extremity, which is occasionally bifurcated.

Its length varies from an inch and a half to fourteen inches, and sometimes it becomes dilated towards its extremity: more commonly, it retains the same diameter throughout. Its structure is similar to that of the vas deferens. Occasionally, it is found unconnected with the epididymis.

The *Vas Deferens*, the excretory duct of the testis, is the continuation of the epididymis. Commencing at the lower part of the globus minor, it ascends along the posterior and inner side of the testis and epididymis, and along the back part of the spermatic cord, through the spermatic canal, to the internal abdominal ring. From the ring it descends into the pelvis, crossing the external iliac vessels, and curves round the outer side of the epigastric artery: at the side of the bladder, it arches backwards and downwards to its base, crossing outside the obliterated hypogastric artery, and to the inner side of the ureter. At the base of the bladder, it lies between that viscus and the rectum, running along the inner border of the vesicula seminalis. In this situation, it becomes enlarged and sacculated; and, becoming narrowed, at the base of the prostate, unites with the duct of the vesicula seminalis to form the ejaculatory duct. The vas deferens presents a hard and cord-like sensation to the fingers; it is about two feet in length, of cylindrical form, and about a line and a quarter in diameter. Its walls are of extreme density and thickness, measuring one-third of a line; and its canal is extremely small, measuring about half a line.

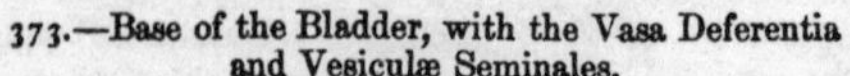

373.—Base of the Bladder, with the Vasa Deferentia and Vesiculæ Seminales.

In *structure*, the vas deferens consists of three coats: 1. An external, or cellular coat; 2. A muscular coat, which is thick, dense, elastic, and consists of two longitudinal, and an intermediate circular layer of muscular fibres; 3. An internal, or mucous coat, which is pale, and arranged in longitudinal folds; its epithelial covering is of the columnar variety.

Vesiculæ Seminales.

The *Seminal Vesicles* are two lobulated membranous pouches, placed between the base of the bladder and the rectum, serving as reservoirs for the semen, and secreting some fluid to be added to that of the testicles. Each sac is somewhat pyramidal in form, the broad end being directed backwards, and the narrow end forward towards the prostate. They measure about two and a half inches in length, about five lines in breadth, and from two to three lines in thickness.

They vary, however, in size, not only in different individuals, but also in the same individual on the two sides. Their *upper surface* is in contact with the base of the bladder, extending from near the termination of the ureters to the base of the prostate gland. Their *under surface* rests upon the rectum, from which they are separated by the recto-vesical fascia. Their *posterior extremities* diverge from each other. Their *anterior extremities* are pointed, and converge towards the base of the prostate gland, where each joins with the corresponding vas deferens to form the ejaculatory duct. Along the inner margin of each vesicula runs the enlarged and convoluted vas deferens. The inner border of the vesiculæ, and the corresponding vas deferens, form the lateral boundary of a triangular space, limited behind by the recto-vesical peritoneal fold; the portion of the bladder included in this space rests on the rectum, and corresponds with the trigonum vesicæ in its interior.

Structure. Each vesicula consists of a single tube, coiled upon itself, and giving off several irregular cæcal diverticula; the separate coils, as well as the diverticula, being connected together by fibrous tissue. When uncoiled, this tube is about the diameter of a quill, and varies in length from four to six inches; it terminates posteriorly in a *cul-de-sac*; its anterior extremity becomes constricted into a narrow straight duct, which joins on its inner side with the corresponding vas deferens, and forms the ejaculatory duct.

The *ejaculatory ducts*, two in number, one on each side, are formed by the junction of the duct of the vesicula seminalis with the vas deferens. Each duct is about three-quarters of an inch in length; it commences at the base of the prostate, and runs forwards and upwards in a canal in its substance, and along the side of the utriculus, to terminate by a separate slit-like orifice upon or within the margins of the sinus pocularis. The ducts diminish in size, and converge towards their termination.

Structure. The vesiculæ seminales are composed of three coats: an *external* or *fibro-cellular*, derived from the recto-vesical fascia; a *middle* or *fibrous coat*, which is firm, dense, fibrous in structure, somewhat elastic, and contains, according to E. H. Weber, muscular fibres; and an *internal* or *mucous coat*, which is pale, of a whitish-brown colour, and presents a delicate reticular structure, like that seen in the gall-bladder, but the meshes are finer. The epithelium is squamous. The coats of the ejaculatory ducts are extremely thin, the outer fibrous layer being almost entirely lost after their entrance into the prostate, a thin layer of muscular fibres and the mucous membrane forming the only constituents of the tubes.

Vessels and Nerves. The *arteries* supplying the vesiculæ seminales are derived from the inferior vesical and middle hæmorrhoidal. The *veins* and *lymphatics* accompany the arteries. The *nerves* are derived from the hypogastric plexus.

The *Semen* is a thick whitish fluid, having a peculiar odour. It consists of a fluid, the liquor seminis, and solid particles, the seminal granules, and spermatozoa.

The *liquor seminis* is transparent, colourless, and of an albuminous composition, containing particles of squamous and columnar epithelium, with oil-globules and granular matter floating in it, besides the above-mentioned solid elements.

The *seminal granules* are round finely-granular corpuscles, measuring $\frac{1}{4000}$th of an inch in diameter.

The *spermatozoa*, or spermatic filaments, are the essential agents in producing fecundation. They are minute elongated particles, consisting of a small flattened oval extremity or body, and a long slender caudal filament. A small circular spot is observed in the centre of the body, and at its point of connection with the tail there is frequently seen a projecting rim or collar. The movements of these bodies are remarkable, and consist of a lashing or undulatory motion of the tail.

Descent of the Testes.

The testes, at an early period of fœtal life, are placed at the back part of the abdominal cavity, behind the peritoneum, in front and a little below the kidneys.

The anterior surface and sides are invested by peritoneum; the blood-vessels and efferent ducts are connected with their posterior surface; and attached to the lower end is a peculiar structure, the gubernaculum testis, which is said to assist in their descent.

The *Gubernaculum Testis* attains its full development between the fifth and sixth months; it is a conical-shaped cord, attached above to the lower end of the epididymis, and below to the bottom of the scrotum. It is placed behind the peritoneum, lying upon the front of the Psoas muscle, and completely filling the inguinal canal. It consists of a soft transparent areolar tissue within, which often appears partially hollow, surrounded by a layer of striped muscular fibres, the Cremaster, which ascends upon this body to be attached to the testis. According to Mr. Curling, the gubernaculum, as well as these muscular fibres, divides below into three processes: the external and broadest process is connected with Poupart's ligament in the inguinal canal; the middle process descends along the inguinal canal to the bottom of the scrotum, where it joins the dartos; the internal one is firmly attached to the os pubis and sheath of the Rectus muscle; some fibres, moreover, are reflected from the Internal oblique on to the front of the gubernaculum. Up to the fifth month, the testis is situated in the lumbar region, covered in front and at the sides by peritoneum, and supported in its position by a fold of that membrane, called the *mesorchium*; between the fifth and sixth months the testis descends to the iliac fossa, the gubernaculum at the same time becoming shortened; during the seventh month, it enters the internal abdominal ring, a small pouch of peritoneum (*processus vaginalis*) preceding the testis in its course through the canal. By the end of the eighth month, the testis has descended into the scrotum, carrying down with it a lengthened pouch of peritoneum, which communicates by its upper extremity with the peritoneal cavity. Just before birth, the upper part of this pouch usually becomes closed, and this obliteration extends gradually downwards to within a short distance of the testis. The process of peritoneum surrounding the testis, which is now entirely cut off from the general peritoneal cavity, constitutes the *tunica vaginalis.**

Mr. Curling believes that the descent of the testis is effected by means of the muscular fibres of the gubernaculum; those fibres which proceed from Poupart's ligament and the Obliquus internus are said to guide the organ into the inguinal canal; those attached to the pubis draw it below the external abdominal ring; and those attached to the bottom of the scrotum complete its descent. During the descent of the organ these muscular fibres become gradually everted, forming a muscular layer, which becomes placed external to the process of the peritoneum, surrounding the gland and spermatic cord, and constitutes the Cremaster. In the female, a small cord, corresponding to the gubernaculum in the male, descends to the inguinal region, and ultimately forms the round ligament of the uterus. A pouch of peritoneum accompanies it along the inguinal canal, analogous to the processus vaginalis in the male; it is called the *canal of Nuck.*

* The obliteration of the process of peritoneum which accompanies the cord, and is hence called the *funicular process*, is often incomplete. For an account of the various conditions produced by such incomplete obliteration (which are of great importance in the pathological anatomy of Inguinal Hernia), the student is referred to the Essay on Hernia, by Mr. Birkett, in 'A System of Surgery,' edited by T. Holmes, vol. iv.

Female Organs of Generation.

THE external Organs of Generation in the female, are the mons Veneris, the labia majora and minora, the clitoris, the meatus urinarius, and the orifice of the vagina. The term 'vulva' or 'pudendum,' as generally applied, includes all these parts.

The *mons Veneris* is the rounded eminence in front of the pubes, formed by a collection of fatty tissue beneath the integument. It surmounts the vulva, and is covered with hair at the time of puberty.

374.—The Vulva. External Female Organs of Generation.

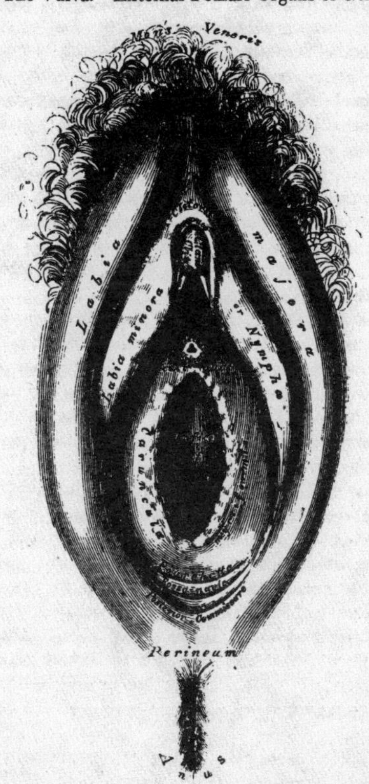

The *labia majora* are two prominent longitudinal cutaneous folds, extending downwards from the mons Veneris to the anterior boundary of the perinæum, and enclosing an elliptical fissure, the common urino-sexual opening. Each labium is formed externally of integument, covered with hair; internally, of mucous membrane, which is continuous with the genito-urinary mucous tract; and between the two, of a considerable quantity of areolar tissue, fat, and a tissue resembling the dartos of the scrotum, besides vessels, nerves, and glands. The labia are thicker in front than behind, and joined together at each extremity, forming the anterior and posterior commissures. The interval left between the posterior commissure and the margin of the anus is about an inch in length, and constitutes the perinæum. Just within the posterior commissure is a small transverse fold, the *frænulum pudendi* or *fourchette*, which is commonly ruptured in the first parturition, and the space between it and the commissure is called the *fossa navicularis*. The labia are analogous to the scrotum in the male.

The *labia minora* or *nymphæ* are two small folds of mucous membrane, situated within the labia majora, extending from the clitoris obliquely downwards and outwards for about an inch and a half on each side of the orifice of the vagina, on the sides of which they are lost. They are continuous externally with the labia majora, internally with the inner surface of the vagina. As they converge towards the clitoris in front, each labium divides into two folds, which surround the glans clitoridis, the superior folds uniting to form the præputium clitoridis, the inferior folds being attached to the glans, and forming the frænum. The nymphæ are composed of mucous membrane, covered by a thin epithelial layer. They contain a plexus of vessels in their interior, and are provided with numerous large mucous crypts which secrete abundance of sebaceous matter.

The *clitoris* is an erectile structure, analogous to the corpora cavernosa of the penis. It is situated beneath the anterior commissure, partially hidden between the anterior extremities of the labia minora. It is an elongated organ, connected to the rami of the pubes and ischia on each side by two crura; the body is short, and concealed beneath the labia; the free extremity, or glans clitoridis, is a small rounded tubercle, consisting of spongy erectile tissue, and highly sensitive. The clitoris consists of two corpora cavernosa, composed of erectile tissue enclosed in a dense layer of fibrous membrane, united together along their inner surfaces by an incomplete fibrous pectiniform septum. It is provided, like the penis, with a suspensory ligament, and with two small muscles, the Erectores clitoridis, which are inserted into the crura of the corpora cavernosa.

Between the clitoris, and the entrance of the vagina, is a triangular smooth surface, bounded on each side by the nymphæ: this is the vestibule.

The orifice of the urethra (*meatus urinarius*), is situated at the back part of the vestibule, about an inch below the clitoris, and near the margin of the vagina, surrounded by a prominent elevation of the mucous membrane. Below the meatus urinarius, is the orifice of the vagina, an elliptical aperture, more or less closed in the virgin, by a membranous fold, the hymen.

The *hymen* is a thin semilunar fold of mucous membrane, stretched across the lower part of the orifice of the vagina; its concave margin being turned upwards towards the pubes. Sometimes this membrane forms a complete septum across the orifice of the vagina: a condition known as imperforate hymen. Occasionally, it forms a circular septum, perforated in the centre by a round opening; sometimes it is cribriform, or its free margin forms a membranous fringe, or it may be entirely absent. It may also persist after copulation. The hymen cannot, consequently, be considered as a test of virginity. Its rupture, or the rudimentary condition of the membrane above referred to, gives rise to those small rounded elevations which surround the opening of the vagina, the *carunculæ myrtiformes*.

Glands of Bartholine. On each side of the commencement of the vagina is a round or oblong body, of a reddish-yellow colour, and of the size of a horse-bean, analogous to Cowper's gland in the male. It is called the *gland of Bartholine.* Each gland opens by means of a long single duct, upon the inner side of the nymphæ, external to the hymen. Extending from the clitoris, along either side of the vestibule, and lying a little behind the nymphæ, are two large oblong masses, about an inch in length, consisting of a plexus of veins, enclosed in a thin layer of fibrous membrane. These bodies are narrow in front, rounded below, and are connected with the crura of the clitoris and rami of the pubes.

they are termed by Kobelt the *bulbi vestibuli*; and he considers them analogous to the bulb of the corpus spongiosum in the male. Immediately in front of these bodies is a smaller venous plexus, continuous with the bulbi vestibuli behind, and the glans clitoridis in front: it is called by Kobelt the *pars intermedia*, and is considered by him as analogous to that part of the body of the corpus spongiosum which immediately succeeds the bulb.

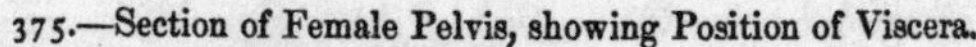
375.—Section of Female Pelvis, showing Position of Viscera.

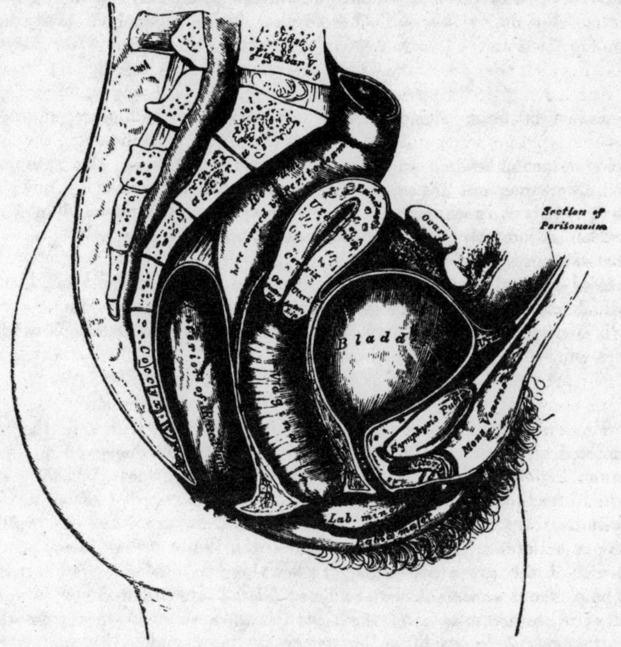

Bladder.

The bladder is situated at the anterior part of the pelvis. It is in relation, *in front*, with the os pubis; *behind*, with the uterus, some convolutions of the small intestine being interposed; its *base* lies in contact with the neck of the uterus, and with the anterior wall of the vagina. The bladder is said to be larger in the female than in the male, and is very broad in its transverse diameter.

Urethra.

The urethra is a narrow membranous canal, about an inch and a half in length, extending from the neck of the bladder to the meatus urinarius. It is placed beneath the symphysis pubis, imbedded in the anterior wall of the vagina; and its direction is obliquely downwards and forwards, its course being slightly curved, the concavity directed forwards and upwards. Its diameter, when undilated, is about a quarter of an inch. The urethra perforates the triangular ligament, precisely as in the male, and is surrounded by the muscular fibres of the Compressor urethræ.

Structure. The urethra consists of three coats: muscular, erectile, and mucous.

The *muscular coat* is continuous with that of the bladder; it extends the whole length of the tube, and consists of a thick stratum of circular fibres.

A thin layer of spongy, erectile tissue, intermixed with much elastic tissue, lies immediately beneath the mucous coat.

The *mucous coat* is pale, continuous, externally, with that of the vulva, and internally with that of the bladder. It is thrown into longitudinal folds, one of which, placed along the floor of the canal, resembles he verumontanum in the male urethra. It is lined by laminated epithelium, which becomes spheroidal at the bladder. Its external orifice is surrounded by a few mucous follicles.

The urethra, from not being surrounded by dense resisting structures, as in the male, admits of considerable dilatation, which enables the surgeon to remove with considerable facility calculi, or other foreign bodies, from the cavity of the bladder.

Rectum.

The rectum is more capacious, and less curved in the female, than in the male.

The *first portion* extends from the left sacro-iliac symphysis to the middle of the sacrum. Its connections are similar to those in the male.

The *second portion* extends to the tip of the coccyx. It is covered in front by the peritoneum, but only for a short distance, at its upper part, and is in relation with the posterior wall of the vagina.

The *third portion* curves backwards, from the vagina to the anus, leaving a space which corresponds on the surface of the body to the perinæum. Its extremity is surrounded by the Sphincter muscles, and its sides are supported by the Levatores ani.

The Vagina.

The vagina is a membranous canal, extending from the vulva to the uterus. It is situated in the cavity of the pelvis, behind the bladder, and in front of the rectum. Its direction is curved forwards and downwards, following at first the line of the axis of the cavity of the pelvis, and afterwards that of the outlet. It is cylindrical in shape, flattened from before backwards, and its walls are ordinarily in contact with each other. Its length is about four inches along its anterior wall, and between five and six inches along its posterior wall. It is constricted at its commencement, and becomes dilated near its uterine extremity; it surrounds the vaginal portion of the cervix uteri, a short distance from the os, and its attachment extends higher up on the posterior than on the anterior wall of the uterus.

Relations. Its *anterior surface* is concave, and in relation with the base of the bladder, and with the urethra. Its *posterior surface* is convex, and connected to the anterior wall of the rectum, for the lower three-fourths of its extent, the upper fourth being separated from that tube by the recto-uterine fold of peritoneum, which forms a *cul-de-sac* between the vagina and rectum. Its sides give attachment superiorly to the broad ligaments, and inferiorly to the Levatores ani muscles and recto-vesical fascia.

Structure. The vagina consists of an external, or muscular coat, a layer of erectile tissue, and an internal mucous lining.

The *muscular coat* consists of longitudinal fibres, which surround the vagina, and are continuous with the superficial muscular fibres of the uterus. The strongest fasciculi are those attached to the recto-vesical fascia on each side.

The *erectile tissue* is enclosed between two layers of fibrous membrane: it is more abundant at the lower than at the upper part of the vagina.

The *mucous membrane* is continuous, above, with that lining the uterus, and below, with the integument covering the labia majora. Its inner surface presents, along the anterior and posterior walls, a longitudinal ridge, or raphe, called the *columns of the vagina*, and numerous transverse ridges, or rugæ, extending outwards from the raphe on each side. These rugæ are most distinct near the orifice of the vagina, especially in females before parturition. They indicate its adaptation for dilatation, and are calculated to facilitate its enlargement during parturition. The mucous membrane is covered with conical and filiform papillæ, and provided with mucous glands and follicles, which are especially numerous in its upper part, and around the cervix uteri.

The Uterus.

The uterus is the organ of gestation, receiving the fecundated ovum in its cavity, retaining and supporting it during the development of the fœtus, and becoming the principal agent in its expulsion at the time of parturition.

In the virgin state it is pear-shaped, flattened from before backwards, and situated in the cavity of the pelvis, between the bladder and rectum; it is retained in its position by the round and broad ligaments on each side, and projects into the upper end of the vagina below. Its upper end, or base, is directed upwards and forwards; its lower end, or apex, downwards and backwards, in the line of the axis of the inlet of the pelvis. It therefore forms an angle with the vagina, since the direction of the vagina corresponds to the axis of the cavity and outlet of the pelvis. The uterus measures about three inches in length, two in breadth at its upper part, and an inch in thickness, and it weighs from an ounce to an ounce and a half.

The *fundus* is the upper broad extremity of the organ: it is convex, covered by peritoneum, and placed on a line below the level of the brim of the pelvis.

The *body* gradually narrows from the fundus to the neck. Its *anterior surface* is flattened, covered by peritoneum in the upper three-fourths of its extent, and separated from the bladder by some convolutions of the small intestine: the lower fourth is connected with the bladder. Its *posterior surface* is convex, covered by peritoneum throughout, and separated from the rectum by some convolutions of the intestine. Its *lateral margins* are concave, and give attachment to the Fallopian tube above, the round ligament below and in front of this, and the ligament of the ovary behind and below both of these structures.

The *cervix* is the lower rounded and constricted portion of the uterus: around its circumference is attached the upper end of the vagina, which extends upwards a greater distance behind than in front.

At the vaginal extremity of the uterus is a transverse aperture, the *os uteri*, bounded by two lips, the anterior of which is thick, the posterior narrow and long.

Ligaments. The ligaments of the uterus are six in number: two anterior, two posterior, and two lateral. They are formed of peritoneum.

The *two anterior ligaments* (*vesico-uterine*), are two semilunar folds, which pass between the neck of the uterus and the posterior surface of the bladder.

The *two posterior ligaments* (*recto-uterine*), pass between the sides of the uterus and rectum.

The *two lateral* or *broad ligaments* pass from the sides of the uterus to the lateral walls of the pelvis, forming a septum across the pelvis, which divides that cavity into two portions. In the anterior part are contained the bladder, urethra, and vagina; in the posterior part, the rectum.

The *cavity of the uterus* is small in comparison with the size of the organ: that portion of the cavity which corresponds to the body is triangular, flattened from before backwards, so that its walls are closely approximated, and having its base directed upwards towards the fundus. At each superior angle is a funnel-shaped cavity, which constitutes the remains of the division of the body of the uterus into two cornua; and at the bottom of each cavity is the minute orifice of the Fallopian tube. At the inferior angle of the uterine cavity is a small constricted opening, the internal orifice (*ostium internum*), which leads into the cavity of the cervix. The cavity in the cervix is somewhat cylindrical, flattened from before backwards, broader at the middle than at either extremity, and communicates, below, with the vagina. Each wall of the canal presents a longitudinal column, from which proceed a number of small oblique columns, giving the appearance of branches from the stem of a tree, and hence the name *arbor vitæ uterinus* applied to it. These folds usually become very indistinct after the first labour.

Structure. The uterus is composed of three coats: an external serous coat, a middle or muscular layer, and an internal mucous coat.

The *serous coat* is derived from the peritoneum; it invests the fundus and the whole of the posterior surface of the body of the uterus; but only the upper three-fourths of its anterior surface.

The *muscular coat* forms the chief bulk of the substance of the uterus. In the unimpregnated state, it is dense, firm, of a greyish colour, and cuts almost like cartilage. It is thick opposite the middle of the body and fundus, and thin at the orifices of the Fallopian tubes. It consists of bundles of unstriped muscular fibres, disposed in layers, intermixed with areolar tissue, blood-vessels, lymphatic vessels and nerves. In the impregnated state, the muscular tissue becomes more prominently developed, and is disposed in three layers: external, middle, and internal.

The external layer is placed beneath the peritoneum, disposed as a thin plane on the anterior and posterior surfaces. It consists of fibres, which pass transversely across the fundus, and, converging at each superior angle of the uterus, are continued on the Fallopian tubes, the round ligament, and ligament of the ovary; some passing at each side into the broad ligament, and others running backwards from the cervix into the recto-uterine ligaments.

The middle layer of fibres presents no regularity in its arrangement, being disposed longitudinally, obliquely, and transversely.

The internal, or deep layer, consists of circular fibres arranged in the form of two hollow cones, the apices of which surround the orifices of the Fallopian tubes, their bases intermingling with one another on the middle of the body of the uterus. At the cervix these fibres are disposed transversely.

The *mucous membrane* is thin, smooth, and closely adherent to the subjacent tissue. It is continuous, through the fimbriated extremity of the Fallopian tubes, with the peritoneum; and, through the os uteri, with the mucous membrane lining the vagina.

In the body of the uterus, it is smooth, soft, of a reddish colour, lined by columnar-ciliated epithelium, and presents, when viewed with a lens, the orifices of numerous tubular follicles arranged perpendicularly to the surface. They are of small size in the unimpregnated uterus, but shortly after impregnation they are enlarged, elongated, presenting a contorted or waved appearance towards their closed extremities, which occasionally dilate into two or three sacculated extremities. The circular orifices of these glands may be seen on the inner surface of the mucous membrane, many of which during the early period of pregnancy are surrounded by a whitish ring formed of epithelium which lines the follicles.

In the cervix, the mucous membrane between the rugæ and around the os uteri is provided with numerous mucous follicles, and glands. The small, transparent, vesicular elevations, so often found within the os and cervix uteri are due to closure of the mouths of these follicles, and their distension with their proper secretion. They were called the *ovula of Naboth.* The mucous membrane covering the lower half of the cervix presents numerous papillæ.

Vessels and Nerves. The *arteries* of the uterus are the uterine, from the internal iliac; and the ovarian, from the aorta. They are remarkable for their tortuous course in the substance of the organ, and for their frequent anastomoses. The *veins* are of large size, and correspond with the arteries. In the impregnated uterus these vessels are termed the *uterine sinuses,* consisting of the lining membrane of the veins adhering to the walls of canals channelled through the substance of the uterus. They terminate in the uterine plexuses. The *lymphatics* are of large size in the impregnated uterus, and terminate in the pelvic and lumbar glands. The nerves are derived from the inferior hypogastric and spermatic plexuses, and from the third and fourth sacral nerves.

The form, size, and situation of the uterus vary at different periods of life and under different circumstances.

In the fœtus, the uterus is contained in the abdominal cavity, projecting beyond the brim of the pelvis. The cervix is considerably larger than the body.

At puberty, the uterus is pyriform in shape, and weighs from eight to ten drachms. It has descended into the pelvis, the fundus being just below the level of the brim of this cavity. The arbor vitæ is distinct, and extends to the upper part of the cavity of the organ.

During, and after menstruation, the organ is enlarged, and more vascular, its surfaces rounder; the os externum is rounded, its labia swollen, and the lining membrane of the body thickened, softer, and of a darker colour.

During Pregnancy, the uterus increases so as to weigh from one pound and a half to three pounds. It becomes enormously enlarged, and projects into the hypogastric and lower part of the umbilical regions. This enlargement, which continues up to the sixth month of gestation, is partially due to increased development of pre-existing and new-formed muscular tissue. The round ligaments are enlarged, and the broad ligaments become encroached upon by the uterus making its way between their laminæ. The mucous membrane becomes more vascular, its mucous follicles and glands enlarged; the rugæ and folds in the canal of the cervix become obliterated; the blood and lymphatic vessels as well as the nerves, according to the researches of Dr. Lee, become greatly enlarged.

After Parturition, the uterus nearly regains its usual size, weighing from two to three ounces; but its cavity is larger than in the virgin state; the external orifice is more marked, and assumes a transverse direction; its edges present a fissured surface; its vessels are tortuous; and its muscular layers are more defined.

In old age, the uterus becomes atrophied, and paler and denser in texture; a more distinct constriction separates the body and cervix. The ostium internum, and, occasionally, the vaginal orifice, often become obliterated, and its labia almost entirely disappear.

Appendages of the Uterus.

The appendages of the uterus are, the Fallopian tubes, the ovaries and their ligaments, and the round ligaments. These structures, together with their nutrient vessels and nerves, and some scattered muscular fibres, are enclosed between the two folds of peritoneum, which constitute the broad ligaments; they are placed in the following order; in front is the round ligament; the Fallopian tube occupies the free margin of the broad ligament; the ovary and its ligament are behind and below the latter.

The *Fallopian Tubes*, or oviducts, convey the ova from the ovaries to the cavity of the uterus. They are two in number, one on each side, situated in the free margin of the broad ligament, extending from each superior angle of the uterus to the sides of the pelvis. Each tube is about four inches in length; its canal is exceedingly minute, and commences at the superior angle of the uterus by a minute orifice, the *ostium internum*, which will hardly admit a fine bristle; it continues narrow along the inner half of the tube, and then gradually widens into a trumpet-shaped extremity, which becomes contracted at its termination. This orifice is called the *ostium abdominale*, and communicates with the peritoneal cavity. Its margins are surrounded by a series of fringe-like processes, termed *fimbriæ*, and one of these processes is connected with the outer end of the ovary. To this part of the tube the name *fimbriated extremity* is applied; it is also called *morsus diaboli*, from the peculiar manner in which it embraces the surface of the ovary during sexual excitement.

Structure. The Fallopian tube consists of three coats, serous, muscular, and mucous.

The *external* or *serous coat* is derived from the peritoneum.

The *middle* or *muscular coat* consists of an external longitudinal and an internal or circular layer of muscular fibres continuous with those of the uterus.

The *internal* or *mucous coat* is continuous with the mucous lining of the uterus, and at the free extremity of the tube with the peritoneum. It is thrown into longitudinal folds in the outer part of the tube, which indicate its adaptation for dilatation, and is covered by columnar ciliated epithelium. This form of epithelium is also found on the inner and outer surfaces of the fimbriæ.

The *Ovaries* (*testes muliebres*, Galen) are analogous to the testes in the male.

They are oval-shaped bodies, of an elongated form, flattened from above downwards, situated one on each side of the uterus, in the posterior part of the broad ligament behind and below the Fallopian tubes. Each ovary is connected, by its anterior margin, to the broad ligament; by its inner extremity to the uterus by a proper ligament, the ligament of the ovary; and by its outer end to the fimbriated

376.—The Uterus and its Appendages. Anterior View.

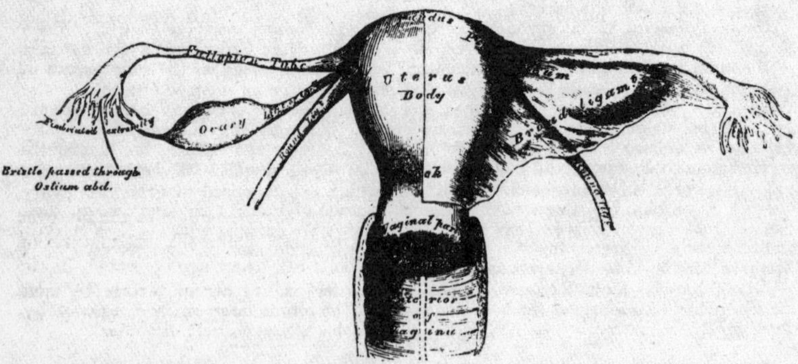

extremity of the Fallopian tube by a short ligamentous cord. The ovaries are of a whitish colour, and present either a smooth or puckered uneven surface. They are each about an inch and a half in length, three-quarters of an inch in width, and about a third of an inch thick; and weigh from one to two drachms. The surfaces and posterior convex border are free, the anterior straight border being attached to the broad ligament.

Structure. The ovary is invested by peritoneum, excepting along its anterior attached margin; beneath this, is the proper fibrous covering of the organ, the *tunica albuginea*, which is extremely dense and firm in structure, and encloses a peculiar soft fibrous tissue, or stroma, abundantly supplied with blood-vessels (fig. 377). Imbedded in the meshes of this tissue are numerous small, round, transparent vesicles, in various stages of development; they are the Graafian vesicles, the ovisacs containing the ova. In women who have not borne children, they vary in number from ten to fifteen or twenty; and in size from a pin's head to a pea; but Dr. Martin Barry has shown, that a large number of microscopic ovisacs exist in the parenchyma of the organ, few of which produce ova. These vesicles have thin, transparent walls, and are filled with a clear, colourless, albuminous fluid.

377.—Section of the Ovary of a Virgin, showing the Stroma and Graafian Vesicles.

378.—Section of the Graafian Vesicle. After Von Baer.

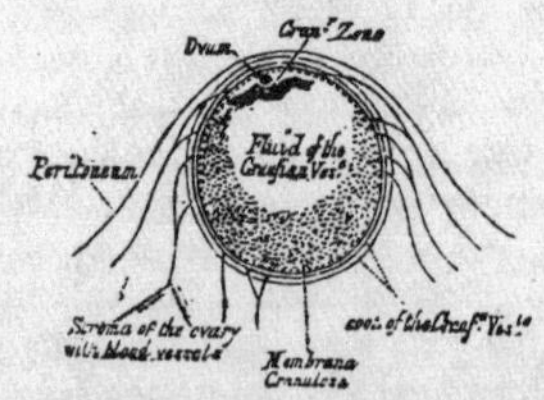

The *Graafian vesicles* are, during their early development, small, and deeply seated in the substance of the ovary; as they enlarge, they approach the surface; and, when mature, form small projections on the exterior of the ovary beneath the peritoneum. Each vesicle consists of an external fibro-vascular coat, connected with the surrounding stroma of the ovary by a network of blood-vessels; and an internal coat, named *ovi-capsule*, which is lined by a layer of nucleated cells, called the *membrana granulosa.* The fluid contained in the interior of the vesicles is transparent and albuminous, and in it is suspended the ovum.

The formation, development, and maturation of the Graafian vesicles and ova continue uninterruptedly from infancy to the end of the fruitful period of woman's life. Before puberty, the ovaries are small, the Graafian vesicles contained in them minute, and few in number; and few, probably, ever attain full development, but shrink and disappear, their ova being incapable of impregnation. At puberty, the ovaries enlarge, are more vascular, the Graafian vesicles are developed in greater abundance, and their ova capable of fecundation.

Discharge of the Ovum. The Graafian vesicles, after gradually approaching the surface of the ovary, burst; the ovum and fluid contents of the vesicles are liberated, and escape on the exterior of the ovary, passing from thence into the Fallopian tube, the fimbriated processes of which are supposed to grasp the ovary, the aperture of the tube being applied to the part corresponding to the matured and bursting vesicle. In the human subject and most mammalia, the maturation and discharge of the ova occur at regular periods only, and are indicated, in the mammalia, by the phenomena of heat or rut; and in the human female, by menstruation. Sexual desire is more intense in females at this period; and if the union of the sexes takes place, the ovum may be fecundated.

Corpus Luteum. Immediately after the rupture of a Graafian vesicle, and the escape of its ovum, the vesicle is filled with blood-tinged fluid; and in a short time the circumference of the vesicle is occupied by a firm, yellow substance, which is probably formed from plasma exuded from its walls. Dr. Lee believes that this yellow matter is deposited outside both the membranes of the follicle; Montgomery regards it as placed between the layers; while Kölliker considers it as a thickening of the inner layer of the outer coat of the follicle. The exudation is at first of a dark brown or brownish-red colour, but it soon becomes paler, and its consistence more dense.

For every follicle in the ovary from which an ovum is discharged, a corpus luteum will be found. But the characters it exhibits, and the changes produced in it, will be determined by the circumstance of the ovum being impregnated or not.

Although there is little doubt that corpora lutea exist in the ovaries after the escape of ova, independent of coitus or impregnation, it appears that the corpus luteum of pregnancy (true corpus luteum) possesses characters by which it may be distinguished from one formed in a follicle, from which an ovum has been discharged without subsequent impregnation (false corpus luteum).

The *true corpora lutea* are of large size, often as large as a mulberry; of a rounded form, and project from the surface of the ovary, the summit of the projection presenting a triangular depression or cicatrix, where the peritoneum appears to have been torn. They contain a small cavity in their centre during the early period of their formation, which becomes contracted, and exhibits a stellate cicatrix during the latter stages of pregnancy. Their vascularity, lobulated or puckered appearance, firm consistence, and yellow colour, are also characteristic marks of true corpora lutea.

False corpora lutea are of small size, do not project from the surface of the ovary, are angular in form, seldom present any cicatrix, contain no cavity in their centre; the material composing them is not lobulated, its consistence is usually soft, often resembling coagulated blood; the yellow matter exists in the form of a very thin layer, or more commonly is entirely wanting. False corpora lutea most frequently result from the effusion into the cavities of the Graafian vesicles of serum or blood, which subsequently undergoes various changes, and is ultimately removed. Dr. Lee states, that in the false corpora lutea the yellow substance is contained within, or attached to, the inner surface of the Graafian vesicle, and does not surround it, as is the case in the true corpora lutea.

In the fœtus, the ovaries are situated, like the testes, in the lumbar region, near the kidneys. They may be distinguished from those bodies at an early period by their elongated and flattened form, and by their position, which is at first oblique, and then nearly transverse. They gradually descend into the pelvis.

The *Ligament of the Ovary* is a rounded cord, which extends from each superior angle of the uterus to the inner extremity of the ovary; it consists of fibrous tissue, and a few muscular fibres derived from the uterus.

The *Round Ligaments* are two rounded cords, between four and five inches in length, situated between the layers of the broad ligament in front of and below the Fallopian tube. Commencing on each side at the superior angle of the uterus, this ligament passes forwards and outwards through the internal abdominal ring, along the inguinal canal to the labia majora, in which it becomes lost. The round ligament consists of areolar tissue, vessels, and nerves, besides a dense bundle of fibrous tissue, and muscular fibres prolonged from the uterus, enclosed in a duplicature of peritoneum, which, in the fœtus, is prolonged in the form of a tubular process for a short distance into the inguinal canal. This process is called the *canal of Nuck.* It is generally obliterated in the adult, but sometimes remains pervious even in advanced life. It is analogous to the peritoneal pouch which accompanies the descent of the testis.

Vessels and Nerves. The *arteries* of the ovaries and Fallopian tubes are the ovarian from the aorta. They anastomose with the termination of the uterine arteries, and enter the attached border of the ovary. The *veins* follow the course of the arteries; they form a plexus near the ovary, the *pampiniform plexus.* The *nerves* are derived from the spermatic plexus, the Fallopian tube receiving a branch from one of the uterine nerves.

Mammary Glands.

The *mammæ*, or breasts, are accessory glands of the generative system, which secrete the milk. They exist in the male as well as in the female; but in the former only in a rudimentary state, unless their growth is excited by peculiar circumstances. In the female, they are two large hemispherical eminences situated towards the lateral aspect of the pectoral region, corresponding to the interval between the third and sixth or seventh ribs, and extending from the side of the sternum to the axilla. Their weight and dimensions differ at different periods of life, and in different individuals. Before puberty they are of small size, but enlarge as the generative organs become more completely developed. They increase during pregnancy, and especially after delivery, and become atrophied in old age. The left mamma is generally a little larger than the right. Their base is nearly circular, flattened or slightly concave, and having their long diameter directed upwards and outwards towards the axilla; they are separated from the Pectoral muscles by a thin layer of superficial fascia. The outer surface of the mamma is convex, and presents, just below the centre, a small conical prominence, the nipple (*mammilla*). The surface of the nipple is dark-coloured, and surrounded by an areola having a coloured tint. In the virgin, the areola is of a delicate rosy hue; about the second month of impregnation, it enlarges, and acquires a darker tinge, which increases as pregnancy advances, becoming, in some cases, a dark brown or even black colour. This colour diminishes as soon as lactation is over, but is never entirely lost through life. These changes in the colour of the areola are of extreme importance in forming a conclusion in a case of suspected pregnancy.

The *nipple* is a cylindrical or conical eminence, capable of undergoing a sort of erection from mechanical excitement. It is of a pink or brownish hue, its surface wrinkled and provided with papillæ, and its summit perforated by numerous orifices, the apertures of the lactiferous ducts. Near the base of the nipple, and upon the surface of the areola, are numerous sebaceous glands, which become much enlarged during lactation, and present the appearance of small tubercles beneath the skin. These glands secrete a peculiar fatty substance, which serves as a protection to the integument of the nipple in the act of sucking. The nipple consists of numerous vessels, which form a kind of erectile tissue, intermixed with plain muscular fibres.

Structure. The mamma consists of gland-tissue; of fibrous tissue, connecting its lobes; and of fatty tissue in the intervals between the lobes. The mammary gland, when freed from cellular tissue and fat, is of a pale reddish colour, firm in texture, circular in form, flattened from before backwards, thicker in the centre than at the circumference, and presenting several inequalities on its surface, especially in front. It consists of numerous lobes, and these are composed of lobules, connected together by areolar tissue, blood-vessels and ducts. The smallest lobules consist of a cluster of rounded vesicles, which open into the smallest branches of the lactiferous ducts; these ducts uniting, form larger ducts, which terminate in a single canal, corresponding with one of the chief subdivisions of the gland. The number of excretory ducts varies from fifteen to twenty: they are termed the *tubuli lactiferi, galactophori.* They converge towards the areola, beneath which they form dilatations, or ampullæ, which serve as reservoirs for the milk, and, at the base of the nipple, become contracted, and pursue a straight course to its summit, perforating it by separate orifices considerably narrower than the ducts themselves. The ducts are composed of areolar tissue, with longitudinal and transverse elastic fibres, and longitudinal muscular fibres: their mucous lining is continuous, at the point of the nipple, with the integument: the epithelium is of the tesselated or scaly variety.

The *fibrous tissue* invests the entire surface of the breast, and sends down septa between its lobes, connecting them together.

The *fatty tissue* surrounds the surface of the gland, and occupies the intervals between its lobes and lobules. It usually exists in considerable abundance, and determines the form and size of the gland. There is no fat immediately beneath the areola and nipple.

Vessels and Nerves. The *arteries* supplying the mammæ are derived from the thoracic branches of the axillary, the intercostals, and internal mammary. The *veins* describe an anastomotic circle round the base of the nipple, called by Haller the *circulus venosus.* From this, large branches transmit the blood to the circumference of the gland, and end in the axillary and internal mammary veins. The *lymphatics* run along the lower border of the Pectoralis major to the axillary glands. The *nerves* are derived from the anterior and lateral cutaneous nerves of the thorax.

The Surgical Anatomy of Inguinal Hernia.

Dissection (fig. 379). For the dissection of the parts concerned in inguinal hernia, a male subject, free from fat, should always be selected. The body should be placed in the supine position, the abdomen and pelvis raised by means of blocks placed beneath them, and the lower extremities rotated outwards, so as to make the parts as tense as possible. If the abdominal walls are flaccid, the cavity of the abdomen should be inflated by an aperture through the umbilicus. An incision should be made along the middle line, from the umbilicus to the pubes, and continued along the front of the scrotum; and a second incision, from the anterior superior spine of the ilium to just below the umbilicus. These incisions should divide the integument; and the triangular-shaped flap included between them should be reflected downwards and outwards, when the superficial fascia will be exposed.

The *superficial fascia* in this region consists of two layers, between which are found the superficial vessels and nerves, and the inguinal lymphatic glands.

The superficial layer is thick, areolar in texture, containing adipose tissue in its meshes, the quantity of which varies in different subjects. Below, it passes over Poupart's ligament, and is continuous with the outer layer of the superficial fascia of the thigh. This fascia is continued as a tubular prolongation around the outer surface of the cord and testis. In this situation, it changes its character; it becomes thin, destitute of adipose tissue, and of a pale reddish colour, and assists in forming the dartos. From the scrotum, it may be traced backwards to be continuous with the superficial fascia of the perinæum. This layer should be removed, by dividing it across in the same direction as the external incisions, and reflecting it downwards and outwards, when the following vessels and nerves will be exposed :—

The superficial epigastric, superficial circumflex iliac, and external pudic vessels; the terminal filaments of the ilio-hypogastric and ilio-inguinal nerves; and the upper chain of inguinal lymphatic glands.

The *superficial epigastric artery* crosses Poupart's ligament, and ascends obliquely towards the umbilicus, lying midway between the spine of the ilium and the pubes. It supplies the integument, and anastomoses with the deep epigastric. This vessel is a branch of the common femoral artery, and pierces the fascia lata, below Poupart's ligament. Its accompanying vein empties itself into the internal saphenous, after having pierced the cribriform fascia.

The *superficial circumflex iliac artery* passes outwards towards the crest of the ilium.

The *superficial external pudic artery* passes transversely inwards across the spermatic cord, and supplies the integument of the hypogastric region, and of the penis and scrotum. This vessel is usually divided in the first incision made in the operation for inguinal hernia, and occasionally requires the application of a ligature.

The veins accompanying these superficial vessels are usually much larger than the arteries: they terminate in the internal saphenous vein.

Lymphatic vessels are found, taking the same course as the blood-vessels: they return the lymph from the superficial structures in the lower part of the abdomen, the scrotum, penis, and external surface of the buttock, and terminate in a small chain of lymphatic glands, three or four in number, which lie on a level with Poupart's ligament.

Nerves. The terminal branch of the ilio-inguinal nerve emerges at the external abdominal ring: and the hypogastric branch of the ilio-hypogastric nerve perforates the aponeurosis of the external oblique, above and to the outer side of the external ring.

The *deep layer of superficial fascia* should be divided across in the same direction as the external incisions, separated from the aponeurosis of the External oblique, to which it is connected by delicate areolar tissue, and reflected downwards and outwards. It is thin, aponeurotic in structure, and of considerable strength. It is intimately adherent, in the middle line, to the linea alba, and below, to the whole length of Poupart's ligament and the upper part of the fascia lata. It forms a thin tubular prolongation round the outer surface of the cord, which blends with the superficial layer, and is continuous with the dartos of the scrotum. From the back of the scrotum, the conjoined layers may be traced into the perinæum, where they are continuous with the deep layer of the superficial fascia in that region, which is attached, behind, to the triangular ligament, and on each side, to the ramus of the pubes and ischium. The connections of this fascia serve to explain the course taken by the urine in extravasation of that fluid from rupture of the urethra: passing forwards from the perinæum into the scrotum, it ascends on to the abdomen, but is prevented extending into the thighs by the attachment of the fascia to the ramus of the pubes and ischium, on each side, and to Poupart's ligament in front, and is prevented from passing on to the buttock by the posterior connections of the perinæal fascia.

379.—Inguinal Hernia. Superficial Dissection.

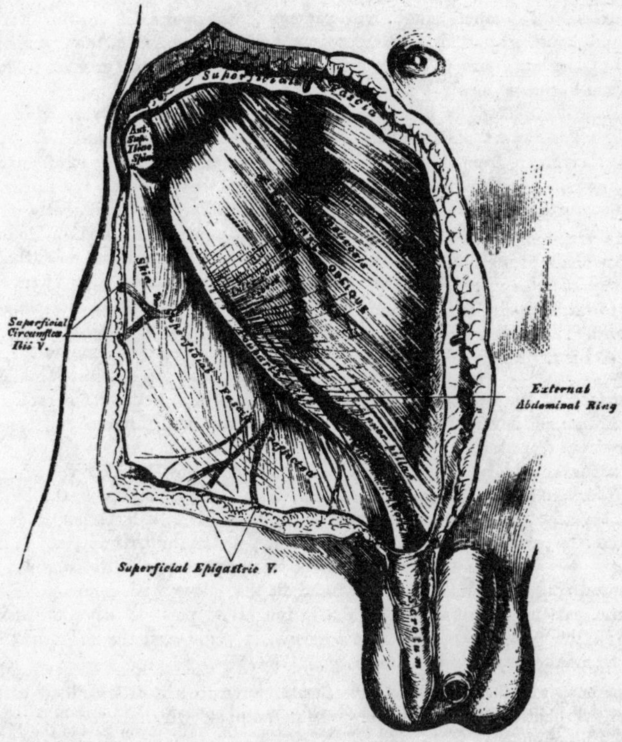

The *aponeurosis of the External oblique muscle* is exposed on the removal of this fascia. It is a thin, strong, membranous aponeurosis, the fibres of which are directed obliquely downwards and inwards. It is attached to the anterior superior spinous process of the ilium, the spine of the pubes, the pectineal line, front of the pubes, and linea alba. That portion of the aponeurosis which extends from the anterior superior spine of the ilium, to the spine of the pubes, is termed Poupart's ligament, or the crural arch; and that portion which is inserted into the pectineal line, is termed Gimbernat's ligament.

Just above and to the outer side of the crest of the pubes, a triangular interval is seen in the aponeurosis of the External oblique, called the *external abdominal ring*, which transmits the spermatic cord in the male, and the round ligament in the female. This aperture is oblique in direction, somewhat triangular in form, and corresponds with the course of the fibres of the aponeurosis. It usually measures from base to apex about an inch, and transversely about half an inch. It is bounded below by the crest of the os pubis; above, by a series of curved fibres, the *intercolumnar*, which pass across the upper angle of the ring so as to increase its strength; and on either side, by the free borders of the aponeurosis, which are called the *columns* or *pillars of the ring*.

The *external pillar*, which, at the same time, is inferior from the obliquity of its direction, is the stronger; it is formed by that portion of Poupart's ligament which is inserted into the spine of the pubes; it is curved round the spermatic cord, so as to form a kind of groove, upon which the cord rests.

The *internal* or *superior pillar* is a broad, thin, flat band, which interlaces with its fellow of the opposite side, in front of the symphysis pubis, that of the right side being superficial.

The external abdominal ring gives passage to the spermatic cord in the male, and round ligament in the female; it is much larger in men than women, on account of the large size of the spermatic cord, and hence the greater frequency of inguinal hernia in men.

The *intercolumnar fibres* are a series of curved tendinous fibres, which arch across the lower part of the aponeurosis of the External oblique. They have received their name from stretching across between the two pillars of the external ring; they increase the strength of the membrane which bounds the upper part of this aperture, and prevent the divergence of the pillars from one another. They are thickest below, where they are connected to the outer third of Poupart's ligament, and are inserted into the linea alba; describing a curve, with the convexity downwards. They are much thicker and stronger at the outer angle of the external ring than internally, and are more strongly developed in the male than in the female. These fibres are continuous with a thin fascia, which is closely connected to the margins of the external ring, and has received the name of the *intercolumnar* or *external spermatic fascia*; it forms a tubular prolongation around the outer surface of the cord and testis, and encloses them in a distinct sheath. The sac of an inguinal hernia, in passing through the external abdominal ring, receives an investment from the intercolumnar fascia.

The finger should be introduced a short distance into the external ring, and then, if the limb is extended and rotated outwards, the aponeurosis of the External oblique, together with the iliac portion of the fascia lata, will be felt to become tense, and the external ring much contracted; if the limb is, on the contrary, flexed upon the pelvis and rotated inwards, this aponeurosis will become lax, and the external ring sufficiently enlarged to admit the finger with comparative ease; hence the patient should always be put in the latter position when the taxis is applied for the reduction of an inguinal hernia, in order that the abdominal walls may be as much relaxed as possible.

The aponeurosis of the External oblique should be removed by dividing it across in the same direction as the external incisions, and reflecting it outwards; great care is requisite in separating it from the aponeurosis of the muscle beneath. The lower part of the Internal oblique and the Cremaster are then exposed, together with the inguinal canal, which contains the spermatic cord (fig. 380). The mode of insertion of Poupart's and Gimbernat's ligaments into the pubes should also be examined.

Poupart's ligament, or the crural arch, extends from the anterior superior spine of the ilium to the spine of the pubes. It is also attached to the pectineal line to the extent of about an inch, forming Gimbernat's ligament. Its general direction is curved towards the thigh, where it is continuous with the fascia lata. Its outer half is rounded, oblique in its direction, and continuous with the iliac fascia. Its inner half gradually widens at its attachment to the pubes, is more horizontal in direction, and lies beneath the spermatic cord.

Gimbernat's ligament is that portion of the aponeurosis of the External oblique which is inserted into the pectineal line; it is thin, membranous in structure, triangular in shape, the base directed outwards, and passes upwards and backwards beneath the spermatic cord, from the spine of the os pubis to the pectineal line, to the extent of about half an inch.

The *triangular ligament* is a band of tendinous fibres, of a triangular shape, which is continued from Poupart's ligament at its attachment to the pectineal line upwards and inwards, behind the inner pillar of the external ring to the linea alba.

The *Internal oblique Muscle* has been described (p. 243). The part which is now exposed is partly muscular and partly tendinous in structure. Those fibres which arise from the outer part of Poupart's ligament are thin, pale in colour, curve downwards, and terminate in an aponeurosis, which passes in front of the Rectus and Pyramidalis muscles, to be inserted into the crest of the os pubis and

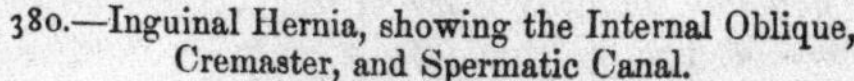

380.—Inguinal Hernia, showing the Internal Oblique, Cremaster, and Spermatic Canal.

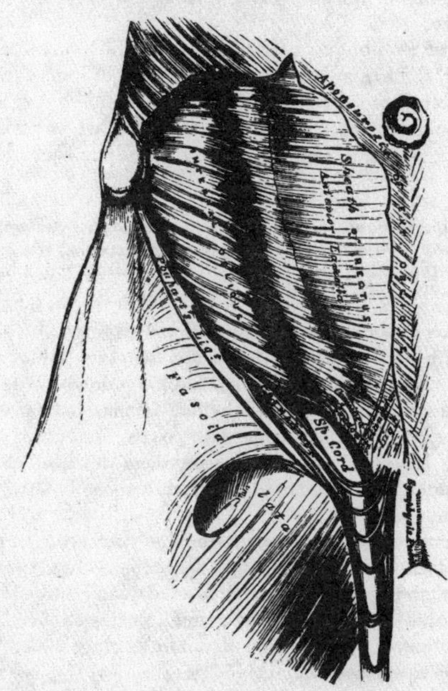

pectineal line, to the extent of half an inch, in common with that of the Transversalis muscle, forming by their junction the conjoined tendon. This tendon is placed immediately behind Gimbernat's ligament and the external abdominal ring, and serves to strengthen what would otherwise be a very weak point in the abdominal wall. When a direct inguinal hernia passes through the external ring, the conjoined tendon usually forms one of its coverings.

The *Cremaster* is a slender muscular fasciculus, which arises from the middle of

Poupart's ligament at the inner side of the Internal oblique, being connected with that muscle, and also occasionally with the Transversalis. It passes along the outer side of the spermatic cord, descends with it through the external ring upon the front and sides of the cord, and forms a series of loops, which differ in thickness and length in different subjects. Those at the upper part of the cord are exceedingly short, but they become in succession longer and longer, the longest reaching down as low as the testicle, where a few are inserted into the tunica vaginalis. These loops are united together by areolar tissue, and form a thin covering over the cord, the *fascia cremasterica.* The fibres ascend along the inner side of the cord, and are inserted by a small pointed tendon, into the crest of the os pubis and front of the sheath of the Rectus muscle.

It will be observed, that the origin and insertion of the Cremaster is precisely similar to that of the lower fibres of the Internal oblique. This fact affords an easy explanation of the manner in which the testicle and cord are invested by this muscle. At an early period of fœtal life, the testis is placed at the lower and back part of the abdominal cavity, but, during its descent towards the scrotum, which takes place before birth, it passes beneath the arched border of the Internal oblique. In its passage beneath this muscle some fibres are derived from its lower part, which accompany the testicle and cord into the scrotum.

It occasionally happens that the loops of the Cremaster surround the cord, some lying behind as well as in front. It is probable that, under these circumstances, the testis, in its descent, passed through instead of beneath the fibres of the Internal oblique.

In the descent of an oblique inguinal hernia, which takes the same course as the spermatic cord, the Cremaster muscle forms one of its coverings. This muscle becomes largely developed in cases of hydrocele and large old scrotal herniæ. No such muscle exists in the female, but an analogous structure is developed in those cases where an oblique inguinal hernia descends beneath the margin of the Internal oblique.

The Internal oblique should be detached from Poupart's ligament, separated from the Transversalis to the same extent as in the previous incisions, and reflected inwards on to the sheath of the Rectus (fig. 384). The circumflex iliac vessels, which lie between these two muscles, form a valuable guide to their separation.

The *Transversalis muscle* has been previously described (p. 246). Its lower part is partly fleshy and partly tendinous in structure; this portion arises from the outer third of Poupart's ligament, and, arching downwards and inwards over the cord, terminates in an aponeurosis, which is inserted into the linea alba, the crest of the pubes, and the pectineal line to the extent of an inch, forming, together with the Internal oblique, the conjoined tendon. Between the lower border of this muscle and Poupart's ligament, a space is left in which is seen the fascia transversalis.

The *inguinal*, or *spermatic canal*, contains the spermatic cord in the male, and the round ligament in the female. It is an oblique canal, about an inch and a half in length, directed downwards and inwards, and placed parallel with, and a little above, Poupart's ligament. It communicates, above, with the cavity of the abdomen, by means of the internal abdominal ring, which is the point where the cord enters the spermatic canal; and terminates, below, at the external ring. It is bounded, in front, by the integument and superficial fascia, by the aponeurosis of the External oblique throughout its whole length, and by the Internal oblique for its outer third; behind, by the conjoined tendon of the Internal oblique and Transversalis, the triangular ligament, transversalis fascia, and the subperitoneal fat and peritoneum; above, by the arched fibres of the Internal oblique and Transversalis; below, by the union of the fascia transversalis with Poupart's ligament. That form of protrusion in which the intestine follows the course of the spermatic cord along the spermatic canal, is called *oblique inguinal hernia.*

The *fascia transversalis* is a thin aponeurotic membrane, which lies between the inner surface of the Transversalis muscle and the peritoneum. It forms part of the general layer of fascia which lines the interior of the abdominal and pelvic cavities, and is directly continuous with the iliac and pelvic fasciæ.

In the inguinal region, the transversalis fascia is thick and dense in structure, and joined by fibres from the aponeurosis of the Transversalis; but it becomes thin and cellular as it ascends to the Diaphragm. Below, it has the following attachments: external to the femoral vessels, it is connected to the posterior margin of Poupart's ligament, and is there continuous with the iliac fascia. Internal to the vessels, it is thin, and attached to the pubes and pectineal line, behind the conjoined tendon, with which it is united; and, corresponding to the point where the femoral vessels pass into the thigh, this fascia descends in front of them, forming the anterior wall of the crural sheath.

381.—Inguinal Hernia, showing the Transversalis Muscle, the Transversalis Fascia, and the Internal Abdominal Ring.

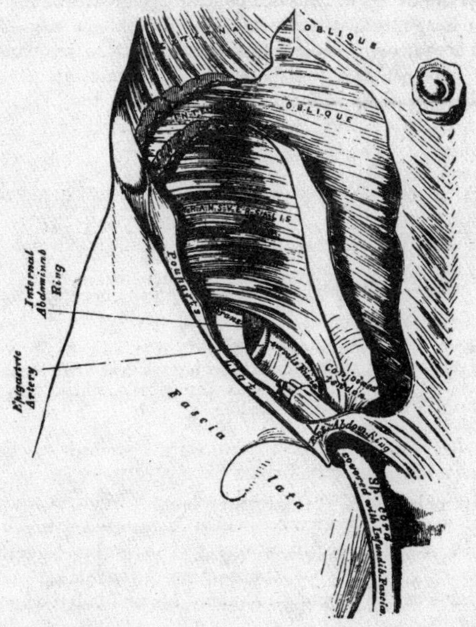

The *internal abdominal ring* is situated in the transversalis fascia, midway between the anterior superior spine of the ilium and the spine of the pubes, and about half an inch above Poupart's ligament. It is of an oval form, the extremities of the oval directed upwards and downwards, varies in size in different subjects, and is much larger in the male than in the female. It is bounded, above, by the arched fibres of the Transversalis muscle, and internally, by the epigastric vessels. It transmits the spermatic cord in the male, and the round ligament in the female, and from its circumference a thin, funnel-shaped membrane, *the infundibuliform fascia*, is continued round the cord and testis, enclosing them in a distinct pouch. When the sac of an oblique inguinal hernia passes through the internal ring, the infundibuliform process of the transversalis fascia forms one of its coverings.

Between the transversalis fascia and the peritoneum, is a quantity of loose areolar tissue. In some subjects it is of considerable thickness, and loaded with adipose tissue. Opposite the internal ring, it is continued round the surface of the cord, forming a loose sheath for it.

The *epigastric artery* bears a very important relation to the internal abdominal ring. This vessel lies between the transversalis fascia and peritoneum, and passes obliquely upwards and inwards, from its origin from the external iliac, to the margin of the sheath of the Rectus muscle. In this course, it lies along the lower and inner margin of the internal ring, and beneath the commencement of the spermatic cord, the vas deferens curving round it as it passes from the ring into the pelvis.

The *peritoneum*, corresponding to the inner surface of the internal ring, presents a well-marked depression, the depth of which varies in different subjects. A thin fibrous band is continued from it along the front of the cord, for a variable distance, and becomes ultimately lost. This is the remains of the pouch of peritoneum which, in the fœtus, accompanies the cord and testis into the scrotum, the obliteration of which commences soon after birth. In some cases, the fibrous band can only be traced a short distance; but occasionally, it may be followed, as a fine cord, as far as the upper end of the tunica vaginalis. Sometimes the tube of peritoneum is only closed at intervals, and presents a sacculated appearance; or a single pouch may extend along the whole length of the cord, which may be closed above; or the pouch may be directly continuous with the peritoneum by an opening at its upper part.

Inguinal Hernia.

Inguinal hernia is that form of protrusion which makes its way through the abdomen in the inguinal region.

There are two principal varieties of inguinal hernia: external or oblique, and internal or direct.

External or *oblique inguinal hernia*, the more frequent of the two, takes the same course as the spermatic cord. It is called *external*, from the neck of the sac being on the outer or iliac side of the epigastric artery.

Internal, or *direct inguinal hernia* does not follow the same course as the cord, but protrudes through the abdominal wall on the inner or pubic side of the epigastric artery.

Oblique Inguinal Hernia.

In oblique inguinal hernia, the intestine escapes from the abdominal cavity at the internal ring, pushing before it a pouch of peritoneum, which forms the hernial sac. As it enters the inguinal canal, it receives an investment from the subserous areolar tissue, and is enclosed in the infundibuliform process of the transversalis fascia. In passing along the inguinal canal, it displaces upwards the arched fibres of the Transversalis and Internal oblique muscles, and is surrounded by the fibres of the Cremaster. It then passes along the front of the cord, and escapes from the inguinal canal at the external ring, receiving an investment from the intercolumnar fascia. Lastly, it descends into the scrotum, receiving coverings from the superficial fascia and the integument.

The coverings of this form of hernia, after it has passed through the external ring, are, from without inwards, the integument, superficial fascia, intercolumnar fascia, Cremaster muscle, infundibuliform fascia, subserous cellular tissue, and peritoneum.

This form of hernia lies in front of the vessels of the spermatic cord, and seldom extends below the testis, on account of the intimate adhesion of the coverings of the cord to the tunica vaginalis.

The *seat of stricture* in oblique inguinal hernia, is either at the external ring, in the inguinal canal, caused by the fibres of the Internal oblique or Transversalis or at the internal ring, more frequently in the latter situation. If it is situated at the external ring, the division of a few fibres at one point of its circumference, is all that is necessary for the replacement of the hernia. If in the inguinal canal, or at the internal ring, it will be necessary to divide the aponeurosis of the External oblique so as to lay open the inguinal canal. In dividing the stricture, the direction of the incision should be directly upwards.

When the intestine passes along the spermatic canal, and escapes from the external ring into the scrotum, it is called *complete oblique inguinal*, or *scrotal hernia*. If the intestine does not escape from the external ring, but is retained in the inguinal canal, it is called *incomplete inguinal hernia* or *bubonocele*. In each of these cases, the coverings which invest it will depend upon the extent to which it descends in the inguinal canal.

There are two other varieties of oblique inguinal hernia: the congenital, and infantile.

Congenital hernia is liable to occur in those cases where the pouch of peritoneum which accompanies the cord and testis in its descent in the fœtus remains unclosed, and communicates directly with the peritoneum. The intestine descends along this pouch into the cavity of the tunica vaginalis, and lies in contact with the testis. This form of hernia has no proper sac, being contained within the tunica vaginalis.

In Infantile hernia, the hernial sac descends along the inguinal canal into the scrotum, behind the pouch of peritoneum which accompanies the cord and testis into the same part. The abdominal aperture of this pouch is closed, but the portion contained in the inguinal canal remains unobliterated. The hernial sac is consequently invested, more or less completely, by the posterior layer of the tunica vaginalis, from which it is separated by a little loose areolar tissue: so that in operating npon this variety of hernia, three layers of peritoneum would require division; the first and second being the layers of the tunica vaginalis, the third the anterior layer of the hernial sac.

Direct Inguinal Hernia.

In direct inguinal hernia, the protrusion makes its way through some part of the abdominal wall internal to the epigastric artery, and passes directly through the abdominal parietes and external ring. At the lower part of the abdominal wall is a triangular space (Hesselbach's triangle), bounded, externally, by the epigastric artery; internally, by the margin of the Rectus muscle; below, by Poupart's ligament. The conjoined tendon is stretched across the inner two-thirds of this space, the remaining portion of the space being filled in by the transversalis fascia.

In some cases the hernial protrusion escapes from the abdomen on the outer side of the conjoined tendon, pushing before it the peritoneum, the subserous cellular tissue, and the transversalis fascia. It then enters the inguinal canal, passing along nearly its whole length, and finally emerges from the external ring, receiving an investment from the intercolumnar fascia. The coverings of this form of hernia are precisely similar to those investing the oblique form.

In other cases, and this is the more frequent variety, the intestine is either forced through the fibres of the conjoined tendon, or the tendon is gradually distended in front of it, so as to form a complete investment for it. The intestine then enters the lower end of the inguinal canal, escapes at the external ring lying on the inner side of the cord, and receives additional coverings from the superficial fascia and the integument. This form of hernia has the same coverings as the oblique variety, excepting that the conjoined tendon is substituted for the Cremaster, and the infundibulum fascia is replaced by a part of the general fascia transversalis.

The *seat of stricture* in both varieties of direct hernia is most frequently at the neck of the sac, or at the external ring. In that form of hernia which perforates the conjoined tendon, it not unfrequently occurs at the edges of the fissure through which the gut passes. In dividing the stricture, the incision should in all cases be directed upwards.

If the hernial protrusion passes into the inguinal canal, but does not escape from the external abdominal ring, it forms what is called *incomplete direct hernia.* This form of hernia is usually of small size, and in corpulent persons very difficult of detection.

Direct inguinal hernia is of much less frequent occurrence than the oblique, their comparative frequency being, according to Cloquet, as one to five. It occurs far more frequently in men than women, on account of the larger size of the external ring in the former sex. It differs from the oblique in its smaller size and globular form, dependent most probably on the resistance offered to its progress by the transversalis fascia and conjoined tendon. It differs also in its position, being placed over the pubes, and not in the course of the inguinal canal. The epigastric artery runs on the outer or iliac side of the neck of the sac, and the spermatic cord along its external and posterior side, not directly behind it, as in oblique inguinal hernia.

SURGICAL ANATOMY OF FEMORAL HERNIA.

The dissection of the parts comprised in the anatomy of femoral hernia should be performed, if possible, upon a female subject free from fat. The subject should lie upon its back; a block is first placed under the pelvis, the thigh everted, and the knee slightly bent, and retained in this position. An incision should then be made from the anterior superior spinous process of the ilium along Poupart's ligament to the symphysis pubis; a second incision should be carried transversely across the thigh about six inches beneath the preceding; and these are to be connected together by a vertical one carried along the inner side of the thigh. These several incisions should divide merely the integument; this is to be reflected outwards, when the superficial fascia will be exposed.

The *superficial fascia* at the upper part of the thigh, consists of two layers, between which are found the cutaneous vessels and nerves, and numerous lymphatic glands.

The superficial layer is a thick and dense cellulo-fibrous membrane, in the meshes of which is found a considerable amount of adipose tissue, varying in quantity in different subjects; this layer may be traced upwards over Poupart's ligament to be continuous with the superficial fascia of the abdomen; whilst below, and on the inner and outer sides of the limb, it is continuous with the superficial fascia covering the rest of the thigh.

This layer should be detached by dividing it across in the same direction as the external incisions; its removal will be facilitated by commencing at the lower and inner angle of the space, detaching it at first from the front of the internal saphenous vein, and dissecting it off from the anterior surface of that vessel and its branches; it should then be reflected outwards, in the same manner as the integument. The cutaneous vessels and nerves, and superficial inguinal glands are then exposed, lying upon the deep layer of the superficial fascia. These are the internal saphenous vein, and the superficial epigastric, superficial circumflexa ilii, and superficial external pudic vessels, as well as numerous lymphatics ascending with the saphenous vein to the inguinal glands.

The *internal saphenous vein* is a vessel of considerable size, which ascends obliquely upwards along the inner side of the thigh, below Poupart's ligament. It passes through the saphenous opening in the fascia lata to terminate in the femoral vein. This vessel is accompanied by numerous lymphatics, which return the lymph from the dorsum of the foot and inner side of the leg and thigh; they terminate in the inguinal glands, which surround the saphenous opening. Diverging from the same point are the superficial epigastric vessels, which run across Poupart's ligament, obliquely upwards and inwards, to the lower part of the abdomen; the superficial circumflexa ilii vessels, which pass obliquely outwards along Poupart's ligament to the crest of the ilium; and the superficial external pudic vessels, which pass inwards to the perinæum and scrotum. These vessels supply the subcutaneous areolar tissue and the integument, and are accompanied by numerous lymphatic vessels, which return the lymph from the same parts to the inguinal glands.

The *superficial inguinal glands* are arranged in two groups, one of which is disposed above and parallel with Poupart's ligament, and the other below the ligament, surrounding the termination of the saphenous vein, and following (occasionally, the course of that vessel a short distance along the thigh. The upper chain receives the lymphatic vessels from the penis, scrotum, lower part of the abdomen, perinæum, and buttock; the lower chain receives the lymphatic vessels from the lower extremity.

382.—Femoral Hernia. Superficial Dissection.

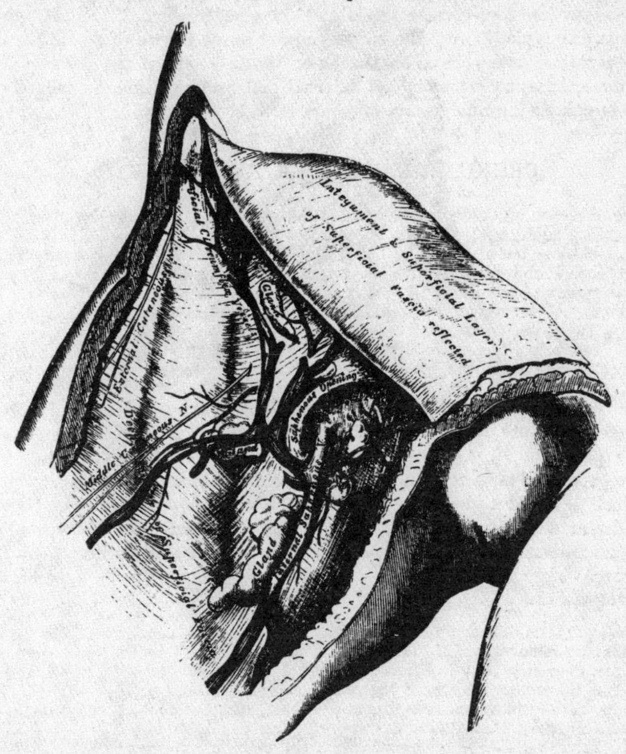

The *nerves* supplying the integument of this region are derived from the ilio-inguinal, the genito-crural, and anterior crural. The ilio-inguinal nerve may be found on the inner side of the internal saphenous vein, the terminal branch of the genito-crural nerve outside the vein, and the middle and external cutaneous nerves more external.

The deep layer of superficial fascia should be divided in the same direction as the external incisions, and separated from the fascia lata; this is easily effected, from its extreme thinness. It is a thin but dense membrane, placed beneath the subcutaneous vessels and nerves, and upon the surface of the fascia lata. It is intimately adherent above to the lower margin of Poupart's ligament, and about one inch below this ligament covers the saphenous opening in the fascia lata, is closely united to its circumference, and is connected to the sheath of the femoral vessels corresponding to its under surface. The portion of fascia covering this aperture is perforated by the internal saphenous vein, and by numerous blood and lymphatic vessels; hence it has been termed the *cribriform fascia*, the openings of these vessels having been likened to the holes in a sieve. The cribriform fascia adheres closely both to the superficial fascia and the fascia lata, so that it is described by some anatomists as a part of the fascia lata, but is usually considered (as in this work) as belonging to the superficial fascia. It is not till the cribriform fascia has been cleared away, that the saphenous opening is seen, so that this opening does not, in ordinary cases, exist naturally, but is the result of dissection. Mr. Callender, however, speaks of cases in which, probably as the result of pressure from enlarged

383.—Femoral Hernia, showing Fascia Lata and Saphenous Opening.

inguinal glands, the fascia has become atrophied, and a saphenous opening exists independent of dissection.* A femoral hernia, in passing through the saphenous opening, receives the cribriform fascia as one of its coverings.

The deep layer of superficial fascia, together with the cribriform fascia, having been removed, the fascia lata is exposed.

The *Fascia Lata*, already described (p. 288), is a dense fibrous aponeurosis, which forms a uniform investment for the whole of this region of the limb. At the upper and inner part of the thigh, a large oval-shaped aperture is observed in it; it transmits the internal saphenous vein and other small vessels, and is called the *saphenous opening*. In order the more correctly to consider the mode of formation of this aperture, the fascia lata in this part of the thigh is described as consisting of two portions, an iliac portion and a pubic portion.

The *iliac portion* of the fascia lata is situated on the outer side of the saphenous opening, covering the outer surface of the Sartorius, the Rectus, and the Psoas

* Anatomy of Femoral Rupture, note on p. 18.

and Iliacus muscles. It is attached externally to the crest of the ilium and its anterior superior spine, to the whole length of Poupart's ligament as far internally as the spine of the pubes, and to the pectineal line in conjunction with Gimbernat's ligament, where it becomes continuous with the pubic portion. From the spine of the pubes, it is reflected downwards and outwards, forming an arched margin, the outer boundary (*superior cornu*) of the saphenous opening. This is sometimes called the *falciform process* of the fascia lata (femoral ligament of Hey); it overlies, and is adherent to, the sheath of the femoral vessels beneath; to its edge is attached the cribriform fascia, and it is continuous below with the pubic portion of the fascia lata by a well-defined curved margin.

The *pubic portion* of the fascia lata is situated at the inner side of the saphenous opening: at the lower margin of this aperture, it is continuous with the iliac portion: traced upwards, it covers the surface of the Pectineus, Adductor longus, and Gracilis muscles; and passing behind the sheath of the femoral vessels, to which it is closely united, is continuous with the sheath of the Psoas and Iliacus muscles, and is finally lost in the fibrous capsule of the hip-joint. This fascia is attached above to the pectineal line, and internally to the margin of the pubic arch. It may be observed from this description that the iliac portion of the fascia lata passes in front of the femoral vessels, the pubic portion behind them; an apparent aperture consequently exists between the two, through which the internal saphenous joins the femoral vein.

The *Saphenous Opening* is an oval-shaped aperture, measuring about an inch and a half in length, and half an inch in width. It is situated at the upper and inner part of the thigh, below Poupart's ligament, towards the inner side, and is directed obliquely downwards and outwards.

Its *outer margin* is of a semilunar form, thin, strong, sharply-defined, and lies on a plane considerably anterior to the inner margin. If this edge is traced upwards, it will be seen to form a curved elongated process or cornu (the superior cornu), or *falciform process of Burns*, which ascends in front of the femoral vessels, and curving inwards, is attached to Poupart's ligament and to the spine of the pubes and pectineal line, where it is continuous with the pubic portion.* If traced downwards, it is found continuous with another curved margin, the concavity of which is directed upwards and inwards; this is the inferior cornu of the saphenous opening, and is blended with the pubic portion of the fascia lata covering the Pectineus muscle.

The *inner boundary* of the opening is on a plane posterior to the outer margin, and behind the level of the femoral vessels; it is much less prominent and defined than the outer, from being stretched over the subjacent Pectineus muscle. It is through the saphenous opening that a femoral hernia passes after descending along the crural canal.

If the finger is introduced into the saphenous opening while the limb is moved in different directions, the aperture will be found to be greatly constricted on

* It is difficult to perceive in the recognised description of these ligaments (Hey's and Burns's), any difference between the two; nor is it clear what structure Mr. Hey really intended to describe. Mr. Gay (on '*Femoral Rupture*,' p. 16) gives very cogent reasons for thinking that the 'deep crural arch' was the structure which Hey had in view. The most recent writer on Femoral Hernia speaks thus while treating of these parts:—'The whole upper edge of the iliac fascia lata is commonly called the "falciform process," whilst its deeper fibres receive the name of "Burns's ligament." Hey's femoral ligament would appear to consist of distinct fibres connected with the inner fold of the iliac fascia, which extend immediately beneath the tendon of the external oblique to the subperitoneal fascia.' (CALLENDER, '*On the Anatomy of the Parts Concerned in Femoral Rupture*,' p. 19, note.) This description of Hey's ligament accords closely with that of the deep crural arch, for the subperitoneal fascia is Mr. Callender's name for the fascia transversalis. Mr. Callender goes on to say,—'The upper border of this (saphenous) opening thus receives by an unfortunate complication, the names of "Falciform process," "Femoral ligament," "Burns's or Hey's ligament." The various divisions of the iliac fascia lata depend in great measure upon the skill of the dissector, and are, in my opinion, artificial.'

extending the limb, or rotating it outwards, and to be relaxed on flexing the limb and inverting it: hence the necessity of placing the limb in the latter position in employing the taxis for the reduction of a femoral hernia.

The iliac portion of the fascia lata, together with its falciform process, should now be removed, by detaching it from the lower margin of Poupart's ligament, carefully dissecting it from the subjacent structures, and turning it aside when the sheath of the femoral vessels is exposed descending beneath Poupart's ligament (fig. 384).

The *Crural Arch*, or *Poupart's Ligament*, is the lower border of the aponeurosis of the External oblique muscle, which stretches across from the anterior superior spine of the ilium, to the spine of the os pubis and pectineal line; the portion corresponding to the latter insertion is called *Gimbernat's ligament.* The direction of Poupart's ligament is curved downwards towards the thigh; its outer half being oblique, its inner half nearly horizontal. Nearly the whole of the space included between the crural arch and innominate bone is filled in by the parts which descend from the abdomen into the thigh. The outer half of the space is occupied by the Iliacus and Psoas muscles, together with the external cutaneous and anterior crural nerves. The pubic side of the space is occupied by the femoral vessels included in

384.—Femoral Hernia. Iliac Portion of Fascia Lata removed, and Sheath of Femoral Vessels and Femoral Canal exposed.

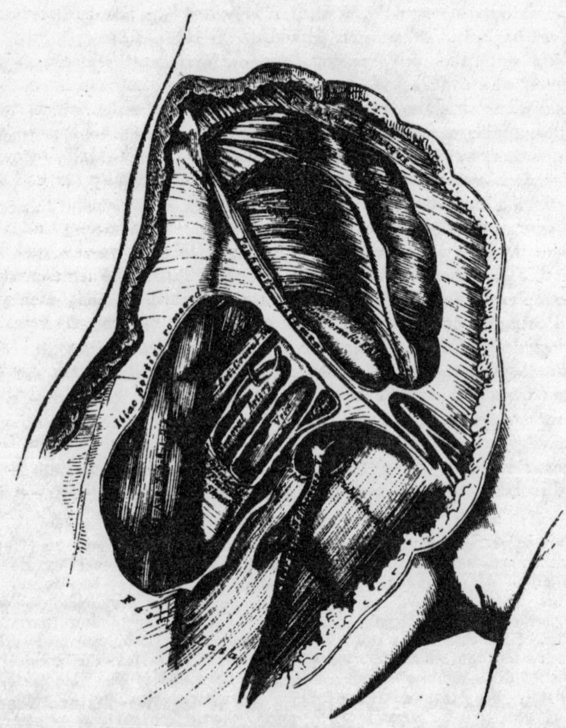

their sheath, a small oval-shaped interval existing between the femoral vein and the inner wall of the sheath, which is occupied merely by a little loose areolar tissue, and occasionally by a small lymphatic gland; this is the crural canal, along which the gut descends in femoral hernia.

Gimbernat's Ligament (fig. 385) is that part of the aponeurosis of the External oblique muscle which is reflected downwards and outwards, to be inserted into the pectineal line of the os pubis. It is about an inch in length, larger in the male than in the female, almost horizontal in direction in the erect posture, and of a triangular form, the base directed outwards. Its *base*, or outer margin, is concave, thin and sharp, lies in contact with the crural sheath, and is blended with the pubic portion of the fascia lata. Its *apex* corresponds to the spine of the pubes. Its *posterior margin* is attached to the pectineal line. Its *anterior margin* is continuous with Poupart's ligament.

Crural Sheath. If Poupart's ligament is divided, the femoral or crural sheath may be demonstrated as a continuation downwards of the fasciæ that line the abdomen, the transversalis fascia passing down in front of the femoral vessels, and the iliac fascia descending behind them; these fasciæ are directly continuous on the iliac side of the femoral artery, but a small space exists between the femoral vein and the point where they are continuous on the pubic side of that vessel, which constitutes the femoral or crural canal. The femoral sheath is closely adherent to the contained vessels about an inch below the saphenous opening, becoming blended with the areolar sheath of the vessels, but opposite Poupart's ligament it is much larger than is required to contain them; hence the funnel-shaped form which it presents. The outer border of the sheath is perforated by the genito-crural nerve. Its inner border is pierced by the internal saphenous vein, and numerous lymphatic vessels. In front, it is covered by the iliac portion of the fascia lata; and behind it is the pubic portion of the same fascia.

Deep Crural Arch. Passing across the front of the crural sheath, and closely connected with it, is a thickened band of fibres, called the *deep crural arch.* It is apparently a thickening of the fascia transversalis, joining externally to the centre of Poupart's ligament, and arching across the front of the crural sheath, to be inserted by a broad attachment into the pectineal line, behind the conjoined tendon. In some subjects, this structure is not very prominently marked, and not unfrequently it is altogether wanting.

If the anterior wall of the sheath is removed, the femoral artery and vein are seen lying side by side, a thin septum separating the two vessels, and another septum separates the vein from the inner wall of the sheath. The septa are stretched between the anterior and posterior walls of the sheath, so that each vessel is enclosed in a separate compartment. The interval left between the vein and the inner wall of the sheath is not filled up by any structure, excepting a little loose areolar tissue, a few lymphatic vessels, and occasionally a lymphatic gland; this is the femoral or crural canal, through which a portion of intestine descends in femoral hernia.

The *crural canal* is the narrow interval between the femoral vein and the inner wall of the crural sheath. It exists as a distinct canal only when the sheath has been separated from the vein by dissection, or by the pressure of a hernia or tumour. Its length is from a quarter to half an inch, and it extends from Gimbernat's ligament to the upper part of the saphenous opening.

Its *anterior wall* is very narrow, and formed by the fascia transversalis, Poupart's ligament, and the falciform process of the fascia lata.

Its *posterior wall* is formed by the iliac fascia and the pubic portion of the fascia lata.

Its *outer wall* is formed by the fibrous septum covering the inner side of the femoral vein.

Its *inner wall* is formed by the junction of the transversalis and iliac fasciæ, which forms the inner side of the femoral sheath, and covers the outer edge of Gimbernat's ligament.

This canal has two orifices: a lower one, the *saphenous opening*, closed by the cribriform fascia; an upper one, the *femoral* or *crural ring*, closed by the septum crurale.

The *femoral* or *crural ring* (fig. 385) is the upper opening of the femoral canal and leads into the cavity of the abdomen.* It is bounded in front by Poupart's ligament and the deep crural arch; behind by the pubes, covered by the Pectineus muscle, and the pubic portion of the fascia lata; internally, by Gimbernat's ligament, the conjoined tendon, the transversalis fascia, and the deep crural arch; externally, by the femoral vein, covered by its sheath. The femoral ring is of an oval form, its long diameter, directed transversely, measures about half an inch, and it is larger in the female than in the male, which is one of the reasons of the greater frequency of femoral hernia in the former sex.

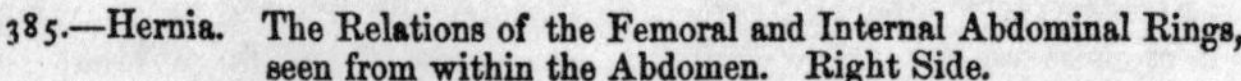

385.—Hernia. The Relations of the Femoral and Internal Abdominal Rings, seen from within the Abdomen. Right Side.

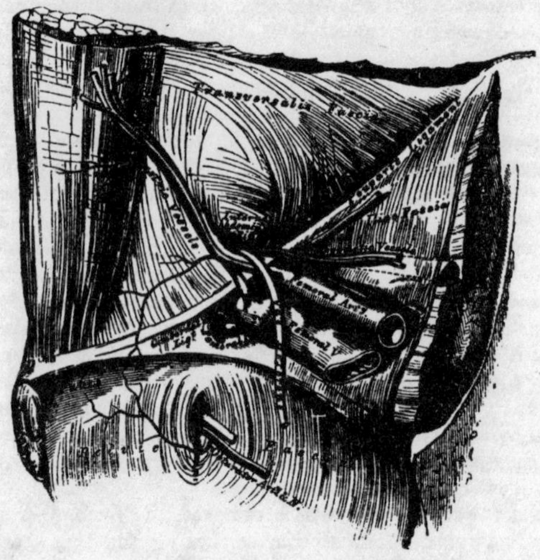

Position of Parts around the Ring. The spermatic cord in the male, and round ligament in the female, lie immediately above the anterior margin of the femoral ring, and may be divided in an operation for femoral hernia if the incision for the relief of the stricture is not of limited extent. In the female, this is of little importance, but in the male the spermatic artery and vas deferens may be divided.

The *femoral vein* lies on the outer side of the ring.

The *epigastric artery*, in its passage inwards from the external iliac to the umbilicus, passes across the upper and outer angle of the crural ring, and is consequently in danger of being wounded if the stricture is divided in a direction upwards and outwards.

The *communicating branch* between the epigastric and obturator lies in front of the ring.

The circumference of the ring is thus seen to be bounded by vessels in every part excepting internally and behind. It is in the former position that the stricture is divided in cases of strangulated femoral hernia.

The *obturator artery*, when it arises by a common trunk with the epigastric, which occurs once in every three subjects and a half, bears a very important relation to the crural ring. In some cases (fig. 386), it descends on the inner side of the external iliac vein to the obturator foramen, and will consequently lie on the outer side of the crural ring, where there is no danger of its being wounded in the operation for dividing the stricture in femoral hernia. Occasionally, however,

* This ring, like the crural canal, is a morbid or an artificial product. 'Each femoral hernia makes for itself (for neither outlet exists in the natural anatomy of the region), a crural canal, and a crural (femoral) ring.'—CALLENDER, *op. cit.*, p. 40.

the obturator artery curves along the free margin of Gimbernat's ligament in its passage to the obturator foramen; it would, consequently, skirt along the greater part of the circumference of the crural canal, and could hardly avoid being wounded in the operation (fig. 387).

Variations in Origin and Course of Obturator Artery.

386.

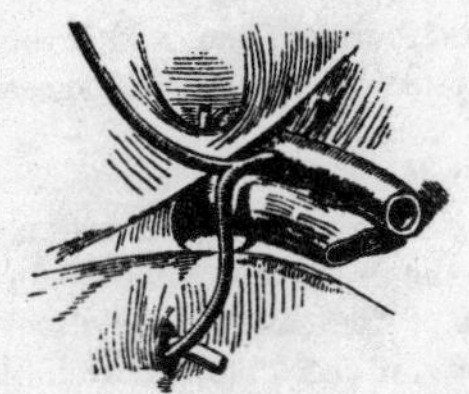

387.

Septum Crurale. The femoral ring is closed by a layer of condensed areolar tissue, called, by J. Cloquet, the *septum crurale.* This serves as a barrier to the protrusion of a hernia through this part. Its upper surface is slightly concave, and supports a small lymphatic gland, by which it is separated from the subserous areolar tissue and peritoneum. Its under surface is turned towards the femoral canal. The septum crurale is perforated by numerous apertures for the passage of lymphatic vessels, connecting the deep inguinal glands with those surrounding the external iliac artery.

The size of the femoral canal, the degree of tension of its orifices, and, consequently, the degree of constriction of a hernia, varies according to the position of the limb. If the leg and thigh are extended, abducted, or everted, the femoral canal and its orifices are rendered tense, from the traction on these parts by Poupart's ligament and the fascia lata, as may be ascertained by passing the finger along the canal. If, on the contrary, the thigh is flexed upon the pelvis, and, at the same time, adducted and rotated inwards, the femoral canal and its orifices become considerably relaxed; for this reason, the limb should always be placed in the latter position when the application of the taxis is made in attempting the reduction of a femoral hernia.

The septum crurale is separated from the peritoneum by a quantity of loose subserous areolar tissue. In some subjects, this tissue contains a considerable amount of adipose substance, which, when protruded forwards in front of the sac of a femoral hernia, may be mistaken for a portion of omentum.

Descent of the Hernia. From the preceding description it follows, that the femoral ring must be a weak point in the abdominal wall; hence it is, that when violent or long-continued pressure is made upon the abdominal viscera, a portion of intestine may be forced into it, constituting a femoral hernia; and the changes in the tissues of the abdomen which are produced by pregnancy, together with the larger size of this aperture in the female, serve to explain the frequency of this form of hernia in women.

When a portion of intestine is forced through the femoral ring, it carries before it a pouch of peritoneum, which forms what is called the *hernial sac*; it receives an investment from the subserous areolar tissue, and from the septum crurale, and descends vertically along the crural canal in the inner compartment of the sheath of the femoral vessels as far as the saphenous opening: at this point, it changes its course, being prevented from extending further down the sheath, on account of the narrowing of the sheath and its close contact with the vessels, and also from the close attachment of the superficial fascia and crural sheath to the lower part of the circumference of the saphenous opening; the tumour is, consequently, directed forwards, pushing before it the cribriform fascia, and then curves upwards on to the falciform process of the fascia lata and lower part of the tendon of the Externa oblique, being covered by the superficial fascia and integument. While the hernia is contained in the femoral canal, it is usually of small size, owing to the resisting nature of the surrounding parts; but when it has escaped from the saphenous opening into the loose areolar tissue of the groin, it becomes considerably enlarged. The direction taken by a femoral hernia in its descent is at first downwards, then forwards and upwards; this should be borne in mind, as in the application of the taxis for the reduction of a femoral hernia, pressure should be directed in the reverse order.

Coverings of the Hernia. The coverings of a femoral hernia from within outwards are peritoneum, subserous areolar tissue, the septum crurale, crural sheath, cribriform fascia, superficial fascia, and integument.*

Varieties of Femoral Hernia. If the intestine descends along the femoral canal only as far as the saphenous opening, and does not escape from this aperture, it is called *incomplete femoral hernia.* The small size of the protrusion in this form of hernia, on account of the firm and resisting nature of the canal in which it is contained, renders it an exceedingly dangerous variety of the disease, from the extreme difficulty of detecting the existence of the swelling, especially in corpulent subjects. The coverings of an incomplete femoral hernia would be, from without inwards, integument, superficial fascia, falciform process of fascia lata, fascia propria, septum crurale, subserous cellular tissue, and peritoneum. When, however, the hernial tumour protrudes through the saphenous opening, and directs itself forwards and upwards, it forms a *complete femoral hernia.* Occasionally, the hernial sac descends on the iliac side of the femoral vessels, or in front of these vessels, or even sometimes behind them.

The *seat of stricture* of a femoral hernia varies: it may be in the peritoneum at the neck of the hernial sac; in the greater number of cases it would appear to be at the point of junction of the falciform process of the fascia lata with the lunated edge of Gimbernat's ligament; or at the margin of the saphenous opening in the thigh. The stricture should in every case be divided in a direction upwards and inwards; and the extent necessary in the majority of cases is about two or three lines. By these means, all vessels or other structures of importance, in relation with the neck of the hernial sac, will be avoided.

* Sir A. Cooper has described an investment for femoral hernia under the name of 'Fascia propria,' lying immediately external to the peritoneal sac, but frequently separated from it by more or less adipose tissue. Surgically, it is important to remember the existence (at any rate the occasional existence) of this layer, on account of the ease with which an inexperienced operator may mistake the fascia for the peritoneal sac, and the contained fat for omentum. Anatomically, this fascia appears to be identical with what is called in the text 'subserous areolar tissue,' the areolar tissue being thickened and caused to assume a membranous appearance, by the pressure of the hernia.

Surgical Anatomy of the Perinæum and Ischio-Rectal Region.

Dissection. The student should select a well-developed muscular subject, free from fat, and the dissection should be commenced early, in order that the parts may be examined in as recent a state as possible. A staff having been introduced into the bladder, and the subject placed in the position shown in fig. 388, the scrotum should be raised upwards, and retained in that position, and the rectum moderately distended with tow.

The space which is now exposed, corresponds to the inferior aperture, or outlet of the pelvis. Its deep boundaries are, in front, the pubic arch and subpubic ligament; behind, the tip of the coccyx; and on each side, the ramus of the pubes and ischium, the tuberosity of the ischium, and great sacro-sciatic ligament. The space included by these boundaries is somewhat lozenge-shaped, and is limited on the surface of the body by the scrotum, in front, by the buttocks behind, and on each side by the inner side of the thighs. It measures, from before backwards, about four inches, and about three in the broadest part of its transverse diameter, between the ischial tuberosities. A line drawn transversely between the anterior part of the tuberosity of the ischium, on each side, in front of the anus, subdivides this space into two portions. The anterior portion contains the penis and urethra, and is called the *perinæum*. The posterior portion contains the termination of the rectum, and is called the *ischio-rectal region.*

Ischio-Rectal Region.

The ischio-rectal region corresponds to the portion of the outlet of the pelvis situated immediately behind the perinæum: it contains the termination of the rectum. A deep fossa, filled with fat, is seen on either side of the intestine, between it and the tuberosity of the ischium: this is called the '*ischio-rectal fossa.*

The *ischio-rectal region* presents, in the middle line, the aperture of the anus; around this orifice, the integument is thrown into numerous folds, which are obliterated on distension of the intestine. The integument is of a dark colour, continuous with the mucous membrane of the rectum, and provided with numerous follicles, which occasionally inflame and suppurate, and may be mistaken for fistulæ. The veins around the margin of the anus are occasionally much dilated, forming a number of hard, pendent masses, of a dark bluish colour, covered partly by mucous membrane, and partly by the integument. These tumours constitute the disease called *external piles.*

Dissection. Make an incision through the integument, along the median line, from the base of the scrotum to the anterior extremity of the anus; carry it round the margins of this aperture to its posterior extremity, and continue it backwards about an inch behind the tip of the coccyx. A transverse incision should now be carried across the base of the scrotum, joining the anterior extremity of the preceding; a second, carried in the same direction, should be made in front of the anus; and a third at the posterior extremity of the gut. These incisions should be sufficiently extensive to enable the dissector to raise the integument from the inner side of the thighs. The flaps of skin corresponding to the ischio-rectal region (fig. 388—2), should now be removed. In dissecting the integument from this region, great care is required, otherwise the External sphincter will be removed, as it is intimately adherent to the skin.

The *superficial fascia* is exposed on the removal of the skin: it is very thick, areolar in texture, and contains much fat in its meshes. In it are found ramifying two or three cutaneous branches of the small sciatic nerve; these turn round the inferior border of the Glutæus maximus, and are distributed to the integument in this region.

388.—Dissection of Perinæum and Ischio-Rectal Region.

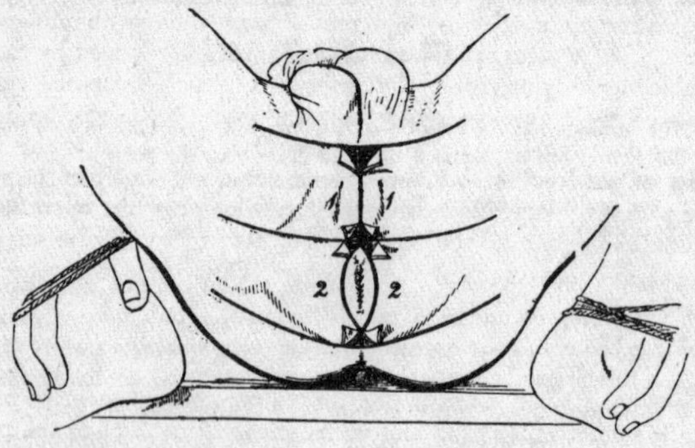

The *External sphincter* is a thin flat plane of muscular fibres, elliptical in shape, and intimately adherent to the integument surrounding the margin of the anus. It measures about three or four inches in length, from its anterior to its posterior extremity, being about an inch in breadth, opposite the anus. It arises from the tip of the coccyx, by a narrow tendinous band; and from the superficial fascia in front of that bone; and is inserted into the tendinous centre of the perinæum, joining with the Transversus perinæi, the Levator ani, and the Accelerator urinæ. Like other Sphincter muscles, it consists of two planes of muscular fibre, which surround the margin of the anus, and join in a commissure before and behind.

Relations. By its *superficial surface*, with the integument; by its *deep surface* it is in contact with the Internal sphincter; and is separated from the Levator ani by loose areolar tissue.

The Sphincter ani is a voluntary muscle, supplied by the hæmorrhoidal branch of the fourth sacral nerve. This muscle is divided in the operation for fistula in ano; and also in some cases of fissure of the rectum, especially if attended with much pain or spasm. The object of its division is to keep the parts at rest and in contact during the healing process.

The *Internal sphincter* is a muscular ring, about half an inch in breadth, which surrounds the lower extremity of the rectum, about an inch from the margin of the anus. This muscle is about two lines in thickness, and is formed by an aggregation of the involuntary circular fibres of the intestine. It is paler in colour, and less coarse in texture, than the External sphincter.

The *ischio-rectal fossa* is situated between the end of the rectum and the tuberosity of the ischium on each side. It is triangular in shape, its base, directed to the surface of the body, is formed by the integument of the ischio-rectal region; its *apex*, directed upwards, corresponds to the point of division of the obturator fascia, and the thin membrane given off from it, which covers the outer surface of the Levator ani (ischio-rectal fascia). Its dimensions are about an inch in breadth, at the base, and about two inches in depth, being deeper behind than in front. It is bounded, *internally*, by the Sphincter ani, Levator ani, and Coccygeus muscles; *externally*, by the tuberosity of the ischium, and the obturator fascia, which covers the inner surface of the Obturator internus muscle; *in front*, it is limited by the line of junction of the superficial and deep perinæal fasciæ: and *behind*, by the margin of the Glutæus maximus, and the great sacro-sciatic ligament. This space is filled with a large mass of adipose substance, which explains the frequency with which abscesses in the neighbourhood of the rectum burrow to a considerable depth.

If the subject has been injected, on placing the finger on the outer wall of this fossa, the internal pudic artery, with its accompanying veins and nerve, will be felt about an inch and a half above the margin of the ischiatic tuberosity, but approaching nearer the surface as they pass forwards along the inner margin of the pubic arch. These structures are enclosed in a sheath formed by the obturator fascia, the pubic nerve lying below the artery. Crossing the space transversely, about its centre, are the inferior hæmorrhoidal vessels and nerves, branches of the pudic; they are distributed to the integument of the anus, and to the muscles of the lower end of the rectum. These vessels are occasionally of large size, and may give rise to troublesome hæmorrhage, when divided in the operation of lithotomy, or of fistula in ano. At the back part of this space may be seen a branch of the fourth sacral nerve; and, at the fore part of the space, a cutaneous branch of the perinæal nerve.

Perinæum.

The perinæal space is of a triangular form; its deep boundaries are limited, laterally, by the rami of the pubes and ischia, meeting in front at the pubic arch; behind, by an imaginary transverse line, extending between the tuberosities of the ischia. The lateral boundaries vary, in the adult, from three inches to three inches and a half in length; and the base from two to three inches and a half in breadth; the average extent of the base being two inches and three-quarters. The variations in the diameter of this space are of extreme interest in connection with the operation of lithotomy, and the extraction of a stone from the cavity of the bladder. In those cases where the tuberosities of the ischia are near together, it would be necessary to make the incisions in the lateral operation of lithotomy less oblique than if the tuberosities were widely separated, and the perinæal space, consequently, wider. The perinæum is subdivided by the median raphe into two equal parts. Of these, the left is the one in which the operation of lithotomy is performed.

In the middle line, the perinæum is convex, and corresponds to the bulb of the urethra. The skin covering it is of a dark colour, thin, freely moveable upon the subjacent parts, and covered with sharp crisp hairs which should be removed before the dissection of the part is commenced. In front of the anus, a prominent line commences, the raphe, continuous in front with the raphe of the scrotum. The flaps of integument corresponding to this space having been removed, in the manner shown in figs. 388—1, the superficial fascia is exposed.

The *Superficial Fascia* consists of two layers, superficial and deep, as in other regions of the body.

The *superficial layer* is thick, loose, areolar in texture, and contains much adipose tissue in its meshes, the amount of which varies in different subjects. In front, it is continuous with the dartos of the scrotum; behind, it is continuous with the subcutaneous areolar tissue surrounding the anus; and, on either side, with the same fascia on the inner side of the thighs. This layer should be carefully removed, after it has been examined, when the deep layer will be exposed.

The *deep layer of superficial fascia* (superficial perinæal fascia) is thin, aponeurotic in structure, and of considerable strength, serving to bind down the muscles of the root of the penis. It is continuous, in front, with the dartos of the scrotum; on either side, it is firmly attached to the margins of the rami of the pubes and ischium, external to the crus penis, and as far back as the tuberosity of the ischium; posteriorly, it curves down behind the Transversus perinæi muscles to join the lower margin of the deep perinæal fascia. This fascia not only covers the muscles in this region, but sends down a vertical septum from its under surface, which separates the back part of the subjacent space into two, being incomplete in front.

In rupture of the anterior portion of the urethra, accompanied by extravasation of urine, the fluid makes its way forwards, beneath this fascia, into the areolar tissue of the scrotum, penis, and anterior and lateral portions of the abdomen; it rarely extends into the areolar tissue on the inner side of the thighs, or backwards around the anus. This limitation of the extravasated fluid to the parts above-named is easy of explanation, when the attachments of the deep layer of the

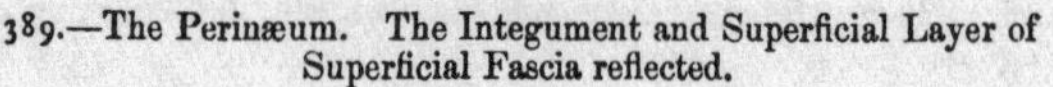

389.—The Perinæum. The Integument and Superficial Layer of Superficial Fascia reflected.

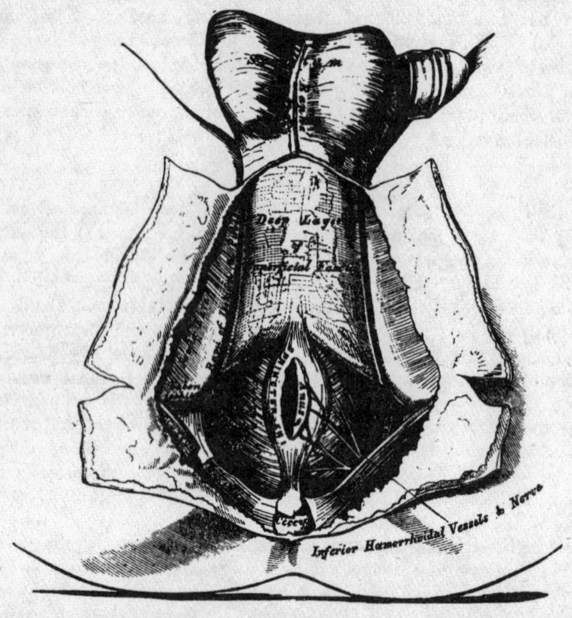

superficial fascia are considered. When this fascia is removed, the muscles connected with the penis and urethra will be exposed; these are, in the middle line, the Accelerator urinæ; on each side, the Erector penis, and behind, the Transversus perinæi.

The *Accelerator urinæ* is placed in the middle line of the perinæum, immediately in front of the anus. It consists of two symmetrical halves, united along the median line by a tendinous raphe. It arises from the central tendon of the perinæum, and from the median raphe in front. From this point, its fibres diverge like the plumes of a pen; the most posterior form a thin layer, which are lost on the anterior surface of the triangular ligament; the middle fibres encircle the bulb and adjacent part of the corpus spongiosum, and join with the fibres of the opposite side, on the upper part of the corpus spongiosum, in a strong aponeurosis; the anterior fibres, the longest and most distinct, spread out over the sides of the corpus cavernosum, to be inserted partly into that body, anterior to the Erector penis; partly terminating in a tendinous expansion, which covers the dorsal vessels of the penis. The latter fibres are best seen by dividing the muscle longitudinally, and dissecting it outwards from the surface of the urethra.

Action. This muscle may serve to accelerate the flow of the urine or semen along the canal of the urethra. The middle fibres are supposed, by Krause, to assist in the erection of the corpus spongiosum, by compressing the erectile tissue of the bulb. The anterior fibres, according to Tyrrel, also contribute to the erection of the penis, as they are inserted into, and continuous with, the fascia of the penis, compressing the dorsal vein during the contraction of the muscle.

The *Erector Penis* covers the unattached part of the crus penis. It is an elongated muscle, broader in the middle than at either extremity, and situated on either side of the lateral boundary of the perinæum. It arises by tendinous and fleshy fibres from the inner surface of the tuberosity of the ischium, behind the

390.—The Superficial Muscles and Vessels of the Perinæum.

crus penis, from the surface of the crus, and from the adjacent portion of the ramus of the pubes. From these points, fleshy fibres succeed, which end in an aponeurosis which is inserted into the side and under surface of the crus penis. This muscle compresses the crus penis, and thus serves to maintain the organ erect.

The *Transverse Perinæi* is a narrow muscular slip, which passes more or less transversely across the back part of the perinæal space. It arises by a smal tendon from the inner and fore side of the tuberosity of the ischium, and, passing obliquely forwards and inwards, is inserted into the central tendinous point of the perinæum, joining in this situation with the muscle of the opposite side, the Sphincter ani behind, and the Accelerator urinæ in front.

Between the muscles just examined, a triangular space exists, bounded internally by the Accelerator urinæ, externally by the Erector penis, and behind by the Transversus perinæi. The floor of this space is formed by the triangular ligament of the urethra (deep perinæal fascia), and, running from behind forwards in it, are the superficial perinæal vessels and nerves, the transverse perinæal artery coursing along the posterior boundary of the space, on the Transversus perinæi muscle.

In the lateral operation of lithotomy, the knife is carried obliquely across the back part of this space, downwards and outwards, into the ischio-rectal fossa, dividing the Transversus perinæi muscle and artery, the posterior fibres of the Accelerator urinæ, the superficial perinæal vessels and nerve, and more posteriorly the external hæmorrhoidal vessels.

The superficial and transverse perinæal arteries are described at p. 395; and the superficial perinæal and inferior pudendal nerves at pp. 540, 542.

The muscles of the perinæum in the female are, the

Sphincter vaginæ.	Compressor urethræ.
Erector clitoridis.	Sphincter ani.
Transversus perinæi.	Levator ani.

Coccygeus.

The *Sphincter Vaginæ* surrounds the orifice of the vagina, and is analogous to the Accelerator urinæ in the male. It is attached, posteriorly, to the central tendon of the perinæum, where it blends with the Sphincter ani. Its fibres pass forwards on each side of the vagina, to be inserted into the corpora cavernosa and body of the clitoris.

The *Erector Clitoridis* resembles the Erector penis in the male, but is smaller than it.

The *Transversus Perinæi* is inserted into the side of the Sphincter vaginæ, and the Levator ani into the side of the vagina. The other muscles are precisely similar to those in the male.

The Accelerator urinæ and Erector penis muscles should now be removed, when the deep perinæal fascia will be exposed, stretching across the front part of the outlet of the pelvis. The urethra is seen perforating its centre, just behind the bulb; and on either side is the crus penis, connecting the, corpus cavernosum with the ramus of the ischium and pubes.

391.—Deep Perinæal Fascia. On the left side, the anterior layer has been removed.

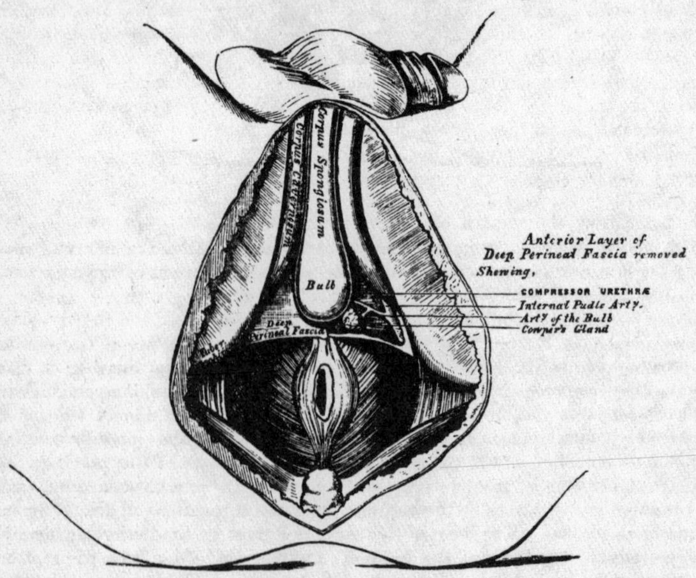

The *Deep Perinæal Fascia* (triangular ligament), is a dense membranous lamina, which closes the front part of the outlet of the pelvis. It is triangular in shape, about an inch and a half in depth, attached above, by its apex, to the under surface of the symphysis pubis and subpubic ligament; and, on each side, to the rami of the ischium and pubes, beneath the crura penis. Its inferior margin, or base, is directed towards the rectum, and connected to the central tendinous point of the perinæum. It is continuous with the deep layer of the superficial fascia behind the Transversus perinæi muscle, and with a thin fascia which covers the cutaneous surface of the Levator ani muscle.

The deep perinæal fascia is perforated by the urethra, about an inch below the symphysis pubis. The aperture is circular in form, and about three or four lines in diameter. Above this is the aperture for the dorsal vein of the penis; and, outside the latter, the pudic nerve and artery pierce it.

The deep perinæal fascia consists of two layers, anterior and posterior: these are separated above, but united below.

The *anterior layer* is continued forwards, around the anterior part of the membranous portion of the urethra, becoming lost upon the bulb.

The *posterior layer* is derived from the pelvic fascia: it is continued backwards around the posterior part of the membranous portion of the urethra, and the outer surface of the prostate gland.

If the anterior layer of this fascia is detached on either side, the following parts are seen between it and the posterior layer: the subpubic ligament above, close to the pubes; the dorsal vein of the penis; the membranous portion of the urethra, and the muscles of the urethra; Cowper's glands and their ducts; the pudic vessels and nerve; the artery and nerve of the bulb, and a plexus of veins.

The *Compressor Urethræ* (constrictor urethræ), surrounds the whole length of the membranous portion of the urethra, and is contained between the two layers of the deep perinæal fascia. It arises, by aponeurotic fibres, from the upper part of the ramus of the pubes on each side, to the extent of half or three quarters of an inch; each segment of the muscle passes inwards, and divides into two fasciculi, which surround the urethra from the prostate gland behind, to the bulbous portion of the urethra in front; and unite, at the upper and lower surfaces of this tube, with the muscle of the opposite side, by means of a tendinous raphe.

Circular Muscular Fibres surround the membranous portion of the urethra, from the bulb in front to the prostate gland behind; they are placed immediately beneath the transverse fibres already described, and are continuous with the circular fibres of the bladder. These fibres are involuntary.

Cowper's Glands are situated immediately below the membranous portion of the urethra, close behind the bulb, and below the artery of the bulb (p. 687).

The *Pudic Vessels and Nerves* are placed along the inner margin of the pubic arch (p. 393).

The *Artery of the Bulb* passes transversely inwards, from the internal pudic along the base of the triangular ligament, between the two layers of fascia, accompanied by a branch of the pudic nerve (p. 395).

If the posterior layer of the deep perinæal fascia is removed, and the crus penis of one side detached from the bone, the under or perinæal surface of the Levator ani is brought fully into view. This muscle, with the triangular ligament in front and the Coccygeus and Pyriformis behind, closes in the outlet of the pelvis.

The *Levator ani* is a broad thin muscle, situated on each side of the pelvis. It is attached to the inner surface of the sides of the true pelvis, and, descending, unites with its fellow of the opposite side to form the floor of the pelvic cavity. It supports the viscera in this cavity, and surrounds the various structures which pass through it. It arises, in front, from the posterior surface of the body and ramus of the pubes, on the outer side of the symphysis; posteriorly, from the inner surface of the spine of the ischium: and between these two points, from the angle of division between the obturator and recto-vesical layers of the pelvic fascia at their under part: the fibres pass downwards to the middle line of the floor of the pelvis, and are inserted, the most posterior fibres into the sides of the apex of the coccyx; those placed more anteriorly unite with the muscle of the opposite side, in a median fibrous raphe, which extends between the coccyx and the margin of the anus. The middle fibres, which form the larger portion of the muscle, are inserted into the side of the rectum, blending with the fibres of the

Sphincter muscles; lastly, the anterior fibres, the longest, descend upon the side of the prostate gland to unite beneath it with the muscle of the opposite side, blending with the fibres of the external sphincter and Transversus perinæi muscles, at the tendinous centre of the perinæum.

The anterior portion is occasionally separated from the rest of the muscle by cellular tissue. From this circumstance, as well as from its peculiar relation with the prostate gland, descending by its side and surrounding it as in a sling, it has been described by Santorini and others as a distinct muscle, under the name of the *Levator prostatæ*. In the female, the anterior fibres of the Levator ani descend upon the sides of the vagina.

Relations. By its *upper* or *pelvic surface* with the recto-vesical fascia, which separates it from the viscera of the pelvis and from the peritoneum. By its *outer* or *perinæal surface*, it forms the inner boundary of the ischio-rectal fossa, and is covered by a quantity of fat, and by a thin layer of fascia continued from the obturator fascia. Its *posterior border* is continuous with the Coccygeus muscle. Its *anterior border* is separated from the muscle of the opposite side by a triangular space, through which the urethra, and, in the female, the vagina, passes from the pelvis.

Actions. This muscle supports the lower end of the rectum and vagina, and also the bladder during the efforts of expulsion.

The *Coccygeus* is situated behind and parallel with the preceding. It is a triangular plane of muscular and tendinous fibres, arising, by its apex, from the spine of the ischium and lesser sacro-sciatic ligament, and inserted, by its base, into the margin of the coccyx and into the side of the lower piece of the sacrum. This muscle is continuous with the posterior border of the Levator ani, and closes in the back part of the outlet of the pelvis.

Relations. By its *inner* or *pelvic surface*, with the rectum. By its *external surface*, with the lesser sacro-sciatic ligament. By its *posterior border*, with the Pyriformis.

Action. The Coccygei muscles raise and support the coccyx, after it has been pressed backwards during defæcation or parturition.

Position of the Viscera at the Outlet of the Pelvis. Divide the central tendinous point of the perinæum, separate the rectum from its connections by dividing the fibres of the Levator ani, which descend upon the sides of the prostate gland, and draw the gut backwards towards the coccyx, when the under surface of the prostate gland, the neck and base of the bladder, the vesiculæ seminales, and vasa deferentia will be exposed.

The *Prostate Gland* is placed immediately in front of the neck of the bladder, around the prostatic portion of the urethra, its base being turned backwards, and its under surface towards the rectum. It is retained in its position by the Levator prostatæ and by the pubo-prostatic ligaments, and is invested by a dense fibrous covering, continued from the posterior layer of the deep perinæal fascia. The longest diameters of this gland are in the antero-posterior direction, and transversely at its base; and hence the greatest extent of incision that can be made in it without dividing its substance completely across, is obliquely outwards and backwards. This is the direction in which the incision is made through it in the operation of lithotomy, the extent of which should seldom exceed an inch in length. The relations of the prostate to the rectum should be noticed: by means of the finger introduced into the gut, the surgeon detects enlargement or other disease of this organ; he is enabled also, by the same means, to direct the point of a catheter when its introduction is attended with much difficulty, either from injury or disease of the membranous or prostatic portions of the urethra.

Behind the prostate is the posterior surface of the neck and base of the bladder; a small triangular portion of this organ is seen, bounded, in front by the prostate gland; behind, by the recto-vesical fold of the peritoneum; on either side, by the vesiculæ seminales and vasa deferentia; and separated from direct contact with the rectum by the recto-vesical fascia. The relation of this portion of the bladder to the rectum is of extreme interest to the surgeon. In cases of retention of urine, this portion of the organ is found projecting into the rectum, between three and four inches from the margin of the anus, and may be easily perforated during life without injury to any important parts; this portion of the bladder is, consequently,

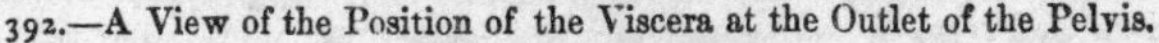

392.—A View of the Position of the Viscera at the Outlet of the Pelvis.

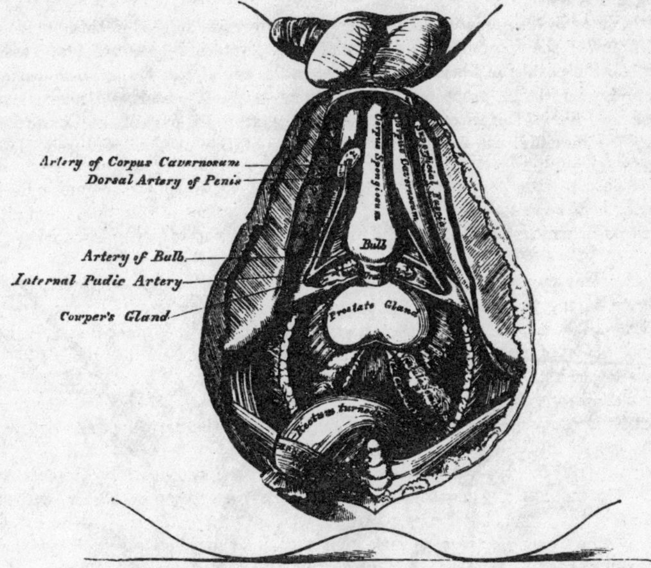

frequently selected for the performance of the operation of tapping the bladder. If the finger is introduced into the bowel, the surgeon may, in some cases, learn the position, as well as the size and weight, of a calculus in the bladder; and in the operation for its removal, if, as is not unfrequently the case, it should be lodged behind an enlarged prostate, it may be displaced from its position by pressing upwards the base of the bladder from the rectum.

Parts concerned in the Operation of Lithotomy. The triangular ligament must be replaced and the rectum drawn forwards so as to occupy its normal position. The student should then consider the position of the various parts in reference to the lateral operation of lithotomy. This operation is performed on the left side of the perinæum, as it is most convenient for the right hand of the operator. A staff having been introduced into the bladder, the first incision is commenced midway between the anus and the back of the scrotum (*i.e.*, in an ordinary adult perinæum, about an inch and a half in front of the anus), a little on the left side of the raphe, and carried obliquely backwards and outwards to midway between the anus and tuberosity of the ischium. The incision divides the integument and superficial fascia, the external hæmorrhoidal vessels and nerves, and the superficial and transverse perinæal vessels; if the fore-finger of the left hand is thrust upwards and forwards into the wound, pressing at the same time the rectum inwards and backwards, the staff may be felt in the membranous portion of the urethra. The finger is fixed upon the staff, and the structures covering it are divided with the point of the knife, which must be directed along the groove towards the bladder, the edge of the knife being carried outwards and backwards, dividing in its course the membranous portion of the urethra, and part of the left lobe of the prostate gland, to the extent of about an inch. The knife is then withdrawn, and the fore-finger of the left hand passed along the staff into the bladder; the staff having been withdrawn, and the position of the stone ascertained, the forceps are introduced over the finger into the bladder. If the stone is very large, the opposite side of the prostate may be notched before the forceps are introduced; the finger is now withdrawn, and the blades of the forceps opened, and made to grasp the stone, which must be extracted by slow and cautious undulating movements.

Parts divided in the operation. The various structures divided in this operation are as follows: the integument, superficial fascia, external hæmorrhoidal vessels and nerves, the posterior fibres of the Accelerator urinæ, the Transversus perinæi muscle and artery (and, probably, the superficial perinæal vessels and nerves), the deep perinæal fascia, the anterior fibres of the Levator ani, part of the Compressor urethræ, the membranous and prostatic portions of the urethra, and part of the prostate gland.

393.—A Transverse Section of the Pelvis; showing the Pelvic Fascia.

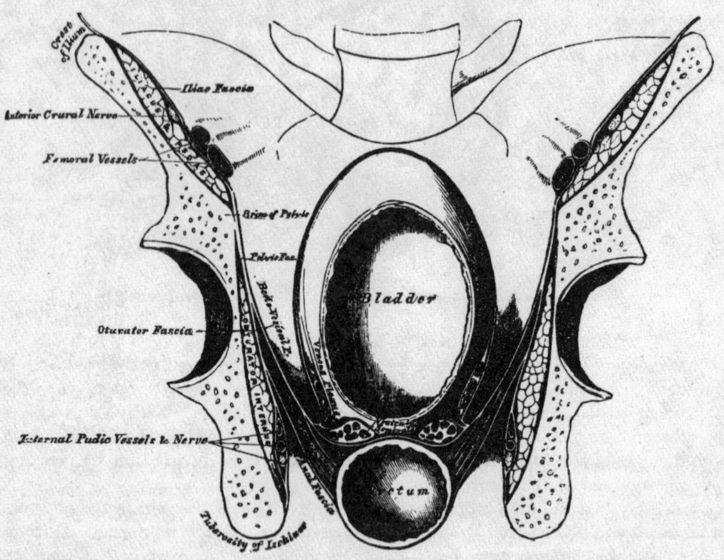

Parts to be avoided in the operation. In making the necessary incisions in the perinæum for the extraction of a calculus, the following parts should be avoided. The primary incisions should not be made too near the middle line, for fear of wounding the bulb of the corpus spongiosum or the rectum; nor too far externally, otherwise the pudic artery may be implicated as it ascends along the inner border of the pubic arch. If the incisions are carried too far forward, the artery of the bulb may be divided; if carried too far backwards, the entire breadth of the prostate and neck of the bladder may be cut through, which allows the urine to become infiltrated behind the pelvic fascia into the loose cellular tissue between the bladder and rectum, instead of escaping externally; diffuse inflammation is consequently set up, and peritonitis from the close proximity of the recto-vesical peritoneal fold is the consequence. If, on the contrary, the prostate is divided in front of the base of the gland, the urine makes its way externally, and there is less danger of infiltration taking place.

During the operation, it is of great importance that the finger should be passed into the bladder *before* the staff is removed; if this is neglected, and the incision made through the prostate and neck of the bladder be too small, great difficulty may be experienced in introducing the finger afterwards; and in the child, where the connections of the bladder to the surrounding parts are very loose, the force made in the attempt is sufficient to displace the bladder up into the abdomen, out of the reach of the operator. Such a proceeding has not unfrequently occurred, producing the most embarrassing results, and total failure of the operation.

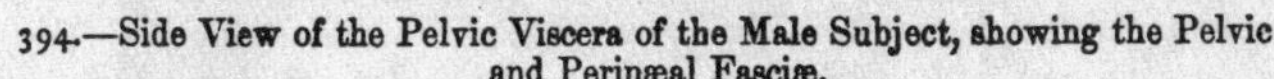

394.—Side View of the Pelvic Viscera of the Male Subject, showing the Pelvic and Perinæal Fasciæ.

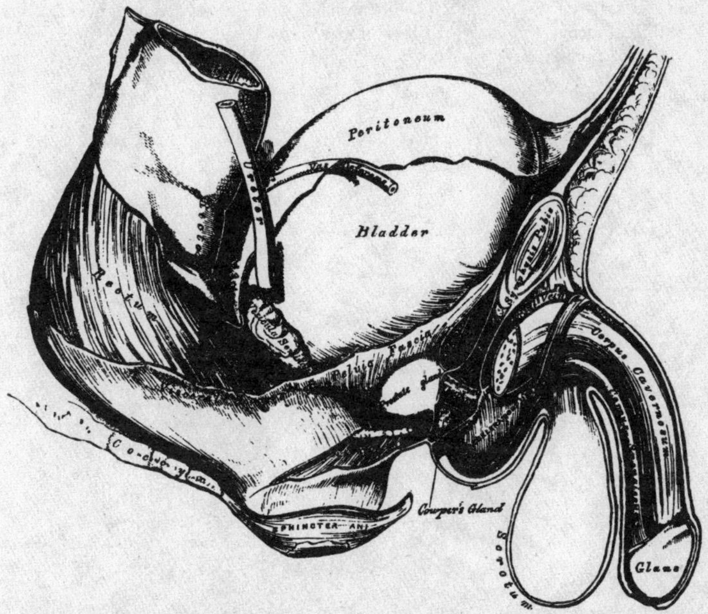

It is necessary to bear in mind that the arteries in the perinæum occasionally take an abnormal course. Thus the artery of the bulb, when it arises, as sometimes happens, from the pudic, opposite the tuber ischii, is liable to be wounded in the operation for lithotomy, in its passage forwards to the bulb. The accessory pudic may be divided near the posterior border of the prostate gland, if this is completely cut across: and the prostatic veins, especially in people advanced in life, are of large size, and give rise, when divided, to troublesome hæmorrhage.

Pelvic Fascia.

The pelvic fascia (fig. 395) is a thin membrane which lines the whole of the cavity of the pelvis, and is continuous with the transversalis and iliac fasciæ. It is attached to the brim of the pelvis for a short distance at the side of the cavity, and to the inner surface of the bone round the attachment of the Obturator internus. At the posterior border of this muscle, it is continued backwards as a very thin membrane in front of the Pyriformis muscle and sacral nerves, behind the branches of the internal iliac artery and vein which perforate it, to the front of the sacrum. In front, it follows the attachment of the Obturator internus to the bone, arches beneath the obturator vessels, completing the orifice of the obturator canal, and at the front of the pelvis is attached to the lower part of the symphysis pubis; being continuous below the pubes with the fascia of the opposite side so as to close the front part of the outlet of the pelvis, blending with the posterior layer of the triangular ligament. At the level of a line extending from the lower part of the symphysis pubis to the spine of the ischium, is a thickened whitish

395.—Pelvic Fascia.

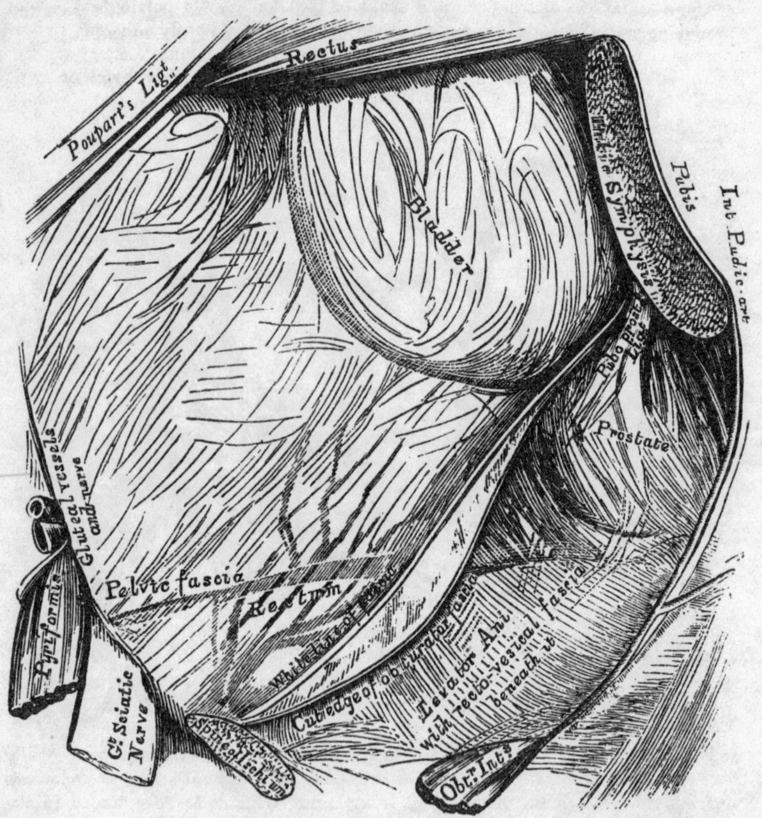

band; this marks the attachment of the Levator ani muscle to the pelvic fascia, and corresponds to its point of division into two layers, the obturator and recto-vesical.

The *obturator fascia* descends and covers the Obturator internus muscle. It is a direct continuation of the pelvic fascia below the white line above mentioned, and is attached to the pubic arch and to the margin of the great sacro-sciatic ligament. This fascia forms a canal for the pudic vessels and nerve in their passage forwards to the perinæum, and is continuous with a thin membrane which covers the perinæal aspect of the Levator ani muscle, called the *ischio-rectal (anal) fascia*.

The *recto-vesical fascia* (visceral layer of the pelvic fascia) descends into the pelvis upon the upper surface of the Levator ani muscle, and invests the prostate, bladder, and rectum. From the inner surface of the symphysis pubis a short rounded band is continued to the upper surface of the prostate and neck of the bladder, forming the pubo-prostatic or anterior true ligaments of the bladder. At the side, this fascia is connected to the side of the prostate, enclosing this gland

and the vesical prostatic plexus, and is continued upwards on the surface of the bladder, forming the lateral true ligaments of the organ. Another prolongation invests the vesiculæ seminales, and passes across between the bladder and rectum, being continuous with the same fascia of the opposite side. Another thin prolongation is reflected round the surface of the lower end of the rectum. The Levator ani muscle arises from the point of division of the pelvic fascia; the visceral layer of the fascia descending upon and being intimately adherent to the upper surface of the muscle, while the under-surface of the muscle is covered by a thin layer derived from the obturator fascia, called the ischio-rectal or anal fascia. In the female, the vagina perforates the recto-vesical fascia, and receives a prolongation from it.

INDEX.

ERRATA.

Page 35, line 14 from bottom, *for* medulla oblongata *read* Pons Varolii.
,, 284, ,, 4, *for* apophysis *read* diaphysis.
,, 334, last line, *for* dental foramen *read* mental foramen.